Look for these symbols next to many of the most interesting, instructive
Problems in *Statistics for
Business and Economics*, Second Editi

 deals with Sales/Market

 deals with Manufacturing/Production

 deals with Finance/Accounting/Economics

 deals with Government/Public Affairs/Public Policy

# STATISTICS FOR BUSINESS AND ECONOMICS

## SECOND EDITION

### JOHN A. INGRAM
Senior Statistician for Metromail Corporation,
a subsidiary of R. R. Donnelley & Sons, Inc.

### JOSEPH G. MONKS
Gonzaga University

**The Dryden Press**
**Harcourt Brace Jovanovich College Publishers**
Fort Worth   Philadelphia   San Diego   New York   Orlando   Austin   San Antonio
Toronto   Montreal   London   Sydney   Tokyo

# THE DRYDEN PRESS SERIES IN MANAGEMENT SCIENCE AND QUANTITATIVE METHODS

*Acquisitions Editor:* Scott Isenberg
*Manuscript Editor:* Cate DaPron
*Production Editor:* Pat Gonzalez
*Art Editor:* Louise Sandy-Karkoutli
*Designer:* Suzanne Montazer
*Production Manager:* Diane Southworth

Cover: © Stephen Huneck "Corporate Structure Secretary" 1990

ISBN: 0-15-583549-1

Library of Congress Catalog Card Number: 91-71628

Printed in the United States of America

# PREFACE

*Statistics for Business and Economics*, Second Edition, remains an introduction to the uses of statistics for understanding data, for the purpose of making meaningful business decisions. Its primary objective is to help students experience decision making based upon patterns observed from data. Since statistics can be used most effectively when data are well understood, a second objective of this book is to help students develop some skills for evaluating the reasonableness of data through data analysis. Careful editing, additions to concept explanations, and more examples and problems have shored up some important topics and thus have met yet another objective, which is to provide a clear and readable presentation of the fundamentals of business and economics statistics. Using computers is optional in *Statistics for Business and Economics*, but computers can be a highly effective tool in the use of statistics. So a fourth objective, available at the instructor's choosing, is to build computer skills with which to circumvent the otherwise time-consuming calculations required for obtaining some statistics.

An ultimate objective is for business students to become users, not simply learners, of statistics. *Statistics for Business and Economics*, Second Edition, offers students what would otherwise require several courses, and other practical experience: the combination of (1) directed study of fundamental concepts and (2) the building of skills for the practical use of statistics. When students learn statistics by studying the chapter and working the varied selection of regular Problems, and then understanding the Computer Applications and working the Computer Problems, they can become consumers of statistics. The authors believe that a coverage of this textbook that includes the Computer Applications and Computer Problems will provide students a reasonable background to become users of statistics in a business or economics work position, or it can be a stepping stone for more advanced (quantitative) studies.

## WHAT IS NEW IN THIS EDITION?

Quite a lot is new! From our experience and that shared with us by others who have used the First Edition, we have made numerous changes and enhancements. In particular, to underscore its relevance in business, we have added new features in four general areas:

Data Analysis  Experiencing and solving difficulties with data is real, even critical, in business work; it is essential that students — and professionals — understand the data before they begin analyzing a quantitative problem. So we have added, to an already generous number of Problems, many Problems that include 20 to 50 observations each. Further, some of these Problems require critical evaluation to assure that the data are both clean and appropriate for a proposed statistical analysis. Our treatment of data analysis includes the most current techniques — box plots; stem-and-leaf displays; graphics, including Pareto charts, engineering diagrams, and flowcharts; data plots with residual analysis; and data frequency displays. Practical data analysis Problems are interspersed throughout the chapters.

**Problem Solving for Real-World Applications**   Each chapter has (1) a Business Case, with questions, that illustrates how key chapter concepts can be used on the job, and (2) Statistics in Action: Challenge Problems, a special section that gives a new perspective to various chapter concepts. Both types of Problems require more thought than the section or chapter Problems. And because they come from our own research and from business journals, magazines, and government publications, the Business Cases and Challenge Problems reproduce actual on-the-job experiences.

**Computer-Assisted Problem Solving**   This feature is optional (except in multiple regression), and the Computer Applications and Computer Problems are color-highlighted and placed at the end of the chapter to make them easy to find. Because Chapter 1 introduces the user-friendly Minitab software along with the data base, and each chapter continues to build usage skills, students have the opportunity to learn computer-assisted statistics problem solving without taking computer courses. The Computer Applications and Computer Problems provide instruction through examples that show program steps, explain program statements, and interpret the statistics outputs. Chapter 5 introduces the SPSS software, and thereafter the text continues to develop experience with both Minitab and SPSS. The Computer Applications and Computer Problems in Chapters 1-18 expose students to a high percentage of the concepts described in the *Minitab Student Edition Handbook*. The SPSS usage covers the same concepts, but is oriented to mainframe computers.

**Fine-Tuned Fundamentals**   Except for adding a new chapter (15) on total quality control, we have left the topic content unchanged. However, emphasis has been added to descriptive statistics and data analysis, and estimation and hypothesis testing have been reorganized for more time-efficient coverage, with more flexibility in topic choices for a first course. The text material is ample to allow options for a second course as well. Here are the specific enhancements made to this Edition:

**Chapter 1**   The Computer Application and Computer Problems promote understanding of the data base, which is U.S. decennial census data that describe family membership, ages, education, and more.

**Chapter 2, 3**   Box plots and stem-and-leaf displays are introduced. Then the emphasis is on data analysis and data cleaning guides.

**Chapter 5**   The Poisson distribution coverage is expanded with another example and more background on using a probability table. New Minitab and SPSS computer setups help students calculate discrete variable probability distributions.

**Chapters 6, 7**   The Statistics in Action: Challenge Problems here provide more work on probability distributions.

**Chapters 8, 9, 10**   The $t$-distribution has been moved from Chapter 10 to Chapter 8 for estimation and to Chapter 9 for hypothesis testing. This gives an early distinction for the use of the $Z$- versus and the $t$-distribution, creating more options for a first course.

**Chapter 11**    Now we display the ANOVA assumptions and computing formulas for the 2-factor equal-numbers factorial design. Also new are examples of how to interpret experiment results when there is a significant interaction effect, plus more small data sets and questions on data analysis.

**Chapter 12**    Both the algebra solution (in the regular sections) and a computer approach (in the Computer Applications and Problems) are given for estimating a simple linear regression model. This allows a smooth transition to the multivariate procedures of multiple regression and the analysis of variance. Also, the First Edition Business Case has been reworked to improve its quality and meaning.

**Chapter 13**    Outliers, leverage points, and influential observations are discussed, thus expanding the analysis of residual data to include more discussion of data analysis.

**Chapter 14**    First Edition Chapters 14 and 15 are combined here to produce an integrated, cohesive coverage of time series and index numbers.

**Chapter 15**    This Total Quality Control chapter is all new. It emphasizes the continuous improvement of system and people processes for improved quality in the production of goods and services.

**Chapter 17**    Again, small data sets of 20 to 50 observations reinforce data analysis. This chapter, especially, shows the benefits of using computers — by including Problems asking for hand calculations, then doing the same calculations with statistical software.

**LEARNING FEATURES IN EVERY CHAPTER**

- Each chapter Introduction includes a pertinent, motivating illustration of statistics being used in business, followed by a statement of purpose for the chapter and an overview of the material that follows.
- Learning objectives describe 4–6 specific abilities students can expect to gain from studying the chapter.
- Examples illustrate fundamental statistics concepts in a business context. An interpretation of the computed results immediately follows each of the 200 Examples in the text.
- Problems are of three kinds. Section Problems within the chapters relate to specific concepts. End-of-chapter Problems reinforce the concepts and enhance analytical skills. Computer Problems, applied to real decision situations, make use of "live" data from the data base. There are 950 regular Problems and 120 more computer-assisted statistics Problems.
- Learning guides are procedures to guide students in solving problems that require more than a few steps. There are learning guides and equation summary tables in virtually every chapter.
- A content-based Summary reviews the key elements of the chapter.
- Key terms, italicized throughout the chapter, emphasize basic concepts. A list of Key Terms following the chapter Summary shows the pages on which each of approximately 300 key terms is defined.
- Business Cases use some of the authors' experiences and other business situations to show how statistics is used in specific decision circumstances.

These 18 experiences allow the student to assume the role of a decision maker; the questions that follow the Business Cases require some critical thinking.

- Statistics in Action: Challenge Problems include a brief description, the problem statement, and the data source for 78 carefully selected Problems from journals, magazines, and government publications. By answering the questions associated with these applications, students are introduced to various professional publications and become aware of the breadth of applications of statistical methods in business.

- The optional Computer Applications build on the textbook's 20,116-record data base and the scenario of a company that is marketing personal computers. (A sample of 100 of the data bank records, and the record layout, appears in Appendix A.) Computer Problems offer students the opportunity to generate computer-assisted solutions to approximately 120 decision questions.

- About 150 of the most interesting and instructive Problems are highlighted by icons:

 Sales/Marketing

 Manufacturing/Production

 Finance/Accounting/Economics

 Government/Public Affairs/Public Policy

- Numerous figures and tables clarify and condense the discussions of important concepts.

- At the end of the textbook are Appendices containing a brief data base, numerous statistical tables, math essentials for statistics, and the answers to the odd-numbered Problems. (The text Examples and Problems were worked using both the Student Edition of Minitab and SPSS, so students can expect these answers to give an accurate check of the results.)

## CONTENT AND FLEXIBILITY

This work covers all areas normally included in the first course in business and economic statistics. The material in Chapters 1–9 provides a foundation for the business "specialty" topics of the remaining nine chapters. Chapters 1–3 concentrate on the terms and methods used to describe business data, and they emphasize data analysis techniques. Chapters 4–6 add more realism to the data environment by introducing structured measures for chance events in probabilities and probability distributions. Chapters 7, 8, and 9 form the basis for classical statistical inference. They are concerned with the use of sampling, estimation, and hypothesis tests in business situations. The remaining nine chapters introduce the most widely used extensions of statistical methods for business decisions. Even for a first course, this allows a selection of topics by the instructor.

Chapter 10 provides a bridge that builds from the Z- and t-statistic inference, including estimation and hypothesis testing in Chapters 8 and 9. Chapter 10 also introduces the chi-square and *F*-distributions that are the foundation for Chapters 11 through 16. Analysis of variance and experimental design (Chapter 11) and chi-square tests (Chapter 16) cover techniques for testing the equality of means and proportions, respectively. Regression and correlation measures describe the nature and degree of association between two variables (Chapter 12) or for more than two variables (Chapter 13). This is extended to time series data and index numbers in Chapter 14. Chapter 15 introduces the procedures of total quality control. Nonparametrical statistics (Chapter 17) are generally not as ''powerful'' as statistics based on a known probability distribution, but they are gaining widespread use in economics, marketing, and management. Finally, Chapter 18 offers interested users the opportunity to extend some basic concepts of probability and statistics into the area of decision theory. This work includes a modern, innovative introduction to Bayesian decision theory, which has numerous applications in accounting and in marketing and operations management.

The text is designed to accommodate either a one-term or a two-term course, and the arrangement and organization of the chapters can be modified to suit a variety of usage patterns. Also, the more difficult material and Problems appear in optional starred sections. Alternative outlines for one- and two-term courses are given in the Instructor's Manual.

## ADDITIONAL LEARNING MATERIALS

Several publications are available to augment the text:

- A Study Guide, developed by Professor A. Thomas Mason of the College of St. Thomas and Professor Dana Schumacher of Iowa State University, for students who seek additional insights and want more study problems.
- An Instructor's Manual containing teaching suggestions, solutions to all Problems in the textbook, and 60 overhead transparency masters.
- New to this edition, a Testbook (also available as a computerized Test Bank in IBM format) with over 500 test questions and their answers.
- Also available from the publisher, on computer disks, the 20,116-record data base and optional data base sample files of 500 records and 100 records.

## ACKNOWLEDGMENTS

The authors extend heartfelt thanks to the many people who have contributed in developing, testing, and building this book. First, we appreciate the sure sense of direction and many helpful ideas provided by Scott Isenberg, our Acquisitions Editor. And we are grateful for the numerous suggestions and questions provided in the reviews submitted by Michael S. Broida, Miami University; Thomas S. Brown, Southwest Missouri State University; Mark Eakin, University of Texas at Arlington; Nicholas R. Farnum, California State University, Fullerton; Ralph C. Gamble, Fort Hays State University; David W. Gerbing, Portland State University; John O. Ifediora, The University of Wisconsin — Platteville; Thomas Johnson, North Carolina State University; Joseph F. Kearney, Davenport University; Eddie M. Lewis, The University of Southern Mississippi; Paul A. Thompson, The Ohio State University; James A. Xander, University of Central Florida; and Fike Zahroon, Moorhead State University.

We offer sincere thanks to those HBJ staff members who have patiently worked with us in the preparation of this book, including Cate DaPron, Manuscript Editor; Pat Gonzalez, Production Editor; Louise Sandy-Karkoutli, Art Editor; Suzanne Montazer, Designer; and Diane Southworth, Production Manager. We are indebted to them for making this a clear presentation of the content; for producing the clean figures and tables; and for giving the final product a most pleasing appearance. But most of all we thank the HBJ team for making our shared work a pleasant, positive experience.

We express our appreciation to Molly Mostel Keys, who processed the solutions, charts, and graphs for the Instructor's Manual. Thanks also to graduate student Lisa Kinney for applying her considerable analytical skills to help organize materials, check problem solutions, and edit the drafts of the manuscript. In particular, we acknowledge the wholehearted support we received from Dean Bud Barnes of Gonzaga University and from Terry Coombes for arranging the personnel assistance and computer facilities to complete this project. Finally, we express sincere gratitude to our wives and families, who have very patiently stood by and encouraged us through this work.

John A. Ingrams
Joseph G. Monks

# CONTENTS

**PART THREE**

# ESTIMATION AND TESTS OF HYPOTHESES    283

CHAPTER 8

CHAPTER 9

CHAPTER 10

CHAPTER 15

**TOTAL QUALITY CONTROL   715**

CHAPTER 16

**CHI-SQUARE TESTS OF FREQUENCIES AND PROPORTIONS   752**

CHAPTER 17

**NONPARAMETRIC STATISTICS   794**

CHAPTER  18

# INTRODUCTION

Some years ago, a multibillion-dollar plan was launched to provide the Pacific Northwest with inexpensive electrical power. Five huge nuclear power plants were to be built. Bonds for projects 1, 2, and 3 were backed by the federal Bonneville Power Administration and bonds for units 4 and 5 were backed by 88 Northwest utilities.

About 10 years later, two of the incomplete projects (4 and 5) were canceled and work on another (1) was suspended. The estimate of expected power demand had been unnecessarily high, and construction costs were skyrocketing. Three years after that, the Washington Public Power System (WPPS) failed to make a $15.2 million monthly interest payment on the bonds for plants 4 and 5 to Chemical Bank of New York. Being short of cash, the WPPS's executive board suspended work on project 3 (which was about two-thirds complete), leaving only project 2. Moody's Investment Service then followed the lead of Standard and Poor's and suspended its ratings on the $6 billion in bonds used to finance units 1, 2, and 3. This was perhaps the *largest individual project bond default* in U.S. history. As shock waves reverberated throughout the business community, utility executives, construction managers, and investment brokers told themselves, "This can't be happening." But it was.

The reasons behind such a gigantic debacle are numerous and complex. Of prime importance was the failure to use good statistical information to support ongoing managerial decisions. Initially there were incorrect estimates of demand. Then came escalating equipment and labor costs, high interest rates, strikes, and a myriad of other managerial problems. Problems surfaced in nearly every business discipline. Marketing estimates were inaccurate, production costs soared out of control, the personnel department faced strikes and labor disputes, legal problems arose, and financing difficulties ultimately proved disastrous. Management lacked timely, accurate statistical information to correct the problems or at least to anticipate and deal with them years earlier. As a result, instead of being rescued through a series of carefully guided, skillful decisions, the project collapsed—to the bewilderment of a host of investors, thousands of employees, and millions of ratepayers in the Pacific Northwest. Ratepayers are still paying for these mistakes—poor management decisions stemming from inadequate information—and will be for years to come.

## 1.1 WHY STATISTICS?

Decision making is the major responsibility of management. Some decisions, like scheduling meetings or planning business trips, are simple. Others, like negotiating the end of a strike, are more complex. All require information. Managers in all types of organizations rely on readily available, accurate information to control current operations and guide future activities. Statistics is concerned with extracting the best information from existing data so that sound business decisions can be made. Figure 1.1 shows the statistical decision-making process. Decisions about such concerns as marketing new products or raising needed capital generally come after careful consideration of information.

**FIGURE 1.1**    The Statistical Decision-Making Process

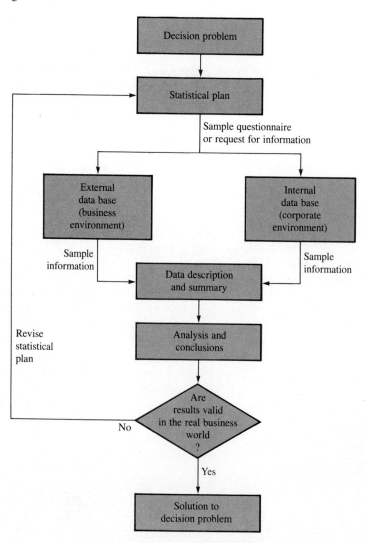

Most *decision problems* stem from questions about current operations or about opportunities to accomplish something better in the future. To solve current problems and direct future activities successfully, managers need a *plan*. And to prepare and implement a plan, they need a source of information, a *data base*. Moreover, they need some valid techniques for extracting useful information from the data base.

Only a portion of the desired data (a sample) is usually needed to reveal logical patterns that exist in it. *Analysis* and *conclusions* about the data can be made from the sample. If the sample data are randomly selected and the methodology is correct, then the results will be valid in the business world. A *solution* to the decision problem will be achieved. All of these elements are vital to good managerial decision making.

## Recognizing the Decision Problem

A manager once remarked, "I have no problems—just opportunities!" Whether managerial action is prompted by problems or by opportunities, the first step is to define clearly the decision situation under study. This means understanding the problem (or opportunity) and stating a goal for correcting it. A full definition of the problem also reveals which factors must be accepted as given and which factors are manageable. Analysts often refer to these factors as uncontrollable and controllable variables. In the WPPS situation, electrical demand was largely uncontrollable (given), but costs, schedules, and labor relations should have been more controllable (manageable).

## The Statistical Plan

Business managers need detailed plans to meet their specific goals. If the plans to market a new product are based upon reliable data, they will more likely result in a successful business strategy.

> **A statistical plan includes identification of a data base, techniques for describing the data, a means of analyzing it, and measures of validity (error) in working with it.**

The *statistical plan* provides a basis for management action.

## The Data Base

Because we regularly use data, all businesses need *data bases*.

> **A data base is a collection of data that is pertinent to the operation of a business. This information is commonly stored on a computer for use in making management decisions.**

Why have car sales dropped over the past quarter? What level of employment can we expect in the building construction industry next year? How can we increase grain production without exceeding our storage limitations? These are all questions that show a need for relevant data. Business, as well as other technical and nontechnical fields, requires data to set goals, to move toward them, and to measure performance.

In business, data from both the internal corporate environment and external sources may be useful for analysis. Data may come from one's own business (such as demand, production, or sales data) or may be developed by outside sources (such as government statistics, stock reports, and consumer statistics). Managers use both *primary data* and *secondary data* as bases for developing statistics to answer questions of why, what, and how.

Data collected by observation, by experimentation, by simulation or from surveys are called primary data. Data from other sources are called secondary data.

The purpose of sampling is to get the best information available without using all the possible data. A sample may be only a small part of the total information that exists, but it can accurately characterize the larger population if it is scientifically selected. Then business managers can be confident that sample information will lead them to valid conclusions.

We have become an information-based society, with computers tracking almost every economic variable and with high-speed printers issuing reams of printouts. Managers often have more data than they can possibly use. A lesser amount of scientifically collected and methodically analyzed data is usually just as informative (and more manageable) than volumes of facts. The point is not just to get more data, but to get the information needed for making responsible business decisions. Computer systems that supply this type of data are referred to as management information systems (MIS) and decision support systems (DSS).

## Data Description and Summary

Data are transformed into useful information via statistical methods and summarization techniques. You will become familiar with the most commonly used statistical description techniques in the next two chapters of this book. When coupled with the data-processing capabilities of computers, these techniques give managers summary measures almost instantaneously. Furthermore, the summary statistics can be collected from virtually any location via telephone or satellite and presented in almost any form (textual, tabular, or graphic). Thus, purchasing managers can receive up-to-the-minute prices for materials, and project managers can quickly estimate the hours of overtime necessary to complete a job.

Beware of determining and presenting statistical results before you have done some analysis of your data. It is always possible that your basic assumptions are false or that the data contain errors due to data editing or misprints. Statistics will not compensate for incorrect data. And a computer simply calculates results, whether the data are right or wrong. Data analysis—checking for reasonable data—is critical to ensure that resulting statistics are derived from the right numbers. Chapter 3 includes some considerations for checking data before making a statistical analysis.

## Analysis and Conclusions

Complex decisions often require managers to project their thinking from current experiences to a new situation. To illustrate, consider a question like this: If we get the government contract, then what level of employment can we expect for next month? To answer such a question managers use well-developed methods of statistical inference. Most of these methods are available on computer in a user-friendly format. But first it is essential to understand the statistics that computer programs perform. This means doing enough hand calculations to gain a good grasp of the methodologies. Because computer results are used extensively in business today, we will be making frequent reference to computer programs in the text and will illustrate some of our analyses with computer printouts.

## An Illustration of the Statistical Decision Process

Whether data are analyzed manually or on computer, a manager needs to have confidence in the analysis. A manager's livelihood depends upon his or her decisions, so the decisions must be good. Example 1.1 illustrates the steps in a typical statistical decision-making situation.

**EXAMPLE 1.1**

There are six steps to consider in a decision-making process that uses statistics.

1. *Decision Situation/Question*   Fashion Fabrics Ltd. has developed a new wrinkle-resistant synthetic for use in women's apparel.

In which two of five major markets—Chicago, New York, Houston, Los Angeles, and Miami—will the new fabric be best received? This question is motivated by management's need to test-market the product before making a commitment to the expense associated with large-volume production and distribution. Management needs information to make a good decision.

2. *Statistical Plan*   The fabric will be test-marketed in randomly selected retail stores in the five market areas. Results will then be projected to estimate the total expected sales in each area. The two markets with the statistically highest volume of sales will be chosen for widespread distribution of the new fabric.

3. *Data Base*   Ten retail outlets will be randomly chosen in each market area. In those ten outlets, advertising, pricing, inventory, and other variables will be kept at the same levels across all areas. Sales data (primary data) will be collected for a three-month period.

4. *Data Description*   Comparative sales data for each market area will be summarized to reveal average weekly sales per 10,000 population of market area. Graphs will be prepared to compare the average and total sales in each area.

5. *Analysis*   Statistical tests will be conducted on the sales data to determine if sales are markedly higher in some areas than in others or if the difference in sales is due simply to chance.

6. *Conclusions/Recommendations*   Using average sales per 10,000 population, management will project annual sales for all five major market areas. Distribution costs will be incorporated and profit projections will be made for each area. The profit projections will be compared with earlier sales figures and evaluated by regional sales managers who have a good grasp of market potential. If any discrepancies exist, they will be resolved at this stage or the plan will be revised.

*Solution*   The two markets with highest total profit potential will be selected for the full-scale marketing program. Then the statistical plan will be developed. With proper implementation, it can provide useful information for a management decision.

## So Why Statistics?

Statistics provides a systematic and logical approach to solving business problems. Done properly, a statistical plan takes the guesswork out of making decisions, especially when only some representative portion of the total information is available.

Our work in statistics emphasizes formulating problems in business (as in other fields) as quantifiable questions, utilizing random samples from a specified data base, and applying the appropriate statistical methods for making decisions. In short, the focus of this book is on using statistics to make business decisions.

This first chapter describes the role of statistics in decision making.

Upon completing Chapter 1, you should be able to

1. define statistics;
2. explain the role of statistics in business decision making;

3. differentiate between decision approaches suitable for no information, some information, and all information situations;
4. distinguish between descriptive and inferential statistics;
5. distinguish between statistical induction and statistical deduction;
6. explain the relationships of sample to population and statistic to parameter;
7. understand one data base by using the variables in it.

This chapter will also give you a general idea of the direction we are taking in the remainder of the book. This textbook has been designed to help you gradually develop a solid foundation upon which to build a useful knowledge of statistics. Once you master some terms and get a feel for the power of statistical inference, we think you'll find statistics interesting—and perhaps even exciting. We guarantee it will be useful when you are in a decision-making capacity.

## 1.2 THE ROLE OF STATISTICS IN BUSINESS DECISION MAKING

Because decision making is such a vital part of business, crucial decisions cannot rest upon hunches or intuition. Competition will not permit that. Nor will the company directors or stockholders. Managers are held accountable for their decisions, even when the data they must use are incomplete or uncertain. In fact, it is precisely because of the uncertainty of information that statistics has become so important to corporate decision makers. Statistics enables decision makers to draw valid conclusions from less than complete information. Through statistics, decisions are made and the chance of a wrong decision is measured.

Figure 1.2 depicts the *information environment* of the business decision maker.

The information environment is the amount of information, ranging from none to complete, that is available to the decision maker.

Visualize the *information environment* as a spectrum ranging from one extreme, in which the decision maker has little or no information, to the other extreme, in which he or she has nearly total information about the situation.

The "all information" extreme means that data are available on every element in the population. The 10-year (decennial) population census is an example of this. The census is used as a basis for the apportionment of federal funds to our cities, so it affects the flow of millions of dollars. Taking a census is costly and time consuming, so businesses prefer not to use a census.

Most business decision making takes place within the "some information" region. For example, a merchant might periodically scan key items on some shelves to determine an appropriate time to restock all of the shelves. This visual sample provides information that can be very meaningful.

A "no information" situation could arise in the marketing of a new product. For example, an inventor may not know whether a new toy will be accepted by the public or not. The inventor has no experience, so there are no data to go on. Yet if the toy is to be marketed, then some level of production must be set.

In many business decision situations, managers might like more information, but due to cost and time constraints they must settle for less. In this case, statistical methods become valuable decision tools.

**FIGURE 1.2**   The Information Environment

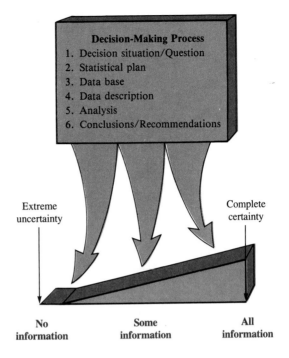

Figure 1.3 lists some of the statistical methods that are useful for making decisions. The figure reflects the same certainty–uncertainty continuum as Figure 1.2, except that the sources of information are now identified. As you can see, larger samples are closer to the "all information" end (lend more certainty) than smaller samples.

Different quantitative procedures apply as the spectrum moves from "all information" to "no information." Under conditions of complete certainty, deterministic mathematical approaches such as calculus and linear programming are appropriate. For example, an accountant who knows all the revenues and assumes all costs are known can compute profits using the formula

$$\text{Profit} = \text{Revenues} - \text{Costs}$$

This is an instance where simple mathematics is appropriate. But the "all information" situations usually do not prevail in decision making, and so they are not the major focus of this book.

Although some situations of extreme uncertainty exist in business decisions, they are not the main thrust of statistics either. Most business decisions must be made on the basis of some information. Such situations are the mainstream of statistics and are our emphasis in this book.

**Decision Situations Using Some Information**

*From production and accounting*   Firms that maintain physical goods usually perform an annual inventory. Historically, druggists, grocers, and many wholesalers have taken a census—a complete count. A recent innovation is the sample inventory. This requires a count of some percentage of the items, along with a projection for a total value. The complete count may be more accurate, but the sample approach can also be accurate, and it takes less time and costs less.

**FIGURE 1.3**   Examples of Decisions That Can Utilize Statistical Methods

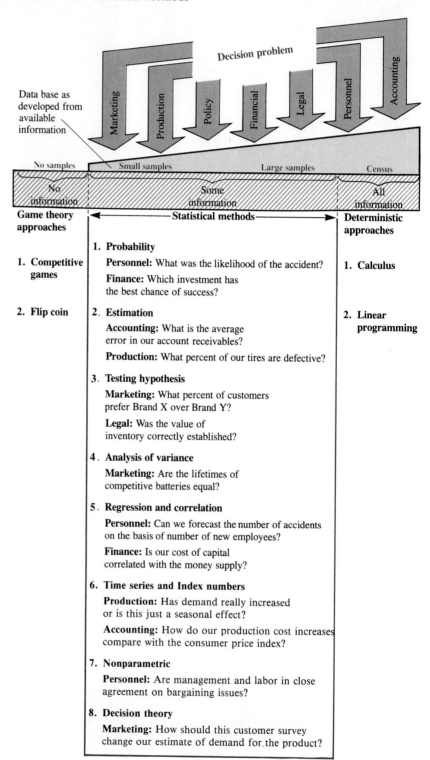

Which approach is better? The answer depends upon costs and the level of accuracy required. In many cases the sample approach will satisfy accounting standards at much lower costs.

*From finance*  When you reach a stable earnings level, you may choose to save for retirement by investing in a mutual fund. Mutual fund managers recognize that it is not feasible to invest funds in every stock on the New York Stock Exchange. Instead they analyze the profit structure of selected firms and then help their clients by offering them a mutual fund with a representative sample of stocks.

*From marketing*  Consider a nationwide overnight delivery business as it first begins to serve a new area. How many delivery routes should it have? Due to the cost of investment in airplanes, trucks, labor, and so on, the firm needs to limit the number of routes yet still be able to deliver the goods.

Decision theory is a statistical approach that allows managers to make decisions based on the existing information. This decision process can begin with a relatively small amount of information. As actual data become available, they can be incorporated into the decision process. Thus managers can use statistics from a few routes to decide how to expand the routes for the successful delivery business.

Statistics provides managers with the tools to answer many vital questions with confidence. Statistical techniques cover a wide range of activity, from production to public relations. Knowledge of statistics is thus becoming more essential to every manager. The major benefit of this knowledge is in helping to make decisions on the basis of some information and in justifying the reasonableness of such decisions.

## 1.3 BASIC CONCEPTS OF STATISTICS

Statistics is concerned with drawing conclusions from data. The validity of the conclusions depends, to some extent, upon the accuracy and comprehensiveness of the data collection effort. Some data may constitute a whole *population*, whereas other data may be only a part, or a *sample*, of the whole population.

> A population is a total collection of all items under study that have one or more specified characteristics.

> A sample is any part of a population.

Thus a population might be the collection of income tax amounts paid last year by all firms in one state. One sample from that population would consist of the tax amounts paid by a dozen firms selected from all those who paid taxes. Such a sample may not be very representative of the total population, but because it is a part of the population, it is a sample. Further, it is only one of many possible samples; some will be better and others will be worse in representing the full population.

One objective of statistics is to obtain samples that are reasonable representations of the parent population. A random sample has the best chance of representing the population. A procedure that assures a random sample is called *random sampling*.

> Random sampling is a procedure that assures every member of the population an equal chance of inclusion in the sample.

The term *statistics* means "coming from data" and denotes a science concerned with the logical treatment of data.

**Statistics is the science of collecting, describing, analyzing, and interpreting data for purposes of making decisions under the condition of some information.**

In a statistics problem, first the population is defined. Then sample data are collected. Analyses are performed to determine what information about the population is contained in the sample. We can then make inferences about the population by projecting the findings from the sample information. Statistics can provide a measure of the reliability or confidence that the inference will lead to a correct decision.

A value used to describe data is called a *parameter* or a *statistic*.

**A parameter is a numerical measure of a population.**

**A statistic is a numerical measure of a sample.**

Consider the business firms in a state as a population that we are trying to understand. Then the average tax paid by all firms in the state is a parameter, whereas the average tax paid by the sample of a dozen firms is a statistic. Parameters are symbolized by Greek letters, and statistics by English letters. Thus the average income tax paid by all firms could be designated by the Greek letter $\mu$ (mu), and the average tax paid by a sample of firms could be designated by $\bar{X}$ (read "$X$ bar").

## Descriptive Statistics

Users of statistics distinguish between descriptive and inferential statistics. Knowledge that comes from observation allows us to describe the data. We refer to data summary measures as *descriptive statistics*.

**Descriptive statistics condense, summarize, or graphically illustrate data. They are derived from observation or measurement of sample data.**

Descriptive statistics includes collection, description, and presentation activities and constitutes the bulk of the "data description" block in Figure 1.1.

Statistical description is used extensively in sales and in business communication, as illustrated by the numerous graphs and charts that appear in business publications such as *Business Week* and *The Wall Street Journal*. Other collections of statistics include summary reports like stock market averages and the consumer price index (CPI). These are applications to summarize or inform rather than to set managerial policy. Chapters 2 and 3 are on description.

## Statistical Inference

The second subdivision—inference—relates to the analysis and interpretation uses of statistics. Inference is often the basis for decisions. The decision rationale typically follows either a deductive or an inductive pattern of reasoning, as depicted in Figure 1.4.

*Deductive inference* is used extensively to develop probabilities where we have some empirical or *a priori* (prior) knowledge about a population and wish to draw conclusions about specific events.

**FIGURE 1.4**   Patterns of Reasoning in Statistical Inference

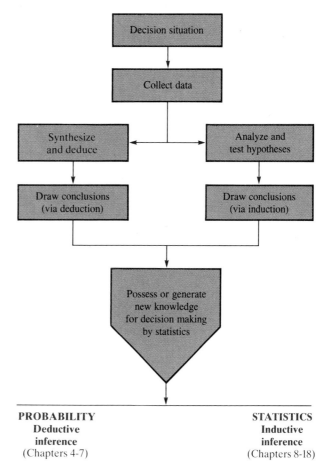

Deductive inference involves reasoning from a general situation to a specific outcome.

*A priori* knowledge is information known before any experiments, tests, or observations.

For example, if we know that 98% of the accounts receivable at a certain bank are usually correct, we would feel there is a high probability that an audit of any single account would show it to be correct. Deductive inference applies a generalization about a known population to a specific sample. It is reasoning from the general to the specific. Methods for assigning a measure of chance to a specific sample through deductive inference are developed in Chapters 4 through 7 as *probability*.

*Statistical inference* involves inductive reasoning from a sample to the bigger population. It is the kind of reasoning in which we infer something about the unknown population on the basis of known sample information. These techniques fit more appropriately in the "Analyze" and "Draw conclusions" blocks on the right-hand side of Figure 1.4.

Statistical (inductive) inference is a generalization about the population that is derived from the information contained in a sample.

This is reasoning from the specific to the general situation. For example, suppose a random sample of 5000 households reveals that 2100 like a new brand of breakfast cereal. The cereal company could draw from this data to infer that 42% of all households would like the new cereal. The actual population proportion—the parameter—will not be known until after the firm makes a large investment and markets the cereal. (Of course, by then it is too late to rescind the marketing effort!)

To assure valid conclusions, statistical inference by induction must be carefully controlled with respect to (1) how the sample data are obtained and (2) how the results are stated. First, the sample must be selected by chance (not purposely) from the population. Second, the risk of an error must be admitted and measured; that is, we must explicitly quantify the chance for a correct decision (called the level of confidence) and the range of results over which this level applies. These two conditions are the statistical insurance for sound management decisions.

Thus the breakfast cereal company can infer with a high degree of confidence, say 95%, that the proportion of the population that would like the new cereal is from, for instance, 37.7% to 46.3%. In saying "95%," they knowingly accept (and admit) a specified risk of 5% of being wrong.

Statistics helps managers draw conclusions about populations by analyzing samples. The managers can then infer the best action for a given situation from the information contained in the sample. This book emphasizes statistical inference in Chapter 8 and following chapters. The primary focus is on the most useful statistical methods for common business situations. These include analysis of variance, regression and correlation, time series, nonparametric statistics, quality control, and decision theory.

## SUMMARY

Decision making is the major function of management, and managers can make better decisions by using statistics. *Statistics* is the science of collecting, presenting, analyzing, and interpreting data for making decisions under conditions of uncertainty. Uncertainty refers to a decision situation in which limited information is available, as opposed to one in which all information or no information is available. Sound decision making with statistics involves the following six steps:

1. defining a clear statement of the problem or opportunity,
2. developing a plan for solving the problem,
3. selecting a data base and drawing a random sample,
4. describing and summarizing the sample results,
5. using statistical methods to analyze the data (while recognizing the risk of error),
6. resolving the problem and recommending a course of action.

These steps, which define a scientific approach to decision making, will be followed throughout this book. Examples and chapter problems are drawn largely from industry to give you a feel for relevant business questions. Since government-derived statistics play a major role in describing corporate business and national economic conditions, a number of problems involve working with a data base that is drawn from a census of the United States.

This first chapter introduced some key terms, as follows: A *population* is the whole collection of elements under study, and a *sample* is any part of the population. *Parameters* describe characteristics of populations, whereas *statistics* describe characteristics of samples.

*Descriptive statistics* stem from empirical (observed) data and often take the form of summary statistics, tables, charts, and graphs. *Inferential statistics* are either *deductive* (if they deduce something about a specific element of the population from a knowledge of the population) or *inductive* (if they infer something about the population from a knowledge of the sample). Since inference is the major thrust in business decisions, it is emphasized (over descriptive statistics) in this book.

One caution: Perhaps the most difficult aspect of making decisions is translating the results of a statistical test to a correct conclusion and then to a decision. For example, suppose you establish a sampling plan for the incoming inspection of television sets. You might require that at least 95% of the sets inspected be in working condition before the lot is accepted. Suppose the first lot passes. You would be using poor judgment to assume, without further inspection, that the next ten carloads are of the same high quality. The error would be due to a fault in logic, not to any shortcomings in statistics. A wise decision maker always gives considerable attention to using good judgment.

We have made an effort to collect and pose business questions that you could experience. You are encouraged to study each problem or example in the chapters that follow as if it were your own decision situation.

Included in Appendix A are some sample data records (PUMS) available from the Bureau of the Census. We will be using these data for some illustrative business applications throughout the textbook. Review Appendix A before proceeding with the Chapter 1 problems. The data base is described further in the computer applications section that follows.

## KEY TERMS

*a priori* knowledge, 11
data analysis
   primary data, 4
   secondary data, 4
data base, 3
descriptive statistics, 10
information environment, 6
parameter, 10
population, 9

random sampling, 9
sample, 9
statistic, 10
statistical inference 12
   deductive, 11
   inductive, 12
statistical plan, 3
statistics, 10

## CHAPTER 1 PROBLEMS

1. The following terms are basic to statistics:

| | | |
|---|---|---|
| statistic | parameter | population |
| sample | descriptive statistics | statistical inference |
| data base | primary data | secondary data |
| statistical plan | decision problem | probability |

Assign the appropriate statistical term to each of the following descriptions.
a. Part of a population  *sample*
b. That subdivision of statistics that primarily concerns data summary  *desc. stat*
c. Data that are derived from your own business or experience  *primary*
d. A problem that requires managerial action  *dec. prob*
e. A numerical value that describes a sample  *statistic*
f. Government statistics, trade or association figures, and other data external to a business  *sec. data*
g. A numerical value that describes a population  *parameter*
h. That branch of statistics using projections from samples to the population  *stat inf.*

i. A total collection of objects, people, or things _population_
j. A subject area within statistics that uses deductive inference to assign a measure of chance to a specific sample outcome _probability_
k. A computer file that embodies the key data for a business _data base_
l. A guide for a logical managerial decision process that includes obtaining reasonable data, describing or analyzing the data facts, and then developing a solution to a decision problem _stat plan_

_char of pop_

2. Explain whether each of the following items describes a statistic or a parameter.
P a. Your grade-point average _s_
   b. The percentage of persons who respond to an opinion survey by agreeing that our president is doing a good job _s_
   c. The average age of all U.S. citizens as reported by the 1990 census _p_
P d. The average income for all persons who are members of the American Medical Association _s_
P e. Sales made by General Foods Corp. in July as an indicator of annual sales _p_

3. Indicate whether a sample or a population is described in each of the following items.
   a. An opinion survey
   b. A biannual agriculture survey
   c. A quarterly stockholders summary of assets and liabilities
   d. The collection of all grades assigned for this class

4. Description and inference are the two broad subareas of statistics. Identify which subarea classification—description or inference—describes each of the following items better.
   a. Computing a baseball player's batting average _d_
   b. Using the first-quarter summary of production to project second-quarter production _i_
   c. A bar chart that displays the corporate revenues for the current year, the preceding year, two years ago, and so on _d_
   d. A survey estimate of corporate year-end assets at somewhere between $1.2 million and $1.4 million _i_
   e. Estimating an assembly worker's weekly output as 1200 units plus or minus 100 units from eight hours of work performance _i_

5. The complete U.S. census describes about 80 million households. Data Services is considering purchasing the A (5%), the B (1%), or the C (0.1%) SAMPLE tapes. What might be the advantage of each sample ( A, B, or C) for the following purposes?
   a. Having some information available for their data analysis work
   b. The cost-efficiency in processing the data, assuming they will use all of the available data

6. Suppose you work for Data Services, Inc., a company that assists other firms by performing a demographic data analysis on their customer files. If Data Services uses the Census Bureau's SAMPLE tapes, is this a primary or secondary data source?

7. How do you think SAMPLES differ from the census with respect to each of the following items?
   a. Count
   b. Accuracy
   c. Use of the results

8. Look at the first three records in Appendix A, Table A.1. Count the number of persons (see columns 12–13) in each record. Sum these numbers. Then divide by the number of records (three). This is an average of the number of persons per household, PPH. Is this a statistic or a parameter? (*Hint*: The national number of PPH is 2.7.)

9. Suppose you were asked to use your calculation in Problem 8 to tell about those records. Would you be using statistics to provide a description or to make an inference?

10. Assume that you do not know the true value of 2.7 persons per household. You are asked to use your number from Problem 8 to estimate the population census value (to make an inference). Is this inference by deduction or by induction? Explain your choice.

11. Many businesses and other agencies use census data. Here are a few illustrations:
    a. The auto industry uses demographics to profile (or classify) households in areas according to affluence, social status, income, and age relative to their customer profiles, so they can concentrate direct marketing where it will be most profitable.
    b. State and municipal governments are allocated federal funding based on population counts.
    c. Land brokers uses national statistics on residential moves to appraise the potential value of land for housing developments.
    Select a business firm that you know something about. Explain how the firm uses (or might use) census data. If you are not familiar with a firm, then consider one of the following areas.

    > *In accounting*   Ernst and Young evaluating media circulation areas (TV news) for advertising their service.
    >
    > *In finance*   Shearson Lehman Brothers establishing housing areas for promotional mailings.
    >
    > *In tax*   H&R Block determining a site location for one of its offices.
    >
    > *In business law*   Evers & Shays estimating employment trends.
    >
    > *In marketing*   L. L. Bean mailing unsolicited catalogs.
    >
    > *In personnel*   Sears estimating housing costs in different locations.
    >
    > *In production*   Hewlett-Packard Corp. seeking a location for a calculator production plant.
    >
    > *In management*   General Electric planning a marketing strategy.

12. A manufacturer of family household consumer products, like Johnson & Johnson, seeks your advice on demographic variables to help them locate their kind of customers (families with young children). From Table A.1 in Appendix A, can you identify five demographics that should interest them? How about ten demographics? Name them.

13. In the early 1980s, Safeway food stores faced a real dilemma. Their operations were running in the red. The problem was not lack of business, but excessive labor expenses. Management decided to meet labor head-on. After extensive negotiations the two sides were at a stalemate. Neither would concede enough for a settlement. Then management decided to collect data on labor costs for stores in all states where Safeway operated. A statistical analysis of wage levels showed that several states (including Iowa and Nebraska) had high labor costs. Management took decisive action. They offered labor a choice: Either accept a final settlement or Safeway would relocate to other states. For the latter option, employees could relocate but under a new wage structure. Labor refused the settlement, and Safeway relocated its operations from the high-labor-cost states to others. Only a percentage of the employees were relocated.

    This scenario describes a management decision problem and its eventual resolution. It encompasses the six steps outlined in the summary section for making sound decisions. Briefly describe the six steps as they apply to this decision situation.

14. Identify a decision problem of your choice from one of the areas in Problem 11. Describe the six steps necessary to resolve the decision situation.

---

The Statistics in Action sections—one for each chapter—are presented so that you can experience real-world data problems in the classroom. Following are some key features of these Problems.

1. A few sentences describe the purpose of the research, give the data source (a reference), and define terms as appropriate.
2. The Problems have been hand-picked to be readable, relevant, and interesting to business and economics students.
3. At first, only simplistic tools are required to analyze the data. As your skills grow, you will analyze the same data with greater discernment, using other, more complex statistical methods.
4. The amount of raw data presented is based on space available in this textbook and the amount provided in the sources. Seeing the data can provide you with insights that otherwise are not possible.
5. The emphasis is on the key concepts that have been covered in the chapter. This way you can gain experience using them with real data to solve decision-making problems.

The Problems come from the business literature and from data fact books. The journal articles are carefully selected to be both meaningful and readable. Many of the articles are available at your university library.

15. Is watching TV becoming the American love affair? For this study a national sample of people kept a one-day diary of their activities. According to the author, "watching TV is the dominant leisure activity of Americans, consuming 40 per cent of the average person's free time as a primary activity." (Primary viewing is when people give TV their undivided attention.) The article includes statistics for the number of weekly primary television-viewing hours for several groupings of people—by age, sex, marital status, and education.

Total Weekly TV Hours

|   |   | Total | Men | Women |
|---|---|---|---|---|
|   |   | 15.0 | 15.7 | 14.4 |
| Age | 18 to 24 | 14.6 | 14.7 | 14.5 |
|   | 25 to 34 | 14.6 | 15.2 | 14.0 |
|   | 35 to 44 | 13.7 | 14.9 | 12.7 |
|   | 45 to 54 | 14.7 | 16.3 | 13.4 |
|   | 55 to 64 | 18.5 | 18.2 | 18.7 |

SOURCE: Robinson, John P., "I Love My TV." *American Demographics* 12, no. 9 (1990): 24–29.

Questions

a. Does there appear to be a direct relationship between hours of viewing and age? Explain.

b. Would you say there is a generation gap between you and your mother's generation, or you and your father's generation, relative to hours of TV viewing? Young women, use the table to compare yourselves to your mothers; young men, to your fathers.

16. What state (or metropolitan area) pays the highest wages? This is a question of interest to many of us. Some relevant statistics are available from the 1990 *Statistical Abstract*.

Labor Force, Employment, and Earnings (No. 669, Average Annual Pay, by State: 1986 to 1988)

| State | Average Annual Pay (Dollars) | | | Percent Change | | State | Average Annual Pay (Dollars) | | | Percent Change | |
|---|---|---|---|---|---|---|---|---|---|---|---|
| | *1986* | *1987* | *1988*[1] | *1986– 1987* | *1987– 1988* | | *1986* | *1987* | *1988*[1] | *1986– 1987* | *1987– 1988* |
| U.S. .... | 19,966 | 20,857 | 21,871 | 4.5 | 4.9 | MO....... | 18,915 | 19,601 | 20,295 | 3.6 | 3.5 |
| | | | | | | MT ...... | 16,085 | 16,438 | 16,957 | 2.2 | 3.2 |
| AL........ | 17,638 | 18,318 | 19,003 | 3.9 | 3.7 | NE ...... | 16,106 | 16,526 | 17,190 | 2.6 | 4.0 |
| AK ...... | 28,442 | 28,008 | 28,033 | −1.5 | .1 | NV ...... | 18,739 | 19,521 | 20,556 | 4.2 | 5.3 |
| AZ........ | 18,870 | 19,610 | 20,383 | 3.9 | 3.9 | NH ...... | 18,303 | 19,414 | 20,749 | 6.1 | 6.9 |
| AR........ | 16,162 | 16,529 | 17,023 | 2.3 | 3.0 | NJ........ | 22,309 | 23,842 | 25,748 | 6.9 | 8.0 |
| CA....... | 21,998 | 23,100 | 24,124 | 5.0 | 4.4 | NM....... | 17,301 | 17,767 | 18,259 | 2.7 | 2.8 |
| CO ...... | 20,275 | 20,830 | 21,472 | 2.7 | 3.1 | NY ...... | 23,200 | 24,634 | 26,347 | 6.2 | 7.0 |
| CT........ | 22,518 | 24,322 | 26,244 | 8.0 | 7.9 | NC ...... | 16,999 | 17,861 | 18,637 | 5.1 | 4.3 |
| DE ...... | 19,637 | 20,764 | 21,977 | 5.7 | 5.8 | ND ...... | 15,778 | 16,157 | 16,508 | 2.4 | 2.2 |
| DC ...... | 27,137 | 28,477 | 30,254 | 4.9 | 6.2 | OH ...... | 19,903 | 20,568 | 21,500 | 3.3 | 4.5 |
| FL........ | 17,680 | 18,674 | 19,520 | 5.6 | 4.5 | OK ...... | 18,345 | 18,615 | 19,098 | 1.5 | 2.6 |
| GA ...... | 18,745 | 19,651 | 20,501 | 4.8 | 4.3 | OR ...... | 18,321 | 18,888 | 19,637 | 3.1 | 4.0 |
| HI ....... | 18,101 | 19,091 | 20,444 | 5.5 | 7.1 | PA........ | 19,403 | 20,408 | 21,485 | 5.2 | 5.3 |
| ID ....... | 16,623 | 17,062 | 17,648 | 2.6 | 3.4 | RI ....... | 17,733 | 18,858 | 20,206 | 6.3 | 7.1 |
| IL ....... | 21,445 | 22,250 | 23,606 | 3.8 | 6.1 | SC........ | 16,603 | 17,280 | 18,009 | 4.1 | 4.2 |
| IN ....... | 19,024 | 19,692 | 20,437 | 3.5 | 3.8 | SD........ | 14,477 | 14,963 | 15,424 | 3.4 | 3.1 |
| IA ....... | 16,598 | 17,292 | 17,928 | 4.2 | 3.7 | TN ...... | 17,661 | 18,501 | 19,209 | 4.8 | 3.8 |
| KS....... | 17,934 | 18,424 | 19,030 | 2.7 | 3.3 | TX........ | 19,934 | 20,463 | 21,130 | 2.6 | 3.3 |
| KY ...... | 17,357 | 18,008 | 18,545 | 3.7 | 3.0 | UT ...... | 17,863 | 18,303 | 18,910 | 2.5 | 3.3 |
| LA........ | 18,290 | 18,707 | 19,330 | 2.3 | 3.3 | VT........ | 16,862 | 17,703 | 18,640 | 5.0 | 5.3 |
| ME....... | 16,326 | 17,447 | 18,347 | 6.9 | 5.2 | VA....... | 18,972 | 19,963 | 21,052 | 5.2 | 5.5 |
| MD....... | 20,121 | 21,324 | 22,500 | 6.0 | 5.5 | WA....... | 19,645 | 20,110 | 20,806 | 2.4 | 3.5 |
| MA ...... | 20,925 | 22,486 | 24,143 | 7.5 | 7.4 | WV....... | 18,402 | 18,820 | 19,341 | 2.3 | 2.8 |
| MI........ | 22,721 | 23,081 | 24,193 | 1.6 | 4.8 | WI........ | 18,202 | 18,890 | 19,743 | 3.8 | 4.5 |
| MN....... | 19,633 | 20,450 | 21,481 | 4.2 | 5.0 | WY ...... | 18,969 | 18,817 | 19,097 | −.8 | 1.5 |
| MS ...... | 15,420 | 15,938 | 16,522 | 3.4 | 3.7 | | | | | | |

[1] Preliminary.

Source: U.S. Bureau of Labor Statistics, *Employment and Wages, Annual Averages 1988*, and USDL News Release 89–415, *Average Annual Pay by State and Industry, 1988*.

SOURCE: U.S. Bureau of the Census, *Statistical Abstract of the United States: 1990* (110th ed.) Washington, D.C.: U.S. Government Printing Office, 1990.

Questions
a. Find the average annual pay for your state in 1988.
b. For 1987, what state had the highest average annual pay?
c. What population does this summary table describe?
d. Is the average annual pay for a U.S. citizen in 1988—$21,871—a statistic or a parameter?
e. Can one state be a sample to describe the United States? yes, but it would be biased

17. Is the 1990 decennial census really accurate? It is well known that the 1990 census, and every earlier decennial census, was subject to undercount. (Undercount produces a number that is lower than the actual value.) Yet accuracy is important because the results are used for political redistricting and for allocation of federal funds.

SOURCES: Savage, Richard I. "Who Counts?" *The American Statistician* 36, no. 3 (1982): 195–200.
U.S. Bureau of the Census. *Census '90 Basics.* Washington, D.C.: U.S. Government Printing Office, 1990.

Question   From your knowledge of the 1990 census, describe two or three actions that might have been taken by the Bureau of the Census to assure as accurate a count as possible.

---

## BUSINESS CASE

*Jamie Gets a Job*

Economic conditions notwithstanding, Jamie Lewis was reporting for work. After four years (plus the last two summers) of meeting university and business school requirements, she had been ready for a vacation. But when the offer came for joining Dexter Building Products as a management trainee, she jumped at the chance. And the salary was pretty good! After all, she had chosen some of the tougher courses, and now it was paying off.

Jamie's new boss, Herman Horton, had promised her an easy first few months. Mr. Horton was in charge of corporate training, and his staff was often called on to do special projects.

Jamie's first day was to be spent getting acquainted with some of the major departments of the company. Their initial visit was with Stephanie Jorgenson, the assistant marketing manager.

**Mr. H.**   Stephanie, I'd like you to meet our new employee, Jamie Lewis.

**Ms. J.**   Pleased to meet you, Jamie. Herman said he was getting some more help—but for the life of me I can't see why! (turning to Mr. Horton) Just kidding, Herm. What will Jamie be doing?

**Mr. H.**   She's a little different from some of the other trainees in that she's not specifically a marketing, finance, or personnel major. She'll be in my staff group for about a year or so—like the previous trainees. By that time we expect she'll find a home somewhere—maybe with you, production, finance, even engineering—who knows?

**Ms. J.**   Really! What are you interested in Jamie?

**Jamie**   I am not really sure yet. I mean, I like lots of things. Just solving problems is the most interesting, I guess.

**Ms. J**   We've got plenty of them! In fact, my day seems to be nothing but one problem after another. Last month our Midwest sales manager grossly overestimated the demand for insulation for the Chicago market area—and it wasn't the first time! I told him to plan for about the same demand as last year, but he had a hunch something big would break. Well, as it turned out, we needed the insulation in Boise instead. With the extra shipping and handling costs, our profits in that product line are shot for this quarter!

**Jamie**   Do you have any historical data on demand?

**Ms. J.**   Oh, yes—certainly. The problem isn't having enough information. It's getting it into some useful form and getting those "crystal ball gazers" in our field sales force to use it.

**Jamie**   Have you ever run any correlations of your insulation demand with any construction indicators, like quarterly building permits, or anything like that?

**Ms. J.**    I'm not exactly sure. Jim Carpenter, our marketing analyst, started on something a couple of years ago. But then he got promoted to a district-level job in Seattle before it was finished. He did leave a bunch of numbers and equations in the file but nothing ever came of it, as I recall. Are you interested in that kind of thing—numbers and computer printouts?

**Jamie**    Not really in numbers per se. But I do like working with them when they're useful. Sometimes they're more meaingful than words.

**Ms. J.**    If I didn't know it was your first day I'd swear you were writing a speech for our executive vice-president, Ms. Courtney. That's just what she's always preaching—the precision of the bottom line.

It's nice to meet you, Ms. Lewis. I took forward to having you with us.

The next visit was in the Personnel Department, with the manager, Tom Crandall. After brief introductions, Mr. Crandall began telling Jamie about some of their current activities.

**Mr. C.**    The hottest thing at the moment seems to be the company-wide wage plan. Ms. Courtney wants a step-type plan with 90% of the employees within plus or minus 20% of the average.

**Jamie**    That's a problem?

**Mr. C.**    It is when the steps are letter graduations—A, B, C, and so on—and 20% for wage group H is quite different from 20% for group A. Oh, we can figure them all out individually, but it takes time to do that. Ms. Courtney wants some kind of quick reference model for collective bargaining negotiations. Sometimes she's got to be able to come up with overall wage costs pretty quickly or the opportunity for a settlement is lost.

**Jamie**    Are you trying to develop an expression for an average?

**Mr. C.**    Well, there's more to it than that, Ms. Lewis. The average changes with each pay step, and she wants some kind of formula to cover them all at once.

**Jamie**    I'm sorry. I meant to say a moving average time series analysis. I was thinking of a least squares regression to summarize the data. Would that do the trick?

**Mr. C.**    Maybe so. I'm not all that familiar with quantitative procedures. Most of my employees are "people" people. We're not really into the equations like the engineers are. I've told Ms. Courtney that this kind of stuff really borders on engineering, but she disagrees. Anyway, we can do without engineers taking over the personnel area. By the way, Jamie, what's your specialty?

**Jamie**    My degree is in business with a major in management science. It emphasized a broad base of management topics that included a solid core of mathematics and statistics.

Following this conversation, Jamie and Mr. Horton stopped in to meet the financial vice-president, Stanley Fairfield. His major area of concern at present was analyzing some corporate financial figures, which appeared to be out of line.

**Jamie**    You mean you haven't been developing your financial model with in-house people?

**Mr. F.**    We thought it would be too complex at first. We had hoped to bring in Arlene Campbell from the controller's office by now. However, the model isn't far enough along, and Arlene's got enough work to keep her busy right where she is.

**Jamie**    If I understand you correctly, you just want to simulate the revenues and costs on a computer so that you can experiment with the effect of different overhead liquidation rates. Is that right?

| | |
|---|---|
| **Mr F.** | That's very good. That's exactly what the model will do when it's finished. Then we can analyze the costs, compute some averages, and so on. |
| **Jamie** | Has anyone tried to model this on a microcomputer? I wonder if a spreadsheet package might work nicely for that. We used it for simulations in an advanced finance class. It's neat because you can watch everything on a terminal and pull off statistical summaries at the touch of a print key. |
| **Mr. F.** | That may be worth exploring. |
| **Mr. H.** | We've got to run, Stan. I promised Jamie I'd take her to lunch at Maxwell's Restaurant—this being her first day and all. |

After lunch, Jamie had some more get-acquainted meetings scheduled. Mr. Horton's assistant, Carol Baker, was to take Jamie to visit with the production, legal, and engineering departments.

When they returned from lunch, Mr. Horton had three messages on his desk. He was to call Stephanie Jorgenson and Stan Fairfield right away, and Tom Crandall was anxious to set up a meeting for tomorrow morning. Mr. Horton knew what the calls were about without even asking. Before he turned Jamie over to Ms. Baker, he drew Ms. Baker aside and whispered quietly, "We do want to keep Jamie in this training program for at least two months, Carol. Don't get her too involved in any one department. Just introductory stuff, you know. Say, she might be interested in walking over to the shipping department. It would take some time getting over and back, but I think you could include that in the afternoon if you hurry a little."

**Business Case Questions**

18. Why did Mr. Horton receive messages from Stephanie Jorgenson, Stan Fairfield, and Tom Crandall?

19. Can an effective manager restrict himself or herself to being a "people person"? Why or why not?

**COMPUTER PROBLEMS**

Definition of Our Data Base

The decennial census is a recurring event that first took place in the United States in 1790. The census was decreed by our Constitution to "determine for the nation, essential information about its people." More recently, the enormous task of compiling census data, which took many months for the 1920 and 1930 censuses, led to some statistical sampling in the 1940 census [3].* By 1970 most of the information from the census came from samples. In 1980 the only complete counts were the count of the people—with limited information on race, the relationship of the people in the household (including marital status, age, family, or other relationships)—and the basic features of the residence. Numerous other variables were collected as sample data.

More detailed information was obtained from samples of 5%, of 1%, and of 0.1% of the population. These data are referred to collectively as the *Public Use Microdata Samples* (PUMS). We have extracted some information from the 5% sample; the data records appear in Appendix A. Additional details about the PUMS are available from [1].

The original Bureau of Census/PUMS survey contains 193 spaces for information about the household (H) followed by 193 spaces available to describe each person (P) within the household. Each computer record includes one (H) description followed by one or more (P) descriptions. Every household description

* Bracketed numbers refer to items listed in "For Further Reading" at the end of the chapter.

includes size, age, value, cost of utilities, and more. Each person is described by age, sex, race, education, and work characteristics. We have selectively reduced the PUMS format to include a record number, followed by 34 positions of household and 57 positions for describing up to six persons in the household. The record layout (with variable names) and descriptions appear in Appendix A. For those records that include fewer than six persons, some positions are coded to blanks. Each record is complete; there are no missing data.

The data base is available in two forms. A sample of 100 records appears in Appendix A. These data are used as much as possible for each chapter's computer problems. A full data base of 20,116 records is available on a computer file from the publishers.

You may well ask how our data base is integrated into this work. Recall that a data base is a collection of information essential to the operation of a business. It often includes accounting information, sales data, production and personnel information, and operational expenditures. The objective is to have one source of information to assist in making management decisions. The data base in this textbook is a collection of economic information useful to many businesses. We use it to illustrate some of the more important statistical concepts through chapter exercises labeled Computer Problems. Although our data base is only a small part of the 5% PUMS sample, it is still substantial. That is why we emphasize the use of computers for problem solving.

Figure 1.5 is a Minitab summary that describes the record layout for the PUMS data base.

**FIGURE 1.5**  A Minitab Computer Run That Explores the Format of the PUMS Data Base

```
MTB >PAPER
MTB > READ 'B:SMPL1.PUM' C1-C29; ←——————————————————————————————————1
SUBC> FORMAT (F5.0,F2.0,F4.0,F2.0,F1.0,F2.0,4F1.0,2F3.0,F4.0,F3.0,F1.0,  &
CONT> F5.0,2(F2.0,F1.0),2F2.0,F1.0,2F2.0,F5.0,F2.0,F1.0,F2.0).
     100 ROWS READ
* 69 blank fields converted to *
```

| ROW | C1 | C2 | C3 | C4 | C5 | C6 | C7 | C8 | C9 | C10 | C11 | C12 |
|-----|-----|-----|------|-----|-----|-----|-----|-----|-----|-----|-----|-----|
| 1 | 181 | 39 | 1680 | 2 | 1 | 2 | 4 | 3 | 2 | 5 | 33 | 27 |
| 2 | 337 | 39 | 80 | 1 | 2 | 6 | 2 | 2 | 2 | 2 | 0 | 30 |
| 3 | 431 | 39 | 1320 | 4 | 1 | 2 | 4 | 3 | 3 | 2 | 78 | 0 |
| 4 | 499 | 39 | 4320 | 1 | 2 | 7 | 2 | 2 | 1 | 1 | 0 | 0 |

· · ·

| ROW | C13 | C14 | C15 | C16 | C17 | C18 | C19 | C20 | C21 | C22 | C23 |
|-----|-----|-----|-----|-------|-----|-----|-----|-----|-----|-----|-----|
| 1 | 100 | 0 | 1 | 13320 | 0 | 0 | 64 | 0 | 1 | 10 | 6 |
| 2 | 0 | 135 | 4 | 9005 | 0 | 1 | 31 | 4 | 1 | 14 | 1 |
| 3 | 197 | 0 | 1 | 18425 | 0 | 0 | 33 | 0 | 1 | 14 | 1 |
| 4 | 0 | 165 | 4 | 3365 | 0 | 1 | 36 | 2 | 1 | 13 | 1 |

· · ·

(*continued*)

**FIGURE 1.5**    (*continued*)

| ROW | C24 | C25 | C26 | C27 | C28 | C29 |
|-----|-----|-----|------|-----|-----|-----|
| 1 | 21 | 40 | 13320 | 1 | 1 | 61 |
| 2 | 52 | 46 | 9005 | * | * | * |
| 3 | 52 | 40 | 18240 | 1 | 1 | 33 |
| 4 | 48 | 42 | 3365 | * | * | * |

. . .

```
MTB > NAME C1 'RECORD' C2 'STATE' C3 'MSA' C4 'PERS' C5 'TENURE' C6 'UNITS'     &
CONT> C7 'BDRMS' C8 'BATHS' C9 'AUTOS' C10 'YRMVDIN' C11 'ELEC' C12 'GAS'       &
CONT> C13 'MOPMT' C14 'MORENT' C15 'HSHDTP' C16 'HSHDINC' C17 'RELATN1'         &
CONT> C18 'SEX1' C19 'AGE1' C20 'MARITL1' C21 'RACE1' C22 'EDN1' C23 'WRKST1'   &
CONT> C24 'WKSWRKD1' C25 'GRSWRKD1' C26 'INDVINC1' C27 'RELATN2'                &
CONT> C28 'SEX2' C29 'AGE2'
MTB > INFO
```

| COLUMN | NAME | COUNT | MISSING |
|--------|------|-------|---------|
| C1 | RECORD | 100 | |
| C2 | STATE | 100 | |
| C3 | MSA | 100 | |
| C4 | PERS | 100 | |
| C5 | TENURE | 100 | |
| C6 | UNITS | 100 | |
| C7 | BDRMS | 100 | |
| C8 | BATHS | 100 | |
| C9 | AUTOS | 100 | |
| C10 | YRMVDIN | 100 | |
| C11 | ELEC | 100 | |
| C12 | GAS | 100 | |
| C13 | MOPMT | 100 | |
| C14 | MORENT | 100 | |
| C15 | HSHDTP | 100 | |
| C16 | HSHDINC | 100 | |
| C17 | RELATN1 | 100 | |
| C18 | SEX1 | 100 | |
| C19 | AGE1 | 100 | |
| C20 | MARITL1 | 100 | |
| C21 | RACE1 | 100 | |
| C22 | EDN1 | 100 | |
| C23 | WRKST1 | 100 | |
| C24 | WKSWRKD1 | 100 | |
| C25 | GRSWRKD1 | 100 | |
| C26 | INDVINC1 | 100 | |
| C27 | RELATN2 | 100 | 23 |
| C28 | SEX2 | 100 | 23 |
| C29 | AGE2 | 100 | 23 |

```
CONSTANTS USED: NONE

MTB > COUNT C27
   COUNT   =     100
    N GOOD =    77  N MISSING = 23
```

**FIGURE 1.5**  (*continued*)

```
MTB > MINIMUM 'RECORD'
   MINIMUM =      181.00
MTB > MAXIMUM 'RECORD'
   MAXIMUM =        19899


MTB > MINIMUM 'STATE'
   MINIMUM =      1.0000
MTB > MAX 'STATE'
   MAXIMUM =      55.000
MTB > MIN 'PERS'
   MINIMUM =      1.0000
MTB > MAX 'PERS'
   MAXIMUM =      12.000
MTB > MEAN 'PERS'
   MEAN    =      2.7300
MTB > STOP
```

5

The following descriptions can help you understand the Computer Applications in subsequent chapters and use the data base with computer packages to solve business decision problems. The Minitab computer run in Figure 1.5 demonstrates how to read a data file and shows how Minitab provides a check that the data have been read as you intended. Finally, it illustrates how you can do some "data snopping" to understand the data.

If you want a printed copy of everything that is displayed on your CRT screen, then at the beginning add the command MTB > PAPER.

1 These data were read from the data disk 'SMPL1. PUM' available from the publisher). If yours is a hard-disk drive system and the file was installed on the hard drive, then read from the C drive.
Notice that the variables are in an F-type FORMAT because all their values are numerical.

Listing a few records for a data set enables us to do a good data check. For example, compare the ROW 2 values against the second record displayed in Table 1 of Appendix A. The values are identical. The entries * for C27–C29 for record 337 indicate that there is no person #2, so too no persons #3 through #6. This is consistent with the value of C4 for variable 4—#persons—which is 1, and with Appendix A, which is blank for Person 2. The data appear to have been read correctly.

2 A NAME is assigned to each variable, C1 through C29. C1, the record number, occupies columns 1–5. C2 is the state code, occupying columns 6 and 7, and so on. See Figure A.1 and Table A.2 in Appendix A for more definitions of assignments. Minitab allows up to eight characters in the name of each variable. The names we give these variables here are the ones we will use throughout the textbook.

3 Notice the assignment of names to the variables (which are displayed in columns C1–C29). Hence the labels COLUMNS and NAMES along with the record COUNT. Only columns C27, C28, and C29 have no values. Blanks occur when there are fewer than six persons on the record.

**4** Commands for data snooping, that is, for viewing specific pieces of data or a statistic (variable description), appear in the Minitab student manual under "Summarizing the Data in Columns." Some of these commands are illustrated here. These displays include COUNT for C27 (RELATN 2), with 77 good records and 23 missing. The MINImum and MAXImum record numbers are 181 and 19899, respectively. Notice that we have provided MIN, MAX, and arithmetic MEAN values for the variable #PERSONS,' but only MIN and MAX values for 'STATE.' This is because STATE is a qualitative rather than quantitative variable, and the mean is not an appropriate descriptor for it.

**5** To close a Minitab session, simply type in the statement MTB > STOP.

We have intentionally not used the data base information in the text examples. You are invited to explore the definitions of these variables further in the Computer Problems section that follows. The Computer Problems, which use the data base information, can become real applications for testing your understanding of key business statistics concepts. They represent problems similar to ones you may encounter in business.

Chapter 1 has laid a foundation for the remainder of the book by describing some first principles of statistics. Similarly, you will need to understand the variables in the data base in order to work the computer problems in the succeeding chapters. Thus you will want to explore the data base file structure as described in the computer problems for this chapter. These problems cover exploring the record layout, viewing the kinds of variables that occur, listing some of the codes for specific variables, and experiencing some computer manipulations required to access the data. The computer problems in Chapter 1 can be answered from the 100 sample records that appear in Appendix A.

## COMPUTER APPLICATION

*Data Base Applications*

These problems will help introduce you to the data in the data base in Appendix A. You may want to key some of this data into your own computer disk files (for microcomputers) or other such storage. First you will need to understand the data. It will be helpful if you first refer to Table A.1 in Appendix A and the variable definitions that follow it.

20. The codes from the household information portion of a PUMS record are listed in Table A.2: Data Dictionary.
    a. Find these values on the listing of record #181 in Appendix A Table A.1.
    b. Use the data dictionary in Appendix A to report a value for each household information code and use this to describe the household. For example, Household type = 1 is a married couple family.

| Record #181 | No. of Baths | 3 | Monthly cost of electricity | 33 |
| Persons 2 | No. of Autos | 2 | Monthly payments to lender | 100 |
| Bedrooms 4 | Yr. moved in | 5 | Household type | 1 |

21. The codes for some variable descriptions on two persons in record #181 are listed in the table.
    a. Find these values on the listing of record #181 in Appendix A.

|          | Relationship | Sex | Age | Highest Year of School | Usual Hours Worked |
|----------|--------------|-----|-----|------------------------|--------------------|
| Person 1 | 00           | 0   | 64  | 10                     | 40                 |
| Person 2 | 01           | 1   | 61  | 10                     | 00                 |

    b. Use the data dictionary (Appendix A) to report value(s) for each of the person codes, and use it to describe these persons in this household.

22. Consider the two *person* descriptions in record #181.
    a. What is the family relationship?
    b. Indicate the sex for each person.
    c. What is the highest year of school attended by the younger person?
    d. Compare the household income against income from all sources for individual persons. Do these numbers agree?

23. Some of the data on each housing record require decoding; that is, the values on the records are not the values one would use for a statistical description. For example, state code 01 represents Alabama and tenure code 1 indicates an owner-occupied housing unit. List those housing-record variables that require decoding for statistical description. (*Hint:* Most will fall into this classification.)

24. Search through the 100 records in Appendix A, beginning (each time) with the first listed record. Indicate the record number that first appears that satisfies each of the following categories.
    a. A household with two children (for a child under 17 years)
    b. A household with household income that exceeds $30,000
    c. A female head of household
    d. A household with four or more persons

25. Key the first two records in Appendix A into a personal computer. Have a classmate ask you to identify codes from the screen for some variables. Have this classmate explain these codes from the data dictionary in Appendix A.

**FOR FURTHER READING**

[1] *Census of Population and Housing 1980: Public Use Microdata Samples Technical Documentation*, Washington D.C.: U.S. Department of Commerce, Bureau of the Census, 1983 and 1987.

[2] Groebner, David F., and Patrick W. Shannon. *Business Statistics: A Decision Making Approach*, 3rd ed. Columbus, MO: Charles E. Merrill, 1989.

[3] Hansen, Morris. "How to Count Better: Using Statistics to Improve the Census." In *Statistics: A Guide to the Unknown*, 2nd ed. Edited by Judith Tanner, F. Mosteller, et al. San Francisco: Holden-Day, 1982.

[4] Iman, Ronald L., and W. J. Conover. *Modern Business Statistics*, 2nd ed. New York: John Wiley, 1989.

[5] Lapin, Lawrence L. *Statistics for Modern Business Decisions*, 5th ed. San Diego: Harcourt Brace Jovanovich, 1990.

[6] Levin, Richard I., and David S. Rubin. *Business Statistics*. Englewood Cliffs, NJ: Prentice-Hall, 1983.

[7] McClave, James T., and P. George Benson. *Statistics for Business and Economics*, 4th ed. San Francisco: Dellen, 1988.

[8] Singer, Judith D., and John B. Willett. "Improving the Teaching of Applied Statistics: Putting the Data Back into Data Analysis." *The American Statistician* 44, no. 3 (1990): 223–230.

# DESCRIPTIVE STATISTICS

# 2

# DESCRIPTIVE
# STATISTICS

## 2.1
## INTRODUCTION

Bill Howe, president of Skyline Developers, called his company's statistician with the following request:

> John, I need to see those maps of Houston's Delta residential area. Bring over those population counts and average age for heads of household too. We need to put some statistics together for the city council meeting tomorrow. You know they're expecting us to make some logical arguments in favor of the Delta Center. We're going to have to come up with a description of the people who live in that area first and we haven't much time!

Bill Howe is using simple statistics to draw conclusions. His purpose is to *describe* a group of people, using certain characteristics as a basis for making some business decisions.

In Chapter 1 we said that statistics is concerned with drawing conclusions from data. More specifically, statistics consists of collecting, presenting, analyzing, and interpreting data for the purpose of making decisions. The simplest form of this process is called *descriptive statistics*.

> Descriptive statistics includes summaries derived from sample data that are used to condense, summarize, or graphically illustrate the data.

The descriptive statistics Bill Howe will use are averages (measures of location), measures of variation, and diagrams that will describe the available data on the residents of Houston's Delta area.

When we use any data for making decisions, we are using statistics. Thus Bill Howe could simply count the new housing starts in Houston's Delta area

**TABLE 2.1** Commonly Used Descriptive Statistics

| Measures of Location | Measures of Variation |
|---|---|
| Mean | Range |
| Median | Mean absolute deviation |
| Mode | Standard deviation |
| Weighted mean | Variance |
| | Coefficient of variation |

and use that number as an informal statistical indicator of the need for a Delta-area shopping center. However, for business decision making it is desirable to formalize such relatively casual procedures so that they are statistically valid. In this chapter we discuss systematic ways of describing and presenting data.

The two principal classes of descriptive statistics discussed in this chapter are *measures of location* and *measures of variation*, as outlined in Table 2.1.

Measures of location—averages—describe data by providing a central tendency (location) value for the data.

Measures of variation describe data by indicating the extent of the differences between the values of a data set.

Measures of location specify a central value (called a measure of central tendency) in a set of data. Measures of variation describe the dispersion or spread of the data.

Upon completing Chapter 2 you should be able to

1. distinguish among the measures of location and measures of variation;
2. compute values for common measures of location, including the arithmetic mean, median, and mode of a set of data;
3. determine the range, mean absolute deviation, standard deviation, and variance as measures of variation for a set of data;
4. compute a coefficient of variation and use it as a measure of relative variation;
5. explain the meaning of Chebyshev's rule and know when it applies;
6. use box plots to describe data.

These statistics are most frequently used for describing business data. Because measures of variation build upon measures of location, we begin with a discussion of location.

## 2.2 MEASURES OF LOCATION

As a measure of location, or central tendency, the average is one of the most widely used descriptive measures of a set of data.

Many averages exist, but only a few are widely used. The appropriate one to use depends upon the intended business application and the nature of the data. The most often used averages are the arithmetic mean, the median, the mode, and a generalization of the arithmetic mean known as a weighted mean.

**Arithmetic Mean**

The average familiar to most people is the *arithmetic mean*. You already know how to find an arithmetic mean. If you live in a cooperative house, for example, you could add up everyone's grocery bills at the end of the month and divide the total amount by the number of residents to find each person's share of the cost.

> The arithmetic mean is a location measure that describes the values in a data set by equal parts of their total. It is computed as the sum of the individual values divided by their number.

To formalize the computational process, we represent the individual data values by $X_i$. Then $X_1, X_2, \ldots, X_i, \ldots, X_n$ represent the values for some variable defined on a data set, and the arithmetic mean is designated as $\bar{X}$ (read "X bar"), where

$$\bar{X} = \frac{X_1 + X_2 + \cdots + X_i + \cdots + X_n}{n}$$

We can write this expression more concisely by showing the sum of $X_1, X_2, \ldots, X_i, \ldots, X_n$ in a different way:

$$\bar{X} = \frac{\sum_{i=1}^{n} X_i}{n}$$

The summation sign $\sum_{i=1}^{n}$ means that the values are summed for a variable—$X$—beginning with a first value, $X_1$, which is indexed as $i = 1, \ldots, n$, where $n$ is the number of observations.* We could use this detailed notation, but for simplicity we omit the subscripts and express the arithmetic mean using less notation.

The *simple arithmetic mean* for a variable $X$ is designated as $\bar{X}$ for sample data and as the Greek letter $\mu$ (pronounced *mu*) for population data.

> For sample data, the simple arithmetic mean is calculated as

$$\bar{X} = \frac{\sum^{n} X}{n} \quad \text{for } n = \text{sample size} \tag{2.1}$$

> For population data, the simple arithmetic mean is calculated as

$$\mu = \frac{\sum^{N} X}{N} \quad \text{for } N = \text{population size} \tag{2.1a}$$

The mean, $\bar{X}$ (or $\mu$), takes a value that depends upon the values for $X$. Because this chapter is about statistical description of sample data, we use $\bar{X}$. Example 2.1 shows one use of the arithmetic mean.

* See "Appendix C: Math Essentials" for summation notation and rules.

**EXAMPLE 2.1**

An investment manager is summarizing the dividend per share paid by one stock over the most recent six quarters. The actual values are $X_1 = \$2.00$, $X_2 = \$3.50$, $X_3 = \$0.40$, $X_4 = \$0.90$, $X_5 = \$2.50$, and $X_6 = \$0.30$. What is the mean dividend per share computed over the six quarters?

*Solution*  Letting $X$ denote the dividend per share, then for $X_1$ through $X_6$ as given,

$$\bar{X} = \frac{\sum X}{n} = \frac{\$2.00 + 3.50 + 0.40 + 0.90 + 2.50 + 0.30}{6} = \frac{\$9.60}{6}$$

$$= \$1.60/\text{share}$$

The mean dividend per share is $1.60. This value is one-sixth of the total per-share dividend ($9.60) paid over the most recent six quarters.

---

The arithmetic mean is also appropriately used to average dollars of sales, ages of insurance policy holders, or scores on exams—in short, to handle data for which the actual values can be meaningfully subdivided into fractional parts of a whole number.

## Median

The *median*, another measure of location, usually does not have the same value as the mean for a given set of data. Instead, it partitions a data set into halves by locating a midpoint value that equals or exceeds the values of 50% of the data. The median is also called the second quartile point.

> The median is a value at the middle position when the data values are arranged in order by size.

Like the arithmetic mean, the median does not have to be one of the original data values. When the number of observations is odd, the median is simply the value in the middle position. For instance, the median of 1, 5, 6, 9, and 20 is 6. When the number of observations is even, the median is a number halfway between the two middle values. Thus for 1, 5, 6, 8, 9, and 20, the median is halfway between 6 and 8, or 7.

An easy way to determine *the position* of the median is to add 1 to the number of observations and divide by 2. That is, the median is at position $(n + 1)/2$. Consider Example 2.2.

---

**EXAMPLE 2.2**

The scores of eight accountants taking a CPA exam were 87, 68, 79, 90, 76, 82, 56, and 81. What was the median score?

*Solution*  There are eight scores. The median score is the value located at position $(n + 1)/2$. Since $(8 + 1)/2 = 4\frac{1}{2}$, the median is halfway between the fourth and fifth scores *as they appear in order of size*. The ordered scores are 56, 68, 76, 79, 81, 82, 87, and 90. Their median is halfway (the mean) between the fourth and fifth scores:

$$\text{Median} = \frac{79 + 81}{2} = 80$$

Thus the median is 80. Half of the accountants (4) had scores above 80; half had scores below 80.

Remember that the median denotes a positional value. When $n$ is even, as in Example 2.2, the median need not take on any of the observed values.

The median is an appropriate average for data that display a natural order. For example, the decennial census of population in the United States records the median age for males in each census area. Since ages are commonly recorded by year of birth, the numbers are readily ordered and the median is an appropriate average.

The method of partitioning a set of data into halves can be extended to fourths (quartiles), tenths (deciles), and hundredths (percentiles). These subdivisions are commonly used for ratings or rankings. A student's class standing, for example, might be recorded as being in the first quartile, the second decile, or even more precisely at a specified percentile. Similar ratings are used for comparing the market shares held by leading brands of soaps, appliances, soft drinks, and other products. Quartiles, deciles, and percentiles are called *quantiles*. They extend the concept of partitioning to indicate an order within parts of four, ten, or one hundred, respectively.

**Mode**

Another common average is the *mode*.

The mode of a set of data is the value that occurs most frequently.

As shown in Example 2.3, a set of data may have no mode, one mode, or several modes.

---

**EXAMPLE 2.3**

Find the mode or modes of the following data sets.
a. 1, 3, 3, 3, 4, 5, 6
b. 1, 3, 3, 3, 4, 4, 4
c. 1, 1, 1, 3, 3, 3, 4, 4, 4
d. brand A, brand A, brand B, brand C

*Solution*   Look for the value that occurs most often in each set.
a. The mode is 3.
b. There are two modes: 3 and 4.
c. There is *no* mode, since no value occurs more frequently than the others.
d. The mode is brand A.

---

The mode is the average that is commonly reported for opinion and market research surveys. The purpose of these surveys is to measure market identity. For instance, a study of brand advertising effectiveness might ask 100 television viewers, "What brand of hair shampoo do you first recall from your viewing of television ads?" The most frequently mentioned brand name is the mode.

The mode is a coarser average, or measure of location, than the mean and median. Because the mode is the most frequently occurring value, it requires only that the data fit into identity classes, such as catalog versus store or telephone purchase. It is necessary neither to order the data, as we must do for the median, nor to scale the data values, as we must do for the mean. Furthermore, it is the only appropriate average for data that cannot be ordered or scaled. Nevertheless, although the mode is often the easiest average to determine, sometimes there is no single most common value. This disadvantage limits its usefulness as an average.

**Measurement Classification Scales**

Let's examine these three common averages a bit further. How would we know which is the most appropriate measure to describe a specific set of data? We should first consider the *level of measurement*. Measurement refers to the process of assigning values to the data.

> The level of measurement of a data set—nominal, ordinal, interval, or ratio—refers to the properties of values assigned to the data. These levels are determined by the properties of order, distance, and a fixed zero.

The properties of order and distance determine the level of measurement. There are four different levels of measurement: nominal, ordinal, interval, and ratio, as illustrated in Table 2.2.

The *nominal* level is the *lowest* level of measurement. No assumptions are made about the values that are assigned to the data. In fact, the values serve only as labels or names of categories, such as blue eyes and male. There is no meaningful order or distance between categories. The mode is an appropriate average for nominal data.

An *ordinal* level of measurement is achieved when it is possible to rank-order all the categories according to some criterion. Collegiate class—freshman, sophomore, junior, senior—is an ordinal measurement. Each category has a unique position relative to the other categories; each category is either lower in value than some categories and higher than other categories or it is one of the extreme categories. However, the characteristic of ordering is the sole mathematical property of this level. It is not possible to determine by *how much* the categories differ (the distance between them). Either the mode or the median is appropriate to describe ordinal data. However, since the median makes use of order, it is the most appropriate descriptor for ordinal data.

In addition to ordering, data having the *interval* level of measurement have the property that the distances between categories are defined by fixed and equal units. Temperature scales are the classical example of interval level measurement. For example, a change from 0°F to 1° is the same *difference* as the change from 30°F to 31°. Therefore, the interval level of measurement has order and distance. However, 0°F does not mean the absence of heat, so there is no fixed zero point.

The *ratio* level is the highest level of measurement; it has the additional property that a zero point is naturally defined. Examples are physical measures, including distances, miles per hour, dollars of sales, and stock sales prices. Data with equal units and a natural zero point allow meaningful ratio comparisons to be made. Thus it is quite reasonable to say that production of 20

**TABLE 2.2** Characteristics of Measurement Scales

| Type of Scale | Example | Data Ordered? | Fixed Distance between Data Points? | Fixed Zero? | Appropriate Measures of Location |
|---|---|---|---|---|---|
| 1. Nominal | Colors (red, blue) | No | No | No | Mode |
| 2. Ordinal | Class (Jr., Sr.) | Yes | No | No | Mode, median |
| 3. Interval | Temp. (°F) | Yes | Yes | No | Mode, median, mean |
| 4. Ratio | Sales ($) | Yes | Yes | Yes | Mode, median, mean |

**FIGURE 2.1** Measurement Levels and the Appropriate Averages

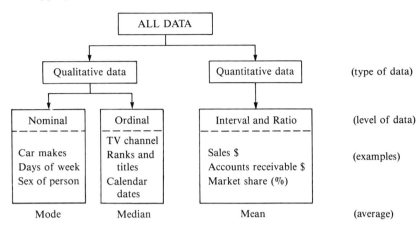

units per hour (ratio level) is twice the production of 10 units per hour. However, it is not reasonable to say that 60°F (interval level) is twice as warm as 30°F. The mode, median, and mean can be used to describe interval and ratio data. However, because it makes use of both ordering and distance properties, the arithmetic mean is the most appropriate average for interval and ratio data.

A simple scheme for levels of measurement is to distinguish variables as *qualitative* and *quantitative*, as shown in Figure 2.1. All quantitative variables have a fixed unit of measurement (indicating order *and* distance). Interval and ratio level data are quantitative. Then the arithmetic mean is normally appropriate. All other variables, such as those describing nominal and ordinal data, are qualitative. Here the median or the mode is appropriate. In determining the appropriate average, always consider the level of measurement of the data. Keep in mind, however, that secondary considerations can sometimes determine the preferred average. For example, consider the need to have an "average" salary in a small company of ten employees. The company president's salary may be three or four (or more) times that of the nearest salary of any other employee. If one were to compute a mean, it could be pulled or skewed toward a larger value than would be representative of the bulk of the employees. The median would be a more appropriate average in this situation. Remember, a lower-level average can always be substituted for a higher-level average. Throughout this textbook we use the word *average* as synonymous with arithmetic mean.* If the median or mode is desired, this will be stated.

**Weighted Mean**

The *weighted mean* is a generalization of the simple arithmetic mean. Each value is given its designated share (or weight) in determining the average. For example, if you buy three shares of a stock you will get (a weight of) three votes at a shareholder's meeting, whereas if I have one share, my vote (weight) is only one; the votes are weighted by the number of shares held. The weighted mean is appropriate when the components being averaged have unequal parts or value, which introduces the need for weights. In an election, where each citizen

*some values in summation are more important than others* [handwritten margin note]

---

* Other averages, such as the geometric mean and the harmonic mean, are also useful in specific situations, such as averaging rates of change in economic variables, but they are outside the scope of this textbook.

**TABLE 2.3**    A Grade Record

| Course | Grade | Grade Points $X$ | Credit Hours $W$ | Weighted Points $(X) \times (W)$ |
|---|---|---|---|---|
| Health | C | 2 | 3 | 6 |
| Political Science | B | 3 | 2 | 6 |
| Statistics | A | 4 | 4 | 16 |
| Russian | A | 4 | 3 | 12 |
| Biology | B | 3 | 4 | 12 |
| Total | | | 16 | 52 |

$$\text{Grade-point average} = \frac{\sum(\text{Weighted points})}{\sum(\text{Credit hours})} = \frac{52}{16} = 3.25$$

has one (equal) vote, the weights are all equal and the weighted mean reduces to the simple arithmetic mean.

The *weighted mean* is a generalization of the simple arithmetic mean. Consider a grade-point average with grades as shown in Table 2.3. Under a 4-point grading system, each credit hour of A work gets 4 points, a B gets 3, a C gets 2, and so forth. To determine the grade-point average, you divide the total weighted points (grade value times credit hours) by the total number of credit hours, or weights.

Although the items being averaged are the student's grade points, each grade point must be weighted by the number of hours per course. Using course hours as weights takes into account the fact that the student spends more time in some classes than in others.

A weighted mean is an arithmetic mean for which each value ($X$) is weighted ($W$) according to some well-defined criterion.

The weighted arithmetic mean is calculated as

$$\bar{X}_w = \frac{\sum XW}{\sum W} \tag{2.2}$$

To compute the weighted arithmetic mean, first answer the question, What is being averaged? The answer is the variable $X$. The relative importance assigned to each value of $X$ is indicated by the weights, $W$. Notice how the values ($X$) and the weights ($W$) are defined in the business context of Example 2.4.

**EXAMPLE 2.4**

As a part of an analysis of the survival of small businesses, the U.S. Small Business Administration wants to estimate the interest rate that a single small business is paying on its loans. The firm selected has three loans outstanding: Loan 1 is for $5000 at a 10% annual rate of interest; loans 2 and 3 are for $2500 and $7500, respectively, both at 15% interest. What *average rate of interest* is the firm paying on its loans? The current rate will be compared with rates for the two previous years—15% and 14.1%. Based on the cost of money, is this firm now in a better or a worse position relative to the two previous years?

*Solution*    The solution involves an average of the rates of interest for three loans of unequal amounts. We therefore use a weighted mean with $X$ = the rates of interest on loans 1, 2, and 3 and $W$ = the respective dollar amount of each loan:

$$\bar{X}_W = \frac{\sum (XW)}{\sum W} = \frac{(0.10 \cdot \$5000) + (0.15 \cdot \$2500) + (0.15 \cdot \$7500)}{\$5000 + \$2500 + \$7500}$$

$$= \frac{\$2000}{\$15,000} = 0.133, \text{ or } 13\tfrac{1}{3}\%$$

This rate of interest is lower than those for the past two years. The firm is in a better cost-of-money position relative to the two previous years.

---

Weighted averages are used extensively for constructing index numbers (discussed in Chapter 14). The economists who construct the Consumer Price Index, for example, weight some widely used items (such as the cost of food, clothing, and shelter) more than others. Business forecasters also make extensive use of weighting by giving more weight to recent demand for goods than to demand for the same goods several years prior.

The problems for this section give you some practice in using the four averages we have described.

**Problems 2.2**

1. Given the following scores, compute the arithmetic mean, the median, and the mode:

   $X_1 = 6$    $X_3 = 8$    $X_5 = 8$    $X_7 = 10$
   $X_2 = 7$    $X_4 = 8$    $X_6 = 9$

2. Compute the mean, the median, and the mode for the six sample values 7, 5, 3, 10, 12, and 7.

3. The personnel department of a firm intends to hire a sales representative on the basis of scores made on four personality and product-knowledge tests. See the table below for the test scores of three candidates. Which average of their own scores—mean, median, or mode—might be *preferred* by Alice, by Bill, and by Clark? Explain.

| Job | Score on Test | | | |
|---|---|---|---|---|
| Candidate | A | B | C | D |
| Alice | 87 | 63 | 72 | 72 |
| Bill | 63 | 86 | 54 | 86 |
| Clark | 81 | 57 | 76 | 74 |

4. Delta Chemical Company has a tuition-reimbursement program for employees who achieve a 3.00 or better scholastic average in a given term. Compute the grade-point average for an employee whose course record is given in the table (assume A = 4, B = 3, and so on). Would this employee get a tuition reimbursement? Try to answer this question (yes or no) before doing any calculations; then check your guess by computing the grade-point average.

| Course | Grade | Hours |
|--------|-------|-------|
| Linear algebra | B | 4 |
| Finance | B | 3 |
| Auditing | C | 4 |
| Inventory control | A | 3 |

5. Of the 500 employees hired by a multinational firm during the past year, 350 are men.
   a. If the mean age of the men is 30.3 years and that of the women is 22.0 years, compute their weighted mean age. Explain why a simple average, $(30.3 + 22.0)/2$, would be inappropriate here.
   b. If the mean age of all men in this firm is 40.0 years, what is the mean age of the 650 previously employed *men* (not those men hired during the past year)? *Hint:* You will need some data from part (a).

## 2.3 MEASURES OF VARIATION

In Section 2.2 we found central values for a data set by determining one of several average values. But an average alone does not adequately characterize any given set of data. A more complete description includes a measure of variation on the differences for the data values that occur.

To illustrate the meaning of variation in data, consider your own spending habits. Although you may have a constant weekly income, you do not necessarily spend the same amount every week. There is variation in your spending. You pay rent each month, car insurance twice a year, and tuition at the beginning of each term. At unpredictable times, you may have to come up with money for medical expenses, clothing, and trips or even new tires for your car. Just as your personal finances show ups and downs, so too do the incomes and expenditures of any business. Measures of variation are used to describe these fluctuations in a data *distribution*.

> A distribution is an assignment of data values or data labels into logical groups.

Chapter 2 is about descriptors—measures of location and of variation—for data distributions. Chapter 3 describes how to construct and interpret *distributions*.

Like averages, measures of variation come in various grades, from coarse ones that describe only the most extreme values (here the greatest and the least differences) to refined ones that describe differences for every observation. Some business data illustrate this point.

Consider the sales made by agents for three insurance firms during the same week. The three distributions in Figure 2.2 have the same mean, $125,000, and yet they look quite different. The sales of Firm B are spread more widely about the mean than those of Firm A. The sales of Firm C exhibit even wider variation, suggesting that salespeople working for Firm C are less likely to be selling amounts near $125,000 per week than are those of Firms A and B. Firm C's sales do not cluster so closely about the mean. Thus we say that sales of Firm C show the greatest spread or variation because its graph has the widest distribution.

**FIGURE 2.2**   Variation in the Weekly Sales of Three Insurance Firms

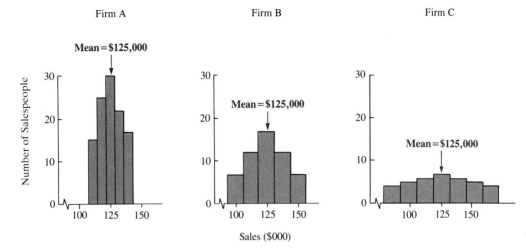

Although all three firms might state the same average number of sales per representative in their annual report to stockholders, that statistic alone would not reveal how the dollar amounts of the sales vary. Three possible choices for measuring that variation are the range, the mean absolute deviation, and the standard deviation.

## Range

The simplest measure of variation is the *range*.

> The range of a data set is the difference between the largest and smallest values in the set.

$$\text{Range } (R) = \text{Largest value} - \text{Smallest value} \qquad (2.3)$$

Example 2.5 and Table 2.4 demonstrate a typical range calculation.

## EXAMPLE 2.5

Table 2.4 shows the weekly sales made by 40 insurance salespeople. What is the range of the sales?

**TABLE 2.4**   Sales ($000) for a Sample of 40 Sales Representatives

| | | | | | | | |
|---|---|---|---|---|---|---|---|
| 75 | 109 | 119 | 123 | 125 | 127 | 133 | 144 |
| 96 | 112 | 120 | 124 | 125 | 128 | 134 | 147 |
| 99 | 114 | 122 | 124 | 126 | 130 | 135 | 152 |
| 102 | 116 | 122 | 125 | 127 | 131 | 137 | 156 |
| 106 | 118 | 123 | 125 | 127 | 131 | 141 | 170 |

*Solution*   Subtract the lowest value from the highest: $170 - 75 = 95$ ($000). Thus the range is $95,000.

As a measure of variation, the range has many business applications. For example, in manufacturing operations the range is often used as a quality-control measure. Consider the steel shafts used to drive the cutter in a common household electric can opener. In a production situation, it is not enough to know that the average diameter of the can opener shafts is satisfactory; the extremes must also be checked to ensure that the process is producing shafts that are neither too small nor too large to work in the appliance. Standards are then set so that a machine operator can quickly identify any shafts that are outside the acceptable range.

**Interquartile Range**

The interquartile range extends the ideas of the range and the median. Recall from our earlier discussion that the median is the second quartile value, $Q_2$ and divides the data set into halves. Similarly, the first quartile, $Q_1$, segments off the lower one-quarter, while the third quartile separates the highest quarter of the data set. The first quartile lies in the position $(n + 1)/4$, and $Q_3$ lies in position $3(n + 1)/4$. A measure that describes the range for the middle 50% of the data values is the *interquartile range*. This measure of variation is defined from the first and third quartile values.

> The interquartile range (IR) describes the difference between the third-quartile and first-quartile values and is the range for the middle 50% of the data values.

$$\text{Interquartile range (IR)} = Q_3 - Q_1 \qquad (2.4)$$

The interquartile range tells the amount of spread in the middle 50% of the data points. Together with the median, the $Q_1$ and $Q_3$ points apportion the data into quarters.

---

**EXAMPLE 2.6**

Returning to Table 2.4, weekly sales for 40 salespeople, what is the range of sales for the middle 50% of the data values?

*Solution*   The sales amounts are already in low to high order, which simplifies finding the $Q_1$ and $Q_3$ values. $Q_1$ lies in the $(n + 1)/4 = (40 + 1)/4 = 10.25$th position. Insofar as there is one unit of distance between the tenth and eleventh values, $Q_1$ is $118 + 0.25(119 - 118) = 118.25$. $Q_3$ lies in the $3(n + 1)/4 = 30.75$th position, so the value we get for $Q_3 = 131 + 0.75(133 - 131) = 132.50$. Then the interquartile range is IR $= 132.50 - 118.25 = 14.25$ ($000). For this sample of 40 records, the middle 50% of weekly sales spans $14,250.

---

The range and interquartile range are of limited use because each measures only the difference for specific values only. However, these are intuitively useful measures, particularly when we can have a graphic representation of them. For now, let's review a more general measure that gives information about the whole data set by measuring differences among all of the values.

**Mean Absolute Deviation**

The mean absolute deviation is another indicator of differences in data values.

> The mean absolute deviation (MAD) describes the mean of the absolute differences of the data values from their arithmetic mean.

We disregard the sign of the differences—that is, whether they are positive or negative—because the differences measure the distance from each value to the mean. This sum would ordinarily equal zero. However, since the distance is nondirectional, the sign is of no consequence: All differences are defined by their absolute (positive) values (indicated by | |). MAD is computed as the sum of the absolute differences between each value and their mean, divided by the number of values, $n$.

$$\text{MAD} = \frac{\sum |X - \bar{X}|}{n} \qquad (2.5)$$

*how far away is avg distrib.*

Businesses use MAD in sales forecasting as a means of measuring and controlling the difference between predicted sales and actual sales. When MAD exceeds a specific limit, the manager responsible for forecasting knows that the forecasts are not on target. In this application, MAD is a measure of the average deviation of the actual demand from a forecast demand.

$$\text{MAD} = \frac{\sum |\text{Actual} - \text{Forecast}|}{n}$$

where $n$ is the number of time periods or data points used to determine the average. In this case, the forecast is assumed to be the "average" value. An illustration is presented in Example 2.7.

---

**EXAMPLE 2.7**

The forecast values and the actual demand for a firm's microwave ovens for June through October are as shown in the table. What is the mean absolute deviation of actual from forecast values? What information does the statistic give the sales manager?

| Month | Actual | Forecast |
|-------|--------|----------|
| June  | 40     | 42       |
| July  | 38     | 45       |
| Aug.  | 50     | 49       |
| Sept. | 45     | 50       |
| Oct.  | 47     | 42       |

*Solution*  Using each forecast value as the mean value,

$$\text{MAD} = \frac{|40 - 42| + |38 - 45| + |50 - 49| + |45 - 50| + |47 - 42|}{5}$$

$$= \frac{2 + 7 + 1 + 5 + 5}{5} = 4$$

The MAD value of 4 provides a gauge of the accuracy of the monthly forecasts. Four is the average (absolute) difference of actual values from the forecast. The sales manager could investigate reasons for poor forecasts in months that have deviations substantially greater than 4.

In using the MAD, we have moved from the rough measure of variation provided by the range to a measure that uses all the data. The MAD is reported in the same units as the data themselves.

Another measure of variation that uses information from all the data is the standard deviation. It is more useful in statistics than the MAD.

## Standard Deviation

A third measure of variation, the *standard deviation*, describes differences among all the values in a set of data and retains the original units.

> The standard deviation (sd) is a measure of variation used to describe the dispersion, or spread, in data. It is an average of squared differences for the data values, taken from their mean, that is standardized by taking the square root.

*[handwritten margin notes: on avg, how far away are observations or how far away from mean to avg. observation]*

The standard deviation is extremely useful because there are numerous (consistent) interpretations about the percent of data values that reside within any specified number of standard deviations. Taking the square root, after squaring, puts the value back in the same units as the original data (such as dollars, hours, and units produced).

Several displays can show how standard deviations are used to describe data. Figure 2.3 displays two data distributions. In each one the top horizontal scale is the original units, whereas the lower scale is in units of standard deviation; then, for both, +1 is the distance of 1 standard deviation. Because its distribution of data values is more dispersed, the graph in part (a) has the larger standard deviation unit. Relative to Figure 2.3(b), Figure 2.3(a), with its larger standard deviation unit, reflects that there is greater variation in those data. Thus the size of a standard deviation unit depends upon the variation or spread of values in the data set. The greater the variation, the larger the standard deviation.

The calculation of the standard deviation uses the differences between every data value and the mean. The differences are individually squared and

**FIGURE 2.3** Units of Standard Deviation

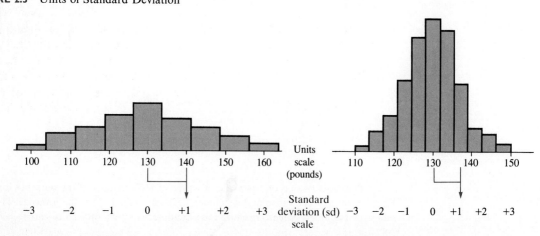

(a) Scale for widespread values          (b) Scale for narrowly spread values then summed, giving a sum of squared differences, or *sum of squares.*\* The sum of squares is then divided by the effective sample size\*\* to produce the *variance.*

> A sum of squares describes the value we obtain by first squaring differences of individual data values from their mean, then summing.

> The variance is a measure of variation we obtain by dividing the sum of squares by the effective sample size.

The population variance $\sigma^2$ (sigma squared) is

$$\sigma^2 = \frac{\sum (X - \mu)^2}{N} \qquad (2.6a)$$

The population standard deviation $\sigma$ (sigma) is

$$\sigma = \sqrt{\sigma^2} \qquad (2.7a)$$

Intuitively, we see that the variance is an average of squared distances from observations to the mean. As data values occur farther from the mean, the average squared distance, or variance, increases. As the data values are more closely aligned, their squared distance shrinks and the variance decreases. The smallest possible value is zero, which occurs in the (rare) case that all of the data have the same value. In the standard deviation, we simply standardize the squared operation of the variance calculation by taking a square root. The variance has squared units, whereas the standard deviation retains the data's units—like dollars or hours.

The *population standard deviation* measures the variability inherent in (all) the data in the population.

The *sample variance*, $s^2$, is also a measure of the average sum of squares. In this case, however, the squared differences are around the sample mean $\bar{X}$ instead of the (true) population mean, $\mu$.

If we were to divide the sum of squares $\sum (X - \bar{X})^2$ by the sample size $n$, we would have a measure of the sample variance. However, it would tend to underestimate the population variance; that is, it would be *biased*. We can remove this bias by dividing the sample sum of squares by $n - 1$, instead of by $n$. In other words, our best estimator of the population variance is the sample variance, as calculated with $n - 1$ in the denominator. The $n - 1$ is referred to as the *degrees of freedom*, the number of free choices for the values of $n$ observations. This corrected form (with $n - 1$) is so widely used that it is commonly referred to as the sample variance $s^2$. The square root of this sample variance

---

\* The differences, before squaring, always satisfy

$$\sum (X_i - \bar{X}) = (X_1 - \bar{X}) + (X_2 - \bar{X}) + \cdots + (X_n - \bar{X})$$
$$= (X_1 + X_2 + \cdots + X_n) - n\bar{X}$$
$$= \sum X_i - n\bar{X} = 0$$

So squaring the differences provides a useful and nonzero result, the sum of squares.

\*\* The effective sample size of a statistic, which is also called degrees of freedom (df), represents the number of free choices for the values of the $n$ observations.

is one of the most useful measures of data variability—the *sample standard deviation, s.*

The sample variance is calculated

$$s^2 = \frac{\sum (X - \bar{X})^2}{(n - 1)} \qquad \text{(2.6)}*$$

Then the sample standard deviation is

$$s = \sqrt{s^2} \qquad \text{(2.7)}*$$

The sample standard deviation, $s$, is calculated the same as $\sigma$, except that the sample standard deviation has as its divisor the sample size minus 1, $n - 1$, whereas the population standard deviation uses $N$. In using sample data, subtract 1 from the observed sample size to provide an effective sample size because the population mean is not known.

The sample variance, $s^2$, with divisor $n - 1$ *is an unbiased estimator of $\sigma^2$ because its average value equals the population value $\sigma^2$.* It is also the best estimator we have of $\sigma^2$. In the same sense, the sample standard deviation, $s$, is a best estimator of $\sigma$—even though it is not an unbiased estimator.

As can be seen from Equation (2.6), the variance of a data set is a function of the amount by which the individual values (collectively) differ from their mean. If all the values in a data set were equal, the sum of squared deviations would be zero and the resultant variance would also be zero. But as differences occur and are squared, the result is a magnification of those differences. Moreover, because the calculation of variance uses squared differences, the resultant variance values are stated in squared units rather than in the original units of the data. For example, if the data units are in dollars, the variance would be in units of dollars squared. However, insofar as the standard deviation is the square root of the variance, it would be stated in the same units as the original data, for example, dollars.

Much of our discussion here will pertain to sample descriptions, so we will use $s^2$ and $s$ rather than $\sigma^2$ and $\sigma$. Keep in mind, however, that $\sigma^2$ and $\sigma$ provide comparable measures for the population.

Example 2.8 illustrates the calculation of a sample data standard deviation. This includes steps for evaluating the mean, the sum of squares, the sample variance, and then the standard deviation.

---

* Equations (2.6) and (2.6a) are often referred to as the definition forms for calculating the sample or population standard deviation. A more efficient, but less intuitive, expression of the standard deviation is

$$s = \sqrt{\frac{n(\sum X^2) - (\sum X)^2}{n(n - 1)}} \qquad \text{(2.6b machine form)}$$

There are other forms as well, but all versions take the general form of

$$\sqrt{\frac{\text{Sum of squares}}{n - 1}}$$

**EXAMPLE 2.8**

A produce manager is concerned about possible extreme variation in the weights of boxes of produce being delivered to the supermarket. If the weights vary extremely or are low on the average, the store's customers will be dissatisfied. As a test, the manager has selected a sample of nine produce boxes. These weigh 29.7, 29.8, 29.9, 29.9, 30.0, 30.0, 30.2, 30.2, and 30.3 pounds. Table 2.5 shows the preliminary calculations. Complete the computation of $\bar{X}$ and $s$. Then identify a range that includes one standard deviation away from $\bar{X}$ for weights of boxes of produce. Do these data display an extreme variation?

**TABLE 2.5**   Weights of Produce Boxes

| Weight $X$ (in pounds) | $(X - \bar{X})$ | $(X - \bar{X})^2$ |
|---|---|---|
| 29.7 | −0.3 | 0.09 |
| 29.8 | −0.2 | 0.04 |
| 29.9 | −0.1 | 0.01 |
| 29.9 | −0.1 | 0.01 |
| 30.0 | 0.0 | 0.00 |
| 30.0 | 0.0 | 0.00 |
| 30.2 | 0.2 | 0.04 |
| 30.2 | 0.2 | 0.04 |
| 30.3 | 0.3 | 0.09 |
| 270.0 | 0.0 | $0.32 = \sum (X - \bar{X})^2$ |

*Solution*   The calculation follows five steps for computing the standard deviation.

Step 1   $\bar{X} = \dfrac{\sum X}{n} = \dfrac{270.0}{9} = 30.0 \text{ lb}$

Steps 2, 3   $\sum (X - \bar{X})^2 = 0.32$   (by Table 2.5)

Step 4   $s^2 = \dfrac{\sum (X - \bar{X})^2}{n - 1} = \dfrac{0.32}{8} = 0.04 \text{ lb}^2$   (0.04 square pounds)

Step 5   $s = \sqrt{0.04} = 0.2 \text{ lb}$   (0.2 pounds)

The numbers in parentheses are a reminder that the variance ($s^2$) has squared units, while the standard deviation ($s$) has the same units, here pounds, as the original variable. One standard deviation above $\bar{X}$ to one standard deviation below $\bar{X}$ gives 29.8 to 30.2 pounds. Except for two boxes, all the individual weights are within 1 standard deviation of the mean. The most extreme weights, 29.7 and 30.3 pounds, are both $1\frac{1}{2}$ standard deviations from the mean. This is not excessive variation.

The importance of the standard deviation as a measure of variation stems from the fact that it has become a base for computing numerous statistical measures. To be most meaningful, the standard deviation should be cited along with the value of the mean. For a known distribution, given the mean and standard deviation values, the decision maker can address problems like this: Estimate how many cases have values between 29.4 and 30.6 pounds. We take up this kind of problem in Chapter 3.

We have described three measures of variation. Of these, the range is least informative because it describes differences for only the two extreme values. On the other hand, mean average deviation and the standard deviation make use of all the data points.

Although the MAD is easier to calculate and understand than the standard deviation, it is difficult to use in sampling theory. Insofar as advanced work in statistics depends on sampling theory, the standard deviation is the most useful measure of variation.

## The Coefficient of Variation

The standard deviation provides us with a measure of the dispersion (variation) of items about their mean. However, the standard deviation units pertain only to the specific set of data under study. Sometimes business analysts want to compare the dispersion of one set of data with that of another, for example, the variation in stock prices for a mutual fund's portfolio A with the variation in prices for portfolio B. A stock market analyst can make such a comparison using the *coefficient of variation V* for each set of data.

> The coefficient of variation $V$ is a measure of *relative variation*. It expresses the standard deviation as a percentage of the arithmetic mean.
>
> The coefficient of variation can be computed for sample data or for population data.

$$V = \frac{s}{\bar{X}}(100) \quad \text{(sample data)} \tag{2.8}$$

or

$$V = \frac{\sigma}{\mu}(100) \quad \text{(population data)} \tag{2.8a}$$

Because $V$ is the ratio of the standard deviation to the mean, the units cancel, so that $V$ is a pure number *without units*. Thus two data sets can have different units, but their coefficients of variation can be compared. Example 2.9 illustrates calculation of a coefficient of variation.

Thus, the coefficient of variation has an advantage over the standard deviation in that the coefficient of variation is independent of the units of measurement. Therefore, it can be used to compare the variation inherent in different types of measure, such as U.S. dollars and Japanese yen. Example 2.9 illustrates such a comparison.

---

**EXAMPLE 2.9**

A multinational firm manufacturing car frames in both the United States and Canada has obtained several frame length measurements from each of its plants to compare the variability inherent within the manufacturing process. The U.S. plant measurements are $\bar{X} = 15.20$ feet and $s_X = 0.05$ feet, whereas the Canadian plant products average $\bar{Y} = 4.53$ meters and $s_Y = 0.02$ meters. Which plant has the more stable process? This question could be of interest for determining which plants are producing to the closest quality standard specifications.

*Solution*   The more stable process will be the one with the smaller coefficient of variation.

$$V_X = \frac{s_X}{\bar{X}}(100) = \frac{0.05}{15.20}(100) = 0.33\%$$

$$V_Y = \frac{s_Y}{\bar{Y}}(100) = \frac{0.02}{4.53}(100) = 0.44\%$$

Because $V_X$ is less than $V_Y$, decide that the U.S. plant has a more stable process.

---

**Problems 2.3**

6. The following measures represent times in seconds required for an auto-matic bottling machine to fill a one-gallon container: 7.3, 7.0, 7.3, 7.0, 6.8, and 7.2.
   a. Compute the mean, variance, and standard deviation for these measures.
   b. Is this sample or population data? What difference would this have on the computations in part (a)? *Suggestion:* Consider that this machine is used for production to fill many one-gallon containers.

7. Assume that the following set of values is a sample from a very large population:
   $$X_1 = 4 \quad X_3 = 9 \quad X_5 = 8 \quad X_7 = 8$$
   $$X_2 = 7 \quad X_4 = 2 \quad X_6 = 10 \quad X_8 = 8$$
   Find each value:
   a. The range and the interquartile range
   b. The average deviation  same as MAD (mean absolute dev.)
   c. The standard deviation
   d. The coeffecient of variation

8. Find the mean and standard deviation for these sample data: 7, −5, 0, 2, −2, 1, 4, 6, 2, and 0.

9. The balance in a checking account for each of the past 12 months (rounded to the nearest $100) is $500, $400, $600, $800, $600, $400, $200, $300, $200, $100, $300, and $400.
   a. Compute the range, the interquartile range, MAD, and $s$.
   b. Now code the amounts by calling $500 just $5 (hundreds). Use $5, $4, $6, $8, . . . , $4 to compute new answers for each answer in part (a). Attach the word *hundreds* to each answer. How does this coding effect the answer?

10. Given the set of scores 201, 203, 206, and 206, compute $\bar{X}$, $s^2$, and $s$. Subtract 200 from each score and again compute $\bar{X}$, $s^2$, and $s$. Compare your answers. Also, compare this type of coding with the coding used in Problem 9.

11. Given the values $X_1 = -8$, $X_2 = -5$, $X_3 = -5$, $X_4 = -4$, and $X_5 = -3$,
    a. compute $\bar{X}$. Then show that $\sum (X - \bar{X}) = 0$.
    b. compute $\sum (X - \bar{X})^2$. Then evaluate $s^2$.
    c. Can $s^2$ be negative? Under what conditions does $s^2 = 0$?

 12. A testing laboratory in Paris has been asked to verify the accuracy of ma-chines used to fill containers of hair spray produced in two different plants of International Chemical Co. A random sample of 40 bottles from an Aus-tralian plant revealed a mean fill quantity of 705.0 ml, with $s = 12.0$ ml.

Another sample of 35 bottles from the firm's Minneapolis plant revealed a mean fill amount of 16.2 oz with $s = 0.9$ oz. Are the machines roughly equivalent in terms of the relative variability of fill amounts, or are the machines at one plant more precise than at the other?

13. An automobile manufacturer produces cars in plants in both Europe and the United States. As part of their quality control, samples of 10 cars from both plants are given mileage tests on local area roads, with results as shown below. However, the manufacturer is concerned that the cars may not have been driven over the same type of terrain and wishes to use the coefficient of variation to compare the variability in mileage from the two locations. Compute the coefficient of variation for the two data sets and draw appropriate conclusions about the comparability of the tests.

| U.S. miles/gal | European km/liter |
|---|---|
| 28.4 | 15.1 |
| 31.4 | 15.5 |
| 30.0 | 14.0 |
| 38.0 | 15.7 |
| 29.2 | 16.8 |
| 35.3 | 13.4 |
| 37.1 | 14.2 |
| 41.0 | 14.4 |
| 30.2 | 15.3 |
| 29.4 | 15.6 |

## 2.4 USES OF DESCRIPTIVE STATISTICS

The mean and the standard deviation are the two statistics most widely used to describe data. Together they provide information about the location (mean) and the dispersion (standard deviation) of a set of data. The fact that the voluminous data contained in a distribution can be summarized in terms of a mean and a standard deviation makes these two statistics especially useful to decision makers. Business analysts can describe populations and make forecasts from these few summary measures, whereas it would be impossible to comprehend or grasp the same meaning from the raw data alone. We probably use summary measures more than we realize. Consider, for example, using your GPA to describe *all* your university coursework, or the window sticker on every new car that states the (average) miles per gallon a driver can expect to get from that vehicle.

### Chebyshev's Rule

A few summary statistics enable us to make inferences. For example, it can be shown mathematically that at least 75% of the values in *any* data distribution are within 2 standard deviation units of the mean. That is, given any set of data, it is impossible to have more than 25% of the observed values greater than a distance 2 standard deviations above or below the mean (in the tails). Similarly, from the same formula, at least 88.9% of all data values will be found always to lie within $+3$ or $-3$ standard deviations of their mean. *Chebyshev's rule* describes the percentage of values that must lie within a given number of standard deviations from their mean.

### CHEBYSHEV'S RULE

For any distribution of values, the percentage of cases within $K$ standard deviations of the mean is *at least*

$$1 - (1/K^2), \quad \text{where } K \text{ is a number greater than 1} \qquad \textbf{(2.9)}$$

Chebyshev's rule is meaningful only for values greater than 1, but it places no restriction on the shape of the distribution (see Figure 2.4); however, the mean and standard deviation must have finite values.

Example 2.10 (page 50) applies Chebyshev's rule to a decision situation.

---

**FIGURE 2.4**  Illustration of Chebyshev's Rule for Any Distribution

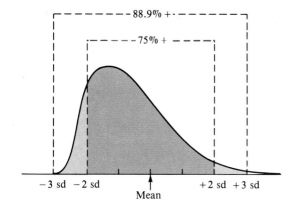

(a) Positively (right) skewed distribution

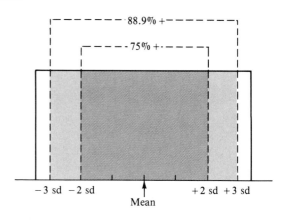

(b) Uniform distribution

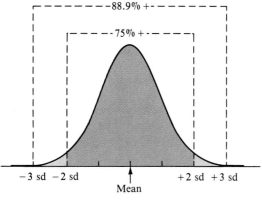

(c) Bell-shaped distribution

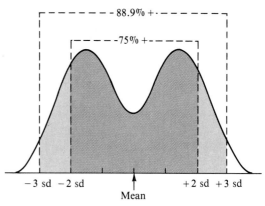

(d) Bi-modal distribution

**EXAMPLE 2.10**

Cedar Woods Furniture Company tries to maintain drying temperatures within a range of 57° to 68°F. A sample of mean daily temperature readings for one of the company's kilns showed $\bar{X} = 62.4°F$ and $s = 3.7°F$. What percentage of the days (at the very least) had temperatures within the optimal range? Is it at least 50%? (This is minimal to assure an economically feasible operation.)

*Solution*    The minimum percentage of days within $K$ standard deviations is $1 - (1/K^2)$, where $K$ is the number of standard deviations above and below the mean, and the value of each standard deviation is 3.7°F.

$$Above \quad K = \frac{68 - 62.4}{3.7} = +1.5 \text{ standard deviations}$$

$$Below \quad K = \frac{57 - 62.4}{3.7} = -1.5 \text{ standard deviations (same)}$$

The percentage within 1.5 sd is at least $1 - (1/K^2)$.

$$Percentage = 1 - (1/1.5^2) = 0.56, \quad \text{or } 56\%$$

This 56% is a minimum; the actual percentage of days could be higher, but it is not less than 56% of all days. Based on the stated condition (of exceeding 50%), the operation is feasible.

Chebyshev's rule is an important mathematical limit, but it is not used extensively in business because both sample and population data tend to be more closely clustered about the mean than Chebyshev's rule recognizes. Using more information about a distribution can give a more accurate description of the data.

### Normal (Bell) Curve Data

The bell-shaped curve of Figure 2.4(c) aptly describes numerous business data distributions. Analysts are quick to take advantage of its occurrence by applying normal curve percentages. Given that this data pattern exists, the bell curve gives a more accurate data description than does Chebyshev's rule.

For a bell curve,

$\pm 1$ standard deviation from the mean includes 68.3% of all the observations.

$\pm 2$ standard deviations from the mean includes 95.4% of all the observations.

$\pm 3$ standard deviations from the mean includes 99.7% of all the observations.

Other patterns and the associated percentages exist, but the bell curve is a common profile for business data. However, Chebyshev's rule has one special property: If we don't know what the distribution is, then we can always resort to Chebyshev's rule.

We do not use Chebyshev's rule to any great extent. However, we will use the normal distribution (bell curve) extensively beginning with Chapter 6. The day-to-day sample readings of a business variable become more meaningful if they can be recognized to fit an overall distribution pattern, such as the normal distribution. Chebyshev's rule and the normal distribution illustrate the dependence of statistical statements upon knowledge of the values of both the mean and the standard deviation.

**Box Plots**

One useful method of describing the pattern of data is a box plot.

> A box plot is a graphical display that shows the central portion (50%) of a data set within a box and the ends or extremes in positional locations outside the box.

Box plots can display the lowest value, the first quartile, the median, the third quartile, and the highest value in the data set. The ends of the plot are identified by dotted lines or "whiskers" that connect the lowest and highest data values to their respective ends of a "box"; hence the name "box plot." The ends of the box, the "hinges," are located at the first ($Q_1$) and third ($Q_3$) quartiles of the data. The distance between the hinges is the interquartile range, IR. The median resides within the box, with its location commonly indicated by a symbol, such as $+$. Essentially, 50% of the data values will reside within the range of the hinges, and another 25% will be outside each hinge of the box.

> An outlier is any observation that has a value markedly large or small relative to the total distribution of values.

Outliers, or extreme values, can produce a marked distortion in the values for some statistics, especially the mean and standard deviation. That is, a single data value can have an undue, undesirable impact on determining the statistic's value. So, identifying possible outliers is an important preliminary step to understanding a data set.

Box plots are also used to identify the values of *outliers* that reside far from the center. Outliers are classified as either mild (outside a hinge) or extreme (far beyond a hinge). A mild outlier resides beyond a hinge a distance that exceeds 1.5 times the interquartile range. An extreme outlier is 3 times the interquartile range, or a greater distance beyond a hinge.

The location of the median ($+$) is indicative of the skewness, or shift from symmetry, for a data distribution. For a symmetric distribution there is zero (no) skew, and the median is equal to the mean. When the distribution is skewed, then the mean is pulled in the direction of the skew, while the median resides in an opposite direction. A box plot describes this. When a distribution is right-skewed, then the median is located in the left half of the box. When a distribution is left-skewed, then $+$ is located in the right half of the box. These cases are illustrated in Figure 2.5. Example 2.11 uses the data from Table 2.4 (repeated on page 52).

---

**FIGURE 2.5**    The Relation of Skewness, the Median ($+$), the Mean, and the Shape

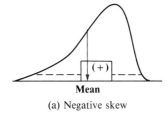

**Mean**

(a) Negative skew

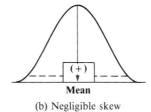

**Mean**

(b) Negligible skew

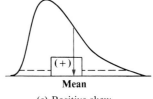

**Mean**

(c) Positive skew

**EXAMPLE 2.11**

What descriptions are available from a box plot for the data values in Table 2.4? The data are sales (in $000) for a sample of 40 insurance sales representatives.

**TABLE 2.4**   Sales ($000) for a Sample of 40 Sales
Representatives

| 75  | 109 | 119 | 123 | 125 | 127 | 133 | 144 |
|-----|-----|-----|-----|-----|-----|-----|-----|
| 96  | 112 | 120 | 124 | 125 | 128 | 134 | 147 |
| 99  | 114 | 122 | 124 | 126 | 130 | 135 | 152 |
| 102 | 116 | 122 | 125 | 127 | 131 | 137 | 156 |
| 106 | 118 | 123 | 125 | 127 | 131 | 141 | 170 |

*Solution*    The descriptions summarized here display the developments for a box plot summary. You can develop the statistics from our earlier work in this chapter.

$$Q_1 \quad \text{(from Example 2.6)} = 118.25 \ (\$000)$$

$$\text{Median} = (125 + 125)/2 = 125.0$$

$$\text{Mean} = \frac{75 + 96 + 99 + \cdots + 156 + 170}{40} = 125.0$$

$$Q_3 = 132.5$$

$$\text{Minimum value} = 75$$

$$\text{Maximum value} = 170$$

In steps, the box plot appears as follows:

The scale and the median:

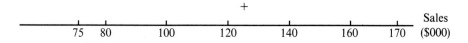

Adding "hinges" (I) at $Q_1$ and $Q_3$ and "whiskers" (----):

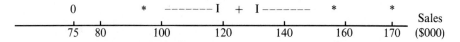

Each whisker is about 1.5 times longer than the span of the interquartile range. Finally, adding three (3) mild outliers, *, and one extreme outlier, 0:

The mild outlier values—96, 156, and 170—are beyond 1.5 times the interquartile range, while the one extreme outlier—75—is more than 3.0 times the length of the interquartile range away from the median.

The box plot shows that the median value is about $125,000. There were three mild outliers (*) at $96,000, $156,000, and $170,000 of sales. One extreme outlier (0) exists at $75,000. The box plot shows that the median is nearly centered, with two outliers in either extreme.

---

A box plot is a simple visual representation that helps us understand a data set. Outlier observations need to be investigated. Are the outlier values data-entered correctly? For the data in Table 2.4, was the $75,000 amount only a partial value, not for the entire week? All reasonable data values should be used. If an incorrect reading or a recording error exists, then the correct value should be inserted and the statistics recalculated. Box plots can be used to identify questionable values.

Computer Problem 48 illustrates the Minitab statements for the box plot on insurance sales in Example 2.6. Try the section Problems for a variety of applications including use of $\bar{X}$, $s$, the bell curve, Chebyshev's rule, and box plots.

**Problems 2.4**

14. The table below gives a few values for $K$ in Chebyshev's rule and also provides the percentage of cases beyond $\pm K$ standard deviations distance from the mean for two of the values.

| $K$ | $\dfrac{1}{K^2}$ | $1 - \dfrac{1}{K^2}$ | Cases Between $\pm K$ | Cases Beyond $\pm K$ |
|---|---|---|---|---|
| 1 | 1 | 0 | at least 0% | at most 100% |
| 2 | $\frac{1}{4}$ | $\frac{3}{4}$ | at least 75% | at most 25% |
| 2.5 | | | | |
| 3 | | | | |
| 3.5 | | | | |
| 4 | | | | |

a. Complete the table.
b. Explain why Chebyshev's rule does not work for $K = 1$.
c. Someone has said that 4 is about the largest value necessary to use Chebyshev's rule. Explain whether you agree with this statement.
d. Explain why the column for cases beyond $\pm K$ has *at most* a percentage (versus an exact, or least, percentage) of the data values.

15. The summary statistics for weekly sales at three computer stores are listed in the table below.
a. For each store, estimate the percentage of weeks with sales below $60,000 or above $180,000 (provide a single number).
b. Compare these percentages for the three stores.
c. Why are the percentages so different?
d. Can Chebyshev's rule be used to answer this question: In what percent of weeks will store 1 have sales less than $100,000? Explain.

| Store | Mean Sales | Standard Deviation |
|-------|------------|--------------------|
| 1     | $120,000   | $10,000            |
| 2     | $120,000   | $20,000            |
| 3     | $120,000   | $60,000            |

16. The graphed profile below shows the closing price (in dollars) of a certain stock over the past 100 weeks. Numbers at the top of the bars represent the number of weeks in which that closing price prevailed. The mean is $10 with standard deviation $1.10. Consider the range data in the table and complete the table. Explain which result, that from Chebyshev's rule or the bell curve, most closely approximates the observed pattern of closing prices for this stock.

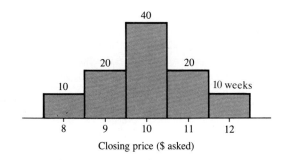

Closing price ($ asked)

| Range | Observed Number of Weeks | Chebyshev | Bell Curve |
|-------|--------------------------|-----------|------------|
| Mean ±1 sd | 80 (%) | at least 0% | 68.3% |
| Mean ±2 sd |        |             |        |
| Mean ±3 sd |        |             |        |

★ 17. The subsection on the normal (bell) curve (page 50) gives the proportion of observations that can be expected under ±1, ±2, and ±3 standard deviations of a bell-shaped distribution. Recognizing the symmetry of the curve,
   a. what percentage of the observations lie within ±2 standard deviations of the mean?
   b. what percentage of the observations lie within 1 standard deviation to the left and 2 standard deviations to the right of the mean?

18. The data below are corporate profit rates (percentages) rounded to one decimal place for the rate of profit on invested capital for 20 firms. Develop a box plot, and describe the median, distribution skew, and outliers.

★ Starred problems are optional.

| Rates of Return on Invested Capital (%) | | | | |
| --- | --- | --- | --- | --- |
| 0.4 | 4.5 | 6.3 | 7.3 | 7.7 |
| 2.1 | 5.0 | 6.6 | 7.4 | 11.6 |
| 2.6 | 5.0 | 6.9 | 7.5 | 12.1 |
| 3.0 | 6.3 | 7.3 | 7.6 | 12.2 |

19. Weekly take-home pay was recorded for a sample of 25 semiskilled factory workers. Use the data to develop a box plot. Indicate minimum, median, and maximum values on the display. Identify any take-home pay amounts that are mild outliers and extreme outliers.

| Weekly Take-home Pay ($) | | | |
| --- | --- | --- | --- |
| 319.12 | 326.81 | 324.79 | 313.48 |
| 331.50 | 348.39 | 337.24 | 326.67 |
| 320.76 | 321.67 | 331.47 | 326.67 |
| 325.42 | 315.38 | 304.12 | 321.19 |
| 333.98 | 340.89 | 327.02 | 327.11 |

## SUMMARY

The most fundamental statistics for describing data are *measures of location* (the arithmetic mean, the median, the mode, and the weighted mean) and *measures of variation* (the range, the mean absolute deviation, the variance, and the standard deviation). These statistics are used to give summaries of a data set.

A business manager regularly encounters much data—often too much to comprehend all at once. Then it becomes imperative to condense that bulk of data into a few pointed statistics. The two key statistics for business data description are an average and a measure of variation.

The *arithmetic mean* is the most commonly used average. It is affected by all values of the data set and has statistical properties that we will find useful later on. The *median* is the middle value. It is not as useful for inference, but it is not influenced by actual values at the ends of the data set. The *mode*, although easy to find (if it exists), is not a widely used measure.

For measures of variation the *range* is easiest to compute, but it does not take into account individual values. The most useful measures of dispersion are the *standard deviation* and the *variance* (square of the standard deviation). The standard deviation reflects information from all members of the data set and yields a measure of error that is widely used for inference purposes.

Statistics are used to describe data sets. The mean and standard deviation are central to Chebyshev's rule and the bell curve, which describe data patterns. The median is useful with box plots to view distribution shape, or skewness, and to identify extreme data values.

The computational forms for the mean and standard deviation reviewed in Table 2.6, are important because they are used to develop many other statistics. The differences in sample versus population computations are subtle, yet important. Note that the population standard deviation computation, $\sigma$, requires the full count $N$ from the population as divisor, whereas the sample standard deviation uses one less than the sample count, $n - 1$, to correct for the use of the estimator $\bar{X}$ for the population mean, $\mu$.

**TABLE 2.6**    Summary of Descriptive Statistics

| | | | |
|---|---|---|---|
| **Means** | Sample data | $\bar{X} = \dfrac{\sum X}{n}$    where $n$ denotes the sample size | **(2.1)** |
| | Population | $\mu = \dfrac{\sum X}{N}$    where $N$ denotes the population | **(2.1a)** |
| | Weighted mean | $\bar{X}_w = \dfrac{\sum XW}{\sum W}$    where $W$ denotes weights | **(2.2)** |
| **Measures of Variation** | Range | $R = \text{Largest value} - \text{Smallest value}$ | **(2.3)** |
| | Interquartile Range | $IR = Q_3 - Q_1$ | **(2.4)** |
| | Mean absolute deviation | $\text{MAD} = \dfrac{\sum |X - \bar{X}|}{n}$    where $|\ |$ denotes absolute (positive) differences | **(2.5)** |
| | Population variance | $\sigma^2 = \dfrac{\sum (X - \mu)^2}{N}$ | **(2.6a)** |
| | Population standard deviation | $\sigma = \sqrt{\sigma^2}$ | **(2.7a)** |
| | Sample variance | $s^2 = \dfrac{\sum (X - \bar{X})^2}{n - 1}$ | **(2.6)** |
| | Sample standard deviation | $s = \sqrt{s^2}$ | **(2.7)** |
| **Measure of Relative Variation** | Coefficient of variation | $V = \dfrac{s}{\bar{X}}(100)$    sample | **(2.8)** |
| | | $V = \dfrac{\sigma}{\mu}(100)$    population | **(2.8a)** |

---

---

## CHAPTER 2 PROBLEMS

20. Several measurements were made on the length of a special coil spring before ship-ment to a customer in Germany. For five measurements using the English system, $\bar{X} = 5$ foot, 3 inches with $s_X = \frac{1}{2}$ inch. An equal number of measurements using the metric system gave $\bar{Y} = 1.70$ meters with $s_Y = 0.01$ meters. Use the coefficients of variation to disclose which measure, English or metric, was more precise.

21. Industrial Products Co. requires their product managers to issue monthly reports on the dollar value of finished products on hand in order to help control any wildly fluctuating inventory investment. Two product managers, Kale and Jones, have

reported their monthly inventory levels, in thousands of dollars, as shown in the table.

a. Compute the standard deviation of investment for each manager.
b. Compute the coefficient of variation for each manager.
c. Discuss the difference in performance for the two managers as indicated by your computations.

*[handwritten annotations:]*

$b).\ V = \frac{\sigma}{X}\ (100)$

$K = \frac{5}{80}\ (100) = 6.25$

$J = \frac{22.3}{80}\ (100) = 27.9$

| Inventory Investment (in $000) | | | |
|---|---|---|---|
| | At the end of month | | |
| | *1* | *2* | *3* |
| Kale | 80 | 75 | 85 |
| Jones | 55 | 98 | 87 |

*[handwritten:]* $\overline{X}$   $K$ 80   $J$ 80   $a)$ $\sigma = 5$   $V =$   $\sigma = 22.3$  $V =$

*[handwritten margin notes: pg .50 example]*

22. The manager of a fast food outlet read in a competitor's report that the mean income of its customers was $27,500 with standard deviation of $3,000. The manager wanted to find out what proportion of her competitor's clientele had incomes between $22,500 and $32,500, since this range covered the incomes of a vast majority of her own clientele. However, the report did not provide this information. What information would Chebyshev's rule provide?

23. The distribution of weekly insurance sales for a sample of $n = 100$ salespeople had mean of $80.0 thousand and standard deviation of $20.0 thousand.
    a. What proportion of these salespeople (as a minimum) sold between $40.0 thousand and $120.0 thousand of insurance?
    b. How many of the salespeople, in absolute numbers, sold more than $120.0 thousand or less than $40.0 thousand of insurance?

*[handwritten margin: $\overline{x} = 12$   med. $\frac{n+1}{2} = 3.5$   $11,$]*

24. Suppose that in four one-hour examinations you receive grades of 78, 86, 82, and 66. Discussing the concepts of mean, median, and mode, select which course "average" you *earned*. Give statistical computations to support your selection.

25. Given the scores $X_1 = 7$, $X_2 = 9$, $X_3 = 13$, $X_4 = 17$, $X_5 = 10$, and $X_6 = 16$, compute $\overline{X}$, the median, the mode, $s^2$, and $s$.

26. A survey of the hourly earnings for workers in three different canneries is summarized in the table below. For each question, indicate the appropriate response, A, B, or C. Use statistics to justify your choice in each part.
    a. Which plant pays the highest *overall* wage rates?
    b. Which has the greatest wage *differentials*?
    c. Which has the most *uniform* (consistent) wage rates?

| Plant | Mean Wage | Standard Deviation |
|---|---|---|
| A | $5.88 | $0.68 |
| B | 6.04 | 0.51 |
| C | 5.65 | 0.48 |

27. Measurements of the time required for an automatic bottling machine to fill a one-gallon milk container are 7.3, 7.2, 7.0, 7.3, 7.0, and 6.8 seconds. If a coefficient of variation of 2.0 (%) or less is required to satisfy minimum operating performance, is this machine functioning within desired specifications? See Problem 6 in Section 2.3 for calculation of $\bar{X}$ and $s$.

28. The following are times from date of purchase until the first service call on eight washing machines: 2.8, 3.0, 2.7, 2.9, 3.0, 3.1, 3.2, and 3.3 years. Compute $\bar{X}$, the median, mode, range, MAD, and $s$.

29. Two sets of data have means of 40 and 50 based on samples of 30 and 20 observations, respectively. Find the mean for the combined set of observations.

30. Suppose you are the payroll manager for a restaurant that employs 150 people. From data on the first week of June of last year, the median wage was $206, the mean wage was $230, and the modal wage was $180. Assuming positions and wages are somewhat stable, except that the overall wage level has increased 6%, how much money will be needed to pay these 150 employees this June?

31. An investor is considering for sale two stocks, A and B. Using the information in the table, discuss which one he should sell. (Assume that he has the same total dollar value invested in each stock and wants to sell the one that has made the best gain relative to its average price.)

| Stock | Average Price | Standard Deviation of Prices | Current Prices |
|-------|---------------|------------------------------|----------------|
| A | $60.0 | $10.0 | $74.50 |
| B | 35.0 | 4.0 | 47.00 |

32. Let $X$ represent the weight of football players in the following statistics on the 11 starting defensive players on each team. Write a short paragraph comparing the weights for these two defensive teams. Use the concepts of arithmetic mean, standard deviation, and coefficient of variation in your discussion.

Culpepper College          State U

$\bar{X} = 180.0 \text{ lb}$          $\bar{Y} = 220.0 \text{ lb}$

$\sum (X_i - \bar{X})^2 = 250.0 \text{ lb}^2$     $\sum (Y_i - \bar{Y})^2 = 62.5 \text{ lb}^2$

33. A small-car manufacturer has taken a random sample of 1600 men from the population of the United States to help establish space requirements for their new compact model. The data revealed a mean height of 69 inches with standard deviation of 3 inches. Use Chebyshev's rule with $K = \pm 3$ to make a statement about the height of all U.S. men on the basis of the preceding sample. Could you make a more exacting statement if you knew that the distribution of height follows a bell curve? Assume that the height of men in the United States follows a bell curve. Then what percentage fall between 63 inches and 75 inches in height?

34. Consider the summary information on two distributions (see table). Indicate which distribution (A, B, both, or neither) satisfies each of the following criterion.
    a. Has a smaller coefficient of variation

b. Has over 50% of the measure above the score 90
c. Has the larger variance

| Statistics | Distribution | |
|---|---|---|
| | A | B |
| Mean | 100 | 90 |
| Median | 90 | 80 |
| Standard deviation | 10 | 10 |

35. The data below are move rates for 40 residential neighborhoods. The national move rate is about 0.20 (that is, 20%). Develop a box plot, and describe the distribution of move rates with mean, median, skewness, and outliers. Are these data consistent with a 20% move rate? It was determined after the fact that the value 0.36 was wrong. The correct value is 0.26. Now answer the questions.

Move Rates

| | | | | | | | |
|---|---|---|---|---|---|---|---|
| 0.10 | 0.15 | 0.16 | 0.18 | 0.19 | 0.21 | 0.23 | 0.26 |
| 0.12 | 0.15 | 0.16 | 0.18 | 0.20 | 0.22 | 0.24 | 0.26 |
| 0.12 | 0.15 | 0.17 | 0.19 | 0.20 | 0.22 | 0.25 | 0.28 |
| 0.13 | 0.16 | 0.17 | 0.19 | 0.21 | 0.22 | 0.25 | 0.30 |
| 0.14 | 0.16 | 0.18 | 0.19 | 0.21 | 0.22 | 0.26 | 0.36 |

36. A list broker's sales were recorded for a 5% random sample of 25 client firms for the first quarter of a fiscal year. Use descriptive statistics and a box plot to describe the distribution for these list sales. Include values for the mean, median, standard deviation, and any outliers. Discuss skewness. Based on your findings and past experience that second-quarter sales are up 12.5% over first-quarter sales, what sales value should the corporate officers expect for the second quarter?

Dollar Sales

| | | | | |
|---|---|---|---|---|
| 2,071.16 | 6,039.48 | 3.985.71 | 6,521.85 | 4,171.45 |
| 2,766.88 | 3,511.28 | 4,496.25 | 3,618.88 | 2,972.08 |
| 3,206.10 | 7,810.78 | 4,514.17 | 2,700.42 | 4,453.44 |
| 2,227.78 | 6,655.96 | 3,059.24 | 7,592.60 | 3,498.85 |
| 2,811.12 | 2,072.27 | 4,538.55 | 5,250.40 | 6,614.03 |

**STATISTICS IN ACTION: CHALLENGE PROBLEMS**

The following problems include applications of statistics that are discussed in this chapter. The journal articles and statistical facts come from diverse areas of business; however, you will see familiar statistics like the mean and standard deviation. Each problem includes background information, and displays the relevant data as it was published. Each problem ends with questions relating to how statistics are used for making business decisions.

37. How aware are consumers of their rights? A sample including 450 questionnaires with 297 (66%) usable responses addressed the awareness of consumer rights in a developing economy. Respondents were judged for awareness on a scale from 1 (low) to 10 (high). Some results appear below.

Raw Scores of Respondent Awareness by Subsamples

| Serial Number | Respondent Category | Subsample Size | Mean Score | Standard Deviation | F-Value |
|---|---|---|---|---|---|
| 1 | High education | 81 | 5.32 | 1.46 | |
| 2 | Moderate education | 111 | 4.85 | 1.32 | 10.64* |
| 3 | Low education | 105 | 4.37 | 1.40 | |
| 4 | 18–24 years old | 112 | 4.71 | 1.42 | |
| 5 | 25–39 years old | 160 | 4.87 | 1.47 | 0.39 |
| 6 | 40–59 years old | 25 | 4.84 | 1.29 | |
| 7 | Above 59 years old | 0 | — | | |
| | | | | | Z-Value |
| 8 | Male | 202 | 4.84 | 1.45 | |
| 9 | Female | 95 | 4.74 | 1.41 | 0.709 |
| 10 | All respondents | 297 | 4.81 | 1.44 | |

\* Significant at the 95% level of confidence.    *Source:* Authors' fieldwork.

SOURCE: Agbonifoh, Bas A., and Pius E. Edoreh. "Consumer Awareness and Complaining Behavior." *European Journal of Marketing* 20, no. 7 (1986): 43–49.

Questions

a. Which group had higher awareness: those 25–39 years old or those 40–59 years old?

b. The standard deviation is largest (1.47) for the group 25–39 years old. Do you agree that a relative large standard deviation is good here? Explain.

38. Rating scales are used to measure people's attitudes to stimuli such as products, advertisements, and opinion surveys. Three types of rating scales were tried in one study: 1 is balanced, 2 are unbalanced, as shown in Table 1.

**TABLE 1**   Three Rating Scales—Descriptor of Category

| Type | (assigned numerical value) | | | | |
|---|---|---|---|---|---|
| Balanced | VG (1) | G (2) | AV (3) | P (4) | |
| Negatively unbalanced | | G (1) | AV (2) | P (3) | VP (4)   AWF (5) |
| Positively unbalanced | EX (1)   VG (2) | G (3) | AV (4) | P (5) | |

Descriptor abbreviations: EX, excellent; VG, very good; G, good; AV, average; P, poor; VP, very poor; AWF, awful.

Using the results of applying these three rating scales, Table 2 shows means and medians for five survey stimuli.

**TABLE 2**   Scale Means and *F*-Values for the Five Stimuli

| | Type of Scale | | | | | | |
|---|---|---|---|---|---|---|---|
| | (*n* = 65) Balanced | | (*n* = 65) Positively Unbalanced | | (*n* = 65) Negatively Unbalanced | | Calculated *F*-Value* |
| Stimulus | *Mean* | *Median* | *Mean* | *Median* | *Mean* | *Median* | *df = 2,192* |
| Carter | 2.78 | 3 | 3.65 | 4 | 2.12 | 2 | 40.67 |
| Koch | 2.98 | 3 | 3.62 | 4 | 2.26 | 2 | 25.10 |
| Subway | 3.83 | 4 | 4.35 | 5 | 3.35 | 3 | 13.22 |
| Police Dept. | 3.40 | 4 | 4.02 | 4 | 2.68 | 2 | 27.02 |
| Prime-time TV | 3.25 | 3 | 3.74 | 4 | 2.77 | 2 | 11.75 |

\* $F(\alpha = 0.01$, deg. of freedom $= 2,192) = 4.72$

SOURCE: Friedman, Hershey H., Yonah Wilamowsky, and Linda W. Friedman. "A Comparison of Balanced and Unbalanced Rating Scales." *The Mid-Atlantic Journal of Business* 19, no. 2 (1981): 1–7.

Question   Explain how the relative values for the means and medians in Table 2 reflect the skew that is established in the unbalanced rating scales of Table 1.

---

## BUSINESS CASE

*Average Isn't Good Enough*

Henry Conroy, the business manager of Cascade City, expected the city council meeting to be controversial, but he wasn't prepared for what Audry Kemp was telling him. Audry was his newly hired administrative assistant, and she was suggesting in a polite way that his recommendation to raise property tax rates from $22 to $32 per thousand dollars of assessed valuation was on shaky ground. Mr. Conroy had put the proposal together on the basis of data from cities in two groups (see table). Group I contained cities of comparable size, and Group II contained cities in the same geographical latitude.

Tax Rate ($ per 1000)

| Group I (same size) | Group II (same latitude) |
|---|---|
| $26, 29, 30, 24, 22, 86, 24, 26, 25, 28 | $37, 14, 55, 36, 42, 15, 56, 12, 15, 38 |
| Mean I = $32 per thousand | Mean II = $32 per thousand |

**Mr. C.**   What do you see as a problem, Audry? Both sets of data support my $32 recommendation, and city government certainly needs more money.

**Mrs. K.**   Oh, I agree about the need! In fact, I think you should make your case on the basis of costs and needs, not on the basis of those statistics.

**Mr. C.**   But both groups have an average tax rate of about $32 per thousand. I consider that pretty convincing. And the council will expect to get some supporting facts.

**Mrs. K.**   Well, I came up with an average of $26 per thousand in Group I. I'm afraid the data in Group II vary so much that our council will not think the mean tax rate is a reliable statistic to support your case.

**Mr. C.**   I agree that Group II includes some strange places, but I had to go outside our region to stay in the same latitude. Where do you get the $26 for the first group? I calculated the average at $32.

**Mrs. K.**   The *arithmetic mean* is $32. But the *median* is $26, and it does appear to be a better measure of location for those data. Don't you think so?

**Mr. C.**   Well, I guess I hadn't given that much thought to the data. That $86 rate at Paradise Palisades does distort the average—I mean the mean—doesn't it? That's a unique spot, you know. Do you think we should drop the Group I data?

**Mrs. K.**   They appear to be good data, Mr. Conroy. Once the $86 figure is deleted, the mean is $26 and the standard deviation is only $2.60. Also, the coefficient of variation is very low, 10% to be exact. I checked it out. Actually, Group I makes a pretty sound case for a rate of $26 per thousand.

**Mr. C.**   How about the Group II data?

**Mrs. K.**   That's just the point, Mr. Conroy. The Group II cities have a very high coefficient of variation. It's almost 53%. The standard deviation, by comparison, is almost $17! Those data are so variable, they hardly produce a precise mean estimate. You're left with a good case for a $26 rate, as I see it.

**Mr. C.**   Maybe this does need a second look. I think we still have a few loose ends to tie up before the council meeting tonight. Let's meet on this again after lunch.

## Business Case Questions

39. When is a median a better measure of central tendency than a mean?

40. On what basis did Mrs. Kemp discard the Group II data? Was that a good decision?

41. Suppose Mr. Conroy could gain access to a much larger sample. What effect would you expect it to have on the following measures?
    a. The mean of Group I    b. The median of Group I
    c. The mean of Group II

## COMPUTER APPLICATION

*Marketing a Product to a Typical American Family*

In Chapter 1 you were introduced to the data base from the PUMS file. In that introduction you identified variables in the data base and investigated the numerical values assigned to some of them. The purpose of this computer application is to obtain some descriptive statistics for a few variables from the data base. The statistics will include the mean, the median, the mode, sample variance, standard deviation, and maximum (largest) and minimum (smallest) values (for calculation of the range).

This application takes the perspective of a firm seeking to market their product to a typical U.S. family. So the variables used here relate to family structure. We begin by first defining a "typical" U.S. family as three or more persons—two parents and one or more children. Although families is not a variable on the data base, we have added a qualification that the age for the head of household be 35 to 50 years. Using these conditions to define a target population will help ensure a high concentration of families. The variables include total household income and householder age.

A large number of computer software packages are available to provide summary statistics. We chose the Minitab package to illustrate some pertinent descriptions (Figure 2.6).

**FIGURE 2.6**  A Minitab Run That Displays Descriptive Statistics and Box Plots for Sample Data from the PUMS Data Base

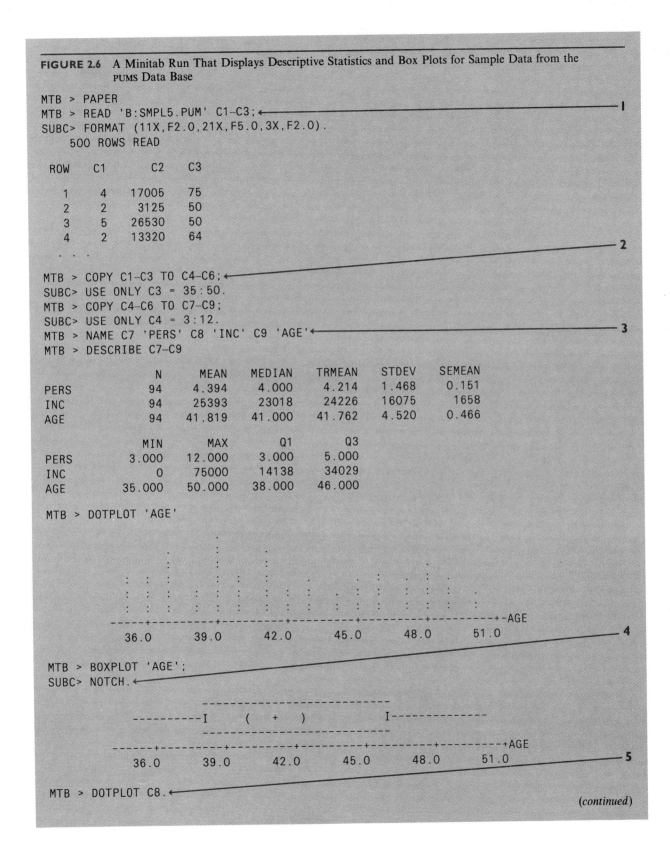

```
MTB > PAPER
MTB > READ 'B:SMPL5.PUM' C1–C3;  ←─────────────────────────────────────────  1
SUBC> FORMAT (11X,F2.0,21X,F5.0,3X,F2.0).
    500 ROWS READ

  ROW    C1        C2     C3

    1     4     17005     75
    2     2      3125     50
    3     5     26530     50
    4     2     13320     64
   . . .
                                                                          2
MTB > COPY C1–C3 TO C4–C6;  ←──────────────────────────
SUBC> USE ONLY C3 = 35:50.
MTB > COPY C4–C6 TO C7–C9;
SUBC> USE ONLY C4 = 3:12.
MTB > NAME C7 'PERS' C8 'INC' C9 'AGE'  ←─────────────────────────────────  3
MTB > DESCRIBE C7–C9

               N      MEAN    MEDIAN    TRMEAN    STDEV    SEMEAN
PERS          94     4.394     4.000     4.214    1.468     0.151
INC           94     25393     23018     24226    16075      1658
AGE           94    41.819    41.000    41.762    4.520     0.466

             MIN       MAX        Q1        Q3
PERS       3.000    12.000     3.000     5.000
INC            0     75000     14138     34029
AGE       35.000    50.000    38.000    46.000

MTB > DOTPLOT 'AGE'

                   .         :
                   .         :       .
          :  .  :  :      :  :  :  .  :  :  :  .
          :  :  :  :  :   :  :  :  :  :  :  :  :  .
          :  :  :  :  :   :  :  :  :  :  :  :  :  :
          :  :  :  :  :   :  :  :  :  :  :  :  :  :
        ----+---------+---------+---------+---------+---------+-AGE
          36.0      39.0      42.0      45.0      48.0      51.0
                                                                          4
MTB > BOXPLOT 'AGE';
SUBC> NOTCH.  ←────────────────────────────

                 -----------------------------
          ---------I    (   +   )         I--------------
                 -----------------------------
        ------+---------+---------+---------+---------+---------+AGE
          36.0      39.0      42.0      45.0      48.0      51.0
                                                                          5
MTB > DOTPLOT C8.  ←───────────────
```

*(continued)*

**FIGURE 2.6**   (*continued*)

```
                         .     .
                  .    :  .:   :.. .
           .   . :. ::: ::   ::: .
          :. :.:::::::: ::: ::..::  .: . . .    . :        :
          +---------+---------+---------+---------+---------+-------INC
          0        15000     30000     45000     60000     75000
```

```
MTB > BOXPLOT 'INC':
SUBC> NOTCH 95.
```

```
                   ----------------
          --------I    (+   )     I----------------            *
                   ----------------
          +---------+---------+---------+---------+---------+-------INC
          0        15000     30000     45000     60000     75000
```

MTB > INFO ←————————————————————————————————————————————— 6

```
COLUMN    NAME    COUNT
C1                 500
C2                 500
C3                 500
C4                 133
C5                 133
C6                 133
C7        PERS      94
C8        INC       94
C9        AGE       94

CONSTANTS USED: NONE

MTB > ERASE C1-C6



MTB > SAVE 'B:DESC1'
 Replace Existing File (Y/N)?

Worksheet saved into file: B:DESC1.MTW
MTB > RETRIEVE 'B:DESC1'
 WORKSHEET SAVED  1/ 1/2000

Worksheet retrieved from file: B:DESC1.MTW
MTB > INFO

COLUMN    NAME    COUNT
C7        PERS      94
C8        INC       94
C9        AGE       94
```

**FIGURE 2.6**    *(continued)*

```
CONSTANTS USED: NONE

MTB > SAVE 'B:DESC1'
 Replace Existing File (Y/N)?

Worksheet saved into file: B:DESC1.MTW
MTB > STOP
```

The Minitab routine provides descriptions for the number of persons, PERS; the household income, INC; and the age for the householder, AGE. For this example the data are restricted to the 500-record PUMS sample and are qualified to households with 3+ persons and age of householder 35 to 50 years.* Some observations are marked with numbers and arrows leading to the program steps. Here is a summary of these descriptions.

1 The data are read from the B-drive, the 'SMPL5.PUM' data file. This assumes a two-disk drive system. Use 'C:SMPL5.PUM' if the data file is stored on a hard-disk (C) drive.

2 Four Minitab statements (COPY–USE–COPY–USE) provide the reduction from the original 500 records to records with AGES 35 through 50 and with number of PERSONS 3 through 12. These qualifications reduce the active file to 94 records.

3 The variables PERSONS–C7, household INCOME–C8, and householder's AGE–C9 are DESCRIBEd with descriptive statistics, including the MEAN, MEDIAN, STandard DEViation, MINimum and MAXimum values, Quartile 1–Q1, and Quartile 3–Q3. (See the Minitab manual for a definition of the TRIMmed MEAN. We do not use it. Also, the standard ERROR of the MEAN, SE MEAN, is used in Chapter 7, and so is defined there). The DOTPLOT, displays the data values for AGE, providing some checks for the descriptive statistics. The MINimum and MAXimum values are 35 and 50. The MEAN—41.819–and MEDIAN—41.000—years values appear to be centrally located for the DOTPLOT data points. The data all appear to be reasonable values.

4 The BOXPLOT for AGE gives additional visual confirmation of the numerics provided at the DESCRIBE step. Q1 = 38.000 and Q3 = 46.000 form the end hinges, marked with the symbols (I). The MEDIAN (+) resides on the age scale, at about 41.00. The whiskers extend to the left and to the right to the MIN = 35.000 and MAX = 50.000 years values, respectively.

5 A DOTPLOT and a BOXPLOT for INCOME again support the descriptive statistics at **3**. The DOTPLOT displays the MINimum ($0), and MAXimum ($75,000) INCOME values. The MEAN ($25,393) and MEDIAN ($23,018) values are central for the INC DOTPLOT display. Minitab flags with a '*' as a mild outlier any value between 1.5 and 3 times the box width beyond either hinge. The box plot for INCOME displays one mild outlier, which is the $75,000 income.

* If you have access to the Minitab software and the 'SMPL5.PUM' data disk available from the publisher, then you can repeat the steps displayed in Figure 2.6 for a computer experience with the data base.

**6** The remaining Minitab commands involve data definition and SAVEing the work file. That is, the conditions at **2** have reduced the active file from 500 to 94 records by imposing the qualifiers 35-to-50-years householder AGE and 3 to 12 PERSONS. The SAVED file—'DESC1.MTW'—contains these 94 records.

It is now possible to describe some of the implications for marketing a consumer product to a typical U.S. family profile. In Figure 2.6 the data for householder age provide a check on the data selection. The maximum age, 50 years, and the minimum age, 35 years, are within the target population requirements. The MEDIAN value is close to, and hence supports, the arithmetic MEAN—41.8 years. By Chebyshev's rule the mean age plus or minus 2 standard deviations, or 28.2 to 55.4 years, includes at least 75% of the sample householder ages (the actual count for this range is 100%). This gives some initial insights into the age distribution for 35- to 50-year-old heads of household for a typical American family.

The statistics for household income are also restricted to the typical American family population. Observe that the MEAN and MEDIAN values are not as close for income as for age. This is as expected because there is greater variation in income values. However, the MEAN is a good descriptor for the ratio level measure of dollars income. The range, or MAXIMUM − MINIMUM = \$75,000, divided by 4 gives a coarse check on the standard deviation value.* Here RANGE/4 = \$18,750 is reasonably close to STD DEV = \$16,075.

In summary, our marketing data includes the following estimates for the typical U.S. family profile.**

1. Three or four people including children, age of head of household 35–50 (prespecified).
2. Mean age of householder is 41.8 years with standard deviation 4.5 years.
3. Mean household income is \$25,393 with standard deviation \$16,075.

These summary statistics are derived from the PUMS data base, reduced by qualifiers from 500 records to a sample of 94 records for the target population.

Although these statistics may appear to be limited, they are sufficient as a base for building some models to market family-oriented products. For example, it would be reasonable initially to target home computers as a family product toward households with head of household aged 35 years and older (to 50 years), aiming for the presence of older children, and with incomes \$25,393 and higher, signaling the ability to buy.

Working the Computer Problems at the end of Chapter 2 will provide you with more applications of computers for obtaining descriptive statistics.

---

* The relationship range/4 ~ standard deviation is justified by Chebyshev's rule because for any distribution *at least* 75% of the values are within $\pm 2$ standard deviations of their mean.

** The SPSS FREQUENCIES procedure provides more summary statistics than are displayed here. We have suppressed the actual FREQUENCIES listing that displays the unique values and their counts of occurrence to provide an example unencumbered of the data.

## COMPUTER PROBLEMS

**Data Base Problems**  The following problems are intended for computer solution. They draw on data from the 100-record PUMS sample. Minitab, SPSS, or SAS software is suggested for consistent use throughout the exercise. However, numerous other statistics software packages can give the same results.

42. Record (a) the household income and (b) the number of persons for a sample of 10 records from the PUMS. Calculate $\bar{X}$, $s$, and the median for these data (preferably do this by computer). Use your data and statistics to demonstrate Chebyshev's rule for $K = 2$ and $3$. For Minitab use DESCRIBE, for SPSS use the DESCRIPTIVES procedure, and for SAS use PROC MEANS in the data step.

43. Using your sample from Problem 42, find the number of persons for each record. Next divide the household income by the number of persons to obtain a per capita income, PCI.
    a. Compute mean and standard deviation of per capita income. (If you're using SPSS, this will require a single COMPUTE-type statement, such as

    ```
    COMPUTE PCI = RND (INCOME/NOPERS)
    ```

    b. How many of your 10 records have per capita income within ($\bar{X} \pm 1$ sd), ($\bar{X} \pm 2$ sd), and ($\bar{X} \pm 3$ sd)?
    c. Compare your findings (i) to Chebyshev's rule and (ii) against the bell-curve percentages. Also compare your findings with those of several of your classmates.

44. Repeat Problem 42 for a sample of 25 records. This time do not hand-check the calculations. What are some advantages of computer calculation of descriptive statistics over hand calculation? What are some disadvantages?

45. Repeat Problem 43 for your sample of 25 records.

46. Use computer runs to check the statistics provided in the business case. This can be done with SPSS by a FREQUENCIES procedure on the values for Group I, then Group II. An alternative is to use another program, such as Minitab. With Minitab, create a worksheet with two columns of data named Group I and Group II, and containing the tax rate data. Use your computer procedure to compute (a) mean, (b) median, (c) standard deviation, and, if it is available, (d) the coefficient of variation. (For example, MTB > DESCRIBE 'GROUP I' gives all but (d).) Compare your numbers with the ones given in the business case.

47. Now obtain a display of the data in Problem 46. For example, in SPSS use

    ```
    FREQUENCIES VARIABLES
       =GROUP I GROUP II/HISTOGRAM/
    ```

    For Minitab, use

    ```
    MTB > DOTPLOT 'GROUP I'.
    ```

    Using your displays and statistics from the previous problem, explain why Mrs. Kemp can say that when the $86 figure is deleted, the Group I data makes a good case for a rate of $26 per thousand. On what statistical basis does she discredit the Group II data?

48. Here is a Minitab routine for a box plot summary of the data in Table 2.4: Weekly sales (in $000) for 40 insurance sales representatives. If you have access to Minitab, key in the program and compare your results with those for Example 2.11.

    ```
    MTB > PAPER
    MTB > SET INTO C1
    DATA>   75  96  99 102 106 109 112 114 116 118
    ```

```
DATA>   119 120 122 122 123 123 124 124 125 125
DATA>   125 125 126 127 127 127 128 130 131 131
DATA>   133 134 135 137 141 144 147 152 156 170
DATA> END
MTB > NAME C1 'SALES'
MTB > PRINT 'SALES'
MTB > DESCRIBE 'SALES'
MTB > DOTPLOT 'SALES'
MTB > HISTOGRAM 'SALES'
MTB > BOXPLOT 'SALES';
SUBC> NOTCH.
MTB > STOP
```

49. The data for corporate profit rates (percentages) are the rate of profit on invested capital for 20 firms. Use a Minitab computer run like the one in the previous problem to develop a box plot. Indicate the values that are outliers and describe the distribution skew as left or right or negligible. Compare your results with those for Problem 18.

| | | | | |
|---|---|---|---|---|
| 6.9 | 2.1 | 3.0 | 0.4 | 7.3 |
| 7.5 | 7.6 | 11.6 | 7.7 | 4.5 |
| 7.4 | 7.3 | 5.0 | 2.6 | 6.6 |
| 6.3 | 12.1 | 12.2 | 5.0 | 6.3 |

## FOR FURTHER READING

[1] Campbell, Steven K. *Flaws and Fallacies in Statistical Thinking*. Englewood Cliffs, NJ: Prentice-Hall, 1974.

[2] Folks, J. Leroy. *Ideas of Statistics*. New York: John Wiley, 1981.

[3] Hamburg, Morris. *Statistical Analysis for Decision Making*, 5th ed. San Diego: Harcourt Brace Jovanovich, 1991.

[4] Hooke, Robert. *How to Tell the Liars from the Statisticians*. New York: Marcel Dekker, 1983.

[5] Ryan, Barbara E., Brian L. Joiner, and Thomas A. Ryan, Jr. *Minitab Handbook*, 2nd ed. Boston: PWS-Kent, 1985.

[6] *SPSS Reference Guide* (version 4.0). Chicago: SPSS, Inc., 1990.

# 3

# DATA SUMMARIZATION

## 3.1 INTRODUCTION

In the past, business decision makers often complained of not getting enough information in time to make the right decisions. One reason for this was that statistics had to be developed by hand. The first business machine—the comptometer—was little more than an abacus. Later, paper-tape adding machines were developed to check for human errors in tabulation.

Today, computers spew out data extremely rapidly, and, barring electrical short circuits or operator error, there is no error in the computations. An entire data set, such as for all tax returns in a tax district, might yield a stack of computer output a foot high. Yet a few pages of summary statistics might serve the tax auditor's needs equally well. Thus the problem becomes one of choosing to print out an entire file of many pages or to print out something less. The decision maker's dilemma has changed from having too little information too late to having too much data that is arriving too quickly to be used effectively.

Our brains can comprehend only a limited amount of numerical data at any one time. Thus it is advantageous to group data and characterize them by summary statistics that we can readily comprehend and use.

In this chapter, our purpose is to learn about alternative ways to develop statistics for large data sets. When data consist of relatively few observations of a variable (up to 20 or 30 values, for instance), we can easily handle them as ungrouped data, as we did in Chapter 2. But when data consist of many observations (like stock market prices), it is advantageous to group them into classes. Grouping results in slightly less accurate calculations, but it usually makes the meaning of the data clearer because it reduces the data classes to a number we can readily comprehend.

Grouped data are data categorized into classes and displayed by means of tables, graphs, or summary measures.

This chapter demonstrates methods for grouping data into frequency distributions and displaying those distributions by means of graphs. Studying graphs can provide insights about data summarization. Examples will show how computer calculations on ungrouped data provide quick and accurate summaries of business data. By comparing ungrouped to grouped data results, we can also appreciate the loss of identity that results from grouping and the subsequent inaccuracy in numerical descriptions. This helps us see the limitations of grouped data.

Upon completing Chapter 3, you should be able to

1. group (original) numerical data into a reasonable number of classes;
2. construct a frequency distribution for numerical data;
3. display grouped data in a histogram and a frequency polygon;
4. calculate numerical measures of grouped data;
5. explain how calculators and computers can simplify statistical calculations.

## 3.2 DISCRETE AND CONTINUOUS VARIABLES

The data used for business decision making are typically gathered by observing some variable of interest. A *variable* is any name or symbol that takes on different values.

A variable is a name or symbol that can take on different values.

For example, a variable might be the level of employment in cities of over 100,000 population. For convenience, we represent each variable by a symbol, such as $X$ or $Y$.

Variables have either *discrete* or *continuous* measure.

A discrete variable can have a countable number of distinct values.

A continuous variable can have more than a countable number of values.

*CATEGORICAL*

The number of shoppers in a particular department store at noon and the number of business failures recorded last year are both *discrete variables*. Here the level of measurement, as described in Chapter 2, can be nominal or ordinal.

A *continuous variable* can assume any value from the possible values on a continuous scale, such as 3.2, 3.21, and 3.214. Here the level of measurement is either ratio or interval. The expected service life of a new appliance and the weight of copper ore shipped from a mine are examples of continuous variables. Copper ore, for example, might be recorded to the nearest ton (21 tons), tenth of a ton (21.4), or even thousandth of a ton (21.432), depending upon the desired accuracy. That is, because the variable is continuous, it could take any value over a given interval (in this example, 21 to 21.432).

As a matter of practicality, continuous variable values must be rounded. We can express continuous variables in the manner of discrete variables by rounding to discrete-sounding values like 21. However, rounding does not make the data discrete.

We distinguish between discrete and continuous variables because we must treat them differently when finding their average and their dispersion; another distinction is made in the next section, which describes procedures for grouping data.

## 3.3 FREQUENCY DISTRIBUTIONS

*Frequency distributions* condense and summarize data by grouping them into classes. Each class then represents a convenient grouping of one or more values of the variable that defines the data.

> A frequency distribution displays the number of times a variable takes on the data values for each of several classes.

Table 3.1 illustrates a frequency distribution of the weekly earnings of 200 employees at the Sunsoya Motorcycle plant. The range of earnings has been divided into seven classes and the distribution reveals the number of employees with earnings that reside in each class. Frequency distributions are referred to as either *absolute* or *relative*. The classification depends on whether the distribution shows the total frequency in each class (absolute) or the ratio of the frequency of each class to the total frequency for all classes (relative). Table 3.1 illustrates both types.

The classes into which data are grouped have four important defining characteristics: *class limits*, *boundaries*, *intervals*, and *midpoints*.

> Class limits are the smallest and largest observed values that can belong to a class.

> Class boundaries are the actual values that separate successive classes or define a (upper or lower) boundary to the class.

*Class limits* are the smallest and largest observed values that go into any class. Because they are the values published in tables and reports, we refer to them as the *stated class limits*. In Table 3.1, the stated limits of the first class are $250 and $299. Stated class limits must accommodate all the data; that is, no data should fall between classes or outside the stated upper and lower class limits. (The original data collected for Table 3.1 were rounded to the nearest dollar to fit within the classes).*

*Class boundaries* are the actual values that divide successive classes and are sometimes referred to as *real class limits*. We need them to compute some summary statistics of grouped data. Class boundaries are halfway between the upper class limit of one class and the lower class limit of the next. They are usually stated in one more decimal place of accuracy than the stated class limits. In Table 3.1 the class boundaries for the second class are $299.50 and $349.50. (These are carried to two decimal places because the variable (cents) is commonly stated to this accuracy.)

The *class interval* and the *class midpoint* are used to calculate statistics from a frequency distribution.

> The class interval is the distance spanned by the boundaries of a class.

> The class midpoint is the arithmetic mean of its class boundaries.

The *class interval* is the difference between the upper and lower boundaries of any given class. For example, the interval for the second class in Table 3.1 is $349.50 − $299.50 = $50.00.

---

\* For example, an earnings of $299.54 was rounded to the nearest $1 and recorded as $300. Earnings exactly halfway between the nearest whole numbers are rounded to an *even* number. Thus $298.50 is rounded down to $298.

**TABLE 3.1**   Frequency Distribution of
Weekly Earnings

| Earnings | Absolute Frequency | Relative Frequency |
|---|---|---|
| $250–299 | 20 | 20/200 = 0.10 |
| 300–349 | 48 | 48/200 = 0.24 |
| 350–399 | 62 | 62/200 = 0.31 |
| 400–449 | 36 | 36/200 = 0.18 |
| 450–499 | 22 | 22/200 = 0.11 |
| 500–549 | 8 | 8/200 = 0.04 |
| 550–599 | 4 | 4/200 = 0.02 |
|  | 200 | 1.00   (100%) |

The *class midpoint*, also called the class mark, is the mean of the boundaries for any given class. For example, the midpoint of the second class in Table 3.1 is ($349.50 + $299.50) ÷ 2 = $324.50.

Five steps are used to construct a frequency distribution like the one in Table 3.1.

### STEPS FOR CONSTRUCTING A FREQUENCY DISTRIBUTION DISPLAY TABLE

1. Array (list) the data values in order by size from lowest to highest (or vice versa).
2. Compute the difference between the smallest and largest values, that is, the range.
3. Divide the range into a convenient number of class intervals of equal size.
4. Count the number of observations in each class to determine the total frequency.
5. Display the class intervals with their frequencies as a frequency distribution table or graph.

In Step 3, either the number of classes or the class interval size must be arbitrarily chosen. It is generally wisest to set the class interval first, because a judicious choice here can simplify later calculations. Intervals of 10, 5, or 2 can be convenient for calculations and so are common in working with grouped data. A good rule of thumb is to select a class interval that allows from 6 to 15 classes. Too many classes can destroy the summary effect of the grouping; too few classes can produce an oversimplification of the data and result in inaccuracies from subsequent calculations.

Classes must contain all the data without any classes overlapping. For example, intervals of 200–300 and 300–400 overlap at 300. Therefore, the intervals should be 200 to <300 and 300 to <400, and so on. For clarity and convenience, classes should be of equal size rather than unequal or open-ended (such as "$550 and higher").

We review the construction of a frequency distribution by grouping some primary data into classes. The variable—speed—is continuous because it is measurable (rather than countable). This fact determines some conditions for the frequency distribution. Example 3.1 demonstrates the construction of a frequency distribution.

---

**EXAMPLE 3.1**

An automobile manufacturer has conducted speed tests on 50 engines of a new design. This testing is essential for quality control to ensure that the product, when mass-produced, will meet the manufacturer's warranty or other advertised standards.

a. Group the data (see table) into classes and construct a frequency distribution of the tested speeds.

b. What is the frequency for engine speeds from 1700 to 1800 revolutions per minute (rpm)?

| | | | | | | | | | |
|------|------|------|------|------|------|------|------|------|------|
| 1716 | 1743 | 1752 | 1737 | 1723 | 1809 | 1758 | 1724 | 1803 | 1787 |
| 1738 | 1680 | 1741 | (1667) | 1690 | 1681 | 1747 | 1779 | 1752 | 1754 |
| 1684 | 1761 | 1766 | 1701 | (1818) | 1752 | 1686 | 1692 | 1760 | 1694 |
| 1700 | 1738 | 1687 | 1754 | 1755 | 1735 | 1785 | 1703 | 1731 | 1729 |
| 1667 | 1813 | 1775 | 1747 | 1723 | 1768 | 1735 | 1749 | 1743 | 1755 |

*Solution*   a. Following the steps just outlined, we array the data in order by size and find that the highest speed is 1818 rpm and the lowest is 1667 (both are circled in the table).* The range is thus $1818 - 1667 = 151$. An arbitrary choice of 20 as a class interval would give 8 classes, since $151 \div 20 = 7.55$. It is often convenient to assume that the limits and boundaries coincide and then state the class intervals so that they include everything under the next class limit. Thus the first class includes all observations from 1660 up to but not including 1680, the second includes 1680 up to but not including 1700, and so forth. See the frequency distribution shown in Table 3.2.

**TABLE 3.2**   Frequency Distribution of Engine Speeds

| Engine Speed (rpm) | Tally | Frequency |
|---|---|---|
| 1660 but less than 1680 | \|\| | 2 |
| 1680 but less than 1700 | ⊬⊦ \|\|\| | 8 |
| 1700 but less than 1720 | \|\|\|\| | 4 |
| 1720 but less than 1740 | ⊬⊦ ⊬⊦ | 10 |
| 1740 but less than 1760 | ⊬⊦ ⊬⊦ \|\|\|\| | 14 |
| 1760 but less than 1780 | ⊬⊦ \| | 6 |
| 1780 but less than 1800 | \|\| | 2 |
| 1800 but less than 1820 | \|\|\|\| | 4 |
| | | $n = 50$ |

b. Table 3.2 provides the essential information to determine that 36 (of 50) engine speeds were from 1700 up to but less than 1800 rpm. That is, 72% of the engine speeds were within the specified range.

---

* Any large data sets will most likely be handled on computer, where sort routines can do this in seconds.

A tally column like the one in Table 3.2 is an aid to placing values into their appropriate classes. However, the tally is generally not shown in a final presentation of the frequency distribution.

Note that the data were recorded to the nearest revolution per minute; the class limits are stated to that same accuracy. Each observation clearly falls into only one class. These data are continuous and have been rounded to whole numbers. Because the class limits and boundaries coincide, there is no gap from the upper stated class limit of one class to the lower stated class limit of the next, as is often the case when discrete data are illustrated.

## 3.4 GRAPHICAL DESCRIPTIONS OF GROUPED DATA

The purpose of graphing is to provide a visual display of data. For example, suppose we were interested in knowing what percentage of employees have weekly earnings higher than $500. Is the manufacturing cost per unit of a product fluctuating by more than $12 from one month to the next? We *could* resolve these questions with the aid of frequency distributions or simply by making counts, but graphing adds another dimension by allowing the decision maker to visualize patterns in the data quickly. Graphs and charts are valuable decision-making tools because they show key ideas and relationships. Marketing representatives frequently organize their sales presentations around a few key graphs.

Computers are, of course, becoming indispensible for displaying data. By using graphics programs, managers can obtain their companies' accounting, inventory, or marketing statistics in seconds. This section discusses some of the graphics most used in business.

Histograms and frequency polygons depict frequency distributions.

**A histogram is a graphical display of a frequency distribution that uses rectangles whose widths represent class intervals and whose heights are proportional to the corresponding frequencies.**

**A frequency polygon is a line graph of a frequency distribution on which the frequency of each class is plotted above its class midpoint.**

A *histogram,* or bar graph, uses rectangular areas to define the frequency for each class. It displays the overall pattern of the data and the frequency of observations in each class. This helps the manager to get an intuitive understanding of the data. Figure 3.1(a) is a histogram of the data from Table 3.1. The class limits of weekly earnings are shown on the horizontal axis, and the areas are proportional to the frequencies of the respective classes.

Figure 3.1(b) is a *frequency polygon,* or line graph, of the same data. The frequencies for each class are plotted above the class midpoints. To give a sense of completeness to the graph, both extremes of the graph are extended to the base, at points that *would be* the adjacent class midpoints. The frequency polygon is an alternative display to a histogram. Observe that both convey the same overall pattern.

Frequency data are sometimes plotted in a *cumulative frequency distribution,* also referred to as an *ogive* (pronounced 'ō-jīv).

**A cumulative frequency distribution, or ogive, is a frequency polygon whose line segments display an increasing (decreasing) graph that is defined by the accumulation (elimination) of frequency class counts.**

**FIGURE 3.1** Graphs of Weekly Earnings

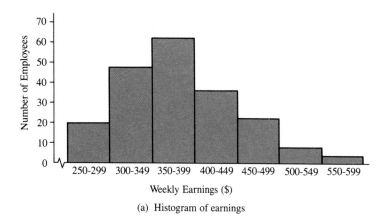

(a) Histogram of earnings

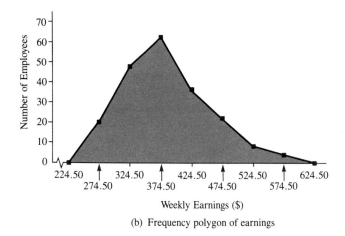

(b) Frequency polygon of earnings

In an ogive the frequencies are accumulated to the class boundaries, as illustrated in Table 3.3 and Figure 3.2.

**TABLE 3.3** Weekly Earnings of Employees—Cumulative Distribution

| Earnings | Frequency | Cumulative Classes Less Than | Cumulative Frequencies |
|---|---|---|---|
| | | less than $249.50 | 0 |
| 250–299 | 20 | less than 299.50 | 20 |
| 300–349 | 48 | less than 349.50 | 68 |
| 350–399 | 62 | less than 399.50 | 130 |
| 400–499 | 36 | less than 449.50 | 166 |
| 450–499 | 22 | less than 499.50 | 188 |
| 500–549 | 8 | less than 549.50 | 196 |
| 550–599 | 4 | less than 599.50 | 200 |
| | $n = 200$ | | |

**FIGURE 3.2**  A Less-Than Cumulative Frequency Distribution (Ogive) for Weekly
Earnings

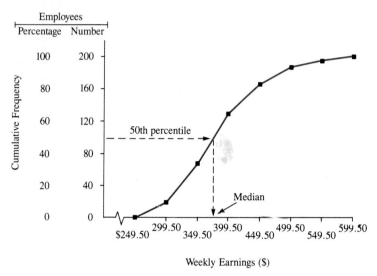

This ogive is a *less-than* frequency distribution because the cumulative
frequency totals plotted at each class boundary show the number of observations that are less than the boundary value. For example, Table 3.3 shows that
130 employees received *less than* $399.50 per week. A *more-than* cumulative dis-
tribution would have the reverse shape, beginning with the initial observation
that all 200 employees earned more than $249.50 per week, 180 employees
earned more than $299.50 per week, and so on.

Cumulative distributions are useful for analyzing financial and production problems. For example, they can show the rates of return available for
various investment amounts and the inventory levels (in excess of median
demand) necessary to limit the risk of running out of production stock. The
median is, of course, the 50th percentile value. For the weekly earnings data in
Figure 3.2, the median appears to be roughly $375. Other percentile values—
the 10th, 25th, 75th, and so on—can be read just as quickly. Simply read from
the scale to the left of the vertical axis in Figure 3.2.

Whether you use a histogram, a frequency polygon, or an ogive depends
upon the situation and your preference. If, for example, the purpose is to display
sales from the perspective of *more than* breaking even against expenses, then
an ogive would be most meaningful. If you want to convey the idea of a smooth
pattern or transition of data, then a frequency polygon is preferred over a histogram. The histogram gives a clear distinction for differences in class counts.
It is often the preferred marketing tool and is used extensively to display accounting data in corporation annual reports.

In the past, businesses frequently employed graphic artists. This has
changed, however, now that computer graphics and desktop publishing software
are available for most standard business needs. In addition, laser printers are
equipped to produce a wide variety of graphics in black and white or in color.

**Problems 3.4**

1. Use the frequency distribution data in the table.
   a. What are the class limits for the first (lowest) class?
   b. What are the class boundaries for the class 3.00–5.99?

c. What is the class midpoint for the class 6.00–8.99?

d. What is the class interval size? *5.995 – 8.99  ≥ 3*

e. If the two most extreme scores were 3.08 and 14.65, what would be the range of the values?

| Class Limits | 3.00–5.99 | 6.00–8.99 | 9.00–11.99 | 12.00–14.99 |
|---|---|---|---|---|
| Frequency | 1 | 4 | 10 | 6 |

2. A sample of 120 spools of kite string produced by a winding machine yields a lowest breaking strength of 3.1 pounds and a maximum strength of 18.6 pounds. Using a class interval of 2.0, calculate the appropriate number of classes and display all class intervals.

*3.0 4.9*
*5.0 - 6.9*

3. For the given scores of a discrete variable, set up a frequency distribution and plot the associated histogram. (*Suggestion:* Use 3 through 7 as the interval for the first class.)

| Scores | 8 | 7 | 15 | 19 | 27 | 17 | 21 | 16 |
|---|---|---|---|---|---|---|---|---|
| | 19 | 24 | 3 | 20 | 17 | 26 | 14 | |

4. A frequency distribution is needed to display the rate of profit on invested capital of 20 firms. The given profit rates (percentages) are rounded to one decimal place. Establish a frequency distribution table and plot the related histogram.

Rates of Return on Invested Capital (%)

| | | | | |
|---|---|---|---|---|
| 6.9 | 2.1 | 3.0 | 0.4 | 7.3 |
| 7.5 | 7.6 | 11.6 | 7.7 | 4.5 |
| 7.4 | 7.3 | 5.0 | 2.6 | 6.6 |
| 6.3 | 12.1 | 12.2 | 5.0 | 6.3 |

5. The data in the table below describe the weekly take-home pay for 20 semi-skilled laborers. Using the classes $300.00–$309.99, $310.00–$319.99, ..., $340.00–$349.99, set up a frequency distribution table and plot the appropriate histogram and frequency polygon. State the class boundaries and class midpoint for the class $330.00–339.99.

Weekly Take-Home Pay ($)

| | | | |
|---|---|---|---|
| 319.12 | 326.81 | 324.79 | 313.48 |
| 331.50 | 348.39 | 337.24 | 326.67 |
| 320.76 | 321.67 | 331.47 | 326.67 |
| 325.42 | 315.38 | 304.12 | 321.19 |
| 333.98 | 340.89 | 327.02 | 327.11 |

6. A check on the number of metal washers per box was made on 25 boxes received by the Pioneer Metal Stove Company (see table).

| Washers per Box | | | | |
|---|---|---|---|---|
| 56 | 76 | 63 | 53 | 35 |
| 47 | 62 | 68 | 49 | 42 |
| 82 | 65 | 45 | 56 | 56 |
| 45 | 46 | 53 | 59 | 57 |
| 63 | 66 | 50 | 67 | 58 |

a. Set up a frequency distribution and plot a histogram. Include labels for each scale and a title for your figure.
b. Establish an ogive (like Figure 3.2) of the percentage of boxes with less than the upper class limits of your distribution. Use your graph to estimate (i) the median, (ii) the "value" (washers per box) that defines the top 10 percentile, and (iii) two values to define a range for the 25th percentile to the 75th percentile.

7. Use the accompanying frequency distribution (see table).

| Class | Frequency | Class | Frequency |
|---|---|---|---|
| 30–39 | 2 | 70–79 | 21 |
| 40–49 | 6 | 80–89 | 7 |
| 50–59 | 4 | 90–99 | 8 |
| 60–69 | 12 | Total | 60 |

a. What are the class boundaries for the class 60–69?
b. How many of the scores are in the range 40–59?
c. If the actual values of the scores in the class 50–59 are 51, 53, 54, and 58, then by how much does the class midpoint differ from their mean?
d. Using information from part (c) as well as the frequency distribution, determine what percentage of the scores are less than 55.
e. Construct a cumulative frequency distribution and use it to estimate the value of the median.

8. The market research department of a clothing manufacturer must select an appropriate fabric for its line of spring fashions in an eastern market area. A real consideration is the climate and how warm the resulting clothes will be. The researchers have obtained historical data on the daily mean temperature for 30 days as shown.

| Daily Mean Temperature Readings (degrees Fahrenheit for $n = 30$ days) | | | | | | | | | |
|---|---|---|---|---|---|---|---|---|---|
| 62 | 71 | 64 | 60 | 62 | 66 | 66 | 62 | 69 | 64 |
| 68 | 67 | 63 | 61 | 65 | 63 | 60 | 58 | 57 | 60 |
| 64 | 56 | 59 | 63 | 64 | 59 | 61 | 61 | 59 | 57 |

a. Arrange the data into a frequency distribution having eight classes and show both the absolute and relative frequencies.
b. Construct a frequency polygon for the data.
c. On how many days was the mean temperature greater than 67°F?
d. On how many days was the temperature 59°F or less?
e. What proportion of the time was the mean temperature in the sixties?
f. What percentage of the days had a temperature of 69°F or more? (*Hint:* This question cannot be answered directly from the frequency distribution, since the grouping process results in a loss of identity of individual values. Try to answer it by using the individual readings, as needed, in conjunction with the frequency distribution.)

## 3.5 DATA ANALYSIS

Although the mean, median, standard deviation, and other statistics described in Chapter 2 can be considered summary measures, data summarization also involves cleaning, condensing, and correcting data.

> Data analysis encompasses procedures for checking data and making corrections to ensure that the data accurately describe the real situation.

Data analysis encompasses an action and a reaction. The action of data analysis is to check that data are meaningful. This includes implementing procedures to identify values that cannot be reasonable due to errors in the data preparation. If it is necessary, the reaction is to make corrections before the data set is used: The data need to be real. Further, it is critical that good data not be changed. There is no place for distortion or misuse of data in statistical decision making.

Chapter 2 included several procedures for data analysis action—for example, box plots to check for outliers plus bell-curve and skewness checks to analyze the distribution pattern or general shape. Also, a levels-of-measurement evaluation is a data analysis action that can help identify the most appropriate average.

In data analysis the reaction most commonly warranted is a simple one. It is to correct any data value found to have been altered. Errors commonly result from incorrect data entry, incorrect copying of data from the original source, incorrect translations, and incomplete or missing data. The reaction can be as simple as editing the data. However, methods for carrying out specific reactions are not normally the concern of statistics textbooks, and so are not covered here. Reactions in statistics include the treatment of outliers (which we will revisit in Chapter 13); the use of assumptions or conditions that make a statistics procedure appropriate (in Chapter 8 and following); data plots in Chapters 12, 13, and 14; control charts in Chapter 15; and, in Chapter 17, the option of using nonparametric statistics when the data are of the interval or ratio measurement level.

Data analysis is timely, and can help you know whether your results are reasonable before you do any statistics. This is especially important when computers are used or data are grouped, since both can give reasonable-looking results whether the data are correct or not. The illustration that follows uses accounting billings for which the data are sales for direct-mail advertising.

**TABLE 3.4**   Accounting Job File Inventory Sequenced by Descending Price per Thousand Names for Six Months Ending July 1

| Sequence Number | Job Number | Client | Date (Mo–Day) | (A) Revenue Amount | (B) Quantity of Names | (A/B) Price per 1000 Names |
|---|---|---|---|---|---|---|
| 1 | N327B | Suburban Rsch | 3–2 | $  47,189 | 131,212 | $360 |
| 2 | Z128A | BILCO | 3–8 | 49,813 | 199,900 | 249 |
| 3 | W001C | NCB | 5–24 | 41,099 | 563,030 | 73 |
| 4 | A124A | Zion Assoc. | 3–15 | 55,198 | 1,000,000 | 55 |
| 5 | H010B | ABC List | 2–11 | 14,775 | 286,900 | 51 |
| 6 | S999C | S&L Banks | 6–21 | 6,014 | 125,100 | 48 |
| 7 | F675K | Media Mart | 2–15 | 10,508 | 228,490 | 46 |
| 8 | G900A | Tyne Labels | 1–28 | 9,000 | 200,000 | 45 |
| ⋮ | | | ⋮ | | | ⋮ |
| 800 | J263B | World Globe | 2–13 | 19,350 | 500,000 | 39 |
| 801 | D986D | ABC Donors | 1–3 | 11,124 | 300,230 | 37 |
| 802 | K243B | US List | 5–7 | 5,145 | 150,000 | 34 |
| ⋮ | | | ⋮ | | | ⋮ |
| 2122 | L921A | Research Inc. | 4–9 | 6,860 | 326,700 | 21 |
| 2123 | Z128B | BILCO | 1–3 | 4,000 | 400,000 | 10 |
| 2124 | Z128C | BILCO | 2–8 | 4,000 | 400,000 | 10 |
| 2125 | Z128D | BILCO | 4–12 | 4,000 | 400,000 | 10 |
| 2126 | Z128E | BILCO | 5–6 | 4,000 | 400,000 | 10 |
| 2127 | Z128F | BILCO | 6–7 | 4,000 | 400,000 | 10 |
| Total | | | | $9,638,246 | 461,043,002 | $ 20.90 |

The data in Table 3.4 show the six-month sales of (hypothetical) XYZ Company by job number and client. The XYZ Company sells lists of names of potential customers to other companies for their use in direct-mail advertising. XYZ had 2127 individual orders for mailing lists during this six-month period. The orders ranged from fewer than one hundred thousand to over one million names and generated a revenue of over $9.6 million for the firm. Table 3.4 includes the quantity of names supplied to each client under each order and the price per 1000 names.

Every data summary warrants some checks. This one is no different. The data include a sequence number, a job number, the client's business name, the date on which the order was received, the job dollars revenue (A), the quantity of names sold (B), and the price per 1000 names sold (A/B). Simple data analysis—editing—is appropriate here. Two jobs require examination based on extreme values for price per thousand.

1. Suburban Rsch shows a price of $360 per 1000 names. This is more than 10 times the overall price per thousand—$20.90. Here it is a common practice to round the Quantity to the nearest 10s position and place a zero in the units position. By this convention the Suburban Rsch quantity appears to not be rounded. A check with the salesperson indicated that the correct quantity was 1,312,120—10 times the number reported. So the correct price per 1000 names was $35.96. This rounds to $36.

**2.** Observation of the BILCO job shows multiple entries, all with the job number Z128_. This suggests a possible miscommunication. Investigation revealed that the job was sold with monthly payments and monthly quotas on quantity. To get a realistic price, we must combine all job numbers Z128_ to compute a job Revenue Amount and a Quantity, then divide, $A/B$, for a realistic Price per 1000 Names. The corrected entries for BILCO are

| | |
|---|---|
| Revenue Amount | $69,813 |
| Quantity | 2,199,900 |
| Price per 1000 | $32 |

With these corrections and the $(A/B)$ total value, the report is ready for a statistical summary.

The error with Suburban Rsch is a case of incorrect data editing, a human error. These can be difficult to find and they often require a review of the original data. The error on BILCO illustrates a logical error in preparing the inventory report. This can be a consequence of two or more persons working on one data summary. When data are prepared by one person, then used by another, there can be omissions or incorrect presentation. The user should check with those who have prepared the data to ensure that the steps were complete and appropriate. Here a statistic—price per 1000 units—helped to identify the existence of problems.

Viewing some of the original data before beginning any analysis is a worthwhile practice. For example, in the Chapter 2 "Computer Application," we listed some of the data prior to a statistical description. This is our regular practice. First, it helps us to understand the data. "Has it been correctly read into the computer?" and "Are all of the data values entered?" are basic questions that need a positive response if any subsequent results are to be valid. Second, statistics software frequently allows grouping of the data, with subsequent summaries. The raw data can be used to check on a few of the group summaries. Third, even with clean (edited) data there can still be extreme values—possible outliers. We can check for those values with box plots. Finally, another exploratory data analysis procedure is the stem-and-leaf display. [11]*

**Stem-and-Leaf Displays**  Histograms and frequency distributions are alternate ways of visualizing an entire data set. Another such concept, the stem-and-leaf display, is best explained by example, as follows.

This example uses the data on sales for 40 insurance sales reps as first presented in Table 2.4. The data are repeated in Table 3.5, where sales are in thousands of dollars ($000).

For example, 75 represents $75,000 of sales. To form a stem-and-leaf display, the numbers are split into two parts. By choice, the two left-most positions will become the stem, while the units position is treated as a fractional part (as the leaf). In stem-and-leaf notation, 75 is expressed as 7 | 5 and represents, here,

---

* Numbers in brackets refer to sources listed at the end of the chapter in "For Further Reading."

**TABLE 3.5** Sales ($000) for a Sample of 40 Sales Representatives

| | | | | | | | |
|---|---|---|---|---|---|---|---|
| 75 | 109 | 119 | 123 | 125 | 127 | 133 | 144 |
| 96 | 112 | 120 | 124 | 125 | 128 | 134 | 147 |
| 99 | 114 | 122 | 124 | 126 | 130 | 135 | 152 |
| 102 | 116 | 122 | 125 | 127 | 131 | 137 | 156 |
| 106 | 118 | 123 | 125 | 127 | 131 | 141 | 170 |

$75 \cdot \$1,000 = \$75,000$ in sales. A similar description for the 40 insurance sales amounts results in the stem-and-leaf display of Figure 3.3.

The stem-and-leaf display provides some visual description of the characteristics of the data. The lowest and highest sales values are, respectively, 75 and 170, in $1000 units. Then the range of sales is $170 - 75 = 95$ ($000), or $95,000. Each stem—the 7, 8, 9, and so on—identifies a group or class, while the number of leaves in the class displays a frequency for data values in that class. For example, stem 7 is a class with frequency one—the data value 75. Observe that stem 8 has no leaves; the frequency is zero. Continuing, the class with stem 9 has two leaves—the values 96 and 99—and the class frequency is two. Thus a stem-and-leaf display is a frequency distribution with specifically defined classes, the stems, and with frequencies, the leaves, that display every data value.

> A stem-and-leaf display is a frequency distribution wherein the summarized data values define a set of classes, or stems, and the number of leaves defines the frequency with which the values are distributed among the classes.

Stem-and-leaf displays and box plots (Chapter 2) are techniques of exploratory data analysis that were introduced by John Tukey [11]. These procedures, partly numerical and partly graphical, focus on using arithmetic and graphic displays to summarize data quickly. Both can be used for statistical description.

The stem-and-leaf diagram is an option in many of the commercial statistics software packages. To illustrate, the next example uses the Minitab soft-

**FIGURE 3.3** Stem-and-Leaf Display for 40 Insurance Sales Amounts (leaf digit unit is $1000)

```
 7 | 5
 8 |
 9 | 6 9
10 | 2 6 9
11 | 2 4 6 8 9
12 | 0 2 2 3 3 4 4 5 5 5 5 6 7 7 7 8
13 | 0 1 1 3 4 5 7
14 | 1 4 7
15 | 2 6
16 |
17 | 0
```

ware to construct a histogram and a stem-and-leaf display. The example also illustrates the point that grouping data causes a loss of identity and can result in inaccuracies for descriptive statistics.

| **EXAMPLE 3.2** |
|---|

The 40 insurance sales ($000) in Table 3.5 can be described with a stem-and-leaf display or as a frequency distribution (a histogram). Compare the alternative displays for these data.

*Solution*   In Figure 3.4, the Minitab software provides the graphic displays from the (1) STEM-AND-LEAF and (2) HISTOGRAM procedures. (The same data were used to illustrate the box plot procedure in Chapter 2.)

**FIGURE 3.4**   Display of 40 Insurance Sales ($000) by the Minitab Stem-and-Leaf and Histogram Procedures

```
  1     7 | 5                        Midpoint   Count
  1     8 |                             80        1    *
  3     9 | 69                          90        0
  6    10 | 269                        100        3    ***
 11    11 | 24689                      110        4    ****
(16)   12 | 0223344555567778          120       10    **********
 13    13 | 0113457                    130       14    **************
  6    14 | 147                        140        4    ****
  3    15 | 26                         150        2    **
  1    16 |                            160        1    *
  1    17 | 0                          170        1    *
```

    (a) Stem-and-Leaf Display            (b) Histogram

The stem-and-leaf display is essentially the same as in Figure 3.3. Here, however, an additional (first) column indicates the cumulative count for the number of leaves on the stem or on a stem nearer to the end of the display (Figure 3.4(a)). So there is one value at 75, no 80s, two values in the 90s, and so on. The number in parentheses (16) is the number of leaves in the median class.

The histogram in Figure 3.4(b) shows a slightly different picture because it uses a different logic for the grouping. It has grouped the data into classes that are defined by the midpoints. The pattern is perceptibly different from that of Figure 3.4(a). Observe that the histogram displays one value (156) for the class with midpoint 160, while the stem-and-leaf display has no values in the 160s.

With any grouping there is a loss of identity, and a loss in accuracy for subsequent statistical calculations. However, the stem-and-leaf distribution has the advantage that the original data values are preserved with the display, making it a useful tool for data analysis.

Throughout this book we have taken precautions to ensure that the data are clean. This is data analysis. Here are some accepted procedures.

### RULES FOR DATA CLEANING AND TOOLS FOR DATA ANALYSIS

1. Whenever they are available, always view some of the original data. Remember that original data are worth asking for.

2. If you are not familiar with the source, show the data to someone who is, then discuss possible problems and potential omissions. Understand the data before you begin using them.

3. Correct any errors arising from the data preparation. Errors can occur during the editing, from data entry, or through miscommunication.

4. Investigate the data. Look for unusual or extreme data values. This procedure can include exploratory data analysis like stem-and-leaf displays and box plots. Other identification for data problems using statistics is in regression (Chapters 12, 13), by time series (Chapter 14), and through sequential plots and control charts (Chapter 15).

5. Use all of the data, or a good sample of them. Be careful how you draw a sample: Our built-in biases can cause us to choose samples that make a point but misrepresent the data. See Chapter 7 for information on random sampling.

6. As a rule, do not change or omit any data without a valid reason.

This section's Problems will give you some practice in viewing real data. The Chapter Problems and Statistics in Action will frequently request data analysis prior to any statistical decision procedures. These Problems regularly use controlled sample sizes of 20 to 50 observations—small enough for you to become acquainted with the data, and a good way to prepare for the Computer Problems sections that use 100, 500, or even more data values.

**Problems 3.5**

9. A sample of 25 mail bags produced the following weights in pounds:

| 35 | 37 | 36 | 34 | 55 |
|----|----|----|----|----|
| 42 | 34 | 52 | 46 | 76 |
| 51 | 43 | 50 | 39 | 43 |
| 37 | 51 | 44 | 45 | 33 |
| 60 | 49 | 41 | 42 | 38 |

a. View these data. Indicate any values you consider extreme.
b. If the mail bags are later recounted and there are actually 26 of them, what steps would you take before computing a mean weight?

10. A machine that is supposed to put three pounds of raisins into a container is suspected of malfunction. The table shows the actual weights as measured on 48 fills (one per hour of shift). What action would you take preceding a statistical evaluation for a malfunction? Here 3.0 pounds is the defined standard. The sampler is the regular quality-control person.

| 2.8 | 3.0 | 2.9 | 2.8 | 2.5 | 2.4 | 2.8 | 2.6 | 2.8 | 2.8 |
|-----|-----|-----|-----|-----|-----|-----|-----|-----|-----|
| 2.6 | 2.8 | 2.6 | 2.9 | 2.6 | 2.9 | 2.7 | 2.9 | 2.7 | 2.7 |
| 2.9 | 2.9 | 2.8 | 2.7 | 3.1 | 2.8 | 2.6 | 2.7 | 2.8 | 2.8 |
| 2.8 | 2.7 | 2.7 | 2.7 | 2.8 | 2.7 | 2.8 | 2.8 | 2.9 | |
| 2.5 | 2.6 | 2.9 | 2.6 | 2.8 | 2.7 | 2.9 | 2.8 | 2.8 | |

a. Is 3.0 pounds a usual value or an unusual value for the 48 sample weights? Give your opinion and reasoning. (A calculation is not expected here.)

b. Someone says these numbers are all too low to be considered real, and it will be necessary to draw another sample and recheck before any statistical analysis is done. Do you agree? Explain.

11. A set of data describes move rates, which are defined as the number of households that move in an area in one calendar year divided by the total number of households in that area. Records come from a sample of areas across the United States. The data are summarized as move-rate percents.

| Area Description | | | Mover's Description | |
|---|---|---|---|---|
| Number of Households | Percent of Households | Cumulative Percent | Move Rate (%) | Sample Cases |
| 1 | 38.2 | 38.2 | 0 | 706 |
| 2 | 18.1 | 56.3 | 20 | 8 |
| 3 | 10.6 | 67.0 | 25 | 13 |
| 4 | 6.8 | 73.8 | 33.3 | 24 |
| 5 | 4.8 | 78.6 | 40 | 6 |
| 6–10 | 11.2 | 89.8 | 50 | 47 |
| 11–20 | 6.1 | 95.9 | 60 | 13 |
| 21–1000 | 4.1 | 100.0 | 66.7 | 15 |
| | | | 80 | 3 |
| | | | 100 | 82 |
| | | | All other | 83 |
| | | | Total | 1000 |

a. Use the "Mover's Description" to find the median and modal values for "Move Rate (%)."

b. The U.S. (average) annual move rate is 20%. The question of real interest is, "Is the procedure used here a good procedure?" Your manager answers, "No, because the areas used have too few households. Bigger area units should be used." Use the data to explain what the manager means. Also, how might the plan be improved to give a more real description for the move rate? Consider the scale of measurement.

12. A grocery store records its daily cash register sales on a computer file. Codes of interest are M–meat, P–produce, and G–groceries.

| Item | Price | Item | Price | Item | Price | Item | Price |
|---|---|---|---|---|---|---|---|
| M | $5.39 | G | $ .03 | P | $3.98 | G | $2.76 |
| M | 7.80 | P | 1.16 | P | 2.37 | G | .42 |
| P | .89 | G | .57 | G | .49 | G | .38 |
| P | 1.30 | G | .92 | G | .49 | P | .01 |
| G | .57 | G | 1.24 | G | 2.46 | G | .72 |

Upon viewing the above listing, the store manager directs you as the assistant manager to close the register and call for service on the machine's price scanner. Why has she taken this action? Assume that the prices include tax and that all numbers were entered by the scanner, not hand-keyed by the grocery checker.

13. A personal ledger is used to check the accuracy of a bank statement, and vice versa.

**Bank Statement**

| Date | Item | Additions | Deductions | Balance |
|------|------|-----------|------------|---------|
| | Beginning Balance | | | $1,520.52 |
| 08–29 | Check 674 | | $ 21.46 | 1,499.06 |
| 08–31 | Check 675 | | 90.00 | 1,409.06 |
| 09–05 | Deposit | $1,240.00 | | 2,649.06 |
| 09–07 | Deposit | 240.00 | | 2,889.06 |
| 09–10 | Check 676 | | 882.87 | 2,006.19 |
| 09–13 | Check 677 | | 18.89 | 1,987.30 |
| 09–20 | Check 678 | | 123.97 | 1,863.33 |
| 09–26 | Interest Payment | | 6.15 | 1,869.48 |

**Personal Ledger**

| Number | Date | Description | Payment | Deposit | Balance |
|--------|------|-------------|---------|---------|---------|
| 673 | 08–19 | Gas Company | $ 20.44 | | $1,520.52 |
| 674 | 08–24 | Electric Company | 21.46 | | 1,499.06 |
| 675 | 08–26 | Sanitary Inc. | 90.00 | | 1,409.06 |
| 676 | 09–04 | Income Tax | 882.87 | | 526.19 |
| | 09–05 | Rents | | $1,240.00 | 1,766.19 |
| | 09–07 | Rents | | 240.00 | 2,006.19 |
| 678 | | Water Dept. (3 mos.) | 123.97 | | 1,882.22 |
| | 09–26 | Interest | | 6.15 | 1,888.37 |

a. There is an obvious difference in the closing balances. What has caused the difference? How did you find the error?
b. Give another example, in which a similar procedure can be used as one check on the correctness of data.

14. A business office has five computer terminals for data editing. The expected production is ten pages of data entered per hour per terminal. Operators stay at the same terminal for the week. All five terminals were kept busy for one week on a single job.
a. One summary is a set of statistics for the week. Write two or three sentences that describe the operator–terminal work and the overall production.

Production Summary—Units Produced per Day

| | | |
|------|------|------|
| Operator–terminal 1 | Mean = 82.0 | Standard deviation = 1.6 |
| Operator–terminal 2 | Mean = 80.0 | Standard deviation = 5.2 |
| Operator–terminal 3 | Mean = 80.0 | Standard deviation = 6.9 |
| Operator–terminal 4 | Mean = 80.0 | Standard deviation = 6.5 |
| Operator–terminal 5 | Mean = 80.0 | Standard deviation = 0.0 |

b. The actual work output, by day, is given in another table. Explain what additional information is available from this raw data that we did not get from the condensed summary statistics in part (a). That is, what is the advantage, if any, of having the actual data? *Note:* You are not expected to calculate the mean and standard deviation. That has already been done.

| Day | Operator–Terminal | | | | |
|---|---|---|---|---|---|
| | 1 | 2 | 3 | 4 | 5 |
| 1 Monday | 80 | 76 | 85 | 74 | 80 |
| 2 Tuesday | 81 | 88 | 82 | 84 | 80 |
| 3 Wednesday | 82 | 75 | 81 | 84 | 80 |
| 4 Thursday | 84 | 82 | 84 | 86 | 80 |
| 5 Friday | 83 | 79 | 68 | 72 | 80 |

15. This table provides rates of return on investment:

| | | | | |
|---|---|---|---|---|
| 6.9 | 2.1 | 3.0 | 0.4 | 7.3 |
| 7.5 | 7.6 | 11.6 | 7.7 | 4.6 |
| 7.4 | 7.3 | 5.0 | 2.6 | 6.6 |
| 6.3 | 12.1 | 12.2 | 5.0 | 6.3 |

a. Use these data to establish a frequency distribution. Or use the one that you developed for this data in Problem 4.
b. Now develop a histogram and a stem-and-leaf display for the data. Define the stem by the units and tens numbers—0, 1, 2, . . . , 12—and the leaves by the tenths place values. What information is provided by the stem-and-leaf display beyond that given by the frequency distribution? (*Note:* To prepare the stem-and-leaf display using Minitab, after reading in the data, use MTB > STEM-AND-LEAF 'INVST'.)
c. Use the stem-and-leaf results to estimate the median and to determine the percent of cases in which investment return is greater than 10%.

16. Develop a stem-and-leaf display for the data on weekly take-home pay as indicated in the table below. Use stems 30, 31, 32, 33, and 34 with a leaf increase of 5. Here is a start:

```
1   30 | 4
2   30 |
3   31 | 3
4   31 | 59
```

Construct the table by hand. If you have access to Minitab or another statistical software package, use it to check your hand-developed stem-and-leaf plot. Use the stem-and-leaf display to determine the median and the inter-quartile range.

| Take-Home Pay ($) | | | |
|---|---|---|---|
| 319 | 326 | 325 | 313 |
| 333 | 348 | 337 | 327 |
| 321 | 320 | 331 | 327 |
| 324 | 315 | 304 | 321 |
| 331 | 340 | 326 | 326 |

## 3.6 NUMERICAL DESCRIPTION OF GROUPED DATA

Now that we have developed procedures for grouping data and for displaying the results, let us turn to methods of summarizing the grouped data. For example, we might wish to make a summary statement concerning the salary of all employees at a large company. We again use the mean, median, mode, and standard deviation to characterize such data. Because we are working with data in which the value of each individual observation has been lost through grouping, these methods give only an approximation of the true values. Nevertheless, the approximation is generally very good, and it becomes even more accurate as the total frequency increases.

Here are the techniques to shorten the calculations for descriptive statistics, including the arithmetic mean, median, mode, and standard deviation, using grouping. Insofar as computers can readily deliver summary statistics for even massive amounts of ungrouped data, these techniques are useful primarily for data that has already been grouped into frequency distributions.

### Mean

Computing the *mean* for grouped data is much like computing the weighted mean as described in Chapter 2. Recall that the weighted mean is equal to the sum of weights times values ($\sum WX$) divided by the sum of weights ($\sum W$). For grouped data, the weights are class frequencies ($f$); class midpoints ($M$) are the values.

The sample mean of a frequency distribution is

$$\bar{X} = \frac{\sum (fM)}{\sum f} \tag{3.1}$$

where    $M$ = class midpoints

$f$ = class frequencies

The assumption underlying Equation (3.1) is that the values of all the data in a class are satisfactorily represented by the class midpoint, $M$. That assumption is more accurate if the class intervals are narrow and if the data in each interval are not consistently clustered at one end. Example 3.3 illustrates the calculation of the mean of a frequency distribution.

EXAMPLE 3.3

Table 3.6 is a frequency distribution of the time it takes a worker to spray paint each of 50 table tops in a furniture plant, and Figure 3.5 is a histogram of the data. What is the mean of the grouped data? The accuracy of the result is important because the result will be used to set a work standard for all employees who perform this task.

**TABLE 3.6** Distribution of Spray Paint Times

| Painting Time (seconds) | Number of Tables $f$ | Class Midpoint $M$ | Class Value $fM$ |
|---|---|---|---|
| 14.00 but less than 16.00 | 4 | 15 | 60 |
| 16.00 but less than 18.00 | 9 | 17 | 153 |
| 18.00 but less than 20.00 | 12 | 19 | 228 |
| 20.00 but less than 22.00 | 10 | 21 | 210 |
| 22.00 but less than 24.00 | 7 | 23 | 161 |
| 24.00 but less than 26.00 | 5 | 25 | 125 |
| 26.00 but less than 28.00 | 3 | 27 | 81 |
| | $\sum f = 50$ | | $\sum (fM) = 1018$ |

**FIGURE 3.5** Histogram of Spray Paint Times

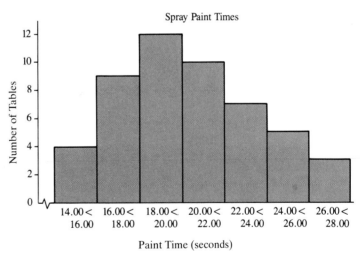

**Solution** By the preliminary calculations in Table 3.6,

$$\bar{X} = \frac{\sum (fM)}{\sum f} = \frac{1018}{50} = 20.36 \text{ seconds}$$

The arithmetic mean thus lies in the class "20 but less than 22 seconds"; the value 20.36 seconds gives an approximation of the exact value. If a more accurate value is warranted, then it should be obtained by averaging the 50 actual times—possibly by computer.

## Median

Because the median (Med) is the positional middle value in an array of data, we can approximate it by accumulating frequencies until one-half of the data are included. When the data are grouped, however, we must estimate the location of the middle value within the *class* that contains the median (called the median class). We can identify the median class by counting. However, an approximation occurs as we assume that the values of the individual observations are equally spaced throughout the class that contains the median.

Given this assumption, the median evaluation begins at the lower boundary of the median class (*L*) and adds a percentage of the class interval. The added percentage is the number of observations in the median class that reside below the median relative to the total number of observations in that (median) class. This adjustment is a percentage of the median class interval, so must be multiplied by the value of the class interval (*i*).

**The median of a frequency distribution is**

$$\text{Med} = L + \left(\frac{n_1}{n_2}\right)i \tag{3.2}$$

with  *L* = the lower *boundary* of the median class

  $n_1$ = **the number of observations in the median class that lie below the median position**

  $n_2$ = **the number of observations in the median class**

  *i* = **the class interval**

When the total number of observations is an odd number, value $n_1$ includes one-half (0.5) of the middle item. Example 3.4 uses the data of the previous example, but asks for calculation of the median.

---

### EXAMPLE 3.4

Find the median for the grouped data of Example 3.3. Then compare this value with the mean (20.36 seconds).

*Solution*   These data include 50 observations, so the median is that value in the array such that 25 values lie on each side. Table 3.6 shows that the first three classes contain 25 observations (4 + 9 + 12) and the last four contain the remaining 25. The median thus lies at the boundary of the third and fourth classes, or at 20.00 seconds. This same conclusion results by using Equation (3.2).

$$\text{Med} = L + \left(\frac{n_1}{n_2}\right)i = 20.00 + \left(\frac{0}{10}\right)2.00 = 20.00 \text{ seconds}$$

where  *L* = the lower class boundary of the median class = 20.00

  $n_1$ = distance into the class = 0

  $n_2$ = number of observations in median class = 10

  *i* = class interval = 2

The mean, 20.36 seconds, is quite close to the median value of 20.00 seconds. The distribution in Figure 3.5 is fairly symmetrical, so the results are reasonable.

We can obtain the same result by working down from an upper boundary of the median class rather than up from the lower boundary.

## Mode

One easy (but coarse) method of approximating the mode of actual data is to use the midpoint of the class that has the greatest frequency. This method gives a mode of 19.00 seconds for the spray paint times in Table 3.6. Remember, however, that since the data are grouped, this is only an estimate. It is possible that no tables were painted in exactly 19 seconds.

For distributions that are reasonably symmetrical, the *mode* can be used to approximate values for the mean and median.

**The mode is approximated from a frequency distribution as the midpoint of the class having the greatest frequency.**

To find the mode quickly, simply observe the midpoint for the class with the greatest frequency. However, if no one class evidences a larger frequency than the others, then there is no approximation for the mode.

## Standard Deviation

The procedure for computing the standard deviation (sd) from grouped data is similar to that using ungrouped data. Here, class midpoints are used to approximate all the values in a class, and the deviation of each class midpoint from the mean is weighted by the frequency of the class.

**The standard deviation approximated from grouped data is\***

$$s = \sqrt{\frac{\sum f(M - \bar{X})^2}{n - 1}} \qquad (3.3)$$

where   $f$ = **class frequency**

$M$ = **class midpoint**

$\bar{X}$ = **sample mean**

$n$ = **total number of observations**

Observe that the equation for grouped data is like that for ungrouped data, Equation (2.6), but with two changes. First, individual values are replaced by their class midpoints. Second, each squared deviation is multiplied by its class frequency to get a total squared difference for all values in the class. The summation is over all classes to give a total sum of squared differences for all the data.

* Table 3.8 in the Summary section includes a "calculation" formula for *s*.

| EXAMPLE 3.5 | Recall that the mean of the time distribution for spray painting table tops was calculated in Example 3.3 to be $\bar{X} = 20.36$ seconds. |
|---|---|

a. Compute the standard deviation for the frequency distribution.
b. What is the span of time for the mean $\pm 2$ standard deviations?

*Solution*   a. The first step is to find the weighted squared deviations by class (114.92, 101.61, ...), as shown in Table 3.7. Add these (to get 531.54); then

$$ s = \sqrt{\frac{\sum f(M - \bar{X})^2}{n - 1}} = \sqrt{\frac{531.54}{49}} = \sqrt{10.85} = 3.29 \text{ seconds} $$

**TABLE 3.7**   Calculation of Standard Deviation of Grouped Data

| Time (seconds) | Number of Tables $f$ | Class Midpoint $M$ | Deviation $(M - \bar{X})$ | Squared Deviation $(M - \bar{X})^2$ | Weighted Squared Deviation $f(M - \bar{X})^2$ |
|---|---|---|---|---|---|
| 14.00 but less than 16.00 | 4 | 15.00 | −5.36 | 28.73 | 114.92 |
| 16.00 but less than 18.00 | 9 | 17.00 | −3.36 | 11.29 | 101.61 |
| 18.00 but less than 20.00 | 12 | 19.00 | −1.36 | 1.85 | 22.20 |
| 20.00 but less than 22.00 | 10 | 21.00 | .64 | .41 | 4.10 |
| 22.00 but less than 24.00 | 7 | 23.00 | 2.64 | 6.97 | 48.79 |
| 24.00 but less than 26.00 | 5 | 25.00 | 4.64 | 21.53 | 107.65 |
| 26.00 but less than 28.00 | 3 | 27.00 | 6.64 | 44.09 | 132.27 |
|  | $\sum = 50$ |  |  |  | $\sum = 531.54$ |

*Note:* $\bar{X} = 20.36$, from Example 3.3.

b. Using Chebyshev's rule, $\bar{X} \pm 2\,sd = 20.36 \pm 2(3.29)$, so that the interval 13.78 to 26.94 includes at least 75% of the values. Note, however, that in this example the summation is over *classes* (whereas for ungrouped data the sum was over the *individual values*).

**Problems 3.6**

17. Given the following frequency distribution, compute the (a) mean, (b) median, and (c) standard deviation. (d) Approximate the mode and (e) calculate the variance.

| Class | Frequency |
|---|---|
| 3–7 | 10 |
| 8–12 | 10 |
| 13–17 | 51 |
| 18–22 | 30 |
| Total | 101 |

18. Given the data array 1.4, 2.1, 2.6, 3.0, 4.5, 5.0, 5.0, 6.3, 6.3, 6.6, 6.9, 7.3, 7.3, 7.4, 7.5, 8.6, 9.7, 10.6, 11.1, 12.3 and an associated frequency distribution,

| Class | Frequency |
|---|---|
| 0–2.4 | 2 |
| 2.5–4.9 | 3 |
| 5.0–7.4 | 9 |
| 7.5–9.9 | 3 |
| 10.0–12.4 | 3 |

compute (a) the mean and (b) the standard deviation. Calculate both from the raw data; then calculate using grouped data procedures. Explain why your answers are slightly different. Which answers are better—from the ungrouped or from the grouped procedures?

19. The hourly wages of a sample of automobile spray painters were collected and organized into a distribution.
    a. Compute $\bar{X}$ and $s$.
    b. Compute the median.
    c. What percent of these painters earn at least $10 per hour? Less than $10 per hour? Between $10 and $14 per hour, including $10 but not $14?

| Hourly Wages | Number of Painters |
|---|---|
| $6 but less than $8 | 5 |
| 8 but less than 10 | 8 |
| 10 but less than 12 | 20 |
| 12 but less than 14 | 10 |
| 14 but less than 16 | 7 |

20. Use the accompanying distribution of scores on a 25-point quiz.
    a. Compute $\bar{X}$, $s^2$, and $s$.
    b. Estimate the median, and the mode.
    c. If a grade of C was given for all scores 8 to 13, inclusive, what percent of the students earned this grade?

| Class (quiz score) | Frequency |
|---|---|
| 5–7 | 2 |
| 8–10 | 10 |
| 11–13 | 26 |
| 14–16 | 32 |
| 17–19 | 8 |
| 20–22 | 2 |
| | $n = 80$ |

21. Weekly accident rates from a sample of selected firms revealed the accompanying numbers of injuries per 1000 worker-hours.
    a. Plot a histogram of the frequency distribution.
    b. Compute values for the mean, the median, and the mode.
    c. Compute the standard deviation and the variance.

| Number of Injuries | Frequency | Number of Injuries | Frequency |
|---|---|---|---|
| 1.5–1.9 | 1 | 4.0–4.4 | 7 |
| 2.0–2.4 | 3 | 4.5–4.9 | 2 |
| 2.5–2.9 | 8 | 5.0–5.4 | 1 |
| 3.0–3.4 | 16 | | $n = \overline{52}$ |
| 3.5–3.9 | 14 | | |

## SUMMARY

We convert uncoded data into *grouped*, or summary, form to increase their convenience for decision making, even though the results are less accurate than if the individual values were used. Grouping makes it easier to calculate statistics such as the mean and the standard deviation.

A *frequency distribution* gives a numerical breakdown of the occurrence of data in various groups called classes. As with ungrouped data, the most commonly used numerical descriptions of grouped data are the mean and the standard deviation. Frequency (grouped) data can also be graphed as *histograms* or *frequency polygons*. Such graphs, including those generated by computers, are used extensively to display sales, payroll, production, market-penetration statistics, and other measures of business vitality.

**TABLE 3.8**   Key Formulas on Data Summarization from Frequency Distributions

### Arithmetic Mean

$$\bar{X} = \frac{\sum (fM)}{\sum f} \qquad \begin{array}{l} f = \text{class frequency} \\ M = \text{class midpoint} \end{array} \tag{3.1}$$

### Median

$$\text{Med} = L + \left(\frac{n_1}{n_2}\right)i$$

$L$ = the lower boundary of the median class

$i$ = the class interval

$n_1$ = the number of observations in the median class that are below the median position

$n_2$ = the number of observations in the median class

(3.2)

### Standard Deviation

Definition formula (sample)

$$s = \sqrt{\frac{\sum f(M - \bar{X})^2}{n - 1}} \qquad \begin{array}{l} \bar{X} = \text{mean} \\ n = \text{total number of observations} \end{array}$$

(3.3)

Calculation formula (sample)

$$s = \sqrt{\frac{n \sum (M^2 f) - (\sum Mf)^2}{n(n - 1)}}$$

*Data analysis* encompasses procedures for checking data and making corrections to ensure that the data accurately describe the real situation. *Stem-and-leaf displays*, *box plots* (Chapter 2), and logical checks on the data are all data analysis procedures. It is important to check the data for reasonableness before you begin any statistical summary.

It is usually advantageous to use a calculator or computer for data summaries if more than a few observations are involved. A calculator is an essential tool in business today, as is the skill of being able to interpret computer printouts. You will get practice by working the computer problems.

Table 3.8 summarizes the formulas developed in this chapter.

## KEY TERMS

| | |
|---|---|
| class boundaries, 71 | median (Med), 90 |
| class interval, 72 | mode, 91 |
| class limits, 71 | standard deviation (sd), 91 |
| class midpoint, 72 | frequency distribution, 71 |
| continuous variable, 70 | frequency polygon, 74 |
| cumulative frequency distribution or ogive, 74 | grouped data, 69 |
| | histogram, 74 |
| data analysis, 79 | stem-and-leaf display, 82 |
| discrete variable, 70 | variable, 70 |
| frequency descriptive measures mean, 88 | |

## CHAPTER 3 PROBLEMS

22. Are the following variables discrete or continuous?
    a. The number of calculators shipped each week by a manufacturer  d
    b. The weight of randomly selected boxes of detergent  c
    c. The amount of room (volume) in a freight car  c
    d. The amount of money invested in various stocks in a portfolio (exact dollars and cents)  d
    e. The maximum daily temperature at a local weather station  c

23. Distinguish between each pair of terms.
    a. Discrete versus continuous data
    b. Raw data versus an array
    c. Class limits versus class boundaries
    d. Histogram versus frequency polygon
    e. Class midpoint versus class mark

24. Given in the table are the classes for three frequency distributions.

| Class | (A) | (B) | (C) |
|---|---|---|---|
| 1st | 10–14 | 10.0–19.9 | 10 but less than 20 |
| 2nd | 15–19 | 20.0–29.9 | 20 but less than 30 |
| 3rd | 20–24 | 30.0–39.9 | 30 but less than 40 |
| 4th | 25–29 | 40.0–49.9 | 40 but less than 50 |

For each of the distributions,
    a. what are the class limits of the first class?  10 & 14
    b. what are the class boundaries of the second class?  14.5  19.5

c. what is the class interval? 14.5 – 19.5

d. what is the midpoint of the second class? (7

25. For the frequency distribution shown in the table, compute each value.

a. Mean  10.25

b. Median

c. Estimate of the mode

d. Standard deviation

e. Variance  square sd

| Class | Frequency |
|-------|-----------|
| 16–21 | 15 |
| 22–27 | 16 |
| 28–33 | 5 |
| 34–39 | 5 |
|       | 41 |

26. A building contractor has just received a shipment of 50 boxes of insulation, which are supposed to weigh 150 pounds each. The contractor prepared a frequency distribution of the actual weights (see table).

| Weight (lb) | No. Boxes |
|-------------|-----------|
| 141.0–143.9 | 12 |
| 144.0–146.9 | 16 |
| 147.0–149.9 | 11 |
| 150.0–152.9 | 5 |
| 153.0–155.9 | 4 |
| 156.0–158.9 | 2 |

a. Express the weights as a relative frequency (percentage) distribution by dividing each class frequency by 50 and then multiplying by 100. The total should be 100%.

b. Plot the relative frequency distribution as a histogram.

c. Compute the mean and median and indicate which is larger.

d. Using the mean, estimate the amount of shortage of insulation (in pounds) and associated cost if it is valued at $2.00 per pound. Assume the contractor has paid for $50 \times 150 = 7500$ pounds.

27. Use the distribution of daily earnings for a sample of 60 steelworkers in Pittsburgh (see table).

| Earnings | Frequency |
|----------|-----------|
| $97.50–97.99 | 11 |
| 98.00–98.49 | 15 |
| 98.50–98.99 | 14 |
| 99.00–99.49 | 11 |
| 99.50–99.99 | 9 |

a. Plot a frequency distribution.

b. Compute the mean.

c. Compute the standard deviation. (*Note:* You may choose to use a calculator for this.)

28. The annual travel expenses for 200 executives are distributed as shown in the table.

| Annual Expenses | Number |
|---|---|
| $10,000 up to 12,000 | 16 |
| 12,000 up to 14,000 | 43 |
| 14,000 up to 16,000 | 98 |
| 16,000 up to 18,000 | 31 |
| 18,000 up to 20,000 | 12 |
| | 200 |

a. Plot a histogram illustrating this distribution and guess the value of the mean.
b. Now compute the mean to see how close you were.

29. The productivity of individual workers at a textile plant varies from 80 units per day for the slowest workers up to almost 130 units for the fastest workers. Production figures for the number of units produced by 200 workers on a given day are shown in the table. Answer these questions in order to analyze costs.

| Units Produced per Worker | Number of Workers $f$ |
|---|---|
| 80–89 | 22 |
| 90–99 | 53 |
| 100–109 | 75 |
| 110–119 | 40 |
| 120–129 | 10 |
| | 200 |

a. Estimate the total number of units produced by these 200 workers.
b. What percentage of the workers produced 90 to 119 units, inclusive?
c. If the workers are paid $.30 per unit, how many received $33.00 or more for this day?

30. Prices paid by a sample of 100 used car buyers at Quick Deal Auto Sales are given in the accompanying table.

| Price (dollars) | Number of Cars |
|---|---|
| 1400 but less than 2000 | 2 |
| 2000 but less than 2600 | 7 |
| 2600 but less than 3200 | 14 |
| 3200 but less than 3800 | 21 |
| 3800 but less than 4400 | 28 |
| 4400 but less than 5000 | 16 |
| 5000 but less than 5600 | 12 |

a. Compute $\bar{X}$, $s^2$, and $s$. (*Suggestion:* Use coding.)
b. What is the total dollar sales on these 100 cars?
c. What percentage of the cars sold for $3800 or over? Under $5000? Between $2600 and $3799?

31. Using the data reprinted here for three pounds of raisin weights, develop a stem-and-leaf plot. Use the classes 2.4, 2.5, 2.6, 2.7, 2.8, 2.9, 3.0, and 3.1. Use the display to compute a median weight. Explain why, in this case, the result is exact and not an estimation.

| | | | | | | | | | |
|---|---|---|---|---|---|---|---|---|---|
| 2.8 | 3.0 | 2.9 | 2.8 | 2.5 | 2.4 | 2.8 | 2.6 | 2.8 | 2.8 |
| 2.6 | 2.8 | 2.6 | 2.9 | 2.6 | 2.9 | 2.7 | 2.9 | 2.7 | 2.7 |
| 2.9 | 2.9 | 2.8 | 2.7 | 3.1 | 2.8 | 2.6 | 2.7 | 2.8 | 2.8 |
| 2.8 | 2.7 | 2.7 | 2.7 | 2.8 | 2.7 | 2.8 | 2.8 | 2.9 | |
| 2.5 | 2.6 | 2.9 | 2.6 | 2.8 | 2.7 | 2.9 | 2.8 | 2.8 | |

32. Here is a stem-and-leaf plot for the number of salespeople who are traveling abroad for a large multinational retail sales corporation. The time frame is 26 weeks. The extremes are 11 and 51 people.

```
1 | 1  2  2  5  5  7  8
2 | 2  2  5  5  7  7  8  9  9
3 | 1  1  3  4  6  6
4 | 0  0  3
5 | 1
```

a. During what percentage of these 26 weeks was the number of salespeople abroad more than 35? 22 or fewer?
b. Now reconstruct the data as a histogram using groups 10–19, 20–29, 30–39, ..., 50–59. Using only the histogram, do you get the same answers as you did for the questions in part (a)?

**STATISTICS IN ACTION: CHALLENGE PROBLEMS**

33. What can be done to improve traditional graphing procedures? William Cleveland gives illustrated suggestions that include full-scale breaks for two-dimensional data plots. For example, Figures 1 and 2 display the same data:

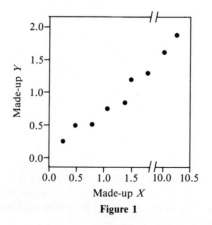

**Figure 1**

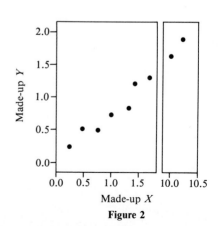

**Figure 2**

Figure 1—with a partial-scale break—can give the false impression that the right-most points continue the nearly straight-line behavior of the points at the left. Figure 2—with a full-scale break—forcefully shows the scale jump (1.5 to 10.0) and discourages mental–visual connections by viewers, or actual connection by users.

SOURCE: Cleveland, William S. "Graphical Methods for Data Preparation: Full Scale Breaks Dot Charts, and Multibased Logging." *The American Statistician* 38, no. 4 (1984): 270–280.

Questions   Search the business journals, periodicals, and data fact books in your library to find an example of a misuse in graphing. Reconstruct the display to show a similar but corrected graph. Discuss the potential confusion that is overcome with your correction. Common misuses in graphing include horizontal to vertical (total) dimensions far off from 1:1 ratio, omission of a lower part of the vertical scale, improper use of relative sizes for charactures, and unlabeled axes.

34. W. E. Hillis discusses the relationship of world population growth to existing trends in the use of wood resources. This article, which encourages resource conservation, is highly informative and uses graphical displays to illustrate a variety of statistics. For example,

The Proportion of Global Population in Developed and Developing Countries for 1980 and 2000, and Their Proportions of Demand for Firewood and Industrial Wood

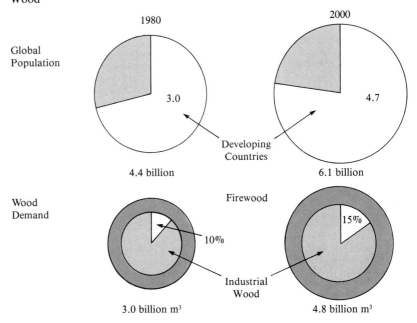

SOURCE: Hillis, W. E. "Forestry, Technology and Society—Their Interdependence." *Journal of Business Administration* 14, nos. 1, 2 (1983–84): 121–142.

Questions   This figure contains some interesting facts, but there is room for clarification. What changes would provide a clearer, more informative display? You may want to make a rough sketch to redisplay your ideas.

35. Conducted six times since 1970, the Virginia Slims Opinion Poll has viewed women's changing status through national surveys. The 1990 poll, conducted by the Roper Organization, found that 57% of women polled considered ideal a lifestyle combining marriage, career, and children. Only 27% would rather marry and have children but no career. A majority of women said they get satisfaction from their job. Response is summarized for one of the questions: "Why do women work?"

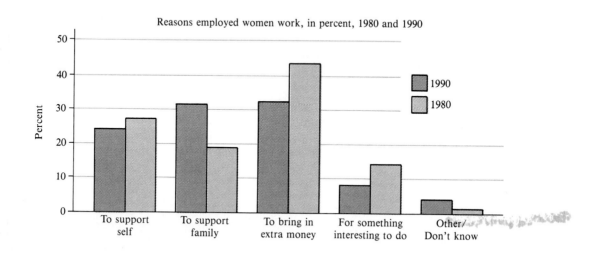

Reasons employed women work, in percent, 1980 and 1990

SOURCE: Townsend, Bickley, and Kathleen O'Neil. "American Women Get Mad." *American Demographics* 12, no. 8 (1990): 26–32.

Questions   Answer these questions from the bar chart.
a. For 1990, about what percent of the women work to support themselves or their family?
b. Is the percent who work to support themselves or their family higher, about the same, or lower than in 1980?
c. In 1990, what was the most common reason the women gave for working? In 1980?
d. Which category—to support self, to bring in extra money, or for something interesting to do—shows the greatest percentage increase over its 1980 percentage?

36. Age and income are the two characteristics of greatest interest in consumer marketing. One summary based on a tabulation of the Census Bureau's March 1985 Current Population Survey includes the following statements. (This study defines "downscale" households as having incomes below $25,000; "middle income," $25,000 to $49,999; and "upscale," $50,000 or higher.)
a. Almost half of American households are downscale.
b. The median age of downscale householders (51) is ten years older than the median for householders with higher incomes ($25,000 and higher).
c. The proportion of people under age 40 and over age 60 is greater among middle-income householders than among upscales.
d. Only 15% of upscale have incomes greater than $100,000.

Questions   Use this table from the article to verify each of the four statements.

Distribution of Households, by Income and Age; Numbers in Thousands

| Age of Householder | Total Households | Less than $5,000 | $5,000 to $9,999 | $10,000 to $14,999 | $15,000 to $19,999 | $20,000 to $24,999 | $25,000 to $29,999 | $30,000 to $34,999 | $35,000 to $39,999 | $40,000 to $44,999 | $45,000 to $49,999 | $50,000 to $74,999 | $75,000 to $99,999 | $100,000 or more |
|---|---|---|---|---|---|---|---|---|---|---|---|---|---|---|
| All ages | 92,830 | 5,737 | 10,006 | 9,516 | 9,126 | 8,184 | 7,891 | 6,984 | 6,414 | 5,373 | 4,265 | 12,455 | 3,966 | 2,911 |
| Under 25 | 5,415 | 711 | 783 | 848 | 792 | 616 | 460 | 389 | 274 | 199 | 95 | 161 | 40 | 47 |
| 25 to 29 | 9,624 | 589 | 786 | 949 | 1,088 | 1,095 | 1,039 | 937 | 839 | 590 | 417 | 994 | 197 | 105 |
| 30 to 34 | 11,300 | 536 | 755 | 874 | 1,062 | 1,121 | 1,216 | 1,088 | 961 | 788 | 627 | 1,736 | 356 | 180 |
| 35 to 39 | 10,513 | 454 | 555 | 763 | 795 | 846 | 935 | 929 | 940 | 820 | 677 | 1,949 | 495 | 356 |
| 40 to 44 | 9,439 | 298 | 459 | 569 | 679 | 670 | 683 | 766 | 708 | 742 | 649 | 2,006 | 709 | 499 |
| 45 to 49 | 7,597 | 310 | 368 | 444 | 469 | 523 | 631 | 597 | 571 | 543 | 487 | 1,570 | 646 | 439 |
| 50 to 54 | 6,420 | 314 | 397 | 411 | 379 | 432 | 504 | 499 | 484 | 449 | 344 | 1,236 | 507 | 465 |
| 55 to 59 | 6,413 | 371 | 509 | 533 | 517 | 490 | 522 | 432 | 451 | 414 | 374 | 1,027 | 418 | 358 |
| 60 to 64 | 6,392 | 413 | 686 | 664 | 742 | 611 | 564 | 465 | 373 | 341 | 245 | 786 | 295 | 206 |
| 65 to 69 | 6,265 | 392 | 1,068 | 1,011 | 815 | 643 | 568 | 344 | 372 | 213 | 126 | 440 | 149 | 124 |
| 70 to 74 | 5,325 | 422 | 1,085 | 960 | 832 | 495 | 395 | 255 | 220 | 138 | 123 | 266 | 71 | 62 |
| 75 and older | 8,125 | 928 | 2,555 | 1,489 | 957 | 640 | 376 | 284 | 223 | 137 | 102 | 283 | 83 | 70 |

SOURCE: Waldrop, Judith. "Up and Down the Income Scale." *American Demographics* 12, no. 7 (1990): 24–30.

## BUSINESS CASE

*Measuring Customer Demand*

Cliff Robertson slammed down the phone. He had lost his third customer in less than a month, and all for the same reason—"insufficient stock." Cliff, marketing manager for a national auto parts firm, was perturbed enough to arrange for a meeting with his company's president, Alex Quinlan, the next day.

By 3 o'clock the next afternoon, Cliff had calmed down some, but he was still upset with Joe Barton, the inventory manager. Mr. Quinlan listened to Cliff's complaint about how Joe "never carried enough transmissions" to supply Cliff's distributors. His argument was that with that kind of service, the company would soon be "losing more transmission business than we're gaining." Cliff implied that any inventory manager who didn't have better business sense ought to be looking for another job.

Mr Quinlan noted that transmissions were only one of several thousand items the company had in inventory, but he agreed it was worth looking into. In defense of Joe Barton he did explain that inventories of all items were kept down in order to limit the cash tied up in inventories. Turning to his intercom, he asked his secretary to get the corporate finance department on the phone. After a couple minutes of conversation he turned back to Cliff.

**Mr. Q.** George tells me we have company guidelines that limit our investment in transmission inventory to $10,000. That sounds reasonable to me. It's designed to provide our customers with Level III service—do you think that's too low?

**Mr. R.** I know our policy manual has something in it about levels, though I'm not exactly sure what Level III is. But I *do* know what $10,000 worth of transmission inventory looks like, and we don't have it—not now anyway. Sometimes maybe, but not when we need it!

**Mr. Q.** Level III means our production people have designed an inventory system to provide our customers with 95% service. Do you feel comfortable with a 95% figure, or is that too low?

**Mr. R.** Ninety-five percent! If we could ship 95% of our transmission demand from stock, I'd be out celebrating. We'd give Barton our "Salesperson of the Year" award.

**Mr. Q.** You mean we're not giving our customers a Level III service now?

**Mr. R.** Well, I know they are really unhappy when we can't ship an order!

Mr. Q.   Let's get some data on the transmission situation and take a close look at it. Can you get together some demand figures for the past year or two?

Mr. R.   Yes, I can have them for you first thing in the morning.

Mr. Q.   Fine. In the meantime, you might go down and have a chat with Joe. He's quite reasonable. I think it might help if you two were better acquainted.

The next morning a table of demand figures (Table 3.9) was on Alex Quinlan's desk when he arrived for work. A note explained that the data covered 2 years plus 16 weeks.

**TABLE 3.9**   Weekly Demand for NR-7 Transmissions (beginning January 1989)

| Week | Demand (number of units) | Week | Demand (number of units) | Week | Demand (number of units) |
|------|------|------|------|------|------|
| 1 | 41 | 41 | 36 | 81 | 24 |
| 2 | 31 | 42 | 28 | 82 | 32 |
| 3 | 27 | 43 | 20 | 83 | 25 |
| 4 | 35 | 44 | 39 | 84 | 8 |
| 5 | 7 | 45 | 15 | 85 | 33 |
| 6 | 21 | 46 | 37 | 86 | 38 |
| 7 | 33 | 47 | 22 | 87 | 17 |
| 8 | 29 | 48 | 54 | 88 | 44 |
| 9 | 39 | 49 | 32 | 89 | 23 |
| 10 | 49 | 50 | 13 | 90 | 11 |
| 11 | 31 | 51 | 26 | 91 | 39 |
| 12 | 26 | 52 | 34 | 92 | 12 |
| 13 | 38 | 53 | 43 | 93 | 42 |
| 14 | 29 | 54 | 9 | 94 | 22 |
| 15 | 15 | 55 | 23 | 95 | 31 |
| 16 | 45 | 56 | 33 | 96 | 29 |
| 17 | 35 | 57 | 28 | 97 | 39 |
| 18 | 46 | 58 | 47 | 98 | 48 |
| 19 | 27 | 59 | 37 | 99 | 36 |
| 20 | 38 | 60 | 29 | 100 | 21 |
| 21 | 14 | 61 | 31 | 101 | 33 |
| 22 | 41 | 62 | 24 | 102 | 3 |
| 23 | 25 | 63 | 45 | 103 | 38 |
| 24 | 47 | 64 | 18 | 104 | 13 |
| 25 | 61 | 65 | 47 | 105 | 39 |
| 26 | 21 | 66 | 26 | 106 | 27 |
| 27 | 36 | 67 | 36 | 107 | 34 |
| 28 | 39 | 68 | 2 | 108 | 67 |
| 29 | 12 | 69 | 48 | 109 | 16 |
| 30 | 35 | 70 | 23 | 110 | 30 |
| 31 | 58 | 71 | 35 | 111 | 21 |
| 32 | 23 | 72 | 15 | 112 | 36 |
| 33 | 45 | 73 | 38 | 113 | 56 |
| 34 | 34 | 74 | 25 | 114 | 42 |
| 35 | 14 | 75 | 49 | 115 | 12 |
| 36 | 42 | 76 | 17 | 116 | 43 |
| 37 | 26 | 77 | 48 | 117 | 33 |
| 38 | 38 | 78 | 40 | 118 | 26 |
| 39 | 31 | 79 | 32 | 119 | 29 |
| 40 | 23 | 80 | 20 | 120 | 17 |

**FIGURE 3.6**    Frequency Distribution and Histogram, Showing Weekly Demand
for Transmissions for 120 Weeks

Additional Data on Weekly Demand for Transmissions—120 Weeks

| | |
|---|---|
| Design service level | 95% |
| Stock level at beginning of each week | 50 transmissions |
| Stock investment (50 units at $200 each) | $10,000 |
| Carrying cost on this investment (@ 20%) | $2000/yr |

Median weekly demand: $\text{Med} = 35 + \dfrac{(-47)}{120}(10) = 31$ transmissions

| Class | $f$ | $d$ | $fd$ |
|---|---|---|---|
| 0 less than 10 | 5 | −3 | −15 |
| 10 less than 20 | 16 | −2 | −32 |
| 20 less than 30 | 33 | −3 | −33 |
| 30 less than 40 | 40 | 0 | 0 |
| 40 less than 50 | 21 | 1 | 21 |
| 50 less than 60 | 3 | 2 | 6 |
| 60 less than 70 | 2 | 3 | 6 |
| | 120 | | −47 |

($f$ = frequency, $d$ = unit deviation)

(a) Frequency Distribution

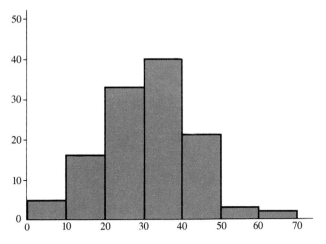

Number of Transmissions Demanded

(b) Histogram

Mr. Quinlan immediately asked his staff assistant, Elaine Sims, to get some additional information on the amount of stock maintained and to summarize and graph the data so that he could visualize it. Elaine suggested that a cumulative distribution would do the best job of illustrating the levels of demand and said she would develop one after preparing a few preliminary figures. After lunch, Elaine returned with a frequency distribution table (Figure 3.6(a)) and a histogram (Figure 3.6(b)) as well as a cumulative frequency distribution table (Figure 3.7(a)) and an ogive (Figure 3.7(b)).

**FIGURE 3.7** Cumulative Distribution and Ogive, Showing Weekly Demand for Transmission for 120 Weeks

| Lower Class Boundary (LCB) | Number >LCB | Percentage >LCB |
|---|---|---|
| 0 | 120 | 100 |
| 10 | 115 | 96 |
| 20 | 99 | 82 |
| 30 | 66 | 55 |
| 40 | 26 | 22 |
| 50 | 5 | 4 |
| 60 | 2 | 2 |
| 70 | 0 | 0 |

(a) Cumulative distribution table

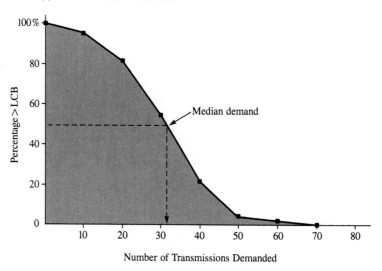

(b) Cumulative distribution graph

The next morning Alex Quinlan called Cliff Robertson to his office to review the results of Elaine's analysis. Cliff promptly mentioned that he had talked with Joe Barton and apologized for flying off the handle, but he repeated that this was a real problem and he felt it important to get to the bottom of it. At that point Mr. Quinlan interrupted.

**Mr. Q.** I think Elaine's graphs are a good summary of the facts, Cliff. The histogram shows that the heaviest demand is from 20 to 40 units per week. Elaine has also made up a cumulative frequency distribution, which shows the percentage of weeks that demand was greater than each of the lower limits in the table. You can see (in Figure 3.7(b)) our median demand—the 50% mark—is just over 30 units. Over the past 2 years, we have begun each week with 50 transmissions in stock. Finally, the additional data she worked up shows that the median weekly demand was 31 units.

**Mr. R.** Yes, I can see that. But it isn't those average weeks that cause me the problem. I would like to advertise that we can ship our orders during the same week that they arrive. Now, that's not asking too much is it?

**Mr. Q.** It is if you're talking about *all* orders. Take a look at Elaine's cumulative distribution on the right. This shows that our weekly demand exceeded 50 units less than 5% of the time—actually it was only 4%. That is, there were only 5 weeks out of the 120 when all orders were not shipped from stock.

**Mr. R.** Hmm. It sure *seemed* like a lot more. But I must admit I kind of suspected that after talking with Joe this morning. I do tend to get pretty wrapped up satisfying my customers, but, of course, Joe and I are both working for the same thing.

**Mr. Q.** Good, I'm glad you said that, Cliff. Your job is to manage your department in the best interests of everyone involved—your customers, your employees, and our stockholders. Much as we'd like to satisfy every customer every time, we can't. But if you can anticipate those large orders and let Joe know, he may be able to work something out for you.

**Mr. R.** Yes, we talked about that. I think it will help.

**Mr. Q.** Did you learn anything else from this episode, Cliff?

**Mr. R.** Well, I always knew it cost a lot to keep those thousands of inventory items in stock. But I didn't realize how much until Joe showed me his computer printout of inventory investment. When you add up the investment in transmissions with that in mufflers, generators, and everything else, it involves a lot of money.

**Mr. Q.** Right. That's why we have to set investment limits. We'd go broke if we tried to give 100% service on everything.

**Mr. R.** I can see your point—and 95% is OK with me. I'll just have to school our salespeople a little better on how our system works now that I know more about it myself. I learned something else too, though. I've learned how to get the facts to back me up—or show me my impression's wrong. Next time I'll be better prepared. Mind if I run some things by Elaine Sims once in a while?

**Mr. Q.** I hope you will. That's what she's here for.

## Business Case Questions

37. Consider Elaine Sims' cumulative distribution for the number of transmissions demanded. In what percentage of the weeks were 45 transmissions demanded?

38. Discuss the 95% design service level. The data showed that 5 of 120 weeks had 50+ transmissions ordered. This is 4%, so 96% had under 50 demands. Why does Mr. Quinlan say the service level is only 95%?

## COMPUTER APPLICATION

*Statistics for Customer Demand*

In this application we look at a solution to the problem presented in the Business Case—maintaining an adequate inventory to meet weekly demand for automobile transmissions. This time we take the viewpoint of the data analyst, Elaine Sims, instead of the marketing manager. Although the sample of 120 weeks worth of sales is not really too large to handle with a calculator or by grouping (the method used in the application), it is large enough to use computer-derived statistics.

We, like Elaine, want to answer the question, How often (that is, with what frequency) does demand exceed the regular inventory supply of 50 units? Other statistics, while relevant, are incidental to this question.

We have reproduced the output we obtained by applying a Minitab computer run to the transmission demand data. This program reduces the data to measure the number of weeks that show any number of transmission units of demand. The data, the Minitab program, and the results appear in Figure 3.8.

**FIGURE 3.8** A Minitab Run That Describes Customer Demand for Model NR-7
Transmissions

```
MTB > PAPER
MTB > READ C1◄──────────────────────────────────────────────────────────1
    120 ROWS READ
MTB > END
MTB > NAME C1 'TRANS'◄────────────────────────────────────────────────────2
MTB > PRINT 'TRANS'
TRANS
    41    31    27    35     7    21    33    29    39    49    31    26    38
    29    15    45    35    46    27    38    14    41    25    47    61    21
    36    39    12    35    58    23    45    34    14    42    26    38    31
    23    36    28    20    39    15    37    22    54    32    13    26    34
    43     9    23    33    28    47    37    29    31    24    45    18    47
    26    36     2    48    23    35    15    38    25    49    17    48    40
    32    20    24    32    25     8    33    38    17    44    23    11    39
    12    42    22    31    29    39    48    36    21    33     3    38    13
    39    27    34    67    16    30    21    36    56    42    12    43    33
    26    29    17

MTB > TALLY 'TRANS';◄──────────────────────────────────────────────────────3
SUBC> ALL.
```

| TRANS | COUNT | CUMCNT | PERCENT | CUMPCT | | TRANS | COUNT | CUMCNT | PERCENT | CUMPCT |
|---|---|---|---|---|---|---|---|---|---|---|
| 2 | 1 | 1 | 0.83 | 0.83 | | 28 | 2 | 49 | 1.67 | 40.83 |
| 3 | 1 | 2 | 0.83 | 1.67 | | 29 | 5 | 54 | 4.17 | 45.00 |
| 7 | 1 | 3 | 0.83 | 2.50 | | 30 | 1 | 55 | 0.83 | 45.83 |
| 8 | 1 | 4 | 0.83 | 3.33 | | 31 | 5 | 60 | 4.17 | 50.00 |
| 9 | 1 | 5 | 0.83 | 4.17 | | 32 | 3 | 63 | 2.50 | 52.50 |
| 11 | 1 | 6 | 0.83 | 5.00 | | 33 | 5 | 68 | 4.17 | 56.67 |
| 12 | 3 | 9 | 2.50 | 7.50 | | 34 | 3 | 71 | 2.50 | 59.17 |
| 13 | 2 | 11 | 1.67 | 9.17 | | 35 | 4 | 75 | 3.33 | 62.50 |
| 14 | 2 | 13 | 1.67 | 10.83 | | 36 | 5 | 80 | 4.17 | 66.67 |
| 15 | 3 | 16 | 2.50 | 13.33 | | 37 | 2 | 82 | 1.67 | 68.33 |
| 16 | 1 | 17 | 0.83 | 14.17 | | 38 | 6 | 88 | 5.00 | 73.33 |
| 17 | 3 | 20 | 2.50 | 16.67 | | 39 | 6 | 94 | 5.00 | 78.33 |
| 18 | 1 | 21 | 0.83 | 17.50 | | 40 | 1 | 95 | 0.83 | 79.17 |
| 20 | 2 | 23 | 1.67 | 19.17 | | 41 | 2 | 97 | 1.67 | 80.83 |
| 21 | 4 | 27 | 3.83 | 22.50 | | 42 | 3 | 100 | 2.50 | 83.33 |
| 22 | 2 | 29 | 1.67 | 24.17 | | 43 | 2 | 102 | 1.67 | 85.00 |
| 23 | 5 | 34 | 4.17 | 28.33 | | 44 | 1 | 103 | 0.83 | 85.83 |
| 24 | 2 | 36 | 1.67 | 30.00 | | 45 | 3 | 106 | 2.50 | 88.33 |
| 25 | 3 | 39 | 2.50 | 32.50 | | 46 | 1 | 107 | 0.83 | 89.17 |
| 26 | 5 | 44 | 4.17 | 36.67 | | 47 | 3 | 110 | 2.50 | 91.67 |
| 27 | 3 | 47 | 2.50 | 39.17 | | 48 | 3 | 113 | 2.50 | 94.17 |
| | | | | | | 49 | 2 | 115 | 1.67 | 95.83 |
| | | | | | | 54 | 1 | 116 | 0.83 | 96.67 |
| | | | | | | 56 | 1 | 117 | 0.83 | 97.50 |
| | | | | | | 58 | 1 | 118 | 0.83 | 98.33 |
| | | | | | | 61 | 1 | 119 | 0.83 | 99.17 |
| | | | | | | 67 | 1 | 120 | 0.83 | 100.00 |
| | | | | | | N= | 120 | | | |

**FIGURE 3.8**  (*continued*)

MTB > DESCRIBE 'TRANS ←————————————————————————— **4**

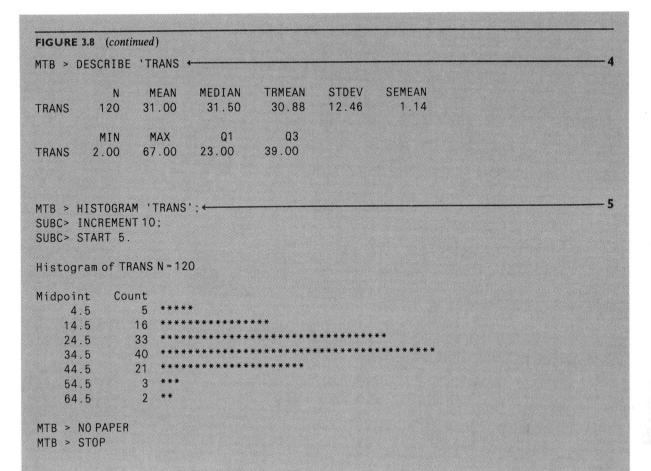

```
                N      MEAN    MEDIAN   TRMEAN    STDEV    SEMEAN
TRANS         120     31.00     31.50    30.88    12.46      1.14


               MIN      MAX        Q1       Q3
TRANS         2.00    67.00     23.00    39.00
```

MTB > HISTOGRAM 'TRANS'; ←—————————————————————————— **5**
SUBC> INCREMENT 10;
SUBC> START 5.

Histogram of TRANS N = 120

```
Midpoint    Count
     4.5        5   *****
    14.5       16   ****************
    24.5       33   *********************************
    34.5       40   ****************************************
    44.5       21   *********************
    54.5        3   ***
    64.5        2   **
```

MTB > NO PAPER
MTB > STOP

The essential points are indicated by numbers and arrows. Following is a summary of this most salient information.

1 The data points are read one value per line. Minitab does not display this.

2 The data are given the name 'TRANS'. Then a PRINT statement reproduces the exact values that were keyed in. This affords a check that the data values are identical to those in Cliff Robert's Table 3.9, which is the data source. The boxed-in numbers are the lowest and highest demand numbers: 2 and 67, respectively.

3 For the TALLY procedure, the COUNT column heads a list of the number of TRANSmission units of demand (NR-7 transmissions) that were experienced for any of the 120 weeks. These vary from 2 to 67, for a range of 65 TRANS units for this discrete variable.

4 The DESCRIBE section includes values for the range and other statistics. You already know about the mean (which is 31.00 here), median, standard deviation, and minimum and maximum values. We defer discussion of the remaining measures because they are not central to our question.

This information is sufficient to show how Elaine Sims obtained the statistic that "weekly demand [for NR-7 transmission units] exceeded 50 units less than 5% of the time—actually it was only 4%." The frequency distribution, or TALLY, in Figure 3.8 shows that the demand COUNT exceeds 50 units (actually 54, 56, 58, 61, 67) for $1 + 1 + 1 + 1 + 1 = 5$ weeks, or 5/120 of all weeks, which is $0.83\% + 0.83\% + 0.83\% + 0.83\% + 0.83\% = 4.12\%$, or, rounded, 4%. This answers the question, How often has demand exceeded the supply?

**5** A final part for Figure 3.8 is a HISTOGRAM that displays the frequencies for various classes (TRANS values) of demand for transmissions. Compare this with the hand-prepared display, Figure 3.6, in the Business Case. The histogram describes the same data with a coding that uses the classes 0 through 9, 10–19, 20–29, ..., 60–69. These classes are identified by their class midpoints: 4.5, 14.5, ..., 64.5, respectively. The conclusion is the same as before. The classes (with midpoints 54.5 and 64.5) that include 50 or more units of demand were experienced for $3 + 2 = 5$ weeks, or 4% of all weeks. The frequency distribution or Minitab TALLY and the histogram provide alternative ways to answer the question.

The histogram in Figure 3.6 adds a visual description about the data. The graph appears somewhat bell-shaped. This graphical display is supported by the statistics that appear in the DESCRIBE procedure. For example, the MEAN demand is 31 units (transmissions). The MEAN and MEDIAN values are quite close: 31.0 and 31.5 units, respectively. The closeness of these measures is one indicator that the distribution is nearly symmetrical. The statistic that directly measures symmetry is SKEWNESS. True symmetry exists only when the skewness coefficient is zero. This distribution (the coefficient would be 0.08) would be considered symmetric.*

Another general descriptor of the shape of a distribution is the KURTOSIS. Again the standard is zero; values above zero indicate a more peaked (narrow) distribution relative to a distribution we will use later as a standard (normal distribution), whereas values below zero indicate a flatter distribution. This data set has a slightly negative value for KURTOSIS ($-0.034$) and is slightly flat compared to the standard normal shape. The standard error, SEMEAN, is a statistic that we discuss in Chapter 7.

This example illustrates the use of a few computer-generated statistics for description and, along with the business application, shows how summary statistics can be used for management decision making.

* The SKEWNESS and KURTOSIS values are not displayed with the Minitab outputs. These values are easy to obtain by the SPSS FREQUENCIES procedure.

## COMPUTER PROBLEMS

*Data Base Applications*

These problems call for data from the 100-record PUMS sample. This requires drawing samples and some programming statements.

39. This extends Problem 42 of Chapter 2 to include more data than is practical for hand calculation of descriptive statistics. Select all 100 records from the PUMS source and obtain a listing on paper of the 100 incomes. Also use computer software to obtain (a) the mean, (b) the median, and (c) the standard deviation for the 100 household incomes.

40. Do the following by hand, that is, by calculator and using grouped data procedures. Group the 100 income values from Problem 39 into six to ten classes, then use grouped-data procedures to calculate these statistics:
    a. Frequency distribution table      b. Mean
    c. Median                            d. Standard deviation
    Explain why your hand-calculated numbers are slightly different from the computer-generated values.

41. Repeat Problem 39, but use data for age of head of household (columns 43–44) and include a frequency distribution table.

42. Repeat Problem 40, but now use the age of head of household. Again make the calculations in (a) through (d) by hand or by calculator and compare your values to those developed on a computer.

43. Select a sample of 50 records from the PUMS file. For each record, display a listing of values for (a) year moved in, (b) bedrooms, and (c) automobiles. Use a computer routine to develop a frequency table for each of parts (a) through (c). Also develop one or more graphical displays for each variable.

44. Use the 40 sales amounts from Table 3.5, repeated here, to develop a stem-and-leaf display for insurance 'SALES'. Minitab users, try these three procedures:

```
1. MTB > STEM-AND-LEAF 'SALES'
2. MTB > HISTOGRAM 'SALES'
3. MTB > STEM-AND-LEAF 'SALES';
   SUBC> INCREMENT = 5;
   SUBC> TRIM OUTLIERS.
```

Use the stem-and-leaf display from Example 3.2 to check your results.

| | | | | | | | |
|---|---|---|---|---|---|---|---|
| 75 | 109 | 119 | 124 | 125 | 130 | 135 | 152 |
| 96 | 112 | 120 | 124 | 126 | 131 | 137 | 156 |
| 99 | 114 | 122 | 125 | 127 | 131 | 141 | 170 |
| 102 | 116 | 123 | 125 | 127 | 133 | 144 | 122 |
| 106 | 118 | 123 | 125 | 128 | 134 | 147 | 127 |

45. Using the data for three pounds of raisins and the Minitab software, establish a stem-and-leaf diagram to check the results for Problem 31. Use

```
MTB > STEM-AND-LEAF 'RASN'
MTB > STEM-AND-LEAF 'RASN';
SUBC> INCREMENT = 0.1;
SUBC> TRIM OUTLIERS.
```

Would you advise the company to advertise 3.0 pounds per fill? Discuss any outliers.

**FOR FURTHER READING**

[1] *Byte* Magazine. McGraw-Hill, Inc., 70 Main Street, Peterborough, NH.

[2] *Compute* Magazine. Compute Publications, Inc., 505 Edwadia Drive, Greensboro, NC.

[3] *Computerworld* Magazine. Computerworld Communications, Inc., 375 Cochituate Road, Framingham, MA (a mainframe weekly).

[4] Dixon, W. J. (ed.) *BMDP Statistical Software Manual.* 1988 ed. Los Angeles: University of California Press, 1985.

[5] *Infoworld* Magazine. Popular Computing, Inc., a subsidiary of Computerworld Communications, Inc., 375 Cochituate Road, Framingham, MA. (A weekly magazine oriented to microcomputers.)

[6] Klecka, William R., Norman H. Nie, and C. Hadlai Hull. *SPSS Primer.* New York: McGraw-Hill, 1975.

[7] Ryan, Barbara F., Brian L. Joiner, and Thomas A. Ryan. *Minitab Handbook*, 2nd ed. Boston: PWS–Kent, 1985.

[8] *SAS (Statistical Analysis System) User's Guide: Statistics.* 1985 ed. Cary, NC: SAS Institute.

[9] Schaefer, Robert L., and Richard B. Anderson. *The Student Edition of Minitab.* Reading, MA: Addison–Wesley, 1989.

[10] *SPSS Reference Guide.* Chicago: SPSS, Inc., 1990.

[11] Tukey, John W. *Exploratory Data Analysis.* Reading, MA: Addison-Wesley, 1977.

# PROBABILITY AND PROBABILITY DISTRIBUTIONS

# CHAPTER

# 4

---

# PROBABILITY

---

## 4.1
### INTRODUCTION

What is the chance that the demand for electric cars will increase over the next year? How likely is it that our new computer will sell well enough for us to break even during its first year on the market? Is there much risk that another pump failure will shut down the power plant? How likely is it that airline mechanics will go on strike in July? The answers to questions like these are often given in terms of the chance, or *probability*, that the event in question will occur.

*Probability* is a measure of how certain we are that an event will take place.

> Probability is a measure of the chance that an event will occur or that a statistical experiment will have a particular outcome.

Decision makers in all businesses use probabilities, even though they do not always formalize them and give them precise numerical values. For example, the manager of a chain of health spas might estimate that there is an even chance that the chain will capture 25% of the market next year. Or the president of a trucking concern might feel that there is an 80% chance the price of oil will increase again next quarter. These measures of chance are expressions of probability. Probabilities give bases for making decisions when the managers are uncertain about the outcomes that will follow from their actions.

There are three kinds of probability, distinguished by the way data are used to determine them: *classical*, *empirical*, and *subjective probabilities*.

> Probabilities based on the inherent characteristics of events are called classical probabilities.

> Probabilities based on a large amount of objective evidence are empirical probabilities.

> Probabilities that use intuitive knowledge or beliefs are subjective probabilities.

Chapter 4 lays a foundation for understanding these three approaches to defining probability. In Section 4.2, we first identify the kinds of events for which probabilities can be measured. In Section 4.3 we consider the quality of the information that is used for each definition of probability. Sections 4.4 through 4.6 illustrate three rules that simplify the calculation of probabilities. Finally, the Business Case illustrates some probability concepts that set the stage for a discussion of probability distributions in Chapters 5 through 7.

**Upon completing Chapter 4, you should be able to**

1. define simple events and combine them into compound events;
2. depict a sample space in terms of a coordinate system or a Venn diagram;
3. define probability and distinguish among classical, empirical and subjective probabilities;
4. state, explain, and apply the basic rules of probability;
5. explain and apply Bayes' theorem;
6. calculate the number of possible outcomes of any event.

## 4.2 SAMPLE SPACES AND EVENTS

Several years ago, the U.S. Department of Justice accused some major electrical manufacturing firms of price fixing. The evidence suggested that competitors had met secretly at times determined by phases of the moon to agree on prices for electrical products to be sold to utilities. There appeared to be a good chance that some executives would go to jail or lose their jobs (or both) if they were found guilty.* However, with every day of the trial, the chances of conviction seemed to change as evidence was introduced or contradicted.

We use this situation to introduce the components of a probability statement. Suppose we wanted to make some statement specifying the chance that a well-defined event will occur. Letting the symbol $P$ represent the word *probability*, we can state probabilities for the two events in Figure 4.1: the probability of being convicted and the probability of both going to jail and losing one's job. (Suppose these are the only two outcomes.)

The basic elements of probability are called *events*. Being convicted is a *simple event*. Going to jail *and* losing one's job is a compound event. Thus a *compound event* is two or more simple events joined together.

A simple event is one that cannot be written as two or more other events; that is, it is a single outcome of a process.

Compound events are collections of two or more simple events.

In this section we distinguish precisely between simple events and compound events. We then explore the role of the connectors (or operators) used to join simple events into compound events.

---

* As it turned out, several executives *were* convicted and jailed, and some of them lost their jobs as well.

**FIGURE 4.1**    Probablity Statements

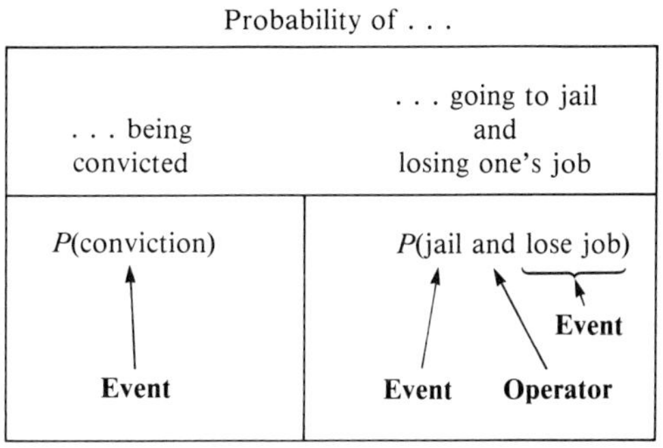

Probability of . . .

| . . . being convicted | . . . going to jail and losing one's job |
|---|---|
| $P(\text{conviction})$ | $P(\text{jail and lose job})$ |

Event

Event    Operator    Event

**Simple Events and Compound Events**

We gather data by observing a *statistical experiment.*

**A statistical experiment is any repeatable process (trials) from which observations of a probability outcome can be obtained.**

The experiment can be as simple as tossing a coin and observing the result or as complex as studying the effect on sales of channeling different amounts of money into advertising. The *essential feature of a statistical experiment is that the outcome of a given trial—whether a single toss of a coin or a one-time advertising compaign—cannot be predetermined.*

Simple events are symbolized by the lowercase letters $e_1, e_2, e_3, \ldots, e_n$. The single trials of an experiment relate to drawing a sample, a process that we considered in Chapter 1. But with probabilities it is essential to identify every conceivable outcome that could occur. These possible outcomes are the simple events of the experiment.

Suppose, for example, that a food warehouser with a central office in Colorado has eight distribution warehouses throughout the western United States. Shipments to the warehouses are made in response to individual warehouse demand, which is variable. The plant inventory-control manager does not know in advance when the orders will arrive, from whom they will come, or how many rail carloads will be requested. If the manager speculates about the destination of the next shipment, the possible simple events are

$$e_1 = \text{next shipment goes to warehouse 1}$$

$$e_2 = \text{next shipment goes to warehouse 2}$$

$$e_3 = \text{next shipment goes to warehouse 3}$$

$$\vdots$$

$$e_8 = \text{next shipment goes to warehouse 8}$$

The manager would be interested in the probabilities of these simple events. If the frequency and size of shipments going to the eight warehouses could be anticipated, the manager could better plan to meet the individual warehouse requirements. The benefits are good service to the warehouses and a minimum of excess inventory at the home plant. This translates to dollars on a profit-and-loss statement, so this type of information is commonly sought.

Managers frequently use probabilities to help them make decisions intended to minimize costs or maximize revenues.

This example illustrates the simple events for one statistical experiment. But to assign numerical probabilities to the events requires a complete set of simple events (that is, all events). A complete set of the possible outcomes in an experiment uses an orderly list or plot of all the simple events.

> The sample space of an experiment is the collection of all the possible simple events for the experiment. The sample space is denoted by $S$.

In the warehouse example, the sample space of outcomes consists of all warehouse locations where the next shipment could go. Then $S = \{e_1, e_2, e_3, \ldots, e_8\}$. An event is any subset or collection of the sample space.

### Rectangular Coordinates

There are two common ways to display sample spaces: the rectangular co-ordinate system defined by $X$- and $Y$-axes, and Venn diagrams. We begin with a single experiment that shows the coordinate system method and illustrates simple events, compound events, and event operators. Example 4.1 illustrates a sample space with simple events.

### EXAMPLE 4.1

We expand the food warehouse example to simple events that describe two locations. Suppose management is interested in the number of rail carload shipments for one month to two warehouses, one in Dallas and the other in Houston. In the past, the number of shipments to Dallas in one month has been 0, 1, or 2. For Houston the number has been 0, 1, 2, or 3. Use rectangular coordinates to demonstrate the sample space of possibilities for the simple events (number of shipments to Dallas, number of shipments to Houston).

**FIGURE 4.2**  Sample Space for Carload Shipment of Food to Two Locations

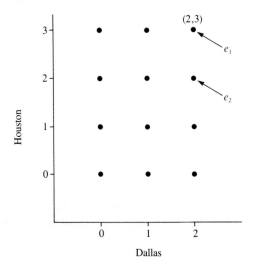

*Solution*   The simple events can be written as pairs: Dallas shipments, Houston shipments. First, we recognize the possible values for each location:

Dallas   0  1  2
Houston  0  1  2  3

Next, we draw a two-dimensional coordinate axis and plot the simple events, as shown in Figure 4.2.

Each simple event in Figure 4.2 is located by a pair of values, one for Dallas and one for Houston. For the simple event $e_1$, the pair (2, 3) indicates that two shipments go to Dallas and three shipments go to Houston. All possible simple events are described by the 12 points displayed in Figure 4.2; it displays the sample space of possibilities for shipments to the two locations, based on past experience.

Events that are defined in two dimensions (variables) can be visualized by ordered pairs. Example 4.1 illustrated simple events in rectangular coordinates. Thus $e_1 = (2,3)$ indicates the label on the horizontal axis is "2" and on the vertical axis is "3."

Often we are more interested in compound events. Figure 4.3 displays three compound events—*A*, *B*, and *C*—on the sample space of Dallas and Houston shipments. Each closed figure includes the points that make up a compound event.

For example, compound event *A*, "both locations get the same number of shipments," is represented by the three ordered pairs (Dallas shipment, Houston shipment): {(0,0), (1,1), (2,2)}. Event *B*, that Dallas gets one shipment, is described by the ordered pairs {(1,0), (1,1), (1,2), (1,3)}. Event *C*, "more shipments go to Dallas than to Houston," is described by {(1,0), (2,0), (2,1)}.

An event (simple or compound) is a restriction on the sample space, and it usually contains fewer than all the points in the space. Using ordered pairs

**FIGURE 4.3**   Three Compound Events

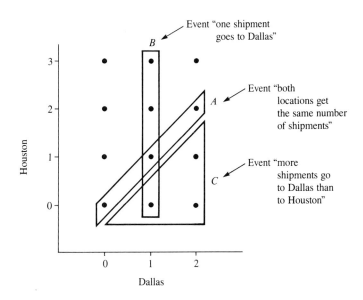

or other labels allows us to designate the elements in the sample space that make up an event.

**Event Operators**

In arithmetic, operators such as +, −, and × are used to combine elements; in probability, *event operators* are used to combine events.

> An event operator is a symbol that indicates a rule for assigning some simple events to a compound event.

There are three common operators: the *union operator*, the *intersection operator*, and the *complement operator*.

> The union operator on events *A* and *B* assigns those simple events belonging to *A*, or to *B*, or to both *A* and *B*, to an event that is the union of *A* with *B*. The symbol ∪ denotes the union operator.

For example, the event "both locations get the same number of shipments *or* Dallas gets one shipment" describes the union of event *A* with event *B* (written *A* ∪ *B*). This is the total of all simple events, or points, contained in the two events. Thus, as Figure 4.4(a) shows, the event *A* ∪ *B* contains six points, so

$$A \cup B = \{(0,0), (1,0),(1,1), (1,2), (1,3), (2,2)\}$$

Note that (1,1) occurs in both events *A* and *B* but is counted only once.

> The intersection operator on events *A* and *B* assigns those simple events (and only those) that belong to both *A* and *B* to an event that is the intersection of *A* with *B*. The symbol ∩ denotes the intersection operator.

---

**FIGURE 4.4**   Comparison of Union and Intersection Operators

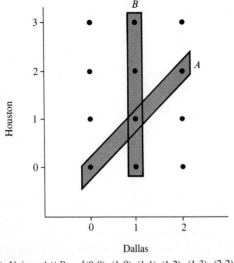

(a)  Union, *A* ∪ *B* = {(0,0), (1,0), (1,1), (1,2), (1,3), (2,2)}

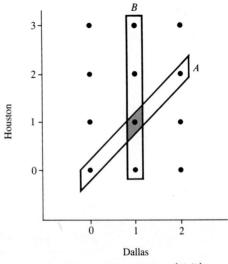

(b)  Intersection, *A* ∩ *B* = {(1,1)}

For example, the event "both locations get the same number of shipments *and* Dallas gets one shipment" (written $A \cap B$) contains only one point (1, 1). We write $(A \cap B) = \{(1, 1)\}$. This is shown as the single point of intersection (overlap) of events $A$ and $B$ shaded in Figure 4.4(b).

A comparison will help distinguish the union and intersection operators. *Union* is an inclusive operator; it includes all simple events that satisfy *either* (one) event as well as those that satisfy both. So the union operator is appropriate when we want $A$ or $B$ or both, that is, at least one of the events. In contrast, *intersection* is an exclusive operator because it excludes all simple events that are not shared; it includes only those descriptions that are in both $A$ and $B$. Figure 4.4 illustrates this distinction for two events that are displayed as $A$ and $B$.

A third useful event operator is the *complement*.

> The complement operator, denoted by $A'$, contains all simple events in the sample space that are not contained in event $A$.

For example, in Figure 4.4 the event that the two locations—Dallas and Houston—do *not* receive the same number of shipments can be defined by nine ordered pairs:

$$A' = \{(0,1), (0,2), (0,3), (1,0), (1,2), (1,3), (2,0), (2,1), (2,3)\}$$

The event $(A \cup B)'$, including the six points outside $A \cup B$ in Figure 4.4(a), is

$$(A \cup B)' = \{(0,1), (0,2), (0,3), (2,0), (2,1), (2,3)\}$$

A useful property of the *complement* is that the union of any event $A$ with its complement $A'$ makes up the entire sample space: $A \cup A' = S$. This fact is used extensively in later developments.

Some event relationships can produce unusual results for the event operators. One case in point is the unusual relationship of *mutually exclusive events*.

> Two events are mutually exclusive if they have no simple events in common, that is, if $A \cap B$ contains no members.

If, in the warehouse example, both locations have the same number of shipments (event $A$ in Figure 4.3), then the number of shipments to Dallas could not exceed the number to Houston (event $C$). So no simple events in $C$ could be in $A$. Thus the events $A$ and $C$ in Figure 4.3 are *mutually exclusive*. In other words, the two events do not overlap; the intersection of $A$ and $C$ is empty. This is symbolized by $A \cap C = \varnothing$, where $\varnothing$ is the *null*, or empty, set. If two events are mutually exclusive, then the probability of both events occurring together is always zero.

**Venn Diagrams**

Earlier we said there are two standard ways to display sample spaces. The use of rectangular coordinates with ordered pairs has already been discussed. The second way, with Venn diagrams, uses the three-event operators union, intersection, and complement. The total area inside each rectangle in Figure 4.5 denotes the full sample space ($S$). Shaded areas indicate the event operators symbolized in each case.

**FIGURE 4.5** Venn Diagrams

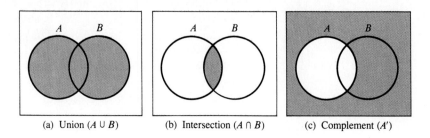

(a) Union $(A \cup B)$        (b) Intersection $(A \cap B)$        (c) Complement $(A')$

In an alternative solution for Example 4.1, we use a Venn diagram to depict the sample space of possibilities for shipments by a food warehouser to two locations.

*Alternative Solution for Example 4.1*

The specific events of interest are

$A$ = both locations get the same number of shipments

$B$ = Dallas gets one shipment

The sample space is displayed in Figure 4.6. It displays the number of points or simple events that make up each event, denoted by $N(\ )$, where the event is written within the parentheses. Then for any event description, the simple events can be counted.

Recall from the warehouse example and Figure 4.3 (page 117) that three points (simple events) satisfied event $A$; thus $N(A) = 3$. Four points satisfied event $B$ and one point satisfied the intersection of event $A$ with event $B$, so $N(A \cap B) = 1$. Finally, six points satisfied the union, so $N(A \cup B) = 6$. You can count the points in Figure 4.6 that satisfy each situation.

The Venn diagram of Figure 4.6 can be expanded to include event $C$ (that Dallas gets more shipments than Houston). See Figure 4.7. Notice that events $A$ and $C$ have no common points. Since they are mutually exclusive, their intersection is empty. In contrast, events $B$ and $C$ are *not* mutually exclusive, since they have one common point.

**FIGURE 4.6**    Counting with Venn Diagrams

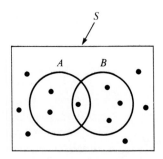

**FIGURE 4.7** A Venn Diagram for Food Warehousing

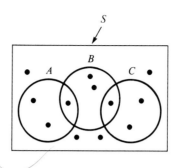

Figure 4.7 enables us to count the number of points, $N(\ )$, in each of several events. For example,

$$N(C) = 3 \qquad N(B \cup C) = 6 \qquad N(B') = 8$$

$$N(B \cap C) = 1 \qquad N(A \cup C) = 6 \qquad N(A \cap C) = N(\varnothing) = 0$$

These counts are central to a definition of probability in the next section. Yet whether they come from a rectangular coordinate display, a Venn diagram, or elsewhere, probability calculations require some enumeration as a base for numerical values.

**Problems 4.2**

1. Let $A = \{1, 2\}$, $B = \{2, 3\}$, $C = \{3\}$, and $S = $ sample space $= \{1, 2, 3\}$. List all descriptions satisfying the following events:
   a. $A \cup B$    b. $C'$    c. $A \cap C$    d. $B \cap C$    e. $(A \cup C)'$

2. For $S = \{1, 2, 3, \ldots, 10\}$, $A = \{2, 4, 6, 9, 10\}$, $B = \{1, 2, 3, 5, 7\}$ and $C = \{5, 10\}$, find each of the following:
   a. $A \cup B$    b. $C'$    c. $A \cap C$    d. $B \cap C$    e. $(A \cup C)'$

3. An experiment consists of tossing two four-sided objects (that is, two four-sided dice) and observing the two numbers from 1, 2, 3, or 4, on the down faces. One object is red and the other is green.
   a. Using a rectangular coordinate system as in Figure 4.2, describe the space. Record the numbers on the red object on the horizontal scale.
   b. Describe in words each of the following events.

   $$A = \{(2,4), (3,3), (3,4), (4,2), (4,3), (4,4)\}$$
   $$B = \{(1,1), (1,2), (1,3), (1,4), (3,1), (3,2), (3,3), (3,4)\}$$
   $$C = \{(2,2), (4,4)\}$$

   c. Describe each of the following by listing all ordered pairs that satisfy the events: $A \cup B$, $B \cap C$, $B'$, $A \cap C'$.
   d. Describe the events in part (c) in words.
   e. Are events $A$ and $B$ mutually exclusive? What about events $B$ and $C$?

 4. The diagram below depicts the outcomes for the inspection by an environmental testing agency of two automobile mufflers from each of two manufacturers, $X$ and $Y$. Scores indicate the number of mufflers that pass inspection.

For example, (2,1), displayed on the graph, indicates the possibility that two of the mufflers from Firm $X$ pass the inspection, whereas one muffler made by Firm $Y$ passes that inspection.

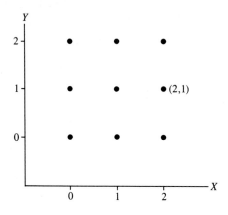

a. List the ordered pairs that satisfy each of the following events.

$A$ = event "exactly one muffler from Firm $X$ passes inspection"

$B$ = event "more mufflers from Firm $X$ than from Firm $Y$ pass inspection"

$C$ = event "the same number of mufflers from Firms $X$ and $Y$ do not pass inspection"

b. Describe each of the following by listing ordered pairs and also by giving statements related to those given in part (a): $A \cup B$, $A \cap C$, $C'$, and $B \cup C$.

c. Count $N(A \cup B)$, $N(A \cap C)$, $N(C')$, $N(B \cup C)$, $N(S)$.

d. Which, if any, of the events in part (b), are mutually exclusive?

## 4.3
## DEFINITIONS OF
## PROBABILITY

Classical probability had its origin in theories about games of chance. Much of the earliest recognized study of probabilities, about 1640, centered around the work of a distinguished French philosopher and passionate gambler named Chevalier de Mère. Many of de Mère's writings concerned games of chance. One question he asked was, Why isn't it favorable to the house to bet that a player will throw at least one double 6 in 24 throws of a pair of dice? De Mère was interested in assessing the outcome of a chance event. A similar type of problem exists in business when, for example, a production manager needs to assess the chance of at least one machine breakdown over the next 24 days of operation. The analysis of both situations requires that we specify probabilities.

Earlier we defined probability as a measure of the chance that an event will occur or that a statistical experiment will have a particular outcome. Probabilities are numbers that can range from 0 to 1 and that describe the chance that a specified event will occur. Zero (0) represents a certainty that an event will not occur, whereas one (1) indicates that the event will occur. As was pointed out in Figure 1.2, probability analysis lies within the "some information" portion of the continuum from no information to all information.

All probability situtations have three characteristics that can be stated as *probability axioms*. These three axioms form the basis for the definitions and rules of probability that follow.

### PROBABILITY AXIOMS CHARACTERISTIC OF ALL PROBABILITY SITUATIONS

*Axiom 1*   The probability of any event ($A$) is a number greater than or equal to zero: $P(A) \geqslant 0$.

*Axiom 2*   The probability of the union of two simple events is equal to the sum of the probabilities of the individual events $P(e_i \cup e_j) = P(e_i) + P(e_j)$.

*Axiom 3*   The sum of the probabilities of all simple events in the sample space must equal one: $P(e_1) + \cdots + P(e_n) = P(S) = 1$.

As stated earlier, the three common approaches to assigning probabilities to events are the (1) classical, (2) empirical, and (3) subjective methods. They differ in the way data are used to determine the probability measure. De Mère's question about throwing a double 6 is an example of *classical* probability. The profitable operation of gambling casinos in Las Vegas and Atlantic City and of state lotteries nationwide is ample evidence of a continued belief in classical probability by the casino owners and game operators. *Empirical* probability, which is based on observations, measures the relative frequency with which an event occurs. For instance, data on the relative frequency of deaths at given ages allow life insurance companies to construct the mortality tables they use to establish insurance premiums. The *subjective* approach uses the assumption that "reasonable" numerical values can be assigned to one's beliefs. With subjective probabilities, logical rules, judgments, or intuition can be used to assign values to these beliefs. For instance, the probability that one firm will market a new electronic product before a competitor does may involve some subjectively determined values. This section explores these three common approaches to defining probability.

**Classical Probability**

Classical probabilities can be determined without a statistical experiment. The earliest applications of classical probability were in gaming situations (tossing coins or dice), where chances could be computed *a priori* (that is, prior to the game). Thus, the probability of getting a head on the toss of a fair coin was reasoned to be 1/2, even without experimental evidence. Early statisticians focused on events having equal chances, like getting heads or tails in tossing a coin. This led to the *classical*, or *equal-likelihood*, definition of *probability*. Likelihood refers to an event's recurrence upon repeated trials of an experiment.

> If a random process can result in any of $N$ equally likely and mutually exclusive outcomes and if $N(A)$ is the number of these outcomes that satisfy the event $A$, then the classical probability of event $A$ is

$$P(A) = \frac{N(A)}{N} \qquad (4.1)$$

Example 4.2, a first example of probabilities, relates to a simple game of chance—tossing dice. We use this game because it is simple and yet conveys important concepts.

**EXAMPLE 4.2**

What is the probability of getting a six on a single toss of a fair six-sided die? *Note:* A practical application is to the percentage of answers that one can expect to get right by guessing at the correct choice for a multiple-choice test where each question has six choices.

*Solution*    Let

$$N = \text{the number of equally likely outcomes } (=6) \text{ (They are equally likely because the die is symmetrical.)}$$

$$N(A) = \text{the number of outcomes that result in a six (that is, the face with six dots) } (=1)$$

Then

$$P(A) = \frac{N(A)}{N} = \frac{1}{6}$$

Note that the numerator is the number of successful outcomes, whereas the denominator is the number of successful outcomes + the number of failures = total outcomes. The probability of getting a six on a single toss is, thus,

$$P(\text{six}) = \frac{\text{Successes}}{\text{Successes} + \text{Failures}} = \frac{1 \text{ six}}{1 \text{ six} + 5 \text{ others}} = \frac{1}{6}$$

So, whether the desired result is a die showing six spots or a correct answer that you got by guessing in a multiple choice test with six options per question, the chance is 1/6. This illustrates Axiom 1: The probability of the event exceeds zero.

Example 4.3 requires summing probabilities over several simple events. This illustrates Axiom 2. Here, by the classical definition, the probability is the same (that is, equally likely) for every simple event.

**EXAMPLE 4.3**

Use the data from Example 4.1 about a food warehouser's distribution of railroad shipments to two locations. The events of interest are

$$A = \text{event "both locations get the same number of shipments"}$$

$$B = \text{event "one shipment goes to Dallas"}$$

$$C = \text{event "more shipments go to Dallas than to Houston"}$$

Use the classical definition of probability to find
a. $P(A)$
b. $P(C)$, $P(B \cup C)$, $P(B')$, $P(B \cap C)$, $P(A \cup C)$, and $P(A \cap C)$

*Solution*    a. Figure 4.8 illustrates the classical probability assignment of 1/12 to each of the 12 possibilities. Using this assignment assumes that each of the 12 possible simple events is *equally likely*. Hence, where $A$ is the event "the two locations get the same value" (this is composed of three simple events), the probability of $A$ is

$$P(A) = \frac{N(A)}{N} = \frac{3}{12} = \frac{1}{4}$$

**FIGURE 4.8**  Sample Space of Equally Likely Events

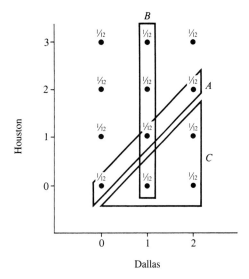

That is, the probability that both locations get none ($P\{(0,0)\}$), one ($P\{(1,1)\}$), or two ($P\{(2,2)\}$) shipments is $P = \{(0,0) \text{ or } (1,1) \text{ or } (2,2)\} = (1/12) + (1/12) + (1/12) = (3/12)$. This solution recognizes that the compound event encompasses a collection of three simple events. Then, by Axiom 2 the probability of the compound event is the sum of the probabilities for these three simple events.

b. By the classical definition, each simple event has probability 1/12. To solve (b) requires finding the probability of several compound events.

$$P(C) = \frac{N(C)}{N} = \frac{3}{12} = 0.25 \qquad P(B \cap C) = \frac{N(B \cap C)}{N} = \frac{1}{12} = 0.08$$

$$P(B \cup C) = \frac{N(B \cup C)}{N} = \frac{6}{12} = 0.50 \quad P(A \cup C) = \frac{N(A \cup C)}{N} = \frac{6}{12} = 0.50$$

$$P(B') = \frac{N(B')}{N} = \frac{8}{12} = 0.67 \qquad P(A \cap C) = \frac{N(A \cap C)}{N} = \frac{0}{12} = 0$$

Check the results by counting the points in Figure 4.8 that satisfy each event and multiplying that number by 1/12. Also, you can observe that Axioms 1, 2, and 3 are maintained in the preceding calculations.

---

The classical definition of probability is limited because in most real situations all possible outcomes are *not* equally likely. Furthermore, in many situations the number of possible outcomes is more than can be counted. Consequently, the *classical definition is limited to those situations in which every simple event has an equal chance of occurrence and the sample space is finite.* As you might suspect, empirical probabilities have a wider application in business.

**Empirical Probability**

When all the outcomes in an experiment are not equally likely, then we may be able to determine the chance of occurrence by gathering empirical evidence, that is, by making observations. This approach leads to the *empirical or relative frequency definition of probability.*

The empirical probability of an event is the value approached by the relative frequency of occurrence as the number of trials of a statistical experiment becomes very large:

$$P(X) = \frac{f(X)}{n} \qquad \text{(4.2)}$$

where  $f(X)$ = frequency of occurrence of event $X$

$n$ = number of trials

For example, the government analyst who wished to find out whether more male than female babies were born in a particular locale might obtain county records on the number of births of each sex for many years. As the analyst looked at more and more records, she would get a better approximation of the true proportion of male infants living in that locale. Suppose the birth data provided the *relative frequency of male* births to total births as follows:

$$\frac{4}{10}, \quad \text{then} \quad \frac{48}{100}, \quad \text{then} \quad \frac{246}{500}, \quad \text{then} \quad \frac{488}{1000}, \dots$$

That is, of the first 10 births recorded, 4 were boys; of the first 100 births, 48 were boys; and so on. At some point, after a sufficiently large number of birth records have been examined, a single value for the relative frequency of male babies can be determined. From the given sequence, the empirical probability of the birth of a boy, $P(B)$, is 488/1000, or approximately 49%. Thus the researcher's answer is that fewer male than female births were recorded in this locale.

With minor modification, our food warehouser example provides another illustration of empirical probabilities. Actually, it is improbable that all the possible pairs of scores depicted in Figure 4.8 would be equally likely. More reasonable measures could be determined from long-term records of shipments to the two locations, as shown in Example 4.4.

---

**EXAMPLE 4.4**

Suppose the food warehouser has operated warehouses in Dallas and Houston for 120 months (that is, 10 years). The relative frequencies of shipments are shown in Figure 4.9.

For example, position (0,0) indicates the event that "both locations get zero shipments," and has a value of 0.01. That is, (Dallas shipments, Houston shipments) = (0,0) occurs once out of every 100 months. Use the empirical probabilities such as $P\{(0,0)\} = 1/100 = 0.01$ and the events from Figure 4.9

$A$ = the two locations get the same number of shipments

$B$ = Dallas gets one shipment

$C$ = Dallas gets more shipments than Houston

to obtain $P(C)$, $P(B \cup C)$, $P(B')$, $P(B \cap C)$, $P(A \cup C)$, and $P(A \cap C)$.

**Solution**   Find the solutions by observing the ordered pairs in Figure 4.9 that define each compound event and then adding the probabilities assigned to these pairs:

$$P(C) = \{(1,0), (2,0), (2,1)\} = 0.03 + 0.08 + 0.13 = 0.24$$

**FIGURE 4.9**  Empirical Probabilities for Warehousing Food at Two Locations

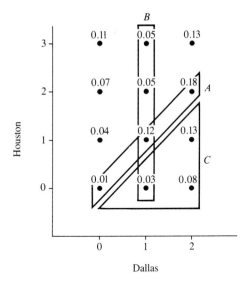

$$P(B \cup C) = P\{(1,0), (1,1), (1,2), (1,3), (2,0), (2,1)\}$$
$$= 0.03 + 0.12 + 0.05 + 0.05 + 0.08 + 0.13 = 0.46$$
$$P(B') = P\{(0,0), (0,1), (0,2), (0,3), (2,0), (2,1), (2,2), (2,3)\} = 0.75$$
$$P(B \cap C) = P\{(1,0)\} = 0.03$$
$$P(A \cup C) = P\{(0,0), (1,1), (2,2), (1,0), (2, 0), (2,1)\} = 0.55$$
$$P(A \cap C) = P[\varnothing] = 0$$

Other probabilities can be calculated using the same source.

A comparison of these results with those from Example 4.3 reveals that the probabilities we determine by the classical definition can be very different from those we determine by using actual data and the empirical probability definition. Empirical data take advantage of past history, so we can expect different results by using past business experience (facts) as a base for probability calculations rather than using the equal-likelihood assumption. Empirical probabilities are preferred because they are based on actual experience.

**Subjective Probability**

The third approach to probability is harder than the other two to condense into an equation because *subjective probabilities* need not be based on either prior knowledge or empirical data.

> The subjective probability of an event is the numerical value of chance assigned to it on the basis of a judgment or a measure of personal belief.

Subjective probabilities need not be based on calculations founded on the laws of nature (like classical probabilities), nor do they require a large collection of data as do empirical probabilities. The only technical requirement is that the likelihood be stated in the same numerical manner as a classical or an empirical probability; the three probability axioms must hold. Example 4.5 demonstrates such a subjective probability.

**EXAMPLE 4.5**

A construction company has contracted to build a chemical plant in Saudi Arabia within nine months, and the firm's comptroller needs to determine the probability that the plant will be completed on time. If the plant is not finished on time, the construction company must forfeit some of its payment for the work. Based upon discussions with purchasing, engineering, and other members of management, the comptroller subjectively estimates that 15% of such construction projects are normally completed in eight months or less. Another 65% of them are completed during the ninth month, and 20% of them take more than nine months to complete. Use these data to derive a subjective probability that they will complete the plant within nine months or less. (This will give the comptroller a measure of the chance that the project will be done on time and an indication of the chance of receiving full payment on their contracted work).

*Solution*    Note that the sum of the percentages for the three outcomes is 100% (15% + 65% + 20%). Based upon his subjective appraisal, the comptroller may assign a probability of 0.15 + 0.65 = 0.80 to the event of "completing the plant in nine months or less."

How do we decide which definition of probability is appropriate in a given situation? Usually it is not difficult. When a situation occurs in which each possible outcome has an equal chance, then the classical (*a priori*) definition applies. Otherwise, we search for empirical data. Nevertheless, many business decisions rely on less tangible judgments, and subjective probabilities play a key role when objective information is unavailable or when events are inherently not repeatable.

**Problems 4.3**

5. Executives of City Power & Light Co. are considering a customer survey on a proposal to raise rates, but they want to avoid surveying many customers who are already concerned about their bills. Of the last 400 inquiries at the customer service center, 120 have been concerned with higher electricity bills. Based on this data, what is the probability that the next inquiry will be concerned with a higher electricity bill?

6. Given $S = \{1, 2, 3, \ldots, 10\}$, $A = \{2, 4, 6, 8, 10\}$, $B = \{1, 2, 3, 5, 7\}$, and $C = \{5, 10\}$, use the classical definition of probability to evaluate each of the following: (a) $P(A \cup B)$, (b) $P(A \cap C)$, (c) $P(C')$, (d) $P(B \cup C)$, (e) $P(S)$.

7. Over the past 100 weeks, the Eastern and Western Divisions of the Chicago Lighting Company have competed vigorously to see which division best meets its weekly sales quota. The frequency with which each division has been under its sales quota, on target, or over its quota is shown in a graph. (0 = Under quota, 1 = On target, 2 = Over quota). Numbers recorded in each space are the number of times out of 100 (the percentage). Let

$A$ = event "the Eastern Division was on target"

$B$ = event "the Eastern Division did better than the Western Division"

$C$ = event "the Eastern and Western Divisions did not perform equally well"

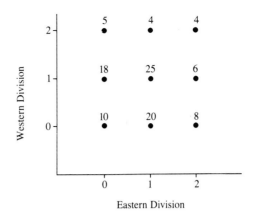

Given this empirical data, evaluate each of the folllowing:
a. $P(A \cup B)$         c. $P(C')$                    (e) $P(S)$
b. $P(A \cap C)$         d. $P(B \cup C)$

8. A container in the safety office of a chemical plant contains nine specially colored caution signs: two white, three black, and four yellow. A helper, who does not know that the colors have a specific meaning, is sent to get one sign. Using the classical definition, compute the probability that it is (a) yellow, (b) either black or white, (c) not white, (d) yellow if you know it is not black.

9. The Reliable Life Insurance Company developed a chart based on the insured lives of many thousand men. Let

$A$ = event "the man is under age 30 when he purchases a policy"

$B$ = event "death occurs after age 50"

$C$ = event "the insured purchases at age 30 or later and his death occurs at age over 50"

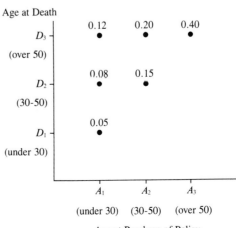

Age at Purchase of Policy

a. List the ordered pairs that describe each event. Attach the prescribed empirical probability to each pair.

b. Compute $P(A)$, $P(C)$, $P(A \cap B)$, and $P(A \cup C)$.

c. Are any of the events $A$ and $B$, $A$ and $C$, or $B$ and $C$ mutually exclusive?

10. A survey of the subjective opinions of managers from 80 firms engaged in mining, milling, and processing nuclear fuel ($U_3O_8$) revealed the given price expectations for the year 2000 (see graph). Let

$$A = \text{event "the price of } U_3O_8 \text{ will be over \$50 per pound"}$$

$$B = \text{event "the price of } U_3O_8 \text{ will be from \$30 to \$50 per pound"}$$

$$C = \text{event "the price of } U_3O_8 \text{ will be less than \$30 per pound"}$$

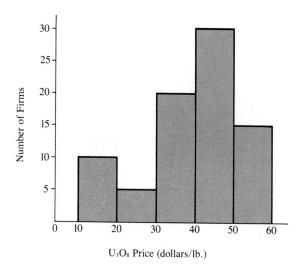

$U_3O_8$ Price (dollars/lb.)

a. Based on the subjective opinions, estimate the probability that in the year 2000 the price will be within each of the classes shown—that is, \$10 but less than \$20, \$20 but less than \$30, and so on.

b. Find $P(A)$, $P(B)$, and $P(C)$.

c. Find $P(A \cup C)$ and $P(B')$.

## 4.4
## PROBABILITY RULES

Three basic operational rules govern calculations of probabilities. Each of the rules is explained and illustrated in the discussion that follows. First, however, we define some terms that are commonly used to describe relationships among probabilities.

### Relationships among Probabilities

The statements in Figure 4.1 express specific relationships between events and the sample space upon which they were defined. The relationships differ, depending upon what (or whether an) event operator is used to connect them. Analysts sometimes refer to event relations as being joint, marginal, or conditional probabilities. We can illustrate the use of these terms by considering the penalties associated with some illegal business behavior, such as insider trading in the stock market.

$$P(\text{Jail and losing job}) \quad \text{is} \quad a \ joint \ probability$$

$$P(\text{Conviction}) \quad \text{is} \quad a \ marginal \ probability$$

$$P(\text{Jail given conviction}) \quad \text{is} \quad a \ conditional \ probability$$

For two events $A$ and $B$, the joint probability is the probability that $A$ and $B$—that is, both—will occur.

The marginal probability for $A$ is the probability that event $A$ will occur.

The conditional probability for $B$ given $A$ is the probability that event $B$ will occur given that event $A$ has already occurred.

Let us explore these definitions further. We can illustrate all three types of probabilities for some business data.

A Denver manufacturer of plastic pipe couplings has shipped 500 couplings to a contractor in Kansas City. The contractor has received a favorable price because some of the couplings deviate more than 0.05 inch from standard pipe sizes and are therefore considered defective. The defective couplings can be used on drain lines, but the contractor can use only good couplings on all pressurized water lines.

The couplings all look alike, but they were produced on three different injection molding machines. The manufacturer has provided the information contained in Table 4.1. This displays the numbers produced by each machine for "good" and then for "defective" couplings. Of the 500 couplings in the shipping container, 450 are good and 50 are technically defective.

Table 4.2 expresses the same data in terms of probabilities. We calculated probabilities by using the empirical (frequency) definition of probability expressed in Equation (4.2). Let's follow through the computation of values for Table 4.2 and use the tables to identify joint, marginal, and conditional probabilities.

We begin with *joint probability*. Suppose the contractor has selected one coupling and wants to know the probability for the event that "the coupling was produced by machine $X$ *and* is good." The sample space consists of the $n = 500$ possible couplings that could be selected. Table 4.1 shows that the frequency of occurrence of "a good coupling (and) produced by machine $X$" is $f = 225$.

**TABLE 4.1**  Counts on Couplings

| Machine | Quality Level | | Total |
| | G (Good) | D (Defective) | |
|---|---|---|---|
| $X$ | $X \cap G$ 225 | 25 | 250 |
| $Y$ | 135 | 15 | 150 |
| $Z$ | 90 | 10 | 100 |
| Total | 450 | 50 | 500 |

**TABLE 4.2**   Probabilities on Couplings

| Machine | Quality Level | | Total |
| --- | --- | --- | --- |
| | G (Good) | D (Defective) | |
| X | P(X ∩ G) 0.45 | 0.05 | P(X) 0.50 |
| Y | 0.27 | 0.03 | 0.30 |
| Z | 0.18 | 0.02 | 0.20 |
| Total | 0.90 | 0.10 | 1.00 |

The probability that "a coupling was produced by machine $X$ *and* is good," denoted by $P(X \cap G)$, is represented by the frequency at the intersection of $X$ and $G$ divided by the total frequency.

$$P(X \cap G) = \frac{f(X \cap G)}{n} = \frac{225}{500} = 0.45$$

Thus the probability of selecting a good coupling produced by machine $X$ is 0.45, as shown in the dark-shaded area of Table 4.2. Notice that the intersection encompasses those simple events (the plastic couplings) that belong *jointly* to the events "being produced by machine $X$" and "being good" ($G$). The probability that an item will simultaneously satisfy two or more events is known as a joint probability. The other joint probabilities in Table 4.2 are computed similarly from counts in the body of Table 4.1; for example, the joint probability of $X$ and $D$ is the frequency at the intersection of $X$ and $D$ divided by the total frequency.

$$P(X \cap D) = \frac{25}{500} = 0.05$$

$$\vdots$$

$$P(Z \cap D) = \frac{10}{500} = 0.02$$

The second relation, a *marginal probability*, is the chance for a single event. Table 4.2 displays this as the aggregate of all those events of a single row (or column) and disregards the cross-classifications involved. Suppose a supervisor is interested *only* in whether a randomly selected coupling was produced on machine $X$, regardless of its perfection or imperfection. We know from Table 4.1 that the frequency of occurrence of couplings produced on machine $X$ is $f(X) = 250$. This is a marginal total, so the (marginal) probability that "a coupling was produced by machine $X$," denoted by $P(X)$, is

$$P(X) = \frac{f(X)}{n} = \frac{250}{500} = 0.50$$

This probability is referred to as a marginal probability; it is computed from values in a margin of the table. The marginal probability takes account of *all* ways in which the designated event can occur. In this example, the probability of a coupling being produced on machine $X$ must include the probabilities of both the good and the defective couplings produced on this machine. The *marginal probability* of $X$ is the sum of joint probabilities of $X$ that lie in the body of Table 4.2. Thus it can also be written

$$P(X) = P(G \cap X) + P(D \cap X)$$
$$= 0.45 + 0.05 = 0.50$$

This second method of computing the marginal probability of $X$ yields the same result as the first; we are simply adding the appropriate cell probabilities from Table 4.2 (that is, 0.45 and 0.05) rather than working with the row total (250/500). In Table 4.2, the marginal value $P(X)$ is thus calculated using the shaded cells in machine $X$ row, and $P(G)$ uses the shaded cells in the Quality Level $G$ column. Here, $P(G) = 0.90$. The other marginal probabilities in Table 4.2 can be identified at the margins of the table in a similar manner.

A third relation is a *conditional probability*, that is, the probability of one event, *given* another. It is symbolized by $P(D|X)$, read "the probability of $D$ given $X$." For example, suppose an inspector wished to establish the defective rate of machine $X$. That is, "given machine $X$, what is the rate of production defectives?" To find this probability the inspector would restrict sampling to couplings produced by machine $X$. Then the probability of a defective coupling is based on the number of defectives produced by machine $X$ relative to that machine's total production. From Table 4.1 we have

$$P(D|X) = \frac{\text{Frequency of } (D \cap X)}{\text{Number of } X} = \frac{P(D \cap X)}{P(X)} = \frac{25}{250} = 0.10$$

*The conditional probability of D given X equals the joint probability of the two events divided by the marginal probability of X;* that is, from Table 4.2,

$$P(D|X) = \frac{P(D \cap X)}{P(X)} = \frac{0.05}{0.50} = 0.10$$

Similarly, the probability of the event "a coupling was produced by machine $Y$, *given* the selection is restricted to good couplings" is

$$P(Y|G) = \frac{P(Y \cap G)}{P(G)} = \frac{0.27}{0.90} = 0.30$$

With these terms in mind, we consider the basic rules of probability. These rules are guidelines for calculating the value for an unknown joint, marginal, or conditional probability when some information is known about other joint, marginal, or conditional probabilities.

**Complement Rule**

One rule for probabilities describes the complement of an event. The *complement of an event* is the remainder of the sample space beyond the event.

**COMPLEMENT RULE**

$$P(A') = 1 - P(A) \tag{4.3}$$

Example 4.6 uses Rule 4.3 to show the complement of an event.

---

**EXAMPLE 4.6**

A carton containing 500 couplings has 450 good units. Use the complement rule to find the probability of the event "one coupling selected at random will be defective."

*Solution*    The couplings are classified as either defective ($G'$) or good ($G$), and these two events constitute the entire sample space, with probability 1. The good couplings are designated as ($G$), which gives

$$P(G) = \frac{450}{500} = 0.90$$

$$P(G') = 1 - P(G) = 1 - 0.90 = 0.10, \quad \text{or } 10\%$$

The probability is 0.90 that any coupling is not defective. Then the probability for a defective is $1.00 - 0.90 = 0.10$.

---

## Multiplication Rule

Probabilities are multiplied to measure the likelihood that one event happens *and* (multiply) another event happens too. Thus, a second rule for combining probabilities is the *multiplication rule*.

> The multiplication rule describes a joint probability as equal to the product of a marginal probability times a conditional probability.

The *multiplication rule* states that the probability of the joint occurrence (intersection) of two events $A$ and $B$, $P(A \cap B)$, is the product of the probability of the first, $P(A)$, times the conditional probability of the second given the first, $P(B|A)$.

**MULTIPLICATION RULE**

$$P(A \cap B) = P(A)P(B|A)$$

or, interchanging events $A$ and $B$,

$$P(A \cap B) = P(B)P(A|B) \tag{4.4}$$

Example 4.7 illustrates the multiplication rule using the data on plastic couplings.

---

**EXAMPLE 4.7**

The data for plastic couplings (from Table 4.1) are repeated here. Use the multiplication rule to find the probability of the event "a coupling selected at random was produced by machine $Y$ and is good."

| Machine | Quality Level | | Total |
|---|---|---|---|
| | G (Good) | D (Defective) | |
| X | $X \cap G$ 225 | 25 | 250 |
| Y | 135 | 15 | 150 |
| Z | 90 | 10 | 100 |
| Total | 450 | 50 | 500 |

*Solution*    By inspection of the table and Equation (4.4),

$$P(Y \cap G) = P(Y)P(G|Y) \qquad \text{where} \qquad P(Y) = \frac{150}{500} = 0.30$$

$$P(G|Y) = \frac{135}{150} = 0.90$$

$$P(Y \cap G) = (0.30)(0.90) = 0.27$$

Thus 27% of the couplings were (both) produced by machine $Y$ and were good.

Because the complete table of frequencies is available, we could have solved Example 4.7 by simply dividing the count at the intersection of $Y$ and $G$ by the total, so that $P(Y \cap G) = 135/500 = 0.27$. However, a complete table is often not available; then probability rules like Equation (4.4) provide the means for a solution.

*A Special Case: Independent Events*

The special case of the multiplication rule that we mentioned earlier relates to *independent events*. So, next we define this term:

> Two events $A$ and $B$ are independent if the occurrence of one has no bearing on the chance of occurrence for the other. Then the conditional probabilities are equal to their corresponding marginal probabilities. So,

$$P(A|B) = P(A) \quad \text{and} \quad P(B|A) = P(B)$$

Consider the business situation we are working with. If the probability of "selecting a good coupling" were not influenced (or conditioned) by whether the selection was restricted to machine $X$, $Y$, or $Z$, then the percentage of defective products would be *independent* of machines.

When statistical independence exists, all conditional probabilities equal their corresponding unconditional (marginal) probabilities. Thus, if $P(B|A) = P(B)$ for events $A$ and $B$, then the events are independent and the multiplication rule $P(A \cap B) = P(A)P(B|A)$ reduces to $P(A \cap B) = P(A)P(B)$, which is another method of confirming independence in a data set. Example 4.8 shows how independence can be ascertained in a problem situation.

---

**EXAMPLE 4.8**

Suppose the manufacturer of couplings wishes to demonstrate the production activity to a client. Given the data on production of couplings, determine whether the quality level is independent of machine type. Is one machine as effective as the other machines in producing good items? If so, a client demonstration could be equally effective on any of the three machines.

*Solution*    If the quality level were independent of the machine, then the probability of obtaining a good coupling would be the same whether it came from machine $X$, $Y$, or $Z$. Then

$P(X \cap G)$ would equal $P(X)P(G)$
$P(Y \cap G)$ would equal $P(Y)P(G)$
$P(Z \cap G)$ would equal $P(Z)P(G)$

The data for the production of couplings indicate that *this is* the case for machine $X$ and quality level $G$:

$$P(X \cap G) = P(X)P(G \mid X) = P(X)P(G)$$

$$= \frac{250}{500} \cdot \frac{225}{250} = \frac{250}{500} \cdot \frac{450}{500} = 0.45$$

The events $X$ and $G$ are independent. Similarly, $P(Y \cap G) = P(Y) \cdot P(G)$, and $P(Z \cap G) = P(Z)P(G)$. So, the quality of any selected coupling is statistically independent of the machine upon which it is made ($X$, $Y$, or $Z$). The probability of a good coupling does not depend upon the choice of machines; all three machines produce 90% good items.

---

Although for the data in Example 4.8 machines and quality level are independent events, this is an unusual relation between events. *Independence is a special relationship that exists between some events, and we cannot expect it to hold for all events.*

**Addition Rule**

A third rule for combining probabilities, the addition rule, involves adding (and subtracting) probabilities. Some probabilities are added, and others subtracted, to measure the likelihood that an event happens *or* (addition) another event happens.

> The addition rule describes the probability for the union of two events as the sum of their marginal probabilities minus their (common) joint probability.

The general *rule of addition* states that the probability of the occurrence of either or both of two events $A$ or $B$ (that is, their union) is the sum of the probability of $A$ plus the probability of $B$ less the probability of their joint occurrence.

**ADDITION RULE**

$$P(A \cup B) = P(A) + P(B) - P(A \cap B) \qquad (4.5)$$

or, when $A$ and $B$ are mutually exclusive events,

$$P(A \cup B) = P(A) + P(B) \qquad (4.5a)$$

Example 4.9 illustrates the general addition rule, whereas Example 4.10 is the special case in which the events are mutually exclusive.

**EXAMPLE 4.9**

Use the addition rule to find the probability for the event "a coupling selected at random was produced by machine $Y$ or is good or it satisfies both." This can be developed from the counts of couplings (Table 4.1) or from probabilities of couplings (see Table 4.2).

|  | Quality Level | | |
|---|---|---|---|
| Machine | G (Good) | D (Defective) | Total |
| X | 225 | 25 | 250 |
| Y | 135 | 15 | 150 |
| Z | 90 | 10 | 100 |
| Total | 450 | 50 | 500 |

*Solution*   The question will be answered with data from Table 4.1, which is repeated here.

$$P(Y \cup G) = P(Y) + P(G) - P(Y \cap G)$$

where   $P(Y) = \dfrac{150}{500} = 0.30 \quad P(G) = \dfrac{450}{500} = 0.90$

and   $P(Y \cap G) = \dfrac{135}{500} = 0.27$

Thus we have $P(Y \cup G) = 0.30 + 0.90 - 0.27 = 0.93$.

We can also find the answer to Example 4.9 by reading probabilities directly from the counts table, Table 4.1. Notice that the 135 good couplings produced on machine $Y$ were included in both the $Y$ total of 150 and the $G$ total of 450. Because these couplings were counted twice (shown as a dark-shaded area), one of the counts is subtracted in the addition rule.

*A Special Case: Mutually Exclusive Events*

A special case of the addition rule applies to mutually exclusive events. You may recall that mutually exclusive events have no common elements, so their intersection is empty. Then $P(A \cap B) = 0$, and the addition rule becomes $P(A \cap B) = P(A) + P(B)$. We illustrate this in Example 4.10 with the coupling data.

**EXAMPLE 4.10**

Use the addition rule to find the probability that a coupling selected at random was produced by machine Y" *or* "was produced by machine Z." That is, what percentage of the production (for machines X, Y, and Z) has come from machine Y or machine Z?

*Solution*

Using the previously developed probability values, the general rule yields this result:

$$P(Y \cup Z) = P(Y) + P(Z) - P(Y \cap Z)$$
$$= 0.30 + 0.20 - 0 = 0.50$$

where $P(Y \cap Z) = 0$ means there is zero chance that any coupling was produced by both machines.

Recognizing that the events of Y, "couplings are produced on machine Y," and of Z, "couplings are produced on machine Z," are mutually exclusive, we use Equation (4.5a):

$$P(Y \cup Z) = P(Y) + P(Z) = 0.30 + 0.28 = 0.50$$

So 50% of the production has come from machine Y or machine Z.

Although Example 4.10 illustrates mutually exclusive events, they are not the usual case. However, it is always safe to begin with the general rule for either addition or multiplication of probabilities because the general rules reduce to special cases when the events are mutually exclusive or independent.

The following list reviews the three basic rules stated at the beginning of this section. As you study each rule, think about the type of probability (joint, marginal, or conditional) and the meaning of the special relationships (independent and mutually exclusive). The problems that follow give ample opportunity for you to apply these rules.

**BASIC RULES OF PROBABILITY**

1. Complement:

$$P(A') = 1 - P(A) \tag{4.3}$$

2. Multiplication:

$$P(A \cap B) = P(A)P(B|A)$$
$$= P(A)P(B) \quad \text{if } A \text{ and } B \text{ are independent} \tag{4.4}$$

3. Addition:

$$P(A \cup B) = P(A) + P(B) - P(A \cap B)$$
$$= P(A) + P(B) \quad \text{if } A \text{ and } B \text{ are mutually exclusive} \tag{4.5}$$

**Problems 4.4**

11. The probability that a small foreign-made car using Super-Go gasoline will get 35 miles or more to the gallon is 0.70. What is the probability the car will get less than 35 miles per gallon? Are the events in question independent, mutually exclusive, or neither? Explain your choice.

12. Given $P(A') = 0.4$, $P(B) = 0.2$, and $P(B|A) = 0.3$, find each of the following.
    a. $P(A)$
    b. $P(A \cap B)$
    c. $P(A \cup B)$
    d. Are the events $A$ and $B$ mutually exclusive? Explain.

13. If the probability that stock $A$ will go up in price is 0.6, the probability that stock $B$ will go up is 0.4, and their actions are independent, what is the probability of each of the following events?
    a. Stock $A$ will not go up in price.
    b. Both $A$ and $B$ will go up in price.
    c. Either $A$ or $B$ (or both) will go up in price.
    d. Neither $A$ nor $B$ will go up in price.
    e. $A$ and $B$ will not both go up in price.

14. The probability that a female customer entering a supermarket will buy jelly is 0.1. The probability that the customer will buy bread if (or given that) she buys jelly is 0.4. The probability is 0.5 that she will buy milk whether or not she buys anything else. What is the probability of each of the following events?
    a. The customer will buy both jelly and bread.
    b. The customer will buy either jelly or milk (or both).

15. $A$ and $B$ are independent events, with $P(A) = 0.4$ and $P(B) = 0.8$. What is the probability of the following events?
    a. $A$ and $B$
    b. $A$ or $B$
    c. neither $A$ nor $B$ (that is, $A' \cap B'$)

16. Given the mutually exclusive events $A$ and $B$, for which $P(A) = 0.5$ and $P(B) = 0.3$, use Venn diagrams or the basic rules of probability to find
    a. $P(A')$         c. $P(A' \cap B)$         e. $P(A' \cup B)$
    b. $P(A \cup B)$   d. $P(A|B)$               f. $P(A \cup B)'$

17. Given $P(A) = 0.4$, $P(B) = 0.6$, and $P(A \text{ and } B) = 0.2$, which of the following are true?
    a. $P(A \text{ or } B) = 1$.
    b. $A$ and $B$ are mutually exclusive. That is, $P(A \cup B) = P(A) + P(B)$.
    c. $P(B|A) = 0.5$.
    d. $A$ and $B$ are independent.

18. Among the members of a certain club, 10% are bankers and 20% have incomes over $90,000 per year. If it is known that 80% of the bankers have incomes over $90,000 per year, what percentage of the club members
    a. have incomes of over $90,000 and are bankers?
    b. either are bankers or have incomes of over $90,000 per year (or satisfy both)?

19. The accompanying data show the status of employees at the Fairfield Plant with respect to their jobs and whether they belong to the company stock plan. For (a), (b), and (c), first indicate the type of probability involved (that is, joint, marginal, or conditional) and then find the computed value of the probability.

|  | (B) Belongs | (B') Does Not Belong | Total |
|---|---|---|---|
| A = Administration | 30 | 20 | 50 |
| M = Marketing | 600 | 200 | 800 |
| P = Production | 100 | 50 | 150 |
| Total | 730 | 270 | 1000 |

a. Probability that an employee does not belong to the plan.
b. Probability that an employee belongs, given that he or she works in marketing.
c. Probability that an employee belongs and works in production.
d. Probability that an employee is either an administrative worker or does not belong.

20. The Flex Furniture Company seeks information about its production. For instance, what is the relative production by shift? By sex?
a. Compute probabilities for each of the following, and indicate what event each calculation describes:

$$P(M \cap 1), P(3|M), P(M \cup 2), \text{ and } P(M|1 \cup 2)$$

b. Does $P(F \cap 2) = P(F)P(2)$? What statistical fact does this indicate?

|  | Shift |  |  |  |
|---|---|---|---|---|
| Worker | 1 | 2 | 3 | Total |
| M | 0.35 | 0.20 | 0.25 | 0.80 |
| F | 0.05 | 0.10 | 0.05 | 0.20 |
| Total | 0.40 | 0.30 | 0.30 | 1.00 |

## 4.5 BAYES' THEOREM

As new information is obtained in a situation, we sometimes need to revise probabilities as part of a probability analysis. We may begin with specific probability estimates and then receive new information from tests, reports, and so on. We then want to use this new information to calculate new probabilities. Bayes' theorem is a probability rule that allows us to calculate these revised probabilities. Its use is primarily associated with subjective probabilities, for reasons that will soon become apparent. Bayes' theorem is a statement of how to calculate conditional probabilities. For example, suppose $\theta$ = the conditioning event "it rains" and $Z$ = the conditioned event "I get wet." Then we would write the joint probability

$$P(\theta \cap Z) = P(\theta) \cdot P(Z|\theta)$$

$$P(\text{It rains } and \text{ I get wet}) = P(\text{It rains}) \cdot P(\text{I get wet}|\text{It rains})$$

Solving for the conditional probability,

$$P(Z|\theta) = \frac{P(\theta \cap Z)}{P(\theta)}$$

**Conditioned**      **Conditioning**
**event**             **event**
**(I get wet)**       **(It rains)**

Similarly,

$$P(\theta|Z) = \frac{P(\theta \cap Z)}{P(Z)}$$

**It rains   I get wet**

This latter expression indicates that the probability of "It rains" given that "I get wet" is equal to the joint probability "It rains and I get wet" divided by the marginal probability "I get wet." We can use this relationship to express Bayes' theorem. Table 4.3 contains the probabilities for the calculation:

$$P(\theta|Z) = P(\text{"It rains } given \text{ that I get wet"})$$
$$= P(\text{"It rains and I get wet"})/P(\text{"I get wet"})$$
$$= P(\theta \cap Z)/P(Z)$$

This result is summarized more formally as

**BAYES' THEOREM**

$$P(\theta|Z) = \frac{P(\theta \cap Z)}{P(Z)} \tag{4.6}$$

In our example, we are finding the probability "It rains" given that "I get wet." That is, Bayes' theorem gives us a way to compute the probability of a preexisting

**TABLE 4.3**   Probability Table

| Conditioned Event | Conditioning Event | | Total |
| --- | --- | --- | --- |
| | *(It rains)* $\theta$ | *(No rain)* $\theta'$ | |
| I get wet, $Z$ | $P(\theta \cap Z)$ | $P(\theta' \cap Z)$ | $P(Z)$ |
| I do not get wet, $Z'$ | $P(\theta \cap Z')$ | $P(\theta' \cap Z')$ | $P(Z')$ |
| Total | $P(\theta)$ | $P(\theta')$ | 1.00 |

condition given a specific event. This preexisting condition may not appear to have any chance associated with it.

Generally, Bayes' theorem is given in an expanded form. $P(\theta \cap Z)$ is replaced by the equivalent (and more familiar) form $P(\theta)P(Z|\theta)$. Sometimes $P(Z)$ is available, but generally $P(Z)$ is the sum of the probabilities $P(\theta \cap Z)$ and $P(\theta' \cap Z)$, since $\theta$ and $\theta'$ are mutually exclusive. We found earlier that the marginal probability of an event equals the sum of the joint probabilities that make up that event. Thus we can write

$$P(Z) = P(\theta \cap Z) + P(\theta' \cap Z)$$

Then, by the rule of conditional probability,

$$P(\theta \cap Z) = P(\theta)P(Z|\theta)$$

and

$$P(\theta' \cap Z) = P(\theta')P(Z|\theta')$$

Therefore, we can rewrite Bayes' theorem in expanded form:

### BAYES' THEOREM IN EXPANDED FORM

$$P(\theta|Z) = \frac{P(\theta)P(Z|\theta)}{P(\theta \cap Z) + P(\theta' \cap Z)}$$

$$= \frac{P(\theta)P(Z|\theta)}{P(\theta)P(Z|\theta) + P(\theta')P(Z|\theta')} \tag{4.7}$$

Bayes' theorem is mathematically correct. Yet many scholars have questioned it since the time it was first introduced in 1876 following the death of the Reverend Thomas Bayes. One reason for the critics' concern is the "inverse," or backward, reasoning that it employs. No one would question the idea of estimating the probability of "I get wet" given that "It rains," $P(Z|\theta)$. But many become skeptical when it comes to computing the probability of a preexisting condition such as $P(\theta|Z)$ = the probability "It rains" given that "I get wet." Either it had rained or it had not—chance is not involved. Classical statisticians look upon the *condition* (in this case "It rains") as a true but unknown population parameter. They feel it is wrong to compute probabilities for the parameter and treat it as though it were random. (Recall that a random variable assumes values as a result of a chance event.)

Since classical statisticians reasoned that a prior condition (for example, "It rains") either does or does not exist, they resisted attaching probability values to it. As a result, it was not until about the mid-1900s that Bayes' theorem received much attention. Since then, however, Bayesian statistics have become widely used in economics and business.

In general, Bayesian approaches have become associated with inverse, subjective, and modified probabilities that result from incorporating experience and judgment into formal probabilities in a systematic manner, that is, with subjective probabilities. More specifically, we use these approaches in Chapter 18 to anticipate the value of market research information prior to deciding what size sample to use. Example 4.11 illustrates the use of Bayes' theorem.

## EXAMPLE 4.11

An oil spill and fire (event) injured two employees and caused substantial environmental damage at a 20-year-old chemical refinery. Although the evidence was destroyed, the insurance investigator is attempting to estimate the chance that the accident was due to the defective condition of a valve.

Let $D$ represent the probability of a "defective valve" and let $A$ describe "an accident occurs." Government safety inspections had previously revealed that $P(D) = 0.20$. Given that a valve is defective, the probability of its failing and causing an accident sometime during the year is 0.4; that is, $P(A|D) = 0.4$. If the valve is not defective, the chance of failure and a resulting accident is reduced to 0.05. Find the probability that the valve was defective, given the recent accident.

*Solution*   For $D$ = "a defective valve" and $A$ = "an accident," we have

$$P(D) = 0.2, P(A|D) = 0.4, \text{ and } P(A|D') = 0.05$$

Then

$$P(D') = 1 - 0.2 = 0.8.$$

We seek $P(D|A)$, which by Bayes' theorem is equal to the joint probability of $D$ and $A$ divided by the marginal probability of $A$. In the expanded form (4.7),

$$P(D|A) = \frac{P(D \cap A)}{P(A)} = \frac{P(D)P(A|D)}{P(D)P(A|D) + P(D')P(A|D')}$$

$$= \frac{(0.2)(0.4)}{(0.2)(0.4) + (0.8)(0.05)} = 0.67$$

Recall that before the accident, the chance was 0.20 that a valve was defective. Now, after the fact, the additional information raises the probability of a defective valve to 0.67. The investigator now has stronger evidence (0.67 probability) that a faulty valve caused this accident.

---

Bayes' theorem is appropriate for this problem because this is a conditional probability and the events of the condition (defective valve and accident) are reversed in time from what we would normally expect. That is, we would commonly seek $P(A|D)$. Also, although the valve was *either* defective or not defective, we are content to associate a probability with each condition.

Note that the denominator of the right-hand side of the Bayes' theorem equation expresses the marginal probability, that is, $P(A)$. When two or more mutually exclusive events are involved, this marginal probability must represent the sum of the appropriate joint probabilities. Thus, for $n$ events, Bayes' theorem becomes

$$P(\theta_1|Z) = \frac{P(\theta_1 \cap Z)}{P(\theta_1 \cap Z) + P(\theta_2 \cap Z) + \cdots + P(\theta_n \cap Z)} \tag{4.8}$$

Problem 24 requires this extended form of Bayes' theorem.

## Problems 4.5

21. A statistics teacher knows from experience that a student who does the homework has a probability of 0.8 of passing, whereas a student who does not do the homework has a probability of 0.2 of passing. If 60% of the

students do the homework, what is the probability that a student who gets a passing grade has done the homework?

22. Suppose the probability is 0.9 that your alarm goes off in the morning. The probability that you will get up in time for your 8:00 class if the alarm goes off is 0.6. The probability that you will get up in time if the alarm does not go off is 0.1. What is the probability that you get up in time for your 8:00 class?

23. At a quality-control inspection station, an inspector might find a material defect, a tolerance variance, or even both. Assume that the probability of finding a material defect is 0.1. If an inspector has already found a material defect, the chance of locating a tolerance variance is 0.8. If no material defect has been found, the probability of detecting a tolerance variance is only 0.2.
    a. What is the chance that a tolerance variance will be detected on a given inspection?
    b. Use Bayes' theorem to compute the probability of finding a material defect, given that a tolerance variance is found.

24. A storeroom contains 80 boxes of fuses, 100 per box. Of these boxes, 20 contain fuses produced by machine $A$, 30 have fuses produced by $B$, and 30 have fuses produced by $C$. The boxes are piled without regard to the machine source. Machine $A$ produces, on the average, 5% defective fuses; machine $B$, 3%; and machine $C$, 2%. If a box is chosen at random and a fuse in it is tested and found to be defective, what is the probability that it was manufactured by machine $A$? By machine $B$? By machine $C$?

25. The chief geologist of Yukon Oil Co. feels that in a certain area of Alaska there is a 30% chance of finding oil; that is, $P(\theta) = 0.3$. She obtains a soil sample ($S$) and determines that the conditional probability of getting such a sample, given that oil was present, was 0.4. Also $P(S|\theta') = 0.5$. Find the revised probability of there being oil, given the sample data.

## 4.6 COUNTING TECHNIQUES

We have defined probability as a measure of the chance that an event will occur or that a statistical experiment will have a particular outcome. An event occurs as the result of selecting an item or combination of items from the total number of items available. For example, an event may be the selection of a worker (an item) for a maintenance job from a crew of eight workers.

It is important that we be able to determine the number of possible events or the number of ways an event can occur, because the classical and empirical approaches (and to some extent even the subjective approach) associate probability with the likelihood that an event will occur on any one trial or selection relative to the total number of possible events. Thus, to assign a probability to an event, we must be able to count, or calculate, both the number of successful events and the total number of possible events. We use three counting techniques, all of which are based on the *fundamental rule of counting*.

### THE FUNDAMENTAL RULE OF COUNTING

If event $A$ can occur in any of $m$ distinct ways and event $B$ can occur in any of $n$ distinct ways, then event $A$ and event $B$ can occur in $m \times n$ ways.

This rule applies when we want to select exactly one of the *m* ways and one of the *n* ways. Each distinct combination of *m* and *n* ways results in a separate event; so the fundamental rule defines the total number of ways the combined events can occur. The fundamental rule of counting is applied in Example 4.12.

---

**EXAMPLE 4.12**

For a group of 60 union and 54 nonunion employees, in how many ways can one union member and one nonunion member be selected for a negotiating team?

*Solution*

There are 60 ways (*m*) of selecting a union employee, and each one of these can be paired with any of the 54 ways (*n*) of selecting a nonunion employee. Using the fundamental rule, the total possibilities are $m \times n = 60 \times 54 = 3240$ ways. Thus there are 3240 ways of choosing a union member and a nonunion member.

---

When each event (for example, selection of a negotiating team) requires not one but two or more items of a type (for example, two or more union members), the fundamental rule no longer applies directly. The number of ways an event can occur can be determined by one of three counting techniques: *multiple choices*, *permutations*, or *combinations*.

Using *n* for the total number of items and *X* for the size of the sample, we express the problem of counting events as "taking *n* items *X* at a time." We denote the number of multiple choices by $_nM_X$, the number of permutations by $_nP_X$, and the number of combinations by $_nC_X$, meaning the number of multiple choices, permutations, or combinations of *n* items when taken *x* at a time. Items are said to be taken *with replacement* when the same item can be counted again. They are taken *without replacement* when any item can be counted only once. For example, if a chairperson and a president are to be selected from a board of directors, the selection would be without replacement unless the same person could fill both positions.

**Multiple Choices**

The total number of outcomes that can arise from selecting two or more items of a type depends on two things: whether a *different order* of outcomes makes a difference—for example, whether *A* followed by *B* constitutes a separate outcome from *B* followed by *A*—and whether *duplication* of outcomes is permissible—for example, whether *A*, *B*, *C*, . . . , are taken with replacement or not. (If duplication is permitted, *AA*, *BB*, . . . , are acceptable.) The appropriate counting technique depends on which of these conditions holds. The examples that follow show how to use the counting techniques and also how to recognize the conditions that determine when to use them.

> **Multiple choices** is a procedure for counting arrangements of items when (1) a different order counts as a distinct outcome (that is, the order of selection is important) and (2) items can be duplicated, (that is, items are taken with replacement).

If there are *n* items and each can occur in *X* different ways, we have

$$\text{Number of ways} = {}_nM_X = \underbrace{n \cdots n \cdots n,}_{X} \quad \text{or} \quad n^X$$

Thus the number of multiple choices is as follows:

**MULTIPLE CHOICES**

$$_nM_X = n^X \qquad (4.9)$$

Example 4.13 illustrates the key conditions of (1) different possible orders and (2) the possible duplication of items. These elements signal a solution that involves multiple choices. Example 4.13 displays a calculation that uses multiple choices for counting.

---

**EXAMPLE 4.13**

The General Manager of Quay's Department Store has authorized funds for a management trainee to represent the store at three professional meetings during the coming year. One meeting concerns managerial accounting, one concerns budgeting, and one concerns internal communications. There are 20 management trainees, and the general manager has no objection to having the same person go to more than one meeting. In how many ways can the trainees be selected for the meetings?

*Solution*    Let $A$, $B$, and $C$ represent the meetings. There are 20 possible choices of a person for meeting $A$. For every one of these 20, there are 20 more choices for meeting $B$. Thus there are $20 \cdot 20$, or 400, different ways of having a trainee or trainees participate in the first two meetings. For every one of these 400 ways there are 20 more choices for meeting $C$, so there are 8000 possible choices. Thus the number of multiple choices of $n$ items (20 trainees) taken $X$ at a time (3 meetings), where order makes a difference and duplication is permitted, is

$$_nM_X = n^X = 20^3 = 8000$$

---

**Permutations**

When we are analyzing *permutations*, duplicates are not counted.

> Permutations is a counting procedure in which different arrangements count as a distinct outcome but the duplication of items is not allowed. The number of permutations of $n$ items taken $x$ at a time is

$$_nP_X = \frac{n!}{(n-X)!} \qquad (4.10)$$

The $n!$ means $n! = n(n-1) \cdot (n-2) \cdots 1$. For example, $5! = 5 \cdot 4 \cdot 3 \cdot 2 \cdot 1 = 120$. If $n = 5$ and $X = 2$,

$$_nP_X = \frac{5!}{(5-2)!} = \frac{5!}{3!} = \frac{5 \cdot 4 \cdot 3 \cdot 2 \cdot 1}{3 \cdot 2 \cdot 1} = 20$$

Example 4.14 illustrates that in counting by permutations (1) the order of the arrangement is important and (2) repeats are not allowed. Then, for example, event $A$ followed by $B$ and event $B$ followed by event $A$ are two distinct outcomes, whereas event $A$ followed by event $A$ and event $B$ followed by event $B$ are not allowed.

---

**EXAMPLE 4.14**

By how many routes can a sales manager visit sales reps in three of four cities (Atlanta, Chicago, Denver, New York) if the order in which he visits the cities makes a difference *and* he does not want to visit a city more than once?

**FIGURE 4.10** One Permutation of Routes between Three of Four Cities

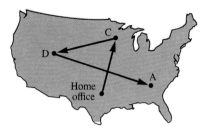

*Solution*   The problem is to select four cities three at a time with no duplications, so counting by permutations is the appropriate technique. Here, $_nP_X = {_4}P_3$.

$$_nP_X = \frac{n!}{(n-X)!} = \frac{4!}{(4-3)!} = \frac{4 \cdot 3 \cdot 2 \cdot 1}{1} = 24$$

Figure 4.10 shows one possible route for the manager's three-city trip, but there are 24 total ways in which the trip could be made! For example, one possible route (permutation) would be Atlanta, Chicago, Denver. This is the top branch of the tree diagram in Figure 4.11, which illustrates the (24) possible permutations.

**FIGURE 4.11**   Permutations of Routes between Three of Four Cities (The heavy line is the route shown in Figure 4.10.)

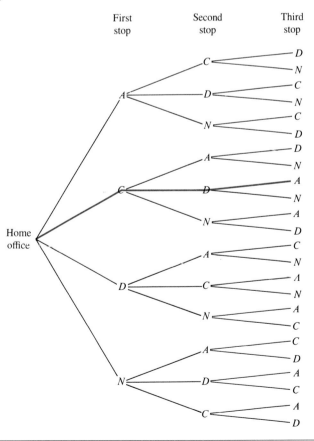

## Combinations

A third counting technique, suggested earlier, is *combinations*.

> **Combinations is a counting procedure in which no duplications are permitted and different orders do not count as distinct outcomes.**

The number of *combinations* may be represented by $\binom{n}{X}$ or $_nC_X$.

### COMBINATIONS

$$_nC_X = \binom{n}{X} = \frac{n!}{X!(n-X)!} \tag{4.11}$$

---

**EXAMPLE 4.15**

In how many ways can two corporate directors be chosen from five executives?

*Solution*    We assume that the order in which the two directors are selected is not important, and that two people are required, so counting by combinations is appropriate.

$$_nC_X = {}_5C_2 = \frac{5!}{2!(5-2)!} = \frac{5 \cdot 4 \cdot 3 \cdot 2 \cdot 1}{2 \cdot 1 \cdot 3 \cdot 2 \cdot 1} = 10$$

Two directors can be chosen from five executives in ten ways.

---

Combinations and permutations seem very similar. However, unlike permutations, combinations consider only the items that make up the event and not the sequence of selection, so a different order of arrangement does not constitute another event. Thus "*A* followed by *B*" and "*B* followed by *A*" are two distinct events in counting permutations, whereas they are the same events when we are counting combinations.

Example 4.16 illustrates the results of applying the three counting techniques to the same collection of items. Notice that the multiple-choice technique, which allows the greatest number of distinguishing characteristics (order, repeats), produces the largest count; the most restrictive technique—combinations—produces the smallest count.

---

**EXAMPLE 4.16**

The personnel department of a small firm employs four persons, Abe, Bob, Cindy, and Dave, whom we shall designate as *A*, *B*, *C*, and *D*, respectively. Two sessions have been announced for training at a microcomputer seminar. The seminar has a morning session on word processing (WP) and an afternoon session on spreadsheet (SS) instruction, so it is possible for one person to attend either one of the sessions or both, depending upon the criteria established by the department manager. Find the number of (a) multiple choices, (b) permutations, and (c) combinations in which a person or persons can be selected to attend the seminars.

*Solution*    a. For multiple choices, duplication is permitted (a person can be chosen to attend both seminars), and different orders are counted (the time of attendance makes a difference). Thus one event is for Abe to attend both sessions (*A*,*A*). Moreover, having Abe attend a morning session and Bob the afternoon session (*A*,*B*) is a different event than having

Bob attend the morning session and Abe the afternoon (B,A). Using the multiple choices rule, we have

$$n^X = 4^2 = 16$$

There are 16 ways to select the participants:

$$(A,A) \quad (B,A) \quad (C,A) \quad (D,A)$$
$$(A,B) \quad (B,B) \quad (C,B) \quad (D,B)$$
$$(A,C) \quad (B,C) \quad (C,C) \quad (D,C)$$
$$(A,D) \quad (B,D) \quad (C,D) \quad (D,D)$$

b. For permutations, different orders are counted but duplication is *not* allowed, so (A,A) is not an acceptable outcome. This is equivalent to saying that no employee may attend both sessions.

$$_nP_X = \frac{n!}{(n - X)!}$$

$$_4P_2 = \frac{4!}{(4 - 2)!} = \frac{4 \cdot 3 \cdot 2 \cdot 1}{2 \cdot 1} = 12$$

There are 12 ways to select the participants. In this case, if Abe attends the WP session and Bob the SS session, this is different from Bob attending the WP and Abe the SS session. But one person may not attend both seminars.

c. For combinations, order makes no difference and duplication is not allowed, so (A,A) is not counted and there is no difference between (A,B) and (B,A).

$$_nC_X = \frac{n!}{X!(n - X)!}$$

$$_4C_2 = \frac{4!}{2!(4 - 2)!} = \frac{4 \cdot 3 \cdot 2 \cdot 1}{2 \cdot 1 \cdot 2 \cdot 1} = 6$$

There are six ways to select the participants. Here the manager just wants to pick two different people, so Abe attending the WP session and Bob doing the SS session is considered the same as Bob doing word processing and Abe doing spreadsheets.

---

Many business applications use combinations, because duplications and different orders are not counted. For example, suppose you are a retailer who has just imported a shipment of 100 down jackets from a supplier in Korea. You are following an inspection plan whereby you inspect ten jackets and accept the shipment if two or fewer of the jackets are defective. If more than two jackets have defects, you ship the entire lot back to Korea. The order in which you select the jackets is not important and inspecting the same jacket more than once is pointless, so counting by combinations is appropriate. But there are about 1.7 trillion possible combinations: $_{100}C_{10} = 1.7 \times 10^{13}$.

Each counting problem has to be evaluated individually. The key to selecting the proper counting technique lies in reading the stated problem carefully to find answers to three questions; these questions appear in the flowchart in Figure 4.12.

**FIGURE 4.12**    A Key to Counting Rules

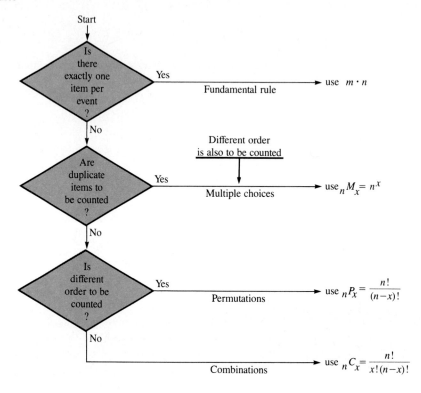

Usually the hardest judgment to make is whether the order of selection is important. You will find a number of problems for developing your decision-making and counting skills in the section Problems.

**Problems 4.6**

26. The distribution manager for a grocery supply center has 12 truck routes to cover but only 10 drivers available on a given day. How many different ways (combinations) are there to assign the 10 drivers to the 12 routes?

27. Find the values of each of the following:
    a. $_{15}M_4$                  b. $_{50}P_2$                  c. $_{10}C_8$
    d. $_{10}M_3$                  e. $_{10}P_3$                  f. $_{10}C_3$

28. You have eight items. In how many ways can you do each of the following?
    a. Make two selections if selections can be repeated (that is, multiple choices)
    b. Select two items if no duplication is permitted (that is, permutations)
    c. Select two items if no duplication is permitted and no different order is permitted (that is, combinations)

29. Seven employees have been recommended to the company president as potential managers to fill two positions in a new plant. In how many ways can the president fill the two positions?
    a. If no duplication and no different order is permitted
    b. If selections can be repeated and different orders are counted as being different

    c. If no duplication is permitted, but different orders are counted as being different

30. In how many ways can four products be advertised on adjacent spaces in a subway station if seven spaces are available? Each product takes one space, and the order is not important.

31. a. How many slates of three officers—a president, a vice president, a secretary—can be formed from a group of ten people? Assume a different person occupies each position.
    b. In how many ways can a committee of five employees be chosen out of a department of nine people?

32. Oregon Insurance, Ltd., plans to put four additional salespeople in the field, two of them somewhere in the eleven Western States, one in the six middle Atlantic states, and one in Minnesota, Wisconsin, Indiana, or Illinois. If the company wants to evaluate the desirability of each possible combination of locations, how many would it have to consider? (*Hint:* Each geographical combination creates a multiple number of joint combinations when coupled with the other.)

## 4.7 SOLVING PROBABILITY PROBLEMS

The following procedure specifies steps for solving probability word problems.

### FIVE STEPS FOR SOLVING PROBABILITY PROBLEMS

1. Identify the problem or read the problem description to classify and bound the problem in a general way. Write the basic events, using literal symbols (with a word description if helpful).
2. Using probability notation, write the given information in the language of statistics.
3. State precisely what is to be found, again in probability notation. (This frequently comes from the sentence that says, "Find the" or "What is" or that ends with a question mark.)
4. Solve the problem by using probability rules, counting techniques, and axioms. Use the "Find the" statement to write a probability rule and use the given information to replace unknowns in the probability rule.
5. Now look back through your solution. Check to see that you have
    a. defined basic events properly,
    b. identified the question correctly,
    c. interpreted the given information correctly,
    d. copied the given numerical values correctly,
    e. taken a logical approach to solving the problem,
    f. obtained a reasonable answer (remember that probabilities must be from 0 to 1, inclusive).

We illustrate this procedure with an application of the multiplication rule.

### EXAMPLE 4.17

At National Industries, 60% of the sales are to foreign customers. Of the foreign customers, 30% are electric utilities. If a market research survey randomly selects one sale for analysis, what is the chance that the sale is to a foreign customer *and* to an electric utility?

*Solution*    1. *Let* $F$ = event "customer is foreign"

$E$ = event "business is electric utilities"

2. *Given*:      $P(F) = 0.60$

$P(E|F) = 0.30$

3. *Find*:   $P(F \cap E)$

4. Apply the multiplication rule:

$$P(F \cap E) = P(F) \cdot P(E|F)$$
$$= (0.60)(0.30) = 0.18, \quad \text{or an } 18\% \text{ chance}$$

5. A *reasonable* value is obtained. That is, 18% of all sales have been to foreign customers that are also electric utilities.

## SUMMARY

Probability is a measure of the chance that an event will occur. How that assignment is made is a function of the information available to the business analyst. An event may be defined before any data are obtained. For example, the percentage of female applicants for a California driver's license may be thought to be 1/2. This assignment uses a classical definition of probability based on *a priori* (before the fact) information, since half of the driving population is assumed to be female and half male. If the analyst has observed or measured the frequency of occurrence over many times, an assignment can use a relative frequency definition. This is common in the manufacturing industry, where product guarantees are established on the basis of historical lifetimes of products manufactured previously. A subjective probability can be assigned on the basis of limited information. One who invests in oil drilling generally uses a subjective probability to assess the chance of an oil strike, because the existence of oil at a particular location is uncertain and there is probably no experience at the location. Increased information (experience) would provide more stable values for assignment of probabilities. Relative frequency or empirical probability and subjective probability are more useful and often more appropriate to managerial decision makers than classical probability.

Several devices, including operational rules for addition, multiplication, and finding the complement, can aid in the calculation of probabilities. In addition, multiple choices, permutations, and combinations provide the counting that is sometimes necessary for determining probabilities. If the number of possibilities is small, a tree or Venn diagram can aid in the counting. Table 4.4 summarizes the concepts in this chapter.

You will use several key concepts from Chapter 4 as you progress through the next chapters; for example, the next three chapters explain how to assign probability to the values of a random variable. This yields a *probability distribution* with associated mean and standard deviation.

## KEY TERMS

addition rule, 136
Bayes' theorem, 141
combinations, 148
complement operator, 119
complement rule, 133
compound event, 115
conditional probability, 131
event operator, 118

fundamental rule of counting, 144
independent events, 135
intersection operator, 118
joint probability, 131
marginal probability, 131
multiple choices, 145
multiplication rule, 134
mutually exclusive events, 119

**TABLE 4.4**    Key Equations for Probability

---

### Classical Probability

$$P(A) = \frac{N(A)}{N} \qquad \text{(4.1)}$$

---

### Empirical Probability

$$P(X) = \frac{f(X)}{n} \qquad \text{(4.2)}$$

---

### Basic Rules for Computing Probabilities

| | | |
|---|---|---|
| *Complement* | $P(A') = 1 - P(A)$ | **(4.3)** |
| *Multiplication* | $P(A \cap B) = P(A) \cdot P(B\|A)$ | **(4.4)** |
| | $\quad = P(A) \cdot P(B) \quad$ if $A$ and $B$ are independent | |
| *Addition* | $P(A \cup B) = P(A) + P(B) - P(A \cap B)$ | **(4.5)** |
| | $\quad = P(A) + P(B) \quad$ if $A$ and $B$ are mutually exclusive | |

---

### Bayes' Theorem

| | | |
|---|---|---|
| *For 2 Events* | $P(\theta\|Z) = \dfrac{P(\theta \cap Z)}{P(Z)} = \dfrac{P(\theta) \cdot P(Z\|\theta)}{P(\theta) \cdot P(Z\|\theta) + P(\theta') \cdot P(Z\|\theta')}$ | **(4.6, 7)** |
| *For n Events* | $P(\theta_1\|Z) = \dfrac{P(\theta_1 \cap Z)}{P(\theta_1 \cap Z) + P(\theta_2 \cap Z) + \cdots + P(\theta_n \cap Z)}$ | **(4.8)** |

---

### Counting Rules

| | | |
|---|---|---|
| *Multiple Choices* | $_nM_X = n^X$ | **(4.9)** |
| *Permutations* | $_nP_X = \dfrac{n!}{(n-X)!}$ | **(4.10)** |
| *Combinations* | $_nC_X = \dfrac{n!}{X!(n-X)!} \quad$ where $n! = n \cdot (n-1) \cdot (n-2) \cdots 1$ | **(4.11)** |

---

---

**CHAPTER 4
PROBLEMS**

33. The Midland Rural Electric board of directors have asked, Who are our customers? For pricing and because of transmission costs, the directors need the percentage of total usage made by commerical accounts in towns outside the county's major city. They have directed the Midland Rural Electric business office to survey the usage of 1000 randomly selected customers. The survey shows 60% of usage is by outlying areas (beyond the county's major city). In outlying areas, 30% of the electricity sales is to commercial customers ($C$). Let $\theta$ denote an outlying area and $\theta'$ be not an outlying area. What part of the total usage is by commercial customers in outlying areas?

34. Given the data in the diagram where the sample space includes 500 observations and constains events $A$, $B$, and $C$,
    a. are events $A$ and $B$ mutually exclusive?
    b. what is the count $N(S)$, $N(A \cup B)$, $N(B \cap C)$, $N(A')$?

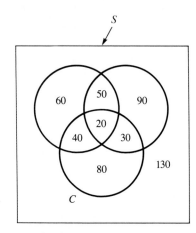

35. Using the figure in the previous problem and the empirical definition of probability, give each of the following probabilities. Leave all answers as fractions.
    a. $P(B \cap C)$                    d. $P(B|A)$
    b. $P(A \cup C)$                    e. $P(A \cup B \cup C)$
    c. $P(A \cap B \cap C)$             f. $P(A \cup B \cup C)'$

36. What is the appropriate name or term for each of the following?
    a. Empty set of events
    b. An event ($A'$) that contains all simple events that are not contained in the event ($A$) itself
    c. Events that have no common simple events
    d. A collection of one or more simple events
    e. A schematic diagram in which the sample space is shown as a rectangle and events (or subsets) of interest are depicted as (sometimes overlapping) circles

37. At a meeting of 60 executives of multinational corporations in Moscow, 50 spoke Russian, 30 spoke English, and 20 spoke both Russian and English.
    a. Depict the sample space of "language spoken" by a Venn diagram.
    b. What is the probability that an executive selected at random spoke Russian but not English?
    c. What is the probability that an executive spoke either English or Russian?
    d. Assume that the selection is restricted to Russian-speaking executives. What is the probability that the executive also speaks English?
    e. Are the two events—"spoke Russian" and "spoke English"—mutually exclusive? Explain.

38. A friend in your class is having difficulty distinguishing when it is appropriate to use permutations, combinations, or multiple choices. Give a brief description of what to look for to distinguish the use of these three counting procedures.

39. A friend in another class wants some way to test for (a) independent events and some way to test for (b) mutually exclusive events. What suggestions can you give for general test rules?

40. A corporation gives its 25-year service employees their choice of a gold watch or a $500 savings bond. Experience shows that the choice has been essentially equal, with half choosing the watch. This year there are ten 25-year service awards. What is the probability that more will choose watches than choose the savings bond? (*Suggestion:* Consider the choices as independent events.)

41. City Hall has surveyed the citizenry for their opinions on a proposal to increase taxes on residential property as a means of securing funds to build a municipal civic center. A summary table was left on the city planner's desk for her assistant to complete and then compute and take the "facts" of this citizens' household survey to the mayor. (*F* = favorable, *F'* = unfavorable, *H* = homeowners, *H'* = renters (non-homeowners).)

| Home Ownership | Opinion | | Percentage of Homeowners |
|---|---|---|---|
| | *F* | *F'* | |
| *H* | | 0.36 | 0.45 |
| *H'* | | | 0.55 |
| Percentage Opinion | 0.40 | 0.60 | 1.00 |

Provide answers to the following questions as a means of analyzing the data in the table.

a. What percentage of those surveyed were favorable and homeowners?

b. What is the probability that an opinion is unfavorable given that the person is a homeowner?

c. What is the probability that a citizen is unfavorable and is a renter?

d. Are the two variables (opinion and home ownership) independent? Might this indicate the possibility of a civic issue?

42. Consider the events of the preceding question (*F* = favorable, *H* = homeowner). Compute the conditional probabilities for (a) $P(F|H)$ and (b) $P(H|F)$. Explain your numerical answers. Do both of these probability calculations make practical good sense? Explain.

43. The production at a furniture- and cabinet-building shop in Georgia requires standards that include milling parts to the correct size and product-finishing with uniform staining. The quality inspector passes finished cabinet products if they meet at least one of these requirements. Given the following table, what percentage of these units have passed inspection? (*C* = milling is correct size, *U* = uniform staining)

| Staining | Milling | | Total |
|---|---|---|---|
| | *C* | *C'* | |
| *U* | 630 | 50 | 680 |
| *U'* | 300 | 20 | 320 |
| Total | 930 | 70 | 1000 |

44. A drugstore chain has 50 delinquent accounts, which should be closed out as bad debt losses. Among these, six have balances in excess of $200. In auditing the firm, an accountant selects five delinquent accounts at random. What is the probability that exactly two of these five accounts will have balances in excess of $200? Set up the expression only; you need not carry out the calculations. (*Hint:* You will want to set up your calculations so the numerator reflects the number of successful outcomes possible and the denominator contains the total number of outcome combinations possible.)

45. Two boxes in a storeroom each contain three AC and three DC relays. One relay is chosen at random from each box. What is each probability that
    a. at least one relay is AC?              b. neither is AC?

46. Given $P(A) = 0.3$, $P(B) = 0.2$, and $P(A|B) = 0.5$, find each value.
    a. $P(A \cap B)$                              d. $P(A' \cup B')$
    b. $P(B|A)$                                   e. $P(A|B')$
    c. $P(A')$                                    f. $P(A'|B')$

47. Given $P(A) = 0.4$, $P(B) = 0.5$, and $P(A|B) = 0.2$, find each value.
    a. $P(A \cap B)$                              b. $P(B|A')$

48. The loan department for a large savings and loan association has experienced 4% defaults on personal loans. Of the defaulted loans, 40% were made to men under age 25. Only 20% of the nondefaulted loans were made to men under age 25. What percent of the bank's personal loans have been made to men who are less than 25 years old?

49. Suppose the probability that an individual is over 45 and earns more than $50,000 a year is 0.15. Moreover, the probability that any individual, regardless of age, makes more than this amount is 0.25. What is the probability that an individual is over 45 given that the person's income exceeds $50,000?

50. The Michigan Molding Co. has just finished production of 1000 two-color legend plates for aircraft cockpits using four machines ($W, X, Y, Z$). After quality-control inspection the plates were graded as shown in the table. ($A$ = acceptable, $B$ = needs repair, $C$ = can be scrapped)

|       | Machine | | | | |
| Grade | $W$ | $X$ | $Y$ | $Z$ | Total |
|-------|-----|-----|-----|-----|-------|
| $A$   | 320 | 80  | 240 | 160 | 800   |
| $B$   | 60  | 15  | 45  | 30  | 150   |
| $C$   | 20  | 5   | 15  | 10  | 50    |
| Total | 400 | 100 | 300 | 200 | 1000  |

a. State, in symbols, the probability that a legend plate selected at random
   (i) was produced by machine $W$.
   (ii) was produced by machine $X$ and needs repair.
   (iii) needs repair given that it was produced by machine $Y$.
   (iv) (restricted to plates produced by machine $W$) is acceptable.
   (v) was either produced by machine $Z$ or can be scrapped.
b. State in words the meaning of each of the following:
   (i) $P(B|Z)$                (ii) $P(A \cap Y)$                (iii) $P(B \cup X)$

c. Compute the probability that a plate selected at random
   - (i) was produced by machine $Z$.
   - (ii) was produced by machine $X$ and is acceptable.
   - (iii) is acceptable, given that it was produced by machine $X$.
   - (iv) is either in need of repair or was produced by machine $W$.
d. Use the addition theorem or multiplication theorem, as appropriate, to compute the following probabilities.
   - (i) $P(Y \cap B)$          (ii) $P(C \cup X)$         (iii) $P(A \cap C)$
e. Determine if the variables (machine and grade) are statistically independent and show why or why not.

51. The following percentages were tabulated in a study of the student body at Delta Senior College.

| Class | Sex M | F | Total |
|-------|------|------|-------|
| $B_1$ | 0.05 | 0.15 | 0.20 |
| $B_2$ | 0.16 | 0.24 | 0.40 |
| $B_3$ | 0.28 | 0.12 | 0.40 |
| Total | 0.49 | 0.51 | 1.00 |

a. Are the subclassifications male, $M$, and graduate, $B_3$, independent? Interpret the meaning of your findings with respect to the two major classifications—sex and class.
b. Compute the probability that a student at Delta is a male if the student was randomly chosen from the junior class.
c. What percentage of the student body is undergraduate women? ($B_1 =$ junior, $B_2 =$ senior, $B_3 =$ graduate)

52. Use the following data to answer the questions. Let

$$R = \text{manager reads } \textit{The Wall Street Journal}, \quad P(R) = 0.55$$

$$R' = \text{manager does not read } \textit{The Wall Street Journal}$$

$$C = \text{manager is a college graduate}$$

$$C' = \text{manager is not a college graduate}, \quad P(C') = 0.40$$

Also, $P(R'|C) = 0.60$.
a. What is the probability that a manager reads *The Wall Street Journal* and is a college graduate?
b. What is the probability that a manager does not read *The Wall Street Journal* given that the manager is a college graduate?
c. What is the probability that a manager does not read *The Wall Street Journal* and is not a college graduate?
d. What is the probability that a manager either does not read *The Wall Street Journal* or is not a college graduate?
e. What types of probability are requested in parts (a), (b), (c), and (d), respectively? Choose from marginal, conditional, and joint probabilities.
f. Are the two variables (readership and education) independent in this population? Show proof.

53. A metallurgist of a mining company in Australia feels there is a 50–50 chance of finding a significant uranium deposit. Results from a test drilling are favorable. However, the probability that the test drilling would give misleading results is 0.3. What should be the metallurgist's revised probability that a significant deposit of uranium exists there?

54. At General Bolt Company, bolts are made by machines $A$ and $B$. Machine $A$ is larger and produces twice as many bolts as $B$ in a given amount of time. Machine $A$ produces 2% defectives and $B$ produces 1% defectives. A bolt is chosen from production at random and found to be defective. What is the probability that this defective bolt was produced by machine $A$? Assume that the two machines are in operation for the same amount of time, and use Bayes' theorem.

**STATISTICS IN ACTION: CHALLENGE PROBLEMS**

55. Who pays for college? One way to answer this question is to split the nation's tuition bill by the age of paying householders—the people who generally pay the bill. Another view is from householder incomes. The data are for 1987.

Tuition Toll*

|  | All Households | Households Paying Tuition | Average Annual Payment |
|---|---|---|---|
| **Age of Householder** | | | |
| All ages | 94,150 | 4,783 | $3,799 |
| Under 25 | 7,811 | 918 | 3,734 |
| 25 to 34 | 21,345 | 1,059 | 2,604 |
| 35 to 44 | 18,747 | 979 | 3,148 |
| 45 to 54 | 13,395 | 1,172 | 5,231 |
| 55 to 64 | 13,080 | 489 | 4,527 |
| 65 to 74 | 11,578 | 148 | 3,623 |
| 75 and older | 8,194 | 14 | 1,688 |
| **Household Income**** | | | |
| All income levels | 81,070 | 4,135 | $3,591 |
| Less than $5,000 | 7,497 | 442 | 3,508 |
| $5,000–$9,999 | 12,490 | 478 | 2,965 |
| $10,000–$14,999 | 10,410 | 304 | 3,396 |
| $15,000–$19,999 | 8,386 | 280 | 2,663 |
| $20,000–$29,999 | 14,175 | 553 | 3,273 |
| $30,000–$39,999 | 10,439 | 522 | 3,392 |
| $40,000–$49,999 | 7,150 | 404 | 2,387 |
| $50,000 and over | 10,524 | 1,147 | 4,841 |

*Note:* Households are Consumer Expenditure Survey (CEX) consumer units.

  * Households paying college tuition and average annual payment, by age and income: 1987; households in thousands.

** The CEX tabulates only complete income reports.

*Source:* Bureau of Labor Statistics, *Consumer Expenditure Survey,* 1987.

SOURCE: Exter, Thomas. "Who Pays for College?" *American Demographics* 12, no. 9 (February 1990): 6.

Questions We will answer the title's question by building the following table. The first column uses the All Households data and the empirical definition of probability. For the second column entries, compute (group) Tuition paid = (Households paying tuition · Average annual payment). Then,

$$\text{Percent of national tuition bill} = \frac{\text{Tuition paid, this group}}{\text{Tuition paid, summed for all groups}}$$

We have started the table. You finish it.

|  | %1 = P(Households) | %2 = P(Tuition paid) |
|---|---|---|
| Age groups, total | 1.00 | 1.00 |
| Younger: under 35 | 0.31 | 0.34 |
| Middle: 35–54 |  |  |
| Older:  over 54 |  |  |
| Household income, total | 1.00 | 1.00 |
| Lower:  under $15,000 |  |  |
| Middle: $15,000–$39,999 |  |  |
| Upper:  $40,000+ |  |  |

56. Use the numbers you developed in Problem 55 to explain who pays for college. (*Suggestion:* Look where %1 is low relative to %2.)

57. A planning and strategy model is designed to determine the market potential, including reach, frequency, and duplication, for two or more extensions $(A, B, C, \ldots)$ as one product line. An example of extensions would be the models of any make of car. The calculations are really just the rules of probability. "Reach" answers the question "What percent of consumers will use (reach) each offering in the product line?"

SOURCE: Miaroulis, George, Valerie Free, and Henry Parsons. "TURF: A New Planning Approach for Product Line Extensions." *Marketing Research* 2, no. 1 (1990): 28–40.

Questions Suppose the product line is flavors for a brand of yogurt, with two product extensions—vanilla and chocolate. Define reach = $P(A \cup B) = P(A) + P(B) - P(AB)$. Calculate reach given $P(A) = 0.25$, $P(B) = 0.30$, and $P(AB) = 0.10$, and display the result as a Venn diagram.

58. Continue to compute the reach if a third product extension—strawberry —adds (to the present numbers) $P(C) = 0.25$, $P(AC) = 0.05$, $P(BC) = 0.05$, and $P(ABC) = 0.01$ and reach = $P(A \cup B \cup C) = P(A) + P(B) - P(AB) - P(AC) - P(BC) + P(ABC)$. Again, display the result with a Venn diagram. Given any number of extensions, what is the largest attainable value for reach?

59. Lewis Carroll, the Reverend Charles Lutwidge Dodgson (1832–1898), was a don at Christ Church and a math lecturer at Oxford University in England. Little has been written about his mathematical creativity; most know him as the author of *Alice in Wonderland* and other literary works for children. In fact, in 1893 Dodgson published 72 probability problems with answers. The article is interesting because

it shows the kinds of questions posed by some of the early thinkers in probability. The problems deal with the basic principles of probability. Three of the formulas that Dodgson used are

$$P(B) = \sum_{i=1}^{n} P(B|A_i)P(A_i) \qquad \text{(i)}$$

$$P(A_j|B) = \frac{P(B|A_j)P(A_j)}{\sum_{i=1}^{n} P(B|A_i)P(A_i)} \qquad \text{(ii)}$$

$$P(A_j|B) = \frac{P(B|A_j)}{\sum_{i=1}^{n} P(B|A_i)} \qquad \text{provided } P(A_1) = P(A_2) = \cdots = P(A_n) = P \qquad \text{(iii)}$$

SOURCE: Senta, E. "Lewis Carroll as a Probabilist and Mathematician." *The Mathematical Scientist* 9, no. 2 (1984): 79–94.

**Questions** Use the values $n = 3$, $P(A_1) = P(A_3) = 1/4$, $P(A_2) = 1/2$, $P(B|A_1) = 1/3$, $P(B|A_2) = 1/6$, and $P(B|A_3) = 1/2$ to compute the probability described by Formula (i). $P(B)$ is the marginal probability for $B$.

60. Formula (iii) requires $P(A_1) = P(A_2) = P(A_3) = P$. With this condition, demonstrate how formula (ii) is changed by some straightforward math to Formula (iii).

***

**BUSINESS CASE**

*Simple Probabilities Can
Reveal Costly Errors*

Jim Reed has been with Bank Systems Co. for only 11 months, but the company is growing so fast that he is already one of the veterans. In its two years of existence, Bank Systems has grown to 380 employees. At present, Jim is acting materials manager and now reports directly to the vice-president for operations.

Last Thursday, Jim had made his first tough decision involving the rejection of some material. As a result, a shipment of 2000 integrated circuits (ICs) was being returned to Chips Unlimited, their supplier in San Jose, California.

Unfortunately, the truck returning the shipment lost its brakes on a hill, crashed, and caught fire. Although the driver escaped, the shipment of ICs was destroyed.

Jim's inspection had shown the shipment to be defective, so he had rejected it, assuming that the supplier still owned it and would be responsible for insuring its safe return. So he didn't instruct his shipping clerk to insure the return shipment.

Chips Unlimited had a different perspective. They assumed that the shipment was good until they received evidence that it wasn't. Furthermore, they now demanded $22,580 in payment for the shipment on the basis of a receipt slip they had from a receiving clerk at Bank Systems. Jim Reed was caught in the middle.

The vice-president for operations, Marcia Hollingsworth, was a graduate in finance but gravitated to operations because she has a knack for handling daily operating problems. When Jim came to her office, Marcia immediately began to collect the facts.

**Ms. H.**    Let's be sure we've got everything straight here, Jim. How many ICs were in that shipment?

**Mr. R.**    Two thousand. But they weren't all good. Some of them had defects.

**Ms. H.**    What kind of defects?

**Mr. R.**    Both mechanical and electrical defects. We had some of each.

**Ms. H.**    Okay. My understanding is that Chips Unlimited guaranteed them to have no more than 2% defectives, is that right?

**Mr. R.**    That's right. But out of 2000 chips, 16 had dimensional problems and 34 had electrical defects. I have the report right here (see Incoming Materials Report).

---

**Incoming Materials Report**

*Date*    12/4

*Item*    Integrated Circuits, IC# 28 B

*Quantity*    2000

*Supplier*    Chips Unlimited, San Jose

*Specifications*    −2% Defective

*Condition*    Mechanical Defects ____16____

Electrical Defects ____34____

*Disposition*    ☐ Accept
☑ Reject

*Reason*    Number of defects exceeds allowable possibilities

---

**Ms. H.**    Do you have any other information on the shipment, Jim? Did Chips Unlimited send along any other information or records?

**Mr. R.**    Just the paperwork that normally accompanies an order. Nothing special, as I recall. You know, there's getting to be so much paperwork with the shipments it would take more time to read it all than to transport the material. Anyway, I've been awfully busy trying to help the boys unload that new shipment of cabinets from Pittsburgh.

**Ms. H.**    Well, let me contact Chips Unlimited. I'll call Susan Grave down in San Jose and see if she can't get me some more information on this shipment. Most firms like this have a pretty good idea about what their shops are turning out. I've known Susan for almost ten years. She may be able to shed some more light on this. I'd sure hate to get stuck with that $22,580 unless we deserve it.

One week later Jim Reed was back in Marcia Hollingsworth's office. Susan Grave, from Chips Unlimited, had researched the defective rates on their production process and forwarded Marcia a table of the most likely composition of the shipment in question (see table on "Mechanical Condition" and "Electrical Condition").

| **Mechanical Condition** | |
|---|---|
| Number out of mechanical tolerance with acceptable voltage (electrical condition) | 2 |
| Number out of mechanical tolerance with excessive voltage (electrical condition) | 13 |
| **Electrical Condition** | |
| Number out of electrical tolerance with acceptable dimensions (mechanical condition) | 22 |
| Number out of electrical tolerance with unacceptable dimensions (mechanical condition) | 13 |

Jim Reed was quick to total the mechanical and electrical out-of-tolerance figures revealed in the report.

**Mr. R.**  Now that I see this report, I do recall seeing something like this in that stack of papers we've been getting. Guess I've been so busy I haven't paid much attention to them. Anyway, it looks like they pretty well confirm what we found, doesn't it?

**Ms. H.**  Well, I'm not sure. Let's put them together in a probability table and find out.

Marcia Hollingsworth sketched out a probability table on a scratch pad on her desk ($E$ = acceptable voltage, $E'$ = voltage out of tolerance):

|  | Mechanical Condition | | |
|---|---|---|---|
| Electrical Condition | Within Tolerance $M$ | Out of Tolerance $M'$ | Total |
| $E$ |  | 2 |  |
| $E'$ | 22 | 13 | 35 |
| Total |  | 15 |  |

**Ms. H.**  (pointing to the 15 and the 35) I guess these are the figures you had focused on.

**Mr. R.**  Yes, that's very close to what we found. We had 16 mechanical defects and 34 electrical ones—same total, as a matter of fact. You know, 50 is more than 2% of 2000. It's $2\frac{1}{2}\%$! That makes our case a little better, doesn't it?

**Ms. H.**  Let's just put your 2000 total in here and see what it does. Since the rows and columns all have to total 2000, that gives us a completed table like this, right?

Marcia Hollingsworth worked out a few subtractions and came up with another probability table:

|  | Mechanical Condition | | |
|---|---|---|---|
| Electrical Condition | Within Tolerance $M$ | Out of Tolerance $M'$ | Total |
| $E$ | 1963 | 2 | 1965 |
| $E'$ | 22 | 13 | 35 |
| Total | 1985 | 15 | 2000 |

**Ms. H.**  Looks to me like we've got 1963 good chips out of 2000. I'd say that's right under 2% defective, wouldn't you?

**Mr. R.**  That's because we inspect for each defect in separate areas.

**Ms. H.**  Well, in that case, you're overlooking the probability of joint defects, aren't you?

**Mr. R.**  I . . . , yes . . . , yes, we definitely have overlooked that possibility. But you suspected it all the time, didn't you?

**Ms. H.**   I didn't know for sure. I just knew that numbers in the probability table have to add up. I guess that comes from a college professor drilling some probability rules into us.

**Mr. R.**   (interrupting) I'm sorry about overlooking those. I wish I . . .

**Ms. H.**   (continuing on) One of them was, "The probability of events $A$ and $B$ equals the probability of $A$ times the probability of $B$ given $A$." In our case, the probability of a good chip would be this:

$$P(E \cap M) = P(E) \cdot P(M \mid E) = \frac{1965}{2000} \cdot \frac{1963}{1965} = 0.982$$

(or, by inspection of Ms. Hollingsworth's table, $P(E \cap M) = 1963/2000 = 0.982$). So, there were 1.8% (that is, 0.018) chips with one or both kinds of defects. That means the data point to an acceptable shipment. I'm afraid we're stuck with the $22,580 this time.

**Mr. R.**   You mean *you're* stuck because of *me*! I'm trying to do a good job, but I've just been so busy. I goofed it up.

**Ms. H.**   You know, Jim, the biggest problem for most managers is that they let the day-to-day business take up the time they should be using for careful analysis and decisions. You should be giving problems like this your attention instead of unloading boxes with the guys. It's nice to keep up the good relationships, but these kinds of mistakes are costly.

**Mr. R.**   Yes, ma'am. I know they are.

**Ms. H.**   We'll absorb this one in our production budget, but we can't stand many more like it.

**Mr. R.**   I get the message. Thanks for being so considerate. It won't happen again.

**Business Case Questions**

61. Suppose the data in the Chips Unlimited example reflected higher figures for voltage out of tolerance (see table). Explain how this would change the outcome that was described in the application. Retain the fixed total of 2000.

|        | $M$ | $M'$ | Total |
|--------|-----|------|-------|
| $E$    |     | 2    |       |
| $E'$   | 30  | 23   | ___   |
| Total  |     |      |       |

62. Suppose that Marcia Hollingsworth believes in data analysis and so directs Jim Reed to "check the facts by reviewing all of the available numbers on that job with the shipping clerk." From this review it was determined that the actual shipment was 1800 units, not the assumed 2000. The other numbers were unchanged. Repeat Marcia Hollingsworth's table and calculations, using 1800 as the total. Does this information support Bank Systems Co.'s case? Explain.

**COMPUTER APPLICATION**

*Empirical Probabilities*

This application continues the marketing study that we began in Chapter 2. That application developed some basic statistics on a target population—families of three or more persons and with householder aged 35–50 years. These statistics included mean and standard deviation for householder ages.

Our interest here is in gaining a better understanding of certain probability characteristics of this population. Specifically, what percentage of the target population has householders aged 41–45 years? This age group is of interest because

**TABLE 4.5**    Calculation of Group Percentages

| 10% Group | Householders Aged 41–45 | Group Count | $[(A)/(B)] \times 100$ Percentage |
|:---:|:---:|:---:|:---:|
| 1 | 115 | 375 | 30.7 |
| 2 | 114 | 364 | 31.3 |
| 3 | 105 | 360 | 29.2 |
| 4 | 110 | 382 | 28.8 |
| 5 | 110 | 347 | 31.7 |
| 6 | 125 | 409 | 30.6 |
| 7 | 101 | 367 | 27.6 |
| 8 | 101 | 366 | 27.6 |
| 9 | 128 | 421 | 30.4 |
| 10 | 123 | 366 | 33.6 |
| Total | 1132 | 3757 | 30.1 |

(1) it covers families with high school- and college-aged youth and (2) it approaches a peak age for higher incomes. These are prime characteristics for marketing products such as personal computers.

We take the view of the classical probabilist and develop an empirical probability to answer this question. This calculation comes from the 20,116 data base records. We can summarize our first 10% sample by having the computer give us a FREQUENCIES count for records with age of householder 41–45 years. Similarly, a second group (#2), representing a second 10% sample, provides another count of householders with age 41–45 years. This process was continued and gives the summary shown in Table 4.5.

Table 4.5 shows that the group percentages seem to vary *around* the total percentage value, which is 30.1%. The group values range from a low of 27.6% to a high of 33.6%.

The classical probabilist would compute the probability for ages 41–45 years as shown in Table 4.6.

Table 4.6 shows that as the sample size increases to a large number (in this case, $n = 3757$ households), the accumulative percentage approaches a stable

**TABLE 4.6**    Calculation of Accumulated Percents

| 10% Group | Householders Aged 41–45 Years | Group Count | Cumulative Percentage | Cumulative Percentage |
|:---:|:---:|:---:|:---:|:---:|
| 1 | 115 | 375 | 30.7 | 30.7 |
| 2 | 114 | 364 | 31.3 | 31.0 |
| 3 | 105 | 360 | 29.2 | 30.4 |
| 4 | 110 | 382 | 28.8 | 30.0 |
| 5 | 110 | 347 | 31.7 | 30.3 |
| 6 | 125 | 409 | 30.6 | 30.4 |
| 7 | 101 | 367 | 27.6 | 30.0 |
| 8 | 101 | 366 | 27.6 | 29.7 |
| 9 | 128 | 421 | 30.4 | 29.8 |
| 10 | 123 | 366 | 33.6 | 30.1 |

value, which we accept as the probability for the event. Here the probability for a household with age of householder 41–45 years from the population with ages 35–50 years is approximated as 30.1%.

This approach of reading the experience of an event is quite common in the insurance industry. Therein, empirical probabilities are used to accumulate traffic accident experience to miles driven, mortality experience in deaths per 1000 persons by sex and age, and birth statistics in number of births per 1000 females to age 44 years. These are all cases in which empirical probabilities are used to determine insurance premiums.

Mail surveys or coupon redemption studies provide the information for manufacturers to measure their market share—the percentage of total sales taken by their products. This computer application is the determination of a potential market.

If the marketers who are targeting home computers as a family product want a broader market than the 30.1% with householders aged 41–45 years, then they could determine probabilities for other ages. More probabilities would be computed as a basis for this decision.

## COMPUTER PROBLEMS

*Data Base Applications*

The following problems are intended for computer-assisted solution. They draw on data from the data base of Appendix A.

63. Using the data base of 20,116 records, select a sample of 100 records and develop a FREQUENCIES count of the number with age of person 1 (the householder) 25 to 35 years. This count can be obtained from a FREQUENCY summary on person-1 ages. A SAMPLE procedure is available in several common software packages (or check with your computer center for a systems "extract" procedure). When you have your data, then ask nine classmates to share their counts. Include their data to complete the table. Use these data to estimate the empirical probability for the event "person 1 (the householder) is 25 to 35 years old."

| | | | Cumulative | |
| | Count of | Total | Count of | Total |
| Sample | 25–35 Years | Count | 25–35 Years | Count |
| --- | --- | --- | --- | --- |
| 1 | | | | 100 |
| 2 | | | | 200 |
| 3 | | | | 300 |
| ⋮ | | | | ⋮ |
| 10 | | | | 1000 |

64. Experience shows that about 50% of all children are girls. Consider those records that have three or more persons. Take a SAMPLE of 1000 records with age for person #3 (requires that positions 80–81 are not blank). To further qualify for children, select only if the field has age 17 years or younger. Then obtain counts for "male" and for "female" (position 79) on those records that qualify. Does your empirical evidence support the classical probability that 50% of all children are girls?

65. What is the probability that a household has three or more autos available for use? Use the count of autos (position 19) for a SAMPLE of 100 households to estimate this probability. Repeat for a new SAMPLE of 200 households. Repeat again, for another SAMPLE of 500 households. Finally, draw a SAMPLE of 1000 households to count the number with three or more autos available. What value would you assign as the probability that a household has three or more autos?

66. What is the probability that the head of household has age in the range 25–35 years? Use the procedure of the preceding problem to answer this question. Compare this with your result for Problem 63 and with 0.242 (actual value).

*Additional Computer Problems*

67. Here is a computer simulation for Minitab users:

```
MTB  > RANDOM 10 OBSERVATIONS INTO C1;
SUBC > INTEGERS 1 TO 2.
MTB  > TALLY C1
```

This simulates a distribution with about half the observations having value 1. Record the COUNT provided for each value 1 and 2. Repeat the three statements, except generate 50 random observations. Again, record the COUNTS for 1 and 2. Repeat again with 100 observations. Does your count of 1s seem to approach 50% of the total? Combine your COUNTS with those from four classmates. Do your results seem to agree with the empirical definition of probability?

68. Repeat the procedures of the preceding problem for Minitab with the following command:

```
MTB  > RANDOM 10 OBSERVATIONS INTO C1;
SUBC > INTEGERS 0 TO 9.
MTB  > TALLY C1
```

Again, consider your COUNT of 1s. Do your results appear consistent with the empirical definition of probability? Explain.

## FOR FURTHER READING

[1] Daniel, Wayne W., and James Terrell. *Business Statistics: Basic Concepts and Methodology*, 4th ed. Boston: Houghton-Mifflin, 1986.

[2] Dixon, Wilfred J., and Frank J. Massey. *Introduction to Statistical Analysis*, 4th ed. New York: McGraw-Hill, 1982.

[3] Freund, John E. *Modern Elementary Statistics*, 6th ed. Englewood Cliffs, NJ: Prentice-Hall, 1984.

[4] Groebner, David, and Patrick Shannon. *Business Statistics*, 3rd ed. Columbus, OH: Charles E. Merrill, 1989.

[5] Johnson, Robert. *Elementary Statistics*, 4th ed. Boston: PWS-Kent, 1984.

[6] Levin, Richard I. *Statistics for Management*, 4th ed. Englewood Cliffs, NJ: Prentice-Hall, 1987.

[7] Naiman, Arnold, and Robert Rosenfeld. *Understanding Statistics*, 3rd ed. New York: McGraw-Hill, 1983.

[8] Ott, Lyman, and David Hildebrand. *Statistical Thinking for Managers*. Boston: PWS-Kent, 1982.

[9] Wonnacott, Ronald, and Thomas Wonnacott. *Introductory Statistics*, 4th ed. New York: John Wiley, 1985.

# 5

## DISCRETE PROBABILITY DISTRIBUTIONS

### 5.1 INTRODUCTION

A favorite TV commercial of recent years depicts the Maytag repairman—just growing old with nothing to do. True, some major appliances are more trouble-free than others, but all appliance manufacturers have to deal with warranties and the likelihood of breakdowns of their products.

Many factors affect a washing machine's operability: the care and precision with which the parts are made and assembled, the type of motor insulation, environmental operating temperatures, frequency of use, load on machine, and so on. Eventually products fail, but it is difficult (if not impossible) to predict whether any given machine will fail before its warranty expires. We say the failure of a product within the warranty period is a "matter of chance" or a "random" event. For the manufacturer, the proportion of machines that fail during any given warranty period could be considered a random variable.

A *random variable* assumes values as the result of a chance event. More specifically,

> A random variable is the function or rule that assigns a numerical value to each possible event for an experiment.

The term *random variable* indicates that there is an underlying probability structure. Chapter 4 included three situations in which probabilities of events could be developed—shipment of chemicals, production of plastic pipe couplings, and emergencies in a chemical refinery. One thing common to all three situations is that the number of possible outcomes can be counted. This means that probabilities could be assigned to distinct outcomes. The ability to count the specific outcomes, in some cases with the assistance of counting rules, identifies these as situations that can be defined by a discrete random variable.

There are two classes of random variables, discrete and continuous. In this chapter we focus upon *discrete random variables*.

A discrete random variable has a countable number of values to which probabilities can be assigned.

A continuous random variable has more values than can be counted.

If, for example, an appliance manufacturer can look at historical data and determine that 4% of its products fail before the warranty period expires, that information could be useful in terms of staffing repair facilities, maintaining inventories of spare parts, etc. Then, for example, if 100 appliances are shipped to a dealer, the likelihood of a discrete number of failures (none, one, two, . . . , up to ten or even twenty) could be estimated. The estimates would stem from probability calculations of a discrete random variable.

This chapter describes the most widely used discrete random variables, beginning with the simplest—the uniform—and progressing to the more complex. Our major interest, however, will be with the binomial random variable because it is used so widely in business.

Upon completing Chapter 5, you should be able to

1. compute uniform, hypergeometric, binomial, and Poisson probability values for discrete random variables;
2. construct discrete random variable probability distribution tables;
3. compute the mean and the standard deviation given a discrete random variable probability distribution;
4. use binomial and Poisson probability tables.

## 5.2 PROBABILITY DISTRIBUTIONS AND EXPECTED VALUES

Chapter 3 described the process of grouping data into frequency distributions, such as the distribution of weekly earnings of employees. Chapter 4 showed that some data tend to conform to identifiable recurring patterns that can be described by probability formulas. For example, the arrival rate of autos at a gas station and the average number of incoming calls to a telephone switching center can be described mathematically by formulas that define discrete random variable probability distributions. Thus, provided there are sufficient empirical data on the relative occurrence of the countable outcomes for a business situation, it is possible to describe a (discrete random variable) probability distribution. Then a discrete random variable takes on various values of $X$ with probabilities specified by its *probability distribution*.

The probability distribution, $P(X)$, for a discrete random variable is an assignment of probability to each of the possible values for the variable.

The expression $P(X)$ denotes the probability for any value $X$ of the discrete random variable. One of the simplest probability distributions is $P(X) = 1/n$, where the discrete random variable $X$ allows $n$ values, each with equal or uniform probability. This is illustrated in Table 5.1 for $n = 6$.

**TABLE 5.1**
The Probability Distribution for
One Roll of a Six-Sided Die

| $X$ | $P(X)$ |
|-----|--------|
| 1 | 1/6 |
| 2 | 1/6 |
| 3 | 1/6 |
| 4 | 1/6 |
| 5 | 1/6 |
| 6 | 1/6 |
| Total | 6/6 = 1 |

The probability distribution for any discrete random variable must meet two conditions: first, that the probability values must be $\geqslant 0$ and $\leqslant 1.0$ for any value of $X$, and second, that when all possible values are summed, their sum must equal 1.

### TWO CONDITIONS FOR A DISCRETE RANDOM VARIABLE PROBABILITY DISTRIBUTION

1. $0 \leqslant P(X) \leqslant 1$   for any value of $X$
2. $\sum P(X) = 1$

Table 5.1 illustrates a discrete probability distribution associated with the roll of a six-sided die. The sample space consists of the six outcomes, or simple events, namely, $S = \{1, 2, 3, 4, 5, 6\}$. The random variable $X$ is the assignment of a label 1 through 6 based on the number of dots on the upturned face. If we assume that each side has an equal chance of ending up, then $P(X)$ can be calculated in accordance with the classical definition of probability, where $P(X) = 1/6$ in each case and $\sum P(X) = 1$.

These results satisfy the conditions for a probability distribution. So Table 5.1 demonstrates a probability distribution for $X =$ the number of dots on the upturned face for one toss of a die. This is an example of a discrete random variable and its probability distribution.

For grouped data we instinctively tend to look for an average. In Chapter 3 the average value of grouped data was called a weighted mean, where the weights correspond to frequencies. Then in Chapter 4 we saw that long-run relative frequencies can be expressed as probabilities. For the average of a random variable, the data values are weighted by the probabilities of events—rather than by relative frequencies—and summed; the resulting value is referred to as the *expected value*.

> The expected value of a discrete random variable is the mean for its probability distribution, and we obtain it by multiplying each value of the random variable by its associated probability and summing the products.

The expected value is simply another type of weighted mean. It represents what we might expect in the long run. Suppose, for example, that historical data reveal that two appliances failed out of the first shipment of 100, then four out of the next 100, and one out of the next 100, and so on. We would expect the number of failures in any shipment of 100 to be a relatively small number, like two, three, or four. If we used the historical data to express the probability of none, one, two, . . . , up to ten or more failures out of 100, we would have a probability distribution, and the mean of that distribution would be the expected value for the number of failures.

Calculation of the weighted mean (Chapter 3) involved $\sum (X \cdot f/n)$. The comparable calculation for a discrete random variable simply replaces the relative frequency ($f/n$) by $P(X)$—a long-run relative frequency or an *a priori* probability. Then the mean or expected value for a discrete random variable is calculated as $E(X)$, by

$$E(X) = \sum [X \cdot P(X)] \qquad \text{(5.1)}$$

Here $X$ represents the values taken by the discrete random variable and $P(X)$ denotes the probabilities associated with the probability distribution for $X$.

Now let us see how a probability distribution is used to determine the expected value, $E(X)$. Example 5.1 extends the probability distribution of Table 5.1, where the probabilities can be determined *a priori*.

## EXAMPLE 5.1

Suppose you must pay $3.75 to roll one die one time. You win as many dollars as there are points that appear on the top side. Would you profit from such a game over the long run?

*Solution*    Here, $X$ is the number of points on the top side, and each point represents $1. We first write the probability distribution of $X$ as shown in the table and then use it to compute the expected value.

| $X$ | $P(X)$ | $[X \cdot P(X)]$ |
|-----|--------|------------------|
| $1 | 1/6 | 1/6 |
| 2 | 1/6 | 2/6 |
| 3 | 1/6 | 3/6 |
| 4 | 1/6 | 4/6 |
| 5 | 1/6 | 5/6 |
| 6 | 1/6 | 6/6 |
|   |   | $21/6 = \sum [X \cdot P(X)]$ |

$$E(X) = \sum [X \cdot P(X)] = 21/6 = 3.50 \quad \text{(or \$3.50)}$$

The result is a $3.75 cost and a $3.50 expected gain, for a net expected loss of $0.25 for each time you roll the die. It would *not* be a profitable game over the long run.

Note that the expected value is the mean of the distribution and is not necessarily a possible value; in Example 5.1 no single roll of a die will have a value of 3.5 points.

Example 5.2 uses empirical data rather than *a priori* information to illustrate the calculation of an expected value.

**EXAMPLE 5.2**

In order to decide how many people were needed on the sales staff, the owner of Uptown Real Estate noted that the company had sold 0, 1, or 2 homes per week 10%, 70%, and 20% of the time, respectively. Based on this record, how many homes can Uptown *expect* to sell in the next week?

*Solution*    For $X$ = the number of homes sold,

| $X$ | $P(X)$ |
|---|---|
| 0 | 0.10 |
| 1 | 0.70 |
| 2 | 0.20 |

Then $E(X) = \sum [X \cdot P(X)] = 0 + 0.70 + 0.40 = 1.10$

In reality, Uptown Real Estate could not sell 1.1 homes in any week. However, over the next 10 weeks, the company can expect to sell $(10)(1.1) = 11$ homes. This number can be a gauge against which the actual number of sales is measured, as well as an indication of the number of salespeople required.

The probability distributions in Examples 5.1 and 5.2 are forms of discrete (random) variable distributions because they involve discrete values and specify all the possible outcomes and the associated probability of each. In Example 5.1, each outcome had an equal probability. This typifies a uniform random variable, which is discussed in the next section.

The material for the remainder of this chapter is arranged to provide a logical development of discrete random variables and their probability distributions. A progressive development of the most common discrete variable distributions—uniform, hypergeometric, binomial, and Poisson—is given in Section 5.3 through 5.5 and 5.7, respectively. Each section includes a rule for assigning probabilities to the distribution of values for the random variable, that is, for forming a probability distribution.* Section 5.6 describes the calculation of distribution properties such as the mean (expected value) and the standard deviation.

**Problems 5.2**

1. During summer weekends, the number of occupied campsites at Lake Wapello has followed the pattern shown:

> 5 campsites occupied with relative frequency 0.10
> 6 campsites occupied with relative frequency 0.15
> 7 campsites occupied with relative frequency 0.30
> 8 campsites occupied with relative frequency 0.15
> 9 campsites occupied with relative frequency 0.25
> 10 campsites occupied with relative frequency 0.05

---

* Some books denote this rule as a probability *mass* function in contrast to a probability *density* function for continuous random variables (in Chapter 6). This distinction is not necessary as long as you understand the concepts covered herein.

Use this information to answer each question.
a. Display the probability distribution.
b. Compute the expected value $E(X)$ of weekend occupancies.
c. Interpret your value for $E(X)$. For example, does this value apply to the Fourth of July weekend occupancies? Explain why or why not.

2. Given the probability distribution shown in the table, compute the mean, $E(X)$. Interpret the meaning of your answer, relative to a small service business, if $X$ = the number of service representatives who are out of the office or on a service call.

| X | P(X) |
|---|------|
| 3 | 0.20 |
| 4 | 0.30 |
| 5 | 0.40 |
| 6 | 0.10 |

## 5.3
## THE DISCRETE UNIFORM DISTRIBUTION

A *discrete uniform random variable* describes a situation in which each of $n$ similar outcomes has an equal chance of occurring.

> For a discrete uniform random variable, each value has equal probability.

### Conditions for a Discrete Uniform Distribution

By definition, $P(X = x_1) = P(X = x_2) = \cdots = P(X = x_n) = 1/n$. Thus the expression for computing discrete uniform probability values $P(X)$ is simply the following:

**THE DISCRETE UNIFORM PROBABILITY RULE**

$$P(X) = \frac{1}{n} \quad \text{for} \quad X = x_1, x_2, \ldots, x_n \tag{5.2}$$

where   $X$ = individual values

$n$ = the number of possible outcomes

$P(X)$ = the uniform probability for every individual value

Figure 5.1 illustrates the uniform distribution that arises from a quality-control sample drawn randomly from one of eight wine casks (I, II, . . . , VIII) in a California winery. (The numbers are not visible to the quality-control inspector.)

The eight casks are all different vintages of a wine, and we want to know the probability of selecting a particular cask (and vintage) for $X$ = the occurrence of any cask type I, II, . . . , or VIII. The identification on each cask is the same. The producer has established a constant quota to assure a high price. So the probability of obtaining any outcome, say cask IV, is $P(X = \text{IV}) = 1/8$.

Chapter 4 methods of calculating probabilities can be used with a known $P(X) = 1/8$, for each of the eight possible values for the random variable. Then,

**FIGURE 5.1**   Discrete Uniform Probability Distribution

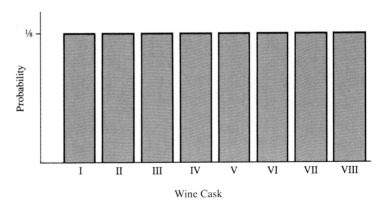

for example, by the addition rule for exclusive outcomes, the probability of obtaining cask I, II, or IV is

$$P(X = \text{I, II, or IV}) = P(\text{I}) + P(\text{II}) + P(\text{IV})$$
$$= \frac{1}{8} + \frac{1}{8} + \frac{1}{8} = \frac{3}{8}$$

This is a simple example, yet it illustrates several generalizations about any
*discrete* variable probability distribution. A first requirement is that all possible
values of the random variable can be identified. Then each value is assigned a
probability by a probability distribution rule. An event is defined in terms of
specific values of the random variable. The event's chance of occurrence is the
summation of the specific probability values. Finally, the mean or expected
value for the (discrete) probability distribution is calculated very much like a
weighted mean in Chapter 3.

**Problems 5.3**

3. A discrete uniform random variable has the probability distribution shown
   in the figure, where $P(X) = 1/4$ for $X = 0, 1, 2,$ or 4. Find each value.
   a. $P(X = 0 \text{ or } 1)$      b. $P(X = 0)$      c. $P(X > 2)$

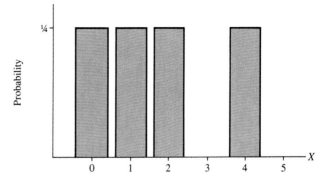

4. A calculator keyboard has ten number keys. From experience, it is known
   that each key has an equal probability of malfunctioning during an automatic life cycle "test to failure." Graph the probability distribution. Let 0, 1,

2, ..., 9 designate the ten number keys. If your firm manufactured this item, how would you respond if a competitor advertised that 5 of 10 users have experienced a malfunction in the number key labeled "2"?

## 5.4 THE HYPERGEOMETRIC DISTRIBUTION

Suppose that a carton containing a dozen fluorescent tubes has two defective tubes. If tubes are selected at random from the carton and are either installed or discarded (but not replaced in the carton), the probability of selecting a defective tube changes each time another tube is selected. Statisticians refer to this type of selection process as *sampling without replacement*. The *hypergeometric random variable* applies to situations in which the random variable describes the number of successes and is finite and in which sample items are randomly drawn without replacement.

> The hypergeometric random variable describes the number of successes in samples drawn at random and without replacement from a small finite population.

### Conditions for a Hypergeometric Probability Distribution

**TWO CONDITIONS THAT DEFINE A HYPERGEOMETRIC RANDOM VARIABLE DISTRIBUTION**

1. Each trial can be reduced to having two possible outcomes—a success or a failure—and the random variable describes the number of trials that result in a success outcome.
2. The outcome at each trial is dependent upon the previous trials.

In simple terms, a hypergeometric probability rule $P(X)$ gives the ratio of the number of successful ways that a specified outcome can occur to the total number of ways that all outcomes can occur. The outcome of a given trial is dependent on the previous trials.

To describe a *hypergeometric probability rule*, let $X$ represent the number of a desired type (successes) in $n$ dependent trials.

### THE HYPERGEOMETRIC PROBABILITY RULE

$$P(X) = \text{Successful ways/Total ways} \qquad (5.3)$$

or

$$P(X) = \frac{{}_aC_x \cdot {}_bC_{n-x}}{{}_{a+b}C_n}, \quad \text{with } X = 0, 1, 2, \ldots, n \qquad (5.4)$$

where   $a$ = the possible number of one type (successes)

$b$ = the number of all other types (nonsuccesses)

$n$ = the number of trials

$X$ = the number of successes in $n$ trials

$P(X)$ = the probability of observing $X$ of this type (successes) in $n$ trials

and the situation indicates that the outcome of each trial is dependent on the outcomes of the previous trials.

Because duplication of outcomes is not permitted and the order in which terms are selected makes no difference, the number of successful ways and the number of total ways can be computed using combinations.

*Successful ways* constitute all combinations of events that can produce outcomes of the desired type, whereas *total ways* include all possible combinations that can produce any outcome (both successful and unsuccessful). Example 5.3 illustrates the calculations.

---

**EXAMPLE 5.3**

Suppose 20 wiring caps remain in a stockroom bin. Of these, 6 are red, 9 are yellow, and 5 are black. If an inventory clerk reached into the bin and indiscriminately picked two wiring caps, what is the probability of getting (a) both red or (b) exactly one red? (c) Construct the probability distribution.

*Solution*    The random variable is the number of red wiring caps $(X)$ in a sample of two caps, which is 0, 1, or 2. Then $a = 6$, the number of red caps, and $b$ (not red) $= 9 + 5 = 14$ other caps. The experiment has $_{a+b}C_n = {}_{20}C_2 = 190$ possible combinations for two wiring caps. It encompasses 190 simple events (total possibilities).

a. The first question requires the evaluation of $P(X = 2)$. Since $_aC_X = {}_6C_2 = 15$ and $_bC_{n-X} = {}_{14}C_0 = 1$, there are $15 \times 1$ successful ways, or

$$P(X = 2) = \frac{({}_6C_2) \cdot ({}_{14}C_0)}{{}_{20}C_2} = \frac{15 \cdot 1}{190} = 0.08$$

The probability is 0.08 that both will be red.

b. For the second question, $X = 1$ red, so $n - X = 2 - 1 = 1$ not red, giving

$$P(X = 1) = \frac{({}_6C_1) \cdot ({}_{14}C_1)}{{}_{20}C_2} = \frac{6 \cdot 14}{190} = 0.44$$

Forty-four percent of all possible random selections would produce 1 red cap.

c. The remaining probability is that $X = 0$:

$$P(X = 0) = \frac{({}_6C_0) \cdot ({}_{14}C_2)}{{}_{20}C_2} = \frac{1 \cdot 91}{190} = 0.48$$

These results for $P(X = 0)$, $P(X = 1)$, and $P(X = 2)$ give the probability distribution shown in Figure 5.2. Because only three values have probabilities, we could have found

**FIGURE 5.2**  A Hypergeometric Probability Distribution

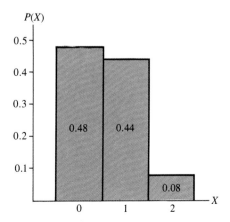

Number of Red Wiring Caps

the probability of zero using the complement rule:

$$P(0) = 1 - [P(1) + P(2)] = 1 - [0.08 + 0.44] = 0.48$$

The distribution shows that it is most likely the clerk would return with either 0 or 1 red cap in a random sample of size two.

---

In many cases you can identify hypergeometric situations only by realizing that sampling is without replacement. For example, the probability of getting a red wiring cap depends on which draw we are talking about. It is a conditional probability that changes with each trial, which means that the outcomes of the draws are dependent. This condition becomes important when we introduce the binomial random variable (in the next section) because there the outcomes of draws *are* independent.

**Problems 5.4**

5. A box of 20 thermostats en route to a building supply firm is dropped in transit and 5 thermostats become inoperable—although the damage is not apparent. If a building contractor selects 4 thermostats from the box, what is the probability that all 4 are operable?

6. In a group of 20 laboratory employees, 4 have some form of blood disease. None is recognized to have the disease. If 10 of the employees are tested, what is the chance that at least 2 will be found to have the disease? Do not determine the entire probability distribution.

7. At the end of the season, a nursery puts 9 untagged apple trees on sale. It is known that 4 of the trees are of the red delicious variety and the rest are Grimes golden. A customer buys 3 of the trees.
   a. What is the probability that 2 are red delicious and 1 is Grimes golden? Assume the types are indistinguishable. (*Note:* The entire distribution is not needed to answer this question.)
   b. Construct the probability distribution for $X$ = the number of red delicious trees in the sample.

8. At the Dravo Co., water meters that have been receipt-inspected are inventoried on blue cards, and those not yet inspected are listed on red cards. At present there are 52 cards, and half are red. If two cards are picked at random, what is each probability?
   a. For draws without replacement, that both will be blue
   b. Without replacement, that both will have the same color (*Hint:* This includes possibilities of two red *or* of two blue.)

9. Ten colored resistors are placed in a closed container. One is green, four are white, three are red, and two are yellow. If someone is mounting these resistors in essentially random order, what is each chance?
   a. That two of the first six mounted are yellow
   b. That the last one placed is red (*Hint:* How many red would be placed among the first nine?)
   *Suggestion:* To answer (a) and (b), consider two groups: (1) those of the type in question and (2) those of all other colors.

## 5.5
## THE BINOMIAL DISTRIBUTION

Many business problems have dichotomous results, meaning that the results fall into one of two possible categories: yes or no, good or bad, and so on. For example, an invoice is either correct or incorrect, consumers either prefer liquid soap or they don't, an automobile starter either works or it doesn't work. When the probabilities associated with the results of repeated trials of a two-outcome experiment are constant, this is a *Bernoulli process*.

> A Bernoulli process consists of repeated (Bernoulli) trials, wherein each outcome is one of two possibilities—a success or a failure—and the probability of a success remains constant for each (Bernoulli) trial.

The random variable, which describes a Bernoulli trial, can have only two possible outcomes: success or failure. The Greek letter $\pi$ is used for the probability of a success. The probability of a failure, then, is $1 - \pi$. For any number of Bernoulli trials, the probability of success $\pi$ at each trial depends on the proportion of successes inherent in the population. It is *not* influenced by what has happened at preceding trials.

As an example, in a Los Angeles electronics plant 60% of the chips produced for electronic calculators are good ($G$), and the remainder are no good ($G'$).

From the given information and for one selection, $P(G) = \pi = 0.6$, so the probability that the next chip is good is 0.6. Then $P(G') = 1 - 0.6 = 0.4$, and the probability that the next chip is bad is 0.4. This simple example illustrates a Bernoulli process for which the number of trials is one. Most frequently, however, we are interested in the outcomes of numerous Bernoulli trials—not just one. To illustrate, we might want to know the probability that exactly three of the next five electronic chips will be good. Assuming that the quality of successive chips is not influenced by the quality of others (that is, assuming that outcomes are independent) and that the probability of a good chip is constant from trial to trial, then the number of successes, $X$, in $n$ Bernoulli trials can be described by the binomial distribution.

### Conditions for a Binomial Probability Distribution

The binomial random variable describes the number of successes in a fixed number, $n$, of independent Bernoulli trials.

The binomial probability distribution is appropriate when

1. the process has Bernoulli trials; each trial has two possible outcomes;
2. the outcome for each trial is independent of all previous trials, so the probability of a success is constant from trial to trial; and
3. the random variable describes the number of successes in a fixed number, $n$, of trials.

Let us expand upon the example about calculator chips. Example 5.4 will enable us to better understand the logic for computing binomial probabilities, and leads to the equation for the binomial probability distribution.

## EXAMPLE 5.4

a. Find the probability of obtaining three good electronic chips in a sample of five when the probability of a good chip is constant at 0.6 and the trials are independent.
b. What are the implications of this finding on the quality of the product?

*Solution*    a. There are $_5C_3$, or ten ways to arrange three good chips in a sample of five. These possible combinations are as shown in the figure below.

Possible Combinations of Three Good Chips in a Sample of Five

| | | | | | | | | | |
|---|---|---|---|---|---|---|---|---|---|
| $G$ | $G$ | $G$ | $G'$ | $G'$ | $G$ | $G'$ | $G$ | $G$ | $G'$ |
| $G$ | $G$ | $G'$ | $G'$ | $G$ | $G'$ | $G$ | $G$ | $G$ | $G'$ |
| $G$ | $G'$ | $G'$ | $G$ | $G$ | $G'$ | $G$ | $G$ | $G'$ | $G$ |
| $G'$ | $G'$ | $G$ | $G$ | $G$ | $G'$ | $G$ | $G'$ | $G$ | $G$ |
| $G$ | $G$ | $G'$ | $G$ | $G'$ | $G$ | $G'$ | $G$ | $G'$ | $G$ |

The first combination consists of three good chips followed by two bad chips. In all selections the probability that any chip is good is the same: $P(G) = 3/5 = 0.6$. Similarly, the probability that a chip is bad is always $P(G') = 0.4$. Thus, for the particular combination $(G, G, G, G', G')$, assuming independence of trials,

$$P(GGGG'G') = P(G) \cdot P(G) \cdot P(G) \cdot P(G') \cdot P(G')$$
$$= [P(G)]^3 \cdot [P(G')]^2$$
$$= (0.6)^3 \cdot (0.4)^2 = 0.0346$$

Since this combination is only one of the ten possible ways of getting three good and two bad chips in five draws, the probability $(0.6)^6 \cdot (0.4)^2$ must be multiplied by the ten possible combinations to yield

$$P(X = 3 \text{ good}) = 10(0.6)^3 \cdot (0.4)^2 = 0.346$$

b. If three good chips are required for "fair quality," the chance of getting them among a selection of five chips is only about 1/3. This may be unacceptable from a quality-control standpoint.

---

Example 5.4 illustrates a long (enumeration) form for computing binomial probabilities. Fortunately, a shortened expression of the binomial probability function is quite easily developed. As suggested in Example 5.4, we use combinations because different orders do not constitute new combinations and duplication is not permitted.

### THE BINOMIAL PROBABILITY RULE

$$P(X|n, \pi) = _nC_X \pi^X (1 - \pi)^{n-X} \quad \text{for } X = 0, 1, \ldots, n \qquad (5.5)$$

where   $n$ = the number of trials

        $X$ = the number of successes (can range from 0 to $n$)

        $\pi$ = the probability of success on any trial
           (a value between 0 and 1)

    $1 - \pi$ = the probability of failure on any trial

**TABLE 5.2**   Probability Distribution for a Binomial Random Variable with $n = 5$, $\pi = 0.6$

|   | $P(X)$ |
|---|---|
| 0 | $P(0) = {}_5C_0(0.6)^0(0.4)^5 = 0.0102$ |
| 1 | $P(1) = {}_5C_1(0.6)^1(0.4)^4 = 0.0768$ |
| 2 | $P(2) = {}_5C_2(0.6)^2(0.4)^3 = 0.2304$ |
| 3 | $P(3) = {}_5C_3(0.6)^3(0.4)^2 = 0.3456$ |
| 4 | $P(4) = {}_5C_4(0.6)^4(0.4)^1 = 0.2592$ |
| 5 | $P(5) = {}_5C_5(0.6)^5(0.4)^0 = 0.0778$ |
|   | Total    1.0000 |

*Note:* Binomial point (single-value) probabilities also appear in Table 5 of Appendix B.

In Equation (5.5), all information necessary to solve the problem is stated in a simple, concise form: The probability sought is $P(X)$, the number of trials is $n$, and the constant probability of success on each trial is $\pi$. Thus, using Example 5.4, for $n = 5$, $X = 3$, and $\pi = 0.6$,

$$P(X = 3 \text{ good} \mid n = 5, \pi = 0.6) = {}_5C_3(0.6)^3 \cdot (0.4)^2 = 0.3456$$

Table 5.2 shows the total probability distribution. Figure 5.3 illustrates two graphic displays for the probability distribution. These are alternative descriptions. Figure 5.3(a) illustrates the probabilities as areas. Figure 5.3(b) better demonstrates the fact that for a discrete random variable the probability exists only at a countable number of points. Thus the probability over any interval consists entirely of the values at the relevant points. However, you may come across either form in business literature.

**FIGURE 5.3**   Binomial Probability Distribution for $(X =)$ the Number of Good Chips, Given $n = 5$, $\pi = 0.6$

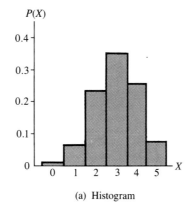

(a) Histogram

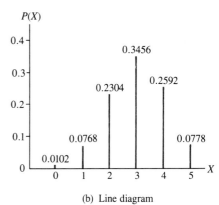

(b) Line diagram

**Binomial Probability Table**

For sample sizes up to 25, binomial probabilities can be taken directly from the probability distribution table (Table 5 in Appendix B). The two parameters $n$ and $\pi$ define the appropriate segment of the table. Simply state the problem as before and go down the left column of the table to the appropriate sample size $n$. Then go across the top row of the table to find the correct value of $\pi$ and read the probability corresponding to the desired value in the $X$ column. Note, for example, for $P(X = x \mid n = 5, \pi = 0.60)$, that Table 5 in Appendix B gives all the same values computed in Table 5.2, beginning with 0.0102 for $P(X = 0)$ and ending with 0.0778 for $P(X = 5)$.

Any difficulty in computing binomial probabilities stems primarily from the exponents in the equation. However, if your calculator is equipped to handle exponents, it may be just as fast to use a calculator as to use a table.

Once you know the random variable is binomial distributed and you have values for $n$ and $\pi$ to define the probability distribution, you can find probabilities. We now introduce a systematic approach to solving binomial probability problems.

#### STEPS FOR SOLVING BINOMIAL PROBABILITY QUESTIONS

1. Identify the question. Then define the random variable $X$ that is being questioned.
2. Determine the particular values of $X =$ the number of successes included in the question, and note $n$ as the number of trials.
3. Be sure that $\pi$ corresponds to the same event as $X$. When $X$ describes the number of something, then $\pi$ is the probability of that outcome.
4. Compute and add the appropriate binomial probabilities (or read them directly from Table B.5 and then find their sum).

These steps are illustrated in Example 5.5.

---

**EXAMPLE 5.5**

Use the data of Example 5.4 to find the probability of obtaining *at least* two good chips in a selection of five.

*Solution*
1. We seek "at least two" good chips in five. The random variable $X$ is the number of good chips.
2. Thus $X = 2, 3, 4,$ or 5 good chips, and $n = 5$.
3. $X$ counts the number of good chips; $\pi = 0.6$ is the probability that any chip is good.
4. Using Table 5.2 or Figure 5.3, we obtain

$$P(X \geqslant 2) = P(2) + P(3) + P(4) + P(5)$$
$$= 0.2304 + 0.3456 + 0.2592 + 0.0778 = 0.9130$$

The probability of at least two good chips in five is 0.9130. There is a relatively strong probability, 0.91, that two or more of the chips will be good.

---

Table 5.2 holds answers to many other probability questions. For example, consider the following. (To read each probability question, substitute for $X$ the phrase "the number of good chips.")

$$P(0 \leqslant X \leqslant 2) = P(X = 0 \text{ or } 1 \text{ or } 2) = P(0) + P(1) + P(2)$$
$$= 0.0102 + 0.0768 + 0.2304 = 0.3174$$

$$P(0 < X \leqslant 2) = P(X = 1 \text{ or } 2) = P(1) + P(2)$$
$$= 0.0768 + 0.2304 = 0.3072$$

$$P(X \leqslant 1.5) = P(0 \leqslant X \leqslant 1.5) = P(X = 0 \text{ or } 1) = P(0) + P(1)$$
$$= 0.0102 + 0.0768 = 0.0870$$

$$P(X > 0) = 1 - P(X = 0) = 1 - 0.0102 = 0.9898$$

$$P(2 < X < 3) = P(\varnothing) = 0 \quad \text{(where } \varnothing\text{—the empty set symbol—is}$$
$$\text{the null or empty event)}$$

Check these results by reading the probabilities from the binomial distribution Table 5 in Appendix B. You can also use the probabilities in Figure 5.3(b) to check these answers.

The binomial random variable is used in business for inventories, in quality control for analyzing production, and in structuring managerial decisions for financial investments. One must be careful to distinguish between situations in which the binomial distribution (with its fixed probabilities and independence) is appropriate and situations in which the hypergeometric (with varying probabilities) applies.

**Problems 5.5**

10. An auditor with a public utility has determined that the probability of randomly selecting a paid-up (good) account is 0.9, so the probability of selecting a delinquent (bad) account is 0.1. He wishes to determine the probability of selecting one delinquent and two paid-up accounts in a random sample of three accounts. Assume independence.
    a. List the possible combinations of two paid-up accounts in a sample of three accounts.
    b. Compute the probability of the first account being delinquent and the second and third accounts being paid up.
    c. Multiply the number of combinations for two paid-up accounts by the probability of each combination (part (b) answer).
    d. Verify your answer to part (c) by finding $P(X = 2 \mid n = 3, \pi = 0.9)$ in the binomial probabilities table—Table 5 in Appendix B.

11. State which conditions required of a probability distribution are violated in each of the following tables.

a.

| X | P(X) |
|---|------|
| 0 | 1/2 |
| 1 | 1/3 |
| 2 | 1/4 |

b.

| X | P(X) |
|---|------|
| 1 | 1/4 |
| 2 | 1/2 |
| 3 | 3/2 |

12. Complete the table to make a probability distribution and then evaluate parts (a), (b), and (c).

| X | P(X) |
|---|------|
| 0 | 1/8 |
| 1 | 2/8 |
| 2 | 3/8 |
| 3 |      |
| Total | |

a. $P(0 \leqslant X \leqslant 2)$      b. $P(0 < X \leqslant 2)$      c. $P(0 < X < 2)$

13. The coefficients of the terms in the expansion of the binomial equation are combination counts. Determine the following:
   a. $_1C_0, \ _5C_0, \ _{10}C_0, \ _1C_1, \ _5C_5, \ _{10}C_{10}$
   b. $_3C_1, \ _4C_1, \ _{20}C_1, \ _3C_2, \ _4C_3, \ _{20}C_{19}$
   c. $_6C_0, \ _6C_6, \ _6C_1, \ _6C_5, \ _6C_2, \ _6C_4$
   Give generalizations for values $(n \geqslant 1)$ of
   d. $_nC_0$ and $_nC_n$
   e. $_nC_1$ and $_nC_{n-1}$
   f. $_nC_X = \ _nC_{n-X}$ for $n \geqslant X$
   Use this result to find $_{12}C_{11}, \ _{15}C_{15}, \ _{18}C_{16},$ and $_{20}C_{15}$.

14. Assuming that the chance for a successful business investment is 3/4 and that investments can be considered independent events, complete the probability distribution for the number of successful business investments among four.

| X | P(X) |
|---|------|
| 0 | $P(0) = (_4C_0)(3/4)^0(1/4)^4 = 1/256 = 0.0039$ |
| 1 | |
| 2 | |
| 3 | |
| 4 | |

15. a. If the probability is 0.1 that a four-person jury in a certain court will contain no women, set up the probability distribution for $X$ = the number of juries that contain no women. Let $n$ = four trials and assume each jury is chosen independently of all others. (*Suggestion:* Use Table 5, Appendix B.)
   b. Use the table you just constructed to determine $P(X \geqslant 2)$.

16. Assuming a binomial situation, use Table 5, Appendix B, to find
   a. $P(3 \leqslant X \leqslant 6)$, for $n = 20$, $\pi = 0.2$
   b. $P(3.6 < X \leqslant 4)$, for $n = 20$, $\pi = 0.7$
   c. $P(X > 0)$, for $n = 25$, $\pi = 0.25$

17. The probability is 0.2 that any one of five hunters will bag the limit on a duck-hunting trip. Assume that they all go separately and that their experi-

ences are independent. What is the probability that
   a. at least four of these hunters will return with their limit?
   b. exactly three will return home with their limit (and the others will not)?
   c. the first three (hunters 1, 2, and 3) each return with their limit and the others do not? (*Hint:* This is similar to the binomial case but is not exactly so because the order is specified.)

18. Experience indicates that along a certain offshore oil drilling area, the chance is 0.1 that any well drilled will be profitable. Assume independence for the drillings.
   a. What is the chance that at least one of four wells will be profitable?
   b. If five wells are drilled, what is the chance that only one well will be profitable?

19. Five percent of all automatic dishwashers of a certain make have defective timers. What is the probability that, in a survey of 15 owners of this make of dishwasher, all will have operable timers?

20. An automatic lathe is out of control and checked for repair if one or more of a sample of five turnings are defective. If 10% of its output is defective, what is the chance that the lathe will be declared out of control after a sample is checked?

---

## 5.6 DESCRIBING DISTRIBUTIONS

Suppose you are the quality-control manager for a large company producing breakfast cereals. A railroad carload of wheat has arrived from Montana, and you must pass judgment on the moisture content. You certainly cannot measure each kernel, so you will probably be content with some sample statistics. First, you may want some estimate of the average (expected value) of moisture. Next, you may want to measure the variability of moisture in the shipment—perhaps by computing a standard deviation. The same measures—mean and standard deviation—are the two most widely used descriptive measures of probability distributions.

   This section illustrates computations for the mean and standard deviation for the binomial and other discrete variable probability distributions. We first take up the mean.

### The Mean (Expected Value)

In Section 5.2, we indicated that the expected value $E(X)$ is the *mean* for a discrete random variable probability distribution. Equation (5.1) is repeated here.

**The mean for discrete variable distributions is**

$$\mu = E(X) = \sum [X \cdot P(X)] \tag{5.1}$$

A simple coin-tossing example will demonstrate the computation. After describing that, we will introduce an equivalent, but computationally shorter, rule for binomial distributions.

**EXAMPLE 5.6**

Let $X$ = the number of heads in $n = 4$ attempts in a coin-tossing experiment. Construct the probability distribution of $X$ and use it to calculate the mean, $\mu$. The random variable is binomial, with $\pi = 1/2$.

*Solution*    The probability distribution appears in Table 5.3. It is symmetrical about the value 2—that is, the mean, the median, and the mode all equal 2. These calculations confirm $\mu = 2$.

**TABLE 5.3**    Calculation of the Mean for a Discrete Random Variable Distribution

| $X$ | 0 | 1 | 2 | 3 | 4 |
|---|---|---|---|---|---|
| $P(X)$ | 1/16 | 4/16 | 6/16 | 4/16 | 1/16 |
| $[X \cdot P(X)]$ | 0 | 4/16 | 12/16 | 12/16 | 4/16 |

$$\mu = \sum [X \cdot P(X)] = 32/16 = 2$$

*Note*: For hand calculation of values from probabilities given as fractions, the least work and the greatest accuracy result from using fractions rather than the equivalent decimal values.

The procedure for finding the mean by Equation (5.1) has only one restriction: The random variable must be discrete. However, calculations become long even for small samples, such as $n = 10$ or 20. Consequently, equivalent but shorter rules have been developed. For the *binomial distribution*, we can compute the mean simply by multiplying the sample size $n$ by the known proportion, $\pi$.*

For binomial distributions

$$\mu = n\pi \tag{5.6}$$

In the coin-tossing experiment, where $n = 4$ tosses and $\pi = 1/2$ (chance of a head), $\mu = 4(1/2) = 2$. This agrees with the result of Example 5.6.

Simplified rules exist for finding the mean in other discrete distributions as well, but they are not covered here. However, the "definition" form of Equation (5.1), $\mu = \sum [X \cdot P(X)]$, works for all cases, so you can always use it.

**Standard Deviation**

Calculation of the *variance and the standard deviation for discrete random variable distributions* follows the same pattern as their calculation in a frequency distribution. Again, relative frequencies are replaced by probabilities. Then, for any discrete random variable with probability function $P(X)$, the variance, $\sigma^2$, is the expected value of the squared differences taken about the distribution mean. An equation for this is as follows.

---

\* This simple result requires substituting Equation (5.5) (binomial probability rule) for $P(X)$ in Equation (5.1):

$$\mu = E(X) = \sum X[_nC_X\pi^X(1 - \pi)^{n - X}]$$
$$= n\pi \sum [_{n-x}C_{X-1} \cdot \pi^{X-1} \cdot (1 - \pi)^{n - 1 - (X - 1)}] = n\pi$$

The variance for discrete variable distributions is

$$\sigma^2 = E[(X - \mu)^2] = \sum [(X - \mu)^2 \cdot P(X)] \tag{5.7}$$

and the standard deviation is

$$\sigma = \sqrt{\sigma^2} \tag{5.8}$$

The *standard deviation* is obtained as the square root of the *variance*. Procedures for calculating the variance change very little from those for calculating the values for a sample variance $(s^2)$ for grouped data in Chapter 3. Recall that $s^2 = \sum f(X - \bar{X})^2/(n - 1)$. For sufficiently large $n$, $f/(n - 1)$ becomes very much like $f/n$ and approximates $P(X)$; hence the rule in Equation (5.7) is quite satisfactory. Example 5.7 demonstrates the calculation of the variance and standard deviation for a discrete random variable distribution.

## EXAMPLE 5.7

Use the data from the coin-tossing experiment to compute the variance and standard deviation of the binomial distribution.

*Solution*

For $\mu = 2$, the calculation of $\sigma^2$ appears in Table 5.4. Since the variance is 1, the standard deviation is also 1.

**TABLE 5.4**   Calculation of the Standard Deviation for a Discrete Random Variable Distribution

| X | P(X) | (X − 2) | (X − 2)² | [(X − 2)² · P(X)] |
|---|------|---------|----------|-------------------|
| 0 | 1/16 | 0 − 2 = −2 | 4 | 4/16 |
| 1 | 4/16 | 1 − 2 = −1 | 1 | 4/16 |
| 2 | 6/16 | 2 − 2 = 0 | 0 | 0 |
| 3 | 4/16 | 3 − 2 = 1 | 1 | 4/16 |
| 4 | 1/16 | 4 − 2 = 2 | 4 | 4/16 |
| Total | 1 | | | $\sigma^2 = [\sum (X - \mu)^2 P(X)] = 1$ |

The calculation in Example 5.7 actually requires one step fewer than computing values for $s^2$ because the division has already been done in the values for $P(X)$.*

As with the mean, a simpler but mathematically equivalent form exists for computing the variance and standard deviation of *binomial* distributions:

**Short calculations for the variance and standard deviation for binomial distributions:**

$$\text{Variance} = \sigma^2 = n\pi(1 - \pi) \tag{5.9}$$

$$\text{Standard deviation} = \sigma = \sqrt{n\pi(1 - \pi)} \tag{5.10}$$

* Note also that we are using *a priori* probabilities $P(X)$, which are inherent in the population distribution. Since we do not need to estimate any values using the sample, the correction $-1$ for sampling bias is not needed in the denominator.

For the coin-tossing example with $n = 4$ and $\pi = 1/2$, the variance is $\sigma^2 = (4)(1/2)(1/2) = 1$ and the standard deviation is $\sigma = 1$. This is the simpler (and therefore preferred) computational form for binomial distributions.

Example 5.8 illustrates an application of a marketing situation.

## EXAMPLE 5.8

The manager of a supermarket chain in Houston, Texas, has developed a contest in which 10 of 5000 customers receive a monetary prize. The winners are selected by chance, and the distribution shown in the table gives the probabilities of a (any) customer receiving the various prize amounts:

| Prize $X$ | Probability $P(X)$ |
|---|---|
| $ 0 | 4990/5000 |
| 100 | 5/5000 |
| 200 | 4/5000 |
| 300 | 1/5000 |

Find (a) the expected cost per customer and (b) the standard deviation of cost per customer to the store.

*Solution*   Using the general form for the mean of discrete variable distributions gives the following solutions:

a. $E(X) = \sum [X \cdot P(X)]$

$$= \left[ 0 \cdot \left( \frac{4990}{5000} \right) \right] + \left[ 100 \cdot \left( \frac{5}{5000} \right) \right] + \left[ 200 \cdot \left( \frac{4}{5000} \right) \right] + \left[ 300 \cdot \left( \frac{1}{5000} \right) \right]$$

$= \$0.32$ per customer

For 5000 customers the expected cost to the store would be $5000 \times \$0.32 = \$1600$.

b. $\sigma = \sqrt{\sum [(X - \mu)^2 \cdot P(X)]}$

$$= \sqrt{ \left[ (0 - \$0.32)^2 \frac{4990}{5000} \right] + \cdots + \left[ (300 - 0.32)^2 \left( \frac{1}{5000} \right) \right] }$$

$= \$7.74$

The average payout would be $0.32 with a standard deviation of $7.74.

The mean and standard deviation values will be used later for probability calculations. For now, it is important to recognize that $\mu$, $\sigma^2$, and $\sigma$ are reasonable extensions to probability distributions of the forms used for calculation of $\bar{X}$, $s^2$, and $s$, respectively, with sample data.

## Problems 5.6

21. Let $X$ be a binomial random variable with $n = 5$, $\pi = 0.4$.
   a. Find the mean, mode, variance, and standard deviation. Use the short forms (Equations (5.9) and (5.10)) for variance and standard deviation calculations for a binomial distribution.

b. Check the values for the mean and variance using the general forms (Equations (5.1) and (5.7)).

22. Records show that 50% of the births at a certain hospital have been boys. For $n = 6$ births and assuming no multiple births, construct the probability distribution table; use this table to find $\mu$, $\sigma^2$, and the mode. See Table 5 in Appendix B. If possible, use a second method to find the values for $\mu$ and $\sigma^2$.

23. Complete the following table for a binomial situation and give values for $n$ and $\pi$:

| $X$ | $P(X)$ | $[X \cdot P(X)]$ | $(X - 0.80)$ | $(X - 0.80)^2$ | $[(X - 0.80)^2 \cdot P(X)]$ |
|---|---|---|---|---|---|
| 0 | 0.36 | 0 | −0.8 | 0.64 | 0.2304 |
| 1 | | | | | |
| 2 | 0.16 | 0.32 | 1.1 | 1.44 | 0.2304 |
| Total | | $\overline{0.80} = \mu$ | | | $\overline{\phantom{0}} = \sigma^2$ |

24. Given that these data are distributed as a discrete uniform variable, with $P(X) = 1/5$ for $X = -4, -2, 0, 1, 2,$
    a. construct the probability distribution table.
    b. using the general forms, find values for $\mu$ and $\sigma^2$.

25. Use the general forms (5.1) and (5.7) to compute values for the population mean and standard deviation for the probability distribution of Problem 7.

26. Explain the difference between the computational procedures for $s^2$ for grouped data and those for $\sigma^2$ for a discrete random variable distribution.

27. Assume a discrete distribution for $X$ = the number of exams given in all university classes, and use the table shown here.
    a. Find values for $\mu$, $\sigma^2$, and $\sigma$.
    b. Give $P(X \leqslant 3)$, $P(X \neq 0)$, and $P(1.3 < X \leqslant 4)$.

| $X$ | 1 | 2 | 3 | 4 | 5 |
|---|---|---|---|---|---|
| $P(X)$ | 0.10 | 0.20 | 0.30 | 0.30 | 0.10 |

28. Evaluate each of the following:
    a. For $X$ binomial with $n = 5$, $\pi = 0.2$, find $\mu$, $\sigma$, $P(X = 4)$.
    b. For $X$ hypergeometric, $a = 2$, $b = 4$, $n = 3$, find $\mu$, $P(X = 1)$, $P(X \geqslant 2)$.
    c. For $X$ uniform, $P(X) = 1/3$, $X = -1, 0, 2$, find $\mu$, $\sigma^2$, $P(X = 0)$.

29. Of the hair driers in a large shipment, 20% are known to be defective. What is the probability that a sample of ten hair driers from this shipment produces eight that are not defective?

30. Find the following:
    a. $P(X = 14 | n = 20, \pi = 0.8)$     c. $P(X = 4 | n = 9, \pi = 0.15)$
    b. $P(X = 0 | n = 10, \pi = 0.1)$

31. A used-car salesman feels that given average-quality cars he should be able to make 1 sale out of every 10 attempts. With really good cars, he expects to sell to 1 out of 5 prospects and with very poor quality cars, only to 1 out of 20. Suppose he makes 1 sale in 15 attempts. What is the probability of such an outcome if the car was
    a. of poor quality?                b. of average quality?

32. Records show that 15% of all shipments of denim jeans from a clothing manufacturer are classified as seconds. If a dozen denim jeans are shipped without inspection, what is the probability that two or more of the items are seconds? Assume independence.

## 5.7 THE POISSON DISTRIBUTION

The binomial, as its name suggests, is a *two-parameter* distribution defined by the parameters $n$ and $\pi$. Given values for these two defining characteristics, we can generate a binomial distribution by Equation (5.5) or by Table 5 of Appendix B. The Poisson variable, another discrete random variable, applies to situations in which the number of times an event can happen is unknown, but the probability is high that in many trials it will happen only a few times or not at all. This distribution is defined by one parameter, $\mu$, the mean rate of the event occurrence (which for this distribution is also the variance). Such infrequent events that occur independently are described by a *Poisson distribution*.

### CONDITIONS FOR THE POISSON DISTRIBUTION

The Poisson process describes the chance occurrence of independent events that happen infrequently but with a specified mean rate.

The Poisson probability distribution is defined by the following conditions:

1. Events occur at some low rate over (continuous) time or space.
2. The outcome for each trial is independent of outcomes for previous trials.
3. It is likely that none to a few events will occur in a small time- (or space-) frame; that is, $P(X = 0)$, $P(X = 1)$, and $P(X = 2)$ are large values compared to $P(X = 10)$, $P(X = 11)$, $P(X = 12)$, . . .

The Poisson distribution is applicable to business situations that require counting unusual events in a fixed space or within a specific time-frame, such as the number of defects in a length of weld on an oil pipeline, the number of automobile accidents per hour along one block in downtown Chicago, or the number of visitors per minute who enter a national park. Each case is appropriately described by a Poisson probability distribution, because each case is based on the average number of events per prescribed unit, that unit being some small dimension of either time or space. The Poisson process gives a discrete, single-parameter distribution that is completely defined by the mean number of occurrences of an event.

### THE POISSON PROBABILITY RULE

$$P(X|\mu) = \frac{\mu^X e^{-\mu}}{X!} \qquad \text{for} \quad X = 0, 1, 2, \ldots \qquad (5.11)$$

where   $\mu$ = the mean number of event occurences

$X$ = the number of events in question

$e$ = the base of natural logarithms

$\doteq 2.71828$

Because the Poisson distribution depends upon $\mu$ only, it is useful for dealing with problems in which (1) an event is a rare occurrence and (2) only an average rate of occurrence is known (or can be estimated). Example 5.9 presents a simplified application of the probability rule.

---

**EXAMPLE 5.9**

Using the Poisson rule, find the probability of zero occurrences when the average is two.

*Solution*    Using Equation (5.11), we can state the probability as

$$P(X = 0 | \mu = 2.0) = \frac{\mu^X e^{-\mu}}{X!} = \frac{2^0 e^{-2}}{0!}$$

Recognizing that any number to the zero power equals one, that is, that $2^0 = 1$ and also that $0! = 1$, simplifies the hand calculation of this particular problem. In addition, if the $e^{-2}$ is moved to the denominator, it becomes $e^{+2}$, so

$$P(X = 0 | \mu = 2.0) = \frac{2^0}{0!e^2} = \frac{1}{1(2.718)^2} = 0.13534$$

---

Although Poisson probabilities are much simpler than binomial probabilities to calculate, the use of tables can sometimes make Poisson computations easier. Table 6 in Appendix B shows some values of $\mu$ and $e^{-\mu}$ that can be used. For example, to compute $P(X = 0 | \mu = 2.0)$ using Table 6, for $\mu = 2.0$, we find $e^{-2} = 0.13534$ and use that directly in the equation. Example 5.10 utilizes table values of $e^{-\mu}$ and asks for the determination of additional probabilities.

---

**EXAMPLE 5.10**

An automatic welding operation in the manufacture of copper tubing typically generates 2.0 minor defects (air pockets) per 1000 foot of tubing. What is the probability that a shipment of 1000 foot of copper tubing to a distributor in Maine will have (a) no defects, (b) one defect, (c) two or more defects? (d) Does the probability of two or fewer defects suggest that the probability distribution of defects can be satisfactorily described by a Poisson process?

*Solution*    We are given $\mu = 2.0$ for $X$ = the number of minor defects in copper tubing that satisfies a Poisson process. Then $P(X) = (2^x e^{-2})/(X!)$ gives the distribution shown in Figure 5.4.

**FIGURE 5.4**  The Poisson Probability Distribution for $\mu = 2.0$

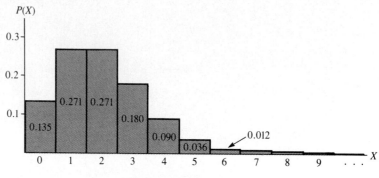

Number of Air Pockets (events)

a.  $P(X = 0) = \dfrac{(2.0)^0 e^{-2}}{0!} = \dfrac{1(0.13534)}{1} = 0.135$

b.  $P(X = 1) = \dfrac{(2.0)^1 e^{-2}}{1!} = \dfrac{2(0.13534)}{1} = 0.271$

c.  $P(X \geqslant 2) = 1 - [P(X \leqslant 1)] = 1 - [P(X = 0) + P(X = 1)]$
    $= 1 - [0.135 + 0.271] = 0.594$

d.  $[P(X = 0) + P(X = 1) + P(X = 2)] = 0.135 + 0.271 + 0.271 = 0.677$
    Yes.

Example 5.10 illustrates the high chance that "a very few events will occur over a short time (or space)," as required by the Poisson experiment. In fact, $P(X \leqslant 3) = 0.677 + 0.180 = 0.857$, so nearly 86% of all (1000 ft) sections of this tubing have three or fewer air pockets! This is consistent with the rare-event Poisson process.

**The Poisson Approximation to Binomial Probabilities**

The Poisson distribution is sometimes used to approximate the binomial when the probability of a success can be considered a rare event. This approximation is most suitable when $n$ is large (say, $n > 20$) and $\pi$ is small ($< 0.05$). It does, however, rely upon independent trials, and it assumes constant probabilities, as does the binomial. Example 5.11 illustrates the different results that we would obtain from using a theoretically correct binomial distribution versus a Poisson approximation.

**EXAMPLE 5.11**

In a precision casting operation, 10% of the finished products require rework. If 20 castings are completed and put into inventory without inspection, what is the probability that two of the castings will require rework? Solve the problem (a) via the exact binomial and (b) via the Poisson approximation.

*Solution*   a.  The binomial distribution gives

$$P(X = 2 \mid n = 20, \pi = 0.10) = 0.285 \quad \text{(Table 5, Appendix B)}$$

b. From the Poisson approximation, $\mu = n\pi = 20(0.10) = 2$. (In other words, 10% of 20, or 2, units are defective items, on the average.)

$$P(X = 2 \mid \mu = 2) = (2^2 \cdot 0.135)/2! = 0.270 \quad \text{(Table B.6)}$$

In Example 5.11 the Poisson approximation yields an answer within about 5% of the theoretically correct binomial, even though the value of $\pi$ was as large as 0.10 and $n$ was as small as 20. For a fixed $\pi$, larger values of $n$ would result in closer approximations.

Binomial tables often go up only to $n = 20$. Above that number it is relatively easy to find binomial probabilities using a computer. As an alternative, the Poisson distribution provides a reasonably good approximation of binomial probabilities as long as $n > 20$ and $\pi < 0.05$.

In sum, the Poisson is the appropriate distribution for rare events, that is, when the probability that the event will occur is very small compared to the large number of times that it could occur (which may be unknown). The event itself is discrete, although it can occur along some continuum. The Poisson distribution can describe the number of accounting errors per report, the number of times in a minute that a receptionist picks up a phone receiver, the number of accidents occurring during a 6-month work period, and many more "rare" situations. It is especially useful in quality control, maintenance, and safety analysis in industrial organizations.

**Problems 5.7**

33. In a lumbermill operation, unusable logs show up at the log-pond on a random basis that can be defined as a Poisson process and at a rate of about two every hour. Find the probability that five unusable logs show up during an hour when a buyer is on tour.

34. A glass company that manufactures "take-home" cola glasses produces an average of 3.4 bubbles (defects) per carton of glasses. The manager of a hamburger chain is considering an order for 100 cartons and wishes to estimate the number of cartons that will have four or more defects. Develop an estimate using the Poisson tables.

35. Use the Poisson distribution to find
    a. $P(X = 5 \mid \mu = 2.8)$        c. $P(X \geqslant 2 \mid \mu = 3.5)$
    b. $P(X = 0 \mid \mu = 4.0)$        d. $P(X = 4 \mid n = 200, \pi = 0.005)$

36. Of the books from a particular publisher, 1% have defective covers. Use the Poisson approximation to find the probability that exactly 10 have defective covers in a lot of 1000.

**SUMMARY**

Chapter 5 initiates our study of *random variables* and their *probability distribution*. A probability distribution can be characterized by its mean $(\mu)$ and standard deviation $(\sigma)$. Equations for these parameters are extensions of equations for frequency distribution calculations in Chapter 3. The probability distributions of *discrete random variables*, treated here, show all possible outcomes, along with the probability for each one. We have identified four principal discrete variable distributions and have computed some associated probabilities to answer business questions for each type.

**TABLE 5.5** Key Equations for Discrete Random Variable Probability Distributions

| Random Variable | Probability Description | Probability Calculation | |
|---|---|---|---|
| Uniform | Equal probability for each $X$ value. One trial is made. | $P(X) = \dfrac{1}{n}$ for $X = x_1, x_2, \ldots, x_n$ | (5.2) |
| Hypergeometric | Probability changes with each trial. Trials are dependent. | $P(X) = \dfrac{{}_aC_x \cdot {}_bC_{n-x}}{{}_{a+b}C_n}$ | (5.4) |
| | | $P(X) = \dfrac{\text{Successful ways}}{\text{Total ways}}$ | (5.3) |
| Binomial | Probability of a success is constant. Trials are independent. | $P(X) = {}_nC_x \pi^X (1 - \pi)^{n-x}$ | (5.5) |
| Poisson | Events occur at some low rate. $P$ (more than a few events) is near zero. Rare event situations. | $P(X) = \dfrac{\mu^X e^{-\mu}}{X!}$ | (5.11) |

| | Probability Distribution Descriptions | | |
|---|---|---|---|
| For all discrete distributions | Mean or expected value | $\mu = E(X) = \sum [X \cdot P(X)]$ | (5.1) |
| | Variance | $\sigma^2 = E(X - \mu)^2 = \sum [(X - \mu)^2 \cdot P(X)]$ | (5.7) |
| | Standard deviation | $\sigma = \sqrt{E(X - \mu)^2} = \sqrt{\sum [(X - \mu)^2 \cdot P(X)]}$ | (5.8) |
| For binomial distributions only | Mean or expected value | $\mu = n\pi$ | (5.6) |
| | Variance | $\sigma^2 = n\pi(1 - \pi)$ | (5.9) |
| | Standard deviation | $\sigma = \sqrt{n\pi(1 - \pi)}$ | (5.10) |

The *binomial* is probably the most widely used discrete random variable. Its use is appropriate when probabilities are constant from one trial to the next, outcomes are limited to two possibilities, and the trials are independent.

Rules appear throughout the chapter to assist you in distinguishing among the binomial, *hypergeometric*, and *Poisson* type situations. The binomial distribution can be approximated by the Poisson when $n > 20$ and $\pi < 0.05$. Tables of point probabilities are presented in Appendix B for binomial and Poisson distributions.

Table 5.5 summarizes the descriptions and equations for computing probabilities for variables that are distributed according to the most commonly encountered discrete distributions. Of these, we will find the most use for the binomial and Poisson distributions.

## KEY TERMS

Bernoulli process and Bernoulli trial, 177
binomial probability distribution
  mean, 184
   variance and standard deviation, 185
binomial random variable, 177
continuous random variable, 168
discrete random variable, 168
discrete uniform random variable, 172
expected value, 169
hypergeometric distribution, 174

hypergeometric random variable, 174
mean for discrete variable probability
  distributions, 183
Poisson distribution, 188
Poisson process, 188
probability distribution, 168
random variable, 167
variance and standard deviation for
  discrete variable distributions, 185

**CHAPTER 5
PROBLEMS**

37. The directors of a new hospital feel there are equal chances that the demand for beds within one year will be 180, 210, 240 or 270. What is the expected value of demand?

38. Empirical data on the number of female births out of groups of five births at St. Steven's Hospital revealed the probability distribution shown below. What is the expected value of female babies in the next five births?

| Number of Girls | $X$ | 0 | 1 | 2 | 3 | 4 | 5 | |
|---|---|---|---|---|---|---|---|---|
| Probability | $P(X)$ | 0.010 | 0.077 | 0.230 | 0.346 | 0.259 | 0.078 | 1.000 |

39. An individual has an opportunity to invest $20,000 with a brokerage firm and is assured that the investment will be made in one of three funds. The firm estimates it has a 30% chance of investing in fund $A$ and a 30% chance for fund $B$, each of which will return $28,000. Fund $C$ would return $32,000. All returns are before investment costs are paid. Compute the *net* expected gain from participating in this venture.

40. A company operating a chain of drugstores plans to open a store in one of two locations, A or B. If successful, location A will show an annual profit of $50,000; if not successful, an annual loss of $8000. If successful, location B is expected to show an annual profit of $55,000; if not successful, an annual loss of $20,000. If the probabilities of success are 3/8 at location A and 2/5 at location B, where should the company open the store? *Hint:* Choose the location that has larger expected profit.

41. Given: $P(X) = 1/5$, $X = -3, 0, 4, 5, 9$; compute $\mu$, $\sigma^2$, and $\sigma$. Find $P(X \leqslant 0)$. Let $X =$ the change in the price of a stock.

42. Given $P(X) = 1/5$, $X = -4, -2, 0, 1,$ and 2. Find
    a. $P(X > 0)$
    b. $P(X \neq 0)$
    c. $P(X < -2 \text{ or } X > 1)$
    d. $P(0 < X < 1)$

43. Given the probability distribution shown in the table, compute the (a) mean, (b) variance, (c) standard deviation.

| $X$ | 4 | 5 | 6 | 7 |
|---|---|---|---|---|
| $P(X)$ | 0.2 | 0.2 | 0.5 | 0.1 |

44. An employment office uses color-coded cards to identify job classifications. A file contains three white, three black, and four green cards. What is each probability, given that two cards are drawn?
    a. Without replacement, both will be green.
    b. With replacement, both will be green.
    c. Without replacement, both will have the same color.

45. A construction firm installing an elevator needs 12 good cable clamps for a job on the twentieth floor. In the past, approximately 10% of the clamps have not been

usable due to prior use. If the supervisor randomly takes 15 clamps (without inspection) from the stockroom, what is the probability he has enough good clamps to complete the installation?

46. A processor in Phoenix produces electronic chips, of which 30% are defective. They are then packed (5000 per box) and sold to an Ohio firm for $700 per box. If a sample of 15 chips is taken at random from one of the boxes, what is the probability that 2 chips will be defective?
    a. Solve the problem using the exact binomial distribution.
    b. Solve using the Poisson approximation.
    c. Why is the Poisson approximation inappropriate here?
    d. What are the values of the mean ($\mu$) and standard deviation ($\sigma$) of the binomial distribution?

47. Patients arrive at the emergency room of Freeway Hospital at an average of 2.8 per hour. The emergency room can accommodate a total of only 6 patients at any one time. Assuming independence of arrivals, what is the probability that more patients will arrive during the next hour than can be accommodated? (*Hint:* Use the Poisson distribution).

48. A new-car dealership finds that, on the average, the paint finish has about 2.3 blemishes per car. What is the probability that the next three cars will be completely free of blemishes? (*Hint:* Requires assumption of independence.)

49. A manufacturer ships unassembled packaging machinery to a distributor. The latest shipment includes 400 bolts, of which 398 are needed for assembly. If the manufacturing process is set so that the probability of a defective bolt is 0.005, what is the probability that any given shipment will *not* have enough good bolts for an assembly?

50. A small-business investment corporation is being formed in Chicago with enough capital to finance nine independent ventures. If the chance of being successful on any one is 25%, what are the corporation's chances of each of the following?
    a. Exactly one successful venture.
    b. At least one successful venture.
    c. Being unsuccessful in all nine ventures.

51. A student does not know the answer to four true–false questions and so must guess. In doing so he may get none, one, two, three, or all four correct.
    a. What is the random variable (in words)?
    b. List the probability distribution for this random variable in table form.
    c. Graph the probability distribution, being sure to label each axis.
    d. If two or more correct answers are required to pass this section of the test and 40 students guess at all questions, how many of them would you expect to pass this portion?

52. An appliance manufacturer estimates it costs $1.00 to correct each of the first two paint imperfections on a refrigerator manufacturing line. For any more than two, the appliance is repainted at a cost of $30.00. However, the mean number of imperfections is only 0.4 per refrigerator. Past data support the assumption that the number of imperfections follows a Poisson distribution. The firm has now received an order for 100 refrigerators and must estimate the cost to correct paint imperfections.
    a. Show the probability distribution of the number of paint imperfections.
    b. What (in words) is the random variable of interest?
    c. Graph the probability distribution. Be sure to label each axis.
    d. What is the estimated total cost to correct the paint imperfections?

**STATISTICS IN ACTION: CHALLENGE PROBLEMS**

53. For a simple coin-toss experiment—say, for three flips of a fair coin—what is the expected longest run of heads? All possible sequences for this simple experiment are HHH, HHT, HTH, THH, HTT, THT, TTH, TTT. There are eight possibilities. The question can be answered with a discrete probability distribution.

| Longest Heads Run | Probability |
|:---:|:---:|
| 0 | 1/8 |
| 1 | 4/8 |
| 2 | 2/8 |
| 3 | 1/8 |

The longest possible run is, of course, three heads. The expected value for the longest heads run distribution is 1: $(0 \cdot 1/8 + 1 \cdot 4/8 + 2 \cdot 2/8 + 3 \cdot 1/8 = 11/8$ or 1, the nearest possible whole number). But suppose the coin is not fair, with, say, $P(H) = 0.1$, so $P(T) = 0.9$. Then

| Longest Heads Run | | Probability |
|:---:|:---:|:---:|
| 0 | $P(TTT) =$ | 0.729 |
| 1 | | 0.243 |
| 2 | | |
| 3 | | |
| | Total | 1.000 |

In the article from which this is taken, Mark Schilling gives a more complicated usage.

SOURCE: Schilling, Mark F. "The Longest Run of Heads." *The College Mathematics Journal* 21, no. 3 (May 1990): 196–207.

Questions   Complete the table for a discrete probability distribution on the longest heads run with $P(H)$ set at 0.1. Use the table to calculate the expected value for this longest heads run distribution.

54. Imagine a production process for the manufacture of computer diskettes in which each batch is three diskettes (not real, but to use our easy calculations). The process is known to produce 10% defectives. If one or more (of three) diskettes are found defective, what would you conclude about the acceptability of the process?

55. Ghosh and McLafferty state, "The evaluation and choice of store locations is one of the most important decisions for retail managers. A good location provides strategic and often monopolistic advantages over competitors, while a poor location can be unprofitable and difficult to correct." These authors discuss the structure of a computer game—COMPLOC—that allows people to apply their knowledge to store

location models in a realistic scenario. We use the concepts to explore the development and use of discrete probability distributions. Some essential notation follows:

$P_{ij}$ = the probability of any individual ($i$) choosing a particular retail store ($j$)—a store location preference

$S_j$ = the relative square feet of floor space for store $j$

$D_{ij}$ = the distance for individual $i$ to travel to store $j$

Assume three store locations, so $j = 1, 2, 3$. Then the formula for the preferred store location, $j$, for an individual $i = 1$

$$P_{1j} = \frac{(S_j/(D_{1j})}{\sum_{j=1}^{3} (S_j/(D_{1j})} \quad \text{for } i = 1, \text{ all individuals in area 1.}$$

*Note:* Usage assumes that all individuals in an area, say a subdivision, are the same distance from each store location.

SOURCE: Ghosh, Avijit, and Sara L. McLafferty. "COMPLOC: A Competitive Store Location Game." *Journal of Marketing Education* (Fall 1984): 38–45.

Questions    Use the following data to establish a probability distribution for individuals in area 1 ($i = 1$) for choosing a retail store location, ($j = 1, 2,$ or 3).

| Store Location, $j$ | Relative Size, $S_j$ | Distance (mi), $D_{1j}$ |
|:---:|:---:|:---:|
| 1 | 1 | 2 |
| 2 | 2 | 2 |
| 3 | 0.5 | 1 |

Use these data to construct a probability distribution for $P_{1j}$. Then give the distribution mode value and its practical meaning.

56. The individuals $i = 2$ for area 2—another subdivision—could shop at this location. Relative to area 2, the statistics are unchanged (same store-location options and relative sizes), except $D_1 = 3$, $D_2 = 3$, $D_3 = 3$.
   a. Based on the earlier criterion of highest $P_{ij}$, what is the best choice of store location for the individuals in area 2? *Suggestion:* Begin by making a new probability distribution.
   b. Since area ($i =$) 1 has two times the population of area ($i =$) 2, assign the weights 2 and 1, respectively. What is the best location ($j = 1, 2,$ or 3) to service both areas? Use the criterion $P_j = \text{wt } 1 \cdot P_{1j} + \text{wt } 2 \cdot P_{2j}$.

57. Friedman and Krausz discuss a use of expected values in financial decision areas of product elimination. Products that produce low income can have negative effects on profits and may require excessive use of managerial expertise. These researchers have used financial portfolio theory to develop an approach for determining which products to eliminate. Assume that if a product is eliminated, its resources are evenly allocated among the remaining products. Three tables are presented.

**TABLE 1**   Rate of Return for Each Product

| Scenario | Prob. | A | B | C | D |
|---|---|---|---|---|---|
| Boom | 0.10 | 20% | 4% | 10% | 5% |
| Recovery | 0.20 | 17 | 6 | 7 | 5 |
| Normal | 0.40 | 12 | 8 | 6 | 5 |
| Recession | 0.20 | 6 | 10 | 5 | 5 |
| Depression | 0.10 | −2 | 20 | −4 | 4 |

**TABLE 2**   Mean, Variance, and Standard Deviation on Rate of Return for Each Product

| | A | B | C | D |
|---|---|---|---|---|
| Mean | 11.20% | 8.80% | 5.40% | 4.90% |
| Variance | 37.56 | 16.96 | 11.64 | 0.09 |
| Standard Deviation | 6.13 | 4.12 | 3.41 | 0.30 |

**TABLE 3**   Weights and Portfolio Statistics

| | Weights ($w_i$) | | | | | | | |
|---|---|---|---|---|---|---|---|---|
| Portfolio | A | B | C | D | Mean | Var. | $\sigma$ | C.V. |
| 1 | 0.25 | 0.25 | 0.25 | 0.25 | 7.58% | 1.94% | 1.39% | 0.18 |
| 2 | 0.33 | 0.33 | 0.33 | 0 | 8.38 | 3.12 | 1.77 | 0.21 |
| 3 | 0.33 | 0.33 | 0 | 0.33 | 8.22 | 0.82 | 0.91 | 0.11 |
| 4 | 0.33 | 0 | 0.33 | 0.33 | 7.10 | 9.98 | 3.16 | 0.45 |
| 5 | 0 | 0.33 | 0.33 | 0.33 | 6.30 | 0.05 | 0.23 | 0.04 |

Table 1 gives scenarios, probabilities, and rate-of-return distributions for four products—A, B, C, and D. Table 2 gives the mean, variance, and standard deviation for the rate of return on each product. Table 3 includes five portfolios (assignments of weights to the four products) and the resulting statistics for overall return rate.

SOURCE: Friedman, Hershey H., and Joshua Krausz. "A Portfolio Theory Approach to Solve the Product Elimination Problem." *The Mid-Atlantic Journal of Business* 24, no. 2 (Summer 1986): 43–48.

Questions    Which product should be considered first for elimination? Table 2 lacks information to answer the question, but it points to products C and D as candidates. In Table 3 we will assume that all mean percents are at an acceptable level. Consider the criterion for elimination: Use that portfolio with the highest mean. Then the choice, portfolio 3, would eliminate product C by virtue of its zero "weight." Indicate which product—A, B, C, or D—would be eliminated under each of these criteria:
a. The lowest variance
b. The lowest standard deviation
c. The lowest C.V.

58. a. How do the authors compute the mean values that appear in Table 3 of Problem 57? Here is a start for computing $E(r_1) = \sum_{i=1}^{n} w_i r_i$. Then

$$\text{Portfolio 1 mean} = P(\text{Boom}) \cdot E(r_1 | \text{Boom})$$
$$+ P(\text{Recovery}) \cdot E(r_1 | \text{Recovery})$$
$$+ \cdots + P(\text{Depression}) \cdot E(r_1 | \text{Depression})$$

Complete the table to show that the mean for Portfolio 1 is 7.58%. Use this procedure to show that the Portfolio 2 mean is 8.38%, and that for Portfolio 4 is 7.10%.

| Scenario | Probability | A | B | C | D | Mean Calculation |
|---|---|---|---|---|---|---|
| Boom | 0.10 | 20% | 4% | 10% | 5% | $0.10[(20 \cdot 0.25) + \cdots + (5 \cdot 0.25)] = ?$ |
| Recovery | 0.20 | | | | | $0.20[(17 \cdot 0.25) + \cdots + (5 \cdot 0.25)] = ?$ |
| Normal | 0.40 | | ⋮ | | | ⋮ |
| Recession | 0.20 | | | | | |
| Depression | 0.10 | −2% | 20% | −4% | 4% | $0.10[(-2 \cdot 0.25) + \cdots + (4 \cdot 0.25)] = ?$ |
| | | | | | | Portfolio 1 mean = ? |

b. From Table 3, considering values for the mean and $\sigma$, one could argue that Portfolio 3 is best, and so eliminate product C. Explain the basis for this argument.

## BUSINESS CASE

*Random ≠ Unknown*

It started out as just another promotional campaign for Market Resources, Inc. (MR Inc.) They had been engaged to design and conduct a national sweepstakes event for Sendit Publishers, a magazine clearinghouse located in Philadelphia. Sandy Holt had masterminded the campaign for MR Inc. and, to all appearances, it was a huge success. She had already reached the goal of attracting 200,000 new subscriptions for Sendit, and some late arrivals were still uncounted. Then an unexpected problem arose.

The problem was with a competing clearinghouse, Subscriber's Service. Subscriber's held a market advantage in the far West and had always fought vigorously to keep it. As soon as Sendit's sweepstakes winners were announced, Subscriber's Service issued a news release stating that it was preparing to file a lawsuit against MR Inc. and Sendit Publishers for deceptive marketing practices.

The situation boiled down to this. MR Inc.'s campaign had been advertised as a "national" sweepstakes that subscribers from *all* states had equal chances of winning. The ten semifinalists were from ten different states: Alabama, California, Florida, Hawaii, Illinois, Maryland, New Jersey, Oregon, South Carolina, and Wisconsin. Subscriber's service's decision to file suit was prompted by the announcement of the two top prizes. The first prize winner was from California, and second prize went to a new subscriber in Hawaii. These two states, along with Washington and Oregon, were the states where Subscriber's Service had the largest market share and where publication of the sweepstakes results was bound to affect its business. Subscriber's Service charged that the final selection was "deliberately determined to injure us in an unfair and deceptive manner." The chance of selecting two western states out of ten semifinalists, they charged, was "virtually nil."

Sandy Holt thought she had followed legitimate procedures in selecting both the semifinalists and the winners, but she now had to persuade her company's attorney, Jonas Miller, and prepare him to fight the suit. So she walked into Jonas's office loaded with files from the campaign.

| Ms. H. | I guess we're in for a good one this time, eh Jonas? |
|---|---|
| Mr. M. | Why? What do you mean, Sandy? |
| Ms. H. | That Subscriber's outfit is trying to pull us into their battle with Sendit. Just because they can't get a good campaign going themselves. Now I suppose we're in for a long battle. |
| Mr. M. | Not at all! I talked to their attorney today. They'll be dropping the suit tomorrow. |
| Ms. H. | Dropping it! What do you mean? |
| Mr. M. | I mean dropping it. After I talked to Marlene I told them they didn't have a leg to stand on. Didn't she tell you? |
| Ms. H. | Marlene? I just hired her two weeks ago. What does she have to do with this? |
| Mr. M. | Well, I called for you yesterday and you were out, so I told her about the case. She looked up the file for me and then said she wanted to do a few quick calculations. So she did, adding a few new words—like *combinatorial* and *hypergeometric*—to my legal vocabulary. She said the chances against getting those two states were about 14 to 1 and it sounded like she knew what she was talking about. Anyway, I asked her to stick around while I called Subscriber's attorney. By the time we threw some facts and figures at him, he was ready to back down. I told him 14 to 1 was nothing like "nil" or 1000 to 1, and that we had full documentation on the whole procedure. |
| Ms. H. | And he took it? |
| Mr. M. | You bet he took it! I told him he was lucky we didn't file a countersuit to recover damages to our name! He agreed. Next time I think he'll be a little more cautious. Marlene seems to be right on top of things, doesn't she? |
| Ms. H. | She's had an outstanding recommendation—and she seems to be good at identifying the key issues in a problem. |
| Mr. M. | And not too bad at solving them! Once she came up with the number of possible combinations and said it was just a matter of defining a probability distribution, I felt like we were on solid ground. You know, I always thought of something random as being unable to pin down—I mean, not really knowable. |
| Ms. H. | You've changed your mind? |
| Mr. M. | According to Marlene, just because something is random doesn't mean it's uncaused or unknowable. We just don't usually bother to figure it out—or maybe can't get enough information to figure it out. But it sure is nice to have people around who understand a random process. |
| Ms. H. | I agree. Let's hope she stays a long while. |

**Business Case Questions**

59. Let $X$ = the number of western states represented in the Sendit Publisher Sweepstakes semifinals. The experiment was to select two top-prize winners from ten semifinalists. What is the probability distribution for this discrete random variable $X$?

60. Confirm the calculation of the probability of 1/15 that the two top prizes would go to semifinalists from western states (California, Hawaii).

## COMPUTER APPLICATION

*Binomial Probabilities*

This extends a general marketing problem that was introduced in Computer Applications of Chapters 2 and 4. This application involves calculating some binomial probabilities and interpreting their meanings. A review of several common statistics software packages indicates that not all include routines for calculating binomial probabilities, so we have provided some commands for a brief routine that appears in the middle of the application.

For marketing home computers it could be important to plan the advertising for the highest possible appeal. The target population is defined as households with 3+ persons and with householders aged 35–50 years. Now the advertising manager asks, Do you want the advertising directed only to a male audience, or should it also include an appeal for women?

Census data demonstrate that women constitute 25% of all heads of household, and that the percentage is growing. This (25%) is a substantial percentage, so the question is timely. But how interested are female heads of household in home computers? Management has decided to survey a random sample of households, for both female and male heads, to get the views of both sexes. Because they seek a fair sample but have a fixed budget—enough dollars for 1000 surveys—the next question was how to assure a reasonable selection of female heads of household in the sample. An initial suggestion was to sample 10 households in each of 100 markets in hopes of reaching a reasonable number of households with a female head. If a response is sought from at least three females in each market, is the sample of 10 households/market reasonable?

We don't need a computer to find the expected value, as $X$ = the number of female heads of household is binomial and $n = 10$, $\pi = 0.25$ are given. Here $\mu = n\pi = 10 \cdot 0.25 = 2.5$ female heads of household. On the average, this procedure gives 2.5 female heads of household per 10 households. Then what is the chance for 3 or more? This question can be answered by computer, and (as you may suspect) the procedure can be used to evaluate other alternative sampling schemes. By the binomial probability rule (Equation (5.5)),

$$P(X \mid n, \pi) = {}_nC_X\pi^X(1 - \pi)^{n - X}$$

For $n = 10$ households, the question requires $P(X \geqslant 3) = P(X = 3$ or 4 or ... or 10), which also equals $1 - \{P(X = 0) + P(X = 1) + P(X = 2)\}$, the latter being a shorter calculation. By a computer solution, this sampling plan gives a 47%+ chance for obtaining three or more female heads of household for (10 responders) each market.

For all of that work we could have found the answer more quickly in Table 5, Appendix B (and we suggest that you do just that to check our computer results). However, the manager for this project wants better than a 47% chance, for that is "like guessing" (near 50–50). So we provide her with other options, as shown in the table below. The decision was made to take 50 samples each of 20 households.

Several Sampling Alternatives, Assuming a Fixed
Total of 1000 Surveys

| Samples | Households ≠ Sample ($=n$) | $\pi$ | $\mu(=n\pi)$ | $P(X \geqslant 3)$ |
|---|---|---|---|---|
| 100 | 10 | 0.25 | 2.50 | 0.4744 |
| 50 | 20 | 0.25 | 5.00 | 0.9088 |
| 40 | 25 | 0.25 | 6.25 | 0.9680 |

This would assure three or more female heads of household for 90.88% of all possible samples. Then $50 \cdot 0.9088 = 45.44$, or 45 (of the 50 samples) are expected to provide 3+ female heads of household; the manager was satisfied.

The application of the binomial probability rule for $n = 20$ and $\pi = 0.25$ allows us to answer numerous questions. We could do a hand calculation using the binomial probability rule (Equation (5.5)); or use a table of binomial probabilities, like Table B.5; or use a computer. Although SPSS does not provide a tailored binomial probability rule, here is an SPSS routine to compute $P(X \geqslant 3) = 1 - \{P(X = 0) + P(X = 1) + P(X = 2)\}$, for $X$ binomially distributed with $n = 20$ and $\pi = 0.25$.

```
COMPUTE N = 20
COMPUTE PI = 0.25
COMPUTE P0 = 1 * (PI**0) * [(1 - PI)**N]
COMPUTE P1 = N * (PI**1) * [(1 - PI)**(N - 1)]
COMPUTE P2 = ((N**N - N)/2) * (PI**2) * [(1 - PI)**(N - 2)]
COMPUTE P3+ = 1 - {P(0) + P(1) + P(2)}
LIST VARIABLES = N PI P0 P1 P2 P3.
```

Then $P(X \geqslant 3) = P(3+) = 1 - \{0.0913\} = 0.9087$.

Minitab offers several options for binomial (and Poisson) probability calculations, including probability (mass) distribution function (PDF) tables and cumulative probability distribution function (CDF) tables, as illustrated in Figure 5.5.

**FIGURE 5.5**  Calculated Binomial Probabilities for $n = 20$, $\pi = 0.25$ by Minitab Probability (mass) Distribution Function (PDF) and from a Cumulative Distribution Function (CDF)

```
MTB > PDF 0;                                                              1
SUBC> BINOMIAL 20 0.25.
       K          P(X = K)                                               2
    0.00            0.0032
MTB > PDF;                                      MTB > CDF;
SUBC> BINOMIAL 20 0.25.                         SUBC> BINOMIAL 20 0.25.


  BINOMIAL WITH N = 20 P = 0.250000             BINOMIAL WITH N = 20  P = 0.250000
       K          P(X = K)                          K   P(X LESS OR = K)
       0            0.0032                           0       0.0032
       1            0.0211                           1       0.0243
       2            0.0669                           2       0.0913
       3            0.1339                           3       0.2252
       4            0.1897                           4       0.4148
       5            0.2023                           5       0.6172
       6            0.1686                           6       0.7858
       7            0.1124                           7       0.8982
       8            0.0609                           8       0.9591
       9            0.0271                           9       0.9861
      10            0.0099                          10       0.9961
      11            0.0030                          11       0.9991
      12            0.0008                          12       0.9998
      13            0.0002                          13       1.0000
      14            0.0000
```

1 The probability for a single value $X = 0$ results from the commands PDF 0 and BINOMIAL $(n =)$ 20 $(\pi =)$ 0.25. The value $P(X = 0) = 0.0032$ is the same as was obtained in the SPSS illustration as $P(0) = 0.0032$.

**2** The CDF table accumulates probabilities as

$$P(X = 0) = P(X = 0) = 0.0032$$

$$P(X \leqslant 1) = P(X = 0) + P(X = 1) = 0.0032 + 0.0211 = 0.0243$$

and so on. This allows a quick solution for the question

$$P(X \geqslant 3) = 1 - P(X \leqslant 2) = 1 - 0.0912 = 0.9088$$

The "point" probability function—the PDF—and the cumulative function—the CDF—are simply options for a solution. For example,

$$\text{by the PDF:} \quad P(X = 3 \text{ or } 4) = P(X = 3) + P(X = 4)$$
$$= 0.1339 + 0.1897 = 0.3236$$

$$\text{by the CDF:} \quad P(X = 3 \text{ or } 4) = P(X \leqslant 4) - P(X \leqslant 2)$$
$$= 0.4148 - 0.0913 = 0.3235$$

Incidentally, the reason why the tables in Figure 5.5 display values through only $K = 14$ and $K = 13$, respectively, is that for $n = 20$, $\pi = 0.25$, the chance for 14 or more successes is zero (to four decimal places).

This application demonstrates that it is not practical to have binomial tables available for every possible application. Moreover, this is a situation for which the Poisson approximation to a binomial distribution is not appropriate (as it defies $\pi < 0.05$), and a computer procedure can give the desired answers. We will learn in the next chapter how to use the normal distribution to approximate binomial probabilities. Nonetheless, the computer approach provides an exact answer for binomial probabilities. It can be more convenient to use a computer, especially when the value of $\pi$ is not given in a binomial probabilities table (like $\pi = 0.33$), or when $n$ exceeds 25. Several of the Computer Problems invite a computer calculation for binomial probabilities.

## COMPUTER PROBLEMS

*Data Base Applications*

The following problems are intended for computer solution. Some draw on data from the data base. A few key statements are given for specific problems in which some programming is required. Or you might want to visit your computer center for some programming assistance.

61. Use the 100-record sample in Appendix A, Table 5 in Appendix B, and a SAMPLE routine or other random numbers procedure to select a random sample of size 20 (from the 100-record PUMS file). Be sure *not* to allow any repeats of sample records. Perform a FREQUENCIES count for column 42—sex of head of household. How many females (code = 1) appeared in your sample? Check with your classmates about how many persons had three or more female heads of household in their sample. How close is this to the probability in the computer application?

Using the 500-record PUMS file and a Minitab routine, we obtained the following 20 samples, where $(X =) 1$ is the code for a female head of household, so $(X =) 0$ denotes a male head of household.

```
MTB > READ 'B:SMPL5.PUM' C1
SUBC> FORMAT (41X,F1.0).
MTB > NAME C1 'SEX1'
MTB > SAMPLE 20 'SEX1' C2
MTB > PRINT C2

C2

0 0 1 1 0 0 0 0 1 0 0 1 0 0 0 0 1 0 0 0
```

Repeating the SAMPLE and PRINT statements nine more times gives

```
sample 2:  0 0 0 1 1 0 0 0 1 0 0 1 0 0 1 1 0 0 0 1
sample 3:  0 0 0 0 0 0 0 0 0 0 0 0 1 0 0 0 0 0 0
sample 4:  1 0 0 0 0 1 0 0 0 1 0 0 1 0 0 0 1 0 0 0
sample 5:  0 0 0 1 0 0 0 0 0 0 0 0 0 0 0 0 0 1 1 0
sample 6:  0 0 0 0 0 0 0 0 1 1 1 0 0 0 1 1 1 0 0
sample 7:  0 1 1 1 0 0 1 0 1 0 0 0 0 0 0 0 0 1 0 1
sample 8:  1 1 1 0 0 0 0 1 0 0 0 0 0 0 0 1 0 1 0 0
sample 9:  0 0 1 1 0 0 0 0 0 0 0 0 0 1 0 0 1 1 0 0
sample10:  0 1 0 0 0 0 0 1 1 0 0 0 0 0 0 0 1 0 0 1 0
```

How close is this to the probability for $P(X = 3 | n = 20, \pi = 0.25)$ that was given in the computer application?

   If you have Minitab available, then repeat the above procedure to determine for yourself $P(X = 3 | n = 20, \pi = 0.25)$. Or you might use Minitab to select two samples, ask four classmates to do the same, and then combine all to determine how many of the ten samples had three or more female heads of household.

*Additional Computer Problems*   62. A common computer software function provides the ability to generate various distributions on random numbers, such as a uniform distribution. We can check the results of a procedure that offers random numbers that display the discrete uniform probability distribution. (This provides a fairly simple application to develop a discrete random variable distribution on a computer.) The following routine, or a modification for your software, should provide a list of numbers with values between 0 and 9. By counting the frequencies and preparing a probability distribution, check whether the numbers satisfy (in your judgment) a discrete uniform distribution, that is, such that $P(X) = 1/10$, $X = 0, 1, 2, \ldots, 9$. The routine is SPSS.

```
NUMBERED
TITLE          UNIFORM RANDOM NUMBERS
INPUT PROGRAM
LOOP #I=1 TO 500
COMPUTE RNDNO=TRUNC(UNIFORM (9.99))
END CASE
END LOOP
END FILE
END INPUT PROGRAM
PRINT/RNDNO
FREQUENCIES VARIABLES=RNDNO/
               STATISTICS=ALL/
```

With Minitab use

```
MTB > RANDOM 500 C1;
SUBC> INTEGER 0 9.
MTB > PRINT C1
```

63. The Minitab system provides simple procedures for computing binomial and Poisson distribution probabilities. If you have access to Minitab, do the following:
   a. Use the statements

   ```
   MTB > PDF;
   SUBC> Binomial n = 5 p = 0.6.
   ```

   to check the probabilities listed in Table 5.2.
   b. Use the statements

   ```
   MTB > PDF;
   SUBC> Poisson Mean = 2.0.
   ```

   to check the Poisson probabilities displayed in Figure 5.4.

   Some computer software does not provide preprogrammed routines to calculate probabilities for binomial, hypergeometric, Poisson and other discrete random variable distributions. Here are some (SPSS) examples for the hypergeometric and Poisson random variables.

64. At the Drovo Co., water meters that have been inspected are inventoried on blue cards, and those not yet inspected are listed on red cards. At present there are 52 cards, and half are red. For two cards picked at random, what is the probability that one card is red and one is blue? The SPSS program segment is

   ```
   INPUT PROGRAM
   COMPUTE COMBN1 = 26/1
   COMPUTE COMBN2 = (52 * 51)/(2 * 1)
   COMPUTE HYPERGEO =
   (COMBN1 * COMBN1)/COMBN2
   END FILE
   END INPUT PROGRAM
   PRINT /COMBN1 COMBN2 HYPERGEO
   EXECUTE
   ```

65. A glass company that manufactures take-home cola glasses produces an average of 3.4 bubbles (defects) per carton of glasses.
   a. What is the probability that any carton contains 0 or 1 bubbles?
   b. For an order of 100 cartons, how many are expected to have 0 or 1 bubbles? *Suggestion:* Since $P(X|3.4) = 3.4^X e^{-3.4}/X!$, use the following program (SPSS) for part (a):

   ```
   TITLE             POISSON PROBABILITY
                        CALCULATION
   INPUT PROGRAM
   COMPUTE NUM1 = ((3.4**0) * (2.71828** - 3.4))
   COMPUTE NUM2 = ((3.4**1) * (2.71828** - 3.4))
   ```

```
COMPUTE DENOM = 1
COMPUTE POISS1 = NUM1/DENOM
COMPUTE POISS2 = NUM2/DENOM
COMPUTE POISSON = POISS1 + POISS2
PRINT POISSON POISS1 POISS2
END FILE
END INPUT PROGRAM
EXECUTE
```

With Minitab, use

```
MTB > PDF;
SUBC> POISSON 3.4.
```

## FOR FURTHER READING

[1] Albright, S. Christian. *Statistics for Business and Economics.* New York: Macmillan, 1987.

[2] Daniel, Wayne W., and James C. Terrell. *Business Statistics.* 4th ed. Boston: Houghton Mifflin, 1986.

[3] Hamburg, Morris. *Statistical Analysis for Decision Making.* 5th ed. Orlando, FL: Harcourt Brace Jovanovich, 1991.

[4] Pearson, E. S., and H. O. Hartley. *Biometrika Tables for Statisticians.* 3rd ed., I, Cambridge, England: Lubrecht and Cramer, 1976.

[5] Ryan, Barbara F., Brian L. Joiner, and Thomas A. Ryan, Jr. *Minitab Handbook,* 2nd ed. Boston: PWS-Kent, 1985.

[6] Snedecor, George W., and William G. Cochran. *Statistical Methods.* 7th ed. Ames, IA: Iowa State University Press, 1980.

# CONTINUOUS PROBABILITY DISTRIBUTIONS

## 6.1 INTRODUCTION

In Chapter 5 we identified two classes of random variables: discrete and continuous. Chapter 5 focused upon the discrete variables, which have a countable number of possible values, and we used counting techniques to evaluate individual point probabilities. In this chapter we address continuous random variables.

**Continuous Variables**

One business application that involves a continuous variable is the setting of time standards for employees to do assigned tasks. For example, suppose a production analyst must set a time standard for a solar panel assembly operation. Suppose further that the standard should be set at a level such that 95% of all employees can meet or better it. Figure 6.1 is a continuous function describing some sample assembly times. The mean time (and mode) here is $\mu = 5$ minutes. Because time is a continuous variable, the level of accuracy of the standard depends upon how many decimal places are used to define the time variable. A value in minutes such that 95% of the area under the curve lies *below* the value should be chosen. Leaving the details of the calculations until later, we can simply observe the percents in the following table.

| If the Standard Is | The Area Below the Curve Is |
|---|---|
| 6 minutes | 84.13% |
| 6.6 minutes | 94.52% |
| 6.64 minutes | 94.95% |
| 6.645 minutes | 95.00+% |

**FIGURE 6.1** A Probability Distribution for a Continuous Random Variable (X)

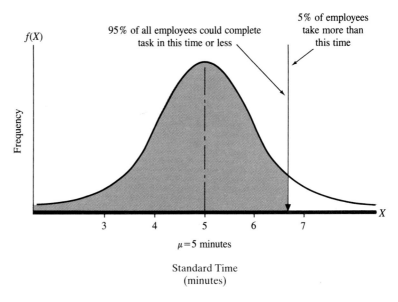

Theoretically, the shaded area in Figure 6.1 could describe an infinite number of possible values, depending upon how precisely the time variable can be recorded. A standard of 6.645 minutes would correspond to the most precise interpretation of Figure 6.1. However, the analyst must select a practical standard that can be measured by reasonably precise instruments and that is appropriate to the problem under investigation. When it is meaningful to differentiate among fractional values, it is common to use decimal numbers with three- or four-place accuracy. This is in contrast to discrete variables, which are specified as whole numbers or as a few well-rounded values.

For discrete variables, accuracy beyond a certain level has little or no effect on probability distributions. To illustrate, it serves just as well to say that a car dealership sold 6 cars last week as to say that it sold 6.0 or 6.00. The probability of selling 6 cars is the same as the probability of selling 6.0 or 6.00.

For continuous variables, however, the level of accuracy does affect the probability distributions. Figure 6.2 shows what happens to the graphs of discrete and continuous variables when successive measurements become increasingly precise. For the discrete variable in Figure 6.2(a), the probability distribution is unchanged as the level of accuracy is increased. For the continuous variable in Figure 6.2(b), the graph of the distribution changes perceptibly as the variable is recorded to higher levels of precision. Thus a teller who services customers at a rate of 1.5 customers per minute serves a greater number of customers than one who serves customers at a rate of 1.4 per minute. The former can serve 90 customers in an hour; the latter, only 84. In this case, one-decimal-place accuracy was actually required for a reasonably clear approximation of the probability distribution. Extension to two-place accuracy sharpens the picture even more, although it does not alter the general distribution pattern.

This chapter deals with continuous variable distributions—and primarily with the normal distribution. We begin with a distribution that lays the foundation for probability calculations for all continuous variable distributions—the continuous uniform distribution. Then we show how to compute probabilities

**FIGURE 6.2**   The Effect of Increased Precision of Measurement on Discrete and
Continuous Random Variable Distributions

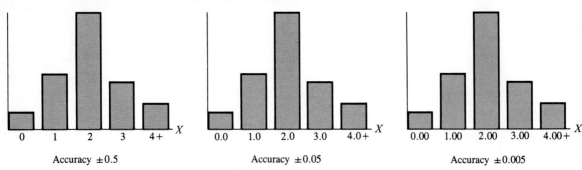

(a)  A discrete variable: *X* describes the number of cars sold

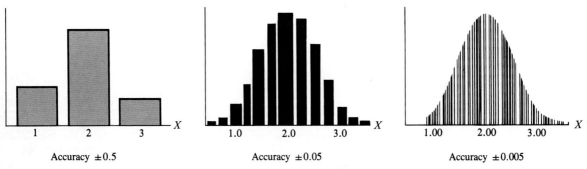

(b)  A continuous variable: *X* describes a customer service rate (customer/min) by a bank teller

using the normal distribution. Section 6.5 concludes the chapter with a discussion of the normal approximation to binomial probabilities.

Upon completing Chapter 6, you should be able to

1. state the conditions required for a continuous probability distribution;
2. compute the mean and variance and find probabilities for any continuous uniform distribution;
3. convert normal random variable values to standard normal deviation ($Z$) scores;
4. compute areas (probabilities) using the standard normal distribution;
5. use the normal approximation to binomial probabilities where it is appropriate.

## 6.2 PROPERTIES OF CONTINUOUS VARIABLE DISTRIBUTIONS

The graph of a continuous (variable) probability distribution can take many shapes. One shape is illustrated by the bell-shaped curve in Figure 6.3. Areas under this curve are representative of the probabilities of a normally distributed random variable $X$. This is because the area between any two points (such as from *a* to *b* on Figure 6.3) depicts the percentage of time when $X$ takes a value between the two points.

**FIGURE 6.3**   The Normal Probability (Density) Curve

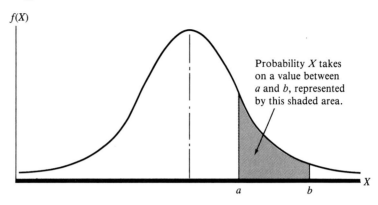

We sometimes refer to the curve of a continuous probability function as a *probability density curve*, or a *probability distribution*.

> The probability density curve, or probability distribution, for a continuous random variable, $X$, is a smooth curve that describes the probability, or area, for any interval of $X$-values.

The total area under the curve must equal 1 (1.00) for it to describe a total probability distribution.

There are other continuous variable distributions besides the bell-shaped curve shown in Figure 6.3. Three examples appear in Figure 6.4. These distributions are also characterized by smooth unbroken curves, which let us find probabilities by making calculations of the area under the curves.

For all continuous variable probability distributions:

1. The probability over any interval must be equal to or greater than zero.
2. The total probability (area) is 1. That is, the area over the range of all possible values, bounded beneath a probability density curve, is 1.

These properties are analogous to the requirements for a discrete variable distribution as defined in Section 5.2.

**The Continuous Uniform Random Variable**

We saw earlier that the discrete uniform random variable has one of the simplest discrete probability distributions. Its continuous variable counterpart is equally simple. The continuous uniform curve is flat. Consequently, finding probabilities is simply a matter of finding the areas of rectangles. Because it is easy to understand, we use the continuous uniform distribution to explain the essential characteristics of *all* continuous probability distributions, including the *mean*, $\mu$; the *standard deviation*, $\sigma$; and the *probability density function*, $f(X)$.

> A probability density function is a rule that can be used to determine how the probability (area) is distributed over intervals defined on the values of the continuous random variable.

**FIGURE 6.4**    Three Continuous Variable Distributions

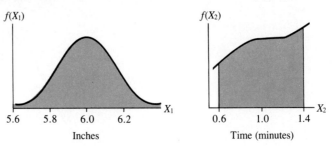

(a) $X_1$: Diameters for plastic spools
    formed on a molding machine

(b) $X_2$: The learning rate with artificial
    intelligence (computer) systems

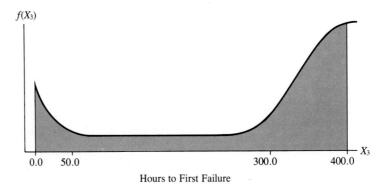

(c) $X_3$: A distribution for describing machine malfunctions

Simply put, a probability density function is a rule that describes a probability density curve.*

The uniform variable probability density function is

$$f(X) = \frac{1}{(b - a)} \quad \text{for } a < X < b, \quad \text{where } b < a \tag{6.1}$$

This equation simply expresses the height of the probability (density) function—$f(X)$—as the ratio of 1 to the range of possible values of $X$. In words, it represents the probability of any subinterval within the range of the $X$-variable. The qualifying restriction $a < X < b$ requires that the variable $X$ lie between $a$ and $b$; otherwise $f(X) = 0$.

A graph of the uniform probability function shows that the distribution is flat throughout the interval from $a$ to $b$ (see Figure 6.5).

That is, for a *continuous uniform random variable*, $X$, the area is the same for subintervals of *equal length* everywhere between point $a$ and point $b$, where $a < X < b$. Otherwise the probability (area) is zero.

> The continuous uniform random variable assigns equal probability
> to intervals of equal length.

---

\* Continuous probability functions will be designated by $f(X)$ rather than $P(X)$, which is used for discrete variable descriptions.

**FIGURE 6.5**   The Continuous Uniform Distribution for $X$ When $b > a$

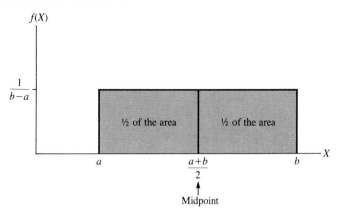

For a continuous uniform variable, probability calculations involve the area within a rectangle of height $1/(b - a)$ over the interval $(a,b)$. The total area gives total probability of 1.0.

$$\left(\frac{1}{b - a}\right) \cdot (b - a) = 1$$

Recall that we determine the area of a rectangle by multiplying its height times its width. This same procedure is used to find the area (and hence the probability) over any interval within $a < X < b$. For example, in Figure 6.5 the area over the interval from $a$ to $(a + b)/2$ is height times width:

$$\left(\frac{1}{b - a}\right) \cdot \left(\frac{a + b}{2} - a\right) = \left(\frac{1}{b - a}\right) \cdot \left(\frac{b - a}{2}\right) = \frac{1}{2}$$

This is reasonable because $(a + b)/2$ locates the median, so that half of all possible values lie within the interval from $a$ to $(a + b)/2$.

Other intervals *within* $(a,b)$ are evaluated in a similar manner. Yet for any interval entirely to the left of point $a$ or entirely to the right of point $b$, there is zero area and, consequently, zero probability. Example 6.1 illustrates these properties for one continuous uniform distribution. These probabilities are just area of rectangles. The values of $a$ and $b$ define the width of the interval of interest. The height, a constant, is given in Equation (6.1) as $1/(b - a)$.

| **EXAMPLE 6.1** | The departure times for trains from New York to Boston are uniform over a four-minute interval, from departure on time to departure four minutes late. The traffic manager who is responsible for maintaining the schedule may be interested in questions such as, Will departure be on time? What are the chances a train will leave over two minutes late? To provide a technical answer to these questions requires a probability structure. The height of the continuous uniform distribution over interval $a = 0.0$ to $b = 4.0$ is |
| --- | --- |

$$\frac{1}{(b - a)} = \frac{1}{(4.0 - 0.0)} = \frac{1}{4}$$

**FIGURE 6.6**   The Continuous Uniform Distribution for $a = 0.0$, $b = 4.0$ Minutes, and $X$ Is Train Departure Time

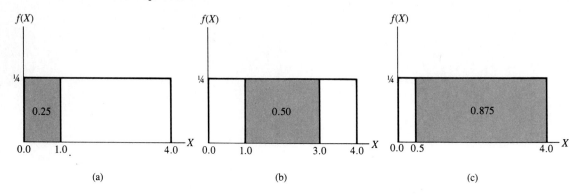

(a)                          (b)                          (c)

The probability distribution is displayed in Figure 6.6.

a. Find $P(X < 1.0)$, the probability that a train will leave less than one minute beyond the scheduled departure time.
b. Find $P(1.0 < X < 3.0)$.
c. Find $P(X > 0.5)$.
d. Find $P(X < 4.0)$.

*Solution*   The desired probabilities are simply areas under the continuous graph of height $f(X) = 1/4$. Multiplying the height times the width gives the answers.

a. $P(X < 1.0) = \dfrac{1}{4}(1.0 - 0) = \dfrac{1}{4}$, or 0.25

b. $P(1.0 < X < 3.0) = \dfrac{3.0 - 1.0}{4} = \dfrac{2}{4}$, or 0.5

c. $P(X > 0.50) = \dfrac{4.0 - 0.50}{4} = \dfrac{3.50}{4} = \dfrac{7}{8}$, or 0.875

d. $P(X < 4.0) = \dfrac{4.0 - 0.0}{4} = 1.0$

**Point Probabilities for Continuous Variables**

We can use Example 6.1 to illustrate a generalization that *the probability of a departure at any instant is zero.* That is, the probability is zero that any time point on Figure 6.6 represents the departure time. How can this be, since all departures are between 0 and 4.0 minutes? This poses a seeming contradiction. The probability of a departure within 4 minutes or less is 1; for a departure from 0 to 2 minutes the probability is 1/2; for a departure from 1 to 2 minutes the probability is 1/4, and so on. Yet the probability for a departure at precisely 1.0 minutes is zero: $P(X = 1.0) = 0$. The time of departure, $X$, is a continuous variable. The key here is the relationship of probability to area.

Suppose the probability $P(X = 1.0)$ is based on computational accuracy of 0.1 minutes. Then 1.0 minutes would be recorded for any departure occurring between 0.95 and 1.05 minutes, giving area and hence probability of $(1.05 - 0.95)/2 = 0.1/2 = 0.05$. Yet with increased accuracy (to the nearest hundredth), the area (or probability) for a departure at 1.00 minutes is reduced

**FIGURE 6.7**   Different Levels of Accuracy in Stating Values of Continuous Variables

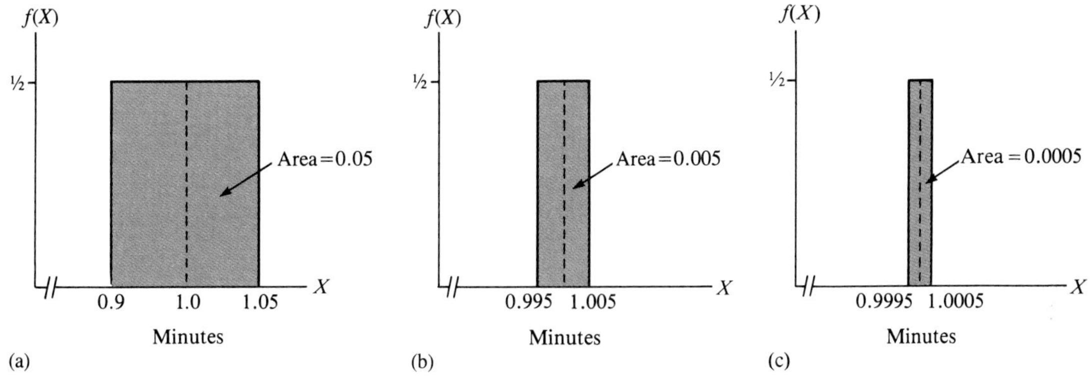

to $(1.005 - 0.995)/2 = 0.01/2 = 0.005$. With even greater accuracy the area is further reduced. In fact, as the interval width approaches zero, the area under the curve also approaches zero. So the probability approaches zero (see Figure 6.7).

This is logical because in the extreme we are trying to find the area of a line. A line, defined as the distance between two points, has only one dimension, length. Since a line has zero width, the area over any point is zero, and so the probability is zero. That is, the probability of departure occurring at any particular point in time is zero. For an interval of time, an area can be calculated, and so the probability is not zero. The conclusion that the probability is zero at any specific point holds true for *all* continuous variable probability distributions.

### DIFFERENCES BETWEEN CONTINUOUS AND DISCRETE VARIABLE PROBABILITY CALCULATIONS

For *discrete random variables* (like the binomial and Poisson), nonzero probability exists at specific points; hence the probability over any interval is the *sum* of probabilities for the points in the interval. The sum over *all* points gives total probability of 1.

For *continuous random variables*, nonzero probability exists only over intervals; the probability at any point is zero. Hence, the probability over an interval is the area defined on that interval by the continuous random variable probability distribution. The total area (and probability) equals 1.

The continuous uniform distribution is unusually simple. Yet it allows us to visualize probability calculations by equating probability to the area beneath a curve and defined over some interval on $X$-values. This concept is essential to dealing with the normal distribution later in this chapter.

**Describing the Continuous Uniform Probability Distribution**

Calculus is needed to derive formulas for the computation of mean, variance, and standard deviation values of many continuous probability distributions. However, without deriving these parameters, we can give expressions for the mean ($\mu$) and variance ($\sigma^2$) for the continuous uniform distribution.

**Mean and variance for any uniform continuous random variable distribution:**

$$\text{Mean} \qquad \mu = \frac{a + b}{2} \qquad\qquad (6.2)$$

$$\text{Variance} \quad \sigma^2 = \frac{(b - a)^2}{12} \qquad\qquad (6.3)$$

Thus for Example 6.1, the mean is

$$\mu = \frac{(0.0 + 4.0)}{2} = 2 \text{ minutes}$$

and the variance is

$$\sigma^2 = \frac{(4 - 0)^2}{12} = 1\tfrac{1}{3} \text{ minutes squared}$$

The mean and variance values are useful descriptors for the discrete uniform distribution, but they are not required for probability calculations. However, many continuous variable distributions require values for $\mu$ and $\sigma$ for the evaluation of probabilities.

---

**EXAMPLE 6.2**

The time in minutes per day that a machinist must wait for material from the stockroom is described by the continuous function $f(X) = 1/15$, where $5 < X < 20$. The waiting time is important because it is nonproductive time for a highly paid craftsman.
a. What is the probability that the machinist will have to wait at most 10 minutes?
b. What is the probability that the machinist will have to wait exactly 10.0 minutes?
c. What is the probability that the machinist will have to wait from 7.5 to 15 minutes?
d. On the average, how long must the machinist wait?

*Solution*    For this example $a = 5$ and $b = 20$, as shown in the graph. The probabilities we seek can be computed as areas under the curve of height 1/15.
    The uniform distribution gives values as follows.
a. $P(X < 10.0) = \text{length} \cdot \text{height} = (10 - 5)/15 = 5/15 = 1/3$ (see shading)

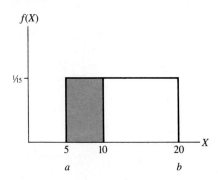

b. $P(X = 10.0) = 0$
c. $P(7.5 < X < 15) = (15 - 7.5)/15 = 1/2$
d. Mean $= (a + b)/2 = (5 + 20)/2 = 12.5$ minutes

Although this discussion has centered on the uniform distribution, the basic idea of associating probabilities with the area under the probability curve applies to *all* continuous random variables. However, many continuous variable distributions do not follow a regular pattern such as a rectangle, a triangle, or a semicircle. Thus computing areas for other random variables is often more difficult than for the uniform case. These situations may require the use of calculus or numerical integration by computer. Another approach is to use tables of already-calculated areas, which is what we shall do as we move on to working with the normal distribution.

**Problems 6.2**

1. For $f(X) = 1/10$ and $-5 < X < 5$,
   a. find $\mu$, $\sigma^2$, and $\sigma$.
   b. compute $P(X < 0)$, $P(X = 1)$, and $P(-2 < X < 3)$.

2. For $f(X) = 1/12$ and $-6 < X < 6$, find $\mu$, $\sigma^2$, and $\sigma$.

3. Consider a discrete uniform probability (function) rule $P(X) = 1/10$, $X = -5, -4, -3, -2, -1, 1, 2, 3, 4, 5$.
   a. Display the probability distribution as a line diagram (see p. 00).
   b. Compute $P(X < 0)$, $P(X = 1)$, and $P(-2 < X < 3)$
   c. Compare the results of parts (a) and (b) with the results of Problem 1.

4. Given the probability distribution shown in the graph, find
   a. the probability function.
   b. values for $\mu$, $\sigma^2$, and $\sigma$.
   c. $P(X \leqslant 2)$, $P(X < 2)$, and $P(X = 2)$.

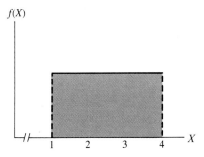

5. The city's highway and roads department must be aware of the flow of traffic on major roads. An especially busy intersection is at the corner of Broadway and Vine Streets. To accommodate the heavier traffic flow on Vine, the light is green on Broadway starting at 12:00 P.M. for one minute. It does not turn green again on Broadway for two minutes. The cycle is repeated throughout the day.
   a. If you come to this intersection by way of Broadway and arrive at 8:02, will the light be green?
   b. What percentage of the time during any given hour will the light be green on Broadway?
   c. If your arrival time at this intersection is a random variable with uniform distribution over the interval 7:59 to 8:06 and you are driving on Broadway, what is the probability that you will *not* have to stop for this light?

## 6.3
## THE NORMAL DISTRIBUTION

The normal distribution is a bell-shaped pattern of variation that was identified by the German mathematician and astronomer Carl Gauss (1777–1855). He observed a recurring pattern of errors in repeated measurements of the same object. For example, suppose a machinist was asked to measure the diameter of 100 similar bolts—one at a time—to an accuracy of 0.0001 inch with a micrometer. Further, suppose that, unknown to the machinist, all measurements were made on the same bolt. Most of the reported values would tend to cluster closely about some central value, with perhaps a few observations of somewhat higher or lower values. Many measurements, such as diameters of bolts produced to the same specification, have an overall bell-shaped pattern of error. This pattern of error, which was initially attributed to chance, is called the *Gaussian*, or *normal, curve.*

As the theory of statistical inference has developed over the past 200 years, the normal distribution has played a key role. There are several reasons for this. First, many empirical data are by nature normally distributed (for instance, the diameters of bolts, weights of packages of foods, or SAT scores). Second, data that are not normally distributed can often be analyzed via the normal distribution, provided an appropriate sampling technique is employed and provided the question is about a mean value or a population proportion. Third, the normal distribution has been made extremely easy to work with by virtue of a coding process that converts any particular normal distribution to a standard normal distribution. (This is sufficiently common that some hand calculators have a normal distribution function programmed into their logic.) Fourth, the normal distribution can be effectively manipulated and used for making inferences.

### The Standard Normal Distribution

Because the normal curve is not a simple polygon, the procedures required to compute areas under the curve are more complicated than those used for the uniform distribution. Finding probabilities for every different normal curve could be a formidable task. Fortunately, we can evaluate areas for all *normal distributions* by making one simple *conversion* to the *standard normal distribution* or *Z-*form. The notation $N(\mu, \sigma^2)$ is a common generalization for a normal distribution with parameters mean $= \mu$ and variance $= \sigma^2$. For the standard normal probability distribution with $\mu = 0$ and $\sigma^2 = 1^2 = 1$, we write $N(0,1)$.

> The normal or Gaussian distribution is a bell-shaped curve with its explicit shape defined by two parameters—its mean value, $\mu$, and standard deviation, $\sigma$.
>
> The standard normal distribution is a continuous, symmetrical, bell-shaped distribution that has a mean of 0 and a standard deviation of 1.

These two parameters—mean and standard deviation—thus define the specific normal distribution.

The normal distribution is a continuous, symmetrical, bell-shaped distribution. Because it is symmetrical, half of the area under the curve lies to the left of the central value and half lies to the right. The mean, median, and mode are all the same central value, which for the standard normal distribution is zero.

Although all normal distributions are characterized by a familiar bell shape, each one is distinguished by the location of its center and by its spread,

**FIGURE 6.8** Three Normal Distributions

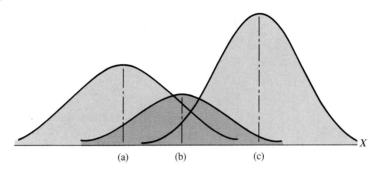

or variation about the mean. Figure 6.8 illustrates three normal distributions with different means and different standard deviations.

Having identified the general character of the distribution, we now describe how to evaluate normal probabilities (areas). For this we use a form of coding to convert the values to the *standard* normal distribution.

The *standard normal score*, or *Z-score*, is a simple coding device for converting any normal distribution to a standard normal distribution.

**The Z-score is a measure of the number of standard deviations of distance that any point resides from the mean.**

We will practice coding some normally distributed random variables to Z-scores shortly, but first some comments about the area (or probability) under the standard normal curve may be helpful.

**Using the Table of Standard Normal Probabilities**

Computation of probabilities by use of an equation is very time-consuming. Fortunately, the probabilities associated with areas under the standard normal curve have been computed and some are in Table B.1 of Appendix B. This table gives probability values over an interval from the center, $Z = 0$, to a given Z-score (see Figure 6.9(a)). Z-scores are located on the left-hand margin, and areas (probabilities) are located in the body of the table (Figure 6.9(b)).

The column headings indicate accuracy for the Z-values to hundreths. In most problems and examples in this textbook, Z-scores will be given to this

**FIGURE 6.9** The Z-Curve and the Table of Standard Normal Probabilities

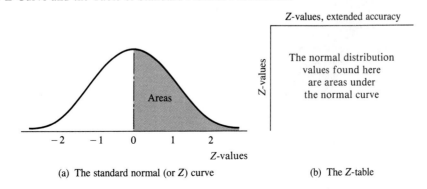

(a) The standard normal (or Z) curve        (b) The Z-table

accuracy. The areas in the table are always for an interval from the mean ($Z = 0$) to some $Z$-value.

Although Table B.1 indicates only positive $Z$-values, there are corresponding negative values because the standard normal distribution is symmetric about zero. Thus the total area for an interval defined by a positive $Z$-value and the same negative value is double the area shown in the table. For example, for $Z = 1$, the table shows $P = 0.3413$. Doubling that number gives 0.6826, or 68.26%. Thus just over 68% of the area of the standard normal distribution is within one $Z$, or one standard deviation unit, from the mean, that is, from $Z = -1$ to $Z = +1$. Table 6.1 gives this area. The distance of two $Z$-scores encompasses the interval from $Z = -2$ to $Z = +2$ and includes 95.44% of the total; and three $Z$-scores, $-3$ to $+3$, include almost all (99.74%) of the area.

**TABLE 6.1**   Examples of the Standard Normal Distribution Areas

| Z-Score | Table Value | Area between $-Z$ and $+Z$ |
|---|---|---|
| 1 | 0.3413 | 68.26% |
| 2 | 0.4772 | 95.44% |
| 3 | 0.4987 | 99.74% |

The examples that follow illustrate the use of Table 1 in Appendix B for determining $Z$-scores and the associated areas under the standard normal curve. They assume the random selection of items from a normally distributed population. (*Note:* The table (B.1) of $Z$-scores is so widely used that it is repeated, for easier access, inside the back cover of this book.)

**EXAMPLE 6.3**

For the standard normal distribution, what percent of the area under the curve lies below $Z = +1.00$?

*Solution*   In Table B.1 we see that the area from the mean to $Z = +1.00$ is 0.3413. Because the distribution is symmetrical, half, or 0.5000, of the area is below the mean ($Z = 0$). So $P(Z < +1.00) = P(Z < 0) + P(0 < Z < 1) = 0.5000 + 0.3413 = 0.8413$, or 84.13%. That is, $Z = +1.00$ defines approximately the 84th percentile for the standard normal curve. Figure 6.10 illustrates the solution, with two shaded areas.

**FIGURE 6.10**   Finding the Area Beneath a Normal Curve for $P(Z < 1.00)$

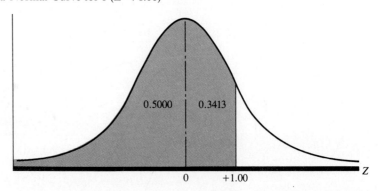

Observe how the shading in Figure 6.10 clarifies the solution to Example 6.3. A visual display is a useful guide for the inclusion or exclusion of certain areas. This can prove extremely helpful in solving problems involving the normal distribution, so use of a sketch is highly recommended. A variety of applications is given in Examples 6.4 through 6.8. They all assume random sampling.

---

| **EXAMPLE 6.4** | What is the probability for a $Z$-score between 0 and $-1.45$? |
|---|---|

*Solution*   From Table B.1, the area for $Z = 1.45$ is 0.4265. Because positive and negative $Z$-values are symmetrical, we know that the area from $Z = -1.45$ to $Z = 0$ is 0.4265. Therefore, $P(-1.45 < Z < 0) = 0.4265$, as shown by the shaded area in the figure.

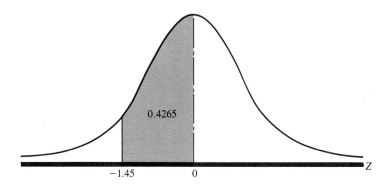

---

| **EXAMPLE 6.5** | What fraction of the total area beneath a normal curve lies to the right of $Z = 2.486$? |
|---|---|

*Solution*   Begin by rounding the $Z$-score to the nearest hundredth: $Z = 2.49$. Table B.1 shows the area 0.4936 for $Z = 2.49$. However, we seek the area to the right of this point, that is, the remainder of the 50% of the area beyond the 0.4936—the dark-shaded area in the figure below. So $P(Z > 2.49) = 0.5000 - 0.4936 = 0.0064$. This is less than 1%, that is, 0.64% of the area.

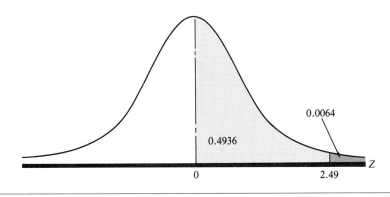

---

Now let's reverse the process by beginning with a given area (probability measure) and finding the associated $Z$-values. We will need to do this to solve many of the problems in later chapters of the book. We obtain $Z$-value by going to the body of Table B.1, locating the area value, and then "realizing" its position to the left and top margins to determine the $Z$-value. Examples 6.6 through 6.8 should clarify the process.

### EXAMPLE 6.6

Find the $Z$-value for which 19.50% of the area is between $Z = 0$ and that $Z$-value.

*Solution*    Going to Table B.1, we find the value 0.1950 in the .01 column of the $Z = 0.5$ row. The left margin gives a $Z$-score accurate to the nearest tenth, 0.5. Reading up to the column heading extends the accuracy to 0.01, so $Z = 0.51$ (dark-shaded area). Since the question does not specify direction above or below $Z = 0$, $Z = -0.51$ is also correct, and the light-shaded area in the figure would also qualify as an answer.

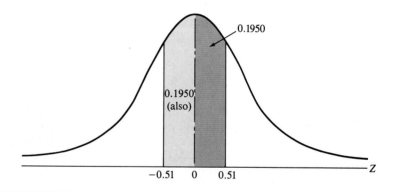

### EXAMPLE 6.7

We seek a $Z$-value for which 75% of the area is to its right. That is, what is the first-quartile value $Q_1$ on the standard normal curve?

*Solution*    Since the area to the left (0.2500) is less than 50%, the corresponding $Z$-value is negative. The area of interest from Table B.1 is $0.5000 - 0.2500 = 0.2500$, though the exact value 0.2500 does not appear in Table B.1. The area value we seek is between 0.2486 and 0.2517. The nearer value is 0.2486 (it is incorrect by $-14$ parts, whereas 0.2517 is incorrect by $+17$ parts), so we read the $Z$-value for area 0.2486. This is $Z = -0.67$. So $P(Z > -0.67) = 0.2486 + 0.5000 = 0.7486$, as shown by the shaded area in the figure. The first-quartile point on the standard normal curve is at $Z = -0.67$.

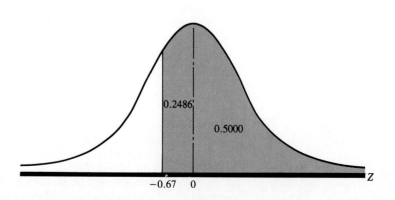

Although it would be nice to get an answer closer to the exact value, to do so requires either interpolation or a more extensive probability table. Furthermore, for most business needs this approximation should be close enough.

Example 6.8 shows a different use of the normal curve.

---

**EXAMPLE 6.8**

Find the value for which the area between $+Z$ and $-Z$ is 0.3400.

*Solution*

Again, symmetry is used. The area of interest is centered at $Z = 0$, so that 0.3400/2, or 17%, of the area resides between 0 and the unknown (positive) $Z$-value. Table B.1 displays 0.1700 as the area corresponding to $Z = 0.44$. This answer is depicted in the figure. (*Note:* This type of question will be asked again when we develop confidence intervals in Section 8.2.)

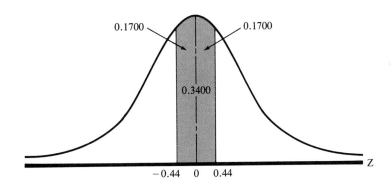

---

With this foundation for determining normal probabilities, we turn now to the method of converting all normal curves to a standard normal form—standard ($Z$) scores.

**Obtaining Z-Scores**

To obtain standard ($Z$) scores, we use a coding technique that converts values for any normal distribution to comparable values on the standard normal scale (where the mean is zero and the standard deviation is one). The letter $Z$ then represents the number of standard deviations of distance that any value of $X$ is from the mean, so it is called a *standard normal* (or $Z$) *score*.

> To convert a random variable value $X$ from any normal distribution with mean $\mu$ and standard deviation $\sigma$ to the standard normal form, simply divide the difference $X - \mu$ by one standard deviation, $\sigma$.

$$Z = \frac{X - \mu}{\sigma} \tag{6.4}$$

> The $Z$-score or standard normal score indicates the number of standard deviations of distance between $X$ and $\mu$.

> The distribution of standard normal scores is called the standard normal distribution.

This conversion allows all normal distributions to be expressed in one form, that for which the probabilities (areas) appear in Table 1 of Appendix B.

**FIGURE 6.11**    Conversion to the Standard Normal Distribution

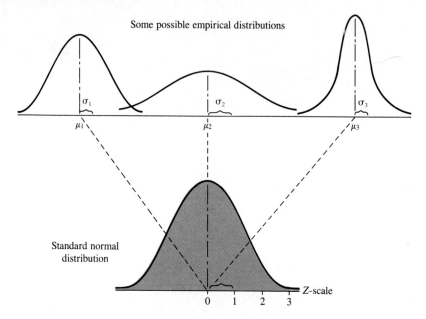

Figure 6.11 symbolizes the conversion. Note that the Z-score holds a special relation to the standard deviation. That is, one unit on the scale of Z equals the distance spanned by one standard deviation. Numerically, when $X - \mu = \sigma$, then $Z = \sigma/\sigma = 1$. Thus Z-values record differences for which the basic unit is the standard deviation. That is why Z is sometimes called a standard normal deviate. The standard normal distribution provides a method for assigning probability based on standard deviation units.

**EXAMPLE 6.9**

An apprentice plumber wants to solder a copper pipe section within an acceptable time in order to qualify as a professional pipe fitter. The requirement is that the joint be completed within one standard deviation of the professional standard time of $\mu = 0.7$ minutes, and $\sigma = 0.2$ minutes. Soldering times are normally distributed.

In practice soldering, the apprentice's time ranges from 0.6 minutes to 1.1 minutes, with standard deviation of 0.2 minutes. Is the apprentice ready for the test?

*Solution*    Compute the number of standard deviations that 0.6 minutes and 1.1 minutes are from $\mu$ by using the expression

$$Z = \frac{X - \mu}{\sigma}$$

The mean $\mu$ is at the center (zero point) on the Z-scale. The 0.6-minute time is one-half unit below the mean, as shown in (a).

$$Z = \frac{0.6 - 0.7}{0.2} = -0.5$$

So the lower limit of the welder's time is satisfactory.

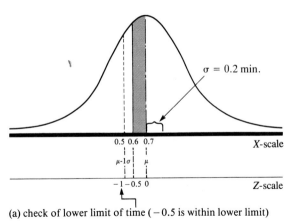

(a) check of lower limit of time ($-0.5$ is within lower limit)

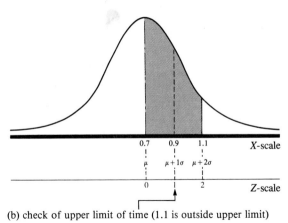

(b) check of upper limit of time (1.1 is outside upper limit)

As demonstrated in (b), the 1.1-minute time is

$$Z = \frac{1.1 - 0.7}{0.2} = +2.0$$

standard deviations above the professional standard and so is not satisfactory. This apprentice needs more practice to assure a time within one standard deviation of the 0.7-minute requirement.

Computing $Z$-scores requires skill, and you can improve rapidly with practice. Section 6.3 offers some skill-building problems that you should complete successfully before continuing to study the applications of normal random variables. Then in Section 6.4 you can concentrate on understanding applications and not be concerned with computing procedures.

**Problems 6.3**

6. Supply the proper term or number as it pertains to the normal distribution.
   a. It is a _____ (continuous, discrete) distribution with total area under the curve equal to _____ (supply number).
   b. Within $\pm 1$ standard deviation of the mean lies _____% of the area.
   c. The probability of a value lying outside 1 standard deviation is _____ (supply number).
   d. The standard normal distribution has a mean of _____ and a standard deviation of _____ (supply numbers).

7. Draw a bell curve showing a display of pertinent $Z$-values and areas as part of your solution for each of the following.
   a. $P(-1.00 < Z < 0.00)$          d. $P(Z < -1.64)$
   b. $P(-1.96 < Z < 1.96)$          e. $P(Z > 1.96)$
   c. $P(1.00 < Z < 2.33)$          f. $P(Z < 1.00 \text{ or } Z > 2.00)$

8. Given the following probabilities (areas), find $Z_0$ for which
   a. $P(-Z_0 < Z < Z_0) = 0.9802$          d. $P(Z < Z_0) = 0.4960$
   b. $P(Z < Z_0) = 0.0505$          e. $P(Z > Z_0) = 0.0038$
   c. $P(Z_0 < Z < 0) = 0.3810$          f. $P(Z < Z_0) = 0.6772$

9. For the following, find the nearest $Z$-value that will satisfy the statement:
   a. 9.68% of the measure is below this $Z$.
   b. 0.4192 is the area between $Z = 0$ and this value.
   c. This is (approximately) the first-quartile value.   *.25 on table*
   d. 92.22% of the area resides above this value.
   e. 99.80% of the measure falls between $\pm Z$.   *9980 ÷ 2*
   *.4990 = 3.08*

10. Find areas (probabilities) for
    a. $Z < 1.326$          d. $-1.364 < Z < 2.41$
    b. $0 < Z < 1.095$       e. $Z < 4.00$
    c. $Z \geqslant -0.163$   f. $Z > 3.09$

11. Find each of the following. Use Table 1, Appendix B, as required.
    a. Find $Z_0$ such that the area to the left of $Z_0$ is 0.5040.
    b. Compute $P(Z > -3.20)$.
    c. Find the area for $(-2.065 < Z)$.
    d. If $P(0.81 < Z < Z_0) = 0.1921$, find $Z_0$.
    e. Find the area between $Z = -1.56$ and $Z = -0.56$.
    f. Which is larger, $P_{60}$ (60th percentile value on $Z$-scale) or $Z = 0.30$? Justify your choice.

12. Sketch a normal distribution with mean 80 and standard deviation 20. Show the $Z$-scale below the $X$-scale indicating $Z$-values of $\pm 1$, $\pm 2$, and $\pm 3$.

13. Given that the distribution of $X$ is normal,
    a. if $\mu = 4$ and $\sigma = 1$, find $P(X > -2)$.
    b. if $\mu = -60$ and $\sigma = 6$, find $P(X \leqslant -54)$.
    c. if $\mu = 10$ and $\sigma = 0.8$, find $P(X \leqslant 5)$.
    d. if $\mu = 2$ and $\sigma = 2$, find $P(-2 < X < 3)$.
    e. if $\mu = 1$ and $\sigma = 2$, find $P(X < -1.5 \text{ or } X > 0)$.

14. For a normal distribution with mean 6.00 and standard deviation 0.50, that is, for $X$ being $N[6.00, (0.50)^2]$, find
    a. $P(X < 6.75)$.
    b. the area to the left of $X = 6.30$.
    c. the measure above $X = 6.30$.
    d. $P(0 < X < 6.00)$.
    e. the median for the probability distribution.

## 6.4
## APPLICATIONS OF THE NORMAL DISTRIBUTION

You are now equipped to use the normal distribution to answer probability questions. Using the procedure for converting normally distributed $X$-variables to $Z$-scores, we can follow a systematic outline of steps for solving normal probability questions.

**Procedures for Solving Normal Probability Distribution Questions**

### SYSTEMATIC PROCEDURE FOR ANSWERING NORMAL DISTRIBUTION PROBABILITY QUESTIONS

1. Identify the question and the random variable.
2. Use the random variable to express the question in probability form.
3. Quantify the given information; use a sketch where useful.

4. Calculate the unknown, using Equation (6.4). This will usually mean calculating a $Z$-value but could mean calculating $\mu$, $\sigma$, or even $X$.
5. Combine the given information using $Z$-scores, a bell curve, and Table 1 in Appendix B to obtain an answer.
6. Recheck the solution—be sure it is reasonable and makes good common sense.
7. State the practical implications of your numerical answer.

In the examples that follow, the notation $N(\mu,\sigma^2)$ indicates a normal distribution with mean $\mu$ and variance $\sigma^2$. In the remainder of this section we use this systematic procedure to answer normal distribution probability questions.

---

**EXAMPLE 6.10**

Suppose a highway patrol officer has clocked many cars in a marked speed zone and has found that the mean speed is 65.0 miles per hour. Assume the standard deviation is 3.0 miles per hour. If the distribution of auto speeds is normal and if any car exceeding 70.0 miles per hour is speeding, what percent of the cars that travel this stretch of road are speeding?

*Solution*   The variable is $X$ = the speed for cars on this stretch of road. The probability distribution is $N(65.0, 3.0^2)$.

Steps 1, 2   The question asks us to find $P(X > 70.0)$ for $X$ = automobile speeds in miles per hour.

Step 3   We use Figure 6.12 to depict the distribution of speeds.

Step 4   Figure 6.12 indicates the given information. We compute the appropriate $Z$-scores.

$$Z = \frac{X - \mu}{\sigma} = \frac{70.0 - 65.0}{3.0} = \frac{5.0}{3.0} = 1.67$$

The appropriate probability (area) from Table B.1 is 0.4525.

**FIGURE 6.12**   Normal Distribution of Car Speeds

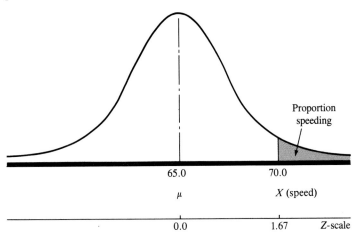

Steps 5, 6   From the standard normal curve diagram, $P(X > 70.0) = P(Z > 1.67) = 0.5000 - 0.4525 = 0.0475$, or 4.75%.

Step 7   We conclude that 4.75% of those who travel this stretch of road are speeding. It is a reasonable answer.

Notice that Figure 6.12 gives a visual representation of the area in question. Procedures and computations then follow logically.

## EXAMPLE 6.11

Union negotiators are asking for a raise of $1.45 per hour. Past union increases have averaged $1.00 per hour with a standard deviation of $0.30 per hour. Assuming that similar conditions prevail now, what is the probability that their increase will be less than $1.45? Assume that the wage increases can be described by a normal distribution.

*Solution*   Although our concern here is to focus upon data that can be analyzed quantitatively, the analyst would always apply his or her experience and common sense to the extent possible. We can analyze the statistical aspect of the problem by letting the variable $X$ = the wage increase; the probability distribution is $N[\$1.00, (\$.30)^2]$.

Steps 1, 2   We want $P(X < \$1.45)$ for $X =$ a wage increase.

Steps 3, 4   Figure 6.13 depicts the distribution of past wage increases. The value in question, $1.45, equates to

$$Z = \frac{1.45 - 1.00}{0.30} = \frac{0.45}{0.30} = 1.50 \text{ standard normal scores}$$

Step 5   From Table B.1, $P(0 < Z < 1.50) = 0.4332$. Thus,

$$P(X < \$1.45) = 0.4332 + 0.5000 = 0.8332$$

Steps 6, 7   There is an 83.3% chance that the union will receive something less than a $1.45-per-hour raise. The "commonsense" interpretation may be that this chance is suffi-

**FIGURE 6.13**   Normal Distribution of Wage Increases

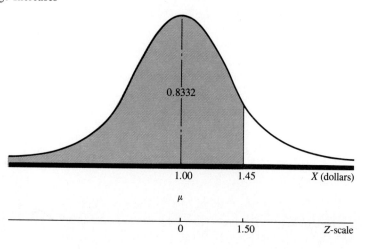

ciently low that the group should request a lower wage increase in order to be assured of getting it.

Example 6.12 differs from this one only in that we are required to determine two probabilities. A diagram, if drawn properly, can assist in the solution.

---

**EXAMPLE 6.12** ◦

Suppose that the union negotiators in Example 6.11 expect to get something less than their request, but hope for a raise of no less than $0.90 per hour. What is the chance they will receive a raise within the interval $0.90–$1.45?

*Solution*  **Step 1**  The random variable remains $X$ = a wage increase, with an $N[\$1.00, (\$0.30)^2]$ distribution.

**Step 2**  This question asks for $P(\$0.90 < X < \$1.45)$.

**Steps 3, 4**  $Z_1 = \dfrac{1.45 - 1.00}{0.30} = 1.50$

$$Z_2 = \frac{0.90 - 1.00}{0.30} = \frac{-0.10}{0.30} = -0.33$$

Dollar amounts are converted to $Z$-scores. Then Figure 6.14 describes the probability as a shaded region under the normal curve.

**Step 5**  Table 1, Appendix B gives $P(-0.33 < Z < 0) = 0.1293$
$\qquad\qquad\qquad\qquad P(0 < Z < 1.50) \qquad\quad = 0.4332$
We seek the sum $\qquad P(-0.33 < Z < 1.50) \quad = \overline{0.5625}$

**Steps 6, 7**  Based on past experience, there is a 56.25% chance that these union members will receive a raise of $0.90 to $1.45 per hour.

---

**FIGURE 6.14**  Normal Distribution of Wage Increases (Again)

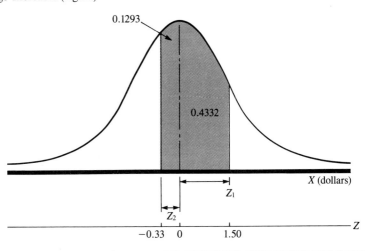

**Applications to Testing: Percentile Ranks and Percentile Values**

Examples 6.10 through 6.12 illustrated a variety of ways to find probabilities associated with the normal distribution, given values for $X$. But often we want to reverse the process: We want to find values of $X$ given a probability value. If we think of probabilities as parts of a total, we can also answer questions relating to percentiles and percentile ranks using the $Z$-distribution. For example, scores for a standardized aptitude test can follow a normal distribution. If the mean value is known, we can answer questions like these: What value is at the 90th percentile? What is the percentile rank for someone who achieved a 480-point score? This approach has been used for scoring the aptitude of students for graduate studies in business.

---

**EXAMPLE 6.13**

The distribution of aptitude scores on a standardized test had a mean of 432 with a standard deviation of 42 points and can be considered normal. Lois scored 480. Determine her percentile rank.

*Solution*    Step 1    The variable $X$ = individual aptitude scores is normal, $N[432, (42)^2]$.

Steps 2, 3    We seek the percentile rank for a score $X = 480$.

Step 4    The raw score is converted to a $Z$-value as follows:

$$Z = \frac{X - \mu}{\sigma} = \frac{480 - 432}{42} = \frac{48}{42} = 1.14$$

Step 5    Percentile = $0.3729 + 0.5000 = 0.8729$.

Steps 6, 7    The score 480 ranks at the 87th percentile (rounded) on this normal distribution. Lois has achieved at a level slightly above 87% of those who took this test.

---

The processes of determining the raw score associated with any percentile is like finding a $Z$-value where we know the probability. An additional step is necessary to convert from the $Z$-score to raw-score ($X$) form.

---

**EXAMPLE 6.14**

Find the value at the 40th percentile, $P_{40}$, for the distribution given in Example 6.13.

*Solution*    A percentage less than 50% indicates a negative $Z$-value. We seek the $Z$-value that gives an area of $0.5000 - 0.4000 = 0.1000$ below the mean. Using the body of Table B.1 and reading the corresponding $Z$-value, we find the area for $Z = 0.25$. But $P_{40}$ is below the mean, so by symmetry $Z = -0.25$. The remaining step involves decoding the $Z$-formula. The situation is depicted in Figure 6.15. Solving the expression $Z = (X - \mu)/\sigma$ for $X$, we have $X = \mu - Z\sigma$.

$$-0.25 = \frac{X - 432}{42}, \quad \text{or} \quad X = 432 - (0.25 \cdot 42) = 421.5$$

The 40th percentile is 421.5. A score of 421 is just below the 40th percentile; 422 is just above the 40th percentile. Approximately 60% of all persons who have taken this test have scored above 422 points.

**FIGURE 6.15**   Finding an Unknown $X$-Value from a Normal Distribution

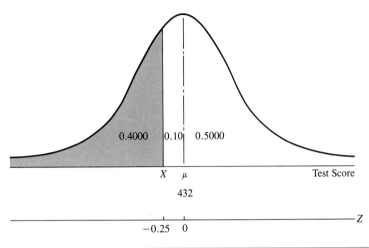

Example 6.14 is a common application of $Z$-scores in education and in psychological testing of employees. A normal distribution is frequently assumed in the guidelines for interpreting the tests. However, when there is a countable number of possible scores, the distribution is actually discrete, but the continuous normal distribution may provide a good *approximation* of the discrete variable probability pattern. It is used because it gives good approximations for many standardized tests with only a fraction of the work that would be required to evaluate the exact discrete probabilities. The (small) error in the normal approximation is ignored here but will be the primary topic of Section 6.5.

**Problems 6.4**

15. Suppose that a restaurant manager tries to maintain the average serving of roast beef at 5.0 ounces with a standard deviation of 0.5 ounces. If, unknown to the management, a standards inspector comes through the line, what is the probability that her serving will weigh less than 4.0 ounces? Assume a normal distribution for serving weights.

16. One-pound boxes of sugar are known to contain an average of 1.03 pounds of sugar with a standard deviation of 0.02 pounds. If the weights of the boxes have a normal distribution, what percent will weigh less than 1.0 pound?

17. The Yard and Garden Company owns a machine that fills bags of lawn fertilizer. The machine is set for a 49.0-pound fill. Fills have a standard deviation of 0.6 pounds and are normally distributed.
   a. For what fraction of the containers will the net weight differ from the mean weight of 49.0 pounds by more than 1.0 pound?
   b. If refilling each reject by hand costs the company an average of 8¢, what is the expected cost of refills per 100 bags produced? Assume that fills between 48.0 and 50.0 pounds are acceptable and that all rejects must be refilled by hand.

18. If a normally distributed variable has a mean of 100 and standard deviation of 10, what percentage of the scores have the following characteristics?

a. Greater than or equal to 100
b. Greater than or equal to 110
c. Between 90 and 110
d. Between 110 and 120
e. Less than or equal to 130

19. The life expectancy of Neverready batteries is normally distributed with a mean of 400 hours and a standard deviation of 120 hours. What is the probability that a battery will last at least 300 hours?

20. The average time required for workers in a furniture plant to assemble a certain piece of furniture is 40.5 minutes with a standard deviation of 9.6 minutes. If the time distribution is normal, what percentage of the furniture pieces will be finished within 60 minutes?

21. Ingots produced at an aluminum plant have a mean weight of 386 pounds with a standard deviation of 20 pounds. Assume the distribution of weights is normal.
    a. What is the percentile rank for a weight of 400?
    b. Find the value of the 60th percentile, $P_{60}$.
    c. Find the values for $P_{25}$ and $P_{75}$.
    d. Which quarter of the distribution includes the weight 396?
    e. Determine the percentile rank for a weight of 370.

22. In a very large business-law class, the 200-point final exam had a mean of 140 with a standard deviation of 18.2 points. A normal distribution was a reasonable approximation to the actual distribution of scores. The instructor decided to grade on the curve and assigned 8% A's, 24% B's, 36% C's, 24% D's, and 8% F's.
    a. Determine the value that separates A and B grades.
    b. What letter grade would be assigned a score of 165?
    c. What is the lowest passing score (lowest D)?
    d. What percent of the class achieved below Laura, whose raw score was 185?

| 6.5 | We discussed binomial variables at length in Chapter 5. Were you to go back |
|-----|-----|
| **THE NORMAL APPROXIMATION OF THE BINOMIAL DISTRIBUTION** | and review those problems, you would find the probability questions were restricted to cases in which the sample size, $n$, was 25 or smaller. This allowed us to use Table 5 in Appendix B. For larger samples we can no longer use that binomial table. The options then are either to use a calculator or computer that can handle the extensive calculations or to use a *normal approximation* to binomial probabilities. |

A normal approximation to binomial probabilities is appropriate when both $n\pi \geqslant 5$ and $n(1 - \pi) \geqslant 5$ are true.

The normal approximation uses the concepts of probability as area under a curve and a continuity correction of $\pm 0.5$ to form normal curve areas for approximating binomial variable probability values.

Suppose we want the probability of nine *or fewer* successes in a binomial experiment of $n = 50$ trials with the fixed chance of success being 0.2. The *exact*

**FIGURE 6.16**   Normal Approximation to Binomial Probabilities

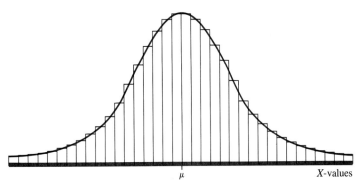

probability is

$$P(X = 0 \text{ or } 1 \text{ or } \ldots \text{ or } 9)$$
$$= {}_{50}C_0(0.2)^0(0.8)^{50} + \cdots + {}_{50}C_9(0.2)^9(0.8)^{41} = 0.4437$$

Unfortunately, calculating the binomial probability values by hand can be a formidable chore. A calculator or computer can make it much easier. However, the normal distribution can also provide a reasonable approximation provided both $n\pi$ and $n(1 - \pi)$ are equal to 5 or more. The normal approximation is quite accurate for $\pi$ close to 1/2, even for small samples of 15 or 20. But its accuracy is less as $\pi$ approaches either 0 or 1. For any value of $\pi$, the approximation improves as $n$ increases. Thus, for $\pi = 0.2$ and a sample of 50, the approximation is reasonable. Moreover, the two values $n\pi = (50)(0.2) = 10$ and $n(1 - \pi) = 40$ satisfy the requirements that the quantities $n\pi$ and $n(1 - \pi)$ both equal or exceed 5.

**The Continuity Correction Factor**

The reason the normal distribution provides a good approximation to binomial probabilities is that the area bounded beneath the binomial probability distribution is approximately equal to the area bounded by the normal curve over a comparable interval. Figure 6.16 illustrates this approximation. The histogram represents exact binomial probabilities, and the smooth (normal) curve approximates this area. Figure 6.17 shows a portion of the exact representation for the binomial distribution with $n = 50$ and $\pi = 0.20$. Because of the normal approximation, the error for this interval of one unit width is the difference in the areas of the shaded triangles (see cutout).

The normal approximation for discrete values uses an interval that covers 0.5 units to either side of the integer value. This correction for calculation of probability as an area is called the *continuity correction*.

> **The continuity correction uses the addition and/or subtraction of 0.5 to the values of a discrete random variable to define an interval for the determination of area (a probability) by a continuous random variable probability distribution.**

The question posed at the beginning of this section provides us with an example.

**FIGURE 6.17** The Normal Approximation at $X = 9$ for a Binomial Distribution with $n = 50$ and $\pi = 0.20$

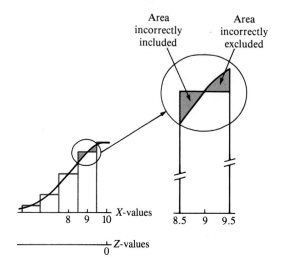

---

**EXAMPLE 6.15**

Approximate the probability of nine *or fewer* successes in a binomial experiment with 50 trials, each having a chance for success of 0.2.

*Solution*

The question asks for $P(X \leqslant 9)$, so the values of interest include $X = 0, 1, \ldots, 9$. The extreme integer value is 9. However, on a continuous scale we should include all values up to the midpoint between 9 and 10. Thus, with the continuity correction we use $(9 + 0.5)$.* Then for $n = 50$ and $\pi = 0.2$, the mean is $\mu = n\pi = (50) \cdot (0.2) = 10$, and $\sigma = \sqrt{n\pi(1 - \pi)} = \sqrt{50(0.2)(0.8)} = \sqrt{8} = 2.83$. The normal approximation form gives us the $Z$-value corresponding to $X = 9.5$.

$$Z = \frac{X - \mu}{\sigma} = \frac{(9 + 0.5) - 10}{2.83} = \frac{-0.50}{2.83} = -0.18$$

Using Table B.1, we have $P(X \leqslant 9) = P(Z < -0.18) = 0.5000 - 0.0714 = 0.4286$. The difference between this and the exact value from a computer calculation is $0.4437 - 0.4286 = 0.0151$. This approximation is reasonably close to the exact value, being off by 1.5 percentage points (which is about a 3% error range). This error can be explained as the difference between the two shaded areas displayed in Figure 6.17.

---

**Steps for Applying a Normal Approximation**

Having defined the essentials for a normal approximation, we now formalize the steps for solving probability questions that can be answered with the help of a normal approximation. Note that the statistical table for this approximation is the same one (Table B.1) we have used for other normal probabilities.

---

\* This includes the interval containing $X = 9$ as a part of the area for the values $0, 1, 2, \ldots, 9$.

### STEPS FOR APPLYING THE NORMAL APPROXIMATION TO BINOMIAL PROBABILITIES

1. Determine the interval of $X$-values of interest. Indicate the integer values that are included in the question.
2. Decide how to apply the *continuity correction*. Either add 0.5 to or subtract 0.5 from the value in question, thus defining the set of values that should be included. (This correction adjusts for the fact that we are now using a continuous distribution to approximate discrete values.)
3. Compute the mean, $\mu = n\pi$, and the standard deviation,

$$\sigma = \sqrt{n\pi(1 - \pi)}$$

4. Compute $Z$-score(s) and proceed as with the exact normal distribution but use the normal approximation form:

$$Z = \frac{(X \pm 0.5) - n\pi}{\sqrt{n\pi(1 - \pi)}} \tag{6.5}$$

In the examples that follow, you should concentrate on (1) recognizing binomial experiments; (2) deciding whether an approximation is necessary, especially when $n$ exceeds 25; and (3) deciding whether to include or exclude (cut off) extreme values by use of the continuity correction.

---

**EXAMPLE 6.16**

Surveys, Inc., a company that performs residential surveys, has found that in 80% of all households, at least one adult member (age 18 or older) is at home between 5 P.M. and 7 P.M. on weekdays. If a random sample of 100 residences is selected and calls are made during this time period, what is the chance that at least one adult will be home in 75 *or more* households in the sample? This number (75) is critical because Surveys, Inc., requires at least 75 adult opinions for the results of a survey to be considered valid. Otherwise the entire survey is scrapped.

*Solution*   We let $X$ = the number of households with one or more adult member at home during this time period. The conditions for a binomial experiment are met. Moreover, the normal approximation is appropriate because $n\pi = (100) \cdot (0.8) = 80$ and $n(1 - \pi) = (100)(0.2) = 20$.

1. The interval of $X$-values of interest is 75 or more, so the probability we seek is $P(X \geqslant 75)$.
2. The continuity correction is subtracted, so that 75 is included with the values larger than 75. The extreme value becomes $(75 - 0.5) = 74.5$.
3. The mean is $\mu = 100 \cdot 0.8 = 80$ with standard deviation $\sigma = \sqrt{100 \cdot 0.8 \cdot 0.2} = 4$.
4. The appropriate $Z$-value is

$$Z = \frac{X - \mu}{\sigma} = \frac{74.5 - 80}{4} = -1.38.$$

This is shown on the $Z$-scale of Figure 6.18.

**FIGURE 6.18**    Normal Approximation to Number of Households

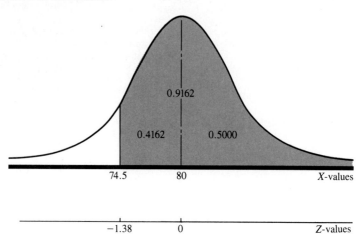

This yields a probability (area) = 0.4162, as shown in Figure 6.18. Therefore, $P(X \geqslant 75) \doteq P[Z > (-1.38)] = 0.4162 + 0.5000 = 0.9162$. There is about a 91.6% chance that an adult would be available to answer questions in each of 75 *or more* of 100 households in this sample. Thus Surveys, Inc., will get, by their definition, adequate information in 91.6% of their surveys. Approximately 8% of the time they will have to repeat their calls to meet the minimum quota.

A phrase such as *or more* or *more than* directs how we convert $X$ to a numerical value that is used to approximate the binomial probability by the normal approximation. In this example, using the continuity correction of $X - 0.5$ rather than simply $X$ added about 2% of the accuracy of the approximation. Using the binomial distribution would give the exact answer, but it requires lengthy calculations.

In the next example we seek a *point binomial probability* using the bell curve. That is, we want the probability for a single value, with the aid of the normal approximation.

---

**EXAMPLE 6.17**

Of the land leases made by oil-drilling companies, 1 in 10 will show a net profit in excess of $100,000. What is the probability that exactly 10 from a selection of 100 leases would show a net profit in excess of $100,000?

*Solution*    The variable is $X =$ the number of land leases that would show a net profit that exceeds $100,000. This can be considered a binomial experiment with $n = 100$ and $\pi = 0.10$. We seek the probability $P(X = 10 | n = 100, \pi = 0.10)$. The normal approximation is appropriate because $n\pi = 10$ and $n(1 - \pi) = 90$.

1. A point probability, $P(X = 10)$, is requested. Recall from our earlier discussion that a point has no area (hence no probability) under a continuous probability density function. The continuity correction factor will enable us to convert the point (location) to a measure of area (density).

2. This approximation requires both subtraction and addition of the correction factor (two values) to form an interval of approximation. That is, the approximated area

**FIGURE 6.19** A Normal Approximation

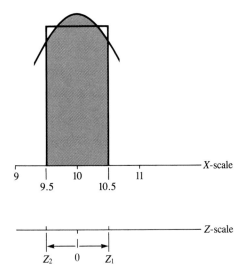

is that on the interval from 9.5 to 10.5. Consequently, two $Z$-values must be computed. The normal approximation is shaded in Figure 6.19.

3. The mean is 10, and the standard deviation is

$$\sigma = \sqrt{n(\pi)(1 - \pi)} = \sqrt{100 \cdot 0.1 \cdot 0.9} = 3$$

*Area*

4. $Z_1 = \dfrac{X_1 - \mu}{\sigma} = \dfrac{(10 + 0.5) - 10}{3} = \dfrac{0.5}{3} = 0.17$    0.0675

$Z_2 = \dfrac{X_2 - \mu}{\sigma} = \dfrac{(10 - 0.5) - 10}{3} = -0.17$    $\underline{0.0675}$

Thus $P(X = 10) \doteq P(9.5 < Z < 10.5) =$    0.1350

This approximate probability describes the shaded area in Figure 6.19. There is approximately a $13\frac{1}{2}\%$ chance that a selection of 100 land leases for drilling oil will produce exactly 10 that would each show a net profit in excess of $100,000.

---

With large $n$, when we make a proper transformation of a binomial distribution to a normal approximation, the resulting distribution has mean $\mu = (n\pi)$ and standard deviation $\sigma = \sqrt{n\pi(1 - \pi)}$. In this form, $\mu$ and $\sigma$ are both expressed as numbers. We could also express them in percentage form by dividing by $n$:

$$\mu_p = \frac{n\pi}{n} = \pi$$

$$\sigma_p = \sqrt{\frac{n\pi(1 - \pi)}{n^2}} = \sqrt{\frac{\pi(1 - \pi)}{n}}$$

We shall use these forms in some of the problems.

In conclusion, the normal approximation has wide use in approximating the probabilities for the binomial and some other discrete distributions. However, it does not give reasonable approximations to all discrete variable distributions and should be used only if the exact distribution is very nearly bell-shaped. Such is the case for binomial distributions when $n\pi \geqslant 5$ and $n(1 - \pi) \geqslant 5$.

**Problems 6.5**

23. For a binomial random variable with $n = 25$ and $\pi = 0.2$, calculate $P(7 \leqslant X \leqslant 9)$ using
    a. the exact binomial probabilities, Table B.5.
    b. the normal approximation with the aid of Table B.1. Compare answers.

24. Answer parts (a) and (b) of Problem 23 assuming a binomial distribution with $n = 25$ and $\pi = 0.3$. Is the approximation closer to the exact value in 23(b) *or* closer in 24(b)? Why?

25. A new insecticide called The Bug Stops Here is being tested by a chemical company. It is designed for an 80% kill of Minnesota mosquitoes on contact. Assuming that this value is correct, what is the probability that a controlled experiment results in a kill of over 85 of a swarm of 100 Minnesota mosquitoes? Use the normal approximation.

26. If 40% of the customers of an auto parts store use credit cards for their purchases, what is the probability that in a random sample of 100 customers, 45 or more use credit cards?

27. A market research firm has found that the probability is about 0.20 that the person buying his or her first painting, print, or sculpture will spend more money than any other buyer. If the dealer handles 100 shows, what is the probability that first-time buyers will spend the most money in 20 or fewer of the shows? Assume that $X$, the number of first-time buyers, follows a binomial probability distribution and all shows have some first-time buyers. Thus we are looking for $P(X \leqslant 20)$.

28. From past experience, management knows that 72% of the company's employees have participated in the company stock plan. What is the probability that 150 of 200 employees in a new plant will participate in the plan?

29. At the Burger-In Cafe, a third of the customers choose the daily special. If, on a given day, the cooks prepared enough food for 160 specials, what is the probability that the demand for the special will exceed the supply? Assume there are 450 customers.

**SUMMARY**

This chapter extends the study of probabilities to questions about *continuous random variables*. We have considered two common cases—the continuous uniform and the normal. For these and other continuous variables, we view probability as the area under a *probability density curve*.

The *uniform distribution* was used to show that continuous variable distributions have area over intervals of nonzero length and that the total area (probability) is one. This can be generalized to all continuous variable distributions.

Because many actual phenomenon are distributed normally, the normal random variable has extensive application in business. Some examples are daily cash dispersements at a bank, the time required for assembly of a machine part, scores on a business-

**TABLE 6.2**   Variables for Summary of Continuous Distributions

**Continuous Uniform**

$$f(X) = \frac{1}{b-a}, \quad a < X < b, \quad b < a \qquad (6.1)$$

Mean $\quad \mu = \dfrac{a+b}{2}$ $\qquad (6.2)$

Variance $\quad \sigma^2 = \dfrac{(b-a)^2}{12}$ $\qquad (6.3)$

**Normal Distribution**

Standard normal conversions $\quad Z = \dfrac{X-\mu}{\sigma}$ $\qquad (6.4)$

Normal approximation to binomial $\quad Z \doteq \dfrac{(X \pm 0.5) - n\pi}{\sqrt{n\pi(1-\pi)}}$ $\qquad (6.5)$

college entrance exam, and stock-price fluctuations about an average market price. The Statistics in Action section displays some data sets with normal distribution.

Table 6.2 summarizes equations for using the continuous uniform and normal distributions.

Calculation of normal probabilities involves coding to a standard normal distribution by $Z$-scores. The $Z$-score simply tells the number of standard deviations of any $X$-value from the population mean. Normal probability values are read directly from the table of standard normal areas, Table 1 in Appendix B.

The normal distribution also provides a useful approximation to the binomial and other discrete variable distributions. It gives a reasonable approximation as long as the (exact) discrete distribution is reasonably symmetric. For the binomial this means that $n\pi \geqslant 5$ and $n(1-\pi) \geqslant 5$. The approximation requires a continuity correction because we are using a continuous variable distribution to approximate a discrete one. However, as $n$ increases, the correction becomes less significant because the error in the approximation approaches zero and becomes insignificant. The normal distribution is the basis for statistical inference that begins with Chapter 7.

| **KEY TERMS** | |
|---|---|
| continuity correction, 231 | probability as area, 209 |
| continuous uniform random variable, 210 | (probability) density function, 209 |
| | probability distribution, 209 |
| distinction between continuous and discrete variable probability calculations, 213 | standard normal distribution, 216 |
| | uniform variable probability density function, 210 |
| normal approximation for binomial probability values, 230 | $Z$-scores, or standard normal scores, 217 |
| normal (Gaussian) distribution, 216 | |

**CHAPTER 6 PROBLEMS**

30. Assume that a population of cotton bales bound for Korea has a normally distributed weight with a mean of 200 pounds and a standard deviation of 10 pounds. Which of the following statements about this population are *false*?
   a. The distribution may not be symmetrical.
   b. We can be absolutely certain that the distribution is not bimodal.

c. Less than 1% of the items would vary from the mean by as much as 30 pounds.

d. Over 95% of the items would fall between 180 and 220 pounds.

31. For $P(X) = 1/6$, $X = -6, -4, -1, 0, 1, 3$,
    a. compute $\mu$.
    b. find $P(X \geqslant -1)$.

32. Assume that the arrival time of the Northern Europe flight to Miami International Airport is uniform over the time interval of 8:05 to 8:15 P.M. What is the latest time at which you should get to the arrival gate to be 80% sure that you will arrive ahead of the flight?

33. The Walk-A-Long Shoe Company produces 160,000 pairs of a given type of shoe each year. Company records show that these shoes have a mean lifetime of 20.0 months with a standard deviation of 2.0 months. Assuming lifetimes are normally distributed, find the probability that a pair of shoes selected at random could be expected to last two years or longer.

34. The diameters of steel axles produced in a plant are normally distributed with a mean of 2.000 inches and a standard deviation of 0.015 inches. What is the probability that an axle selected at random will have a diameter of more than 1.970 inches?

35. Assume a normal distribution with the given mean and standard deviation:
    a. If $\mu = 0$ and $\sigma = 1$, find $P(Z > 1.06)$.
    b. If $\mu = -2$ and $\sigma = 2.5$, find $P(-1 < X < -0.5)$.
    c. For $N(0,1)$, find $Z_0$ such that $P(Z < Z_0) = 0.9821$.
    d. For $N(0,1)$, find $Z_0$ such that $P(-Z_0 < Z < Z_0) = 0.2358$.
    e. If $\mu = 1010$ and $\sigma = 200$, evaluate $P(X > 1300)$.
    f. If $\mu = 0$ and $\sigma = 2$, find the 70th percentile value.

36. Apples from a large shipment average 8.0 ounces in weight with a standard deviation of 1.6. What proportion of the fruit weighs between 6.6 and 9.4 ounces? Assume a normal distribution of weights.

37. The average score on a nationwide 200-point civil service test was 170.0 with a standard deviation of 10.0. The distribution of scores is considered normal.
    a. What is the percentile rank for a score of 175?
    b. Determine the value for the 45th percentile.

38. Assume that the heights of men applying for U.S. Forest Service training are normally distributed with a mean of 68.2 inches and $\sigma = 1.5$ inches. Past experience indicates that 20% will be classified as short, 55% as medium height, 20% as tall, and 5% as very tall. What is the tallest height in the medium class? How would a candidate who is 5 feet 9 inches tall be classified?

39. For the members of a union, income is normally distributed with $\mu = \$32,000.00$ and $\sigma = \$1,000.00$. Find the proportion of members who make under $\$30,000.00$.

40. Castparts, Inc., produces castings that have a mean weight of 2.3 pounds and a standard deviation of 0.3 pounds. Each is 1.5 inches high. Assume that the weights can be measured to any desired degree of accuracy and that the distribution of weights can be reasonably described with a normal curve.
    a. What percent of the castings have a weight greater than or equal to 3.0 pounds?
    b. Below what weight will the lightest 20% of the castings be found?

41. If $n = 36$ and $\pi = 0.50$, use the normal approximation to the binomial to find the
    a. mean.
    b. standard deviation of the approximating distribution.

42. A production process turns out roofing shingles that are normally 29% blemished. A random sample of 310 shingles was taken during an inspection of the plant by a

large homebuilder. What would be the
a. expected number of blemishes?
b. standard deviation?
Use a normal distribution.

43. The manufacturer of a new drink, Slim Cola, is conducting a taste-test against their own well-established diet cola. If there is essentially no difference in taste, what is the probability that at most 60 of 100 tasters would prefer the new cola? (*Hint:* Use $\pi = 1/2$.)

44. A manufacturer knows that 10% of the power units for one make of Citizens' Band (CB) radio burn out before their guarantee has expired. What is the probability that a merchant who has sold 100 such CB radios might be asked to replace at least 20 of them?

45. A grocery wholesaler has learned that in a railroad car of peaches 20% of the boxes are spoiled and cannot be sold.
    a. If she agrees to accept a randomly selected shipment of 400 boxes, what is the probability that 90 or more of the boxes will be spoiled?
    b. If she pays $1.00 per box for 400 boxes, what is her expected gross profit assuming she can sell the good boxes for $4.00 each? Assume the $1.00 cost includes labor.

46. A soft-drink manufacturer bottles 5000 drinks per day with a machine that produces normally distributed weights. State law requires that no more than 1% of the bottles can have contents below the advertised weight (12 fluid ounces). If the standard deviation of the content weight distribution is historically 0.2 fluid ounces, what should be the value of the setting of the filling machine if the manufacturer wants to just meet minimum requirements?

47. Ten percent of the fire bricks baked in an oven turn out to be defective in some way. Find the probability that exactly two will be defective in a sample of ten:
    a. Via the binomial distribution
    b. Via the Poisson distribution
    c. Via the normal approximation

**STATISTICS IN ACTION: CHALLENGE PROBLEMS**

48. The ability to predict and control the output of each production step is a key to sustaining both quality and quantity in manufacturing. James Krupp defines the essential conditions for a manufacturing process to be operating satisfactorily:

> The process capability study must answer two questions simultaneously:
> Is the process stable, i.e., are the results of the specific process being studied predictable?
> Is the process in control, i.e., do the predicted results of the process fall within the performance or dimensional range specified on the engineering drawing?

Being "in control" requires that the product measure, or count, follow a normal distribution.

SOURCE: Krupp, James A. G. "Process Capability: One Element of Zero Inventories." *Journal of Production and Inventory Management* 28, no. 3 (1987): 17–21.

Another "control" condition involves the limits of variability for the data and is discussed in Chapter 15.

Questions   The data below come from an assembly production. The data are 48 samples of the number of "good" (nondefective) units produced per minute of work. The samples were collected at 10-minute work intervals during an 8-hour work shift.

| Sample | Good Items | Sample | Good Items | Sample | Good Items | Sample | Good Items | Sample | Good Items | Sample | Good Items |
|--------|------------|--------|------------|--------|------------|--------|------------|--------|------------|--------|------------|
| 1 | 68 | 9 | 75 | 17 | 65 | 25 | 67 | 33 | 66 | 41 | 61 |
| 2 | 66 | 10 | 67 | 18 | 64 | 26 | 65 | 34 | 60 | 42 | 63 |
| 3 | 63 | 11 | 64 | 19 | 72 | 27 | 66 | 35 | 65 | 43 | 58 |
| 4 | 65 | 12 | 66 | 20 | 67 | 28 | 66 | 36 | 63 | 44 | 60 |
| 5 | 64 | 13 | 64 | 21 | 68 | 29 | 64 | 37 | 59 | 45 | 60 |
| 6 | 70 | 14 | 62 | 22 | 65 | 30 | 69 | 38 | 63 | 46 | 62 |
| 7 | 63 | 15 | 65 | 23 | 65 | 31 | 64 | 39 | 62 | 47 | 52 |
| 8 | 69 | 16 | 65 | 24 | 64 | 32 | 64 | 40 | 62 | 48 | 55 |

Complete the following histogram plot. Then discuss the reasonableness of a normal distribution. (*Suggestion:* Compare the observed percent of cases within $\pm 1$ sd, $\pm 1.5$ sd, $\pm 2$ sd, $\pm 3$ sd from the mean against theoretical values from Table B.1. Use 64.2 for the mean and 3.9 for the standard deviation.)

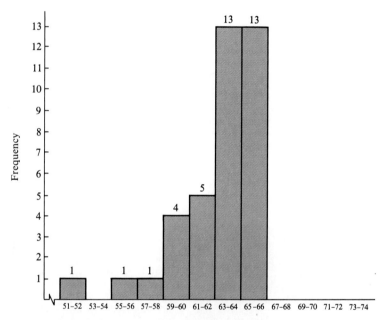

Number of good (non-defective)
units produced per minute

49. A normal distribution is a requirement, but is not enough to assure that a manufacturing process is satisfactory. Consider the sample number as a time measure in Problem 48. Plot the data with "Sample Number" on the horizontal and "Number of Good Items Produced" on the vertical axis. What is happening to the production as time progresses?

50. Some data are normally distributed, some are not. The following displays are Minitab stem-and-leaf plots* for three data sets taken from the *Statistical Abstract*. Use the procedures suggested in Problem 48 to determine if a normal distribution is reasonable in each case. Compare the observed percent of cases within $\pm 1$ sd, $\pm 1.5$ sd, $\pm 2$ sd, $\pm 3$ sd from the mean against normal distribution values from Table B.1.

* Example 3.3 describes the notation used in Minitab stem-and-leaf displays.

| Percent of the female population 16 years + who are in the labor force, by state, 1988 (from Table 629, p. 381) | | | Percent of the male population 16 years + who are in the labor force, by state, 1988 (from Table 629, p. 381) | | | Live births—rate per 1000 population, by state, 1988 (from Table 83, p. 64) | | |
|---|---|---|---|---|---|---|---|---|
| 1 | 41 | 1 ← 41.1% | 1 | 63 | 2 ← 63.2% | 1 | 12 | 0 ← 12.0 births / 1000 pop. |
| 1 | 42 |  | 1 | 64 |  | 5 | 13 | 2679 |
| 1 | 43 |  | 1 | 65 |  | 19 | 14 | 01445556778889 |
| 1 | 44 |  | 1 | 66 |  | (17) | 15 | 00122223455556788 |
| 1 | 45 |  | 1 | 67 |  | 15 | 16 | 1259 |
| 1 | 46 |  | 1 | 68 |  | 11 | 17 | 1146 |
| 1 | 47 |  | 1 | 69 |  | 7 | 18 | 2228 |
| 1 | 48 |  | 1 | 70 |  | 3 | 19 |  |
| 2 | 49 | 7 | 4 | 71 | 668 | 3 | 20 |  |
| 5 | 50 | 889 | 6 | 72 | 58 | 3 | 21 | 1 |
| 7 | 51 | 36 | 11 | 73 | 13377 | 2 | 22 | 0 |
| 7 | 52 |  | 17 | 74 | 133399 | 1 | 23 |  |
| 10 | 53 | 359 | 21 | 75 | 7899 | 1 | 24 |  |
| 12 | 54 | 56 | 25 | 76 | 4555 | 1 | 25 |  |
| 14 | 55 | 29 | (6) | 77 | 001367 | 1 | 26 |  |
| 18 | 56 | 0245 | 20 | 78 | 0113344679 | 1 | 27 |  |
| 21 | 57 | 004 | 10 | 79 | 1345679 | 1 | 28 |  |
| 24 | 58 | 469 | 3 | 80 | 3 | 1 | 29 |  |
| (7) | 59 | 0145789 | 2 | 81 | 1 | 1 | 30 |  |
| 20 | 60 | 002223559 | 1 | 82 | 4 | 1 | 31 | 3 |
| 11 | 61 | 5688 |  |  |  |  |  |  |
| 7 | 62 | 48 |  |  |  |  |  |  |
| 5 | 63 | 03 |  |  |  |  |  |  |
| 3 | 64 | 1 |  |  |  |  |  |  |
| 2 | 65 |  |  |  |  |  |  |  |
| 2 | 66 | 17 |  |  |  |  |  |  |
| Mean = 57.9, sd = 4.7 | | | Mean = 76.4, sd = 3.25 | | | Mean = 16.0, sd = 2.8 | | |

SOURCE: U.S. Bureau of the Census. *Statistical Abstract of the United States: 1990* (110th ed.). Washington, D.C., 1990.

51. In Problem 50 the data for "live births" have one extreme value: 31.3 for Washington, D.C. The remaining values are for the 50 states. As a matter of data analysis, discuss the possibility of excluding the Washington, D.C. value from the data set.

## BUSINESS CASE

*Everything's Normal*

Jim Lockhart is quality control manager of Duotronix, an electronics producer in the Silicon Valley near San Jose. On a recent morning he put his name and reputation on the line. He guaranteed a defense contractor that a large shipment of circuit boards would have mounting holes of the proper diameter, even though there was no time to inspect them. The circuit boards were already packaged and shipped (late). If not enough of them fit, a space project would be delayed six weeks. Duotronix might lose its other NASA contracts, and Lockhart might lose his job.

Duotronix had shipped 605 circuit boards. Unless 600 were satisfactory, the whole shipment would be returned. The critical item was a 1.000-inch-diameter mounting hole. If it was more than 0.010 out of tolerance (either + or −), the board would not fit.

Beth Tyle, Duotronix production manager, stopped by Lockhart's office after lunch to discuss the problem. She had just received a phone call from C. H. Mitchell, the vice president of marketing. Beth had assured Mr. Mitchell that the shipment would meet specifications, even though she wasn't totally convinced of it.

**Ms. T.**   Now we're both out on a limb on this thing, Jim. We both know we've had problems with that mounting dimension in the past. Mitchell knows it too. Are you sure you're right? I was just going on what you said yesterday.

**Mr. L.**   No, I'm not sure!

**Ms. T.**   Well, this is a fine time to tell me. You said yesterday that the order was probably OK.

**Mr. L.**   I didn't really say it was probably OK. I said that by all probabilities, it should be OK. There's a difference.

**Ms. T.**   The floor is yours. Please explain.

**Mr. L.**   Well, look. I've done enough inspections of those mounting holes to know the diameters follow a certain pattern. It's called a normal distribution. Furthermore, the out-of-tolerance holes follow a random pattern. That is, they're unpredictable—you can't tell when they occur.

**Ms. T.**   Seems to me that's worse. Wouldn't it be better if we knew when they were going bad?

**Mr. L.**   Well, it would seem so, but since we don't know what causes them, we treat the error amount as random. Do you know that those holes have mean diameter of 1.002 inches with a standard deviation of 0.003 inches?

**Ms. T.**   You mean they aren't 1.000 inches on the average? I thought that's exactly what they're supposed to be!

**Mr. L.**   They're supposed to be within a tolerance. The standard deviation is a measure of spread, like a range. The type of distribution is normal, and since I know the mean and the standard deviation, I have a pretty good feel for what those diameters are. The way I figure it, there is practically no chance that the holes will be undersize. Furthermore, the probability of getting any mounting holes larger than 1.010 inches is about 0.004. Just look at this sketch and my calculations (see Figure 6.20).

$$Z = \frac{X - \mu}{\sigma} = \frac{1.010 - 1.002}{0.003} = 2.67$$

Using some probability tables, we can compute $P(Z > 2.67) = 0.5000 - 0.4962 = 0.0038$. If slightly less than 0.4 of 1% of the boards exceed 1.010 inches, in 605 boards that is only

$$605(0.0038) = 2.3 \text{ boards}$$

---

**FIGURE 6.20**   Normal Distribution of Mounting Hole Diameters

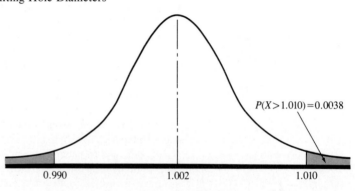

$P(X > 1.010) = 0.0038$

0.990          1.002          1.010

X (diameter) in inches

So I figure about $2\frac{1}{2}$ boards are defective. I can be 100% off (that is, 5 boards defective) and we're still OK. I don't like playing this close any better than you do. But I figure in this case the risk was worth it. The odds are on our side.

**Ms. T.**    Well now, I don't know if I feel better or worse! You mean you can really operate like that?

**Mr. L.**    Beth, when you're in quality control, everything is a risk. I just prefer better odds than this, that's all.

Beth Tyle answered her phone the next day.

**Mr. L.**    Beth, I heard from our contractor—four bad ones.

**Ms. T.**    Congratulations!

**Mr. L.**    Not really! I had a 100% safety factor but I used up about 70%. Kind of scary, isn't it?

**Business Case Questions**

52. What is Mr. Lockhart's concept of randomness? In particular, how does he distinguish a "random" error from any other kind of tolerance situation?

53. What kind of evidence might Mr. Lockhart have had to "know the type of distribution"? That is, how might he have known the hole diameters were normally distributed? *inspections, history, sampling*

54. Suppose the process was producing holes with a mean diameter of 1.004 inches instead of 1.002 inches. Could the shipment still be expected to meet the specifications? (*Hint:* Compute the percent of the boards that could be expected to exceed 1.010 inches if $\mu = 1.004$.) *find z value, find % outside 1.004*
    *z = 2.    13.79 app*

55. Another customer who received 2000 circuit boards from the batch, with mean 1.002 inches and standard deviation 0.003 inches, indicates that the shipment is not acceptable, because it contains 13 boards beyond their tolerance. What is this company's tolerance for the board mounting-hole diameters?

## COMPUTER APPLICATION

*A Normal Distribution*

The normal random variable has extensive application in business. The household income variable from the PUMS data base provides one example. Figure 6.21 is a SPSS setup that gives us two displays.

**FIGURE 6.21**    A SPSS Computer Run That Gives Two Frequency Distributions for the PUMS Household Income Variable

**(Input)**

```
                                                              1
1  NUMBERED
2  TITLE          PUMS INCOME SUMMARY
3  DATA LIST      FILE = INDATA/                              2
4                 PERSONS 12-13 HSHDINC 35-39 AGE1 43-44

5  SET BLANKS = -1
6  SELECT IF      ((PERSONS GE 3) AND ((AGE1 GT 34) AND (AGE1 LT 51)))    3
7  COMPUTE        INCGP = HSHDINC
8  RECODE         HSHDINC (75000 THRU HI = 75000)
9  RECODE         INCGP (LO THRU 0 = 1) (1 THRU 700 = 2)
```

*(continued)*

**FIGURE 6.21**    *(continued)*

```
10                          (701 THRU 2300 = 3) (2301 THRU 3400 = 4)
11                          (3401 THRU 4600 = 5) (4601 THRU 5300 = 6)
12                          (5301 THRU 8800 = 7) (8801 THRU 11200 = 8)
13                          (11201 THRU 11600 = 9) (11601 THRU 14000 = 10)
14                          (14001 THRU 16000 = 11) (16001 THRU 24900 = 12)
15                          (24901 THRU 34500 = 13) (34501 THRU 37500 = 14)
16                          (37501 THRU 41500 = 15) (41501 THRU 42600 = 16)
17                          (42601 THRU 47000 = 17) (47001 THRU 50400 = 18)
18                          (50401 THRU 52600 = 19) (52601 THRU 55200 = 20)
19                          (55201 THRU 58100 = 21) (58101 THRU 58500 = 22)
20                          (58501 THRU 62800 = 23) (62801 THRU 70000 = 24)
21                          (70001 THRU 74000 = 25) (74001 THRU 75000 = 26)
22                          (75001 THRU HI = 27)
23  FREQUENCIES    VARIABLES = INCGP/
24                 FORMAT = DOUBLE NEWPAGE/
25                 STATISTICS = MEAN MODE MAX STDDEV RANGE SUM
                               MEDIAN VARIANCE SKEWNESS MIN/
```

**(Output)**

INCGP

| Value Label | Value | Frequency | Percent | Valid Percent | Cum Percent |
|---|---|---|---|---|---|
| | 1.00 | 18 | .5 | .5 | .5 |
| | 2.00 | 19 | .5 | .5 | 1.0 |
| | 3.00 | 37 | 1.0 | 1.0 | 2.0 |
| | 4.00 | 37 | 1.0 | 1.0 | 3.0 |
| | 5.00 | 38 | 1.0 | 1.0 | 4.0 |
| | 6.00 | 38 | 1.0 | 1.0 | 5.0 |
| | 7.00 | 183 | 4.9 | 4.9 | 9.8 |
| | 8.00 | 162 | 4.3 | 4.3 | 14.2 |
| | 9.00 | 26 | .7 | .7 | 14.9 |
| | 10.00 | 170 | 4.5 | 4.5 | 19.4 |
| | 11.00 | 192 | 5.1 | 5.1 | 24.5 |
| | 12.00 | 955 | 25.4 | 25.4 | 49.9 |
| | 13.00 | 940 | 25.0 | 25.0 | 74.9 |
| | 14.00 | 185 | 4.9 | 4.9 | 79.9 |
| | 15.00 | 188 | 5.0 | 5.0 | 84.9 |
| | 16.00 | 51 | 1.4 | 1.4 | 86.2 |
| | 17.00 | 134 | 3.6 | 3.6 | 89.8 |
| | 18.00 | 85 | 2.3 | 2.3 | 92.0 |
| | 19.00 | 36 | 1.0 | 1.0 | 93.0 |
| | 20.00 | 36 | 1.0 | 1.0 | 94.0 |
| | 21.00 | 41 | 1.1 | 1.1 | 95.0 |
| | 22.00 | 1 | .0 | .0 | 95.1 |
| | 23.00 | 33 | .9 | .9 | 96.0 |
| | 24.00 | 37 | 1.0 | 1.0 | 96.9 |
| | 25.00 | 20 | .5 | .5 | 97.5 |
| | 26.00 | 95 | 2.5 | 2.5 | 100.0 |
| | Total | 3757 | 100.0 | 100.0 | |

This computer run relates to the marketing problem that targets households with 3+ persons and with age of head of household 35–50 years. The frequencies we have run on the 20,116 PUMS data base require some explanation.

**FIGURE 6.22**    A Histogram on Household Incomes for 3+ Person Households
with Head of Household Aged 35–50 Years

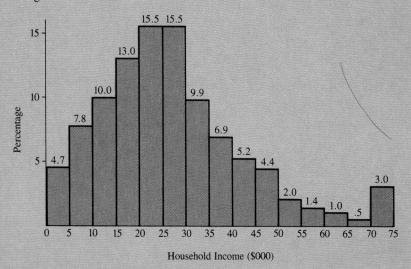

1  NUMBERED indicates that the user wants the program statements to be numbered on the paper output. The TITLE is simply for user recognition. DATA
   LIST indicates the name of the file where the data are stored and the file
   positions for the variables (PERSONS, HSHDINC, and AGE1).

2  SET BLANKS is a procedure for identifying any missing (blank) numeric data.
   SELECT IF restricts the file to those records (3757 of the 20,116) that have greater
   than or equal (GE) 3 persons and householder with age 35 to 50.

3  The statement COMPUTE INCGROUP = HSHDINC and the subsequent RECODE
   INCGP provide the data to answer some marketing questions. (Actually this
   requires two computer runs. A first run uses FREQUENCIES VARIABLES = HSHDINC.
   This is the basis for Figure 6.22. Since it produced a 100+ page output, the
   results are not displayed here! The first run also is the basis for the RECODE
   INCGP (income groups). The second run was FREQUENCIES VARIABLES = INCGP
   and it is displayed here.)

These results are summarized in Table 6.3 on page 246.

Marketing managers want to be able to select households at various percentages of the population. For example, what household income level includes
the highest 20% of the target population? Suppose the top 20% can afford and will
have home computers if they want. What is the household income range for the
next highest 20% income households?

Rather than answer questions singly, a manager may first want to check
whether a normal distribution is appropriate. If so, then the normal distribution
and Table B.1 can be used to answer many questions about the householder income variable. Figure 6.22 suggests a bell-shaped curve. But is the distribution for
these 3757 incomes adequately described by a normal curve?

To answer this question, we calculate some selected normal distribution
percentages and compare them to the actual values. For example, the mean plus
or minus two standard deviations spans $26.93 \pm (2 \times 15.71)$, or $0 to $58.35

**TABLE 6.3** Distribution for the Target Population for PUMS Data Household Incomes by Selected Incomes

| Income ($000) | Cumulative Distribution (%) | Income ($000) | Cumulative Distribution (%) |
|---|---|---|---|
| $0.0 and lower | 0.5 | 37.5 and lower | 79.9 |
| 0.7 and lower | 1.0 | 41.5 and lower | 84.9 |
| 2.3 and lower | 2.0 | 42.6 and lower | 86.2 |
| 3.4 and lower | 3.0 | 47.0 and lower | 89.8 |
| 4.6 and lower | 4.0 | 50.4 and lower | 92.0 |
| 5.3 and lower | 5.0 | 52.6 and lower | 93.0 |
| 8.8 and lower | 9.8 | 55.2 and lower | 94.0 |
| 11.2 and lower | 14.2 | 58.1 and lower | 95.0 |
| 11.6 and lower | 14.9 | 58.5 and lower | 95.1 |
| 14.0 and lower | 19.4 | 62.8 and lower | 96.0 |
| 16.0 and lower | 24.5 | 70.0 and lower | 96.9 |
| 24.9 and lower | 49.9 | 74.0 and lower | 97.5 |
| 34.5 and lower | 74.9 | 75.0 and lower | 100.0 |

| | | | | | |
|---|---|---|---|---|---|
| Mean | 26.93 | St Dev | 15.71 | Kurtosis | 1.12 |
| Median | 25.00 | Variance | 246.94 | Maximum | 750.00 |
| Mode | 75.00 | Skewness | 0.98 | Minimum | 0.00 |

*Note:* The statistics (Mean, Median, and so on) come from the SPSS frequencies on HSHDINC.

(thousand). By Table 6.3 this range includes about 95.1% of the distribution. By normal distribution theory and Table 1 in Appendix B, two standard deviations from the mean encompass 95.44% of the data. This is very close. Other comparatives are summarized in Table 6.4.

With these data a marketing manager could be satisfied that a normal distribution reasonably describes this PUMS data. So the manager can answer the following questions—and many related ones—by using Table 1, Appendix B and the normal curve probability distribution. This is not a perfect fit, but it illustrates one way that empirical data can be used to identify a reasonable probability distribution model.

The marketing manager's specific questions are answered for $X$ = household income.

**TABLE 6.4** Dispersion of Household Income Data

| Standard Deviations | Range ($000) | Actual Percent | Normal Curve Theory (Table B.1) |
|---|---|---|---|
| ±1.0 | 11.22–42.64 | 72.8% | 68.26% |
| ±1.5 | 3.36–50.50 | 89.1 | 86.64 |
| ±2.0 | 0.0 –58.35 | 95.1+ | 95.44 |
| ±3.0 | 0.0 –74.06 | 97.5 | 99.74 |

1. The manager asks, What income $X_0$ defines the top 20% of income? By Table 1, Appendix B, $Z_0 = +0.84$ is the 80th percentile, so $Z_0 = +0.84$ gives $X_0 = 40.13$. By normal distribution the top 20% of incomes are approximately $40,000 and higher.

2. The manager's second question is, what is the income range for the 60th to 80th percentiles for income? $P(Z > Z_0) = 0.20$, so find $Z_0$, then $X_0$. $P(Z > Z_1) = 0.40$; find $Z_1$, then $X_1$. Then $X_0 - X_1 =$ range of the 60–80 percentiles for the income distribution $P(Z > Z_0) = 0.40$ requires $Z_0 = 0.255$, so $X_0 = 30.9$, or approximately $31,000. Incomes from (approximately) $31,000 to $40,000 include the 60th to 80th percentiles for the household income distribution.

   Some comments are appropriate. First, these data are truncated at both extremes $0 and $75,000. This is a matter of the data source and is not a choice for some secondary data sources. For example, the 3.0% for income range 70–75 ($ thousand) is forced by the truncation. This includes incomes that are much larger than $75,000. However, the normal curve is a reasonable approximation. Second, Chapter 16 includes a statistical test for the goodness of fit for a normal curve to the data, so a more exact method is available.

   The Computer Problems offer more applications to PUMS data.

## COMPUTER PROBLEMS

*Data Base Problems*

These problems are related to the Chapter 6 Computer Application.

56. Use the mean and standard deviation as reported in Table 6.3 along with Table 1, Appendix B, to complete the following table.

| Income ($000) | Cumulative Distribution (%) |
|---|---|
| 13.7 and lower | 20.0 |
| ___ and lower | 25.1 |
| | 50.0 |
| ⋮ | 75.1 |
| | 89.9 |
| ___ and lower | 100.0 |

*Note:* 13.7 is calculated as follows:

$$Z = \frac{X - \text{Mean}}{\text{Standard deviation}}$$

$$-0.84 = \frac{X - 26.93}{15.71} \quad \text{gives} \quad X = 13.7$$

Compare your answers to the incomes in Table 6.3.
*Suggestion:* You might want to use a simple computer program to calculate your $X$-values.

57. Develop a computer frequency distribution for the PUMS file of 20,116 records for households with 3+ persons with age of head of household 35–39 years. Here is a skeleton routine (SPSS).

```
SELECT IF    ((PERSONS GT 2) AND (AGE GT 34)
             AND (AGE LT 40))
COMPUTE      INCOME = TRUNC((HSHDINC + 50)/100)
FREQUENCIES  VARIABLES = INCOME/
             STATISTICS = ALL/
```

Use the results of the frequency distribution to construct a table like Table 6.3.

58. A table for the description in Problem 57 can be checked by a computer run (as suggested in Problem 57). Use the data in the table to construct a graph similar to Figure 6.22. Is the pattern close to a normal curve?

| Income ($000) | Cumulative Distribution (%) | Income ($000) | Cumulative Distribution (%) |
|---|---|---|---|
| 0 and lower | 0.4 | 40 and lower | 88.0 |
| 5 and lower | 4.9 | 45 and lower | 92.1 |
| 10 and lower | 13.2 | 50 and lower | 94.2 |
| 15 and lower | 24.4 | 55 and lower | 95.2 |
| 20 and lower | 39.4 | 60 and lower | 96.2 |
| 25 and lower | 57.5 | 65 and lower | 97.0 |
| 30 and lower | 73.3 | 70 and lower | 97.6 |
| 35 and lower | 82.6 | 75 and lower | 100.0 |

1351 cases

59. Here is a Minitab routine that gives values for a histogram that approximates a normal curve for the household income data that are displayed in Figure 6.22.

```
MTB > SET C1
DATA>  2.5  7.5 12.5 17.5 22.5 27.5 32.5 37.5
DATA> 42.5 47.5 52.5 57.5 62.5 67.5 72.5
DATA> END
MTB > PDF C1, PUT INTO C2;
SUBC> NORMAL 26.93, 15.71.
MTB > PRINT C1 C2
```

The output is calculated heights of the normal curve ( = C1) at each class midpoint value ( = C2). Construct a histogram from your data and visually compare this to Figure 6.22.

60. How closely does the household income data of Figure 6.22 follow a normal distribution (mean = 26.93, standard deviation = 15.71)? Continue the Minitab program of the preceding problem with

```
MTB > CDF 5;
SUBC> NORMAL 26.93, 15.71.
```

This should give $-1.40$ (a $Z$-score), 0.0814 (a cumulative probability). Repeat the above (2) statements for 10, then 15, ..., through 75 to obtain 15 cumulative probabilities. Use your data and the percentages in Figure 6.22 to complete the table.

| Income ($000) | Cumulative Actual Percent (Figure 6.22) | Normal Distribution ($\times$ 100) |
|---|---|---|
| 0 | 0.0 | 0.0 |
| 5 | 4.70 | 8.14 |
| 10 | 12.50 | 14.06 |
| ⋮ | | |
| 75 | | |

*Note:* You will need to accumulate the percents in Figure 6.22 to obtain the entries 4.70, 12.50, .... How closely does the household income data follow this normal distribution, $N(26.93, 15.71^2)$?

**FOR FURTHER READING**

[1] Anderson, David R., Dennis J. Sweeney, and Thomas A. Williams, *Statistics for Business and Economics*, 4th ed. St. Paul, MN: West, 1989.

[2] Fruend, John E., and Ronald Walpole. *Mathematical Statistics*, 4th ed. Englewood Cliffs, NJ: Prentice-Hall, 1987.

[3] Mason, Robert D. *Statistical Techniques in Business and Economics*, 7th ed. Homewood, IL: Irwin, 1990.

[4] Richards, Larry E., and Jerry L. LaCava. *Business Statistics: Why and When?* 2nd ed. New York: McGraw-Hill, 1983.

[5] Ryan, Barbara F., Brian L. Joiner, and Thomas A. Ryan, Jr. *Minitab Handbook*, 2nd ed. Boston: PWS-Kent, 1985.

[6] *SPSS Reference Guide* (version 4.0). Chicago: SPSS, Inc., 1990.

# SAMPLING DISTRIBUTIONS

## 7.1 INTRODUCTION

Many of us are familiar with the A. C. Nielsen company, which produces television program ratings. The Nielson ratings have a good deal to say about whether a TV program will be awarded prime time or whether it will be dropped. Data for these decisions are obtained from a sample of viewers. Sampling is also used to assess voter opinions, obtain weather readings, make traffic counts, and take periodic readings of heartbeat, temperature, and blood pressure. Sampling is also a part of our everday consumer buying. Supermarkets frequently offer small samples of food to entice customers to try their products. *The purpose of sampling is to obtain a description of population characteristics without observing all of the population values.*

Sampling gives us statistical measures and saves time and money. So you should not be surprised to find that it plays a major role in business statistics.

### Finite and Infinite Populations

Because of costs, the population size is an important consideration in deciding whether to sample or to take a complete census. Population sizes are classified as either *finite* or *infinite*.

A population is finite if its elements can be counted or enumerated—1, 2, 3, . . . , $N$, where $N$ is a whole number greater than or equal to 1.

An infinite population is one that is not finite.

A population is finite if all of its elements can be counted; otherwise it is infinite.* The population of the United States is large but finite. Because a com-

---

* Any population is considered infinite if it is *countably infinite*. That is, elements of the population can be given labels 1, 2, 3, . . . , 1000, . . . , 1,000,000, . . . , ∞ (∞ stands for infinity). This means that although we might envision assigning a counting label to each element, the number of elements is so great that it would be physically impossible to assign a value to each one.

plete count of such a large population is costly, the U.S. population census—a complete count—is taken only once each 10 years. Sampling is carried out between censuses. The set of all numbers is an infinite population because it is impossible to count all numbers.

Businesses rely heavily on sampling. For example, manufacturers of soap, breakfast cereal, and toothpaste frequently distribute samples of their products as a means of gauging consumer response. Manufacturers must limit the number of samples, however, because sampling costs can run from $0.50 to $20 or more per sample! Time is also critical, because competitors are frequently quick to copy a popular product. Data from a sample can be collected more quickly than data from the full population, and with reasonable planning sample data can provide much information. Thus it is practical for businesses to use samples. But because the sampling procedures are different, we distinguish finite from infinite populations.

## Inference from Samples

Generally, the purpose in drawing a sample is to *infer* from a sample value, or statistic, something about a population value, or parameter. This notion of *inductive inference* is the counterpart of the deductive inference we used in Chapters 4 through 6. In those chapters we assumed some knowledge of the underlying probability distribution and then *deduced* statements about the probability of specific events or elements of the distribution.

The focus in Chapter 7 is on statistical (inductive) inference. That is, we shall use specific information from a sample to *infer* something about the general population from which the sample was drawn. This is the fundamental approach that underlies inferential statistics. In this chapter we show how a specific sampling procedure, random sampling, enables us to infer something about the mean, a proportion, or other characteristics of the population under study. In particular, we shall be assuming the use of simple random sampling. In Chapter 1 we saw that simple random sampling was a procedure that assured every member of the population an equal chance of being included.

In this chapter we elaborate on this concept and on some of its many applications in business.

Upon completing Chapter 7, you should be able to

1. define simple random sampling;
2. select a simple random sample;
3. recognize what a sampling distribution is and how it differs from other types of probability distributions;
4. explain the central limit theorem;
5. make statistical statements that rely upon knowledge of the central limit theorem.

## 7.2 SIMPLE RANDOM SAMPLING AND RELATED SAMPLING METHODS

This section introduces two classes of sampling and five well-known sampling procedures. Then we focus upon the one procedure that is best for our work, which is random sampling.

Recall that a statistic describes some characteristic of a sample, whereas a parameter describes some characteristic of a population. The primary difference is that a parameter is determined from the total distribution—that is, by a census—and therefore has a fixed value. A statistic, however, is dependent upon the chance occurrence of the items in the sample selected from a population, and so may change from sample to sample.*

**Using Sample Statistics as Estimators**

One objective in statistical inference is to make reasonable projections from a statistic to a related parameter. Statistics used in this way are called *estimators*. Common estimators include $\bar{X}$ for $\mu$, $s$ for $\sigma$, and $X/n$ (relative frequency) for $\pi$ (the population proportion). A *point estimate* is the value obtained from applying an estimator to a data sample.

> An estimator is a statistic used to estimate a population parameter.
>
> A point estimate is the numerical value of the estimator.

An example should help to distinguish between estimators and estimates. Suppose a marketing study is to be taken to estimate the true proportion of consumers who prefer the company's brand of cologne over all competing brands. An estimator would be the relative frequency of preference, $X$, in a sample of consumers, $n$. That is, $p = X/n$ is the estimator for $\pi$, the population proportion of consumers who prefer the company's brand over all other brands. If $X = 32$ consumers prefer the product and there are $n = 100$ consumers in the sample, then the point estimate of the population proportion who prefer the company's brand is 32/100, or 0.32.

Estimation is one of the two major areas in inference described in subsequent chapters. To obtain meaningful estimates, we must conduct the sampling carefully to assure that the sample data do not possess any inherent bias.

The purpose of sampling is to make reasonable estimates from a statistic to a parameter. But how does one ensure that the projections will be reasonable? For example, suppose an industrial engineer wants to estimate the number of garments completed by textile workers in a factory. Would it be better to count the output of one day or to take samples at randomly selected hours throughout the week? The first technique might produce an incorrect estimate because production tends to vary with the day of the week. So a sample of one day's output has a potential for bias. The latter approach, which checks production at random times, is more likely to result in samples that represent the actual rate of production. Random samples can give estimates that are unbiased and reliable.

Let us turn now to two widely recognized approaches to sampling.

**Two Classes of Sampling**

Samples are classified either as (1) *probability samples* or as (2) *judgment* (nonprobability) *samples*.

> A probability sample is a sample chosen in a way that assures every unit in the population a nonzero chance of being included.
>
> A judgmental sample is a sample chosen on the basis of someone's opinion of which elements of the population should be included in the sample.

---

* Under traditional classical statistics, parameters are fixed. Recall, however, that in Bayesian analysis the parameter is viewed as though it were a random variable. This is one of the major extensions of the Bayesian approach and has proved to be a useful analytical tool for decision making under uncertainty, as we shall see in Chapter 18.

The probability approach results in a sample that is representative of the population to the extent that nonrepresentative samples have small probabilities of being selected. For example, if the true mean weight of bags of flour is 50 pounds (with $\sigma = 2$ pounds), the likelihood of obtaining a nonrepresentative mean as low as 45 pounds (from a sample of, say 30 bags) would be very small. The probabilities of obtaining various sample results (e.g., 45 pounds, 51 pounds) are typically not equal in probability sampling. But if the population parameter were known beforehand, the probabilities of obtaining specific sample results could be calculated prior to the actual sampling activity. In this sense, probability sampling affords analysts a measure of how well or how poorly they can expect a sample statistic to describe the parameter. On the other hand, judgmental sampling can result in considerable savings in time and money—although it *may* result in biased samples and some incorrect estimates and conclusions.

**Probability Sampling**

*Probability sampling* means that every item in the population has *some identifiable probability* of being selected for a sample. Common ways of assigning probability include (1) probability proportional to size (pps sampling) and (2) equal probability for every item (random sampling). For example, pps sampling could be used by a car dealership to inventory parts in their parts storage. The pps approach would permit the company to vary the sample proportions according to cost or some other criteria. For example, a high percentage of car engine blocks would probably be counted. Less accuracy in dollar value might be ascribed to cotter pins. A very large error in the number of cotter pins would be required to offset the inaccuracy in cost of one engine block.

Pps sampling could be used to inventory machine parts, where the probability of being included is proportional to a unit's value. However, a more widely used technique is (simple) random sampling.

**Simple Random Sampling**

Random sampling is the best known of the probability sampling techniques, partly because it assures each element an equal chance of being included in the sample. Many sampling applications involve units that are reasonably uniform or have equal importance. An example is the PUMS sample of the U.S. population to which we referred in Chapter 1 and which is our data base. Many firms use the PUMS sample rather than looking at the full census because it is a very large *random sample* of our population.

> Simple random sampling is any procedure that guarantees an equal chance of inclusion for every item in the population at a first selection and equal chance for all remaining items at each succeeding selection.

This condition is easily visualized with *finite populations*. For example, suppose you have a box of $N$ tickets, where each ticket represents a distinct inventory item. If these tickets are thoroughly mixed and a single ticket is drawn, the selection represents a random sample of one. The chance of selection, $1/N$, is the same for every ticket. The remaining tickets could again be mixed and then a second ticket chosen. By repeating this process $n$ times, you would select a random sample of $n$ tickets (representing items). This process might be used, for example, if $n = 5$ items are to be chosen from $N = 100$ items to estimate the total value of an inventory. This can be formalized: The probability for any specific sample of $n$ items selected at random and without replacement from a finite population of size $N$ is

$$P(\text{Any sample of } n \text{ from a population of size } N) = \frac{1}{{}_N C_n} \qquad (7.1)$$

Notice that when $n = 1$, then $P(\text{Any sample of 1}) = 1/{}_N C_1 = 1/N$. Probability calculations concerned with random sampling from finite populations rely on computing the number of combinations. Samples are generally drawn without replacement.

The illustration of drawing tickets randomly from a box constitutes sampling without replacement and without regard to order. We can use the combinatorial procedure of Chapter 4 to compute the number of combinations available. A simple random sample of $n = 5$ from a population of 100 could be chosen in any one of ${}_N C_n$ ways:

$$_{100}C_5 = \frac{100!}{5! \; 95!} = \frac{100 \cdot 99 \cdot 98 \cdot 97 \cdot 96}{5!} = 75{,}287{,}520 \text{ ways}$$

Recognizing that a specific set of 5 tickets is one of the 75,287,520 possible combinations, we can make a probability statement, using Equation (7.1), that

$$P(\text{Any random sample of 5 from 100}) = \frac{1}{75{,}287{,}520} = 0.0000000133$$

Numbers like this help explain why so many states are finding that their lotteries are such good money-makers.

The condition of randomness depends upon the process by which sample items are drawn. It does not depend upon the underlying probability distribution of the variable. To ensure randomness, the sampling procedures must guarantee that all elements of the population have an equal chance of being selected. For samples of $n < N$, random sampling is the most reliable way to produce a sample representative of the population because, in the long run, every sample of $n$ has an equal chance of being chosen.

For *infinite populations*, a random sample is chosen by independent selection of items from the distribution of an infinite number of possible observations. For example, suppose a management analyst wished to take a random sample to estimate the time spent by tellers serving customers at a drive-in bank. Note that time values, being ratio-level data, have an infinite number of possibilities. We estimate service time by using a random numbers source to specify $n$ service points in time. This process yields a random sample of $n$ independently determined observations from the distribution of an infinite number of possible outcomes—in this case, points in time.

## Other Approaches to Probability Sampling

In addition to pps and simple random sampling, other methods of probability sampling include systematic sampling, stratified random sampling, and cluster sampling.

*Systematic sampling* is commonly used instead of random samples in production processes, business surveys, and for selecting items from files.

> For systematic sampling, elements of the population are categorized in some way (such as alphabetically or numerically) and a random starting point is selected. Then every $N$th item of the categorized population is included until the sample size $n$ is satisfied.

For example, a systematic sampling procedure might begin with a randomly chosen start (from 001 to 100) as selected from an assembly line, a computerized file, or a customer list. After the first item is selected, then, for example, every 100th item might be checked. This is often thought to be more convenient and operational than simple random sampling. Systematic sampling usually assures a good cross-sectional representation of the population. An exception occurs when the $N$th item coincides with a cycle or seasonal pattern. For example, we would *not* systematically use only December sales information to develop a profile for annual retail sales. December sales might be well above the annual average and thereby produce an upward bias in a sales forecast.

*Stratified random sampling* is sometimes used to reduce costs while still ensuring that the major groups in the population are appropriately represented. Stratification simply means that like units are grouped together. Then random sampling is performed within each group.

> For stratified random sampling, elements of the population are first grouped into relatively homogeneous subgroups, or strata. Then a random sample is selected from each stratum (subgroup).

For example, in a wage survey of personnel records, employees might be stratified by job (engineers, production workers, or salespeople) before random sampling is begun. In doing this, we seek to make the strata as internally homogeneous as possible by recognizing that the wages of engineers are likely to be different from those of production workers or sales representatives.

Stratification thus partitions the population so that there is less variability within each stratum than in the population at large. With less variability in a stratum, the sample size requirements are reduced and there is a potential for cost saving. On the other hand, some additional cost is involved in the actual work of stratifying the population. A shortcoming of this method can be the inability to identify good (homogeneous) strata and to determine their proportions in the total population.

*Cluster sampling* is another probability sampling technique. Here, the target population (the subject of the study) is segmented into mutually exclusive subgroups, each representing the entire population. A random sample of the subgroups is then selected to provide estimates of the population values.

> For cluster sampling, elements of the population are first grouped into primary subgroups, or clusters, each of which should be representative of the entire population. Then a random sample of clusters is selected to provide the required data.

If the estimates are calculated directly from the subgroups, the process is referred to as single-stage cluster sampling. Alternatively, one may further divide the subgroups into smaller groups and select randomly from them. This is two-stage cluster sampling.

Cluster sampling is typically less expensive (per observation) than simple random sampling. However, it requires a larger sample for the same amount of variation. To be effective the clusters should be homogeneous relative to other clusters, and internally as heterogeneous as the population. The objective is to form groups of clusters that are small images of the target population. Localizing the sample units to relatively few groups or areas can produce substantial cost reduction for the fieldwork of obtaining the sample.

The procedures for calculating summary statistics of stratified and cluster samples and projecting them onto their populations are slightly more complex than those used for simple random samples. You may wish to consult a textbook on sampling if you have occasion to use one of these sampling methods.

## Nonprobability Sampling

The second major type of sampling is *judgment*, or *nonprobability sampling*.

**For judgment sampling the sample items are deliberately chosen to satisfy a condition or belief.**

With judgment sampling, the sample elements are deliberately chosen to satisfy a given criterion. The condition may be a *quota*, or it can be based simply on *convenience*. Sometimes a judgment sample is *purposeful* in that specific items are consciously included. This is the case for the items that make up the consumer price index.

Judgment samples are usually cheaper than probability samples of comparable size. However, *they are not statistically reliable*, and there is no accepted way to assess their validity. The degree of "nonrepresentativeness" of probability samples can be computed. With judgment samples this is not the case. Judgment samples can be nonrepresentative because of poor or biased judgment. If the judgment is good, then the sample may be representative; if not, the sample results will not accurately reflect the full population.

Our analysis in the remainder of the book will rest upon the assumption that random samples are used to obtain the data. Probability samples ensure a known, calculable chance that any item in the population will be in the sample. This is a necessary prerequisite for the statistical inference that follows.

## Problems 7.2

1. Let a population of size 4 be defined by the values $-1, 2, 3$, and 4. Assuming random samples of $n = 2$, there are $_4C_2 = 6$ pairs: $(-1,2)$, $(-1,3)$, $(-1,4)$, $(2,3)$, $(2,4)$, and $(3,4)$. Determine, but do not singly identify, the number of distinct samples of $n = 2$:
   a. From $N = 40$     b. From $N = 400$

2. For a certain infinite population, 20% of the probability resides on the interval 2.0 to 3.0. What is the probability that a sample of two chosen at random from this population will give each of the following?
   a. Both values on this interval
   b. At least one on this interval

3. A recent study indicates that in a certain firm 400 persons voted in the last union election. There were 642 employees eligible to vote, and the mean age of those who voted was 28.4 years.
   a. For the population of all employees eligible to vote, identify and estimate as many parameters as you can.

b. Could the sample of those who voted be considered a random sample of all the employees eligible to vote? Explain.

4. Suppose you wanted to make a survey of the safety condition of cars in your county. If there are 150,000 cars registered in the county, how many random samples of size 100 are possible? (Just set up the expression; do not compute a final answer.)

5. Mr. Baker's television is not operating properly, and since he has just moved to this locality, he does not know which repair shops give satisfactory service. The Better Business Bureau has rated the firms in this area as shown in the table ($S$ = satisfactory, $U$ = unsatisfactory).

a. If Mr. Baker selects a firm at random from the telephone directory, what is the chance that the firm selected is rated satisfactory by the Better Business Bureau?

| Firm | Service | Firm | Service | Firm | Service |
|------|---------|------|---------|------|---------|
| 1 | S | 6 | S | 11 | S |
| 2 | U | 7 | S | 12 | U |
| 3 | S | 8 | U | 13 | U |
| 4 | U | 9 | U | 14 | S |
| 5 | S | 10 | S | 15 | S |

b. If he narrows his choice by selecting a simple random sample of three from which to obtain repair estimates, what is the probability that at least one will be rated satisfactory?

6. Consider the following figures as a population of sales values ($000): 20, 56, 78, 30, 63, 89, 36, 64, 89, 36, 67, 96, 42, 82, 102. The following categories are proposed for grouping the units:

Group 1    20, 56, 63, 89, 89
Group 2    30, 42, 64, 78, 102
Group 3    36, 36, 67, 82, 96

a. Is this grouping more conducive to stratified sampling or to cluster sampling? Give reasons for your choice.
b. Using every third value from the sales values, select an $N$th ($N = 3$) systematic sample from a random start (start at 1, 2, or 3; it doesn't matter). Use your results to explain why systematic sampling should not be used when $N$ equals or nearly equals the period in cyclical data. (*Note:* These data are periodic repetitive, having a period of three values.)

## 7.3 SAMPLING DISTRIBUTIONS

As a business analyst you will recognize that most jobs have a few key elements. This is true in statistics, also. One of the key elements in statistics is the relationship between random sampling and probabilities. This section explores that relationship and defines the important probability rule that follows from it.

**FIGURE 7.1**    The Population Distribution for a Discrete Uniform Random Variable

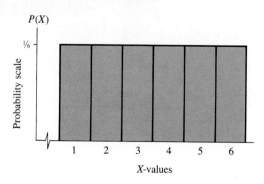

In the last section we observed that in a finite population of size $N$, there exist $_NC_n$ possible simple random samples, each of size $n$. It seems logical to ask what kinds of values we would observe for statistics from these different samples. Also, is it possible to determine some regularity of chance of occurrence for the various values obtained for statistics from one sample? We will answer these questions for the case in which the statistic is the sample mean. Then, for example, inferences about mean production, average income, expenses, and the like are similarly obtained. We begin by introducing the concept of a *sampling distribution*.

**Requirements for a Sampling Distribution**

A sampling distribution is the probability distribution for a statistic. Its description includes (1) all possible values that can occur for the statistic and (2) the probability of each value or each interval of values for a given sample size.

To illustrate this concept, assume we have a population composed of six values, each with an equal probability. The probability distribution is

$$P(X) = 1/6 \quad \text{where} \quad X = 1, 2, 3, 4, 5, 6$$

This distribution is discrete and uniform with mean $\mu = (1 + 6)/2 = 3.5$ and standard deviation $\sigma = 1.71$. The population distribution appears in Figure 7.1.

Now assume that we draw random samples of three items from this distribution and record the mean, $\bar{X}$, for each. By the rule for random samples, there are $_6C_3 = 20$ distinct samples of three observations each. These 20 possible samples appear in Table 7.1.

Thus, a first sample might contain the values 1, 2, and 3; a second, 1, 2, and 4; a third, 1, 2, and 5; and so on. These samples would have means $\bar{X}_1 = 2$, $\bar{X}_2 = 2\frac{1}{3}$, and $\bar{X}_3 = 2\frac{2}{3}$, respectively. Each of the 20 sample combinations has a chance of 1/20 of occurring. *This is a new and important concept. We no longer think of the sample mean as a single value but rather as a random variable that can take on different values.*

The distribution of the 20 sample means for random sampling with $n = 3$ is displayed in Figure 7.2(b). It illustrates several properties that indicate $\bar{X}$ is a suitable statistic for estimating $\mu$, which is 3.5. First, from Figure 7.2(b), the mean of the sample $\bar{X}$ values is also 3.5. Under random sampling the mean of the

**TABLE 7.1**   The Sampling Distribution of the Mean, Random Sampling ($n = 3$, $N = 6$)

| Possible Samples | $\bar{X}$ | Frequency | $P(\bar{X})$ |
|---|---|---|---|
| (1, 2, 3) | 2.00 | 1 | 0.05* |
| (1, 2, 4) | 2.33 | 1 | 0.05 |
| (1, 2, 5), (1, 3, 4) | 2.67 | 2 | 0.10 |
| (1, 2, 6), (1, 3, 5), (2, 3, 4) | 3.00 | 3 | 0.15 |
| (1, 3, 6), (1, 4, 5), (2, 3, 5) | 3.33 | 3 | 0.15 |
| (1, 4, 6), (2, 3, 6), (2, 4, 5) | 3.67 | 3 | 0.15 |
| (1, 5, 6), (2, 4, 6), (3, 4, 5) | 4.00 | 3 | 0.15 |
| (2, 5, 6) (3, 4, 6) | 4.33 | 2 | 0.10 |
| (3, 5, 6) | 4.67 | 1 | 0.05 |
| (4, 5, 6) | 5.00 | 1 | 0.05 |
| | | 20 | 1.00 |

\* $0.05 = 1/20$

$\bar{X}$-distribution is the same as the mean of the $X$ (population) distribution.\* This suggests that under simple random sampling, $\bar{X}$ is an *unbiased estimator* of $\mu$.

> An unbiased estimator has for its mean a value that equals the population parameter value.

Second, compared to the population distribution, the sampling distribution shows a marked clustering of values near $\mu = 3.5$ (see Figure 7.2). Thus, even though a single sample will give a value for $\bar{X}$ other than 3.5, most samples have a mean ($\bar{X}$) value reasonably close to $\mu$.

**The Law of Large Numbers**

As a way of exploring the effect of increased sample size on the values of the sample mean, we will look at sampling distributions for random samples of sizes

---

**FIGURE 7.2**   Sampling Distribution for the Mean $\bar{X}$, $n = 3$, $N = 6$

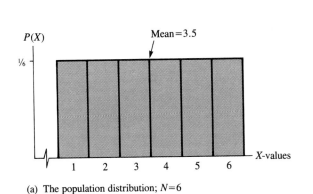

(a) The population distribution; $N=6$

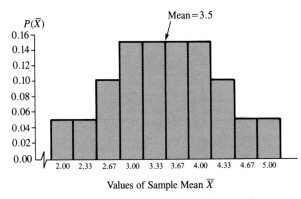

(b) The sampling distribution for samples of size $n=3$

\* The expected value of the sample means ($\bar{X}$'s) is designated as $E(\bar{X})$. Statistical theory shows that with random sampling, the sampling distribution for $\bar{X}$ has the same mean as the parent ($X$) distribution. That is, under random sampling $E(\bar{X}) = E(X) = \mu$.

3, 4, 5, and 6. The concentration of mean values about $\mu = 3.5$ is heightened as the sample size increases. For example, for our parent distribution, there are $_6C_4 = 15$ distinct random samples of size 4 and each has a probability of 1/15. Continuing, there are $_6C_5 = 6$ samples of five; each has probability 1/6. But there is only one sample of the whole, $_6C_6 = 1$; that is, a census of $n = N = 6$ with probability 6/6 = 1. These distributions are recorded for $n = 3, 4, 5,$ and 6 in Figure 7.3. Note that all of the distributions are centered about the population mean, $\mu = 3.5$. How does the distribution of $\bar{X}$-values change as the sample size increases? It becomes more tightly clustered about as $n$ increases toward $N(=6)$. This is characteristic of simple random sampling. Larger samples generally produce closer estimates. This has been extended to a *law of large numbers*, one interpretation of which follows.

### LAW OF LARGE NUMBERS

**For random sampling with a large sample size, the sample mean has a high probability of taking on a value near the population mean.**

Note that the present example is not a large sample. But Figure 7.3 indicates how the larger sample size resulted in a greater cluster of mean values near the population mean. This concentration extends to the extreme where $n = N$, so there is only $_NC_N = 1$ sample. Then the sampling distribution becomes the single census value $\mu$.

**FIGURE 7.3**  Sampling Distributions for the Mean for Sample Size $n = 3, 4, 5, 6$ from a Uniform Distribution with $P(X) = 1/6$ for $X = 1, 2, \ldots, 6$

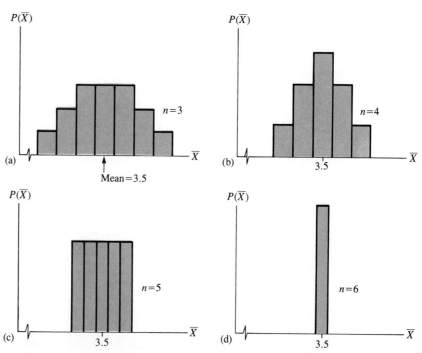

**TABLE 7.2**  Symbols for Means and Standard Deviations

| | Population | Sample | Sampling Distribution of Means |
|---|---|---|---|
| Mean | $\mu$ | $\bar{X}$ | $\mu_{\bar{x}} = \mu$ |
| Standard deviation | $\sigma$ | $s$ | $\sigma_{\bar{x}} = \dfrac{\sigma}{\sqrt{n}}$ |
| | | | $n < 0.05N$  when $n$ is finite |

## The Sampling Distribution of $\bar{X}$ under Random Sampling

Figure 7.3 also demonstrates the character of the mean and standard deviation for the distribution of the means. Table 7.2 gives symbols for these expressions.

We learned earlier that the expression for calculating the standard deviation of a discrete variable is

$$\sigma = \sqrt{\frac{\sum (X - \mu)^2}{N}}$$

A similar expression exists for calculating the standard deviation for $\bar{X}$-values. Since the sampling distribution of $\bar{X}$ consists entirely of values of the sample means, the squared deviations are the squares of the difference between the individual sample mean values $\bar{X}$ and $\mu$, or $(\bar{X}_i - \mu)^2$. Summing and then dividing by the number of samples and taking the square root gives us the standard deviation of the sampling distribution.

The standard deviation of the sampling distribution of the means $\sigma_{\bar{x}}$ is more commonly known as the *standard error of the mean*.

**The standard error of the mean, denoted by $\sigma_{\bar{x}}$, is the standard deviation of the sampling distribution of sample means.**

Intuitively, it represents the dispersion of sample means ($\bar{X}$'s) about the population mean $\mu$. Equation (7.2) gives the calculation formula for random sampling.

$$\sigma_{\bar{x}} = \sqrt{\frac{\sum (\bar{X}_i - \mu)^2}{_NC_n}}, \quad \text{which is equivalent to } \frac{\sigma}{\sqrt{n}} \tag{7.2}$$

This holds for finite populations with random sampling and for $n/N < 0.05$ (up to where sample sizes are at most 5% of the population size). Some statisticians designate $\sigma$ as $\sigma_x$ to distinguish it from $\sigma_{\bar{x}}$ because they are standard deviations for different variables.

## The Relation of Standard Error to Sample Size

As the sample size becomes larger, the difference between the sample mean and the population mean should decrease. Thus with larger sample sizes the error due to using a sample decreases. Figure 7.3 displayed this result. In fact, the standard error goes to zero as the sample size ($n$) approaches the population

**TABLE 7.3**   The Sampling Distribution of the Mean for Random Sampling ($n = 4$, $N = 6$)

| Possible Samples | $\bar{X}$ | Frequency | $P(\bar{X})$ |
|---|---|---|---|
| (1, 2, 3, 4) | 2.5 | 1 | 0.067* |
| (1, 2, 3, 5) | 2.75 | 1 | 0.067 |
| (1, 2, 3, 6), (1, 2, 4, 5) | 3.0 | 2 | 0.133 |
| (1, 2, 4, 6), (1, 3, 4, 5) | 3.25 | 2 | 0.133 |
| (1, 2, 5, 6), (1, 3, 4, 6), (2, 3, 4, 5) | 3.50 | 3 | 0.200 |
| (1, 3, 5, 6), (2, 3, 4, 6) | 3.75 | 2 | 0.133 |
| (1, 4, 5, 6), (2, 3, 5, 6) | 4.0 | 2 | 0.133 |
| (2, 4, 5, 6) | 4.25 | 1 | 0.067 |
| (3, 4, 5, 6) | 4.50 | 1 | 0.067 |
| | | 15 | 1.000 |

\* $0.067 = 1/15$

size ($N$).* For infinite populations the error goes to zero as the sample size becomes "very large." In either case the limiting value for $\sigma_{\bar{X}}$ is zero. Then the sample mean equals the population mean.

Figure 7.3 illustrated the effect of increased sample size on the shape of the sampling distribution. As $n$ increases, $\sigma_{\bar{X}}$ decreases. We can illustrate this numerically by using the definition formula for $\mu$ and $\sigma$ but changing $X$ to $\bar{X}$:

$$\mu_{\bar{X}} = \sum [\bar{X}_i \cdot P(\bar{X}_i)] \qquad \text{for all } \bar{X}_i \tag{7.3}$$

$$\sigma_{\bar{X}} = \sqrt{\sum (\bar{X}_i - \mu_{\bar{X}})^2 \cdot P(\bar{X}_i)} \tag{7.4}$$

Returning to the data from Figure 7.1, we calculate values for a uniform distribution with random samples of size $n = 4$. These calculations appear in Table 7.3. This provides the essential information for calculating $\mu_{\bar{X}}$ and $\sigma_{\bar{X}}$ when $n = 4$ and $N = 6$ under random sampling.

$$\mu_{\bar{X}} = \left[ 2.50 \frac{1}{15} \right] + \left[ 2.75 \cdot \frac{1}{15} \right] + \cdots + \left[ 4.50 \cdot \frac{1}{15} \right] = \frac{52.50}{15} = 3.50$$

$$\sigma_{\bar{X}}^2 = \left[ (2.50 - 3.50)^2 \cdot \frac{1}{15} \right] + \left[ (2.75 - 3.50)^2 \cdot \frac{1}{15} \right]$$

$$+ \cdots + \left[ (4.50 - 3.50)^2 \cdot \frac{1}{15} \right] = \frac{4.3750}{15} = 0.2917$$

$$\sigma_{\bar{X}} = \sqrt{0.2917} = 0.54$$

Computed values of $\sigma_{\bar{X}}$ are given for $n = 3$ (using Table 7.1 data), for $n = 4$ (above), and on through $n = N = 6$, for purposes of comparing the standard

---

\* Although Equation (7.2) does not become zero when the sample approaches the population size, it does become zero when a finite population correction factor is applied. This is discussed next and is defined in Equation (7.5).

errors for alternative random sample sizes (see table below). As $n$ increases, the standard error, $\sigma_{\bar{X}}$, decreases.

| $n$ | 3 | 4 | 5 | 6 ($=N$) |
|---|---|---|---|---|
| $\sigma_{\bar{X}}$ | 0.76 | 0.54 | 0.37 | 0 |

**A Finite Population Correction Factor**

The preceding relation between sample size and standard error shows that for larger sample sizes, the $\bar{X}$ values are more closely clustered about $\mu$. Increasing the sample size to $N = 6$ should, of course, reduce the standard error to zero. To see the basis for this statement, let us return to the common expression for standard error as presented in Equation (7.2).

$$\sigma_{\bar{X}} = \frac{\sigma}{\sqrt{n}}$$

As $n$ (in the denominator) gets larger, the standard error of the mean approaches zero. However, it reaches zero only if $n$ is *very large* (or approaches $N$ for finite populations). Nothing in Equation (7.2) accounts for the fact that the population may be finite and therefore limit the size of $n$. To take account of this, a correction factor must be applied to the calculation of the standard error when the sample constitutes a significant portion—5% or more—of the whole.

For finite populations, where $n$ is 5% or more of $N$, a *finite population correction factor*, fpc, should be applied in computing the standard error.

> The finite population correction factor, fpc, adjusts the standard error to most accurately describe the amount of variation. It should be applied whenever the sample size is 5% or more of the population size. The fpc is $\sqrt{(N-n)/(N-1)}$.

The fpc factor ensures that as the sample size ($n$) approaches the population size ($N$), the standard error, $\sigma_{\bar{X}}$, does in fact approach zero. The corrected equation for the standard error of the mean thus becomes

$$\sigma_{\bar{X}} = \frac{\sigma}{\sqrt{n}} \sqrt{\frac{N-n}{N-1}} \quad \text{for } n \geqslant 0.05N \qquad (7.5)$$

**EXAMPLE 7.1**

Given a population of $n =$ six values (1, 2, 3, 4, 5, and 6), the standard deviation is $\sigma = 1.71$. For samples of $n = 6$ (so, here, $n = N = 6 =$ the population size), the standard error of the mean using Equation (7.2) is

$$\sigma_{\bar{X}} = \frac{\sigma}{\sqrt{n}} = \frac{1.71}{\sqrt{6}} = 0.70 \quad \text{(is not correct)}$$

Theoretically it should be zero. Compute the corrected value of $\sigma_{\bar{X}}$.

*Solution*    Applying the fpc (using Equation (7.5)) gives the same result as using Equation (7.4):

$$\sigma_{\bar{x}} = \frac{\sigma}{\sqrt{n}}\sqrt{\frac{N-n}{N-1}} = \frac{1.71}{\sqrt{6}}\sqrt{\frac{6-6}{6-1}}$$

$$= 0.70(0) = 0 \quad \text{(is correct)}$$

The fpc is not necessary if a sample is less than 5% of the population. Also, where the population is infinite, we assume that the sample is less than 5% of the population and ignore the fpc.

**Problems 7.3**

7. Assuming a random sample of size 100 from a very large population with a mean $\mu = 50$ meters and a standard deviation $\sigma = 20$ meters, find values for the mean and standard deviation for the distribution of the sample means. That is, determine values for $\mu_{\bar{x}}$ and $\sigma_{\bar{x}}$.

8. We say that under random sampling $\bar{X}$, the sample mean, is an "unbiased estimator of the population mean." What is meant by this statement?

9. What is the difference between $\sigma$, $\sigma_X$, and $\sigma_{\bar{x}}$ in sampling theory? (*Hint:* To what distribution of values does each refer?)

10. Given a population of $N = 6$ with the discrete uniform distribution shown in the table, let $X$ = weekly contribution to United Fund.
    a. Find the mean of the contributions for all possible random samples of size 2, assuming the sample is without replacement.
    b. Find the mean of the sample means of part (a).
    c. Compute the standard error of the mean under random sampling.
    d. What percentage of the individual values are $2.50 to $4.50, inclusive? What percentage of the means are in this same interval?

| Employee | X | P(X) |
|----------|------|------|
| A | $1 | 1/6 |
| B | 2 | 1/6 |
| C | 3 | 1/6 |
| D | 3 | 1/6 |
| E | 4 | 1/6 |
| F | 5 | 1/6 |

11. Work parts (a), (b), and (c) of Problem 10 but use $n = 3$. Compare your results.

12. A simple random sample of 100 items is taken from an infinite population. To how many items must the sample be *increased* in order to reduce by one-half the standard error of the sample means?

13. What relationship exists between the population standard deviation, the standard error of the mean, and the sample size?

14. Why can the population mean be estimated more closely by a sample mean with an increased sample size than by a mean from a smaller sample size?

## 7.4 THE CENTRAL LIMIT THEOREM

We have seen that random sampling is an accepted technique for selecting samples. Within this framework, the sampling distribution of the means can be used to make statements about a population mean. In this section, we focus on the specific type of distribution formed by sample means, $\bar{X}$'s.

Figure 7.3 gives us a clue to an appropriate distribution for $\bar{X}$. Although the sampling distributions in Figure 7.3 are discrete and are drawn with steps, there is a tendency toward a concentration of values near the center. This peaking of values about the mean is not unique to any one experiment. In fact, with sufficiently large samples it occurs for the distribution of means from every parent distribution. As the sample size increases, the distribution of the sample means approaches the shape of the normal curve, no matter what the distribution of the original variable! This is stated in a fundamental theorem of statistics, the *central limit theorem*.

### CENTRAL LIMIT THEOREM

For simple random samples from any distribution with finite mean and variance, as *n* becomes increasingly large, the sampling distribution of the sample means is approximately normally distributed.*

This theorem should not seem unusual since we saw numerous normal-like distributions in Chapter 6, including wage distributions, automobile speeds, some test scores, and physical characteristics such as heights. It would seem natural to expect the distribution of $\bar{X}$-values to be normal if the original distribution of $X$-values is itself normal. We have also seen that binomial distributions tend to be normally distributed when $n$ is large if neither $\pi$ nor $(1 - \pi)$ is close to 0.

What is perhaps surprising is that for a population with any distribution shape—triangular, uniform, and so on—the distribution of $\bar{X}$-values from these distributions tends to be bell-shaped (see Figure 7.4 on page 266.)

For sufficiently large random samples, the sampling distribution for $\bar{X}$-values will be normally distributed with mean $E(\bar{X}) = \mu_{\bar{X}} = \mu$ and standard deviation $\sigma_{\bar{X}} = \sigma/\sqrt{n}$.

### Applications of the Central Limit Theorem

Because the sampling distribution is normally distributed, we use an approximate form of the standard normal deviate $Z$ for probability calculations. The only restriction is that the sample size be large, that is, more than 30 observations.** Then

$$Z = \frac{\bar{X}_i - \mu_{\bar{X}}}{\sigma_{\bar{X}}} \qquad (7.6)$$

* In the case of a census (100% sample), $\bar{X}$ goes to a single value, $\mu$, the population mean. The central limit theorem actually applies to sums of random variables and has a broader application than just individual means. However, because it is commonly stated in terms of means, we follow that convention here.

** The central limit theorem applies to samples of any size *if* the parent distribution is normal. However, when the population distribution is *not* normally distributed, sample sizes must sometimes be large in order for the sampling distribution to be normally distributed.

**FIGURE 7.4**    Illustration of the Central Limit Theorem Showing Parent Distribution
Forms

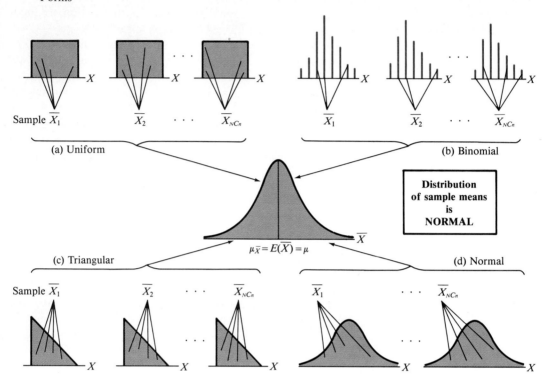

That is, the standard normal deviate $Z$ is distributed normally with mean
0 and variance 1, $[N(0,1)]$. This form compares the deviation of $\bar{X}$-values from
$\mu$ to the size of the standard unit of deviation, now on the scale of $\bar{X}$-values.
We use it when we are seeking the probabilities of "average" values. The following examples show some diverse applications.

---

**EXAMPLE 7.2**

Suppose the (true nationwide) average starting salary for graduates in finance is $28,100
with a standard deviation of $450. What is the probability that a random sample of 100
finance graduates would have a mean salary of $28,000 or more?

*Solution*    The distribution of individual salaries is unknown. It is not essential, but we do know
its mean ($\mu = \$28,100$) and standard deviation ($\sigma = \$450$). The question concerns $\bar{X}$, the
mean for samples of 100 incomes.

1. We seek $P(\bar{X} \geqslant \$28,000)$.
2. Using the rule, we get

$$Z = \frac{\bar{X} - \mu}{\sigma/\sqrt{n}} = \frac{\$28,000 - \$28,100}{\$450/\sqrt{100}} = \frac{-\$100}{\$450/10} = -2.22$$

3. Table B.1 gives $P(-2.22 < Z < 0) = 0.4868$. This is shown as the dark-shaded
area to the left of the mean in the graph.

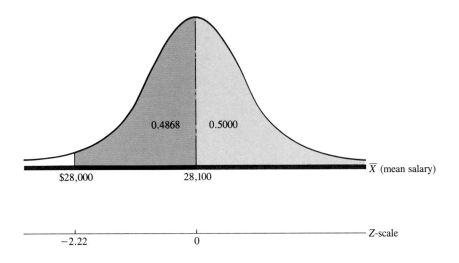

4. For samples of size $n = 100$, adding the probability associated with the sample means > \$28,000 (light-shaded area) gives

$$P(\bar{X} > \$28,000) = 0.4868 + 0.5000 = 0.9868$$

There is a 98.68% chance that the mean salary for a random sample of 100 accounting graduates would exceed \$28,000.

---

Note that the standard error of the sampling distribution in Example 7.2 is \$45.00 (not \$450). *A common error is to forget to divide the standard deviation by the square root of the sample size.*

Example 7.3 points out that the standard error must be "sized" when we are working with normal probability problems.

**EXAMPLE 7.3**

The experience of a telephone company has been that the average amount owed on overdue accounts is \$22.00 with a standard deviation of \$1.50. The distribution of these amounts is essentially normal. What is the probability that (a) the mean for a random sample of 100 overdue accounts exceeds \$22.50 and (b) a single overdue account will show a value in excess of \$22.50? (Obviously the better audit check would be to sample more than one account.)

*Solution*   a. The variable is $\bar{X}$ = mean (or average) for samples of 100 accounts.

1. We seek $P(\bar{X} > \$22.50)$.
2. $Z = \dfrac{\bar{X} - \mu}{\sigma/\sqrt{n}} = \dfrac{\$22.50 - \$22.00}{\$1.50/\sqrt{100}} = \dfrac{\$0.50}{\$0.15} = 3.33$
3. $P(\bar{X} > \$22.50) = P(Z > 3.33) \doteq 0.0000$ because this exceeds the largest entry in Table B.1.
4. There is essentially zero chance for a mean value in excess of \$22.50 as long as random samples of $n = 100$ are used. The standard error (size of a deviation unit) for this situation is $\$1.50/\sqrt{100} = \$0.15$. See Figure 7.5(a).

**FIGURE 7.5**   Comparison of Deviation Units When the Variable Is $\bar{X}$ or $X$

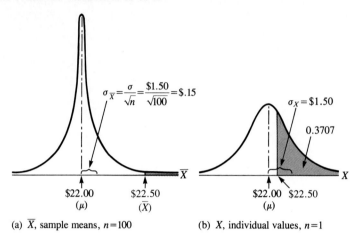

(a) $\bar{X}$, sample means, $n=100$          (b) $X$, individual values, $n=1$

b. The variable is $X$ = individual amounts owed. This becomes a normal distribution problem of the type we worked with in Chapter 6.

1. We want $P(X > \$22.50)$ for $X$ normally distributed.
2. The first $Z$-form (from Chapter 6) gives

$$Z = \frac{X - \mu}{\sigma} = \frac{\$22.50 - \$22.00}{\$1.50} = \frac{\$0.50}{\$1.50} = 0.33$$

3. $P(X > \$22.50) = P(Z > 0.33) = 0.5000 - 0.1293 = 0.3707$ (See Figure 7.5(b).)
4. Over 37% of the individuals with overdue accounts owe more than $22.50. Observe that the standard deviation, here $1.50, is ten times that of the standard error of the $\bar{X}$-distribution, $0.15.

---

Example 7.3 shows that it is critically important to determine which random variable—$\bar{X}$ or $X$—is under question. The first question asks for the mean or average; the latter refers to the individual. This choice can greatly alter the probability, as we just saw in the solutions to parts (a) and (b).

---

**EXAMPLE 7.4**

An energy shortage has forced airlines to seek larger passenger loads as a means of holding down ticket prices. If flights between Altanta and Chicago average 120 passengers with a standard deviation of 12, what is the probability that 64 flights (considered a random sample) would average 122 to 125 passengers, inclusive?

*Solution*   The variable is $\bar{X}$ = average number of passengers. We want to find $P(122 \leqslant \bar{X} \leqslant 125)$, as indicated in Figure 7.6.

For $\bar{X} = 125$:

*Table Area*

$$Z = \frac{\bar{X} - \mu}{\sigma/\sqrt{n}} = \frac{(125 - 120)}{12/\sqrt{64}} = \frac{5}{12/8} = \frac{5}{1.5} = 3.33 \qquad 0.5000$$

**FIGURE 7.6**   Sampling Distribution of Number of Airline Passengers

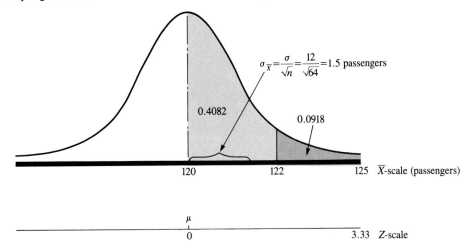

For $\bar{X} = 122$:

$$Z = \frac{(122 - 120)}{12/\sqrt{64}} = \frac{2}{1.5} = 1.33 \qquad\qquad 0.4082$$

Then,

$$P(122 \leqslant \bar{X} \leqslant 125) \quad \text{by subtraction} = \qquad\qquad \overline{0.0918}$$

Thus the probability that 64 flights would average 122–125 passengers is 0.0918, the area shown in the dark-shaded portion of Figure 7.6

---

The preceding examples are just a few applications of the central limit theorem. It will appear again, many times, in the remaining chapters because it is one of the most important rules in statistics.

**Problems 7.4**

15. On the average, containers of frozen vegetables in storage lose 6.5 grams in weight with a standard deviation of 1.8 grams. What is the probability that a random sample of 36 containers will have a mean loss of weight in excess of 7.0 grams during storage?

16. Flashlight bulbs manufactured by a batch process have a mean lifetime of 600.0 hours with a standard deviation of 45.0 hours. Samples of 36 bulbs are taken from a batch. What percentage of these samples will have a mean lifetime of 585.0 to 620.0 hours?

17. Given that $X$ is normally distributed with a mean of 30 and standard deviation of 8, calculate the probability that the sample mean, $\bar{X}$ based on a random sample of size 30, will be less than 32.

18. At a savings and loan association, the average amount for short-term loans to university students is $255.00 with a standard deviation of $12.50. The distribution of amounts lent is normal.

a. What percentage of the (individual) loans are for $200.00 or less?
b. What is the probability that 25 loans, considered a random sample, will have a mean over $220.00? (*Note:* The central limit theorem applies even though *n* is less than 30, because we started from a normally distributed parent distribution).

19. Suppose that individuals eating at a pancake restaurant spend, on the average, $4.24 for pancakes with a standard deviation of $0.15.
    a. What percentage of the customers spend over $4.30 for pancakes? Assume a normal distribution of amounts spent.
    b. What is the probability that a random sample of 100 customers will spend an average of $4.20 to $4.30?

20. The mean height for a population of 10,000 college men is 69.2 inches with a standard deviation of 2.5 inches.
    a. Find the probability that in a random sample of 100 college men, the mean height will be greater than 69.9 inches.
    b. How many of these men are over 6 feet tall? Assume a normal distribution for heights.

---

**SUMMARY**

Chapter 7 is short but important. Statistical inference relies upon *random sampling*, wherein each element of the population has an equal chance of being chosen. If all possible samples of a given size were taken from a finite population and we made a list of the sample means obtained along with the respective probabilities for each, we would have a sampling distribution of the means. The mean of the sampling distribution, $\mu_{\bar{x}}$, equals the population mean, $\mu$, and the standard deviation, $\sigma_{\bar{x}}$, equals $\sigma/\sqrt{n}$. The standard deviation of a sampling distribution of means is called the standard error of the mean.

The central limit theorem is perhaps the most important rule in classical statistics. It affirms the normality of sampling distributions for means of large samples ($n > 30$) when random sampling is used. This is true regardless of the distribution from which

---

**TABLE 7.4**   Key Equations for Sampling Distributions and the Central Limit Theorem

**Probability of a Sample *under Random Sampling***

$$P(\text{Any sample of } n) = \frac{1}{{}_NC_n} \qquad (7.1)$$

**Sampling Distribution of the Mean**

| | | |
|---|---|---|
| Mean | $\mu_{\bar{x}} = \sum[\bar{X}_i \cdot P(\bar{X}_i)] \qquad \text{for all } \bar{X}_i$ | (7.3) |
| Standard error of the mean | $\sigma_{\bar{x}} = \sqrt{\dfrac{\sum(\bar{X}_i - \mu_{\bar{x}})^2}{{}_NC_n}} = \dfrac{\sigma}{\sqrt{n}}$ | (7.2) |
| | $= \dfrac{\sigma}{\sqrt{n}}\sqrt{\dfrac{N-n}{N-1}} \qquad \text{for } n \geqslant 0.05N$ | (7.5) |

**Central Limit Theorem**

| | | |
|---|---|---|
| Z-Scores | $Z = \dfrac{\bar{X}_i - \mu_{\bar{x}}}{\sigma_{\bar{x}}}$ | (7.6) |

the sample is drawn. This tremendous observation has opened the door to widespread use of statistical sampling in practically all phases of government and private industry.

The significance of this theorem will become more apparent as we move through the next chapters in exploring the methods of statistical inference.

Table 7.4 illustrates the more important formulas that were introduced in Chapter 7.

## KEY TERMS

central limit theorem, 265
cluster sampling, 255
estimate, 252
estimator, 252
finite (versus infinite) population, 250
finite population correction factor, 263
judgment sampling, 256
law of large numbers, 259

probability sampling, 252
sampling distribution, 258
simple random sampling, 253
standard error of the mean, 261
stratified random sampling, 255
systematic sampling, 255
unbiased estimator, 259

## CHAPTER 7 PROBLEMS

21. Distinguish among stratified, random, convenience, and purposeful samples. Which is classified as a probability sample?

22. Distinguish between $\sigma_x$ and $\sigma_{\bar{x}}$. Use sketches of the normal distribution to illustrate your answer.

23. Read the following statement: "If the standard deviation of annual retail sales per store is the same for Middleberg (20,000 population) and for a city of 2 million, the standard error of the mean for a simple random sample of 100 stores would be approximately the same for both cities." Is this statement true or false? Explain.

24. Mountain Airlines has determined that the mean weight of baggage each customer transports is 32.0 pounds and the population standard deviation is 2.1 pounds. If a random sample of 49 passengers is taken from the population (considered infinite), what is the probability that the *sample mean* will exceed 32.9 pounds?

25. The mean weight of bags of fertilizer is known to be 82.0 pounds, and $\sigma = 4.0$ pounds. If a sample of $n = 64$ is randomly drawn from the population (considered infinite), what is the probability that the *sample mean* will exceed 82.5 pounds?

26. In Chapter 6 (Problem 19) we had the following problem: The life expectancy of Neverready batteries is normally distributed with a mean of 400 hours and a standard deviation of 120 hours. What is the probability a battery will last at least 300 hours?
    a. Solve this problem. (The answer is the same).
    b. If a scout purchases three dozen batteries for a camp-out, what is the probability the average life of the 36 batteries exceeds 360 hours?
    c. Sketch the distributions used in parts (a) and (b) and label them with the correct parameters.

27. This question concerns the dependence of the standard error of the mean upon the sample size.
    a. Assume samples from an infinite population. By what fraction is the standard error of the mean reduced when the sample size is increased from 60 to 240?
    b. Suppose a population consists of only $N = 20$ items and has a population mean of 187 and standard deviation of 10. What would be the standard error of the mean for samples of $n = 9$ drawn at random?

**STATISTICS IN ACTION: CHALLENGE PROBLEMS**

28. Data for sales (millions of dollars) for food stores by state is one indicator for retail sales potential. The U.S. Bureau of Labor Statistics compiles summary data on an annual basis.

Retail Trade—Sales ($ millions) for Food Stores

| U.S. 301,847 | | | | | |
|---|---|---|---|---|---|
| North East | | West North Central | | West South Central | |
| Me | 1,809 | Mn | 4,890 | Ar | 2,494 |
| NH | 1,903 | Ia | 3,288 | La | 5,428 |
| Vt | 840 | Mo | 5,820 | Ok | 3,646 |
| Ma | 8,075 | ND | 681 | Tx | 21,291 |
| RI | 1,179 | SD | 735 | | |
| Ct | 4,503 | Ne | 1,672 | | |
| | | Ks | 2,803 | | |
| Mid Atlantic | | South Atlantic | | Mountain | |
| NY | 21,296 | De | 901 | Mt | 1,025 |
| NJ | 11,119 | Md | 6,113 | Id | 1,132 |
| Pa | 14,422 | DC | 591 | Wy | 540 |
| | | Va | 8,049 | Co | 4,276 |
| | | WV | 2,143 | NM | 1,667 |
| | | NC | 7,850 | Az | 4,983 |
| | | SC | 4,093 | Ut | 1,900 |
| | | Ga | 7,709 | Nv | 1,537 |
| | | Fl | 17,002 | | |
| East North Central | | East South Central | | Pacific | |
| Oh | 12,977 | Ky | 4,287 | Wa | 6,206 |
| In | 6,062 | Tn | 5,654 | Or | 3,328 |
| Il | 12,454 | Al | 4,476 | Ca | 36,495 |
| Mi | 9,922 | Ms | 2,761 | Ak | 912 |
| Wi | 5,337 | | | Hi | 1,572 |

*Source:* U.S. Bureau of the Census, *1987 Census of Retail Trade*, Table 1358, Geographic Area series, RC87-A-1 to 52. Washington, D.C.: U.S. Government Printing Office.

*Source: Statistical Abstract of the United States 1990* (110th ed.). Washington, D.C.: U.S. Department of Commerce, Census Bureau, 1990.

Questions    What is the mean sales by state for food stores? This is a (rare) case in which the population mean is available. It is $301,847/51 = 5918.6$ million dollars.

How good is a random sample estimate from, say, five states (a 10% sample)? This question is answered in the next four Problems.

Use Table B.7, Random Numbers, to draw a random sample without replacement of five states. Indicate the five states along with their food store sales.

29. Calculate the mean food store sales for your sample of five states. Is this bigger than, smaller than, or "close to" the state average—5918.6 million dollars?

30. Given $\sigma = 6656$ million dollars as the standard deviation in (state) food store sales, compute the standard error of the mean for samples of $n = 5$ states from $N = 51$.

31. Use the result of the previous problem and $\mu = 5918.6$ million dollars to locate your sample mean within $\pm 1$, $\pm 2$, or $\pm 3$ standard errors of the mean distance from $\mu$.

32. One of the major considerations in sampling is whether to select with or without replacement of the chosen items. These approaches are different. Specifically, with replacement, sampling permits an item to appear in the sample more than once. Bill Williams illustrates the differences by sampling eight discs, one for each of eight taxpayers, considered a population (for the purpose of demonstration with a few values).

| Taxpayer | 1 | 2 | 3 | 4 | 5 | 6 | 7 | 8 |
|---|---|---|---|---|---|---|---|---|
| Tax Paid ($y) | 60 | 72 | 68 | 94 | 90 | 102 | 116 | 130 |

Then $\bar{Y} = \$91.5$, $N = 8$, and

$$\sigma^2 = (1/N) \sum (y - \bar{Y})^2 = 515.75 \quad \text{and}$$

$$S^2 = (1/(N-1)) \sum (y - \bar{Y})^2 = 589.41$$

A random sample of two is drawn either (a) without replacement or (b) with replacement. Mr. Williams shows (in Table 10.1, page 106, of his book) that the sample mean is unbiased for sampling with replacement and indicates that the difference in procedures—with versus without replacement—is in the variance for each sampling distribution. That table is reproduced here.

## With Replacement Sampling of Taxpayers*

| Taxpayers Selected | Sample Mean | Taxpayers Selected | Sample Mean | Taxpayers Selected | Sample Mean | Taxpayers Selected | Sample Mean |
|---|---|---|---|---|---|---|---|
| 1,1 | 60.0 | 2,1 | 66.0 | 3,1 | 64.0 | 4,1 | 77.0 |
| 1,2 | 66.0 | 2,2 | 72.0 | 3,2 | 70.0 | 4,2 | 83.0 |
| 1,3 | 64.0 | 2,3 | 70.0 | 3,3 | 68.0 | 4,3 | 81.0 |
| 1,4 | 77.0 | 2,4 | 83.0 | 3,4 | 81.0 | 4,4 | 94.0 |
| 1,5 | 75.0 | 2,5 | 81.0 | 3,5 | 79.0 | 4,5 | 92.0 |
| 1,6 | 81.0 | 2,6 | 87.0 | 3,6 | 85.0 | 4,6 | 98.0 |
| 1,7 | 88.0 | 2,7 | 94.0 | 3,7 | 92.0 | 4,7 | 105.0 |
| 1,8 | 95.0 | 2,8 | 101.0 | 3,8 | 99.0 | 4,8 | 112.0 |

| Taxpayers Selected | Sample Mean | Taxpayers Selected | Sample Mean | Taxpayers Selected | Sample Mean | Taxpayers Selected | Sample Mean |
|---|---|---|---|---|---|---|---|
| 5,1 | 75.0 | 6,1 | 81.0 | 7,1 | 88.0 | 8,1 | 95.0 |
| 5,2 | 81.0 | 6,2 | 87.0 | 7,2 | 94.0 | 8,2 | 101.0 |
| 5,3 | 79.0 | 6,3 | 85.0 | 7,3 | 92.0 | 8,3 | 99.0 |
| 5,4 | 92.0 | 6,4 | 98.0 | 7,4 | 105.0 | 8,4 | 112.0 |
| 5,5 | 90.0 | 6,5 | 96.0 | 7,5 | 103.0 | 8,5 | 110.0 |
| 5,6 | 96.0 | 6,6 | 102.0 | 7,6 | 109.0 | 8,6 | 116.0 |
| 5,7 | 103.0 | 6,7 | 109.0 | 7,7 | 116.0 | 8,7 | 123.0 |
| 5,8 | 110.0 | 6,8 | 116.0 | 7,8 | 123.0 | 8,8 | 130.0 |

$$E(\bar{y}) = 91.5 = \bar{Y}$$

$$E(\bar{y} - \bar{Y})^2 = \frac{1}{64} \left[ (60 - 91.5)^2 + (66 - 91.5)^2 + \cdots + (130 - 91.5)^2 \right]$$

$$= 257.88 = \frac{515.75}{2} = \frac{\sigma^2}{n}$$

* 2 taxpayers from 8; 64 equally likely ordered pairs; unit listed first is selected first.

SOURCE: Williams, Bill. *A Sampler on Sampling*. New York: John Wiley, 1978.

Questions   For sampling without replacement, this table is changed only by exclusion of the repeats: 1,1 60; 2,2   72; 3,3 68; 4,4 94; 5,5 90; 6,6 102; 7,7 116; 8,8 130. There remain $64 - 8 = 56$ equally likely ordered pairs. Calculate $E(\bar{y})$ for random sampling without replacement (so excluding the eight repeats), using the table. State a conclusion.

33. Also, for sampling without replacement, calculate $E(\bar{y} - \bar{Y})^2$.

34. On page 109 of his book, Mr. Williams gives a formula for sampling without replacement: $\text{var}(\bar{y}) = (1 - n/N)(S^2/n)$. Use $S^2 = 589.41$, calculated above, and this formula to find a value for $\text{var}(\bar{y})$. Compare your value against the answer to the preceding problem. Which procedure, sampling *with* or *without* replacement, gives the larger variance? Why?

35. Give two examples for which you would prefer sampling with replacement and two for which sampling without replacement is preferred.

36. Use

$$(N - 1)S^2 = \sum (y - \bar{Y})^2$$
$$N\sigma^2 = \sum (y - \bar{Y})^2$$

to show that, for sampling without replacement,

$$\left(\frac{N - n}{N - 1}\right)\frac{\sigma^2}{n} = \left(1 - \frac{n}{N}\right)\frac{S^2}{n}$$

---

## BUSINESS CASE

*Bob Sharpe Gains Acceptance with a Sample Inventory*

Bob Sharpe had just joined Hanford Tractor Company as a junior partner, and he was anxious to make a meaningful contribution to management. His major responsibility was being the plant manager. The first step, he felt, was to gain the acceptance of the line employees.

Bob's initial weeks were devoted to learning about the organizational structure. In doing this he talked with most of the regular employees. The one troublesome area for everyone was the "annual inventory headache." The company had computerized its parts file, so parts were generally ordered and on hand as needed. However, there was an annual December month-long work detail, in which all service personnel took turns counting the inventory. This extra-hours work detail had been a source of employee complaints for several years. Bob resolved to do something about the situation. This year Hanford Tractor would do its inventory using statistical sampling, rather than counting every single item by hand!

As a first step Bob called in a local statistician, Betty Ladd. The parts section manager, Dennis Duval, was also present at their first meeting. After some preliminary discussion they sat down to map out a plan for sampling the inventory.

**Mr. S.**   There are over 13,000 different machine parts in the storage bins. What we need is a way to avoid counting the quantity of every part for the December inventory.

**Ms. L.**   Let's see, how much time did you spend for a full-count inventory last year?

**Mr. D.**   I remember that well. We worked a four-person crew for the six weekends before Christmas. So that would amount to 48 person-workdays.

**Ms. L.**   Was that effective?

**Mr. D.**   Well … the auditors were satisfied and our people liked the time-and-a-half overtime pay. But it really is not good for morale. In the long run it costs the company big dollars.

**Ms. L.** Oh! How so?

**Mr. D.** Our people work one or two weekends, then get burned out. The hours are long, so they get tired and there's no enthusiasm left for work during the regular week. There's a real upswing in days absent for sickness—and client relations go downhill.

**Mr. S.** Apparently the situation has degenerated. This year several people have already volunteered that they'd just as soon not earn the overtime. They want to spend the time with their families. These are signs that it's time for a change. That's why we invited you, Betty. Do you think you can help us?

**Ms. L.** Well, I'd be inclined to think we might reduce that extra workload by a factor of 2 or more. Suppose we begin by discussing some special needs for a sample inventory. We will need some computer support. Are you using the computer for any of your regular inventory work?

**Mr. S.** Actually, we use it quite a lot. Our parts holdings are each identified by a code number that is stored on the computer. This code includes a value for each part and the quantity on hand.

**Mr. D.** And every night we feed it a report of the quantity for each part used for service or the quantity that was sold outright that day. Then each Monday we get a computer update for part quantities. In fact, there is an order automatically placed the night the quantity of any part goes below a prespecified level.

**Ms. L.** Then you have a foundation for our inventory needs. But you may still want to hire a programmer to computerize the sampling procedure.

**Mr. S.** Betty, about the sampling plan. I've heard you talking about multicluster sample designs and ratio estimates. Sounds fancy. However, we need to keep in mind that the plan should be easy to implement. Also, our auditors have to approve the plan before any action can be taken. Those accountants will need to understand where we're going. Can you give them something reasonably straightforward?

**Ms. L.** (reflecting) I can. It will probably be a stratified random sampling procedure. It may take a couple of sessions to develop a plan. But the three of us should be able to work out the details. Then we can go for the accountants' approval.

Several weeks later the company's tax auditor Tom Adler come to review the sample inventory plan. Bob and Betty had already supplied the auditors with an outline of the plan. Now they wanted a formal presentation.

**Mr. S.** You know that we have spent considerable time in past years preparing the year-end parts inventory. Our proposal is a way to do the job more time-efficiently yet provide results that will be acceptable for your auditing standards.

**Mr. A.** We are aware that other firms have used sampling to perform an inventory. The procedure is acceptable for tax purposes if it's within certain standards. In the past you have maintained a complete inventory with 2% error. Will your plan achieve this?

**Ms. L.** Our proposal is to preset inventory standards to a maximum 3% error with a 95% statistical reliability. These conditions, along with our stratified random sampling procedure, will allow a determination of the number of parts to sample in order to assure a result that will range within $97 to $103 for an actual holding of $100.

**Mr. A.** But doesn't the 95% reliability mean there is a 5% chance you will be wrong?

**Ms. L.** Yes. Our sampling procedure will assure an error, of $\pm 3\%$, for 95% of all possible sample readings. We will assume that the single reading we take will be a good one. Our chances could be increased to 98%, but this would require taking a bigger sample. Does 95% seem too low?

**Mr. A.**    That number is satisfactory. We understand that anytime you take a sample there is a chance for error.

**Mr. S.**    This is so. But also there can be error in taking a complete count. For example, I understand that last year we measured some sheet metal parts up to three times. We finally took the average reading because the measurements were all different!

**Mr. A.**    Another question is what you mean by a stratified sample. How will you stratify and what advantage has it over random sampling for getting a good reading?

**Ms. L.**    We will assign parts to classes in accordance with their unit price. The classes might be under $0.10, $0.10–$0.15 per unit, ... , $20–$50 per unit, and so on. This allows determination of the variability in part value (unit price × number of units) by class. This in turn leads to a determination of how many parts to sample in each class in order to combine the values to a single number with no more than 3% error. Stratification should increase our sampling efficiency. It provides better results for a fixed sample size as compared to random sampling.

**Mr. S.**    Obviously, we would take more parts from the higher price classes because these will show the largest variations.

**Ms. L.**    That's right. In fact, the sample will likely require counts of *every item* that has a unit price over $100, or $200.

**Mr. A.**    But where does random sampling come in?

**Mr. S.**    We've planned one computer routine to compute the statistics, like class means and standard deviations, to get the counts for part-class sample requirements. Another program is required to generate the random numbers for selecting machine parts within each class. Random sampling within classes assures every part a fair chance of being selected for inspection.

**Mr. A.**    That's fair enough. But there's one more critical point. Once you get a result, how do you know whether the result is within the 3% tolerance? What is your standard?

**Ms. L.**    That becomes automatic. Our inventory standard is the weekly computer listing. This report is made independently of our inventory. There will be a physical count of what is found in the bins for our sample, then that is raised to a projected total. . . .

**Mr. A.**    Let me guess the rest. Then you compare the sample projected against the regular weekly report to see if the two agree within 3%?

**Ms. L.**    Yes. Provided such agreement exists, the weekly total will be the number for taxation purposes. Of course the weekly inventory is corrected for any discrepancies that are found through bin counts.

**Mr. A.**    And if not?

**Ms. L.**    If not, we simply go back to the bins for more samples until the 3% agreement is reached.

After two more short sessions, the auditor agreed to the sampling plan.

The sample inventory was taken in December. The computer derived sample requirement was just over 30% of the total parts. The counts were made in 4 days, in comparison to 12 days for the previous year. And the weekly computer listing agreed to within not 3% but 2% of the sample projected value. The procedure worked, with a projected value $2 in error for every $100 in parts holdings. The auditors accepted the results, and Bob Sharpe gained the acceptance of the line employees.

**Business Case Questions**

37.  Because the machine parts were "code numbered" with a quantity recorded on computer, an option for sampling would have been to use systematic sampling on code numbers. Suppose the part items were numbered sequentially 1 to 13,562 with 1 being #1 flat washers, 2 being #3 flat washers, ... , 13,561 being 12-foot scraper

blades, 13,562 being 15-foot scraper blades. Explain some pros and cons for systematic sampling, here, in place of the stratified random sampling.

38. Why would judgment sampling *not* be appropriate in this (parts-inventory) situation?

39. Cluster sampling was briefly considered for this job. However, the logical clusters, in terms of (1) the sequential numbering system and (2) the physical location of parts storage, were, for example, washers and cotter pins, belts and pulleys, tractor wheels and scrapers, and so on. Explain why cluster sampling would not be a viable alternative here.

## COMPUTER APPLICATION

*Stratified Random Sampling*

This application uses the data from a national marketing analysis on individual household expenditures for food. These data are not available on the PUMS data base and are included here instead. The specific question for the study was, What is the mean expenditure for food purchases for home consumption (excluding meals eaten away from home)?

Logically, the cost for food is positively related to the number of persons in a household; generally, the greater the number of persons, the greater the amount of money spent on food. So the marketers decided to group or stratify households on the variable persons per household.

Stratification is a way of grouping like units together. Then random sampling is performed within each group, or stratum (Section 7.2). This is a probability sampling procedure. Stratification is better than simple random sampling to the extent that there is less variability within each stratum than if a simple random sample were taken from the total. For stratification to be economically advantageous, the groups need to be homogeneous (alike) in the amounts spent for food purchases and heterogeneous (different) across the groups.

The available information included (1) a distribution of the U.S. households by number of persons per household and (2) statistics from a *preliminary sample of size 50* from each person per household group. These statistics are displayed in Table 7.5.

In comparison to simple random samples, the result of a meaningful stratified sampling is either a smaller variation in the estimates for a specified sample size, or a smaller sample size requirement for a given amount of variability. This analysis used a fixed sample size of 1000 with stratified random sample selection to ensure a smaller variation in the estimates from alternative samples. (This differs

**TABLE 7.5**   Statistics for a Stratification on Persons per Household for Estimating Monthly Mean Dollars for Household Food Expenditures in the United States

| Persons-per-Household Groups | (1) Households (millions) | (2) Statistics from a Preliminary Sample | |
|---|---|---|---|
| | | $n_i$ | $s_i$ |
| 1 ($N_1 =$) | 19.95 | ($n_1 =$) 50 | $20.462 |
| 2 | 26.89 | 50 | 29.622 |
| 3 | 15.13 | 50 | 30.962 |
| 4 | 13.59 | 50 | 36.830 |
| 5 + ($N_5 =$) | 9.84 | ($n_5 =$) 50 | 42.556 |
| | Total   85.40 | 250 | $38.489 |

from the preliminary sample of $5 \cdot 50 = 250$, which is used to obtain variance estimates before allocation of the 1000 sample records.)

Is the plan a reasonable one? The preliminary statistics relate to this question. First, consider the appropriateness of stratification. These data indicate reasonable homogeneity within groups. This is intuitive, because more persons in a household require more food. Also, if these were not homogeneous within groups, then some of the group standard deviations would be large relative to the total. However, by Table 7.5, only the group-5 standard deviation exceeds the total standard deviation, and group-1 through -3 deviations are substantially less. The stratification is reasonable.

Because the total sample is preset at 1000, the objective of stratifying is to allocate the sample to produce the best possible result. Then, from Table 7.5, group 1 with $N_1 = 19.95$ is close to the average for total households with a quite low standard deviation (20.462). This should require a small sample. In contrast, group 2 has high household count and a greater standard deviation. Relative to group 1, group 2 warrants a relatively high sample count. So, the allocation of the 1000 sample records to the groups should be directly proportional to (1) the total household count and (2) the larger standard deviation estimates. Groups like group 2, with higher counts and larger standard deviations, require a larger allocation from the total sample.

The allocation that ensures the least variance uses data from Table 7.5 and the procedure

$$ n_i = \left( \frac{N_i s_i}{\sum N_i s_i} \right) n \qquad i = 1, 2, 3, 4, 5 $$

where    $N_i = $ the U.S. total households for group $i$

$s_i = $ the preliminary sample standard deviations

$n = 1000$, the prespecified total sample

$n_i = $ the sample allocated to group $i$

The calculations give

| | | |
|---|---|---|
| Group 1 | $n_1 =$ | 158 |
| Group 2 | $n_2 =$ | 308 |
| Group 3 | $n_3 =$ | 180 |
| Group 4 | $n_4 =$ | 192 |
| Group 5 | $n_5 =$ | 162 |
| Total sample | | 1000 households |

Again, the objective in stratifying is to provide sample requirements that will decrease the sampling variability and so improve the chance for selecting a "good" sample. This allocation requires $n_1 = 158$ households, $n_2 = 308$ households, and so on. (Recall from the previous discussion that, by logic, relative to group 1, group 2 requires a relatively high sample count.)

You might be surprised that the total sample requirement is only 1000 households to estimate 85,400,000 households. This sample requirement is only

(1000/85.4 million) · 100 = 0.12%, or 1/8 of 1% of the population! However, by random sampling theory*

$$Z = \frac{\bar{X} - \mu}{\sigma/\sqrt{n}} \text{ gives error,} \quad E = |\bar{X} - \mu| = \frac{Z\sigma}{\sqrt{n}}, \quad \text{or} \quad n = \left(\frac{Z\sigma}{E}\right)^2$$

Then, for 0.95 probability, $Z = +1.96$; for error, $E$, of $\pm\$2.50$ and estimating $\sigma$ by $s = 38.489$, from Table 7.5, we obtain

$$n = \left(\frac{1.96 \cdot \$38.489}{\$2.50}\right)^2 = (30.175)^2 = 911$$

Since the analysis uses stratified sampling and the total sample is 1000 households, the estimate (for a 0.95 probability) will have a sampling error of less than $\pm\$2.50$. Surprisingly small as the number may seem, comparable sample size is used for Roper and Gallup polls and other national probability sampling estimates.

* This is a calculation for estimation. More definition for this appears in Section 8.3.

## COMPUTER PROBLEMS

*Data Base Applications*

The following problems are most easily performed using the SAMPLE command with SPSS, or the random numbers statements in BMDP or other similar random sampling procedure. Also, it is possible to draw numbers from 1 to 20,116 from a random numbers table and then pull samples from a remote terminal without using a special software procedure. Check at your computer center for a procedure. The sample of 100 records in Appendix A is insufficient data for the following problems. However, you can use the tape file of the 20,116 PUMS records that is available on floppy disks.

40. a. Use a sample procedure to select 20 random samples, each consisting of 10 records from the PUMS data file. For each sample, compute and print the mean value for household income. Be sure that each sample includes 10 records. Print the 20 means and 20 standard deviations. (You will be able to use these statistics in a Chapter 8 computer problem). There is no need to print the 20 · 10 = 200 sample incomes. Also, see Problem 42.
    b. Repeat, but now select 20 random samples, each of 50 records. Again print the 20 means and 20 standard deviations. These data will also be used for a computer problem in Chapter 8.
       Here is a SPSS setup to draw one sample.

```
1  NUMBERED
2  TITLE PUMS SAMPLES
3  DATA LIST FILE = INDATA/      ──── Refers to the 20,116-record PUMS data file
4       INCOME 35 — 39           ──── Include PERSONS 12–13 SEX1 42 AGE1 43–44 to obtain
                                       data for Problems 42, 43, and 44
5  SET BLANKS = -1
6  TEMPORARY                     ──── Is not required for one sample, but is for two or more
7  SET SEED = 17955              ──── A random number, 20,116 or smaller, from Table B.7
8  SAMPLE 50 FROM 20116
9  DESCRIPTIVES VARIABLES = INCOME/  ──── If line 4 has them, then include here INCOME PERSONS
                                          SEX1 AGE1/
```

```
10      STATISTICS = MEAN STDDEV
11                  RANGE SUM MIN MAX
12 FINISH
```

This SPSS routine gives the following outputs:

```
VARIABLE INCOME

        MEAN    18073    RANGE    63280        MINIMUM    1205
        STDDEV  12520    SUM      903630       MAXIMUM    64485

VALID OBSERVATIONS -    50              MISSING OBSERVATIONS -       0
```

For SPSS users, see the *SPSS Reference Guide* [5]* for a description of any unfamiliar program statements. Also, to obtain all 20 samples in one computer run, repeat 20 times the block of statements 6 through 11, each time with a new random number "seed" from Table B.7. You will need the 20 means and 20 standard deviations that are output by the DESCRIPTIVES procedure. Save these results as well, because they can be used in the Computer Problems in Chapters 8 and 9.

c. Use your statistics from parts (a) and (b), possibly with the aid of a computer FREQUENCY distribution, to demonstrate the law of large numbers. That is, plot your 20 sample means for part (a) samples of $n = 10$, then for part (b) samples of $n = 50$. Which distribution has its 20 sample means positioned closer to the (PUMS) mean $\mu = \$19,700$?

41. Explain how your results in Problem 40(b) demonstrate the central limit theorem.
   a. Do this by computing $\bar{\bar{X}} = \sum \bar{X}/20$. Then $\sigma_{\bar{x}}$ is approximated by Equation (7.2) with $\bar{\bar{X}}$ used in place of $\mu$.
   b. Compute the table using $Z = (\bar{X} - \bar{\bar{X}})/\sigma_{\bar{x}}$ with $\bar{X}$ taking the values for each of the 20 means.

| $Z \cdot \sigma_{\bar{x}}$ | $\bar{\bar{X}} \pm Z\hat{\sigma}_{\bar{x}}$ | Normal Probabilities | Observed Percentage (number of means ÷ 20) |
|---|---|---|---|
| $\pm 1\sigma_{\bar{x}}$ | 17.83–21.91 | 0.6826 | |
| $\pm 2\sigma_{\bar{x}}$ | | 0.9544 | |
| $\pm 3\sigma_{\bar{x}}$ | | 0.9974 | |

c. Explain your results for part (b).

42. Repeat Problem 40, but for age of head of household. That is, complete parts (a)–(c) for the variable age of person 1 (columns 43–44). Use $\mu = 47.6$ years as the (PUMS) mean age for heads of household. *Suggestion*: Use the same samples as for Problem 40, but include age of person 1.

* Numerals in brackets refer to items cited in "For Further Reading" at the end of the chapter.

43. Select 20 random samples each of 50 households from the PUMS data and calculate, for each,

$$p = \text{the percentage of heads of household}$$
$$\text{who are female}$$

This can be done by a coding so that $X = 1$ for female head of household (if PUMS record, column 42 = 1) and otherwise $X = 0$. Then run a frequency count on the variable $X$ for each of your 20 samples.
   a. Plot a histogram of your $p$ values.
   b. Define $\pi$ and give one point estimate for it.
   c. What is the theoretical distribution that describes the random variable $X$? If your computer program provides it, include the standard error value with the outputs. This can be used in a Chapter 8 Computer Problem.

44. Select 20 random samples each of 50 households from the 20,116 PUMS file and calculate for each sample: $\bar{X} = $ the mean number of persons per household (pph). This can be obtained from the SPSS DESCRIPTIVES procedure for PERSONS (columns 12–13) SUM statistic divided by 50.
   a. Plot a histogram of your $\bar{X}$ values.
   b. Define $\mu$ and give several point estimates for its value.
   c. Explain how the central limit theorem can be used to answer probability questions about $\bar{X}$, and give two requirements for using the central limit theorem.

## FOR FURTHER READING

[1] Mansfield, Edwin. *Statistics for Business and Economics*, 3rd ed. New York: Norton, 1987.
[2] Mason, Robert D. *Statistical Techniques in Business and Economics*, 7th ed. Homewood, IL: Irwin, 1990.
[3] Slonim, Morris J. *Sampling*. New York: Simon & Schuster, 1971.
[4] Williams, Bill. *A Sampler on Sampling*. New York: Wiley, 1978.
[5] *SPSS Reference Guide* (version 4.0). Chicago: SPSS, Inc., 1990.

# ESTIMATION AND TESTS OF HYPOTHESES

# 8

## ESTIMATION

When you look back over the past ten years, it's really shocking to see what's happened to this industry. But we've not only survived, we've grown and prospered.

> Ronald W. Allen, CEO, Delta Air Lines
> (*Forbes* Magazine, Oct. 29, 1990, p. 36)

In the ten years prior to 1990, Delta Air Lines rose from the nation's sixth-largest U.S. carrier to its third largest. By 1990, Delta was expanding at the fastest rate in its 61-year history—taking on about 90 new pilots and 160 new flight attendants per month—while other airlines were cutting expenses in anticipation of a recession.

How could an airline dare expand while competitors such as USAir and Pan Am were laying people off? Several factors pointed to an exceptionally strong ability of management to anticipate uncertainty and adjust to it in a systematic, logical manner. Other airlines, such as United, had to deal with contentious unions and face the possibility of work stoppages. Delta maintained good communications within its workforce, paid its workers 21% more than the industry average, and benefited from high productivity and good customer service. While other airlines based major decisions on the 3%-per-year growth rate of travel within the United States, Delta's information system alerted management to the 17%-per-year growth rate of passengers into and out of the United States. So Delta organized its domestic network to feed passengers into its expanding transatlantic and Pacific routes.

Delta management has been careful to collect accurate information, for it must be used to deal with the same uncertainties faced by competitors. Delta managers know about market shares, airport capacities, fuel prices, and little things like customer complaints. The data tell them that every cent-per-gallon rise in fuel prices costs Delta $20 million a year. But with the newest equipment of the six major airlines (average age of 8.6 years per plane) they also have a fuel-efficient fleet. The company has had the fewest consumer complaints of any major carrier for the past 16 years. And while its management personnel maintain a cautious hands-on approach, their strategic planning process also alerts them to the ups and downs of business cycles and allows them to capitalize on opportunities without getting overly optimistic or pessimistic.

Delta Air Lines is not the first firm that has learned to cope successfully with major uncertainties. Virtually every major business decision calls for some venture into the unknown. Estimates must be made of upcoming costs of fuel and other supplies, of labor expenses, of what demand can be expected to be, and of how customers feel about current levels of service. What is the true value of our inventory? How much time does it really take to fill customer orders? Do our packages actually contain the product weights they are advertised to contain? Is our advertising directed to the right market segment?

Fortunately, many of the business decisions about products to produce, prices to charge, and the amount of inventory to stock can be made in a thoughtful and reliable manner. And even though there is less information than the manager would like, he or she can have some measure of confidence in the result if the decision process being followed is undertaken in a systematic and professional manner. This chapter introduces some statistical procedures designed to do just that: help make good decisions on the basis of limited, or sample, information.

Introducing high-tech products, forecasting sales, and managing inventory levels are not simple tasks, but they are not unusual assignments. Management must routinely make decisions without full information. Some decisions can be made using readily available data; others require that new information be collected. In both cases, the proper use of statistical techniques can help maximize the number of good decisions and minimize the number of poor ones—with the hoped-for result of orderly corporate growth rather than layoffs and general decline. No firm relishes the prospect of having to lay off employees, but many firms are forced to do just that.

Sometimes decisions must be made on the spot. For the most part, however, management is a planned and systematic process. Managers must assess a decision situation and apply their knowledge and experience to develop the best course of action. Theory allows them to make *statistical inferences* about a population from information that is contained in a sample.

> Statistical inference uses sample data and statistical procedures (1) to estimate population parameters or (2) to test whether a population parameter is likely to satisfy some specified value.

Statistical inference includes two procedures we shall study: estimation (in this chapter) and hypothesis testing (in the next). Both techniques rely upon sample data, which often comes from survey information or the firm's own data base (primary data). However, it can come from elsewhere in the business environment (secondary data).

Anyone can make an estimate or a guess about the value of something. Statistical *estimation* is different. It is more refined and scientifically credible.

> Estimation is the process of using a statistic to infer about the value of an unknown parameter with a specified level of certainty.

By following statistical procedures we know how close we should be to the value being estimated (expressed as an interval) and how likely we are to be correct (expressed as an amount of confidence). For example, a manufacturer of computer disks might want to know how many times the disks can be run on a disk drive before they wear out. By collecting usage data from a representative sample of disks, a manufacturer can make an inference about the average lifetime of all disks of that type. Or a marketing manager may wish to estimate the proportion of video-cassette-recorder (VCR) owners who plan to buy two or more video films within the next month. In either case the statistical estimate will include a measure of statistical error. Then a properly chosen sample will enable a manager to make a valid inference about the much larger market segment.

In *hypothesis testing* we establish some claim, or hypothesis, about the value of a population parameter. Then we collect sample evidence to substantiate or refute the hypothesis. Statistical procedures allow us to draw a conclusion about the hypothesis and state the chance (risk) that our conclusion is wrong.

> Hypothesis testing is the use of sample data to make a decision either to support or to refute a statement about a parameter value or about a population distribution.

For example, suppose a pharmaceutical firm feels that a newly developed drug is effective in arresting an allergy. The firm would develop a hypothesis concerning the cure rate for the drug. Maybe the hypothesis would claim that the drug was effective in at least 75% of the allergy cases. Then the hypothesis would be tested, perhaps first by treating animals, but eventually by treating a sample of people who have the allergy. The sample results would either support or refute the hypothesis. The firm could quantify this further by stating the risk that they will accept of being wrong (such as 1% or 5%) when making the claim. Such a procedure was used for testing the Salk polio vaccine before it was accepted for nationwide distribution.

In many respects, estimation and hypothesis testing constitute the heart of statistical inference. The concepts in Chapters 8 and 9 will be used throughout the remainder of this book. For now, however, you will avoid potential confusion if you keep one distinction in mind: Estimation begins with sample statistics (sample means or proportions), whereas hypothesis testing starts with a hypothesis about a population parameter. Both use sample data in helping people make business decisions. We will define these terms more precisely later, but for now you can assume that the larger the sample, the better the decision information. However, larger samples are also more expensive. Therefore, managers usually have to be satisfied with less than total information, and the conclusions will not be as certain as they might like them to be.

This chapter uses sample statistics to infer something about population values. We work with both measurable (continuous) and countable (discrete) data. Our main effort is to learn the procedures for estimating either the population mean or the population proportion. Procedures for estimating the difference in two population means or two population proportions are included in Section 8.6. We also consider how to select an appropriate sample size. This chapter uses large-sample techniques ($n > 30$) and relies upon the central limit theorem. The $t$-distribution is introduced, then used when the original data are normally distributed and the population variance is not known. Chapter 10 discusses both estimation and testing when smaller samples ($n \leqslant 30$) are used.

Upon completing Chapter 8, you should be able to

1. explain the principles underlying confidence interval estimation;
2. make a confidence interval estimate of a population mean, $\mu$;
3. use the $t$-table as appropriate to estimate means;
4. make a confidence interval estimate of a population proportion;
5. determine the appropriate sample size for interval estimation of means and proportions;
6. estimate the difference between two population means or two population proportions.

## 8.2 ESTIMATING A POPULATION MEAN WHEN THE STANDARD DEVIATION IS KNOWN

Suppose you have a "homegrown" business of supplying watermelons to supermarkets. You have determined that the best time to harvest a field of melons is when the average sugar content reaches a specified level. You estimate this time by plugging representative melons to get 1 cubic inch of fruit from the inside of each and then doing a chemical analysis for sugar content. When your sample suggests the melons are just right for shipment, you have them picked and sent off in railroad cars to your customers. You are estimating a population parameter from a sample statistic. The cost is high for an inaccurate estimate. If the melons are too green or too ripe, you may be out of business next year! Thus it is very important to obtain an accurate estimate of the average sugar content in the whole field.

### Properties of Point Estimators

In using data on the average sugar content of a sample ($\bar{X}$) of watermelons to make a judgment about the average sugar content of the whole field, or population, $\mu$, you are using estimation. Estimates usually take one of two forms: point estimates or interval estimates.

*Point estimation* is a process of obtaining a single value from a sample to approximate a population parameter whose true value is unknown.

An estimator is a statistic that is used to estimate a population parameter.

A point estimate is a specific single value calculated from an estimator and used to estimate the value of an unknown population parameter.

The sample mean, $\bar{X}$, is an *estimator* of the parameter, $\mu$, and the value of $\bar{X}$ from one sample is an *estimate* of the true population mean. The mean sugar content of a sample of watermelons is an example of a point estimate. A point estimate can be useful if it is reasonably close to the population, or true value. With many possible sample values, it is unlikely that any random sample mean will in fact *equal* the population mean. Unfortunately, time and cost considerations usually restrict us to a limited number of observations. (Consider the potential loss in sales if 30% of the melons are plugged!) Averaged together, the individual observations for one sample make up one point estimate.

The merits of a point estimation procedure are generalized in a set of four statistical properties: (1) unbiasedness, (2) efficiency, (3) consistency, and (4) sufficiency.

## FOUR PROPERTIES OF POINT ESTIMATORS

1. **Bias** describes the location of an estimator. An estimator is unbiased if the mean of its sampling distribution equals the parameter value being estimated. Otherwise the estimator is biased.
2. **Efficiency** is a measure of the amount of variation in the sampling distributions for alternative estimators of a parameter. For two estimators and a sample size, *n*, the *more efficient estimator* is the one having a sampling distribution with a smaller standard error.
3. **Consistency** describes how the sampling distribution of an estimator concentrates around the parameter value as the sample size, *n*, increases. A *consistent estimator* is more reliable for larger samples.
4. **Sufficiency** is a measure of the use of information. A *sufficient estimator* uses all the information about a parameter that is available in the sample.

These four properties are required for what is termed a *best point estimator* of a parameter. All parameters do not necessarily have this ideal, or best, point estimator. However, with random sampling, $\bar{X}$ is the best estimator of $\mu$. Let us see why.

---

**EXAMPLE 8.1**

Explain why under random sampling $\bar{X}$ is the "best" estimator for the population mean, $\mu$.

*Solution*

To be a "best" point estimator requires satisfaction of four conditions—unbiased, most efficient, consistent, and sufficient.

1. From Chapter 7 we know that $E(\bar{X}) = \mu$. (See, for example, Figure 7.2 and Table 7.3). Under random sampling, $\bar{X}$ is an *unbiased estimator* for $\mu$.
2. For normal distributions, the median is also an unbiased estimator for $\mu$! However, the sample mean is a more efficient estimator because its standard error is smaller than is that of the sample median. In fact, $\bar{X}$ has the smallest possible standard error for the estimators of $\mu$. [1]* It is, therefore, the *most efficient estimator* for $\mu$.
3. By the *law of large numbers* (in Chapter 7), the standard error of the mean decreases as the sample size increases. That is, the larger the sample is, the closer, on average, the sample mean is to $\mu$. This makes $\bar{X}$ a *consistent estimator*.
4. Finally, the sample mean is a *sufficient estimator* because it uses all the information relevant to the estimation of $\mu$ that is contained in the sample. For the purpose of estimating $\mu$, there is no additional information (beyond $\bar{X}$, in the sample) to be gained by specifying the individual values of the sample or indicating their order of occurrence.

---

* Numerals in brackets refer to items cited in "For Further Reading" at the end of the chapter.

Thus the sample mean, $\bar{X}$, is the "best" statistical estimator for $\mu$ because, under random sampling, it clearly satisfies the four requirements: unbiased, most efficient, consistent, and sufficient.

We can also identify a best estimator for samples from dichotomous distributions, where $X$ is the number of successes in $n$ Bernoulli trials. Under random sampling $p = X/n$ is unbiased, most efficient, consistent, and sufficient for estimating $\pi$. Therefore, the sample proportion $p$ is a best estimator of $\pi$. The Computer Application illustrates that $p$ is a consistent estimator for $\pi$ with a specific example.

We noted that it is not always possible to obtain a best estimator for a parameter. For example, consider $s$ as an estimator of $\sigma$. You will recall that when calculating the sample variance, $s^2$, we divide by $n - 1$ rather than by $n$. For random samples from any infinitely large population with known $\sigma$, $s^2$ (with divisor $n - 1$) is an unbiased estimator of $\sigma^2$. However this (*unbiased*) property does not transfer to $s$ as estimator for $\sigma$; $s$ is a biased estimator of $\sigma$. Still, $s$ has other properties that make it as good as any other estimator for $\sigma$, so we use it.

We use *best estimators* if they are available. Otherwise, we use estimators that statistical theory shows are better than any others.

## Error of Estimation

In estimation we want to use a statistic that can estimate a parameter with the least possible error. This translates into using an estimator with a sampling distribution that gives sample values close to the parameter value.

> The error of estimation for a given point estimate is the difference between the sample estimate and the parameter value, ignoring the sign of the result. The smaller the error, the more precise the resulting inference.

Point estimation of a parameter has some serious deficiencies. First, a specific point estimate is very likely to be wrong. Second, it provides no assessment of the probability that a single point estimate value is reasonably close to the parameter value. On the other hand, interval estimation for means makes use of the central limit theorem (Chapter 7) to provide a probability value for the error of estimation, $E$.

You will recall that with normal distributions or for $n > 30$ the distribution of sample means is normal. Table 8.1 displays the general form used for interval estimates, here for the mean, $\mu$, and with known standard deviation, $\sigma$.

Thus the central limit theorem can be used to determine the probability, or a measure of confidence, that $\bar{X}$ falls within any specified error distance from the mean. We use these aids—the central limit theorem, the measure of error, and the probability of observing a chosen size for the error (confidence)—to overcome the shortcomings of point estimation. By using interval estimation procedures we can estimate a population mean and find the probability, or level of confidence, that the estimate is within a specified distance (error) from the actual population mean, even though the population mean is unknown.

**TABLE 8.1**  The General Form of Interval Estimates on the Mean, $\mu$

| Interval Around Mean, $\mu$ | Probability of Containing $\bar{X}$ |
|---|---|
| $\mu - \left(1\dfrac{\sigma}{\sqrt{n}}\right)$ to $\mu + \left(1\dfrac{\sigma}{\sqrt{n}}\right)$ | 0.6826 |
| $\mu - \left(2\dfrac{\sigma}{\sqrt{n}}\right)$ to $\mu + \left(2\dfrac{\sigma}{\sqrt{n}}\right)$ | 0.9544 |
| $\mu - \left(3\dfrac{\sigma}{\sqrt{n}}\right)$ to $\mu + \left(3\dfrac{\sigma}{\sqrt{n}}\right)$ | 0.9974 |

**Interval Estimation Procedure**

By adding and subtracting any appropriate number from the point estimate, we can form an interval of values within which the population parameter lies, with some specified probability. The standard procedure for interval estimation of $\mu$ is to define the interval using the sample mean plus and minus some multiple of the standard error. The interval is defined by the two endpoints $\bar{X} - Z\sigma_{\bar{X}}$ and $\bar{X} + Z\sigma_{\bar{X}}$. A good *interval estimate* is one that has a high probability of containing the population parameter.

The definition for confidence interval estimates follows.

> An interval estimate is a span of values that should include an unknown parameter value. It is defined by two end values.

> Confidence is a number between 0 and 100 that reflects the probability (multiplied by 100) that the interval estimate will include the parameter value. We normally want high confidence (probability), in the range of 90% to 99.7%.

We form a confidence interval by subtracting and adding the bound on the error of estimation to the point estimate to obtain two end values—a lower confidence limit, LCL, and an upper confidence limit, UCL. The *bound on the error of estimation*, what is subtracted or added, is ($Z \cdot$ Standard error).

The logic for making confidence interval estimates is based on the centrally located point estimate. We can illustrate this logic by referring to Figure 8.1. This is simply another application of the central limit theorem. Here the parameter is $\mu$ and the estimator is $\bar{X}$. We describe the error for estimating $\mu$ at a 95% confidence level. (*Note:* The computation procedure is summarized following this discussion.)

Figure 8.1 assumes a variable that is normally distributed. Then 95% of the sample means reside within the interval $Z = -1.96$ to $Z = +1.96$. Thus we could observe that $P(-1.96 < Z < +1.96) = 0.95$. We replace $Z$ with its equivalent form $(\bar{X} - \mu)/\sigma_{\bar{X}}$, so

$$P\left(-1.96 < \frac{\bar{X} - \mu}{\sigma_{\bar{X}}} < +1.96\right) = 0.95$$

**FIGURE 8.1**    Establishing a 95% Confidence Interval for $\mu$

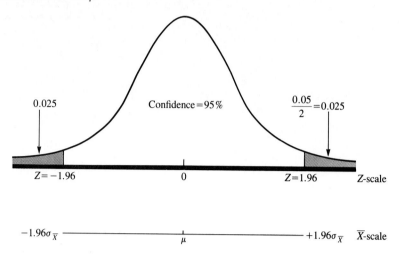

We obtain a solution by "pivoting" on, and isolating, the parameter—here, $\mu$. This approach is called the *pivotal method* for establishing confidence intervals.

$$P(-1.96\sigma_{\bar{X}} < \bar{X} - \mu < +1.96\sigma_{\bar{X}}) = 0.95$$

$$P(\bar{X} - 1.96\sigma_{\bar{X}} < \mu < \bar{X} + 1.96\sigma_{\bar{X}}) = 0.95$$

so    $$\underbrace{\bar{X} - (1.96\sigma_{\bar{X}})}_{\substack{\textbf{Lower confidence} \\ \textbf{limit, LCL}}} < \mu < \underbrace{\bar{X} + (1.96\sigma_{\bar{X}})}_{\substack{\textbf{Upper confidence} \\ \textbf{limit, UCL}}}$$

That is, we have found an interval around $\bar{X}$, the point estimate, that is highly likely (95% confidence) to contain the population parameter $\mu$. This procedure requires large random samples of $n > 30$ with known $\sigma$ or assumes that the sample statistic $\bar{X}$ is normally distributed from a population with known $\sigma$. The standard error of the mean is $\sigma_{\bar{X}} = \sigma/\sqrt{n}$.

The illustration in Figure 8.1 specifies the confidence as being 95%. This is generalized to a $[(1 - \alpha)100]\%$ confidence interval.

A $[(1 - \alpha)100]\%$ confidence interval has the form

Point estimate $\pm Z_{\alpha/2} \cdot$ Standard error

For $[(1 - \alpha)100]\% = 95\%$, then, $1 - \alpha = 0.95$ and $\alpha = 0.05$, or 5%. Since the confidence interval is symmetric, centered at $Z = 0$, the $Z$-values $\pm 1.96$ each remove $\alpha/2 = 0.05/2 = 0.025$, or 2.5%, of the area from either tail of the $Z$-distribution. We generalize this as $\pm Z_{\alpha/2}$: In Figure 8.1, $Z_{0.025} = 1.96$ and $-Z_{0.025} = -1.96$.

We can form confidence interval estimates by "pivoting" around parameters, including $\mu$, $\pi$, $\sigma^2$, and the ratio of variances $(\sigma_1^2/\sigma_2^2)$. The appropriate statistics related to these parameters utilize the $Z$-, $t$-, chi-square, and $F$-distributions, respectively. These are probability curves for sampling distributions that relate an estimator to a parameter. The form of each distribution is introduced as we discuss the related parameter (and its estimator) in the text. We begin by using the procedure for making confidence interval estimates of $\mu$. This uses the normal ($Z$) distribution.

**Confidence Intervals for $\mu$**

Before proceeding further, let us summarize the steps in using sample data to make a confidence interval estimate of a population mean $\mu$.

> **PROCEDURE FOR MAKING CONFIDENCE INTERVAL ESTIMATES OF $\mu$ FOR A NORMALLY DISTRIBUTED POPULATION WITH KNOWN VARIANCE AND $n > 30$.**
>
> 1. Select a random sample of size $n$ and calculate the sample mean $\bar{X}$.
> 2. Calculate the standard error of the mean $\sigma_{\bar{x}} = \sigma/\sqrt{n}$.
> 3. Establish confidence limits of a width $\pm(Z_{\alpha/2} \cdot \sigma_{\bar{x}})$ about the sample mean $\bar{X}$. The $Z_{\alpha/2}$ represents the confidence coefficient and reflects the amount of confidence desired (for example, $Z_{\alpha/2} = 1.96$ for 95% confidence). The area outside the interval is divided into the two tails with $\alpha/2$ area in each tail. Then
>
> $$\text{LCL} = \bar{X} - (Z_{\alpha/2} \cdot \sigma_{\bar{x}}) \qquad (8.1)$$
>
> $$\text{UCL} = \bar{X} + (Z_{\alpha/2} \cdot \sigma_{\bar{x}})$$
>
> 4. Interpret the confidence interval in this manner: *—explain what of this says*
> "By using this procedure and sample size, 95% of all samples will include the population mean, $\mu$."

---

**EXAMPLE 8.2**

*parametric*
*stats sample*
*random*
*normal distrib.*

The production manager of a frozen-foods packaging plant must maintain strict controls over package weights in order to satisfy state licensing requirements. She now wishes to estimate the mean weight of restaurant packs of frozen corn at a confidence level of 95%. A simple random sample of 100 packs reveals a mean weight of 10.1 pounds. The packaging machine tolerances are such that the process standard deviation is $\sigma = 0.50$ pounds. Find the 95% confidence limits.

*Solution*    The procedure for making confidence interval estimates gives direction to the solution.

Step 1   $\bar{X} = 10.10$ pounds

*std error*

Step 2   $\sigma_{\bar{x}} = \dfrac{\sigma}{\sqrt{n}} = \dfrac{0.50}{\sqrt{100}} = 0.05$ pounds

Step 3   For 95% confidence, the interval on the $Z$-scale must go from $-1.96$ to $+1.96$ (see figure). Thus,

$$\text{LCL} = \bar{X} - (Z_{\alpha/2} \cdot \sigma_{\bar{x}}) = 10.1 - (1.96 \cdot 0.05) = 10.0$$

$$\text{UCL} = \bar{X} + (Z_{\alpha/2} \cdot \sigma_{\bar{x}}) = 10.1 + (1.96 \cdot 0.05) = 10.2$$

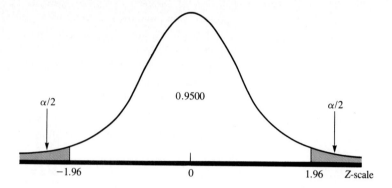

Step 4    We have confidence that the mean weight of the packs of frozen corn is a number within the interval 10.0 to 10.2 pounds. This confidence (95%) is placed on the procedure we followed in making the interval estimate.

Now let us analyze Example 8.2. First, we did not know the true mean weight of the process packs—nor do we yet know it. Our knowledge about a mean value is limited to the value of one random sample mean ($\bar{X} = 10.1$ pounds). But there are perhaps thousands of other sample means as well. However, from the central limit theorem we do know that if $n > 30$, the $\bar{X}$'s are normally distributed. This means that they cluster about the population mean, $\mu$, in a normal curve shape, as shown in Figure 8.2(a). Using the central limit theorem, we know by *deduction* that 95% of all sample means for random sam-

**FIGURE 8.2**    Inference from the Central Limit Theorem

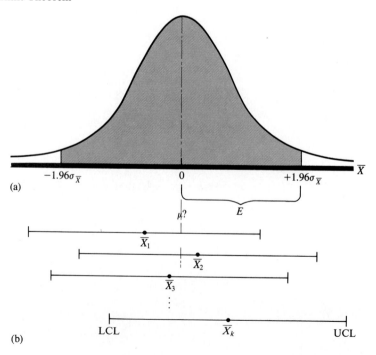

ples of size $n$ lie within an interval with width $-1.96\sigma_{\bar{X}}$ to $+1.96\sigma_{\bar{X}}$. By deduction, we know that 95% of the $\bar{X}$'s are within $1.96\sigma_{\bar{X}}$ distance of $\mu$. The width of the interval is $E = 1.96\sigma_{\bar{X}} - (1.960_{\bar{X}}) = 2(1.96\sigma_{\bar{X}})$. The confidence interval is indicated by the shaded area in Figure 8.2(a).

Suppose we were to establish an interval of width $\pm(1.96 \cdot \sigma_{\bar{X}})$ about all the *sample means* in Figure 8.2(a). Using the same size interval, we learn from *induction* that 95% of the specific intervals $\bar{X} \pm (1.96\sigma_{\bar{X}})$ will include $\mu$.

Now for the inferential step. We do not, of course, establish such intervals for all sample means. We typically use only one of the many ($k$) possible sample means (and resulting intervals) shown in Figure 8.2(b). However, because we are using the same size interval ($\pm 1.96\sigma_{\bar{X}}$ in this case), we can infer that 95% of the $k$ cases would yield intervals that include $\mu$. Conversely, there is a 5% chance that any interval chosen at random excludes $\mu$. We don't know whether any given interval includes the population mean or not. By establishing a 95% confidence interval about our one sample mean ($\bar{X} = 10.1$ pounds in Example 8.2), we are simply *following a procedure*, whereby 95% of all possible intervals obtained in this way will include the population parameter. We would expect to be correct 95% of the time, so we expect that our sample interval is one of those that includes $\mu$.

---

## EXAMPLE 8.3

An operations analyst for an airline company has been asked to develop a fairly accurate estimate of the average refueling and baggage handling time at a foreign airport. The estimate is to be used to arrange flight schedules so that the airline has a very high percentage of "on-time" departures. Suppose the analyst randomly samples refueling and baggage handling times on 36 flights and finds they average 24.2 minutes. The population standard deviation is 4.2 minutes. Establish a 90% confidence interval estimate.

*Solution*  The figure depicts the sampling distribution that includes 90% of the area beneath the normal curve and is centered at $Z = 0$.

$$CL = \bar{X} \pm (Z_{\alpha/2} \cdot \sigma_{\bar{X}})$$

where     $\bar{X} = 24.2$ minutes

tabulated $Z_{0.05} = 1.645$   (for 0.4500 area; see figure below)
Z (one you look up)

std error $\sigma_{\bar{X}} = \dfrac{\sigma}{\sqrt{n}} = \dfrac{4.2}{\sqrt{36}} = 0.7$ minutes

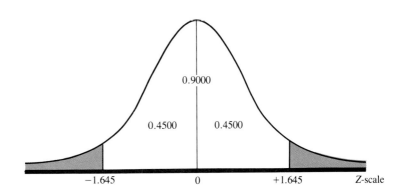

Therefore,

$$LCL = 24.2 - (1.645 \cdot 0.7) = 23.0 \text{ minutes}$$

$$UCL = 24.2 + 1.2 = 25.4 \text{ minutes}$$

The analyst should report an interval of 23.0 to 25.4 minutes at the 90% confidence level.

Suppose that in Example 8.3 the flight scheduler balks at the analyst's qualification of being only 90% sure. The analyst could offer more confidence simply by changing to a higher $Z$-level. But this change would cause another concern. Intuitively, we expect to trade confidence for precision. That is, for a fixed sample size, higher confidence means lowered precision (a longer interval), and vice versa. Example 8.4 illustrates this point.

Find the 99.7% confidence interval estimate for the data in Example 8.3. Retain the sample size at 36 units.

*Solution*    Relative to a lower confidence, the 99.7% confidence interval has wider limits on the $Z$-scale. This is illustrated in the figure below. For 99.7% confidence, $Z_{\alpha/2} = 3.0$.

$$CL = \bar{X} \pm (Z_{\alpha/2} \cdot \sigma_{\bar{X}})$$
$$= 24.2 \pm (3.0 \cdot 0.7) = 22.1 \text{ to } 26.3 \text{ minutes}$$

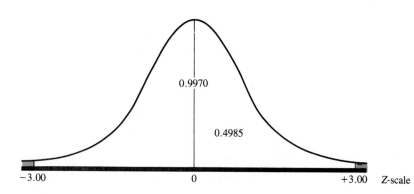

This wider interval is shown in Figure 8.3. Now the analyst could report an interval of 22.1 to 26.3 minutes at the 99.7% confidence level. For a fixed sample (here, 36), a higher confidence requires a wider confidence interval.

Prior to leaving this section, we emphasize two points. First, although we have assumed $\sigma$ is known, sometimes we have to use $s$ as an estimator of $\sigma$. Then our estimated standard error is $s/\sqrt{n}$. This requires another sampling distribution, which is described in the next section. Second, before sampling, we have a probability situation in which the sample mean is considered a random variable. We can properly speak of the probability that the sample will take on a value in some specific interval. After sampling, when an interval is established

**FIGURE 8.3** Higher Confidence for a Fixed Sample Size Requires Wider Intervals (More Error)

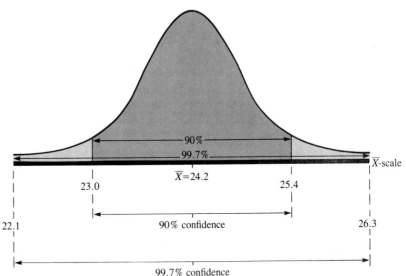

about the sample mean, there is no probability involved (unless we take a Bayesian approach of viewing the population parameter as a random variable). The interval about the sample mean either includes the true mean $\mu$ or it doesn't. This is why *we speak of confidence rather than of probability. The statement refers to the process, not the individual result.* Similarly, it is incorrect to say the population mean falls in the confidence interval. The population mean is already there—the interval either encompasses it or it doesn't.

Managers typically should define confidence levels prior to collecting the sample data; otherwise they could set the level to arbitrarily show wider or narrower intervals. Additionally, if all other factors are set (that is, the confidence limits, $\bar{X}$, $\sigma_{\bar{x}}$ and $n$), one can reverse the calculation and find out specifically how much confidence can be put into certain claims.

**Problems 8.2**

1. A simple random sample of 625 high-school students revealed a mean TV-watching time of 110 minutes per day with $\sigma = 50$ minutes per day. What is the 99.7% confidence interval estimate of the mean time the high-school students in this area spent watching TV?

2. A sample of 100 ingots at an aluminum plant revealed a mean weight of 602 pounds, and the standard deviation is known as $\sigma = 20$ pounds. Compute the 95% confidence interval estimate and carefully write out a statement expressing your conclusion.

3. A market research study was made in a community of 20,000 families to establish the average expenditure per month for gasoline, oil, and other motor fluids. If, in a random sample of 100 families, the average expenditure was $42.13 with a population standard deviation of $6.50, estimate the average monthly expenditure for motor fluids in the community. Use a 95% confidence interval level.

4. In their last contract the average increase in weekly salary for a random sample of 49 workers was $35.00. The standard deviation was $\sigma = \$5.60$. Find a 90% confidence interval for the true average weekly salary increase for all such workers.

5. A random sample of 64 cars passing a checkpoint on a certain highway showed a mean speed of 60.0 miles per hour. The standard deviation of 15.0 miles per hour is known from earlier speed tests, $\sigma = 15.0$ mph.
   a. Establish a 95% confidence interval for the mean speed of cars at this checkpoint. Interpret the meaning of this interval.
   b. Establish a 99% confidence interval using the same data. Now compare the widths of the two intervals.

6. While performing a stimulus–response experiment, a psychologist found that a random sample of 49 railroad employees required an average of 6.20 seconds to respond to one stimulus. The standard deviation in response time for experiments of this type is 1.00 second. Establish a 94% confidence interval of the true time required for railroad employees to react to this stimulus.

7. A cost accountant in a job shop selected a random sample of 50 customer orders in an attempt to estimate the average number of engineering changes made before the orders are shipped. Data are shown below.
   a. Give point estimates for the mean, $\mu$, and the standard deviation, $\sigma$, of the number of changes.
   b. Construct a 95% confidence interval estimate for the mean. Let $\sigma = 1.0$.

| Number of Changes | Frequency (number of orders) |
|:---:|:---:|
| 0 | 25 |
| 1 | 14 |
| 2 | 7 |
| 3 | 4 |
| Total | 50 |

## 8.3 ESTIMATING A POPULATION MEAN WHEN THE STANDARD DEVIATION IS NOT KNOWN

Trade associations frequently publish statistics such as average life, endurance, reliability, and other measures of the usefulness of consumer products. To a limited extent, these statistics can be used to assess and test the claims of producers. However, they are usually not sufficient to meet the needs of the production and marketing departments of manufacturers. Firms must often collect and analyze their own primary data.

But sampling and testing can be expensive. In addition, some types of tests damage, wear out, or even destroy the products. As an economic consequence, many business situations must be analyzed using small samples. This section develops the procedures for making interval estimates about averages when the information comes from small samples. In making this inference we assume the observations come from a normally distributed population.

**The t-Statistic and t-Sampling Distribution**

The sampling distribution for means from small samples from a normal distribution with unknown variance is the *Student's t-distribution.*

The Student's *t*-statistic is

$$t = \frac{\bar{X} - \mu}{s_{\bar{X}}} \tag{8.2}$$

where $s_{\bar{X}} = \dfrac{s}{\sqrt{n}}$

Assumptions: The random variable $X$ is normally distributed.
The population standard deviation is not known.

As noted earlier, the *t-distribution* has pronounced similarities to the *Z-distribution*. The Z- and t-statistics are calculated in the same way; both distributions have a mean of zero. The Z-distribution has a standard deviation equal to one, but the t-distribution has a standard deviation greater than one. Both Z and t are smooth bell-shaped probability curves. In contrast to the fixed appearance of the standard normal, the flatness or peakedness of the t-distribution is determined by the sample size. Relative to the Z-curve, Figure 8.4 displays the distance on the t-scale for several t-curves to cover a 95% centered area. Observe that as the sample size, n, grows larger to n > 30, the t-distribution approaches the Z as a limiting shape.

The percentage points for a t-distribution are derived from calculus. For our use, the important distinction is that the t-values in the table decrease to Z-values as the sample size becomes large. Larger samples provide less error

---

**FIGURE 8.4**   Plots of the *t*-Distribution for Different Sample Sizes

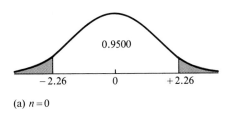

(a) $n = 0$

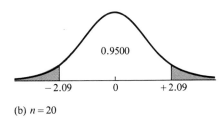

(b) $n = 20$

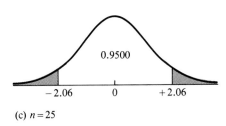

(c) $n = 25$

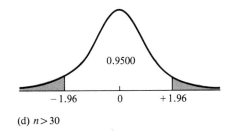

(d) $n > 30$

and more useful information. This idea is captured in a simple concept called *degrees of freedom.*

> The degrees of freedom (df) are a measure of the amount of usable information in a sample and are related to the sample size.

The degrees of freedom reflect the fact that some of the information is lost when we calculate a variance using the statistic $\bar{X}$ in place of the parameter $\mu$. The result cannot be expected to be as good as it would be if $\mu$ were known and used.

All *t*-statistics have an associated df; but the df depends upon the explicit *t*-form. The *t*-statistic (8.2) has $n - 1$ df. For all *t*-statistics, df can be observed as the divisor in the expression of the standard deviation. In Equation (8.2), $s = \sqrt{\sum (X - \bar{X})^2/(n - 1)}$ has a divisor of $n - 1$. One parameter, $\mu$, is estimated (from $\bar{X}$), so one is subtracted from the sample information. This leaves the divisor as df $= n - 1$.

The *t*-table (Table B.2 in Appendix B) is constructed somewhat differently from the Z-table (Table B.1 in Appendix B). Percentage points for the *t*-distribution appear in the body of the table, and probabilities are in the column headings. Since the flatness or peakedness of the *t*-distribution depends on the sample size (which is described in degrees of freedom), df heads the leftmost column in the table. For now we will use df $= n - 1$ and consider only single-sample procedures. A portion of Table B.2 is displayed in Table 8.2. The *t*-table is an opposite description compared to the Z-table, which describes areas in the center of the distribution. Table 8.2 displays the area in an extreme—that is, a tail area—of the *t*-distribution. This is represented by the shaded area, where $\alpha$ takes a value 10%, 5%, or other. The column headings 0.10, 0.05, ..., and 0.005 denote the percentage of area that is in a tail and beyond the tabular *t*-value. Then, for example, for a *t*-distribution defined by 14 df, 10% of the area is to the right of $t = 1.345$. By symmetry the percentages are for either

**TABLE 8.2**    A Segment of the *t*-Table (Appendix B, Table 2)

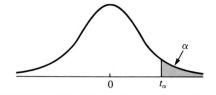

| Degrees of Freedom | $t_{0.100}$ | $t_{0.050}$ | $t_{0.025}$ | $t_{0.010}$ | $t_{0.005}$ |
|---|---|---|---|---|---|
| 1 | 3.078 | 6.314 | 12.706 | 31.821 | 63.657 |
| 2 | 1.886 | 2.920 | 4.303 | 6.965 | 9.925 |
| ⋮ | ⋮ | ⋮ | ⋮ | ⋮ | ⋮ |
| 14 | 1.345 | 1.761 | 2.145 | 2.624 | 2.977 |
| 15 | 1.341 | 1.753 | 2.131 | 2.602 | 2.947 |
| ⋮ | ⋮ | ⋮ | ⋮ | ⋮ | ⋮ |
| 29 | 1.311 | 1.699 | 2.045 | 2.462 | 2.756 |
| inf | 1.282 | 1.645 | 1.960 | 2.326 | 2.576 |

tail, but since the t-distribution
are negative. So for df = 14, anothe
tion is to the left of $t = -1.345$. Som
Table B.2.

---

### EXAMPLE 8.5

Find the appropriate t-value for a 95% confidence
from a normal distribution.

*Solution*   By the symmetry of the t-distribution, one-half of the 0.95
center, which is at the value zero (0). The balance, or $1 - 0.9$
into two parts. So the $\alpha$-value in Table 8.2 is $0.05/2 = 0.025$. For $n$
15, we read $t_{0.025} = 2.145$. Thus 2.5% of the area falls above $+2.14$
is below $-2.145$, while 95% is in between.

---

### EXAMPLE 8.6

Find the t-value for a 95% confidence interval using a sample of size $n = 31$.

*Solution*   For $n = 31$ (df $= \infty$) with 95% of the area in the center of the distribution, the t-percent
age points are $t_{0.025,\infty} = 1.960$. That is, $\alpha/2 = 0.025$, so 2.5% of the area is in each tail
(beyond $-1.960$ or $+1.960$).

---

We can check the t-percentage points for the two preceding examples by
comparison. Our numbers are reasonable because, with $n > 30$, the values for
$t_\infty$ are $-1.960$ and $1.960$, respectively. That is, for df $=$ infinity, the t-distribution becomes the Z. You will recognize the numbers $-1.96$ and $1.96$ as the
percentage points on the Z-scale for a 95% confidence interval. Also, this limiting
form was displayed in Figure 8.4.

**Estimation Procedures**   The t-statistic is widely used because many random variables are normally distributed but the distribution variance is unknown. Then, for sample sizes under
31, the sample means are distributed as a Student's t-distribution. The procedures for estimation are the same as those outlined in Section 8.2, except that
(1) t-values replace Z-values and (2) the sample standard deviation, $s$, replaces
$\sigma$. Upper and lower confidence interval forms are given in Equation (8.3).

> Confidence intervals for the means of random samples of size $n \leqslant 30$
> from a normal distribution use LCL $< \mu <$ UCL with
>
> $$\text{LCL} = \bar{X} - (t_{\alpha/2} \cdot s_{\bar{X}}) \qquad (8.3)$$
> $$\text{UCL} = \bar{X} + (t_{\alpha/2} \cdot s_{\bar{X}})$$
>
> where $t_{\alpha/2}$ denotes $\alpha/2$ probability in each tail of the t-distribution
> with df $= n - 1$ and $s_{\bar{X}} = s/\sqrt{n}$.

These upper and lower confidence limits are consistent with the *pivotal
method* procedure described in Section 8.2.

Establishing confidence intervals using Equation (8.3) requires the calculation of $\bar{X}$ and $s$. These sample calculations are often most efficiently done

(rotated text, partially obscured)
has a central value of zero, lower-tail $t$-values
10% of the area bounded by the $t$-distribu-
examples will help illustrate the use of

interval using a sample of size 16

sides on either side of the
$= 0.05$, is split equally
16 and df $= n - 1 =$
and another 2.5%

l samples it is sometimes easier to use a hand cal-
calculations, the next example includes the extra
f $\bar{X}$ and $s$ values in the interval estimation of a

onal fresh-produce broker and receives air cargo ship-
d the world. Most produce is priced by the pound, so
oming produce weights accurately. The air transports
an estimate a shipment weight by using a sample mean
craft haul 100 containers). Suppose a random sample
nanas had weights of 560, 540, 570, 550, 570, 560, and
stribution of container weights, find a 90% confidence
veight of the containers. Then estimate the total cargo
ents.

ibution requirements lead us to the $t$-statistic proce-
and $s$. We let $X = $ the weight in 100 pounds for an

| $X$ (100 pounds) | $(X - \bar{X})$ | $(X - \bar{X})^2$ |
|---|---|---|
| 5.4 | −0.2 | 0.04 |
| 5.5 | −0.1 | 0.01 |
| 5.6 | 0.0 | 0.00 |
| 5.6 | 0.0 | 0.00 |
| 5.7 | 0.1 | 0.01 |
| 5.7 | 0.1 | 0.01 |
| 5.7 | 0.1 | 0.01 |
| Totals   39.2 | 0 | 0.08 |

$$\bar{X} = \frac{\sum X}{n} = \frac{39.2}{7} = 5.6(100) \quad \text{or 560 pounds}$$

$$s = \sqrt{\frac{\sum (X - \bar{X})^2}{n - 1}} = \sqrt{\frac{\sum (X - 5.6)^2}{7 - 1}} = \sqrt{\frac{0.080}{6}}$$

$$= 0.115$$

Then $t_{\alpha/2, n-1} = t_{0.05,6} = 1.94$.

$$\text{CL} = \bar{X} \pm t_{\alpha/2} s_{\bar{X}}$$

$$\text{LCL} = 5.60 - 1.94 \cdot 0.115/\sqrt{7} = 5.52 \ (100) \text{ pounds}$$

$$\text{UCL} = 5.60 + 1.94 \cdot 0.115/\sqrt{7} = 5.68 \ (100) \text{ pounds}$$

That is, the 90% confidence interval estimate of the mean weight of Jamaican
banana shipping containers is between 552 and 568 pounds. A transport that hauls 100
containers should be carrying between 55,200 and 56,800 pounds

Our next example compares the numerical results of estimation with a small sample using the $t$-formula versus using a large sample, $n > 30$, and the $Z$-formula with known standard deviation.

---

**EXAMPLE 8.8**

The Texas Chemicals Company has developed a new liquid solution that is supposed to increase the useful lifetime of auto batteries. In a test on 16 randomly selected batteries, the average lifetime was increased by 11.00 months with a standard deviation of 6.00 months. Assuming that the increase in lifetimes is a normal random variable, establish a 99% confidence estimate of the true increase in lifetime.

a. What increase in months of lifetime can reasonably be expected from the use of this solution?

b. Compare the results with $n = 16$ and the $t$-statistic against using a random sample of $n = 64$ and the $Z$-statistic.

*Solution*   a. The $t$-statistic is again used, and for 99% confidence with 15 df we have $t_{0.005,15} = 2.95$. This yields confidence limits of $11.00 - (2.95 \cdot 6.00/\sqrt{16}) < \mu < 11.00 + 4.42$, so that LCL = 6.58 months and UCL = 15.42 months. The true mean increase in lifetime affected by this new liquid is estimated to be between 6.58 and 15.42 months at 99% confidence. The company can safely advertise a 6-month extension of lifetime with the use of their new liquid battery solution.

b. Estimates based on the $t$-statistic have larger errors than do those based on a $Z$-statistic because less information (a smaller sample) is used. In the preceding example, if we had obtained approximately the same mean, $\bar{X} = 11.00$, from a random sample of $n = 64$ batteries, and, for this situation, $\sigma$ also is 6.00, the error for a 99% confidence estimate would be $E = Z\sigma/\sqrt{n} = (2.58 \cdot 6.00)/\sqrt{64} = 1.94$. This is less than one-half the previous error of 4.42 units. Yet, we must pay a price for the better estimate; reducing the error necessitates a larger sample at an increased cost. But sometimes we may not be able to afford the added cost!

---

Having looked briefly at the estimation of means, we turn next to choosing the appropriate random sample size. The section Problems include skill exercises using the $t$-distribution, Table B.2, and interval estimation with small samples, $n \leqslant 30$, from normal distributions.

**Problems 8.3**

8. Assuming we wish to estimate the mean, with small samples, use the $t$-table to find the $t$-value for
   a. $n = 24, \alpha = 0.05$        d. $n = 25, \alpha = 0.01$
   b. $n = 15, \alpha = 0.05$        e. $n = 25, \alpha = 0.05$
   c. $n = 5, \alpha = 0.05$         f. $n = 25, \alpha = 0.10$

9. Answer each of the following questions concerning the use of the $t$-table (assume a normal distribution for $X$).
   a. Find the tabular $t$-value for a confidence interval on $\mu$ based on $n = 22$, with $\alpha = 0.05$.
   b. A confidence estimate of $\mu$ based on $n = 19$ observations uses tabular $t = \pm 2.10$. What is the confidence level?
   c. For $n = 11$, determine tabular $t$ for a 90% confidence interval.

10. Discuss the similarities and differences in the $t$- and $Z$-probability distributions.

*how do you know when to divide (2)* (handwritten margin note)

11. Write one or two sentences to explain each of the following:
    a. How you would find the 95th percentile using the $t$-table for a $t$-distribution defined by 17 df?
    b. How you would locate the correct $t$-value for estimation of $\mu$ with df $= 12$ and $\alpha = 0.05$?
    c. Why is the $t$-table so incomplete—for instance, why doesn't it contain $t$-values for probability levels 0.04, 0.06, 0.072, and so on?

12. Information provided by 24 food-service caterers in a certain city indicates that the mean age for their employees is 20.0 years with a standard deviation of 2.6 years. Use this information to develop a 95% confidence interval estimate of the average age for employees in food-service work in this city. Assume a normal distribution.

13. Twenty-five cans of dog food were randomly sampled for cereal content, with the result that $\bar{X} = 6.0$ ounces and $s = 0.05$ ounces of cereal. Find the 99% confidence limits for $\mu$. Assume a normal distribution for can weights.

14. A firm having 4000 employees took a random sample of 25 in order to estimate their average annual state tax payment with 95% confidence. If the results were $\bar{X} = \$640.00$ and $s = \$20.00$, what is the best interval estimate? Assume tax payments follow a normal distribution.

15. In a study of the effectiveness of a reducing diet by a consumer protection agency, the following weight-gain figures were obtained from a random sample of 10 men aged 50–60. Note that a minus sign indicates a loss of weight. Given measures of $-3$, $-7$, $-1$, $0$, $+3$, $+1$, $-1$, $-6$, $-10$, and $-6$ pounds, estimate the true mean loss effected by this diet at a 90% confidence level. Assume a normal distribution of weight loss.

## 8.4 CHOOSING THE SAMPLE SIZE

One of the first questions the analyst faces in estimation is, How large should the random sample be? Since the sample size depends on the level of confidence desired, the first step in determining the sample size is to specify the confidence level. The confidence level and the sampling variability establish the limits for the confidence interval. In determining sample size we refer to these limits as the allowable bound of error, $E$.

The fundamental relation used to determine sample size for estimation of the mean is, again, the $Z$ formula from the central limit theorem. Figure 8.5 illustrates the *bound of error* relation ($E$) for 95% confidence. We generalize this for $[(1 - \alpha) \cdot 100]\%$ confidence.

> The bound of error, $E$, is the size of the maximum error that will be tolerated in estimation of the parameter. This establishes an interval of estimation as the interval of values from the point estimate minus $E$ to the point estimate plus $E$, so the bound of error, $E$, is one-half the width of the $(1 - \alpha)$ confidence interval.

Figure 8.5 displays the relationship that is expressed in the preceding definition.

Because $\mu$ is unknown and is being estimated, interval estimates are centered at $\bar{X}$. Using (1) the $Z$-value associated with the level of confidence desired and (2) a measure of the sampling variation gives $\pm E = Z\sigma_{\bar{x}}$. So, the bound of

**FIGURE 8.5** Bound of Error, *E*, Relations

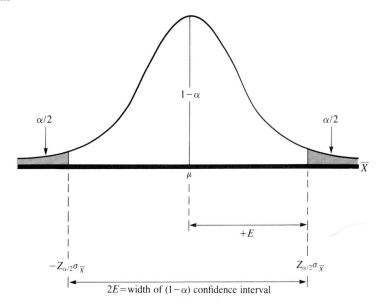

error equals confidence times sampling error ($\sigma_{\bar{X}}$). Then, replacing $\sigma_{\bar{X}}$ by $\sigma/\sqrt{n}$ gives $\pm E = Z\sigma/\sqrt{n}$. Using some algebra we can solve for *n*:

$$n = (Z\sigma/E)^2$$

These results are summarized in Equations (8.4) and (8.5).

### BOUND-OF-ERROR (*E*) AND SAMPLE-SIZE (*n*) CALCULATIONS FOR ESTIMATING $\mu$

$$\text{Bound of error} \quad E = Z\sigma_{\bar{X}} = \left(\frac{Z_{\alpha/2}\sigma}{\sqrt{n}}\right) \qquad (8.4)$$

$$\text{Sample size} \quad n = \left(\frac{Z\sigma}{E}\right)^2 \qquad (8.5)$$

where   $n$ = the random sample size ($> 30$)

   $E$ = the bound of error of estimation or, equivalently, one-half the width of the confidence interval

   $Z$ = the $Z_{\alpha/2}$-value for the level of confidence where $\alpha$ is equally divided between the two tails of the distribution

   $\sigma_{\bar{X}} = \sigma/\sqrt{n}$, the standard error of the mean, which is the measure of sampling variability.

Note that to determine the sample size for a specified level of confidence requires a value for $\sigma$. If $\sigma$ is not known, then it must be estimated, perhaps by analysis of another sample or perhaps by a sample value obtained in a very similar experiment. The value for $E$ is specified by the decision maker. Thus, in sample-size considerations there are four elements to consider. Knowing or specifying any three items allows us to find the fourth.

## ITEMS TO CONSIDER IN SAMPLE-SIZE COMPUTATION

1. $n$     the random sample size;
2. $Z_{\alpha/2}$   the indicator of the confidence level;
3. $\sigma$     the measure of dispersion in the original distribution;
4. $E$     the maximum amount of difference allowable between the point estimate and the parameter value. This is the bound of error and is equal to one-half the confidence interval width.

In the following example we seek the sample size, $n$, after finding or evaluating the other three factors.

---

**EXAMPLE 8.9**

The Chamber of Commerce in a college town is anxious to estimate the average amount of money that university students spend in the local economy in any given year. If the amount is high enough, they will promote advertising that should appeal to students. The researcher would have to specify values for both the confidence level, a $Z$-value, and the bound of error, $E$, in this situation. Suppose these were set at 95% confidence (so $Z = 1.96$) and $E = \pm\$50.00$. The standard deviation of dollars spent by individual students, $\sigma$, is known from previous studies and is $\sigma = \$256.00$. How many students should be surveyed?

*Solution*   The sample size calculation uses Equation 8.5 with the given information:

$$n = \left(\frac{Z\sigma}{E}\right)^2 = \left(\frac{1.96 \cdot \$256.00}{\$50.00}\right)^2 = (10.04)^2 = 100.7$$

In this case, the calculated sample size (100.7) is not a whole number. So we round up to the next larger whole number. Rounding up, 100.7 becomes 101—so 101 students should be surveyed.

---

When a sample size calculation gives other than a whole number, then round up. This will ensure that the specified confidence can be achieved within the maximum error specified.

Better estimates mean higher confidence, a narrower interval, or both. Example 8.10 shows that for fixed confidence (here, 95%), an improved estimate in terms of reduced interval width requires an increased sample size.

---

**EXAMPLE 8.10**

Suppose that in Example 8.9 a Chamber of Commerce official wanted less error—assume the manager wants an estimate for the mean amount spent by these students with a maximum error of only $25.00. How many students should be surveyed to provide an estimate that is twice as precise as the first one?

*Solution*   The answer is more than the previous value. Assuming the same confidence, 95%, and again using $\sigma = \$256.00$, we have

$$\sigma = \$256.00$$

$$n = \left(\frac{Z\sigma}{E}\right)^2 = \left(\frac{1.96 \cdot \$256.00}{\$25.00}\right)^2 = 402.8$$

Rounding up gives $n = 403$, the number of students who should be surveyed.

Relative to Example 8.9, Example 8.10 asks for a better estimate in the sense of less error, so we must pay for that accuracy by taking a larger sample. In fact, cutting the bound of error in half produced a fourfold increase in sample size.

A continuation of Example 8.10 enables us to observe the effect of requiring a higher probability level while maintaining the original bound of error.

## EXAMPLE 8.11

Suppose we want 99% confidence ($Z_{0.005} = 2.58$) and $E = \pm\$50.00$ in the estimate of average amount spent by the university students of Example 8.9. That is, compared to the original specifications, we want increased confidence while the allowable error is unchanged. Now what is the required sample size?

*Solution*    Again the sample size must be increased.

$$n = \left(\frac{Z\sigma}{E}\right)^2 = \left(\frac{2.58 \cdot \$256.00}{\$50.00}\right)^2 = (13.21)^2 = 174.5$$

Rounding up gives $n = 175$, the required sample size.

To generalize these three examples, we can achieve better estimates in the sense of increased confidence, decreased bound of error, or both by taking a larger random sample. This may be expensive, however. For example, it might cost $0.50 to interview each student. Then 101 interviews (Example 8.9) would cost $50.50, whereas 404 interviews (Example 8.10) would cost $202. The latter cost might well be prohibitive for a marketing class project. Anytime an estimate is sought, a balance must be found between cost, confidence, and bound on the error of estimation.

We have observed that the spread of the $t$-distribution is a function of the sample size; $t$-scores are based on degrees of freedom, which for our use here are $n - 1$. In addition, the sample size, $n$, influences the standard error as $s_{\bar{X}} = s/\sqrt{n}$. To evaluate the problem of estimating an appropriate sample size, we consider the *bound-of-error* relation described in Equation (8.6).

**The bound of error for estimating a normal distribution mean, with unknown variance, is equal to a $t$-statistic value times the standard error of the mean.**

$$\text{Bound of error} = (t\text{-value}) \cdot (s_{\bar{X}}) \tag{8.6}$$

$$E = ts_{\bar{X}} = t_{\alpha/2,\, n-1}(s/\sqrt{n})$$

Note that the sample size is used both to specify the $t$-value and to calculate the standard error. Thus, finding the sample size that is needed to limit the error to a specified amount appears to require knowing the $t$-value first. But the $t$-value depends on $n$. This is circular, so a direct estimate of a small sample size cannot be made from the error relationship.

Example 8.12 illustrates the steps of an estimation project. You will find that the sample size is determined, then the sample is drawn, and, finally, calculations on the sampled units are used to estimate the mean and related total amounts.

**EXAMPLE 8.12**

For tax purposes, a sample inventory is to be made of the warehouse holdings for a canned-foods wholesaler. Generally, items are stored in cases of 3 dozen cans. The warehouse currently contains 24,500 such cases. An estimate of the average price per case is to be projected to an estimate for the total value of canned foods in the warehouse through multiplication of the mean by $N = 24,500$. The allowable error for estimating the mean price is $0.05 per case. A 95% confidence level is specified, and a normal distribution for sample mean price per case is reasonable.

*Solution*

First we must find an appropriate sample size. Random sampling is used. The sample specifications are 95% confidence and $E = \pm \$0.05$/case. The standard deviation in price per case, $\sigma$, is $0.84. Since the sample size is large, $t_\infty = 1.96$ and the required sample is determined by

$$n = \left(\frac{Z \cdot \sigma}{E}\right)^2 \doteq \left(\frac{1.96 \cdot \$0.84}{\$.05}\right)^2 = (32.93)^2 = 1084.2$$

So, inspection of $n = 1085$ cases will give an interval estimate that satisfies the specified allowable error and confidence values. This amounts to a $(1085/24,500) \cdot (100) = 4.4\%$ random sample. The sample size is critical, since every selected case must be inspected. This takes work-time, which means added cost.

Next, 1085 sample cases are chosen using invoice numbers with the aid of a random numbers table. Assume that, upon inspection, the sample cases show an estimated average cost per case of $\bar{X} = \$17.28$. From sample specifications this gives 95% confidence limits of $17.28 − $0.05 to $17.28 + $0.05, or $17.23 to $17.33 per case.

A point estimate of the total inventory value is then

$$N \cdot \text{Mean} = N\bar{X} = 24,500 \cdot \$17.28 = \$423,360$$

Considering all possible intervals from the different possible samples using this sampling procedure, we find that 95% will include the parameter value. Again, we assume this interval includes the correct (but unknown) parameter value. Here the error of estimation is at most $(24,500) \cdot (\$0.05) = \$1225$, with 95% confidence. If the population value is in this interval, then it is probably no more than $1225 above to $1225 below the point estimate of $423,360.

---

In Example 8.12, we calculated the sample size. Next, we drew the sample and then made the estimate. This demonstrates the statistics calculations, the numbers part of a survey project. Problem 58, a problem from [8], gives a more comprehensive example of the steps in a sample survey.

Another consideration is the use of the finite population correction (fpc) factor, as discussed in Chapter 7. Example 8.11 suggested a 4.4% sample. Had this been 5% or larger, the finite correction factor should have been applied. The benefits are high levels of confidence and low error but with reduced sample.

**Problems 8.4**

16. A California company canning olives has found that the standard deviation of drained weights per can in a production lot is consistently 0.60 ounces. How large a sample of these cans would have to be weighed for management to have 95% confidence that the error made in using $\bar{X}$ to estimate the average weight is less than 0.15 ounce?

17. Suppose that the random sample of $n = 101$ suggested in Example 8.9 is of students from your college or university. If the sample showed $\bar{X} =$

$1300.00, what would you estimate as the mean amount spent by students from your campus in the local economy? Remember the example asks for a 95% confidence level. Use $\sigma = \$280.00$.

18. We wish to know the average number of miles driven per day by all the truckers in a certain area. If there are 1000 truckers in this area, what size random sample is required to estimate the mean with 95% confidence if the error is estimated to be at most $\frac{1}{4}$ mile? Use $\sigma = \frac{3}{4}$ mile.

19. A normal distribution has $\sigma = 20.0$. How large a sample must one take so that the 99% confidence interval will not be more than 5.0 units wide? Recall that length UCL $-$ LCL $= 2E$.

20. A random sample of families is selected to estimate the mean monthly income in a certain area. The standard deviation is approximately $40.00. Use 95% confidence.
   a. For error, $E$, to be within $10, how large a sample should be selected?
   b. If sufficient funds are available to take a random sample of $n = 100$, determine $E$, the bound on the error of estimation for this sample size.

21. A supplier of crushed rock and a road construction firm are seeking an equitable way to measure loads of rock hauled to the job site without having to weigh each one (gross weight $-$ net weight $=$ tare, or load weight). The trucks are all of 20-yards capacity and a single loading operator is used, so conditions are fairly uniform. The firms agree to take a random sample of load weights sufficient in number to estimate the average weight, $\mu$, to within $\pm 0.1$ tons. Assume each yard of this rock weighs 1.25 tons.
   a. For a 95% confidence level, how many loads should be weighed? Use $\sigma = 0.6$ ton.
   b. If the random sample of the size you suggest in part (a) produced $\bar{X} = 25.2$ tons, establish a 95% confidence estimate of $\mu$.
   c. Using information from part (b) and the knowledge that a total of 3000 loads were hauled for this job, make a point estimate for the total amount of rock (in tons) hauled.

22. In an attempt to determine the wearability of a certain line of tires, a manufacturer decides, primarily because of costs, to road-test 20 sets of tires ($20 \cdot 4 = 80$ tires total).
   a. What maximum error can we expect in estimating the mean life of all tires in this line? Use 95% confidence and assume $\sigma = 2000$ miles.
   b. Suppose the maximum error obtained in part (a) is to be cut in half, while the confidence and standard deviation values remain unchanged. Assuming random sampling, how many sets of tires (a set equals four tires) should be tested?
   c. Answer the question in part (a), except now use 94% confidence. Explain the difference between your answers for parts (a), (b), and (c).

## 8.5
## ESTIMATING A POPULATION PROPORTION

Estimation of population proportions is probably the most common estimate made for public use. It includes voter preference polls, inquiry into interest rates, percent of unemployment, proportions for cost indexes, and, not least, sports and weather percentages. Although we commonly see a single-valued point estimate of percentages, an error estimate is necessary if the results are to be used for statistical projection. This need is recognized by major polling agencies such as Gallup and Roper. The National Weather Service also broadcasts confidence levels with its weather warnings.

The basis for estimating a population proportion is the binomial distribution, because it is characterized by a dichotomous classification of events (those *for* versus those *against*, *good* versus *defective*, and so on) with fixed percents. The binomial distribution applies to experiments with discrete, countable outcomes. However, estimates of the value of $\pi$ are typically made only from *large* samples. (The limitations of small samples are apparent; for example, a sample of only $n = 5$ would limit $p$, the point estimator, to values of 0.00, 0.20, 0.40, 0.60, 0.80, or 1.00.) For large samples, with $np \geqslant 5$ and $nq \geqslant 5$, the binomial distribution is approximately the same as the normal distribution. Therefore, when we use large samples, we can use the normal approximation as the appropriate distribution for making an inference about the population proportion. That is, with proper conditions, binomial experiments can use normal distribution procedures to answer estimation questions.

Chapter 7 demonstrates that the distribution of sample means ($\bar{X}$'s) is normal for samples of size $n > 30$. This same principle, embodied in the central limit theorem, also holds for sample proportions, $p$'s.

As depicted in Figure 8.6, the distribution of sample proportions also tends to be normal. However, this may require a larger sample. In fact, as seen in Chapter 6, as long as $np \geqslant 5$ and $nq \geqslant 5$, the distribution of $p$ will be nearly a normal curve. Some analysts prefer to stipulate further that the sample size be at least 100, although samples of less than 100 clearly show normal curve patterns. That is, the larger the sample, the better the normal approximation to a binomial distribution. Even when the binomial population is skewed, such as when $p$ is close to 0.1 or 0.9, the sampling distribution of $p$ will be approximately normally distributed when the sample is sufficiently large.

In working with sample means, we established confidence limits as $\bar{X} \pm (Z_{\alpha/2} \cdot \text{Standard error of the mean})$. For proportions, we use $p$ and $\sigma_p$, the *standard error of proportion* in $p \pm Z_{\alpha/2} \cdot \sigma_p$. And $\sigma_p$ is dependent upon the true population proportion, $\pi$.

**For $X$ binomially distributed the standard error of proportion is**

$$\sigma_p = \sqrt{\frac{\pi(1 - \pi)}{n}} \qquad (8.7)$$

---

**FIGURE 8.6** Similarity of Distributions of Means and Proportions for Large Samples

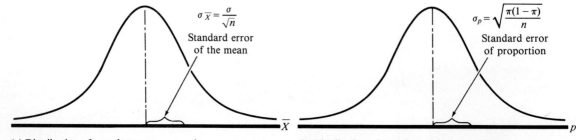

$\sigma_{\bar{X}} = \dfrac{\sigma}{\sqrt{n}}$
Standard error of the mean

$\sigma_p = \sqrt{\dfrac{\pi(1 - \pi)}{n}}$
Standard error of proportion

(a) Distribution of sample means = normal

(b) Distribution of sample proportions = normal

For large $n$, the sampling distribution of proportions is normally distributed with mean $\mu_p = \pi$ and standard error of proportion, $\sigma_p$. Then, since $\pi$ is unknown, we substitute the sample proportion without any significant loss of accuracy.

The estimated standard error of proportions is

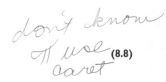

$$\hat{\sigma}_p = \sqrt{\frac{pq}{n}} \qquad (8.8)$$

*don't know*
*if use caret*

*Note:* The caret (^) is used in statistics to denote an estimator. Some texts also use the symbol $s_p$ for the estimated standard error of proportion. We will consistently use $\hat{\sigma}_p$.

**Confidence Limits for $\pi$**

When we use the standard error of proportion, the confidence limits for estimating $\pi$ are as follows.*

Confidence limits for estimating $\pi$ are $p \pm Z_{\alpha/2}\hat{\sigma}_p$:

$$\text{LCL} = p - Z_{\alpha/2}\hat{\sigma}_p \qquad (8.9)$$

$$\text{UCL} = p + Z_{\alpha/2}\hat{\sigma}_p$$

These limits hold for $np \geqslant 5$ and $nq \geqslant 5$ and where $p = X/n$ is used as the estimator for $\pi$ in evaluating the standard error. As with confidence limits for estimating means, the point estimator (in this case, $p$) is adjusted by the addition or subtraction of a bound on the error of estimation. Here, that bound of error $E$ is calculated as $E = Z_{\alpha/2} \cdot \hat{\sigma}_p$.

Following are some examples of estimation of population proportions. Keep in mind that proportions and the corresponding percentages are equivalent, except for placement of the decimal. One must simply be consistent in using either decimal proportions or percentages—the equations hold true in either case.

---

**EXAMPLE 8.13**

A political candidate has retained a polling agency to determine her voter strength. At the time of its most recent poll, the agency found that 225 of 400 registered voters would vote for this candidate. Assuming the agency took a random sample, what percent of the electorate favored this candidate at the time of this poll?

*Solution*    Using a 95% confidence level, we can estimate $\pi$ as

$$\text{CL} = p \pm (Z_{\alpha/2} \cdot \hat{\sigma}_p)$$

*don't know*

where

$$p = \frac{X}{n} = \frac{225}{400} = 0.56$$

---

* $Z = (p - \pi)/\sqrt{pq/n}$ is the sampling distribution for our estimation with samples $n \geqslant 100$. See Chapter 16 for inference on frequencies and proportions with smaller samples.

Therefore,

*estimating π (population)*

*95% sure for 51-61% of electorate will vote for candidate*

$$q = 1 - p = 1 - 0.56 = 0.44$$

$$\hat{\sigma}_p = \sqrt{\frac{pq}{n}} = \sqrt{\frac{(0.56)(0.44)}{400}} = 0.025$$

$$\text{LCL} = 0.56 - (1.96 \cdot 0.025) = 0.51$$

$$\text{UCL} = 0.56 + (1.96 \cdot 0.025) = 0.61$$

At the time of this polling, between 51% and 61% of the voters favored this candidate. There is a 95% confidence level for the procedure used to make this estimate.

Statistical estimation is widely used as the basis for election projections. With samples of only 1500 to 2500 potential voters, scientific polls can accurately predict national results for 70 million or more voters. However, the pollsters use techniques that are more sophisticated than simple random sampling.

A second example shows the diversity of applications of statistical estimation for proportions.

## EXAMPLE 8.14

The personnel manager of a large firm wants to find the percentage of employee absences per day in order to determine a standard against which to gauge periods of excessive absenteeism. His sample is the record of absenteeism for a random selection of 500 employee-days taken from work records over the past several years. The data reveals $X = 30$ absences for the sample of $n = 500$ employee-days. Establish a 90% confidence interval.

*Solution*   The procedure is the same as before. Here $\pi$ = the (unknown) proportion of employee-day absences. Since

$$p = \frac{X}{n} = \frac{30}{500} = 0.06 \quad \text{and} \quad Z = 1.645$$

then

$$\text{CL} = p \pm (Z_{\alpha/2} \cdot \hat{\sigma}_p) = 0.06 \pm \left( 1.645 \sqrt{\frac{0.06(1 - 0.06)}{500}} \right)$$

$$= 0.06 \pm 0.02 = (0.04, 0.08)$$

Therefore, 90% confidence limits are $4\% < \pi < 8\%$.

The personnel manager can use 4% to 8% as a gauge of the usual percentage of absences. Then, for example, an absence rate of 10% on a given day would indicate excessive absenteeism.

Examples 8.13 and 8.14 have shown two applications for establishing confidence limits for a population proportion, $\pi$. Next, we view sample size calculations for estimating a population proportion.

**Sample-Size Considerations**

Suppose, in Example 8.14, the personnel manager wanted an estimate with less error. For example, he might want the bound on the error (which was approximately 2%) reduced to 1%. Determining sample size in this situation uses a familiar procedure; it requires only a new computing form. Sample size in binomial experiments is determined using the following relation:

### BOUND-OF-ERROR (E) AND SAMPLE-SIZE (n) CALCULATIONS FOR ESTIMATING $\pi$

Bound-of-error    $E = Z_{\alpha/2}\hat{\sigma}_p = Z \cdot \sqrt{\dfrac{pq}{n}}$    (8.10)

Sample-size    $n = \dfrac{Z^2 \cdot pq}{E^2}$    (8.11)

where    $\hat{\sigma}_p$ = the estimated standard error of proportion

Equation (8.11) is used in Example 8.15 for a sample size calculation.

### EXAMPLE 8.15

What sample will be necessary for the personnel manager of Example 8.14 to estimate the true percentage of absenteeism ($\pi$) for his firm to within 1% with 90% confidence?

*Solution*    The confidence level is unchanged, but this requires less error than in the preceding example. Logically this sample must exceed the last one, which was $n = 500$. This can serve as one check. Example 8.14 provides the estimates $p = 0.06$ and $q = 0.94$ for $\pi$ and $(1 - \pi)$, respectively.

Using Equation (8.11) for the sample size calculation gives

$$n = \frac{Z^2 pq}{E^2} = \frac{(1.645)^2 \cdot (0.06) \cdot (0.94)}{(0.01)^2} = 1526.2$$

The manager needs 1527 randomly chosen employee-day records to estimate $\pi$ to within $\pm 1\%$ at 90% confidence.

Examples 8.14 and 8.15 illustrate the fundamental principle that one can get better sampling results by taking a larger (random) sample. Of course, the best results come from taking a sample of the whole population, a census. In most cases, however, this extreme is not economically feasible, so the analyst must accept some error and less than 100% confidence in the result.

As Example 8.16 shows, costs play a large part in the determination of sample size for some projects.

### EXAMPLE 8.16

The advertising manager for a manufacturer of self-teaching language cassettes wants to know the market share (percentage of total consumer sales) for the company's language cassettes. The company has allocated $3000 for this project, and the manager wants an estimate to within $\pm 2\%$ of its actual market share. Overhead expenses for the project are $500, and each consumer interview costs $1. A confidence of 95% is required. Can their market share be estimated from a random sample of language cassette customers within the required precision for $3000?

Solution  This example involves two issues. First, how large a sample can the manager afford to buy? Since Sample funds = Total allocation − Overhead, or $3000 − $500 = $2500, then at $1 per interview, he can afford $n = \$2500/\$1 = 2500$ interviews. The second requirement is that the bound on error of estimation be within $\pm 0.02$. Because no prior estimate is available for $\pi$, we let $p = 0.50$. (We'll explain why momentarily.) Can these requirements be met with $n = 2500$, for $p = 0.50$? Using 95% confidence gives

$$E = (1.96) \cdot \sqrt{\frac{(0.5)(0.5)}{2500}} = (1.96) \cdot \left(\frac{0.5}{50}\right) = \pm 0.02$$

The desired precision can just be met. The consumer advertising section can support its survey.

---

Note that in Example 8.16, $p = 0.5$ seemed to be an arbitrary choice. A better initial value would be the market share for the previous quarter. Of course, a previous value is not available if the product is new. But, "Why the $p = 0.5$ value?" can now be answered.

## Estimating n When p Is Unknown

The value $p = 0.5$, when used in sample size computations, leads to the largest (that is, the safest) sample size that might be required for specified confidence (Z-value) and error (E). In other words, had we used any value other than 0.5 the sample size would have been less. Recall that the equation is $n = Z^2 pq/E^2$. Then, for example,

if $p = 0.5$ and $q = 0.5$, then $pq = 0.25$
if $p = 0.4$ and $q = 0.6$, then $pq = 0.24$
if $p = 0.8$ and $q = 0.2$, then $pq = 0.16$

and so on. Thus, 0.5 gives the largest possible value for $pq$. Then, by Equation (8.11) and for specified values of confidence (Z) and error (E), this translates to the largest sample requirement. So, $p = 0.5$ should be used only if no better value is available. One alternative is to obtain an initial estimate of $\pi$. This might come from the scientific results of a closely related study.

## Estimating the Number of Successes

One final note pertains to the use of the normal approximation for estimation of $\pi$. When we converted from percentages to the number of successes in Chapter 6, we computed the standard deviation as $\sigma_X = \sqrt{n\pi(1 - \pi)}$. This standard deviation is effectively a *standard error of the number of successes*, which we might more appropriately designate here as $\sigma_{np}$. We then computed probabilities for discrete *numbers*—for example, $P(75 \text{ or more} \mid n = 100, \pi = 0.80)$. In this chapter we have been working almost wholly with *proportions*, but the two concepts are completely interchangeable. A proportion is simply a ratio of some part to the whole sample. Thus,

(Proportion)  ·  (Number in sample) = (Number of successes)

$\left(\dfrac{X}{n}\right)$      times      $(n)$      =      $X$

Example 8.17 illustrates the relationship.

| EXAMPLE 8.17 | A sample of 400 is taken from a population and 80 defectives are found. Compute the standard error in terms of (a) proportions and (b) numbers and (c) show their equivalence. |
| --- | --- |

*Solution*   a. Proportion of defectives:

b. Number of defectives:

$$\hat{\sigma}_p \doteq \sqrt{\frac{pq}{n}}$$

$$\hat{\sigma}_{np} \doteq \sqrt{npq}$$

$$= \sqrt{\frac{(0.20)(0.80)}{400}}$$

$$= \sqrt{(400)(0.2)(0.8)}$$

$$= 0.02$$

$$= 8$$

c. The equivalence is demonstrated from the fact that

$$\left(\frac{X}{n}\right) \cdot n = X$$

$$0.02 \times 400 = 8$$

Thus, a standard error of proportion $\hat{\sigma}_p$ multiplied by the sample size ($n$) gives a standard error of the number of successes, $\hat{\sigma}_{np}$.

$$\hat{\sigma}_p \cdot (n) \doteq \left(\sqrt{\frac{pq}{n}}\right) \cdot (n) = \sqrt{\frac{pqn^2}{n}} = \sqrt{pqn} = \hat{\sigma}_{np}$$

For Example 8.17, this gives $\hat{\sigma}_p \cdot n = 0.02 \cdot 400 = 8 = \hat{\sigma}_{np}$.

The following problems include some that require you to compute both the *proportion* of successes and the *number* of successes.

**Problems 8.5**

23. Indicate whether the following are true or false.
    a. $Z_{\alpha/2}$ is a measure of deviation from the mean in terms of standard errors.
    b. Estimates of the value of $\pi$ are usually made with samples of fewer than 30.
    c. Whereas for large samples the distribution of sample means is normal, the distribution of sample proportions usually follows the uniform distribution.
    d. We introduce significant error if we try to substitute $p$ for $\pi$ in computing a standard error of proportion.
    e. Increasing the sample size decreases the value of the standard error of proportion (other factors held constant).

24. Of 100 potential customers surveyed, 50% favored a new product.
    a. Estimate the value of the standard error of proportion.
    b. Provide a 95% confidence interval estimate of the true population proportion in favor.

25. In a corporate survey of employee attitudes a sample of 900 contained 576 people who gave an identical response to a certain stimulus word.
    a. What is the point estimate of the percentage of identical response?
    b. What is the standard error of proportion for this estimate? That is, find $\hat{\sigma}_p$.

   c. With a probability of 0.95, what can you say about the error, $E$, of your estimate in part (a)? That is, what is the size of $E$?

   d. Using the normal approximation, find a 95% confidence interval for the true proportion.

26. A national employment agency, in an attempt to estimate the percentage of college graduates who take a job in their primary area of training, found that 640 in a random sample of 1000 had done so. Estimate, with 94% confidence, the true percentage who take jobs in their specialties.

27. What sample size should be taken in order to estimate $\pi$ to within $\pm 2\%$ with 98% confidence? Assume $p$ is about 0.20.

28. Of 1000 people treated with a new drug, 200 showed an allergic reaction. With 90% confidence, estimate the proportion of the population that would show an allergic reaction to the drug.

29. A professional society wants to estimate the percentage of its applicants who pass a state auditor's exam. In a random sample of 120 applicants who took the exam over the past three years, 90 passed and received their licenses. Estimate the true percentage with 98% confidence.

30. During a production study, a skilled technician is to be observed at randomly chosen times during a workday in order to estimate within $\pm 0.05$ and with 95% confidence the proportion of time that she is productive.

   *385* a. What is the largest sample size required under the foregoing restrictions?

   *138.3* b. If a study in a similar company resulted in $p = 0.90$, could you use this

   *yes* information to reduce the sample size? To how much?

31. The advertising department of a national firm wishes to find the market share (percentage of total consumer sales) for their brand. They have $2000 for this project and want an estimate to within $\pm 2\%$ of their actual market share. Overhead expenses for the project are $400, and each household interview costs $1.00. In the last quarter (three-month period) the firm's market share was 25%. A confidence of 95% is requested.

   a. Can this proportion be estimated to the desired specifications within the given cost constraints?

   b. Assume the firm provided sufficient funds to take the sample computed in (a); the sample showed that $X = 400$ customers favored their brand. Estimate their current market share. Again, use 95% confidence.

   c. What sample would be required to estimate the market share within $\pm 1\%$, which is half the error given above? (Retain the 25% market share estimate and 95% confidence to compute $n$.)

   d. Assuming a fixed overhead of $500, how much would this sample cost?

32. Samples of 400 pages are randomly selected from a computer listing of a telephone directory compilation that has a proportion defective of 0.10. A defective page is one that has one or more misspelled words or one or more keying errors.

   a. Find the standard error of proportion, $\hat{\sigma}_p$.

   b. Compute the standard error of the number $\hat{\sigma}_{np}$.

   c. Show how we can obtain (b) more directly from (a) by multiplying by $n$.

# 8.6 ESTIMATING DIFFERENCES— LARGE, INDEPENDENT SAMPLES WITH KNOWN STANDARD DEVIATIONS

Up to this point, we have concerned ourselves with estimating a single population mean or population proportion. However, managers are frequently interested in comparing the parameters of two populations to find out by how much they differ. For example, how different is the average cost of living in Dallas from the average cost of living in Honolulu? What is the difference in highway miles per gallon for one make of car versus another? How does the proportion of rejects produced during a day shift compare with the proportion produced on swing shift? We can estimate the difference between two *independent samples* by following a procedure similar to that used earlier in the chapter. We must, however, make some adjustments to accommodate the sampling distribution for the difference in (a) two sample means or (b) two proportions.

The assumptions underlying the estimation of the difference in means (or proportions) of two populations are straightforward.

**ASSUMPTIONS FOR ESTIMATING DIFFERENCES**
$$\mu_1 - \mu_2 \ (\text{OR} \ \pi_1 - \pi_2)$$

1. Both samples—$n_1$ and $n_2$—are randomly drawn from their respective populations.
2. The samples are independent of each other. The results of one sample in no way affect the other.
3. The sample size must be sufficiently large that
   a. the distribution of the difference in sample means, $\bar{X}$'s, is normal (or $n_1$ and $n_2 > 30$) for $\bar{X}_1 - \bar{X}_2$, or
   b. the distribution of the difference in sample proportions, $p_1 - p_2$, is approximately normal (or $n_i p_i > 5$, $n_i q_i > 5$, $i = 1, 2$).
4. The variances of the two populations are known (for means) or can be estimated from the (large) random samples (for proportions).

Chapter 10 considers the changes necessary to accommodate small sample sizes and unknown population variances.

## Confidence Intervals for $\mu_1 - \mu_2$

We know from the central limit theorem that for large samples the distribution of sample means is normal. The same holds true for differences between sample means, that is, for $\bar{X}_1 - \bar{X}_2$. For independent random samples of sizes $n_1$ and $n_2$ drawn from two populations with known variances, the distribution of $(\bar{X}_1 - \bar{X}_2)$ values is normal with a mean equal to the population difference $(\mu_1 - \mu_2)$. The *standard error of the difference between two means* is designated $\sigma_{\bar{X}_1 - \bar{X}_2}$.

The standard error of the difference between two means for independent large random samples is

$$\sigma_{\bar{X}_1 - \bar{X}_2} = \sqrt{\frac{\sigma_1^2}{n_1} + \frac{\sigma_2^2}{n_2}} \tag{8.12}$$

Equation (8.12) expresses the standard error of the difference between two means as the square root of the sum of the error in the first population, $\sigma_1^2/n_1$, and that in the second, $\sigma_2^2/n_2$. The standard error of the difference requires independence to ensure that there is no common variance (covariance) that might alter the value.

To find a $[(1 - \alpha) \cdot 100]\%$ confidence interval, we add and subtract the bound of error for the difference between the two sample means, which is $E = Z_{\alpha/2} \cdot \sigma_{X_1 - X_2}$. The limits for the interval take the familiar form of

$$\text{Point estimate} \pm [Z_{\alpha/2} \cdot (\text{Standard error})]$$

**Confidence interval for the difference in means for independent large random samples:**

$$\text{CL} = (\bar{X}_1 - \bar{X}_2) \pm \left( Z_{\alpha/2} \cdot \sqrt{\frac{\sigma_1^2}{n_1} + \frac{\sigma_2^2}{n_2}} \right) \qquad (8.13)$$

Example 8.18 demonstrates the use of Equation 8.13.

---

**EXAMPLE 8.18**

Teletex Communications is closing a plant in Dallas and offers to move its employees to a new facility in Honolulu. However, an employee union is also insisting on a cost-of-living adjustment for food. To determine the cost difference, the Teletex personnel manager has obtained published data describing the prices for a random selection of foods, including milk, sugar, peanut butter, orange juice, and ground beef.

The data are summarized in the table. The personnel manager wishes to develop an interval estimate of the difference in average costs, so that she can work within an acceptable range to negotiate an adjustment for cost-of-living with the company. Estimate the cost difference at 90% confidence.

|                              | Honolulu          | Dallas            |
|------------------------------|-------------------|-------------------|
| Mean cost of food items      | $\bar{X}_1 = \$2842$ | $\bar{X}_2 = \$2323$ |
| Known standard deviations    | $\sigma_1 = \$337$   | $\sigma_2 = \$218$   |
| Sample sizes used, $n$       | $n_1 = 31$        | $n_2 = 36$        |

*Solution*    For 90% confidence, $Z_{\alpha/2} = Z_{0.05} = 1.645$.

$$\text{CL} = (\bar{X}_1 - \bar{X}_2) \pm Z_{\alpha/2} \sqrt{\frac{\sigma_1^2}{n_1} + \frac{\sigma_2^2}{n_2}}$$

$$\text{CL} = (\$2842 - \$2323) \pm 1.645 \sqrt{\frac{\$337^2}{31} + \frac{\$218^2}{36}}$$

$$= \$519 \pm \$116 \quad \text{or } \$403 \text{ to } \$635$$

Therefore,

$$\$403 < (\mu_{\text{Honolulu}} - \mu_{\text{Dallas}}) < \$635$$

The personnel manager can be 90% confident that the procedure reasonably describes the difference in mean food cost for these selected items. Something between $403 and \$635 would (with a high probability) be a reasonable cost-of-living adjustment for average food costs in Honolulu (in excess) over Dallas, so the manager can reasonably negotiate a value within this range.

---

## Confidence Intervals for $\pi_1 - \pi_2$

With sufficiently large samples, the sampling distribution of the difference in sample proportions is also approximately normally distributed. This distribution has an estimated *standard error of difference in proportions* $\hat{\sigma}_{p_1 - p_2}$ that is estimated from the sample proportion values $p_1$ and $p_2$.

The standard error for the difference in proportions is

$$\hat{\sigma}_{p_1 - p_2} = \sqrt{\frac{p_1 q_1}{n_1} + \frac{p_2 q_2}{n_2}} \tag{8.14}$$

We estimate the *confidence interval for the difference in the population proportions* $(\pi_1 - \pi_2)$ from the difference in sample proportions $(p_1 - p_2)$ by adding and subtracting $(Z_{\alpha/2} \cdot$ standard error).

Confidence limits are defined for the difference in proportions:

$$CL = (p_1 - p_2) \pm \left( Z_{\alpha/2} \cdot \sqrt{\frac{p_1 q_1}{n_1} + \frac{p_2 q_2}{n_2}} \right) \tag{8.15}$$

Here is an example.

---

### EXAMPLE 8.19

Sue Spencer was recently promoted to quality-control manager of Space Propulsion Systems, Inc. She had been on the new job less than one week when her boss asked her to explain why some customers appeared to be receiving a higher proportion of defective castings than others. Sue suspected a difference in percent defective by shift, and she knew that the production from each shift went to a different location. So, she had a quality-control inspector randomly sample 320 castings produced on a day shift and 200 produced on the swing shift.

   The results were 45 defectives in the day-shift sample and 18 in the swing-shift sample. Estimate the difference in percent defective using (a) a 95% level of confidence and (b) an 80% level of confidence. What answer does this provide to Sue for her boss?

*Solution*   a.  For 95% confidence,

$$Z_{\alpha/2} = Z_{0.025} = 1.96$$

$$p_1 = \frac{45}{320} = 0.141$$

$$p_2 = \frac{18}{200} = 0.090$$

$$CL = (p_1 - p_2) \pm \left( Z_{\alpha/2} \cdot \sqrt{\frac{p_1 q_1}{n_1} + \frac{p_2 q_2}{n_2}} \right)$$

$$= (0.141 - 0.090) \pm \left( 1.96 \cdot \sqrt{\frac{(0.141)(0.859)}{320} + \frac{(0.090)(0.910)}{200}} \right)$$

$$= 0.051 \pm 1.96(0.028) = 0.051 \pm 0.055$$

$$= (-0.4\%, 10.6\%)$$

For 95% confidence there may be a small negative difference, or it may be as much as +10.6%, with the day shift having the higher defective rates.

b. At 80% confidence, $Z_{0.10} = 1.28$, so

$$CL = 0.051 \pm (1.28 \cdot 0.028) = 1.5\% \text{ to } 8.7\%$$

If Sue is willing to risk a 20% chance of being wrong, she could say the defective rate on the day shift is 1.5% to 8.7% *higher* than on swing shift. If Sue is willing to risk only a 5% chance of being wrong, she could say the defective rate on the day shift is from 0.4% *lower* to 10.6% *higher* than on the swing shift.

---

Chapter 9 extends the concepts of differences to hypothesis tests. More often than not, the population standard deviations are not known. And it may not be reasonable, even if they are known, to assume equal population standard deviation values. Finally, the two samples may not be statistically independent. Such cases are discussed in Chapter 10, Extensions of Estimation and Hypothesis Testing. Firms competing for sales frequently compare their products' costs and performances with those of their competitors. For these comparisons, an analysis of paired differences is sometimes appropriate over an analysis of differences for individual means. Again, this case is covered in Chapter 10.

**Problems 8.6**

33. A plant in Tokyo uses both robots and humans on its production lines. Samples of the activities reveal that the average production for a human operator is 112.2 units per hour with standard deviation $\sigma = 8.0$ units. Comparable values for the robot-controlled line are 108.6 and 6.0 units per hour. Assume the samples were independently selected for $n = 36$ hourly observations each. Estimate the true difference in mean production for the human versus robot-controlled lines. Interpret your conclusion. Does this test cast doubt on the efficiency of these robots? Let $\alpha = 0.05$.

34. Much has been written about the number of accidents on U.S. highways. But how good a job is done in educating our young drivers? In one area, a comparison of driving ability was made by using random samples of 100 young drivers who had taken a driver education course and an equal number of others who had not. Each group was given the same driving test. Their average test scores and standard deviations are listed in the table. Does this measure indicate higher test achievement for those with driver education training? Use a 90% confidence interval to support your answer.

|  | Mean Test Score | Standard Deviation |
|---|---|---|
| Driver education | 95 | 10 |
| No driver education | 85 | 15 |

35. Can teenagers distinguish by taste the differences between Coke and Pepsi? Some places—grocery stores, lunch counters, and even state fairs—have

booths where you can perform your own taste tests. Suppose 65 of 100 Iowa teenagers in a sample at the Iowa State Fair and 140 of 200 Nebraska teens at the Nebraska State Fair can distinguish Pepsi from Coke by taste.
a. Estimate with 95% confidence the difference in percentage of Iowa–Nebraska teens who can distinguish between the two brands.
b. Someone has said that Nebraska teens have a more discriminating sense of taste than Iowa teens. Does your calculation in part (a) support this for discrimination of Coke versus Pepsi? Explain.

36. Crystal figurines must be flawless to sell at a collector's price. A firm in Vermont has developed a process to produce cast figurines that they feel are equal in quality to a brand made in Germany. They have been invited to test their new process on a batch against a batch prepared by the old process. The results are shown in the table. Estimate with 95% confidence the difference in percent at collector's quality for the two processes. May the Vermont company develop an advertising program to say their process is superior? Explain.

|                          | Method | |
|                          | Old | New |
| ------------------------ | --- | --- |
| Collector quality        | 140 | 125 |
| Below collector quality  | 20  | 15  |
| Total figurines          | 160 | 140 |

37. An expert in physical fitness has classified each of 500 college men as being either athletic or nonathletic. Among the 300 considered athletic, 48 have a continuing program for exercise. For the nonathletic, only 22 had ongoing exercise. Estimate the difference in percent of college men who have an ongoing program by comparing the percentages of the two groups, that is, 48 of the 300 considered athletic versus 22 of the 200 considered nonathletic. Let $[(1 - \alpha) \cdot 100]\% = 95.44\%$.

38. The time needed to broadcast a television commercial can be reduced by means of an electronic process called *time compression*. This process enables networks to broadcast a 30-second commercial in 24 seconds. No shift occurs in voice pitch, and viewers are not aware that commercials have been altered. Sponsors are concerned about how recall scores are affected by time compression relative to their normal-speed counterparts.

James MacLachlan and Michael Siegel have reported one study on 131 undergraduate students at State University College, Oneonta, New York.* The viewing was a recorded portion of the television program "60 Minutes." The segment covered a possibly fraudulent cancer care clinic in California. Two commercial breaks were dubbed into the program. The students were randomly divided into two groups. Those in group one (control)

* MacLachlan, James, and Michael Siegel. "Reducing the Costs of TV Commercials by Use of Time Compressors." *Journal of Marketing Research* XVII (Feb. 1980): 52–57.

saw four normal commercials; those in group two (treatment group) saw the same four commercials in time-compressed form plus two additional commercials. After seeing the taped show, students filled out a brief questionnaire about their reactions to it. Two days later they answered unaided and aided recall questions about the commercials. The table summarizes the percentage recalling each ad from unaided recall. Can shorter commercials be effective? Establish 95% confidence intervals for differences in the percentage of people recalling each commercial (that is, four confidence intervals). Were the shorter commercials effective for recall by these subjects?

| Commercial | Normal Version ($n = 57$ viewers) | | Compressed Version ($n = 74$ viewers) | |
|---|---|---|---|---|
| | *Percentage of People Recalling* | | *Percentage of People Recalling* | |
| Dupont Lucite paint | 17.5% | 10 | 23.0% | 17 |
| Allstate Insurance | 26.3 | 15 | 43.2 | 32 |
| Western Union | 5.4 | 3 | 6.8 | 5 |
| Dentine Gum | 8.8 | 5 | 10.8 | 8 |
| Total | | 33 people | | 62 people |

## SUMMARY

*Estimation* is the process of approximating values for unknown parameters. Point estimates are nearly always in error, but *confidence interval* estimates can be made with full knowledge of the percentage of times we can expect to include the correct value. Estimation from large samples relies upon knowledge of the central limit theorem. This estimation is reasonable not because of some special secret information, but because of the statistical procedures followed.

The Student's *t*- and the *Z*-distributions have numerous similarities. Both are bell-shaped and center around a mean value of zero. However, the *t-distribution* depends on the sample size, *n*, whereas the *Z*-distribution does not. The standard deviation for the *t*-distribution is greater than one. With larger samples the *t*-distribution becomes more pointed and approaches the form of the *Z*-distribution. Also, as *n* becomes increasingly larger, the standard deviation approaches the value 1. The appropriate form of distribution to use, *Z*- or *t*-statistic, depends upon whether $\sigma$ is known (use *Z*) or if the distribution is normal with unknown population standard deviation (use the *t*-statistic).

We can check for the reasonableness of tabular *t*-values by comparing the entry from Table B.2 against the *Z*-value (that is, $t_\infty$) at the same $\alpha$-level. Because the *t*-distribution is based on smaller samples, *t*-values are larger than comparables *Z*-scores. Wider intervals are required for *t* than for *Z* for the same confidence.

Estimation forms based on the two statistics are quite similar, but the *t*-form yields wider confidence intervals. With increased sample size, the *t*-distribution approaches the *Z*. So larger samples give more precise (narrower) confidence intervals. Because *t*-scores are a function of sample size, calculations of *n* cannot be made directly from the *t*-distribution as they can from the *Z*. This is a limitation of the *t*-statistic.

*Confidence intervals for means* and *proportions* are developed similarly in that each is centered on a sample statistic (either $\bar{X}$ or *p*). Both the distribution of sample means and the distribution of sample proportions are normal (for large random samples). And the confidence limits bound the error (*E*) by the amount of the confidence coefficient (*Z*) times the appropriate standard error. The computation of standard error differs, of course, because $\bar{X}$ and *p* are different statistics.

The *estimation of differences for two means or differences on* two *proportions* follows a similar procedure, except we use the difference (rather than a single sample value) and take both samples into account when calculating the *standard error of the difference.*

The determination of sample size is similar for means and proportions. For estimation of means, we multiply $Z_{\alpha/2}$ by the standard deviation and then divide by the desired error $E$. We square the result and round up to the nearest whole number. For proportions, we square the $Z_{\alpha/2}$ value, multiply it by $p$ and by $q$, and then divide by $E$ squared.

Statistical estimation incorporates a sequence of steps. It begins with specification of a confidence level. This, along with the standard deviation and an indication of the allowable error, $E$, is the input necessary to determine a sample size. Next the random sample is collected. Then the sample is summarized to provide a point estimate for a mean or population proportion. This is expanded to a confidence interval by adding and subtracting a designated number of standard errors to the point estimate to obtain a *confidence interval* of values.

Table 8.3 summarizes the estimation equations appearing in this chapter.

**TABLE 8.3**  Key Equations for Estimation

| Distribution and Parameters | Conditions | Distribution | Estimation | Sample Size |
|---|---|---|---|---|
| $\mu$ — $n > 30$ large | $\sigma$-known | $Z = \dfrac{\bar{X} - \mu}{\sigma/\sqrt{n}}$ | $\bar{X} \pm (Z\sigma/\sqrt{n})$  (8.1) | $n = \left(\dfrac{Z\sigma}{E}\right)^2$ (8.5) |
| | $\sigma$-unknown | $t = \dfrac{\bar{X} - \mu}{s/\sqrt{n}}$  (8.2) | $\bar{X} \pm ts/\sqrt{n}$  (8.3) | |
| $n \leq 30$ small (normal dist.) | $\sigma$-known | $Z = \dfrac{\bar{X} - \mu}{\sigma/\sqrt{n}}$ | $\bar{X} \pm Z\sigma/\sqrt{n}$ | |
| | $\sigma$-unknown | $t = \dfrac{\bar{X} - \mu}{s/\sqrt{n}}$ | $\bar{X} \pm ts/\sqrt{n}$ | |
| $\pi$ | $np \geqslant 5, nq \geqslant 5$ | $Z = \dfrac{p - \pi}{\sqrt{\dfrac{pq}{n}}}$ | $p \pm Z\sqrt{\dfrac{pq}{n}}$  (8.9) | $n = \dfrac{Z^2 pq}{E^2}$ (8.11) |
| | other | (see Chapter 16) | | |
| $\mu_1 - \mu_2$ — large indep. samples | $\sigma_1, \sigma_2$ known | $Z = \dfrac{(\bar{X}_1 - \bar{X}_2) - (\mu_1 - \mu_2)}{\sqrt{\dfrac{\sigma_1^2}{n_1} + \dfrac{\sigma_2^2}{n_2}}}$ | $(\bar{X}_1 - \bar{X}_2) \pm Z\sqrt{\dfrac{\sigma_1^2}{n_1} + \dfrac{\sigma_2^2}{n_2}}$  (8.13) | |
| | $\sigma_1, \sigma_2$ unknown | (see Chapter 10) | | |
| small samples (normal dist.) | | (see Chapter 10) | | |
| $\pi_1 - \pi_2$ — large indep. samples | | $Z = \dfrac{(p_1 - p_2) - (\pi_1 - \pi_2)}{\sqrt{\dfrac{p_1 q_1}{n_1} + \dfrac{p_2 q_2}{n_2}}}$ | $(p_1 - p_2) \pm Z\sqrt{\dfrac{p_1 q_1}{n_1} + \dfrac{p_2 q_2}{n_2}}$  (8.15) | |
| | small samples | (see Chapter 16) | | |

## KEY TERMS

best estimator, 289
bound on the error of estimation, 291
confidence interval, 291
  for means, 301
  for proportions, 311
confidence vs. probability, 297
consistency, 289
degrees of freedom, 300
efficiency, 289
error of estimation, 290
estimation, 286
estimator, 288

hypothesis testing, 287
interval estimate, 291
pivotal method, 292
point estimate, 288
properties of point estimators, 289
standard error of the difference
  in means, 317
  in proportions, 319
statistical inference, 286
sufficiency, 289
Student's $t$-statistic, 299
unbiased, 289

## CHAPTER 8 PROBLEMS

39. Which of the following statements best describes the meaning of the 95% confidence interval for a population mean?
    a. There is a 0.95 probability that the mean is a correct point estimate.
    b. The probability is 0.95 that the population mean will fall within the specified interval.
    c. The sample mean is a random variable that is accurate 95% of the time.
    d. If additional samples of the same size were taken and the 95% confidence interval established in the same manner, we would expect to bound the population mean about 95% of the time.
    e. The sample mean is at least 95% as large as the population mean but does not exceed the population mean.

40. As distinguished from the normal distribution, indicate whether the following are true or false for the $t$-distribution.
    a. Symmetrical
    b. More dispersed values than the normal
    c. No relation to the sample size
    d. Zero mean and standard deviation 1.
    e. For a fixed confidence level and population standard deviation, the only way to reduce a confidence level is to increase the standard error.

41. Assume that a random sample of 20 fuses is drawn from a box of 200 in order to establish a confidence interval for the true mean current (in amps) required to cause failure. Indicate whether the following statements are true or false.
    a. The $t$-distribution should be used instead of $Z$.
    b. The number of degrees of freedom should be 19.
    c. A 95% confidence interval using $t$ would be wider than a 99% confidence interval using $t$.
    d. To use the $t$-distribution requires that the variable $X$ = the electric current required to cause failure is normally distributed.
    e. The confidence interval width would be decreased if a smaller sample were used.

42. What is the 99.7% confidence interval for the population proportion of successes if a sample of 600 had 60% successes?

43. A firm has two million outstanding shares of stock and management wishes to estimate the mean number of shares held per stockholder and be 99% confident their estimate is within five shares of the population mean. How large a sample should they take if the standard deviation is 25 shares per stockholder?

44. A manufacturer thinks a certain process produces 8% defectives and wishes to take a sample large enough to estimate the accuracy to within $\pm 3\%$ with 95.5% confidence. How large a sample should be taken?   $328$

45. A union vote on whether to accept a contract is expected to be *very* close. If union leaders wish to gain a preliminary feeling for how the union will vote and they want 95% confidence that their estimate is within $\pm 5$ percentage points, compute the (maximum) sample size required.

46. Department of Labor officials estimate the unemployment percentage in Riverton to be somewhere between 10% and 20%. They wish to take a sample that will permit them to estimate the true percentage within $\pm 1\%$ and want 95.5% confidence in their estimate. To obtain this precision, how large a sample must they take?

47. Given that $\bar{X} = 20.0$, $\sigma = 4.0$, $n = 36$, and $X$ is normally distributed:
    a. Compute the 95% confidence limits for $\mu$.
    b. What sample size would be required to give one-half the maximum error, $E$, indicated in part (a)?

48. For a sample of 25 textile mill workers in a New England community, the mean weekly wage was $400 with a standard deviation of $s = \$40$. What are the 98% confidence limits for the true mean wage in this community? Assume the wages are normally distributed.

49. For over two years, doctors at Sacred Heart Medical Center have used a fixed dose of a certain drug to bring about an average increase in the pulse rate of 10.0 beats per minute. A group of nine patients given the same dose from a new shipment of this drug showed increases of 16, 15, 14, 10, 8, 12, 13, 20, and 9 beats per minute. Assume a normal distribution. The standard deviation is not known. Establish a 95% confidence interval estimate for $\mu$ based on this sample data.

50. A student council candidate found that in a random sample of 100 students, 54 favored him over the other candidates.
    a. Using the normal distribution, construct a 90% confidence interval for $\pi$. Define $\pi$.
    b. How large a sample must he take to be sure that his error was no larger than $\pm 0.04$ with a confidence of 90%?

51. A random sample of households is selected in order to estimate the mean monthly income in a certain area. The standard deviation is $40.00. Use 95% confidence.
    a. For error, $E$, to be within $10, how large a sample should be selected?
    b. If sufficient funds are available to get a random sample of $n = 100$, how large an error, $E$, should be expected?

52. How large a sample should you take to estimate the average number of customers per day at a certain bakery? The standard deviation is known to be about 20. We want to be 90% sure that our estimate is within 5 of the true mean.

53. Suppose that you are in charge of a poll to determine how many students favor an exam week. If you wish to estimate the true proportion of students favoring an exam week to within 5% with 90% confidence, how large a sample should you take?

54. Auditors of the Drapo Co. took a random sample of 25 of its 200 market representatives in order to estimate their mean travel expenses with 95% confidence. If the results were $\bar{X} = \$860.00$ and $s = \$40.00$, what is the best interval estimate? Round to the nearest 50 cents. Assume a normal distribution.

55. A random sample of 16 lumber mill workers revealed an average wage of $315.00 per week with $s = \$9.60$ per week. Find the 95% confidence limits for the true average wage $\mu$. Assume a normal distribution of the population.

56. A random sample of 100 lamps taken from a very large shipment contains 18 with imperfections.
    a. What is the possible size of our error if we estimate this proportion of imperfect lamps as being 0.18? Use a level of confidence of 0.98 and $p = 0.18$.
    b. Construct a 98% confidence interval for the actual percent of imperfect lamps in this shipment.

57. Find a 95% confidence interval for the mean based on the data listed in the frequency table. Assume an approximately normal population distribution. First you will need point estimates for $\mu$ and $\sigma$. Use $\bar{X}$ and $s$.

| Class | Frequency |
|---|---|
| 10 but less than 20 | 6 |
| 20 but less than 30 | 18 |
| 30 but less than 40 | 24 |
| 40 but less than 50 | 12 |
| Total | 60 |

**STATISTICS IN ACTION: CHALLENGE PROBLEMS**

58. Sometimes you can't see the forest for the trees. Chapters 7 and 8 explore some fairly complicated details in the statistics of sampling, yet a business sampling project involves more than determining sample size and calculating point and interval estimates for a parameter. In his book *Sampling in a Nutshell*, Morris Slonim outlines some other steps in a survey project, including defining the sampling universe, specifying exactly what information is being sought in the survey, reviewing the literature to ensure that the information is not already available in a publication, and more. Slonim's chapter "Steps in a Sample Survey" (II) outlines key considerations for anyone who intends to perform, or use, sample surveys. Here is an excerpt from the book.

> The data you amass for the "brass" should be timely, accurate, and as inexpensive as possible. No reader should be startled by now to learn that scientific, or probability, sampling can very often provide data with these commendable attributes. Thus, whenever one is confronted with the problem of producing vital information for his chiefs he should certainly give at least a passing thought or two to the possibilities of sampling. He will find that sampling appeals to the "wheels" if they get what they want and save time and money in the process.
>
> If it appears that a sample might do the trick, one should, as a first step, determine as precisely as possible the population, or universe, to be surveyed. It might consist of all of a company's accounts receivable unreceived as of a certain date, the total number of railroad ties with knots, or the number of rakes who have taken grass widows to lawn parties in the past month.
>
> The next step is to set up a sampling "frame." This is a list or file of all the units in the statistical universe. Sometimes this is far from easy to accomplish. In one study the universe consisted of all the military pay

records in the Air Force. At any given date a number of these records were being transported overseas or back by their owners and were not available for sample selection in any Finance Office. Thus the frame of sampling units did not include all the units in the population under study. This is often the case in an actual study, and unless the differences between frame and universe data are negligible, a bias may be present in the sample results.

A very important step that should not be overlooked is to set down in terms of utmost clarity exactly what information is being sought in the survey. Unless the item or items of information are specified precisely, the reporting forms or questionnaires may yield something quite different from what was actually wanted. In a project designed to obtain the size distributions of Air Force clothing items, it was necessary to stipulate that the size of garment actually needed by the airman rather than the size purchased be reported; for many a size-7 cap has enveloped a size-6 7/8 cranium at some Air Force bases because no 6 7/8s were available.

Before proceeding too far with the survey, it is always advisable to check whether the information sought has already been found by someone else. In a large organization it is not unusual to find that the information one seeks or a substitute suitable for the purpose at hand has already been unearthed by another branch or unit. Obviously, no survey at all is even quicker and cheaper than a small sample survey. It is an excellent idea, too, to make sure that every single item of data listed is absolutely necessary to accomplish the aim of the survey. With the exception, perhaps, of stilts for a serpent there is nothing more useless or more ridiculous than a mass of figures collected at great travail, added, multiplied, divided by the cube root of $\pi$ and converted (by a \$350-an-hour electronic computer) to homogenized index numbers—that have no bearing on the problem.

It is necessary at this point that the desired degree of precision be specified by the users of the sample data. We noted earlier that the precision can be expressed in terms of permissible error (tolerance) and acceptable risk (confidence). Obviously, these will depend on the purpose at hand. For example, one could tolerate a larger percentage error in an estimate of the unpopped kernels per ladle of popcorn than would be acceptable in calculating the average lethal dose of samples of strychnine.

It is now time to investigate the relative efficiency of various sample designs for the degree of precision specified. This is the technical phase of the survey and requires the efforts of highly skilled sampling technicians. By examination of past data and from test samples, it is usually possible to compare the cost of several different sampling methods for obtaining the specified precision. The most efficient sampling method can then be selected, although the time element and other administrative considerations must be taken into account. These would include such items as availability of maps, enumerators, data-processing equipment, etc. In some instances the amount of money available for a sample survey is stipulated in advance. In such instances the sampling expert investigates alternative sample designs and comes up with an administratively feasible one that will provide the maximum precision for the money available. If this precision is too low to be acceptable to the users, they must either relax their requirements, raise additional cash, or drop the matter entirely.

It is advisable next to give some thought to the questionnaire or reporting form, if any such item is to be used in the survey. It is important to devise the form or questionnaire so as to obtain correct answers to the questions asked. Clarifying instructions are usually in order, as anyone can testify who has seen some responses to very simple queries. In cases where the survey is to make use of enumerators, instructions for training these inquisitive folk must be developed.

In our own experience, a small-scale pretest of reporting forms, questionnaires and instructions has proved inexpensive and indispensable. Regardless of how thoroughly the job was done or how carefully all loopholes had been plugged, every pretest has come up with its fair share of surprises. The final forms and instructions have always been much improved as a result of the pretest. In addition, the small-scale test has often been used to obtain data regarding the relationship and distribution of the factors under study, and these have been used in developing better sample methods for the survey. In this connection, it should be noted that the steps in a survey as outlined here are far from firm or in fixed order. Often they overlap, and in any case they should be performed in whatever fashion is logical for the purpose at hand.

When we have reached this stage, the time has come to conduct the survey as planned. The sampling units are selected, the data collected, the figures tabulated, the tabulations audited, the analysis conducted and the results interpreted and reported to management. Compressing these important steps into one sentence tends to conceal the fact that much remains to be done even after the development of the sampling method has been completed. The scientific sample approach is merely a way of getting information quickly, cheaply and reliably, within predetermined limits of precision. It may be worth while to repeat that we use the word "precision" instead of "accuracy" because precision refers to the degree of similarity between sample results and results from a 100 per cent count *obtained in the identical manner*.

It is a good idea, in presenting the results of a sample survey to management, to explain in as simple terms as possible the limits of reliability of the sample results. The U.S. Census Bureau was one of the first to do this, and many others have followed suit. It is only fair, in presenting sample results, to furnish the user an idea of how far he can trust them.

SOURCE: For Problems 58–60—Slonim, Morris J., ed. *Sampling in a Nutshell*. New York: Simon & Schuster, 1971 (pp. 19–23).

Questions   This application can help you understand some key steps for planning a sample survey:

1. Identify the universe and establish a frame.
2. Determine the specific information that is being sought.
3. Evaluate specific sample precision requirements.
4. Choose a sample design.
5. Prepare the survey form or questionnaire.
6. Pretest the questionnaire form and field procedures.
7. Perform the survey.
8. Present the results.

Use two or three sentences to define each of these eight steps and to point out a few cautions in performing them.

59. A distributor of containers of mixed nuts regularly checks that the company is meeting its advertising claim of "No more than 50 percent of the nuts, by count, are

peanuts." For this they inspect filled containers and count the number of peanuts and the number of other nuts. The nut mix contains Brazil nuts (B), cashews (C), almonds (A), filberts (F), and peanuts (P). Use the following sample of 50 nuts to establish a 95% confidence interval on the proportion of peanuts.

| Nut Number | Type | Weight (mg) | Nut Number | Type | Weight (mg) | Nut Number | Type | Weight (mg) |
|---|---|---|---|---|---|---|---|---|
| 1 | B | 63 | 18 | F | 20 | 35 | P | 8 |
| 2 | B | 61 | 19 | F | 18 | 36 | P | 8 |
| 3 | B | 57 | 20 | F | 17 | 37 | P | 8 |
| 4 | B | 56 | 21 | P | 11 | 38 | P | 8 |
| 5 | C | 35 | 22 | P | 10 | 39 | P | 8 |
| 6 | C | 34 | 23 | P | 10 | 40 | P | 8 |
| 7 | C | 34 | 24 | P | 9 | 41 | P | 8 |
| 8 | C | 32 | 25 | P | 9 | 42 | P | 8 |
| 9 | C | 31 | 26 | P | 9 | 43 | P | 8 |
| 10 | C | 28 | 27 | P | 9 | 44 | P | 8 |
| 11 | A | 34 | 28 | P | 9 | 45 | P | 8 |
| 12 | A | 32 | 29 | P | 9 | 46 | P | 8 |
| 13 | A | 31 | 30 | P | 8 | 47 | P | 8 |
| 14 | A | 30 | 31 | P | 8 | 48 | P | 7 |
| 15 | A | 28 | 32 | P | 8 | 49 | P | 7 |
| 16 | A | 28 | 33 | P | 8 | 50 | P | 5 |
| 17 | F | 21 | 34 | P | 8 | | | |

60. a. Estimate with 95% confidence the mean weight for random samples of 50 mixed nuts taken from this process. Use the data from the preceding problem and $\bar{X} = 18.76$ mg, $s_{\bar{X}} = 2.21$ mg. Assume the distribution of weights for samples of 50 nuts is normal.

   b. The nut types—Brazil, cashews, almonds, filberts, peanuts—are easily distinguished by size, shape, and color. Discuss the feasibility of using stratified random sampling. Explain the procedures you would use to draw the sample and to calculate an estimate for the *mean weight* for samples of 500 nuts.

## BUSINESS CASE

*I'm an Accountant*

Peg Lambert had never really thought about it before—and frankly she felt she could still be quite happy in her ignorance. Except her boss, Grover Mitchell, wasn't happy. Grover Mitchell was the manager of Internal Auditing for VISTACARD, a national credit-card firm. It seems that discrepancies were surfacing between the company accounting department's month-end summary and some customer claims. And Grover Mitchell was getting some complaints from his boss.

Being fresh out of college, Peg Lambert didn't know what had happened in the past. She only knew she had turned down two higher-paying jobs to be with VISTACARD because she wanted an auditing position—and she got a discount on personal travel. But she hadn't planned on being a statistician. Nevertheless, Mr. Mitchell had dumped this on her for a first assignment.

**Mr. M.**   Peg, your first assignment is certainly main line accounting—auditing monthly accounting reports, to be more exact.

**Ms. L.**   What kind of procedures do you have?

**Mr. M.**  That's just it, Peg. You see, we don't really have a procedures manual for this job yet. As a matter of fact, we haven't had the time to design an auditing system yet. That will be your first assignment. We need to know if the charges that appear on the various account numbers are correct and truly reflective of the past month's activity. The bulk of them are okay, but we're finding more and more customers challenging our records. And more often than not, we've had to give in. Our losses are on the verge of skyrocketing. And the front office wants monthly statements audited so we can find out what is going on.

**Ms. L.**  This sounds like a statistical sampling problem to me. Are you sure you want to put me on this? Statistics is not my strongest suit.

**Mr. M.**  Peg, this is *our* problem! I don't put labels on problems. And we don't compartmentalize people around here. Give it a shot. Look at some alternatives. I'm sure you'll come up with something. You've had some statistics, haven't you?

Peg investigated several auditing procedures over the next two weeks. The most accurate method would be to gather data by an independent source and keep a separate record. The account balances could be compared to the accounting monthly reports. However, such duplication would be terribly expensive and would probably disrupt the regular accounting work flow.

Another approach would take selected samples of the incoming charges at random intervals and match them with the resultant billing information sent to the customers. This could then be extrapolated to a monthly balance. This is the approach that Peg decided to pursue. She decided to call it the input/output method.

Since the input/output takes several forms, the question arose as to what forms of data would be most easily counted. Peg also needed enough accounts for a representative sample. She investigated the input/output documents.

1. *Keyboard input/output:* This proved to be an elusive source for data because of the wide variety (and location) of data that was currently keyed onto the customer purchase records. The data base, though complete, would be nonuniform and so would make for cumbersome calculations.

2. *Microfiche:* Peg also learned that some data were kept on microfiche. These items could easily be counted, but only selected (major) items were included. They did not represent the large number of items that could appear on the charge cards. This was incomplete as a sampling frame (base).

3. *Computer printouts (from branch locations):* Computer printouts from branch locations contained lists of items that could be counted. However, upon further study Peg found that only about half of the branch offices had a common computerized reporting system, and many of them reported on an irregular basis.

4. *Transaction format:* A little further investigation revealed that an information system utilizing disk storage on personal computers for randomly selected customers was being implemented on a trial basis. This data base would be geographically representative of the customer base and would include the total information desired in an easily identifiable form. Peg would simply have the information (already on disk) transmitted to a receiver in her office, and that could be done over a phone line.

The hard part was over. Peg arranged to obtain the disk information for the past month (September). It turned out to be almost a 10% sample of customer charges. The charges totaled $7,487,328 as determined from 1364 records, each of which contained 100 customers. The computer produced the totals in short order and also revealed a standard deviation of $2312.50 per record (of 100 customers).

Peg now had the empirical input data from the customer charges with which to establish her audit. She felt a 98% confidence level would be appropriate as a starter. That would yield a confidence interval, as determined from her sample, that should include the actual resulting monthly charge 98% of the time. Her sample data yielded a mean of

$$\bar{X} = \frac{\$7,487,328}{1364} = \$5489.20 \text{ per 100 customers}$$

From this she established a confidence interval estimate of

$$CL = \$5489.20 \pm \left[ (2.33)\left(\frac{\$2312.50}{\sqrt{1364}}\right) \right]$$

$$= \$5343.30 \text{ to } \$5635.10 \text{ per 100 customers}$$

Next, she obtained the output data—that is, the complete customer charge report. It was readily available and constituted a good standard of comparison. The average per 100 customers was $5502.12, well within the audit interval.

Peg then wrote up a preliminary report and stopped by Mr. Mitchell's office to let him review it prior to final typing.

**Mr. M.**  How are you doing on the project, Peg?

**Ms. L.**  You were right, Mr. Mitchell. It really wasn't the statistics that was the problem. It was the data base to use. It took me almost two weeks to find a reliable data base.

**Mr. M.**  I thought that would be the case. What did you come up with?

**Ms. L.**  Well, the transaction reporting system being installed by the management information systems group will work just fine. It gives us a good sample of customer charges. The data are already computerized and we can get a quick estimate of the upper and lower confidence limits. I ran a 98% confidence interval for the September data and then compared it with the actual billing information. If the actual amount lies within the confidence interval, that's a good basis for saying the billing figure is correct. If not, a more detailed auditing check should be carried out.

**Mr. M.**  Sounds fine! How did the September figures turn out?

**Ms. L.**  Oh, fine—right within the interval.

**Mr. M.**  Great.

**Ms. L.**  Mr. Mitchell, about those errors in the past ... when I was going through the data entered on one of the keyboards, I noticed a curious thing. In several cases the numeral 2 was not being recorded in the correct location. When the last digit to be keyed in is a 2, it sometimes does not come out in the right column. The people in data entry say it's no problem since they all know about it and watch for it, but I thought I'd just mention it to you anyway. I didn't plan to mention it in the report though—do you think I should?

**Mr. M.**  Ah ... well, that won't be necessary, Peg. I think maybe that keyboard will be out of service by the time your report is finished.

**Ms. L.**  Which brings me to the final question: When do you need my completed report?

**Mr. M.**  How about tomorrow noon?

**Ms. L.**  O.K. See you then.

**Business Case Questions**

61. Explain the potential nonrepresentation of monthly accounts for each auditing source in the example, including (1) keyboard I/O, (2) microfiche, (3) branch office computer printouts, and (4) computerized transaction format.

62. The complete customer charges report for September showed a mean of $5502.12. Is this amount within the limits for a 95% confidence interval estimate for Peg Lambert's sample? Indicate one advantage and one disadvantage of a 98% confidence interval over a 95% confidence interval estimate.

63. Explain in two or three sentences how Ms. Lambert performed a test audit on the VISTACARD September customer charges.

## COMPUTER APPLICATION

*Interval Estimation of $\pi$, with Sample Size Considerations*

This application continues the marketing study that we began in Chapter 2. In the Chapter 4 application we explored the question, What percentage of the target population—families of 3+ persons with householder aged 35–50 years—is householders aged 41–45 years?

This is an unusual situation in that we have more than one sample available to estimate a population value: $\pi$ = the percentage of 3+ person families with head of household 35 to 50 years old, and with head of household 41 to 45 years old. Our total value (30.1%) is a composite from the PUMS file, so 30.1% is a parameter, $\pi$. Our objective is to explore confidence interval estimation of $\pi$ by demonstrating the results from six alternative samples. Here is an SPSS program that provides the essential data.

```
 1 NUMBERED ←─────────────────────────────────────────────────────── 1
 2 TITLE       SAMPLES FOR PERCENT 41 TO 45 YEARS OLD
 3 DATA LIST      FILE = INDATA/
 4                    PERSONS  12–13
 5                    INCOME   35–39
 6                    AGE1     43–44 ─────────────────────────────── 2
 7 SET BLANKS = -1
 8 SELECT IF     ((PERSONS GE 3) AND (AGE1 GE 35) AND (AGE1 LE 50)) ── 3
 9 COMPUTE       AGE4145 = AGE1
10 RECODE        AGE4145 (41 THRU 45 = 1) (ELSE = 0) ──────────────── 4
11 DESCRIPTIVES  VARIABLES = AGE4145/
12               STATISTICS = MEAN SEMEAN/
13 TEMPORARY ──────────────────────────────────────────────────────── 5
14 SET SEED = 582 ←
15 SAMPLE 38 FROM 3757
16 DESCRIPTIVES  VARIABLES = AGE4145/
17               STATISTICS = MEAN SEMEAN/
18 TEMPORARY
19 SET SEED = 725
20 SAMPLE 75 FROM 3757
21 DESCRIPTIVES  VARIABLES = AGE4145/
22               STATISTICS = MEAN SEMEAN/
23 TENMPORARY
24 SET SEED = 2206
25 SAMPLE 150 FROM 3757
26 DESCRIPTIVES  VARIABLES = AGE4145/
27               STATISTICS = MEAN SEMEAN/
28 TEMPORARY
29 SET SEED = 883
30 SAMPLE 301 FROM 3757
31 DESCRIPTIVES  VARIABLES = AGE4145/
32               STATISTICS = MEAN SEMEAN/
```

```
33 TEMPORARY
34 SET SEED = 2271
35 SAMPLE 601 FROM 3757
36 DESCRIPTIVES   VARIABLES = AGE4145/
37                STATISTICS = MEAN SEMEAN/
38 TEMPORARY
39 SET SEED = 1817
40 SAMPLE 1202 FROM 3757
41 DESCRIPTIVES   VARIABLES = AGE4145/
42                STATISTICS = MEAN SEMEAN/
43 FINISH
```

The outputs are condensed into a single summary to avoid repeating some print labels.

---

```
VARIABLE AGE4145
   MEAN   .301 (=π)   S. E. MEAN   .007      POPULATION

      MEAN   .368      S. E. MEAN   .079   SAMPLE 1   n =     38
      MEAN   .267      S. E. MEAN   .051   SAMPLE 2   n =     75
      MEAN   .353      S. E. MEAN   .039   SAMPLE 3   n =    150
      MEAN   .312      S. E. MEAN   .027   SAMPLE 4   n =    301
      MEAN   .328      S. E. MEAN   .019   SAMPLE 5   n =    602
      MEAN   .284      S. E. MEAN   .013   SAMPLE 6   n =   1202
```

---

Several of the SPSS program statements are noteworthy. See the SPSS manual [9] for more details on specific statements.

1 Lines 1 through 7 of the program are data setup. NUMBERED gives program step line numbers on the output. The DATA LIST tells the operating system that the data set is on a file INDATA and, for PERSONS INCOME AGE 1, tells it the data positions on the data file. SET BLANKS is a data analysis flag to code blank data.

2 Line 8, a SELECT IF statement, restricts the data usage from the original 20,116 records to those 3757 records for which the number of PERSONS is 3+ and for which AGE1, the householder age, is 35 to 50 years. This is the population being analyzed.

3 Lines 9 and 10 produce a recoded variable—AGE4145—that has value one (1) when AGE1 is 41, 42, 43, 44, or 45 years, and otherwise has value zero (0). AGE4145 is a Bernoulli random variable.

4 Lines 11 and 12 give SPSS DESCRIPTIVES inputs to summarize the population of $N = 3757$ records. The output summary includes its MEAN = $\pi$ = 0.301.

5 The remainder, Lines 13 through 42, define six samples of increasing size. SET SEED = # provides unique random samples, and SAMPLE # FROM 3757 defines the required sample size through a user-supplied random number, #, taken from Table B.7. For illustration, the sample sizes are approximately $37.5, 2 \cdot 37.5, \ldots, 6 \cdot 37.5$—that is, 38, 75, 150, 301, 601, and 1202 records. FINISH, line 43, signals the end of the SPSS program statements.

The (6) sample results are used individually to provide 95% confidence interval estimates for $\pi$ (which is purposely known for our illustration: $\pi = 0.301$). SAMPLE 1, which uses the sample standard deviation, calls for a $t$-statistic. Since $n = 38$, then $df = 38 - 1 = 37$ in Table B.2 requires a $t$-value = 1.96. Since the SAMPLE 2, 3, 4, 5, 6 sizes increase, this value—1.96—is also used with the subsequent (larger) samples. The confidence intervals for $\pi$ are defined by $p \pm (1.96\sqrt{pq/n})$. Table 8.4 displays the calculated 95% confidence intervals.

The point estimates in Table 8.4 are different. They range from 0.267 to 0.368. The point-estimate plots on the confidence intervals illustrate that with increasing sample size the point estimates are moving closer to the population value, 0.301. This can also be observed by the decrease in error (that is, a narrower interval) that occurs with increased sample size. This illustrates that under random sampling the sample proportion, $p$, is a *consistent estimator* for $\pi$.

The relation of error in estimation to sample size is displayed in Table 8.3. Consider the standard error values for

$$
\begin{array}{lll}
n = \ \ 38 & \text{Standard error} = 0.079 & \\
n = 150, \ \doteq \ 4 \cdot 38 & \text{Standard error} = 0.039, & = (1/2) \cdot 0.079 \\
n = 601, \ \doteq 16 \cdot 38 & \text{Standard error} = 0.019, & = (1/4) \cdot 0.079
\end{array}
$$

The point is that there is a direct (reciprocal) relation between the sample size and the error of estimation. When the sample size is increased by 4 times, the error is reduced to 1/2; by 16 times, the error is reduced to 1/4 the original value, and so on. This observation can be generalized.

*Rule:* For random sampling with large samples ($n > 30$), increasing the sample size by a factor of $K$ will result in reducing the error of estimation by a factor of $1/\sqrt{K}$.

For example, increasing the sample size from 38 to $16 \times 38 \doteq 601$ resulted in a confidence-interval size reduction by $1/\sqrt{16} = 1/4$. Indeed, one way to get a better estimate is to increase the random sample size. Another way to show this reduction (in Table 8.4) is with bounds that decrease as the sample size increases.

**TABLE 8.4**  Sample Results for Estimates of the Proportion of Householders Who Are 41–45 Years Old from Among Those Who Are 35–50 Years Old

| Sample Size | Point Estimate | Standard Error | Bounds for 95% Confidence Intervals | Proportion 41–45 Years Old |
|---|---|---|---|---|
| 38 | 0.368 | 0.079 | 0.213–0.523 | (--------------------•----------------------) |
| 75 | 0.267 | 0.051 | 0.167–0.367 | (----------------•---------------) |
| 150 | 0.353 | 0.039 | 0.277–0.429 | (--------•------) |
| 301 | 0.312 | 0.027 | 0.259–0.365 | (---------•----) |
| 601 | 0.328 | 0.019 | 0.291–0.365 | (---•----) |
| 1202 | 0.284 | 0.013 | 0.259–0.310 | (---•---) |

Proportion 41–45 Years Old scale: 0.16  0.20  0.24  0.28  0.32  0.36  0.42  0.46  0.50  0.54

Population 0.301 ————————————————→ $\pi = 0.301$

Of course, managers want a bigger sample—more information—for making their decisions. But this has some limits. First, a bigger sample requires more money and more time to obtain the results. Because managers have many decisions to make and budget restraints to stay within, it is usually most feasible to use moderate-sized samples. Second, the Chapter 7 Business Case demonstrates a situation for which it is best to take a sample; and for bigger samples, the benefits in better estimates can diminish as the sample size increases beyond a point.

Remember, to a reasonable point of sampling, larger samples can provide better estimates. The computer problems provide more applications for determining a reasonable sample.

## COMPUTER PROBLEMS

*Data Base Applications*

64. Use your 20 samples from Chapter 7, Computer Problem 40(b), to develop (i) point and (ii) 95% confidence interval estimates for $\mu$ = mean household income by computing each of the following.
   a. Use the sample mean and standard deviation from your sample #1 in Chapter 7, Problem 40 to establish a 95% confidence interval estimate for $\mu$ = the mean household income.
   b. Repeat as in part (a) to establish a 95% confidence interval from each of your 20 samples. Determine how many of your intervals include $\mu$. (This should be near 95%.) Use $\mu$ = $19,700.

65. Repeat Problem 64, but now estimate $\mu$ = the mean age for heads of household. For this, use your sample means and standard deviations for 20 random samples selected in Chapter 7, Problem 42. How many of your (20) intervals include $\mu$ = 47.6 years? Use $Z$ = 1.96 for 95% confidence interval estimates.

66. Again use the approach of Problem 64, now to estimate $\pi$ = the percentage of heads of household who are female. Problem 43 in Chapter 7 requests the essential data for your calculations, including the standard error of proportion. Now determine how many of your 95% confidence intervals include $\pi$ = 0.278.

67. Select a random sample of 50 records. For each record, extract the count for persons per household (positions 12–13). Then calculate a point estimate

$$\bar{X} = \text{the mean of persons per household}$$

as the sum on the 50 counts of persons divided by 50. Continue to estimate, with 90% confidence, $\mu$ = the true mean of persons per household. The PUMS national number is 2.74. Does your interval include this value?

68. Select a random sample of 50 PUMS records. For each record use the weeks worked for person 1 (columns 51–52) to obtain the statistics $\bar{X}$, standard deviation, and (if available) standard error. Develop a 95% confidence interval estimate for $\mu$ = the mean of weeks worked by heads of household (person 1). The mean is 34.4 weeks. Does your interval include this (true) value?

69. The following is one random sample of 50 incomes from the 20,116-record PUMS file (as in Chapter 7, Computer Problem 40).

```
INCOME
 20115  21580  16840  11215  11940  17410  21010  22630   9005
 38510  17360   3995  32780  18820   5405  28115   5565  75000
  5020  28825  12365   7515   4880  16170  36800  20770  15010
 20010  75000  75000  21010  16815  11985  14005  39420   4050
  3515  17690   4710  16505  42060   2580   6150  56015  10240
 28010   5605   6775  28015   9005
```

It has $\bar{X} = \$20,776$ and $s = \$18,032$.

a. Another computer run on the total (20,116-record) PUMS file gave $\sigma = \$14,931$. Minitab statements to establish a 95% confidence interval for $\mu =$ the mean household income with $\sigma = \$14,931$ (known) are

```
MTB > SET IN C1
DATA>←——————— enter the (50) given incomes
DATA> END
MTB > NAME C1 'INCOME'
MTB > PRINT 'INCOME'
MTB > DESCRIBE 'INCOME'
MTB > ZINTERVAL 95% confidence assuming sigma =14931
      for the data in C1
MTB > SAVE 'B:INF1'
```

Minitab users set up and run this program. Interpret your confidence interval.

b. Now repeat part (a), except use the TINTERVAL procedure in place of the ZINTERVAL one.

```
MTB > TINTERVAL 95% confidence using the data in C1
```

Again, interpret the confidence interval. Then explain why the confidence interval in part (b) is wider than that obtained in part (a).

70. Here are data for the age of householder, AGE1, for a random sample of 50 from the PUMS 20,116-record file. The total PUMS file has $\sigma = 17.74$ years.

```
AGE1
 76  56  44  28  33  74  30  27  31  41  26  53  54
 31  57  40  77  30  23  43  24  74  79  65  52  37
 57  35  32  35  44  30  66  63  38  36  69  34  68
 33  42  70  52  53  55  44  69  39  56  21
```

This sample has $\bar{X} = 46.92$ and $s = 16.69$.

a. Use a Minitab procedure similar to that in Problem 69, with the AGE1 data, 90% confidence and the MTB > ZINTERVAL procedure to estimate AGE1 for the 20,116-record PUMS file.

b. Repeat part (a), except use the Minitab TINTERVAL procedure. Explain why this 90% confidence interval is narrower than the one for part (a).

*Note:* Save this data for use in a Computer Problem in Chapter 9.

71. Use the data in Example 8.7 to establish a 90% confidence interval on the mean weight of shipping cartons. Compare your (LCL, UCL) interval with that given in the example. *Notes:* For Minitab use the TINTERVAL procedure.

Let

```
MTB > SET
DATA> 5.4, 5.5, 5.6, 5.6, 5.7, 5.7, 5.7
DATA> END
MTB > TINTERVAL 90 PERCENT C1
```

For SPSS users the FREQUENCIES procedure gives the statistics required for a confidence interval estimate, or a *t*-test. For the Example 8.7 check, read in the data, then use

```
FREQUENCIES VARIABLES = pounds (5.4, 5.7)/
             STATISTICS = MEANS SEMEAN/
```

In this case you are required to hand-calculate the interval.

---

## FOR FURTHER READING

[1] Freund, John E., and Ronald Walpole. *Mathematical Statistics*, 4th ed. Englewood Cliffs, NJ: Prentice-Hall, 1987.

[2] Gallup Opinion Index. Princeton, NJ: American Institute of Public Opinion, January 1976.

[3] Groebner, David F., and Patrick W. Shannon. *Business Statistics: A Decision Making Approach*, 3rd ed. Columbus, OH: Charles E. Merrill, 1989.

[4] Hamburg, Morris. *Statistical Analysis for Decision-Making*, 5th ed. Orlando, FL: Harcourt Brace Jovanovich, 1991.

[5] Larson, Harold. *Introduction to Probability Theory and Statistical Inference*, 3rd ed. New York: John Wiley, 1982.

[6] MacLachlan, James, and Michael Siegel. "Reducing the Cost of TV Commercials by Use of Time Compressors." *Journal of Marketing Research* XVII (Feb. 1980): 52–57.

[7] Snedecor, George, and William Cochran. *Statistical Methods*, 7th ed. Ames, IA: Iowa State University Press, 1980.

[8] Slonim, Morris J. *Sampling in a Nutshell*. New York: Simon and Schuster, 1971.

[9] *SPSS Reference Guide* (version 4.0). Chicago: SPSS, Inc., 1990.

[10] Zuwaylif, Fadil H. *Applied Business Statistics*, 2nd ed. Menlo Park, CA: Addison-Wesley, 1984.

# HYPOTHESIS TESTING

## 9.1
## INTRODUCTION

The vice president of marketing for Sporthaus, a national chain of sporting goods stores, is considering having a nationwide promotional campaign. She has obtained a random sample of 500 households from the mailing list of a direct-mail advertising firm. The firm claims the list has a high proportion of families with members in the 15- to 35-year-old age range. But the Sporthaus vice president wants assurance of this before authorizing funds to use the list for a large national mailing. She will use the national list only if there is some assurance that 80% or more of households include someone in the 15-to-35 age group. If you were a market analyst for Sporthaus, it could be your job to determine whether the list satisfies the age requirement. How would you decide? By the time you finish this chapter you should be able to answer this question by using principles of hypothesis testing.

The Sporthaus illustration is a decision question about a population characteristic—a question that can be structured as a test of hypotheses. A *hypothesis* is a belief about a population, or state of nature.

This chapter is concerned with testing hypotheses and making decisions about population values. The process of using statistics to form a decision about a population parameter is called statistical *hypothesis testing*.

> Hypothesis testing is a process of using sample data and statistical procedures to decide whether to reject or not reject a hypothesis (statement) about a population parameter value (or about its distribution characteristics).

The kinds of hypotheses we will test are familiar ones about means and percentages. Sporthaus's vice president is concerned about a percentage. Other examples of percentage-type questions: Does the incumbent governor have the support of over 50% of the electorate? Are Apple computers favored by 30% of all microcomputer users? Other tests concern means: Is the mean SAT score of entering freshmen 10 points higher than it was last year? What salary can you expect when you enter the job market? Are average salaries in your field above \$25,000? Are they below \$35,000? The applications of hypothesis testing are many and varied.

Hypothesis testing helps managers decide what products to produce, what marketing strategies to follow, and how to assess the validity of their advertising claims. Court cases are even decided on the basis of statistical tests. Hypothesis testing is a powerful tool for decision making.

Our approach to hypothesis testing involves identifying the decision problem, formulating an appropriate hypothesis, choosing a statistical test to resolve it, gathering relevant data, and conducting the test. In testing hypotheses, we use sample statistics to compare with "hypothesized" population values. Then, based upon the data available, the hypothesis test results direct us to a best course of action.

In this chapter you will learn how to use hypothesis tests to help resolve business decision questions.

Upon completing Chapter 9, you should be able to

1. explain the principles underlying hypothesis testing;
2. structure a business decision situation about means or proportions into the form of a test of hypotheses;
3. apply systematic testing procedures;
4. interpret hypothesis test results and draw conclusions.

This chapter does not introduce any new statistical measures. It does, however, make use of many of the statistical tools you studied in earlier chapters.

## 9.2 PROCEDURES FOR HYPOTHESIS TESTING

In this section, we develop techniques for making decisions by testing hypotheses. For example, assume you are a marketing trainee for a firm producing moped bikes. Suppose you are given the responsibility for developing an advertising campaign for your company's product. You may wish to advertise that the average gas mileage obtained for your bikes is over 125.0 mi/gal, but you need some test data to support this claim before you advertise it. This section outlines a step-by-step procedure to accomplish such tests.

The procedures for statistical testing are an application of the familiar scientific method.

### STEPS FOR CONDUCTING A HYPOTHESIS TEST

1. Define the question to be tested and formulate a hypothesis (and alternatives) for stating the problem.
2. Establish a procedure for testing the hypothesis and a criterion for deciding when to reject it.
3. Collect sample data (evidence) upon which to test the hypothesis and calculate the sample statistic value.

4. Based on your sample findings, use statistical procedures to *decide whether to reject or accept the hypothesis* (or revise the original alternatives and repeat Steps 2 through 4).*

Example 9.1 illustrates the application of this method and introduces the key statistical terms used in testing.

---

**EXAMPLE 9.1**

The marketing manager of a company manufacturing a type of gas-powered moped bike has done some research into competitors' moped bikes and has found that most mopeds get less than 125 miles per gallon (mi/gal) of gasoline. He now wants to advertise the good gasoline mileage of his company's vehicle. In fact, he wants to state, "Our mopeds will average over 125 miles per gallon of gasoline." The question: Is the claim of over 125.0 mi/gal (average) valid?

*Solution*      The logical alternatives are "yes" (the mopeds *do* average over 125.0 mi/gal) and "no" (the mopeds do *not* average over 125.0 mi/gal), but neither completely answers our question. However, we will use the question to illustrate the four steps in the solution procedure (listed previously).

---

**Step I: Define the Problem and Formulate the Hypotheses**

Having first defined the problem, we can develop an initial hypothesis and an *alternative* hypothesis to test. The initial hypothesis is also referred to as the null hypothesis. *Null* means *no change*, and the *null hypothesis* is typically a statement of no (statistical) difference from some predetermined or accepted standard.

In Example 9.1, the null hypothesis is a statement that the mean gas mileage is not different from that of other mopeds; that is, the firm's mopeds get no more than 125.0 mi/gal. This is symbolized by $H_0$: $\mu = 125.0$ mi/gal. However, since the firm's moped is designed to give a high gas mileage, the alternative of concern is that their mean gas mileage is statistically more than 125.0 miles per gallon. This is symbolized by $H_A$: $\mu > 125.0$ miles per gallon. If the manufacturer's claim is valid, sample evidence will cause rejection of the *null hypothesis* and will substantiate the higher-mileage claim.

> A hypothesis is a statement of belief used as a basis for evaluating population values or probability distributions.
>
> The null hypothesis is a statement of no difference or no change from a hypothesized value (or distribution) and is symbolized by $H_0$.
>
> An alternative hypothesis is a statement that a difference exists or that a change has taken place, and is symbolized by $H_A$. It is a logical complement of $H_0$.

Several considerations help establish the test alternatives. First, the *null hypothesis*, $H_0$, structures the test procedure. It is a standard or starting point

---

* Strictly speaking, we may not "accept" a hypothesis because sample evidence never "proves" a hypothesis is true. Data can only be insufficient to disprove it. Thus many researchers conclude "we cannot reject" rather than "we accept" a hypothesis. We shall use the term *accept* only because we feel it lends more intuitive understanding to the methodology and is appropriate for decision-making purposes.

for establishing the test. Thus the null hypothesis includes a statement of equality of the parameter to a standard or predetermined value. It is the norm on which we must rely unless evidence is strong enough to reject the norm.

The *alternative hypothesis*, $H_A$, is a statement of inequality ($<$, $\neq$, or $>$). In the present test, the manufacturer's claim will be supported by a sample reading of somewhat more than 125.0 mi/gal. The manufacturer hopes that the alternative hypothesis will be substantiated. For this reason *the alternative hypothesis is sometimes called the test or research hypothesis.*

If other reasonable alternatives exist, these can also be tested against the null hypothesis. The null hypothesis remains unchanged as a standard against which to test reasonable alternatives one at a time. The ranges for the null and alternative values are stated in such a way that (1) rejecting the null hypothesis suggests that the *alternative* hypothesis is more reasonable and (2) not rejecting the null hypothesis implies that the *null* hypothesis is more reasonable. We might decide in favor of $H_A$ because the evidence contradicts $H_0$, or we might decide in favor of $H_0$ because the evidence supports it. Thus it is possible from our analysis to cast doubt on $H_0$, but neither $H_0$ nor $H_A$ can be "proven" in a causal sense.

Hypotheses should be based upon the researcher's prior knowledge of the problem being investigated. Some knowledge of the problem is important to formulate the hypothesis properly and in a way that will allow unbiased data to support or refute it. Working backward (by using specific data to formulate a general hypothesis after the data have been analyzed) could cause the researcher to accept hypotheses that the data were not designed to test. The hypotheses should be established before any data are collected.

**Step 2: Establish the Test Criterion**

Having set the null and alternative hypotheses, we look next for a test procedure and criterion. The procedure will be scientific and statistical. Moreover, the same format is followed in nearly all statistical tests.

The test criterion is based upon the amount of difference that can reasonably be expected between the hypothesized null population value ($\mu_0$) and the observed sample value ($\bar{X}$). We express this difference in units of standard error, or $Z$ units—much as we did in estimation. For example, suppose the moped manufacturer road-tests a random sample of 49 bikes and finds a mean of $\bar{X} = 126.4$ mi/gal and a sample standard deviation of 4.9 mi/gal. Is the difference between 125.0 mi/gal and 126.4 mi/gal a *significant difference*?

> A significant difference occurs if the difference between the hypothesized (null) value and the sample statistic value is too large to be attributed to chance. A significant difference strongly suggests that the null hypothesis is *not* true.

A difference small enough to be attributed to chance—the oddity of the particular sample chosen—is termed a *nonsignificant difference*. Our test then will be to decide if the null value of 125.0 mi/gal and the sample value of 126.4 mi/gal are significantly different. That is, is the 1.4 mi/gal difference just a result of this particular sample of 49 mopeds?

**Risk of Error**

The path to determine statistical significance leads to the statement of *test errors* and the related *risks* (see Table 9.1). Because our conclusions are based

**TABLE 9.1**    Classification of Test Errors and the
Related Risks in Statistical Testing

| Decision (based on sample evidence) | True Condition | |
|---|---|---|
| | $H_0$ Is True | $H_0$ Is False |
| Reject $H_0$ | Type 1 error* | No error |
| Do not reject $H_0$ | No error | Type 2 error** |

* The $\alpha$-risk is the probability of making a type 1 error:

$$\alpha = P(\text{Type 1 error}) = P(\text{Reject } H_0 | H_0 \text{ is true})$$

** The $\beta$-risk is the probability of making a type 2 error:

$$\beta = P(\text{Type 2 error}) = P(\text{Accept } H_0 | H_0 \text{ is false}).$$

on sample information, we concede some risk of getting an improbable sample and making a decision error. Two possible errors exist: *type 1* and *type 2*.

A type 1 error is rejecting the null hypothesis when it is actually true.

A type 2 error is accepting the null hypothesis when it is actually false.

Note that a *type 1 error* can be committed only if the null hypothesis is true. In Example 9.1 we could commit a type 1 error only if all mopeds of this model did acutally average 125.0 mi/gal or less. If we rejected this (true) fact, we would conclude (erroneously) that this model averages more than 125.0 mi/gal. The risk here is one of faulty advertising, which could result in dissatisfied customers, a bad dealer image, even a damaging lawsuit. Manufacturers cannot afford to make such errors, so they typically want to keep the risk of this error very low. This risk is called the *producer's risk*, or the $\alpha$-risk.

A *type 2 error* results from (erroneously) accepting that the average mileage for this model of bike equals or falls below 125.0 mi/gal when, in fact, the overall average is really above 125.0 mi/gal. The risk here is one of forcing the firm to lower its claim on fuel economy, thereby reducing the marketability of their product. This would damage their competitive position, so that risk must be considered in the business decision. This risk is called the *consumer's risk* because it can be the risk of accepting a product when it does not meet the hypothesized specifications. This is also called the $\beta$-risk.

The foregoing discussion describes the test risks, but it does not consider their quantification. However, the *risks* must be quantified for statistical testing. The risks are quantified as the *probability of making the error*, either type 1 or type 2. Each is described through a conditional probability statement (see Table 9.1).

The structure of a conditional statement for either risk (type 1 or type 2) is

$$\text{Risk} = P(\text{Decision about } H_0 | \text{True condition})$$

Although both risks are important in conducting statistical tests, we concentrate on the α-risk first. With nearly complete information, both risks could be made quite small; but for practical reasons—time and cost considerations—complete information is rarely available. With sample information we except some chance of error, but we hope to limit it to a small amount. Because the α-risk is generally set at a low value, such as 1% or 5%, a test criterion is established according to the α-level.

Since $\alpha = P(\text{Reject } H_0 | H_0 \text{ is true})$, then $1 - \alpha = P(\text{Accept } H_0 | H_0 \text{ is true})$. This suggests a test procedure: For a 5% risk of rejecting a true null hypothesis, set $\alpha = 0.05$. Hence,

$$\alpha = 0.05 = P(\text{Reject } H_0 | H_0 \text{ is true}) = P(Z > 1.645 | \mu = 125.0)$$

The test is on the reasonableness of the result, given the assumption ($H_0$) that the mean mileage does not exceed 125.0 mi/gal for this model of moped. This is illustrated in Figure 9.1, which describes the distribution of sample means.

Suppose the true mean were 125.0 mi/gal. Then, by the central limit theorem, the sample means would be distributed about the value 125.0 in a normal distribution. Thus there is a 95% chance that a random sample mean would yield a Z-score of 1.645 or less. Figure 9.1 shows that 5% of all possible random sample means will give Z-scores of more than $+1.645$. (This comes from the Z-distribution, Table 1 in Appendix B.) So there is only a 5% chance we will get a sample with $\bar{X}$ so large that we will decide erroneously that $\mu > 125.0$ when in fact $\mu = 125.0$ mi/gal. Our test criterion (initial value) is thus a Z-score

**FIGURE 9.1**   Test Conditions for an Upper-Tailed Test, with
$\alpha = 0.05 = P(\text{Reject } H_0 | H_0 : \mu = 125.0)$

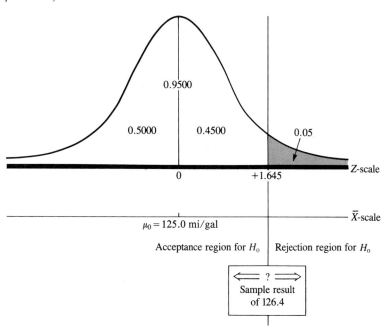

of 1.645. We refer to Figure 9.1 as a *one-tailed* test because the $\alpha$ region of rejection for $H_0$ is all in one tail—the upper tail—of the distribution.

**Step 3: Collect Data and Calculate the Sample Statistic Value**

Step 3 of our procedure calls for collecting sample data and computing a test value. Recall that the random sample of 49 mopeds revealed an average mileage of $\bar{X} = 126.4$ mi/gal. Assume $\sigma = 4.9$ mi/gal. Because we are working with sample means, the central limit theorem provides the appropriate sampling distribution. Our *test statistic*, Z, is a measure of this sampling distribution.

> A test statistic describes the variation in a sampling distribution. The test statistic for testing hypotheses about $\mu$, with $\sigma$ known, and for $n > 30$, is
>
> $$Z = \frac{\bar{X} - \mu}{\sigma_{\bar{X}}} \qquad (9.1)$$

The Z formula provides a meaningful comparison for the sample mean to the null hypothesis value, $\mu_0 = 125.0$. The Z-value calculated from this sample is the sample statistic value:

$$\text{Calculated } Z = \frac{\bar{X} - \mu_0}{\sigma/\sqrt{n}} = \frac{126.4 - 125.0}{4.9/\sqrt{49}} = \frac{1.4}{0.7} = 2.0$$

where $\mu_0$ denotes the value specified by the null hypothesis, $H_0: \mu = \mu_0$.

**Step 4: Accept or Reject the Null Hypothesis: Draw Conclusions**

Since the calculated value of $Z = 2.0$ is greater than the test criterion of $Z = 1.645$, the sample does not support the null hypothesis; that is, the assumption $\mu = 125.0$ mi/gal is not supported because the sample mean exceeds the test value. If the true mean $\mu$ were as low as 125.0, we would obtain Z-values greater than 1.645 only 5% of the time. The sample evidence suggests that moped bikes of this model do average over 125.0 mi/gal. The manufacturer now has statistical evidence at the 5% level of significance with which to support its claim.

**A Summary of Steps in Hypothesis Testing**

In summary, the steps in applying the one-tailed test of hypothesis are as follows:

1. Define the problem and formulate a null and an alternative hypothesis, stated so that accepting the null automatically means rejecting the alternative. For example,

$$H_0: \mu = 125.0 \text{ mi/gal} \quad (\text{where } 125.0 = \mu_0)$$

$$H_A: \mu > 125.0 \text{ mi/gal}$$

2. Establish a criteria for deciding whether to accept or to reject the null hypothesis. We typically set this test by specifying an $\alpha$-risk. (Other alternatives are presented in Section 9.5.) A greater-than alternative leads to a one-tailed, upper-rejection-region test, whereas a less-than alternative leads to a lower-tailed rejection-region test. For example, if $\alpha = 0.05$, the

test criterion or critical $Z$-value is 1.645. The test criterion for an upper-tailed test would be

> Test: Reject $H_0$ if the $Z$-value calculated from the sample is greater than 1.645.

3. Draw a random sample and compute a sample mean. Then compute the $Z$-value. In our case,

$$Z = \frac{\bar{X} - \mu_0}{\sigma/\sqrt{n}} = 2.0$$

4. Accept or reject the null hypothesis based on a comparison of the observed $Z$-value from the sample and the test criterion, or critical $Z$-value. Express the conclusion in a clear and straightforward way. In Example 9.1,

> Decision: Reject $H_0$ since $2.00 > 1.645$; the true mean mileage exceeds 125 mi/gal at the 5% level of risk.

Rather than accept a null hypothesis as a fact, researchers say that based on the sample evidence, we cannot reject the null hypothesis at the given $\alpha$-level. This implies two things. *First*, it recognizes the possibility of a type 1 error. *Second*, the classical theory of testing tells us that sample information cannot prove a null hypothesis. Thus if we conclude that we cannot reject $H_0$ based on the sample evidence, we admit the possibility of still other viable alternative values, which can then be tested.

The $\alpha$-risk level should be set by management but will be specified in our examples here. Since the $\alpha$-risk describes a probability measure, it can take on any value from 0 to 1, inclusive. But one of the low-risk values—0.01, 0.025, 0.05, or 0.10—is commonly used. The choice of $\alpha$-risk level depends on many factors, a major one being the cost of committing a type 1 error.

The $\beta$-risk level is also an important consideration. We can reduce the $\alpha$-risk at the expense of increasing the $\beta$-risk (and vice versa). Also, $\beta$ is not a single number. Its value depends on how wrong $H_0$ is. *One way to reduce both risks simultaneously is to increase the sample size.*

Note that *the claim being questioned (or researched) is typically quantified as the alternative hypothesis.* It expresses some deviation from no change—the null hypothesis. Rejecting the null hypothesis supports the claim.

An inability to reject the null hypothesis means the sample evidence does not support the researcher's claim. If the sample value deviates significantly from the hypothesized (null) value, the researcher can put statistical credence in the claim. Users of statistics often refer to the $\alpha$-level as the *level of significance* of a test.

> The level of significance of a test is the $\alpha$-level or risk of rejecting the null hypothesis when it is true.

Thus, if $\alpha = 0.05$ we refer to this as *a 5% $\alpha$-level*, and an associated test is a 5% (level) test.

We have now covered the essentials of hypothesis testing. When using this procedure, be sure to state each step very clearly in terms of the particular problem you are solving. *Communicating your methodology is often more important than communicating an answer.*\*

The major portion of what follows concerns problems for which scientific testing procedures can aid in decision making.

**Problems 9.2**

1. The employees in a national company have complained through their union negotiators that they are being underpaid. The mean wage for all employees in this company is hypothesized to be about $14.75 per hour, but union leaders would like to have sample evidence to show that it is not that high. State suitable $H_0$ and $H_A$, making the alternative a statement that would lend support to the workers' claim for higher wages.

2. What precautions can be taken to reduce the risk of making (a) a type 1 error? (b) a type 2 error?

3. Using $\alpha = 0.025$ and the test "Reject $H_0$ if calculated $Z < -1.96$," complete the decision problem that you started in Problem 1. Suppose a random sample of 49 employees from the company showed a mean wage of $14.54. Assume $\sigma = \$0.84$.

4. Based on a sample, the Allied Plastics Corporation wants to test the hypothesis that shipments of plastic pipe meet specifications for compressive strength.
   a. The two types of error in this situation are (1) the producer's risk and (2) the consumer's risk. Describe the error, either type 1 or 2, that would most logically result in a greater loss for the producer.
   b. What can be done to minimize the producer's risk?
   c. What would you do if you wanted to reduce the consumer's risk without increasing the producer's risk?

 5. The director of operations for a large hospital is concerned with the effectiveness of a recently implemented procedure for admitting cardiac (heart attack) patients.
   a. What null hypothesis is she testing (if she is committing a type 1 error) when she erroneously decides that the average time required to admit a cardiac patient exceeds 5 minutes?
   b. What null hypothesis is she testing (if she is committing a type 2 error) when she says erroneously that the average is 5 or more minutes?

6. A manufacturer of transducers claims that his company's products meet specifications of $\mu = 2.500$ with tolerance limits of $\pm 0.050$. A random sample of 300 units revealed a mean of 2.493. The standard deviation $\sigma = 0.040$. State the null and alternative hypotheses to test whether the true mean is actually 2.500. All units are pounds per square inch.

---

\* We cannot overemphasize the importance of sound judgment in deciding when a test gives reasonable results. For example, for sufficiently large samples, very small differences can be significantly different. This does not mean, however, that such differences are really important or consequential in a particular situation. See source [8] in "For Further Reading" at the end of Chapter 9 for more explanation on statistical versus actual significance.

# 9.3
# TESTS OF MEANS:
# ONE POPULATION

Tests of means for a single population are subdivided according to whether $\sigma$ is known. First we consider the cases in which (1) $\sigma$ is known and the random variable is normally distributed or (2) $\sigma$ is known and $n > 30$. For the latter case the central limit theorem makes the $Z$-distribution appropriate. For the former case $Z$-scores are appropriate by normal distribution theory. When $\sigma$ is unknown, we can substitute $s$ for $\sigma$ and, as long as the variable is normally distributed, use the $t$-statistic for inference on means. This case is covered later in this section.

## Tests of Means When the Standard Deviation Is Known

This section illustrates the testing procedures and interpretation of tests of means. Three types of test alternatives are discussed:

1. Less than ($<$)
2. Does not equal ($\neq$)
3. Greater than ($>$)

It is essential first to determine whether a test should be one- or two-tailed. If a one-tailed test is to be used, then we must go on to state whether it is to have a greater-than ($>$) or a less-than ($<$) rejection region. In all cases, we use sample data and a statistic to decide either (1) to reject or (2) to accept the null hypothesis based upon whether the statistic (1) falls into the rejection region or (2) does not fall into the rejection region.

For a two-tailed test (where $H_A$ reads $\neq$), we want two test criteria, or critical $Z$-values. These values are chosen to define the region of rejection, or the $\alpha$-level of the test. Thus one critical value cuts off $\alpha/2$ probability in the upper tail (extreme), and the other cuts off $\alpha/2$ in the lower extreme of the sampling distribution.

Situations in which we would reject the null hypothesis only when the statistic is larger than some critical (sampling distribution) value require an upper one-tailed test. Then the rejection of $H_0$ occurs only in the upper region of the sampling distribution. A lower one-tailed test results in rejection of the null hypothesis if the sample statistic falls in a lower region of the sampling distribution.

The alternative hypothesis helps structure the test relation, so it is important to determine a proper alternative. This will require careful reading of each question and assessment of the situation. We illustrate each of the three types of alternative hypothesis ($<$, $\neq$, and $>$) next. Section 9.6 further discusses one-tailed versus two-tailed tests.

## A Less-Than Alternative Hypothesis Test

Our first test uses a lower-tailed rejection region. In it, the decision maker hopes to show that a new assembly time is better (lower) than the existing standard. The alternative hypothesis is that the mean is less than some (null) value, which is symbolized as $\mu_0$.

## EXAMPLE 9.2

An industrial engineer at the National Airframe assembly plant has developed a new jig that he feels will shorten a standard 5-minute assembly time without causing any hardship on the employees. In a dry run using the new jig, a random sample of 36 units required an average of 4.85 minutes for assembly. The standard deviation of assembly times is known to be 0.50 minutes. Using the sample times, test whether the new jig has reduced the mean time to less than 5 minutes. Use the $\alpha = 0.025$ significance level.

**FIGURE 9.2**   Lower-Tailed Test of $H_0: \mu = 5.0$

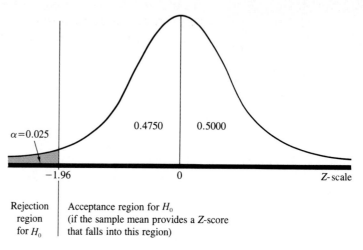

Rejection region for $H_0$ | Acceptance region for $H_0$ (if the sample mean provides a Z-score that falls into this region)

*Solution*   The solution uses the four steps for making a hypothesis test. The first step is key because it defines the question and sets the rejection region. The test is *lower-tailed* because the engineer is testing to see if these data display a *reduced* time. Thus the alternative hypothesis is stated as *less than*.

1. Does the new jig reduce mean time for assembly to less than 5 minutes?

$$H_0: \mu = 5.0 \text{ minutes} \qquad H_A: \mu < 5.0 \text{ minutes}$$

This requires a one-tailed (lower) rejection region, as shown in Figure 9.2.

2. $\alpha = 0.025$, $n = 36$. The Z-value associated with 0.025 $\alpha$-risk is 1.96. Since this is a lower, one-tailed test, the test criterion, or critical Z, is $-1.96$.

Test: Reject $H_0$ if calculated $Z < -1.96$.   (the test value)

The rejection region of the sampling distribution is shown as the shaded area in Figure 9.2.

3. Using the sample data and the central limit theorem, we convert the mean of the sample into a Z-score.

$$\text{Calculated } Z = \frac{\bar{X} - \mu_0}{\sigma_{\bar{X}}} = \frac{4.85 - 5.00}{0.5/\sqrt{36}} = \frac{-0.15}{(1/12)} = -1.80$$

4. Decision: We accept $H_0$ at the $\alpha = 0.025$ level of significance, because the sample calculated Z-value ($-1.80$) is more positive than the critical test value of $-1.96$. We conclude that the new jig does not reduce time to less than 5.0 minutes.

Since there is more than a 0.025 probability for obtaining a sample mean reading at or below 4.85 minutes, we attribute the difference from 5.0 minutes as being due to chance. In other words, it is a peculiarity of this sample. The evidence is not sufficient to conclude that (statistically) the new jig has reduced the mean time to below 5.0 minutes.

## A Greater-Than Alternative Hypothesis Test

This example employs an upper-tailed rejection region. As you might expect, a few key words again signal the type of test. One common situation stems from testing for an *increase*. Phrases such as *bigger than* and *more* also indicate an upper-tailed test.

### EXAMPLE 9.3

A manufacturer of premixed dog foods is considering offering bonuses to its sales representatives as an incentive for more sales. Forty-nine of the salespersons were randomly chosen to operate under a commission plus a bonus system for one year. Sales prior to the issue of bonuses averaged $15,500 per distributorship with a (known) standard deviation of $2940. During the trial period, sales to the same distributors averaged $16,470 (after adjustment for inflation in prices and for the cost of the bonuses). Based on the criterion of increased sales, is the increase in sales statistically significant for $\alpha = 0.01$?

*Solution*  For $\mu =$ the average sales to distributors, the sampling distribution takes the form shown in Figure 9.3.

1. Does the addition of a bonus result in higher sales?

$$H_0: \mu = \$15,500 \qquad H_A: \mu > \$15,500 \quad \text{one-tailed (upper) test}$$

2. $\alpha = 0.01$, $n = 49$

   Test: Reject $H_0$ if calculated $Z > 2.33$.

   The rejection region is identified as the shaded region in Figure 9.3.

3. The calculated value is

$$Z = \frac{\bar{X} - \mu_0}{\sigma/\sqrt{n}} = \frac{\$16,470 - \$15,500}{\$2940/\sqrt{49}} = 2.31$$

4. Decision: Do not reject $H_0$, because $2.31 < 2.33$.

   We conclude from this test that because the bonuses have not resulted in higher sales they are not worthwhile.

**FIGURE 9.3**  Upper-Tailed Test of $H_0: \mu = \$15,500$

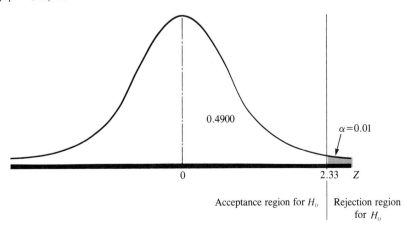

In Example 9.3, the $\alpha = 0.01$ is a rather stringent test. Had $\alpha$ been specified as 0.05, the decision would have been a clear rejection of the null hypothesis, for the test criterion would have been $Z > 1.645$. So the specification of $\alpha$ is important.

It is a fairly common practice to report research findings significant at the 0.10 or 0.05 or even the 0.01 level. To avoid any criticism of making the data fit the result, *it is* also *important to set the significance level ($\alpha$) for a hypothesis test before any data are analyzed.* In doing this, be sure to set a realistic $\alpha$-level.

### The Does-Not-Equal Alternative Hypothesis Test

Another example uses a two-tailed test, which divides the rejection region into two equal parts. This approach is appropriate when the question concerns equal versus unequal results.

**EXAMPLE 9.4**

Golf balls produced in a New Jersey plant are supposed to average 1.750 inches in diameter. A random sample of 900 golf balls from the production of a single machine gives $\bar{X} = 1.754$ inches. Let $\sigma = 0.050$ inches. Is the machine within acceptable adjustment? Use $\alpha = 0.05$.

*Solution*    Here $\mu$ = the mean diameter of golf balls made by this machine. As this question illustrates, one must often use logical considerations about the problem to define the test. Logically, golf balls either too big or too small would be an indication that the machine is outside the limits of proper adjustment. The alternative hypothesis is two-tailed.

1. $H_0: \mu = 1.750 \qquad H_A: \mu \neq 1.750$
2. $\alpha = 0.05, n = 900$

Test: Reject $H_0$ if $|\text{calculated } Z| > 1.96$.

The rejection region of the sampling distribution is shown as the shaded area in Figure 9.4.

**FIGURE 9.4**   A Two-Tailed Test of $H_0: \mu = 1.750$

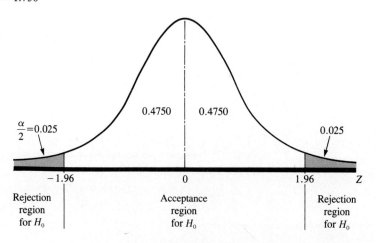

3. The calculated value is

$$Z = \frac{\bar{X} - \mu_0}{\sigma/\sqrt{n}} = \frac{1.754 - 1.750}{0.050/\sqrt{900}} = \frac{0.004}{0.050/30} = 2.40$$

4. Decision: Reject $H_0$ at $\alpha = 0.05$ significance level, since the $Z$-value of 2.40 is greater than 1.96.

The machine needs adjustment; the mean is not 1.750 inches.

Because the calculated $Z$-value is $> +1.96$, the diameters appear to be too large. Whether this is a production or marketing problem depends upon the practical significance of the engineering specifications of the product. A diameter averaging 0.004 inches larger than the specifications may well be within an acceptable quality control tolerance even though it is a statistically significant difference from the specifications.

In many cases of exploratory research the test alternative is *does not equal*. If the null hypothesis is rejected, a more discriminating test—that is, one-tailed—is sometimes made. New sample data should be used in each succeeding test. Thus, for this last situation, another sample might be drawn and the revised hypothesis $H_A$: $\mu > 1.75$ inches tested.

Each example in this section illustrated a different type of alternative hypothesis. The assembly time example illustrated the less-than (lower-tailed) alternative test. The "bonus" sales commission example used an upper-tailed (greater-than) alternative test. The does-not-equal alternative was depicted in the example about production of golf balls and is called a two-tailed test because the null hypothesis is rejected if the sample $Z$-value falls in either tail.

In each of these three examples the *claim of change is expressed in the alternative hypothesis*. Also, the alternative hypothesis identifies the rejection region as being either the left or lower tail ($<$); the right or upper tail ($>$); or both tails of the distribution ($\neq$). The distinction of a one-tailed versus a two-tailed test is a matter of the business hypothesis being tested. If one's experience allows a directional test, then a one-tailed test is appropriate. If there is no basis for suspecting a higher or lower value, then a two-tailed alternative is appropriate.

The end result of a hypothesis test is a process of deductive elimination of possibilities. Hypothesis testing does not prove a result or justify a parameter value. If the null hypothesis is rejected, then one can only conclude (and state) that the test discredits the null value. The random sample supports other values as more likely. When the null hypothesis is not rejected, then the sample supports the null value. Nonzero $\alpha$- and $\beta$-risks remind us that there is always some chance that either decision can be wrong.

**The Relation of Hypothesis Testing and Confidence Interval Estimation**

The preceding example affords us an opportunity to note an important relationship between hypothesis testing, as just described above, and confidence interval estimation, as discussed in Chapter 8. In Chapter 8 we took careful note of the fact that estimates were centered around sample means, $\bar{X} \pm Z\sigma_{\bar{X}}$. The area inside the normal curve between $+Z\sigma_{\bar{X}}$ and $-Z\sigma_{\bar{X}}$ was referred to as the *level of confidence*.

**FIGURE 9.5**   Structural Differences Between Hypothesis Testing and Confidence
Interval Estimation

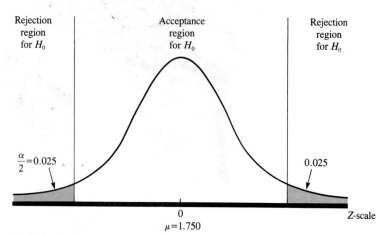

(a)  Testing the hypothesis $H_o$ : $\mu = 1.750$ at 0.05 level of significance

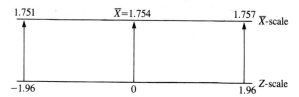

(b)  95% confidence interval estimation

Testing, on the other hand, is concerned with statistical evidence as to whether or not a population parameter has a certain hypothesized value. Tests are made of parameters, and the region of acceptance surrounds the *hypothesized population value—not a sample value*. Also, in testing, the area under the normal curve *outside* the acceptance region is referred to as the *level of significance* of the test. This area, which is commonly 1%, 2.5%, 5% or 10%, is generalized as $\alpha$.

Figure 9.5 illustrates the relationship between estimation and testing in terms of Example 9.4. Since $H_0$: $\mu = 1.750$ inches is rejected, we no longer have a valid estimate of the population mean. It is essentially unknown, but we can (if we want to) make a new estimate of it. This, of course, is the job of confidence interval estimation. A 95% confidence interval estimate would be $1.754 \pm 1.96$ $(0.050/\sqrt{900})$, or 1.751 to 1.757 inches.

It is important to distinguish the terms associated with testing from those of estimation. Remember, we express *confidence* in our estimation procedure, but tests are conducted at a specified *level of significance*. Confidence levels are typically in the range of 90% to 99.7%, whereas levels of significance are usually set at 1% to 5%, or possibly 10%.

## t-Tests of Means for Normal Distributions with Unknown Standard Deviation

Having looked briefly at the form for testing the mean with $\sigma$ known and at the relation between confidence interval estimation and hypothesis testing, we turn next to hypothesis testing when $\sigma$ is not known. The procedures for testing introduced earlier in Chapter 9 remain the foundation for hypothesis tests of the mean when the population standard deviation is unknown. However, for inference for small samples ($n \leqslant 30$) to be valid, the random variable must be normally distributed. The sampling distribution is the $t$-statistic that was described in Equation (8.2).

The Student's $t$-statistic is

$$t = \frac{\bar{X} - \mu}{s/\sqrt{n}} \tag{9.2}$$

Assumptions: The random variable $X$ is normally distributed.
The population standard deviation is not known.

---

**EXAMPLE 9.5**

The manager of a major airlines' customer courtesy unit knows that the average tension on the release for overhead storage shelves on the airplane should be 5.0 pounds or more. He does not wish to make any adjustments unless the tension is significantly less than 5 pounds. He randomly inspects 25 shelf releases and finds that the tension averages 4.7 pounds with a standard deviation of 0.2 pounds. Can he conclude that the average tension (in the population of overhead shelf releases) is less than 5.0 pounds? Use $\alpha = 0.025$ and assume the individual measures of tension are normally distributed.

*Solution*

The supervisor seeks a course of action that will avoid any discomfort to passengers. The parameter is $\mu =$ the mean tension on the shelf releases. Is the average tension less than 5 pounds?

1. $H_0$: $\mu = 5.0$ pounds      $H_A$: $\mu < 5.0$ pounds      $\alpha = 0.025, n = 25$
2. Test: Reject $H_0$ if calculated $t < t_{0.025,24} = -2.06$
3. Calculated $t = \dfrac{\bar{X} - \mu}{s/\sqrt{n}} = \dfrac{4.7 - 5.0}{0.2/\sqrt{25}} = -7.5$
4. Decision: Reject $H_0$ at $\alpha = 0.025$, since $-7.5 < -2.06$.

This sample indicates that the average tension is clearly less than 5.0 pounds. The tension should be increased.

---

Note that in Example 9.5 we used the same procedures as the large-sample test, only the test statistic was changed from a $Z$- to a $t$-statistic with $n - 1$ df and the test statistic value is read from Table B.2. This example was a one-tailed test, so all the rejection region was on one side of the distribution.

The next example, in addition to displaying a $t$-test, illustrates an equivalent way of computing the sample standard deviation. Bear in mind that this is just another alternative; the earlier procedure would work too.

---

**EXAMPLE 9.6**

One measure of the need for preventive maintenance is the work stoppages with which managers must contend. The following are downtimes (work stoppages in hours) for a brewery during nine randomly selected months: 18, 5, 9, 10, 13, 7, 2, 11, and 6. Assuming that downtime hours per month are normally distributed, test the claim that this plant has been nonoperable an average of less than 10 hours per month. Use $\alpha = 0.025$.

*Solution*

The sampling distribution for this test is shown in Figure 9.6. The parameter being tested is $\mu =$ the true mean downtime per month.

$H_0$: $\mu = 10$ hours per month      $H_A$: $\mu < 10$ hours per month

$\alpha = 0.025$   (shaded in Figure 9.6)

**FIGURE 9.6**   Distribution for Downtimes, in Hours, at a Brewery

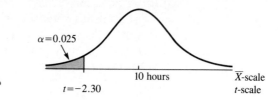

$\alpha = 0.025$

10 hours    $\overline{X}$-scale

$t = -2.30$    $t$-scale

Test: Reject $H_0$ if calculated $t < t_{0.025,8} = -2.31$.

$$\overline{X} = \frac{18 + 5 + \cdots + 6}{9} = \frac{81}{9} = 9$$

$$s^2 = \frac{n \sum (X)^2 - (\sum X)^2}{n(n-1)} = \frac{9(18^2 + 5^2 + \cdots + 6^2) - (81)^2}{(9)(8)} = 22.5$$

Therefore,

$$s = \sqrt{22.5} = 4.74$$

$$\text{Calculated } t = \frac{\overline{X} - \mu}{s/\sqrt{n}} = \frac{9 - 10}{4.74/\sqrt{9}} = \frac{-1}{1.58} = -0.63.$$

The value $-0.63$ is to the right of the test value $(-2.31)$ shown in Figure 9.6. Accept $H_0$. Since calculated $t = -0.63 >$ test $t = -2.31$, the mean downtime is *not* less than 10 hours per month.

---

The computational formula used for $s^2$ gives results equivalent to the definitional form used earlier. Also recall from Chapter 7 that whenever the sample constitutes 5% or more of the population, the finite population correction factor $\sqrt{(N-n)/(N-1)}$ should be applied to the calculation of the standard error $s_{\overline{x}}$, just as it is to $\sigma_{\overline{x}}$. (We assume that the brewery in Example 9.6 has been in operation for many months, so no correction was made.)

In conclusion, inference concerning a population mean is much the same with small samples coming from a normal distribution with unknown $\sigma$ as it is with large samples, except that a $t$-statistic is used. A summary of the formulas for confidence interval estimation and hypothesis testing appears in Table 9.4 at the end of the chapter.

**Problems 9.3**

7. A behavioral scientist has designed an instrument to measure the tolerance score of persons engaged in stressful activities. By the end of a labor-management negotiation session involving 36 persons, the average tolerance score, as measured on a standardized test, had decreased by 12 points. Is the claim justified that this experience will decrease the tolerance level by more than 10 points? Use $\alpha = 0.05$ and $\sigma = 2$. Find each of the following items.
   a. $H_0$ and $H_A$                    c. Value of calculated statistic
   b. Test procedure and test value      d. Decision and conclusion

$$z = \frac{\bar{x} - \mu_0}{\sigma/\sqrt{n}}$$

8. Candies manufactured by Cherrie-Chocolate Candies Co. are supposed to weigh at least 2.00 ounces, and they have a standard deviation of 0.10 ounce. If a sample of size 100 has an average weight of 1.98 ounces, can we conclude that the average weight is less than 2.00 ounces? Let $\alpha = 0.10$. Find each of the following items.
   a. $H_0$ and $H_A$
   b. Rejection region
   c. Value of calculated statistic
   d. Decision and conclusion

9. The mean valuation for single-family homes in a large city is assessed (by the county assessor) to be $88,950. A city council member has questioned whether the valuation for homes is realistic, so a random sample of 36 homes is taken. The sample mean is $92,460 and standard deviation is $5200. Does the sample data support the assessor's value? Use $\alpha = 0.05$ and complete parts (a)–(d) as in Problems 7 and 8.

10. The Longer Life Lightbulb Co. (3L Co.) says its bulbs last 2000 hours or more. The standard deviation is 200 hours. A government agency wishes to test the 3L Co. statement against the alternative hypothesis that the mean life is less than 2000 hours for $\alpha = 2\frac{1}{2}\%$ level. Suppose 100 bulbs are tested and the mean life is only 1975 hours. State all the steps of the test, including the appropriate conclusion.

11. A government economist is interested in whether food costs would be substantially reduced if families would raise their own gardens. In a planned test, a random selection of 49 gardening families, revealed that each spent an average of $300 per month for food (including the cost of planting and caring for the garden). Is this expenditure statistically less than expenditures for similar-sized families who did not raise gardens if their monthly food-bill average is $320 with population standard deviation $50? Use $\alpha = 0.05$.

12. The manager of the Aerospace Insurance Company cafeteria wants to prepare for, on the average, about the number of dinners that will be served. From past experience she thinks that about 525 dinners will generally be needed. If a random sample of 36 days gives a mean of 519, does the sample evidence agree with the manager's experience or is it significantly different? Use $\alpha = 0.05$ and let $\sigma = 24$.

13. Assuming a one-tailed (upper) test of hypothesis of the mean, use the $t$-table to find the test $t$-value for
   a. $n = 24$, $\alpha = 0.05$
   b. $n = 15$, $\alpha = 0.05$
   c. $n = 5$, $\alpha = 0.05$
   d. $n = 25$, $\alpha = 0.01$
   e. $n = 25$, $\alpha = 0.05$
   f. $n = 25$, $\alpha = 0.10$

14. Answer parts (a)–(e) of Problem 13 for a two-tailed test of hypothesis, assuming normal distributions with unknown standard deviation.

15. Answer each of the following questions concerning the use of the $t$-table:
   a. Find the tabular $t$-value for a lower-tailed test of $H_0$: $\mu = 10$ based on $n = 22$, with $\alpha = 0.05$.
   b. For a two-tailed $t$-test, $H_0$: $\mu = -4$, based on $n = 11$ and with $\alpha = 0.10$, determine tabular $t$.
   c. An upper-tailed test of a mean used $n = 17$ normally distributed observations and test value $t = +2.58$. What is the $\alpha$-risk level?

16. A food-packaging machine is supposed to fill packages to 16.0 ounces with a standard deviation of 0.10 ounces. A sample of $n = 25$ packages revealed a mean weight of $\bar{X} = 16.06$ ounces. Test the hypothesis that the machine is properly adjusted at the 5% level. Assume that the weights follow a normal distribution.

17. A manufacturer of elevator cables claims its product has a tensile strength of 4000 pounds. A test of the strength of ten of those cables produced a mean of 3910 pounds and standard deviation of 90 pounds. Test the manufacturer's claim using $\alpha = 0.01$. Make a one-tailed test and assume a normal distribution.

18. A manufacturer of personal computers claims that its computers have an average defect-free life of 3 years or more. A random check of three computers produced failure times of 2.4, 2.9, and 2.8 years, giving $\bar{X} = 2.7$ and $s = 0.26$ years.
    a. Is this evidence sufficient to contradict the manufacturer's claim? Use $\alpha = 0.05$ and assume a normal distribution of lifetimes.
    b. Does your conclusion agree with your intuition? If not, explain what further testing you might do before making a final conclusion. Consider, but do not evaluate, the $\beta$-risk potential.

## 9.4 TESTS OF PROPORTIONS

Tests of proportions follow the same pattern developed earlier for testing means. Recall that the statistic used is the $Z$-statistic.

> The generalized sampling distribution for testing hypotheses of means or of proportions for large samples is
>
> $$Z = \frac{\text{Statistic} - \text{Parameter}}{\text{Standard error}} \tag{9.3}$$

### The Sampling Distribution

With proportions, $\pi$, $Z$-tests are restricted to the use of sufficiently large samples such that $np \geqslant 5$ and $nq \geqslant 5$. Then, for binomially distributed populations, the sampling distribution is approximately normally distributed.

> For $np \geqslant 5$, $nq \geqslant 5$, the sampling distribution for testing hypotheses about the *proportion* of successes is
>
> $$Z = \frac{p - \pi}{\sigma_p} \tag{9.4}$$
>
> or, for the *number* of successes, $X$, it is
>
> $$Z = \frac{np - n\pi}{\sigma_{np}} \tag{9.5}$$

The standard error is expressed in terms of proportions, $\sigma_p$, or numbers, $\sigma_{np}$, depending upon whether the test concerns the proportion of successes or the number of successes, respectively. Since tests of hypotheses are structured in a way that favors the null hypothesis (that is, requires significant evidence in order to reject the null hypothesis), we extend this rationale to the computation of the standard error. That is, *we use hypothesized values of $\pi$ rather than sample values of p to compute the standard error.*

The standard error for the sampling distribution of *proportions* is

$$\sigma_p = \sqrt{\frac{\pi(1-\pi)}{n}} \tag{9.6}$$

or, for the *number of successes*, is

$$\sigma_{np} = \sqrt{n(\pi)(1-\pi)} \tag{9.7}$$

The parameter value (for $\pi$) is that specified in the null hypothesis. For example, for $H_0: \pi = 0.4$, the value 0.4 replaces $\pi$ throughout the Z-form.

We reemphasize that the value used to calculate the standard error in Equations (9.6) and (9.7) differs from the standard error calculation used for estimation in Chapter 8. In hypothesis tests we assume that the null value is the most reasonable value for $\pi$. So we use the null value for $\pi$ to compute $\sigma_p$ in (9.6) or $\sigma_{np}$ in (9.7). In estimation, we used the sample value, $p$, for the unknown value of $\pi$. This procedural difference is due to the difference in the philosophy of hypothesis testing and interval estimation. Remember, confidence intervals start with sample statistics (hence the use of $p$), whereas hypothesis tests begin with the hypothesized population value (hence the use of $\pi$).

## Tests of Proportions

The structure for tests of proportions is the same as for tests of means. The major difference is the form of the Z-statistic. This section contains no new theory, but we give several examples that illustrate tests involving the true proportion of success, $\pi$, in binomial situations. Common applications include tests concerning voter opinion, percentage of defective parts on a production line, and consumer preferences concerning marketable goods.

---

**EXAMPLE 9.7**

Security Windows Corp. has experienced an 8% defective rate on its safety lock settings. Each day a random sample of 100 windows is selected for quality control inspection. Suppose a sample with 15% defective lock settings is obtained. Should production be stopped to adjust the manufacturing process? Test at $\alpha = 0.02$.

*Solution*    Let $\pi =$ the true percentage of defective safety lock settings being used. Because we are concerned with the possibility of too many defective settings, the alternative hypothesis is a greater-than statement. The sampling distribution is shown in Figure 9.7. The test rejection region appears as the shaded area.

1. $H_0: \pi = 0.08$    $H_A: \pi > 0.08$
2. $\alpha = 0.02, n = 100$

   Test: Reject $H_0$ if calculated value is $Z > 2.05$.

3. Calculated

$$Z = \frac{p - \pi_0}{\sigma_p} = \frac{0.15 - 0.08}{\sqrt{\frac{(0.08)(0.92)}{100}}} = 2.58$$

where $\pi_0$ denotes the value specified by the null hypothesis, $H_0: \pi = \pi_0 (= 0.08)$.

$CI = P \pm z \left( \sigma_p^{\wedge} \right)$

**FIGURE 9.7**   One-Tailed Test of Proportions at $\alpha = 0.02$ Level

$CI = .15 \pm 2.05 \left( .0271 \right)$

$.15 \pm .0556$

$.0944 \angle \text{of} \angle .2056$

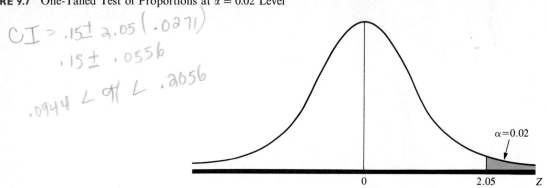

$\alpha = 0.02$

0          2.05          Z

**4.** Decision: Reject $H_0$ at the 0.02 $\alpha$-level, because $2.58 > 2.05$.

The percentage of defective safety lock settings has increased significantly. Based on this criterion, the manufacturing process should be stopped and adjustments made.

Example 9.8 illustrates an evaluation on a matter of common interest at most university campuses: student opinions.

**EXAMPLE 9.8**

In a sample survey conducted at a large private university, 400 randomly selected students were asked if they were in favor of having quiet hours in the dorms. There were 186 students favoring quiet hours. Is there sufficient evidence to indicate that less than half of the students on this campus favor quiet hours in the dorms? Test $H_0: \pi = 0.50$ against $H_A: \pi < 0.50$.

*Solution*   Since the $\alpha$-risk level is not specified, we are at liberty to make our own choice, say, $\alpha = 0.05$. Let $\pi =$ the percentage of students at this university who favor quiet hours. We hypothesize that $\pi = 0.50$, so that the sample result must be significantly less than 50% before we accept the alternative that only a minority favor quiet hours.

**1.** Do less than half the students favor quiet hours?

$$H_0: \pi = 0.50 \qquad H_A: \pi < 0.50$$

**2.** $\alpha = 0.05$, $n = 400$, $p = \dfrac{186}{400} = 0.465$, $\sigma_p = \sqrt{\dfrac{0.5(0.5)}{400}} = 0.025$

Test: Reject $H_0$ if the calculated $Z$-value is $< -1.64$.

**3.** The calculated value is

$$Z = \frac{p - \pi_0}{\sigma_p} = \frac{0.465 - 0.500}{0.025} = -1.40$$

**4.** Decision: We cannot reject $H_0$ at $\alpha = 0.05$, because $-1.40 > -1.64$.

Although a minority in this sample favor quiet hours (that is, 46.5%), we cannot conclude that the percentage of students at this university who favor quiet hours is under 50%. The sample evidence does not discredit the null hypothesis at the $\alpha = 0.05$ level. We conclude that student opinion is about equally divided on this issue; opinion is not sufficiently one-sided to reveal any concensus.

*Note:* This question could have been solved via number (counts) rather than proportions. With that approach, the mean number would be hypothesized as $\mu = n\pi_0 = 400 \cdot 0.50 = 200$ with standard error

$$\sigma_{np} = \sqrt{n\pi(1 - \pi)} = \sqrt{400(0.50 \cdot 0.50)} = 10$$

Then the calculated value is

$$Z = \frac{np - n\pi_0}{\sigma_{np}} = \frac{186 - 200}{10} = -1.40$$

This solution is identical to the calculation based on the proportion of successes.

As with means, a confidence interval estimate could be made to get a better feeling for the true value of the proportion of students who favor quiet hours. For this we use a 98% confidence interval. It is of the form $(p \pm Z \cdot \hat{\sigma}_p)$. Recall again that confidence intervals rely on sample estimates (rather than hypothesized values), so the estimated standard error of proportion would be

$$\hat{\sigma}_p = \sqrt{\frac{pq}{n}} = \sqrt{\frac{(0.465)(0.535)}{400}} = 0.025$$

A 98% confidence interval then becomes

$$(\text{LCL, UCL}) = 0.465 \pm [2.33(0.025)] = (0.41, 0.52)$$

so $0.41 < \pi < 0.52$. This particular interval estimates the true proportion, $\pi$, of students who favor quiet hours is between 0.41 and 0.52. Opinion is split, and since the interval includes values less than 0.50 as well as some above 0.50, there is no general consensus for or against the issue. Remember, this interval used only one of many possible samples, and some of the sample intervals may lie above the population value and some below. Yet 98% of all sample intervals computed in this way will contain the true value. In assuming this is a good interval, we have only a 2% chance that our conclusion is wrong.

The next example illustrates that you should always read a problem carefully to identify the question, that is, the test (alternative) hypothesis. In this case the phrase *significantly greater than* is the key to the alternative statement.

**EXAMPLE 9.9**

The market research section of a national distributor wants to determine whether college students can distinguish their cola from other brands just by its taste. An experiment was designed in which 200 randomly chosen college students were each given four small glasses of cola, only one of which contained their favorite brand. After tasting all four drinks, 64 correctly identified their brand. Are the results significantly greater than would be expected by chance? Use $\alpha = 0.025$.

*Solution*    In deciding what hypothesis to test, we begin by assuming equal chance of selection for each of the four glasses. That is, if an individual cannot actually identify his or her cola by taste, the chance is 25% for identifying it by guessing, so $H_0$: $\pi = 0.25$. An ability to discern the drink by taste from among the different brands can be inferred if statistically more than 25% (some number above 25 per 100 tasters) of the students identify their favorite brand. Then, for $\pi =$ the true percentage of college students who can identify their favorite brand from the other three by its taste, we have $p = 64/200 = 0.32$.

1. $H_0$: $\pi = 0.25$      $H_A$: $\pi > 0.25$
2. $\alpha = 0.025$, $n = 200$

   Test: Reject $H_0$ if the calculated value is $Z > 1.96$ (since this is a one-tailed test).

3. The calculated value is

$$Z = \frac{p - \pi_0}{\sigma_p} = \frac{0.32 - 0.25}{\sqrt{\dfrac{(0.25)(0.75)}{200}}} = \frac{0.07}{0.03} = 2.33$$

   where $\pi_0 = 0.25$ is the value specified for $\pi$ by the null hypothesis.
4. Decision: Reject $H_0$ at $\alpha = 0.025$ since $2.33 > 1.96$.

This test supports the conclusion that college students, in general, have some ability to identify their favorite brands of cola by taste alone. It suggests that more than 25% of all college students have the ability to identify their favorite brand in competition with the three other brands. This finding is significant at the 0.025 level (that is, it would occur by chance less than or equal to $2\frac{1}{2}\%$ of the time), so it indicates that college students can distinguish their favorite brand by taste alone.

---

**Problems 9.4**

19. During a study of employee morale, a random sample of 900 employees were given a test on attitude toward working overtime without pay. The test showed 576 were against such overtime work. Test the hypothesis that 65% of the employees are against working overtime without pay. Use $\alpha = 0.04$.

20. A random sample of 100 men and 100 women voters in a city with 10,000 registered voters showed 73 men and 69 women favored a business and occupations (B&O) tax. Test the hypothesis that 70% of the registered voters in this community favor a B&O tax. Use $\alpha = 0.075$ and assume the registered voters are 50% men and 50% women.

21. A pharmaceutical firm claims to have developed a medicine that will, with 80% effectiveness, reduce the stress on office workers. The medicine was used to treat a random sample of 400 office workers who had a threshold level of stress. The stress was reduced in 310 of the subjects. Test the pharmaceutical company's claim at $\alpha = 0.075$.

22. A supermaket will change its policy from giving trading stamps to having weekly drawings for free groceries if the change is favored by a majority of its customers. In a survey of 200 randomly selected customers, 110 indicated they were in favor of the change. What should the supermarket do?

State the statistical hypotheses; then show your calculations and the critical value. Test at the significance level of 0.05 and state practical conclusions.

23. In a sample survey conducted at a National Inventor's Convention, 100 randomly selected patrons were asked if they favored a change from the English to the metric system. Fifty-seven answered yes and 43 answered no. Is there sufficient evidence to conclude that a majority of the patrons were in favor of such a change? State the hypothesis to be tested, test at the 0.05 level of significance, show calculations, and state your conclusions.

24. The management of San Lando Kennels Racetrack is considering a Ladies' Day event with free admission every Monday for women aged 18 or older. If, for a random selection of 220 admissions, 60 are women aged 18 or older, test the hypothesis that over 25% of Ladies' Day admissions would be free. Use $\alpha = 0.025$. (*Note:* We will assume that Ladies' Day is similar to any other day in the proportion of women in attendance aged 18 or older.)

*see notes*

*.77766*

## 9.5 TESTING CONSIDERATIONS

Because hypothesis testing can be very useful for making decisions, a good conceptual understanding is important. Our earlier discussion introduced the procedures for making hypothesis tests. This section and the next will provide additional processes for testing. In this section we examine alternative ways to perform tests, including (1) *p*- or probability values, (2) tests based on preset $\alpha$- and $\beta$-risk levels, and (3) the use of operating characteristic and power curves. Section 9.6 gives more details on the meaning of hypothesis tests.

### *p*-Values: An Alternative Test Structure

Our approach to hypothesis testing thus far has been a decision-based procedure. At the beginning of Section 9.2 we outlined four steps for this procedure. After formulating the null and alternative hypotheses, the next step is to establish a procedure for testing the null hypothesis and a criterion for deciding whether to accept or to reject it. This entails selecting an $\alpha$-value, drawing a sample, and then performing the test. The decision is based on whether the sample statistic value is greater than the (preset $\alpha$) test criterion. If so, the sample results cast doubt on the null hypothesis, and so it is rejected.

The *p*-value approach provides an alternative means for making the same decisions. The *p-value* is the smallest value of $\alpha$ for which a test result is considered statistically significant.

> The smallest level of significance at which the null hypothesis can be rejected is called the *p*-value.

Here is an example that illustrates testing with *p*-values.

---

### EXAMPLE 9.10

The market research manager of the cola manufacturer in Example 9.9 asks, "Can we provide a stronger test than was initially made?" (That is, an $\alpha$-risk of 0.025.) The question is whether more than one in four college students could identify their favorite cola by taste alone. Two hundred randomly chosen students were involved in the test.

*Solution*    The null and alternative hypotheses are as before:

$$H_0: \pi = 0.25 \qquad H_A: \pi > 0.25$$

**FIGURE 9.8**    *p*-Value for a Test of Hypothesis

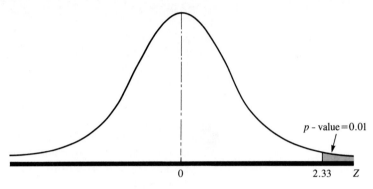

From the sample data, the calculated $Z$-value was

$$Z = \frac{0.32 - 0.25}{0.03} = 2.33 \qquad \textbf{(from Example 9.9)}$$

Referring to Table 1 in Appendix B, we find that the probability corresponding to a $Z$-value of 2.33 is 0.4901. Therefore, the lowest value for the significance is $0.5000 - 0.4901 = 0.0099 = 0.01$. That is, the $p$-value for this test is 0.01. This is shown as the shaded area in Figure 9.8.

The null hypothesis would be rejected for a test value of 0.01 or any larger value, including 0.025, 0.05, 0.10, or higher.

The $p$-value represents the probability of observing a sample outcome more contradictory to the null hypothesis than the observed sample. Thus the smaller the probability, or $p$-value, the greater the weight of evidence for rejecting the null hypothesis.

Observe that the $p$-value is a number derived from the test sample data. It is the likelihood of obtaining a sample result if the null hypothesis is true. The $p$-value (0.01) in Example 9.10 is quite small. So this casts much doubt on the null hypothesis that $\pi = 0.25$.

The proponents of the $p$-value approach favor its flexibility. Whereas the traditional approach forces a decision (either to accept or to reject the null hypothesis), the $p$-value procedure gives the manager an objective measure of the weight of evidence for rejecting the null hypothesis. If, for example, the manager in Example 9.10 requires a $p$-value less than 0.05, 0.025, or even 0.01, this would result in rejecting $\pi = 0.25$ in favor of $\pi > 0.25$. The traditional data analyst will counter such an argument that $\alpha$ should be specified before, not after, the hypothesis test is conducted.

In some research there may be good reason for deriving a $p$-value from the sample data and allowing others to reach their own conclusions. On the other hand, this procedure can permit a decision maker to set an $\alpha$-level that simply justifies a prior decision. This tendency toward manipulation of data to suit prior conclusions can be self-serving and counterproductive, so $p$-values should be used with care. Nevertheless, the $p$-value approach to hypothesis testing is a valid alternative to traditional level-of-significance testing, and both procedures are used in business research literature. Many computer packages for testing hypotheses provide *p*-values (without their being specifically requested.)

**α- and β-Risk Considerations: Determining β-Risk**

Although we have given evaluation of α- and β-risks only limited discussion thus far, they are important considerations. For example, one might think it desirable to use a low α-value, say, in the order of 0.01, to reduce the chance of rejecting a correct (null) hypothesis. This could be false security, for the α- and β-risks have an inverse relation: For a given sample size, as α decreases, the β-risk increases and in a disproportionately quick manner. That is, for a given sample size, if the risk of rejecting a true hypothesis (α) is reduced, then the risk of accepting a false hypothesis (β) is increased. Taking a larger sample is one sure way to reduce both risks at the same time. However, since many situations can justify only a limited sample, you should be fully aware of both the α- and β-risk levels in order to gain the best protection against both possible wrong conclusions. The relationship between α- and β-risk is illustrated in Figure 9.9.

**FIGURE 9.9**   Illustration of the β-Risk for Various Alternative Values of the Mean, $\mu_A$, with $\beta$ = Shaded Area as Part of Total Area Beneath the Curve

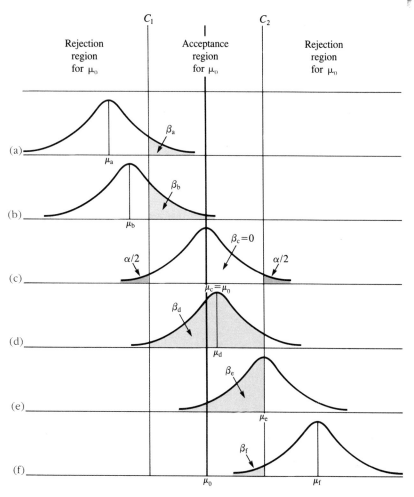

Figure 9.9(a) displays the probability of a type 2 error, the $\beta$-risk, as the shaded area labeled $\beta_a$. The $\beta$-risk is the area in the acceptance region. This is the shaded area in the center frame of Figure 9.9(a); it is a small area relative to the total area beneath this normal curve. That is, when the alternative value of $\mu_A$ is a far distance from $\mu_0$, the $\beta$-risk is a small value.

Observe that in going from Figure 9.9(a) to 9.9(b) and then to 9.9(d) the acceptance region area and $\beta$-risk increase. But as $\mu_A$ is further away from $\mu_0$ (Figure 9.9(e)) and still further away in Figure 9.9(f), then the $\beta$-risk diminishes. Simply, Figure 9.9 shows that as $\mu_A$ approaches the value of $\mu_0$, it becomes increasingly difficult to decide which $\mu_0$ or $\mu_A$ is the correct value. So, for a fixed $\alpha$-level test, the $\beta$-risk increases as $\mu_A$ is closer to $\mu_0$ and decreases as $\mu_A$ moves further away from $\mu_0$.

Figure 9.9(c) is unique for this set because it depicts the possibility that $\mu_A = \mu_0$. Then the $\beta$-risk is zero, because if $\mu_0 = \mu_c$, then $\mu_c$ is the true mean and $\beta_c = P(\text{Accept } H_0 | \mu = \mu_c) = 0$. We qualify this point because it relates to a break point at the value $\mu_0$ in the probability curve for $\beta$-values. This is demonstrated by the fact that no (zero) area is shaded in the acceptance region for $H_0$ in Figure 9.9(c). Figure 9.9(c) also displays the $\alpha$-risk identification and shows that this situation represents a two-tailed test with $\alpha$-(level) risk of rejecting the null hypothesis. Incidentally, although Figure 9.9 displays a test of means, the same structure is used for all normal distribution tests of proportions—for differences in means, and more.

There are many possible alternative values in evaluating the $\beta$-risk; for example, with the alternative hypothesis $H_A: \mu > 10$ there are infinitely many values greater than 10. However, only one value is used for any $\beta$-value calculation. You might have noticed the critical limits—$C_1$ and $C_2$—denoted in Figure 9.9. We find the numerical value for the $\beta$-risk (chance of a type 2 error) at any one value simply by computing the ratio of the shaded area between $C_1$ and $C_2$ relative to the total area beneath the normal curve. Here $C$—that is, as $C_1$ and $C_2$—represents the *critical limits* of $Z$ standard errors distance from $\mu_0$. Now we can define the calculation of *critical limits*.

Critical limits are the values corresponding to the specified $\alpha$-levels at (and beyond) which the null hypothesis can be rejected. They are

|  | Lower-tailed test | Two-tailed test | Upper-tailed test |  |
|---|---|---|---|---|
| Means | $C_1 = \mu_0 - Z_\alpha \sigma_{\bar{x}}$ | $C_1 = \mu_0 - Z_{\alpha/2} \sigma_{\bar{x}}$ $C_2 = \mu_0 + Z_{\alpha/2} \sigma_{\bar{x}}$ | $C_1 = \mu_0 + Z_\alpha \sigma_{\bar{x}}$ | (9.8) |
| Proportions | $C_1 = \pi_0 - Z_\alpha \sigma_p$ | $C_1 = \pi_0 - Z_{\alpha/2} \sigma_p$ $C_2 = \pi_0 + Z_{\alpha/2} \sigma_p$ | $C_1 = \pi_0 + Z_\alpha \sigma_p$ | (9.9) |

The usual test procedure is to fix $\alpha$ at a low value. Then the size of the $\beta$-risk relates to (1) the distance of the critical values from the null value and (2) the sample size. We illustrate the determination of the $\beta$-risk for a single alternative hypothesized value in the test of a population mean.

**EXAMPLE 9.11**

The advertising manager of a large television station has used 10 minutes per day per household as the average rate that people spend watching television ads. He wants to know whether people are spending less time watching ads than previously. This could adversely affect the station's major source of revenues, which is from spot ads. A recent monitor by time meters of 100 randomly selected households showed a mean of 9.5 minutes per day. The population standard deviation is 4.0 minutes. Let $\alpha = 0.025$.

a. Identify (in words) the question and the type 1 and type 2 errors.
b. Compute the critical limit, $C_1$, for a lower-tailed test of the hypothesis $H_0$: $\mu = 10$, $H_A$: $\mu < 10$.
c. Compute the value of the type 2 error if the true mean rate were 9.5 minutes per day. (Notice that this is one value in the range of the alternative hypothesis of $\mu < 10$.)

*Solution*

a. The question is, Are viewers spending less than 10 minutes per day watching television ads?

A type 1 error would be to reject the null hypothesis when it is true and to conclude that people watch fewer ads, although they really watch 10 minutes or more of ads per day.

A type 2 error would be to accept the null hypothesis when it is false and to conclude that people still watch at least 10 minutes per day of ads, although they really watch less.

b. The critical limit is determined from Equation (9.8) as

$$C_1 = \mu_0 - Z\sigma_{\bar{x}} \qquad \text{where} \quad Z_\alpha = 1.96$$
$$= 10.0 - 1.96(0.4)$$
$$= 9.2 \text{ minutes per day} \qquad \sigma_{\bar{x}} = \frac{\sigma}{\sqrt{n}} = \frac{4.0}{\sqrt{100}} = 0.4$$

See Figure 9.10.

**FIGURE 9.10** Type 2 Error for $H_0$: $\mu = 10$ When the True Mean Is 9.5 Minutes

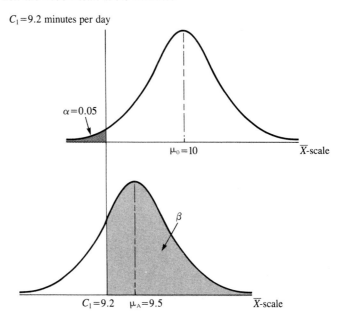

c. A type 2 error is committed if $H_0$: $\mu = 10$ is accepted and the true mean is really less than 10. For example, suppose the true mean is 9.5 minutes. Then a large percentage of the sample distribution lies above 9.2, as can be seen from the shaded portion of the sketch. This shaded area represents the probability that an observed sample mean will be greater than 9.2 (critical $C_1$) and represents the $\beta$-risk for the specific alternative value $\mu = 9.5$. The area under the curve between the critical limit (9.2) and 9.5 is determined by

$$Z = \frac{9.2 - 9.5}{0.4} = -0.75$$

Therefore,

$$P(-0.75 < Z < 0) \quad 0.2734$$

$$\text{Add for area to right of 9.5} \quad \underline{0.5000}$$

$$P(\text{Accepting } H_0 | \mu = 9.5) \quad 0.7734 \quad (\beta\text{-risk for } \mu_A = 9.5)$$

So the probability of the type 2 error is 0.7734 when the true mean is 9.5. That is, $\beta = 0.7734$ when $\mu_A = 9.5$.

---

In Example 9.11, the 9.5 minutes is just one alternative value that the manager might want to check. Because the manager does not know the true mean, he can take the most reasonable (alternative) values and calculate the associated $\beta$-risk for each one. If, for example, he used $\mu_A = 9.2$, another value in the range of the alternative hypotheses, then the probability of accepting the null hypothesis (the $\beta$-risk) would be reduced to 0.500, since $Z = (9.2 - 9.2)/0.4 = 0$ gives $P(\text{Accept } H_0 | \mu = 9.2) = 0 + 0.500$. Given other values of $\mu$, probabilities can be computed for the related (alternative) $\beta$-risk. If these probabilities are computed and graphed, the result is referred to as an *operating characteristic* (OC) curve. Managers use the operating characteristic curve to make decisions about the most reasonable production levels and the quality level that can be expected in manufactured products.

## The Operating Characteristic Curve and Power of a Test

An OC curve gives information about the potential strength of a hypothesis test. Conceptually this is an extension of Table 9.1, which is simply a classification of errors in testing. A classical test result is either (1) to reject the null hypothesis or (2) to not reject (accept) the null hypothesis. The plot of probabilities for (2) under alternative values of the parameter ($\mu_A$) is an *OC curve* if the vertical axis shows the *probability of acceptance*. Insofar as $\beta$ = the probability of accepting the null hypothesis (when it is false), the OC-curve values are graphed directly as $\beta$-values. If the vertical axis shows the *probability of rejecting* the null hypothesis, the plot of probabilities is called a *power curve*.

> An operating characteristic (OC) curve is a graph showing the probability of accepting $H_0$ for various parameter values from the range of the alternative hypothesis.
>
> A power curve is a graph showing the probability of rejecting $H_0$ for various parameter values from the range of the alternative hypothesis.

**FIGURE 9.11**    OC and Power Curves for Three Categories of Hypothesis Tests with $H_0: \mu = \mu_0$

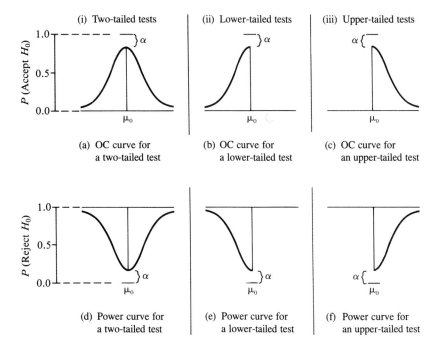

(a) OC curve for a two-tailed test

(b) OC curve for a lower-tailed test

(c) OC curve for an upper-tailed test

(d) Power curve for a two-tailed test

(e) Power curve for a lower-tailed test

(f) Power curve for an upper-tailed test

Figure 9.11 displays the OC curves and power curves associated with (i) two-tailed, (ii) lower-tailed, and (iii) upper-tailed hypothesis tests.

Because $P(\text{Reject } H_0 | H_0 \text{ is false}) + P(\text{Accept } H_0 | H_0 \text{ is false}) = 1$, power $= 1 - \beta$ when $H_0$ is false. If we view this relation as $P(\text{Reject } H_0 | H_0 \text{ is false}) = 1 - P(\text{Accept } H_0 | H_0 \text{ is false})$, then power-curve values are complements of OC-curve probabilities. The null hypothesis value $\mu = \mu_0$ is a special case. For the OC curve, $\mu = \mu_0$ is the highest point on the curve, so it has the highest value for $P(\text{Accept } H_0)$. Similarly, the power curve achieves a lowest value when $\mu = \mu_0$, and that value is $\alpha = P(\text{Reject } H_0)$.

Figure 9.11(a) is really a combination of the parts illustrated in Figure 9.9(a)–(f). That is, for a two-tailed test of hypothesis, the graph of $P(\text{Accept } H_0)$ increases as $\mu$ approaches $\mu_0$; $\beta$-risk defies definition when $\mu = \mu_0$, for then the null hypothesis is not false. Finally, as $\mu$ takes on bigger values, it becomes easier to discern when $\mu > \mu_0$, and the OC curve approaches zero. And, conversely, there is more power (via higher values on the power curve) as $\mu$ takes values further from $\mu_0$. The power curve in Figure 9.11(d) is simply the OC curve in Figure 9.11(a) turned upside down. Since power is the probability to reject a hypothesis, the lowest point on the power curve is for $\mu = \mu_0$. This makes good sense because we want the least possible probability to reject if, in fact, $\mu = \mu_0$. The lower-tailed or upper-tailed OC and power curves in Figure 9.11(b)–(f) are logical extensions of the two-tailed schemes.

Because hypothesis tests very often result in rejection of $H_0$, it can be useful to view a power curve to determine how quickly the test provides a relatively high probability for the rejection (of $H_0$). A strong test should have a relatively high probability of rejecting the null hypothesis when it is false. So when the real value of the parameter ($\mu$) is not close to the hypothesized value ($\mu_0$), we would like the probability of rejecting $H_0$ to be near 1.0. For values of $\mu$ closer to $\mu_0$, the probability drops until it gets to the value of $\alpha$ at $\mu_0$, where the null hypothesis is true and the probability of rejecting $H_0$ is $\alpha$.

For a two-tailed hypothesis test, $\mu_0$ is the only point at which the null hypothesis is true, so the probability of rejection climbs as $\mu$ deviates further from $\mu_0$. For one-tailed tests, the power drops to $\alpha$ (alpha) as $\mu$ approaches from left to right for a lower-tailed test or from right to left for an upper-tailed test. The hypothesis test procedures in this book have high power, meaning the power curves in Figure 9.11(d)–(f) move quickly to a high value (near 1.0) for values of $\mu$ near $\mu_0$. A power curve with high $P(\text{Reject } H_0)$ for values near $\mu_0$ is insurance toward a safe test conclusion.

Example 9.12 illustrates the construction of an operating characteristic curve, using the television viewing times introduced in the previous example.

---

## EXAMPLE 9.12

Develop an OC curve for the preceding test. That is, graph the probabilities for accepting $H_0$ for mean times for spot ads ranging from 8.0 to 10.0 minutes. Recall that $H_0: \mu = 10.0$ and $H_A: \mu < 10.0$ minutes.

*Solution*   One calculation appears in Example 9.11. There $H_A: \mu = 9.5$ gives $\beta$ (for $\mu = 9.5$) = 0.7734. A second possibility is $H_A: \mu = 9.2$. Then $\beta$ (for $\mu = 9.2$) = $P(\text{Accept }_2 | \mu = 9.2)$ = $P(Z < (9.2 - 9.2)/0.4) = P(Z < 0) = 0.5000$. The $\mu$-values, along with $\beta$'s for alternative values of $\mu$, are shown in the table.

| $\mu$-Value | 8.0 | 8.5 | 9.0 | 9.2 | 9.5 | 10.0 | 10.5 |
|---|---|---|---|---|---|---|---|
| $P(\text{Accept } H_0 | \mu)$ | 0.00 | 0.04 | 0.31 | 0.50 | 0.77 | 0.975 | 1.00 |

The case in which $\mu = \mu_0 \, (= 10.0)$ gives the complement to the $\alpha$-risk ($1 - 0.975 = 0.025$). This can serve as a check of your calculation. The resulting OC curve is displayed in Figure 9.12.

**FIGURE 9.12**   An OC Curve for $H_0: \mu = 10$

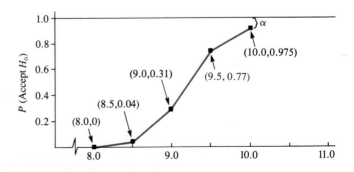

Figure 9.12 shows that the probability of accepting $H_0$ increases toward $1 - \alpha$ as the value for $\mu$ approaches $\mu_0 = 10.0$. Values for $\mu > 10$ are of less interest, and the probability of acceptance for those values approaches 1.0. Compare this to the general form of an OC curve for lower-tailed tests in Figure 9.11. The profile is the same.

## Sample-Size Considerations in Quality Control

The methodology in Example 9.12 allowed us to determine the probability of a type 2 error once the $\alpha$-level was specified. Sometimes both $\alpha$- and $\beta$-levels are specified. (*Note:* In order to specify $\beta$, we must also designate a specific alternative value for the parameter.) If both risks are to be limited, the sample size must be allowed to fluctuate accordingly. This requires a *sampling plan.*

A sampling plan consists of specific procedures, based on predefined $\alpha$- and $\beta$-risks, that define (1) the sample size and (2) the critical value(s) for performing a test of hypothesis.

---

**EXAMPLE 9.13**

Suppose, for the business situation described in Example 9.11, that the manager has not tested $H_0$: $\mu = 10$ minutes but proposes to take a sample that will satisfy the following conditions.

1. He believes that people watch spot ads an average of 10 minutes per day. He sets the $\alpha$-risk at $\alpha = 0.025$ of rejecting this null hypothesis if, in fact, it is true. That is,

$$\alpha = 0.025 \quad \text{when } H_0: \mu = 10.0 \quad (\text{therefore, } Z_\alpha = -1.96)$$

2. If people actually watched only 9.0 minutes per day, he would set the $\beta$-risk at only 0.10 of incorrectly accepting $H_0$: $\mu = 10.0$. That is,

$$\beta = 0.10 \quad \text{when } \mu = 9.0 \quad (\text{therefore, } Z_\beta = 1.28)$$

Find the sample size required to test this null and alternative hypothesis.

*Solution* We know $\sigma = 4$ minutes per day but do not know $\sigma_{\bar{x}}$ because $n$ is unknown. The solution consists of finding an $n$ that will exactly satisfy the $\alpha$ and $\beta$ requirements. We accomplish this by establishing two equations for the critical value $C_1$ and solving them simultaneously. The first equation describes a relation on the $\alpha$-condition; the second gives a condition for the $\beta$-risk requirement. These two conditions are depicted in the two sampling distributions of Figure 9.13, where the $\alpha$- and $\beta$-risks are shaded. From the condition on $\alpha$-risk, and using Equation (9.8), we get

$$C_1 = \mu_0 - \frac{Z_\alpha \sigma}{\sqrt{n}} = 10.0 - \frac{1.96(4)}{\sqrt{n}}$$

With similar logic for the $\beta$-risk requirement,

$$C_1 = \mu_A + \frac{Z_\beta \sigma}{\sqrt{n}} = 9.0 + \frac{1.28(4)}{\sqrt{n}}$$

Since $C_1 = C_1$,

$$10.0 - \frac{1.96(4)}{\sqrt{n}} = 9.0 + \frac{1.28(4)}{\sqrt{n}}$$

**FIGURE 9.13**    Display for Determining Sample Size for Given $\alpha$-Risk and $\beta$-Risk

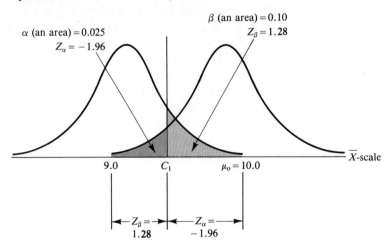

when    $\sqrt{n} = 5.12 + 7.84 = 12.96$

$$n = (12.96)^2 = 168$$

and the sample size requirement is $n = 168$.

We can then find the critical limit $C_1$ by substituting $n = 168$ into either equation:

$$C_1 = 10.0 - \frac{1.96(4)}{\sqrt{168}} = 9.4 \text{ minutes per day}$$

The sample size and the critical (acceptance) limit constitute the essential ingredients of a decision rule for this test.

> Decision rule for Example 9.13: Take a random sample of 168 television viewers and find the sample mean advertising-viewing time per day. If $\bar{X} > 9.4$ minutes per day, accept the hypothesis of an average rate of 10 minutes per day. Otherwise reject the hypothesis and conclude that spot ad viewing averages less than 10 minutes per day.

---

The $\alpha$- and $\beta$-risks for this rule are known in advance. So, if the mean is $\mu = 10.0$ minutes per day, sample means of less than 9.4 minutes will occur $2\frac{1}{2}\%$ of the time (but that is acceptable). On the other hand, if the true mean time is 9.0 minutes, the sample mean will be equal to or greater than 9.4 minutes for 10% of all samples of size $n = 168$. Then the decision rule will lead to an erroneous conclusion of accepting the null hypothesis 10 times in 100. That too is acceptable.

The cost of drawing a sample can be a very real consideration; suppose the cost of obtaining a sample of 168 television viewers is excessive. Then the manager will have to agree to higher $\alpha$- or $\beta$-risks (or both) or else abandon his project.

**$\alpha$- and $\beta$-Risks in t-Tests**    This procedure used preset $\alpha$- and $\beta$-risks to determine a sample size and a critical value for accepting or rejecting the null hypothesis. The critical value, or test criterion, was a value for $\bar{X}$ rather than a $Z$-score because $\bar{X}$ is more understandable to workers and inspection personnel. This test procedure is commonly used in manufacturing and production activities for quality control purposes, and the decision rule is called a *sampling plan*. This procedure is often referred to as an *acceptance sampling* procedure for evaluating the quality level of inputs to or outputs from a production process.

Careful consideration should be given to $\beta$-values in $t$-tests, just as in $Z$-tests. $\beta$-risk calculations help confirm the benefit of having larger sample sizes. Something is lost with smaller samples; namely, for a prespecified $\alpha$-value there is an increase in the $(\beta)$ risk of accepting a false null hypothesis.

Calculation of the $\beta$-risk for reasonable alternative values of the parameter should be a standard research procedure following an "Accept $H_0$" decision. However, values are less accessible from $t$- than from $Z$-tables. This is because $t$-values are usually tabulated for only a limited number of percentage points. If, for example, an "Accept $H_0$" decision also produces a $\beta$-risk of over 50% for the most reasonable alternative, then the decision is probably wrong. Of course, a safer procedure—one with lower $\beta$-risk—results from using larger samples. It can be shown that the tests we have discussed thus far have the smallest $\beta$-risk possible for given $\alpha$ and $n$.

## Sample Size for Proportions

The procedure for finding an appropriate sample size and a decision rule when working with proportions is like that for means, except that $\pi$ replaces $\mu$ and the standard error of proportion $\sqrt{\pi(1-\pi)/n}$ replaces the standard error of the mean, $\sigma/\sqrt{n}$. Again, the solution is to set up two equations for the common critical limit and solve these equations for $n$. Then the critical limit is obtained by substitution of $n$ into either of the two equations.

## Problems 9.5

25. What differences exist for the $p$-value approach to hypothesis testing versus the traditional decision-based procedure for
    a. the statement of null and alternative hypotheses?
    b. establishing the test standard?
    c. the calculated statistic?
    d. the statements of conclusion and findings?

26. The premixed dog food sales example (9.3) used $n = 49$ and calculated $Z = 2.31$ for $H_A$: $\mu > \$15,500$. Use this information to choose the level of significance (from among 0.05, 0.025, 0.015, 0.1) at which the null hypothesis can be rejected.

27. Does a minority of the students in your student dorms favor a designated time period for quiet hours? This question was asked in Example 9.8. The test used $H_A$: $\pi < 0.50$ and, for $n = 400$ respondents, showed calculated $Z = -1.40$. What is the smallest value from among 0.01, 0.025, 0.05, and 0.10 at which this calculated $Z$ would be judged significant? That is, choose a $p$-value for this test.

28. Test the following hypotheses concerning the price obtained for a particular model of new car.

$$H_0: \mu = \$12,000 \qquad H_A: \mu > \$12,000$$

A sample of $n = 100$ randomly selected sales provides calculated $Z = 1.89$. Perform the test and state your conclusions by
a.  the traditional decision approach, with $\alpha = 0.05$.
b.  the $p$-value procedure.
Compare your conclusions by the two approaches.

29.  We are testing the hypothesis that the true mean weight of concrete beams is 200 pounds. We have found the standard error to be 2 pounds and have established rejection limits at 194 and 206 pounds. If the true (unknown) mean is actually 204 pounds, what is the probability of a type 1 error? Assume a normal distribution.

30.  We are testing the hypothesis that the true mean weight of concrete beams is 200 pounds. We have found the standard error to be 2 pounds and have established rejection limits at 194 and 206 pounds. If the true (unknown) mean is actually 204, what is the probability of a type 2 error? Assume a normal distribution.

31.  Employees in the shipping department of Spears Stores, Inc., claim that, on the average, they fill each order in 14 minutes or less. To test this claim, you (a management analyst) randomly sample 400 orders to test $H_0: \mu = 14.0$. You find an average completion time of 15.0 min with $\sigma = 10.0$ minutes.
a.  Use a 0.05 level of significance and test $H_0: \mu = 14.0$.
b.  What is the critical limit $(C)$, in terms of minutes, for the decision rule associated with this hypothesis?
c.  If the true (unknown) mean is actually 14.5 minutes, what is the probability of a type 2 error?

32.  A carload of 20,000 lead sheets has been received by a shipyard, which uses them for radiation shielding. Each sheet is supposed to weigh *at least* 50 pounds, and it is known that the process producing them has a standard deviation of 5 pounds. An inspection agreement is reached whereby the supplier's risk of type 1 error is limited to 0.045. On the other hand, if the true mean weight of the sheets in the carload is only 47 pounds, the shipyard wants only a 5% chance of accepting the shipment. What is the critical value for the sample mean (i.e., the borderline weight) that will satisfy these conditions?

33.  A firm's quality control manager wishes to find an appropriate sample size such that

(1)  if $\pi = 0.10$ defective (as he hypothesizes), then $\alpha = 0.05$.
(2)  if $\pi$ is as bad as 0.20, then $\beta$ should be 0.05.

Using the $H_0: \pi = 0.10$, $H_A: \pi \neq 0.10$, set up equations for the critical limit and determine the sample size required.

34.  A producer and a consumer have agreed upon a sampling plan for the purchase of some construction materials. The plan calls for drawing a random sample of $n = 40$ parts and testing the tensile strength under $H_0: \mu = 80,000$ pounds per square inch (psi). If the sample mean is $\leqslant 72,000$ psi, the shipment is to be rejected. Assume $\sigma = 10,000$ psi.

a. Compute the probability of accepting the shipment if the (true) popu-
tion value is 60,000 psi, 65,000 psi, 70,000 psi, 72,000 psi, 80,000 psi, and
90,000 psi.

b. Draw the OC curve.

c. What is the chance ($\alpha$-risk) of rejecting the lot if the population mean
is 80,000 psi?

35. The quality control supervisor at a grass seed plant is testing the rye grass
content of an incoming shipment of seed. Each bag is supposed to contain
at least 200 pounds of rye grass, and the population standard deviation is
$\sigma = 10.6$ pounds. The sampling plan calls for a random sample of 50 bags
and rejection of the shipment if the mean weight of rye grass is less than
198 pounds.

a. What is the $\alpha$-risk of rejecting the lot if the true mean weight of the rye
grass is 200 pounds?

b. Given the sampling plan ($n = 50$, accept the lot if $\bar{X} \geq 198$), what is the
probability of accepting the lot if the true weight is 195 pounds? 197
pounds? 198 pounds? 200 pounds? 201 pounds?

c. Use the data from (b) to construct an OC curve for the sampling plan.

36. Construct a power curve from the information in Problem 35. Remember
that power is 1 minus the probability of acceptance. (*Hint:* Recall that a
power curve is simply an upside-down OC curve.) What is the power for
a true mean weight of 195 pounds? 200 pounds? 201 pounds?

37. A test of hypothesis uses $H_0$: $\mu = 3.0$ with $H_A$: $\mu > 3.0$ and the following
test procedure: Reject $H_0$ if $\bar{X} > 3.2$ for $n = 100$ randomly selected observations. Explain why it would be unnecessary to compute the power for
$\mu < 3.0$.

## 9.6
## THE MEANING OF
## HYPOTHESIS TESTS

Because hypothesis testing is used throughout the remainder of this textbook,
we pause here to solidify an understanding of what *testing* means. A test of
hypothesis contains four basic elements:

1. A sampling/probability distribution
2. Alternatives for action
3. A decision rule
4. Identification of risks and quantification of the loss or penalty for an
erroneous action

The task in hypothesis testing is to choose the correct probability distribution
for the situation and the most reasonable action from the alternatives. Then a
decision rule is established that will minimize the risk of an incorrect action.
Now that you have a basis for making hypothesis tests, this section can provide
you with a greater understanding of hypothesis testing.

### The Structure of
### Hypothesis Tests

The framework for hypothesis tests requires some criterion for the decision
maker to either reject or accept the null hypothesis. The classical approach is
to make the decision about a population by using sample information. This requires identifying a statistic to be used to test the parametric value in question
and a sampling distribution that relates the parameter and the statistic.

A test begins with the structuring of alternatives. Selecting the null and alternative hypotheses is a matter of where one wants to place the burden of proof. In Example 9.1, the burden was placed on the moped bike manufacturer to verify a higher mileage. The analyst felt the greater risk was in deciding to favor too high a mileage claim. Consequently, the null hypothesis was structured to require a high sample value to reject, and the $\alpha$-risk was kept low at 5%. Had the greater concern been for accepting too low a mileage value, then the test would have been structured in the opposite direction. In either case, *it is necessary to retain the equality in the null hypothesis.*

A test partitions the possible values for the statistic into (1) an acceptance region and (2) a rejection region for the null hypothesis. The moped bike example used a one-tailed test with the rejection region for higher-mileage values. So the acceptance region included values of the statistic ($\bar{X}$) that, within the tolerance of sampling error, make the null values (here $\mu = 125$ mi/gal) more acceptable than those under the alternative hypothesis $\mu > 125$ mi/gal. (See Figure 9.14.)

The remainder of the $\bar{X}$ space (possibilities) favors the alternative hypothesis, and so it defines a rejection region for the null hypothesis. That is, a hypothesis test begins with the assumption that the null hypothesis is true. Then, if the sample evidence tells us that this assumption is not supportable, we reject $H_0$.

You might well ask, "How different from the null value of equality must the statistic be to cause us to reject the null hypothesis? For example, in Figure 9.14, would a reading of 126 mi/gal cause rejection, or is something larger required? The answer to this helps to structure a decision rule. Assuming $\alpha$ is specified, it depends on three considerations:

1. The amount of information available, that is, the sample size
2. Whether the test is directional (one-tailed) or nondirectional (two-tailed)
3. The $\beta$-risk

---

**FIGURE 9.14**   The Space of Possible Values for the Sample Mean

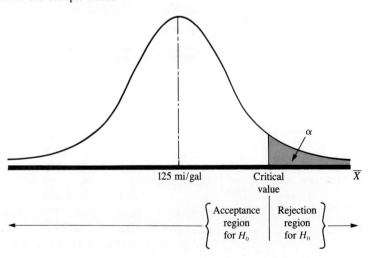

As long as the sample is sufficiently large ($n > 30$), or the distribution is normal with $\sigma$ known, we can use the sampling distribution of the mean and the central limit theorem.

An analyst may use the value of the sample mean directly as the basis for accepting or rejecting the null hypothesis. However, it is common to use a Z-value to fix a probability level for incorrectly rejecting the null hypothesis. So (1) the sampling distribution of the Z-statistic and (2) the $\alpha$-risk establish the decision rule. Nevertheless, the use of Z-scores rather than $\bar{X}$ to establish the decision rule is a matter of convenience. In tests of proportions, the percent defective ($p$) is commonly used in production control testing in lieu of Z-scores.

## Comparing One-Tailed against Two-Tailed Tests

As you may have noted above, one-tailed tests are sometimes referred to as *directional tests* and two-tailed as *nondirectional tests*. Whether a test is directional or nondirectional depends upon the objective of the test and upon the information that is available at the onset of the test. For example, a one-tailed alternative test would be suggested if you had an office copy machine about which there were complaints of too many defective copies. If, however, you were comparing two machines for purchase, it would be most appropriate to know if one performed differently (better or worse) than the other. Then a two-tailed test would be appropriate.

For the same $\alpha$-risk, one- and two-tailed tests show something different. Figure 9.15 illustrates the difference.

Stating a directional alternative hypothesis, such as $<$ or $>$, rather than a does-not-equal ($\neq$) two-tailed alternative indicates an understanding of the most reasonable alternative values. If the null value is rejected, a one-tailed test gives the analyst additional information and certainty about the location of the parameter. Not only is the rejection region undivided, by not being split into two tails, but also this results in a shorter distance between the null-hypothesized parameter and the rejection limit (critical Z). Thus, with one-tailed tests a smaller deviation from the null value leads to rejection of the null hypothesis. Since we are interested in increasing our ability to reject $H_0$ when it is false, using information about the direction of change actually improves the

**FIGURE 9.15**  Comparison of One-Tailed and Two-Tailed Tests

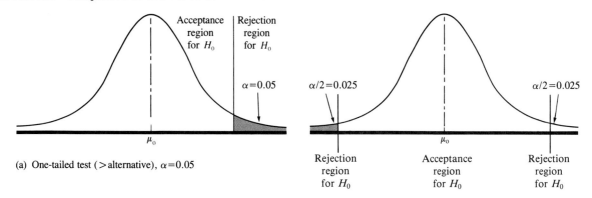

(a) One-tailed test ($>$ alternative), $\alpha = 0.05$

(b) Two-tailed test ($\neq$ alternative), $\alpha = 0.05$

**FIGURE 9.16**   Relating $\alpha$- and $\beta$-Risks

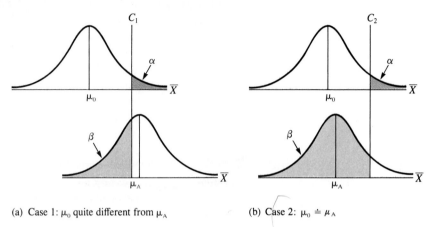

(a) Case 1: $\mu_0$ quite different from $\mu_A$          (b) Case 2: $\mu_0 \doteq \mu_A$

test by increasing its power without increasing $\alpha$. Basically this results from recognizing that the probability of getting a sample mean in one of the tails is virtually zero. In this sense we say that a one-tailed test is more sensitive than the comparable two-tailed test. If the information warrants, it is better to choose a one-tailed test.

**Sample Considerations Relative to $\alpha$- and $\beta$-Risks**

A final condition of hypothesis tests depends upon the importance given to minimizing the $\alpha$- and $\beta$-risks simultaneously. Consider two cases for which $\mu_0$ and $\mu_A$ are the best approximations for the population mean. The test is an upper one-tailed test defined by $H_A: \mu > \mu_0$. This is illustrated in Figure 9.16.

The $\alpha$- and $\beta$-risks can be considered as defining areas under the curve. These range from 0 to 1. In Case 1, in which $\mu_A$ and $\mu_0$ are markedly different (which, of course, we cannot know), there would be little effect on $\beta$ no matter how small the $\alpha$-value. That is, for $C_1$ values further to the right, $\alpha$ decreases and therefore $\beta$ increases. Yet the placement of $C_1$ would, within reasonable limits, be of little consequence. Both $\alpha$- and $\beta$-risks are small (areas) simply because of the large discrepancy between the true parameter value and the hypothesized parameter value. In Case 2, in which this discrepancy is slight, $\mu_0 \doteq \mu_A$, but unknown, the location of $C_2$ does matter. Moving $C_2$ slightly to the left produces a sharp increase in the $\alpha$-risk level, and moving it to the right substantially increases the $\beta$-risk. Since we never know the true parameter value (if we did, we would not have to make a statistical test), we cannot know which situation exists. Therefore, as we showed in Section 9.5, increasing the sample size is the acceptable alternative to shifting $C_2$ to either the left or the right.

**9.7
TESTS OF
DIFFERENCES FOR
TWO POPULATIONS**

Statistical tests often pertain to the difference between two population means or between two proportions. Is there a difference in the average lifetime of 60-watt lightbulbs produced by Sylvania and General Electric? Is the proportion of voters who favor right-to-work laws different in Florida and California? Do the Scholastic Aptitude Test (SAT) scores differ for students applying to Ivy League versus Pac-10 universities?

**TABLE 9.2**   Equations for Differences

|  | Means | Proportions |
|---|---|---|
| Statistic: Sample difference | $\bar{X}_1 - \bar{X}_2$ | $p_1 - p_2$ |
| Parameter: Hypothesized difference | $\mu_1 - \mu_2$ | $\pi_1 - \pi_2$ |
| Standard error of difference | $\sigma_{\bar{X}_1 - \bar{X}_2} = \sqrt{\dfrac{\sigma_1^2}{n_1} + \dfrac{\sigma_2^2}{n_2}}$ | $\sigma_{p_1 - p_2} \doteq \sqrt{\dfrac{p_1 q_1}{n_1} + \dfrac{p_1 q_2}{n_2}}$ |

Tests of the difference in population means ($\mu_1 - \mu_2$) or population proportions ($\pi_1 - \pi_2$) use the normal distribution theory described earlier. Recall from Chapter 8 that for *two independent and large random samples with known variances,* the differences between two sample means (or two sample proportions) will be normally distributed. Under these conditions tests involving the hypothesized difference use the standardized form

$$Z = \frac{\text{Statistic} - \text{Parameter}}{\text{Standard error of statistic}}$$

where the statistic, parameter, and standard error are those for the difference of values rather than for single values. Appropriate equations are summarized in Table 9.2.

**Tests of $\mu_1 - \mu_2$**

For tests of the difference in means from large samples, use the test statistic

$$Z = \frac{(\bar{X}_1 - \bar{X}_2) - (\mu_1 - \mu_2)}{\sigma_{\bar{X}_1 - \bar{X}_2}} \tag{9.10}$$

where   $\sigma_{\bar{X}_1 - \bar{X}_2} = \sqrt{\dfrac{\sigma_1^2}{n_1} + \dfrac{\sigma_2^2}{n_2}}$

Assumptions:
   The sample sizes, $n_1$ and $n_2$, both exceed 30, or the distributions for $X_1$ and $X_2$ are normal.
   The random samples are independent.
   The standard deviations are known.

This statistic is normally distributed with a mean of zero and a standard deviation of one if both $n_1 > 30$ and $n_2 > 30$. Frequently we hypothesize that there is no difference in the population means, so $\mu_1 = \mu_2$ or, alternatively, $\mu_1 - \mu_2 = 0$. However, we can test a hypothesis that the mean difference is a specific value just as easily.

**EXAMPLE 9.14**

Two work methods are being compared for assembly of sports equipment. For a random sample of 64 employees working by the traditional method, the mean production rate was 68.8 units. This population standard deviation is 5.2 units. For a random sample of 80 employees who used a new method, the mean production rate was 70.5 units. The standard deviation for the new method is 5.6 units. Test the hypothesis that the new method results in a statistically higher production mean at $\alpha = 0.025$. If the new method yields a statistically higher rate, it will become the standard procedure.

**Solution**  This is a one-tailed test for which we hypothesize the difference in mean performance $\mu_1 - \mu_2 = 0$ and, alternatively, $\mu_1 - \mu_2 > 0$. We shall let $1 =$ the new method and $2 =$ the traditional method.

1. Does the new method result in higher productivity?

$$H_0: \mu_1 = \mu_2 \quad (\text{or } \mu_1 - \mu_2 = 0)$$
$$H_A: \mu_1 > \mu_2 \quad (\text{or } \mu_1 - \mu_2 > 0)$$

2. $\alpha = 0.025$

Test: Reject $H_0$ if calculated $Z > 1.96$.

3. Then

$$\sigma_{\bar{X}_1 - \bar{X}_2} = \sqrt{\frac{\sigma_1^2}{n_1} + \frac{\sigma_2^2}{n_2}} = \sqrt{\frac{(5.6)^2}{80} + \frac{(5.2)^2}{64}} = 0.90$$

$$Z = \frac{(\bar{X}_1 - \bar{X}_2) - (\mu_1 - \mu_2)}{\sigma_{\bar{X}_1 - \bar{X}_2}} = \frac{(70.50 - 68.80) - 0}{0.90} = \frac{1.70}{0.90}$$

$$= 1.89$$

4. Decision: We cannot reject $H_0$ of no difference, because $1.89 < 1.96$. Therefore, there is no statistical difference in production performance for the new method versus the traditional method.

The evidence does not support a higher performance for those using the new method. This, of course, is at the $\alpha = 0.025$ level, which means we are willing to be wrong only 2.5% of the time. Since the production rate is not statistically increased, we must use other considerations in choosing a procedure.

---

An equivalent way to state the null hypothesis, $H_0: \mu_1 - \mu_2 = 0$, may be preferred over $H_0: \mu_1 = \mu_2$, since the former statement clearly indicates that the true mean difference in Step 3 is zero. In fact, differences other than zero can be hypothesized. For example, $H_0: \mu_1 - \mu_2 = +2$ might be a hypothesis that the average age for workers at the Brookdale plant (1) exceeds that for workers at the Waltham plant (2) by two years.

**Tests of $\pi_1 - \pi_2$**

Earlier, to test a hypothesis about a single population proportion, we used the hypothesized proportion, $\pi$, to compute the standard error. When we are testing the difference between two proportions, two cases arise, depending upon the null hypothesis. The most common test (which we shall illustrate) is a test that both proportions are equal. This would be stated as $H_0: \pi_1 = \pi_2$, or as $H_0: \pi_1 - \pi_2 = 0$. In this case the best estimate for $\pi_1 - \pi_2$ is a weighted value of the total number of successes $(X_1 + X_2)$ divided by the total sample $(n_1 + n_2)$; that is,

$$P_{wtd} = \frac{X_1 + X_2}{n_1 + n_2}$$

Then the test of the difference in proportions $H_0: \pi_1 - \pi_2 = 0$ uses the sampling distribution defined by

$$Z = \frac{(p_1 - p_2) - (\pi_1 - \pi_2)}{\sigma_{p_1 - p_2}} \quad \text{where} \quad n_1 p_1 \geqslant 5, n_1 q_1 \geqslant 5 \qquad (9.11)$$

$$\text{and} \quad n_2 p_2 \geqslant 5, n_2 q_2 \geqslant 5$$

with

$$\sigma_{p_1 - p_2} = \sqrt{P_{wtd} \cdot q_{wtd}\left(\frac{1}{n_1} + \frac{1}{n_2}\right)} = \sqrt{\frac{P_{wtd} q_{wtd}}{n_1} + \frac{P_{wtd} q_{wtd}}{n_2}} \qquad (9.12)$$

and

$$P_{wtd} = \frac{X_1 + X_2}{n_1 + n_2}, \quad q_{wtd} = 1 - P_{wtd}$$

**Assumptions: The samples are independent random samples. The samples are large, commonly 100 or more.**

With Example 9.15, we illustrate a case for which the hypothesized difference is zero.

---

**EXAMPLE 9.15**

A distributor of computer products has major outlets in several cities, including Denver and Seattle. The company president is considering entering into an exclusive dealership arrangement with a manufacturer of fax machines. He wants to know whether potential customers in Denver and Seattle have the same preference for the Faxspeed II. Random samples of customers reveal the data shown in Table 9.3. Test the hypothesis that there is no difference in the proportion preferring the Faxspeed II in Denver and Seattle. Use a 2% level of significance.

**Solution**

1. $H_0: \pi_1 - \pi_2 = 0$

   $H_A: \pi_1 - \pi_2 > 0$

2. Test: Reject $H_0$ if calculated $Z > 2.05$, for $\alpha = 0.02$. (*Note:* This is a one-tailed test and both samples are large enough to permit use of the normal distribution.)

**TABLE 9.3** Customer Preference for Fax Machines

| Customer Preference | Denver | Seattle |
|---|---|---|
| Faxspeed II | 62 | 46 |
| Other brands | 47 | 38 |
| No preference | 111 | 96 |
| Total sample | 220 | 180 |

3. $P_{\text{wtd}} = \dfrac{X_1 + X_2}{n_1 + n_2} = \dfrac{62 + 46}{220 + 180} = 0.270$   where   $X_1 = 62$

$$X_2 = 46$$

$$\sigma_{p_1 - p_2} \doteq \sqrt{P_{\text{wtd}}q_{\text{wtd}}\left(\frac{1}{n_1} + \frac{1}{n_2}\right)} = \sqrt{(0.27)(0.73)\left(\frac{1}{220} + \frac{1}{180}\right)}$$

$$= 0.045$$

Since $p_1 = 62/220 = 0.282$ and $p_2 = 46/180 = 0.256$,

$$Z = \frac{(p_1 - p_2) - (\pi_1 - \pi_2)}{\sigma_{p_1 - p_2}} = \frac{(0.282 - 0.256) - 0}{0.045} = 0.58$$

4. Decision: We cannot reject $H_0$ of no difference, because $0.58 < 2.05$.

The evidence suggests that customer preference for the Faxspeed II is consistent (no difference) in Denver and Seattle for an $\alpha = 0.02$ test of hypothesis.

---

In situations for which the hypothesized difference between two proportions is some nonzero value, $d_0$ (for example, $H_0: \pi_1 - \pi_2 = 0.15$ and $H_A$: $\pi_1 - \pi_2 > 0.15$), we use the following equation to estimate $\sigma_{p_1 - p_2}$.

> Tests of the difference in proportions $H_0: \pi_1 - \pi_2 = d_0 \neq 0$ use the sampling distribution defined by (9.11) with $\hat{\sigma}_{p_1 - p_2}$ the estimator for the standard error of difference in proportions:
>
> $$\hat{\sigma}_{p_1 - p_2} = \sqrt{\frac{p_1 q_1}{n_1} + \frac{p_2 q_2}{n_2}} \tag{9.13}$$

We do not form a weighted proportion in this case because we are operating under the assumption that the proportions are truly different, not accidentally different because of sampling error. Aside from the estimate of $\sigma_{p_1 - p_2}$, other testing procedures are unchanged. For Equations (9.11) and (9.13), we determine $p_1$ as $X_1/n_1$ and $p_2$ as $X_2/n_2$. In both cases, we substitute $\pi_1 - \pi_2$ with its hypothesized value $d_0$ whether this value ($d_0$) is zero or not. It is noteworthy that in Chapter 8 estimation of $\pi_1 - \pi_2$ uses the expression (9.13) for the standard error, because there we do not assume equal proportions.

In this section we have discussed testing means for independent samples with known standard deviations. Chapter 10 includes a test of means for "paired differences" when the samples are not independent. Other procedures are required when the samples are independent but the standard deviations are not known and the random variables come from normal distributions. This is also covered in Chapter 10. Chapter 11 extends the inference to three or more means. Inference on proportions is extended in Chapter 16.

The following problems give a selection of tests on differences in means or in proportions for large, independent, random samples.

**Problems 9.7**

38. You are testing $H_0: \mu_1 - \mu_2 = 0$ with random samples of $n_1$ and $n_2$ both $> 30$. Possible conditions are (i) $\sigma_1$ and $\sigma_2$ are known and (ii) $\sigma_1$ and $\sigma_2$ are

not known and must be estimated. Explain the difference in your test, for $\alpha = 0.05$, in the following items.
a. The test statistic
b. Calculation of the standard error of difference in means
c. The level of significance
d. The interpretation of conclusions

39. There are two distinct cases for testing the difference in proportions for large samples. Explain the differences in performing a test (a) $H_0$: $\pi_1 - \pi_2 = 0$ versus a test (b) $H_0$: $\pi_1 - \pi_2 = 0.02$. Assume each test uses a two-tailed alternative hypothesis with $\alpha = 0.05$.

40. Test whether there is a difference in the weight of plywood sheets from two mills. Thirty-six random samples each reveal average weights of 67.0 pounds per sheet from mill 1 and 65.0 pounds from mill 2. The standard deviations are $\sigma_1 = 4$ pounds and $\sigma_2 = 3.2$ pounds, respectively.
a. Use the 0.01 level of significance to test.
b. At what $p$-value is the difference significant?

41. The marketing manager of a home products firm would like to know if her company's brand of dish soap was priced lower than the competition's price on a nationwide basis. Fifty samples of the company's brand (A) and 40 from the competition (B) were randomly selected from major food stores. The average price was $1.24 for brand A and $1.31 for B. Use a standard deviation of $0.14 for A and $0.16 for B. Test whether the price of brand A soap is significantly lower than that of brand B at the $\alpha = 0.05$ level.

42. Computronics has two manufacturing plants (A and B) that produce plastic molded cases for computer keyboards. The quality control manager thinks there is no difference in the proportion of defective cases from the two plants but asks that a sample be taken to confirm this. One hundred twenty cases are randomly selected from each plant. Plant A yields 8 defective cases and plant B has 11. Test whether there is a difference in the proportion defective. Use $\alpha = 0.10$.

43. The standard (old) method for testing the acid content of grapefruit is under criticism. This is critical because the fruit is picked somewhat green, food coloring is injected into the peel, and then the fruit is shipped to super-markets throughout the United States. If the fruit has excessive acid content, it will *not* be well accepted, resulting in high consumer complaints and low sales. A new method is available to determine readiness to pick, and it will be adopted if it improves over 10% on the current rate of selecting ripe (low acid) fruit.

    Use the statistics in the table to test $H_A$: $\pi_{new} - \pi_{old} > 0.10$, where $\pi$ = the proportion of grapefruit with acceptable (low) acid content. Use $\alpha = 0.05$ and state conclusions.

| Method | | New | Old |
|---|---|---|---|
| $X_i$, $i = 1, 2$ | Number ripe | 365 | 400 |
| $n_i$ | Total | 400 | 500 |

**SUMMARY**

*Statistical testing* is a useful technique for making business decisions. The testing procedure is as follows:

1. Make a clear statement of the hypotheses.
2. Specify the test, including the
   a. $\alpha$-risk,
   b. test statistic,
   c. null rejection value.
3. Draw a random sample and calculate a $Z$-score (or a $t$-score).
4. Make the decision. State a practical conclusion that supports either the null or the research hypothesis.

For means with random samples of $n > 30$, we can make use of the central limit theorem, using the $Z$-distribution to test hypotheses about the population mean. (The same procedure can work for tests of means if $\sigma$ is not known and the distributions are normal, but this uses the $t$-statistic.) For proportions, a sufficiently large sample is one for which $np > 5$ and $nq > 5$.

A common procedural problem is setting up the alternative (test) hypothesis. This is a quantification of the statement of the decision problem. The test can appear as a question or in the form "Test that . . ." Either form will satisfy the following criteria:

1. Identify the parameter being tested.
2. Imply a relation ($<$, $>$, or $\neq$).
3. Specify a value.

Put these together and you have the *alternative hypothesis.*

Establishing the *test procedure* can be difficult. Remember that the test states the same relation ($<$, $>$, or $\neq$) as the alternative hypothesis. For example, if the alternative hypothesis were $H_A$: $\mu < -1.6$, we would reject the null hypothesis if the calculated $Z$-value is *less than* the corresponding (negative of the) $Z$-value from the table. We would then conclude that the mean is $< -1.6$. This is an example of a lower-tailed test.

To determine the $Z$-value for the test, remember that a lower-tailed ($<$) test requires a $Z$-value that puts the total value of $\alpha$ in the lower tail; for an upper-tailed ($>$) test, the $\alpha$-value is totally in the upper tail. For a two-tailed test ($\neq$), the $Z$-values are such that $\alpha/2$ is in each tail. An alternative approach to setting a predetermined $\alpha$ (and therefore $Z$) value is to report the *p*-value, the smallest value at which the sample statistic is significantly different from the hypothesized value of the parameter.

Interpretation of test findings requires critical thinking. If the data lead you to reject $H_0$, then the conclusion is simply to state the alternative hypothesis in words. If the data lead you to accept $H_0$, then it will suffice to say *cannot reject*, stating the null hypothesis. In any case, the preassigned ($\alpha$-risk) *level of significance* should be included in a discussion of the findings. This tells the reliability of the result. Otherwise the *p*-value of the test statistic should be reported.

Two test risks were considered. A low *$\alpha$-risk* guards against incorrect rejection of the null hypothesis. The *$\beta$-risk* concerns accepting a false null hypothesis. A good test requires keeping both errors low. This may mean taking a substantial sample, which can be expensive and time-consuming. A reasonable balance will moderate both errors and the cost reflected in sample size. The *operating characteristic* and *power curves* give us a graphic interpretation of the $\alpha$- and $\beta$-risks. The *OC curve* shows the probability of accepting $H_0$ for various values of the parameter, and the power curve shows the probability of rejecting $H_0$ for these same values. The curves are complements to probability 1.

Tests for means for a single population use a $Z$- or $t$-statistic, depending on whether the standard deviation, $\sigma$, is known. The Z-distribution and Table B.1 are used for tests on means when (1) $\sigma$ is known and the distribution of $X$ is normal, or when (2) $\sigma$ is known and the random sample size is larger than 30. The $t$-distribution and Table B.2 are used for $\sigma$ not known and $X$ normally distributed. Otherwise, the test procedures are the same.

Our discussion included tests of means and differences in means for large (independent) samples, of $n > 30$. Tests of proportions for $np \geqslant 5$, $nq \geqslant 5$ include tests of $H_0: \pi = $ constant and, for two populations, the difference in proportions. Table 9.4 summarizes the relevant test statistics.

**TABLE 9.4** Key Equations for Hypothesis Testing

| | | Inferences | |
|---|---|---|---|
| Definition | Assumptions | Estimation (Ch. 8) | Testing (Ch. 9) |
| One-sample test | Means, $\sigma^2$ known, $n > 30$, or normal distribution | **Confidence limits** (LCL, UCL) $= \bar{X} \pm Z \dfrac{\sigma}{\sqrt{n}}$  (8.1) | **Test statistic** $H_0: \mu = \mu_0$ $Z = \dfrac{\bar{X} - \mu_0}{\sigma/\sqrt{n}}$  (9.1) |
| | Means, $\sigma^2$ unknown, any sample size normal distribution | (LCL, UCL) $= \bar{X} \pm t \dfrac{s}{\sqrt{n}}$  (8.3) | $H_0: \mu = \mu_0$ $t = \dfrac{\bar{X} - \mu_0}{s/\sqrt{n}}$  (9.2) |
| | Proportions, $np > 5$, $nq > 5$ | (LCL, UCL) $= p \pm Z \sqrt{\dfrac{pq}{n}}$  (8.9) | $H_0: \pi = \pi_0$ $Z = \dfrac{p - \pi_0}{\sqrt{\dfrac{\pi_0(1 - \pi_0)}{n}}}$  (9.4) (9.6) |
| Two independent samples | Difference in means, independent samples: $n_1 > 30$ and $n_2 > 30$, or normal distributions | (LCL, UCL) $= (\bar{X}_1 - \bar{X}_2) \pm Z \sqrt{\dfrac{\sigma_1^2}{n_1} + \dfrac{\sigma_2^2}{n_2}}$  (8.13) | $H_0: \mu_1 - \mu_2 = 0$ $Z = \dfrac{(\bar{X}_1 - \bar{X}_2) - (\mu_1 - \mu_2)}{\sqrt{\dfrac{\sigma_1^2}{n_1} + \dfrac{\sigma_2^2}{n_2}}}$  (9.10) |
| | Difference in proportions, independent samples: $n_1 p_1 > 5$ $n_1 q_1 > 5$ and $n_2 p_2 > 5$ $n_2 q_2 > 5$ | (LCL, UCL) $= (p_1 - p_2) \pm Z \sqrt{\dfrac{p_1 q_1}{n_1} + \dfrac{p_2 q_2}{n_2}}$  (8.15) $\pi_1 = \pi_2$ (known), (no estimate req'd) $\pi_1 \neq \pi_2$ (or relation not known), use $p_1 = \dfrac{X_1}{n_1}, \quad p_2 = \dfrac{X_2}{n_2}$ | (a) $H_0: \pi_1 - \pi_2 = 0$ use $p_{\text{wtd}} = \dfrac{X_1 + X_2}{n_1 + n_2}$ $Z = \dfrac{(p_1 - p_2) - (\pi_1 - \pi_2)}{\sqrt{\dfrac{p_{\text{wtd}} q_{\text{wtd}}}{n_1} + \dfrac{p_{\text{wtd}} q_{\text{wtd}}}{n_2}}}$  (9.11), (9.12) (b) $H_0: \pi_1 - \pi_2 = d_0$ use $p_1 = \dfrac{X_1}{n_1}$ for sample 1 $p_2 = \dfrac{X_2}{n_2}$ for sample 2 |

## KEY TERMS

alternative hypothesis, 340
critical limits C, 364
error (type 1, type 2), 342
hypothesis (null, alternative), 340
hypothesis testing, 338
level of significance, 345

operating characteristic (OC)
  curve, 366
power curve, 366
*p*-value, 361
risk of error ($\alpha$, $\beta$), 369
significant difference, 341
test statistic, 344

## CHAPTER 9 PROBLEMS

44. A drug (tablet) is supposed to weigh 10.0 grams with $\sigma = 2.0$ grams. If each tablet in a sample of 100 of the tablets has an average weight of 9.0 grams, can we say that the average weight is statistically less than 10.0 grams? Use $\alpha = 0.025$.

45. A national study has determined that in the mid 1960s the average life span for U.S. males was 66.8 years. If a random sample suggests the average had become 67.7 by the early 1990s, would this evidence a significant increase in mean life span? Use $\alpha = 0.05$ and $\sigma_{\bar{x}} = 0.4$ years.

46. A machine is set to produce no more than 0.10 defective parts when properly adjusted. After this machine had been in operation for some time, a sample of 100 pieces was tested. Fifteen pieces were found to be defective.
    a. Is there evidence, at the 5% level, that the machine needs readjustment?
    b. Using your hypothesis, describe the possible error that is controlled at 5% risk.

47. Barten Fish Products Co. claims that 95% of their run-of-the-sea fish sticks are cod. A recent survey by U.S. FDA inspectors showed that 12 of 140 packages were not cod. Test the validity of the firm's claim at the 10% level of significance against the alternative that 95% is not the correct percentage. State all steps of your test explicitly and clearly.

48. From experience, we know that 5% of certain articles produced in a leather goods factory are defective. A newly hired worker, who has produced 600 of these articles, made 42 defective pieces.
    a. Does this cast doubt on the person's ability to perform the job? Suppose, as the supervisor, that you transferred the worker for quality reasons. Make a test to justify your action, $\alpha = 0.05$.
    b. State the risk involved if the previous finding is "Reject $H_0$." Use everyday language.

49. The Nurry Brothers Corp. markets mixed nuts containing peanuts, cashews, almonds, and so on. Due to a record-keeping error, a large container has mixed nuts in unknown proportions. If, upon inspection, a random sample of 100 nuts yields 40 peanuts, can we safely say that among all of the nuts, less than half are peanuts? Design your test so that if exactly half were peanuts, you would make a wrong conclusion 2.5% of the time.

50. A tire recapping shop is attempting to estimate the average defect-free life in miles, for their product.
    a. If a random sample of 121 tires showed a mean defect-free life of 12,327 miles, estimate with 95% confidence the true average defect-free miles. Let $\sigma = 2,112$ miles.
    b. Would the shop be fairly safe in placing a 12,000-mile warranty on its product? Assume the normal distribution and make a one-tailed test using $\alpha = 0.025$.

51. A real estate sales instructor feels that her students average 85 or better on the state licensing examination. An examiner who wishes to test this claim surveys records of

49 of the instructor's former students. These records show $\bar{X} = 81$. Test the instructor's claim at $\alpha = 0.05$. Let $\sigma = 12.0$.

52. Indicate whether the following are true or false.
    a. Both the $Z$- and $t$-distributions have means of zero and standard deviations of 1.
    b. In hypothesis testing, $\beta$ is the maximum probability of committing a type 1 error.
    c. If $\alpha$ is fixed, then as the sample size is increased, $\beta$ is reduced for all values of the parameter in the region where $H_0$ is false.
    d. The $t$-distribution is theoretically correct for a one-sample test of $H_0$: $\mu =$ some value $\mu_0$ when $\sigma$ is unknown and $X$ is normally distributed.

53. Suppose scores 102, 98, 104, 105, 108, 106, 104, 99, 95, and 99 were obtained by ten employees of an international firm on an imports/exports regulations test. Test the research hypothesis that the population mean score exceeds 100. Let $\alpha = 0.05$ and assume a normal distribution of scores.

54. Charlie's supervisor is concerned about the amount of time Charlie spends on assembly jobs. Charlie thinks he averages less than 50 minutes for each job. As part of the investigation, the supervisor reviewed the records of eight jobs and recorded the times: 39, 35, 42, 50, 51, 40, 42, and 53 minutes. Using this data, which gives $s = 6.5$ min, can Charlie justify a claim of averaging under 50 minutes? Use $\alpha = 0.025$. Assume a normal distribution.

55. A sample of 16 insurance policies is selected to test the hypothesis that the mean value of policies is \$16,500. The sample shows $\bar{X} = \$15{,}990$ and $s = \$400$. What is the calculated $t$-score? Assume a normal distribution.

56. A package-filling device is set to fill cereal boxes with a (mean) weight of 20.10 ounces of cereal per box. The production supervisor suspects that the machine may be overfilling the boxes, thereby increasing material costs. A random sample of 25 boxes is weighed in order to test the claim that the machine is operating as it should against the alternative that it is overfilling. The sample mean turns out to be 20.251 ounces and $s = 0.05$ ounces. Is the machine overfilling the packages (at the 5% level)? Assume a normal distribution for the fills.

57. A manufacturer of elevator cable advertises that its cable has a mean breaking strength of 120,000 pounds. A laboratory test for 20 cables showed a mean of only 117,000 pounds with a standard deviation of 10,000 pounds. Test the manufacturer's claim at a 0.05 level of significance against the alternative that the cable strength is less than 120,000 pounds. Assume a normal distribution.

58. For over two years, doctors at Sacred Heart Medical Center have used a fixed dose of a certain drug to bring about an average increase in the pulse rate of 10.0 beats per minute. A group of nine patients given the same dose from a new shipment of this drug showed increases of 16, 15, 14, 10, 8, 12, 13, 20, and 9 beats per minute. Assume a normal distribution and a sample standard deviation of 4.0 beats per minute.
    a. Is there evidence that the reaction to this shipment of the drug is different from the past response? Use $\alpha = 0.05$.
    b. Establish a 95% confidence interval estimate for $\mu$ based on this sample data.

59. Electricar, Inc., manufactures electric car batteries that it claims have a mean life of 30 months. A research organization tests ten batteries and finds a mean of 25 months with standard deviation of 5 months. Test the firm's claim at a 1% level of significance against the alternative that the mean life $\neq$ 30 months. State all steps of your test procedure clearly. What assumptions are required for your test?

60. Given: $H_0$: $\mu = 195$ pounds, $H_A$: $\mu > 195$ pounds. For $n = 100$ and $\sigma = 16$ pounds, what is the $\beta$-risk when the true mean is $\mu = 200$ pounds? Use a critical limit of $C = 197$ pounds.

61. A company has purchased a large quantity of steel wire. The supplier claims the wire has a mean tensile strength of 80 pounds or more. The company wishes to test a sample of 64 pieces against an alternative hypothesis that the mean strength is less than 80 pounds. They find $\bar{X} = 79.1$. Let $\sigma = 5.6$ pounds. Use a one-tailed test at a 0.05 level of significance and state your steps.
   a. Should the claim be accepted?
   b. What is the $\beta$-risk if the true mean strength is really 78 pounds?
      *Note:* Use the $\alpha = 0.05$ level to set a value for the critical limit $C$.

62. A metals plant produces castings that are thought to have an average weight of 320 pounds with a standard deviation of 20 pounds. For samples of $n = 100$, the standard error of the mean is 2.0 pounds and decision rule limits are set so as to hold the probability of a type 1 error to 0.05 (which is divided into two tails). What is the probability of a type 2 error if the mean weight is really 315 pounds?

63. A tourist shop chain store recorded per-store mean sales of $326 per day during the summer season a year ago. Management has hired you to test (at a 0.05 level) the null hypothesis that mean daily sales during this year's season are equal to last year's against the alternative that they are greater this year. Daily sales during the summer are normally distributed with a standard deviation of $10.
   a. State the null and alternative hypotheses.
   b. State an expression for the critical limit $C$ (in dollars using $\mu$, $Z$, $\sigma$, and $\sqrt{n}$) such that the probability of a type 1 error will be 0.05 when the population mean is $326.
   c. State an expression for the critical limit $C$ for which the probability of type 2 error will be 0.10 when the population mean is $330.
   d. Equate the expressions from parts (b) and (c) to find the sample size $(n)$ and the critical limit $(C)$ for the desired test.

64. A Denver manufacturer advertises that its nylon fishing line will support a constant load of 20 pounds. A sports association collects 100 randomly selected samples, tests them, and finds that the average constant load they will support is 19.5 pounds. $\sigma = 2.0$ pounds.
   a. Test the manufacturer's claim at a 10% level of significance against the alternative that the mean $\neq 20$ pounds.
   b. How significant is the difference determined in part (a)?
      (*Note:* This requires determining the area outside the calculated $Z$-value and doubling it in order to include both tails.)
   c. Provide a 95% confidence interval estimate of the true population mean strength.

65. In the production plant of a concrete block manufacturer, the standard deviation of the weight of blocks is known to be $\sigma = 1.5$ pounds, and the mean is supposed to be 5.0 pounds. A sample of 100 blocks was taken to test the hypothesis that the true mean weight was 5.0 pounds against $H_A$: $\mu \neq 5.0$ pounds. If the test was designed to yield a probability of type 1 error of 0.0455, what is the probability of type 2 error if the true mean is
   a. 5.3 pounds?
   b. 4.4 pounds?
   c. How large should the sample be if the probability of acceptance of the hypothesis is to be limited to 0.005 if the true mean is 4.5 pounds, and the probability of type 1 error is 0.0455 if the true mean is 5.0 pounds?
      (*Note:* This corresponds to a two-tailed test.)

66. Plastic Boats Corp. advertises that its boats have an average impact strength of 10,000 psi or more. Recently a trade association randomly selected and tested segments from 64 boats and found the mean impact strength to be 9800 psi. Use $\sigma = 800$ pounds per square inch.

a. Test the firm's claim at a 5% level of significance against the alternative that the mean strength is less than 10,000 psi.
b. How significant is the difference determined in part (a)?
   (*Note:* This requires determination of the area outside the calculated $Z$-value.)
c. What is the minimum strength (i) in $Z$ units and (ii) in psi that the sample mean must exhibit if the trade association is to accept the claim at this 5% level? Use $n = 64$.

## STATISTICS IN ACTION: CHALLENGE PROBLEMS

What is the practical significance of statistical significance testing? Sawyer and Peter present several issues concerning the use of statistical hypothesis testing procedures.

Their intensive but readable article complements our discussion and provides a broadening perspective of what hypothesis testing is. The following questions relate three common misconceptions about hypothesis tests. In each one you are asked to make a choice and then give reasons for your choice.

SOURCE: For Problems 67–69—Sawyer, Alan G., and J. Paul Peter. "The Significance of Statistical Significance Tests in Marketing Research." *Journal of Marketing Research* 20 (May 1983): 122–133.

67. Choose either (a) or (b) as the true statement and explain your choice.
   a. A *p*-value, $p = 0.05$, means there is a probability of only 0.05 that the observed results were caused by chance.
   b. A *p*-value of 0.05 means that if the null hypothesis is true, then the chance is 5 in 100 of getting a mean difference this large or larger.

68. Choose either (a) or (b) as the true statement and explain your choice.
   a. Rejecting the null hypothesis at a predetermined *p*-level, say, 0.05, supports the inference that sampling error is an unlikely explanation of the results, but it does not make the alternative hupothesis valid.
   b. A predetermined *p*-value of 0.05 means its complement, $1 - 0.05 = 0.95$, is the probability that the alternative hypothesis is true.

69. Choose either (a) or (b) as the true statement and explain your choice for a test of hypothesis on a mean from a normally distributed population with known standard deviation.
   a. Given the same $\alpha$-level, the probability for rejecting the null hypothesis is the same for a random sample whether the sample size is $n = 20$ or $n = 100$.
   b. For the same $\alpha$-value, rejecting a null hypothesis from a sample of 100 gives a higher probability of a valid rejection, versus rejecting with a sample of 20.

70. Martilla and Garvey suggest that once the sampling and statistical test are determined (random sampling, $H_0: \mu_1 = \mu_2$, $\sigma$'s unknown), statistical significance depends on three factors:

   **1.** The size of the difference between the two means
   **2.** The sample size
   **3.** The standard deviations of the two samples

SOURCE: Martilla, John A., and Dan W. Carvey. "Four Subtle Sins in Marketing Research." *Journal of Marketing* 39 (Jan. 1975): 8–15.

Questions   Use the table below, reproduced from page 11 of Martilla and Garvey's article, to explain the effects of these three factors on obtaining a statistically significant difference. Comparison 1 is a starting place; Comparison 2 relates to a change in means; Comparison 3 involves increased sample size; and Comparison 4 considers the precision, that is, the size of the standard deviations.

Hypothetical Average Weekly Unit Sales Comparisons for Two Brands*

| Analysis | Brand A | Comparison 1 Brand B | Comparison 2 Brand B | Comparison 3 Brand B | Comparison 4 Brand B |
|---|---|---|---|---|---|
| Mean | 85 | 87 | 88 | 87 | 87 |
| Sample size | 100 | 100 | 1000 | 100 | 100 |
| Standard deviation | 8 | 12 | 12 | 12 | 6 |
| Computed Z-value | | 1.38 | 2.08 | 2.26 | 2.00 |
| Level of statistical significance | | N.S. | 0.05 | 0.05 | 0.05 |

* Computations are made using a two-tailed Z-test, with a value of 1.96 required for statistical significance at the 0.05 level.

## BUSINESS CASE

*A Case of Fair Pricing**

By the time Tom McKee phoned his attorney's office, the situation looked almost hopeless. He had already signed the agreement to purchase the Northside Supermarket, and it looked like he had been taken! He had agreed on the price of $53,421 for the 6018 types of items in inventory on the shelves. Unfortunately, he hadn't checked the marked prices until this morning, when he unlocked the door as the new owner. But after a quick check he knew the gimmick. Sure the quantity in inventory was accurately reported. The trouble was, it was overpriced!

Now the problem was how to prove it was overpriced, in a very short time. His attorney said he would need evidence in order to bring suit, but to inventory every item in the store now would mean keeping the doors closed for three days. He couldn't afford to do that. On the other hand, if he started selling the goods tomorrow, his evidence would be going out the door. He needed to come up with a solution now—or the case was lost.

Tom saw that his evidence would have to come from a sample. He grabbed a pencil and paper and started down the aisle, picking typical items and recording the prices. He went from Kellogg's Corn Flakes to Van Camp's Pork and Beans and even to paper clips in order to get a good representative sample of the prices.

After sampling prices from 100 items, Tom felt the data were sufficient (also, he was using up selling time). Later, he went over the data with his lawyer. The 100 items had a price total of $105.60. Tom had brought along a price list from the local grocery distribution center, which revealed that the "competitive standard" for the same 100 items was $102.50. So the mean for the difference in prices was ($105.60 − $102.50)/100 = $0.031. Tom had also calculated a standard deviation for the differences in prices for the individual items. That number was $s = \$0.125$. The attorney was convinced that Tom had a good case against the former owner, but, just for insurance, he invited a business professor at the local university to check it over for him. The next week the attorney received a letter from the professor (see letter).

## Business Application Questions

71. Explain why Professor Knight used a greater-than alternative hypothesis, rather than a does-not-equal alternative, for his test.

* This is from a case brought to one of the authors, but the names and figures have been modified to protect the interests of the parties involved.

Mr. Dan Murphy, Attorney
First National Building
City

Dear Mr. Murphy:

I have done some preliminary analysis of the data you left me last week. These data were summarized to the difference of prices (competitive standard and Northside) for a sample of $n = 100$ items, and can be described as follows.

Sample mean of price differences = $0.031 per item

Sample standard deviation of price differences = $0.125

I conducted a test at a 5% level of significance to determine whether the sample data from Northside Supermarket support the hypothesis that the per-item average price difference (from a competitive standard grocer) is $0. It can be summarized as follows:

1. $H_0$: $\mu = \$0$
   $H_A$: $\mu > \$0$
2. $\alpha = 0.05$
3. Test statistic: $t$: Reject if calculated $t > 1.645$ $(= t_\infty)$
4. Calculation:

$$t = \frac{\bar{X} - \mu_0}{s_{\bar{X}}} \quad \text{where} \quad s_{\bar{X}} = \frac{s}{\sqrt{n}} = \frac{\$0.125}{\sqrt{100}} = \$0.012$$

$$= \frac{\$0.031}{\$0.012} = 2.58$$

5. Conclusion: Since $2.58 > 1.645$, reject the hypothesis that the per-item average price difference is $0. This difference exceeds $0, so the Northside prices for the 100 items exceed the $1.025 per-item average price from the standard grocers.

The assumptions underlying the above conclusions are that a random sample of $n = 100$ items was taken from a population consisting of 6018 types of items (e.g., Crisco Oil, Zee Napkins, Best Foods Mayonnaise, etc.) and that readings and recordings were accurately made. From a statistical standpoint, the sample size was adequate. My understanding is, however, that sampling was done on a "walk-through" basis. Because the location of shelf items in a grocery store stems from a conscious marketing strategy, this type of sampling may result in obtaining a nonrepresentative sample. In the data you provided me, the ordering of prices on the list does not appear to be random (as measured by a nonparametric runs test).

Had your client used some sort of random number selection device, based upon the inventory number of the items, I believe you would have a solid case. However, due to the questionable sampling procedure I could not pass judgment of the representativeness of the items in the sample. Sorry.

Very truly yours,

*John A. Knight*

John A. Knight, Professor

72. Describe the type 1 and type 2 errors that could occur in this test of hypothesis. Which error, type 1 or type 2, would be most critical to Tom McKee's business? Explain.

73. Use the sample values and the hypothesis test calculations plus Table 2 in Appendix B to determine a *p*-value for this test (this assumes the α-value was not given). Would the use of a *p*-value contribute anything to this test? Why or why not?

74. Explain, in layman's language, the problem that John Knight indicates makes this sampling questionable and invalidates the test.

## COMPUTER APPLICATION

This application extends the marketing problem of targeting households with 3 + persons and with the head of household 35 to 50 years old. For that population of 3757 PUMS records, the mean income is $\mu = \$26,936$ with $\sigma = \$15,715$. So the population mean is known.

Our objective is to demonstrate the meaning of the α-risk, using samples from the population. We have selected 100 random samples, each of $n = 100$ records, from the $N = 3757$-record population. (The Chapter 8 Computer Application shows a SPSS routine and the DESCRIPTIVES procedure for doing this, except here $n = 100$ and requires 100 random samples.)

The 100 sample means (dollars) are shown in Table 9.5.
An $\alpha = 0.05$ test of hypothesis for

$$H_0: \mu = \$26,936 \quad \text{versus} \quad H_A: \mu \neq \$26,936$$

uses a two-tailed $Z$-test defined by "Reject $H_0$ if |calculated $Z$| > 1.96."

**TABLE 9.5**  Sample Mean Income for 100 Random Samples, $\bar{X}$'s

| Sample | $\bar{X}$ | Sample | $\bar{X}$ | Sample | $\bar{X}$ | Sample | $\bar{X}$ | Sample | $\bar{X}$ |
|---|---|---|---|---|---|---|---|---|---|
| 1 | 26378 | 21 | 23099 | 41 | 24961 | 61 | 28889 | 81 | 28086 |
| 2 | 27442 | 22 | 25483 | 42 | 26047 | 62 | 24990 | 82 | 30205 |
| 3 | 27073 | 23 | 26001 | 43 | 27410 | 63 | 27924 | 83 | 27514 |
| 4 | 27057 | 24 | 26988 | 44 | 26327 | 64 | 26810 | 84 | 26240 |
| 5 | 28130 | 25 | 28967 | 45 | 27772 | 65 | 25298 | 85 | 24817 |
| 6 | 26165 | 26 | 28963 | 46 | 26194 | 66 | 26988 | 86 | 27115 |
| 7 | 26875 | 27 | 27454 | 47 | 27232 | 67 | 27560 | 87 | 29334 |
| 8 | 25102 | 28 | 25354 | 48 | 27227 | 68 | 21968 | 88 | 27119 |
| 9 | 27876 | 29 | 27967 | 49 | 27776 | 69 | 27965 | 89 | 25477 |
| 10 | 24116 | 30 | 25906 | 50 | 27306 | 70 | 29212 | 90 | 27960 |
| 11 | 26507 | 31 | 28635 | 51 | 25946 | 71 | 26466 | 91 | 26940 |
| 12 | 27419 | 32 | 26357 | 52 | 28662 | 72 | 27697 | 92 | 29522 |
| 13 | 27722 | 33 | 24750 | 53 | 25217 | 73 | 26615 | 93 | 29050 |
| 14 | 25935 | 34 | 25678 | 54 | 29526 | 74 | 26253 | 94 | 29626 |
| 15 | 27697 | 35 | 26595 | 55 | 30069 | 75 | 28414 | 95 | 27283 |
| 16 | 24496 | 36 | 28036 | 56 | 26729 | 76 | 27110 | 96 | 28260 |
| 17 | 27081 | 37 | 28857 | 57 | 26209 | 77 | 27392 | 97 | 25968 |
| 18 | 25210 | 38 | 27649 | 58 | 27283 | 78 | 29022 | 98 | 26762 |
| 19 | 29653 | 39 | 27095 | 59 | 26489 | 79 | 28014 | 99 | 26764 |
| 20 | 28143 | 40 | 29700 | 60 | 28625 | 80 | 26378 | 100 | 29630 |

In this case we know the answer! $\mu = \$26,936$. And we know that for $\alpha = 0.05$, on the average, 5 of 100 random samples will erroneously reject this true null hypothesis. So, how many of our samples would erroneously lead to the conclusion "Reject the (true) null hypothesis"? The answer is, any that produce

$$P\left(\frac{\bar{X} - 26,936}{15,715/\sqrt{100}} \middle| \mu = \$26,936\right) < -1.96$$

or

$$P\left(\frac{\bar{X} - 26,936}{15,715/\sqrt{100}} \middle| \mu = \$26,936\right) > 1.96$$

The calculation is easily performed by a Minitab run:

```
MTB > SET INTO C1
DATA> ←───────────────────── key in the 100 means from Table 9.5
DATA> END
MTB > NAME C1 'INCMEAN'
MTB > PRINT 'INCMEAN'
MTB > LET ZMINC = 10 * ((C1 - 26936)/15715)    Z-scores for
                                                mean incomes
MTB > PRINT 'ZMINC'
MTB > SAVE 'B:INC'
MTB > STOP
```

The resulting $Z$-scores are displayed in Table 9.6. You can easily observe those that exceed $+1.96$ or are smaller than $-1.96$. There are four of them.

**TABLE 9.6**   $Z$-Scores

```
ZHINC
 -0.36     0.32     0.09     0.08     0.76    -0.49    -0.04
 -1.17     0.60    -1.79    -0.27     0.31     0.50    -0.64
  0.48    -1.55     0.09    -1.10     1.73     0.77    -2.44
 -0.92    -0.59     0.03     1.29     1.29     0.33    -1.01
  0.66    -0.66     1.08    -0.37    -1.39    -0.80    -0.22
  0.70     1.22     0.45     0.10     1.76    -1.26    -0.57
  0.30    -0.39     0.53    -0.47     0.19     0.19     0.53
  0.24    -0.63     1.10    -1.09     1.65     1.99    -0.13
 -0.46     0.22    -0.28     1.07     1.24    -1.24     0.63
 -0.08    -1.04     0.03     0.40    -3.16     0.65     1.45
 -0.30     0.48    -0.20    -0.43     0.94     0.11     0.29
  1.33     0.69    -0.36     0.73     2.08     0.37    -0.44
 -1.35     0.11     1.53     0.12    -0.93     0.65     0.00
  1.65     1.35     1.71     0.22     0.84    -0.62     0.11
 -0.11     1.71
```

This result (four samples would lead to "Reject $\mu = \$26{,}936$") is quite consistent with $\alpha = 0.05$ because for all possible random samples of 100 records, 5% would erroneously reject this (true) null hypothesis. Our 100 samples included four that would reject it. This (4%) is very close to the overall rate (5%).

Through $\alpha$ we define the long-term expected results, considering the possibility of many (sample) repetitions of that procedure. Thus, we did not interpret $\alpha$ for the result of any one sample. In practice, however, it is often practical to draw only one sample. Then the sample result becomes the single expression of the procedure. We assume that it is a good expression for the procedure, so we determine from it whether to reject or not reject the null hypothesis. In doing this, with preset $\alpha = 0.05$, we will incorrectly reject a true null hypothesis only 5 times in 100.

Problem 83 invites you to use the $Z$-values in Table 9.6 to evaluate the results for these 100 samples for $\alpha = 0.10$.

## COMPUTER PROBLEMS

*Data Base Applications*

Problems 75 through 80 can be answered with data from the 100-record PUMS sample in Appendix A. However, if it is available, the 20,116-record PUMS data base is a better source.

75. Select a random sample of 50 households from the PUMS data base. Use this to test the hypothesis that $\mu =$ mean household income exceeded \$19,000. Assume a normal distribution of incomes and test at $\alpha = 0.05$.

76. Select a random sample of 10 records from the PUMS data base. Use this to test the hypothesis that $\mu =$ mean household income exceeds \$19,000. Assume a normal distribution of incomes, and test at $\alpha = 0.05$. Compare this test with the preceding problem on (a) the test statistic, (b) the size of the standard error, and (c) the test conclusion. (d) Discuss, but do not calculate, the relative $\beta$-risk.

77. Select another random sample of 10 records from the PUMS data base. Use the age for person #1—the head of household—to test the hypothesis that the mean age for person #1 is 47.0 years. Assume a normal distribution for person #1 ages. Let $\alpha = 0.05$ and use a two-tailed alternative hypothesis.

78. What percentage of all the heads of household on the PUMS data base are males ($\pi$)? Test $H_0$: $\pi = 0.70$ against $H_A$: $\pi \neq 0.70$ using a random sample of 50 records from the PUMS data base. Use $\alpha = 0.10$ and state practical conclusions.

79. Draw a random sample of 50 PUMS records and record for each person #1 the number of weeks worked. Let $\pi =$ the proportion who worked 30 weeks or more. Test $H_0$: $\pi = 0.75$ against $H_A$: $\pi < 0.75$. Use $\alpha = 0.05$ and state your conclusions.

80. For the sample in the preceding problem, record the age of the head of household (person #1). Use this to test $H_0$: $\mu - 40 = 6$ years against $H_A$: $\mu - 40 \neq 6$ years. Perform this test by the following methods.
   a. The classical decision approach, with $\alpha = 0.05$
   b. The $p$-value approach
   Explain the difference in your conclusions.

Problems 81 through 84 use data that are provided in the problem or data from the Computer Application.

81. Here is a Minitab routine to retrieve the file INF1 (saved from Problem 69 in Chapter 8) and then perform a ZTEST of significance, using the *p*-value. Assuming a two-tailed test, state your test conclusion. Recall that this is a test on $\mu = \$19,698$ for the 20,116 PUMS-record population and that $\sigma = \$14,931$ is known.

```
MTB > RETRIEVE 'B: INF1'
MTB > PRINT 'INCOME'

INCOME
   20115     21580     16840     11215     11940     17410     21010     22630      9005
   38510     17360      3995     32780     18820      5405     28115      5565     75000
    5020     28825     12365      7515      4880     16170     36800     20770     15010
   20010     75000     75000     21010     16815     11985     14005     39420      4050
    3515     17690      4710     16505     42060      2580      6150     56015     10240
   28010      5605      6775     28015      9005
```

```
MTB > ZTEST OF MU = 19698 ASSUMING SIGMA = 14931 USING THE DATA IN C1;
SUBC> ALTERNATIVE = 0.


TEST OF MU = 19698.000 VS MU N.E. 19698.000
THE ASSUMED SIGMA = 14931
                  N       MEAN     STDEV    SE MEAN      Z    P VALUE
INCOME           50   20776.400 18031.631  2111.562    0.51     0.61
```

82. Use the data from Problem 70 in Chapter 8, repeated here, to test the hypothesis that the mean age for heads of household in the 20,116-PUMS file is 47.58 years. Use a *t*-test with $\bar{X} = 46.92$ and $s = 16.69$. Minitab users: After entering the data, use

```
MTB > TTEST OF MU = 47.58 USING THE DATA IN C1;
MTB > ALTERNATIVE = 0.
```

Perform a statistical test of hypothesis and state your conclusion. Let $\alpha = 0.05$.

```
AGE1
   76    56    44    28    33    74    30    27    31    41    26    53    54
   31    57    40    77    30    23    43    24    74    79    65    52    37
   57    35    32    35    44    30    66    63    38    36    69    34    68
   33    42    70    52    53    55    44    69    39    56    21
```

83. Use the *Z*-scores from the Chapter 9 Computer Application to explain how this expression of 100 random samples demonstrates the meaning of an $\alpha = 0.10$ test of hypothesis. Remember that in this case the population mean is known to be $\mu = \$26,936$ and a two-sided alternative hypothesis is used.

84. These data are the age means for 100 random samples, each of 100 records, from the population of 3757 PUMS records defined by 3+ persons and householder aged 35 to 50.

AGEMEAN

| | | | | | | | | |
|---|---|---|---|---|---|---|---|---|
| 42.52 | 42.29 | 42.28 | 42.78 | 42.41 | 41.80 | 41.34 | 41.39 | 42.98 |
| 41.63 | 41.46 | 42.32 | 41.60 | 42.29 | 42.13 | 40.87 | 41.45 | 42.19 |
| 41.96 | 41.88 | 41.79 | 41.57 | 42.18 | 42.01 | 41.97 | 41.57 | 42.02 |
| 41.66 | 41.28 | 42.79 | 42.16 | 42.59 | 41.09 | 42.07 | 41.09 | 41.61 |
| 41.88 | 41.52 | 42.29 | 42.05 | 42.47 | 41.62 | 42.11 | 41.86 | 41.94 |
| 41.49 | 41.87 | 42.99 | 42.61 | 41.69 | 41.57 | 42.25 | 42.50 | 41.76 |
| 42.95 | 42.77 | 41.69 | 41.93 | 41.49 | 41.56 | 41.90 | 41.98 | 41.50 |
| 42.63 | 41.62 | 42.01 | 41.67 | 42.01 | 41.28 | 42.42 | 41.67 | 41.70 |
| 42.56 | 42.25 | 41.73 | 41.37 | 41.44 | 41.55 | 42.31 | 42.52 | 41.19 |
| 41.58 | 41.51 | 42.27 | 41.88 | 40.96 | 42.02 | 42.97 | 42.40 | 41.97 |
| 41.82 | 42.12 | 42.70 | 42.41 | 42.19 | 40.58 | 41.69 | 41.61 | 42.45 |
| 42.11 | | | | | | | | |

Use this data and the procedure outlined in the Computer Application to explain the meaning of $\alpha = 0.05$ for a test of hypothesis on $H_0$: $\mu = 41.91$ against $H_A$: $\mu \neq 41.91$ years. *Suggestion:* Use $\sigma = 4.57$ years (known) and a Minitab computer run to calculate $Z$-scores; then interpret your results.

**FOR FURTHER READING**

[1] Book, Stephen A., and Mark J. Epstein. *Statistical Analysis in Business.* Glenview, IL: Scott Foresman, 1982.

[2] Conover, W. J., and Ronald L. Iman. *Modern Business Statistics*, 2nd ed. New York: John Wiley, 1989.

[3] Dunnett, Charles W. "Drug Screening," in *Statistics—A Guide to the Unknown*, 3rd ed., edited by Judith Tanur et al. Pacific Grove, CA: Brooks-Cole, 1989.

[4] Lapin, Lawrence. *Statistics for Modern Business Decisions*, 5th ed. San Diego: Harcourt Brace Jovanovich, 1990.

[5] McClave, James T., and P. George Benson. *Statistics for Business and Economics*, 4th ed. San Francisco: Dellen, 1988.

[6] Pfaffenberger, Roger C., and James H. Patterson. *Statistical Methods for Business and Economics*, 3rd ed. Homewood, IL: Irwin, 1987.

[7] Presby, Leonard. *Compstat: Solving Statistical Problems by Microcomputer.* New York: Random House, 1984.

[8] Sawyer, Alan G., and J. Paul Peter. "The Significance of Statistical Significance Tests in Marketing Research." *Journal of Marketing Research* 20 (May 1983): 122–133.

# 10

# EXTENSIONS OF ESTIMATION AND HYPOTHESIS TESTING

Suppose you are the operations manager of an advertising firm that distributes large quantities of advertising pieces as inserts in magazines. Your supplier ships the inserts to you in cartons that are supposed to contain 1500 pieces with an average weight of 68.5 pounds. Your concern is to check against an undercount because one of your clients, a sportswear manufacturer, has questioned whether you are distributing the required number of advertisements. If the quota is not met, you are short-changing your customer, and your firm is subject to a stiff fine. You weigh five cartons from a large shipment on the loading dock and obtain weights of 67.3, 68.7, 64.3, 67.1, and 68.2 pounds. Should you allow the supplier to unload the shipment or refuse it due to insufficient weight?

This situation involves the use of a small sample to infer about the mean weight of the cartons in the shipment. It uses the $t$-statistic and associated probability tables. Recall from Chapters 8 and 9 that the $t$-distribution is similar to the standard normal ($Z$-) distribution but affords greater variability. The distribution shape becomes more like the standard normal curve as the sample size is increased.

In Chapters 8 and 9, we studied estimation and hypothesis testing procedures for making inferences about population means from large samples, and from normal distributions with known variance. We did not require normality in every case because the central limit theorem ensured that we would have normal sampling distributions as long as we used large samples of $n > 30$. Nor did we consider inference concerning the variance. Now that the foundation is laid, it is time to develop the procedures for making inferences about differences in population means, using samples of 30 or fewer observations from normal distributions but with $\sigma^2$ unknown. This chapter describes inference about means for $n \leqslant 30$, and for variances for any size sample. Both situations require that the random variable have a normal distribution. Inference on means when the distribution is not normal is deferred to Chapter 17.

We will discuss three probability distributions in this chapter. First is the $t$-distribution, which is used for inference on differences of means for normal distributions with unknown $\sigma^2$'s. Second is the chi-square (pronounced "kī-square") distribution. Chi-square is useful for inference about the variance of a single population. Again, a normally distributed random variable is an essential condition. Third are the $F$-distribution and the associated $F$-statistic for inference about the variances from two populations. All three distributions assume that the initial variables are normally distributed. The $t$-statistic relates to small samples, whereas the chi-square and $F$ statistics can apply to any sample size. Each one relates in a different way to the normal probability distribution and has a distinctive probability distribution with tables that appear in Appendix B. One commonality is that all three are defined by a measure of the sample sizes called degrees of freedom, or df. The distributions themselves are (mathematically) related, but that relationship goes beyond the scope of our inquiry here.

Upon completing Chapter 10, you should be able to

1. distinguish decision problems that require use of the $t$- rather than the $Z$-distribution;
2. look up and use values from the $t$-, chi-square, and $F$-distribution tables;
3. make decisions about means by using paired observations to control extraneous differences;
4. make inferences about differences in means for independent samples from normal distributions;
5. test and estimate the values for the variances of one or of two distributions.

The concepts and inference procedures build upon those used in Chapters 8 and 9. However, we begin with different assumptions and use different statistical distributions, so we need new formulas.

## 10.2 INFERENCE CONCERNING MEANS FROM TWO POPULATIONS: PAIRED OBSERVATIONS

In Chapter 9 we focused on one-sample tests. We turn now to the use of the $t$-distribution in two-sample tests. These include tests concerning (1) paired observations and (2) comparisons of two population means using independent random samples. Both situations require the earlier assumptions of normal distributions with unknown standard deviations. Paired observations are considered here, using paired-differences procedures, so the latter case is covered in Section 10.5.

Paired differences are appropriate when the business variable being tested experiences two environments. Examples are productivity for a data analyst with versus without computer-assisted calculations, the efficiency of a copy machine before and after maintenance, or the accuracy of two identical machines under the same work conditions (same operator, run time, speed, etc.). Pairing is an attempt to eliminate known but unwanted variances so that the resulting test response can focus on the actual difference under study. We remove un-wanted or extraneous variations from the test calculations by pairing the data. Then the resulting test becomes more definitive of the business question under investigation.

Did you take the ACT, SAT, or other test for admission to your college? Some persons take these tests several times. How much improvement in a score might be expected for students who take an ACT test two times? Is the difference really significant? The response to this inquiry would come from a paired differ-ence test. Like sample means, the differences (or sums) of normally distributed variables are also normally distributed. And the resultant difference is a random variable. This enables us to use the single sample $t$-distribution for inference about *paired differences*.

> The paired-differences procedure matches like or related units. This restricts the variability measure to the error among pairs—and removes other, extraneous variations.

The calculations use differences between pairs, denoted by $X_d = X_A - X_B$. Here A and B can denote two measurements for one individual. For example, the first and second scores on an ACT test would constitute a pair of scores. Or A and B could represent the two ACT scores by different individuals who have been "paired" on the basis of IQ, comparable high-school training, age, sex, locale, and the like. The paired-differences test is only one procedure for controlling sources of variation to assure stronger inference. Related and more-comprehensive procedures will be described in Chapter 11.

Business situations in which paired differences tests are used include com-parisons of industrial processes or machines, before and after comparisons of workers, analysis of the monthly performance levels of competing products, testing and retesting in job training, and tests of similar accounts to find ac-counting errors. The calculations can often be performed with hand calculators.

The procedure for paired-difference testing is relatively straightforward. The random variable is $X_d =$ the difference in values of the matched pairs. Then, for small samples taken from normal distributions, the *sampling distribution for mean differences*, $\bar{X}_d$, is distributed according to a $t$-distribution.

The sampling distribution for the means, $\bar{X}_d$, of paired differences is

$$t_{n-1} = \frac{\bar{X}_d - \mu_d}{s_d/\sqrt{n}} \tag{10.1}$$

where   $n =$ the number of pairs

$X_d =$ the individual difference of pairs $= X_A - X_B$

$\bar{X}_d =$ the sample mean of differences $= \dfrac{\sum X_d}{n}$

$s_d =$ the standard deviation of paired differences

$$= \sqrt{\frac{\sum (X_d - \bar{X}_d)^2}{n-1}}$$

Assumption: The difference is normally distributed with unknown standard deviation.

Equation (10.1) is very much like (9.2). Both use the same $t$-tables, and the calculation procedures are quite similar. However, Equation (10.1) requires calculation of $X_d = X_A - X_B$ from the original observations. The df is also calculated differently. In Equation (10.1) the df = the number of pairs $- 1$. Equation (10.1) is appropriate for dependent (as opposed to independent) samples, so that the matching, pairing, or repeated measuring of the variables yields measures that reflect real differences among the pairs, and excludes other nonessential variations. A first example, illustrative of a before–after experiment, demonstrates both estimation and hypothesis testing procdures. It makes use of the sampling distribution for paired differences.

---

**EXAMPLE 10.1**

A supervisor of a cosmetic packaging operation wants to estimate the increase in production for workers who are given work methods training. She selects a random sample of ten workers. The worker's output rate is measured first as he or she begins a new job and then two weeks after the end of the training (see table). Test the hypothesis that the average output has increased by less than or equal to 30 units per period against the alternative that the increase is greater than 30 units per period. Assume that the distribution of differences is normal.

*Solution*    We apply the paired differences test with $\mu_d$ = the true mean increase in units.

| Worker | $X_A$, After Training | $X_B$, Before Training | $X_d$ | $(X_d - \bar{X}_d)$ | $(X_d - \bar{X}_d)^2$ |
|--------|-----------------------|------------------------|-------|---------------------|------------------------|
| 1 | 51 | 15 | 36 | 2 | 4 |
| 2 | 63 | 21 | 42 | 8 | 64 |
| 3 | 55 | 20 | 35 | 1 | 1 |
| 4 | 68 | 36 | 32 | −2 | 4 |
| 5 | 48 | 12 | 36 | 2 | 4 |
| 6 | 38 | 9 | 29 | −5 | 25 |
| 7 | 54 | 17 | 37 | 3 | 9 |
| 8 | 73 | 42 | 31 | −3 | 9 |
| 9 | 49 | 26 | 23 | −11 | 121 |
| 10 | 57 | 18 | 39 | 5 | 25 |
| | | Totals | 340 | 0 | 266 |

Mean   $\bar{X}_d = 34$

$$H_0: \mu_d = 30 \text{ units} \qquad H_A: \mu_d > 30 \text{ units}$$

Insofar as the level of significance is not specified, we shall assume $\alpha = 0.05$.

Test: Reject $H_0$ if calculated $t > t_{0.05,9} = 1.83$.

$$\bar{X}_d = 34, \qquad s_d = \sqrt{\frac{\sum (X_d - \bar{X}_d)^2}{n-1}} = \sqrt{\frac{266}{10-1}} = 5.44$$

$$\text{Calculated } t = \frac{\bar{X}_d - \mu_d}{s_d/\sqrt{n}} = \frac{34 - 30}{5.44/\sqrt{10}} = \frac{4}{1.72} = 2.33$$

Decision: Reject $H_0$ since $2.33 > 1.83$.

This suggests an average increase of more than 30 units for posttraining over pretraining productivity at the $\alpha = 0.05$ level. We can gain additional information by setting a confidence interval estimate about the sample mean increase. Thus, for 90% confidence,

$$\text{CL} = \bar{X}_d \pm t_{0.05,9}(s_d/\sqrt{n})$$

$$(\text{LCL, UCL}) = 34 \pm 1.83(1.72) = 34 \pm 3.15 \text{ units}$$

$$= 30.85 \text{ to } 37.15 \text{ units}$$

The true mean improvement is estimated to be within 31 to 37 units of production.

---

A second example is designed to compare the cleaning power of two types of soap by measuring how well each does in cleaning a dozen types of uniformly soiled cloths. Because some fabrics are inherently brighter than others, the test uses matched swatches of cloth to block out any other differences in the pairs. This way the researcher is assured that there is no preference given to either treatment (in this case two laundry soaps). But the tests are dependent in that each set of data (each measure of cleanliness) stems from a common (uniformly soiled) carrier. This yields an effect similar to the before and after training described in the previous example.

## EXAMPLE 10.2

For purposes of advertising, a market researcher would like to determine if one laundry soap is better than another with respect to brightness of the wash. The researcher takes swatches of 12 different types of cloth, each uniformly soiled, cuts each piece in half, and by random choice washes one part in laundry soap A and the other in soap B. The brightness of the resulting washes is then measured with a special meter. The resulting paired observations appear in the table. Is there a significant difference in the brightness induced by the two laundry soaps? Use a level of significance of $\alpha = 0.05$.

*Solution*   We assume that the sample differences come from a normal distribution. Then our null hypothesis is one of "no difference."

$$H_0: \mu_d = 0 \qquad H_A: \mu_d \neq 0 \qquad \alpha = 0.05$$

Test: Reject $H_0$ if $|\text{Calculated } t| > t_{0.025,11} = 2.20$.

The data, with differences and $t$-calculation, are listed in the table:

| Swatch | 1 | 2 | 3 | 4 | 5 | 6 | 7 | 8 | 9 | 10 | 11 | 12 |
|---|---|---|---|---|---|---|---|---|---|---|---|---|
| $X_A$ | 5 | 4 | 4 | 1 | 6 | 3 | 4 | 5 | 5 | 3 | 6 | 4 |
| $X_B$ | 1 | 3 | 4 | 2 | 4 | 3 | 2 | 3 | 6 | 5 | 3 | 2 |
| $X_d = X_A - X_B$ | 4 | 1 | 0 | −1 | 2 | 0 | 2 | 2 | −1 | −2 | 3 | 2 |

$$\bar{X}_d = \frac{\sum X_d}{n} = \frac{12}{12} = 1$$

$$s_d = \sqrt{\frac{\sum (X_d - \bar{X}_d)^2}{n-1}} = \sqrt{\frac{(4-1)^2 + (1-1)^2 + \cdots + (2-1)^2}{12-1}}$$

$$= \sqrt{\frac{48}{11}} = 2.09$$

$$\text{Calculated } t = \frac{\bar{X}_d - \mu_d}{s_d/\sqrt{n}} = \frac{1 - 0}{2.09/\sqrt{12}} = 1.66$$

Decision: Accept $H_0$ of no difference because $1.66 < 2.20$.

These data evidence no statistical difference in brightness at the $\alpha = 0.05$ level for the fabrics washed in these two laundry soaps.

---

As stated earlier, you should always be aware of the possibility of a type 2 error when the conclusion is to accept $H_0$, especially for small samples. Depending upon what true measure of difference exists, there is some chance we are wrong in concluding there is "no difference."

The paired-differences test is one of a class of statistical designs that use grouping techniques to improve test precision. The paired procedure is appropriate for two-sample tests of the mean when the responses are from normal distributions but are in some way dependent or related in a pairwise manner. In fact, provided the distribution of the differences is normal, the original distributions do not have to be normal. However, normal parent distributions guarantee a normal distribution for the $X_d$'s.

**Problems 10.2**

1. An Iowa meat company obtained figures from a random sample of ten hogs for a two-week trial on a new feed ration (see table). Using $\alpha = 0.01$, test to determine if the ration produces a statistically significant gain of weight. Assume a normal distribution of weight changes and use $H_0: \mu_d = 0$.

|         | Weight (pounds) | |
| :---: | :---: | :---: |
| Subject | Before | After |
| 1 | 142 | 145 |
| 2 | 149 | 156 |
| 3 | 137 | 138 |
| 4 | 150 | 149 |
| 5 | 155 | 152 |
| 6 | 160 | 161 |
| 7 | 128 | 128 |
| 8 | 147 | 153 |
| 9 | 164 | 174 |
| 10 | 136 | 142 |

2. A Cooperative Trade Commission is checking for equal pricing of the same type of meals (dinners) served at two locations of a restaurant chain. Assuming normal distributions, test the hypothesis, at $\alpha = 0.05$, that average prices are higher at the second location. Also establish a 95% confidence interval for the true mean difference. The numbers in the table include state taxes.

| | Cost at Location | |
|---|---|---|
| Meal | A | B |
| 1 | $14.72 | $14.67 |
| 2 | 14.81 | 15.12 |
| 3 | 13.90 | 15.01 |
| 4 | 15.16 | 14.93 |
| 5 | 14.56 | 14.98 |
| 6 | 14.13 | 14.82 |
| 7 | 14.55 | 14.56 |
| 8 | 14.38 | 15.28 |

3. A systems specialist has studied the workflow of clerks, all doing the same processing job. She has designed a new workflow layout for a clerical work-station and wants to compare average production for the new method with production for the older method. Nine clerks are available for the test. After ample familiarization with the new station, each clerk is assigned a common task. See the table for the number of units processed by each clerk. The order of stations is randomized. Assuming the necessary assumptions are met, test the hypothesis that the new station facilitates a faster work rate. Use $\alpha = 0.025$.

| Clerk | 1 | 2 | 3 | 4 | 5 | 6 | 7 | 8 | 9 |
|---|---|---|---|---|---|---|---|---|---|
| Old rate | 66 | 76 | 65 | 73 | 84 | 82 | 72 | 71 | 74 |
| New rate | 71 | 90 | 72 | 80 | 90 | 91 | 68 | 76 | 80 |

4. In a service test on wear, two brands of jogging shoes were tested by 12 different women. The figures in the table give wear time in months. The women serve to identify pairs. To assure against a bias, the order of wearing brands was made by random assignment for each woman. Assume measures are normally distributed and test the hypothesis of equal wear for the two brands. Use $\alpha = 0.05$.

| Woman | 1 | 2 | 3 | 4 | 5 | 6 | 7 | 8 | 9 | 10 | 11 | 12 |
|---|---|---|---|---|---|---|---|---|---|---|---|---|
| Brand A | 7 | 9 | 8 | 6 | 5 | 7 | 6 | 5 | 4 | 6 | 3 | 4 |
| Brand B | 8 | 10 | 8 | 5 | 3 | 9 | 8 | 5 | 5 | 9 | 5 | 7 |

# 10.3 INFERENCE CONCERNING A SINGLE POPULATION VARIANCE

The tests discussed thus far in this chapter have assumed unknown variance. We now consider some inferential procedures concerning a population variance. More often than not in business we do not know the true value for $\sigma^2$. However, we can usually determine a reasonable value by estimation or by testing procedures.

The variance can be of interest in its own right. For example, a machine that stamps lines on paper must show reasonably uniform results. Otherwise

**FIGURE 10.1**   The Relation Between the Standard Normal and Chi-Square ($\chi_1^2$)
Distributions $[N(0,1)]^2 \sim \chi_1^2$

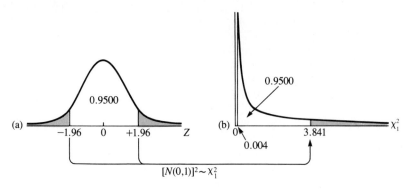

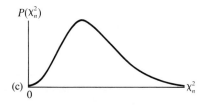

buyers will avoid using the product. In secretarial work, a key to efficiency is one's consistency, or lack of variance, in typing or word-processing activities. And a stock broker should be aware of the variability in a stock's daily prices before recommending an investment. Thus variance considerations are important to many business decisions.

## The Chi-Square Distribution and Tables

The sampling distribution for inference about a population variance relates $\sigma^2$ to its estimator $s^2$. Freund and Walpole [2]* show that this relation is described by a chi-square ($\chi^2$) distribution. This requires that the original random variable have a normal distribution. Then the chi-square distribution is the accumulation of squared deviations from independent normal random variables. The relation is illustrated in Figure 10.1.

The relation between normal and chi-square percentile values is $\chi^2 = (Z)^2$. A chi-square percentile with 1 df equals the associated $Z$-value squared. For example in Figure 10.1(a) two values, $-1.96$ and $+1.96$, are related by $(\pm 1.96)^2$ to 3.841 in Figure 10.1(b). Understanding this relation can help you understand the construction of the chi-square probability distribution, Table B.3.

Consider the entries for Table B.3 for df $= 1$.

For $\alpha = 0.95$, $\chi_{0.95,1}^2 = 0.004$ in Table B.3(a)
For $\alpha = 0.05$, $\chi_{0.05,1}^2 = 3.841$ in Table B.3(b)

---

* Numbers in brackets refer to items listed in "For Further Reading" at the end of the chapter.

**TABLE 10.1**   A Segment of the Chi-Square Table (Appendix B, Table B.3)

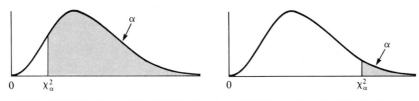

| | Upper-tail area $= \alpha$ | | | | | Upper-tail area $= \alpha$ | | | | |
|---|---|---|---|---|---|---|---|---|---|---|
| df | 0.995 | 0.99 | 0.95 | 0.90 | 0.75 | 0.25 | 0.10 | 0.05 | 0.01 | 0.005 |
| 1 | — | — | 0.004 | 0.016 | 0.102 | 1.323 | 2.706 | 3.841 | 6.635 | 7.879 |
| 2 | 0.010 | 0.020 | 0.103 | 0.211 | 0.575 | 2.773 | 4.605 | 5.991 | 9.210 | 10.597 |
| 3 | 0.072 | 0.115 | 0.352 | 0.584 | 1.213 | 4.108 | 6.251 | 7.815 | 11.345 | 12.838 |
| 4 | 0.207 | 0.297 | 0.711 | 1.064 | 1.923 | 5.385 | 7.779 | 9.488 | 13.277 | 14.860 |
| ⋮ | | | | | | | | | | |
| 29 | 13.121 | 14.257 | 17.708 | 19.768 | 23.567 | 33.711 | 39.087 | 42.557 | 49.588 | 52.336 |
| 30 | 13.787 | 14.954 | 18.493 | 20.599 | 24.478 | 34.800 | 40.256 | 43.773 | 50.892 | 53.672 |
| 40 | 20.707 | 22.164 | 26.509 | 29.051 | 33.660 | 45.616 | 51.805 | 55.758 | 63.691 | 66.766 |
| ⋮ | | | | | | | | | | |

If we can think of the squaring as folding over the normal curve about the value zero, then chi-square percentage values are the squares of the corresponding $N(0, 1)$ values. For example, $Z = +1.96$ and $-1.96$ both go into $(\pm 1.96)^2 = 3.841$. That is, the area from $Z = 0$ to $Z = +1.96$ *and* that from $Z = 0$ to $Z = -1.96$ are both included in the area for the region from $\chi_1^2 = 0$ to $\chi_1^2 = 3.841$. You will recall that $Z = \pm 1.96$ apportions off $2 \times (0.5000 - 0.4750) = 0.05$ of the area outside the center of the $Z$-distribution. This is precisely the area— 0.05—recorded at the top of the column in Table B.3 for $\chi_{0.05,1}^2 = 3.841$ as an upper-tailed percentage point. Other chi-square distribution percentages are defined similarly.

Mathematical statistics can be used to show that with random sampling, two, three, or more variables that are independently drawn from a normal (0,1) distribution will, when each is squared, sum to provide a chi-square distribution defined by two, three, or more degrees of freedom. The variable values will mix so that the shape of a chi-square distribution tapers off from the $\chi_1^2$ and appears for $n > 1$ as in Figure 10.1(c).

Table 10.1 displays the shape of the chi-square curve and includes some percentage points from both tails.

Observe that the lower-tail percentage points are numbers that begin near zero. These appear in the upper left corner of Table 10.1(a). Meanwhile, upper-tail percentage points appear to the right and bottom of Table 10.1(b). These grow quite large as a result of squaring very large $Z$-values. This explains why the chi-square distribution is skewed to the right.

The percentage points in Table 10.1(a) relate to lower confidence limits and hypothesis tests that have a "less-than" alternative hypothesis.

Then the test criterion $P(\chi^2 < \chi^2_\alpha) = 1 - \alpha$ is the probability that a calculated $\chi^2$-value (from sample data) is less than a tabled value. Otherwise the percentage points $\chi^2_\alpha$ in Table B.3 represent upper-tail percentages as $P(\chi^2 > \chi^2_\alpha) = \alpha$. We illustrate with an example.

---

**EXAMPLE 10.3**

Identify the value for $\chi^2_\alpha$ that cuts off 5% of the area in the upper tail of a chi-square distribution defined by 1 df. Relate this value to the associated standard normal distribution.

*Solution*    From Table 10.1(b), for df = 1 and $\alpha = 0.05$ we obtain $\chi^2_{0.05,1} = 3.841$. This value of 3.841 relates to the standard normal distribution. For df = 1, the $\chi^2$-value is simply the square of the value from a standard normal variable. So $\sqrt{\chi^2_{0.05,1}} = \sqrt{3.841} = 1.96$. Thus chi-square, $\chi^2_{0.05,1}$, comes from the square of $Z_{0.025} = \pm 1.96$ (that is, from two tails).

---

**The Sampling Distribution for a Single Population Variance**

The *sampling distribution for the variance from a normal distribution* can now be defined as a chi-square distribution.

> The sampling distribution for the variance of a normally distributed variable is a chi-square distribution. The chi-square variable is
>
> $$\chi^2_{n-1} = (n-1)s^2/\sigma^2 \qquad (10.2)$$
>
> Assumption: The random variable, $X$, is normally distributed.

For interval estimation of $\sigma^2$ we rely on the pivotal method, with chi-square in Equation (10.2) as the pivotal statistic. For a specific $\alpha$-value, $P(\chi^2_{1-(\alpha/2)} < \chi^2 < \chi^2_{\alpha/2}) = 1 - \alpha$. Then, for (10.2) as the pivotal statistic,

$$\chi^2_{1-(\alpha/2)} < \frac{(n-1)s^2}{\sigma^2} < \chi^2_{\alpha/2}$$

gives

$$\frac{(n-1)s^2}{\chi^2_{\alpha/2}} < \sigma^2 < \frac{(n-1)s^2}{\chi^2_{1-(\alpha/2)}}$$

In this last expression, the end values are the LCL and UCL of the interval estimate. Thus the equation gives us the limits for a $[(1 - \alpha) \cdot 100]\%$ confidence interval estimate of $\sigma^2$. To estimate $\sigma$, we simply take the square roots of the limits on the variance.

Tests of hypothesis about a population variance use the same chi-square distribution. In a testing situation, we first specify a value for $\sigma^2$. Then Equation (10.2) is used to obtain a calculated chi-square value that is compared to the tabled value.

Upper-tail alternative hypotheses use $\chi^2_\alpha$ and Table B.3(b) to define the rejection region; lower-tailed tests require $\chi^2_{1-\alpha}$ values from Table B.3(a). Estimation and two-tailed alternative tests use $\chi^2_{1-(\alpha/2)}$ (lower) and $\chi^2_{\alpha/2}$ (upper) percentage points.

**EXAMPLE 10.4**

A mail advertiser receives its mail brochures packaged in cartons. The quantities can be reasonably measured by the carton weights. The following are weights in pounds of a sample of five cartons from a large lot shipped by the regular supplier.

| Carton | 1 | 2 | 3 | 4 | 5 |
|--------|--------|--------|--------|--------|--------|
| Weight (pounds) | 1497.7 | 1493.7 | 1453.7 | 1443.8 | 1438.5 |

Specifications call for a mean of 1470.0 pounds with variance of no more than 4% of the mean weight. Excessive variation is a concern because of potential spillage loss (split cartons from overfills) and potential under- or overfilling of work orders. Is this shipment within the contract specifications for weight variance? Let $\alpha = 0.05$.

*Solution*   Since the contract specifies weight variance no greater than 4% of mean weight, or $0.04 \cdot 1470.0 = 58.8$ pounds, we test

$$H_0: \sigma^2 = 58.8 \text{ pounds against } H_A: \sigma^2 > 58.8 \text{ pounds}$$

Test: Reject if calculated $\chi^2 > \chi^2_{0.05,4} = 9.488$.

Calculated $\chi^2 = (n-1)s^2/\sigma^2$

where $s^2 =$ variance calculated from 5 sample values

$\qquad = 792.80$

$\sigma^2 =$ assumed variance $= 58.8$

Calculated $\chi^2 = (4)(792.80)/58.8 = 53.93$

Decision: Reject $H_0$ since calculated $\chi^2 > \chi^2_{0.05,4} = 9.488$.

Conclusion: The sample suggests that the weight variance is beyond the contract specifications for $\alpha = 0.05$.

We summarize the use of $\chi^2$ for inference about variances from a single normally distributed population:

**Inference about a single population variance for a normally distributed variable:**

*Hypothesis testing:*   $H_0: \sigma^2 = \sigma_0^2$ uses the sample statistic
$$\chi^2_{n-1} = (n-1)s^2/\sigma_0^2$$

**for $H_A: \sigma^2 > \sigma_0^2$   compare against tabular $\chi^2_\alpha$**

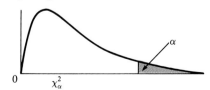

for $H_A$: $\sigma^2 < \sigma_0^2$   compare against tabular $\chi^2_{1-\alpha}$

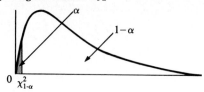

for $H_A$: $\sigma^2 \neq \sigma_0^2$   compare against tabular $\chi^2_{1-(\alpha/2)}$ and $\chi^2_{\alpha/2}$

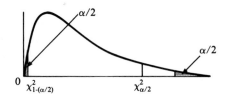

$$\textit{Estimation:}\quad \text{LCL} = (n-1)s^2/\chi^2_{\alpha/2}$$
$$\text{UCL} = (n-1)s^2/\chi^2_{1-(\alpha/2)} \tag{10.3}$$

The chi-square distribution is also used for inference concerning proportions. This application relates to the extension of tests on proportions to multinomial (versus binomial) situations. Since this is a rather extensive topic, and coverage is deferred to Chapter 16, the purpose of the chi-square exposition in this chapter is to ensure that you have the tools for estimation and testing hypotheses about population variances while we are still working in the context of estimation and testing. In Chapter 16, we will use Table B.3 for more inference and give additional intuitive development on the chi-square distribution.

## 10.4 INFERENCE CONCERNING THE VARIANCES OF TWO POPULATIONS

In earlier situations where we tested the equality of two means, we assumed that the two population variances were equal. This was a condition for the sampling distribution for inference about means of independent samples from two normally distributed populations. Then, before making the test on means, it is necessary to confirm that the two variances really are equal. This section is about the $F$-distribution and testing for equal variances. In the next section we can consider equal variances as a condition for testing for equal means for two independent samples.

Sometimes we may wish to compare two variances for other reasons. For example, two machines may both produce bolts that have an average diameter of $\frac{1}{2}$ inch. Yet if the variation in bolt diameters from one machine is much greater than that from the other, the output may be unusable. Tests of equal variance will reveal the problem. Another example comes from the financial markets. When you compare two stocks for investment, the one with greater variability in price is likely to be a more risky investment. That is one reason why the Dow-Jones averages reported in our daily newspapers include daily highs and lows as well as average stock prices.

## The $F$-Distribution and Tables

The sampling distribution for inference about the equality of two population variances is described by the $F$-distribution. The $F$-distribution form assumes that independent samples are taken from normally distributed populations having variances $\sigma_1^2$ and $\sigma_2^2$. As with other tests, sample values are used to estimate population values. In this case the sample variances $s_1^2$ and $s_2^2$ are then used to estimate the ratio of the population variances $\sigma_1^2$ and $\sigma_2^2$. Given that $n_1$ and $n_2$ are the respective sample sizes, the (theoretical) *sampling distribution for the ratio of two variances* is then distributed as $F$, where

> **The $F$-distribution is a sampling distribution of the ratio of two variances.**

$$F_{n_1-1,\,n_2-1} = \frac{s_1^2/\sigma_1^2}{s_2^2/\sigma_2^2} = \frac{\sigma_2^2 s_1^2}{\sigma_1^2 s_2^2} \qquad (10.4)$$

Under $H_0: \sigma_1^2 = \sigma_2^2$, 
$$= s_1^2/s_2^2 \qquad (10.5)$$

Assumptions:
1. **The distributions are both normal.**
2. **The samples are independently selected.**

For the most common application, where $\sigma_1^2 = \sigma_2^2$, then $\sigma_1^2$ and $\sigma_2^2$ cancel in Equation (10.4), leaving us with the more conventionally stated sampling distribution for the ratio of two sample variances, Equation (10.5).

Thus under $H_0: \sigma_1^2 = \sigma_2^2$ we can view the $F$-statistic as a ratio of two independent sample variances. And we develop the probability distribution using chi-square distributions—although the mathematical development of it is not so intuitive as for chi-square. (See [2]). For our purposes, we shall simply note that the $F$-distribution is skewed to the right with much the same shape as chi-square, and it does not allow negative values.

Two sets of degrees of freedom are associated with $F$, one for the numerator variance ($df_1 = n_1 - 1$) and one for the denominator variance ($df_2 = n_2 - 1$). Table 10.2 illustrates $F$-distribution percentage points for selected $df_1$ and $df_2$

**TABLE I0.2**  A Segment of the F-Distribution Table (Appendix B, Table B.4)

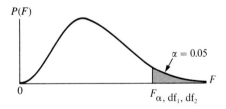

**Numerator Degrees of Freedom df$_1$**

| df$_2$ \ df$_1$ | 1 | 2 | 3 | 4 | $\cdots$ | 120 | inf |
|---|---|---|---|---|---|---|---|
| 1 | 161.45 | 199.50 | 215.71 | 224.58 | $\cdots$ | 253.25 | 254.31 |
| 2 | 18.513 | 19.000 | 19.164 | 19.247 | | 19.487 | 19.496 |
| 3 | 10.128 | 9.5521 | 9.2766 | 9.1172 | | 8.5494 | 8.5264 |
| 4 | 7.7086 | 6.9443 | 6.5914 | 6.3882 | | 5.6581 | 5.6281 |
| $\vdots$ | | | | | | | |
| 120 | 3.9201 | 3.0718 | 2.6802 | 2.4472 | $\cdots$ | 1.3519 | 1.2539 |
| $\infty$ | 3.8415 | 2.9957 | 2.6049 | 2.3719 | $\cdots$ | 1.2214 | 1.0000 |

Denominator Degrees of Freedom df$_2$

values when $\alpha = 0.05$. Because each $\alpha$-level table requires a two-dimensional display to show the percentage points, we have elected to include only the ones for $\alpha = 0.10, 0.05, 0.025$ and $0.01$. (See Appendix B, Table 4; that is, Table B.4).

Notice that because the $F$-distribution is skewed to the right, for low df the (upper-tail) entries are quite large. For example, $F_{0.05,1,1} = 161.45$. Thus samples of $n_1 = n_2 = 2$ require a huge $F$-value (161.45) to show a statistical difference at $\alpha = 0.05$, which is reasonable for such small information. Yet for moderate samples like $n_1 = 5$ and $n_2 = 4$, the criterion is much smaller: $F_{0.05,4,3} = 9.1172$.

Our use of the $F$-distribution is restricted to upper-tail percentage points. This is because when the two variances are equal, the ratio of sample variances should have a value near 1. If the variances are quite different, then the ratio of the *larger* variance divided by the *smaller* variance will produce a large $F$-value. So it is a common practice to structure tests to require that the variance in the numerator will always be equal to or larger than that in the denominator. This way we can manage with upper-tail distribution percentages as given in Table B.4. Here is an example using the table.

---

## EXAMPLE 10.5

A variance ratio test using independent samples of $n_1 = 5$ and $n_2 = 4$ revealed a calculated $F$-statistic value of 8.6. Does this calculated value exceed the $\alpha = 0.05$ upper-tail percentage point from the $F$-distribution table?

*Solution*    The appropriate entry from Table 10.2 (or Table B.4) has $df_1 = 5 - 1 = 4$. Similarly, $df_2 = 4 - 1 = 3$. The value for $F_{0.05,4,3}$ is 9.1172. Therefore, the calculated $F$-value of 8.6 does not exceed tabular $F$. So the two variances can be considered equal at the 5% level of significance.

---

## Estimation and Testing the Equality of Two Variances

As in all preceding cases of interval estimation, the limits for a confidence interval are developed by the pivotal method. Equation (10.4) defines the pivotal statistic. For a specified $\alpha$-value, $P(F_{1-(\alpha/2)} < F < F_{\alpha/2}) = 1 - \alpha$. Then substituting Equation (10.4) as the pivotal statistic gives

$$F_{1-(\alpha/2)} < \frac{\sigma_2^2 s_1^2}{\sigma_1^2 s_2^2} < F_{\alpha/2}$$

and

$$\frac{s_2^2}{s_1^2 F_{\alpha/2}} < \sigma_1^2/\sigma_2^2 < \frac{s_2^2}{s_1^2 F_{1-(\alpha/2)}} \quad \text{or} \quad \text{LCL} < \sigma_1^2/\sigma_2^2 < \text{UCL}$$

This provides the limits for a $[(1 - \alpha)100]\%$ confidence interval estimate of the variance ratio.

Because $F$-tables require a great deal of space and testing requires only $F_\alpha$ values, we have elected to include $F$-tables for only the upper-tailed percentage points, the $F_\alpha$-values. Since confidence interval calculations also require $F_{1-\alpha}$-values, these are obtained through the relation $F_{1-\alpha,df_1,df_2} = 1/F_{\alpha,df_2,df_1}$.

Hypothesis tests for the equality of two variances are stated as $H_0: \sigma_1^2 = \sigma_2^2$. Then, under the null hypothesis, $\sigma_2^2/\sigma_1^2 = 1$ and Equation (10.4) reduces to Equation (10.5). Our inclusion of only upper-tail $F$-percentage points tables poses no problem for testing a hypothesis when the alternative is one-tailed. Thus for $H_A: \sigma_1^2 > \sigma_2^2$, we simply compare the sample variance ratio, $s_1^2/s_2^2$, to the appropriate tabular $F$-value in Table B.4 for the specified $\alpha$-risk level.

The two-tailed alternative, $H_A: \sigma_1^2 \neq \sigma_2^2$, requires an adjustment, however. Here we must identify the population with the larger *sample* variance as population 1, and the other as population 2. By always placing the larger sample variance in the numerator, we force the ratio, $F = s_1^2/s_2^2$, to be compared against the upper-tail percentage points of the $F$-distribution. These results are summarized in Figure 10.2. Part (a) illustrates the null hypothesis of equal variances. Part (b) illustrates a one-tailed test in which the rejection region is identified by an $\alpha$-risk. Part (c) illustrates a two-tailed alternative hypothesis, so the rejection region is designated by $\alpha/2$.

Hypothesis testing of population variances is illustrated with an example.

---

**FIGURE 10.2** Procedures for Testing the Equality of Two Population Variances

(a) Tests of hypothesis for the equality of two population variances use $H_0: \sigma_1^2 = \sigma_2^2$.

$$\text{Calculated } F = s_1^2/s_2^2 = \frac{\text{Larger variance}}{\text{Smaller variance}} \qquad \textbf{(10.5 repeated)}$$

where $s_1^2$ is the larger numeric value.

(b) For $H_A: \sigma_1^2 > \sigma_2^2$, reject $H_0$ if calculated $F > F_{\alpha, n_1 - 1, n_2 - 1}$

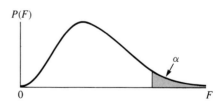

(c) For $H_A: \sigma_1^2 \neq \sigma_2^2$, reject $H_0$ if calculated $F > F_{\alpha/2, n_1 - 1, n_2 - 1}$

Assumptions:
1. Both populations are normally distributed.
2. The two samples are selected at random and are independent.

**EXAMPLE 10.6**

The mail advertiser in Example 10.4 has decided to test its regular supplier package weights against those of another supplier (supplier 2). The weights (in pounds) for a sample of five cartons packed by the new supplier are displayed below along with the weights for five cartons shipped by the regular supplier.

| Carton No. | 1 | 2 | 3 | 4 | 5 |
|---|---|---|---|---|---|
| Supplier 1 (regular) | 1497.7 | 1493.7 | 1453.7 | 1443.8 | 1438.5 |
| Supplier 2 (new) | 1467.9 | 1487.4 | 1456.8 | 1461.1 | 1470.5 |

Both suppliers are tested against the contract specifications that the mean weight should be 1470.0 pounds with a variance no greater than 4% of the weight. Are the cartons shipped by the new supplier (supplier 2) less variable in weight than those shipped by the regular supplier? Let $\alpha = 0.05$.

*Solution*    To qualify as a second source, supplier 2 is required to substantiate a lower variance in shipping carton weights than supplier 1. So $H_0: \sigma_1^2 = \sigma_2^2$ and $H_A: \sigma_1^2 > \sigma_2^2$. Then the test uses the $F$-distribution, and we reject the $H_0$ if calculated $F > F_{0.05,4,4} = 6.39$. From the data given, we have

$$s_1^2 = \frac{\sum (X_1 - \bar{X}_1)^2}{n_1 - 1} = 792.80 \qquad s_2^2 = \frac{\sum (X_2 - \bar{X}_2)^2}{n_2 - 1} = 138.23$$

$$\text{Calculated } F = \frac{\text{Sample variance for supplier 1}}{\text{Sample variable for supplier 2}} = \frac{792.80}{138.23} = 5.74$$

Decision: Supplier 2's cartons are not statistically less variable than those of supplier 1 at the $\alpha = 0.05$ level of significance. There is not sufficient reason to switch suppliers based on the results of this test.

We have viewed several applications for inference concerning variances for one or two populations. The chi-square statistic will be used extensively in Chapter 16; the $F$-distribution is central to Chapter 11, where we take up the analysis of variance; and the test for equality of two variances is key in the next section. Use the following problems to confirm your understanding of inference concerning variances. Again, we have placed the important forms in Table 10.3.

**Problems 10.3 and 10.4**

5. Indicate the appropriate values from Table B.3 for
   a. $\chi_\alpha^2$ for $P(\chi^2 > \chi_\alpha^2) = 0.05$ with df $= 6$.
   b. $\chi_{1-\alpha}^2$ for $P(\chi^2 < \chi_{0.95,10}^2) = 0.05$.
   c. the 99th percentile on the chi-square distribution defined by df $= 19$.

6. Indicate the appropriate values from Table B.4 for
   a. $F_{\alpha,df_1,df_2}$ for $\alpha = 0.05$, df$_1 = 8$, df$_2 = 15$.
   b. the test statistic value for $H_0: \sigma_1^2 = \sigma_2^2$, $H_A: \sigma_1^2 > \sigma_2^2$ for samples $n_1 = 12$ and $n_2 = 17$ and $\alpha = 0.05$. (*Note:* Insofar as Table B.4 does not have 11 df for numerator values, you may wish to approximate the table value.)
   c. the 90th percentile on the $F$-distribution defined by df$_1 = $ df$_2 = 8$.

7. Complete the following inference that requires use of Table B.3.
   a. A test of hypothesis of $H_0$: $\sigma^2 = 2$ against $H_A$: $\sigma^2 > 2$ uses a sample of $n = 21$ to calculate $s^2$. What is the chi-square test value for $\alpha = 0.10$?
   b. Given $s^2 = 48/(9 - 1)$ for a small random sample from a normal distribution, use $1 - \alpha = 0.95$ to establish LCL and UCL for $\sigma^2$.
   c. Extend the result of part (b) to a 95% confidence interval estimate of $\sigma$.

8. Explain why we use the $1 - \alpha/2$ percentage points for an $\alpha$-level hypothesis test for equal variances against a "does not equal" alternative.

9. For independent samples of $n_1 = n_2 = 9$ from two normally distributed populations, we find $s_1^2 = 25.25$ and $s_2^2 = 35.61$. Test the hypothesis of equal population variances against a does-not-equal alternate hypothesis. Use $\alpha = 0.05$.

10. Some hand-woven ponchos are made in a plant outside Mexico City. Estimate, with 95% confidence, the variance in time required to weave a poncho. A random sample of 12 times yielded a sample variance of 26.9 minutes squared.

11. An economy-minded shopper wants to buy those groceries that are most greatly reduced in price. Consider the prices in the table to be a random sampling of six weeks. Is the variation statistically greater for generic beans as opposed to generic corn? Assume normal distributions of prices and let $\alpha = 0.10$.

|  |  | | Week | | | |
| --- | --- | --- | --- | --- | --- | --- |
| Item | *1* | *2* | *3* | *4* | *5* | *6* |
| Beans | $0.35 | $0.44 | $0.29 | $0.39 | $0.44 | $0.32 |
| Corn | 0.43 | 0.41 | 0.36 | 0.35 | 0.46 | 0.39 |

12. Local union leaders are debating what salary conditions to bargain for with management. The trade standard is that the variance in union salaries should not exceed 100 dollars squared. If this value is exceeded, then the union leaders feel they should bargain for internal equities; otherwise they should try for uniform increases for all members. Assume that the wages are normally distributed and that a random sample of 15 union salaries shows a variance of 108 dollars squared. Which choice should they take? Let $\alpha = 0.05$.

## 10.5 INFERENCE ON MEANS FROM TWO POPULATIONS: INDEPENDENT SAMPLES

In this section we focus upon the *sampling distribution for the difference in means from independent samples* (as opposed to the dependent samples of Section 10.2).

> The sampling distribution for the difference in means, $\bar{X}_1 - \bar{X}_2$, for independent samples of $n_1 \leqslant 30$ and $n_2 \leqslant 30$ is

$$t_{n_1 + n_2 - 2} = \frac{(\bar{X}_1 - \bar{X}_2) - (\mu_1 - \mu_2)}{\sqrt{(s_w^2/n_1) + (s_w^2/n_2)}} \tag{10.6}$$

where $s_w^2$ is a pooled estimator for the common variance and is the weighted mean of the sample variances

$$s_w^2 = \frac{(n-1)s_1^2 + (n_2 - 1)s_2^2}{n_1 + n_2 - 2}$$

Assumptions:
1. The samples are each of size 30 or less.
2. The samples are independently selected from normally distributed populations.
3. The distributions have a common, but unknown, variance, $\sigma^2$.

Observe that the pooled variance, $s_w^2$, is the weighted mean for the two sample variances, $s_1^2$ and $s_2^2$. The weights are the respective sample sizes, each less one. This yields an effective total sample size of $n_1 + n_2 - 2$, which is the degrees of freedom: $df = n_1 + n_2 - 2$. The pooled variance does *not* require equal sample sizes. But *if the sample sizes are equal*, the calculation of $s_w^2$ is reduced to $s_w^2 = (s_1^2 + s_2^2)/2$, that is, the sample average of the variances.

The next example extends the test for equal variances—just studied in Section 10.4—to a test for equal means.

**EXAMPLE 10.7**

The suppliers in Example 10.5 were found to be sending cartons whose weights were of statistically equal variance. This is important because the shipper will charge a stiff penalty for cartons that are overweight. A final requirement is for the supplier to be on target for the required mean weight. The variability could be quite small, but if the mean were excessive, the mail advertiser would be charged for overweight cartons! Test the carton weights for equal means using $\alpha = 0.05$.

*Solution*   The data for the five samples each of supplier carton weights are repeated here.

| Carton No. | 1 | 2 | 3 | 4 | 5 |
|---|---|---|---|---|---|
| Supplier 1 (regular) | 1497.7 | 1493.7 | 1453.7 | 1443.8 | 1438.5 |
| Supplier 2 (new) | 1467.9 | 1487.4 | 1456.8 | 1461.1 | 1470.5 |

The test for equal means uses Equation (10.6), with an appropriate pooled variance from Example 10.6. Then, under the null hypothesis $H_0: \mu_1 = \mu_2$ and using Equation (10.6), we get

$$\text{Calculated } t = \frac{(1465.5 - 1468.7) - 0}{\sqrt{465.52\left(\frac{1}{5} + \frac{1}{5}\right)}} = -0.24$$

where $s_w^2 = \frac{4 \cdot 792.80 + 4 \cdot 138.23}{5 + 5 - 2} = 465.52$

with $s_1^2 = 792.80$ and $s_2^2 = 138.23$ from Example 10.6.

Decision: We cannot reject the hypothesis of equal means for
$|\text{Calculated } t| < t_{0.05,8} = 1.86.$

Both means appear to be below the target weight, 1470 pounds, and so are acceptable. We will need other criteria to decide on the choice of a supplier.

There are many business applications for Equation (10.6). For example, the two means being compared might represent two market areas, the production rates of workers on similar jobs, or the yield rates under different production processes. Our next example illustrates both hypothesis testing and estimation for independent samples.

**EXAMPLE 10.8**

A timber company with forests in the East and the South wishes to compare timber yields in the two areas. If the difference is substantial, that knowledge will influence their plans for expansions to new timber lands. They have randomly selected $n = 9$ one-acre parcels to be logged in each geographical area and have carefully measured the lumber produced from each. The mean units of lumber produced from the East ($\bar{X}_1$) and the South ($\bar{X}_2$) are given next, along with variances in production for each area.

| East | South |
|---|---|
| $\bar{X}_1 = 40.33$ units | $\bar{X}_2 = 36.54$ units |
| $s_1^2 = 25.25$ | $s_2^2 = 35.61$ |

We will assume that both yields are normal and have a common, but unknown, variance. Is there sufficient evidence to conclude a significant difference in mean yield in these two sections of the country? Let $\alpha = 0.05$.

*Solution*   For $\mu_1$ = mean yield in the East and $\mu_2$ = mean yield in the South, this is a two-tailed test in which we are simply looking for any significant difference ($+$ or $-$). The sample sizes are $n_1 = 9$ and $n_2 = 9$.

$$H_0: \mu_1 - \mu_2 = 0 \qquad H_A: \mu_1 - \mu_2 \neq 0 \qquad \alpha = 0.05$$

Test: Reject $H_0$ if $|\text{Calculated } t| > t_{\alpha/2, \, n_1 + n_2 - 2} = t_{0.025,16} = 2.12.$

A first requirement is to test for equal variances. Since calculated $F = s_1^2/s_2^2 = 35.61/25.25 = 1.41 < 3.44 \, (= F_{0.05,8,8})$ we can proceed to test for equal means. Then, for

$$s_w^2 = \frac{(n_1 - 1)s_1^2 + (n_2 - 1)s_2^2}{n_1 + n_2 - 2} = \frac{(9 - 1)/(25.25) + (9 - 1)(34.61)}{(9 + 9 - 2)}$$

$$= 29.93*$$

---

* Because $n_1 = n_2 = 9$, this calculation can be simplified to $(s_1^2 + s_2^2)/2 = (25.25 + 34.61)/2 = 29.93$. However, the (equivalent) solution as given also demonstrates the more general calculation for cases in which $n_1 \neq n_2$.

we obtain

$$\text{Calculated } t = \frac{(\bar{X}_1 - \bar{X}_2) - (\mu_1 - \mu_2)}{\sqrt{\dfrac{s_w^2}{n_1} + \dfrac{s_w^2}{n_2}}} = \frac{(40.33 - 36.54) - 0}{\sqrt{(29.93/9) + (29.93/9)}} = \frac{3.79}{2.58} = 1.47$$

Decision: Do not reject $H_0$.

We must accept $\mu_1 = \mu_2$ if we are willing to make a type 1 error only 5% of the time. From this evidence, the firm must conclude there is no significant difference in the yield from these two sections of the country.

Although they have accepted the hypothesis of no difference, the probability of a type 2 error is likely quite high. So the timber company may, in fact, be accepting a false null hypothesis. Unless the value of $\beta$ is calculated, the firm has no measure of the risk they are taking in accepting that there is no difference in the means. As a practical matter, if the company plans to continue to log the areas, they might consider increasing the sample size and making a retest.

In the preceding example we could establish a confidence interval estimate of the true difference, as a more definitive measure of the difference in average yield. A confidence estimate appropriate to this situation uses Equation (10.7).

The [(1 − α) · 100]% confidence limits for estimating differences in means from small independent samples from normal distributions with a common, but unknown, variance is

$$(\text{LCL, UCL}) = (\bar{X}_1 - \bar{X}_2) \pm \left[ (t_{\alpha/2,\, n_1 + n_2 - 2}) \sqrt{\frac{s_w^2}{n_1} + \frac{s_w^2}{n_2}} \right] \qquad (10.7)$$

**EXAMPLE 10.9**

Compute a 95% confidence interval estimate for the data of the timber yields in the previous example.

*Solution*   Using Equation (10.7) with

$$\bar{X}_1 - \bar{X}_2 = 3.79, \qquad \sqrt{\frac{s_w^2}{n_1} + \frac{s_w^2}{n_2}} = 2.58, \qquad \text{and } t_{0.025,16} = 2.12$$

gives

$$\text{LCL} = 3.79 - [2.12 \cdot (2.58)] = -1.68$$
$$\text{UCL} = 3.79 + [5.47] = 9.26,$$

Thus CL $= -1.68 < (\mu_1 - \mu_2) < 9.26$ units.

Conclusion: The difference in average yield per acre in the West is as much as 1.68 units *less*, to as much as 9.26 units *more*, than for that in the South. Insofar as the range spans zero, there is not a clear difference in yield.

A second example using the *t*-form of Equation (10.6) adds the requirement of computing values for $s_1^2$ and $s_2^2$. This example also illustrates a case wherein the null hypothesis projects a nonzero difference.

**EXAMPLE 10.10**

A buyer who deals in quarter horses is interested in the relative ages of horses in two areas—the Northeast and the Far West—at the time they are first sold. Having been in the business for some years, his experience makes him believe generally that in the Northeast, quarter horses at first sale are one year older than quarter horses at first sale in the Far West. Test his belief using the age-at-first-sale data in the table, which represents a random sample of 11 first sales in the Northeast and 14 in the Far West. Assume normal distributions, common but unknown variances, and let $\alpha = 0.025$.

| 1<br>Northeast Region | 2<br>Far West Region |
|---|---|
| 3  3  4  5 | 2  3  3  4 |
| 5  5  6  6 | 4  4  4  4 |
| 7  8  8 | 4  4  4  5 |
| | 5  6 |

*Solution*   For $\mu_1$ = mean age of quarter horses at first sale in the Northeast, this is a one-tailed test of the hypothesis

$$H_0: \mu_1 - \mu_2 = 1 \text{ year} \qquad H_A: \mu_1 - \mu_2 > 1 \text{ year} \qquad \alpha = 0.025$$

Test: Reject $H_0$ if calculated $t > t_{0.025, 23} = 2.07$.

Summary statistics (calculations are requested in Problem 15):

$$\bar{X}_1 = 5.45 \text{ years} \qquad \bar{X}_2 = 4.00 \text{ years} \qquad s_w^2 = 1.85 \text{ years squared}$$

Calculated

$$t = \frac{(\bar{X}_1 - \bar{X}_2) - (\mu_1 - \mu_2)}{\sqrt{(s_w^2/n_1) + (s_w^2/n_2)}} = \frac{(5.45 - 4.00) - 1}{\sqrt{(1.85/11) + (1.85/14)}}$$

$$= \frac{0.45}{\sqrt{0.3003}} = 0.82$$

Decision: Do not reject $H_0$ since calculated $t = 0.82 < 2.07$.

We conclude that the mean age of quarter horses at first sale in the Northeast is *not* more than one year older than the mean age of those in the Far West.

We are unable to show, statistically, that the mean age at first sale in the Northeast is over one year more than in the Far West if we are willing to be wrong only 2.5% of the time. Because the sample size is small, we necessarily accept a high risk of making a type 2 error.

In Example 10.10 we used an upper-tailed test. This was a matter of convenience. With some adjustments we could have used a lower-tailed test ($H_0: \mu_2 - \mu_1 = -1$, $H_A: \mu_2 - \mu_1 < -1$). The conclusion would be unaltered. There is, however, one important precaution concerning the one-tailed test. The difference of sample means must be used in an order consistent with the test hypothesis. Example 10.10 used the sequence $\mu_1 - \mu_2$, so we must follow with $\bar{X}_1 - \bar{X}_2$. Using $\bar{X}_2 - \bar{X}_1$ with the preceding hypotheses would give a grossly incorrect calculated $t$-value.

The samples for the last example have been assumed to be independent. That is, the first sale of any quarter horse in the Northeast had no influence on the first sale of one in the Far West, and vice versa. Inference concerning the difference of means for dependent samples from normal distributions was discussed earlier. See [3] for inference when the variances cannot be considered equal but the samples are independently drawn from two normal distributions. In Chapter 17 the Mann–Whitney $U$-test provides a nonparametric test when the distributions are not normal. The following problems will provide a check on your understanding of the calculations and procedures described in this section. The forms for confidence interval estimation and hypothesis tests from this section are also included in Table 10.3 at the end of the chapter.

**Problems 10.5**

13. Two division managers of the International Chemical Company want to compare performance levels for employees in their respective plants. Times to complete similar jobs are sampled for 18 workers at plant 1 (Mexico) and 12 workers at plant 2 (Brazil) with the following results:

$$\text{(Mexico)} \quad \bar{X}_1 = 76.2 \text{ minutes} \quad s_1^2 = 18.6 \quad (n = 18)$$

$$\text{(Brazil)} \quad \bar{X}_2 = 83.1 \text{ minutes} \quad s_2^2 = 38.2 \quad (n = 12)$$

Is there a significant difference in performance on this measure for workers in these two plants at $\alpha = 0.05$? Assume common variances and normal distributions.

14. The word *cold* is an acronym for "chronic obstructive lung disease," which includes various ailments. Suppose a chemist develops a drug called SOS, which she claims will give people longer relief from colds than will other drugs. To test this claim, a sample of 36 people with colds are given the drug and subsequently find relief for an average of 8.6 hours with a standard deviation of 3.0 hours. Thirty-six others take a different cold medicine for which the mean relief is 7.2 hours with a standard deviation of 4.0 hours. Assume that the relief time follows a normal distribution.
    a. Is the chemist's claim justified at $\alpha = 0.05$?
    b. Is the chemist's claim justified at $\alpha = 0.025$?

15. Perform the calculations suggested in the quarter-horse sales example (10.10). That is,
    a. compute $s_1^2$ and $s_2^2$.
    b. Use your answers from part (a) to check $s_w^2$. Check the other calculations in that solution as well.

16. A training director in a large hospital contends that the orderlies who take her public relations course perform better on the job than do those not receiving this training. Of 30 recently hired orderlies, 15 were randomly selected to receive the training. The other 15 received no special job training. Six months later, on-the-job evaluations produced the following productivity indexes. For the trained group, $\bar{X}_1 = 84.63$, $s_1 = 3.5$. For the

orderlies without this training, $\bar{X}_2 = 81.45$, $s_2 = 4.0$. Test the director's claim using $\alpha = 0.05$. Assume common variances and that both distributions are normal.

## SUMMARY

Chapter 10 continues the inference on population means that was started in Chapters 8 and 9. There we introduced the ideas of estimation and hypothesis testing. The inference (using a $Z$-statistic) was on means for normal distributions with known variance or on normal distributions with unknown variance (then using a $t$-statistic). Chapter 10 extends the inference on means to include cases for (1) paired differences and (2) differences

**TABLE 10.3**   Key Equations for Inference in Extensions of Estimation and Hypothesis Testing (Variable—or Difference in Variables—Normally Distributed)

| Inference Subject and Type | Conditions | Testing Hypotheses | Estimation | Component Calculations |
|---|---|---|---|---|
| **Means**<br>One sample<br>(from Chapters 8 and 9) | normal distribution<br>$\sigma^2$ unknown<br>$n \leqslant 30$<br>use $t$-distribution | $H_0: \mu = \mu_0$<br><br>$statistic: t = \dfrac{\bar{X} - \mu_0}{s/\sqrt{n}}$ | $\text{CL} = \bar{X} \pm t_{\alpha/2}\dfrac{s}{\sqrt{n}}$ | |
| Paired differences<br>(dependent samples) | differences are normally distributed<br>$\sigma^2$ unknown<br>$n \leqslant 30$<br>use $t$-distribution | $H_0: \mu_d = \mu_0$<br><br>$statistic: t = \dfrac{\bar{X}_d - \mu_0}{s_d/\sqrt{n}}$ | $\text{CL} = \bar{X} \pm t_{\alpha/2}\dfrac{s_d}{\sqrt{n}}$ | $X_d = X_A - X_B$<br><br>$s_d^2 = \dfrac{\sum(X_d - \bar{X}_d)^2}{n-1}$ |
| Differences (two)<br>Independent samples | common $\sigma^2$, unknown<br>$n_1, n_2 \leqslant 30$<br>normal distributions | $H_0: \mu_1 - \mu_2 = 0$<br>$statistic:$<br><br>$t = \dfrac{(\bar{X}_1 - \bar{X}_2) - (\mu_1 - \mu_2)}{\sqrt{\dfrac{s_w^2}{n_1} + \dfrac{s_w^2}{n_2}}}$ | $\text{CL} = (\bar{X}_1 - \bar{X}_2) \pm t\sqrt{\dfrac{s_w^2}{n_1} + \dfrac{s_w^2}{n_2}}$ | $\text{df} = n_1 + n_2 - 2$<br><br>$s_w^2 = \dfrac{(n_1-1)s_1^2 + (n_2-1)s_2^2}{n_1 + n_2 - 2}$ |
| **Variances**<br>One sample | normal distribution<br>$\sigma^2$ unknown<br>$n = $ any value | $H_0: \sigma^2 = \sigma_0^2$<br><br>$statistic: \chi^2 = \dfrac{(n-1)s^2}{\sigma_0^2}$ | $\text{CL} = \dfrac{(n-1)s^2}{\chi_{\alpha/2}^2}, \dfrac{(n-1)s^2}{\chi_{1-\alpha/2}^2}$<br><br>(for estimation of $\sigma^2$) | |
| (two)<br>Independent samples | normal distribution<br>$\sigma^2$ unknown<br>$n = $ any size | $H_0: \sigma_1^2 = \sigma_2^2$<br><br>$statistic: F = \dfrac{\sigma_2^2 s_1^2}{\sigma_1^2 s_2^2}$ | $\text{CL} = \dfrac{s_2^2}{s_1^2 F_{\alpha/2}}, \dfrac{s_2^2}{s_1^2 F_{1-\alpha/2}}$<br><br>$\left(\text{for estimation of } \dfrac{\sigma_1^2}{\sigma_2^2}\right)$ | *Note for testing:*<br>$H_A: \sigma_1^2 > \sigma_2^2$<br>use $F_{\alpha, n_1-1, n_2-1}$<br>$H_A: \sigma_1^2 \neq \sigma_2^2$<br>use $F_{\alpha/2, n_1-1, n_2-1}$<br><br>Both use $F_{\text{calc}} = \dfrac{s_1^2}{s_2^2}$<br><br>(where $s_1^2 > s_2^2$). |

between two means for independent samples with a pooled estimate for the common variance. Also, a discussion of inference on variances introduces the *chi-square distribution* ($\chi^2$), *for a single population variance*, $\sigma^2$, and the *F-distribution for comparing the ratio of two variances* ($\sigma_1^2/\sigma_2^2$).

*Paired difference tests* presume the matched units are too much alike to give independent readings. This is usually obvious. The comparison can be on the same individual (or machine or process) before and after some treatment. Or it can use matched pairs. Paired-differences inference compares changes in samples on the basis of before and after maintenance, no treatment versus treatment, or matched pairs. Pairing is a means of controlling nonessential variations to achieve a precise measure of the test variation. The computations for testing and estimation for differences in dependent samples are similar to the single-sample *t*-procedures.

Tests for the *difference between means for independent samples* assume the calculation of a common pooled variance. We have explored procedures for inference about a single population variance. This uses the chi-square distribution. The test for common variances for two populations uses the *F*-distribution. The statistics for inference about two independent sample means, for differences in means, and for variances appear in Table 10.3. The assumptions are included along with testing and estimation outlines.

## KEY TERMS

chi-square distribution, 404
confidence intervals for differences in means, independent samples, 415
F-distribution, 407
paired-differences procedure, 397
sampling distribution for differences in means, independent samples, 412

sampling distribution for equality ratio of two variances, 407
sampling distribution for a normal distribution variance, 404
sampling distribution for the mean of paired differences, 397

## CHAPTER 10 PROBLEMS

17. Two personal computers are rated on efficiency, which is the percent of time that the system was processing for a fixed set of tasks. An efficiency rating of 100 is possible.

| | Efficiency | |
|---|---|---|
| Task | *For PC 1 (%)* | *For PC 2 (%)* |
| 1 | 98 | 96 |
| 2 | 91 | 93 |
| 3 | 94 | 89 |
| 4 | 92 | 85 |
| 5 | 94 | 95 |
| 6 | 95 | 90 |
| 7 | 97 | 94 |
| 8 | 92 | 98 |

Is one personal computer more efficient than the other? Assume a normal distribution for the difference in efficiencies and use $\alpha = 0.05$.

18. Thirty students were tested before *and* after taking a math course. The average increase in their scores was 8 points with $s = 16$ points. Using a significance level of 0.01, test the hypothesis that the students have improved from the first test to the

second test. Under what circumstances would a type 2 error occur? Assume a normal distribution of differences.

19. The table below lists two separate ratings (1 and 2) by a consumer panel on the taste of each of nine soft drinks, on a preference scale of 0 to 10.0 (highest). Using $\alpha = 0.05$, test the claim of one producer that, in general, the first rating is higher. Assume a normal distribution.

| Drink No. | Rating #1 | Rating #2 | Drink No. | Rating #1 | Rating #2 |
|-----------|-----------|-----------|-----------|-----------|-----------|
| 1 | 5.0 | 3.8 | 6 | 5.1 | 5.4 |
| 2 | 4.6 | 4.2 | 7 | 4.3 | 4.2 |
| 3 | 6.1 | 6.2 | 8 | 5.4 | 4.7 |
| 4 | 3.8 | 3.2 | 9 | 4.6 | 4.3 |
| 5 | 6.1 | 5.4 | | | |

20. How large a sample should we take to estimate the average number of emergency calls per week to a hospital? The standard deviation is known to be about 20. We want to be 90% sure that our estimate is within 5 of the true mean. Assume the number of calls is normally distributed and that $n$ will exceed 30.

21. A manager at Reed Electrical Company has decided to test supervisors on their communications improvement over the span of six months of participation in a company-sponsored education program.
    a. Assuming necessary test conditions are met, test at $\alpha = 0.025$ that the average increase in communication score is over 30 points. The results are for a random selection of 16 supervisors. Scores are 100 points maximum.

Communication Scores

| Supervisor | Beginning of Program | End of Program | Supervisor | Beginning of Program | End of Program |
|------------|----------------------|----------------|------------|----------------------|----------------|
| 1 | 55 | 89 | 9 | 66 | 98 |
| 2 | 62 | 86 | 10 | 57 | 88 |
| 3 | 41 | 76 | 11 | 66 | 100 |
| 4 | 70 | 93 | 12 | 61 | 97 |
| 5 | 36 | 74 | 13 | 72 | 98 |
| 6 | 59 | 92 | 14 | 50 | 88 |
| 7 | 56 | 85 | 15 | 68 | 94 |
| 8 | 39 | 69 | 16 | 43 | 86 |

    b. Estimate, with 95% confidence, the true average increase in communications-skills score for supervisors at this plant.

22. Use the following summary information for independent samples from two normal distributions to test
    a. $H_0: \mu_1 = \mu_2$ against a two-tailed alternative. Use $\alpha = 0.10$.
        $n_1 = 10 \quad \bar{X}_1 = 16 \quad s_1 = 7$
        $n_2 = 12 \quad \bar{X}_2 = 13 \quad s_2 = 5$
    b. Is it reasonable to assume that these population variances are statistically equal?

23. Listed below are summary statistics concerning the time (in minutes) required by two assembly teams to complete a series of simulated assemblies in an electrical manufacturing plant. Test the claim that Team 1 takes an average of over 5 minutes longer to do the job. Assume that all essential test conditions are met. Use $\alpha = 0.10$.

| Team | Mean | Standard Deviation | $n$ |
|------|------|--------------------|-----|
| 1 | 23.0 | 4.2 | 15 |
| 2 | 15.0 | 3.7 | 17 |

24. The work in a lawn-spraying service is fairly simple and requires a high percentage of semi- and unskilled laborers. Should the workers be hired (1) on a hit-or-miss basis such as a quick interview or (2) on the basis of the results of a job test? Both methods were tried for several months, and a record was kept of the productivity in lawns sprayed per day. A random selection of workers gave the results displayed in the following table. Using $\alpha = 0.025$, test the hypothesis that the average production is higher for those selected by the job test.

| Productivity (Number of Lawns per Day) | |
|----------------------------------------|----------------------|
| Hit-or-Miss Selection | Selection by Test |
| 4 5 5 8 7 4 3 5 6 | 6 8 8 6 5 9 7 |

Assume normal distributions and use $\sum (X_1 - \bar{X}_1)^2 = 19.56$ and $(n_2 - 1)s_2^2 = \sum (X_2 - \bar{X}_2)^2 = 12.00$, to estimate the common but unknown variances.

25. In a management development conference designed to test the time required to solve a complex decision problem, 35 managers using direction-set 1 required an average of $\bar{X}_1 = 57$ minutes with $s_1^2 = 64$ minutes squared. Thirty-two managers each using direction-set 2 took $\bar{X}_2 = 54$ minutes with $s_2^2 = 36$ minutes squared. Assume independence of trials, normal distributions, and use a 5% $\alpha$-risk.
    a. Test for equal variances, $\alpha = 0.05$. (That is, is it reasonable to pool these variances?)
    b. Test whether mean times required to solve the problem are significantly different.

26. A financial advisory firm wants to know whether one of its portfolios is yielding a higher return on investments. This test controlled on the dollars invested and on timing. Portfolio 1 was investments in stocks; Portfolio 2 was investments in bonds. The response was the return on investment, a percent.

| Control | Portfolio 1 (%) | Portfolio 2 (%) |
|---------|-----------------|-----------------|
| 1 | 7.5 | 7.0 |
| 2 | 10.3 | 9.8 |
| 3 | 9.4 | 9.1 |
| 4 | 8.9 | 8.8 |
| 5 | 8.0 | 7.3 |
| 6 | 7.6 | 7.7 |
| 7 | 8.4 | 8.2 |
| 8 | 9.6 | 9.2 |

Test for an equal return on investment for the two portfolios against the alternative that stocks yield a higher return. Let $\alpha = 0.05$ and assume that the difference values follow a normal distribution.

27. Two accounting methods are compared that claim to reduce the actuarial error that occurs in $10,000 worth of insurance premiums, that is, the overcharge in premium rates to ensure a desired profit. If a "better" procedure gives a statistically lower overcharge, is (either) one a better method? Use $\alpha = 0.05$ and assume the differences for the product classes follow a normal distribution.

| Product Class* | Method 1 | Method 2 |
|---|---|---|
| 1 | 0.75 | 0.60 |
| 2 | 1.40 | 1.31 |
| 3 | 1.51 | 1.55 |
| 4 | 0.77 | 0.73 |
| 5 | 0.63 | 0.55 |
| 6 | 0.99 | 0.92 |
| 7 | 1.36 | 1.28 |
| 8 | 1.03 | 0.99 |
| 9 | 1.37 | 1.39 |
| 10 | 1.31 | 1.21 |

\* The product classes are defined by customer's age and sex.

28. An industry-wide study compared the absentee rates under two sick-benefits plans. Companies were chosen at random to use either Plan 1 or Plan 2 for one year. Then the rates of absenteeism for the two plans were compared. Ten companies used Plan 1, and another 10, by random choice, used Plan 2. The numbers are person-days per 100 employees and are absences for one year.

| Plan 1 | 3.44 | 3.81 | 3.41 | 3.26 | 3.69 | 3.46 | 3.39 | 3.55 | 3.64 | 3.33 |
|---|---|---|---|---|---|---|---|---|---|---|
| Plan 2 | 2.96 | 2.55 | 3.22 | 2.88 | 2.67 | 2.94 | 2.62 | 2.71 | 3.14 | 2.77 |

The standard deviations for Plan 1 and Plan 2 are, respectively, 0.172 and 0.222, and the population distributions are normally distributed. Answer the following.
a. Test the hypothesis of equal population variances. Use $\alpha = 0.05$.
b. Provided your answer to (a) shows it is appropriate, test the research hypothesis $\mu_1 > \mu_2$. Again, use $\alpha = 0.05$. Is one plan statistically better? State your recommendations for a plan.

29. Are college men or college women more informed on business finance? For a random selection of students in a large general finance course, the following scores were achieved on a finance, banking, and investments test prepared by the instructor. The data are scores out of a possible 100 points.

| Men's scores | 83 | 82 | 66 | 70 | 79 | 63 | 82 | 66 | 70 | 92 | 75 | 85 | 73 | 81 |
|---|---|---|---|---|---|---|---|---|---|---|---|---|---|---|
| Women's scores | 74 | 82 | 86 | 66 | 92 | 97 | 70 | 73 | 75 | 91 | 84 | 89 | 84 | 88 |

Assume the distributions are normal, for independent samples, and use $\alpha = 0.05$ to test the following.

a. Given that the standard deviation for men's scores is 8.51 and for women's scores is 9.23, test for equal variances.

b. Based on your results for part (a), test for equal population means.

c. Establish a 95% confidence interval on the difference in means for men's scores minus women's scores.

For the population of these samples, are college men or college women or neither more informed on the items of this business finance test?

30. Salaries are compared for presidents at private and public larger-enrollment universities. The data are random samples of eight salaries, each rounded to the nearest $1000.

| Private | 134 | 128 | 139 | 136 | 133 | 130 | 140 | 137 |
| Public | 133 | 139 | 134 | 142 | 145 | 150 | 145 | 140 |

Assume that each sample comes from a normal distribution and that the two samples are independent. Then use $\alpha = 0.10$ in the following questions.

a. Test the hypothesis of equal population variances, given $s_1^2 = 4.21$, $s_2^2 = 5.76$.

b. If the test in part (a) is "Do not reject $H_0$," then proceed to test the research hypothesis that large-public-university presidents are paid more than presidents at private universities. Test at $\alpha = 0.10$.

c. Establish a 90% confidence interval on the difference for public minus private large-university mean salaries. Does this result provide more information than part (b)? If so, what is the added information?

**STATISTICS IN ACTION: CHALLENGE PROBLEMS**

31. Janssen and Schutz have used statistics to evaluate the performance of a professional hockey team. (The reason for including this non-business application is that it displays very good use of procedure. They are aware of, and display proper consideration for, assumptions; accordingly, they use timely statistics.) The data, from Table 1, page 276, of the article, relates to a severe slump experienced by the Edmonton Oilers in the middle of the 1983–84 National Hockey League season.

Means and Standard Deviations of the Two Samples

| Average (per game) | Normal Play (N = 24) | | Slump (N = 6) | |
|---|---|---|---|---|
| | $\bar{X}$ | $s$ | $\bar{X}$ | $s$ |
| Team points* | 1.79 | 0.59 | 0.33 | 0.82 |
| Goals-for | 5.88 | 2.42 | 3.17 | 3.54 |
| Goals-against | 3.54 | 1.64 | 6.33 | 2.94 |

* Two points are awarded for a win, one for a tie.

Relative to tests on means, the authors indicate (on page 277) that "Having investigated variability in performance we are now in a position to test the differences in average performance rates. $t$-tests for differences between means, with pooled estimates of the variance, were significant beyond the 0.05 level for team points, $t = 5.03$, as well as for goals-for, $t = 2.33$. With respect to goals-against, the $t$-test for differences between means is inapplicable due to the inequality of the population variances."

SOURCE: Janssen, Christian T. L., and R. W. Schutz. "Six Odd Games: An Analysis of the Midseason Slump of the Edmonton Oilers." *Journal of Recreational Mathematics* 17, no. 4 (1984–1985): 275–281.

**Questions** Use the table data to check the authors' statements about the goals-for data. That is, perform an $F$-test for equal variances. Then, provided the variances test is not statistically different for $\alpha = 0.05$, continue to pool the variances and test for equal means. Perform the tests and state your conclusions.

32. Again use data from the table by Janssen and Schutz, this time to check the authors' statements about goals-against. That is, as appropriate, perform the $F$-test for equal variances, followed by a ($t$-)test for equal means. As before, use $\alpha = 0.05$.

33. Friedman and Friedman discuss "More Subtle Sins in Marketing Research." Of interest here is the error of performing a univariate statistical test multiple times, on the same set of data, when it is more appropriate to perform a multivariate test one time. Specifically, the Business Case, following, concerns a comparison of the mean buying power for householders in four trading zones. It is possible to test by comparing a zone-1 mean against a zone-2 mean, then the zone-1 mean against a zone-3 mean, and so on, with $t$-tests. However, the authors point out that when the $\alpha$-level of the individual $t$-tests is fixed at, say, 0.05, the $\alpha$-level of the whole study, the experiment-wise error rate, is not 0.05. It can be substantially larger!

Alpha is the probability of rejecting the null hypothesis when it is true. For a single $t$-test this can be fixed at, say, 0.05. But if several $t$-tests are performed on the same data, the probability of rejecting at least one of the null hypotheses when it is true increases with the number of tests. The true experiment-wise error rate can be as high as $1 - (1 - \alpha)^P$, where $P$ is the number of $t$-tests, if the individual measures are all independent. Their Table 1 includes the following.

Maximum Experiment-wise Error as a Function
of the Number of Univariate Tests, $P$

| $P$ | $\alpha = 0.05$ |
| --- | --- |
| 1 | 0.050 |
| 3 | 0.143 |
| 5 | 0.226 |
| 10 | 0.401 |
| 15 | 0.537 |
| 20 | 0.642 |
| 100 | 0.994 |

For 15 individual comparisons, then, with independence, there is over a 50% chance that one or more true null hypothesis will be (falsely) rejected. This is not a good procedure. The business application provides an alternative.

SOURCE: Friedman, Hershey H., and Linda W. Friedman. "More Subtle Sins in Marketing Research." *The Mid-Atlantic Journal of Business* 20, no. 1 (Winter 1981–1982): 21–31.

Questions   Let's check a few of Friedman and Friedman's experiment-wise (maximum) error rates. For ($P =$) three tests, it is possible to incorrectly reject all three, any two, or only (any) one with probability $(0.05)^3 + 3 \cdot (0.05)^2 \cdot (0.95) + 3 \cdot (0.05) \cdot (0.95)^2 = 0.1436$ or, by Friedman and Friedman, $1 - (1 - 0.05)^3 = 0.1436$. What theoretical probability distribution is in effect? (*Hint:* It was key in Chapter 5.)

34. Provided your calculator has sufficient decimal places of accuracy, use the formula in Problem 33, $1 - (1 - 0.05)^P$, to check the (maximum) experiment-wise error rates for $P = 5$, 10, and 15 given in the preceding problem.

---

## BUSINESS CASE

*Gene Stewart Develops Supermarket Site Location Statistics*

His first assignment with B&Q National Grocers could have been a case study situation from Professor Easley's market research course. Gene Stewart had observed several site location strategies in his classes, but this was the real thing! He was assigned by Jim Rogers, the vice president of marketing, to analyze and help determine a new store site in the Dallas market.

Gene recognized that a site choice requires evaluation of numerous factors. Key ones include

1. availability of a site with good access and high traffic flow;
2. a favorable market environment in a trading zone not already saturated with grocery outlets;
3. more than one acceptable site, to provide B&Q's buyers with some bargaining power for acquisition of a site location;
4. consumer buying-power potential adequate to ensure a favorable ratio of revenues to costs.

The first three factors had been researched by other personnel on Mr. Roger's staff, resulting in identification of four possible zones. The last factor (4) was assigned to Gene.

Gene analyzed the research, including Sales and Marketing Management's *Survey of Buying Power*. The reports available to him covered the metropolitan statistical areas (MSA's), including Dallas. Unfortunately, none of the reports got as specific as identifying trading zones within a MSA.

An alternative was to collect primary data; Gene elected to perform a market-basket analysis. Mr. Rogers agreed to budget $5000 for development of a market-basket questionnaire and for the necessary field surveys.

Gene's plan was to survey a random sample of 20 householders in each trading zone. In-depth personal interviews with the person most responsible for buying food in each household would disclose the actual items that person was planning to buy. These shoppers were asked to prepare a complete grocery shopping list each week. To assure minimal bias in timing, they would be interviewed by random assignment.

| Trading Zone | Adjusted Mean Buying Power | Standard Deviation | Sample Size |
|---|---|---|---|
| 1 | $128 | $26 | 20 |
| 2 | 136 | 32 | 20 |
| 3 | 91 | 30 | 20 |
| 4 | 112 | 41 | 20 |

Prices for all items used the area retail standard scale of food prices. An adjustment was made to the mean expenditure to reflect the average number of persons in that trading zone. Once he had this information in hand, the rest was up to Gene.

Gene's quantitative analysis textbook had a section labeled "multiple comparisons" [5]. Referring to this, he determined that his objective was to compare all the trading zone means. But he had to compare them two at a time. That meant $_4C_2 = 6$ $t$-tests. The text said there was a (correct) way to do this, one that would ensure an overall experiment-wise $\alpha$-risk level. The procedure was Fisher's least significant difference (LSD) method.*

Gene proceeded with a test, setting his $\alpha$-risk at 5%:

$H_0$: Mean expenditures are equal for the four trading zones.

$H_A$: Mean expenditures differ for two or more trading zones.

The LSD statistic for an $\alpha$-risk test is

$$\text{LSD} = t_{\alpha/2}\sqrt{s_w^2\left(\frac{1}{n_i} + \frac{1}{n_j}\right)} \qquad \text{where} \quad i = \text{trading zone } i$$
$$j = \text{trading zone } j$$

The value $s_w^2$ is a within-trading-zones pooled variance, and $t_{\alpha/2}$ has df $= (n_1 - 1) + (n_2 - 1) + \cdots + (n_4 - 1)$. Any mean difference that exceeds the LSD number is judged to be statistically significant at an overall 5% $\alpha$-level.

Gene calculated for $n_1 = n_2 = n_3 = n_4 = 20$:

$$s_w^2 = \frac{(20 - 1)26^2 + (20 - 1)32^2 + (20 - 1)30^2 + (20 - 1)41^2}{(20 - 1) + (20 - 1) + (20 - 1) + (20 - 1)} = 1{,}070.25$$

and then computed

$$\text{LSD} = 1.96\sqrt{1070.25\left(\frac{1}{20} + \frac{1}{20}\right)} = \$20.28.$$

What remained was to array the four means from low to high and apply the LSD comparison. Any two zones with a difference in means of over $20.28 are significantly different, as shown in Figure 10.3, so Gene elected to use a rule that any two means with less than the LSD value are underscored by an unbroken line (part (b) of figure). This holds for areas 1 and 2 displayed in Figure 10.3(b), since $136 − $128 = $8 and $8 is less than $20.28. Because they are *not* significantly different, areas 1 and 2 have a common underline in part (c), as do areas 1 and 4 in part (c). However, because areas 2 and 4 differ by $136 − $112 = $24, which exceeds $20.28, these areas do not have a common underline in Figure 10.3.

---

* The Fisher LSD procedure as applied here assumes that a preliminary test of the hypothesis $H_0$: $\sigma_1^2 = \sigma_2^2 = \sigma_3^2 = \sigma_4^2$ is not rejected. By Hartley's test for homogeneity of variances, described in [5],

$$\text{Calculated } F = \frac{s_{\max}^2}{s_{\min}^2} = \frac{41^2}{26^2} = 2.49$$

which is less than test $F_{0.05,4,19} = 2.90$. So the variances are statistically equal and can be pooled. Also, it requires that the hypothesis $H_0$: $\mu_1 = \mu_2 = \mu_3 = \mu_4$ be rejected. This is true for our data. The test, which uses an $F$-ratio statistic, is discussed in the next chapter.

**FIGURE 10.3**   An Example of Multiple Comparisons Using the Least Significant
Differences Procedure

| | Area | 3 | 4 | 1 | 2 |
|---|---|---|---|---|---|
| (a) | Mean | $91 | $112 | $128 | $136 |

| | Area | | | 1 | 2 |
|---|---|---|---|---|---|
| (b) | Mean | | | $128 | $136 |

| | Area | 3 | 4 | 1 | 2 |
|---|---|---|---|---|---|
| (c) | Mean | $91 | $112 | $128 | $136 |

Gene concluded, with an overall $\alpha$-level of 5%, that the significant differences in mean buying power by trading zones are

$$2 > 4 \quad 2 > 3 \quad 1 > 3 \quad 4 > 3$$

Based on this market-basket analysis, zone 2 had the highest mean. It was statistically greater than all other zone means except 1. The zone-3 mean was statistically less than the (3) other means.

Gene's report stated that zone 2 should be considered as having the strongest buying power potential, followed by zone 1. Zone 3 was not in contention. This information would be evaluated by Mr. Rogers and other management along with the results of the research on the other key factors for a site choice.

**Business Case Questions**

35. State briefly the statistical nature of the problem faced by Gene Stewart.

36. What role is played by the $s_w^2$-value in the Fisher LSD procedure?

37. How was Gene Stewart able to draw a conclusion about the relative position of more than two means when he was only able to compare the means two at a time?

**COMPUTER APPLICATION**

*Tests with t- and F-statistics*

We again consider the marketing problem that targets households with 3+ persons and with head of household 35–50 years old. In Chapter 6 we established that, for 3757 records, the normal curve is a reasonable distribution for household incomes for that target universe.

A timely question is whether the average household income is statistically equal for households (with 3+ persons) with head of household (a) 35–39 years versus (b) 40–50 years. If there is a difference, then these marketers can use that knowledge to segment the population from the more affluent to the less affluent age groups. From Chapter 10, this points to testing for the difference in population means. We have already established the required condition of normal distribution for the 35–39-year-old group in Computer Problems 57 and 58 in Chapter 6. We will assume that normal distribution holds for the incomes for both of these groups.

The data in Table 10.4 were obtained for a random sample of 50 records from the PUMS file. Remember that this target population is restricted to households with 3+ persons and head of household 35–50 years old. The data are split

**TABLE 10.4**  Incomes for Two Groups: (1) Householder Aged 35–39 Years and (2) Householder 40–50 Years

| Number | Incomes for Group 1 (35–39) | Incomes for Group 2 (40–50) | Number | Incomes for Group 1 (35–39) | Incomes for Group 2 (40–50) |
|---|---|---|---|---|---|
| 1 | $20,010 | $33,065 | 15 | $46,495 | $12,005 |
| 2 | 44,510 | 41,810 | 16 | 32,010 | 8,005 |
| 3 | 9,760 | 20,005 | 17 | 20,285 | 21,165 |
| 4 | 7,125 | 35,720 | 18 | 24,130 | 26,590 |
| 5 | 17,010 | 41,275 | 19 | 50,315 | 44,005 |
| 6 | 17,165 | 31,575 | 20 | 48,010 | 19,850 |
| 7 | 10,520 | 31,140 | 21 | 18,630 | 16,510 |
| 8 | 45,915 | 38,665 | 22 | 17,535 | 18,285 |
| 9 | 28,015 | 27,015 | 23 | | 44,615 |
| 10 | 75,000 | 17,285 | 24 | | 51,015 |
| 11 | 38,600 | 59,265 | 25 | | 28,135 |
| 12 | 22,395 | 41,985 | 26 | | 29,915 |
| 13 | 30,915 | 24,510 | 27 | | 25,215 |
| 14 | 43,590 | 75,000 | 28 | | 41,920 |

by the age of householder, AGE1, into two groups: 35–39 years old and 40–50 years old.

From observation it is not clear which sample group has the higher mean. However, the incomes are quite widespread for both groups. The condition of equal variances is another assumption. A test for equal variances precedes the *t*-test for equal means. The SPSS software [8] gives information for both tests in the T-TEST procedure. Here is a routine to produce the essentials for testing for (1) equal variances and then, provided it is appropriate, (2) equal means using Equation (10.6):

```
1  NUMBERED
2  TITLE      PUMS FILE TTEST EXAMPLE
3  DATA LIST FILE= INFILE/
4             PERSONS 12-13 INCOME 35-39 AGE1 43-44
5  SET BLANKS = -1
6  SELECT IF   ((PERSONS GE 3) AND (AGE1 GE 35) AND (AGE1 LE 50))←————————1
7  COMPUTE         AGECD = AGE1
8  RECODE          AGECD (35 THRU 39 = 1) (ELSE = 2)
9  SET SEED = 2566 ←————————————————————————————————————————2
10 SAMPLE 50 FROM 3757
11 T-TEST GROUPS = AGECD(1,2)/VARIABLES = INCOME ←————————————3
12 LIST VARIABLES    AGECD AGE1 INCOME/FORMAT = NUMBERED/
13                   CASES FROM 1 TO 50
14 FINISH
```

A few comments will identify the essential procedures for this test. For a more complete description, and for additional outputs, see the *SPSS Guide* [8]. The data are read, as before, from the 20,116-record PUMS data base.

**1**  The SELECT IF statement qualifies the selection to the target population, which includes 3757 records. The COMPUTE and RECODE statements convert householder age—AGE1—to AGECD (a group), either 1 (35–39 years) or 2 (40–50 years).

**2** The choice for a SEED number is arbitrary, and it can come from Table B.7—Random Numbers. The SEED number needs to be a value 3757 or smaller.

**3** The T-TEST procedure allows tests for two, three, or more independent sample means, but taken two at a time. LIST VARIABLES provides a check that the data were read into the computer as intended. Because the data appear in Table 10.4, they will not be repeated here.

The T-TEST summary appears in Table 10.5. Observe that the outputs include a calculated F-VALUE and the $p$-value, for testing $H_0: \sigma_1^2 = \sigma_2^2$, assuming a two-tailed test of hypothesis. Of course, there is a summary for the $t$-test on $H_0: \mu_1 = \mu_2$. It also includes the group standard deviations for income by group and the group sample means. We will not use the SEPARATE VARIANCE ESTIMATE $t$-test.

**TABLE 10.5** A SPSS $t$-Test for Equal Mean Incomes for Heads of Household in which Group 1 is 35–39 and Group 2 is 40–50 Years, Both Restricted to Households with Three or More Persons

```
T-TESTS FOR INDEPENDENT SAMPLES OF AGECD

GROUP 1 — AGECD EQ 1.00
GROUP 2 — AGECD EQ 2.00
```

| | | | | | | | POOLED VARIANCE ESTIMATE | | | SEPARATE VARIANCE ESTIMATE | | |
| VARIABLE | NUMBER OF CASES | MEAN | STANDARD DEVIATION | STANDARD ERROR | F VALUE | 2-TAIL PROB. | T VALUE | DEGREES OF FREEDOM | 2-TAIL PROB. | T VALUE | DEGREES OF FREEDOM | 2-TAIL PROB. |
|---|---|---|---|---|---|---|---|---|---|---|---|---|
| INCOME | | | | | | | | | | | | |
| GROUP 1 | 22 | 30360.9091 | 16874.283 | 3597.609 | 1.30 | .511 | -.44 | 48 | .661 | | | |
| GROUP 2 | 28 | 32340.5357 | 14781.225 | 2793.389 | | | | | | -.43 | 42.06 | .666 |

$$S_w^2 = (15696.940)^2$$

The test for equal variances uses the two sample standard deviations

$$H_0: \sigma_1^2 = \sigma_2^2 \quad \text{versus} \quad H_A: \sigma_1^2 \neq \sigma_2^2$$

Because GROUP 1 has the larger standard deviation, it is placed in the numerator for the calculated $F$-statistic:

$$\text{Calculated } F = s_1^2/s_2^2 = \frac{(16,874.283)^2}{(14,781.225)^2} = 1.30$$

For our two-sided alternative hypothesis, this calculated $F$ has a $p$-value of 0.511. The two variances are not statistically different. It is appropriate to continue with a $t$-test for equal means.

The $t$-test for independent samples with a POOLED VARIANCE ESTIMATE gives the essential test information for testing $H_0: \mu_1 = \mu_2$ against $H_A: \mu_1 \neq \mu_2$.

$$\text{Calculated } t = \frac{30,360.91 - 32,340.54}{\sqrt{s_w^2(1/22 + 1/28)}} = -0.44$$

where $s_w^2 = \left( \dfrac{21 \cdot 16{,}874.283 + 27 \cdot 14{,}781.225}{22 + 28 - 2} \right)^2 = (15{,}696.94)^2$

Again, the calculated statistic is not significant for an $\alpha = 0.05$ test. From Table 10.5, the $p$-value is 0.661.

These tests indicate that the mean income for heads of household aged 35–39 years does not differ, statistically, from the mean income for those 40–50 years old when both are qualified to households with 3+ persons. The marketers may conclude that if they split the target population by age of householder 35–39 versus 40–50, then they cannot expect to obtain higher/lower income segments. This does not discount the possibility that some as yet-untested age range could provide a higher income population segment.

Here the assumptions were valid for a $t$-test of equal means. However, in cases in which the assumptions of normality or constant variances are not met, the nonparametric Mann–Whitney $U$-test in Chapter 17 provides a viable procedure. The Mann–Whitney test, however, compares medians rather than means. Chapter 11 introduces the procedures for testing the equality of means for three or more independent random samples.

## COMPUTER PROBLEMS

38. The following procedures can be used to check the paired differences $t$-test in Example 10.1. Using Minitab,

```
MTB > READ C1 C2
DATA> 51 15
DATA> 63 21
        ⋮
DATA> 57 18
DATA> END
MTB > NAME C1 'AFTER' C2 'BEFORE'
MTB > NAME C3 = 'DIFF'
MTB > LET 'DIFF' = 'AFTER' - 'BEFORE'
MTB > TTEST 30 'DIFF'
```

If you are using SPSS, read the data as cases for two variables—AFTER and BEFORE. Then use

```
COMPUTE BEFORE1 = BEFORE + 30
T-TEST PAIRS = AFTER WITH BEFORE1
```

Or use another procedure to calculate a paired differences $t$-value. When your $t$-value agrees with that in Example 10.1, then proceed to use similar logic to check Example 10.2.

39. A test of hypothesis for two independent samples from normal distributions can use the data in Example 10.10 to test $H_0: \mu_1 - \mu_2 = 1$ year versus $H_A: \mu_1 - \mu_2 > 1$ year. Read the data as was done in the previous problem except that, for group 2 (Far West), add 1 to each value (that is, $\mu_1 = \mu_2 + 1$). For Minitab, use

```
MTB > TWOSAMPLE-T 'NOEAST' BY 'FARWEST';
SUBC> POOLED.
```

For SPSS read the data in as a single variable FRSTSALE with the Northeast data preceding the Far West data. Then

```
COMPUTE NUMBER = 1
IF ($CASENUM GT 11) NUMBER = 2
T-TEST GROUPS = NUMBER(1,2)/VARIABLES
              = FRSTSALE
```

This shows another way to read in the data. Or use another procedure to calculate an independent-samples $t$-value. Compare your calculated $t$-value to the value 0.82 given in Example 10.10.

40. Use the $t$-test procedure for dependent samples, a computer program that is available to you, and the data in Problem 2 to test the hypothesis of equal pricing for the dinners served at two locations of a chain restaurant. Use a $p$-value testing approach. Here is a setup for Minitab users. Let $H_0$: $\mu_d = 0$ and $H_A$: $\mu_d > 0$.

```
MTB > READ C1 C2
DATA> ◄───────────────────────── Read in eight data pairs.
MTB > LET C3 = C2 - C1
MTB > NAME C1 'LOCA' C2 'LOCB' C3 'LOCDIFF'
MTB > PRINT C2 C1 C3
MTB > TTEST 'LOCDIFF';
SUBC> ALTERNATIVE = +1.
```

41. Use a $t$-statistic procedure and the data in Problem 4 to compare the wear for Brand A and Brand B women's jogging shoes. Let $\alpha = 0.05$, and compare your calculated $t$-value with that for Problem 4. What is your test conclusion? *Note:* For Minitab users, the essential Minitab statements are the same as given in the previous problem but with the appropriate data.

```
MTB > TTEST 'BRNDDIF';
SUBC> ALTERNATIVE = 0.
MTB > TINTERVAL 95% 'BRNDDIF'
```

42. The following can be used to check the independent-samples $t$-test in Problem 24. Using Minitab,

```
MTB > READ C3 C4
DATA> 4 1
DATA> 5 1
    ⋮
DATA> 7 2
DATA> END
MTB > NAME C3 'LAWN' C4 'GROUP'
MTB > PRINT 'LAWN' 'GROUP'
MTB > TWOT 97.5% 'LAWN' 'GROUP';
SUBC> ALTERNATIVE = +1.
SUBC> POOLED.
```

The data are read one value at a time, followed by a group number, because there are unequal numbers in the groups. Then use the TWOT procedure.

43. Recall Problem 29, about who is better informed on general business finance— college men or college women. Use the data from that problem and a computer

routine of your choice to test the hypothesis of equal means. Begin with a test for equal variances and, if appropriate, follow with a $t$-test for equal means for independent samples from normally distributed populations. (The normality is assumed.) Use a $p$-value for your test determination. Also, establish a 95% confidence interval on the difference of the population means. For Minitab users, after reading in the data in pairs, use

```
MTB > TWOSAMPLE 95% 'MEN' 'WOMEN';
SUBC> ALTERNATIVE = 0;
SUBC> POOLED.
```

The TWOSAMPLE procedure is appropriate when there are equal numbers in the two groups. This produces both a $t$-test and a 95% confidence interval. Compare this with your results for Problem 29.

## FOR FURTHER READING

[1] Fisher, R. A. *The Design of Experiments*, 9th ed. New York: Hafner Publishing, 1974.

[2] Freund, John E., and Ronald E. Walpole. *Mathematical Statistics*, 4th ed. Englewood Cliffs, NJ: Prentice-Hall, 1987.

[3] Iman, Ronald L., and W. J. Conover. *Modern Business Statistics*, 2nd ed. New York: John Wiley, 1989.

[4] Olson, Charles L., and Mario J. Picconi. *Statistics for Business Decision Making*. Glenview, IL: Scott Foresman, 1983.

[5] Ott, Lyman. *An Introduction to Statistical Methods and Data Analysis*, 3rd ed. Boston: PWS-Kent, 1988.

[6] Pearson, E. S. *A Statistical Biography of William Sealy Gossett*. Cambridge, England: Cambridge University Press, 1990.

[7] Schaefer, Robert L., and Richard B. Anderson. *The Student Edition of Minitab*. Reading, MA: Addison-Wesley, 1989.

[8] *SPSS Reference Guide* (version 4.0). Chicago: SPSS, Inc., 1990.

[9] Steel, Robert G. D., and James H. Torrie. *Principles and Procedures of Statistics: A Biometrical Approach*, 2nd ed. New York: McGraw-Hill, 1980.

[10] *The Survey of Buying Power-Data Service 1984*. New York: Sales and Marketing Management, 1984.

# STATISTICS FOR BUSINESS DECISIONS

# 11

# ANALYSIS OF VARIANCE AND EXPERIMENTAL DESIGNS

Boise Cascade, DiGiorgio, Georgia Pacific, and Weyerhauser all compete in the building materials market. In these firms, as in others, the company presidents are concerned about the firm's position in the market. The president of a firm must be aware of the firm's corporate sales in wood products. And in order to make statements about market share, the president may have to compare the dollar sales of competitors to the company's own.

As chief executive officer (CEO) of a company, the president is expected to make precise statements about how effectively the firm is using the capital provided by the stockholders. But that responsibility does not stop with the CEO. Nearly every manager, whether involved with production, marketing, finance, personnel, or any other area, has some need for making valid comparisons. For example, a medical insurance administrator may have to establish the amount of coverage the firm will allow for semiprivate rooms in hospitals in three adjoining cities. Are the mean rates charged in the three cities close enough to be considered the same, or are they really different? Management can answer this question by using sample data and a statistical test to compare the mean rate in the three cities.

The same test procedure can be used to compare average costs in construction industries, or the mean performance for different work groups in a manufacturing environment. Advertisers can compare the average service lives of competitive products. Supermarket managers can compare the prices charged by competing wholesalers by using the average price for a standard market basket of goods. Many purchasing departments keep computerized records of average delivery times from different suppliers. Then they can automatically channel orders to their best suppliers. All these situations concern management decision situations that can be treated by comparisons of the means for two or more populations. Related decision questions are analyzed in this chapter.

**Experimental Design Concepts**

Business data comparisons can involve more than two population means, so we must have statistical procedures for inference about the means of three or more populations. We assume that a quantitative measure is made of the response variable under study and that the responses in each population are normally distributed and independent of those for the other groups. The response variable can be interval or ratio scale.

A planned, or designed, experiment is used for tests of three or more group means. The plan is always established before any data are collected. The statistical plan, including the selection of variables and instructions for making tests, is called an *experimental design.*

An experimental design is a plan that specifies

1. the experiment test units, called experimental units (eu's);
2. how the test variables will be assigned to the eu's;
3. what measurements will be taken to define the response;
4. the response test variables (called treatments) and the control variables that will be considered; *factor for the effect,*
5. a model for testing the treatment variables to determine their level of significance in explaining variations in response;
6. the data analysis approach that is to be used to perform the testing.

By making a plan or statistical design, we can model and analyze experiment results to determine whether there are significant differences in the mean response for some treatment populations.

The theory of designed experiments originated in the agricultural sciences, where people have long been concerned with how such factors as different soil types, fertilizers, or amounts of moisture affect plant growth. Many of the terms in experimental design come from that discipline.

Experiments are conducted to determine whether selected variables influence mean crop yields or whether the variations are just chance occurrences. For an agriculture experiment, a test variable could be brands of fertilizer. This variable is often called a treatment, factor, or effect. The subjects of the experiment are called the *experimental units,* or eu's

An experimental unit (eu) is a person or subject upon which a response measurement is made. *Observation*

For an agribusiness crop study, the experimental units may be one-acre plots of soil; for a marketing test, the eu's may be individual consumers.

For our purposes, all testing will assume random assignment of treatment variables to the eu's. This ensures that there is no systematic bias, so that the experiment is fair and subsequent tests can be valid.

In a designed experiment, accurate and timely measurements are made for a well-defined response. The variations in response are determined and then are attributed to either (1) differences due to treatment groups (explained differences) or (2) residual error (unexplained differences). We seek a design in which the treatments explain a high percentage of the total variability in the response variable. The remaining variability, a smaller portion, is attributed to unexplained error. Otherwise, a relatively high measure for unexplained variability indicates a weak design. Our objective is to find those variables and an experimental design that will do the best possible job of explaining differences in response—that is, to find strong treatment variables. So we consider several alternative designs.

One of the simplest experimental plans is the completely randomized design, or CRD. For this design the treatment groups are randomly assigned to the experimental units. For example, a CRD experiment would result from a direct mail charity appeal where a suggested gift amount—$1, or $5, or $10— is randomly assigned to the households on a mailing list. The purpose of our test would then be to determine whether the suggested amount made a difference in the amount received. Other, more complicated designs could explain more about variations in charitable gift-giving and possibly result in more gift dollars. The CRD will be defined more formally shortly. We will also look at two other common designs: (1) the randomized complete block design (Section 11.5) and (2) factorial designed experiments (Section 11.6).

## Data Analysis Procedures

The data analysis approach to testing for equal group means uses some statistics that we studied in Chapter 10. There we used the normal and $t$-distributions for comparing the means for samples from two populations. Now we use the $F$-distribution for testing the equality of two or more population means. From Chapter 10 the $F$-statistic is defined as the ratio of two variance estimators: Now we use the $F$-distribution in the form

$$F = \frac{\text{Variance estimate among treatment groups}}{\text{Variance estimate within treatment groups}}$$

We structure a test for the equality of means of treatment groups by dividing the estimated variance *among* groups by that estimated for *within* groups. If the population means are approximately equal, the two estimates will, on average, be nearly equal. This will result in a low value for the $F$-(variance ratio) statistic. So a low $F$-value supports a hypothesis of equality among the population means. In contrast, if the group means vary substantially from one to the other, the resulting $F$-ratio statistic can be much larger than 1. A large $F$-ratio value will cause us to conclude that the population means are significantly different and therefore are not equal.

Performing the $F$-ratio test for equal treatment group means is simplified by a systematic procedure called *analysis of variance*.

> Analysis of variance, ANOVA, is an algebraic procedure for partitioning the total response variance into explained (treatment) and unexplained (error) parts, and it establishes a structure for reporting the results by an $F$-ratio test.

The processes of an $F$-variance ratio test and the analysis of variance for systematic calculations are universal and apply to many designs. The key is to find treatment variables and a design that will identify sources explaining substantial variation in the response.

If, by testing, we determine that the group means are significantly different, then we can use estimation procedures to determine which treatment populations are different, and estimate by how much their means differ. This estimation process is an extension of the earlier $t$-statistic confidence intervals to (multiple) comparisons for differences among three or more group means.

Upon completing Chapter 11, you should be able to

1. explain the principles underlying the analysis of variance (ANOVA), including the meanings of treatment (explained) and error (unexplained) variation;

2. determine whether the conditions in a business situation warrant making an $F$-ratio test for equal population means;

3. apply the ANOVA procedure to data classified according to a single treatment variable and interpret the meaning of the calculated $F$-value;

4. explain how the ANOVA procedures can be extended to two-variable classifications and distinguish between blocking and factorial experiments;

5. develop and interpret confidence intervals for the multiple comparisons of differences in group means;

6. use standard computer software packages for one-way and two-way ANOVA summaries for testing the equality of two or more group means.

This chapter explains three experimental designs. But to apply these designs, it is helpful first to understand the basic terminology that is used in testing for equal population means.

**Problems 11.1**

1. A business experiment is designed to compare the production from two machines that seal cans of a cola. Of a test sampling including 1000 cans, half are chosen at random to be capped by machine A. The remaining 500 cans are sealed by machine B. This test is repeated 10 times. The test will use the difference in the time required to seal the cans. Identify each of the following.
   a. The experimental test units (the eu's)
   b. The test variables
   c. The response measure
   d. The statistical test question

2. The hourly exams in a business class contain some elements of an experimental design. Indicate whether each of the following designed-experiment elements is included in course hourly exams.
   a. Experimental units. Name the subjects or eu's.
   b. A test measurement. What measure is assigned to each eu?
   c. A mixed or random assignment of test variables to the eu's. Are test questions assigned at random to the students?
   d. A measure of the significance of variations in the response. What report is generally given about students' test scores? From your answers to (a)–(d), can you conclude that hourly testing is a designed experiment? Explain.

3. Explain why it is not possible to have a negative value for

$$F = \frac{\text{Variance estimate among treatment means}}{\text{Variance estimate within treatment means}}$$

4. A completely randomized design is one for which the treatment groups are randomly assigned to the experimental units. Consider an experiment wherein several accountants from three firms are assigned a similar task. Some accountants will use a computer for this task, others will use calculators, and others will do the computations by hand. Explain how this experiment satisfies being a completely randomized design. If some accountants (eu's) are not trained in using computers, then can this be a CRD experiment? Explain your answers.

## 11.2 IDENTIFYING THE STRUCTURE OF A ONE-WAY DESIGNED EXPERIMENT

Limiting comparisons to two groups at a time is not always sufficient. For example, suppose the Federal Aviation Administration (FAA) wishes to analyze the labor costs per passenger-mile for air carriers flying from Boston to Chicago. This information could be used to assess air fares. Although they could use the *t*-statistic (or *Z*) to compare two means at a time (for example, United versus TWA), their results would be restricted to the relative costs of the two airlines being compared.

The *F*-statistic offers a more comprehensive approach. It would allow the FAA to compare at one time the average costs for all airlines flying the Boston-to-Chicago route. Related tests would also enable the FAA to single out which airlines deviated substantially from the others in cost. This section develops the data structure for testing the equality of means for two, three, or more business populations.

### The Data Structure for One-Way Designed Experiments

At this point we introduce some test data. For this, we elect to use the weight of sheets of plywood purchased from a manufacturer as the response variable, as shown in Table 11.1. Suppose a contractor has just been awarded the construction contract for a large domed stadium and wants to choose the strongest plywood sheeting available. Because the plywood is exposed, it needs to be strong and able to withstand extreme weather conditions. A characteristic of

**TABLE 11.1**   Weight (in pounds) of Plywood Sheets Classified According to the Wood Preservative Treatment

|        | Treatment 1 | Treatment 2 | Treatment 3 |
|--------|-------------|-------------|-------------|
|        | 65          | 64          | 68          |
|        | 68          | 65          | 70          |
|        | 68          | 66          | 70          |
|        | 70          | 68          | 74          |
|        | 70          | 72          | 74          |
|        | 71          | 74          | 76          |
| Totals | 412         | 409         | 432         |
| Means  | 68.6667     | 68.1667     | 72.0000     |

plywood strength is its weight. This reflects not only the kind of wood but also the wood preservative treatment (the treatment groups). Suppose the manufacturer used a standard wood, and three wood preservative processes are available: (1) surface stain, (2) pressure spray sealer, (3) liquid injection treatment. The first two procedures use a surface application, whereas the third penetrates the wood to fill it with a liquid preservative. This can affect the weight.

Let us consider the $F$-ratio test of means in the context of the plywood data. How alike are the weights for plywood from the three wood preservative treatments? Do they differ substantially from one process (treatment population) to another, or are the differences in the 18 weights simply due to chance?

What response structure is required to conclude that the treatments do, in fact, have different means? Alternatively, what results would convince the contractor that the data are simply three independent samples from the same population (with the same population mean)? The data in Figure 11.1 can be used to distinguish these two extremes.

The two data results appear to be quite different. We compare them by evaluating the weight variation *among*, or across, treatment groups relative to the variation *within* the individual treatments. We shall first discuss this comparison in concept, and then follow with some procedures to quantify and calculate differences in treatment means.

Figure 11.1(a) shows treatments with quite different means, compared to individual weights that are quite close within the treatment groups. This case would seem to indicate three treatments that have different means.

In contrast, Figure 11.1(b) shows weights that seem to occur at random, with a mix of high, middle, and low weights for every treatment. The population means here could be quite close. Relative to the variation of weights *within* the treatment groups, the variation *among* the treatment means is judged to be a small value. So Figure 11.1(b) actually shows three samples of weights that come from one population. Their means appear to be the same value.

**FIGURE 11.1**    Illustration of Alternative Data Structures for a One-Way Designed Experiment

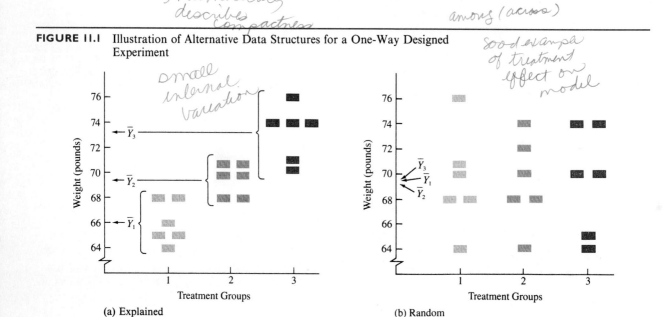

(a) Explained

(b) Random

Figure 11.1 illustrates the extreme conditions. The data in Table 11.1 do not fit either situation exactly. However, our next effort is to define procedures that will allow us to determine if these data are simply three independent samples coming from one population or are unique and coming from two or more treatment populations that have different means.

**Problems 11.2**

5. Some data displays that appear later in this chapter are listed in tables (a) and (b).
   a. Do the number of responses received from the three methods of mailings (table (a)) appear to be different (your judgment)? Explain.
   b. Answer part (a) for the data on the power rating of batteries.

| | Mailing Method | | |
| --- | --- | --- | --- |
| | *1* | *2* | *3* |
| | 16 | 9 | 12 |
| Number of | 19 | 7 | 6 |
| responses | 18 | 8 | 10 |
| received | 15 | 5 | 7 |
| | 17 | 6 | 15 |

(a) Responses from three methods of equal count mailings

| | Power rating of | | |
| --- | --- | --- | --- |
| | *Brand A* | *Brand B* | *Brand C* |
| | 5 | 5 | 5 |
| | 7 | 5 | 6 |
| | 8 | 6 | 7 |
| | 8 | 7 | 7 |
| | 10 | 7 | 8 |
| | 10 | 8 | 9 |
| Means | 8.0 | 6.3 | 7.1 |

(b) Power ratings of batteries

6. How do the data displayed in Problem 5(a) illustrate the results for a one-way designed experiment?

7. Which means from Figure 11.1(b) appear different?

## 11.3 ONE-WAY DESIGNED EXPERIMENTS: THE COMPLETELY RANDOMIZED DESIGN

This section develops the procedures for testing the equality of means for two, three, or more business populations. Figures 11.1(a) and 11.1(b) illustrate two extremes for the data in modeling a single-treatment, or one-way, *completely randomized design experiment*.

A completely randomized design, CRD, is the random assignment of $T$ treatment levels to the experimental units so that $n_1$ units receive treatment 1, $n_2$ receive treatment 2, . . . , and $n_T$ receive treatment $T$.

The CRD is a random assignment of the treatments to the experimental units. The two sources of response variation identified are (1) due to treatment differences and (2) due to random error. The random error measures the variability in response resulting from all sources other than the treatment variable.

The response values are denoted by $Y_{ij}$, for $i = 1, 2, \ldots, T$ treatment groups with $j = 1, 2, \ldots, n_i$ replications per group. (A replication is another or additional observation.) Hence $Y_{ij}$ represents the $j$th observation under treatment $i$, and the error or unexplained sources are described by $\varepsilon_{ij}$.

**The Model with Assumptions and Hypotheses**

The elements of a designed experiment are summarized in a linear additive model. If we let $Y_{ij}$ be the observed response, we can state the design model for the *completely randomized design* as follows:

### MODEL FOR COMPLETELY RANDOMIZED DESIGN

**(with either equal or unequal sample sizes)**

$$Y_{ij} = \mu + t_i + \varepsilon_{ij} \quad \text{with} \quad i = 1, 2, \ldots, T$$
$$j = 1, 2, \ldots, n_i$$

where    $\mu$ = an overall mean

   $t_i$ = treatments variable effect

   $\varepsilon_{ij}$ = random error

with    $i$ = treatment level

   $j$ = observation number

   $n_i$ = the number of repeats (replicates) receiving treatment $i$

   ($n_i = n$ when the sample sizes are equal)

This model recognizes the differences that arise from the treatment groups as one source for explaining variations in the response. If a real treatment effect exists, the sample response values will be consistently higher in some groups than in others. Then the statistical test will show a significant difference in the response values for some of the treatment means. This possibility was illustrated in Figure 11.1(a). Conversely, if the treatments induce no real differences in response, then the response values will be more dispersed. Each treatment group will have a similar share of high, middle, and low response values. Figure 11.1(b) illustrates this case.

The CRD model requires the following assumptions for a hypothesis test of equal treatment means.

### ASSUMPTIONS

1. The responses come from independent random samples. (The outcome of one sample in no way affects the results of another.)
2. Each sample is selected from a normal treatment population, so the errors are normally distributed.
3. The population means are $\mu_i$, and the variance is a constant value, denoted by $\sigma^2$.*

### MODEL

$$Y_{ij} = \mu + t_i + \varepsilon_{ij}$$

---

* Section 10.4 describes the procedures for testing the equality of variances assumption. See source [4] for testing for equal variances on three or more populations. (Bracketed numerals refer to items in "For Further Reading" at the end of the chapter.)

The model describes a response $Y_{ij}$ as the sum of three terms. The term $\mu$ denotes an overall mean that is unknown but is a constant. The term $t_i$ denotes an effect due to the treatment population $i$; $t_i$ is an unknown constant for each treatment group $i$. Finally, the term $\varepsilon_{ij}$ denotes a random error for the $j$th observation from treatment population $i$. Because the $\varepsilon_{ij}$ are normally and independently distributed with *mean zero*, $E(Y_{ij}) = \mu + t_i$. So the response values come from populations with means $(\mu + t_i)$. Because $t_i$ can be negative, zero, or positive, the mean for any population $i$ can be less than, equal to, or greater than the overall mean. This, then, establishes the null hypothesis as $H_0: \mu_1 = \mu_2 = \cdots = \mu_T$. Or if we use the model with

$$\sum_{i}^{T} t_i = 0*$$

this is equivalent to $H_0: t_1 = t_2 = t_3 = \cdots = t_T = 0$.

If $H_0$ is true, then the $(T)$ populations have equal means, and by assumption 3 the variances are equal. Then it follows that the treatment groups are random samples from one population, $N(\mu, \sigma^2)$.

The alternative hypothesis is $H_A$: At least two means are not equal. If the null hypothesis is rejected in favor of $H_A$, then the $t_i$ are not all zero and there are treatment differences, so the populations have different (unequal) means.

## Partitioning Total Variance for an F-Variance Ratio Statistic

We return to the data in Table 11.1. Which case (Figure 11.1(a) or 11.1(b)) exists here? That is, are the weights *affected by* the wood preserving process (case 1) or is plywood strength, measured by weight, *independent of* the wood-preserving process (case 2)?

To answer this question we need an understanding of estimation of the variance *among* treatment groups, symbolized by MST, and of estimation of the variance *within* groups, called mean squared error, or MSE. The ratio of these estimators defines an $F$-ratio statistic for testing equal treatment population means. Figure 11.2 and Table 11.2 help explain these concepts.

We begin with a condition that the treatment group responses have equal variance, $\sigma^2$, and come from normal distributions. Provided we can substantiate equal means, then the treatment groups are simply random samples from a single population, $N(\mu, \sigma^2)$.

To begin, Table 11.2 displays response values, $Y_{ij}$; treatment group totals, $Y_{i\cdot}$; and treatment group means $\bar{Y}_{i\cdot}.$** One estimator for the (assumed common) population variance compares each treatment mean value to $\bar{Y}_{\cdot\cdot}$, the overall mean. The differences are squared: $(\bar{Y}_{i\cdot} - \bar{Y}_{\cdot\cdot})^2$. Since a mean represents all the

---

* This is a condition of the *fixed-effects model*, which is also called model 1. This relates to $T$ populations represented by the sample data. An alternative, the *random effects model*, or model 2, considers the $T$ populations as a random sample of size $T$ from a bigger population of treatments. In the fixed effects model, inference is limited to those treatments that occur in the experiment. We consider only fixed-effects models here.

** A dot $(\cdot)$ in the subscript indicates a summation over that level. Then

$$\bar{Y}_{1\cdot} = \sum_{j=1}^{n_1} Y_{1j}/n_1$$

indicates a sum over the replicates for treatment variable 1.

**FIGURE 11.2** Display of the Data Structure for a One-Way Designed Experiment with $Y_{ij}$ = Weight of Plywood Sheets from Different Wood Preservative Processes

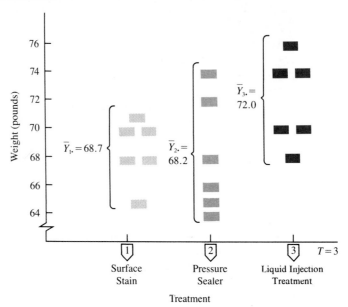

responses for a treatment, each squared difference is multiplied by the sample size for that group. We then sum the squared differences over all treatment groups to display the total for differences *among* the treatment groups. We secure the variance estimate by dividing the "among" total by the number of groups minus one. This estimated variance calculation is a comparison among (across) the groups. So it is called the *mean square among treatments*, or MST.

$$\text{MST} = \sum n_i (\bar{Y}_i. - \bar{Y}..)^2 / (T - 1) \tag{11.1}$$

**TABLE 11.2** Display of the Data for a One-Way Designed Experiment

| Treatment ($i =$) | Surface Stain ①  | Pressure Sealer ② | Liquid Injection Treatment ③ | Totals ($T = 3$) |
|---|---|---|---|---|
| $Y_{ij}$ = Weight of plywood sheets from different wood preservative processes | | | | |
| | $Y_{11} = 65$ | $Y_{21} = 64$ | $Y_{31} = 68$ | |
| | $Y_{12} = 68$ | $Y_{22} = 65$ | $Y_{32} = 70$ | |
| | $Y_{13} = 68$ | $Y_{23} = 66$ | $Y_{33} = 70$ | |
| | $Y_{14} = 70$ | $Y_{24} = 68$ | $Y_{34} = 74$ | |
| | $Y_{15} = 70$ | $Y_{25} = 72$ | $Y_{35} = 74$ | |
| | $Y_{16} = 71$ | $Y_{26} = 74$ | $Y_{36} = 76$ | |
| Total weight | $Y_1. = 412$ | $Y_2. = 409$ | $Y_3. = 432$ | $Y.. = 1253$ |
| Mean weight | $\bar{Y}_1. = 68.6667$ | $\bar{Y}_2. = 68.1667$ | $\bar{Y}_3. = 72.0000$ | $\bar{Y}.. = 69.6111$ |
| Number in sample | $n_1 = 6$ | $n_2 = 6$ | $n_3 = 6$ | $\sum_{i=1}^{3} n_i = 18$ |

By Table 11.2 for the plywood data,

$$\text{MST} = \frac{[(6) \cdot (68.6667 - 69.6111)^2 + \cdots + (6) \cdot (72.0000 - 69.6111)^2]}{3 - 1}$$

$$= 26.056$$

The name "mean square" indicates that the calculation uses sample data rather than total population values.

> The mean square among treatments, MST, is a sample estimator for $\sigma^2$ that compares the mean response for treatment groups to the overall sample mean. So the source of variation being measured is *among treatments*. It has $T - 1$ independent terms, where $T$ is the number of treatment groups.

A second, and independent, estimator for $\sigma^2$ uses differences for the response values within a treatment group from their treatment means; it is a *within*-treatments estimator for $\sigma^2$. The deviations for treatment 1 are for $Y_{11}$ from $\bar{Y}_1.$, $Y_{12}$ from $\bar{Y}_1.$, ..., and $Y_{1n}$ from $\bar{Y}_1..$ Computing a variance, $s_1^2$, within this treatment group is precisely what we did in Chapter 2. Similarly, $s_1^2, s_2^2, \ldots, s_T^2$ are all calculated. These sample variances are summed and then averaged over all $(T)$ treatment groups, giving a mean squared error, MSE.*

$$\text{MSE} = \frac{(n_1 - 1)s_1^2 + (n_2 - 1)s_2^2 + \cdots + (n_T - 1)s_T^2}{(n_1 - 1) + (n_2 - 1) + \cdots + (n_T - 1)}$$

$$= \frac{\sum_{j=1}^{n_1} (Y_{1j} - \bar{Y}_1.)^2 + \sum_{j=1}^{n_2} (Y_{2j} - \bar{Y}_2.)^2 + \cdots + \sum_{j=1}^{n_T} (Y_{Tj} - \bar{Y}_T.)^2}{(\sum n_i) - T}$$

$$\text{(11.2a)}$$

From Table 11.2,

$$\text{MSE} = \frac{\begin{bmatrix} (65 - 68.6667)^2 + \cdots + (71 - 68.6667)^2 \\ + (64 - 68.1667)^2 + \cdots + (74 - 68.1667)^2 \\ + (68 - 72.0000)^2 + \cdots + (76 - 72.0000)^2 \end{bmatrix}}{18 - 3} = 10.144$$

Since 1 df is used to estimate the population mean for each treatment group, each $s^2$ contains $n_i - 1$ independent values. So this within-groups estimator for $\sigma^2$ has $(\sum_i^T n_i) - T$ independent values, or degrees of freedom. The result is a within-treatments estimator for the population variance, called the *mean squared error within treatments*, or MSE.

> The mean squared error within treatments, MSE, is an estimator for $\sigma^2$ that compares the response for eu's with the same treatment against their treatment mean. Thus the source of the variations being measured is within treatments and is considered to be due to random error.

* For the case of equal sample sizes for all treatments, this expression is simplified to MSE = $(s_1^2 + s_2^2 + \cdots + s_T^2)/T$.

Because all the responses within a comparison have the same treatment, this is a measure of random error. (This assumes that the treatment variable is the single appreciable source for describing variations in the response.) These results are summarized as follows:

### AN F-RATIO STATISTIC

$$H_0: \mu_1 = \mu_2 = \mu_3 = \cdots = \mu_T$$

1. *Mean square among treatments*, MST (estimator 1 for $\sigma^2$):
   This estimator of $\sigma^2$ compares each treatment group mean ($\bar{Y}_i.$) to the overall mean ($\bar{Y}..$).

$$\text{MST} = \frac{\sum_i^T n_i(\bar{Y}_i. - \bar{Y}..)^2}{T-1} \tag{11.1}$$

2. *Mean squared error*, MSE (estimator 2 for $\sigma^2$):
   This estimator of $\sigma^2$ is a pooled variance of values ($Y_{ij}$) from their respective treatment group means ($\bar{Y}_i.$).

$$\text{MSE} = \frac{\sum^T (\sum^{n_i} (Y_{ij} - \bar{Y}_i.)^2)}{(\sum n_i) - T} \tag{11.2b}$$

or

$$\text{MSE} = \frac{(n_1 - 1)s_1^2 + (n_2 - 1)s_2^2 + \cdots + (n_T - 1)s_T^2}{\sum^T (n_i) - T}$$

where   $T$ = the number of treatment groups

$n_i$ = the number of eu's assigned to each group

$Y_{ij}$ = response for the individual experimental units

$\bar{Y}_i.$ = mean response for treatment $i$

$\bar{Y}..$ = the overall mean

Then, the *F*-statistic for a test of equal treatment means is

$$F = \frac{\text{MST}}{\text{MSE}} \tag{11.3}$$

for comparison against tabular $F_{\alpha, T-1, (\sum^T n_i) - T}$.

The MST and MSE calculations summarized above are for situations in which the treatment sample sizes are all equal. The *F*-ratio statistic is $F = \text{MST}/\text{MSE}$. For the plywood data, $F = 26.056/10.144 = 2.57$. Figure 11.2 shows the data for this case in which the question concerns whether the plywood data represents samples from different populations, or three samples from a single population. The question has essentially been answered. Example 11.1 summarizes our findings.

---

**EXAMPLE 11.1**

A building contractor wants to select the best possible plywood sheeting. Insofar as the size, type, and aging of the wood are uniformly the same, a key indicator of quality is the weight of the sheets after a wood-preserving treatment is applied. Since the plywood sheets are randomly selected for the assignment of treatments, we shall use a CRD model to test for equal treatment mean weights. The data from Figure 11.2, repeated in Table 11.3, are used to test whether the (mean) weights of sheets from wood-preserving processes 1, 2, and 3 are equal. Let $\alpha = 0.05$.

*Solution*  Here,

$$H_0: \mu_1 = \mu_2 = \mu_3 \qquad H_A: \text{At least two means are not equal.}$$

*(treatment groups −1)*

Test: Reject $H_0$ if calculated $F > 3.68$  $(F_{0.05,2,15})$.

Table 11.3 shows a summary analysis for a CRD experiment. By Equation (11.1),

$$\text{MST} = \frac{[(6) \cdot (68.6667 - 69.6111)^2] + \cdots + \overbrace{[(6) \cdot (72.0000 - 69.6111)^2]}^{A}}{3 - 1}$$

$$= 26.056$$

The calculation of MST is illustrated in the block A from Table 11.3. You can follow the calculation by tracing the items at A in Table 11.3 to the comparable elements in the MST calculation.

**TABLE 11.3**  A Summary Analysis for a Completely Randomized Experiment

| (i =) | Weight of Plywood Sheets from Different Wood-Preservative Processes | | | |
| | Surface Stain 1 | Pressure Sealer 2 | Liquid Injection Treatment 3 | Totals |
|---|---|---|---|---|
| | 65 | 64 | 68 | |
| | 68 | 65 | 70 | |
| | 68 | 66 | 70 | |
| | 70 | 68 | 74 | |
| | 70 | 72 | 74 | |
| | 71 | 74 | 76 | |
| Total weight, $Y_{i\cdot}$ | 412 | 409 | 432 | 1253 |
| Mean weight, $\bar{Y}_{i\cdot}$ | 68.6667 | 68.1667 | 72.0000 | 69.6111 |
| Number in sample, $n_i$ | 6 | 6 | 6 | 18 |

(To B)  (To A)

By Equation (11.2),

$$\text{MSE} = \frac{\begin{bmatrix} (65 - 68.6667)^2 + \cdots + (71 - 68.6667)^2 \\ + (64 - 68.1667)^2 + \cdots + (74 - 68.1667)^2 \\ + (68 - 72.0000)^2 + \cdots + (76 - 72.0000)^2 \end{bmatrix}}{3(6-1)} = 10.144$$

Calculation of MSE is illustrated with the block B from Table 11.3. Again, to see the calculation pattern, trace the items at B in Table 11.3 to the comparable elements in the MSE calculation. There, observe the math that is used to combine the data. The divisor in MSE (also in MST) indicates the count of usable data points (also called df).

$$\text{Calculated } F = \frac{\text{MST}}{\text{MSE}} = \frac{26.056}{10.144} = 2.57 \quad \text{(as before)}$$

Since the calculated $F$ (2.57) is less than $F_{0.05, 2, 15} = 3.68$, from Table B.4, the sample data show no significant difference in plywood weights for the three wood-preserving processes. This test does not provide a clear choice of a wood-preserving process. However, these data display somewhat higher values for Process 3—the liquid treatment—over the other treatment means. That is, 72.0000 appears higher (heavier) than 68.6667 and 68.1667 pounds. Because our intuition suggests a different conclusion, we will re-examine this test in a later section.

---

Equations (11.1) and (11.2) provided variance estimates in Example 11.1, followed by an $F$-ratio test using (11.3). This process is rather clumsy. In the next section we outline an ANOVA procedure to systematize the calculations for making the $F$-ratio test and simplify the calculation of MST and MSE.

**Problems 11.3**

8. An investment situation resulted in the following $F$-ratio statistic calculation:

$$F = \frac{\text{SS for treatment groups}/(T - 1)}{\text{Error SS}/((\sum n_i) - T)} = 5.83$$

For numerator df $= 2$ and denominator df $= 12$, indicate whether each of the following statements is true or false:
a. The null hypothesis was $\mu_1 = \mu_2 = \mu_3$.
b. There were 16 observations in all.
c. There were three treatment groups.
d. The null hypothesis can be rejected at the 0.01 level.
e. The null hypothesis can be rejected at the 0.05 level since $5.83 > 3.89$.

9. Assume that there were four treatment levels, with five repeats for each. (The experiment requires 20 eu's, with random assignment of 5 eu's to each treatment.) The $F$-test is on the hypothesis $H_0: \mu_1 = \mu_2 = \mu_3 = \mu_4$ with calculated $F = 12.04$. With what $F$-table value should 12.04 be compared if $\alpha = 0.01$ was specified? State the appropriate test conclusion.

10. An experiment on 5 groups, with 12 observations randomly assigned per group, showed $F = \text{MST}/\text{MSE} = 44.1/17$. What can you conclude for a test

of the equality of means at $\alpha = 0.05$? For $\alpha = 0.025$, what can you conclude? Why should the $\alpha$-risk level be set before any statistical calculations are made? Can you find the $p$-value for this test from the $F$-tables (Table 4 in Appendix B)? Explain.

11. An industrial engineer is experimenting with three methods for filling food packages on a food processing line. She selects 15 workers (eu's) and randomly assigns 5 to each method. Numbers in the table represent cartons of food processed per day by the workers under each method. Use the procedures of this section and $\alpha = 0.025$ to test whether the average production levels under the three methods are equal.

| Cartons of Food Packaged by the Workers | | | | | |
|---|---|---|---|---|---|
| Method 1 | 16 | 19 | 18 | 15 | 17 |
| Method 2 | 9 | 7 | 8 | 5 | 6 |
| Method 3 | 12 | 6 | 10 | 7 | 15 |

12. A marketing group is studying the effectiveness of a television commercial. Ten subjects are randomly assigned to a control group and ten others to an experimental group (see table). The control group is immediately asked to recall an advertisement. The members of the experimental group are asked to recall it several hours later. The response is a measure of recall time in seconds. Assume these samples are from normal distributions. (You are invited to test the assumption of equal variances.)

   a. Test $H_0: \mu_C = \mu_E$ using the $t$-test for independent samples with equal but unknown variances. Use $\alpha = 0.05$.
   b. Use the $F$-ratio test developed in this section to answer part (a). Compare your results. Does calculated $t^2$ = calculated $F$ for this experiment?

| Treatment | Control | Experimental |
|---|---|---|
| Mean | 13 | 15 |
| $s^2$ | 3.2 | 4.6 |

13. The table shows the production (in units) for each of five machinists using the exact same equipment. Calculate $F$ and test at $\alpha = 0.05$ whether there is a significant difference in the mean production rate among at least two of these machinists. Assume these are samples from normal distributions of individual production with equal variances.

| Art | Brenda | Carol | Dan | Edwin |
|---|---|---|---|---|
| 171 | 161 | 139 | 199 | 172 |
| 148 | 156 | 126 | 125 | 182 |
| 176 | 184 | 152 | 155 | 204 |

14. An auto parts distributor handles batteries from three suppliers. He questions whether to continue carrying all three brands. As a test, six batteries of each brand were connected to a load; each was tested every hour for a minimum power potential (enough power to start a car). By this test (see table), do the batteries sold by suppliers $A$, $B$, and $C$ have equal life? Use $\alpha = 0.05$. The overall mean is 7.111.

|        | Supplier | | |
|--------|:-----:|:-----:|:-----:|
|        | *A* | *B* | *C* |
|        | 5 | 5 | 5 |
|        | 7 | 5 | 6 |
|        | 8 | 6 | 7 |
|        | 8 | 7 | 7 |
|        | 10 | 7 | 8 |
|        | 10 | 8 | 9 |
| Totals | 48 | 38 | 42 |
| Means  | 8.000 | 6.333 | 7.000 |

15. This problem was presented in Chapter 10, where a *t*-test was requested; now solve the problem with an *F*-ratio test: The training director in a large hospital asserts that orderlies who take a public relations (PR) course perform better on the job than do those not receiving this training. Of 72 most recently hired orderlies, 36 were randomly selected to take the PR training. The others did not have this course. The productivity measures for the on-the-job evaluations six months after hiring were as shown in the following table.

|                    | Treatment Groups | |
|--------------------|:-----:|:-----:|
|                    | *PR Training* | *No PR Training* |
| Mean $\bar{Y}_i$   | 84.63 | 81.45 |
| Standard deviation | 3.50  | 4.00  |

    a. Test the director's claim using $\alpha = 0.05$. Does your conclusion agree with Problem 16 in Chapter 10?

    b. Check that the assumption of equal variances is valid, $\alpha = 0.05$.

16. Persons were randomly assigned to take a cold medicine, either SOS or another. The statistics (see table) summarize the relief, in hours, from cold symptoms for persons (the eu's) who took SOS, versus the relief for persons who took the *other* cold medicine.

| | Treatment | |
|---|---|---|
| | *SOS* | *Other* |
| Mean | 8.6 hours | 7.2 hours |
| Variance | 9 hours squared | 16 hours squared |
| $n$ | 36 | 36 |

a. Test $H_0: \mu_{SOS} = \mu_{other}$ against $H_A: \mu_{SOS} \neq \mu_{other}$ by an $F$-ratio test. Use $\alpha = 0.05$.

★b. Compare to the conclusions for Problem 14, Chapter 10.

★17. Equation (11.2) for equal sample sizes, $n_i = n$, gives us the following expressions:

$$(1) \; MSE = \frac{s_1^2 + s_2^2 + \cdots + s_T^2}{T} \quad \text{and}$$

$$(2) \; MSE = \frac{\sum\limits^{T} \left( \sum\limits^{n_i} (Y_{ij} - \bar{Y}_{i.})^2 \right)}{T(n-1)}$$

Let $T = 3$ and $n = 4$. Using

$$s_1^2 = \frac{\sum\limits^{4} (Y_{1j} - \bar{Y}_{1.})^2}{4-1}, \ldots, s_3^2 = \frac{\sum\limits^{4} (Y_{3j} - \bar{Y}_{3.})^2}{4-1}$$

show that (1) can be changed into (2).

18. The data in Example 11.1 provide the numbers to illustrate the equality of expressions (1) and (2) in Problem 17. Display the calculations from these data to show that (1) = (2).

## 11.4 THE ANALYSIS OF VARIANCE PROCEDURE

The *analysis of variance* (ANOVA) is an algebraic procedure for structuring the calculation of variance estimates.

> The analysis of variance (ANOVA) is an algebraic procedure for partitioning the total variation into treatment (explained) and error (unexplained) parts, and it sets a structure for reporting the results in terms of an $F$-ratio test.

The ANOVA process apportions the total response sum of squared deviations, TSS, into treatment (among groups), SST, and error (within groups), SSE, components. The total SS is calculated as

$$TSS = \sum^{T} \sum^{n_i} (Y_{ij} - \bar{Y}_{..})^2$$

★ Star indicates supplementary material for enhanced (optional) course work.

Then the treatments SS is

$$\text{SST} = \sum^{T} n_i (\bar{Y}_{i.} - \bar{Y}_{..})^2$$

Error SS is obtained by subtraction:

$$\text{SSE} = \text{TSS} - \text{SST}$$

Similarly, degrees of freedom are split:

Total df = Treatment df + Error df

$$\left(\sum^{T} n_i\right) - 1 = (T - 1) + \left[\left(\sum^{T} n_i\right) - T\right]$$

A mean square is the appropriate sum of squares divided by the corresponding degrees of freedom. Treatments mean square is $\text{MST} = \text{SST}/(T - 1)$, and error mean square is $\text{MSE} = \text{SSE}/[(\sum^{T} n_i) - T]$. *The ANOVA simply provides a procedure for the systematic calculation of the mean squares.*

Table 11.4 gives a summary of the ANOVA calculations and an associated $F$-test on the equality of treatment means. The MST, MSE, and $F$-statistic ($= \text{MST}/\text{MSE}$) use Equations (11.1) through (11.3).

Table 11.4 defines the same procedure we used in Example 11.1. But the ANOVA format structures the calculations into a tabular display. Although we

**TABLE 11.4** ANOVA Procedure for a One-Way Analysis of Variance with $F$-Test for Equal Treatment Means

| Source of Variation | df | Sum of Squares | Mean Squares | |
|---|---|---|---|---|
| Among treatments | $T - 1$ | SST | $\text{MST} = \dfrac{\text{SST}}{(T - 1)}$ | |
| Error (within treatments) | $\left(\sum^{T} n_i\right) - T$ | SSE | $\text{MSE} = \dfrac{\text{SSE}}{\left(\sum^{T} n_i\right) - T}$ | (11.4) |
| Total | $\left(\sum^{T} n_i\right) - 1$ | TSS | | |

with $\text{TSS} = \sum_i^{T} \sum_j^{n_i} (Y_{ij} - \bar{Y}_{..})^2 = \sum_i^{T} \sum_j^{n_i} Y_{ij}^2 - \text{CT}$

$\text{SST} = \sum_i^{T} n_i (\bar{Y}_{i.} - \bar{Y}_{..})^2 = \sum_i^{T} \left(\dfrac{Y_{i.}^2}{n_i}\right) - \text{CT}$  $\text{CT} = \left(\dfrac{\sum\sum Y_{ij}}{\sum n_i}\right)^2$

*what is CT?*

$\text{SSE} = \text{TSS} - \text{SST}$

**$F$-Ratio Test for Equal Treatment Means**

Hypotheses: $H_0: \mu_1 = \cdots = \mu_T (= \mu)$ (all treatment means are equal)
$H_A$: Some treatment means are not equal.

Test statistic: Calculated $F = \text{MST}/\text{MSE}$  **(11.3) repeated**

Test procedure: Reject $H_0$ if calculated $F > F_{\alpha,\, T-1,\, (\Sigma^T n_i) - T}$.
Otherwise, accept $H_0$.

have not emphasized the condition of constant variance, the test for equal variances precedes any test for equal means.

Example 11.2 uses the ANOVA procedures as outlined in Table 11.4, and displays them in a computer listing format.

---

**EXAMPLE 11.2**

Here we illustrate the ANOVA and test for equality of means as in Example 11.1. The test compares weights for plywood sheets produced by three preservative processes. The variance calculations for (among) treatment mean square and error mean square use the ANOVA process described in Table 11.4. The display, giving only the most pertinent outputs, is from an SPSS computer printout.

*Computer Solution*    Table 11.5, which is a computer listing, shows descriptive statistics on the response for three processes. The means—68.6667, 68.1667, and 72.0000—are the same as presented earlier in Table 11.1. The ANOVA and $F$-statistic calculation are discussed next.

The ANOVA calculations are defined in Table 11.4, with $F = \text{MST}/\text{MSE}$ defined by Equations (11.1) through (11.3). The calculations are done by computer using the data in Table 11.3, with $\bar{Y}.. = 69.6111$.

$$\text{TSS} = [(65 - 69.611)^2 + (68 - 69.6111)^2 + \cdots + (76 - 69.6111)^2]$$

$$= 204.278$$

$$\text{SST} = \text{value from Example 11.1} = 52.112$$

$$\text{SSE} = \text{TSS} - \text{SST} = 204.278 - 52.112 = 152.166$$

The degrees of freedom are

$$\text{Treatments df} = 3 - 1 = 2$$

$$\text{Error df} = 18 - 3 = 15$$

$$\text{Total df} = (3 \cdot 6) - 1 = 17$$

**TABLE 11.5**  Summary Statistics and ANOVA for the Plywood Data (SPSS output)

Description of Subpopulations
Criterion (response) Variable Plywood Sheet Weights
Broken Down by (treatment variable) Preservative Process

| Variable | Code | Value Label | Sum | Mean | Std Dev | Variance | N |
|----------|------|-------------|-----|------|---------|----------|---|
| Process | 1 | surface stain | 412.0000 | 68.6667 | 2.1602 | 4.6667 | (6) |
| Process | 2 | pressure sealer | 409.0000 | 68.1667 | 4.0208 | 16.1667 | (6) |
| Process | 3 | liquid injection | 432.0000 | 72.0000 | 3.0984 | 9.6000 | (6) |
| Total cases=18 | | | | | | | |

Analysis of Variance and F-Statistic Calculation

| Source | DF | Sums of Squares | Mean Square | F | Significance |
|--------|----|-----------------|-------------|---|--------------|
| Treatments | 2 | 52.112 | 26.056 | 2.568 | 0.1098 |
| Error | 15 | 152.166 | 10.144 | | |
| Total | 17 | 204.278 | | | |

Then

$$MST = (SST/\text{Treatments df}) = 26.056$$

$$MSE = (SSE/\text{Error df}) = 10.144$$

so that calculated $F = 26.056/10.144 = 2.568$. These results are summarized with an analysis of variance and $F$-statistic calculation in Table 11.5.

The interpretation uses the level of significance, or $p$-value, approach. Our test was to reject $H_0$ if the $p$-value were less than 0.05. But the $p$-value is 0.1098, so the difference in mean weights is *not* significant.

The random assignment of treatments to experimental units, such as the random selection of people, machines, or plywood sheets (the experimental units) for treatments groups, is essential to this design.

Because the examples in subsequent sections use computer outputs, you need to understand the following four procedures for testing hypotheses in designed experiments.

### STEPS FOR A SYSTEMATIC SOLUTION TO HYPOTHESIS TESTS IN DESIGNED EXPERIMENTS

1. The layout of the ANOVA table that displays degrees of freedom and sums of squares for recognized sources of variation and for error (unexplained sources)
2. The calculations to determine df and SS entries
3. The development of mean squares to estimate the true variance explained by each source of variation
4. Calculation of the $F$-ratio statistic and determination of an appropriate $F$-test value from Table B.4 for each explained source of variation.

The conclusion for a test of hypothesis is either to reject or to not reject equal means. In the former case, more testing (Section 11.7) can identify which means are statistically different. In the latter case, the testing is complete; the treatment groups have means that are not statistically different. Example 11.3 displays the steps with calculations for an experiment with unequal numbers of replicates for the treatment groups.

**EXAMPLE 11.3**

A national retail chain department store has arranged to give merchandise-pricing training to its employees. Three groups of sales employees are selected at random to experience the training. The three types of training are (1) hands-on training under the supervision of an experienced pricing manager, (2) video instruction in merchandise pricing, and (3) training by reading a pricing manual and guide. Upon completion of their instruction, the individuals are given a practical test in pricing. The highest possible score is 400 points. Due to illnesses, vacations, and so on, the test produced unequal test groups of 7, 5, and 6 persons (see table). Is there a statistical difference in the scores under the three methods? Use $\alpha = 0.01$. This is an example with unequal numbers of replicates for the treatment groups. Although the sample sizes are smaller than would normally be encountered in practice, they enable us to illustrate the procedure more clearly.

| | Group (Treatment) | | | |
|---|---|---|---|---|
| | *1* (*Hands-On*) | *2* (*Video Instruction*) | *3* (*Pricing Manual*) | Totals |
| | 396 | 383 | 368 | |
| | 397 | 380 | 369 | |
| | 398 | 387 | 366 | |
| | 400 | 385 | 366 | |
| | 398 | 378 | 364 | |
| | 395 | | 368 | |
| | 383 | | | |
| Totals ($Y_i.$) | 2767 | 1913 | 2201 | 6881 |
| Means ($\bar{Y}_i.$) | 395.286 | 382.600 | 366.833 ($\bar{Y}.. =$) | 382.278 |
| $n_i$ | 7 | 5 | 6 | 18 |

**Solution**   This situation requests a test for equal means for individuals who were trained under different methods to perform pricing activities. The test is

$$H_0: \mu_1 = \mu_2 = \mu_3 \qquad H_A: \text{At least one } \mu_i \neq \text{others.}$$

The model is a CRD because (1) constant variance and (2) independent normal distributions for the three populations are reasonable assumptions here and because the assignment of treatments to experimental units is random.

**Step 1**   The ANOVA includes sources of variation for (1) among (treatment) groups, (2) error (within groups), and (3) total. The ANOVA display, including degrees of freedom, sums of squares, and mean squares for each *source* of variation, appears in Table 11.6.

The total sum of squares, TSS, is the sum of the squared deviations of each individual value, $Y_{ij}$, from the grand mean, $\bar{Y}...$ The treatment sum of squares, SST, is the sum of squared differences for each group mean from the overall mean times the number of observations in the group. Then the error sum of squares is obtained by subtraction: SSE = TSS − SST.

**Step 2**   The calculation of sums of squares and degrees of freedom uses procedures already discussed.

$$\text{TSS} = \sum\sum (Y_{ij} - \bar{Y}..)^2 = [(396 - 382.28)^2 + (397 - 382.28)^2 + \cdots + (368 - 382.28)^2] = 2877.611$$

or

$$\text{TSS} = \sum\sum Y_{ij}^2 - \text{CT} = (396^2 + 397^2 + \cdots + 368^2) - \frac{(396 + 397 + \cdots + 368)^2}{18} = 2877.611$$

$$SST = \sum n_i(\bar{Y}_{i.} - \bar{Y}_{..})^2$$
$$= [7(395.29 - 382.28)^2 + 5(382.60 - 382.28)^2 + 6(366.83 - 382.28)^2]$$
$$= 2616.149$$

or

$$SST = \frac{\sum Y_{i.}^2}{n_i} - CT = \frac{2767^2}{7} + \frac{1913^2}{5} + \frac{2201^2}{6} - \frac{(6881)^2}{18} = 2616.149$$

$$SSE = TSS - SST = 2877.611 - 2616.149 = 261.462$$

Total df = Treatments df + Error df:

$$18 - 1 = (3 - 1) + (7 + 5 + 6 - 3) = 2 + 15$$

**Step 3**    Mean squares are, by definition, a sum of squares divided by the associated df. The results appear in Table 11.6. MST = 1308.075 and MSE = 17.431.

**Step 4**    The calculated $F$-value is, by Equation (11.3), $F = MST/MSE = 1308.075/17.431 = 75.04$.

A summary for the four steps is the ANOVA and $F$-test in Table 11.6 by a popular statistics program for the Macintosh. In this output, SST corresponds to the among-groups sum of squares and SSE to the within-groups sum of squares. The calculated $F$ of 75.04 is substantially larger than the $F_{0.01,2.15} = 6.36$. So it is significant at $p < 0.0001$.

We conclude that the pricing test scores are statistically different for these three groups. The within-group scores are quite uniform. This has induced a low error mean square, which has resulted in a large $F$-ratio statistic (75.04), with a $p$-value less than 0.0001. So the conclusion is very reasonable—there is much more variance across the groups (explained) than there is inherently (unexplained) within the groups. The means

**TABLE 11.6**    Microcomputer Summary of an ANOVA and $F$-Test

| Group | Frequency | Mean |
|---|---|---|
| Hands-on | 7 | 395.286 |
| Video instruction | 5 | 382.600 |
| Pricing manual | 6 | 366.833 |

One-Way ANOVA 3 Groups ANOVA TABLE

| Source | DF | Sums of Squares | Mean Square | F-test |
|---|---|---|---|---|
| Among groups | 2 | SST = 2616.149 | 1308.075 | 75.044 |
| Error (within groups) | 15 | SSE = 261.462 | 17.431 | $p < 0.0001$ |
| Total | 17 | 2877.611 | | |

indicate that performance scores were ranked high to low in the order of hands-on experience, video instruction, and pricing manual. Which of these are statistically different means is a topic for a later section.

---

The designs discussed up to this point have one-way or single-treatment sources for explaining the response variations. Generally speaking, we measure the *efficiency of a design* by the MSE. We seek a small SSE with an appropriate error df. In this sense the CRD is straightforward, but relative to other designs it can be inefficient. That is, it may be possible to identify more treatment sources that would explain substantial response variations, so would greatly reduce the error sum of squares, but only slightly reduce the error df. The MSE that results can reverse a test conclusion! For example, the conclusion in Example 11.1 (of no significant difference in treatment mean weights) may actually be a consequence of an inefficient statistical design.

**Problems 11.4**

19. The plant manager at Lakeside Industries wants to determine if work absences relate to the shift an employee works. Workers are randomly assigned to shifts—day, swing, or graveyard—at the beginning of a month. A random selection of 24 workers from each shift gave the following number of days absent over a six-month period. We assume absences in each shift approximate a normal distribution, and the assumption of equal variances is reasonable. Moreover, others have found that total SS = 1104.90. The test is for equal numbers of absences in the three shifts at $\alpha = 0.05$. Two summaries appear in the following table. One is the response totals and means for the (3) treatment groups. The second is an ANOVA table. The first summary is complete; the second is not.
    a. Complete the following ANOVA table.
    b. State the null and alternative hypotheses for an $F$-ratio test of means.
    c. Determine the critical $F$-value for $\alpha = 0.05$, df = 2 and 69.
    d. Calculate the $F$-ratio statistic, test $H_0$, and state your conclusion.

**Summary 1**   Absences over 6 months for 24 workers from each of 3 shifts

| | Shift (Treatment) | | | Overall (3) Shifts |
|---|---|---|---|---|
| | Day | Swing | Graveyard | |
| Total | 82 | 112 | 43 | 237 |
| Mean | 3.42 | 4.67 | 1.49 | 3.29 |

**Summary 2**   An ANOVA partition of total SS and df (incomplete)

| Source of Variation | df | Sum of Squares | Mean Squares |
|---|---|---|---|
| Treatments | $3 - 1 = 2$ | | |
| Error | | 1004.79 | |
| Total | 71 | 1104.90 | |

20. The procedure for ANOVA is used in conjunction with hypothesis tests of which of the following four procedures?
    a. $H_0: \mu_1 = \mu_2$, using $Z$
    b. $H_0: \mu_1 = \mu_2 = \mu_3 = \cdots = \mu_T$, using $F$
    c. $H_0: \pi_1 = \pi_2 = \pi_3$, using $t$
    d. $H_0: \mu_1 - \mu_2 = 0$, using $t$

21. Using $\alpha = 0.01$ and the information in the table below, complete a test of $H_0: \mu_1 = \mu_2 = \mu_3 = \mu_4$ versus $H_A$: At least one of the $\mu_i \neq$ the others. Assume a normal distribution for treatment groups. State your conclusions.

| | Treatment Group | | |
|---|---|---|---|
| *1* | *2* | *3* | *4* |
| 13 | 7 | 15 | 11 |
| 15 | 10 | 18 | 10 |
| 20 | 13 | 21 | 15 |

| Source of Variation | df | Sum of Squares | Mean Squares | F |
|---|---|---|---|---|
| Among treatments | | | | |
| Error | | 76 | | |
| Total | | | | |

22. The data in the table represent the mileage (in thousands of miles) obtained from five randomly selected tires from each of three different brands. Assuming a normal distribution of tire life, complete the test for equality of means using $\alpha = 0.05$ and state your summary in the form of an ANOVA table.

| | Brand | | |
|---|---|---|---|
| | *A* | *B* | *C* |
| | 30 | 35 | 28 |
| | 25 | 42 | 28 |
| | 31 | 38 | 30 |
| | 38 | 39 | 24 |
| | 36 | 31 | 25 |
| Means | 32 | 37 | 27 |

| Source | df | Sum of Squares |
|---|---|---|
| Treatment | | |
| Error | 12 | |
| Total | | 450 |

23. Complete the analysis of variance table; then test $H_0: \mu_1 = \mu_2 = \mu_3$ at $\alpha = 0.05$. State your conclusions.

| Treatment Group | $n_i$ | Mean |
|---|---|---|
| 1 | 6 | 104.4 |
| 2 | 5 | 146.8 |
| 3 | 7 | 155.4 |

| Source | df | SS | Mean Square | F |
|---|---|---|---|---|
| Treatment Groups | | | | |
| Error | | | | |
| Total | | 20,642.61 | | |

24. A sales manager feels that there is a substantial difference in total business obtained depending on the sales region. This is a real concern because salespeople get a fixed-percentage commission of sales. Given the random selection of a monthly intake for three regions (see table), what can you conclude? Use $\alpha = 0.05$ and TSS = 136.50. Assume normal distributions with equal variances

| Sales by Region ($0,000) | | |
|---|---|---|
| *1* | *2* | *3* |
| $25.95 | $24.65 | $20.14 |
| 20.93 | 28.37 | 30.61 |
| 21.80 | 28.27 | 21.17 |
| 25.66 | 27.25 | 21.46 |
| Totals $94.34 | $108.54 | $93.38 |

25. Explain the differences between a one-way, completely randomized design with *equal* numbers versus a CRD with *unequal* numbers for each of the following.
    a. The model
    b. Calculation of degrees of freedom
    c. Sum of squares for treatments
    d. Calculation of error sum of squares

26. The data in the table were recorded in a test for equal means that compared four treatments and had TSS = 265. Assume that the necessary test assumptions are met and, using $\alpha = 0.01$, test for significant differences between treatment group means. Observe that this experiment contains unequal sample sizes.

| Treatment Group | Sample Size | Mean |
|---|---|---|
| A | 4 | 50.0 |
| B | 4 | 45.3 |
| C | 3 | 41.7 |
| D | 4 | 47.3 |

The next two problems were presented in Chapter 10 for a test of the equality of two treatment group means. Repeat the test, now using an $F$-statistic. Compare your conclusions with Problems 13 and 24 in Chapter 10.

27. The statistics in the table summarize performance levels for workers in two divisions of the multinational International Chemicals Company (ICC). The times to complete a standardized task were measured for a random sample of 18 workers in Mexico (1) and for another random sample of 12 workers in Brazil (2). The response is the time required to complete the task.

|  | Location | |
|---|---|---|
|  | Mexico | Brazil |
| Mean | 76.2 min | 83.1 min |
| Variance | 18.6 min$^2$ | 38.2 min$^2$ |
| $n_i$ | 18 | 12 |

a. Test for equal variances, $\alpha = 0.05$.
b. Test for a significant difference in the mean response in performance for the ICC workers in Mexico and Brazil, $\alpha = 0.05$.

To accommodate the unequal numbers, use

$$\text{MSE} = \frac{(n_1 - 1)s_1^2 + (n_2 - 1)s_2^2}{n_1 + n_2 - 2}$$

28. The data in the table relate the performance for lawn-care service workers who are hired based on either (1) hit-or-miss quick interviews or (2) a job-skills test. The response measure is productivity: the number of lawns serviced per day from three months of the firm's work-service record. Assume that a CRD is appropriate. Use the data and an $F$-ratio statistic to test $H_0: \mu_1 = \mu_2$ with $\alpha = 0.025$. Does this procedure and test allow you to conclude that $\mu_1 < \mu_2$? Explain. Use MSE $= 2.25$, $\bar{Y}_{..} = 6.00$.

| Group 1 | 4 | 5 | 5 | 8 | 7 | 4 | 3 | 5 | 6 |
|---|---|---|---|---|---|---|---|---|---|
| Group 2 | 6 | 8 | 8 | 6 | 5 | 9 | 7 | | |

29. The test is $H_0: \mu_1 = \mu_2$ against $H_A: \mu_1 \neq \mu_2$. The samples in the table are from normally distributed treatment groups with

$$\bar{Y}_1. = 16 \quad s_1^2 = 28 \quad n_1 = 13 \quad \bar{Y}.. = 14.70$$

$$\bar{Y}_2. = 13 \quad s_2^2 = 7 \quad n_2 = 10 \quad MSE = \frac{12 \cdot 28 + 9 \cdot 7}{21} = 19.0$$

| Source | df | Sum of Squares | Mean Square | F |
|--------|-----|----------------|-------------|------|
| Treatment | 1 | 50.87 | 50.87 | 2.67 |
| Error | 21 | | 19.00 | |
| Total | 22 | | Tabular $F_{0.05,1,21} = 4.32$ | |

Conclusion: The treatment means are statistically equal for $\alpha = 0.05$. Do you agree with this conclusion? Explain.

30. A consumer-research course assignment is to develop a comparative test for the prices of groceries at three supermarkets—Safeway, A&P, IGA—in the Chicago market. Treat this as a CRD, with response being the price of a fixed basket of groceries. The treatments are three chain supermarkets; the eu's are the weekly baskets of items specified for price checks. Explain how you would do the following tasks to assure a valid test.
    a. Specify the items for price check
    b. Control potential outside variables that could bias the results
    c. Perform all essential statistical tests (give alternate action for the possible alternate test conclusions)
    d. Use more store sites (replicates) to obtain better test precision

31. A CRD with unequal numbers is established for the power usage of five brands of microwave ovens. The random samples include 12, 8, 7, 11, and 10 units for the five brands (see table). Something is wrong with the ANOVA and $F$-calculations. Explain.

| Source | df | Sum of Squares | Mean Square | F |
|--------|-----|----------------|-------------|------|
| Treatments | 4 | 15,638 | 3909.50 | |
| Error | 43 | 32,971 | 766.77 | Calculated $F = 5.10$ |
| Total | 47 | 47,069 | | $F_{0.05,4,47} = 2.61$ |

## 11.5 THE RANDOMIZED COMPLETE BLOCK DESIGN

Thus far we have analyzed designs that have only one variable of classification. For example, we have asked, Does plywood from different wood preservative treatment processes have the same average weight? In another experiment we asked, Are pricing test scores for sales trainees for a national retail chain department store equal for three groups—(1) hands-on training, (2) video instruction, and (3) pricing manual? When the experimental units are randomly

**TABLE 11.7** Weight (in pounds) of Plywood Sheets Classified Two Ways—According to (1) Preservative Process and (2) Grade

| Grade | Wood Preservative Process | | |
| --- | --- | --- | --- |
| | Surface Stain (1) | Pressure Sealer (2) | Liquid Injection Treatment (3) |
| Economy | 65 | 64 | 70 |
| Standard | 68 | 68 | 74 |
| Best | 71 | 72 | 76 |

selected from their respective treatment-group populations, then we can perform statistical tests for the equality of the population means.

It can be advantageous to classify a response by two or more sources for explaining response variations. For example, Table 11.1 classified the weights of plywood sheets in one way. In Table 11.7, the plywood sheets are classed by grade in addition to being classified by the preservative treatment.

The grade of plywood is (1) economy, (2) standard, or (3) best. This *two-way classification* lends itself to testing several hypotheses, including both (1) equality of weights among preservative processes and (2) equality of weights among grades. This suggests two separate tests:

1. $H_0: \mu_1 = \mu_2 = \mu_3$   (no difference of weights among preservative treatment processes)
2. $H_0: \mu_E = \mu_S = \mu_B$   (no difference of weights among grades)

If the grade classification were ignored, then SS for grades is a part of the error SS and the first hypothesis is just what we have already tested. Test 2 could be viewed as a one-way test for grades (quality) of plywood if the preservative source were deleted. However, a two-treatments design yields more information than is contained in two separate one-way tests. This is because the cross-classification provides a better isolation of the error, or unexplained variations. This more accurate measure of random error can increase the test precision through a reduced error mean square.

**Definition of The Randomized Complete Block Design**

We can increase the efficiency of a design by adding variables that explain previously unexplained response variations. One situation involves a randomized block design. This two-way experiment uses (1) a treatment variable and (2) a *blocking variable*.

> A blocking variable is a variable that is an inherent characteristic of the experimental units.

Examples of blocking variables for humans (as eu's) are age, sex, IQ, and height. In this type of design the experimental units are assigned purposely, not at random, to blocks; within each block there is a random assignment of treatments to eu's. The simplest case is the paired-differences model discussed in Chapter 10. This model can be generalized.

A randomized complete block design (RBD) for comparing $T$ treatments in $B$ blocks has treatments randomly assigned to the experimental units within blocks. Each treatment appears once in each block, so an RBD experiment requires $B \times T$ experimental units.

The primary objective of blocking is to remove from the unexplained source those response variations explained by the block variable. A second objective of blocking is to make the eu's within blocks as homogeneous as possible. For example, in an experiment in which the response is dollars of income, block 1 may contain experimental units that are a young age; block 2, all of a mid-age, and block 3, an older age. This type of grouping can help improve the test precision.

The randomized complete block design model extends the CRD model by adding a term for "blocks" that is another source for explaining response variations.

### MODEL FOR A (TWO-WAY) RANDOMIZED COMPLETE BLOCK DESIGN, RBD

$$Y_{ij} = \mu + t_i + b_j + \varepsilon_{ij} \qquad i = 1, 2, \ldots, T$$
$$j = 1, 2, \ldots, B$$

where   $i$ = treatment variable levels

$j$ = block variable levels

$\varepsilon_{ij}$ = unexplained error that represents all sources of response variation except treatments and blocks

*Assumptions:*
1. The responses are independent and normally distributed within each block.
2. The treatment and block effects are additive (so the model shows additive terms), with distribution means $\mu + t_i + b_j$.
3. The variance is constant for all treatment by block combinations.

This model and its assumptions are an extension of the completely randomized design. Two added conditions apply to randomized block designs. First, we know before any testing that the variable used for blocking is a known (inherent) source of response variation. So, in the plywood illustration, we accept that various grades of plywood have different weights. This is a given; otherwise we would use another design. Second, we do not make distribution assumptions on the block variable; it is a known source of substantial response variation.

Treatments are randomly assigned to those experimental units within block 1, again (separately) to those within block 2, and so on. The assignment requires $B \times T$ experimental units. So, for example, an experiment with 5 treatments and 8 blocks requires $8 \cdot 5 = 40$ experimental units. Though more eu's are desirable as a means for reducing the MSE and improving test precision, the high number of observations can be an expensive requirement for a RCB design.

The $F$-ratio test for treatment differences takes the same form as before: MST/MSE. However, relative to a CRD, blocking can provide a more precise

**FIGURE 11.3** Comparison of the Partition of (Total) Response Variation by the Completely Randomized Design (CRD) and by the Randomized Complete Block Design (RBD)

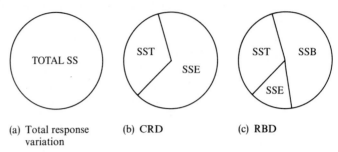

(a) Total response variation  (b) CRD  (c) RBD

test for real treatment differences by a substantial reduction in SSE coupled with a small reduction in error df. See Figure 11.3.

As in most forms of statistical analysis, we gain precision by explaining away the statistical error. As Figure 11.3 suggests, *any improvement from adding block variables is simply the result of removing a substantial portion of the previously unexplained SSE, at the expense of a small reduction in error degrees of freedom.* This reduces the MSE from that for the completely randomized design. So blocking can increase test precision and improve the estimate of true error variance, $\sigma^2$. Even more complex models with more treatments can sometimes do a better job of isolating the explained variations.

We are now equipped to make an $F$-test for equal treatment means for the randomized complete block model. An ANOVA structures the mean square calculations.

Calculations for SS and df are an extension of the CRD. By the design, one eu occurs in every treatment by block combination. Then SST requires calculation on $T$ data points, SSB on $B$ (a number) data points, and SSE uses $B \cdot T$ numbers. Table 11.8 on page 465 gives the details for SS and df calculations as Equation (11.5).

We use the plywood data from Table 11.7 to illustrate a randomized complete block design test.

## EXAMPLE 11.4

Previous analysis in Examples 11.1 and 11.2 used a CRD design to compare the weights for samples of plywood sheets produced by three preservative treatments. Those results indicated "no significant difference" in mean weights for $\alpha = 0.05$. Do the sample data in Table 11.7 with a randomized complete book design support this earlier conclusion? The data are repeated here from Table 11.7:

| | Wood Preservative Process | | |
|---|---|---|---|
| Grade | *(1)* | *(2)* | *(3)* |
| Economy | 65 | 64 | 70 |
| Standard | 68 | 68 | 74 |
| Best | 71 | 72 | 76 |

**TABLE 11.8**   The Two-Way ANOVA and $F$-Test for a Randomized Block Design

| Source | df | SS | MS | |
|---|---|---|---|---|
| Treatments | $T - 1$ | SST | MST $(= \text{SST}/(T - 1))$ | |
| Blocks | $B - 1$ | SSB | MSB | |
| Error | $(B - 1)(T - 1)$ | SSE | MSE | **(11.5)** |
| Total (corrected) | $BT - 1$ | TSS | | |

$$n = BT$$

$$\text{CT} = \frac{(\text{Total of all observations})^2}{n} = \frac{\left(\sum_j \sum_i Y_{ij}\right)^2}{BT}$$

$$i = 1, 2, \ldots, T$$

$$j = 1, 2, \ldots, B$$

$$\text{TSS} = \text{Total sum of squares} = \sum_i \sum_j (Y_{ij} - \bar{Y}_{..})^2$$

$$= \text{SS of all observations} - \text{CT}$$

$$= \text{SS}(Y) = \sum_i \sum_j Y_{ij}^2 - \text{CT}$$

$$\text{SST} = \text{SS for treatments} = \sum_i^T B(\bar{Y}_{i.} - \bar{Y}_{..})^2$$

$$= (\text{SS for treatments, each summed across all block levels}) - \text{CT}$$

$$= \frac{\sum_i^T Y_{i.}^2}{B} - \text{CT} \quad \checkmark$$

$$\text{SSB} = \text{SS for blocks} = \sum_j^B T(\bar{Y}_{.j} - \bar{Y}_{..})^2$$

$$= (\text{SS for blocks, each summed across all treatment levels}) - \text{CT}$$

$$= \frac{\sum_j^B Y_{.j}^2}{T} - \text{CT}$$

$$\text{SSE} = \text{TSS} - \text{SST} - \text{SSB} = \text{Unexplained variations}$$

$$F = \frac{\text{MST}}{\text{MSE}} = \frac{\text{SST}/(T - 1)}{\text{SSE}/(BT - B - T + 1)} \qquad \textbf{(11.6)}$$

*Solution*   The calculations are displayed to test $H_0: \mu_1 = \mu_2 = \mu_3$ against the alternative that at least two treatment means are different. The essential calculations for the two-way ANOVA are as follows:

**1.** Data analysis:

$$\sum \sum Y_{ij}^2 = 65^2 + 68^2 + \cdots + 76^2 = 43{,}946$$

$$\text{CT} = \frac{(65 + 68 + 71 + 64 + 68 + 72 + 70 + 74 + 76)^2}{3 \cdot 3}$$

$$= 43{,}820.444$$

$$\text{TSS} = 43{,}946 - 43{,}820.444 = 125.556$$

$$43877.33 \qquad 131,632$$

$$SST = \frac{(204)^2 + (204)^2 + (220)^2}{3} - CT$$

$$= 56.889 \quad \text{(preservative process)}$$

$$SSB = \frac{(199)^2 + (210)^2 + (219)^2}{3} - CT = 66.889 \quad \text{(grade)}$$

2. The ANOVA and $F$-ratio statistic test: The analysis of variance calculations described in the table use Equation (11.5), and the $F$-ratio test is by Equation (11.6).

| Source | df | SS | MS |
|---|---|---|---|
| Preservative process (treatments) | $3 - 1 = 2$ | 56.889 | 28.444 |
| Grade (blocks) | $3 - 1 = 2$ | 66.889 | 33.444 |
| Error | $9 - 3 - 3 + 1 = 4$ | 1.778 | 0.444 |
| Total | $9 - 1 = 8$ | 125.556 | |

For $H_0: \mu_1 = \mu_2 = \mu_3$ versus $H_A$: At least one $\mu_i \neq$ others:

$$\text{Calculated } F = \frac{28.444}{0.444} = 64.0 \quad (> F_{0.05,2,4} = 9.94)$$

3. Conclusions: This test shows a highly significant difference in weights for some wood preservative processes. By observation of the means,

| Process (treatment) | 1 | 2 | 3 |
|---|---|---|---|
| Mean Weight | 68.0 lb | 68.0 lb | 73.3 lb |

process 3 is providing heavier-weight plywood sheets. These data and the RCB design results do not support the conclusions of Examples 11.1 and 11.2. By this test one of the preservative processes (#3) gives plywood with a higher mean weight.

In Example 11.4 grade was used as a blocking variable in the wood-weight test. (Grade was a good choice for a blocking variable because, if tested, it would explain a significant amount of variation in the response. Plywood grade is an inherent (known) source of differences in weight.

Several enhancements can improve the present experiment. First, we question the use of only one observation or replicate for each two-way cell (blocks by treatments). This has given an error df of only 4, which is minimal for a reliable estimate of $\sigma^2$ by means of the MSE. One improvement would be to identify a variable for blocking that has more than three levels. A second possibility is to introduce more treatment variables with replicates for the treatment combinations. Replication does, however, introduce another component into the model. This component is the two-way effect called *interaction*. Interaction is the combined effect of multiple treatment variables on the eu's. It is different than the sum of the individual effects.

## Considerations for Designed Experiments

To illustrate the value of a two-way classification, let us study two hypothetical situations. First, *suppose* that, unknown to the analyst, the plywood sheets used in the surface stain process were all economy grade, the plywood used in the pressure sealer process was all standard grade, and the liquid-injection-treatment plywood sheets were all best grade. This is illustrated in Table 11.9.

Now assume that a single treatment test is conducted across preservative processes on the hypothesis $H_0$: $\mu_1 = \mu_2 = \mu_3$. You may want to cover the grade categories on the left to visualize this better. Suppose further that the difference in means appears to be statistically significant. This could lead to an incorrect conclusion. The means could differ simply because best-grade plywood is inherently heavier than standard or economy grades—not because there is a difference in weights due to the wood preservative process. The test is biased; the design is poor because in it weight can be attributed only to the preservative process . . . and this contains a bias.

Either of these one-way tests could lead to an incorrect conclusion. So it is important to design the experiment to get valid results, which requires including the treatment variables that have a real effect on variations in the response. Nonetheless, it is not always possible to identify and control (measure) all the important treatment variables in an experiment. Then randomization is the best insurance against biases due to the omitted variables. *Randomization* is achieved by the random assignment of treatments to experimental units or to blocks. Randomization is an essential part of all designed experiments.

Consider another potential problem. In a one-way design, all unexplained variations are incorporated into the error component. Then, if the error sum of squares is an inflated value, this results in a smaller-than-actual calculated *F*-value. Sometimes this situation leads to a false conclusion to reject the null hypothesis. A low calculated *F*-value, sometimes less than 1.0, can indicate an inefficient design.

**TABLE 11.9**  Hypothetical Sources of Variations in Plywood Weight

| Grade | Wood Preservation Process | | |
|---|---|---|---|
|  | *Surface Stain (1)* | *Pressure Sealer (2)* | *Liquid Injection Treatment (3)* |
| Economy | All observations of economy grade from Process 1 | | |
| Standard | | All observations of standard grade from Process 2 | |
| Best | | | All observations of best grade from Process 3 |

Adding a blocking variable will remove error sum of squares; this may increase the test precision. However, Example 11.4 illustrates a situation in which the block variable includes too few (3) levels for the given number of treatments (also 3). In our example this enhancement (blocking) reduced the error mean square but retained a minimal error df ($=4$). This choice for a block variable did not provide sufficient error df. Blocking may achieve an efficient design and a reliable estimate for error variance ($\sigma^2$), but this requires error df $= (B - 1)(T - 1)$ greater than 4!

Because additional treatment variables may remove substantial response variations from unexplained ones, a way potentially to achieve a more efficient design is to add more treatment variables. More treatment variables will remove previously unexplained SS, but they also remove degrees of freedom. Another technique for improving the discrimination of the test is to increase the number of observations. *Replication*, which brings in more data by repeating the random assignment of treatments to eu's, can reduce the error mean square and at the same time maintain a high error df, resulting in a more efficient design.

This section introduced blocking as one way to improve upon the efficiency of a one-way CRD design. By using a blocking variable, we recognize and then remove extraneous variations from the experiment. The result can be a reduced MSE. Then we pointed out the importance of randomization as a means of eliminating potential biases and ensuring reasonable sample estimates for population means.

The next section discusses another option—the use of more than one treatment variable. That discussion incorporates the concept of replication as a means for reducing the error MS and still retaining a reasonable error df.

The bottom line is that business decision situations for testing means require some checks such as blocking, randomization, and replication to ensure an efficient design and reasonable test conclusions.

**Problems 11.5**

32. A designed experiment was treated as a RCB design with the ANOVA summary (see table). Discuss the effectiveness of the blocks variable. Then explain how a blocking variable should differ from a treatment variable.

| Source of Variation | df | Sum of Squares | Mean Squares |
|---|---|---|---|
| Treatments | 3 | 492 | 134 |
| Blocks | 4 | 180 | 45 |
| Error | 32 | 960 | 30 |

33. Given is the following information for a designed experiment using a randomized complete block design:

$H_0: \mu_1 = \mu_2 = \mu_3$

Treatments $df_1 = 2$ and Treatments mean square $= 24$

Error mean square $= 3.2$ with 22 degrees of freedom

a. Test for significant differences in the (3) treatment means, $\alpha = .05$.

b. Consider a completely randomized design experiment having the same treatments $df_1 = 2$ and treatments mean square = 24 but with error mean square = 4.1 and $df_2 = 24$. Use numbers to explain why the RCB design in (a) gives more precision than the CRD design indicated here.

34. A test for the equality of means used samples of ten each from three populations with, respectively, $s_1^2 = 24$, $s_2^2 = 21$, and $s_3^2 = 19$. The treatment variable is brands of cola sold in vending machines. Let $\bar{Y}_1. = 3.2$, $\bar{Y}_2. = 3.8$ and $\bar{Y}_3. = 2.9$, all in thousands of dollars.

a. State the appropriate test hypothesis for a test of equal population means.

b. Build an ANOVA table.

c. Test the hypothesis (part a) using $\alpha = 0.05$.

35. The table below is a summary for an experiment that compares the average price for cases of three brands of cream cheese. Perform a test for equal means with $\alpha = 0.05$.

| Treatment Level | Treatment Mean Price |
|---|---|
| On sale | $25.75 |
| Regular retail | 38.67 |
| Wholesale | 18.75 |

| Source of Variation | df | Mean Squares |
|---|---|---|
| Brand (treatment) | 2 | 53.583 |
| Display location | 4 | |
| Error (unexplained) | 29 | 13.415 |
| Total | 35 | |

36. Explain what number indicates that the table below may be an inefficient design.

ANOVA

| Source of Variation | df | Mean Squares |
|---|---|---|
| Treatments | 2 | 21 |
| Blocks | 2 | 28 |
| Error | 4 | 3 |
| Total | 8 | |

37. An incentive sales plan is under evaluation by a national retail sales firm. The treatment comparison is straight salary (nonincentive) against 1% commission (incentive plan). The analysis is on dollar sales as the response variable, with sales "divisions" as a block variable. Divisions is a logical blocking variable as, for example, appliances are high-priced goods with a high commission; clothing sales are more volume sales with low per-unit commission. Records were kept on eight salespeople (eu's) under each plan. Within each of eight sales divisions, one person was randomly selected for

no incentive and another for an incentive sales position. The table is a summary of mean incomes and the ANOVA for a RCB design. Complete the ANOVA table and perform an $F$-ratio test for equal means for the sales plans. Let $\alpha = 0.05$. State your conclusions.

| | Sales Plan | |
|---|---|---|
| | Nonincentive | Incentive |
| Means | $13,194 | $14,647 |

| Source | df | SS | MS |
|---|---|---|---|
| Sales plan (treatments) | 1 | 1,821,307 | 1,821,307 |
| Division (block) | 7 | 5,593,620 | |
| Error | 7 | 1,142,756 | |
| Total | 15 | | |

## 11.6 FACTORIAL DESIGNS AND INTERACTION

In some situations, an analyst may want to look at two or more treatments on the same eu's. There is a design similar to the randomized block design that considers two or more treatment variables simultaneously. Each one is tested for significance against the MSE. This is called a two-way *factorial design*.

> A factorial design in two or more treatment variables assigns experimental units at random to treatment group combinations.

Here we discuss the two-factor (completely random) factorial design with equal numbers of replicates for every combination of treatment groups. This is called a complete factorial design. See [4] for variations of this design.*

Replication—that is, more than one eu at each given treatment group combination—allows us to test for additive treatment effects. The occurrence of nonadditivity in treatments is called *interaction*.

> Two treatment variables have an interaction effect if the difference in mean response for the levels of one treatment variable is not constant across the levels of the second treatment variable. Then the treatment effects are not simply additive.

By definition, interaction is the lack of simple additivity for treatment variables. For example, the combined physiological effect of alcohol and drugs on a person may be greater than the individual additive effects of each chemical if used separately. To illustrate, if under a given level of a drug the time to react (like applying the brake on a car) is 15 seconds, whereas under a given level of alcohol the time to react is 10 seconds, then under the same levels of drug and alcohol together, the time to react might be 30 seconds rather than $15 + 10 = 25$ seconds. The difference, 5 seconds, measures an (added) effect for drugs with alcohol. In the presence of an interaction, the treatment effects are not simply additive.

* Ott [4] generalizes a $k$-way classification of the data wherein all factor level combinations of the ($k$) treatment variables in the experiment have equal numbers.

When an interaction effect exists, the interpretation of the effect of the treatment variables on the eu's is not straightforward. Luckily, we can measure these interaction effects and identify whether there is (1) a nonsignificant interaction, which results in a straightforward interpretation of the treatment effects, or (2) a significant interaction, which necessitates a cautious interpretation of the treatment effects.

**Two-Factor Factorial Designs Including Replication and Interaction**

A factorial design with two treatment variables differs from the randomized complete block design, wherein one variable (plywood grade) was a block variable. As an inherent source of substantial response variation, the block variable is not manipulated. In contrast, in a factorial experiment, the treatment variables can be chosen and their levels then applied in any combination to any experimental unit. We note, then, that the factorial design affords a random assignment of treatment combinations to the experimental units. This distinguishes a factorial designed experiment from the RCB designed experiment.

We consider only the simplest factorial case, wherein the same (an equal) number of each treatment-level combination is applied to the experimental units. This identifies a factorial designed experiment with equal numbers of *replicates*. Other factorial designs are quite common, but require additional definition and are more computational, so are not discussed here. See [4] for other designs.

The discussion of two-variable factorial designs includes repeat observations in order to generalize the model. Including replicates enhances the reliability of our estimate of the error variance. Further, it can provide data for a measure of the effect of an interaction between treatment variables. Because we have already seen the other key elements for a factorial design experiment—treatments, randomization, and replication—a discussion of interaction is appropriate.

How does interaction affect a design and the model? We illustrate the presence or absence of interaction for a design with two-treatment variables. From Figure 11.4 we can view an interaction effect as some deviation from parallel plots. If the lines remain always (nearly) parallel, then an interaction effect is negligible. Otherwise, an interaction effect exists. Then the effect of the treatments on the response becomes unclear. Figure 11.4(a) displays mean response plots in which there is no interaction of the treatments. Figures 11.4(b) and 11.4(c) display two kinds of interaction effect. Figure 11.4(b) is a crossover

**FIGURE 11.4** Possibilities for Interaction in a 3 · 2 (levels) Two-Treatment Factorial Designed Experiment

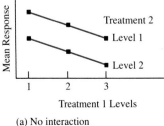

(a) No interaction

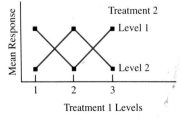

(b) An interaction expressed in a reversal of response (crossover)

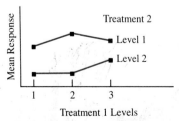

(c) An interaction expressed in the extent of response (non parallelism) effect, while Figure 11.4(c) has an interaction due to the nonparallelism. For example, in Figure 11.4(b) the mean response at level 1 of Treatment 2 exceeds the mean response at level 2 for levels 1 and 3 of Treatment 1, but is a lower response at level 2 of Treatment 1. The response relationship changes for different levels or combinations of the treatments. This "crossover" is one kind of interaction.

The two-factor factorial design model includes treatments and interaction effects as sources for explaining response variation. The factorial model includes a term for interaction effects, as well as terms to denote two treatments.

### MODEL FOR TWO-WAY FACTORIAL DESIGNS WITH EQUAL REPLICATION

$$Y_{ijk} = \mu + \alpha_i + \beta_j + (\alpha\beta)_{ij} + \varepsilon_{ijk} \qquad i = 1, 2, \ldots, A$$

$$j = 1, 2, \ldots, B$$

$$k = 1, 2, \ldots, n$$

(replicates)

where $\mu$ = an overall mean

$\alpha_i$ = treatment effect for treatment 1

$\beta_j$ = treatment effect for treatment 2

$(\alpha\beta)_{ij}$ = a (combined) interaction effect for the two treatments

$\varepsilon_{ijk}$ = random error unique to an experimental unit

In the factorial design model, the interaction term reflects a net effect on the response after the effects of (1) an overall mean and (2) the treatments.

Assumptions:

1. The combinations of treatment levels are randomly assigned to the eu's.
2. The observations in each combination of treatment levels are random samples from independent populations with distribution means: $\mu + \alpha_i + \beta_j + (\alpha\beta)_{ij}$.
3. The random error for each treatment combination is normally distributed with mean zero and a constant variance, $\sigma^2$.

The hypothesis tests follow a specific order, first testing for significant interaction effects. That is, the test on $H_0$: $(\alpha\beta)_{ij} = 0$ versus $H_A$: Not all $(\alpha\beta)_{ij} = 0$ is used to determine if all treatment means can be reduced to the form $\mu + \alpha_i + \beta_j$, that is, if the interaction effect is ignorable.

If the interaction effect is not significant, then $(\alpha\beta)_{ij} = 0$ for all treatment-level combinations. This condition gives two directives. First, the treatment effects are considered additive. This means that a relation on the levels of one treatment will remain or hold true across all levels of the second treatment variable. (Remember Figure 11.4(a), which shows near-parallel response graphs for zero interaction.) Then the "main effect" treatments are universally interpretable. Second, a zero-interaction effect signals that the component is random, and so can be considered a part of the random error. Then the interaction sum of squares and degrees of freedom are added or "pooled" with random error to provide more data, and thus to provide a more precise estimate of the unknown error variance, $\sigma^2$.

When the interaction between two treatments is a significant effect, this means that a change in one treatment gives a change in response at some levels of the second treatment that differs from the response at other levels. This situation is viewed in Figure 11.4(b) as a crossover, or as nonparallel mean plots as displayed in Figure 11.4(c). Here, because the response relationship changes, a separate interpretation is required for the meaning of treatment (main) effects depending upon the combination of treatment levels. This situation is illustrated in Example 11.6.

If all interaction effects are found to be not significant, it becomes appropriate to test for significant treatment "main effects." These tests are

$$H_0: \alpha_i = 0 \qquad H_A: \text{Not all } \alpha_i = 0 \qquad i = 1, 2, \ldots, A$$

This null hypothesis states that all levels of the $\alpha$-treatment variable have statistically equal response. Similarly,

$$H_0: \beta_j = 0 \qquad H_A: \text{Not all } \beta_j = 0 \qquad j = 1, 2, \ldots, B$$

where the null hypothesis, if true, indicates that the levels of the $\beta$-treatment variable all have the same response effect. For those cases in which

$$(\alpha\beta)_{ij} = 0 \text{ for all combinations of } i = 1, 2, \ldots, A \text{ and } j = 1, 2, \ldots, B$$

and     $\alpha_i = 0$ for all levels $i = 1, 2, \ldots, A$

and     $\beta_j = 0$ for all levels $j = 1, 2, \ldots, B$

with the condition of a normal response distribution, the data come from a single population having mean $\mu$ and variance $\sigma^2$ (unknown). From studying other designs, we have seen that the tests use the $F$-distribution. This is computational, but it is simplified by an analysis of variance that appears in Table 11.10.

**TABLE 11.10**  The $F$-Test and Two-Way ANOVA for a Two-Factor Factorial Experiment with Equal Numbers of Replicates

| Source of Variation | df* | Sums of Squares | Mean Squares |
|---|---|---|---|
| Treatment 1 | $A - 1$ | SSA | MSA = SSA/$(A - 1)$ |
| Treatment 2 | $B - 1$ | SSB | MSB = SSB/$(B - 1)$ |
| Interaction | $(A - 1)(B - 1)$ | SSAB | MASB = SSAB/$(A - 1)(B - 1)$ |
| Explained | $AB - 1$ | SST | MST = SST/$(AB - 1)$ |
| Error | $AB(n - 1)$ | SSE | MSE = SSE/$AB(n - 1)$ |
| Total (corrected) | $ABn - 1$ | TSS | |

* Requires equal replicates in all cells.

The *F*-ratio tests use the appropriate explained source (treatment 1 or 2, interaction, or explained) mean square divided by an error mean square, MSE. Here is an outline of the sum of squares calculations:

$$\text{TSS} = \text{SST} + \text{SSE}$$

Total SS = Explained SS + Error (unexplained) SS

Then

$$\text{SST} = \text{SSA} + \text{SSB} + \text{SSAB}$$

Explained SS = Treatment 1 SS + Treatment 2 SS + Interaction SS

Computing formulas for the parts use a correction term,

$$CT = \frac{(\text{Total of all observations})^2}{ABn} = \frac{\left( \sum_i \sum_j \sum_k Y_{ijk} \right)^2}{ABn} = \frac{(Y...)^2}{ABn}$$

where   $i = 1, 2, \ldots, A$     $j = 1, 2, \ldots, B$     $k = 1, 2, \ldots, n$

and the dot subscripts indicate a summation over that factor.

$(65 - 69.611)^2$

$$\text{TSS} = \text{Total sum of squares} = \sum_i \sum_j \sum_k (Y_{ijk} - \bar{Y}...)^2$$

$$= \text{SS of all observations} - CT$$

$$= \text{SS}(Y) = \sum_i \sum_j \sum_k Y_{ijk}^2 - CT$$

$$\text{SST} = \text{Explained sum of squares} = \sum_i \sum_j (\bar{Y}_{ij.} - \bar{Y}...)^2$$

$$= \frac{\sum_i \sum_j Y_{ij.}^2}{n} - CT$$

where $Y_{ij.}$ is the response total for the *n*-values (replicates) in cell $i, j$ and $\bar{Y}_{ij.}$ denotes the associated cell means.

$$\text{SSE} = \text{TSS} - \text{SST}$$

$$= \text{Error, or unexplained, SS}$$

$$\text{SSA} = \text{Treatment 1 sum of squares}$$

$$= \frac{\sum_i Y_{i..}^2}{Bn} - CT$$

$$\text{SSB} = \text{Treatment 2 sum of squares}$$

$$= \frac{\sum_j Y_{.j.}^2}{An} - CT$$

$$\text{SSAB} = \text{Interaction sum of squares}$$

$$= \text{SST} - \text{SSA} - \text{SSB}$$

**TABLE 11.11** Data Base for a $3 \cdot 3$ Two-Factor Factorial Designed Experiment with Two Replicates

| Exterior Finish (Treatment 2) | | Wood Preservative Process (Treatment 1) | | | |
| --- | --- | --- | --- | --- | --- |
| | | Stain (1) | Pressure Sealer (2) | Liquid Injection Treatment (3) | Exterior Finish Totals |
| Stain only | | 65 | 64 | 70 | |
| | | 68 | 66 | 68 | |
| | Cell totals | 133 | 130 | 138 | 401 |
| | Cell means | 66.5 | 65.0 | 69.0 | 66.833 |
| Paint | | 68 | 68 | 74 | |
| | | 70 | 65 | 70 | |
| | Cell totals | 138 | 133 | 144 | 415 |
| | Cell means | 69.0 | 66.5 | 72.0 | 69.167 |
| Plastic coating | | 70 | 72 | 76 | |
| | | 71 | 74 | 74 | |
| | Cell totals | 141 | 146 | 150 | 437 |
| | Cell means | 70.5 | 73.0 | 75.0 | 72.833 |
| Wood preservative Totals | | 412 | 409 | 432 | $(Y_{..})$ 1253 |
| | | 68.667 | 68.167 | 72.000 | $(\bar{Y}_{..})$ 69.611 |

The key to an ANOVA summary is the partition of total SS and degrees of freedom into their component parts. Interaction is a new concept. The SS for interaction explains variation that otherwise would be unexplained, so it derives from the error term. Then, if statistical tests show that the interaction effect is not significant, it will be collapsed, or pooled, with the error.

Except for interaction, the components of this model are familiar from earlier described designs. To better understand the meaning of interaction, we extend the plywood weights to two replicates for a two-way factorial design. As Table 11.11 shows, treatment 1 is the wood preservative process and treatment 2 is the exterior finish for appearance. The finishes are (1) stain only, (2) paint, or (3) a plastic coating.

For the data in Table 11.11, saying that an interaction exists for preservative process by exterior finish means that the weights for different preservative processes do *not* remain in the same relative order for the different exterior finishes. (Remember the discussion relative to parallel lines for Figure 11.4.) Graphs of cell means that show lines with different slopes or lines that cross indicate potential interactions.

Figure 11.5 provides a visual check for interaction on plywood treatments. The nearly parallel graphs for paint and stain finishes display what appears to be negligible interaction effect for preservative process and finish. That is, the differences in mean weight for these two finishes are nearly constant across wood preservative process 1, across process 2, and then across process 3. This parallelism suggests that the stain and paint finishes contribute very little measure to an interaction effect.

**FIGURE 11.5** Visual Check for Interaction for Wood Preservative Process with Type of Finish (The data points in the figure are the cell means from Table 11.11.)

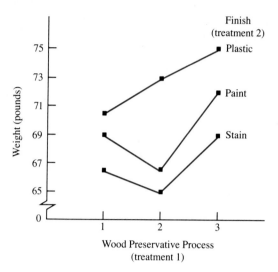

In Figure 11.5 the graph for "plastic" finish is not parallel to the graphs for "paint" and "stain" finishes. This apparent interaction effect brings up the following question: Is the distortion from parallelism slight enough (insignificant) to permit us to conclude that there is a constant difference in mean weight for the three preservative processes across *all* finishes?

**EXAMPLE 11.5**

Using the data in Table 11.11 for a $3 \cdot 3$ two-factor factorial experiment with two replicates, test the following null hypotheses:

1. There is no significant interaction effect between the preservative process (treatment 1) and the finish (treatment 2) processes in mean response.
2. There is no significant effect on mean plywood weights that is attributable to the preservative, treatment 1.
3. There is no significant effect on mean plywood weights that is attributable to the finish, treatment 2.

*Solution* The calculations use data from Table 11.11. The ANOVA is presented in Table 11.12.

$$CT = \frac{(Y...)^2}{3 \cdot 3 \cdot 2 \;\; 18} = \frac{(1253)^2}{18} = 87{,}222.722$$

$$TSS = (65^2 + 68^2 + \cdots + 74^2) - CT = 204.278$$

$$SST = \frac{(133^2 + 130^2 + \cdots + 150^2)}{2} - CT = 176.778$$

$$SSE = 204.278 - 176.778 = 27.500$$

**TABLE 11.12** The ANOVA Table for a Two-Factor Factorial Experiment with Interaction (SPSS output)

| SOURCE OF VARIATION | SUM OF SQUARES | DF | MEAN SQUARE | F | SIG OF F |
|---|---|---|---|---|---|
| MAIN EFFECTS | 161.889 | 4 | 40.472 | 13.245 | .001 |
|     FINISH | 109.778 | 2 | 54.889 | 17.964 | .001 |
|     PRESERV | 52.111 | 2 | 26.056 | 8.527 | .008 |
| | | | | | |
| 2-WAY INTERACTIONS | | | | | |
|     FINISH    PRESERV | 14.889 | 4 | 3.722 | 1.218 | .368 |
| | | | | | |
| EXPLAINED | 176.778 | 8 | 22.097 | 7.232 | .004 |
| | | | | | |
| RESIDUAL *(error, unexplained)* | 27.500 | 9 | 3.056 | | |
| | | | | | |
| TOTAL | 204.278 | 17 | 12.016 | | |

$$SSA = \frac{(412^2 + 409^2 + 432^2)}{6} - CT \qquad = 52.111$$

$$SSB = \frac{(401^2 + 415^2 + 437^2)}{6} - CT \qquad = 109.778$$

$$SSAB = 176.778 - 52.111 - 109.778 \qquad = 14.889$$

A first test compares the explained source with the MSE for an $F$-test of significance for

$H_0$:    None of the explained variance sources—preservative, finish, or interaction—is significant in describing the response—weight. That is, the EXPLAINED sources are all nonsignificant.

$H_A$:    One or more of the effects, including interaction, are significant.

[*Note:* This test is optional but has the advantage that if it shows that the explained sources (collectively) are not significant, then the tests for (1) interaction followed by tests for (2) preservatives and (3) finish treatments are not required.]

    The $p$-value approach is used for this and the remaining tests. This test for significant EXPLAINED variation shows calculated $F = 7.23$ ($= 22.097/3.056$) with a $p$-value of 0.004. This is highlighted in Table 11.12. Therefore we conclude that one or more of the sources explain real differences for the plywood sheet weights, and the testing continues. (Had we not rejected $H_0$, the testing would stop and we would conclude that the data are a sample from one population.) So, in this case individual tests are required for significant interaction and treatment effects.

1. The interaction effect is tested first. It is not significant, as the calculated $F = 1.218$ has a $p$-value of 3.68, which exceeds 0.05. Thus it becomes appropriate to test the treatments. This result appears in Table 11.12.

    For the test on treatments, we first combine the nonsignificant interaction SS with error SS, and divide by their combined degrees of freedom, to obtain a

**TABLE 11.13**   An ANOVA Table for a Two-Factor Factorial Experiment with Interaction Pooled with Error

| SOURCE OF VARIATION | SUM OF SQUARES | DF | MEAN SQUARE | F | SIG OF F |
|---|---|---|---|---|---|
| MAIN EFFECTS | 161.889 | 4 | 40.472 | 12.412 | .000 |
|    FINISH | 109.778 | 2 | 54.889 | 16.834 | .000 |
|    PRESERV | 52.111 | 2 | 26.056 | 7.991 | .005 |
| EXPLAINED | 161.889 | 4 | 40.472 | 12.412 | .000 |
| RESIDUAL | 42.389 | 13 | 3.261 | (=MSE(2)) | |
| TOTAL | 204.278 | 17 | 12.016 | | |

more precise estimate for the unknown population variance. This is MSE(2) in Table 11.13*:

$$\text{MSE}(2) = \frac{14.889 + 27.500}{4 \quad + \quad 9} = \frac{42.389}{13} = 3.261$$

2. The calculated $F$-statistic for testing

$H_0$: There is no significant effect on weight due to the preservative
is $F = 7.991 \ (= 26.056/3.261)$, which by Table B.4 has a $p$-value 0.005.

There are significant differences in mean weight that can be attributed to the different levels of preservative—stain only, pressure, or liquid treatment.

3. Finally, the test for treatment 2, finish, has the null hypothesis

$H_0$: There is no significant effect on weight due to the finish.

The calculated $F$-statistic is $F = 16.834$, which for $df_1 = 2$, $df_2 = 13$ has a $p$-value $< 0.001$ and so is significant. Some differences in weights can be attributed to the choice of finish—stain only, paint, or plastic coating.

---

Because the interaction effect was not significant, differences attributable to the treatment effects are universal. Then it becomes appropriate, for example, to apply multiple-comparison tests for the effect of the individual treatment levels on the plywood sheet weights. This is done in the next section.

The preceding example illustrates the case in which the interaction is not significant. However, the more challenging case for a factorial experiment arises when the interaction is statistically significant. The next example illustrates this.

---

* Generalized for a 2-factor factorial experiment, with treatments 1 and 2 having $A$ and $B$ levels, respectively, and with equal numbers of replicates, $n$,

$$\text{MSE (2)} = \frac{\text{SSAB} + \text{SSE}}{(A-1)(B-1) + AB(n-1)}$$

is a mean square error estimate that incorporates a non-significant (by testing) interaction component.

**EXAMPLE 11.6**

What effect do (1) the product sold and (2) the sales region have on the commission paid to specialized (auto, life, health, home mortgage) insurance salespeople for a large-structured national insurance firm? This question was posed by the firm's corporate board, who had on hand the random sample data that appear in Table 11.14. The response is the annual sales commission in thousands of dollars.

*2 treatments, region, product*

**TABLE 11.14** Sales Commissions ($000) for Specialized Insurance Salespeople Working for a National Company

| Product | | Region | | | | Means |
|---------|-------|------|------|------|------|-------|
| | | *A* | *B* | *C* | *D* | |
| 1 | | 39.3 | 41.6 | 38.8 | 42.9 | |
| | | 37.7 | 42.7 | 37.2 | 39.3 | |
| | | 40.6 | 38.9 | 39.1 | 40.5 | |
| | Means | 39.2 | 41.1 | 38.4 | 40.9 | 39.9 |
| 2 | | 41.5 | 38.4 | 40.2 | 38.9 | |
| | | 39.7 | 37.7 | 41.1 | 38.1 | |
| | | 38.4 | 40.1 | 40.9 | 39.2 | |
| | Means | 39.9 | 38.7 | 40.7 | 38.7 | 39.5 |
| 3 | | 40.6 | 40.3 | 37.2 | 43.6 | |
| | | 39.8 | 38.8 | 38.4 | 42.1 | |
| | | 41.3 | 39.6 | 37.0 | 44.5 | |
| | Means | 40.6 | 39.6 | 37.5 | 43.4 | 40.3 |
| | Total Means | 39.9 | 39.8 | 38.9 | 41.0 | 39.9 |

*is interaction effect*

Analyze these data and advise the corporate board on whether the company should continue with specialized sales assignments that earn product- and region-based commissions. Is this practice of commission payments equitable for the total sales staff—over 4000 salespeople—from which this sample was randomly selected?

*Solution*

The marginal means in Table 11.14 show minimal differences for commissions based on the product sold. There may be some differences by region. But is it reasonable to make such blanket statements? The data are quite mixed. A data plot will give some indication of a possible interaction effect. That plot, in Figure 11.6, uses the mean response for the 12 cells.

By Figure 11.6 it is clear that an interaction effect exists. The question becomes, Is there a way to provide a meaningful discussion about either of the treatments? An SPSS ANOVA produced the summary in Table 11.15.

Table 11.15 indicates a significant two-way interaction as well as significant differences for the REGION treatment, because both $p$-values are $<0.05$. Differences in commissions vary by region according to the product that is sold.

Because there is a significant interaction, differences by REGION are described in parts. This becomes an objective data analysis that uses information from the ANOVA table (11.15) and Figure 11.6. These summaries are descriptive and are not based on statistical significance tests.

**FIGURE 11.6**   A Plot of Sales Commission Means ($000) for a Possible Interaction
of Product by Region for Insurance Salespeople

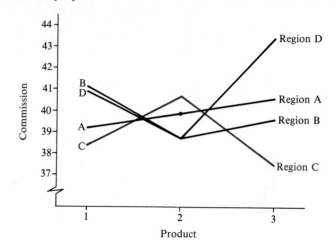

**TABLE 11.15**   An ANOVA for the Factorial Experiment on Insurance Sales
Commissions

| SOURCE OF VARIATION | SUM OF SQUARES | DF | MEAN SQUARE | F | SIG OF F |
|---|---|---|---|---|---|
| MAIN EFFECTS | 24.002 | 5 | 4.800 | 3.200 | .024 |
| PRODUCT | 3.376 | 2 | 1.688 | 1.125 | .341 |
| REGION | 20.627 | 3 | 6.876 | 4.584 | .011 |
| 2-WAY INTERACTIONS | | | | | |
| PRODUCT   REGION | 57.073 | 6 | 9.512 | 6.341 | .000 |
| EXPLAINED | 81.076 | 11 | 7.371 | 4.914 | .001 |
| RESIDUAL | 36.000 | 24 | 1.500 | | |
| TOTAL | 117.076 | 35 | 3.345 | | |

1. Region D shows a high mean commission for products 1 and 3, but has relatively
   low commission for product 2.
2. The commissions for region C show a pattern that is the reverse of region D
   commissions, being the highest for product 2 and quite low for products 1 and 3.
3. The region B commissions were relatively high for those salespeople who sell
   product 1; otherwise, for regions A and B the commissions are reasonably close.

Provided the plan were workable for the salespeople, it would be reasonable to develop
a rotation system for individual salespeople in regions C and D who sell product 1 and,
similarly, a rotation in regions C and D for those who sell product 3. This should level
out their commissions. Since there is not a logical rotation for the salespeople in region
B who sell product 1, it seems best to leave them with some commission advantage.

Lowering their commissions is not an option. Otherwise, the commissions appear reasonably level, so no other changes are suggested.

---

From what we saw in Example 11.6, a significant interaction can be likened to a "bull in a China shop." Both require some sorting out to put the pieces back together. With a significant interaction, it can help to start sorting by viewing a manageable amount of the data. In this respect the table of cell means followed by a plot like Figure 11.6 can give some indication of how to put the pieces together. The ANOVA is central to determining which treatment effects can show response differences.

We turn next to a problem that was mentioned with the CRD comparisons to determine individually which pairs of treatment means are significantly different. The procedures will apply to completely random, randomized block, and factorial design experiments. But first, some exercises over the preceding material are in order.

**Problems 11.6**

The following information applies to Problems 38–41.

J. B. Wilkinson, J. Barry Mason, and Christie H. Paksoy used a three-way factorial design in "Assessing the Impact of Short-Term Supermarket Strategy Variables" (*Journal of Marketing Research*, XIX, February 1982: 72–82). The design was a 3 by 3 by 2 factorial with treatment factors of price, display, and advertising. Four products were treated separately as experimental units. These were (1) Camay soap (bath size), (2) White House apple juice (32 oz), (3) Mahatma rice (1 lb), and (4) Piggly Wiggly frozen pie shells (2/pkg).

The response measure, unit sales per period, was defined operationally as sales from Wednesday noon until Saturday at 9 P.M. The experiment was conducted in one store of the Piggly Wiggly supermarket chain. Two replicates were taken for each of the $18 (= 3 \times 3 \times 2)$ treatment-level combinations for a total of 36 response values. Factor level descriptions are as follows:

*Regular price:* the current warehouse recommended retail price

*Cost price:* the cost to the supermarket

*Reduced price:* an amount halfway between regular price and cost price

*Normal display:* a display based on the stock manager's definition of regular shelf-space allocation

*Expanded display:* two times the area used for normal display

*Special display:* the regular shelf space plus special display, either (a) within the aisle or (b) at the end of aisle

*Advertising:* product name and price in the supermarket's Wednesday newspaper advertisement

*No advertising:* exclusion from the Wednesday newspaper advertisement

38. One ANOVA table displays the results for unit sales of Camay soap. Use residual MS with 18 df to test for a significant three-way interaction (test 1) and for significant two-way interactions (tests 2, 3, and 4). Use $\alpha = 0.05$ for these four tests. State your conclusions.

| Source of Variation | Sum of Squares | df | Mean Square |
|---|---|---|---|
| Main effects | 26,926.222 | 5 | 5,385.244 |
| Price | 5,219.056 | 2 | 2,609.528 |
| Display | 21,578.722 | 2 | 10,789.361 |
| Advertising | 128.444 | 1 | 128.444 |
| Two-way interactions | 6,465.889 | 8 | 808.236 |
| Price/display | 4,310.111 | 4 | 1,077.528 |
| Price/advertising | 1,685.056 | 2 | 842.528 |
| Display/advertisting | 470.722 | 2 | 235.361 |
| Three-way interactions | | | |
| Price/display/advertisting | 4,222.778 | 4 | 1,055.694 |
| Explained | 37,614.889 | 17 | 2,212.641 |
| Residual | 6,135.996 | 18 | 340.889 |
| Total | 43,750.884 | 35 | 1,250.025 |

39. The diagram shows mean sales for Camay soap for price level by display type. This figure displays the two-way interaction for price with display. For $\alpha = 0.05$ this interaction is statistically significant.

   a. How many levels are illustrated for price? How many for display?

   b. What type of interaction (deviation from parallel) occurs for expanded versus regular display with the levels of price? *Suggestion:* See Figure 11.4 for display of types including parallel (no interaction), crossover, or nonparallel.

   c. What interaction occurs for special versus regular display over the levels of price?

   d. With the existence of a significant interaction, it can be unsafe to simply interpret the MAIN EFFECTS (if also statistically significant) as $A > B > C$. For example, a crossover could contradict this. The data in Problem 38 indicate significant interactions. From this figure, is it fair to say that the special display outperforms the expanded and regular displays? Explain.

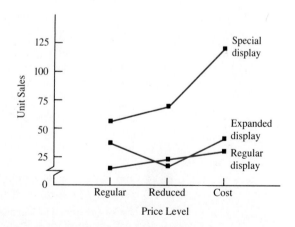

40. Another ANOVA table summarizes variations in unit sales when the eu's are packages of pie shells. Use the data to test three hypotheses.
   a. For prices: $H_0$: $\mu_{reg} = \mu_{red} = \mu_{cost}$
      $H_A$: At least one $\mu \neq$ others
   b. For display: $H_0$: $\mu_{reg} = \mu_{exp} = \mu_{spec}$
      $H_A$: At least one $\mu \neq$ others
   c. For advertising: $H_0$: $\mu_{adv} = \mu_{no\,adv}$
      $H_A$: $\mu_{adv} \neq \mu_{no\,adv}$
   Begin by testing all interactions for significance, $\alpha = 0.05$. Combine with the residual source all interactions that are not significant.

| Source of Variation | df | Sum of Squares | Mean Square | Calculated F |
|---|---|---|---|---|
| Main effects | 5 | 13,719.361 | 2,743.872 | 22.79 |
| Price | 2 | 3,283.909 | 1,641.955 | 13.64 |
| Display | 2 | 9,918.204 | 3,959.102 | 41.19 |
| Advertising | 1 | 208.112 | 208.112 | 1.73 |
| Two-way interactions | 8 | 1,474.810 | 184.351 | 1.53 |
| Price/display | 4 | 1,156.863 | 289.216 | 2.40 |
| Price/advertising | 2 | 109.149 | 54.574 | 0.45 |
| Display/advertising | 2 | 119.986 | 59.993 | 0.50 |
| Three-way interactions | 4 | 488.040 | 122.010 | 1.01 |
| Price/display/advertising | 4 | 488.040 | 122.010 | 1.01 |
| Explained | 20 | 18,664.073 | 933.204 | 7.75 |
| Residual (error) | 15 | 1,805.816 | 120.388 | |
| Total | 35 | 20,469.889 | 584.854 | |

41. The ANOVA table and graph support a significant two-way interaction for price by display on unit sales for Mahatma Rice, $\alpha = 0.05$. Check this by
   a. testing $H_0$: No interaction for price by display.
   b. explaining the interaction effect relative to the levels of price by the levels of display.

| Source | df | Mean Square | F |
|---|---|---|---|
| Price/display interaction | 4 | 266.958 | 4.49 |
| Price/advertising | 2 | 53.583 | 0.90 |
| Display/advertising | 2 | 172.583 | 2.90 |
| Price/display/advertising | 4 | 24.792 | 0.42 |
| Explained | 17 | 842.647 | 14.18 |
| Residual | 18 | 59.444 | |
| Total | 35 | 439.857 | |

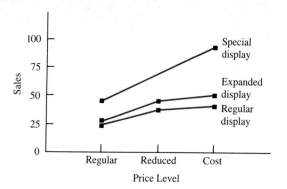

42. For the data in Example 11.6, lets say someone has incorrectly pooled the (significant) interaction SS and df with Error. How has this altered the tests on the treatments—product and region? Why has it altered these tests? To answer these questions, you will need to review the ANOVA table in Example 11.6 and compare the MSE there with the MSE value below.

| SOURCE OF VARIATION | SUM OF SQUARES | DF | MEAN SQUARE | F | SIG OF F |
|---|---|---|---|---|---|
| MAIN EFFECTS | 24.002 | 5 | 4.800 | 1.547 | .205 |
| PRODUCT | 3.376 | 2 | 1.688 | .544 | .586 |
| REGION | 20.627 | 3 | 6.876 | 2.216 | .107 |
| EXPLAINED | 24.002 | 5 | 4.800 | 1.547 | .205 |
| RESIDUAL | 93.073 | 30 | 3.102 | | |
| TOTAL | 117.076 | 35 | 3.345 | | |

43. A table summarizes ($Y =$) units of production, which are the number of engines installed in chassis at a major automobile assembly plant during one workday. (*Note:* Partial installations were included.) Assume that the conditions are met for a $3 \cdot 2$ factorial experiment with three replicates.

| Model | Robot Line 1 | 2 | 3 | 4 | Mean |
|---|---|---|---|---|---|
| 1 | 22.33 (3) | 17.00 (3) | 21.67 (3) | 23.67 (3) | 21.17 (12) |
| 2 | 19.33 (3) | 14.67 (3) | 17.33 (3) | 21.00 (3) | 18.08 (12) |
| 3 | 24.00 (3) | 21.33 (3) | 23.00 (3) | 23.33 (3) | 22.92 (12) |
| Mean | 21.89 (9) | 17.67 (9) | 20.67 (9) | 22.67 (9) | 20.72 (36) |

The numbers in the table are means based on the number of observations listed below the mean in parentheses. The ANOVA table follows:

| SOURCE OF VARIATION | SUM OF SQUARES | DF | MEAN SQUARE | F | SIG OF F |
|---|---|---|---|---|---|
| MAIN EFFECTS | 274.056 | 5 | 54.811 | 22.171 | .000 |
|    MODEL | 143.722 | 2 | 71.861 | 29.067 | .000 |
|    ROBOT LINE | 130.333 | 3 | 43.444 | 17.573 | .000 |
| 2-WAY INTERACTIONS | | | | | |
|    MODEL   ROBOT LINE | 23.833 | 6 | 3.972 | 1.607 | .189 |
| EXPLAINED | 297.889 | 11 | 27.081 | 10.954 | .000 |
| RESIDUAL (ERROR) | 59.333 | 24 | 2.472 | | |
| TOTAL | 357.222 | 35 | 10.206 | | |

Plot the cell means as we did for Example 11.6. Do you believe there is a significant interaction effect? Now test the null hypothesis of no interaction and, if appropriate, test for significant treatment effects—robot line number and model. Use $\alpha = 0.05$ for all tests and state your conclusions.

44. These data are from the same assembly situation as in the preceding problem, except the data were taken one month later, and line 4 is not operating in this experiment.

| | Robot Line | | | |
|---|---|---|---|---|
| Model | *1* | *2* | *3* | Mean |
| 1 | 20.50 | 29.50 | 23.00 | 24.33 |
| | (4) | (4) | (4) | (12) |
| 2 | 17.25 | 26.75 | 15.00 | 19.67 |
| | (4) | (4) | (4) | (12) |
| 3 | 13.00 | 24.25 | 18.50 | 18.58 |
| | (4) | (4) | (4) | (12) |
| | 16.92 | 26.83 | 18.83 | 20.86 |
| Mean | (12) | (12) | (12) | (36) |

Again, the numbers in the table are means with the number of observations indicated in parentheses.

| SOURCE OF VARIATION | SUM OF SQUARES | DF | MEAN SQUARE | F | SIG OF F |
|---|---|---|---|---|---|
| MAIN EFFECTS | 888.111 | 4 | 222.028 | 108.502 | .000 |
|   A-MODEL | 224.056 | 2 | 112.028 | 54.747 | .000 |
|   B-ROBOT LINE | 664.056 | 2 | 332.028 | 162.258 | .000 |
| 2-WAY INTERACTIONS | | | | | |
|   MODEL    ROBOT LINE | 72.944 | 4 | 18.236 | 8.912 | .000 |
| EXPLAINED | 961.056 | 8 | 120.132 | 58.707 | .000 |
| RESIDUAL (ERROR) | 55.250 | 27 | 2.046 | | |
| TOTAL | 1016.306 | 35 | 29.037 | | |

Proceed as in Problem 43 to plot the cell means, and indicate from your plot whether you think there is a significant interaction effect. Follow this with tests of hypothesis, as appropriate. Then discuss the relation of robot line and car model to the (response) number of engines installed.

# 11.7 ESTIMATION OF TREATMENT MEANS AND MULTIPLE COMPARISONS

Although it is informative to know that at least one treatment mean is significantly different from the rest, it may be equally important to determine *which* ones are different and *what* are the amounts of the differences. The problem for the estimation of treatment differences is to develop a procedure for simultaneously comparing three or more group means.

In Chapter 10, we discussed an expression for estimating a single treatment mean: $\bar{Y}_i. \pm (t_{\alpha/2} \cdot \sqrt{\text{MSE}/n})$. Similarly, a $[(1 - \alpha) \cdot 100]\%$ confidence interval for the difference in two treatment means is

$$(\bar{Y}_1 - \bar{Y}_2) \pm \left[ t_{\alpha/2} \sqrt{\text{MSE}\left(\frac{1}{n_1} + \frac{1}{n_2}\right)} \right]$$

You will recognize some similarities in these forms. Here we have independent samples from two treatment populations. Whether the design is CRD, RBD, or other, the test uses the error mean square to estimate treatment differences: The estimate for (assumed) common variance is MSE from the appropriate ANOVA. The df for the *t*-statistic is that of the error (residual) source of variation.

Remember, a test conclusion that does not reject the null hypothesis of $\mu_1 = \mu_2 = \cdots = \mu_T$ implies that the treatment means are *not* significantly different. In this case, there is no need to estimate differences. Otherwise the estimation problem is not as simple as comparing two treatment means, then two more, and so on. Indeed, every mean must be simultaneously estimated against every other mean to yield a ranking for all of the means. There is need to make all pairwise comparisons among $T$ treatment means.

This situation is not as simple as when we compare just two population means. For example, suppose the comparisons are among $T = 4$ population means. Then it is possible to make $T - 1 = 3$ independent comparisons. However, there are six possible pairwise comparisons among the four population means—1 with 2, 1 with 3, 1 with 4, 2 with 3, and so on. If we make each of these six comparisons using a $t$-test with $\alpha = 0.05$, then finding the overall error rate is more involved.

The problem of estimation in designed experiments is to establish a procedure that will assure the desired simultaneous confidence level for all $_TC_2$ comparisons.

> Simultaneous confidence intervals are confidence intervals that reflect true differences across all treatment comparisons with some fixed error rate.

This is not an easy problem, and the solution is beyond the scope of this textbook. Yet several approaches are available that control the overall error rate at a desired level. Some of these procedures, including ones by Fisher, by Tukey, by Student-Newman-Keuls, and by Scheffé are called *multiple comparisons* [4].

> Multiple comparisons are the simultaneous assessment of the confidence estimates of differences among treatment means for all $_TC_2$ possible two-at-a-time comparisons.

A multiple-comparison approach ensures (1) that all pairwise comparisons are made and (2) that a prespecified error rate is maintained for all the comparisons. A major difference among the prominent procedures is the error rate that each one controls. We have chosen the Scheffé procedure, which controls an *"experiment-wise"* error rate.

> In multiple-comparison procedures an experiment-wise error rate, $\alpha$, means that, considering all possible comparisons, the probability of having one or more comparisons falsely declared significant is $\alpha$, where $0 \leqslant \alpha \leqslant 1$.

Relative to the other procedures mentioned, the Scheffé approach is more conservative in detecting significant differences among pairs of population means.

The Scheffé procedure [7] for establishing simultaneous confidence intervals with experiment-wise error rate of $\alpha$ is summarized as follows:

### SCHEFFÉ'S PROCEDURE FOR ESTIMATING SIMULTANEOUS CONFIDENCE INTERVALS

The following confidence intervals have experiment-wise $[(1 - \alpha) \cdot 100]\%$ levels of confidence:

$$(\bar{Y}_1 - \bar{Y}_2) \pm \left[ \sqrt{(T - 1)F_\alpha} \cdot \sqrt{\text{MSE}\left(\frac{1}{n_1} + \frac{1}{n_2}\right)} \right]$$

$$(\bar{Y}_1 - \bar{Y}_3) \pm \left[ \sqrt{(T - 1)F_\alpha} \cdot \sqrt{\text{MSE}\left(\frac{1}{n_1} + \frac{1}{n_3}\right)} \right] \tag{11.7}$$

$$\vdots$$

$$(\bar{Y}_{T-1} - \bar{Y}_T) \pm \left[ \sqrt{(T - 1)F_\alpha} \cdot \sqrt{\text{MSE}\left(\frac{1}{n_{T-1}} + \frac{1}{n_T}\right)} \right]$$

where MSE and $df_2$ for $F_\alpha$ $(= F_{\alpha, T-1, df_2})$ are defined by the error source for the appropriate design, and $T$ is the number of treatment groups.

When we consider the Scheffé approach as a testing procedure, *if the range of the confidence interval includes both minus and plus values, then we accept the hypothesis of "no difference" in the two means.* This is equivalent to simultaneous two-tailed tests of hypothesis with $\alpha =$ the specified (experiment-wise) error rate. Here is an example.

**EXAMPLE 11.7**

In a previous example we concluded that WEIGHTS were not equal for plywood sheets for three wood PRESERVATIVE PROCESSES. We further concluded that the resulting weights of the liquid treatment process 3 are heavier for all FINISHES than are the weights obtained through process 1 or process 2. These conclusions were based on observation, not testing, so they require objective justification. Justify these earlier conclusions at $\alpha = 0.05$, for experiment-wise $[(1 - \alpha) \cdot 100]\% = 95\%$ simultaneous confidence interval estimates.

*Solution*

Here is the essential information for Scheffé's multiple comparisons. From the ANOVA table, 11.13, on page 478,

| Source of Variation | SS | df | MS | |
|---|---|---|---|---|
| Adjusted residual | 42.389 | 13 | 3.261 | (= MSE) |

From a summary of the sample means,

Wood Preservative Process

| | Mean Weight (pounds) | | |
|---|---|---|---|
| | Surface Stain (1) | Pressure Sealer (2) | Liquid Injection Treatment (3) |
| Treatment mean | 68.66 | 68.16 | 72.00 |

The calculations use Equation (11.7) for simultaneous 95% confidence interval estimates with $F_{0.05,2,13} = 3.81$.

for $\quad (68.66 - 68.16) \pm (\sqrt{(3 - 1)3.81} \cdot \sqrt{3.261(\frac{1}{6} + \frac{1}{6})}) = 0.50 \pm 2.88$

for $\quad (72.00 - 68.66) \pm (2.88) \qquad\qquad = 3.34 \pm 2.88$

for $\quad (72.00 - 68.16) \pm (2.88) \qquad\qquad = 3.84 \pm 2.88$

The 95% simultaneous confidence intervals for the difference in means are

for     $\mu_2 - \mu_1$:   $-2.38 < \mu_2 - \mu_1 < +3.38$

for     $\mu_3 - \mu_1$:   $+0.46 < \mu_3 - \mu_1 < +6.22$

for     $\mu_3 - \mu_2$:   $+0.96 < \mu_3 - \mu_2 < +6.72$

The difference in mean response for mean (2) and mean (1) can be either negative or positive because that interval has both negative and positive values. Thus it is possible that process 2 can provide either a lighter or a heavier plywood on the average than process 1. With experiment-wise 95% confidence we conclude there is *no* statistical difference for $\mu_2$ and $\mu_1$.

Since the intervals comparing $\mu_3$ (separately) with $\mu_1$ and $\mu_2$ are positive throughout, $\mu_3$ exceeds $\mu_1$, and $\mu_3$ exceeds $\mu_2$ with experiment-wise error rate 5%. This confirms the earlier statement (Example 11.4 in Section 11.5) that the liquid treatment (process 3) produces the heaviest plywood on average in these three wood preservative processes. These results are summarized as multiple comparisons.

| Wood Preservative Process | 2 | 1 | 3 |
|---|---|---|---|
| Mean weight of plywood | 68.16 | 68.66 | 72.00 |

The common underline for 1 and 2 indicates that, statistically, $\mu_1 = \mu_2$ for experiment-wise $[(1 - \alpha)100]\% = 95\%$ confidence. The disconnected line for process 3 shows that $\mu_3$ is significantly greater than $\mu_1$ and $\mu_3$ is significantly greater than $\mu_2$. For example, the 95% (experiment-wise) confidence interval on $\mu_3 - \mu_1$ indicates that $\mu_3$ can be as little as 0.46 pounds heavier than $\mu_1$ to as much as 6.22 pounds heavier than $\mu_1$.

---

The Scheffé comparisons are simplified for equal numbers of replicates. As before, for $n_1 = n_2 = \cdots = n_T$ all the confidence intervals in Equation (11.7) have the same bound of error ($\pm 2.88$ in the example). With unequal numbers in the treatment samples, the bound of error varies somewhat, depending on the two sample sizes for the treatment comparison. Otherwise the procedure is the same for the common experimental designs.

The problems that follow will provide you the opportunity to practice the techniques used in setting simultaneous confidence intervals for the differences in means. This requires, first, rejecting a hypothesis of equal means for treatments defined in one of the common experimental designs.

**Problems 11.7**

45. An important marketing problem for manufacturers is how to persuade intermediaries to buy, effectively merchandise, and sell their products. As products become more complex, with uncertainty about product attributes of performance, the buyers will look more toward personal communications with the manufacturer. The several channels available to the manufacturer to influence intermediaries include trade shows, journals, direct mail, and personal selling. Gregory Upah reports one such study from a telephone survey of 141 store managers of retail hardware stores in Illinois (*Journal of Business Research* 11 (1983): 107–126). One hypothesis was that these

retailers would judge personal selling by manufacturer representatives to be more important as a source of information when product complexity is high than when it is low. The design was a one-way design in high–low complexity with replicates. The use of repeated measures was intended to hold constant any factors relating to the uniqueness of each retail firm— for example, their number of sales employees (this is akin to blocking). See the incomplete ANOVA table.

a. State null and alternative hypotheses for a test of means for the complexity variable. Calculate an $F$-value and perform the test, $\alpha = 0.05$.

b. Is a Scheffé-type multiple-comparison approach necessary to determine individual treatment differences? Why or why not?

| Source | df | Mean Square |
|---|---|---|
| Complexity | 1 | 142.86 |
| Subjects | 140 | |
| Error | 139 | 3.72 |
| Total | 280 | |

| Treatment Levels | Low Complexity | High Complexity |
|---|---|---|
| Mean importance ratings | 3.73 | 5.18 |
| Visits by manufacturer salespeople | $n = 141$ | $n = 141$ |

46. The Pace Real Estate Appraisal firm employs three appraisers. Currently, the firm has one opening and two appraisers. The company evaluated a recent applicant for the position by having him (separately) appraise five properties and then comparing his values to the appraisals made by the firm's two current appraisers (see table).

| | Appraisals ($000) | | |
|---|---|---|---|
| Property | Applicant | Joe Pace | Joan Cox |
| 1 | 109.8 | 107.2 | 105.6 |
| 2 | 92.5 | 93.6 | 91.5 |
| 3 | 77.5 | 75.9 | 76.8 |
| 4 | 56.9 | 55.1 | 54.7 |
| 5 | 38.3 | 37.3 | 36.9 |

Consider this a randomized complete block experimental design with "property" as blocks.

a. Do the data provide sufficient evidence to indicate a significant difference, $\alpha = 0.05$, among the mean appraisals for the three appraisers? *Suggestion:* If you have the option available, consider obtaining the ANOVA by computer.

b. If part (a) shows it is appropriate, then perform Scheffé's procedure to test for pairwise differences.

c. Based on your findings, is the job applicant making appraisals that are, on average, statistically different from those made by the firm's two appraisers?

47. The State Department of Roads seeks a brand of paint that will produce the yellow center no-passing stripe most economically. Six paint brands are tested on five road segments representing various conditions of road wear (see table). The same equipment and workers are used to paint the stripes, and the response is the linear feet of stripe (road segment) covered by one gallon of paint. Consider this an RCB-designed experiment with six treatment levels (paint brands) and five block levels (road segments). The mean square error is 19.20.
    a. Test for a significant difference in the mean surface coverage for the six paint brands for $\alpha = 0.05$.
    b. Perform Scheffé's procedure to determine the significant pairwise differences using a 95% simultaneous confidence level. *Note:* This can require constructing up to $_6C_2 = 15$ confidence intervals, but a judicious choice of the order in testing can appreciably reduce this number.
    c. Based on the single consideration of surface coverage, what brand or brands are best?

| Paint Brand | 1 | 2 | 3 | 4 | 5 | 6 |
|---|---|---|---|---|---|---|
| Mean (feet) | 8.3 | 21.6 | 12.4 | 11.7 | 16.3 | 13.3 |

48. In Problem 40 all (two- and three-way) interactions were combined to form an adjusted residual (as described in Example 11.5) with 27 df. Use this, the associated MSE, and the treatment means in the table below to develop Scheffé's multiple comparisons for Piggly Wiggly pie shells. Use $\alpha = 0.05$. Note that $n_{cost} = n_{reg} = n_{red} = 12$.

| Treatment (price) | Treatment Means |
|---|---|
| Regular | 25.75 |
| Reduced | 32.75 |
| Cost | 43.67 |

49. Based on the results for Problem 41, on Mahatma rice, continue as is appropriate to
    a. establish 95% simultaneous confidence interval estimates on differences in mean sales for prices on
        (i) $\mu_{cost} - \mu_{red}$
        (ii) $\mu_{cost} - \mu_{reg}$
        (iii) $\mu_{red} - \mu_{reg}$
        Use $n_{cost} = n_{reg} = n_{red} = 12$.
    b. Continue to display the multiple comparisons in the table. Review the figure in Problem 41, which displays the two-way interaction for price

with display. In view of this significant interaction effect, how do you interpret the multiple comparisons requested here?

| Treatment (price) | Treatment Means |
|---|---|
| Regular | 28.17 |
| Reduced | 44.42 |
| Cost | 48.83 |

## SUMMARY

Making inferences about the equality of means from two, three, or more populations requires some specific assumptions. For samples from two normal distributions, we can use a $Z$-, a $t$-, or an $F$-statistic. For testing the equality of means for three or more normally distributed populations, only the $F$-statistic applies. The $F$-test requires that the populations have equal variances.*

The ANOVA format is commonly used to report the results of a variance partition. It also can help to organize the calculations for an $F$-variance ratio statistic. For a *completely randomized design*, the ANOVA identifies sums of squares and df for response variation sources. The calculations are based on additive parts:

$$\text{Total SS} = \text{SS treatments} + \text{SS error}$$

$$\text{Total df} = \text{Treatments df} + \text{Error df}$$

The treatment mean squares (MST = SS treatments/Treatments df) and mean squares for error (MSE = SS error/Error df) give independent estimates of the true error variance in response. If the mean square for differences among treatment means, MST, is small relative to MSE, then $F = \text{MST/MSE}$ will, on average, be a small number. Thus a low calculated $F$-value indicates that the samples come from the same normal (treatment) population. A large value for the sample $F$-ratio evidences unequal mean response for the treatment groups. Then we would reject a hypothesis of equal treatment means, and conclude that the treatment groups represent populations with different means. Calculations differ slightly for equal versus unequal size treatment groups.

Multiple-factor experimental designs use two or more treatment variables to define response variations. The SS error is reduced with each additional treatment variable if the treatment variable is a bonified source of variation. Similarly, in an RBD experiment each additional blocking variable will reduce the error sum of squares. These refinements can result in a more efficient design. Replicates of experimental units for every treatment-by-treatment combination increase the error df, which results in more information with which to estimate the true error variance. Replication for two-factor designs also allows us to measure the interaction effect of treatment variables. If a significant interaction exists, then it is difficult to interpret differences in treatment means.

A significant $F$-value can indicate that two or more treatment means are different. Multiple pairwise comparison procedures, like Scheffé's procedure, allow us to establish simultaneous confidence intervals, from which we can estimate *which* treatment means are statistically unequal and *by how much*.

Because they are extensive, most experimental design computations are done on computers. In practice, even the one-way ANOVA is usually developed on a computer. Numerous software packages are available for the ANOVA calculations on mainframe and personal computers. We briefly discuss the options available on several such packages,

---

* Section 10.4 has a test for equal variances for two populations. See [4] for testing on three or more populations.

including SPSS, BMDP, SAS, and Minitab. The choice of software depends largely on its availability on your computer system. Computer packages are generally quite dependable and substantially utilize the procedures that are described in this chapter.

Table 11.16 is a summary table of the key Chapter 11 formulas. Otherwise the formulas are clearly displayed throughout the chapter as Equations (11.1) through (11.7).

**TABLE 11.16**   Key Equations for Analysis of Variance and Experimental Designs

ANOVA **and** $F$**-test for completely randomized designs, CRD**

| Source of Variation | df | Sum of Squares | Mean Squares |
|---|---|---|---|
| Among treatments | $T - 1$ | SST | $\mathrm{MST} = \dfrac{\mathrm{SST}}{T-1}$ |
| Error (within treatments) | $\left(\sum\limits^{T} n_i\right) - T$ | SSE | $\mathrm{MSE} = \dfrac{\mathrm{SSE}}{\left(\sum\limits^{T} n_i\right) - T}$    **(11.4)** |

| | Total $\left(\sum\limits^{T} n_i\right) - 1$ | TSS | |

where   $\mathrm{TSS} = \sum\limits_{i}^{T}\sum\limits_{j}^{n_i} (Y_{ij} - \bar{Y}..)^2$    $\bar{Y}.. = \dfrac{\sum\limits_{j}\sum\limits_{i} Y_{ij}}{n}$

$\mathrm{SST} = \sum\limits_{i}^{T} n_i(\bar{Y}_{i.} - \bar{Y}..)^2$    $\mathrm{SSE} = \mathrm{TSS} - \mathrm{SST}$    and    $n = \sum\limits_{i} n_j$

$H_0: \mu_1 = \mu_2 = \cdots = \mu_T$
$H_A$: At least two means are not equal.

Calculated $F = \dfrac{\mathrm{MST}}{\mathrm{MSE}}$    with    $\mathrm{df} = T - 1, \left(\sum\limits^{T} n_i\right) - T$

ANOVA **and** $F$**-test for randomized complete block design, RBD**

| Source of Variation | df | Sum of Squares | Mean Squares |
|---|---|---|---|
| Treatments | $T - 1$ | SST | $\mathrm{MST} = \dfrac{\mathrm{SST}}{T-1}$ |
| Blocks | $B - 1$ | SSB | $\mathrm{MSB} = \dfrac{\mathrm{SSB}}{B-1}$    **(11.5)** |
| Error | $(B-1)(T-1)$ | SSE | $\mathrm{MSE} = \dfrac{\mathrm{SSE}}{(B-1)(T-1)}$ |

| | Total   $n-1$ | TSS | |

$n = BT$

where   $\mathrm{TSS} = \sum\limits_{j}\sum\limits_{i} (Y_{ij} - \bar{Y}..)^2$

$\mathrm{SST} = \sum\limits_{i} B(\bar{Y}_{i.} - \bar{Y}..)^2$

$\mathrm{SSB} = \sum\limits_{j} T(\bar{Y}_{.j} - \bar{Y}..)^2$

$\mathrm{SSE} = \mathrm{TSS} - \mathrm{SST} - \mathrm{SSB}$

*(continued)*

**TABLE 11.16**   (*continued*)

$H_0: \mu_1 = \mu_2 = \cdots = \mu_T$
$H_A$: At least two means are not equal.

$$\text{Calculated } F = \frac{\text{MST}}{\text{MSE}} \quad \text{with df} = T - 1, [(B-1)(T-1)] \qquad (11.6)$$

ANOVA **and** *F*-**test for a two-factor factorial experiment with equal numbers of replicates**

| Source of Variation | df* | Sum of Squares | Mean Squares |
|---|---|---|---|
| Treatment 1 | $A - 1$ | SSA | $\text{MSA} = \text{SSA}/(A-1)$ |
| Treatment 2 | $B - 1$ | SSB | $\text{MSB} = \text{SSB}/(B-1)$ |
| Interaction | $(A-1)(B-1)$ | SSAB | $\text{MSAB} = \text{SSAB}/(A-1)(B-1)$ |
| Explained | $AB - 1$ | SST | $\text{MST} = \text{SST}/(AB-1)$ |
| Error | $AB(n-1)$ | SSE | $\text{MSE} = \text{SSE}/AB(n-1)$ |
| Total (corrected) | $ABn - 1$ | TSS | |

where

$$\text{TSS} = \sum_i \sum_j \sum_k (Y_{ijk} - \bar{Y}_{...})^2 = \left( \sum_i \sum_j \sum_k Y_{ijk}^2 \right)^2 - \text{CT}$$

$$\text{SST} = \sum_i \sum_j (\bar{Y}_{ij\cdot} - \bar{Y}_{...})^2 = \frac{\left( \sum_i \sum_j Y_{ij\cdot}^2 \right)^2}{n} - \text{CT}$$

$$\text{SSA} = \sum_i (\bar{Y}_{i\cdot\cdot} - \bar{Y}_{...})^2 = \frac{\left( \sum_i Y_{i\cdot\cdot}^2 \right)}{Bn} - \text{CT}$$

$$\text{SSB} = \sum_j (\bar{Y}_{\cdot j\cdot} - \bar{Y}_{...})^2 = \frac{\left( \sum_j Y_{\cdot j\cdot}^2 \right)}{An} - \text{CT}$$

$$\text{SSAB} = \text{SST} - \text{SSA} - \text{SSB} \quad \text{CT} = \frac{\left( \sum_i \sum_j \sum_k Y_{ijk} \right)^2}{ABn}$$

$$\text{SSE} = \text{TSS} - \text{SST}$$

The *F*-ratio test for interaction (performed first) or treatment 1 or treatment 2 uses the appropriate mean square divided by MSE. For interaction not significant, use

$$\text{MSE}(2) = \frac{\text{SSAB} + \text{SSE}}{(A-1)(B-1) + AB(n-1)}$$

for tests on the treatments.

**Scheffé procedure for estimating simultaneous confidence intervals**

$$(\bar{Y}_1 - \bar{Y}_2) \pm \left( \sqrt{(T-1)F_\alpha} \cdot \sqrt{\text{MSE}\left( \frac{1}{n_1} + \frac{1}{n_2} \right)} \right)$$

$$(\bar{Y}_1 - \bar{Y}_3) \pm \left( \sqrt{(T-1)F_\alpha} \cdot \sqrt{\text{MSE}\left( \frac{1}{n_1} + \frac{1}{n_3} \right)} \right) \qquad (11.7)$$

$$\vdots$$

$$(\bar{Y}_{T-1} - \bar{Y}_T) \pm \left( \sqrt{(T-1)F_\alpha} \cdot \sqrt{\text{MSE}\left( \frac{1}{n_{T-1}} + \frac{1}{n_T} \right)} \right)$$

where MSE comes from the ANOVA for the appropriate design and has $\text{df}_2$, with $F_\alpha = FX,\ _{T-1, \text{df}_2}$.

* Requires equal replicates.

**CHAPTER 11
PROBLEMS**

50. A designed experiment used a completely randomized design with the null hypothesis $H_0$: $\mu_1 = \mu_2 = \mu_3$ using the 0.05 $\alpha$-level. If there were eight observations for treatment level 1, nine for level 2, and six for level 3, the calculated $F$ should be compared to what numerical (test) value?

51. A test for the equality of means is to be conducted at the 1% level. Samples of 20 each are taken from three populations with, respectively, $s_1^2 = 48$, $s_2^2 = 42$, and $s_3^2 = 38$. Assume a CRD test.
    a. State the appropriate hypothesis (you may assume equal variances).
    b. Indicate the test degrees of freedom.
    c. Provide the test value.
    d. If calculated $F = 3.69$, would it be appropriate to proceed on to estimating the true pairwise differences for the (3) population means? Explain.

52. Data were collected on the mean time before failure on four brands of Watt stereo receivers. Perform the test for equal means given this ANOVA. Indicate the design type (CRD, RCB, or other) along with your conclusions for $\alpha = 0.05$. Consider the error df. What suggestions could you make for a more precise experimental design?

| Source | df | SS | MS |
|---|---|---|---|
| Brand (treatment) | 3 | 36 | |
| Blocks | 2 | 36 | |
| Error | 6 | 24 | |
| Total | 11 | | |

53. A random sample of five batteries was selected from each of three different brands. The lifetimes in hours ($X$) were recorded for each individual battery (see table). Determine whether the mean lifetimes of the different brands are statistically different at the 0.05 level. Report your results in an ANOVA table.

| Brand A | | Brand B | | Brand C | |
|---|---|---|---|---|---|
| $X$ | $(X - \bar{X}_A)^2$ | $X$ | $(X - \bar{X}_B)^2$ | $X$ | $(X - \bar{X}_C)^2$ |
| 40 | 0 | 50 | 0 | 60 | 0 |
| 30 | 100 | 50 | 0 | 60 | 0 |
| 40 | 0 | 50 | 0 | 60 | 0 |
| 60 | 400 | 60 | 100 | 70 | 100 |
| 30 | 100 | 40 | 100 | 50 | 100 |
| 200 | 600 | 250 | 200 | 300 | 200 |
| | $\bar{X}_1 = 40$ | | $\bar{X}_2 = 50$ | | $\bar{X}_3 = 60$ |

54. An ANOVA for a CRD experiment with three treatment groups has been started in the table below. Complete the table and test $H_0: \mu_1 = \mu_2 = \mu_3$ assuming equal sample sizes, $n_1 = n_2 = n_3 = 9$, normal distributions, and equal variances. Use $\alpha = 0.05$.

| Source | df | Sum of Squares | Mean Squares | $F$ |
|---|---|---|---|---|
| Treatment | | 198 | | |
| Error | | | | |
| Total | | 392 | | |

55. Consider the listed results from two different production methods used in the processing of rubber. The measurements are for elastic strength and they come from normal distributions.

| | Treatment | |
|---|---|---|
| | $A$ | $B$ |
| Units in sample | 64 | 100 |
| Sample mean | 1.70 | 1.81 |
| Standard deviation | 0.50 | 0.70 |

a. Test $H_0: \sigma_1^2 = \sigma_2^2$ against $H_A: \sigma_1^2 \neq \sigma_2^2$ with $\alpha = 0.05$.
b. Do the two methods produce different products with respect to elastic strength? Test $H_0: \mu_A = \mu_B$ with $\alpha = 0.05$. (*Note:* You should answer this question only if your conclusion to part (a) was "Do not reject $H_0$.")

56. Random samples of three brands of gasoline gave the performance scores shown in the table. Brands were randomly assigned to cars, and the other conditions of the experiment were maintained as uniformly as possible. Assume normal distributions and constant variance.
a. Test for equal mean performance, using $\alpha = 0.05$.
b. Set a 95% confidence interval for the mileage for brand C (only). Use $s_C = 2.16$.

| Brand | Mileage to Nearest Mile | | | | Total |
|-------|----|----|----|----|-------|
| A | 27 | 30 | 25 | 26 | 108 |
| B | 21 | 20 | 22 | 21 | 84 |
| C | 27 | 31 | 30 | 32 | 120 |

57. Productivity measures were compared for a random selection of workers in a Kansas City factory. Group 1 had no formal training; group 2 had one month of on-the-job training; and group 3 had one month of technical-school training. Test for equal achievement using $\alpha = 0.025$ and total SS = 14,820.0. Assume normal distributions. Use 77.5 as the overall mean. (*Suggestion:* Begin by establishing an ANOVA table.)

| $i$ | Group | Mean Productivity Score | $n_i$ |
|-----|-------|-------------------------|-------|
| 1 | No formal training | 72.4 | 16 |
| 2 | One month on the job | 76.8 | 15 |
| 3 | One month of technical | 84.2 | 14 |

58. Is purchasing power comparable for households in five midwestern states? The data in the table summarize the findings from a survey, where age was used as a block variable. This recognizes that various age groups have different levels of purchasing power index (the response). Given total SS = 3556.91 and age (block) SS = 3516.67, test for a significant treatment effect across states. Use $\alpha = 0.025$.

| Age Group | State | | | | |
|-----------|----------|--------|-----------|----------|----------|
|           | Illinois | Iowa   | Minnesota | Missouri | Nebraska |
| 19–35 | 77.9 | 76.3 | 75.3 | 78.1 | 75.4 |
| 36–45 | 92.1 | 88.9 | 86.3 | 88.5 | 87.2 |
| 46–64 | 109.3 | 107.1 | 106.4 | 108.6 | 105.2 |
| 65+ | 73.6 | 73.8 | 74.5 | 73.6 | 72.3 |

59. Assuming that there is a significant treatment effect for states, go on from Problem 58 to determine an ordering from low to high for purchasing power across the five states. That is, perform Scheffé's multiple comparisons at 97.5% simultaneous confidence. Otherwise, indicate that there is no need for the estimation of mean pairwise differences.

60. The table below is a partially completed ANOVA and $F$-ratio test on weekly sales for an inexpensive calculator. The treatment variable is retail outlets—drugstores, discount houses, department stores, and office supply stores. The outlets are all chain stores in a large metropolitan area. Because retail sales can depend on store location, six shopping centers were chosen for the experiment. One retail outlet of each type was used in each store location. Complete the table and make the $F$-test for equal treatment means $\alpha = 0.025$. State practical recommendations.

| Source | df | Sum of Squares | Mean Squares | F |
|---|---|---|---|---|
| Outlets (treatment) | | | | |
| Shopping center (blocks) | | 447.5 | | |
| Residual | | 148.5 | | |
| Total | 23 | 896.0 | — | |

61. The mean weekly sales for the four outlets in Problem 60 are (a) $580, (b) $612, (c) $604, and (d) $599. Go on to determine *which* outlet types have produced statistically higher dollar sales. You will need to use some information from Problem 60. Use 97.5% confidence.

Problems 62 through 64 reintroduce data from Wilkinson, Mason, and Paksoy used in Problems 38–41.* The response measure is unit sales per period for a grocery display, where a period is from Wednesday noon until Saturday 9 P.M. Other definitions appear before the Section 11.6 problems. The problems below relate to the product White House apple juice. The table summarizes the ANOVA on unit sales (the response) for White House apple juice. Note that all interactions were absorbed into an adjusted residual because all of the interactions are not significant for $\alpha = 0.05$.

| Source of Variation | df | Sum of Squares | Mean Squares |
|---|---|---|---|
| Explained | 5 | 7,086.972 | 1,417.394 |
| Price | 2 | 2,842.722 | 1,421.361 |
| Display | 2 | 3,982.889 | 1,991.444 |
| Advertising | 1 | 261.361 | 261.361 |
| Adjusted residual | 30 | 2,773.999 | 92.467 |
| Total | 35 | 9,860.971 | 281.742 |

**Price**

| Treatment | Treatment Means |
|---|---|
| Regular | 28.17 |
| Reduced | 44.42 |
| Cost | 48.83 |

**Display**

| Treatment | Treatment Means |
|---|---|
| Regular | 30.75 |
| Expanded | 35.58 |
| Special | 55.08 |

**Advertising**

| Treatment | Treatment Means |
|---|---|
| No advertising | 37.78 |
| Advertising | 43.17 |

62. Test at $\alpha = 0.05$ for significant (a) price, (b) display, and (c) advertising treatment effects. Show calculated F-values and state test conclusions. If it is appropriate for

* Wilkinson, J. B., J. Barry Mason, and Christie H. Paksoy. "Assessing the Impact of Short-Term Supermarket Strategy Variables." *Journal of Marketing Research* 19 (Feb. 1982): 72–86.

your tests, then use multiple-comparison procedures to determine statistically significant differences in units sold attributable to the "Price" treatment conditions—regular, reduced, or cost. Use $\alpha = 0.05$ and indicate the best pricing option.

63. Again, use multiple-comparison procedures to determine any significant differences in units sold attributable to the "Display" treatment conditions—regular, expanded, special. Use $\alpha = 0.05$ and indicate if there is a single best choice for display.

64. A third treatment variable—"Advertising"—has only two levels, so you can perform an $F$-test for an individual comparison from the ANOVA. Test for equal mean units sold under "Advertising" versus under "No advertising." Use $\alpha = 0.05$. Apply the results from this and the two preceding problems to suggest a best combination of price, display, and advertising to achieve the highest unit sales for White House apple juice.

65. Ahmed Abdel-Halim investigated 229 supervisor–subordinate pairs (eu's) in "Effects of Task and Personality Characteristics on Subordinate Responses to Participatory Decision Making—PDM," *Academy of Management Journal* 3 (1983): 477–84. The sample was employees from a large retail drug company in the midwest. One response measure, job performance, was a sum on five ratings of performance made by each supervisor. The treatments—(1) PDM, (2) need for independence, and (3) task repetitiveness—were dichotomized as above or below the median for each subject. The test uses a replicated $2 \times 2 \times 2$ factorial designed experiment. The ANOVA and $F$-tests are summarized.

| Source of Variation | df | SS | Calculated $F$ |
|---|---|---|---|
| Participative decision making (PDM) | 1 | 153.35 | 12.29 |
| Need for independence (NI) | 1 | 12.38 | <1 |
| Task repetitiveness (TR) | 1 | 5.40 | <1 |
| PDM × NI | 1 | 6.43 | <1 |
| PDM × TR | 1 | 1.64 | <1 |
| NI × TR | 1 | 5.35 | <1 |
| PDM × NI × TR | 1 | 50.16 | 4.02 |
| Residual (error) | 221 | 2758.77 | |
| Total | 228 | | |

a. Check the calculations for the first and last calculated $F$ in the table. Then compare these to the appropriate tabled $F$-value for $\alpha = 0.05$.

b. Why or why not interpret a significant (main) effect for PDM? That is, can we conclude that a significant difference exists in job performance (in this firm) for supervisor–subordinate pairs with higher versus with lower PDM scores?

66. Personnel records from four large utilities were researched in a study concerned with the number of years of service prior to retirement. A random sample of 10 retiree records from a phone company showed mean service of 16 years with standard deviation 7 years. Comparable statistics for 10 electric power company retiree records were mean 11 years and standard deviation 6 years; for a random sample of 9 utility gas retirees the mean was 18 years, with standard deviation 8 years; and for water service utility, 13 retirees had a mean of 22 years and standard deviation of 10 years. Test whether the difference between means is significant for $\alpha = 0.05$. Recall that, for unequal numbers,

$$s_w^2 = \frac{\sum^{T} (n_i - 1)s_i^2}{\left( \sum^{T} n_i - T \right)} \quad (= \text{MSE})$$

STATISTICS IN
ACTION:
CHALLENGE
PROBLEMS

67. Martin et al. have analyzed the 16 possible cell combinations in a $2^4$ factorial (4 variables, each with two levels) design in which the authors randomly assigned every cell 125 subjects, for a total of 2000 participants in a university mail survey. Four response inducement techniques were compared for their relative contributions to response rate improvements and to the per-response cost of each (not discussed here). The four inducements were

| Factor 1 | Prenotification | $P1$ = Prenotification letter used |
| | | $P0$ = Prenotification letter not used |
| Factor 2 | Follow-up | $F1$ = Follow-up letter used |
| | | $F2$ = Follow-up letter not used |
| Factor 3 | Personalization | $Pe1$ = Name used in letter |
| | | $Pe0$ = Letter addressed to "Occupant" |
| Factor 4 | Return postage | $R1$ = Stamped reply envelope |
| | | $R0$ = Business envelope, no stamp |

A summary table of the research results appears below:

Results of an Analysis of Variance of the Four Factors That Affect Mail Survey Response Rates

| Factor Main Effects | df | ANOVA Sum of Squares | Mean Square | $F$-Value |
|---|---|---|---|---|
| $P1$ | 1 | 12.64 | 12.64 | 73.69[a] |
| $R1$ | 1 | 0.01 | 0.01 | 0.07 |
| $Pe1$ | 1 | 0.22 | 0.22 | 1.29 |
| $F1$ | 1 | 0.88 | 0.88 | 4.9[a] |
| Interactive effects between sets of factors taken two at a time | | | | |
| $P1,R1$ | 1 | 0.06 | 0.06 | 0.35 |
| $P1,Pe1$ | 1 | 0.68 | 0.68 | 3.99[a] |
| $R1,Pe1$ | 1 | 0.01 | 0.01 | 0.07 |
| $P1,F1$ | 1 | 0.31 | 0.31 | 1.82 |
| $R1,F1$ | 1 | 0.02 | 0.02 | 0.14 |
| $Pe1,F1$ | 1 | 0.01 | 0.01 | 0.07 |
| Interactive effects between sets of factors taken three at a time | | | | |
| $P1,R1,Pe1$ | 1 | 0.18 | 0.18 | 1.05 |
| $P1,R1,F1$ | 1 | 0.31 | 0.31 | 1.82 |
| $P1,Pe1,F1$ | 1 | 0.18 | 0.18 | 1.05 |
| $R1,Pe1,F1$ | 1 | 0.00 | 0.00 | 0.00 |
| Interactive effects between all four factors | | 0.00 | 0.00 | 0.00 |
| Total model Response rate = $f(P1,F1,Pe1,F1)$ | 15 | 15.50 | 1.033 | 6.02[a] |
| Error | 1984 | 340.32 | 0.1715 | |
| Corrected total | 1999 | 355.82 | | |

[a] $p < 0.001$.

The authors include these points in their summary of the findings in the table above.

1. Of the four factors, prenotification and follow-up displayed significant main effects.

2. Personalization and stamped reply envelopes, while important, did not exert significant treatment effects.

3. Of the eleven interaction effects only one, $P1 \cdot Pe1$, was significant.

4. The results suggest that prenotification and follow-up are the preferred inducements, aside from any consideration of costs.

5. Personalization was also advantageous because by a significant interaction, when both personalization and prenotification were employed, the overall response rate was higher than when only the latter inducement was used.

SOURCE: Martin, Warren S., W. Jack Duncan, Thomas L. Powers, and Jesse C. Sawyer. "Costs and Benefits of Selected Response Inducement Techniques in Mail Survey Research." *Journal of Business Research* 19 (1989): 67–79.

**Questions**   For each of the statements 1 through 5, indicate (a) the sources of variation being described, (b) the null hypotheses, (c) the calculated *F*-value, (d) an estimated *p*-value, and (e) whether the statement is based on a test of hypothesis or other. Discuss whether you would pool any of the interaction sums of squares and df with the "Error."

68. How are individuals influenced by the physical attractiveness of the people displayed in printed advertising? In this research, communicator attractiveness is operationally defined as the degree to which a person's face is pleasing to observe, as determined through a consensus of judges. This study used college students, like you, as the "receivers," or judges, and numerous ad mockups to obtain the ratings. The statistical summary is presented here from Table 2, page 239, of the journal article.

*final*

Analysis of Variance Using Trustworthiness, Expertise, and Liking Scores as Criteria

| Source of Variation | df | Trustworthiness (dependent var.) MS | F | PrF | Expertise MS | F | PrF | Liking MS | F | PrF |
|---|---|---|---|---|---|---|---|---|---|---|
| Receiver gender (A) | 1 | 0.11 | 0.08 | 0.785 | 0.90 | 0.41 | 0.521 | 0.70 | 1.08 | 0.299 |
| Communicator gender (B) | 1 | 2.45 | 1.63 | 0.203 | 0.70 | 0.32 | 0.571 | 0.70 | 1.08 | 0.299 |
| Communicator physical attractiveness (C) | 3 | 14.28 | 9.49 | 0.001 | 6.60 | 3.06 | 0.028 | 13.89 | 21.41 | 0.001 |
| A × B | 1 | 0.61 | 0.41 | 0.524 | 3.40 | 1.56 | 0.213 | 0.08 | 0.12 | 0.729 |
| A × C | 3 | 2.55 | 1.70 | 0.169 | 3.89 | 1.78 | 0.151 | 0.23 | 0.35 | 0.788 |
| B × C | 3 | 2.33 | 1.48 | 0.219 | 0.64 | 0.29 | 0.832 | 0.06 | 0.10 | 0.963 |
| A × B × C | 3 | 0.40 | 0.25 | 0.860 | 3.54 | 1.62 | 0.185 | 1.17 | 1.80 | 0.146 |
| Explained | 15 | 4.10 | 2.72 | 0.001 | 3.28 | 1.50 | 0.102 | 3.17 | 4.89 | 0.001 |
| Residual | 304 | 1.51 | | | 2.18 | | | 0.65 | | |
| Total | 319 | 1.63 | | | 2.24 | | | 0.77 | | |

**Cell Means of Main Effects for Physical Attractiveness: Trustworthiness, Expertise, and Liking Scores as Criterion**
Communicator Physical Attractiveness

| | No-Photo | Low | Moderate | High |
|---|---|---|---|---|
| Trustworthiness | 3.84 | 3.31 | 3.71 | 4.34 |
| Expertise | 3.66 | 3.20 | 3.45 | 3.88 |
| Liking | 4.06 | 3.66 | 4.16 | 4.65 |

The experimental design was a $2 \cdot 2 \cdot 4$ factorial design with three treatments: (1) sex of the receiver, (2) sex of the communicator, and (3) physical attrativeness level of the communicator. The three response variables were 7-point scaled measures assessing the receivers' evaluations of the communicator's (1) trustworthiness and (2) expertise, and (3) the receivers' liking for the communicator if they were to meet. The data were responses, randomly selected to achieve equal cell sizes of 20 subjects per "experimental condition," or 320 responses for 20 receivers in each of 16 experimental conditions (treatments combinations).

SOURCE: Patzer, Gordon. "Source Credibility as a Function of Communicator Physical Attractiveness." *Journal of Business Research* 11 (1983): 229–241.

Questions    Use data from the table above, *p*-values, and the ANOVA procedures for testing in a factorial design with the dependent variable Trustworthiness:
a. to test for significant interaction effects; then, if appropriate,
b. to test for significant treatment effects; then, if appropriate,
c. to test, using multiple-comparison procedures, for mean differences in perceived trustworthiness for No-photo (as control), Low, Moderate, or High level of communicator physical attractiveness.
Give one or two statements that summarize the results of your tests. Note that the multiple comparisons have equal group sizes.

69. Use data from the table in Problem 68, *p*-values, and the ANOVA procedures for testing in a factorial design with the dependent variable Expertise:
a. to test for significant interaction effects; then, if appropriate,
b. to test for significant treatment effects; then, if appropriate,
c. to test, using multiple-comparison procedures, for mean differences in perceived expertise for No-photo (as control), Low, Moderate, or High level of communicator physical attractiveness.
Give one or two statements that summarize the results of your tests.

70. Use data from the table in Problem 68, *p*-values, and the ANOVA procedures for testing in a factorial design with the dependent variable Liking:
a. to test for significant interaction effects; then, if appropriate,
b. to test for significant treatment effects; then, if appropriate,
c. to test, using multiple-comparison procedures, for mean differences in perceived expertise for Non-photo (as control), Low, Moderate, or High level of communicator physical attractiveness.
Give one or two statements that summarize the results of your tests.

---

## BUSINESS CASE

*Setting a Real Estate Profile*

Tony Tyrone's sales staff needed a boost. Business had been down for the past two months, with only four new listings and six sales. As president of Real Estate One (REO) Sales, the pressure was on Tony to get business moving. What could he do? He'd thought about it almost constantly for the past few days. Finally, he called in the Company Vice President and Comptroller, Carla Hines.

**Mr. T.**    Carla, you know we've had a real selling slump for the past two months. The staff is really down. We've got to do something.

**Ms. H.**    You're right, Tony. In fact, I am afraid that if we don't see a turnaround in business this month that we will lose some of our best salespeople. Then we are really in trouble! Do you have any ideas?

**Mr. T.**    Yes, I do. You know that we've had our best results selling in the prestige listings market. If we can associate REO Sales with higher-priced homes, this would invite more expensive listings. They could take fewer listings and still have a good business. What do you think?

**Ms. H.**    (Reflecting) We'd be taking a big risk. If we specialize, that will discourage old friends with lower-priced homes from taking listings with REO. But if you are serious, we'll need to be able to back up any statements we make. Can REO Sales honestly claim to handle higher listings, and are there enough higher-priced listings available for us to specialize?

**Mr. T.**    What you're saying is, Can we afford to specialize in an already limited and highly competitive market? Well, I don't know. Maybe we should get some data as an information base before deciding. What do you think about the idea?

**Ms. H.**    Well, obviously we've got to do something, and soon. Your idea is worth investigating. I'll have Jason Worth get numbers for the last three months from the Board of Realtor's Multiple Listing sheets. Your daughter Marcia is at the university. Do you suppose her marketing training would provide her with enough statistics to make a fair test?

**Mr. T.**    We'll see. I'll check with Marcia. If she can't handle it, then maybe one of her professors will! Go ahead and get the data.

Marcia was able to provide a statistical analysis based on Carla Hine's multiple listings sales numbers. These were summarized by the listing agency (below). These listings seemed to satisfy a random sample of asking prices from the past year. Marcia suggested an $F$-ratio test to compare the average asking price for REO listings against price listings for the other realtors in the city. The standard for comparison was the price variation *within* real estate firms. After making a table of means and sample variances (see page 504) Marcia was equipped to perform the tests. Her first test confirmed equal variances. Next, her calculations for an $F$-ratio test of $H_0$: $\mu_C = \mu_T = \mu_V$ resulted in a

$$\text{Calculated } F = (495.85/3)/(483.06/27) = 9.24$$

This was significant for $\alpha = 0.05$ (tabular $F_{0.05, 3, 27} = 2.96$).

Summary of Multiple Listing Real Estate Sales

| Clayton Realtors | | Taylor Associates | | Village Realty | | Real Estate One (REO) Sales | |
|---|---|---|---|---|---|---|---|
| *Multiple Listing Number* | *Asking Price* | *Multiple Listing Number* | *Asking Price* | *Multiple Listing Number* | *Asking Price* | *Multiple Listing Number* | *Asking Price* |
| 14610 | $85,400 | 14606 | $101,500 | 14607 | $94,000 | 14608 | $105,000 |
| 14613 | 91,900 | 14612 | 94,900 | 14611 | 97,500 | 14609 | 105,000 |
| 14615 | 93,500 | 14616 | 98,500 | 14617 | 91,900 | 14614 | 99,500 |
| 14619 | 87,000 | 14621 | 99,500 | 14618 | 95,500 | 14620 | 98,200 |
| 14624 | 85,900 | 14623 | 91,000 | 14622 | 98,900 | 14625 | 100,000 |
| 14631 | 95,500 | 14627 | 94,500 | 14626 | 88,900 | 14630 | 95,500 |
| 14635 | 94,900 | 14632 | 98,900 | 14628 | 101,500 | 14633 | 108,000 |
| | | | | 14629 | 88,900 | 14636 | 104,100 |
| | | | | 14634 | 98,900 | | |

Table of Means and Sample Variances

| Agency | Clayton Realtors | Taylor Associates | Village Realty | Real Estate One (REO) |
|---|---|---|---|---|
| Mean asking price    ($ thousands) | 90.59 | 96.97 | 95.11 | 101.91 |
| Sample variance    ($ millions$^2$) | 19.11 | 13.17 | 20.51 | 17.90 |
| (New) listings | 7 | 7 | 9 | 8 |

Marcia then tested REO's mean price for new listings against each competitor by Scheffé's test, with all prices in thousands of dollars ($000).

| | Clayton | Village | Taylor | REO |
|---|---|---|---|---|
| 90.59($000) | | 95.11 | 96.97 | 101.91 |

REO had the highest mean listing price—statistically higher than all competitors except Taylor Associates! Mr. Tyrone and Ms. Hines were going over the summary sheets with Marcia.

**Marcia**    You see, Dad, where the line is broken at Taylor Associates?

**Mr. T.**    Yes. What does that mean?

**Marcia**    The test shows that your company has higher listings than all *but* Taylor Associates.

**Ms. H.**    That makes sense. We regularly compete with Taylor for the better listings in town, but does this mean that we can't claim prestige listings?

**Mr. T.**    No, but we'd better acknowledge the statistics on our advertising. We'll simply use an indefinite article—*a*—instead of *the* in our ads. REO Sales is *a* prestige listings real estate company. We can get the point across just as well.

Ms. Hines and Mr. Tyrone discussed the matter further. Considering their consistent high price plus the fact that REO was getting over 25% of all new listings, they concluded that they should initiate a "higher price" advertising slogan. Following a brief presentation, the members of the Board of Associates for REO Sales concurred and the "higher price" approach was adopted. The result was increased sales and an improved position in the real estate market; and Marcia was on her way to becoming a practicing business analyst!

**Business Case Questions**

71. Use the mean asking prices, sample variances, and listings counts to establish a one-way ANOVA for the variance analysis of real estate companies on home sales price.

72. Calculate an $F$-value and test for equal asking prices (the response) for the four real estate companies (treatment levels). Use $\alpha = 0.05$ for your test and compare with the given results.

73. Validate the Scheffé procedure summary by computing the $_4C_2 = 6$ pairwise simultaneous confidence intervals using a 95% confidence level. Again, compare your results with those displayed in the application.

---

**COMPUTER APPLICATION**

Statistical Software for Designed Experiments with an Example

The application displays the SPSS ANOVA procedure for the two-factor factorial designed experiment in Example 11.5. The response measure is the weights for plywood sheets, each treated by random selection with one of the nine combinations of three (3) wood preservative processes by three (3) finishes for a $3 \cdot 3$ two-factor factorial experiment. The hand calculations and interpretations appear in Example 11.5.

```
 1 NUMBERED
 2 TITLE SPSS ANALYSIS FOR PROBLEM 11.5                    I
 3 DATA LIST    FILE = INFILE/
 4              WEIGHT 1-2 PRESERV 4 FINISH 6
 5 SET BLANKS = -1
 6 LIST  VARIABLES = WEIGHT PRESERV FINISH/
 7           FORMAT = NUMBERED/CASES FROM 1 TO 18     2
 8 ANOVA VARIABLES = WEIGHT BY FINISH, PRESERV(1,3)/
 9           MAXORDERS = 2/STATISTICS = MEAN          3
10 ANOVA VARIABLES = WEIGHT BY FINISH, PRESERV(1,3)/
11           MAXORDERS = NONE/STATISTICS = MEAN
12 FINISH
```

A few brief comments will describe some of the key elements of this SPSS setup.

I The appearance of the data input is displayed in the variable LIST output, following. WEIGHT, the response variable, is in positions 1–2, and so on. Observe that there are two replicates for each combination of PRESERVative by FINISH. For example,

65 1 1     and     68 1 1

are two replicates for PRESERVative at level 1 and for FINISH at level 1, respectively.

2 The first ANOVA is for the response—WEIGHT—on PRESERVative and FINISH. MAXORDERS = 2 allows a two-variable interaction as the highest-order interaction. This is a *first-order interaction* for PRESERVative with FINISH. The levels of the factors are required to be integer values; Here both are 1-2-3. Otherwise, we would specify separate ranges for PRESERVative and FINISH. The STATISTICS = MEAN is an option that gives a table of cell means of WEIGHT within the combined categories of PRESERVative and FINISH.

3 The second ANOVA differs from the first only in that the interaction term is excluded by MAXORDERS = NONE. Then the interaction sums of squares and degrees of freedom are combined, or pooled, into the RESIDUAL (error) sum of squares and df, respectively.

The two ANOVA tables and the variable LIST follow. See Example 11.5 for a summary of the results.

Variable LIST

|    | WEIGHT | PRESERV | FINISH |
|----|--------|---------|--------|
| 1  | 65     | 1       | 1      |
| 2  | 68     | 1       | 1      |
| 3  | 68     | 1       | 2      |
| 4  | 70     | 1       | 2      |
| 5  | 70     | 1       | 3      |
| 6  | 71     | 1       | 3      |
| 7  | 64     | 2       | 1      |
| 8  | 66     | 2       | 1      |
| 9  | 68     | 2       | 2      |
| 10 | 65     | 2       | 2      |
| 11 | 72     | 2       | 3      |
| 12 | 74     | 2       | 3      |
| 13 | 70     | 3       | 1      |
| 14 | 68     | 3       | 1      |
| 15 | 74     | 3       | 2      |
| 16 | 70     | 3       | 2      |
| 17 | 76     | 3       | 3      |
| 18 | 74     | 3       | 3      |

NUMBER OF CASES READ: 18

Table of CELL MEANS

| | | PRESERV 1 | PRESERV 2 | PRESERV 3 | |
|---|---|---|---|---|---|
| FINISH | 1 | 66.50 ( 2) | 65.00 ( 2) | 69.00 ( 2) | 66.83 ( 6) |
|        | 2 | 69.00 ( 2) | 66.50 ( 2) | 72.00 ( 2) | 69.17 ( 6) |
|        | 3 | 70.50 ( 2) | 73.00 ( 2) | 75.00 ( 2) | 72.83 ( 6) |
|        |   | 68.67 ( 6) | 68.17 ( 6) | 72.00 ( 6) | 69.61 ( 18) |

ANOVA, Including a Two-Way (First-Order) Interaction Source of Variation

| SOURCE OF VARIATION | SUM OF SQUARES | DF | MEAN SQUARE | F | SIG OF F |
|---|---|---|---|---|---|
| MAIN EFFECTS | 161.889 | 4 | 40.472 | 13.245 | .001 |
|    FINISH | 109.778 | 2 | 54.889 | 17.964 | .001 |
|    PRESERV | 52.111 | 2 | 26.056 | 8.527 | .008 |
| 2-WAY INTERACTIONS | | | | | |
|    FINISH   PRESERV | 14.889 | 4 | 3.722 | 1.218 | .368 |
| EXPLAINED | 176.778 | 8 | 22.097 | 7.232 | .004 |
| RESIDUAL (ERROR) | 27.500 | 9 | 3.056 | | |
| TOTAL | 204.278 | 17 | 12.016 | | |

ANOVA, with Interaction Pooled into RESIDUAL

| SOURCE OF VARIATION | SUM OF SQUARES | DF | MEAN SQUARE | F | SIG OF F |
|---|---|---|---|---|---|
| MAIN EFFECTS | 161.889 | 4 | 40.472 | 12.412 | .000 |
|    FINISH | 109.778 | 2 | 54.889 | 16.834 | .000 |
|    PRESERV | 52.111 | 2 | 26.056 | 7.991 | .005 |
| EXPLAINED | 161.889 | 4 | 40.472 | 12.412 | .000 |
| RESIDUAL (ERROR) | 42.389 | 13 | 3.261 | | |
| TOTAL | 204.278 | 17 | 12.016 | | |

The overall conclusion is that the combination of preservative 3 (liquid treatment) and finish 3 (plastic coating) gives the heaviest average weight for plywood sheets. Computer Problem 80 gives a Minitab setup for this experiment. Other problems use the SPSS ONEWAY procedure.

The analysis of designed experiments is extensively computerized, because it is most convenient to perform the ANOVA calculations by computer. Below, we describe the procedures available from some common commerical software packages: SPSS, BMDP, SAS, and Minitab. We review what each offers for the design models and ANOVA calculations that appear in Chapter 11.

**1.** SPSS [9] has two procedures that encompass the material of this chapter: ANOVA and ONEWAY. The SPSS format includes an overview and operational procedures with annotated examples. The content is descriptive rather than symbols-oriented.

ANOVA incorporates calculations for the most common designs, including the completely random design and a factorial design allowing up to five interaction levels. The ANOVA output includes the calculated $F$ and the significance level of $F$ ($p$-value) for each effect, using error mean square (MSE) as the divisor. Procedure ONEWAY affords alternative multiple-comparison procedures by the RANGES subcommand.

2. BMDP [1] has multiple procedures that relate to this chapter. For beginners, 7D allows both one- and two-way factorial analyses. The discussion covers definition of terms, examples of printouts, sums of squares, calculated forms, and test statistics, and is generally symbolic rather than using word descriptions. The coverage also includes multiple-comparison procedures.

3. SAS [5] procedure ANOVA affords analysis of variance for balanced data from a wide variety of experimental designs. This package uses a semiprogramming approach with model and output specifications declared by keyword statements. Display examples, including CRD, randomized complete block, and two-factor factorial experiments illustrate the essential data inputs, procedure ANOVA setup statements, and the ANOVA results. Several multiple-comparison procedures are included.

4. Minitab offers three commands for analysis of variance: two for one-way analyses (ONEWAY, AOVONEWAY) and the other (TWOWAY) for both RCB and two-factor factorial designs with equal numbers of replicates. These are not as sophisticated as the mainframe procedures of SPPS and others. For example, there is not a multiple-comparison test. Yet the procedures are quite easy to use, and each procedure gives the essential outputs for data checks, ANOVA displays, and $F$-tests. The Minitab *Student Edition* [6] descriptions are clear, and they provide definitions by example. Also, we have provided the essential setups for Minitab solutions for a one-way, an RCB, and a two-factor factorial design with replicates in Problems 78 through 80.

All four—SAS, BMDP, SPSS, and Minitab—*as well as other* software offer flexible procedures for ANOVA analysis. Your choice may well depend upon availability. Once you have used one of these packages, you will find it fairly easy to use others.

## COMPUTER PROBLEMS

*Data Base Applications*

74. Following is the household income recorded for random samples of ten records from each of three Metropolitan Statistical Area (MSA) markets (positions 8–11): 1 = Chicago (MSA #1600), 2 = New York (MSA #5600), 3 = Los Angeles-Long Beach (MSA #4480). (*Note:* Chicago MSA has 622 records, New York has 869 records, and LA-Long Beach has 684 records in the 20,116 record PUMS file.)

| Chicago (1) | New York (2) | Los Angeles-Long Beach (3) |
|---|---|---|
| $15,735 | $ 7,410 | $13,005 |
| 9,085 | 16,035 | 4,655 |
| 8,690 | 6,730 | 28,075 |
| 15,005 | 255 | 17,135 |
| 29,020 | 5,885 | 11,445 |
| 21,410 | 9,480 | 4,295 |
| 12,740 | 36,010 | 16,005 |
| 21,005 | 925 | 14,205 |
| 12,005 | 16,795 | 32,140 |
| 17,260 | 18,005 | 3,810 |

Use the completely randomized design and $\alpha = 0.05$ to test for equal means. Can you make any suggestions about how to design this test better? (*Suggestion:* Consider that the size of the error mean square is large. How might it be made smaller?)

75. Would you agree that more wage earners in a household should mean more income? Suppose we wanted a randomized complete block design to test Problem 74. (Do not calculate). Set out the df for an ANOVA table for three treatments (three markets) by three blocks (wage earners: 0, 1, or 2+) RCB design model. Discuss the precision of this plan (design) relative to that in Problem 74. (*Suggestion:* What is the error df?)

76. Here is an example of a poorly planned experiment. This problem is an exercise in data analysis. The data are a random sample of 40 household records from the PUMS file, including number of persons (positions 12–13) and monthly cost of electricity (positions 21–23)—the response variable.

Monthly Cost of Electricity Based on Number of Persons in Household

| | 1 Person | 2 Persons | 3 Persons | 4 Persons |
|---|---|---|---|---|
| | $ 65 | $75 | $50 | $140 |
| | 0 | 46 | 55 | 50 |
| | 9 | 45 | 16 | 7 |
| | 35 | 20 | 70 | 10 |
| | 42 | 15 | 24 | 0 |
| | 20 | 49 | 25 | 0 |
| | 40 | 20 | | 75 |
| | 20 | 15 | | 80 |
| | 33 | | | 125 |
| | 9 | | | 55 |
| | 125 | | | 30 |
| | | | | 35 |
| | | | | 50 |
| | | | | 13 |
| | | | | 30 |
| Means | $ 36.18 | $36.52 | $40.0 | $ 46.67 |
| $n$ | 11 | 8 | 6 | 15 |

First, test $H_0: \mu_1 = \mu_2 = \mu_3 = \mu_4$ for equal cost of electricity. Let $\alpha = 0.05$. After your test, consider that the $0 amounts were all for renter households. In fact, *most* of the lower electricity costs were for renter households. Explain how this has affected your test of hypothesis. Suggest the action you would take to get a valid test for the cost of electricity.

*Additional Computer Problems*

77. Use the data of Example 11.3 on $Y$ = merchandise pricing test scores and a computer program of your choosing to check the ANOVA and F-test calculations given in the example.

Minitab users, try

```
MTB > NAME C1 = 'PRICING' C2 = 'GROUP'
MTB > READ 'PRICING' 'GROUP'
DATA> 396 1
DATA> 397 1
        ⋮
DATA> 368 3
DATA> END
MTB > PRINT 'PRICING' 'GROUP'
MTB > ONEWAY ON DATA 'PRICING', 'GROUP'
```

This output includes individual 95% confidence intervals for the means. These are *not* multiple comparisons.

78. Here is a SPSS computer-run setup for the data in Example 11.3, which has

```
Response = Score on a practical pricing test
Group (treatment) levels    1 = Hands-On Training
                            2 = Video Instruction
                            3 = Pricing Manual
```

The data are read according to the DATA LIST format in the routine.

```
1 NUMBERED
2 TITLE SPSS ONEWAY ANALYSIS FOR PROBLEM 11.3
3 DATA LIST    FILE = INFILE/
4              PRICING 1–3 GROUP 5
5 SET BLANKS = -1
6 LIST VARIABLES = PRICING GROUP/CASES FROM 1 TO 18
7 ONEWAY    PRICING BY GROUP(1,3)/FORMAT = LABELS/
8           RANGES = SCHEFFE(.05)/STATISTICS = ALL
```

The data and the ANOVA summary appear in Example 11.3. Here is a summary of the Scheffé tests. (Asterisks indicate pairs of groups significantly different at the 0.05 level.)

| | | | Group | |
|---|---|---|---|---|
| Mean | Group | 3 | 2 | 1 |
| 395.2857 | 1 | | * | * |
| 382.6000 | 2 | | | * |
| 366.8333 | 3 | | | |

Use the data from Example 11.3 and employ Scheffé's procedure for estimating simultaneous confidence intervals–Equation (11.7)—to check the SPSS results. Interpret the meaning of the Scheffé results. If you have access to the SPSS software, run the program to check these results. Problem 77 gave the setup for Minitab users.

79. Use the data in Problem 58 on $Y =$ index of purchasing power and a randomized complete block (RCB) computer routine to test for a significant treatment effect across states. Let age be the block variable.

For Minitab, use

```
MTB > NAME C1 = 'PURCHASE'  C2 = 'AGE'  C3 = 'STATE'
MTB > READ 'PURCHASE' 'AGE' 'STATE'
DATA> 77.9 1 1
DATA> 76.3 1 2
       ⋮
DATA> 72.3 4 5
DATA> END
MTB > TABLE 'AGE' BY 'STATE';
SUBC> MEAN 'PURCHASE'.
MTB > LPLOT 'PURCHASE' BY 'AGE', CODE FOR 'STATE'
MTB > TWOWAY 'PURCHASE' 'AGE' 'STATE';
SUBC> ADDITIVE.
```

80. The Table 11.11 data and Example 11.5 illustrate a two-factor design. Check the ANOVA summary and $F$-test results by using any computer routine. The response is weights, and the treatments are (1) wood preservative and (2) exterior finish.

For Minitab, try

```
MTB > NAME C1 = 'WEIGHT' C2 = 'FINISH' C3 = 'PRESERV'
MTB > SET 'WEIGHT'
DATA> 65 68 64 66 70 68 . . . 76 74
DATA> END
MTB > SET 'FINISH'
DATA> 1 1 1 1 1 1 2 2 2 2 2 2 3 3 3 3 3 3
DATA> END
MTB > SET 'PRESERV'
DATA> 1 1 2 2 3 3 1 1 2 2 3 3 1 1 2 2 3 3
DATA> END
MTB > TABLE 'FINISH' 'PRESERV';
SUBC> DATA 'WEIGHT'.
MTB > TABLE 'FINISH' 'PRESERV';
SUBC> MEAN 'WEIGHT'.
MTB > TWOWAY 'WEIGHT' 'FINISH' 'PRESERV'
```

**FOR FURTHER READING**

[1] Dixon, W. J., ed. *BMDP Statistical Software Manual*, vol. 1. Berkeley, CA: University of California Press, 1990.

[2] Dunn, Olive, and Virginia Clark. *Applied Statistics: Analysis of Variance and Regression*, 2nd ed. New York: John Wiley, 1987.

[3] Kirk, Roger. *Experimental Design*, 2nd ed. Belmont, CA: Brooks-Cole Publishing, 1982.

[4] Ott, Lyman. *An Introduction to Statistical Methods and Data Analysis*, 3rd ed. Boston: PWS-Kent, 1988.

[5] *SAS (Statistical Analysis System) Users Guide: Statistics*, version 5 ed. Cary, NC: SAS Institute, 1985.

[6] Schaefer, Robert L., and Richard B. Anderson. *The Student Edition of Minitab.* Reading, MA: Addison-Wesley, 1989.

[7] Scheffé, Henry. *The Analysis of Variance.* New York: John Wiley, 1959.

[8] Snedecor, George, and W. C. Cochran. *Statistical Methods,* 7th ed. Ames, IA: Iowa State University Press, 1980.

[9] *SPSS (Statistical Package for the Social Sciences) Reference Guide.* Chicago: SPSS, Inc., 1990.

[10] Wonnacott, Thomas, and Ronald Wonnacott. *Introductory Statistics for Business and Economics,* 4th ed. New York: John Wiley, 1990.

# REGRESSION AND CORRELATION

## 12.1
### INTRODUCTION TO REGRESSION AND CORRELATION APPLICATIONS

It was already 9:30 and the vice president of the Seattle bank had to notify New York of the bank's decision by 10:00 A.M. Should they go ahead with the proposed sale of securities or not? The right decision would certainly ease tensions and help pull them through the current crisis. But a mistake of this magnitude could threaten the bank's own financial stability! And one major Northwest bank was already in trouble.

The big question concerned the direction of short-term interest rates. Would they rise, remain the same, or fall over the next few weeks? The vice president realized that interest rates were highly dependent upon the money supply.

Then a secretary very alertly passed along last Friday's money-supply report from the Federal Reserve. With a few strokes on the keyboard, the new data became part of an existing data base, and a statistical modeling projection emerged on the terminal—in color. The message was clear: Don't sell the securities. New York had a decision via fax by 9:45 A.M.

The bank vice president in this opening scenario used a regression model that characterized the relationship between the supply of money in the economy ($X$) and the interest rate ($Y$). The vice president's model reflected the fact that when the Federal Reserve Board increases the supply of money, the cost of money (the interest rate) tends to fall. By entering a current value for money supply, the vice president was able to predict the action of the interest rate. In general, financial situations are not as simple as suggested here, but this illustrates the kinds of relationships that are analyzed in simple linear regression and correlation.

### What Are Regression and Correlation?

*Regression* is a statistical procedure for establishing the relationship between two or more variables. It involves (1) determining whether or not the variables are related in some way, (2) describing that relationship in an equation, and (3) using that resulting equation to predict values of one variable (the dependent variable) on the basis of another (the independent variable).

Simple regression utilizes one independent variable ($X$) to predict the value of the dependent variable ($Y$). If the relationship between the two variables can be accurately described by an equation of the form $Y = a + bX + e$, then the model is called a simple linear regression model.

A familiar example of the use of a simple linear regression model is to predict freshman fall enrollment at a university from the number of high school diplomas awarded in the previous spring. In business an appliance marketing manager might use regression to predict sales on the basis of the level of disposable income in the economy. In construction industries a building materials dealer may forecast demand from information on the number of housing starts in the area. In manufacturing, an automobile assembly line production manager might use regression to show how the cost of a new car is related to the number of hours of training given assembly-line workers.

*Correlation* is a measure of the degree of association between two variables. Correlation analysis yields a summary statistic, called the correlation coefficient, that measures the linear association between two variables. If the variables are perfectly linearly related, the correlation coefficient will be either $+1$ or $-1$; if they are not linearly related, the correlation coefficient will be zero. Correlation values between $-1$ and $0$ or $0$ and $1$ reflect varying degrees of a linear relationship.

Marketing analysts can use correlations to identify variables, such as age, sex, and income, that are closely related to buying on impulse. They can plan their marketing strategy to capitalize on the known buyer attributes. In retailing, the manager of a highly seasonal product might expect sales to be correlated with the outdoor temperature and would schedule his inventory according to the season.

Regression and correlation are described in this chapter, and distinctions between the two procedures are developed.

Upon completing Chapter 12, you should be able to

1. identify business situations in which a regression model might be useful;
2. identify the dependent and independent variables used in regression;
3. graph the two variables and use the graph to verify that their relationship can be described by a simple linear regression model;
4. use the least squares procedure to find the best linear relation;
5. test whether a proposed linear model is statistically meaningful;
6. use a linear regression model to predict values of a dependent variable from values of an independent variable;
7. compute and interpret the correlation coefficient for data on two variables.

Nearly all the computational techniques used in this chapter are available on mainframe and personal computer packages. The structure and terminology we use is designed to enable you to understand and apply the most popular types of software packages available. The applications sections in this chapter give you an introduction to computer calculations. However, an intensive discussion of the notation and terminology specific to statistical packages is reserved for Chapter 13.

## 12.2 THE SIMPLE LINEAR REGRESSION MODEL

The bank vice president in the opening example assumed that interest rates could be accurately predicted by a simple linear regression model that had money supply as the one independent variable. However, we know that interest rates are also influenced by other variables, such as the opportunities available for the use of funds and the demand from foreign investors. Unless the effects of these other variables are explicitly recognized, we simply acknowledge them as error and assume they are random. (If these additional independent variables were introduced into the model, we would have a multiple regression model.) The complete regression model then consists of an equation (including a random error variable) along with any essential assumptions or conditions. The model assumptions are discussed in Section 12.4, where they are first used.

A *simple linear regression model* is a linear equation relating a dependent variable, denoted by $Y$, to an independent variable, $X$, and allowing for random error, $\varepsilon$.

> A simple linear regression model is an equation that describes a dependent variable in terms of an independent variable plus random error.

$$Y = \beta_0 + \beta_1 X + \varepsilon$$

Components : Deterministic + Random

The deterministic, or mathematical, portion of the simple regression model describes a straight line. This line is defined by two numbers: (1) $\beta_0$, the *intercept* of the line with the $Y$-axis, and (2) $\beta_1$, the *slope* of the line. These model parameters are illustrated on Figure 12.1.

**FIGURE 12.1** A Data Plot with a Simple Regression (Conceptual)

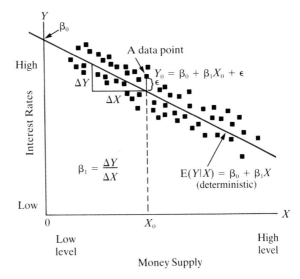

The random *error* component of the simple regression model, $\varepsilon$, describes the difference of the data point from the deterministic value. This error is illustrated in Figure 12.1 as a vertical distance from each point to the regression line. We prefer a model for which this error is small, because then the data points are close to the regression line. Then the model is said to have a good fit to the data.

For the money-supply example, we would plot the data using rectangular coordinates of $X$ as the money supply and $Y$ as the interest rate. If the simple linear model is a good fit, then the plot of the data points will show a strong linear pattern. The regression model would have a negative slope with only small error values. Of course, before a linear model is accepted, all the requisite assumptions must be satisfied. Then reasonable predictions of interest rates, the dependent variable, can be made.

**FIGURE 12.2** Graphs of Common Regression Relations for Paired Data

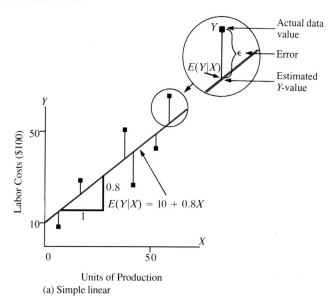

(a) Simple linear

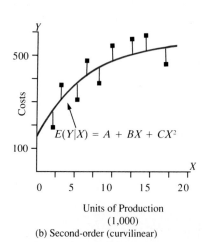

(b) Second-order (curvilinear)

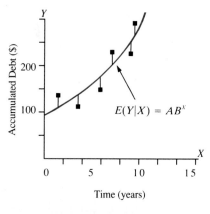

(c) Exponential (nonlinear)

Figure 12.2 illustrates graphs for the simple linear model as well as for several other regression models. The constant slope is the distinctive feature of the simple linear regression graph (Figure 12.2(a)).

Other regression models include higher-order additive terms like $X^2$ or $X^3$, as in the parabola in Figure 12.2(b). Still other data are described by an exponential model like the one illustrated in Figure 12.2(c). This nonlinear model uses a product of components. Debt accumulation by compound interest is one application of an exponential model. Only simple linear models are considered in this chapter.

Regression uses sampling to develop an estimated linear model of a relationship inherent in the population: $Y = \beta_0 + \beta_1 X + \varepsilon$. The deterministic part of the model, called the *linear regression*, describes the expected, or average, value of $Y$ for given $X$-values.

> The simple linear regression is the expected value for $Y$ given a value for $X$, $E(Y|X) = \beta_0 + \beta_1 X$. From sample data the *estimated simple linear regression* is $\hat{Y} = b_0 + b_1 X$.

When we work with sample data, the estimator for $E(Y|X)$ is denoted by $\hat{Y}$. Then the simple linear regression line, as estimated from the sample data, is $\hat{Y} = b_0 + b_1 X$. Fitting an estimated regression line to sample data means determining values for $b_0$ and $b_1$ that are the best possible estimators for $\beta_0$ and $\beta_1$. Then, for a specified $X$ value, $E(Y|X)$ is estimated by $\hat{Y}$.

We introduce a business problem that can be resolved through regression modeling. These data will be used to illustrate several concepts, so they are used throughout the chapter.

---

**EXAMPLE 12.1**

Walt Disney World, near Orlando, Florida, has become a key factor in the economy of the entire central Florida area. Table 12.1 gives some airline passenger booking information that may be useful for business planning in the area. Since airline reservations

**TABLE 12.1**  Hotel Occupancy and East Coast Airlines Passengers

| East Coast Passengers (000) (X) | Occupancy (%) (Y) |
| --- | --- |
| 65.689 | 40 |
| 71.593 | 41 |
| 53.653 | 48 |
| 70.177 | 49 |
| 74.974 | 73 |
| 85.626 | 74 |
| 84.575 | 68 |
| 58.046 | 51 |
| 72.809 | 63 |
| 87.607 | 75 |
| 85.380 | 70 |
| 50.611 | 38 |

**FIGURE 12.3**   A Data Scattergram with Freehand Regression Line for Percentage
of Hotel Occupancy ($Y$) on East Coast Passenger Bookings ($X$)

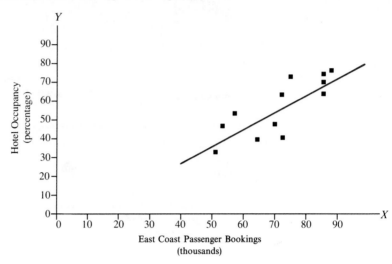

are generally made in advance of lodging reservations, management feels that airline
bookings might be used to predict hotel occupancy.

To develop a reasonable linear model will require values for the slope and $Y$-intercept. The two variables are

> $X$ = thousands of passengers booked for airline flights by East Coast
> Airlines to Orlando International Airport

> $Y$ = occupancy rate for selected Walt Disney World area hotels

Is a linear regression model based on East Coast passenger booking counts ($X$) reasonable for predicting hotel occupancy rates ($Y$) for this area?

*Solution*   To see if a linear model is reasonable, we begin with a data plot of the sample data
from Table 12.1. The plot appears in Figure 12.3. These data appear as a scatter diagram,
or *scattergram*, on $XY$-coordinates. A scatter diagram is simply a graph of the data
points in two-dimensional space that helps to show if any relationship is evident. (Many
computer programs provide scatter diagrams as a part of their regression analysis.)

By visual inspection the data points appear to be somewhat in a line. So a line
has been sketched on the data.\* If we can justify that this estimated linear model gives
a satisfactory fit, then it might be used for predictions of occupancy rates for Walt
Disney World hotels.

Example 12.1 presents several questions. First, by what rationale can a
straight line be drawn, and what are the characteristics of this line? Second,
does this sample estimated model describe hotel occupancy rates that are good
enough to be used for actual predictions? These questions will be answered
after the following set of exercises.

\* For a quick summary of rectangular coordinates and linear equations see Appendix C: Math
Essentials for Statistics.

**Problems 12.2**

1. Construct a graph for the line described by $Y = \frac{1}{2}X + 1$.

2. Find the slope and $Y$-intercept for the pictured line. Write a linear equation. (See Appendix C for a rule for computing the slope of a line.)

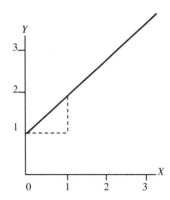

3. The following scores are a random sample from a much larger class. Plot the data on rectangular coordinates (as in Problem 2) and then make a visual freehand plot of the linear relation. Estimate the $Y$-intercept and the slope.

| Placement score, $X$ | 93 | 59 | 85 | 78 | 60 | 90 | 72 | 73 | 60 | 85 |
| Economics grade, $Y$ | 95 | 64 | 80 | 87 | 64 | 88 | 74 | 74 | 80 | 84 |

4. For the given pairs of values,
   a. answer the questions given in Problem 3.

| Time (coded), $X$ | $-3$ | $-2$ | $-1$ | 0 | 1 | 2 | 3 |
| Production, $Y$ | 1 | 2 | 3 | 5 | 8 | 11 | 12 |

   b. Is the slope positive or negative? Explain. (See Appendix C on linear and curvilinear equations.)

5. Use the given data plots.
   a. Which of graphs (a), (b), and (c) has an approximately linear relation with negative slope?
   b. Which appears to show a fairly good linear statistical relation?
   c. For the data set in graph (c), which is a more likely relation:
      $Y = a + bX + \varepsilon$ or $Y = c + dX + fX^2 + \varepsilon$. Why? (Here, $a$, $b$, $c$, $d$, and $f$ are generalizations for some unspecified numbers. Recall Figure 12.2.)

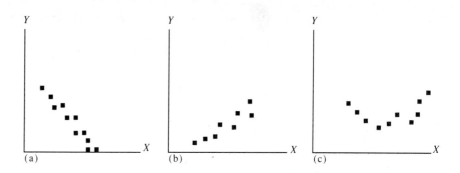

## 12.3
## LEAST SQUARES
## ESTIMATION OF $\beta_0$
## AND $\beta_1$

The "best" regression line for sample data will result in the least possible error in describing that data. Hence the error component is analyzed to define the least total error, and from this analysis the best estimators for $\beta_0$ and $\beta_1$ are determined. The error is measured by the difference between actual response value ($Y$) and the estimated response value ($\hat{Y}$) as determined from the sample regression equation. This difference is designated by the English letter $e$, where $e = Y - \hat{Y}$; it is called a *residual* to indicate that it comes from sample data. In contrast, the *error* for the population data and the model is $\varepsilon$.

The least squares procedure assures a good fit to the sample data in that it defines a line with the sum of squared residual distances for the data points that is a minimum. The least squares estimation of $\beta_0$ and $\beta_1$ can now be outlined.

### Procedures for Least Squares Estimation of $\beta_0$ and $\beta_1$

A mathematics procedure—calculus—is used to minimize the sum of squared errors for *least squares estimation* of $\beta_0$ and $\beta_1$. Symbolically, this procedure minimizes the sum of squared errors, SSE, where

$$\text{SSE} = \sum_{}^{n} [Y - \hat{Y}]^2 = \sum_{}^{n} [Y - (b_0 + b_1 X)]^2$$

The minimization results in computation formulas for $b_1$, the sample estimator of $\beta_1$, and $b_0$, the estimator for $\beta_0$.

> The least squares estimation procedure minimizes SSE to determine values for $b_0$ and $b_1$ that ensure a best fit for the estimated regression line to the sample data points.

The least squares estimation results in two equations called *normal equations*.

$$\sum Y = nb_0 + b_1 \sum X$$
$$\sum XY = b_0 \sum X + b_1 \sum X^2 \tag{I}$$

Substitution of values for $\sum Y$, $\sum X$, $\sum XY$, and $\sum X^2$ from sample data allows solution of the Equations (I) values for $b_0$ and $b_1$.

The least squares procedure defines a line that is a best fit in that it is as close to *all* the data points as possible. In addition, the sample estimators $b_0$ and $b_1$ are unbiased and efficient estimators for $\beta_0$ and $\beta_1$, respectively. Neter,

Wasserman, and Kutner [4]* give the mathematical basis for the least squares estimation and a development of the normal equations.** The solutions for the $Y$-intercept and slope estimators $b_0$ and $b_1$ are as follows:

**LEAST SQUARES ESTIMATORS FOR $\beta_0$ AND $\beta_1$**

$$Y\text{-intercept} \qquad b_0 = \bar{Y} - b_1\bar{X} \qquad \qquad \text{(12.1)}$$

$$\text{Slope} \qquad b_1 = SS(XY)/SS(X) \qquad \qquad \text{(12.2)}$$

where 
$$SS(Y) = \sum (Y - \bar{Y})^2 = \sum Y^2 - [(\sum Y)^2/n]$$
$$SS(X) = \sum (X - \bar{X})^2 = \sum X^2 - [(\sum X)^2/n]$$
$$SS(XY) = \sum [(X - \bar{X})(Y - \bar{Y})] = \sum XY - [(\sum X \sum Y)/n]$$

The notations $SS(X)$ and $SS(XY)$ are the same as in ANOVA (Chapter 11). That is, the symbols represent the sum of squared deviations about $\bar{X}$ and about $(\bar{X},\bar{Y})$, respectively. Total variation, $SS(Y)$, is defined similarly.

Example 12.1 displayed the estimated regression line for one set of data. Using the data from Table 12.1 and Equations (12.1) and (12.2), we can now demonstrate the calculation of $b_0$ and $b_1$ values.

---

**EXAMPLE 12.2**

Use the least squares estimation procedure and data from Table 12.1 to estimate the slope and $Y$-intercept values for a linear regression of occupancy rate $(Y)$ on East Coast Airlines passenger counts $(X)$.

*Solution*   This example requires calculation of values for $b_0$, and $b_1$. The data from Table 12.1 for

$X$ = thousands of passengers booked for East Coast Airlines flights to Orlando, Florida

$Y$ = occupancy rate (%) for selected Walt Disney World area hotels

| $X$ | 65.689 | 71.593 | 53.653 | 70.177 | 74.974 | 85.626 |
|---|---|---|---|---|---|---|
| $Y$ | 40 | 41 | 48 | 49 | 73 | 74 |
| $X$ | 84.575 | 58.046 | 72.809 | 87.607 | 85.380 | 50.611 |
| $Y$ | 68 | 51 | 63 | 75 | 70 | 38 |

give

$$\sum X = 860.7400 \qquad \sum XY = 51{,}155.0720 \qquad n = 12$$
$$\sum Y = 690 \qquad \sum X^2 = 63{,}546.5941$$

---

* Bracketed numbers refer to items listed in "For Further Reading" at the end of the chapter.

** The minimization is an application of calculus. See [1] for an algebra solution to the least squares estimators $b_0$, $b_1$ for $\beta_0$ and $\beta_1$.

Then,

$$SS(XY) = \sum (XY) - \frac{\sum X \sum Y}{n} = 1{,}662.5220$$

$$SS(X) = \sum (X^2) - \frac{(\sum X)^2}{n} = 1{,}807.1485$$

and

$$b_1 = \frac{1{,}662.5220}{1{,}807.1485} = 0.9200$$

$$b_0 = \frac{690}{12} - \left(0.9200 \cdot \frac{860.740}{12}\right) = -8.4879$$

The estimated regression equation is $\hat{Y} = -8.4879 + 0.9200X$. To check the reasonableness of the equation and the amount of error for estimating one data point, let $X = 65{,}689$ East Coast passengers, the first observation. Then, $\hat{Y} = -8.4879 + 0.9200(65.689) = 51.9$, or 52% occupancy. The residual associated with this actual data point $X = 65.689$ is $e = Y - \hat{Y} = 40 - 52 = -12\%$.

---

The least squares estimated regression has several properties that are useful for data calculations:

1. The least squares regression line passes through the point $(\bar{X}, \bar{Y})$.
2. The sum of the errors (residuals) is zero (i.e., $\sum e = 0$).
3. The sum of the (error)$^2$ is a minimum value.

To plot the regression line, we can use the points $(\bar{X}, \bar{Y})$ and $(0, b_0)$. The point $(\bar{X}, \bar{Y})$ is near the middle of the regression line, and the point $(0, b_0)$ is plotted where $X = $ zero (when the range of the independent variable includes zero).

$\sum e = 0$ is used principally by computer programs to check for calculation errors. Data plots that include the residual values can also help to determine how well the estimated regression line describes the data.

**Using Data Plots to Assess the Estimated Regression Equation**

Several of the properties just discussed can be illustrated by example.

---

**EXAMPLE 12.3**

Plot the least squares regression line on the data scatter diagram for % hotel occupany ($Y$) and East-Coast Airlines passenger bookings ($X$). Also display the residuals to the line $\hat{Y} = -8.4879 + 0.9200X$.

*Solution*    Figure 12.4(a) shows that one point on the estimated regression line imposed on the sample is $(\bar{X}, \bar{Y}) = (71.1, 57.5)$. A visual check about whether $\sum (\text{error})^2$ is a minimum can also be made on Figure 12.4(a).

Figure 12.4(b) illustrates the residuals plotted against $X$. The residuals occur somewhat at random about the line $\bar{e} = 0$, so this does not discredit a linear regression in $X$.

**FIGURE 12.4**   Graphical Checks on an Estimated Linear Regression for Walt Disney
World Area Data

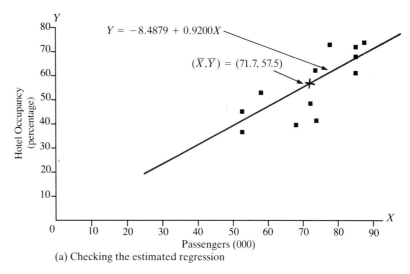

(a) Checking the estimated regression

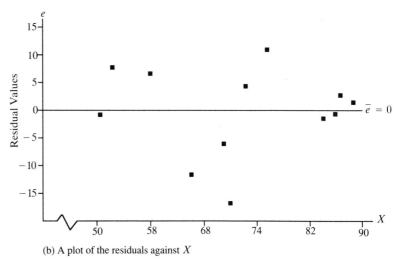

(b) A plot of the residuals against $X$

The property that the residuals sum to zero is one check on the calculations for $b_0$ and $b_1$. Another check is the regression line imposed on the scattergram ($XY$ data plot) in Figure 12.4(a). The estimated regression appears to fit
the data reasonably well. Because the line lies close to the data points, the calculations of $b_0$ and $b_1$ are within reason. Plot 12.4(b) shows a dispersion of the
residual values; there is not a linear pattern. A simple linear regression appears
reasonable; but additional independent variables might reduce the unexplained
variations.

The following problems offer you an opportunity to try these concepts.

**Problems 12.3**

6. The given table contains values for two variables, $X$ and $Y$.
   a. Plot the data. Does a linear regression appear reasonable? Make a freehand plot of a regression line.

b. Establish the estimated regression of $Y$ on $X$. Use Equations (12.1) and (12.2) for the least squares line of best fit.

| Week (coded), $X$ | $-2$ | $-1$ | 0 | 1 | 2 |
|---|---|---|---|---|---|
| In-plant accidents, $Y$ | 0 | 1 | 0 | 1 | 3 |

7. The following sample gives values of $X =$ hours of continuous processing and $Y =$ computer down-time in minutes.

| $X$ | 4 | 5 | 6 | 7 | 8 | 9 | 10 |
|---|---|---|---|---|---|---|---|
| $Y$ | 4 | 6 | 8 | 12 | 18 | 24 | 26 |

a. Use the least squares procedure to estimate the regression of $Y$ and $X$, assumed to be linear, given that

$$\sum (X^2) = 371, \quad \sum (XY) = 798, \quad \bar{X} = 7, \quad \bar{Y} = 14$$

b. Does the point with coordinates (7,14) fall on your regression line? Why or why not?
c. What is the estimated value of $Y$ given $X = 5.5$?

8. Use the data from Problem 3 to find the estimated regression coefficients $b_0$ and $b_1$ in $\hat{Y} = b_0 + b_1 X$. Use $\sum X = 755$; $\sum Y = 790$; $\sum (XY) = 60{,}627$; $\sum (X^2) = 58{,}477$; $\sum (Y^2) = 63{,}338$. Compare these values with your first (freehand) estimate. Recall that $n = 10$.

9. a. Plot the given data and make a freehand estimated regression line.
   b. Use the given information to develop an estimated regression equation of the form $\hat{Y} = b_0 + b_1 X$ by evaluating $b_0$ and $b_1$. Now impose the least squares regression line on your figure. Compare this to your answer in Problem 4.

| $X$ | $-3$ | $-2$ | $-1$ | 0 | 1 | 2 | 3 |
|---|---|---|---|---|---|---|---|
| $Y$ | 1 | 2 | 3 | 5 | 8 | 11 | 12 |

10. An industrial psychologist has initiated the following plan aimed at reducing the percentage of absenteeism for workers in a large manufacturing plant. Every workday (five days per week), each worker who arrives on time is given a playing card from a standard deck of cards. On Friday the worker in each section with the best poker hand wins a bonus of $25.00. Suppose that after five weeks the results were as follows:

| Week, $X$ | 1 | 2 | 3 | 4 | 5 |
|---|---|---|---|---|---|
| Percent absent, $Y$ (%) | 7 | 6.5 | 7 | 6.5 | 6 |

a. Plot the data. Does there appear to be a reasonably strong linear relation?

b. Using the following summary statistics, compute estimates for the regression coefficients and plot your estimated regression.

$$\sum (X^2) = 55 \qquad SS(X) = \sum (X - \bar{X})^2 = 10.0$$

$$\sum (Y^2) = 218.5 \qquad SS(Y) = \sum (Y - \bar{Y})^2 = 0.7$$

$$\sum (XY) = 97 \qquad SS(XY) = \left[ \sum (XY) - \frac{\sum X \sum Y}{n} \right] = 2.0$$

c. Use your regression equation to estimate the percent absent for the fifth week. Compare the estimated and observed values.

d. Does the approach seem to result in reduced absenteeism? Consider the sign and size of the slope for your regression.

11. The solution of Equations (12.1) and (12.2) is essentially a problem from algebra of the simultaneous solution of linear equations. Solve the following for $b_0$ and $b_1$:

$$5 = b_0 + 2b_1$$
$$6 = 3b_0 + 4b_1$$

12. Using the results from Problem 6, draw your estimated regression line on the same diagram with the data plot and your freehand guestimated regression line. Compare the freehand and the least squares regression lines. Which one gives a better fit to the data?

13. a. Using the results from Problem 7, draw your estimated regression line on the same diagram with a plot of the data and your freehand "guestimated" regression line. Compare the freehand and least squares regression lines. Which one gives a better fit to the data?

b. The normal equations for simple linear regression are

$$\sum Y = nb_0 + b_1 \sum X$$
$$\sum XY = b_0 \sum X + b_1 \sum X^2$$

For this problem,

$$98 = 7b_0 + 49b_1$$
$$798 = 49b_0 + 371b_1$$

Solve these two equations to find values for $b_0$ and $b_1$. Compare your answers with those for Problem 7, part (a).

14. The residuals are given for a linear regression on the data in Problem 7. Plot the residual values on the vertical axis against the $X$-values on the horizontal axis. Discuss the advisability of including an $X^2$ term in this model. Also see the data plot in Problem 7.

| $X$ | 4 | 5 | 6 | 7 | 8 | 9 | 10 |
|-----|-----|-----|-----|-----|-----|-----|-----|
| Residuals | $-4.46$ | $-4.62$ | $-4.77$ | $-2.92$ | 0.92 | 4.77 | 4.61 |

15. For $X =$ time (coded) and $Y =$ production, the data in Problem 4 give a least squares estimated regression $\hat{Y} = 6 + 2.0X$. Use the $X$-values given in Problem 4 to determine values for $\hat{Y}$. Then complete the table below.

| $X$ | $-3$ | $-2$ | $-1$ | 0 | 1 | 2 | 3 |
|-----|-----|-----|-----|-----|-----|-----|-----|
| $Y$ | 1 | 2 | 3 | 5 | | 11 | 12 |
| $\hat{Y}$ | 0 | 2 | | | 8 | | |

Compute values for $e = Y - \hat{Y}$ and plot these values on the vertical axis against $X$-values on the horizontal axis of rectangular coordinates. Does a simple linear regression appear to give a good fit to this data?

## 12.4
## THE STANDARD ERROR OF ESTIMATE: AN ESTIMATOR OF $\sigma^2$

We have *assumed* that a simple linear regression model can accurately describe business data. But that assumption should be tested before the model is used for making any predictions. This section introduces some procedures for testing whether a regression equation is accurate in its description. If the equation is not accurate, it is meaningless and should not be used.

Certain assumptions must be satisfied in order to make valid inference via regression. These assumptions pertain to and limit the error that results when actual data points deviate from the least squares line of best fit. The deviation is embodied in what is called the *standard error of estimate*. After the assumptions underlying the simple regression model are considered, the standard error of estimate is used to evaluate the significance of a regression line.

### Assumptions for Inference about Regression

A major objective of regression is to predict values of the dependent variable. If these predictions are to be valid, then four assumptions should be met.

**THE SIMPLE LINEAR REGRESSION MODEL WITH ASSUMPTIONS**

$$Y = \beta_0 + \beta_1 X + \varepsilon \tag{I}$$

**FIGURE 12.5** The Assumptions for a Linear Regression

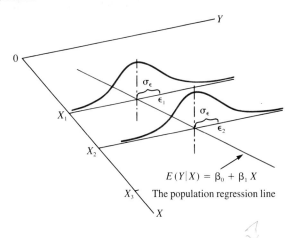

where

1. the mean response is a linear function. That is, $E(Y|X) = \beta_0 + \beta_1 X$;
2. the $\varepsilon$ for a given fixed value of $X$ is normally distributed, $N(0, \sigma^2)$;
3. the variance is the same for all values of $X$: $\text{Var}(\varepsilon) = \sigma^2$;
4. the errors ($\varepsilon$'s) are independent of each other.

The model assumptions are illustrated in Figure 12.5. The regression value is the expected value of $Y$ for any given value of $X$, denoted by $E(Y|X)$. By assumption 1, the regression relation is a line defined by $E(Y|X) = \beta_0 + \beta_1 X$. Note that the errors are about the line $E(Y|X)$ according to assumption 2 that the errors are normally distributed. When this is so, the bell curve (for $\varepsilon$) is centered at the regression line, and the positive and negative errors cancel out. Assumption 3 simply points to a constant variance of $\sigma^2$ for all $X$-values. This means that the spread is the same regardless of the $X$-value used. Assumption 4 states that the errors are independent. For any observation the error for the value of $Y$ is unaffected by the error for any other observation.

**The Standard Error of Estimate**

Having developed an estimated regression equation, we might ask, "Just how strong is the model?" and "How close do observed $Y$-values come to the values predicted by the estimated regression line?" To answer these questions, we compute the variance of points about the regression line. This measure (which we now develop) is the sample estimator for $\sigma^2$.

One of the regression assumptions is that the data points have a constant variance about the population regression line. The sum of squared differences of observed $Y$-values from the corresponding sample regression line values, $\hat{Y}$, is denoted by $\text{SSE} = \sum (Y - \hat{Y})^2$. As we might expect, if SSE is divided by the number of degrees of freedom, the mean squared error, MSE, is obtained. We use all $n$ of the sample data points to calculate SSE. However, 2 degrees of freedom are lost because both the population slope and $Y$-intercept values must be estimated by sample statistics. Therefore, there are $n - 2$ degrees of freedom. Hence, the mean squared error, $s_e^2$, is $\text{SSE}/(n - 2)$.

By taking the square root of $s_e^2$, we obtain the *standard error of estimate*, $s_e$. The standard error of estimate is a widely used term, and is really nothing more than an estimate of the *standard deviation of the data points about the regression line*, or the standard deviation of the error term.

### MEAN SQUARED ERROR (AN ESTIMATOR FOR $\sigma^2$)

$$\text{MSE} = \hat{\sigma}^2 = s_e^2 = \text{SSE}/(n-2) = \frac{\text{SS}(Y) - b_1 \, \text{SS}(XY)}{n-2} \qquad (12.3)$$

where $\quad \text{SSE} = \sum (Y - \hat{Y})^2 = \text{SS}(Y) - b_1 \, \text{SS}(XY) \qquad (12.3\text{a})$

$$\text{SS}(Y) = \sum (Y - \bar{Y})^2 = \sum (Y^2) - (\sum Y)^2/n$$

$$\text{SS}(XY) = \sum (X - \bar{X})(Y - \bar{Y}) = \sum (XY) - [(\sum X)(\sum Y)/n]$$

The standard error of estimate is the standard deviation of differences of the data points taken about the regression line.

$$s_e = \sqrt{\frac{\text{SSE}}{n-2}} \qquad (12.4)$$

The standard error of estimate is a simple measure that can be used to evaluate the reliability of the estimated regression equation.

---

### EXAMPLE 12.4

a. What are the residuals ($e$) to hotel occupancy rates ($Y$) based on the estimated regression on East Coast Airlines passengers, $\hat{Y}$? (Refer to Table 12.1.)
b. Use the residuals to develop values for $s_e^2$ and $s_e$. These are used for inference on occupancy rates in Walt Disney World area hotels.

*Solution*    a. Recall the calculation of the first residual value in Example 12.1. All 12 residuals appear here for $\hat{Y} = -8.4879 + 0.9200X$.

| | | | | | | |
|---|---|---|---|---|---|---|
| $Y$ | 40 | 41 | 48 | 49 | 73 | 74 |
| $\hat{Y}$ | 51.94 | 57.38 | 40.87 | 56.07 | 60.49 | 70.29 |
| $e = Y - \hat{Y}$ | $-11.94$ | $-16.38$ | 7.13 | $-7.07$ | 12.51 | 3.71 |
| $Y$ | 68 | 51 | 63 | 75 | 70 | 38 |
| $\hat{Y}$ | 69.32 | 44.91 | 58.49 | 72.11 | 70.06 | 38.07 |
| $e = Y - \hat{Y}$ | $-1.32$ | 6.09 | 4.51 | 2.89 | $-0.06$ | $-0.07$ |

b. We obtain SSE by summing the squared residuals, $\sum (Y - \hat{Y})^2$. Then,

$$s_e^2 = \frac{\text{SSE}}{n-2} = \frac{\sum (Y - \hat{Y})^2}{n-2}$$

$$= \frac{(-11.94)^2 + (-16.38)^2 + \cdots + (-0.07)^2}{12 - 2} = 75.0$$

$$s_e = \sqrt{75.0} = 8.66$$

A check calculation uses

$$s_e^2 = \frac{\text{SSE}}{(n-2)} = \frac{\text{SS}(Y) - b_1\,\text{SS}(XY)}{n-2}$$

$$= \frac{2279 - (0.9200)(1662.5220)}{12 - 2} = 74.95$$

This confirms the earlier calculation to within procedural rounding differences.

The accuracy of the SSE calculation is important in regression inference. First, all error measures for testing and for prediction intervals incorporate $s_e$. Second, it is most common for $s_e$ to be a small value, often near zero. So at least three decimal places should be retained at all steps in the calculation.

**Model Checks**

The residual ($e$) values, along with $s_e$, provide the essentials for ensuring that the assumptions for regression are satisfied. Checks on the assumptions include the following:

1. Determining the percent of $e$-values in each interval $\pm 1s_e$, $\pm 2s_e$, $\pm 3s_e$ as an evaluation of the assumption that the error, $\varepsilon$, is normally distributed.
2. Listing the sign, $+$ or $-$, for the residuals in their original sequence. If the signs appear to occur at random, this suggests that the $\varepsilon$'s are independently distributed.

Other useful properties are defined by the residuals:

3. Calculation of $\sum e$, which should equal zero, gives a check on the calculation of $b$-values.
4. Plots of the residuals, such as $e$ versus $\hat{Y}$ or $e$ versus $X$, can provide leads for steps toward improving the model.

Example 12.5 illustrates these ideas.

**EXAMPLE 12.5**

Are the assumptions of a linear regression reasonable for the data on hotel occupancy and East Coast Airlines passengers? Use information from Example 12.4 (a) to check the model assumptions and (b) to explore possible improvements in the regression description.

*Solution*

a. We check the reasonableness of the model assumptions by evaluating the data, as suggested in model checks 1 and 2.

Assumption I   There is a normal distribution for the errors, $\varepsilon$. From Example 12.4, $\sum e = (-11.94) + (-16.38) + \cdots + (-0.07) = 0$. Then $\bar{e} = \sum e/12 = 0$, and from Example 12.4, where $s_e = 8.66$, we have the values shown in the following table. The actual values are distributed reasonably close to those in a normal distribution. Therefore, the assumption of a normal distribution of errors appears to be acceptable. (With $n = 12$ data points, it would be unusual if actual and theoretical values were identical. But these are quite close!)

| Interval | Actual | Theoretical |
|----------|--------|-------------|
| $\bar{e} \pm s_e = \pm 8.66$ | $9/12 = 75\%$ | $68.3\%$ |
| $\bar{e} \pm 2s_e = \pm 17.32$ | $12/12 = 100\%$ | $95.4\%$ |
| $\bar{e} \pm 3s_e = \pm 25.98$ | $100\%$ | $99.7\%$ |

**Assumption 2**   The $\varepsilon$'s are independently distributed. Again we use the residual values from Example 12.4. Here it suffices to display either a plus sign or a minus sign for each $e$-value: $-, -, +, -, +, +, -, +, +, +, -, -$. Since these signs are mixed, which would suggest a random distribution, the assumption of random errors appears to be reasonable. This question of randomness is confirmed in Chapter 17 when a nonparametric statistic test for randomness in data is developed.

b. What information for possible improvements in the model is provided by the residuals? For this we look to data plots that include $e$-values. These plots can demonstrate whether the residuals contain more information that can be used to improve the description of response. For a strong linear relation the residuals should fall in a horizontal band centered around zero.

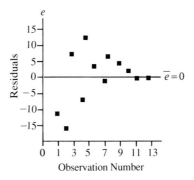

This plot does not show an obvious strong linear pattern.

---

Example 12.5 gave descriptive procedures for checking regression assumptions. Similar techniques are used in Chapter 13 to investigate adding independent variables to form a multiple $X$-variable regression model. Chapters 16 and 17 contain hypothesis-testing procedures for testing goodness of fit for actual data to a theoretical distribution (Section 16.3) and for testing the assumption of random residuals (Section 17.3).

**Problems 12.4**

16. In research conducted among 202 families, an equation of the form $\hat{Y} = 50 + 10X$ was developed, where $\sum Y = 635.6, \sum Y^2 = 4400$, and $\text{SS}(XY) = 80$. Here $X =$ number of persons in the family and $Y =$ weekly expenditure for food. Find $s_e$.

17. Find $s_e$ given the following data: $b_0 = 20, b_1 = 0.20, n = 102, \sum (Y - \hat{Y})^2 = 20{,}000, \sum (Y - \bar{Y})^2 = 42{,}000$.

18. Two regression models were tested on a single data set. One model (1) resulted in $s_e = 0.25$ while the second model (2) showed $s_e = 0.50$. Recognizing

that these results are from sample data, compare the worth of the two models in terms of the *relative variance* displayed for each of them. In terms of least variance, which model is better and by how much?

19. Use the residual values given in the table to compute $s_e$ for that data. Check your result with Equation (12.3).

$$s_e^2 = \frac{SS(Y) - b_1\, SS(XY)}{n-2}$$

given $SS(Y) = 116$, $b_1 = 2.0$, $SS(XY) = 56$.
*Note:* This data is from Problem 4.

| X | -3 | -2 | -1 | 0 | 1 | 2 | 3 |
|---|---|---|---|---|---|---|---|
| Y | 1 | 2 | 3 | 5 | 8 | 11 | 12 |
| $\hat{Y}$ | 0 | 2 | 4 | 6 | 8 | 10 | 12 |

20. Market researchers believe that if potential customers can be convinced to try a product, there is an identifiable *conversion ratio* describing how likely they are to be repeat buyers. One method of motivating the first purchase is by a special coupon offer. The given data summarize the results of a model to predict the response to a trial coupon offer. The data show, from six sample mailings (called *waves*), both the predicted and the actual percentage of people who responded.

| Sampling | Actual | Predicted | Residual Values |
|---|---|---|---|
| | — Percent Response Comparison between Actual and Predicted Trial | | |
| Wave 1 | 22.6% | 21.5% | -1.1% |
| Wave 2 | 19.6 | 22.6 | +3.0 |
| Wave 3 | 28.4 | 24.9 | -3.5 |
| Wave 4 | 20.7 | 22.5 | +1.8 |
| Wave 5 | 23.5 | 22.3 | -1.2 |
| Wave 6 | 22.1 | 23.0 | +0.9 |

Use $s_e = 2.64$ and the residual values to confirm the assumptions for a regression analysis by doing the following steps.

a. Determine the percent of the (6) residual values within $\bar{e} \pm 1s_e$, $\bar{e} \pm 2s_e$, $\bar{e} \pm 3s_e$. Compare to a normal distribution. (*Note:* $n = 6$ may be too small a sample for realistic conclusions.)

b. Plot the residuals in the sequence of the waves 1 through 6. Does there appear to be a systematic pattern in the residuals or a random scatter indicative of a good fit?

# 12.5
# TESTING THE
# REGRESSION MODEL

A procedure for testing the model for a statistically significant relation can now be developed. This procedure, not surprisingly, is a test of the slope of the (hypothesized) regression line, $\beta_1$. Here the size of the residual error $s_e^2$ is a key to determining the strength of an assumed linear regression.

## Hypothesis Test of $\beta_1$

The essential question in regression analysis is, "Does the model help to make better predictions?" The answer is yes if it describes a significant relationship between the independent and dependent variables. A significant relationship is supported or is negated by a test of the slope of the regression line.

Consider the two regression lines depicted in Figure 12.6. For Case 1, the regression line is flat; the slope is zero. That the variable $X_1$ does not predict $Y$ is evidenced in that the estimated regression, $\hat{Y}$, is essentially a constant. A constant is easy to explain (and compute), so a regression line seems unnecessary because $X$ is of no help in predicting $Y$. In contrast, Case 2 illustrates a significant slope. Here $\beta_1 \neq 0$, so $X_2$ can be used to predict $Y$ by a linear regression.

This is the framework for conclusions based on hypothesis tests of $\beta_1$. Neter, Wasserman, and Kutner [4] show that a $t$-statistic is the appropriate sampling distribution for testing $H_0: \beta_1 = 0$ against $H_A: \beta_1 \neq 0$. The sampling distribution for this test is as follows:

**SAMPLING DISTRIBUTION FOR $b_1$**

$$t = \frac{b_1 - \beta_1}{s_{b_1}}$$    (12.5)

with

$$s_{b_1}^2 = \frac{s_e^2}{\text{SS}(X)} \quad \text{and} \quad \text{df} = n - 2$$

where the model assumptions appear with the simple linear regression model (I). The sample size minus the number of model parameters that are estimated determines the number of degrees of freedom. Both the slope and $Y$-intercept

FIGURE 12.6    Two Regressions with Sample Data

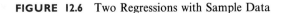

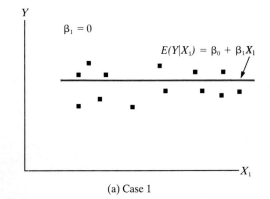

(a) Case 1

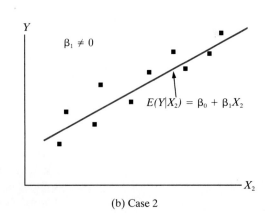

(b) Case 2

have been estimated, so df $= n - 2$. For sample data, the estimated variance about the true slope, $s_{b_1}^2$, replaces the population $\sigma_{b_1}^2$.

Tests on the value of $\beta_1$ follow precisely the procedures outlined in Chapter 9. The sampling distribution in Equation (12.5) provides the test statistic. The test begins with the null hypothesis of no significant slope—that is, the null hypothesis states that the slope is zero. The alternative hypothesis states that the slope is not equal to zero. A two-tailed test is used. A rejection of the null hypothesis provides (statistical) evidence of nonzero slope and supports the conclusion of a meaningful linear regression.

| **EXAMPLE 12.6** | A linear equation was developed earlier relating hotel occupancy ($Y$) to the number of East Coast Airlines passengers ($X$). Is the simple linear regression significant at a level of $\alpha = 0.05$? If so, it would be reasonable to use the model to predict vacancy rates for hotels in the Walt Disney World area. |

*Solution*  The test is $H_0$: $\beta_1 = 0$ versus $H_A$: $\beta_1 \neq 0$. The essential statistics are

$$b_1 = 0.9200 \quad s_e^2 = 75.0000 \quad SS(X) = 1807.1485 \quad n = 12$$

Test: Reject $H_0$ if calculated $|t| > t_{0.025,10} = 2.228$.

Here,

$$s_{b_1} = \sqrt{\frac{s_e^2}{SS(X)}} = \sqrt{\frac{75.0000}{1807.1485}} = 0.2037$$

so

$$\text{Calculated } t = \frac{b_1}{s_{b_1}} = \frac{0.9200}{0.2037} = 4.52$$

Decision: Reject $H_0$.

Conclusion:  We conclude that the East Coast Airlines passenger counts provide a meaningful prediction of Walt Disney World area hotel occupancy rates. This estimated simple linear regression *can* give meaningful predictions.

The next step in regression inference is to make the prediction. However, caution is necessary. For example, if East Coast Airlines went on strike (which is possible), its passenger count would be zero and the estimated regression would project a negative occupancy rate! This is, of course, an impossible value. Such a use of the regression model is invalid because a zero passenger count is far outside the range of passenger counts that was used to develop the estimated regression.

A second consideration for Example 12.6 is the possibility of improving the predictions by adding more independent variables to the model. For example, would predictions be better if other air carriers' counts were included? Would it be helpful to incorporate passenger counts for ground traffic that stops at Florida visitor centers? Chapter 13 explains how to improve predictions by including more independent ($X$) variables.

The statistical concepts that are developed here enable us to make additional inference about the slope of the regression line. This section concludes with a discussion of the estimation of $\beta_1$.

**Confidence Interval Estimation of $\beta_1$**

Having rejected the hypothesis of zero slope, we next make an interval estimate of the true slope. This follows the logic of the *pivotal method* for confidence interval estimation described in Chapter 8. The pivotal statistic, a *t*-statistic, is defined by Equation (12.6).

**CONFIDENCE INTERVAL ESTIMATION OF $\beta_1$**

$$(b_1 - ts_{b_1}) < \beta_1 < (b_1 + ts_{b_1}) \tag{12.6}$$

The data for Walt Disney World area hotel occupancy provides an illustration.

---

**EXAMPLE 12.7**

For the Walt Disney World area data, $\beta_1$ = the true increase in hotel occupancy rate at selected hotels per 1000 East Coast Airlines passengers. What is the 95% confidence interval estimate of $\beta_1$?

*Solution*    We use the confidence interval procedure in Equation (12.6) and $s_{b_1} = 0.2037$ from Example 12.6. Then, under the assumptions for inference in regression and for $n = 12$,

$$(b_1 - ts_{b_1}) < \beta_1 < (b_1 + ts_{b_1})$$

$$(0.920 - [2.228 \cdot 0.2037]) < \beta_1 < (0.920 + [2.228 \cdot 0.2037])$$

$$0.47\% < \beta_1 < 1.37\%$$

Conclusion:   The rate of increase in percent occupancy at Walt Disney World area hotels is somewhere between $+0.47$ and $1.37$ (%) per 1000 East Coast passengers.

---

This section started with the hypothesis that $\beta_1 \neq 0$. The hypothesis was rejected on the basis of a two-tailed rejection region. A $[(1 - \alpha) \cdot 100]\%$ confidence interval can be used for the same determination. A confidence interval that includes zero tells us that the slope of the regression line might very well be zero. This is equivalent to using a two-tailed hypothesis test and finding that $\beta_1 = 0$. Conversely, for a meaningful regression at the $\alpha$-risk level, the $[(1 - \alpha) \cdot 100]\%$ confidence interval for $\beta_1$ excludes zero, so we conclude that $\beta_1 \neq 0$.

It is reasonable to expect a business analyst to indicate the accuracy of prepared estimates. The preceding example illustrates the error in estimating $\beta_1$ as a result of using $b_1$. Similarly, regression predictions are subject to sampling error. The next section extends point predictions to confidence intervals for average regression values to predictions of an estimated response range for individual values.

**Problems 12.5**

21. The table shows sales, $Y$, for each of eight salespeople. Years of sales experience are also recorded for each.

| X (years) | 6 | 3 | 2 | 4 | 1 | 6 | 3 | 5 |
|-----------|---|---|---|---|---|---|---|---|
| Y ($000) | 8 | 5 | 2 | 3 | 3 | 9 |   | 6 |

a. Given $\bar{Y} = 5$, find the value missing from the table.

b. Estimate $\beta_1$, the mean increase in sales for each additional year of experience. Use $s_{b_1} = 0.27$ and 95% confidence. Develop a point estimate; then place bounds of error on your estimate by computing the upper and lower confidence limits for the slope.

22. a. Plot the data given in the table. Does a linear regression appear meaningful?

| X | −2 | −1 | 0 | 1 | 2 |
|---|----|----|---|---|---|
| Y | 0 | 1 | 0 | 1 | 3 |

b. The data above are repeated from Problem 6. Use the values for $b_0$ and $b_1$ found there (or recalculate) and $s_{b_1} = 0.28$ to test $H_0: \beta_1 = 0$ against $H_A: \beta_1 \neq 0$. Use $\alpha = 0.05$ and consider that assumptions for inference are met. State your conclusions.

23. Suppose you are the production manager for a firm at which productivity for each worker can be described as the number of units of output, $Y$, for $\hat{Y} = 400 + 2X$, where $X$ is years of experience in this trade. Let $n = 25$ and $s_{b_1} = 0.71$. Write two or three sentences in response to each question.

a. Is there a meaningful linear relation between units produced and years of experience? (*Note:* The given information is sufficient to make this test.)

b. If there is a meaningful linear relation, estimate the increase in units produced for each additional year of experience. Give bounds and explain their meaning.

c. If a linear relation is not meaningful, describe the steps you would take to discern a meaningful relation between $Y$ and $X$.

24. For the equally spaced data in the table, compute a 95% confidence interval for estimation of $\beta_1$. Use $\hat{Y} = 6 + 2X$ and $s_{b_1} = 0.17$. (*Note:* These data were also considered in Problem 9.)

| X | −3 | −2 | −1 | 0 | 1 | 2 | 3 |
|---|----|----|----|---|---|---|---|
| Y | 1 | 2 | 3 | 5 | 8 | 11 | 12 |

25. A market research firm supplies names of potential customers to a list broker, who sells the lists of names to noncompetitive customers for use in direct mail advertising. The market research firm and the brokers share in the revenue from a sale. In one case the broker sold a list containing the names of antique collectors to an agency that specializes in continental

tours. The regression analysis on 46 records gave $b_1 = 1.08328$ and $s_{b_1} = 0.25713$. Test the hypothesis that $\beta_1 = 0$ for $\alpha = 0.05$. Can customer names of antique collectors be used to predict participation in continental tours?

## 12.6 INFERENCE BASED ON THE REGRESSION LINE

A regression equation is used to predict values of the dependent variable from the values of an independent variable. This involves predicting either (1) the regression value, $E(Y|X)$, which is an average, or (2) an individual response value, $Y$.

Whether interest lies in the *mean response* or in an *individual prediction* depends upon the purpose of the regression. If the prediction of a response that will occur on the average (in the long run) is desired, the regression value is used. In contrast, inference about a specific individual response requires an additional component associated with the individual's unique variation.

To illustrate, suppose the owner of a pizza chain runs a coupon advertisement in a local newspaper. The owner of the stores is most interested in the area-wide average number of coupons that are redeemed. Yet the manager of a single store would surely be more interested in that individual location. Both situations, average response and individual response, are considered here.

Point estimation is the same for an individual response as for the average response. Both estimates are based on the model in Equation (12.7).

$$Y = \underbrace{\beta_0 + \beta_1 X}_{\substack{\text{Mean} \\ \text{response}}} + \underbrace{\varepsilon}_{\substack{\text{Unique response} \\ \text{(to the individual)}}} = E(Y|X) + \varepsilon \qquad \text{(12.7)}$$

Since $E(\varepsilon) = 0$, then the point estimates have identical expected values:

|  FOR  |  FOR  |
| population mean response | individual response |

$$E(Y|X) = \beta_0 + \beta_1 X \qquad E(Y_{\text{indiv}}) = \beta_0 + \beta_1 X \qquad \text{(12.8)}$$

USE
sample estimates defined by

$$\hat{Y} = b_0 + b_1 X \qquad \text{AND} \qquad \hat{Y} = b_0 + b_1 X$$

Both require assigning a value for the independent variable ($X$), thereby calculating a predicted response. The point estimator for both is $\hat{Y} = b_0 + b_1 X$.

**Interval Estimation: Confidence and Prediction Intervals**

There is a difference in the *error* associated with these predictions. This difference relates to the variances of individuals about an average response because individuals have an additional (unique) source of variation beyond the average. The difference is calculated from the relation in Equation 12.7.

$$Y_{\text{indiv}} = E(Y|X) + \varepsilon$$

The variances are additive so that Individual response variance = Variance due to regression + Random error variance. That is,

$$\text{Variance } (Y_{\text{indiv}}) = \text{Variance } (E(Y|X)) + \text{Variance } \varepsilon$$

This relation is estimated from the sample data by

$$s^2_{\text{indiv}} = s^2_{\hat{Y}} + s^2_e$$

Since $s^2_e$ is zero or larger, the estimated variance for an individual prediction, $s^2_{\text{indiv}}$, equals or exceeds that for average response, $s^2_{\hat{Y}}$. This relation is generalized for establishing *confidence limits for the estimation of mean response* and *prediction limits for the estimation of individual response*. The narrower limits in Figure 12.7 are the upper and lower confidence limits, UCL and LCL, that describe possible variations in mean response. These limits identify the $[(1 - \alpha) \cdot 100\%]$ confidence limits for a regression line.

The fact that individual response varies more than mean response is illustrated by the wider interval for individuals in Figure 12.7. To summarize:

- Confidence limits are for estimation of mean response, $E(Y|X)$ (dark-shaded area in Figure 12.7).
- Prediction limits are for estimation of an individual's response, $Y$ (light-shaded area) and provide a wider interval.

> A $[(1 - \alpha) \cdot 100]\%$ *confidence interval for mean response* adjusts the estimated regression value (a point estimate) for variation in the estimated mean response at the $(1 - \alpha)$ probability level.
>
> A $[(1 - \alpha) \cdot 100]\%$ *prediction interval for individual response* adjusts the point estimate by the variability in estimated mean response *plus* the unique variation for the individual at $(1 - \alpha)$ probability.

The *only* distinction for the calculation of confidence intervals (mean response) versus prediction intervals (individuals) is in the variance calculation. The variance for estimated (mean) regression incorporates variations due to the slope and the $Y$-intercept estimates. The variance for individual response estimates *adds* variations due to individual differences. The sampling distributions are summarized next.

---

**FIGURE 12.7**   Comparison of Confidence Limits for *Mean Response* (Narrower) with Prediction Limits for *Individual Response* (Wider)

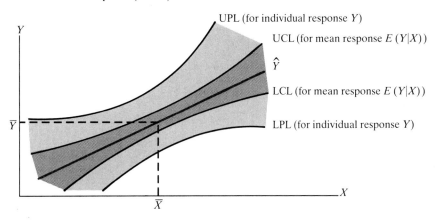

**ESTIMATED SAMPLING DISTRIBUTIONS
FOR (1) MEAN AND (2) INDIVIDUAL RESPONSE**

(1) For mean response,            (2) For individuals,

$$t = \frac{\hat{Y} - E(Y|X)}{s_e \sqrt{\dfrac{1}{n} + \dfrac{(X - \bar{X})^2}{SS(X)}}} \quad (12.9) \qquad\qquad t = \frac{\hat{Y} - Y}{s_e \sqrt{1 + \dfrac{1}{n} + \dfrac{(X - \bar{X})^2}{SS(X)}}} \quad (12.10)$$

Both have df $= n - 2$.

**ESTIMATION INTERVALS**

Confidence limits on $E(Y|X)$

$$(\text{LCL, UCL}) = \hat{Y} \pm \left( t \cdot s_e \sqrt{\frac{1}{n} + \frac{(X - \bar{X})^2}{SS(X)}} \right) \qquad (12.11)$$

Prediction limits on $Y$

$$(\text{LPL, UPL}) = \hat{Y} \pm \left( t \cdot s_e \sqrt{1 + \frac{1}{n} + \frac{(X - \bar{X})^2}{SS(X)}} \right) \qquad (12.12)$$

The estimation of true variances $\sigma_{\hat{Y}}^2$ and $\sigma_{\text{indiv}}^2$ results from the approximation of $\sigma^2$ by the sample estimator $s_e^2$.

**Predicting Mean Response**

Example 12.8 illustrates the use of *confidence limits* for the estimation of a mean response.

---

**EXAMPLE 12.8**

Use the model developed in Example 12.2. If the East Coast Airlines passenger count were 70,000, compute the 95% confidence interval estimate of mean occupancy for the selected Walt Disney World area hotels. (Consider the assumptions for inference to be reasonably satisfied.)

*Solution*    Using the regression equation developed previously, we have for $X = 70.0$ (thousand) passengers,  $\hat{Y} = -8.4879 + 0.9200(70.0000) - 55.9121$,  and  $s_e = \sqrt{75.0000} = 8.6603$. Then

$$s_{\hat{Y}} = s_e \sqrt{\left( \frac{1}{n} + \frac{(X - \bar{X})^2}{SS(X)} \right)}$$

$$= 8.6603 \sqrt{\left( \frac{1}{12} + \frac{(70 - 71.7283)^2}{1807.1485} \right)} = 2.5247$$

Also, for 95% confidence and $12 - 2 = 10$ df, $t = 2.228$, so a 95% confidence interval on $E(Y|X)$ is

$$\hat{Y} \pm t s_e \sqrt{\frac{1}{n} + \frac{(X - \bar{X})^2}{SS(X)}} = 55.9121 \pm [(2.228)(2.5247)] = 55.9 \pm 5.6$$

Conclusion:   $50\% < E(Y|70,000) < 62\%$. The estimated average hotel occupancy rate is between 50% and 62% for a month with 70,000 East Coast Airlines passengers.

**Predicting Individual Response**

This procedure is easily extended to illustrate the prediction of individual values. Example 12.9 illustrates this by posing a question for a business analyst.

**EXAMPLE 12.9**

A marketing analysis of 40 sales markets used regression modeling to estimate the demand for a breakfast cereal. The data base was derived from returns to an advertising campaign involving newspaper coupons. The research department then compiled these statistics:

$$\hat{Y} = \$65,000 + \$40X$$

where    $Y$ = dollar sales per day for March

$X$ = the number of coupons returned per day

and $\beta_1 \neq 0$ was concluded from a preliminary statistical test. Also, $s_e = \$2857$. For $X = 300$ coupons/day,

$$\sqrt{\frac{1}{n} + \frac{(X - \bar{X})^2}{SS(X)}} = 0.0858$$

a. Find a 95.5% confidence interval estimate of average annual sales for all markets with 300 coupons redeemed per day. An estimated $18,000 gross sales is required to achieve a suitable profit level. Is an *overall* suitable profit level achieved if 300 coupons are redeemed per day?
b. Because an individual has more variation than a mean response, the greater benefit (or loss) from using coupons rests with the individual market. What profit interval is expected for the Rochester, Minnesota, market, which achieved 300 coupons per day? The research department has calculated

$$\sqrt{1 + \frac{1}{n} + \frac{(X - \bar{X})^2}{SS(X)}} = 1.0037$$

*Solution*    a. The point estimate for $E(Y \mid X) = 300$ is $\hat{Y} = \$6500 + \$40(300) = \$18,500$, and we can use the Z-distribution because $n > 30$. For 95.5% confidence,

$$LCL = \hat{Y} - \left( Zs_e \sqrt{\frac{1}{n} + \frac{(X - \bar{X})^2}{SS(X)}} \right)$$

$$= \$18,500 - [2(\$2857)(0.0858)]$$

$$= \$18,500 - \$490 = \$18,010$$

$$UCL = \hat{Y} + \left( Zs_e \sqrt{\frac{1}{n} + \frac{(X - \bar{X})^2}{SS(X)}} \right) = \$18,500 + \$490$$

$$= \$18,990$$

$$\$18,010 < \text{Average sales given 300 coupons} < \$18,990$$

Conclusion (a)    Sales average from $18,010 to $18,990 for markets in which 300 coupons are redeemed per day. The product would be profitable to the company if marketed at this level. There is 95.5% confidence that this procedure has produced an interval that includes the true average sales.

b. The estimated sales remain at $\hat{Y} = \$18,500$. But the prediction limits are different. The expected profit interval for the individual market–Rochester, Minnesota, is

$$\hat{Y} \pm Z s_e \sqrt{1 + \frac{1}{n} + \frac{(X - \bar{X})^2}{SS(X)}} = \$18,500 \pm [(2)(\$2857)(1.0037)]$$

$12,765$ (LPL) $<$ Individual market sales $<$ $24,235$ (UPL)

Conclusion (b)   The point estimate remains at \$18,500. However, this *prediction* interval extends from \$12,765 to \$24,235. It is over ten times as wide as the comparable *confidence* interval for an average market at this coupon-redemption level!

Overall conclusion:   In this situation, the manufacturer has reasonable assurance of a profitable distribution. This does not hold for the individual markets. Sales at the individual market level show a substantial variation. Some individual markets may have sales below a marginal profit level.

---

As expected, the last example displayed individual prediction limits that were substantially wider than the confidence limits for the regression. This increase stems from the substantial inherent variability due to individual error that is not measured by the regression.

Observe in Figure 12.7 how the confidence intervals (dark-shaded) flare out as $X$ takes values further away from $\bar{X}$. Predictions have a different accuracy depending on how far $X$ is from $\bar{X}$. Thus the error for a confidence (or prediction) interval differs only in the calculation of $(X - \bar{X})^2$ for alternative values of $X$. The most accurate predictions are at $X = \bar{X}$, and the error increases for $X$-values further away from $\bar{X}$. This is another reason for restricting predictions to the range of $X$-values used for developing the estimated regression model.

**Problems 12.6**

26. In trying to relate sales of an agricultural chemical (tons) to average daily temperature in a given region, company analysts used 102 observations to derive the expression $\hat{Y} = 320 + 8X$, where $X$ is the maximum weekly temperature in degrees Fahrenheit. In addition, $\sum (Y - \bar{Y})^2 = 490,000$.
   a. Find the standard deviation of regression.
   b. Develop a 90% confidence interval estimate of the average tons sold when the maximum weekly temperature is $X = 100°F$. Let

   $$\sqrt{\frac{1}{n} + \frac{(X - \bar{X})^2}{\sum (X - \bar{X})^2}} = 0.1$$

   c. Predict the sales on an individual week when the temperature goes to $X = 100°F$. Let

   $$\sqrt{1 + \frac{1}{n} + \frac{(X - \bar{X})^2}{\sum (X - \bar{X})^2}} = 1.0$$

27. A regression equation is $\hat{Y} = 12 + 8X$, where $n = 100$ and $s_e = 4$. Develop a 99.7% *confidence interval* estimate for the average response of $Y$ when $X$

takes on a value of 10. Let

$$\sqrt{\frac{1}{n} + \frac{(X - \bar{X})^2}{\sum (X - \bar{X})^2}} = 0.1$$

28. Standard Oil is mailing two letters to its credit-card holders, each letter offering a different product for sale. The first mailing is an offer involving compact disks and the second involves silver-plated jewelry. The company wishes to predict the percentage of people in a ZIP code area who will buy the product offered on the second mailing based on percent who purchased in the first mailing.

| Product | Percentage of Cardholders Who Purchased in Each Area | | | | | | | | | | |
|---|---|---|---|---|---|---|---|---|---|---|---|
| | *A* | *B* | *C* | *D* | *E* | *F* | *G* | *H* | *I* | *J* | *K* |
| Compact disks, *X* | 2.65 | 2.80 | 2.21 | 2.39 | 1.81 | 2.30 | 1.64 | 1.97 | 1.45 | 1.65 | 1.28 |
| Silver jewelry, *Y* | 4.16 | 3.93 | 3.78 | 3.41 | 3.36 | 3.25 | 3.16 | 2.91 | 2.85 | 2.79 | 2.23 |

An estimated regression is $\hat{Y} = 1.26 + 0.99X$, where $s_e = 0.30$.
a. Give a point estimate of the average percent purchase of silver for areas with 2.50% purchase of compact disks.
b. Expand this to a 95% confidence estimate of the true average percent who purchase silver from all areas with a 2.50% purchase of compact disks.
c. Assume 2.50% of the people in area *M* purchased compact disks. Establish a 95% prediction interval for the response to the second mailing for this area.

29. The Universal Microchips Company employs substantial numbers of people for work in sorting, assembling, and packing small units of its merchandise. In an attempt to improve the selection of its employees, Universal is evaluating a dexterity test, which is supposed to predict an individual's production rating. The top rating is 100 and the lowest rating is 1. The test was given to 25 employees who are currently working at Universal and for whom a production rating was given. The available statistics include the following sums.

$$\sum X = 1373 \quad \sum Y^2 = 117{,}690 \quad Y = \text{production rating}$$
$$\sum Y = 1656 \quad \sum X^2 = 84{,}113 \quad X = \text{dexterity test score}$$
$$\sum XY = 99{,}013$$

Also, $\hat{Y} = 15.16 + 0.93X$, with $s_e = 4.485$.
a. What is the expected production rating for a test score of 60? Give both (i) a point estimate and (ii) a 95% confidence interval.
b. Susan Ladd was among those tested at Universal. Her dexterity test score was 60. Does the estimated prediction interval include her actual score, for 95% probability? Her production rating is 75.

30. The accounting department at State U. has established a linear regression for standard college entrance exam scores ($X$) and first-year grade average ($Y$) for its accounting majors. Available statistics include

$$\hat{Y} = 0.751 + 0.00435X \quad s_e = 0.194 \quad \bar{X} = 540 \quad SS(X) = 22{,}000 \quad n = 88$$

Answer each question, keeping in mind the distinction between interval estimates for individuals and interval estimates for the regression.

a. As a senior majoring in accounting, you are asked to advise incoming students how they, individually, can have the most success in choosing a major. What success in first-year accounting would you predict for an individual who has just scored 500 on the college entrance exam?

b. As an academic student recruiter, you visit numerous high schools to talk about the university programs. In a group session the following question is asked: "What success would you expect for the (average) new student who scores 500 on the college entrance exam?" How would you answer?

31. Given $s_e = 0.36$, $s_{\hat{y}} = 0.20$, $s_{indiv} = 1.20$, and $n = 100$, a 95% interval estimate on $E(Y|X)$ gives the result shown in the figure. Something is wrong. What is it?

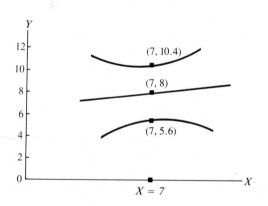

## 12.7 CORRELATION ANALYSIS

Suppose you have graphed a data set and you feel that a meaningful linear relation exists. You use the least squares procedure to get the "best" estimators for slope and $Y$-intercept and write the regression model equation. Next you want to know if the model can provide meaningful predictions. You test the hypothesis that $\beta_1 = 0$ to answer that question. Now you might ask, "To what extent can one variable be used to describe another?" That is, how *close* is the linear relation of the two variables? This question concerns correlation, which is our next topic for discussion.

### Coefficient of Determination

Correlation is a measure of association. To understand correlation, we first need to know how association is measured. The technique is to split the total sum of squares for $Y$ into a regression sum (SSR) plus an error sum (SSE).

$$\text{Total SS} = \begin{bmatrix} \textbf{Sum of squares} \\ \textbf{for regression} \end{bmatrix} + \begin{bmatrix} \textbf{Sum of squares} \\ \textbf{for error} \end{bmatrix} \qquad \text{(12.13)}$$

$$\text{SS}(Y) = \text{SSR} + \text{SSE}$$

$$\Sigma(Y - \bar{Y})^2 = \Sigma(\hat{Y} - \bar{Y})^2 + \Sigma(Y - \hat{Y})^2$$

SSR, the sum of squares for regression, is a new term, but SSE is a familiar concept. The interrelations are illustrated in Figure 12.8.

The objective in regression modeling is to explain as much as possible of the total sum of squared deviations of the dependent variable $\text{SS}(Y)$. This $\text{SS}(Y)$ represents the sum of squared deviations of points $(Y)$ from their mean, $\bar{Y}$. The SSR represents the variations in $Y$ that are explained by the regression model, and SSE describes the unexplained variations.

Relative to the $\text{SS}(Y)$, when SSR is large than SSE is small, and vice versa. Figure 12.8 displays two extremes.

**Case I**   In Figure 12.8, Case 1, SSE (unexplained) is small, in that $\text{SSE}/\text{SS}(Y)$ is close to zero. Then $X$ and $Y$ are highly associated. The data points are close to the regression line and the correlation is high.

**Case 2**   SSE is large, approaching $\text{SS}(Y)$. Then $X$ and $Y$ have a low association. In Figure 12.8, Case 2, the data are widely dispersed about the regression line and the linear correlation is weak.

**FIGURE 12.8**   Measuring the Correlation for Two Variables, Where "Explained" Contributes to the Sum of Squares for Regression (SSR); "Unexplained" Contributes to Error (SSE); and "Total" Contributes to $\text{SS}(Y)$

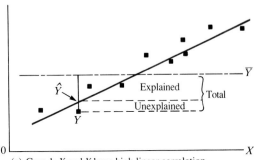

(a) Case 1: $X$ and $Y$ have high linear correlation

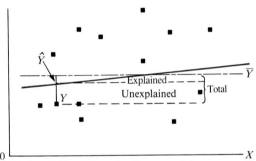

(b) Case 2: $X$ and $Y$ have low linear correlation

In Case 1, high values for $Y$ are directly related to high values for $X$, middle values for $Y$ relate to middle-size $X$, and low $Y$ values related to low $X$'s; or there is a reversal. Because the data lie close to the regression line, the correlation is *strong*. It is also *positive* (a direct relation), not *negative* (a reversal). In contrast, Case 2 illustrates a weak linear correlation. Here, sometimes a high value of $Y$ corresponds to a high value of $X$, sometimes a high value of $Y$ corresponds to a middle or low value of $X$.

The *coefficient of determination* is a measure for the degree of association between two variables. It is a ratio that compares the sum for regression (explained) variations to the total variations.

> The coefficient of determination, $r^2$, gives the percentage of total variations in $Y$ that is explained by the linear regression with the $X$-variable.

$$r^2 = \frac{\text{SSR}}{\text{SS}(Y)} = \frac{\text{SS}(Y) - \text{SSE}}{\text{SS}(Y)} = 1 - \frac{\text{SSE}}{\text{SS}(Y)} \qquad (12.14)$$

A sum of squares must equal or exceed zero. By definition, $\text{SS}(Y)$ equals or exceeds SSE. These conditions provide us with sufficient information for establishing bounds on the size of $r^2$. Refer again to Figure 12.8.

In **Case 1**: SSE, the unexplained, is much smaller than $\text{SS}(Y)$, and $\text{SSE}/\text{SS}(Y)$ is near 0. Then $r^2 = 1 - (\text{SSE}/\text{SS}(Y)) \doteq 1 - 0$, so $r^2$ approaches 1. Here the data points reside close to the regression *line*.

In **Case 2**: SSE approaches the size of $\text{SS}(Y)$. Then

$$r^2 = 1 - \frac{\text{SSE}}{\text{SS}(Y)} \doteq 1 - 1$$

so $r^2$ approaches 0. Here the data are scattered and generally do not fall close to the regression line. So, $0 \leqslant r^2 \leqslant 1$ defines the range of values for a coefficient of determination.

We rarely encounter the extreme values (of 0 and 1) in calculating $r^2$. Interest is thus in interpreting values between 0 and 1. Suppose that the following coefficients of determination between the stated variables are given:

$$\text{(Sales and years of experience)} \qquad r^2 = 0.4$$

$$\text{(Sales and hours worked)} \qquad r^2 = 0.8$$

Then years of experience are associated with, or explain, 40% of the variations in sales. Similarly, hours worked are associated with, or explain, 80% of the variations in sales. The association between sales and hours worked is twice as strong as the association between sales and years of experience. It may be that other variables (for example, advertising expense) have an even higher association. Part of the regression/correlation modeling process is to find the best variable to use in the regression model and the degree of association that can be directly related to regression. Just how correlation and regression are related is the next topic.

**A Coefficient of
Correlation**

The coefficient of correlation is a widely used measure of association that is derived from the coefficient of determination.* The coefficient of correlation is also used to answer the question of how close the linear relation is for two variables. It was first derived by the English scientist Karl Pearson and is sometimes called Pearson's product moment *coefficient of correlation.*

> The coefficient of correlation (Pearson) is a measure of the amount of variation that is shared by two variables. It is often used as an indicator of the strength of a linear relation.**

$$r = \pm \sqrt{\frac{SSR}{SS(Y)}} = \frac{SS(XY)}{\sqrt{SS(X) \cdot SS(Y)}} \qquad \text{(12.15)}$$

The sign of $r$ in Equation (12.15) is the same as that for the slope of the associated linear regression. If the regression line has a positive slope, then $Y$ increases as $X$ increases and the correlation between the variables has a positive $(+)$ coefficient. If the regression line has a negative slope, then $Y$ decreases as $X$ increases and the correlation between the variables is negative $(-)$. The first expression in Equation (12.15) requires us to supply the sign for $r$, either $+$ or $-$. The second expression automatically gives the sign from the calculation of $SS(XY)$. The sign can be checked for agreement with the direction of slope for the associated regression line.

The name *correlation* is appropriate because $r$ is a measure of co- (for common) relationship. The symbol for the correlation in the population is $\rho$ (rho). In this text, the work is primarily with samples, so $r$, the (sample) estimated coefficient of $\rho$, is used.

Recognizing that $r = \sqrt{r^2}$, we see that the bounds for $\rho$ are, simply, $\sqrt{0} = 0$ and $\pm\sqrt{1} = \pm1$. So $-1 \leqslant r \leqslant +1$. In fact, for $r = \pm1$, both give $r^2 = +1$, so *a correlation of $r = -1$ represents an equal amount of explained variation as a correlation of $r = +1$. Consequently, $r = +1$ and $r = -1$ are equally strong.* The sign, whether plus or minus, simply indicates the direction of the association. Similarly, $r = -0.6$ signals a relation as strong as the one indicated by $r = +0.6$. Also, $r = -0.8$ indicates a slightly stronger linear relation than $r = +0.79$.

Figure 12.9 illustrates values of $r$ ranging from the extremes of $-1$ (part (a)) to $+1$ (part (f)).

Parts (c) and (d) display a weak, or nonexistent, linear relation. Hence $r = 0$, and the graph is a line with zero slope. Though not linear, the data in

---

\* The coefficient of correlation ($r$) is the square root of the coefficient of determination ($r^2$).

$$r^2 = \frac{SSR}{SS(Y)} = \frac{\left(\frac{SS(XY)}{SS(X)}\right) \cdot SS(XY)}{SS(Y)} = \frac{[SS(XY)]^2}{SS(X) \cdot SS(Y)} \qquad \text{Then } r = \sqrt{r^2}.$$

\*\* A more convenient expression for hand-calculation is

$$r = \frac{\sum XY - \frac{\sum X \sum Y}{n}}{\sqrt{\left(\sum X^2 - \frac{(\sum X)^2}{n}\right)\left(\sum Y^2 - \frac{(\sum Y)^2}{n}\right)}}$$

**FIGURE 12.9**  A Display of Linear Correlations over the Range $-1$ to $0$ to $+1$

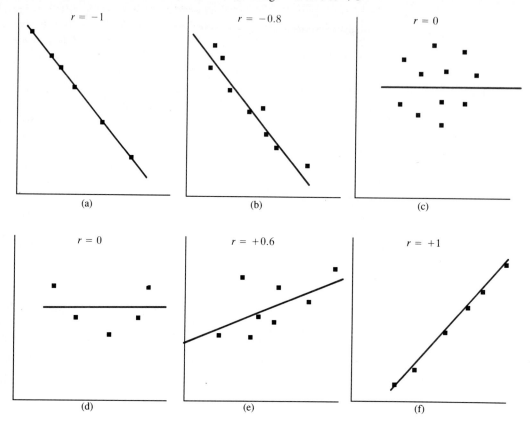

part (d) hold a strong pattern; there is a strong curvilinear relation between $X$ and $Y$. The coefficient of correlation measures the strength of the relationship between variables for a simple linear regression. It does not discern the strength of the relationship for other regression models.

---

**EXAMPLE 12.10**

We have developed an estimated regression for hotel occupancy rate in the Walt Disney World area from passengers of East Coast Airlines. What is the strength of that relation?

*Solution*   Earlier calculations provide the following values:

$$\text{SS}(XY) = 1662.5220 \quad \text{SS}(X) = 1807.1485 \quad \text{SS}(Y) = 2279$$

Then,

$$r = \frac{\text{SS}(XY)}{\sqrt{\text{SS}(X) \cdot \text{SS}(Y)}} = \frac{1662.5220}{\sqrt{1807.1485 \cdot (2279)}} = 0.8192$$

The linear correlation is 0.8192.

---

*Note*: In Example 12.10 the correlation is positive. Many correlations are negative. It can be reassuring to check the sign, whether $+$ or $-$, for the calculation. Several checks are available. First, the correlation coefficient and the estimated slope *always* have the same sign. (Recall Example 12.2, where $b_1 = +0.9200$).

Also, when $r$ is positive, the scatter plot of data points generally moves upward to the right. For a negative $r$, the data points generally move downward to the right. Finally, for $r = 0$, the data points are nearly flat, or else they are scattered with no apparent linear pattern. These quick checks can be helpful when you calculate or interpret correlations.

What is a strong correlation? Is 0.6 or 0.8 strong, or must it be higher yet? Several approaches exist to answer this question. One is to calculate $r^2$ and then interpret from it the *percent of total variation (in Y) that is associated with (or explained by) the regression on X*. Second, numerous computer software packages provide a level of statistical significance for correlations (for example, see *SPSS Reference Guide* [8]). Such checks can be useful. But what is deemed a "close" association in one field (marketing) may be different in another (finance). Research that reports correlations frequently includes the significance level for the association, such as 1% or 5%.

**Inference about the Correlation Coefficient**

Under the assumptions for the linear regression model (I) and with $H_0$: $\rho = 0$, a $t$-statistic provides the sampling distribution for inference concerning $\rho$. There are $n - 2$ degrees of freedom because $\bar{X}$ estimates $\mu_X$ and $\bar{Y}$ estimates $\mu_Y$.

Tests of hypothesis of $\rho = 0$ for $r$ use*

$$t = \frac{r - \rho}{\sqrt{(1 - r^2)/(n - 2)}} \quad \text{with df} = n - 2 \qquad (12.16)$$

With these conditions, a test for *zero* correlation is equivalent to a test for zero slope in simple linear regression.

---

**EXAMPLE 12.11**

Again, referring to the data of Example 12.1, is there a significant linear correlation between the hotel occupancy rate ($Y$) and the number of East Coast Airlines passengers ($X$)? Use $\alpha = 0.05$, and $r = 0.8192$.

*Solution*   This is a test for zero correlation.

$$H_0\text{: }\rho = 0 \qquad H_A\text{: }\rho \neq 0$$

Test: Reject $H_0$ if calculated $|t| > t_{0.025,10}$, which $= 2.228$.

$$\text{Calculated } t = \frac{r - 0}{\sqrt{(1 - r^2)/(n - 2)}} = \frac{0.8192}{\sqrt{(1 - 0.8192^2)/(12 - 2)}} = 4.52$$

---

* With a conclusion $\rho = 0$, a confidence interval for $\rho$ is

$$r \pm (t_{\alpha/2}\sqrt{(1 - r^2)/(n - 2)}) \qquad (12.17)$$

But of most interest is estimation of $\rho$ for the *general case*, $\rho \neq 0$. Then, for large $n$ (common use is $n \geqslant 25$), an appropriate sampling distribution for testing $H_0$: $\rho = \rho_0 (\neq 0)$ is

$$Z = (Z' - \rho')/\sigma(Z') \qquad (12.18)$$

where $\quad Z' = \dfrac{1}{2}\log_e\!\left(\dfrac{1+r}{1-r}\right) \quad$ and $\quad \sigma(Z') = 1/\sqrt{n - 3}.$ $\qquad (12.19)$

Then an interval estimate for $\rho$ uses $Z' \pm (Z_{\alpha/2} \cdot \sigma(Z'))$.

Conclusion:  $\rho \neq 0$, because calculated $t = 4.52 > 2.228$. There is a significant linear correlation between $Y$ and $X$. (Note that this test is an alternative to the test for a meaningful regression for hotel occupancy (Example 12.6), where the test was $H_0$: $\beta_1 = 0$. In both cases we arrived at the same calculated $t$-value: 4.52. There is a significant linear regression for $Y$ on $X$.)

---

Computer software packages (including SAS, SPSS, and BMDP) provide significance levels for interpreting the strength of correlations. Significance is meaningful when the sample is a reasonable size, but it becomes meaningless for very large samples. In such samples even chance associations can be significant because for large $n$, the divisor in Equation (12.16) is controlled by $n - 2$ rather than by $1 - r^2$. So the calculated $t$-value is insensitive to the $r$-value. For example, with a sample size of $n = 902$ and a "low" correlation coefficient of $r = 0.10$,

$$\text{Calculated } t = \frac{0.10}{\sqrt{[1 - (0.10)^2]/(902 - 2)}} = 3.02$$

This is statistically significant at the $\alpha = 0.001$ level. But on a scale of 0 to 1, the 0.10 is a weak correlation, explaining only 1% of the variance. Therefore, for sample sizes larger than $n = 100$, compare your correlations with similar research in the literature. Practical usefulness is described by the strength of your correlation compared to that achieved by similar research.

**Correlation Aside from Regression**

It is possible to obtain significant correlations simply because the sample is large. Even for small samples, significant correlations can occur by chance. Significant correlations that do not appear strong (either by a low $|r|$ or because of experience in that particular area) can indicate an inappropriate regression model.

Regression and correlation procedures are used concurrently for many business problems. However, each stems from different conditions. The sampling distribution appropriate for correlation is different from Equation (12.16). Correlation inference requires that the two variables are *random variables with normal distributions*.

#### DISTINCTIONS BETWEEN REGRESSION AND CORRELATION

1. Regression models have independent and dependent variables, whereas in correlation all variables have equal status.
2. In regression, the dependent variable, denoted by $Y$, is assumed to be random while the independent variable, $X$, has fixed (nonrandom) values; in correlation all variables are assumed to be random.
3. In regression, the dependent variable needs to be normally distributed, but in correlation all variables are assumed to be normally distributed. That is, they form a joint normal distribution.

Although this chapter distinguishes between the two processes, it covers only principles and applications of regression and correlation. The correlation model for variables that are jointly normally distributed is, however, discussed by others (see Daniel and Terrell [2]).

Correlations have an important role in business decisions quite beyond regression. The following example illustrates one application.

**EXAMPLE 12.12**

The consumer research department of an advertising firm must select variables to measure the success of a direct-mail effort to Los Angeles County. The sample includes 7508 census records for this geography. Data are given in a two-way (matrix) display of correlations for a selection of demographics. Which pairs of demographic variables, if any, hold a strong correlation?

(*Note:* A typical correlation matrix contains 1.0000s on the diagonals and repeats the off-diagonal numbers symmetric about the diagonal. Below, the off-diagonal numbers were deleted to simplify the display.)

| Variable | Description | Variable (by number) 1 | 2 | 3 | 4 | 5 | 6 | 7 |
|---|---|---|---|---|---|---|---|---|
| 1 | Persons per household | 1.000 | 0.9011 | −0.0317 | 0.6605 | 0.8877 | 0.4535 | −0.3352 |
| 2 | Percent population less than 18 years | | 1.0000 | −0.1836 | 0.5274 | 0.7836 | 0.4584 | −0.4081 |
| 3 | Male median age 25+ | | | 1.0000 | 0.2702 | 0.2013 | −0.2363 | 0.2289 |
| 4 | Percent single-family dwellings (SFDUS) | | | | 1.0000 | 0.8013 | 0.0177 | 0.0956 |
| 5 | Percent households, families | | | | | 1.0000 | 0.2221 | −0.1157 |
| 6 | Percent households, Hispanic | | | | | | 1.0000 | −0.6380 |
| 7 | Median school years | | | | | | | 1.0000 |

Correlation for percent households Hispanic and median school years

*Solution* Sometimes simply reordering the entries in a matrix to group those that exhibit higher correlations can help to identify high correlations that are otherwise hidden. Reordering with the original numbers gives a block of highly correlated variables.

A Matrix Display of Demographic Variable Correlations

| Variable | Description | Variable (by number) 1 | 2 | 4 | 5 | 3 | 6 | 7 |
|---|---|---|---|---|---|---|---|---|
| 1 | Persons per household | 1.0000 | 0.9011 | 0.6605 | 0.8877 | −0.0317 | 0.4535 | −0.3352 |
| 2 | Percent population less than 18 years | | 1.0000 | 0.5274 | 0.7836 | −0.1836 | 0.4584 | −0.4081 |
| 4 | Percent SFDUS | | | 1.0000 | 0.8013 | 0.2702 | 0.0177 | 0.0956 |
| 5 | Percent families | | | | 1.0000 | 0.2013 | 0.2221 | −0.1157 |
| 3 | Male median age 25+ | | | | | 1.0000 | −0.2363 | 0.2289 |
| 6 | Percent Hispanic households | | | | | | 1.0000 | −0.6380 |
| 7 | Median school years | | | | | | | 1.0000 |

Conclusion:   In the table, variables 1, 2, 4, and 5 in the shaded area are all reasonably highly correlated. Any one could give a fair expression to represent the others. (See the descriptions.) Similarly, the percent of Hispanic households (6) and median school years (7) have a high inverse correlation. Percent of households with families (5), male median age of 25+ years (3), and median school years (7) constitute a reduced set of low-correlating demographic variables.

---

Example 12.12 identifies correlations that would allow us to reduce the number of variables preliminary to a decision analysis. This is referred to as using a correlation matrix for *variable elimination, or screening.* One use of a screen is in direct-mail advertising to eliminate numerous redundant census variables. Those variables that remain after screening have lower correlations. Correlations are used in Chapter 13 to eliminate *X*-variables that are highly correlated and contain redundant information. This can eliminate confusion in the *X*-variables and so lead to a more reliable linear model. It is another form of data analysis.

Another use of a screen is to determine surrogate variables that are most easily calculated. For example, with a chemical production process that is fairly constant, managers could use production time as a surrogate measure to establish reasonable shutdown times without measuring the exact quantities of chemicals produced. This would reduce the expense of measuring output while still being somewhat efficient. The use of production time as a surrogate for the quantity of production might be recognized from a high correlation for these variables.

## Correlation and Causation

If companies can determine what causes variables to change, what causes production shutdowns, and why sales go up or down, then these factors might possibly be controlled and operations improved. For example, who would not enjoy the benefits of knowing what causes the fluctuations in stock market prices?

In seeking explanations for business events, analysts often attempt to explain the underlying causes of a response. But consider the word *cause*. One thing is a cause of another thing if a change in the first *produces* a change in the second thing. In contrast, two things are correlated or associated if changes in the two tend to *accompany* one another. We have discussed one variable as being correlated, that is, related to or described by another. But *strong correlation* alone *does not prove causation.* Any attempt to discover a causal relation requires (1) checking that the variables are directly associated, (2) verifying that the assumed cause precedes the assumed effect, and (3) eliminating all other plausible alternative explanations as a cause.

Proving a cause-and-effect relationship is somewhat analogous to proving a hypothesis. Proof requires the elimination of all other plausible alternatives. Yet the elimination of nonplausible causes (by low correlation or by showing that an assumed cause follows an assumed effect) can help the business manager to eliminate nonessential variables in a decision problem.

Smoking and the incidence of lung cancer for U.S. males is a case in which the idea that correlation "approximates" a cause is accepted by many [7]. But *this is the exception rather than the rule.* The conclusion that smoking is a cause of lung cancer is a consequence of many years of government research and testing. Statistics tend to support a causal relationship for both men and women. However, it takes very extensive studies like [7] to prove or disprove causal relationships such as those between smoking and lung cancer.

Again, correlation alone does not prove causation. As long as any alternatives remain, causation is impossible to verify. But it is possible to refute causation. Therefore, one should be content in determining if and how two business variables are related. This information, in terms of regression models and correlations, can be quite useful for decision-making in business even if causation cannot be established.

**Problems 12.7**

32. For the following data, $r = 1$.    For the next data, $r = -1$.

| X | 2 | 6 | 4 | 7 |
|---|---|---|---|---|
| Y | -15 | 41 | 13 | 55 |

| X | 5 | 20 | 25 |
|---|---|---|---|
| Y | 42 | 12 | 2 |

33. These data were obtained in a correlation study: $n = 10$, $\bar{X} = 4$, $\bar{Y} = 7$, $\sum Y^2 = 890$, $\sum (XY) = 180$, $\sum X^2 = 260$. Compute the coefficient of correlation. Interpret the result using $r^2$.

34. The following data indicate yields of potatoes in pounds and amounts of fertilizer in pounds used by six individuals in a competition. The objective was to see how many pounds of potatoes could be grown from one seed potato.
   a. Plot the data. Does a strong correlation appear to exist?
   b. Compute $r$. Interpret your result in terms of $r^2$.

| Yield, Y (in pounds) | 2 | 4 | 6 | 6 | 4 | 2 |
|---|---|---|---|---|---|---|
| Fertilizer, X (in pounds) | 0.5 | 1 | 2.5 | 4.5 | 4.5 | 6 |

35. In two related research studies, one person got $r = -0.65$, whereas the second found $r = 0.58$. Which is the stronger linear relation and by how much?

36. A study of New White detergent surveyed 15 test markets to determine the relation of two variables to sales. The first variable, $X_1$, gave expenditures for television advertising in thousands of dollars. The second, $X_2$, was the unit price of the product. Sales, $Y$, were measured as the proportion or share of the market held by New White. Given:

$$[n \sum (X_1 Y) - \sum (X_1) \sum (Y)] = 11.18 \qquad [n \sum (Y^2) - (\sum Y)^2] = 0.88$$
$$[n \sum (X_2 Y) - \sum (X_2) \sum (Y)] = -0.21 \quad [n \sum (X_1^2) - (\sum X_1)^2] = 300$$
$$[n \sum (X_2^2) - (\sum X_2)^2] = 0.12$$

Compute the correlation of $Y$ with $X_1$ and the correlation of $Y$ with $X_2$. Which variable, $X_1$ or $X_2$, gives the stronger description of sales?

37. Use the information from Problem 6 to find a value for $r$. Interpret $r$ by using $r^2$.

38. Use information and calculations from Problem 9 to find a value for $r$. See the computation note with Example 12.10 that suggests several checks for a reasonable value for $r$.

39. Suppose someone ran a correlation analysis between age and score (100-point scale) in a college course. If the results show $r = +1.00$, is this sufficient indication to conclude that the older persons will get higher grades in this course? Explain.

40. For each of the graphs, indicate those options from among (a)–(e) below that you would try. One or more options should be appropriate for each data plot.
    a. Test the hypothesis $H_0: \beta_1 = 0$ against $H_A: \beta_1 \neq 0$.
    b. Test for a higher-order polynomial regression.
    c. Computer $r$ as a (point) estimate of the linear correlation.
    d. Gather more data before attempting any analyses.
    e. Compute an estimate for the true slope.

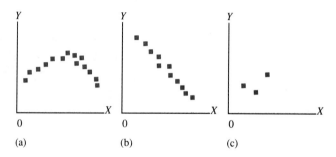

(a)           (b)           (c)

41. Indicate which options from among (a)–(e) in Problem 40 you would try for data plots A and B.

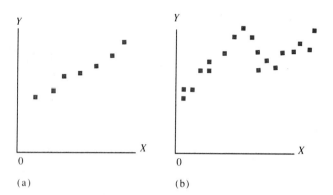

(a)           (b)

42. An extremely important and expensive promotional tool used in industrial marketing is personal selling. In personal selling a major expense is the cost related to turnover of sales personnel. In one study, A. Paresuraman and C. Futrell (*Journal of Business Research* 11, no. 1 (March 1983): 33–48) examined the interrelationships among pharmaceutical salespeople. Their

sample encompassed a mail survey of a national pharmaceutical manufacturer's sales force. Usable responses were obtained from 216 of the 347 salespeople surveyed. One finding concerns demographic variable correlations. Discuss the meaning for the correlations on these variables.

|  | Correlations with | | |
| --- | --- | --- | --- |
| Demographic Variable | *D1* | *D2* | *D3* |
| Age (D1) | 1.000 | | |
| Tenure (D2) | 0.819 | 1.000 | |
| Education Level (D3) | −0.109 | −0.015 | 1.000 |
| Income (D4) | 0.289 | 0.473 | 0.163 |

43. The data below show two years worth of the monthly sales of a soft drink (expressed as a percentage of average monthly sales) and average monthly temperature.

| Month | | Sales (percent of average month) | Temperature (°F) |
| --- | --- | --- | --- |
| Year 1 | J | 55.9 | 41.6 |
|  | F | 63.2 | 49.0 |
|  | M | 87.8 | 55.0 |
|  | A | 112.9 | 72.2 |
|  | M | 122.0 | 76.1 |
|  | J | 134.0 | 84.1 |
|  | J | 131.4 | 83.6 |
|  | A | 118.4 | 83.3 |
|  | S | 105.0 | 75.3 |
|  | O | 89.3 | 65.1 |
|  | N | 86.6 | 51.6 |
|  | D | 74.1 | 48.6 |
| Year 2 | J | 58.4 | 37.5 |
|  | F | 79.1 | 50.1 |
|  | M | 102.9 | 56.2 |
|  | A | 99.6 | 61.4 |
|  | M | 126.4 | 74.5 |
|  | J | 130.0 | 80.3 |
|  | J | 130.4 | 81.8 |
|  | A | 118.4 | 78.7 |
|  | S | 105.0 | 75.3 |
|  | O | 89.3 | 63.0 |
|  | N | 86.6 | 57.6 |
|  | D | 74.1 | 47.6 |

a. Compute the sample coefficient of determination between monthly sales and temperature. Explain the meaning of your calculated value.

b. Do higher temperatures *cause* higher sales? Justify your answer both (i) with logic and (ii) by your calculation in (a).

c. Test $_2$: $\rho = 0$ versus $H_A$: $\rho \neq 0$ at $\alpha = 0.05$. What does this tell you about the prediction of average sales from the average monthly temperature?

## SUMMARY

Regression modeling is a useful statistical technique that identifies logical relationships. Such knowledge enables managers to do a better job of estimating such variables as sales, production costs, and investment results.

Simple linear regression is used to predict values of a dependent variable from knowledge of an independent variable. This chapter discussed some key questions related to developing a simple linear regression model.

1. *How can we verify that a simple linear model is appropriate?* A scatter diagram, a visual inspection of the data plot, is used first. This graph tells whether a linear relation appears to exist. (The chapter covers simple linear relations only.)

2. *What is a linear model?* Statistical models in business are often based on a business hypothesis, which is basically one's idea of how selected business and economic forces interact. A simple linear regression model is a linear equation that graphs these forces as a line. It relates a business response, or dependent variable, to an independent variable, but allows for error in the relation.

3. *What process is used to develop a model (equation)?* The *least squares* approach is used to estimate the parameters (slope and intercept) of the regression model. The objective of the least squares method is to identify the line that passes closest to all the sample data points. We define closeness by minimizing the squared (error) distances of all data points from the line. This makes the total error as small as possible.

4. *When is an estimated regression line statistically meaningful?* A regression line is statistically significant if it passes a *t*-test of the slope. This test compares the estimated slope against the zero slope of a line that would result if the two variables were unrelated. A significant *t*-value means that the linear regression gives a meaningful description of the data. That is, the regression equation will give better estimates of the dependent variable ($Y$) than would result from estimating all response values with a constant, $\bar{Y}$.

5. *What does the coefficient of correlation tell us about the linear relation?* The coefficient of correlation, $r$, measures the strength of the linear regression. It takes on values between $\pm 1$ and $-1$. A more intuitive measure is the coefficient of determination ($r^2$), which describes the percent of total variations in response that are associated with, or explained by, the linear model. A low coefficient of determination, near zero, is a signal that the model is not appropriate. (In this case, other models should be evaluated.) A high coefficient of determination, with $r^2$ approaching 1, suggests that the model is good for making predictions. However, even $r^2 = 1$ does not indicate cause and effect, although it does mean a strong linear relation exists.

6. *When can a linear model be used for prediction?* If a linear model is meaningful and if the correlation is strong, then predictions of the dependent variable are appropriate. Predictions can be about either an *individual response* or an *average response*. The main difference is that individuals also have a unique variation about the average response, so the error of estimation is larger for individuals.

The answers to these questions give managers a basis for formulating regression models to describe relationships that they hypothesize might exist in the business world. Part of the modeling process involves testing the models with data to see whether they are statistically valid. These ideas will be extended to more sophisticated models in Chapter 13. Table 12.2 shows key formulas for regression and correlation.

**TABLE 12.2** Key Equations for Regression and Correlation

## Least Squares Estimators for $\beta_0$, $\beta_1$

$$b_0 = \bar{Y} - b_1 \bar{X} \qquad (12.1)$$

$$b_1 = SS(XY)/SS(X) \qquad (12.2)$$

and

$$SS(Y) = \sum Y^2 - [\sum Y)^2/n]$$
$$SS(X) = \sum X^2 - [\sum X)^2/n]$$
$$SS(XY) = \sum XY - [\sum X \sum Y/n]$$

## Mean Square Error

$$s_e^2 = \frac{SSE}{n-2} \qquad (12.3)$$

$$SSE = SS(Y) - b_1 \, SS(XY)$$

## Standard Error of Estimate

$$s_e = \sqrt{\frac{SSE}{n-2}} \qquad (12.4)$$

## Inference Concerning the Slope

For hypothesis tests

$$t = \frac{b_1 - \beta_1}{s_{b_1}} \quad \text{with} \quad s_{b_1}^2 = s_e^2/SS(X) \qquad (12.5)$$

Confidence intervals

$$(b_1 - t s_{b_1}) < \beta_1 < (b_1 + t s_{b_1}) \qquad (12.6)$$

## Inference for (1) Mean and (2) Individual Response

For hypothesis tests:
(1) for mean response, $E(Y|X)$

$$t = \frac{\hat{Y} - E(Y|X)}{s_e \sqrt{\dfrac{1}{n} + \dfrac{(X - \bar{X})^2}{SS(X)}}} \qquad (12.9)$$

(2) for individual response, $Y$

$$t = \frac{Y - \hat{Y}}{s_e \sqrt{1 + \dfrac{1}{n} + \dfrac{(X - \bar{X})^2}{SS(X)}}} \qquad (12.10)$$

For estimation:
of $E(Y|X)$

(LCL, UCL)

$$= \hat{Y} \pm \left( t \cdot s_e \sqrt{\frac{1}{n} + \frac{(X - \bar{X})^2}{SS(X)}} \right) \qquad (12.11)$$

of $Y$

(LPL, UPL)

$$= \hat{Y} \pm \left( t \cdot s_e \sqrt{1 + \frac{1}{n} + \frac{(X - \bar{X})^2}{SS(X)}} \right) \qquad (12.12)$$

## Coefficient of Determination

$$r^2 = \frac{SSR}{SS(Y)} = 1 - \frac{SSE}{SS(Y)} \qquad (12.14)$$

where $SS(Y) = SSR + SSE$

## Coefficient of Correlation

$$r = \sqrt{\frac{SSR}{SS(Y)}} = \frac{SS(XY)}{\sqrt{SS(X)\,SS(Y)}} \qquad (12.15)$$

## Inference Concerning $\rho$

For hypothesis tests:
$H_0: \rho = 0$

$$t = (r - \rho)/\sqrt{(1 - r^2)/(n - 2)} \qquad (12.16)$$

$H_0: \rho = \rho_0 \, (\neq 0) \qquad Z = (Z' - \rho_0)/\sigma(Z') \qquad (12.18)$

For estimation:
with $\rho = 0$

$$CL = r \pm (t_{\alpha/2} \cdot \sqrt{(1 - r^2)/(n - 2)}) \qquad (12.17)$$

or the general case, $\rho \neq 0$

$$CL = Z' \pm (Z_{\alpha/2}/\sigma(z')) \qquad (12.19)$$

with $Z' = \dfrac{1}{2} \log_e \left( \dfrac{1 + r}{1 - r} \right)$, $\sigma(Z') = 1/\sqrt{n - 3}$

An analysis of variance for regression appears in the Computer Application section.

**KEY TERMS**

coefficient of correlation, 545
coefficient of determination, 544
confidence interval for mean
  response, 537
estimated regression, 517
least squares estimation, 520
mean squared error, 527

normal equation, 520
prediction interval for individual
  response, 537
simple regression model, 515
standard error of estimate, 528
variable screening, 550

**CHAPTER 12
PROBLEMS**

44. Which, if any, of the following statements concerning the value of $b_2$ in the regression equation is correct?
  a. It shows the height of the regression line where $X = 0$.
  b. It may be either a positive or a negative amount.
  c. It is used for estimating the average $Y$-value associated with a given $X$-value.
  d. It always has the same sign as the coefficient of correlation.

45. In a regression study, the equation $\hat{Y} = -2.0 + 8.6X$ was obtained. What does this indicate?
  a. There is a negative relationship between the two variables.
  b. The coefficient of correlation will be negative.
  c. If $X$ is 10, the value of $\hat{Y}$ is 84.
  d. If $r$ is positive, the value for $r^2$ will be larger than $r$.
  e. The computation is in error, since $b_0$ cannot be negative.

46. Assume that the variance in the $Y$-variable is 100 and $s_e = 6.0$. Find $r$.

47. The sum of squares for the $Y$-variable, SS($Y$), was found to be 3.00. Which of the following must be true?
  a. SSR > SSE
  b. SSE $\leqslant 3.0$
  c. $r^2 = 0.80$
  d. $b_0$ and $b_1$ must be positive numbers.

48. Examine the given diagram. Explain what is right and what is wrong with the following statements.
  a. $r \doteq -0.98$
  b. The linear regression of $Y$ on $X$ is $\hat{Y} = 7 + 1X$.

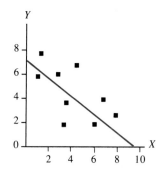

49. In a regression study, $Y = $ net profit, $X_1 = $ percentage of market sales, and $X_2 = $ labor costs. If the correlation coefficient between $Y$ and $X_1$ is $r_{YX_1} = 0.46$ and if $r_{YX_2}^2 = 0.3136$, which variable, $X_1$ or $X_2$, gives stronger linear description? By how much?

50. Consider the display of observed data points and the estimated regression line. Something has gone wrong in the calculations. At what point should the math be rechecked?

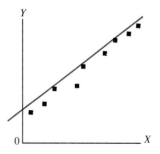

 51. A market research analyst is trying to establish the dependency of industrial shipments upon advertising expenditures of her manufacturing company. She has randomly selected 121 weeks and found their advertising expenses and shipments during the week following each. Results are summarized, where shipments are in units of railroad cartons and advertising expense is in hundreds of dollars. Also, $b_0 = 30$, $b_1 = 2$, and $s_e = 22$.

a. Which of the variables is the dependent variable?

b. State the simple linear regression equation.

c. In what units is the standard deviation of regression expressed?

d. What is the value of the measure commonly known as the standard error of estimate?

e. Make a point estimate of the number of cartons shipped during weeks following a $2000 advertising expenditure.

f. Make a 95% confidence interval estimate of the average (mean) number of cartons shipped during weeks following a $2000 advertising expenditure. Let $\sum (X - \bar{X})^2 = 30,000$.

g. Predict the individual number of cartons to be shipped during the week of July 24, when $500 had been spent on advertising during the previous week. Use $Z = 2$.

| Week | Advertising Expense ($00) $X$ | Shipments (cartons) $Y$ |
|---|---|---|
| 1 | 5 | 40 |
| 2 | 0 | 20 |
| 3 | 20 | 60 |
| — | — | — |
| ⋮ | — | — |
| | — | — |
| 120 | 0 | 20 |
| 121 | 30 | 80 |
| Totals | 1210 | 6050 |
| Mean | 10 | 50 |

52. For the paired observations shown in the table below, $X$ = time from the initial startup and $Y$ = thousands of units produced.
    (*Note:* $Y = -1$ indicates a new lot of 1000 units that were spoiled.)
    a. Plot the data on an $XY$-coordinate system.
    b. Without making any calculations, place your (freehand) estimate of the regression line on the graph. Is the slope positive, negative, or nearly zero?
    c. The regression line, estimated by least squares, is $\hat{Y} = -1.1 + 0.7X$. Plot this line on your graph. How close does this line come to the one that you made for part (b)?

| $X$ | 0 | 2 | 3 | 4 | 6 |
|-----|---|---|---|---|---|
| $Y$ | −1 | 0 | 1 | 2 | 3 |

53. Using the following data, answer parts (a), (b), and (c) of Problem 52, except you will have to develop the equation for the least squares regression line.

| $X$ | 0 | 2 | 3 | 4 | 6 |
|-----|---|---|---|---|---|
| $Y$ | 3 | 2 | 1 | 0 | −1 |

54. Give a short answer of one or two sentences for each of the following questions.
    a. What is the essential difference between a situation in which $r = -0.7$ and one in which $r = +0.7$? Can you make better predictions in one case than in the other? Explain.
    b. An advertising agency selected a random sample of 250 families to study the age relation between husband and wife. For the sample, $r = 0.88$; explain the practical meaning of this value.

55. The following table indicates the number of graduates from a small college who went directly into a job. Assume that the class size was about the same for all years.

| Year | 1 | 2 | 3 | 4 | 5 | 6 |
|------|---|---|---|---|---|---|
| Number of persons | 121 | 138 | 115 | 162 | 160 | 174 |

Plot the data. Does a linear trend appear reasonable? Estimate the trend line by a least squares analysis.

56. For the accompanying sample data, assume that the regression of $Y$ on $X$ is linear, and establish the regression equation. Check for reasonableness of the equation by plotting the data.

$X$ = the number of full-time machine operators

$Y$ = daily production of stenciled T-shirts (thousands, rounded).

| X | 1 | 2 | 4 | 6 | 8 | 9 | 11 | 14 |
|---|---|---|---|---|---|---|----|----|
| Y | 1 | 2 | 4 | 4 | 5 | 7 | 8  | 9  |

*(handwritten above table: $x$  16  36  64  81  121  196)*

Given:

$$\sum X^2 = 524 \qquad n \cdot \sum (X^2) - \left(\sum X\right)^2 = 1056$$

$$\sum (XY) = 364 \quad n \cdot \sum (XY) - \sum (X) \sum (Y) = 672$$

57. A company selling appliances is attempting to estimate the number of microwave ovens it will sell per week on the basis of the number of replies to an advertisement for a "free" microwave cookbook. For a sample of $n = 102$ weeks, the following data were gathered.

$$b_0 = \quad 25 \quad \sum (Y - \hat{Y})^2 = 22{,}500$$

$$b_1 = 0.10 \quad \sum (Y - \bar{Y})^2 = 45{,}000$$

a. Develop a point estimate of the number of ovens sold per week when 80 replies were received for the free cookbook.
b. Find 95% confidence limits for the mean number of ovens sold per week when 80 replies were received. Let $s_{\hat{Y}} = s_e\sqrt{(1/n) + [(X - \bar{X})^2/SS(X)]} = 1.5$.
c. Find the value of the coefficient of determination, $r^2$.
d. Explain the meaning of your $r^2$-value.

58. Compute the coefficient of correlation for the values given below. Now code the data using $X' = X/10$ and $Y' = Y/100$. Again compute $r$. Do you think any other coding could be done to simplify computing $r$?

| X | 20 | 50 | 70 | 40 | 20 |
|---|----|----|----|----|----|
| Y | 300 | 200 | 100 | 300 | 400 |

59. Ten students obtained these exam grades in a marketing course (see table). Compute the coefficient of correlation for these data given that $\sum (XY) = 47{,}777$; $\sum (X^2) = 47{,}394$; and $\sum (Y^2) = 48{,}609$. Interpret the meaning of your value by using $r^2$.

| Student | Midterm | Final | Student | Midterm | Final |
|---------|---------|-------|---------|---------|-------|
| 1 | 65 | 61 | 6 | 65 | 71 |
| 2 | 91 | 89 | 7 | 72 | 78 |
| 3 | 54 | 62 | 8 | 43 | 35 |
| 4 | 80 | 71 | 9 | 57 | 62 |
| 5 | 51 | 62 | 10 | 92 | 89 |

60. Given:

$$n \sum X^2 - (\sum X)^2 = 28$$

$$\hat{Y} = -2.45 + 1.89X$$

$$s_e = 0.40$$

$$\left[ n \sum (XY) - \sum X \sum Y \right] = 53 \quad n = 7$$

$$\left[ n \sum (Y^2) - (\sum Y)^2 \right] = 102 \quad \bar{X} = 7$$

a. Compute $r$, the Pearson coefficient of correlation.
b. Test $H_0: \beta_1 = 0$ against $H_A: \beta_1 \neq 0$ with $\alpha = 0.10$. State your conclusion.

61. Coastal Timber Co. replants its forests after logging operations. The firm wishes to develop a mathematical relationship that will provide estimates of the heights of young trees on the basis of age. A random sample of five trees revealed the following data, with age in years and height in feet. $\hat{Y}$ = the regression estimated value of the dependent variable.
a. Using the normal equations, develop a regression equation that best fits the data.
b. Provide a point estimate of the height of 4-year-old trees.
c. What percent of the variation in tree heights is explained by the age of the tree?

| Age $X$ | Height $Y$ | $X^2$ | $Y^2$ | $XY$ | $\hat{Y}$ | $(Y-\hat{Y})$ | $(Y-\hat{Y})^2$ | $(Y-\bar{Y})^2$ |
|---|---|---|---|---|---|---|---|---|
| 3 | 9 | 9 | 81 | 27 | 9.0 | .0 | .0 | 0 |
| 1 | 5 | 1 | 25 | 5 | 4.8 | +.2 | .04 | 16 |
| 2 | 7 | 4 | 49 | 14 | 6.9 | +.1 | .01 | 4 |
| 5 | 14 | 25 | 196 | 70 | 13.2 | +.8 | .64 | 25 |
| 4 | 10 | 16 | 100 | 40 | 11.1 | −1.1 | 1.21 | 1 |
| 15 | 45 | 55 | 451 | 156 | | 0.0 | 1.9 | 46 |

**STATISTICS IN ACTION: CHALLENGE PROBLEMS**

62. Kenneth M. Johnson describes the use of statistical modeling to target new markets for a national hotel chain. His point is that statistical models can be powerful aids for deciding how to expand into new markets, but that models do have limitations. And, he says, the final assessment must be made not by numbers, but by managers consulting with their staff.

Johnson gives a clear, brief overview of a decision problem that uses regression modeling. This article has no formulas, symbolism, or language that is peculiar to statistics. It provides an example of modeling from a management perspective. (You will need to get this article from your university library to answer Problems 62–64).

What are the steps in a project that uses modeling as a tool for selecting new markets? Johnson begins with a description of the decision problem and why it fits the framework of modeling. Next he describes the steps for building a model. These steps include selecting the data sources, variables considered as predictors, considerations for reducing 61 independent variables to 8, and scoring the model for an individual market—Traverse City, Michigan. Finally, evaluation of the modeling results is described in the sections HOW GOOD IS THE MODEL? and PUTTING THE MODEL TO WORK.

This article covers the steps for the statistical decision-making process as outlined in Chapter 1, and summarized in our Figure 1.1. The following table is a list of those steps. Also, we present some section headings and key statements from Johnson's article. Read the article, then assign the statements to the sequence of our statistical decision-making process.

| Statistical Decision-Making Process | Statement or Section Heading |
|---|---|
| 1. Decision problem | A. PICKING MARKETS (section heading) |
| 2. Statistical plan | B. HOW GOOD IS THE MODEL? and PUTTING THE MODEL TO WORK (two sections) |
| 3. External and internal data bases | C. The statistical analysis began after Welcome researchers had this data in hand. |
| 4. Data description and summary | D. This step is number crunching and was not detailed in the article. |
| 5. Analysis and conclusions | E. The next step was data collection. |
| 6. Test if results are valid in the real business world | F. WHAT BOOSTS REVENUE? (a section heading) |
| 7. Solution to decision problem (using the modeling results) | G. Welcome Hotels wants to move into the middle-sized markets, but which markets will be best? |

SOURCE: Johnson, Kenneth M. "Modeling to Find Markets." *American Demographics* 7, no. 11 (Nov. 1985): 40–44.

63. Use information from the journal article by Kenneth M. Johnson, described in the preceding problem, to answer the following two questions.
    a. One source for data was corporate records on the hotel chain's 100 existing middle-sized markets. List three other sources used and the kind of data for each.
    b. Why was only one of the variables—(1) area household income or (2) household purchasing power—considered in the model?

64. In HOW GOOD IS THE MODEL? (page 43) Johnson indicates that techniques other than statistics are more appropriate in evaluating the quality of a model in real settings. What is Johnson's rationale for this statement, and what "other technique" was used for a quality assessment?

65. The subjects for a study by Laura Otis were 42 students enrolled in a summer-session class. Their ages ranged from 17 to 50 years, with a mean age of 30 years. Many of the people were public-school teachers. The author is careful to point out some limitations of her findings by citing consistencies and differences from related research. Otis found that personality measures did not appear to play a significant role in individual food selections. But she did find, unexpectedly, that the subject's age was significantly correlated with willingness to try unusual foods: $r = +0.47$, $p < 0.01$ (page 743). That is, the older participants were more adventurous than the younger subjects. Some of this discussion warrants a closer look.

SOURCE: Otis, Laura. "Factors Influencing the Willingness to Taste Unusual Foods." *Psychological Reports* 54 (1984): 739–745.

The results relative to age (pages 743 and following) are interesting and are worth investigating for a better understanding of testing $H_0: \rho = 0$. Specifically, use the formula 12.16 and $r = +0.47$, with $n = 42$ to calculate, under $H_0: \rho = 0$, $Z = (r - 0)/\sqrt{(1 - r^2)/(n - 2)} = ?$ Use this and Table B.1 (normal curve) to confirm the $p$-value given in the journal article ($p < 0.01$). What are your findings?

66. S. P. Raj used regression modeling to explore the relationship between (a) a product's brand share of users and (b) its percent loyal customers. Using Simmon's data, a syndicated research on consumer purchase habits covering 1000 brands in 86 product classes, Raj found that brands with a larger share of users have proportionately larger fractions of loyal buyers. The results can be used to allocate resources between increasing the brand's consumer base of users and the developing brand user loyalty. The discussion is descriptive and clear, and it includes understandable summaries, including regression coefficients, $r^2$, sample sizes. One of the models that Raj discusses is

$$\text{Brand user share}_b = \beta_0 + \beta_1 \cdot \text{BL}_b^2 + e_b \tag{1}$$

where BL denotes the percent of user brand loyalty ($0 \leqslant \text{BL} \leqslant 100\%$) for a product brand "b." This is a linear regression in the parameters, the $\beta$'s, but it is not a straight-line function. However, the math is unchanged, except $X = \text{BL}_p^2$, so these numbers are just squared. You should feel comfortable with the concepts discussed herein, and you may find some useful general business results.

The study centers on analyzing a relationship between brand user share and brand loyalty. User share is a function of brand loyalty. For example, Table 2 on page 55 displays product P4 household goods for 138 items, including napkins, paper cups, aluminum foil, and more. Then, Table 3 (page 56) gives estimates for parameters $b_0 = 0.60$ for $\beta_0$ and $b_1 = 0.008$ for $\beta_1$ with $r^2 = 0.60$ for the model 1.

SOURCE: Raj, S. P. "Striking a Balance between Brand Popularity and Brand Loyalty." *Journal of Marketing* 49, no. 1 (Winter 1985): 53–59.

Use the model (1) and the estimated regression coefficients to establish an estimated regression equation for User share = Function (brand loyalty). Check your results by constructing a graph for regression estimated percent brand user share versus (by your equation) 10%, 20%, 30%, ... , 90%, 100% brand user loyalty values. Note that the vertical axis for (estimated) brand user share can have values ranging from zero to 100%. Is your regression graph linear? Explain.

---

## BUSINESS CASE

*Wally's Used Cars*

Wally Sanders not only worked full-time managing his car lot, but cars were his hobby and seemed to be his major interest in life. At 29 years old, he probably knew more about the history of automobiles than anyone in Riverton. He knew that the first recorded car was a 2-foot-long steam-powered model built by a Jesuit priest in Belgium in 1668, and that it wasn't until 1885 that a gasoline-powered car was successfully driven (at top speed of 8 to 10 miles per hour), in Germany. Plus he knew that the fastest car has been the Budweiser rocket, at 739 miles per hour. Ask him a simple question and you'd get a full response. How many people drive cars in the United States? Wally would respond, "My guess there are almost 150 million drivers in the U.S., and they average about 200 miles a week."

Wally liked to call his lot a "vintage car specialty shop" that featured "exceptional late-model cars." He handled mostly better foreign cars, Corvettes, T-Birds, and Mustangs. Most were really in top condition for their age. His business came primarily from college-age people, or the "sports," as Wally called them.

Wally was honest about his products. He emphasized the good and openly admitted the bad, for he intended to stay in Riverton. However, with waning sales, he felt a need to review his marketing strategy. What could he do to convince potential customers of the extra value he was offering? That's when he noticed Jack Baker, a professor at the local university, drive by in his 1965 Plymouth.

Wally gave Jack 20 minutes to get to the campus and get settled in his office.

**Wally**  (on the phone) Hello, Jack, this is Wally Sanders—from Wally's Auto Sales. I haven't talked to you in months, but I just saw you drive by my lot. How's everything?

**Jack**  Fine, Wally. And the old Plymouth is running great! What can I do for you?

**Wally**  I need some advice about how to improve my sales efforts. I'm getting enough potential customers to the lot, but I need to make more sales. Once the customers are here, we kind of inundate them with promotional things. I think they're going away overwhelmed by all the features we emphasize and then they never come back.

**Jack**  Yes, I can understand. Do you talk about mileage, cost, durability, and things like that?

**Wally**  You bet. I tell them they can't expect to get 156 miles per gallon like that three-wheeled "California Commuter." But for older cars, the mileage is good. And they may not hold up like a 1957 Mercedes that went over a million miles, but they're generally pretty durable.

**Jack**  Wally, you know I'm a statistician, not a marketing wizard. So why me?

**Wally**  Well, I thought . . . you being a numbers man . . . I mean, I can appreciate the value of numbers. But you work with them all the time. So I hoped I might hire you to take a look at some of our sales data and just tell me what you think.

**Jack**  Well . . . why not! I'm always game for a consulting fee. How about 4:00 P.M. next Tuesday?

At their meeting Wally explained his circumstance. He'd sold cars to many young people in the area. In fact, maybe he'd sold too many. It was getting to the point where every time he brought out his *Blue Book* of numbers, a potential customer would be turned off. Wally needed a new pitch, one that would be attractive to the college crowd.

**Jack**  Wally, I've read your ads! No tricks, no gimmicks, just solid cars at Wally's! Give me some facts. Do you keep any numbers?

**Wally**  Sure! We've got lots of data, especially on the Corvettes. Also, we've been on computer for about two years now.

**Jack**  That sounds like a good start. What stats do you keep?

**Wally**  We've got age of the car, its mileage, and the number of features. We push up to eight features: clean body (i.e., little rust), stereo radio, power equipment, white sidewalls, stick shift, air conditioning, convertible top, and chrome parts.

**Jack**  (Laughing) Stop! That's plenty! Can you get me the numbers for this season—say, for the past six months? How many Corvettes have you sold?

**Wally**  Just a minute. (He went to a personal computer and keyed in several entries.) Here, Jack, see for yourself. The number is 16 Corvettes sold, and this tells you the features of each one. Do you want to take the stats with you?

**Jack**  Yes, and 16 is a reasonable sampling. Can you get me the selling price, the car's age, its mileage, and the number of features up to eight for each car?

**Wally**  Sure, give me a second on this electronic wizard. You know, I think I could get real interested in working with these computers some day.

**Jack** Looks to me like you already have a fair beginning. Your little computer may be able to help you more than I can. Anyway, I'll stop back later this week.

*A few days later, Jack Baker returned with some papers in hand.*

**Jack** Wally, I've summarized some of your numbers. Here are your data and my statistics. You can see I've organized my thinking and put together some figures. You'll need to understand the bottom line to develop a statistically valid, yet convincing, sales pitch.

Summary of Jack Baker's Analysis

```
Correlations (Corvette sales)

price by mileage      -0.82
price by car's age    -0.77
price by extras        0.58

Regressions

DEPENDENT VARIABLE...PRICE (selling)
VARIABLE ENTERED...MILEAGE
MULTIPLE R  0.82386    ANALYSIS OF VARIANCE   DF   SUM OF SQUARES   MEAN SQUARE      F
R SQUARE    0.67875       REGRESSION           1      1.98153        1.98153     29.5800
STAN. ERROR 0.25882       RESIDUAL            14      0.93784        0.06699

DEPENDENT VARIABLE...PRICE
VARIABLE ENTERED...AGE OF CAR
MULTIPLE R  0.77430    ANALYSIS OF VARIANCE   DF
R SQUARE    0.59955                            1      1.75030        1.75030     20.9604
STAN. ERROR 0.28897                           14      1.16907        0.08351

DEPENDENT VARIABLE...PRICE
VARIABLE ENTERED...EXTRAS
MULTIPLE R  0.57626                     DF
R SQUARE    0.33208                      1      0.90225        0.90225      6.2621
STAN. ERROR 0.37958                     14      2.01712        0.14408
```

Together they looked over a summary of Jack's analysis. (See computer output in the table.)

**Wally** I recognize the terms, but you may have to help me with some meanings. I suspected you would make selling price the *dependent* variable. But I'm not clear on why the VARIABLE ENTERED space indicates MILEAGE.

**Jack** Actually, I first used a little software program that allows me to work with a number of variables, but I decided we should limit ourselves to working with only one independent variable for now. So I did a few of what we call simple linear regressions. And I ultimately chose MILEAGE because it has the biggest number (highest correlation). The minus sign there simply means that your selling price has been best for low-mileage Corvettes. Or, to put this another way, high mileage on a Corvette turns off a buyer, so you have to lower the price. Right?

**Wally** As a matter of fact, I try to avoid handling any cars that have excessive—or even any more than average—miles.

**Jack** Well, after looking at the different features (variables) you mentioned last week, I concluded that the car's mileage gave me the best indication of selling price. We'll use that

MILEAGE regression (pointing to the paper) for now and ignore the rest. Notice that with MILEAGE as the independent variable, the *F*-value is the highest, or best. Its value is *F* = 29.58.

**Wally**    (pointing) What does this number (−0.82) mean?

**Jack**    Simply that the variability in mileage and price are statistically related. Low mileage doesn't necessarily cause high prices, but the two go hand in hand.

**Wally**    I'm impressed with your numbers, Jack, and so far it sounds reasonable. But how do I use it to sell?

**Jack**    Again, I'm not an expert on marketing. But you may want to focus your efforts on the variables that are most significant.

Continue to use the *Blue Book* as a rough guide. However, when a customer comes in, don't bring it out to draw attention to the price. After going over the several features, move the customer's attention to your strong point and the statistically significant feature, which is lower mileage. It's logical that lower-mileage cars give more problem-free miles.

And here's the clincher! Your statistics show that lower-mileage Corvettes have higher sales prices. That's a fact from your data and my calculations. You can honestly indicate a higher resale potential for low mileage.

**Wally**    Hey, you may have something there! We often mention low mileage, but I was never able to say anything solid about it. I'll give this a try for a while. If it works, I'll be back for some more advice.

**Jack**    Don't worry, Wally. I'll be stopping by in two weeks just to satisfy my own curiosity.

**Business Case Questions**

67. Discuss the sampling of records on 16 car (Corvette) sales. More specifically,
    a. is a sample of 16 records a big enough sample for this analysis? Explain.
    b. what assumptions are necessary for Jack Baker to apply a regression analysis on Wally's sales data?

68. Jack indicated that the car's mileage showed the strongest simple regression on selling price. Use the summary table to answer the following questions.
    a. Use the R SQUARE values for (i) mileage, (ii) age of car, then (iii) extras (each) on price to explain why mileage is the strongest variable for predicting selling price.
    b. The *F*-value for mileage is $1.98153/0.06699 = 29.58$. Use the relation that $t^2_{df} = F_{1,df}$, so $t = \pm\sqrt{F}$, to calculate a *t*-statistic value. Then test $H_0: \beta_1 = 0$ for average selling price $= \beta_0 + \beta_1 \cdot$ mileage.

---

**COMPUTER APPLICATION**

*An Analysis of Variance for Simple Linear Regression*

The correlation and regression data summaries in the Chapter 12 Business Case may seem familiar yet not be entirely clear. Some of the notation, such as correlations and ($r^2$), and standard error (for regression), is developed in this chapter. Other notation, including the analysis of variance notation, was first introduced in Chapter 11. Our objective here is to briefly introduce the notation and procedures used in multiple regression modeling for Chapter 13. This discussion includes the equivalence to the procedures in Chapter 12 for regression and correlation using *t*-statistics versus the *F*-statistic in Chapter 13.

The hypothesis testing uses an analysis of variance for regression that is displayed in Table 12.3.

The test for a meaningful linear regression is a direct application of the regression summary in Table 12.3 because under $H_0: \beta_1 = 0$ we have used

$$t_{n-2} = (b_1 - 0)/\sqrt{s_e^2/\text{SS}(X)} \qquad (12.5)$$

**TABLE 12.3** An ANOVA for Regression in Simple Linear Regression

| Source of Variation | df | Sum of Squares | Mean Square |
|---|---|---|---|
| Regression | 1 | SSR | SSR/1 (=MSR) |
| Residuals | $n-2$ | SSE | $s_e^2 = $ SSE/$n-2$ (=MSE) |
| Total | $n-1$ | SS($Y$) | |

and

$$SS(Y) = SSR + SSE \tag{12.20}$$

The form for testing $H_0$: $\beta_1 = 0$ against $H_A$: $\beta_1 \neq 0$ using the preceding ANOVA summary is

$$F = \frac{\text{Regression mean square}}{\text{Residual mean square}} = \frac{\text{SSR}/1}{s_e^2} = \frac{\text{MSR}}{\text{MSE}} \tag{12.21}$$

with 1 and $n-2$ df.

$$r^2 = \frac{\text{Regression sum of squares}}{\text{Total sum of squares}} = \frac{\text{SSR}}{\text{SS}(Y)} \tag{12.15}$$

where

$$s_e^2 = \text{SSE}/(n-2) \tag{12.4}$$

1. Squaring both sides of (12.5) and simplifying the right-hand side gives

$$t_{n-2}^2 = b_1^2 \cdot \text{SS}(X)/s_e^2 = b_1 \cdot \text{SS}(XY)/s_e^2$$

or, from Table 12.3,

$$= \frac{\text{SSR}}{1} \bigg/ \frac{\text{SSE}}{(n-2)}$$

where the regression has divisor 1, its df, because there is one $X$-variable in the model, $Y = \beta_0 + \beta_1 X + \varepsilon$.

2. Applying the additive property of sums of squares to simplify calculations gives

$$\text{SSE} = \text{SS}(Y) - b_1 \cdot \text{SS}(XY)$$

3. From the theory of statistics [4], $t_{n-2}^2 = F_{1,n-2}$.

4. Applying (1) and (2) to (3) gives an $F$-test for a linear regression based on Table 12.3 ANOVA calculations.

$$F_{1,n-2} = t_{n-2}^2 = \frac{[b_1 \cdot \text{SS}(XY)]/1}{[(\text{SS}(Y) - (b_1 \cdot \text{SS}(XY))]/(n-2)} \tag{12.21}$$

$$= \frac{\text{SSR}/1}{\text{SSE}/(n-2)} = \frac{\text{SSR}/1}{s_e^2} = \frac{\text{MSR}}{\text{MSE}}$$

**TABLE 12.4**  Ages for Married Householder (AGE1) and Spouse (AGE2) for a Random Sample of 50 Records from the PUMS File

| AGE1 | AGE2 | AGE1 | AGE2 | AGE1 | AGE2 | AGE1 | AGE2 | AGE1 | AGE2 |
|---|---|---|---|---|---|---|---|---|---|
| 41 | 42 | 31 | 32 | 33 | 38 | 24 | 22 | 58 | 52 |
| 55 | 53 | 43 | 40 | 56 | 55 | 71 | 64 | 41 | 37 |
| 42 | 41 | 30 | 21 | 59 | 52 | 32 | 29 | 58 | 48 |
| 59 | 57 | 55 | 42 | 69 | 61 | 35 | 29 | 73 | 76 |
| 79 | 83 | 54 | 47 | 51 | 51 | 42 | 42 | 36 | 33 |
| 24 | 22 | 44 | 43 | 33 | 33 | 28 | 23 | 25 | 23 |
| 35 | 33 | 57 | 50 | 47 | 40 | 36 | 34 | 30 | 29 |
| 52 | 41 | 24 | 32 | 39 | 36 | 29 | 25 | 57 | 40 |
| 40 | 34 | 73 | 75 | 55 | 55 | 30 | 30 | 31 | 23 |
| 50 | 49 | 41 | 34 | 62 | 61 | 53 | 51 | 49 | 46 |

With this background we can view an example with computer calculations for testing the hypothesis of a meaningful linear regression, evaluating the coefficient of correlation (by Equation (12.15)), and, if appropriate, estimating the regression equation.

This example uses data from the PUMS file for a regression of a spouse's age predicted from the age of a known married householder. A random sample of 50 (see Table 12.4) was selected from the 12,200 (of 20,116 PUMS-file) records with MARITAL, column 65, equals 1 (married). An interest could be to send a specialty magazine subscription mailing to the household, for example, about homemaking, gardening, hobbies, or other.

By inspection the data appear to be highly positively correlated. We can expect a significant linear regression. The following is a SPSS regression subroutine that gives the ANOVA and a plot of the data.

```
 1  NUMBERED
 2  TITLE    SPSS REGRESSION RUN USING PUMS AGE1 AND AGE2 DATA
 3  DATA LIST FILE = INFILE/
 4              AGE1 43-44  PERSON2 60-61  AGE2 63-64
 5  SET BLANKS = -1
 6  SELECT IF (PERSON2 EQ 1)
 7  SET SEED = 1962
 8  SAMPLE 50 FROM 12200
 9  REGRESSION VARIABLES = AGE1  AGE2/
10          DEPENDENT = AGE2/ METHOD = ENTER AGE1/
11          SCATTERPLOT = (AGE1 AGE2) (AGE1 *RESID) (AGE1 *PRED)
12          (AGE2 *RESID) (AGE2 *PRED) (*RESID *PRED)
13  LIST VARIABLES = AGE1 AGE2/FORMAT= NUMBERED/ CASES FROM 1 TO 50
14  PLOT   HSIZE = 40/VSIZE =20/TITLE 'PLOT OF MARRIED COUPLE AGES'/
15         PLOT = AGE2 WITH AGE1
16  FINISH
```

A few comments are appropriate to explain the key commands for the REGRESSION and PLOT procedures. (Computer Problem 72 gives the setup for a Minitab computer regression on this data. Problem 76 asks you to check our calculations.)

l The data are read in, with PERSON2 (the relationship of person #2 to the head of household) used to identify a spouse, which requires PERSON2=1.

**FIGURE 12.10** A SPSS PLOT for AGE2, the Response Variable, against AGE1

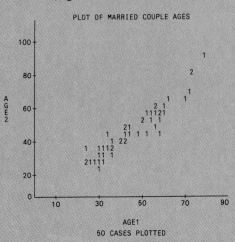

PLOT OF MARRIED COUPLE AGES
50 CASES PLOTTED

2 The SEED number is an arbitrary choice from Table B.7, Random Numbers, but is restricted to a value not bigger than 12,200, which is the total number of records that have a spouse.

3 In REGRESSION the order of the substatements is critical. See the *SPSS Manual* [8] for other options. The SCATTERPLOT statement is listed here for your awareness. This can give very useful data analysis information, and is used extensively in Chapter 13, but is not displayed for this example.

4 LIST VARIABLES enables you to check that the data are entered correctly. The PLOT procedure provides a display of the data, to help you get a first, intuitive feeling about the correlation and regression slope (whether + or −) and a sense of how strong the linear regression is (by how closely the data points fall in a straight-line pattern).

**TABLE 12.5** An SPSS Summary for Regression of Spouse Age, AGE2, on the Age of Householder for a Random Sample of 50 PUMS File Records

```
EQUATION NUMBER 1    DEPENDENT VARIABLE..   AGE2

BLOCK NUMBER  1.  METHOD:  ENTER      AGE1

VARIABLE(S) ENTERED ON STEP NUMBER 1..     AGE1

MULTIPLE R           .95028    ANALYSIS OF VARIANCE
R SQUARE             .90303                   DF     SUM OF SQUARES     MEAN SQUARE
ADJUSTED R SQUARE    .90101    REGRESSION       1        9353.06765      9353.06765
STANDARD ERROR      4.57419    RESIDUAL        48        1004.31235        20.92317

                  (Calculated) F =     447.01954      SIGNIF F =  .0000

------------------ VARIABLES IN THE EQUATION ------------------


VARIABLE            B           SE B       BETA         T      SIG T
AGE1           .954958      .045167    .950281     21.143     .0000
(CONSTANT)   -1.194187     2.151058                 -.555     .5814

END BLOCK NUMBER   1    ALL REQUESTED VARIABLES ENTERED.
```

A summary for the REGRESSION ANOVA and the data PLOT are presented in Table 12.5 and Figure 12.10.

There is a highly significant linear relation for spouse age—AGE2—regressed on AGE1—age of the householder. This is evidenced by the computer CALCULATED $F = 447.02$ (= MSR/MSE = 9353.06765/20.92317) with highly SIGNIF F = .0000. The $p$-value is less than 0.001. This result is supported by the quite high correlation value, MULTIPLE R, of 0.95028, which is the square root of MULTIPLE R SQUARE = $\sqrt{0.90303} = \sqrt{\text{SSR/TSS}} = \sqrt{((9353.06765)/(9353.06765 + 1004.31235))}$. (This is a simple linear correlation. The computer listing is set to accomodate a "MULTIPLE" number of $X$-variables). ADJUSTED R is explained in Section 13.4. The STANDARD ERROR of estimate $s_e^2$ is calculated as MSE = $\sqrt{20.92317} = 4.574$.

The VARIABLES IN THE EQUATION provides the slope estimate—$B_1$ for AGE1—and the AGE2 intercept estimate—$B_0$ for CONSTANT. The regression model is estimated $\widehat{\text{AGE2}} = -1.1942 + 0.9550 \cdot \text{AGE1}$. Then, for example, for AGE1 = 41, the first data value in Table 12.4, the predicted value is $\widehat{\text{AGE2}} = -1.1942 + 0.9550(41) = 38$. The residual, or error, for this individual prediction is AGE2 − AGE2 = 38 − 42 = −3, an underprediction by three years. The "$T$" value is our $t_{n-2}$. This gives an alternative test statistic to the $F$-test, since $(t_{48}^2) = (21.143)^2 = 447.0195 = $ calculated $F$.

The regression summary table (12.3) is an analysis of variance for regression that simply extends the ANOVA procedures of Chapter 11 to include cases in which both the $X$ and the $Y$ can be continuous variables. This is the foundation for the calculations, and testing, in Chapter 13 for multiple $X$-variable regression. Much of the analysis then becomes a matter of understanding and using a computer-developed regression summary.

## COMPUTER PROBLEMS

*Data Base Applications*

69. Using the PUMS data base, select a random sample of 50 household records for owner households. These are identified by "Payment to lender is not $0." For your sample records, obtain values for $Y$ = total monthly payments to lender (positions 27–30) and $X$ = household income (positions 35–39). Plot these data on $XY$-coordinates. If possible, use a scattergram plot program for the data plot. By your judgment of the data plot, does a simple linear regression appear to be reasonable? The following data are for one random sample from the PUMS file.

| Obs. | Pay | Income | Obs. | Pay | Income | Obs. | Pay | Income | Obs. | Pay | Income | Obs. | Pay | Income |
|------|-----|--------|------|-----|--------|------|-----|--------|------|-----|--------|------|-----|--------|
| 1 | 95 | 16505 | 11 | 83 | 29415 | 21 | 365 | 30175 | 31 | 125 | 14005 | 41 | 600 | 30510 |
| 2 | 60 | 24035 | 12 | 95 | 9875 | 22 | 90 | 30825 | 32 | 298 | 29650 | 42 | 110 | 29415 |
| 3 | 456 | 10830 | 13 | 249 | 18750 | 23 | 337 | 36880 | 33 | 204 | 15910 | 43 | 790 | 10210 |
| 4 | 521 | 34010 | 14 | 1440 | 65210 | 24 | 502 | 57715 | 34 | 55 | 31715 | 44 | 167 | 4005 |
| 5 | 255 | 22305 | 15 | 99 | 31460 | 25 | 575 | 25510 | 35 | 126 | 31010 | 45 | 102 | 16950 |
| 6 | 661 | 38445 | 16 | 75 | 4310 | 26 | 469 | 20005 | 36 | 175 | 5515 | 46 | 525 | 16215 |
| 7 | 372 | 18410 | 17 | 74 | 32010 | 27 | 350 | 41610 | 37 | 311 | 31935 | 47 | 12 | 28015 |
| 8 | 165 | 11830 | 18 | 324 | 24795 | 28 | 265 | 58565 | 38 | 300 | 22005 | 48 | 614 | 32305 |
| 9 | 365 | 39515 | 19 | 293 | 62145 | 29 | 185 | 34815 | 39 | 305 | 18815 | 49 | 275 | 47075 |
| 10 | 468 | 16720 | 20 | 100 | 31010 | 30 | 146 | 27975 | 40 | 284 | 55140 | 50 | 750 | 59415 |

Inspection of the data indicates that observations 14 and 47 appear "unusual." However, these values cannot be discredited, and so are used.

70. Use a software program available at your university to estimate a linear regression for $Y$ = total monthly payments to lender on $X$ = household income. Use the $n = 50$ records from Problem 69 to obtain as many of the following as you can with the software.
    a. The estimated regression equation, that is, values for $b_0$, $b_1$
    b. The linear correlation value for $Y$ and $X$
    c. A test of $H_0: \beta_1 = 0$ against $H_A: \beta_1 \neq 0$, for $\alpha = 0.05$
       Is monthly payment to lender linearly related to household income for $\alpha = 0.05$?
    If (a)–(c) are not available preprogrammed, then obtain the appropriate sums, sums of squares, etc., by computer. Then complete (a)–(c) using a calculator.

71. Provided a significant regression exists in Problem 70, use your estimated regression equation with $X$ = \$15,000 to obtain a 95% confidence interval on $\beta_1$.

72. Select a random sample of the ages of 25 couples (50 numbers). This selection requires that the person-record include spouse (positions 60–61) code = 01. List the ages as pairs. Also display the data on $XY$-coordinates. Is the linear correlation for spouses' ages positive, negative, or near zero? Guess whether a linear relation is statistically significant for $\alpha = 0.05$. The following data are one random sample from the PUMS file.

| Pair | AGE1 | AGE2 | Pair | AGE1 | AGE2 | Pair | AGE1 | AGE2 | Pair | AGE1 | AGE2 | Pair | AGE1 | AGE2 |
|------|------|------|------|------|------|------|------|------|------|------|------|------|------|------|
| 1 | 53 | 50 | 6 | 22 | 23 | 11 | 44 | 37 | 16 | 47 | 36 | 21 | 58 | 58 |
| 2 | 47 | 50 | 7 | 71 | 70 | 12 | 61 | 51 | 17 | 23 | 22 | 22 | 71 | 68 |
| 3 | 60 | 70 | 8 | 40 | 38 | 13 | 27 | 32 | 18 | 77 | 65 | 23 | 29 | 27 |
| 4 | 24 | 28 | 9 | 52 | 54 | 14 | 37 | 37 | 19 | 41 | 35 | 24 | 40 | 50 |
| 5 | 66 | 60 | 10 | 24 | 26 | 15 | 39 | 38 | 20 | 43 | 40 | 25 | 74 | 71 |

For Minitab users, here is a run setup for this problem and the next:

```
MTB > READ C1-C2
DATA> ←————— read in the data pairs (AGE1   AGE2)
DATA> ←
    ⋮
DATA> END
MTB > NAME C1 'AGE1' C2 'AGE2'
MTB > PRINT 'AGE1' 'AGE2'
MTB > PLOT 'AGE2' 'AGE1'
MTB > BRIEF 1
MTB > REGRESS 'AGE2' ON 1 'AGE1'
MTB > CORRELATION 'AGE1' 'AGE2'
MTB > STOP
```

SPSS users can use the setup in the Computer Application, but substitute the data that are given here. Again, this will provide information for this and the next problem.

73. Calculate the linear regression and correlation descriptors—$b_0$, $b_1$, and $r$—for Problem 72. Let $X$ = householder's age and $Y$ = spouse's age. Perform a test of hypothesis $H_0: \rho = 0$ against $H_A: \rho \neq 0$ for $\alpha = 0.05$. Decide whether $\rho = 0$ or $\rho \neq 0$.

*Additional Computer Problems*

74. Use a computer routine to test for a significant linear relation for the data in Problem 7

where  $X$ = hours of continuous processing

and  $Y$ = computer downtime in minutes

for

| $X$, hours | 4 | 5 | 6 | 7 | 8 | 9 | 10 |
|---|---|---|---|---|---|---|---|
| $Y$, downtime | 4 | 6 | 8 | 12 | 18 | 24 | 26 |

Use $\alpha = 0.05$ and any appropriate software. For Minitab users, after entering the data and assigning hours ($X$) in C1 and downtime ($Y$) in C2, use

```
MTB > BRIEF 1
MTB > REGRESS 'DOWNTIME' ON 1 'HOURS'

MTB > CORRELATION 'DOWNTIME' 'HOURS'
```

For SPSS users, with the data read in, use

```
REGRESSION VARIABLES = DOWNTIME  HOURS /
         STATISTICS = DEFAULTS /
                DEP = DOWNTIME /
             METHOD = ENTER  HOURS
     LIST  VARIABLES = DOWNTIME  HOURS /
                    CASES FROM 1 TO 7
```

Or use any other software that includes simple linear regression.

75. Repeat the procedure outlined for Problem 74, but use the variables

$X$ = years of retail sales experience (EXPER)

$Y$ = thousands of dollars in sales (SALES)

and the data

| $X$ | 6 | 3 | 2 | 4 | 1 | 6 | 3 | 5 |
|---|---|---|---|---|---|---|---|---|
| $Y$ | 8 | 5 | 2 | 3 | 3 | 9 | 4 | 6 |

Let $\alpha = 0.05$, and use any available statistics software.

76. Use the data in Table 12.4 and a computer program of your choice to check the results given in the Computer Application. SPSS users can use the routine that appears in the Computer Application. For a Minitab run, use the program statements given in Problem 72. Do your results agree with those given in the Computer Application?

**FOR FURTHER READING**

[1] Cureton, Edward E. Letter to the editor. *The Statistician* 25 (June 1972): 54–55.
[2] Daniel, Wayne W., and James C. Terrell. *Business Statistics*, 4th ed. Boston: Houghton-Mifflin, 1986.
[3] Morrison, Donald F. *Applied Linear Statistical Methods*. Englewood Cliffs, NJ: Prentice-Hall, 1983.
[4] Neter, John, William Wasserman, and Michael H. Kutner. *Applied Linear Regression Models*, 3rd ed. Homewood, IL: Irwin, 1990.
[5] Ott, Lyman. *An Introduction to Statistical Methods and Data Analysis*, 3rd ed. Boston: PWS-Kent, 1988.
[6] Schaefer, Robert L., and Richard B. Anderson. *The Student Edition of Minitab*. Reading, MA: Addison-Wesley, 1989.
[7] *Smoking and Health: Report of the Advisory Committee to the Surgeon General*. Washington D.C.: U.S. Dept. of HEW, Public Health Service, Health Services and Mental Health Administration, 1964.
[8] *SPSS Reference Guide* (version 4.0). Chicago: SPSS, Inc., 1990.

# CHAPTER

# 13

## MULTIPLE LINEAR REGRESSION

## 13.1 INTRODUCTION

The vice president of operations of the 80-year-old Pennsylvania steel company was facing another crisis—this time, a cash flow problem. Last year the board of directors had granted him a substantial amount of capital to purchase new equipment, which was supposed to improve productivity. But labor negotiations had not gone well, fuel prices were higher, and some major orders the firm had planned on went instead to firms in Brazil and Korea. Now, as the next board meeting was approaching, the prospects for profit improvement again looked dim. Each time one problem was solved, another seemed to take its place.

The problems faced by this company—high energy costs, escalating wages, and foreign competition—are not unusual, nor are such problems restricted to steel companies. The variables that affect a business are often intimately related. Energy costs, productivity, labor agreements, interest rates, and competitive prices all affect the financial status of a firm and each other. For example, if interest rates are very high, firms will be under pressure to keep inventories low in order to minimize the amount of cash invested in carrying extra stock. But low inventories may mean lost sales and decreased cash flow.

In Chapter 12 we studied how to describe the relationship that exists between two business variables. But most realistic business situations are much more complex than that. Most business situations, like those suggested above, involve several variables. And a change in one variable can have interactive effects on the others. For example, an increase in interest rates can lead to a reduction in finished goods inventories, which can also affect sales, warehouse space cost, and a myriad of other factors.

This chapter allows us to analyze the multivariable situations in much the same way as we analyzed the simple linear regressions in the previous chapter. Multiple regression and correlation are understandably among the most valuable and widely used statistical techniques in business today. Many firms would be unable to do the analysis and market research studies necessary for their survival without utilizing multiple regression and correlation techniques.

Multiple linear regression involves the use of two or more independent variables to predict the value of one dependent variable. For example, in the auto manufacturing industry, profits are a function not only of the labor and material costs but also of overhead costs, interest rates, gas prices, buyer preference, and a host of other variables. In equation form we could depict this as

$$\text{Profits} = f(\text{mfg costs, overhead, interest paid, and so on})$$

Multiple regression often gives us a more realistic description of the business situation, and it typically yields a stronger prediction than simple linear regression. This is because two, three, or more independent variables will very likely explain significantly more of the variation in the dependent variable than one variable will explain.

We saw earlier that the simple correlation coefficient is squared to measure the closeness, or strength, of a simple linear regression. This concept extends directly to a multiple coefficient of correlation, $R$, as well as to $R^2$. Nearly all regression computer-software packages include multiple-$R^2$ as an indicator of linear model strength.

Multiple regression is used for data analysis when simple regression analysis does not adequately explain the variations in a response variable. Because it is an extension of simple regression, the discussion in this chapter parallels that of Chapter 12. However, Chapter 13 relies upon the use of computer outputs to display the more lengthy multiple regression calculations. Because you are likely to do most any multiple regression problem on a computer, we will concentrate on understanding and interpreting common regression outputs.

Upon completing chapter 13, you should be able to

1. extend simple linear regression to a more general multiple linear model and
   a. estimate the model coefficients—$\beta_0, \beta_1, \beta_2, \ldots, \beta_k$;
   b. test the model for significance;
   c. make interval estimates of the $\beta$-values;
2. select an appropriate linear model from numerous independent variables;
3. use a multiple regression model to predict expected response and individual response with interval estimates;
4. use residual data to
   a. test the adequacy of regression assumptions;
   b. determine when you have achieved a reasonable multiple regression equation;
5. use regression (computer) printouts with understanding and recognize that some interpretations require the expertise of a skilled statistician.

The examples and problems in the chapter make frequent reference to computer outputs. Although the outputs are illustrated, you may want to enhance the material by using your own computer regression routines. The references at the end of this chapter also provide more information on multiple regression.

## 13.2 THE MULTIPLE REGRESSION MODEL

As an introduction to a multivariate linear model, consider these two models:

Model (1) $\quad Y = \beta_0 + \beta_1 X_1 + \beta_2 X_2 + \cdots + \beta_k X_k + \varepsilon$

Model (2) $\quad Y = \beta_0 + \beta_1 X_1 + \beta_2 X_2 + \beta_{12} X_1 X_2 + \beta_{11} X_1^2 + \beta_{22} X_2^2 + \varepsilon$

Model (1) is multivariate because it contains two or more independent variables. Its coefficients are linear because no coefficients appear with an exponent or as a product of or divided by another coefficient. Finally, model (1) is linear in the independent variables because each appears only in the first power and not as a product of or divided by another independent variable. Model (1) plots as a plane surface, where $k$ is the number of independent variables defining that surface. The regression model displayed in Figure 13.1 (on page 576) has $k = 2$ independent variables and is the simplest of any multiple regression linear model.

In model (2) the independent variables appear with exponents and as products. The coefficients, however, are linear. The plot of this model would look like a three-dimensional surface with mountains and valleys. Since the terms $(X_1 X_2)$, $X_1^2$, and $X_2^2$ are second order in $X$ but the coefficients are linear, model (2) is called a second-order linear model. To illustrate this, we consider the stock of used cars that often fill our car lots. The price of a used car (dependent variable) might be linearly related to the mileage $(X_1)$, car's age $(X_2)$, worth of accessories $(X_3)$, and square of the number of previous owners $(X_4)$. This is a second-order (linear) model because the coefficients are linear but the independent variables are at most squared. Although regression models can become quite complex, we will show that many can be expressed as first-order linear models like Model 1.

### The Multiple Regression Model

Our emphasis is on *first-order linear models.*

> The multiple linear regression model (general case) is a *first-order linear model* for a dependent variable with $k$ ($> 1$) independent variables that can be written as
>
> $$Y = \beta_0 + \beta_1 X_1 + \beta_2 X_2 + \cdots + \beta_k X_k + \varepsilon \qquad \text{(13.1)}$$

The model in Equation (13.1) identifies a response variable $Y$ as being linearly related to two or more independent variables $X_i$ and a random error term. The regression equation, which excludes random error, is described by

$$E(Y \mid X_1, X_2, \ldots, X_k) = \beta_0 + \beta_1 X_1 + \beta_2 X_2 + \cdots + \beta_k X_k$$

This defines a first-order linear regression surface in $k$ independent variables.*

The *multiple regression model* is a logical extension of the simple regression case. Extending from one independent variable to two translates to extending

---

* We can transform (rescale) linear models of second- $(X^2)$ and higher-order $(X)$ terms into first-order linear models by employing substitute variables. For example we can let $X_3 = X_1^2$. In this way the linear models of order one and higher can be defined by one model. And computer programs can provide these transformations at the touch of a few keys.

from a line in the two dimensions, $X$ by $Y$, to a plane cutting three dimensions, $X_1$, $X_2$, and $Y$. This plane defines a regression surface.

> The multiple linear regression model for exactly *two* independent variables relates a dependent variable to two independent variables plus random error.

$$Y = \underbrace{\beta_0 + \beta_1 X_1 + \beta_2 X_2}_{} + \underbrace{\qquad \varepsilon \qquad}_{} \tag{13.2}$$

Components:            Deterministic   + Random error

with   $Y$   symbolizing dependent variables
       $X$'s symbolizing independent variables
       $\varepsilon$   symbolizing random error

We describe the model assumptions later in this section. For *two* independent variables, the deterministic components define a plane surface in *three* dimensions. In Figure 13.1 the model coefficients $\beta_0$, $\beta_1$, and $\beta_2$ describe the plane of best fit to the data. $\beta_0$ is the height above the $X_1X_2$ plane at which the regression plane cuts the $Y$-axis. This is labeled $\beta_0$ on Figure 13.1. $\beta_1$ is the slope of a line that is formed where the regression plane intersects the $X_1Y$ axis. $\beta_2$ has a similar definition in the $X_2Y$ plane. These extend the simple linear regression $Y$-intercept and slope to an intercept and two slopes. Because the regression equation describes the average or expected response, it defines the regression plane $E(Y|X_1,X_2)$ that best fits the data points.

The random error ($\varepsilon$) for a least squares fitted equation is estimated by the *sample data errors*, called *residuals* (*e*). In Figure 13.1 the residuals are described

---

**FIGURE 13.1**    A Three-Dimensional Scattergram with a Regression Plane
                   (Conceptual)

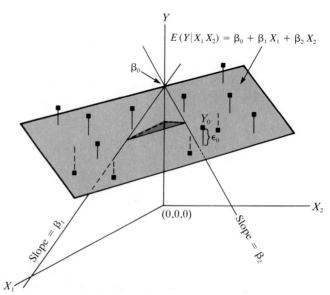

by the vertical distance from each observed response $(Y)$ and the regression plane $E(Y|X_1,X_2)$. These residuals are a base for testing the model assumptions and for making predictions. The standard deviation of the residual values is used for testing and for establishing prediction and confidence intervals. This *standard error of estimate* is central to all inference about the model.

The model assumptions are an extension from simple regression.

### ASSUMPTIONS FOR THE MULTIPLE LINEAR REGRESSION MODEL

1. The errors $(\varepsilon)$ are normally distributed with mean $E(\varepsilon) = 0$ and variance Var $(\varepsilon) = \sigma^2$.
2. The errors are statistically independent. Thus the error for any value of $Y$ is unaffected by the error for any other $Y$-value.
3. The $X$-variables are linear additive (i.e., can be summed).

As before, the model assumptions refer principally to the errors as the basis for inference. Assumptions 1 and 2 are the same as for simple linear regression. Assumption 3 means that the models that we consider can be defined by Equation (13.1).

Having defined the multiple linear regression model and recognized its assumptions, we move next to a procedure for estimating the model parameters, the $\beta$-values. But first are some problems to check your understanding.

**Problems 13.2**

1. Each of the following describes a multiple linear regression in two independent variables. Use the given information to find values for $\beta_0$, $\beta_1$, and $\beta_2$ in the model $E(Y|X_0,X_1,X_2) = \beta_0 + \beta_1X_1 + \beta_2X_2$.
   a. A regression plane that passes through the origin and forms a line $Y = 2X_1$ with the $X_1Y$ axes, and $Y = 3X_2$ with the $X_2Y$ axes
   b. A regression plane that appears as shown in the $X_1Y$ and $X_2Y$ planes

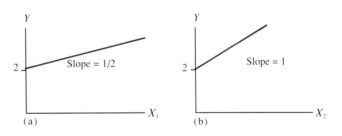

2. We can transform the relation $Y = 3 + 4X + 2X^2 + X^4 + \varepsilon$ to a first-order linear model, Equation (13.1), by setting $X = X_1, X^2 = X_2, X^4 = X_3$. This gives $\hat{Y} = 3 + 4X_1 + 2X_2 + X_3$. Which of the following *cannot* be changed to a first-order linear model?
   a. $Y = -2 + 0.9X + \varepsilon$
   b. $Y = \dfrac{3}{X} + \varepsilon, \quad X \neq 0$
   c. $Y = 17 + 142X^2 - 12X^3 + 0.10X^5 + \varepsilon$
   d. $Y = 3^X X^2 \varepsilon$

3. A multiple linear regression model is described by

$$Y = 3 + 0.5X_1 + 2X_2 + \varepsilon$$

where   $X_1$ = new investments ($100,000)

$X_2$ = weekly payroll ($10,000)

$Y$ = profits

a. Indicate the linear regression equation for $E(Y|X_1X_2)$.
b. Give a meaning for each number in the model.

4. Sometimes, linear models include qualitative $X$-variables. Then one model can be used to display more than one (simple) linear regression. This is used, for example, when two regression lines are parallel (have the same slope), but have different $Y$-intercepts. Consider

$$Y = 12 + X_1 + 3X_2 + \varepsilon$$

where   $X_1$ = 0 for not college graduates

= 1 for college graduates

$X_2$ = years in present position

$Y$ = income ($000)

Let   $X_1 = 0$,   then $E(Y|X) = 12 + 3X_2$;     or

$X_1 = 1$,   then $E(Y|X) = 13 + 3X_2$

a. Plot these lines on rectangular coordinates.
b. How are the graphs similar? How are the graphs different?

## 13.3
## REGRESSION BY THE LEAST SQUARES FITTED PLANE

To estimate the parameters of a regression model, we use an extension of the least squares estimation procedure as developed in Chapter 12. The sample analog for the model described by Equation (13.1) is

$$Y = b_0 + b_1X_1 + b_2X_2 + \cdots + b_kX_k + e$$

The estimation uses sample data to find estimator $b$'s for the $\beta$'s that will minimize the residuals, the $e$'s. Here the least squares procedure minimizes the sum of squares for error, SSE. If we let $\hat{Y}$ equal the estimated value of $Y$, we can express SSE as

$$\text{SSE} = E(Y - \hat{Y})^2 = E[Y - (b_0 + b_1X_1 + b_2X_2 + \cdots + b_kX_k)]^2$$

Solving a series of simultaneous equations provides formulas for the estimate of the $Y$-intercept ($b_0$) and for the estimators of the slopes ($b_1, b_2, \ldots, b_k$). The resulting ordinary *least squares estimation* solutions provide a best fit for an estimated regression surface. By least squares theory, the estimators $b_0$, $b_1$,

$b_2, \ldots, b_k$ are the "best" estimators available, because they minimize SSE and are unbiased for estimating $\beta_0, \beta_1, \ldots, \beta_k$.*

The least squares point estimation procedure minimizes the SSE to obtain values for $b_0, b_1, b_2, \ldots, b_k$. This ensures a best fit of the regression surface to the response data points. The $b$-values are the solutions from solving the set of normal equations:

$$\sum Y = b_0 n + b_1 \sum X_1 + b_2 \sum X_2 + \cdots + b_k \sum X_k$$

$$\sum X_1 Y = b_0 \sum X_1 + b_1 \sum X_1^2 + b_2 \sum X_1 X_2 + \cdots + b_k \sum X_1 X_k$$

$$\sum X_2 Y = b_0 \sum X_2 + b_1 \sum X_1 X_2 + b_2 \sum X_2^2 + \cdots + b_k \sum X_2 X_k$$

$$\vdots$$

$$\sum X_k Y = b_0 \sum X_k + b_1 \sum X_1 X_k + b_2 \sum X_2 X_k + \cdots + b_k \sum X_k^2$$

$$\text{(13.3)}$$

Computers can handle the extensive calculations required to solve for the $b$-values.

Whenever another $X$-variable is included, a new model is being formulated. Then all coefficients have to be reestimated. Consider a comparison of two regression models, (I) and (II).

**(I)** A simple linear regression: $Y = \beta_0 + \beta_1 X + \varepsilon$

**(II)** A multiple linear regression containing two $X$-variables:
$Y = \beta_0 + \beta_1 X_1 + \beta_2 X_2 + \varepsilon$

The normal equations for (I) are outlined in (13.3) by the dark-shaded area and those for (II) by all shaded areas. These equations produce, for model (I), $b_0 = \bar{Y} - b_1 \bar{X}_1$, and for model (II), $b_0 = \bar{Y} - b_1 \bar{X}_1 - b_2 \bar{X}_2$. To estimate the $\beta$-values for each case, model (II) requires more data inputs than does model (I), and so usually the $b_0$'s and $b_1$'s of the two models are not equal. Including more $X$-variables necessitates the solution of more equations from (13.3). The calculations soon become tedious. We assume that you have access to a computer for the problems, so we shall concentrate on the use and meaning of the solutions rather than on calculations.

**Illustrating the Estimated Regression Coefficients**

The data in Table 13.1 will be used for estimating the $Y$-intercept and $\beta$-coefficients for one linear regression model. The dependent variable is a USE SCORE on the cosmetics used by female respondents to a direct-mail advertising campaign in the Philadelphia market. This cosmetics USE SCORE has an interval scale, with values that range from 0 to just over 2.00.

Table 13.1 has column headings that designate the independent variables by a number. Because these variables require individual definitions, an abbreviated name and a short description appear, for each item, at the end of the table.

The *independent* variables warrant some definition. They include two types: (a) individual household variables and (b) demographics. Each individual

---

* Minimizing SSE to obtain the Normal Equations (13.3) is a straightforward application of the calculus. See [5] for a more complete development. (Bracketed numbers refer to items listed in "For Further Reading" at the end of the chapter.)

**TABLE 13.1** Cosmetics USE SCORES and Values on 15 Independent Variables

| Dependent (Cosmetics USE SCORE) | Independent Variables | | | | | | | | | | | | | | |
|---|---|---|---|---|---|---|---|---|---|---|---|---|---|---|---|
| | (1) | (2) | (3) | (4) | (5) | (6) | (7) | (8) | (9) | (10) | (11) | (12) | (13) | (14) | (15) |
| 1.10 | 3 | 1 | 1 | 6 | 33 | 81 | 100 | 345 | 49 | 88 | 652 | 32 | 96 | 96 | 38 |
| 1.55 | 1 | 1 | 1 | 9 | 39 | 95 | 100 | 367 | 52 | 72 | 608 | 26 | 26 | 71 | 42 |
| 0.80 | 9 | 1 | 1 | 6 | 49 | 64 | 100 | 282 | 50 | 64 | 699 | 43 | 33 | 95 | 36 |
| 1.35 | 3 | 1 | 1 | 8 | 22 | 73 | 100 | 328 | 50 | 84 | 615 | 23 | 32 | 75 | 30 |
| 0.70 | 4 | 0 | 0 | 6 | 52 | 85 | 99 | 325 | 48 | 67 | 713 | 23 | 36 | 100 | 25 |
| 2.05 | 9 | 1 | 1 | 6 | 21 | 98 | 100 | 317 | 48 | 99 | 652 | 19 | 35 | 82 | 48 |
| 1.10 | 1 | 0 | 1 | 9 | 9 | 48 | 99 | 309 | 51 | 58 | 648 | 28 | 20 | 85 | 47 |
| 1.30 | 9 | 1 | 1 | 8 | 27 | 76 | 99 | 328 | 53 | 92 | 655 | 32 | 27 | 98 | 32 |
| 0.85 | 9 | 1 | 1 | 8 | 27 | 76 | 99 | 328 | 53 | 92 | 655 | 32 | 27 | 98 | 32 |
| 0.00 | 4 | 0 | 1 | 12 | 13 | 38 | 99 | 260 | 50 | 69 | 631 | 33 | 31 | 76 | 19 |
| 1.55 | 5 | 1 | 1 | 14 | 18 | 81 | 99 | 383 | 50 | 74 | 656 | 26 | 31 | 69 | 31 |
| 1.60 | 9 | 1 | 1 | 5 | 47 | 98 | 98 | 318 | 39 | 98 | 673 | 31 | 46 | 92 | 28 |
| 1.35 | 9 | 1 | 1 | 10 | 19 | 68 | 99 | 378 | 51 | 73 | 628 | 40 | 27 | 73 | 43 |
| 0.60 | 9 | 1 | 1 | 12 | 12 | 47 | 99 | 320 | 55 | 50 | 656 | 43 | 22 | 85 | 22 |
| 0.00 | 1 | 1 | 1 | 3 | 23 | 81 | 100 | 253 | 55 | 38 | 590 | 34 | 31 | 61 | 32 |
| 0.40 | 9 | 1 | 0 | 7 | 26 | 52 | 99 | 249 | 52 | 82 | 714 | 26 | 14 | 118 | 15 |
| 0.80 | 9 | 0 | 0 | 4 | 15 | 61 | 100 | 241 | 50 | 46 | 581 | 22 | 40 | 80 | 25 |
| 0.70 | 1 | 0 | 1 | 6 | 38 | 57 | 99 | 292 | 46 | 52 | 605 | 21 | 42 | 80 | 38 |
| 0.50 | 3 | 0 | 1 | 5 | 20 | 58 | 100 | 301 | 49 | 54 | 579 | 11 | 37 | 80 | 32 |
| 0.50 | 9 | 1 | 1 | 7 | 23 | 54 | 100 | 309 | 50 | 89 | 610 | 19 | 31 | 93 | 38 |
| 0.55 | 3 | 1 | 1 | 5 | 39 | 62 | 100 | 281 | 47 | 81 | 596 | 18 | 44 | 89 | 32 |
| 1.35 | 9 | 1 | 1 | 15 | 24 | 84 | 100 | 375 | 47 | 95 | 631 | 24 | 42 | 79 | 36 |
| 1.00 | 9 | 1 | 1 | 6 | 17 | 71 | 99 | 356 | 49 | 100 | 708 | 24 | 38 | 74 | 33 |
| 1.60 | 9 | 1 | 0 | 19 | 8 | 90 | 100 | 387 | 48 | 95 | 581 | 22 | 40 | 65 | 31 |
| 0.80 | 5 | 1 | 1 | 5 | 48 | 77 | 99 | 273 | 43 | 78 | 676 | 27 | 43 | 109 | 16 |
| 2.10 | 9 | 1 | 1 | 5 | 30 | 91 | 99 | 360 | 51 | 96 | 704 | 31 | 30 | 101 | 23 |
| 1.60 | 9 | 1 | 1 | 5 | 38 | 86 | 99 | 308 | 48 | 82 | 713 | 24 | 36 | 89 | 32 |
| 1.10 | 9 | 1 | 0 | 5 | 38 | 86 | 99 | 308 | 48 | 82 | 713 | 24 | 36 | 89 | 32 |
| 0.70 | 4 | 0 | 1 | 8 | 19 | 49 | 97 | 297 | 45 | 70 | 695 | 24 | 36 | 87 | 21 |
| 0.80 | 3 | 1 | 1 | 10 | 14 | 43 | 98 | 271 | 50 | 44 | 662 | 22 | 28 | 61 | 28 |
| 0.07 | 7 | 0 | 1 | 2 | 58 | 37 | 99 | 262 | 37 | 82 | 797 | 10 | 51 | 102 | 13 |
| 0.65 | 1 | 0 | 1 | 2 | 52 | 71 | 95 | 285 | 40 | 66 | 805 | 22 | 45 | 118 | 28 |
| 0.00 | 2 | 0 | 1 | 2 | 36 | 31 | 98 | 270 | 49 | 71 | 708 | 32 | 35 | 95 | 11 |
| 0.50 | 7 | 1 | 1 | 3 | 31 | 47 | 99 | 279 | 48 | 88 | 713 | 17 | 36 | 99 | 24 |
| 1.60 | 4 | 1 | 1 | 5 | 19 | 85 | 99 | 379 | 51 | 100 | 598 | 10 | 37 | 90 | 34 |
| 0.40 | 4 | 0 | 1 | 7 | 14 | 65 | 100 | 236 | 48 | 72 | 655 | 26 | 35 | 77 | 12 |
| 1.60 | 1 | 1 | 1 | 7 | 40 | 84 | 99 | 356 | 43 | 99 | 661 | 18 | 43 | 86 | 33 |
| 1.60 | 1 | 1 | 1 | 4 | 25 | 89 | 100 | 353 | 48 | 95 | 614 | 10 | 37 | 88 | 43 |
| 1.85 | 2 | 1 | 1 | 4 | 25 | 89 | 100 | 353 | 48 | 95 | 614 | 10 | 37 | 88 | 33 |
| 0.00 | 6 | 0 | 1 | 8 | 28 | 34 | 100 | 307 | 49 | 58 | 664 | 27 | 38 | 108 | 10 |
| 0.40 | 2 | 0 | 1 | 8 | 28 | 34 | 100 | 307 | 49 | 88 | 664 | 27 | 38 | 108 | 10 |
| 0.85 | 5 | 1 | 1 | 10 | 21 | 56 | 98 | 322 | 51 | 89 | 669 | 24 | 32 | 86 | 25 |
| 0.85 | 1 | 0 | 0 | 8 | 26 | 68 | 99 | 308 | 50 | 64 | 679 | 26 | 33 | 96 | 20 |
| 1.60 | 1 | 1 | 0 | 8 | 6 | 65 | 98 | 347 | 58 | 94 | 644 | 13 | 21 | 83 | 45 |
| 0.70 | 2 | 1 | 1 | 3 | 49 | 70 | 97 | 213 | 39 | 76 | 754 | 14 | 43 | 128 | 19 |
| 0.85 | 9 | 1 | 0 | 4 | 22 | 81 | 100 | 302 | 53 | 64 | 657 | 20 | 30 | 86 | 37 |
| 0.00 | 1 | 0 | 1 | 2 | 47 | 48 | 97 | 281 | 46 | 39 | 750 | 21 | 32 | 108 | 21 |
| 0.85 | 9 | 1 | 1 | 2 | 71 | 64 | 99 | 246 | 46 | 70 | 727 | 23 | 44 | 111 | 29 |
| 0.40 | 5 | 1 | 1 | 2 | 71 | 64 | 99 | 246 | 46 | 70 | 727 | 23 | 44 | 111 | 29 |
| 1.60 | 5 | 1 | 1 | 8 | 18 | 87 | 100 | 348 | 51 | 100 | 634 | 19 | 31 | 90 | 41 |

**TABLE 13.1** (*continued*)

---

### Dependent Variable

Cosmetics kit USE SCORE, including use of lipgloss, eyeshadow, blush, and a brand loyalty measure.

---

### Independent Variables

#### (a) Individual Household Level Variables

(1) LOR — length of residence, in years, at this address
(2) DU — dwelling unit size: single household (1) or other (0)
(3) TITLE — the title of the individual listed for this address in a telephone directory: Mr. (1) or else (0)

#### (b) Demographic or Area Level Variables

(4) %FEMHD — % of households with a female head of household
(5) %AUTOS2P — % of households with two or more autos
(6) %OWNER — % of occupied units that are owner-occupied
(7) %WHITE — % of households with race of head of household being Caucasian
(8) SOCIAL — a socioeconomic status measure for an area
(9) MMA — male median age, for males 25 years and older, in an area
(10) PCTSFDU — % of households that are single-family dwellings
(11) MEDHV — median home value for the area
(12) %CLERSLS — % of employed persons 16 years or older, employed in clerical or sales work
(13) %POP18M — % of population that is 18 years old or younger
(14) INCINDX — income index, an area median income indexed to a regional median
(15) %EMPLFEM — % of households with one or more employed females

---

variable gives a value for an individual household. These individual variables are numbered 1, 2, and 3 in Table 13.1 for length of residence (LOR), dwelling size (DU), and title code (TITLE), respectively. For example, a telephone directory listing at Lincoln, Nebraska, since 1978 is John Ingram, 510 Birchwood Drive. In 1991 this household had LOR of 14 years, DU size of one, and TITLE code Mr. These variables are associated with individual households.*

A demographic variable describes a characteristic for numerous households. Demographic variables are usually expressed by the Census Bureau as percents, medians, or means. The variables numbered (4) through (15) in Table 13.1 are census-defined demographics. For example, demographic number (6), %OWNER, is the ratio of the number of owner occupied to the total number of houses in an area. SOCIAL, number (8), is a composite variable that encompasses demographics on occupation levels and on education.

We use these variables to develop an estimated regression for the cosmetics use data.

---

* The LOR variable is ordinal scale. It takes values 1, 2, 3, . . . years length of residence. The variables DU and TITLE are used here as category or type codes so are nominal scale. These variables are coded as indicator or dummy variables. For example, for the DU variable the values are coded or dummied in, with 1 denoting single-family dwelling units (SFDU) and 0 denoting 2+ (multiple) family dwellings (MFDU). Because these codes are not used to indicate an order, they might serve as well if interchanged. So they are indicator or dummy codes and the variables are called *dummy variables*. Section 13.7 includes a discussion of dummy variables.

The completely randomized design model in Chapter 11 could use codes 0 and 1 for a 2+ group experiment. Then a computer procedure could apply the multiple linear regression model to estimate parameters and make tests. In fact, this is a basis for generalized linear models in computer software packages, including those discussed in [4] and [10].

**EXAMPLE 13.1**

A manufacturer produces a cosmetics kit that contains lipgloss, eye shadow, and blush. The manufacturer has taken a random sampling of 50 households in the Philadelphia area in order to develop a market profile of household characteristics for cosmetics kit users. The firm's objective is to derive an expression for predicting an individual household product USE SCORE ($Y$) for the cosmetics kit. This is the sum of scores on the quantity of (any) lipgloss, plus eye shadow, plus blush, plus a value for brand loyalty. The independent variables are 15 demographic and individual household measures. However, for purposes of illustration, our analysis is limited to two independent variables: (1) the percent of households that are OWNER-occupied, and (2) a socioeconomic status measure, SOCIAL. Use the data from Table 13.1 to develop a multiple regression equation to predict $Y$ from these two independent variables.

*Solution*    The data from Table 13.1 were entered into a computer according to an SPSS format [11]. Some output from the regression procedure appears in Table 13.2. (The table provides more information than is needed here, as some terms are introduced now and will be used later for inference. This type of output is common for numerous statistical regression packages.)

In Table 13.2 we have shaded the portion that displays point estimates for the $Y$-intercept (CONSTANT) and the two slope values. Note that the E-03 following one of the values indicates that the number contains three more decimal places to the left. For example, 5.62515E-03 is 0.00562515.

The estimated regression equation is of the form $\hat{Y} = b_0 + b_1 X_1 + b_2 X_2$. By the statistics in the shaded area of Table 13.2, the regression equation is $\hat{Y} = -1.98000 + (0.01732 \cdot \%\,\text{OWNER}) + (0.00563 \cdot \text{SOCIAL})$. To understand this expression better, we consider estimates for the constant ($\beta_0$), and the coefficient parameters $\beta_1$ and $\beta_2$.

$b_0 = -1.98000$ means that the estimated regression plane intersects the $Y$-axis (cosmetics use) 1.98 units *below* the $X_1 X_2$ plane.

**TABLE 13.2**    SPSS Output (Modified) for an Analysis of Cosmetics Use Scores for $n = 50$ Households

DEPENDENT VARIABLE _____ Cosmetics USE SCORE

INDEPENDENT

VARIABLES ENTERED _____ % OWNER,  SOCIAL

| | | ANALYSIS OF VARIANCE | DF | SUM OF SQUARES | MEAN SQUARE |
|---|---|---|---|---|---|
| MULTIPLE R | .86350 | REGRESSION | 2 | 11.90880 | 5.95440 |
| R SQUARE | .74564 | RESIDUAL | 47 | 4.06244 | .08643 |
| ADJUSTED R SQUARE | .73482 | F = 68.88892 | | SIGNIF F = | .00 |
| STANDARD ERROR | .294000 | | | | |

VARIABLES IN THE EQUATION

| VARIABLE | B | SE B | BETA | CORREL | F |
|---|---|---|---|---|---|
| % OWNER | .01732 | 2.6680E-03 | .55890 | .78183 | 42.137 |
| SOCIAL | 5.62515E-03 | 1.1289E-03 | .42904 | .71944 | 24.830 |
| (CONSTANT) | -1.98000 | | | | 43.299 |

$b_1 = 0.01732$ indicates that the estimated *increase* in cosmetics kit USE SCORE is 0.01732 for each increase of 1% in OWNER residences, with SOCIAL held constant.

$b_2 = 0.00563$ means the USE SCORE is increased by 0.00563 for each unit of increase in SOCIAL score, holding % OWNER constant.

If these coefficients result in a good model, then the manufacturer can take advantage of the fact that areas high in owner-occupied housing, along with high SOCIAL scores, identify good prospects for cosmetics kit purchase.

How useful is this regression? We are not yet equipped to answer this question. But the answer relates to the residuals, or amounts by which actual values deviate from predicted ones. The residuals, when squared and summed, constitute the residual mean square error, which is the basis for inference in regression. We shall make use of these values later in the chapter.

**Problems 13.3**

5. Establish a relation of $Y$ (sales in $000) to $X_1$ (advertising expenditures in $000) and $X_2$ (hundreds of worker-hours of labor). For this, use an estimated multiple linear regression equation based on your solutions for $b_0, b_1, b_2$.

**(I)**
$$189 = 8b_0 + 36b_1 + 33b_2$$
$$1057 = 36b_0 + 204b_1 + 173b_2$$
$$1013 = 33b_0 + 173b_1 + 173b_2$$

which gives

**(II)**
$$44 = 3b_0 + 31b_1$$
$$-2184 = 196b_0 - 504b_1$$

a. Use (II) to solve for $b_1$.
b. Next use $b_2 = 5$ and your value for $b_1$ and any equation in (I) to find $b_0$.
c. Establish the estimated regression. Then interpret your values for $b_0, b_1, b_2$ in $\hat{Y} = b_0 + b_1X_1 + b_2X_2$.

6. A district manager for the circulation department of a newspaper wants to compare the circulation on paper routes, $Y$, with total service complaints, $X_1$, and carrier turnover, $X_2$, for the routes. The relationship will be used to plan for improved service. The data are displayed for 20 periods in $Y =$ mean size of route, $X_1 =$ service complaints, and $X_2 =$ carrier turnovers.

| Period | 1 | 2 | 3 | 4 | 5 | 6 | 7 | 8 | 9 | 10 |
|--------|-----|-----|----|-----|----|----|-----|-----|----|----|
| $Y$    | 133 | 115 | 79 | 118 | 70 | 75 | 151 | 119 | 58 | 97 |
| $X_1$  | 68  | 14  | 12 | 23  | 52 | 8  | 34  | 18  | 17 | 47 |
| $X_2$  | 4   | 3   | 1  | 1   | 5  | 1  | 5   | 2   | 2  | 3  |

| Period | 11 | 12 | 13 | 14 | 15 | 16 | 17 | 18 | 19 | 20 |
|--------|-----|-----|-----|----|-----|-----|----|----|-----|----|
| $Y$    | 140 | 115 | 132 | 58 | 145 | 155 | 75 | 53 | 137 | 66 |
| $X_1$  | 51  | 17  | 8   | 22 | 32  | 20  | 9  | 6  | 35  | 25 |
| $X_2$  | 4   | 1   | 3   | 1  | 1   | 2   | 4  | 1  | 3   | 4  |

    a. Use the method of least squares to establish an estimated linear regression for $Y$ on $X_1$ and $X_2$. This is most easily done on a computer. If you use a calculator, then use Equations (13.3) with $b_1 = 0.675$, $b_2 = 0.033$.

    b. Interpret your values for $b_0, b_1, b_2$. Then estimate $Y$ for $(X_1, X_2) = (47,3)$.

7. For a business research experiment, the Normal Equations (13.3) are

$$569 = 22b_0 + 1{,}132b_1 + 62.94b_2$$

$$32{,}603 = 1{,}132b_0 + 75{,}570b_1 + 3{,}356.20b_2$$

$$1{,}679.50 = 62.94b_0 + 3{,}356.20b_1 + 185.1656b_2$$

Provide answers for each of the following.

    a. Give values for $n, \sum Y, \sum X_1 X_2, \sum X_2 X_1$.

    b. Display the related normal equations for a simple linear regression of $Y$ on $X_1$. *Suggestion:* Eliminate from (13.3) all terms that include $X_2$. The remainder gives the desired equations.

    c. What is the value of $b_0$ for $\hat{Y} = b_0 + b_1 X_1 + b_2 X_2$? It will greatly simplify your work to use $b_1 = 0.146072$ and $b_2 = 6.756339$ (these were obtained from a computer run).

8. The management of a quick-service food shop wants to study the relation, assumed to be linear, of $Y$ on $X_1$ and $X_2$ (defined next) during the peak hours of 5–8 P.M. For a random selection of four work periods (one in each of four months) with $Y =$ sales in \$100 per hour, $X_1 =$ number of actively working employees and $X_2 =$ cost for supplies for one month ago on this day (in \$000). The relevant Equations (13.3) are

$$20 = 4b_0 + 20b_1 + 8b_2$$

$$101 = 20b_0 + 106b_1 + 39b_2$$

$$39 = 8b_0 + 39b_1 + 18b_2$$

The data are

| $Y$ | 5 | 5 | 6 | 4 |
|-----|---|---|---|---|
| $X_1$ | 5 | 4 | 6 | 5 |
| $X_2$ | 2 | 1 | 2 | 3 |

    a. Given $b_0 = 60/11$, find values for $b_1$ and $b_2$. Interpret the meaning of each $b_0, b_1$, and $b_2$ value.

    b. Explain why inference concerning $\beta_0$ would have no meaning in this situation.

9. A business consultant specializes in advising corporate executives whose firms are experiencing declining profits. Her experience with other businessses has allowed her to develop an equation for the expectation of the time ($Y$) in quarters for a company using her consulting to show increasing profits.

Her formula is

$$\hat{Y} = 2.48 + 1.14X_1 - 0.50X_2$$

where   $X_1$ = quarters since the company last showed a real increase in profits

$X_2$ = the number of years the chief administrative officer has held that position

a. Interpret each of the coefficients $b_0$, $b_1$, $b_2$.
b. Use this formula to estimate the time, in years, to show a profit, assuming that the company uses this service and that $X_1 = 6$ while $X_2 = 10$.

10. Using the model $Y = \beta_0 + \beta_1 X + \beta_2 X^2 + \varepsilon$, show that the system of Normal Equations (13.3) becomes

$$\sum Y = b_0 n + b_1 \sum X + b_2 \sum X^2$$
$$\sum (XY) = b_0 \sum X + b_1 \sum X^2 + b_2 \sum X^3$$
$$\sum (X^2 Y) = b_0 \sum X^2 + b_1 \sum X^3 + b_2 \sum X^4$$

*Suggestion: Let $X_1 = X$ and $X_2 = X^2$. Establish the normal equations for $X_1$ and $X_2$. Then replace $X_1$ with $X$ and $X_2$ with $X^2$.*

11. A regression model describing $Y$ = the price of vans (\$000) uses $X$ = age as the single independent variable. The model is $Y = \beta_0 + \beta_1 X + \beta_2 X^2 + \varepsilon$.
a. The sales of 16 vans at a single dealership provided the following statistics: $\bar{Y} = 11.25$ (\$000), $\bar{X} = 5$, $\sum X^2 = 548$, $b_0 = 21.100$, $b_1 = -3.395$. Use this information and the equations in Problem 10 to find $b_2$.
b. Use your estimated regression to determine a value for $Y$ given $X = 3$. Explain what your value means.

## 13.4 DESCRIBING THE REGRESSION MODEL WITH COMPUTER GENERATED STATISTICS

Table 13.2 includes outputs common to many commerical computer software packages. These outputs include multiple $R$, $R$-square ($R^2$), adjusted $R^2$ ($R_a^2$) the model standard error, and an analysis of variance with calculated $F$ for a significance test of the regression model, the $\beta$-coefficient estimates ($b$-values), standardized $\beta$-coefficients (BETA), correlations, and calculated $F$-statistics (or $t$-values) for inference about the individual $\beta$-coefficients. Most of these items are extensions from Chapter 12, and we will use them in regression inference in this chapter. However, adjusted $R^2$ and $\beta$-coefficients are new, so we offer an explanation of them now.

### Adjusted $R^2$

An estimated equation usually does not fit the population as well as it fits the sample from which it was derived. Adjusted $R^2$ attempts to correct the $R^2$ value so that it more closely reflects the fit of the model to the population of all response values.

From Chapter 12, recall that $R^2 = 1 - (SSE/TSS)$. Now, $R^2$ cannot decrease when another $X$-variable is added to the regression equation because SSE cannot increase with more $X$ variables, whereas TSS is fixed for a given data set. Then, sometimes we can make $R^2$ large simply by including a large number of $X$-variables. A procedure that adjusts $R^2$ for an increase in the number of $X$-variables is adjusted $R^2$, or

$$R_a^2 = 1 - \left[\left(\frac{n-1}{n-k}\right)\frac{\text{SSE}}{\text{TSS}}\right]$$

This *adjusted coefficient of multiple determination* may decrease when another $X$-variable is included. Adjusted $R^2$ will decrease if the decrease in SSE is more than offset by the loss of 1 df from $n - k$. So, the adjusted $R^2$ has become a popular statistic that is frequently displayed for computer regression summaries.

## Beta Coefficients

In addition to providing the slope ($b$-) values, multiple regression programs usually give a standardized value as well.

These standardized regression coefficients, or *betas*, are defined by $\text{beta}_k = b_k \cdot (s_{X_k}/s_Y)$. That is, the $b$-value is multiplied by the ratio of standard deviations, resulting in a coefficient that is adjusted for the relative dispersion of the $X_k$- and $Y$-variables with all other independent variables held constant. The result is that *beta weights can be directly compared*, since they have the same units of measurement, whereas the $b$-values often have different units.

Sometimes these beta coefficients are interpreted as showing, for instance, that %OWNER has a greater impact than SOCIAL on USE SCORES because beta (%OWNER) = 0.55890 exceeds beta (SOCIAL) = 0.42904 (from Table 13.2). However, it is not wise to interpret the betas rigidly as reflecting the relative importance of a set of independent variables, because adding more independent variables into the regression equation changes the $b$-values and hence the betas. Also, the betas are affected by correlations among the independent variables because the unstandardized regression coefficients are affected by the other $X$-variables in the model. We will use these concepts later when we discuss inference in regression.

## On Notation and Interpreting the Estimated Regression Coefficients

We have described the multiple regression model coefficients as a $Y$-intercept value (constant) and two or more slope estimators. The calculated sample value $b_0$ estimates the population value $\beta_0$, the elevation at which the regression surface intersects the $Y$-axis. When $X_1, X_2, \ldots, X_k$ include the origin within the range of the sample data, then $b_0$ has a meaningful physical interpretation. Otherwise, it is simply used to balance the $b_1, b_2, \ldots, b_k$ values and provide reasonable response predictions.

The values for $b_1, b_2, \ldots, b_k$ estimate the slopes of the regression surface projected into the $X_1 Y, X_2 Y, \ldots, X_k Y$ planes, respectively. Figure 13.2 displays conceptual expressions for $k = 2$ independent variables.

Note that Figure 13.2 introduces a new notation for the estimated slope coefficients. This denotes a coefficient, $b$, with two subscripts separated by a dot, which is read as "holding fixed." Thus $b_{1 \cdot 2}$ means the estimated slope in the $X_1 Y$ plane (for example, %OWNER-USE). The subscript 2 indicates that independent variable 2 (SOCIAL) was used in the data that developed this slope estimate but that the calculated value assumes the variable 2 is held constant, so that it has no effect on the $X_1 Y$ slope value. In contrast, the $b_1$ in a simple linear regression denotes the slope for $X_1$ with $Y$, since there are no other $X$-variables.

**FIGURE 13.2**  Interpreting the Regression Coefficients in an Estimated Multiple
Linear Regression, $k = 2$

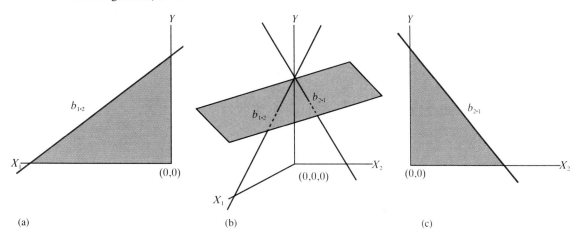

(a)                                                   (b)                                                   (c)

The slope estimators $b_1$ and $b_{1 \cdot 2}$ generally take on different values. For example, if the USE SCORES are regressed against % OWNER for the simple linear regression model of Chapter 12, we get

$$b_1 = 0.02423$$

Instead, in the multiple regression model, we have

$$b_{1 \cdot 2} = 0.01732 \quad \text{(from Table 13.2)}$$

The difference in values is *not* due to computer roundoff. It is due to a difference in models and the subsequent calculation procedures, which use and provide different information. In the simple linear regression model, the $b_1$ coefficient represents the increase expected in cosmetics USE SCORE, with a unit change in the % OWNER occupied. No other variables are considered to affect cosmetics use. In the multiple regression model, the independent variable SOCIAL is considered to affect cosmetics use and is therefore added to the equation. The $b_{1 \cdot 2}$-value describes the same expected increase as $b_1$, but the variable SOCIAL is held constant. Thus $b_{1 \cdot 2}$ isolates the effect of % OWNER on the USE SCORES by removing the influence of the SOCIAL variable.

It is possible, but highly unlikely, that $b_1$ and $b_{1 \cdot 2}$ will have equal values. For $b_1$ to equal $b_{1 \cdot 2}$, $X_1$ and $X_2$ must have zero correlation. However, independence for two $X$-variables is the exception rather than the rule. Conversely, when $b_1$ and $b_{1 \cdot 2}$ have quite different values, it is possible that $X_1$ and $X_2$ are highly correlated. This can indicate high multicollinearity in the model. Multicollinearity will be defined more specifically later, but for now think of it as an interdependence among the independent variables.

The notation $b_{1 \cdot 2}$ was introduced to demonstrate that the multiple regression equation coefficients define something different from simple linear regression coefficients, the $b_1$. However, as more independent variables are included, the notation soon becomes cumbersome. For example, the slope coefficient of $X_1$ with five other variables held constant would be designated $b_{1 \cdot 23456}$.

When working with multiple regression models, writers of textbooks and journal articles do not always explicitly designate all the subscripts describing variables that are held constant. Thus it is important to interpret $b_1$ to be $b_1$, or $b_{1 \cdot 2}$, and so on, in keeping with the model under discussion. We use the abbreviated notation, so you will need to remember the change as independent variables are added to a model.

**Problems 13.4**

12. An analysis performed on executive salaries for multinational firms included the following variables.

$$Y = \text{profit incentives in shares granted}$$

$$X_1 = \text{the executive's age}$$

$$X_2 = \text{the executive's number of years with the firm}$$

From experience we know that both $X_1$ and $X_2$ are positively correlated with $Y$. Based on the Normal Equations (13.3), someone has obtained the following: $b_0 = 20.193$, $b_1 = +2.217$, $b_{1 \cdot 2} = -1.918$. Profit shares are in 100 units. Something in the analyst's numbers seems illogical. Use these variables to explain the difference between $b_1$ and $b_{1 \cdot 2}$.

13. a. For the data in Problem 5, relate the units for each of the coefficients $(b_1, b_2)$ to the respective $X$-variable, either $X_1$ or $X_2$. What clues do these units give about the relative strength of $X_1$ and $X_2$ as predictors for sales?

   b. Suggest a conversion that would make the $b$-coefficients, $b_1$ and $b_2$, directly comparable for relative predictive strength on sales.

14. Explain the differences in the least squares regression estimation of $\beta$-coefficients for a simple versus a multiple regression model with respect to
   a. the information used to make the estimates.
   b. the assumptions.
   c. the feasibility of hand-calculator calculations.

15. Table 13.2 shows $R^2 = 0.74564$ and adjusted $R_a^2 = 0.73482$. Here $R^2 > R_a^2$. Use $R^2 = 1 - (\text{SSE/TSS})$ and $R_a^2 = [1 - ((n-1)/(n-k))(\text{SSE/TSS})]$ to explain why $R^2 \geqslant R_a^2$. What is the relation when $k = 1$?

## 13.5 THE STANDARD ERROR OF ESTIMATE IN MULTIPLE REGRESSION

We have assumed that a multiple linear regression model can be used to predict a business response variable. However, we should always test whether the model is statistically valid before using it to make any predictions. This section extends the standard error of estimate to the multiple regression model.

The assumptions underlying multiple regression models were summarized in Section 13.2. An important assumption is that the error variance is constant regardless of what values the independent variables take on. The error variance is estimated as the mean square error, MSE, of the sample data. Insofar as the MSE is used in all tests and interval estimates, this assumption (of constant variance) is essential to our inference. Further, to conduct those tests and make the interval estimates, we use the $t$- and $F$-statistics, which assume that the errors are independent and normally distributed.

At this point, you may think there are so many constraints that regression modeling appears useless, that no clear-cut applications exist. However, multiple regression models *are* widely used in business. And they are used even though all the assumptions described above are not completely satisfied. This is because some of the assumptions can be relaxed without significant effect in certain situations. It is good to be aware of what consequences can result if the assumptions are not met, so we will point out some of these effects in a later section. Finally, we note that diagnostics, such as data plots and statistics describing the errors (residuals), can provide clues for improving the regression model. These topics are introduced in Section 13.7, after we discuss the basics for multiple regression inference.

## The Standard Error of Estimate Defined

As with simple linear regression models, the *standard error of estimate*, SE, is the measure of error used with multiple regression models. The SE is, in turn, calculated from the *mean square error* (MSE), which is an estimate of the population error variance, $\sigma^2$.

> Mean Square Error (MSE) for multiple regression is a single, comprehensive measure of the reliability of the estimated regression:
>
> $$MSE = SSE/[n - (k + 1)] \tag{13.4}$$
>
> with SSE computed from a model with $k$ independent variables.
>
> The Standard Error of estimate is
>
> $$SE^* = \sqrt{MSE} \tag{13.5}$$

The divisor for MSE is easily remembered as the sample size minus the number of model parameters being estimated. The number of parameter estimates is one for each of the $k$ independent variables plus one for the constant ($Y$-intercept).

The size of the MSE affects all regression inference. Specifically, the smaller the MSE, the more precise the estimates of the $\beta$'s, the narrower the confidence intervals on $Y$ and $E(Y|X_1 X_2 \cdots X_k)$, and the more reliable the predictions. *The mean square error provides a single measure to evaluate the reliability of the estimated regression.* The next example illustrates the gain in reliability for one situation.

**EXAMPLE 13.2**

Two models, summarized in Table 13.3, are explored for estimating the cosmetics kit USE SCORE. The first model has one independent variable; the other has two. Compare the MSEs for the two estimated regressions.

* For a regression in $X_1$, $X_2$ this reduces to an algebraic equation:

$$SE = \sqrt{MSE} = \sqrt{\frac{SSE}{n-3}} = \sqrt{\frac{\sum (Y - \hat{Y})^2}{n-3}}$$

$$SE = \sqrt{\frac{\sum Y^2 - b_0 \sum Y - b_1 \sum X_1 Y - b_2 \sum X_2 Y}{n-3}}$$

for $\hat{Y} = b_0 + b_1 X_1 + b_2 X_2$

For three or more $X$-variables, it is most convenient to use computer results.

**TABLE 13.3**  SPSS Output (Modified) Displaying Summaries for Cosmetics Kit Use Score Modeling

```
VARIABLES(S) ENTERED ON STEP NUMBER 1..      % OWNER


MULTIPLE R            .78183   ANALYSIS OF VARIANCE    DF   SUM OF SQUARES   MEAN SQUARE
R SQUARE             .61126   REGRESSION               1         9.76260      9.76260
ADJUSTED R SQUARE    .60316   RESIDUAL                48         6.20864       .12935
STANDARD ERROR       .35965

                              F =      75.47619       SIGNIF F =   .0000


                 VARIABLES IN THE EQUATION

VARIABLE            B        SE B       BETA        T      SIG T

% OWNER        .024227    .002789    .781832     8.688    .0000
(CONSTANT)    -.711544    .196276               -3.625    .0007


VARIABLES(S) ENTERED ON STEP NUMBER 2..      SOCIAL


MULTIPLE R            .86350   ANALYSIS OF VARIANCE    DF   SUM OF SQUARES   MEAN SQUARE
R SQUARE             .74564   REGRESSION               2        11.90880      5.95440
ADJUSTED R SQUARE    .73482   RESIDUAL                47         4.06244       .08643
STANDARD ERROR       .29400

                              F =      68.88892       SIGNIF F =   .0000


                 VARIABLES IN THE EQUATION

VARIABLE            B        SE B       BETA        T      SIG T

% OWNER        .017319    .002668    .558905     6.491    .0000
SOCIAL         .005625    .001129    .429041     4.983    .0000
(CONSTANT)   -1.980003    .300904               -6.580    .0000
```

*Solution*   Table 13.3 incorporates two displays from an SPSS computer listing. The mean square error MSE = 0.12935 (shown shaded) results from using one independent variable, %OWNER. At Step 2, with the addition of the SOCIAL variable, the MSE = 0.08643 (also shaded). Thus, adding a second independent variable reduces MSE by 33%. This means that the regression equation at Step 2 will give better predictions for the cosmetics kit USE SCORE.

In Example 13.2, the inclusion of a second variable increased the precision of our results. Whether this is a significant gain is a matter for hypothesis testing. The MSE can also be used to visually *check* for a statistically improved model. The technique for doing this uses a plot by regression degrees of freedom (df) against MSE. The result is called a *scree plot*.

**A scree plot is a plot of df against MSE and shows when MSE levels off or becomes flat with the addition of more independent variables.**

It is used as a visual check when the inclusion of additional variables adds little or no more useful information to the model.

See [2] for discussion of the use of scree plots, including a MSE plot.

**Problems 13.5**

16. Table 13.3 displays two regression summaries. Explain the relation between STANDARD ERROR and MEAN SQUARE for RESIDUALS in each summary. Specifically, demonstrate that

$$\sqrt{\text{MEAN SQUARE FOR RESIDUAL}} = \text{STANDARD ERROR}$$

17. Again we use Table 13.3. A statistical test shows that the addition of SOCIAL (at Step 2) improves the regression description on USE SCORE (at Step 1). Quantify the gain in increased precision by a comparison: (RESIDUAL 2 MEAN SQUARE/RESIDUAL 1 MEAN SQUARE) · 100. What does this tell about the scatter of observations about the regression?

18. A summary shows a decreasing RESIDUAL MEAN SQUARE as more $X$-variables enter the model.

| Model | $X$-Variables | df | Mean Square |
|-------|---------------|-----|-------------|
| 1 | 1 $X$-variable | 1 | |
| | Residuals | 23 | 403.00 |
| 2 | 2 $X$-variables | 2 | |
| | Residuals | 22 | 102.70 |
| 3 | 3 $X$-variables | 3 | |
| | Residuals | 21 | 107.06 |

Of the values displayed, the lowest mean square for residuals is Model 2. Use this information to explain which model (1, 2, or 3) you would choose. Explain your choice.

19. One possible problem with the residual mean square evaluation, as in Problem 18, might be that a plateau is reached and testing stops short before the plateau is recognized. Subsequently, an improved model can be overlooked. Consider again the data in Problem 18, extended here.

| Model | $X$-Variables | df | Mean Square |
|-------|---------------|-----|-------------|
| 3 | 3 $X$-variables | 3 | |
| | Residuals | 21 | 107.06 |
| 4 | 4 $X$-variables | 4 | |
| | Residuals | 20 | 86.39 |
| 5 | 5 $X$-variables | 5 | |
| | Residuals | 19 | 84.65 |

a. Make a scree plot of the residual mean squares (MSE) through the model with five $X$-variables.

b. Explain how a scree plot can provide information to help you choose a model.

## 13.6 TESTING THE REGRESSION MODEL

As with simple regression, the critical question for multiple regression inference is also, "Does the regression model have predictive meaning?" To test this in simple linear regression, we hypothesized $H_0$: $\beta_1 = 0$ and used a $t$-statistic on the sampling distribution of $b_1$. This was described as equivalent to the $t$-statistic for testing zero correlation. Then, under the assumptions for regression inference, $t^2 = F = (\text{SSR}/1)/\text{MSE}$ has an $F$-distribution with 1 and $n - 2$ degrees of freedom.

> **Mean Square for Regression (MSR) is a measure of the average sum of squares explained by the variables in the model, MSR = SSR/k, where $k$ = the number of independent variables.**

We can now extend this to a general procedure for testing for zero slopes in a multiple linear regression model.

The test for a meaningful multiple linear regression model is a test that at least one of the $X$-variables gives a significant description of response. The null hypothesis is that all $\beta$'s equal zero. Alternatively, *at least one* of the $\beta$'s does not equal zero. We use the $F$-statistic for testing and extend the sampling distribution to the case for $k$ independent variables. Recall that $t^2 = F$, so you should expect a single-tailed rejection region. The test is whether there is a multiple linear relation between $Y$ and the set of $X$-variables $X_1, X_2, \ldots, X_k$. It is defined by

$$H_0: \beta_1 = \beta_2 = \cdots = \beta_k = 0$$

$$H_A: \text{Not all } \beta_i \text{ equal } 0 \qquad i = 1, 2, \ldots, k$$

This uses the test statistic $F = \text{MSR}/\text{MSE}$. Table 13.4 details the ANOVA procedure for making this test.

The multiple linear regression has associated multiple $R$- and $R^2$-values, where the definitions are extensions of the simple regression concepts.

**TABLE 13.4**  An ANOVA for Regression for Testing $H_0$: $\beta_1 = \beta_2 = \cdots = \beta_k = 0$

The $F$-Ratio Statistic

$$F_{k,n-(k+1)} = \frac{\text{SSR}/k}{\text{SSE}/(n-(k+1))} = \frac{\text{Regression MS}}{\text{Residual MS}} = \frac{\text{MSR}}{\text{MSE}} \qquad (13.6)$$

**An ANOVA for Regression with $F$-Statistic**

| Source | df | Sum of Squares | Mean Square | $F$ |
|---|---|---|---|---|
| REGRESSION | $k$ | SSR | MSR | MSR |
| RESIDUALS | $n-(k+1)$ | SSE | MSE | $\overline{\text{MSE}}$ |
| TOTAL | $n-1$ | TSS($=$ SS $(Y)$) | | |

The coefficient of multiple determination, $R^2$ (or $R^2_{Y\cdot 1,2\ldots k}$), describes the percentage of total variation in $Y$ that is associated with $X_1, X_2, \ldots,$ and $X_k$.

$$R^2 = \frac{\text{SSR}}{\text{TSS}} = 1 - \frac{\text{SSE}}{\text{TSS}} \qquad (13.7)$$

The coefficient of multiple correlation is

$$R = +\sqrt{\frac{\text{SSR}}{\text{TSS}}} \qquad (13.8)$$

Because $R$ is the correlation between observations and the regression plane $(Y)$, $R$ is always positive and ranges from 0 to 1. Thus $0 \leqslant R^2 \leqslant 1$.

Also, $R^2$ increases with the addition of $X$-variables. You can see this increase in Table 13.3, where the $R^2$ change is $0.74564 - 0.61126 = 0.13438$ with the addition of SOCIAL to the regression equation. Whether the improvement is significant is a matter for hypothesis testing.

---

**EXAMPLE 13.3**

Are one or more of the SOCIAL and %OWNER variables significant in describing the cosmetics USE SCORES? Use $\alpha = 0.01$.

*Solution*   The question is whether either of the variables or both have a nonzero slope. The solution follows the usual steps for a hypothesis test.

1. Table 13.3, Step 2, displays the regression output of interest. The test is for a significant linear regression using two $X$-variables.

$$H_0: \beta_1 = \beta_2 = 0 \qquad H_A: \text{At least one of } \beta_1, \beta_2 \neq 0.$$

The most pertinent parts of Table 13.3 are repeated here.

```
DEPENDENT VARIABLE _____ (Cosmetics Kit) USE SCORE
VARIABLES ENTERED _____ % OWNER, SOCIAL

                          ANALYSIS OF              SUM OF      MEAN
                          VARIANCE       DF        SQUARES     SQUARE

MULTIPLE R    .86350      REGRESSION      2        11.90880    5.95440
R SQUARE      .74564      RESIDUAL       47         4.06244     .08643
                          F=68.88892               SIGNIF F=.00
```

2. Test: Reject $H_0$ if calculated $F > F_{0.01,2,47} \doteq 5.10$.
3. Calculated $F = 68.89$.
4. Conclusion: Calculated $F$ exceeds test $F$. Also, "SIGNIF F = .00" indicates that the calculated $F$ is significant with a $p$-value smaller than 0.01. We reject $H_0$ for $\alpha = 0.01$ and conclude that *one or both* of %OWNER and SOCIAL variables are significantly related to the cosmetics kit USE SCORE.

The procedure used in Example 13.3 can be used to test a linear regression model with any number of $X$ variables. If our conclusion is to accept $H_0$, then none of the variables has predictive meaning and we discard the model. Otherwise, we test for those variables from among $X_1, X_2, \ldots, X_k$ that are useful for making the best predictions. This relates to regression model building procedures, which are described in Section 13.7.

Another question is, "What is the relative strength in describing the response for those independent variables that enter a regression equation?" This is our next topic.

## Inference Concerning the Individual $\beta$-Coefficients

Inference about the individual $\beta$-coefficients uses procedures similar to ones defined in Chapter 12. Under the assumptions for the multiple regression model, page 577, the sampling distribution is a $t$-distribution.

> The sampling distribution for inference concerning the individual $X$-variables in a multiple linear regression model is
>
> $$t_{n-(k+1)} = \frac{b_i - \beta_i}{s_{b_i}} \qquad i = 1, 2, \ldots, k \qquad (13.9)$$
>
> where    $s_{b_i}$ = standard error of $b_i$*

The degrees of freedom $n - (k + 1)$ reflect the multiple linear regression model with $k$-independent variables. With some additional discussion we could compute the standard error of $b_i$ (see [8]). However, this is regularly available from a regression computer listing, so we use the standard error of $b_i$, or SE $B$, from computer-generated results.

The next example illustrates the inference concerning individual coefficients for the multiple regression on cosmetics USE SCORE. This includes (1) 95% confidence estimates and (2) hypothesis tests on the $\beta$'s.

---

**EXAMPLE 13.4**

Several relative comparisons are made for %OWNER and SOCIAL as descriptors for the cosmetics kit USE SCORES. The summary is from Table 13.3, Step 2, some of which is repeated in Table 13.5. This displays statistics needed to test the individual $\beta$-coefficients. And Table 13.5 also displays numbers for calculating the upper and lower limits for the 95% confidence interval on each $\beta$.
a. Develop separate 95% confidence interval estimates for the $\beta$-coefficients.
b. Within the two-$X$-variable model, test %OWNER, then SOCIAL, for significance in the regression model. Let $\alpha = 0.01$.

*Solution*    This problem requires an evaluation of each variable as a predictor for cosmetics kit USE SCORES, $Y$. Table 13.5 provides the information essential for our answers.
a. Use $B \pm (t \cdot \text{SE } B)$ with $t_{0.025,47} \doteq 1.96$ for 95% confidence estimates.

* For a regression in $X_1$ and $X_2$, this can be expressed as

$$s_{b_1} = \frac{\text{SE}}{\sqrt{\text{SS}(X_1) \cdot (1 - r_{12}^2)}} \qquad \text{and} \qquad s_{b_2} = \frac{\text{SE}}{\sqrt{\text{SS}(X_2) \cdot (1 - r_{12}^2)}}$$

and SE = the standard error of estimate (13.5).
For three or more $X$-variables, it is most convenient to use computer results.

**TABLE 13.5**  Summary Statistics for Inference on the Individual $\beta$-Values

```
DEPENDENT VARIABLE _____ Cosmetics USE SCORE
VARIABLES ENTERED _____ % OWNER and SOCIAL   RESIDUAL DF = 47

                  VARIABLES IN THE EQUATION

VARIABLE               B       SE B       BETA         T    SIG T
% OWNER          .017319    .002668    .558905     6.491    .0000
SOCIAL           .005625    .001129    .429041     4.983    .0000
(CONSTANT)     -1.980003    .300904               -6.580    .0000
```

For $b_1$ = Slope for % OWNER:

Confidence Interval = $0.01732 \pm [(1.96)(.00267)] \doteq 0.0120$ to $0.0227$

For $b_2$ = Slope for SOCIAL:

Confidence Interval = $0.00563 \pm [(1.96)(0.00113)] \doteq 0.00335$ to $0.00790$

b.  The tests for significance can be either one- or two-tailed and use the $t$-statistic $t_{n-(k+1)} = (b_i - \beta_i)/s_{b_i}$ from Equation (13.9).

Tests for $H_A$: $\beta_i > 0$ use the critical test value $t_{0.01,47} \doteq 2.33$

For % OWNER:   Calculated $t = \dfrac{0.01732}{0.002668} = 6.49$

with SIG T $= 0.000$ in Table 13.5. The % OWNER coefficient is significant with $p$-value $< 0.001$.

For SOCIAL:   Calculated $t = \dfrac{0.005625}{0.001129} = 4.98$

with SIGT $= 0.000$. The SOCIAL variable coefficient is significant with $p$-value $< 0.001$. Again the calculated values are very close to the $T$-values generated by the computer program. Both % OWNER and SOCIAL are statistically significant at $\alpha = 0.01$.

## Beta Coefficients

The preceding example contains one more useful point. That is, because % OWNER and SOCIAL have different units, we cannot compare the descriptive strength for these two independent variables simply by comparing their $b$-coefficients. Here SOCIAL scores are in thousands, while % OWNER is scaled 0 to 100%. The BETA, or standardized scores, equalize the scales for all $X$-variables. So *the relative order of strength in describing response is observed in the sizes of the BETA coefficients.* In the example, % OWNER is stronger in describing cosmetics USE SCORE than was SOCIAL, because the beta coefficient for % OWNER (0.55890) is larger than the beta value for SOCIAL (0.42904). (Remember the betas can change as the number of $X$-variables is changed in the equation. This is because the relative contribution of each $X$-variable is affected by all the variables in the model.)

**Predicting Expected and Individual Response**

Having developed a meaningful regression equation, we can use it to predict either the average value or an individual response. The point estimate is the same for both. We obtain it simply by substituting the specific values for $X_1$, $X_2, \ldots, X_k$ into the estimated regression equation and solving for the dependent variable, $Y$. However, the interval estimates are different. *Prediction intervals*, for individuals, are wider than the corresponding *confidence intervals* for the estimation of average response. The argument is similar to that presented for the simple regression case in Chapter 12.

The distinction for average versus individual response is a reflection of the respective models,

$$E(Y|X_1X_2\cdots X_k) = \beta_0 + \beta_1 X_1 + \beta_2 X_2 + \cdots + \beta_k X_k \qquad \text{for average response}$$

and
$$Y = \beta_0 + \underbrace{\beta_1 X_1 + \beta_2 X_2 + \cdots + \beta_k X_k}_{\textbf{Mean response}} + \underbrace{\varepsilon}_{\textbf{+ Unique (to the individual)}} \quad \text{for individuals}$$

Since, by assumption, $E(\varepsilon) = 0$, the point estimates are identical. However, only the variance for individuals includes a "unique" component of error. This results in a greater width in the individual estimate than exists for estimates of the average. The sampling distributions are summarized here.

### ESTIMATED SAMPLING DISTRIBUTIONS: (1) MEAN AND (2) INDIVIDUAL RESPONSE

In multiple linear regression
where the $t$-statistics have df $= n - (k + 1)$:

(1) **Mean Response**  $\quad t = \dfrac{\hat{Y} - E(Y|X_1X_2\cdots X_k)}{s_{\hat{Y}}}$  (13.10)

$s_{\hat{Y}}^2$ = estimated variance of mean response

$s_{\hat{Y}}$ = standard error of mean (predicted) response

(2) **Individual Response**  $\quad t = \dfrac{\hat{Y} - Y_{\text{indiv}}}{s_{\text{indiv}}}$  (13.11)

$s_{\text{indiv}}^2 = s_{\hat{Y}}^2 + \text{MSE}$

$s_{\text{indiv}}$ = standard error of the individual response

Interval estimates are

(1) **Mean response confidence interval** for $E(Y|X_1X_2\cdots X_k)$:

$$(\text{LCL, UCL}) = \hat{Y} \pm (t \cdot s_{\hat{Y}})$$  (13.12)

**(2) Individual response prediction interval for $Y_{indiv}$:**

$$(\text{LPL}, \text{UPL}) = \hat{Y} \pm t\sqrt{s_{\hat{Y}}^2 + \text{MSE}} \qquad (13.13)$$

Relative to simple linear regression, the calculations for the estimated variances in the mean and individual response models are computationally more difficult. They are best handled with matrices. Since we have elected to avoid the complexity of matrix calculations, we will not give computation formulas. However, you should understand that the variance of mean response includes contributions from each $\beta$-value. In addition, the variance of individual response *adds* a component for unique variations. Thus, the variance for individual response values exceeds that for mean response values.

Ott [8] provides a development of the variance forms that requires the use of matrices. However, the statistical analysis system (SAS) software provides ready-made calculations for both prediction and confidence limits. Computer Problem 65 contains data from an SPSS run that allows you to use Equations (13.12) and (13.13) to check the calculations from this SAS output.

---

**EXAMPLE 13.5**

The estimated regression equation for cosmetics kit USE SCORES, $\hat{Y} = -1.98000 + (0.01732 \,\% \text{OWNER}) + (0.00563 \cdot \text{SOCIAL})$, was found to be statistically significant in Example 13.3. Use this result and the following SAS output to establish (a) 99% confidence and (b) 99% prediction limits for a USE SCORE with ($\%$ OWNER, SOCIAL) = (81, 345). This individual record is Case 1 in Table 13.1.

*Solution*    Table 13.6 includes only the essentials for interval estimates from a SAS computer listing for the General Linear Models, GLM, procedure [10]. Then, for the first observation in Table 13.6 with ($\%$ OWNER, SOCIAL) = (81, 345) and reading from the table, (a) 99% confidence limits for the mean are (LCL, UCL) = (1.2135, 1.5135), and (b) 99% prediction limits for this individual are (LPL, UPL) = (0.5061, 2.1669).

---

Table 13.6 displays confidence and prediction limits for several observations. Observe that some *lower* confidence limits are negative numbers. This is not physically possible, since the actual USE SCORES were zero or positive. However, Table 13.6 shows the actual computer output, which is based solely

**TABLE 13.6**    SAS Output Displays for 99% Confidence Limits for Mean Response and 99% Prediction Limits for Individual Response (2 Outputs Combined)

| OBS. | OBS. VALUE | PRED. VALUE | LOWER 99% CL FOR MEANS | UPPER 99% CL FOR MEANS | LOWER 99% PL INDIVIDUALS | UPPER 99% PL INDIVIDUALS |
|---|---|---|---|---|---|---|
| 1 | 1.1000 | 1.3635 | 1.2135 | 1.5135 | .5601 | 2.1669 |
| 2 | 1.5500 | 1.7297 | 1.5165 | 1.9430 | .9122 | 2.5150 |
| 3 | .8000 | .7147 | .5823 | .8471 | -.0859 | 1.5150 |
| 4 | 1.3500 | 1.1293 | 1.0072 | 1.2515 | .3307 | 1.9280 |
| ⋮ | | | ⋮ | | | ⋮ |
| 48 | .8500 | .5122 | .3023 | .7221 | -.3045 | 1.3289 |
| 49 | .8500 | .5122 | .3023 | .7221 | -.3045 | 1.3289 |
| 50 | 1.6000 | 1.4843 | 1.3162 | 1.6524 | .6772 | 2.2913 |

upon computational procedures. The *real* lower limits are zero. This discrepancy serves as a good reminder that one should always assess the answer to a problem in the context of good judgment and report results accordingly. Chapter Problem 65 invites you to check the reported results by using a multiple regression program. Related 99% probability intervals are displayed for several other cases. Remember that the prediction interval is always wider than the associated confidence interval.

This completes our discussion of the basic inference for multiple linear regression models. An understanding of this material should allow you to use the common commercial regression printouts. The remainder of this chapter will be useful if you need to develop your own models for a "best" selection from a set of independent variables.

**Problems 13.6**

20. The following information relates to the table below:

$$Y = \text{annual sales for the ABC Company, \$ millions}$$

$$X_1 = \text{advertising expenses, \$ millions}$$

$$X_2 = \text{comparative price index on the major product,}$$
$$\text{for current year relative to the 1976–85 average}$$

```
DEPENDENT VARIABLE _____ SALES
VARIABLES ENTERED _____ ADVERTISING, PRICE INDEX
```

|                | | ANALYSIS OF | | SUM OF | MEAN |
|----------------|------|-------------|----|----------|----------|
| MULTIPLE R     | .988 | VARIANCE    | DF | SQUARES  | SQUARES  |
|                |      |             |    |          |          |
| R SQUARE       | .977 | REGRESSION  | 2  | 44.9800  | 22.4900  |
|                |      | RESIDUAL    | 7  | 1.0500   | .1500    |

```
VARIABLES IN THE EQUATION
```

| VARIABLE    | B     | BETA    | SE B  |
|-------------|-------|---------|-------|
| ADVERTISING | .740  | 1.0513  | .1063 |
| PRICE INDEX | -.212 | -.3384  | .2992 |
| CONSTANT    | 4.710 |         |       |

a. Test the hypothesis $H_0: \beta_1 = \beta_2 = 0$ against $H_A$: At least one $\beta_i \neq 0$. Use $\alpha = 0.05$.

b. Now test the individual coefficients (Test 1) $H_0: \beta_1 = 0$, $H_A: \beta_1 \neq 0$, and (Test 2) $H_0: \beta_2 = 0$, $H_A: \beta_2 \neq 0$. Again, use $\alpha = 0.05$.

c. From the results of (a) and (b), what model would you choose from

$$Y = \beta_0 + \beta_1 X_1 + \beta_2 X_2 + \varepsilon$$

$$Y = D_0 + D_1 X_1 + \varepsilon$$

$$Y = C_0 + C_2 X_2 + \varepsilon$$

Explain.

21. A computer regression analysis provides the following information:

| Source | SS | df | Mean Square | F |
|---|---|---|---|---|
| Regression | 27.7258 | 3 | MSR | MSR/MSE |
| Residuals | 0.7433 | 8 | MSE | $F_{0.10,3,8} = ?$ |

| Variable | B | Std Error of B |
|---|---|---|
| $X_1$ | 5.8256 | 0.1525 |
| $X_2$ | $-8.0010$ | 1.0800 |
| $X_3$ | 1.3176 | 0.0804 |
| Constant | $-48.1001$ | |

a. Complete the ANOVA for regression and test $H_0$: $\beta_1 = \beta_2 = \beta_3 = 0$ for $\alpha = 0.10$.
b. Compute multiple $R$ and $R^2$.
c. Establish a 90% confidence interval estimate for $\beta_2$.

22. A color photograpy developing firm is explaining factors that relate to the color intensity of its film, measured on a scale from 0 to 10.0. Critical factors appear to relate to the exposure time (hundredths of a second) and the relative humidity in the photo processing lab. For convenience, the data were coded using

$$X_1 = \text{exposure time minus } 5.5$$

$$X_2 = \text{relative humidity minus } 0.3$$

The regression analysis shows a significant multiple regression on $X_1$ and $X_2$ (both coded), estimated by

$$\hat{Y} = 5.50 + 1.33X_1 - 8.01X_2$$

Other statistics include MSE $= 0.2812$, $n = 16$, and $s_{\hat{Y}}$. The estimated variance on mean response is 0.3586. Using this data, answer the following.
a. Establish 95% confidence limits for the average color intensity when exposure time is 7.0 and relative humidity is 0.42.
b. Develop the comparable 95% prediction limits for the photo now being developed if exposure time and relative humidity are the same as in part (a).

23. The owner of a sheet metal shop is concerned about the warp strength of metal sheets used for lightweight race car bodies. A high warp strength is desirable, because this means high resistance to force without bending. The percentage of steel, $X_1$, and the percentage of nonmetallic filler, $X_2$, are important factors in determining warp strength, $Y$. A model developed from 25 test sheets is $\hat{Y} = 9.1286 + 0.2030X_1 - 0.0708X_2$. Use this along with MSE $= 0.4400$ and $s_{\hat{Y}} = 0.3333$ to find answers to the following when $(X_1, X_2) = (20, 60)$.

    a. 98% confidence limits on the average warp strength

    b. 98% prediction limits for the warp strength on a single sheet with 60% filler and 20% steel

    c. If NASCAR safety rulings require a warp strength in excess of 8.0, does the average value for $(X_1, X_2) = (20,60)$ satisfy the standard? How about the individual sheets?

24. What is wrong with the following regression summary?

```
DEPENDENT VARIABLE _____ Y
VARIABLES ENTERED _____ X₁, X₂, X₃

                                         SUM OF
                                 DF      SQUARES

MULTIPLE R    .7328   REGRESSION    3      8055
R SQUARE      .5370   RESIDUALS    12      6948
```

Calculated $F = 8055/6948 = 1.159$

Tabular $F_{0.05,3,12} = 3.49$

Conclusion: There is not a significant linear relation for $Y$ on $X_1$, $X_2$, $X_3$.

25. An estimated linear regression for a parcel delivery service is described as follows.

$$\hat{Y} = 1441.32 - 4.29X_1 + 18.52X_2$$

where

    $Y$ = ($) claim liability on lost and damaged items

    $X_1$ = total parcels picked up in Council Bluffs

    $X_2$ = total parcels picked up in Omaha

    a. Estimate the average claim liability for $(X_1, X_2) = (1300,250)$.

    b. This is a case where the units of $X_1$, $X_2$ are identical, so the $b_1$, $b_2$ coefficients are directly comparable. Interpret the practical meaning for this model.

26. A marketing analyst was given the responsibility to develop a linear model for

    $Y$ = product usage of a brand of bath soap

based on

    $X_1$ = relative urbanization of the area in %

and

    $X_2$ = education level of the head of household

Two competing models are summarized:

(1) $\hat{Y} = 28.8415 + 13.0128X_2$
(2) $\hat{Y} = 26.9388 + 1.5732X_1 + 6.9932X_2$

We also know that the correlation for $X_1$ and $X_2$ is 0.6838. Discuss the values for $b_2 = 13.0128$ and $b_{2 \cdot 1} = 6.9932$. Are you surprised that they are quite different? Explain.

27. For Problem 26, the analysis of variance for regression on the model including $X_1$ and $X_2$ is as follows.

| SOURCE | DF | SUM OF SQUARES | MEAN SQUARES | F |
|--------|----|----|----|----|
| Regression | 2 | 938.5317 | 469.2659 | 8.26 |
| Residuals | 6 | 340.9058 | 56.8176 | |

a. What is the value for the standard error of estimate?
b. Is the regression of $X_1$ and $X_2$ on $Y$ meaningful at $\alpha = 0.05$?

28. A controlled experiment was made to determine the relation of two gasoline additives, $X_1$ and $X_2$, on gas mileage. Test standards, including randomization of runs with different additives, and use of a single test vehicle guarantee valid results. You can assume that a meaningful regression is estimated by $\hat{Y} = 19.82 - 0.57X_1 + 1.77X_2$. Use this estimated regression along with MSE $= 2.1360$, $s_{\hat{Y}}^2 = 0.5933$, and $n = 9$ to estimate the following items for $X_1 = 1$ unit, $X_2 = 2$ units.
a. The average mileage with 95% confidence limits
b. The mileage for a next run with $(X_1, X_2) = (1,2)$. Give 95% prediction limits.

29. The following data concern $Y = $ sales, $X_1 = $ labor expenditures, and $X_2 = $ advertising expenditures (all in tens of thousands of dollars).

| $X_1$ | 4 | 4 | 7 | 7 | 9 | 12 | 12 | 9 |
|-------|---|---|---|---|---|----|----|---|
| $X_2$ | 1 | 1 | 2 | 2 | 5 | 8 | 8 | 5 |
| $Y$ | 7 | 8 | 12 | 11 | 17 | 20 | 22 | 24 |

a. Use these data and a regression program available at your computer center to develop an ANOVA and test for a meaningful linear regression of $X_1$ and $X_2$ on $Y$-sales. Use $\alpha = 0.05$.
b. If the regression is meaningful, use it to (point) estimate sales for a period wherein $X_1 = 9$ and $X_2 = 5$. Then compare to the actual value $Y = 17$ and compute the residual.

## ★13.7
## REGRESSION
## MODEL-BUILDING:
## A COMPUTER
## APPROACH

Our illustrations thus far have been restricted to the inclusion of only two independent variables. But the multiple linear regression model allows for any number of independent variables. And most computer packages for multiple regression will accept a large number—20 or more—of independent variables. So now we turn to the full problem of regression modeling in the context of available computer programs.

Suppose you are the president of the Chamber of Commerce for a city in a booming tourist area. You regularly receive inquiries about tourism, including traffic and the availability of lodging, food, transportation, and related services. Your clientele expects you to have ready answers. We consider just one aspect of tourism, the availability of lodging. Suppose you want to predict the percent hotel occupancy for your area.

Many variables can influence hotel occupancy rate ($Y$), including ($X_1$) the amount of auto traffic entering the area, ($X_2$) air traffic levels, ($X_3$) mean temperatures, ($X_4$) the cost of gasoline, ($X_5$) lodging rates relative to nearby areas, ($X_6$) precipitation, and ($X_7$) the amount of alternative lodging (campgrounds, condos, trailers). You could develop a multiple regression model to predict $Y$. Which of the many available independent variables should you include in your model?

This situation is rather common in business, in that many variables compete to describe a dependent variable. The objective is to identify those variables that can best describe a response and then use them in a regression model to make predictions. We begin by discussing procedures for selecting the independent variables.

### Step 1: Choosing the Variable Selection Procedure

Because there is no one optimal procedure, we explore several options for selecting the independent variables. One option is adding the variables into the model one at a time. Another is to begin with all of the known $X$-variables and eliminate the least predictive ones. Some of the most common procedures are forward-selection (also called forward-stepping), stepwise-regression, backward-elimination, and all-regressions. Several other procedures are available in the commercial software. Here is a brief introduction.

*All-regressions* is the best procedure in that it allows the user to consider every possible combination of $k$ independent variables and thus to identify the statistically strongest combination. However, there is one major drawback to all-regressions: the large number of possibilities for models. Their number is the possible combinations of $k$-variables: $_kC_0 + {}_kC_1 + {}_kC_2 + \cdots + {}_kC_{k-1} + {}_kC_k = 2^k$. For example, for $k = 1$, there are only $_1C_0 + {}_1C_1 = 2$ possibilities: either exclude $X$ or include $X$. For $k = 2$, four possibilities exist. For $k = 10$ independent variables, there are 1024 possible combinations, so there are 1024 possible linear models to choose from. Can you imagine reviewing 1024 computer tables just to choose the best one? In practice, all-regressions selections generally are not practical because of limited analyst-time and computer-time. So other options are more frequently used.

What are the alternatives for variable selection? Procedures other than all-regressions require decision rules so that the computer can screen the vari-

★ Star indicates supplementary material for enhanced (optional) coursework.

able combinations and display only a few of the best ones. These generally use correlations or other measures of the reduction in the error sum of squares (SSE) to add in (forward-selection) or to remove (backward-elimination) independent variables.

The main approaches to independent variable selection are (1) *all-regressions*, (2) *forward-selection*, (3) *backward-elimination*, and (4) *stepwise-regression*. We consider the major aspect of these modeling alternatives in the following discussion.

### DESCRIPTION OF FOUR COMMON VARIABLES SELECTION PROCEDURES

1. The all-regressions procedure estimates the multiple linear regression equation, considering all available independent variables and any combination down to the single strongest $X$-variable. Because the calculations are extensive, this option is not always available in commercial software packages.

2. Forward-selection regression allows the $X$-variables to enter the equation one at a time. At each step the next variable tested for entry is the one that explains the most from previously unexplained response variations. The test criterion usually centers on reducing the residual error sum of squares (SSE).

3. Backward-elimination begins with all available independent variables entered in the equation. Then variables are removed one at a time. At each step, the $X$-variables that are still in the equation are evaluated for removal. The one that induces the least loss in explained variations is tested for removal.

4. Stepwise-regression is a forward-building procedure. The strongest $X$-variable is tested for first entry. When there are independent variables in the equation, the variable that adds the least to the collective regression sum of squares is tested for removal. If a variable is removed, then the equation is recomputed using the remaining $X$-variables, and each of the $X$-variables in the model is examined for removal. Once no more variables need to be removed, then all $X$-variables not in the equation are examined one at a time for entry. When a variable is entered, then all variables in the model are tested for removal. The process terminates when (1) no variables in the model need to be removed *and* no variables out of the model qualify to enter or when (2) a preset maximum number of steps or other test control is reached.

Substantial discussion on the merits and shortcomings of the regression modeling alternatives is available. We will consider these four approaches in the discussion and problems.

**Step 2: Testing X-Variables: The Extra Sum of Squares Principle**

The appropriate number of variables to include in a regression model depends upon how useful each independent variable is in explaining the variability of the dependent variable. Most computerized regression programs either start with one independent variable and add others in a step-by-step fashion or start with all available independent variables and delete the unnecessary ones until an optimal number is reached. The resultant model will include $X$-variables as long as those variables explain a significant amount of previously unexplained response variation. One criterion that tells whether a variable explains sufficient response ($Y$) variations to be included in a multiple linear regression model is known as the *extra sum of squares principle*, also called *conditional sum of squares*.

> The extra sum of squares principle for testing whether a variable $X_k$ should be included in a linear regression model that already contains $X_1, X_2, X_3, \ldots, X_{k-1}$ uses
>
> $$
> \begin{aligned}
> F_{1,\,n-(k+1)} &= F(X_k \mid X_1, X_2, \ldots, X_{k-1}) \\
> &= \frac{\text{SSE}(X_1 X_2 \cdots X_{k-1}) - \text{SSE}(X_1 X_2 \cdots X_{k-1} X_k)/1}{\text{SSE}(X_1 X_2 \cdots X_{k-1} X_k)/n - (k+1)} \\
> &= \frac{(\text{Reduction in SSE for } X_k \text{ after } X_1 X_2 \cdots X_{k-1})/1}{\text{MSE for the model including } k\text{-variables}} \\
> &= \frac{\text{MSR}(X_k \mid X_1 X_2 \cdots X_{k-1})}{\text{MSE}(X_1 X_2 \cdots X_k)}
> \end{aligned}
> \qquad (13.14)
> $$

Adding more $X$-variables to a model either decreases or maintains the SSE value. So the numerator part of Equation (13.14) is either zero or a positive number. A real improvement in explained variations is evidenced by a significant reduction in SSE for the inclusion of an $X$-variable. The test for a significant reduction in unexplained sums of squares compares the calculated $F$ in (13.14) against a table $F$-value with 1 and $[n - (k+1)]$ degrees of freedom. The procedure is readily extended to tests for entering two or more variables at one step.

We illustrate the forward-selection procedure first. It provides a direct application of the extra sum of squares principle. The example continues the Chapter 12 analyses of selected hotel occupancy rates in the Walt Disney World area. The data from Chapter 12 are repeated in Table 13.7. The time variable,

**TABLE 13.7** Data on Hotel Occupancy Rate, Airline Passengers (in hundreds), and Time

| Observation | Hotel Occupancy Rate OCC | $X_1$ Delta DELTA | $X_2$ East Coast EAST | $X_3$ Southern SOUTH | $X_4$ Time TIME |
|---|---|---|---|---|---|
| 1 | 40 | 345.26 | 656.89 | 133.40 | −11 (Oct.) |
| 2 | 41 | 479.45 | 715.93 | 148.60 | −9 (Nov.) |
| 3 | 48 | 427.81 | 536.53 | 175.92 | −7 |
| 4 | 49 | 423.05 | 701.77 | 155.60 | −5 |
| 5 | 73 | 413.77 | 749.74 | 178.31 | −3 |
| 6 | 74 | 574.86 | 856.26 | 177.08 | −1 |
| 7 | 68 | 463.57 | 845.75 | 178.24 | 1 (Apr.) |
| 8 | 51 | 379.83 | 580.46 | 168.10 | 3 |
| 9 | 63 | 376.08 | 728.09 | 159.58 | 5 |
| 10 | 75 | 384.14 | 876.07 | 164.41 | 7 |
| 11 | 70 | 399.93 | 853.80 | 172.35 | 9 |
| 12 | 38 | 285.44 | 506.11 | 135.65 | 11 (Sept.) |

$X_4$, is in months with origin at month 6. Increments are $\frac{1}{2}$ month; so each month adds $+2$ to the value for the previous month.

---

**EXAMPLE 13.6**

From among the four independent variables in Table 13.7, which combination of two independent variables gives the best statistical model for predicting hotel occupancy? A hotel developer wants a model to estimate occupancy for a period when the values for (Delta, East Coast, Southern, time) are (460, 850, 175, and 1). Using the forward-selection procedure, determine a "best" regression equation for two $X$-variables. Use $\alpha = 0.05$.

*Solution*  The regression summary is by the Minitab microcomputer software [9]. We display only those parts of the summary that are needed for the solution in Table 13.8. These statistics include a correlation matrix and three analysis of variance summaries. This display is a forward-selection procedure because we test only to enter variables, and once entered, a variable is not allowed to be removed.

*Correlations:* In Table 13.8 the first column in the correlation matrix displays the simple correlations for each independent variable with the dependent variable OCC, occupancy rate. The correlation 0.819 being the largest in column 1 indicates that EAST is the strongest independent variable for predicting occupancy rate by a linear regression model. The shaded portion emphasizes the correlations for DELTA, EAST, and SOUTH passenger variables, which are $r = 0.465$ or larger. If highly correlated variables enter an equation, then the model would be suspected of suffering from multicollinearity, a topic we shall take up later in the chapter. As indicated by the last row of correlations, time of the year has relatively low correlation with the other independent variables.

*Variable selection by the forward-selection procedure:* Correlations can indicate much about variable relations, but testing determines which variables enter a regression equation. The variable selection for this modeling requires three steps, 1–3 (marked as subtables 13.8.1 through 13.8.3, but coming from one computer run).

Table 13.8.1 shows $R^2$ (EAST, OCC) = 67.1%. R-sq is shaded. In Chapter 12, Example 3, we found that a simple regression on $X_2$, East Coast passengers, was viable for making predictions of hotel occupancy rate ($Y$). This is confirmed in Table 13.8.1 at Step 1, designated by ① on the computer listing, by a calculated $F = \text{MSR/MSE} = 1529.5/75.0 = 20.39$ (which is equivalent to $t^2 = (4.52)^2 = 20.43$ from Example 12.6). East Coast passenger counts provide a meaningful regression description of occupancy rate. The significance of the EAST variable is confirmed by its $p$-value = 0.000.

Summaries that appear in *Table 13.8.1, Step 1*, include a regression equation, $t$-statistics, $R^2$-adjusted, and a residuals analysis. These items will be discussed after the forward-selection modeling presentation is completed.

**TABLE 13.8**  Descriptive Statistics for a Forward-Selection of Independent Variables from a Minitab Software Summary

```
CORRELATIONS
MTB > CORRELATION 'OCC' 'DELTA' 'EAST' 'SOUTH' 'TIME'
```

|         | Y<br>OCC | $X_1$<br>DELTA | $X_2$<br>EAST | $X_3$<br>SOUTH |
|---------|------|-------|------|-------|
| DELTA   | .428 |       |      |       |
| EAST    | .819 | .537  |      |       |
| SOUTH   | .766 | .586  | .465 |       |
| TIME $X_4$ | .373 | -.370 | .155 | .116  |

**TABLE 13.8.1** Forward-Selection Procedure, Step 1

REGRESSION OF OCCUPANCY RATE ON EAST COAST PASSENGERS ($X_2$)
MTB > BRIEF output level is 3
MTB > REGRESS 'OCC' on 1 variable 'EAST'

The regression equation is
OCC = -8.5 + 0.0920 EAST

| Predictor | Coef | St.Dev | t-ratio | p |
|---|---|---|---|---|
| Constant | -8.49 | 14.82 | -.57 | .579 |
| EAST | .09200 | .02037 | 4.52 | .000 |

$s_e$ = 8.658          R-sq = 67.1%          R-sq(adj) = 63.8%

ANALYSIS OF VARIANCE

| SOURCE | DF | SS | MS |
|---|---|---|---|
| Regression | 1 | 1529.5 | 1529.5 |
| Error | 10 | 749.5 | 75.0 |
| Total | 11 | 2279.0 | $F_{calc}$ = 20.39    p = .000 |

(Residuals analysis)

| Obs. | OCC | Fit | St.Dev Fit | Residual | St.Resid |
|---|---|---|---|---|---|
| 1 | 40.00 | 51.94 | 2.79 | -11.94 | -1.46 |
| 2 | 41.00 | 57.38 | 2.50 | -16.38 | -1.98 |
| 3 | 48.00 | 40.87 | 4.45 | 7.13 | .96 |
| 4 | 49.00 | 56.07 | 2.52 | -7.07 | -.85 |
| 5 | 73.00 | 60.49 | 2.59 | 12.51 | 1.51 |
| 6 | 74.00 | 70.29 | 3.78 | 3.71 | .48 |
| 7 | 68.00 | 69.32 | 3.62 | -1.32 | -.17 |
| 8 | 51.00 | 44.91 | 3.74 | 6.09 | .78 |
| 9 | 63.00 | 58.49 | 2.51 | 4.51 | .54 |
| 10 | 75.00 | 72.11 | 4.09 | 2.89 | .38 |
| 11 | 70.00 | 70.06 | 3.74 | -.06 | -.01 |
| 12 | 38.00 | 38.07 | 4.97 | -.07 | -.01 |

①

*Table 13.8.2, Step 2*, results indicate that adding the variable ($X_2$) SOUTH after EAST results in a combined $R^2$ (EAST, SOUTH) = 86.0%. This combination produces an increase of $(86.0 - 67.1)/67.1 = 0.282$, or 28.2% more explained variations over the equation in EAST, alone. This is a substantial increase. It suggests that the Southern Airlines variable might enter the linear model.

**TABLE 13.8.2** Forward-Selection Procedure, Step 2

REGRESSION OF OCC RATE ON EAST COAST ($X_2$) AND SOUTHERN ($X_3$)
MTB > BRIEF output level is 3
MTB > REGRESS 'OCC' on 2 variables 'EAST' 'SOUTH'

The regression equation is
OCC = -61.5 + 0.0663 EAST + 0.440 SOUTH

**TABLE 13.8.2**   (*continued*)

| Predictor | Coef | St.Dev | t-ratio | p |
|-----------|------|--------|---------|---|
| Constant | -61.49 | 18.30 | -3.36 | .008 |
| EAST | .06633 | .01582 | 4.19 | .002 |
| SOUTH | .440 | .1262 | 3.49 | .007 |

s = 5.952        R-sq = 86.0%          R-sq(adj) = 82.9%

ANALYSIS OF VARIANCE

| SOURCE | DF | SS | MS |
|--------|-----|-----|-----|
| Regression | 2 | 1960.15 | 980.08 |
| Error | 9 | 318.85 | 35.43 |
| Total | 11 | 2279.00 | |

| SOURCE | DF | SEQ SS |
|--------|-----|--------|
| EAST | 1 | 1529.47 |
| SOUTH | 1 | 430.68 |

| Obs. | OCC | Fit | St.Dev Fit | Residual | St.Resid |
|------|-----|-----|------------|----------|----------|
| 1 | 40.00 | 40.79 | 3.73 | -.79 | -.17 |
| 2 | 41.00 | 51.39 | 2.43 | -10.39 | -1.91 |
| 3 | 48.00 | 51.52 | 4.32 | -3.52 | -.86 |
| 4 | 49.00 | 53.54 | 1.88 | -4.54 | -.80 |
| 5 | 73.00 | 66.71 | 2.52 | 6.29 | 1.17 |
| 6 | 74.00 | 73.24 | 2.73 | .76 | .14 |
| 7 | 68.00 | 73.05 | 2.71 | -5.05 | -.95 |
| 8 | 51.00 | 50.99 | 3.11 | .01 | .00 |
| 9 | 63.00 | 57.03 | 1.78 | 5.97 | 1.05 |
| 10 | 75.00 | 68.97 | 2.95 | 6.03 | 1.17 |
| 11 | 70.00 | 70.99 | 2.58 | -.99 | -.18 |
| 12 | 38.00 | 31.78 | 3.87 | 6.22 | 1.38 |
| Total | | | | .00 | |

②

In the next example we test the Southern Airline passenger counts for addition to a regression model that already includes East Coast passengers.

**EXAMPLE 13.7**

Use the extra sum of squares principle to test for inclusion of SOUTH after EAST passengers for a multiple linear regression on occupancy rate. That is,

$$\text{Test:} \quad H_0: \beta_{S \cdot E} = 0 \quad \text{against} \quad H_A: \beta_{S \cdot E} \neq 0 \quad \text{Let } \alpha = 0.05.$$

*Solution*   By the extra sum of squares principle,

$$\text{Calculated } F \text{ (SOUTH after EAST)} = \frac{\text{MSR (SOUTH after EAST)}}{\text{MSE (SOUTH, EAST)}} = \frac{430.65}{35.43} = 12.2$$

Tabular $F_{0.05,1,9} = 5.12$

We construct an analysis of variance from Tables 13.8.1 and 13.8.2.

| SOURCE | DF | SS | MS | F |
|---|---|---|---|---|
| Regression EAST, SOUTH | 2 | 1960.15 | | |
| Regression EAST ($X_2$) | 1 | 1529.50 | | |
| Regression SOUTH after EAST ($X_3$ after $X_2$) | 1 | 430.65 | 430.65 | 12.2 |
| Error for $X_1, X_2$ in the model | 9 | 318.85 | 35.43 | |
| Total | 11 | 2279.00 | | |

So the SOUTH passengers variable adds significant description to the model for OCCupancy rate that already includes the EAST passengers variable. This is consistent with the substantial increase in $R$-square already noted. Similarly, the error mean square of 35.43 in Table 13.8.2 is a substantial reduction from the MSE = 75.0 in Table 13.8.1. The regression on EAST ($X_2$) and SOUTH ($X_3$) variables is estimated by

$$\widehat{\text{OCC}} = -61.490 + 0.066X_2 + 0.440X_3$$

From Table 13.8.3 the strongest combination of three independent variables for this set may be to add DELTA ($X_1$) to EAST ($X_2$) and SOUTH ($X_3$). The Obs. 6 value will be discussed later with graphic displays.

**TABLE 13.8.3**  Forward-Selection Procedure, Step 3

```
REGRESSION OF OCC RATE ON EAST COAST (X₂), SOUTHERN (X₃) AND DELTA (X₁)
MTB > BRIEF output level is 3
MTB > REGRESS 'OCC' in 3 variables 'EAST' 'SOUTH' 'DELTA'

The regression equation is
OCC = -63.9 + .0782 EAST + .560 SOUTH - .0621 DELTA
```

| Predictor | Coef | St.Dev | t-ratio | p |
|---|---|---|---|---|
| Constant | -63.86 | 15.11 | -4.23 | .003 |
| EAST | .07818 | .01402 | 5.58 | .000 |
| SOUTH | .5602 | .1164 | 4.81 | .000 |
| DELTA | -.06209 | .02707 | -2.29 | .051 |

s = 4.904     R-sq = 91.6%     R-sq(adj) = 88.4%

ANALYSIS OF VARIANCE

| SOURCE | DF | SS | MS |
|---|---|---|---|
| Regression | 3 | 2086.62 | 695.54 |
| Error | 8 | 192.38 | 24.05 |
| Total | 11 | 2279.00 | |

| SOURCE | DF | SEQ SS |
|---|---|---|
| EAST | 1 | 1529.47 |
| SOUTH | 1 | 430.68 |
| DELTA | 1 | 126.47 |

| Obs. | OCC | Fit | St.Dev Fit | Residual | St.Resid |
|---|---|---|---|---|---|
| 1 | 40.00 | 40.80 | 3.07 | -.80 | -.21 |
| 2 | 41.00 | 45.60 | 3.22 | -4.60 | -1.24 |

**TABLE 13.8.3**   *(continued)*

| | | | | | |
|---|---|---|---|---|---|
| 3 | 48.00 | 50.08 | 3.62 | -2.08 | -.63 |
| 4 | 49.00 | 51.91 | 1.70 | -2.91 | -.63 |
| 5 | 73.00 | 68.96 | 2.30 | 4.04 | .93 |
| 6 | 74.00 | 66.60 | 3.67 | 7.40 | 2.27R* |
| 7 | 68.00 | 73.34 | 2.23 | -5.34 | -1.22 |
| 8 | 51.00 | 52.11 | 2.61 | -1.11 | -.27 |
| 9 | 63.00 | 59.12 | 1.72 | 3.88 | .85 |
| 10 | 75.00 | 72.89 | 2.97 | 2.11 | .54 |
| 11 | 70.00 | 74.62 | 2.65 | -4.62 | -1.12 |
| 12 | 38.00 | 33.98 | 3.33 | 4.02 | 1.12 |

* R denotes an Obs. with a large St.Resid.

3

The forward-selection procedure extends the extra sum of squares principle to a test for entering $X_1$ *after* $X_2$ and $X_3$ in Example 13.8.

## EXAMPLE 13.8

Test, using $\alpha = 0.05$, for the entry of DELTA ($X_3$) into the regression model for OCCUPANCY rate ($Y$) that already contains EAST ($X_2$) and SOUTH ($X_3$). Determine a best estimated regression model from ones including $X_1$, $X_2$, and $X_3$. Use this equation and (Delta, East Coast, Southern) = (460, 850, 175) to make a point estimate for the occupancy rate.

*Solution*

$$H_0: \beta_{\text{D} \cdot \text{ES}} = 0 \qquad H_A: \beta$$

An analysis of variance (from Tables 13.8.2 and 13.8.3) table displays the calculations required for an $F$-test. Calculated $F$ (DELTA after EAST, SOUTH) = 126.47/24.05 = 5.26. $F_{0.05,1,8} = 5.32$. DELTA ($X_1$) does not enter this regression model for $\alpha = 0.05$. This is consistent with the $p$-values in Table 13.8.3, which show $p = 0.051$ for DELTA. Both the $F$-test and the $p$-value indicate that DELTA just misses entering the model.

| Source | DF | SS | MS | F |
|---|---|---|---|---|
| Regression $X_1$, $X_2$, $X_3$ | 3 | ⎡2086.62 | | |
| Regression $X_2$, $X_3$ | 2 | ⎢1960.15 | | |
| Regression $X_1$ after $X_2$, $X_3$ | 1 | ⎢ 126.47 | 126.47 | 5.26 |
| Error | 8 | +⎢ 192.38 | 24.05 | |
| Total | 11 | =⎣2279.00 | | |

Using the estimated regression equation in Table 13.8.2, at Step 2, we conclude that the best model as estimated by the forward-selection procedure is

$$\hat{Y} = -61.490 + 0.066X_2 + 0.440X_3$$

Therefore, the estimated occupancy rate when EAST ($X_2$) = 850 and SOUTH ($X_3$) = 175 is

$$\hat{Y} = -61.490 + 0.066(850) + 0.440(175) = 71.9\%$$

**TABLE 13.9** Alternatives for a Variable Selection Criterion

| $X$-Variables in Model | $R^2$ | $R^2$-Adj | MSE |
|---|---|---|---|
| EAST $(X_2)$ | 67.1% | 63.8% | 75.0 |
| East Coast, SOUTH $(X_3)$ | 86.0 | 82.9 | 35.43 |
| East Coast, Southern, DELTA $(X_1)$ | 91.6 | 88.4 | 24.05 |
| East Coast, Southern, Delta, TIME $(X_4)$ | 92.6 | 88.3 | 24.16 |

The following discussion provides additional checks on the reasonableness of this equation.

Any computer-based solution for a regression equation requires some user decisions. One requirement is the criterion for selecting the best subset of variables. Three common methods include Mallow's Cp criterion, multiple $R^2$, and adjusted $R^2$. All these methods relate to the percent of variations in the dependent variable that are explained by a set of $X$-variables. Table 13.9 gives an initial view of several measures. The $R^2$-values in Table 13.9 come from the multiple correlations, $R$, squared in Table 13.8.

Table 13.9 displays several informative patterns. First, $R^2$ displays a *growth rate* that declines with more $X$-variables. Second, for three variables $R^2$-adj peaks and MSE is a minimum. This is close to the model with two $X$-variables as determined by the forward-selection procedure. And $R^2$-adjusted can decrease, whereas $R^2$ is nondecreasing. For more discussion on rules for selecting a model, see [5] or [7].

**Step 3: Checks on the Adequacy of the Model and Data Analysis**

The *residuals*, or remainder of variations of the observed values from the regression response surface, are the deviations $Y - \hat{Y}$. Graphical displays of these *residual plots* provide us with an assessment of (1) how well the assumptions are satisfied and (2) how adequately the model fits the data.

> A residual plot is a scatter diagram of the residuals, $e = Y - \hat{Y}$, as plotted against a variable or characteristic of interest.

Some rather common residual patterns appear in Figure 13.3. An indication of a good fit for the model to the data is when the scatter diagram for the data displays the data points as dispersed close to, but scattered at random about, the regression plane. Then $\hat{Y} \doteq Y$ for all of the data points.

Another expression that the linear model is appropriate is a flat residual data plot, as in Figure 13.3(a). If the plot is tight around $\bar{e} = 0$, with no systematic pattern to be positive or negative, then most variations in $Y$ have been described by the regression surface, and the fit is good.

If the residual plot appears like Figure 13.3(b), then an improvement in describing $Y$ may be possible through the inclusion of higher-order terms like $X^2$ or $X^3$ for a curvilinear regression. The residual plot that appears as Figure 13.3(c) identifies an important variable that has been left out of the model. Later we will consider a situation in which a time-related variable has been omitted.

The pattern in Figure 13.3(d) shows the residual values increasing as the $X$-values increase. This megaphone effect suggests the existence of a nonconstant or *heteroscedastic error variance*.

**FIGURE 13.3**  Common Residual Plot Displays

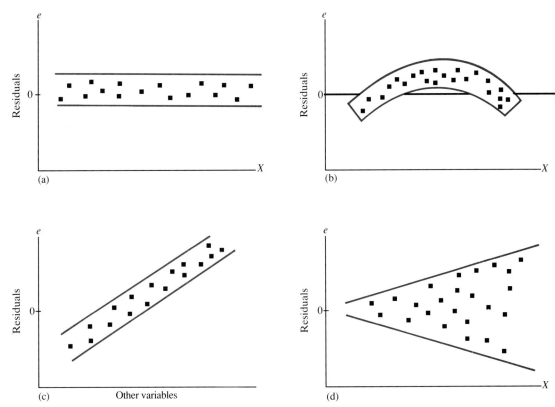

A heteroscedastic (nonconstant) error variance means the assumption of constant variance does not hold for ordinary linear regression.

Applying appropriate weights by a weighted least squares procedure can usually correct this situation and ensure a constant variance [1].

The hotel occupancy data can be used to illustrate several residual plots. These data plots afford us checks on the assumptions and a view of what another $X$-variable could add to the description. Figure 13.4 demonstrates some of these checks.

Some of the insights provided by a residual plot include answers to the following questions.

a. Do the assumptions appear reasonable for (1) normality, (2) constant error variance, and (3) a linear response surface?

b. Will one of the variables (e.g., $X_1$, $X_2$, or other) explain a substantial amount of the currently unexplained response variations?

**EXAMPLE 13.9**    What information is available from the residuals to hotel occupancy rates based on East Coast and Southern passenger variables? Table 13.10 includes occupancy rates from Table 13.7 and estimated occupancy plus residuals from Table 13.8.2.

**FIGURE 13.4**    Residual Plots for a Regression of Hotel Occupancy Rates on East
Coast and Delta Passenger Variables

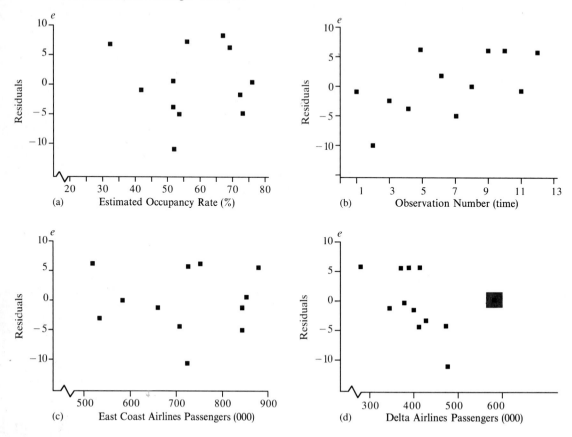

The residuals $e = Y - \hat{Y}$ are differences resulting from using the regression
on East Coast ($X_2$) and Southern passengers ($X_3$) to estimate the observed occupancy
rates ($Y$). The data plots in Figure 13.4 illustrate patterns for these residuals along
with other variables from Table 13.5. By the forward-selection procedure, estimated
occupancy is $\hat{Y} = 61.500 + 0.0663X_2 + 0.440X_3$.

a. Plot the residuals against estimated occupancy rate, $\hat{Y}$.

b. Plot the residuals against the observation number (time).

**TABLE 13.10**    Residuals for Occupancy Rate ($Y$) Predicted by Two Independent Variables ($X_2$) and ($X_3$)

| Obs. | Occupancy Rate $Y$ | Estimated Occupancy $\hat{Y}$ | Residuals $e$ | St. Resid $Z_e$ | Obs. | Occupancy Rate $Y$ | Estimated Occupancy $\hat{Y}$ | Residuals $e$ | St. Resid $Z_e$ |
|---|---|---|---|---|---|---|---|---|---|
| 1 | 40 | 40.79 | −0.79 | −0.17 | 7 | 68 | 73.05 | −5.05 | −0.95 |
| 2 | 41 | 51.39 | −10.39 | −1.91 | 8 | 51 | 50.99 | 0.01 | 0.00 |
| 3 | 48 | 51.52 | −3.52 | −0.86 | 9 | 63 | 57.03 | 5.97 | 1.05 |
| 4 | 49 | 53.54 | −4.54 | −0.80 | 10 | 75 | 68.97 | 6.03 | 1.17 |
| 5 | 73 | 66.71 | 6.29 | 1.17 | 11 | 70 | 70.99 | −0.99 | −0.18 |
| 6 | 74 | 73.24 | 0.76 | 0.14 | 12 | 38 | 31.78 | 6.22 | 1.38 |

c.  Plot the residuals against the number of East Coast passengers, $X_2$.

d.  Plot the residuals against the number of Delta passengers, $X_1$.

These plots, in Figure 13.4, provide information to answer two questions:

1.  Do the assumptions appear reasonable for a linear model?

2.  Will the Delta passengers (variable) reduce substantially the response variations?

*Solution*

1.  Plots (a), (b), and (c) in Figure 13.4 relate to question (1). Plot (a) shows residuals plotted against occupancy rate. It displays minimal heteroscedasticity, thus giving us no reason to question the assumption of a constant variance. Plot (b) displays a recurrence of positive, then negative, residuals in the order of occurrence for the observations. The reasonably flat plot (c) of residuals against East Coast passengers gives us an indication that higher-order $X_2$ terms are not necessary.

    These graphics support a linear equation as being reasonable for hotel occupancy regressed on East Coast passengers and Southern passengers.

2.  For our second question, Figure 13.4(d) shows residuals graphed against Delta passenger counts. The plot is linear with the exception of one point, which is highlighted. This is Observation 6 in Table 13.8.2. Its residual value indicates some irregularity. But the original observation in Table 13.7 shows that this value is unusual. This $X$-observation (574.86) is much higher than the other values for Delta, and should be checked to ensure that it is correct and not a typo, misinformation, or other. Table 13.8.3 shows that Observation 6 has a large standardized residual.

---

The challenge in identifying model disturbances is that they do not occur alone. That is, a residual pattern can show both nonconstant variance and nonlinearity. Can you visualize such a pattern? Of course, there is no sure way to identify all real disturbances visually. A good approach is to study a variety of siutations, so you can recognize the most characteristic patterns. Several problems that follow give you an opportunity to develop your ability to recognize some of these patterns. Some statistical packages display plots on standardized variables. The overall patterns would not change.

The display in Figure 13.4(d) presents an interesting possibility. It appears that the one unusual value could impact the whole set of Delta passenger values. Unless they are actually missing values, there are ways to identify, then treat, unusual data values.

**Detecting Unusual Observations**

Chatterjee and Hadi [3] describe three interrelated conditions under which individual data points can have an undue impact on the regression model: as outliers, as high-leverage points, and as influential observations. They are defined as follows:

> An outlier is an observation for which the standardized residual is large compared to the other standardized residuals.

> A high-leverage point is an observation whose $X$-value (or $X$-coordinates for multiple predictors) is in some sense far from the rest of the data.

> An influential observation exerts excessive influence on the fitted regression equation compared to other observations in the data set.

These concepts are reasonably well defined in the statistics literature. We begin with some base rules for identifying each of these conditions.

Outlier: By normal distribution theory there is less than a 0.05 chance that the absolute value of a standardized residual will be 2.0 or larger. For example, Minitab displays outliers with a notation to the effect that $R$ denotes an observation with a large standardized residual; that is, using $(Y - \hat{Y})$ in standardized form gives $Z_e = (Y - \hat{Y})/s_e$. When $Z_e$ is 2.0 or larger, the data point is considered an outlier.

High-leverage point: This relates to the ability of the regression equation to predict the response for a specified $X$-value (for a specific set of $X_1, X_2, X_3, \ldots, X_k$ values in multiple regression). From Chapter 12, Equation (12.11), the error for prediction at $X = X_i$ depends on $(1/n) + [(X_i - \bar{X})^2/\text{SS}(X)]$ for a prediction of $\hat{Y}_i = b_i + b_1 X_i$. Clearly the error increases as $X_i$ is different from $\bar{X}$. That is, if you are asked "Does the regression predict the data well?" the appropriate answer is, "The fit is better for $X_i$ near $\bar{X}$. This was the point of Figure 12.7. In contrast to residuals, which involve the $Y$-values, the evaluation of high-leverage points depends entirely upon the $X$-values.

If we let $h_i = (1/n) + [(X_i - \bar{X})^2/\text{SS}(X)]$, then theorists have shown that $\sum^n h_i = p$, where $p$ is the number of parameters in the regression model. In simple linear regression, $p$ has value two: one for the slope plus one for the $Y$-intercept. Then, on average, any one of the $n$ observations would contribute $h_i = p/n$ by [3]. Then any observation for which $h_i > 2p/n$ is defined a high-leverage point: $X = X_i$ is a high leverage point if $(1/n) + [(X_i - \bar{X})^2/\text{SS}(X)] > 2p/n$, which requires $X_i$ to be far removed from $\bar{X}$. The extension to multiple regression is beyond the scope of this work. See [3] or [6].

Influential observation: Qualification as an influential observation is admittedly subjective in that the "influence" can be upon any one of several characteristics of the regression model. We will concentrate on influence for the regression model parameters—the slope and $Y$-intercept. Because influential observations can be outliers or high-leverage points, or possibly both, Figure 13.5 illustrates three data points, each having some of these conditions.

First we consider point A in Figure 13.5 (excluding points B and C, which will be considered in turn). Considered with the other data points (except B and C), point A is an outlier and also an influential observation. It is an outlier because it has a large residual, $Z_e > 2$. And it is an influential observation be-

**FIGURE 13.5** A Data Set That Illustrates an Influential Observation, an Outlier, and a High-Leverage Point

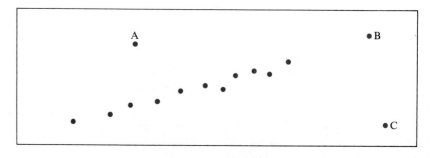

cause its inclusion will change the $Y$-intercept, and possibly the slope value. Point A is not a high-leverage point, because its $X$-value is intermediate to the other $X$-coordinates.

Among the data, point B (without A or C) is a high-leverage point because it is far beyond the other observations. Since point B is in line with the other points, it will not greatly influence the slope or $Y$-intercept values. However, point B will have a positive effect on the $(Y - \hat{Y})$ values, and so provide a reinforcement of the regression for the other data points.

Observation C (without A or B) is all three—a high-leverage point, an outlier, and influential. The display symptom for point A (without B or C) is as an outlier, so the diagnostic information is in the residual analysis; whereas for point B (without A or C) the residuals are small, but the point is far removed, so a check for an unusual point is the $h_i$-statistic. Then point C (without A or B) is extreme in all three conditions. Myers [6], Chapter 8, gives a good coverage of "Influence Diagnostics."

In Example 13.9 we identified one observation for Delta that had an unusually large $X$-value (574.86). This illustrates a case in which the data are in question and, if included, might distort the regression model. In this case the Delta variable does not enter the multiple regression model on $Y$ = hotel occupancy rate, so our interest is in data analysis for a simple linear regression model. Figure 13.6 shows the data.

Determining whether the data point in question is an outlier is not obvious. It requires calculating the regression line, then computing its standardized residual. The Minitab output in Problem 40 provides the needed data. Similarly, the problem gives data for the $h$ statistic to test for a high-leverage point. Is it an influential observation? Block out the point in your mind. What happens to the resulting regression line? To answer this with statistics, Problem 40 gives the regression line based on 11 data points—excluding the point in question. This provides a gauge for the amount that the slope and $Y$-intercept values would change.

**FIGURE 13.6** Data for a Simple Linear Regression of $Y$ = % Occupancy on $X$ = Delta Passengers

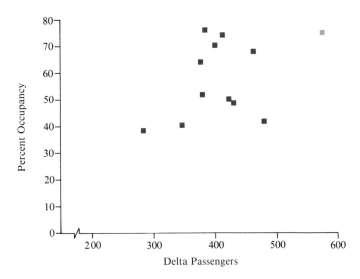

We close with a word of caution. Computer packages like Minitab provide cautionary displays. However, *the objective in data analysis is to retain data so long as they are meaningful*. In fact, a high-leverage point, like point A in Figure 13.5, can be most informative because it describes the data beyond the range of other $X$-values. It can extend the data. In comparison, the combination that produces the greatest influence is a data point that contains a large $h_i$ value and also has a relative large residual (like point $C$ in Figure 13.5). Problems 40 and 41 give practice in detecting unusual observations.

## Step 4: When the Assumptions Do Not Hold

Throughout Chapters 12 and 13, we have emphasized the underlying assumptions that assure us that the regression models are meaningful. We now review these assumptions, along with some descriptive checks and inferential tests upon them.

> Regression assumptions: The errors are independent and normally distributed with constant variance.

This statement encompasses three assumptions: The errors (1) are independent, (2) are normally distributed, and (3) have a constant variance. What procedures exist for checking these assumptions? Also, what statistical procedures are used when the assumptions do not hold? These questions are discussed here.

Analyzing the residuals, like analyzing any other random variable, requires calculation of mean and standard deviation estimates. For a good fit the residual mean, $\bar{e}$, should be near zero, with the residual plot flat with no systematic deviation about the line $\bar{e} = 0$. Then a *check* for normality compares the percent of residuals within $\pm k$ standard deviations distance from the mean against 68%, 95.4%, and 99.7% for $k = \pm 1$, $\pm 2$, and $\pm 3$, respectively. This is only a check; it is not a test. One inferential test uses the $\chi^2$ statistic to test for goodness of fit to a normal distribution. See Chapter 16.

Example 13.19 described a graphic check for nonconstant variance, with suggested corrective action. See [6] [7] for procedures for weighted least squares regression.

Example 13.10 gives a check that the residuals are normally distributed.

---

**EXAMPLE 13.10**

Example 13.7 included data on the residual values plotted against the estimated regression for hotel occupancy rates on East Coast and Southern passengers. The data plot appears in Figure 13.4(a). Is the fit close enough to a bell curve to conclude that the residuals follow a normal distribution?

*Solution*   This answer is based on description, not on inference, but it will substantiate an obvious result. The data in Table 13.8.2 give $\bar{e} = \sum e/12 = 0.00$ and $s_e \, (= \hat{\sigma}_e) = 5.952$. These statistics and the residuals that are displayed in Table 13.8.2 give checks on the assumption that the residuals follow a normal distribution.

| $k$ | $0.00 \pm (k \cdot 5.95)$ | Percent of Cases | |
|---|---|---|---|
| | | *Actual* | *Normal Distribution* |
| 1 | $-5.95$ to $5.95$ | 58% $(=\frac{7}{12})$ | 68% |
| 2 | $-11.90$ to $11.90$ | 100 | 95.4 |
| 3 | $-17.85$ to $17.85$ | 100 | 99.7 |

Considering the small number of data points (12), the residuals seem to follow a normal distribution.

A normal probability plot of the residuals is another way to detect nonnormality. If the residuals are normally distributed, then a plot on normal probability paper will be essentially a straight line. See [8] for an example.

**Assumption—
The Errors Are
Independently
Distributed**

Our model assumes that the errors (the $\varepsilon_i$) are independent, zero-correlated random variables. However, it is common for the residuals of business and economic data to be positively correlated *in* time. Then the error terms are described as being *autocorrelated*.

> Autocorrelation measures the co-variation in residuals induced by a correlation over time.

Omitted variables often relate to time. For example, we used the number of passengers on East Coast flights and Southern flights as independent variables to predict hotel occupancy rates. Tourism in the Disney World area has seasonal highs and lows, so hotel occupancy can have a time dependence. Many people travel to this area by car. Should the model also consider tourist ground traffic counts by month? This omission from the model could cause the residuals to be positively autocorrelated because the ground traffic count is correlated with hotel occupancy rates over time. A high autocorrelation can signal the existence of missing but strong independent variable(s). If there is a significant omission, a residual plot against the missing variable would display a strong pattern.

What problems occur with autocorrelated residuals? A major difficulty is that the residual error estimate, MSE, can *underestimate* the true error variance. There is distortion in the resulting tests and in the estimation of $\beta$'s. Then inference using the model would be invalid.

What techniques are available for analyzing autocorrelated errors? We discuss three: *First*, a data plot of the residuals against time will display any linear association. For example, Figure 13.4(b) is a display for the hotel occupancy rates at different points in time. This indicates minimal autocorrelation because the residual plot is nearly horizontal and appears random. *Second*, one can perform a runs test for randomness of residuals about $\bar{e}$. This procedure is outlined in Section 17.3. *Third*, one can assume a model that includes error over time and incorporate this time effect in evaluating the error variance. Durbin and Watson have devised a test by which to identify the presence of a time effect [7].

**The Durbin–Watson
Statistic**

The Durbin–Watson statistic affords a test for a significant effect due to autocorrelated errors. In its simplest form the procedure assumes that the residuals include a component for autocorrelation as $\varepsilon_t = \rho_{auto}\varepsilon_{t-1} + \mu_t$, where $\rho_{auto}$ is the autocorrelation over time; $t$ and $t-1$ denote successive time periods; and $\mu_t$ is random error, which is normal with a mean of zero and variance of $\sigma^2$. Under the null hypothesis, $H_0$: $\rho_{auto} = 0$, the *Durbin–Watson statistic* is as follows, where $d_L$ and $d_U$ are lower and upper bounds of the test.

The Durbin–Watson statistic, $D$, and a test for autocorrelation

$$D = \frac{\sum\limits_{2}^{n} (e_t - e_{t-1})^2}{\sum\limits_{1}^{n} e_t} \tag{13.15}$$

Test:  If $D < d_L$, conclude $\rho_{auto} > 0$.
If $D > d_U$, conclude $\rho_{auto} = 0$.

The test is inconclusive if $d_L \leqslant D \leqslant d_U$.

When the data are positively autocorrelated as $\rho_{auto} > 0$, then $e_t$ will tend to track $e_{t-1}$ so that adjacent terms have close values. Then the differences in residuals $e_t - e_{t-1}$ are generally small and the $D$-statistic takes a value less than $d_L$, Durbin and Watson have obtained lower and upper bounds $d_L$ and $d_U$ for alternative sample sizes and for different numbers of independent variables. These appear in Table B.10.

If the Durbin–Watson statistic is *not* significant, evidenced by a calculated $D$-value $> d_U$ and indicating $\rho_{auto} = 0$, then the model should *exclude* the component for autocorrelation. Then the modeling procedure becomes the ordinary least squares regression. Otherwise, the model and subsequent inference should use a time series model as described in Chapter 14. SAS, BMDP, and SPSS each provide the Durbin–Watson statistic in a regression procedure and as do personal computer programs such as Minitab and StatView.

---

**EXAMPLE 13.11**

The Durbin–Watson statistic is an output for the regression summary of hotel occupancy rate on East Coast passengers and time. Note that the calculated $D$-value of 1.51 appears in the residual information summary. The data points are few enough, however, for us to hand-calculate $D$ using Equation (13.15). Use $\alpha = 0.05$ to determine whether the data on hotel occupancy evidence a significant autocorrelation effect.

Given the Durbin–Watson statistic = 1.51, test for a significant autocorrelation.

*Solution*   This question asks for a test for zero autocorrelation, so

$$H_0: \rho_{auto} = 0 \qquad H_A: \rho_{auto} > 0$$

$D = 1.51$ by the computer summary

Table B.10 shows $d_L = 0.95$, $d_U = 1.54$ for $n = 12$ and $k = 2$ variables.
Conclusion: Since $D = 1.51 < d_U = 1.54$, then $\rho_{auto} = 0$. So this test is inconclusive by $d_L < 1.51 < d_U$.

---

Other procedures for handling autocorrelation are discussed in Chapter 14.

We have considered the assumptions on the errors ($\varepsilon$), including independence and normal distribution, $N(0,\sigma^2)$. The remaining assumption is that the $X$-variables combine in a linear additive model.

**Regression assumption: The model is a multiple linear regression that can be expressed as $Y = \beta_0 + \beta_1 X_1 + \cdots + \beta_k X_k + \varepsilon$ .**

How do we determine if the model adequately fits the data? For a single $X$-variable, a scatter diagram may help to identify an inappropriate model. With two $X$-variables, the $X_1$ by $X_2$ by $Y$ three-dimensional plot may be difficult to visualize. With three $X$-variables, a four-dimensional plot is not possible. Sometimes plots of residuals from the regression can identify a lack of fit, but a test for lack of fit provides a validation.

A statistical test for lack of fit requires more than one response for some individual (simple linear) or combination of (multiple linear) $X$-variables. The test is based on a split of the SSE into (1) pure experimental error (SSE-pure) and (2) lack-of-fit error (SSE). The SSE-pure is derived from differences in response for identical $X$-values or combinations of $X$-values. The balance of SSE goes to lack of fit. The test uses an $F$-statistic that builds upon these two sums of squares. If the $F$-test for lack of fit is significant for $\alpha$, then the model apparently is inadequate. Otherwise, an $F$-test for lack of fit that is not significant gives some credence to making inference based on the fitted regression model. Then both the pure-error and the lack-of-fit terms can be used to estimate $\sigma^2$. Then $F$-tests have the usual MSE value, which combines the two. See [5] or [7] for an illustration.

**Condition—The X-Variables Are Not Significantly Correlated**

High intercorrelation between $X$-variables, called *multicollinearity*, confuses the evaluation of the relative influence of the $X$-variables.

> Multicollinearity, or intercorrelation, exists when the independent variables are correlated among themselves.

The $X$-variables will often contain some correlation, so some multicollinearity usually exists.

With high multicollinearity the sampling variances of the coefficients may be excessive. The result is unreliable estimates for the $\beta$-coefficients. Then, for example, the addition of a few more observations can produce substantial shifts in estimated regression coefficient values. Hypothesis tests on the individual $\beta$'s can be erroneous, too.

If the (multiple) regression model is to be used only for prediction of a $Y$-value, the existence of multicollinearity may not present a problem. However, if interest lies in the relative influence of the various regression coefficients ($\beta$-values), erroneous conclusions may result. That is, a model with high multicollinearity may be found significant in describing a response variable. This would normally cause us to conclude that at least one $\beta_i \neq 0$. Yet with high multicollinearity it is possible at the same time for all the independent variables to be nonsignificant by simple $t$-tests. This seeming inconsistency is a consequence of the overlap in what the variables are measuring. The combined effect of several variables is accurately measured in inference relative to $\beta_1 = \beta_2 = \cdots = \beta_k = 0$. However, the effect of *each* $X$-variable on the full-model SSE cannot be accurately assessed; the intercorrelations can mask the contribution of each $X$-variable to the SSE. So when each ($X$-) variable is tested alone, the result is a bigger-than-actual error variance. This results in a small calculated $t$-value, which may lead to an incorrect conclusion that the associated individual $\beta$-value is equal to zero.

Several corrective actions are available to adjust for multicollinearity. We mention four. One possibility is to drop one or several $X$-variables from among those that are collinear. This was our tack in the illustration including the Delta Airlines information. The result is reduced standard errors for the independent variables that remain in the model. With reduced multicollinearity, the least squares estimation of regression coefficients becomes more reasonable. This procedure is not without problems, however. First, there remains no direct information for the deleted variables. Also, the size of the coefficients for the remaining $X$-variables is affected by the correlated independent variables that remain in the model. That is, a regression coefficient reflects only a partial effect on the response variable, $Y$, conditioned by whatever other correlated $X$-variables are in the model.

Statistics software packages provide a limited assessment of multicollinearity. A control is through the use of *tolerance* restrictions. A tolerance restriction in regression programs can keep out high correlates. See [11] for a definition of *tolerance*. Pretest screening of $X$-variables from correlation matrices can also be helpful. For example, the 0.537 correlation for East Coast and Delta in Table 13.8 is an indication that these variables contain substantial overlap.

*Ridge regression* is another method for alleviating multicollinearity problems. It uses a modification of the least squares estimation of regression coefficients. This procedure gives biased estimators for the regression coefficients that tend to have small standard errors. This is not all bad, because ridge regression estimators can have a high probability of being close in estimating the parameter value. A discussion of the methodology is beyond the scope of this work; see [7].

Other methods for dealing with multicollinearity include regression with principal components, in which independent variables that have low correlation are formed as additive combinations of the original $X$-variables. Another approach is Bayesian regression, wherein prior information about the regression coefficients is incorporated into the estimation procedure. For more on these topics see [1].

## Dummy Variables—A Coding Device for Qualitative Independent Variables

One way to accommodate qualitative (nominal scale) independent variables is to introduce *dummy*, or *indicator*, *variables* that represent varying conditions or states of nature.

> Dummy or indicator variable coding uses an assignment of 1 if the observation is a specified classification type; otherwise the observation is assigned the value 0.*

An example is coding dichotomous variables such as single-family versus multiple-family dwelling size in Table 13.1. Some qualitative variables warrant more than two classifications. For example, personal titles can be described in five classes, by Mr., Mrs., Miss, Ms., and other. But this requires coding more than one dummy variable.

Dummy variables provide a means for including qualitative variables in a regression model. Yet using dummy variables also introduces several potential problems. First, the dummy variables can expand the number of variables in

---

* Other dummy coding exists. For simplicity we use one common type.

the regression equation. This can grossly increase the computer run time and it can expand the number of possible models to test. Second, if improperly used, dummy variables can induce the extreme case of multicollinearity. You will recall that our application using dwelling type, DU, used one variable less than the number of categories, that is, $2 - 1 = 1$ dummy variable. This was necessary to avoid a situation of high multicollinearity.

We illustrate the use of dummy variables with several examples. The salient features will appear in the discussion.

## EXAMPLE 13.12

In Example 13.2 the cosmetics USE SCORES were described by a regression on % OWNER and SOCIAL variables. An alternative, although a weaker description, includes the SOCIAL and dwelling unit, DU, variables. In Table 13.11 we repeat the essential data from Table 13.1.

Earlier testing in Example 13.2 demonstrated a positive ($\beta$-coefficient) relation between cosmetics use and home ownership, % OWNER. We would expect that persons in single-family households, where home owners concentrate, would have higher USE SCORES than residents of multiple-family dwellings. We can test this contention by introducing a dummy variable, $X_2$, for dwelling size. If the residence is a single-family household, $X_2 = 1$; if a multiple-famliy dwelling, $X_2 = 0$. SOCIAL is quantitative; it does not require coding. How does the use of dummy variable coding for DU provide evaluation of two qualitative classes on a response?

**TABLE 13.11**   Data on Cosmetics USE SCORE ($Y$), Dwelling Size Code ($X_2$), and SOCIAL Score ($X_8$)

| $Y$ | $X_2$ | $X_8$ | $Y$ | $X_2$ | $X_8$ |
|------|------|------|------|------|------|
| 1.10 | 1 | 345 | 2.10 | 1 | 360 |
| 1.55 | 1 | 367 | 1.60 | 1 | 308 |
| 0.80 | 1 | 282 | 1.10 | 1 | 308 |
| 1.35 | 1 | 328 | 0.70 | 0 | 297 |
| 0.70 | 0 | 325 | 0.80 | 1 | 271 |
| 2.05 | 1 | 317 | 0.07 | 0 | 262 |
| 1.10 | 0 | 309 | 0.65 | 0 | 285 |
| 1.30 | 1 | 328 | 0.00 | 0 | 270 |
| 0.85 | 1 | 328 | 0.50 | 1 | 279 |
| 0.00 | 0 | 260 | 1.60 | 1 | 379 |
| 1.55 | 1 | 383 | 0.40 | 0 | 236 |
| 1.60 | 1 | 318 | 1.60 | 1 | 356 |
| 1.35 | 1 | 378 | 1.60 | 1 | 353 |
| 0.60 | 1 | 320 | 1.85 | 1 | 353 |
| 0.00 | 1 | 253 | 0.00 | 0 | 307 |
| 0.40 | 1 | 249 | 0.40 | 0 | 307 |
| 0.80 | 0 | 241 | 0.85 | 1 | 322 |
| 0.70 | 0 | 292 | 0.85 | 0 | 308 |
| 0.50 | 0 | 301 | 1.60 | 1 | 347 |
| 0.50 | 1 | 309 | 0.70 | 1 | 213 |
| 0.55 | 1 | 281 | 0.85 | 1 | 302 |
| 1.35 | 1 | 375 | 0.00 | 0 | 281 |
| 1.00 | 1 | 356 | 0.85 | 1 | 246 |
| 1.60 | 1 | 387 | 0.40 | 1 | 246 |
| 0.80 | 1 | 273 | 1.60 | 1 | 348 |

*Solution*    The data are plotted in Figure 13.7. Irrespective of the SOCIAL SCORE, households in single-family dwellings regularly have higher USE SCORES than do multiple-family dwelling residents. Dwelling size adds description to the SOCIAL variable. Without the DU variable, USE SCORES would tend to be underestimated for single-family dwellings and overestimated for multiple-family dwellers. A single regression line that ignores dwelling size would lie between that labeled SFDU's and that labeled MFDU's on Figure 13.7.

The dummy variable, $X_2$, makes an adjustment for elevation (level of the $Y$-response); on average, USE SCORE is higher for SFDU's than for MFDU's. The regression model is

$$Y = \beta_0 + \beta_2 X_2 + \beta_8 X_8 + \varepsilon \tag{I}$$

where   $X_2$ = single- or multiple-family dwelling (1 or 0)

$X_8$ = SOCIAL score

Then, for single-family dwellings, with $X_2 = 1$, Equation (I) is

$$Y = (\beta_0 + \beta_2) + \beta_8 X_8 + \varepsilon \quad \text{for single family; or else } (X_2 = 0)$$

$$Y = \beta_0 \quad\quad + \beta_8 X_8 + \varepsilon \quad \text{for multiple-family-dwelling residents}$$

**FIGURE 13.7**   A Data Plot for USE SCORES and a Dummy Variable Estimated Regression

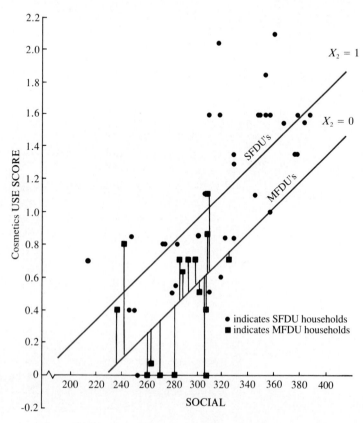

The model (I) assumes the same slope, $\beta_8$, for both the single-family and the multiple-family dwellings. Then the dummy variable affects only the elevation, that is, the intercept parameter. This difference is shown in Figure 13.7.

---

The preceding decision demonstrates that one dummy variable can provide evaluation of two qualitative classes on a response USE SCORE, and this is expressed by one equation (I).

An alternative for a qualitative variable with $k = 2$ classes is to fit two separate regression equations—one to the single-family and one to the multiple-family dwelling data points. Here, however, in the latter case we would be fitting a regression model to only 15 data points. This could result in a relatively unstable standard error. When we use dummy variables in model (I), the error variance for coefficient estimates and testing uses all 50 observations. This gives more information for calculating the MSE, which ensures more precise inference about the model parameters.

There were two classes for the dwelling-unit variable. How do we expand this to variables with more than two classes? For example, suppose we consider seasonal effects of spring, summer, fall, and winter. This seasons variable has four classes. If we arbitrarily assign summer as base, then three dummy variables define all (4) possibilities:

$$X_1 = 1 \text{ or } 0, \text{ for spring or other, respectively}$$

$$X_2 = 1 \text{ or } 0, \text{ for fall or other}$$

$$X_3 = 1 \text{ or } 0, \text{ for winter or other}$$

Summer is defined by $X_1 = 0$, $X_2 = 0$, and $X_3 = 0$. Then a regression model would require three dummy variables to define the seasons variable.

To generalize the preceding for a qualitative variable with $k$ classes, the regression model requires $k - 1$ dummy variables. You might well ask why we don't simply use $k$ dummy variables. If we did, then the least squares estimates for the $\beta$'s could not be found. This is because any one of the dummy variables could be written as a combination of the remaining $k - 1$ dummy variables. The result is that the normal equations do not have a unique solution. So using all ($k$) possible categories as dummy variables would induce multicollinearity in its most extreme form.

Example 13.13 illustrates the case for a variable with $k > 2$ classes. This procedure can be used in our regression models in Chapter 14 to include a seasonal measure in a time series.

---

**EXAMPLE 13.13**

The marketing department suggests that the cosmetics USE SCORES could be linear for quarterly sales patterns. From experience, the marketing people know that the over-the-counter sales are highest in the spring months, with lows in the summer and moderate levels in the fall and winter; that is, actual sales exhibit discernible seasonal variations. How could we extend the model in Example 13.12 to include a quarterly seasonal effect?

*Solution*    Three dummy variables will account for a quarterly seasonal effect:

$$X_1 = \begin{cases} 1 \text{ if month is during spring quarter} \\ 0 \text{ if some other month} \end{cases}$$

$$X_2 = \begin{cases} 1 \text{ if month is during fall quarter} \\ 0 \text{ if some other month} \end{cases}$$

$$X_3 = \begin{cases} 1 \text{ if month is during winter quarter} \\ 0 \text{ if some other month} \end{cases}$$

Summer quarter is, by default, designated as the base with codes $X_1 = 0$, $X_2 = 0$, $X_3 = 0$. Then the regression model for cosmetics USE SCORE $(Y)$, regressed on SOCIAL $(X_4)$ and a seasonal measure, is

$$Y = \beta_0 X_0 + \beta_1 X_1 + \beta_2 X_2 + \beta_3 X_3 + \beta_4 X_4 + \varepsilon \qquad \textbf{(II)}$$

with $X_0 = 1$ (always). The model (II) becomes

| | |
|---|---|
| for spring | $Y = (\beta_0 + \beta_1) + \beta_4 X_4 + \varepsilon$ |
| for summer | $Y = \beta_0 \qquad\quad + \beta_4 X_4 + \varepsilon$ |
| for fall | $Y = (\beta_0 + \beta_2) + \beta_4 X_4 + \varepsilon$ |
| for winter | $Y = (\beta_0 + \beta_3) + \beta_4 X_4 + \varepsilon$ |

Thus the coefficients $\beta_1$, $\beta_2$, $\beta_3$ (for the three dummy variables) indicate a shift up or down from the summer-quarter base. It is reasonable to use equal numbers of observations for each quarter. However, this is not a requirement.

---

In summary, it is important to check for valid assumptions and to take corrective action when the assumptions do not hold. We defined multicollinearity and then described four means of reducing the confusion due to correlated independent variables. Finally, we described the use of dummy variables. However, this is only an introductory coverage of these topics. For more depth, see the chapter references. And, if you experience a regression model with invalid assumptions, then it is advisable to consult with an experienced statistician.

**Problems 13.7**

30. The plots are (a) a scattergram (original data) and (b) residuals from a simple linear regression in time, $t$. The production supervisor says these plots indicate that the production curve, $P$, is best estimated by a second-degree linear model in time, $P = \beta_0 + \beta_1 t + \beta_2 t^2 + \varepsilon$. Do you agree? Explain.

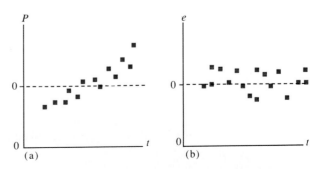

31. For each of the residual plots shown in graphs (a) and graphs (b), indicate the pattern that is displayed and give either a corrective action or another model to test.

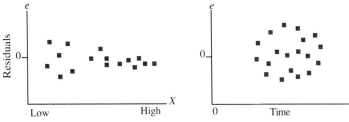

(a) Residuals are to $\hat{Y} = b_0 + b_1X$, where $X$ is a variable other than time.

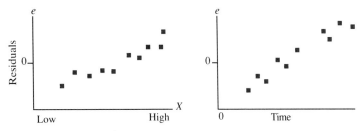

(b) Residuals are to $\hat{Y} = C_0 + C_1X + C_2X^2$. Again, $X \neq$ time.

32. The data below relate to an estimated regression for $Y =$ oversized rivet holes on $X =$ number of repairs in overhauling a commercial airline jet hull. Use the data (i) to make residual plots (vertical axis) against repairs, $X$, and then (ii) to make them against the sample numbers.

| Sample | 1 | 2 | 3 | 4 | 5 | 6 | 7 | 8 |
|---|---|---|---|---|---|---|---|---|
| Repairs, $X$ | 22 | 26 | 21 | 28 | 33 | 32 | 33 | 27 |
| Residuals | −3.1 | −1.3 | 2.2 | 4.2 | 4.7 | 0 | 7.7 | 2.0 |

| Sample | 9 | 10 | 11 | 12 | 13 | 14 | 15 |
|---|---|---|---|---|---|---|---|
| Repairs, $X$ | 34 | 35 | 30 | 37 | 33 | 13 | 12 |
| Residuals | −11.6 | 2.1 | −6.4 | 2.6 | −3.3 | 3.5 | −3.2 |

a. Use (i) to check the adequacy of the model $Y = \beta_0 + \beta_1X + \varepsilon$ against $Y = \beta_0 + \beta_1X + \beta_2X^2 + \cdots + \beta_kX^k + \varepsilon$. Is there need for $X^2, X^3, \ldots$?

b. Use the residual values to evaluate $e$ and $\hat{\sigma}_e$. Use these numbers to check the reasonableness of the assumption of normal errors. Include a comparison of actual expected values for $\bar{e} \pm 1\hat{\sigma}_e$, $\bar{e} \pm 2\hat{\sigma}_e$, $\bar{e} \pm 3\hat{\sigma}_e$, where $s_e = \hat{\sigma}_e$.

33. Establish a best linear model of up to two $X$-variables for $Y =$ sales ($000) on $X_1 =$ advertising expenditures ($000) and $X_2 =$ worker-hours of labor, using either (i) or (ii) below. Use the forward-selection procedure and $\alpha = 0.05$ to choose a best model. Do this by first testing $H_0: \beta_2 = 0$ against $H_A: \beta_2 \neq 0$. If you reject $H_0: \beta_2 = 0$, then test for a significant improvement by adding $X_1$ into the model, i.e., test $H_0: \beta_{1 \cdot 2} = 0$.

(i) The estimate is $\hat{Y} = -3.81 + 6.65X_2$. (*Note:* $r_{Y2} > r_{Y1}$, so ignore $X_1$ alone.)

| Source | df | SS | MS |
|---|---|---|---|
| $X_2$ | 1 | 1635.57 | 1635.57 |
| Residuals | 6 | 571.93 | 95.32 |

(ii) The estimate is $\hat{Y} = -6 + 2X_1 + 5X_2$.

| Source | df | SS | MS |
|---|---|---|---|
| $X_1$ and $X_2$ | 2 | 1645.00 | 822.50 |
| Residuals | 5 | 562.50 | 112.50 |

34. A wholesale distributor seeks a descriptive measure of its sales, $Y$ (in $000), to several retailers. The information includes $X_1$ = number of sales calls made in one year to each retail location, $X_2$ = estimated annual sales for each retailer (see summary statistics table).
   a. What special condition is indicated by the correlation $(X_1X_2) = 0$? How does this affect the hypothesis tests of $\beta_1$ and $\beta_2$?

SUMMARY STATISTICS

| Variable | Mean | St. Dev | Y | $X_1$ | $X_2$ |
|---|---|---|---|---|---|
| | | | | Correlations | |
| Sales, Y | 2.333 | 5.8949 | 1.0000 | .6366 | .7053 |
| Sales Calls, $X_1$ | 10.0000 | 4.3301 | | 1.0000 | .0000 |
| Retailer Sales, $X_2$ | 7.0000 | 3.9686 | | | 1.0000 |

| Source | DF | Sum of Squares | Mean Squares |
|---|---|---|---|
| $X_2$ | 1 | 138.2900 | 138.2900 |
| Residuals to $X_2$ | 7 | 139.7100 | 19.9586 |
| $X_1$, $X_2$ | 2 | 268.2907 | 134.1454 |
| Residuals to $X_1$, $X_2$ | 6 | 9.7093 | 1.6182 |

   b. Use the forward-selection procedure to select a best linear regression from among the following:

$$Y = \beta_0 + \varepsilon \qquad\qquad Y = D_0 + D_1X_2 + \varepsilon$$
$$Y = C_0 + C_1X_1 + \varepsilon \qquad Y = F_0 + F_1X_1 + F_2X_2 + \varepsilon$$

   Let $\alpha = 0.05$.

35. The Durbin–Watson test for autocorrelated errors uses the following:

    Reject $H_0$ if $D < d_L$.
    Accept $H_0$ if $D > d_U$.
    Otherwise, the test is inconclusive.

    Given the values $d_L = 1.08$ and $d_U = 1.36$ ($\alpha = 0.05$) for a simple linear regression for $Y =$ physical dexterity score on $X =$ alcohol level in blood. The data gives $D = 1.31$ for $n = 15$ individuals who were available for a clinical test.
    a. State your conclusions for the Durbin–Watson test.
    b. Suggest further action to determine the presence of autocorrelation.

36. A forensic scientist, investigating a recent upward trend in the crime rate throughout the United States, has studied the arrest rate per 100,000 population, $Y$, as associated with several variables:

    $X_1 =$ area population

    $X_2 =$ percent of households with annual income below poverty level

    $X_3 =$ unemployment rate

    Data are from a hypothetical sample of 20 cities, each exceeding one-half million population. Models are being compared. Identify the linear model, with up to three $X$-variables, that gives a best fit to the data. Let $\alpha = 0.05$ and use a forward-selection procedure. *Note:* To conserve space, we have listed only the essential information. Begin by testing $H_0: \beta_i = 0$, $i = 1, 2,$ or $3$. You choose.
    a. $Y$ regressed on $X_1$ and $X_2$ (see Summary 1)

Summary 1

| Source | df | Sum of Squares |
|---|---|---|
| $X_1$ | 1 | 1,308.34 |
| $X_2$ after $X_1$ | 1 | 9.46 |
| Residuals $X_1, X_2$ | 17 | 537.40 |

b. $Y$ regressed on $X_1$ and $X_3$ (see Summary 2)
c. $Y$ regressed on $X_2$ and $X_3$ (see Summary 3)

Summary 2

| Source | Sum of Squares |
|---|---|
| $X_3$ | 1,360.14 |
| $X_1$ after $X_3$ | 35.35 |
| Residuals $X_1, X_3$ | 459.71 |

Summary 3

| Source | Sum of Squares |
|---|---|
| $X_3$ | 1,360.14 |
| $X_2$ after $X_3$ | 117.23 |
| Residuals $X_2, X_3$ | 377.82 |

d. $Y$ regressed on $X_1$, $X_2$, and $X_3$ (see Summary)

Summary for 1, 2, 3

| Source | df | Sum of Squares |
|---|---|---|
| Regression on $X_1$, $X_2$, $X_3$ | 3 | 1,507.18 |
| Residuals to $X_1$, $X_2$, $X_3$ | 16 | 348.03 |

37. Members of a community school board are trying to evaluate the effects of educational resources on student performance. They have information on

$Y$ = Stanford achievement test scores

$X_1$ = per-pupil expenditure, dollars

$X_2$ = proportion of teachers with master's degree or higher

$X_3$ = pupil–teacher ratio

The sample was 25 sixth-grade classes in a metropolitan area. See the summary of the ANOVA. Use this information with $\alpha = 0.05$ and the forward-selection procedure to determine which variables of $X_1$, $X_2$, $X_3$ should be included in a linear model describing $Y$. You can assume that $X_1$ is the strongest single predictor and that $X_1 X_2$ is the strongest of the two variable combinations.

| Source | df | Sum of Squares |
|---|---|---|
| Regression on $X_1$ | 1 | 18,953.04 |
| Residuals to $X_1$ | 23 | 9,269.19 |
| Regression on $X_1$, $X_2$ | 2 | 25,963.07 |
| Residuals to $X_1$, $X_2$ | 22 | 2,259.16 |
| Regression on $X_1$, $X_2$, $X_3$ | 3 | 25,974.00 |
| Residuals to $X_1$, $X_2$, $X_3$ | 21 | 2,248.23 |
| Total | 24 | 28,222.23 |

38. Recall that $R^2$ = sum of squares for regression divided by total sum of squares. Use this to calculate $R^2$-values for the models

$$Y = \beta_0 + \beta_1 X_1 + \varepsilon$$

$$Y = C_0 + C_1 X_1 + C_2 X_2 + \varepsilon$$

$$Y = D_0 + D_1 X_1 + D_2 X_2 + D_3 X_3 + \varepsilon$$

in Problem 37. Then complete the table. Use the $R^2$ change values to substantiate your choice for a linear model in Problem 37.

| Variables in Model | $R^2$ | $R^2$ Change |
|---|---|---|
| $X_1$ | 0.6716 | 0.6716 |
| $X_1 X_2$ | | |
| $X_1 X_2 X_3$ | | |

39. A partial summary of a forward-selection regression analysis on the hotel's occupancy rate data follows. Use this information along with $\alpha = 0.05$ to confirm the hypothesis tests in Examples 13.8 and 13.9.

```
                          MULTIPLE REGRESSION

DELETION OF MISSING DATA

NUMBER 1  DEPENDENT VARIABLE . .  OCC

 BLOCK NUMBER 1.  METHOD: FORWARD

(S) ENTERED ON STEP NUMBER 1..  EAST (X₂)

R              .81922    ANALYSIS OF VARIANCE      DF       SUM OF SQUARES     MEAN SQUARE
               .67111
R SQUARE       .63823    REGRESSION                 1          1529.47006      1529.47006
ERROR         8.65754    RESIDUAL                  10           749.52994        74.95299
                         F =   20.40572                   SIGNIF F = .0011

(S) ENTERED ON STEP NUMBER 2 . .  SOUTH (X₃)

R              .92741    ANALYSIS OF VARIANCE      DF       SUM OF SQUARES     MEAN SQUARE
               .86009
R SQUARE       .82900    REGRESSION                 2          1960.15400       980.07700
ERROR         5.95209    RESIDUAL                   9           318.84600        35.42733
                         F =   27.66443                   SIGNIF F = .0001

                          MULTIPLE REGRESSION

NUMBER 1  DEPENDENT VARIABLE . .  OCC

(S) ENTERED ON STEP NUMBER 3 . .  DELTA (X₁)

R              .95686    ANALYSIS OF VARIANCE      DF       SUM OF SQUARES     MEAN SQUARE
               .91559
R SQUARE       .88393    REGRESSION                 3          2086.62310       695.54103
ERROR         4.90379    RESIDUAL                   8           192.37690        24.04711
                         F =   28.92410                   SIGNIF F = .0001

                              SUMMARY TABLE

 MULTR    RSQ   ADJRSQ   F(EQN)    SIGF   RSQCH     FCH   SIGCH        VARIABLE   BETAIN   CORREL
 .8192  .6711   .6382   20.406    .001   .6711   20.406   .001   IN:   EAST      .8192    .8192
 .9274  .8601   .8290   27.664    .000   .1890   12.157   .007   IN:   SOUTH     .4911    .7660
 .9569  .9156   .8839   28.924    .000   .0555    5.259   .051   IN:   DELTA    -.3127    .4276
```

40. Figure 13.4(d) is a data plot for occupancy rate ($= Y$) and Delta passenger counts ($= X$) that can relate to a simple linear regression. The 12 data appear in Table 13.7. Here is a Minitab summary for a regression that includes all 12 observations, including the data point in question, $Y = 74$, $X = 574.86$.

```
MTB > READ C1 C2
      12 ROWS READ
MTB > END
MTB > NAME C1 'OCC' C2 'DELTA'
MTB > PRINT C1 C2
MTB >
      ROW     1      2      3      4      5      6      7      8      9     10     11     12
      OCC    40     41     48     49     73     74     68     51     63     75     70     38
      DELTA 345.26 479.45 427.81 423.05 413.77 574.86 463.57 379.83 376.08 384.14 399.93 285.44
MTB > PLOT 'OCC' 'DELTA'
```

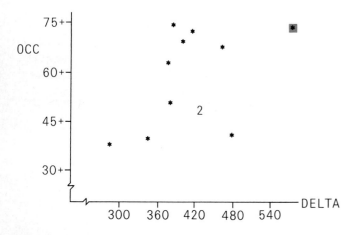

```
MTB > BRIEF output level 3
MTB > REGRESS 'OCC' on 1 predictor in 'DELTA'

The regression equation is
OCC = 22.5 + 0.0849 DELTA

Predictor      Coef      Stdev    t-ratio        p
Constant      22.46      23.76       0.95    0.367
DELTA       0.08489    0.05675       1.50    0.166

s = 13.65      R-sq = 18.3%   R-sq(adj) = 10.1%

Analysis of Variance

SOURCE        DF        SS        MS       F        p
Regression     1     416.6     416.6    2.24    0.166
Error         10    1862.4     186.2
Total         11    2279.0
```

| Obs. | DELTA | OCC | Fit | Stdev.Fit | Residual | St.Resid |
|---|---|---|---|---|---|---|
| 1 | 345 | 40.00 | 51.77 | 5.50 | −11.77 | −0.94 |
| 2 | 479 | 41.00 | 63.16 | 5.46 | −22.16 | −1.77 |
| 3 | 428 | 48.00 | 58.78 | 4.03 | −10.78 | −0.83 |
| 4 | 423 | 49.00 | 58.37 | 3.98 | −9.37 | −0.72 |
| 5 | 414 | 73.00 | 57.59 | 3.94 | 15.41 | 1.18 |
| 6 | 575 | 74.00 | 71.26 | 10.01 | 2.74 | 0.30X* |
| 7 | 464 | 68.00 | 61.81 | 4.88 | 6.19 | 0.49 |
| 8 | 380 | 51.00 | 54.70 | 4.36 | −3.70 | −0.29 |
| 9 | 376 | 63.00 | 54.39 | 4.46 | 8.61 | 0.67 |
| 10 | 384 | 75.00 | 55.07 | 4.26 | 19.93 | 1.54 |
| 11 | 400 | 70.00 | 56.41 | 4.01 | 13.59 | 1.04 |
| 12 | 285 | 38.00 | 46.69 | 8.23 | −8.69 | −0.80 |

*X denotes an Obs. whose X-value gives it large influence.

Note that for this display the Minitab output has rounded the Delta values. For these data, $SS(X) = 58,002.25$.

Another Minitab summary is the regression, now based on the 11 observations, that excludes the data point in question, $Y = 74$, $X = 574.86$.

```
MTB > BRIEF output level 3
MTB > REGRESS 'OCC' on 1 predictor in 'DELTA'
```

The regression equation is
OCC = 28.8 + 0.0683 DELTA

| Predictor | Coef | Stdev | t-ratio | p |
|---|---|---|---|---|
| Constant | 28.83 | 33.66 | 0.86 | 0.414 |
| DELTA | 0.06827 | 0.08388 | 0.81 | 0.437 |

$s_e = 14.32$    R-sq = 6.9%    R-sq(adj) = 0.0%

Analysis of Variance

| SOURCE | DF | SS | MS | F | p |
|---|---|---|---|---|---|
| Regression | 1 | 135.9 | 135.9 | 0.66 | 0.437 |
| Error | 9 | 1846.1 | 205.1 | | |
| Total | 10 | 1982.0 | | | |

| Obs. | 1 | 2 | 3 | 4 | 5 | 7 | 8 | 9 | 10 | 11 | 12 |
|---|---|---|---|---|---|---|---|---|---|---|---|
| DELTA | 345 | 479 | 428 | 423 | 414 | 464 | 380 | 376 | 384 | 400 | 285 |
| OCC | 40 | 41 | 48 | 49 | 73 | 68 | 51 | 63 | 75 | 70 | 38 |
| Fit | 52.40 | 61.56 | 58.03 | 57.71 | 57.07 | 60.47 | 54.76 | 54.50 | 55.05 | 56.13 | 48.31 |
| St.Resid | −0.96 | −1.74 | −0.75 | −0.65 | 1.17 | 0.60 | −0.28 | 0.63 | 1.47 | 1.02 | −1.05 |

Note that Observation 6 is intentionally omitted. $n = 11$ observations remain.

Use the given information to determine if Observation 6, which is $(X,Y) = (574.86,74)$, qualifies for any of the following:

a. high-leverage point

b. outlier

c. influential observation

41. These data, introduced in Problem 29, include $Y$ = sales, $X_1$ = labor costs, and $X_2$ = advertising costs. A first summary uses all eight data points. The Minitab outputs include an estimated regression equation, an analysis of variance for regression, and a summary, including standardized residuals and data plots. Note that this extends the concepts of Problem 40, from a simple linear to a multiple regression in two independent variables. You will not be able to apply the $h_i$ statistic because that form applies only to a single $X$-variable. However, whether the point in question is high-leverage is intuitive.

| ROW | SALES | LABOR | ADVT |
|-----|-------|-------|------|
| 1 | 7 | 4 | 1 |
| 2 | 8 | 4 | 1 |
| 3 | 12 | 7 | 2 |
| 4 | 11 | 7 | 2 |
| 5 | 17 | 9 | 5 |
| 6 | 20 | 12 | 8 |
| 7 | 22 | 12 | 8 |
| 8 | 24 | 9 | 5 |

```
MTB > BRIEF output level 3
MTB > REGRESS Y in C5 on 2 independent variables in C6 C7
```

The regression equation is
SALES = 3.26 + 1.03 LABOR + 0.90 ADVT

| Predictor | Coef | Stdev | t-ratio | p |
|-----------|------|-------|---------|------|
| Constant | 3.261 | 7.129 | 0.46 | 0.667 |
| LABOR | 1.034 | 1.755 | 0.59 | 0.581 |
| ADVT | 0.898 | 1.869 | 0.48 | 0.651 |

$s_e$ = 3.481     R-sq = 79.6%     R-sq(adj) = 71.4%

Analysis of Variance

| SOURCE | DF | SS | MS | F | p |
|--------|----|----|----|----|----|
| Regression | 2 | 236.27 | 118.14 | 9.75 | 0.019 |
| Error | 5 | 60.60 | 12.12 | | |
| Total | 7 | 296.88 | | | |

| SOURCE | DF | SEQ SS |
|--------|----|--------|
| LABOR  | 1  | 233.47 |
| ADVT   | 1  | 2.80   |

| Obs. | LABOR | SALES | Fit   | Stdev.Fit | Residual | St.Resid |
|------|-------|-------|-------|-----------|----------|----------|
| 1    | 4.0   | 7.00  | 8.29  | 2.41      | -1.29    | -0.52    |
| 2    | 4.0   | 8.00  | 8.29  | 2.41      | -0.29    | -0.12    |
| 3    | 7.0   | 12.00 | 12.29 | 2.41      | -0.29    | -0.12    |
| 4    | 7.0   | 11.00 | 12.29 | 2.41      | -1.29    | -0.52    |
| 5    | 9.0   | 17.00 | 17.06 | 1.31      | -0.06    | -0.02    |
| 6    | 12.0  | 20.00 | 22.85 | 2.19      | -2.85    | -1.05    |
| 7    | 12.0  | 22.00 | 22.85 | 2.19      | -0.85    | -0.32    |
| 8    | 9.0   | 24.00 | 17.06 | 1.31      | 6.94     | 2.15R*   |

*R denotes an Obs. with a large St.Resid.

MTB > PLOT 'SALES' 'LABOR'

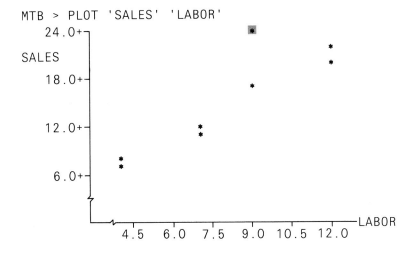

MTB > PLOT 'SALES' 'ADVT'

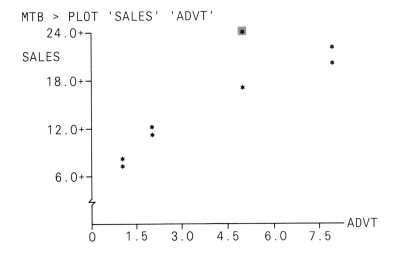

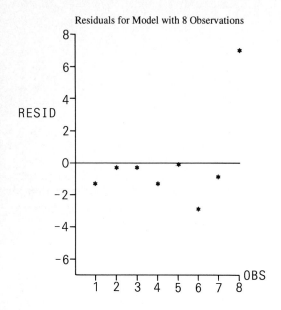

Residuals for Model with 8 Observations

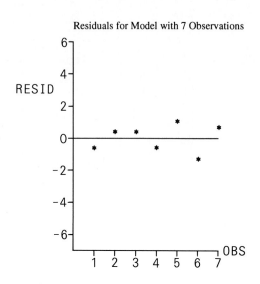

Residuals for Model with 7 Observations

A second summary that excludes Observation 8 follows.

```
MTB > BRIEF output level is 3
MTB > REGRESS Y in C5 on 2 independent variables in C6 C7


The regression equation is
SALES = 2.52 + 1.10 LABOR + 0.693 ADVT


Predictor     Coef      Stdev     t-ratio        p
Constant     2.523     2.157       1.17      0.307
LABOR       1.1025    0.5305       2.08      0.106
ADVT        0.6926    0.5655       1.22      0.288


s = 1.052   R-sq = 97.9%    R-sq(adj) = 96.8%


Analysis of Variance

SOURCE        DF        SS         MS         F         p
Regression     2    202.43     101.21     91.45     0.000
Error          4      4.43       1.11
Total          6    206.86


SOURCE        DF    SEQ SS
LABOR          1    200.77
ADVT           1      1.66
```

| Obs. | LABOR | SALES | Fit | Stdev.Fit | Residual | St.Resid |
|---|---|---|---|---|---|---|
| 1 | 4.0 | 7.000 | 7.626 | 0.736 | -0.626 | -0.83 |
| 2 | 4.0 | 8.000 | 7.626 | 0.736 | 0.374 | 0.50 |
| 3 | 7.0 | 12.000 | 11.626 | 0.736 | 0.374 | 0.50 |
| 4 | 7.0 | 11.000 | 11.626 | 0.736 | -0.626 | -0.83 |
| 5 | 9.0 | 17.000 | 15.909 | 0.428 | 1.091 | 1.14 |
| 6 | 12.0 | 20.000 | 21.294 | 0.697 | -1.294 | -1.64 |
| 7 | 12.0 | 22.000 | 21.294 | 0.697 | 0.706 | 0.90 |

Use the given information to determine if Observation 8—(sales, labor, advertising) = (24, 9, 5)—can be considered a high-leverage point, an outlier, or an influential observation (include the possibility of multiple conditions). Explain your choices. Now compare the predicted values for the two models: These are the data in the columns headed "Fit."

## SUMMARY

This chapter extends the simple linear regression model to a general *first-order multiple linear regression*. The components of this model are a deterministic part and random error. The deterministic component is physically described by a plane or surface in $k + 1$ dimensions, representing $k$ independent variables. The residuals (sample errors) provide a base for the inference, including predictions and checks on the model. The model assumptions are an extension of the simple regression case.

*Least squares estimation* provides the point estimators for the $\beta$-coefficients. The interpretation of the $\beta$ estimates includes a $Y$-intercept and slopes in the $X_i Y$ planes. The $b_1, b_2, \ldots, b_k$ determination in multiple regression recognizes that other $X$-variables contribute something to the description about the response. These are *not* the same values as simple regression $b$'s, because additional information is used for developing the slope estimators in multiple regression.

The variance estimate that is based on the accumulation of residual error, called mean square error (MSE), is the key to inference in multiple regression. Testing incorporates a comparison of *explained variations* in a *mean square for regression*, MSR, to the *unexplained variations* for a response, the *mean square error*, as $F = \text{MSR/MSE}$. This provides a test statistic for the collective description on $Y$ for a full set of $k$ ($> 1$) independent ($X$-) variables. The individual $X$-variables are tested for significance with a $t$- or an $F$-statistic. Predictions are ascribed to either (1) *expected (average) response* or (2) *individual response*. The predictions on individual response carry the larger error.

Computer-assisted calculations are a must for a regression modeling problem with even two independent variables. The several alternatives available for identifying a best set of independent variables include *forward-selection, backward-elimination, stepwise*, and *all-regressions* procedures. A choice of procedure may depend upon what software options are available.

Computer-developed residual data plots can provide useful diagnostics. These include graphic displays to check how well the regression equation fits the sample data as well as to validate the model assumptions. Plots of residuals against untested variables can disclose potential strong predictors.

Regression is a widely used tool in business and economic modeling. Applications described in the chapter include modeling cosmetics USE SCORES from individual and demographic variables, predicting hotel occupancy in the Walt Disney World area, and describing selling features for used cars (Business Case, pp. 644–47). Whatever your application, you will probably want to develop regression models on a computer. Table 13.12 displays some key equations developed in this chapter.

**TABLE 13.12** Key Equations for Multiple Regression

Testing $H_0: \beta_1 = \beta_2 = \cdots = \beta_k = 0$

The ANOVA for regression with $F$-statistic

| Source | df | SS | MS |
|---|---|---|---|
| Regression | $k$ | SSR | MSR |
| Error | $n - (k + 1)$ | SSE | MSE |

$$\text{Calculated } F_{k,(n-k+1)} = \frac{\text{SSR}/(k)}{\text{SSE}/(n-(k+1))} = \frac{\text{MSR}}{\text{MSE}} \qquad (13.6)$$

$$R^2 = \frac{\text{SSR}}{\text{TSS}} = 1 - (\text{SSE/TSS}) \qquad (13.7)$$

Inference on individual $X$-variables

$$t_{n-(k+1)} = \frac{b_i - \beta_i}{s_{b_i}} \qquad (13.9)$$

Mean response confidence intervals

$$\text{for } E(Y \mid X_1 X_2 \cdots X_k) = \hat{Y} \pm (t \cdot s_{\hat{Y}}) \qquad (13.12)$$

Individual response prediction interval

$$\text{for } Y_{\text{indiv}} = \hat{Y} \pm t \cdot \sqrt{s_{\hat{Y}}^2 + \text{MSE}} \qquad (13.13)$$

The extra sum of squares principle

$$\text{for } H_0: \beta_{X_k \mid X_1 X_2 X_3 \cdots X_{k-1}} = 0$$

uses

$$F_{1,n-(k+1)} = \frac{\text{SSR}(X_k \mid X_1 X_2 \cdots X_{k-1})/1}{\text{MSE}(X_1 X_2 \cdots X_k)} \qquad (13.14)$$

## KEY TERMS

adjusted $R^2$, 585
analysis of variance for regression, 607
autocorrelated errors, 617
beta or standardized regression
  coefficients, 586, 595
coefficient of multiple correlation,
  $R$, 593
coefficient of multiple determination,
  $R^2$, 593
confidence intervals in multiple
  regression, 596
dummy variable coding, 620
Durbin–Watson statistic, 617
extra sum of squares principle, 604
heteroscedastic error variance, 611
high-leverage point, 613, 614
influential observation, 613, 614

least squares (point) estimation, 578, 579
mean square error (MSE), 589
mean square for regression (MSR), 592
multicollinearity of $X$-variables, 619
multiple linear regression model, 575
outlier in regression, 613, 614
prediction intervals in multiple
  regression, 596
regression modeling procedures
  all-regressions, 603
  backward-elimination, 603
  forward-(stepping)-selection, 603
  stepwise-regression 603
residual plots, 610
scree plot, 590
standard error of estimate, 577

42. A linear model with the dependent variable $Y$ = age of marathon race winner is estimated by $\hat{Y} = 2.1 + 3.2X_1 - 1.5X_2 + 5.3X_3$. The ANOVA information is supplied.
    a. What is the sample size?
    b. Compute $R^2$ and $R$.
    c. Test $H_0: \beta_1 = \beta_2 = \beta_3 = 0$ against $H_A$: At least one $\beta_i \neq 0$, for $\alpha = 0.05$.
    d. Assume $X_2$ and $X_3$ have different units. Does the 5.3 coefficient of $X_3$ mean that $X_3$ is the stronger predictor of $Y$ than $X_2$, which has a coefficient of 1.5? Explain.
    e. How would an increased sample, to $n = 25$, improve this inference?

| Source | df | Sum of Squares | Mean Squares |
|--------|----|----------------|--------------|
| Regression | 3 | 332.46 | 110.82 |
| Residuals | 6 | 86.26 | 14.38 |

43. A multiple linear regression model is estimated by $\hat{Y} = 14 - 2X_1 - 0.5X_2$ with MSE = 8.5000 and 8 df for the error.
    a. Develop a point estimate for $E(Y|X_1 = 1, X_2 = 8)$.
    b. Test the hypothesis $H_0: E(Y|1,8) = 10$. Use $\alpha = 0.10$ and a two-tailed alternative. Use $s_{\hat{Y}} = 3.0332$.

44. A collegiate testing service is developing a new accounting aptitude test. Related measures are used to validate the test for a sample of 14 subjects. Two validating measures are $X_1$ = score on a senior-level accounting final and $X_2$ = ACT senior aptitude score. The following table summarizes the explained SS relative to tests of hypotheses for the forward-selection of a best model in $X_1$, in $X_2$, or in both. Use $\alpha = 0.05$ to determine a best model.

| Source | df | Sum of Squares | Test |
|--------|----|----------------|------|
| Regression on $X_1X_2$ | 2 | 52,237.027 | a. $H_0: \beta_1 = \beta_2 = 0$ |
| Regression on $X_1$ | 1 | 48,713.530 | b. $H_0: \beta_1 = 0$ |
| $X_2$ after $X_1$ | 1 | 3,523.498 | c. $H_0: \beta_2 = 0$ |
| Residuals to $X_1X_2$ | 11 | 25,588.848 | |

45. The table displays residuals to actual prices for used cars of a single make. The regression is a simple linear model on the car's age.

| Age $X$ | Residual $e$ | Age $X$ | Residual $e$ |
|---------|--------------|---------|--------------|
| 1 | 3.46 | 2 | 2.34 |
| 4 | −1.88 | 3 | −0.77 |
| 10 | 1.42 | 7 | −2.23 |
| 2 | −0.66 | 5 | −1.00 |
| 5 | −3.00 | 5 | −2.00 |
| 6 | −0.12 | 3 | 1.23 |
| 8 | − 0.34 | 6 | −2.12 |
| 1 | 2.46 | 12 | 3.29 |

a. Calculate $\bar{e}$. Then use $s_e = \hat{\sigma}_e = 2.10$ to check the assumption of normally distributed residuals.

b. Plot residuals (horizontal axis) against age. Does this data plot indicate that $Y = \beta_0 + \beta_1 X + \beta_2 X^2 + \varepsilon$ would probably give an improved statistical description?

46. Let $Y$ = chemical production yield of household ammonia cleaner, and

$$X_1 = \text{rate of chemical operation}$$

$$X_2 = \text{temperature of the process-cooling bath}$$

$$X_3 = \text{concentration of the absorption chemicals}$$

Use the provided information to do an all-regressions analysis. We use the minimum $\text{MSE}_p$ criterion for finding the best model from regressions. Here $p$ denotes the number of $X$-variables in the regression. We seek the set of independent variables that makes $\text{MSE}_p$ the smallest, or for which $\text{MSE}_p$ is so close to the minimum that adding more $X$-variables appears not worthwhile. (The procedure is subject to the user's judgment.)

| Variables Entered | Source | df | Sum of Squares | $R^2$ |
|---|---|---|---|---|
| $X_1$ | Regression | 1 | 1750.12 | 0.8458 |
|  | Residuals | 19 | 319.12 |  |
| $X_2$ |  | 1 | 1586.09 | 0.7665 |
|  |  | 19 | 483.15 |  |
| $X_3$ |  | 1 | 330.80 | 0.1599 |
|  |  | 19 | 1738.44 |  |
| $X_1 X_2$ |  | 2 | 1880.44 | 0.9088 |
|  |  | 18 | 188.80 |  |
| $X_1 X_3$ |  | 2 | 1760.10 | 0.8506 |
|  |  | 18 | 309.14 |  |
| $X_2 X_3$ |  | 2 | 1594.18 | 0.7704 |
|  |  | 18 | 475.06 |  |
| $X_1 X_2 X_3$ | Regression | 3 | 1890.41 | 0.9136 |
|  | Residuals | 17 | 178.83 |  |

a. Complete the table. (For example, MSE $(X_1) = 319.12/19 = 16.796$.)

| $p$ | $X$ | $\text{MSE}_p$ |
|---|---|---|
| 0 | none | 103.462 |
| 1 | $X_1$ | 16.796 |
| 1 | $X_2$ |  |
| 1 | $X_3$ |  |
| 2 | $X_1, X_2$ |  |
| 2 | $X_1, X_3$ |  |
| 2 | $X_2, X_3$ |  |
| 3 | $X_1, X_2, X_3$ |  |

b. Now plot $MSE_p$ (vertical axis) against $p$. Use this scree plot to project the best all-regressions model.

47. Use the forward-selection procedure to determine the best regression model, $\alpha = 0.05$, for the data summary in Problem 46. Compare to your all-regressions solution.

48. Use the backward-elimination procedure and summaries in Problem 46 to determine the best regression model, $\alpha = 0.05$. Compare this to your (i) $MSE_p$ solution and (ii) forward-selection models (Problems 46 and 47).

 49. The following information relates economic data relative to energy usage in 22 geographic areas:

$$Y = \text{gasoline consumption (gallons)}$$

$$X_1 = \text{kilowatt hours of electricity}$$

$$X_2 = X_1^2$$

$$X_3 = \text{total energy costs, all fuels (in dollars)}$$

$$X_4 = \text{total energy expended (BTUs)}$$

Use the information in the table and the procedure outlined in Problem 46 to project the variables from among $X_1$ through $X_4$ for a best all-regressions linear model. Let $\alpha = 0.05$. Explain why your choice for a combination of variables is reasonable. Consider the scree plot.

| $p$ | Variables in Model | Error df | SS for Residuals | $p$ | Variables in Model | Error df | SS for Residuals |
|---|---|---|---|---|---|---|---|
| 0 | none | 21 | 2618.81 | 2 | $X_2, X_3$ | 19 | 868.18 |
| 1 | $X_1$ | 20 | 2579.92 | | $X_2, X_4$ | 19 | 46.40 |
| | $X_2$ | 20 | 2618.79 | | $X_3, X_4$ | 19 | 231.98 |
| | $X_3$ | 20 | 868.51 | 3 | $X_1, X_2, X_3$ | 18 | 868.14 |
| | $X_4$ | 20 | 264.72 | | $X_1, X_2, X_4$ | 18 | 40.98 |
| 2 | $X_1, X_2$ | 19 | 2138.76 | | $X_1, X_3, X_4$ | 18 | 41.73 |
| | $X_1, X_3$ | 19 | 868.25 | | $X_2, X_3, X_4$ | 18 | 46.30 |
| | $X_1, X_4$ | 19 | 43.99 | 4 | $X_1, X_2, X_3, X_4$ | 17 | 40.05 |

50. A regression analysis provides the data in the table.

| | | Source | df | Sum of Squares | Mean Squares |
|---|---|---|---|---|---|
| Multiple $R$ | 0.9861 | Regression | 2 | 2641.010 | 1320.505 |
| $R$ Square | 0.9724 | Residuals | 10 | 74.761 | 7.476 |

| | Variables in the Equation | | |
|---|---|---|---|
| Variable | $B$ | SE $B$ | $T$ |
| Var 1 | 1.440 | 0.1384 | 10.40 |
| Var 2 | $-0.614$ | 0.0486 | 12.63 |
| Constant | 103.097 | | |

a. Does the multiple linear regression using Var 1 and Var 2 provide a significant description of the variations in the response variable? First guess; then confirm by using a hypothesis test with $\alpha = 0.01$.

b. Are both Var 1 and Var 2 needed for a best model? Consider the $t$-values (T) displayed. What danger exists in your conclusion (i) if these variables are highly intercorrelated or (ii) if an important independent variable has been omitted from the model?

51. An earlier problem requested a point estimate for a model in which

$$Y = \text{time, in (three-month) quarters, for a company to realize an increase in profits}$$

$$X_1 = \text{quarters since the last real increase in profits}$$

$$X_2 = \text{number of years the chief executive officer has held that position}$$

Given $X_1 = 6$, $X_2 = 10$, and $\hat{Y} = 2.48 + 1.14X_1 - 0.50X_2$, extend this point estimate to a 95% interval estimate (for df = 100) as follows:

a. Find confidence limits, given that the SE of mean response is 0.82 for these values of $X_1$, $X_2$.

b. Find prediction limits, given that the SE of individual response is 1.96 for these values of $X_1$, $X_2$.

Explain how the intervals differ and why this would be important to you if your company were in this situation.

---

**STATISTICS IN ACTION: CHALLENGE PROBLEMS**

52. Paul Sommers' article is a thing of beauty. He states, on page 37, "Although beauty pageant statistics command widespread attention (at least one evening each year), unfortunately the subject has not been examined systematically and objectively." Mr. Sommers gives an interesting application of multiple regression to evaluating the possibility of "biases," that is, preferences, in the judging of Miss USA and Miss Universe competitions.

There are three competitions in a Miss USA–Miss Universe pageant that the judges grade: personality interview, swimsuit, and evening gown. Following the appearance of each semifinalist in any one of the three competitions, each judge records a score, ranging from a low of 1 to a perfect score of 10. An average score, reflecting the sentiments of the entire panel, is then flashed on the TV screen. Based on these data it is possible to offer some judgment of the judging. The purpose of Sommers' research was to test for bias preference according to (1) hair color, (2) order of appearance, and (3) region.

Mr. Sommers concludes that there *were* preferences in judging at the 1980 and 1981 contests. He cites the following general equation:

$$S(\text{score}) = a + (b_1 \cdot \text{PERS}) + (b_2 \cdot \text{SWIM}) + (b_3 \cdot \text{EVENGOWN})$$
$$+ (b_4 \cdot \text{BLOND}) + (b_5 \cdot \text{FIRST6}) + (b_6 \cdot \text{REGION}) \tag{1}$$

Then, Equation (2a) (page 35) provides the data for an analysis of possible preferences in the 1980 Miss USA contest judging. (The values in parentheses below (2a) are $t$-scores for testing the individual coefficients.)

$$S = \begin{array}{cccc} 7.889 & + (0.328 \cdot \text{PERS}) & + (0.089 \cdot \text{SWIM}) & + (0.324 \cdot \text{BLOND}) \\ (39.16) & (1.52) & (0.41) & (1.37) \end{array}$$

$$\begin{array}{cc} - (0.247 \cdot \text{FIRST6}) & + (0.471 \cdot \text{REGION}) \\ (-1.04) & (2.66) \end{array} \tag{2a}$$

$$R^2 = 0.276 \qquad \text{df} = 30$$

SOURCE: Sommers, Paul D. "Testing for Bias in Beauty Pageant Judging." *Journal of Recreational Mathematics* 16, no. 1 (1983–1984): 33–39.

Questions    Use both the appropriate $t$-score and $\alpha = 0.05$ to test the hypothesis $H_0$: There was a preference for BLONDS in this judging. Use a two-tailed alternative hypothesis.

53. Continue to test individually (with $t$-tests) each of the independent variables in Equation (2a) of Problem 52 for significance. If some of these variables are not significant by your $t$-tests, then why might they still be included in the regression model?

54. Dockweiler and Willis have performed a regression analysis to model the junior- and senior-year achievement of entering accounting majors at one university. Their Table 1 gives correlations, with significance levels, for up to 11 independent variables with the dependent variable. Here, the variables are defined below the table.

**TABLE I**    Pearson-Product Moment Correlation Coefficients of Independent Variables with Subsequent GPA in the Accounting Curriculum

| Independent Variables | Students | | |
|---|---|---|---|
| | *Locals* | *Transfers* | *All* |
| $X_1$ | 0.649*** | 0.404** | 0.612*** |
| $X_2$ | 0.105* | −0.102 | −0.048 |
| $X_3$ | 0.576*** | 0.383** | 0.548*** |
| $X_4$ | 0.628*** | 0.404** | 0.590*** |
| $X_5$ | −0.044 | 0.141 | 0.014 |
| $X_6$ | 0.353*** | | |
| $X_7$ | 0.301*** | | |
| $X_8$ | 0.167** | | |
| $X_9$ | 0.208*** | | |
| $X_{10}$ | 0.215*** | | |
| $X_{11}$ | 0.233*** | | |

  * Significant at the 0.10 level.
 * Significant at the 0.01 level.
*** Significant at the 0.001 level.

$Y$ = GPA earned while enrolled in the undergraduate accounting curriculum (junior and senior years)

$X_1$ = GPA earned prior to entering the undergraduate accounting curriculum, excluding grades earned in introductory accounting courses

$X_2$ = total hours of college credit completed prior to entering the undergraduate accounting curriculum

$X_3$ = grade earned in the first introductory accounting course

$X_4$ = grade earned in the second introductory accounting course

$X_5$ = age of student upon entering the undergraduate accounting curriculum

$X_6$ = high school rank (HSR)

$X_7$ = MCET score, Form B

$X_8$ = MMPT score, Form C-2

$X_9$ = SCAT, Series II, Form 1-C, verbal score

$X_{10}$ = SCAT, Series II, Form 1-C, quantitative score

$X_{11}$ = SCAT, Series II, Form 1-C, composite score

SOURCE: Dockweiler, Raymond C., and Carl G. Willis. "On the Use of Entry Requirements for Undergraduate Accounting Programs." *The Accounting Review* 3 (July 1984): 496–504.

Questions  Use the data in Table 1 to identify the four strongest independent variables for predicting $Y$ = junior- and senior-year accounting GPA (a) for local students, (b) for transfer students, and (c) for all students.

55. Table 2 from Dockweiler and Willis (page 499) provides summaries for stepwise regression analyses. Use these summaries to explain the following:
   a. For local students, why does the variable $X_6$ not enter the model yet $X_1$, $X_3$, and $X_4$ all enter the model?
   b. The variable $X_5$ = student's age upon entering the undergraduate accounting program is significant for the model for transfer students. By Table 2, do higher values for this variable increase or decrease the predicted GPA, or can you tell from the given information? Explain.

**TABLE 2**  Results of Stepwise Regression Analysis

| Independent Variables | Local Students | | | Transfer Students | | | All Students | | |
|---|---|---|---|---|---|---|---|---|---|
| | Multiple Correlation | $R^2$ Change | F-Value | Multiple Correlation | $R^2$ Change | F-Value | Multiple Correlation | $R^2$ Change | F-Value |
| $X_1$ | 0.649 | 0.421 | 63.24**** | 0.404 | 0.163 | 9.07** | 0.612 | 0.374 | 79.06**** |
| $X_4$ | 0.741 | 0.128 | 45.69**** | 0.529 | 0.116 | 5.74* | 0.709 | 0.129 | 53.61**** |
| $X_3$ | 0.757 | 0.023 | 14.20*** | 0.537 | 0.009 | 0.79 | 0.725 | 0.023 | 14.37*** |
| $X_2$ | 0.762 | 0.008 | 3.61 | 0.566 | 0.032 | 2.89 | 0.728 | 0.005 | 2.90 |
| $X_5$ | 0.764 | 0.003 | 1.79 | 0.602 | 0.042 | 4.47* | 0.729 | 0.000 | 0.18 |
| $X_8$ | 0.765 | 0.001 | 1.16 | | | | | | |
| $X_{10}$ | 0.766 | 0.002 | 0.44 | | | | | | |
| $X_6$ | 0.766 | 0.000 | 0.05 | | | | | | |
| $X_9$ | 0.766 | 0.000 | 0.02 | | | | | | |
| $X_7$ | 0.766 | 0.000 | 0.00 | | | | | | |
| $X_{11}$ | 0.766 | 0.000 | 0.00 | | | | | | |
| Maximum Values | 0.766 | 0.586 | | 0.602 | 0.362 | | 0.729 | 0.531 | |

  * Significant at the 0.05 level.
 ** Significant at the 0.01 level.
 *** Significant at the 0.001 level.
 **** Significant at the 0.0001 level.

56. Friedman and Friedman point out the advantage of cross-validation of a model. That is, a researcher starts with a dependent variable and a set of independent variables and, after examining the data, obtains a "best" estimated regression equation—say, by stepwise regression. Cross-validation develops a regression equation using only a part (say, two-thirds) of the data, chosen at random. Then the regression equation is applied to the remainder (one-third of the data), to predict values for the dependent variable.

    To illustrate, the authors consider a hypothetical multiple regression equation to explain sales at different branches of a fast-food chain (the dependent variable) by the independent variables: pedestrian traffic near a branch, auto traffic near a branch, area population surrounding a branch, and number of competitors in the neighborhood of the branch. The example suggests using 60 randomly selected branches to develop a model, with a remaining 30 branches as the validation sample—set aside for testing the model. The results are displayed below (from page 27, Table 2).

**TABLE 2**   Hypothetical Regression Analysis: Branch Sales of a Fast-Food Chain

|  | Regression Coefficients Showing How Sales Depend on Regression Variables |
|---|---|
| **Analysis Sample** | |
| Regression Variables | |
| Pedestrian traffic | 0.45 |
| Vehicular traffic | 0.89 |
| Area population | 0.21 |
| Amount of competition | −0.46 |
| Computer $R^2$ | 0.94 |
| Computed $F$-ratio statistic | 23.50 (df = 4, 55) |
| **Validation Sample** | |
| Computed $R^2$ | 0.72 |
| Computed $F$-ratio statistic | 14.60 (df = 4, 25) |

SOURCE: Friedman, Hershey H., and Linda W. Friedman. "More Subtle Sins in Marketing Research." *The Mid-Atlantic Journal of Business* 20, no. 1 (Winter 1981): 21–31.

Questions   Discuss the comparative Computed $R^2$ values for the Analysis Sample and the Validation Sample. Compare the Computed $F$-ratio statistic for the Analysis Sample versus the same for the Validation Sample. What problems can occur in accepting a model without performing a validation?

57. For the data in Friedman and Friedman's Table 2, explain what the authors mean when they say (on their page 22) that an adjusted $R$ (adjusted for the indicated number of degrees of freedom) would be even lower. The suggestion is that the $R$ computed using the analysis sample, i.e., without cross-validation, is "artificially inflated." Recall that

$$R^2 = 1 - \frac{\text{SSE}}{\text{TSS}}$$

and

$$R_a^2 = 1 - \left[ \frac{(n-1)\text{SSE}}{(n-k)\text{TSS}} \right]$$

---

**BUSINESS CASE**

*Wally Wants to Sell More Cars*

Several weeks had passed since Wally Sanders began using Jack Baker's suggestions for improving sales at his car lot. Then Wally was back on the phone.

**Wally**    Jack, this is Wally Sanders. Sales have been better than usual since you gave that pitch. I'd like to hear the rest of your story.

**Jack**    Good, Wally. I'm glad to learn that you've tried my suggestions. It will take me a few days to make some computer runs. Then I could stop over in about a week—say, on Wednesday?

**TABLE 13.13**   The First Step for a Multiple Regression Analysis on Car Prices

```
                    MULTIPLE REGRESSION—FIRST STEP

NUMBER OF CASES                                          16
TOTAL # OF VARIABLES                                     2
F TO ENTER                                              1.0
F TO REMOVE                                             1.0


NUMBER OF PREDICTOR
VARIABLES ENTERED                                        1
MULTIPLE R SQUARE                                       .67875
STD ERROR OF EST                                        .25882


                      ANALYSIS OF VARIANCE

REGRESSION

   DEGREES OF FREEDOM 1                                  1
   SUM OF SQUARES                                       1.98153
   MEAN SQUARE                                          1.98153


RESIDUAL

   DEGREES OF FREEDOM 2                                 14
   SUM OF SQUARES                                        .93784
   MEAN SQUARE                                           .06699
   F RATIO                                             29.58


VAR         ID#              NAME                  COEFFICIENT

 1           2        MILEAGE (1000 MILES)           -.14367
                      CONSTANT ($000)              18.81266
```

At the appointed time Jack appeared at Wally's lot with a bundle of papers that was bigger this time than the last.

**Wally**  (Joking) Hi, Jack. Say, much more of this and we'll have to get you a wheelbarrow.

**Jack**  (Looking about) Seriously, Wally, something is happening here. Your inventory appears to be low.

**Wally**  Yes, in fact I've sold quite a few cars this week. It's got to be your pitch. . . . Let's get down to business. I've got three more Corvettes coming in from Atlanta this afternoon.

**Jack**  We'll need a place to set out my sheets. I have several regression printouts from one of our personal computers.

Wally clears a sales desk, and they spread out the papers. (See Tables 13.13 through 13.15.)

**TABLE 13.14**   The Final Step for a Multiple Regression on Car Prices

```
                      MULTIPLE REGRESSION—FINAL STEP

NUMBER OF CASES                                              16
TOTAL # OF VARIABLES                                         4
F TO ENTER                                                  1.0
F TO REMOVE                                                 1.0


NUMBER OF PREDICTOR
VARIABLES ENTERED                                            3
MULTIPLE R SQUARE                                          .68638
STD ERROR OF EST                                           .27622


                        ANALYSIS OF VARIANCE

REGRESSION

    DEGREES OF FREEDOM 1                                    3
    SUM OF SQUARES                                       2.00381
    MEAN SQUARE                                           .66794


RESIDUAL

    DEGREES OF FREEDOM 2                                   12
    SUM OF SQUARES                                        .91557
    MEAN SQUARE                                           .07630
    F RATIO                                               8.75
```

| VAR | ID# | NAME | COEFFICIENT |
|---|---|---|---|
| 1 | 2 | MILEAGE | -.12078 |
| 2 | 3 | EXTRAS | .02795 |
| 3 | 1 | CAR AGE | -.01153 |
|  |  | CONSTANT | 18.60499 |

**TABLE 13.15** Summary Table for a Regression on Car Prices

SUMMARY TABLE

| # OF PRED VARIABLES | VAR ID# | NAME | MULT R SQUARE |
|---|---|---|---|
| 1 | 2 | MILEAGE | .67875 |
| 2 | 3 | EXTRAS | .68547 |
| 3 | 1 | CAR AGE | .68638 |

.67875 ⟍ 0.7% CHANGE
.68547 ⟍ 0.09%
.68638 ⟋

CORRELATIONS

| | | |
|---|---|---|
| PRICE BY CAR AGE | | -.77430 |
| BY MILEAGE | | -.82356 |
| BY EXTRAS | | .57626 |
| CAR AGE BY MILEAGE | | .91768 |
| BY EXTRAS | | -.63263 |
| MILEAGE BY EXTRAS | | -.62151 |

**Jack** Look at the correlation matrix (in Table 13.15) that I built from several computer runs. Notice the largest correlation there?

**Wally** That would be 0.91768 (shaded in Table 13.15) for a car's age by its mileage. That's a high number. Is that good?

**Jack** Well, it tells quite a lot. The number squared ($0.91768^2$) is about 0.84. This means that about 84% of the time, mileage and a car's age are giving the same information. They are highly correlated, or multicollinear. It means, don't overdo a good thing. Your sales pitch should use one but not both. I suggest mileage, because you deal mostly with a market of young buyers who've likely already decided they want a certain make of car, such as a Corvette. A low-mileage Corvette is a real plus. You should lean on the car's age only if the mileage is high.

**Wally** How about the extras? Do they make a statistical difference in the price?

**Jack** Actually, the statistics show they have no appreciable association with increase in the price, beyond the description by mileage. But they could be important to keep your customer's attention. Look at the numbers I've added to the Summary Table—0.7% and 0.09%.

**Wally** Let's see, you're saying that extras *add* less than 1% (actually 0.7%) to the definition of higher prices (in Table 13.15)? And a car's age adds even less? Jack, I don't believe it!

**Jack** Well, that's what the numbers show. But you're paying me to interpret them. So, first off we've already said that either (1) mileage or (2) a Corvette's age can be your primary selling point. Depending on the car and its mileage, you should emphasize one—mileage or age—but not both. After this, you *should* point out the extras. But don't expect buyers to pay for these. Your experience shows that they won't! The extras could be a holding feature, though, like frosting on the cake.

**Wally** (After a minute of reflecting) You know, you just made me realize that I'm pushing *too* much. If a buyer sees what he likes, then it doesn't take much convincing to make the sale. But I've been trying to grandstand age and low mileage, then low mileage and age. Also, since I'm not in a dealership, I will stop pushing all the great extras. Buyers will just want more. I'll slip in the extras if I sense they're becoming indecisive—or to cap a sale.

**Jack** Right, Wally. I might have helped you some, but mostly I think the statistics helped you sort out what was really important in your sales pitch. You believe in your product—that really makes the sales. Emphasizing some facts, with discretion, can't hurt. Now, if you've got my check, then I'll be on my way.

**Wally** Jack, it's been a pleasure. (Hands him the check) By the way, now that you have some extra cash, could I interest you in this nice 'vette? (Pats the car) It has unusually low mileage (chuckle).

**Jack** Wally?

**Wally** What?

**Jack** Call me if your sales go down again.

**Business Case Questions**

58. Use statistics provided in Table 13.13 to write the estimated regression equation for price $(Y)$ on mileage $X_2$. Use this equation and a mileage value to 10 to (point) estimate the selling price for a Corvette with 10,000 miles.

59. Given that the total SS is 2.91937 and regression SS = 2.00114 for the variables MILEAGE and EXTRAS in the equation, use the available information and a forward-selection procedure to determine which variables from MILEAGE, EXTRAS, and CAR AGE to include in a regression model to estimate car PRICE. You will need to use information from Tables 13.13 through 13.15, the extra sum of squares principle, and $\alpha = 0.05$.

## COMPUTER PROBLEMS

Because this chapter is based on the computer resolution of multiple regression models, no specific section is designated as computer applications. The Problems that follow provide ample coverage of problem-solving with the use of computer calculations.

*Problems from Sources Other Than PUMS*

The computer problems for this chapter are a continuation of the data analysis within the chapter, so they do not use the PUMS data from our data base. Chapter 12 provides regression problems on the PUMS data.

60. We've begun a *forward-selection* solution for an equation describing occupancy rate variables from among DELTA, EAST, SOUTH, and TIME as $X$-variables. Complete the testing and compare your selection of independent variables with the selection in Example 13.8.

Step 1: Test for entry—EAST, $H_0$: $\beta_{EAST} = 0$; $H_A$: $\beta_{EAST} \neq 0$. So (DELTA, SOUTH, TIME) SS adds to ERROR SS.

|  | df | Sum of Squares | Mean Square |
|---|---|---|---|
| Regression | 1 | 1529.50 | 1529.50 |
| Residuals | 10 | 749.50 | 74.95 |

Calculated $F = 20.4$. Because (Calc $F$) $= 20.4 > F_{0.05,1,10} = 4.96$, the significance level is $F < 0.05$. So, retain EAST. Compare this result with Example 13.6.

Step 2:   Test for entry—SOUTH after EAST, $H_0$: $\beta_{\text{SOUTH} \cdot \text{EAST}} = 0$. So (DELTA, TIME) SS adds to ERROR SS.

|  | df | Sum of Squares | Mean Square |
|---|---|---|---|
| Regression | 2 | 1960.15 | 980.08 |
| Residuals | 9 | 318.8 | 35.43 |

Use this and data from the table in Step 1 to perform the calculations by the extra sum of squares principle to test for the entry of SOUTH after EAST. Continue using $\alpha = 0.05$ and compare your results with Example 13.7.

Step 3:   Given the following regression summary,

|  | df | Sum of Squares | Mean Square |
|---|---|---|---|
| Regression | 3 | 2086.62 | 695.54 |
| Residuals | 8 | 192.38 | 24.05 |

for the three variables EAST, SOUTH, and DELTA, continue to test for the entry of DELTA after EAST and SOUTH. Compare your results with Example 13.8.

61. We have begun a backward-elimination solution for a best OCCUPANCY rate model using variables from DELTA, EAST, SOUTH, and TIME. Again, perform the hypothesis tests to identify a best equation. This approach begins with a full model and tests this for statistical significance. Use the extra sum of squares principle to calculate the appropriate sum of squares, for different subsets of $X$-variables, to test. Continue testing and dropping variables until you reach a first conclusion of Reject $H_0$. Then keep all variables that are still in the model, including the variable being tested. Again, let $\alpha = 0.05$.

Step 1:   For $H_0$: $\beta_D = \beta_E = \beta_S = \beta_T = 0$, $H_A$: At least one $\beta_i \neq 0$.

|  | df | Sum of Squares | Mean Square |
|---|---|---|---|
| Regression (models) | 4 | 2109.86 | 527.47 |
| Errors | 7 | 169.14 | 24.16 |

Calculated $F = 21.83$     $p$-value $< 0.05$*
Conclusion: At least one $\beta_i \neq 0$.

* $p$-value $< 0.05$ as Calc $F = 21.83 > F_{0.05, 4, 7} = 4.12$.

Steps 2, 3, and 4:   Use the following to test successively for elimination of TIME (first), then DELTA, then SOUTH. Remember, you should stop with a first conclusion of Reject $H_0$.

|  | df | Sum of Squares |
|---|---|---|
| Regression | 3 | 2086.62 |
| Residuals | 8 | 192.38 |
| Regression | 2 | 1960.15 |
| Residuals | 9 | 318.85 |
| Regression | 1 | 1529.50 |
| Residuals | 10 | 749.50 |

Indicate tabular $F$-values for each test and specify the variables that are retained in your final model.

62. The Summaries (See Table 13.16) provide sums of squares for testing all-regressions for the hotel occupancy data in Table 13.7. This output can be obtained from SPSS, Minitab, or other software. For example, by Minitab the Model 1 statistics come from

    ```
    MTB > REGRESS C1 on 1 independent variable in C2
    ```

    where C1 has the data for ($Y=$) the occupancy rate, and $C_2$ holds values for DELTA. This or another routine gives substantially more output than is displayed here.

**TABLE 13.16**   ANOVA for a Regression Summary

| Model | Regression DF | Regression SS | Error DF | Error SS | INDEPENDENT VARIABLES | | | |
|---|---|---|---|---|---|---|---|---|
| 1 | 1 | 416.60 | 10 | 1862.40 | DELTA | | | |
| 2 | 1 | 1529.50 | 10 | 749.50 | EAST | | | |
| 3 | 1 | 1337.10 | 10 | 941.90 | SOUTH | | | |
| 4 | 1 | 317.30 | 10 | 1961.70 | TIME | | | |
| 5 | 2 | 1529.96 | 9 | 749.04 | DELTA | EAST | | |
| 6 | 2 | 1338.70 | 9 | 940.30 | DELTA | SOUTH | | |
| 7 | 2 | 1161.90 | 9 | 1117.10 | DELTA | TIME | | |
| 8 | 2 | 1960.15 | 9 | 318.85 | EAST | SOUTH | | |
| 9 | 2 | 1671.25 | 9 | 607.75 | EAST | TIME | | |
| 10 | 2 | 1523.40 | 9 | 755.60 | SOUTH | TIME | | |
| 11 | 3 | 2086.62 | 8 | 192.38 | DELTA | EAST | SOUTH | |
| 12 | 3 | 1718.52 | 8 | 560.48 | DELTA | EAST | TIME | |
| 13 | 3 | 1577.59 | 8 | 701.41 | DELTA | SOUTH | TIME | |
| 14 | 3 | 2078.23 | 8 | 200.77 | EAST | SOUTH | TIME | |
| 15 | 4 | 2109.86 | 7 | 169.14 | DELTA | EAST | SOUTH | TIME |
|  | 11 | 2279.00 | (= TOTAL DF and TOTAL SS) | | | | | |

Referring to the data in Table 13.16, use the information to perform an all-regressions analysis. Start by selecting the strongest $X$-variable, EAST for testing. Next select the strongest set of two $X$-variables—EAST, SOUTH—for testing. Then select the strongest set of three $X$-variables for testing. Finally, select all four $X$-variables. Testing uses the extra-SS principle and should stop when the extension to another level does not produce a statistical gain for $\alpha = 0.05$. Explain how this procedure is different from the forward-selection used in Problem 60.

63. Use a software package available at your computer center to perform a *stepwise-regression* analysis on the hotel occupancy rate data. As before, let $\alpha = 0.05$ and compare your results with Example 13.8 and Problem 60.

64. Use the cosmetics kit USE SCORES and data in Table 13.1 for a computer-assisted stepwise-regression analysis. Perform your tests at $\alpha = 0.01$ and then compare your determination of a best model against Examples 13.1–13.4.

65. Estimated cosmetics kit USE SCORES are defined by
$\hat{Y} = -1.98000 + (0.0173 \cdot \% \text{OWNER}) + (0.00563 \cdot \text{SOCIAL})$. Use

$$t_{47,0.005} \doteq 2.60 \qquad (\text{for } \alpha = 0.01)$$

$$\text{MSE} = 0.08643$$

$$s_{\hat{y}} = *\text{SEPRED values}$$

where *SEPRED is the usual standard error for predicted $(\hat{Y})$ values (following along with Equations (13.12) and (13.13) to complete the table). If you have available regression software that gives prediction limits and confidence limits, then use it along with the data from Table 13.1 to check these results by a complete computer solution.

| CASE | SCORE | *PRED | *SEPRED | PREDICTION LPL UPL | CONFIDENCE LCL UCL |
|---|---|---|---|---|---|
| 1 | 1.10 | 1.3635 | 0.0559 | | |
| 2 | 1.55 | 1.7297 | 0.0794 | | |
| 3 | 0.80 | 0.7147 | 0.0493 | | |
| 4 | 1.35 | 1.1293 | 0.0455 | | |
| ⋮ | | | | | |
| 48 | 0.85 | 0.5122 | 0.0782 | | |
| 49 | 0.40 | 0.5122 | 0.0782 | | |
| 50 | 1.60 | 1.4843 | 0.0626 | | |

66. The data in the table below show $Y =$ product usage of brand $Z$ bath soap as it relates to

$$X_1 = \text{relative urbanization of the area}$$

$$X_2 = \text{education level of the head of household}$$

$$X_3 = \text{age of spouse (if no spouse, use the age of head of household)}$$

| Variables in Model | Regression SS | DF | Residual SS | DF |
|:---:|:---:|:---:|:---:|:---:|
| None | 0 | 0 | 1279.4375 | 8 |
| $X_1$ | 822.4924 | 1 | 456.9451 | 7 |
| $X_2$ | 754.6416 | 1 | 542.7959 | 7 |
| $X_3$ | 505.6873 | 1 | 773.7502 | 7 |
| $X_1X_2$ | 938.5317 | 2 | 340.9058 | 6 |
| $X_1X_3$ | 859.9795 | 2 | 419.4580 | 6 |
| $X_2X_3$ | 1079.3416 | 2 | 200.0960 | 6 |
| $X_1X_2X_3$ | 1081.5735 | 3 | 197.8640 | 5 |

Use the forward-selection procedure with $\alpha = 0.05$ to determine a best regression model from those available.

67. Use the all-regressions procedure and the data in Problem 66 to determine a best regression model from those available. Again, let $\alpha = 0.05$. Is this the same or a different model from that obtained by forward-selection?

68. Here is a Minitab run for a regression of $Y =$ sales on $X_1 =$ labor expenditures, and $X_2 =$ advertising expenditures, all in \$10,000 units. The data are from Problem 29.

Step

```
1. MTB > READ C1 C2 C3
        READ 8 ROWS
2. MTB > END
3. MTB > NAME C1 'SALES' C2 'LABOR' C3 'ADVT'
4. MTB > PRINT C1 C2 C3

   ROW    SALES    LABOR    ADVT

    1       7        4       1
    2       8        4       1
    3      12        7       2
    4      11        7       2
    5      17        9       5
    6      20       12       8
    7      22       12       8
    8      24        9       5
5. MTB > CORRELATE C1 C2 C3

          SALES    LABOR
   LABOR   0.887
   ADVT    0.884    0.971
6. MTB > BRIEF output level 3
7. MTB > REGRESS Y in C1 on 2 independent variables in
   C2 C3
```

```
The regression equation is
SALES = 3.26 + 1.03 · LABOR + 0.90 · ADVT

Predictor    Coef    Stdev    t-ratio       p
Constant    3.261    7.129       0.46    0.667
LABOR       1.034    1.755       0.59    0.581
ADVT        0.898    1.869       0.48    0.651

s_e = 3.481    R-sq = 79.6%    R-sq(adj) = 71.4%

Analysis of Variance

SOURCE        DF        SS        MS       F        p
Regression     2    236.27    118.14    9.75    0.019
Error          5     60.60     12.12
Total          7    296.88

SOURCE        DF    SEQ SS
LABOR          1    233.47
ADVT           1      2.80

Obs    LABOR    SALES      FIT    Stdev.Fit    Residual    St.Resid
  1      4.0     7.00     8.29         2.41       -1.29       -0.52
  2      4.0     8.00     8.29         2.41       -0.29       -0.12
  3      7.0    12.00    12.29         2.41       -0.29       -0.12
  4      7.0    11.00    12.29         2.41       -1.29       -0.52
  5      9.0    17.00    17.06         1.31       -0.06       -0.02
  6     12.0    20.00    22.85         2.19       -2.85       -1.05
  7     12.0    22.00    22.85         2.19       -0.85       -0.32
  8      9.0    24.00    17.06         1.31        6.94        2.15R*
```

*R denotes an Obs. with a large St.resid.

At Step 7, the order of the independent variables—C2, then C3—is important because the Minitab output displays the SEQ SS (sequential SS) in the regression analysis of variance for the first variable named. Then Step 5, the correlation matrix, provides essential information that C2 is a stronger predictor than C3 (as $r_{YX_1} = 0.887$ is larger than $r_{YX_2} = 0.884$); hence C2 should precede C3 at Step 7.

Use $\alpha = 0.05$ and the given information to determine a best model for a regression of sales, $Y$, on labor and/or advertising. If you have Minitab available, then run

```
MTB > REGRESS 'SALES' on 1 variable 'LABOR'
```

and use this to help define a best model.

69. If you have Minitab available, run the regression for sales in C1 on two independent variables (labor) in C2 and (advertising) in C3. You can use the (7) Minitab statements in Problem 68 for this. Then add the following statements:

```
MTB > PLOT C1 C2
MTB > PLOT C1 C3
MTB > DELETE 8 C1-C3
```

```
MTB > PRINT C1-C3
MTB > BRIEF output level 3
MTB > REGRESS Y in C1 on 2 independent variables in
      C2 C3
MTB > STOP
```

Use this information to choose a regression equation, based on either eight data points or seven data points. Compare this with your results for Problem 41.

70. For Minitab users, here is a routine to regress occupancy rate on Delta for the data given in Table 13.7.

```
MTB > READ C1 C2
DATA> 40 345.26
     ⋮
DATA> 38 285.44
DATA> END
MTB > NAME C1 'OCC' C2 'DELTA'
MTB > PRINT C1 C2
MTB > PLOT C1 C2
MTB > BRIEF output level 3
MTB > REGRESS 'OCC' on 1 predictor in 'DELTA'
MTB > DELETE 6 C1-C2
MTB > PRINT C1-C2
MTB > PLOT C1 C2
MTB > BRIEF output level 3
MTB > REGRESS 'OCC' on 1 predictor in 'DELTA'
MTB > STOP
```

a. Is Observation 6 a high-leverage point? an outlier?
b. What is the estimated model with 12 observations? Plot this equation on the first PLOT graph. What is the estimated model with 11 observations? Plot this equation on the second PLOT graph. Compare $p$-values for the DELTA coefficient for the two models. Compare $R$-sq values for the two models. Is Observation 6 an influential observation?
c. Do the findings for parts (a) and (b) support your conclusions for Problem 40?

## FOR FURTHER READING

[1] Belsley, David, Edwin Kuh, and Roy Welsch. *Regression Diagnostics: Identifying Influential Data and Sources of Multicollinearity.* New York: Wiley, 1980.
[2] Chatfield, Christopher, and A. J. Collins. *Introduction to Multivariate Analysis.* London: Chapman and Hall, Ltd., 1980.
[3] Chatterjee, Samprit, and Ali S. Hadi. *Sensitivity Analysis in Linear Regression.* New York: Wiley, 1988.
[4] Dixon, Wilfred J., ed. *BMDP Statistical Software*, 1981 ed. Berkeley, CA: University of California Press, 1981.
[5] Draper, Norman R., and Harry Smith. *Applied Regression Analysis*, 2nd ed. New York: Wiley, 1981.
[6] Myers, Raymond H. *Classical and Modern Regression with Applications.* Boston: Duxbury, 1986.
[7] Neter, John, W. Wasserman, and M. Kutner, *Applied Linear Regression Models*, 3rd ed. Homewood, IL: Irwin, 1990.

[8] Ott, Lyman. *An Introduction to Statistical Methods and Data Analysis*, 2nd ed. Belmont, CA: Duxbury, 1984.

[9] Ryan, Barbara F., Brian Joiner, and Thomas A. Ryan, Jr. *Minitab Handbook*, 2nd ed. Boston: Duxbury, 1985.

[10] *SAS (Statistical Analysis System) User's Guide: Statistics*, 1982 ed. Cary, NC: SAS Institute, Inc., 1982.

[11] *SPSS User's Guide*, 2nd ed. New York: McGraw-Hill, 1986.

[12] Wonnacott, Thomas H., and R. M. Wonnacott, *Introductory Statistics for Business and Economics*, 4th ed. New York: Wiley, 1990.

# 14

# TIME SERIES
# AND INDEX NUMBERS

Get ready for higher prices at your department store and your supermarket! The Labor Department's latest Producer Price Index showed a relatively strong increase of 3.2% over the past six months, with increases in the prices of both intermediate and finished goods. Analysts attribute the increase to higher energy prices, a weakening of the dollar, and higher prices for imported goods.

Meanwhile, industrial production climbed only slightly last month, and the Consumer Confidence Index, which uses 1985 as a base year of 100, stood at 102.5, vs. 78.1 a year ago. With the nation's mills and factories now operating at over 83% of capacity, an upward pressure on prices is expected over the next quarter. Economists predict a 0.7% increase in the consumer price index next month, which is equivalent to an annual trend rate of 8.4%. If the increase does materialize, bankers fear the Federal Reserve will view this as inflationary and exert tighter controls on credit, resulting in a hike in interest rates. This could be a blow to comptrollers who have not accurately forecast their financial needs for the next quarter.

## 14.1
## INTRODUCTION

Today, you can pick up any one of several business newspapers or periodicals and find discussions like the preceding one. They concern prices, capacity, employment, and other statistics of every day concern to business managers. Most often, the data are couched in a *time frame* (e.g., month, quarter, year). Very often they use *index numbers*, such as price, productivity, and money market indices, to summarize and convey information about the economy.

The purpose of this chapter is to present some techniques for analyzing business and economic data within the concept of a time dimension. We begin with a definition of a time series and follow with a discussion of the classical approach to time series modeling. Our emphasis here will be on finding the most useful measures of extracting trend and seasonal information from data. Then we briefly explore how time series and other econometric methods can be used in a forecasting context. The latter part of the chapter offers a brief introduction to the types, construction, and uses of index numbers.

Upon completing Chapter 14, you should be able to

1. express the trend and seasonal components of a time series in mathematical terms;
2. extract cyclical variations from time series data;
3. use time series data for forecasting;
4. compute and interpret basic types of indexes;
5. use index numbers to determine the "real" change in economic variables.

## 14.2 COMPONENTS OF A TIME SERIES

Although much of our existence seems to be time-dependent, *time has no direct casual effect* on economic activity. It is simply the carrier of influencing factors such as employment, productivity, and weather-related events. Nevertheless, business managers sometimes disregard the specific cause–effect relationships and lump many factors together under the guise of time. They do this because it is both convenient and useful for decision-making purposes. Then the activities of some business indexes, such as GNP, can be considered a *time series*.

A time series is a set of regular observations of some variable over time.

The observations of a time series consist of data points that may be tabulated or graphed to illustrate how the series changes with time. The graph of the unemployment rate in the United States in Figure 14.1 illustrates a typical time series. We might observe that the unemployment rate has fluctuated from a low of just under 5% to about 10% during the time period of the graph. The figure simply reports the unemployment changes over time—it does not explain the changes. However, some sources (for example, *Survey of Current Business*) do explain some of the underlying causes for changes such as this.

**FIGURE 14.1**   Unemployment Rate in the United States

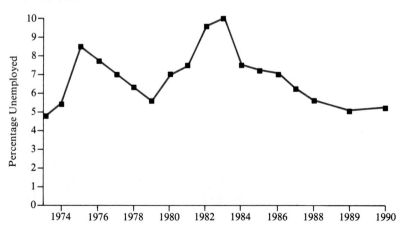

In an attempt to abstract more information from a time series, analysts often *decompose* it, or break it down into component parts. The classical approach to time series analysis results in a breakdown to the homogeneous and distinguishable components of *trend* (*T*), *cyclical* (*C*), *seasonal* (*S*), and *random* (*R*) *variation.*

> Trend is a pattern of movement that describes the long-term change in a time series. It is often characterized by a smooth graph.

> Seasonal variation describes a repetitive pattern that recurs within a year, quarter, or other short term. Seasonal variation usually follows a regular pattern and is often expressed as an index number.

> Cyclical variations (e.g., business cycles) are long-term patterns that are repetitive but can be irregular in size or duration or both. Cycles are affected by business and economic conditions and are more difficult to define than trend or seasonal effects.

> Random (or irregular) variations are changes not accounted for by trend, seasonal, or business cycle effects. These are the least predictable variations and relate to chance or unexplained events, so we refer to them as *error.*

Each of the components of a time series describes a different effect. Trend usually has the longest effect; random, the shortest. The better we can isolate these factors, the more useful the time series becomes for forecasting the future on the basis of the past.

The *classical model* for a time series links the components in a percentage relationship, in which the aggregate value of the data is a multiplicative function of the components.

> The classical model for a time series is

$$Y = T \cdot C \cdot S \cdot R \tag{14.1}$$

> where   $Y$ = actual series value
>
>   $T$ = trend value
>
>   $C$ = cyclical component
>
>   $S$ = seasonal variation
>
>   $R$ = random error or irregular variations

The actual value $Y$, is usually stated in terms of units, such as dollars, tons, or number of items. The trend value ($T$) is also in units, whereas $S$, $C$, and $R$ are most often indexes, or percentages.

The objective of time series analysis is to understand past action of a variable by isolating its components in a time-related manner. Organizations can make use of this understanding to forecast and prepare for future levels of business activity.

Analysts use the classical model of a time series to isolate the components by a method called *decomposition,* which refers to the breaking out of components from the actual values. A trend profile is first determined. Then the actual $Y$ is divided by the trend ($Y/T$) to leave the seasonal, cyclical, and random variations. Next, seasonals are determined, and finally cyclical variations are estimated. After the components are defined and estimated, they can be used to forecast a future series value. This application assumes that the random error is sufficiently small to not appreciably distort the forecast.

More advanced approaches to time series analysis use higher mathematics. For example, econometric models are more complex than the model described in Equation (14.1). Nevertheless, the basic components are the same—trend, seasonal, cyclical, and random variations. There is an essential difference, however. The econometric approaches also model the random component. Moreover, by incorporating a measure of random error, econometric models have some valuable statistical properties that extend beyond those of simple time series. The forecast can then include a point estimate plus a measure of the error that exists in the forecast value. The end result is better forecasts plus a measure of the reliability of the forecast.

**Problems 14.2**

1. Explain which component of the classical time series model is described by each of the variations (a) through (d).
   a. The long-term growth in income as a person approaches middle age
   b. The increased level in grocery retail sales that is experienced near Easter, July 4, Thanksgiving, and Christmas
   c. The variations in our economy that result in somewhat regular patterns of recession, depression, recovery, and economic boom
   d. The uncommon variations in your personal budget that can send you to the bank to correct an overdraft on your account

2. A business has experienced March 1992 sales revenues of $200 thousand. If this relates a trend value of $180 thousand, a cyclical value of 1.10, and an irregular variation of 1.00, estimate the seasonal index value.

3. Using a time line, explain the time frame difference between seasonal variations and cyclical variations.

4. The classical approach is one model for the changes described in time series data. Another approach would be an additive model like $Y = T + C + S + R +$ constant. Give one logical advantage for an additive model. Give one advantage for the classical model

$$Y = T \cdot C \cdot S \cdot R$$

**14.3
CLASSICAL TIME
SERIES: TREND BY
LEAST SQUARES
PROCEDURES**

We have described the classical time series model as including components for trend, seasonal, business cycle, and random variations. The next two sections discuss trend analysis. Section 14.5 covers seasonal variations and random variations. We then turn to an analysis of business cycles and the use of the classical model for forecasting.

> The trend of a time series is a long-term movement arising from basic factors that change only gradually over time.

*Trends* may be due to demographic factors, consumption of resources, growth of technologies, and social phenomena such as family patterns or health-care trends.

A plot of the data points is a useful starting point for analyzing a trend. The graph may take a multitude of shapes and forms. However, many economic

**TABLE 14.1** Data for Investment in Production Equipment

| Year | Investment ($ millions | Year | Investment ($ millions) |
|------|------------------------|------|-------------------------|
| 1982 | 5  | 1988 | 25 |
| 1983 | 7  | 1989 | 31 |
| 1984 | 10 | 1990 | 31 |
| 1985 | 20 | 1991 | 38 |
| 1986 | 15 | 1992 | 40 |
| 1987 | 13 |      |    |

series exhibit a gradually increasing trend, and a linear (first-order) representation of the trend is often satisfactory for business decisions. Nevertheless, a graph is useful to confirm that a linear trend line will provide a good fit to the data.

Several methods of fitting a trend line to a series of data are used in business and government. We shall discuss the (1) freehand, (2) least-squares, and (3) moving-average methods. The freehand and moving-average methods simply portray or describe the data, whereas the least-squares (regression) method describes the data *and* provides a trend equation.

Table 14.1 and Figure 14.2 depict investment data for a firm that produces castings for jet engines. We will use these data to illustrate the three methods of estimating a trend.

**Freehand Method**

A freehand trend line is a line drawn through the data points in an attempt to equalize the amount of deviation above and below the line. In Figure 14.2, one possibility would be to draw a straight line through the two points marked **X**. The freehand line is satisfactory for identifying general patterns, so it can be useful for a preliminary data analysis. However, the location of the trend line becomes subjective, and two people could very well choose different lines. The least squares techniques (used in regression) provide a more objective approach.

**FIGURE 14.2**  Investment in Production by Casting Supply Company, 1982–1992

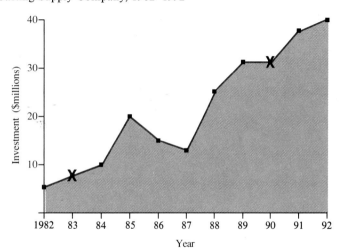

**Trend by Least Squares**

Many business and economic series satisfy the assumption of constant change. Then a simple *linear trend* can be adequately described by the linear regression model $Y = \alpha + \beta X_t + \varepsilon$. For this model, the estimated trend is $\hat{Y} = a + bX_t$, and the trend, $T$, is simply the estimated value $\hat{Y}$. Because the model assumes a dependency over time, $X_t$ represents a time variable. The intercept and slope coefficients are estimated by $a$ and $b$. We can determine these easily by using the same *normal equations* we use for simple linear regression.

Normal equations for least squares trend line:

$$\sum Y = na + b \sum X_t$$

$$\sum X_t Y = a \sum X_t + b \sum X_t^2$$

*Special situation:* If the data can be arranged so that $\sum X_t = 0$, then the solutions for the intercept (a) and the slope (b) values are reduced to

$$\sum Y = na \qquad \text{or} \quad a = \sum Y/n = \bar{Y} \qquad\qquad \text{(14.2)}$$

$$\sum X_t Y = b \sum X_t^2 \quad \text{or} \quad b = \frac{\sum X_t Y}{\sum X_t^2}$$

When working with time series data using hand calculations, we often code the time variable so that $\sum X_t = 0$. This results in easier solutions for $a$ and $b$ with no loss in generality. To code the data, simply designate the middle of the time period as the origin—for instance, *1987 = 0* (origin) in Figure 14.2. Then the equal number of plus and minus periods on each side results in coded values that sum to zero. Thus 1985 takes on a coded value of $-2$, 1986 becomes $-1$, 1987 is 0 (the origin), 1988 is assigned the value $+1$, and so on.

---

### EXAMPLE 14.1

The management of a firm that produces jet engine castings seeks a reasonable projection for their equipment investments in 1996. Use the data in Table 14.1 to develop a linear trend equation.

*Solution*

The data are displayed in Table 14.2. Totals are provided for estimating the model coefficients. For a linear trend the equation takes the form $\hat{Y} = a + bX_t$:

$$a = \frac{\sum Y}{n} = \frac{235}{11} = 21.3636 \qquad b = \frac{\sum X_t Y}{\sum X_t^2} = \frac{394}{110} = 3.5818$$

---

**TABLE 14.2**   Millions of Dollars Investment in Production and Capital Expansion for a Castings Firm, 1982–1992 with 1987 as Origin, 1987 = 0

| Year | 1982 | | 1984 | | 1986 | | 1988 | | 1990 | | 1992 | Total |
|---|---|---|---|---|---|---|---|---|---|---|---|---|
| $X_t$, coded years | −5 | −4 | −3 | −2 | −1 | 0 | 1 | 2 | 3 | 4 | 5 | 0 |
| $Y$, investment ($ millions) | 5 | 7 | 10 | 20 | 15 | 13 | 25 | 31 | 31 | 38 | 40 | 235 |
| $X_t^2$ | 25 | 16 | 9 | 4 | 1 | 0 | 1 | 4 | 9 | 16 | 25 | 110 |
| $X_t Y$ | −25 | −28 | −30 | −40 | −15 | 0 | 25 | 62 | 93 | 152 | 200 | 394 |
| $X_t^2 Y$ | 125 | 112 | 90 | 80 | 15 | 0 | 25 | 124 | 279 | 608 | 1000 | 2458 |

where the summations are taken from Table 14.2. Therefore, the forecasting equation is

$$\hat{Y} = 21.4 + 3.6X_t \quad (1987 = 0, \; X_t = \text{years}, \; Y = \$ \text{ millions})$$

The values in parentheses after the equation are vital to the equation and constitute its *signature*. The trend forecast for 1996, which is nine years after 1987, is

$$\hat{Y} = 21.4 + 3.6(9) = \$53.8 \text{ million}$$

Example 14.1 has illustrated the calculations for a linear trend by the least squares procedure. This can be extended to include an $X_t^2$ term for a quadratic trend.

**Quadratic Trend**

At this point it is useful for us to review some alternative models for describing the trend. In practice, some series, such as population growth and expenditures on technology, tend to follow nonlinear patterns as depicted in Figure 14.3(b), (c), and (d).*

We shall use the next example to illustrate fitting one nonlinear curve, the quadratic model as in Figure 14.3(b). Then, with some direction from the chapter references, you will be equipped to extend the processes to any of the models in Figure 14.3.

* See Linear and Curvilinear Equations in Appendix C for background information on quadratic and other nonlinear trend curves.

**FIGURE 14.3**   Forms of Trend Equations

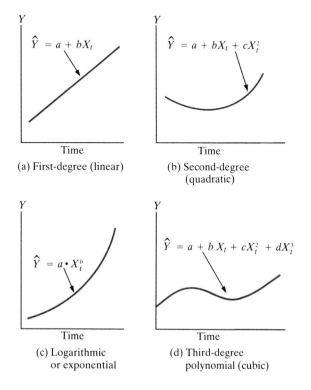

(a) First-degree (linear)   $\hat{Y} = a + bX_t$

(b) Second-degree (quadratic)   $\hat{Y} = a + bX_t + cX_t^2$

(c) Logarithmic or exponential   $\hat{Y} = a \cdot X_t^b$

(d) Third-degree polynomial (cubic)   $\hat{Y} = a + bX_t + cX_t^2 + dX_t^3$

The normal equations for an estimated quadratic polynomial model, or a second-degree curve, $\hat{Y} = a + bX_t + cX_t^2$ as in Figure 14.3(b), extends the least squares solution beyond a linear trend to

1. $\sum Y = \quad na + b \sum X_t + c \sum X_t^2$
2. $\sum X_t Y = a \sum X_t + b \sum X_t^2 + c \sum X_t^3$
3. $\sum X_t^2 Y = a \sum X_t^2 + b \sum X_t^3 + c \sum X_t^4$

These equations are one case of the normal equations for a general linear model as described in Chapter 13, Equations (13.3).

The solutions of these equations for $a$, $b$, and $c$ values are simplified if we use a coding such that $\sum X_t = 0$. This can extend to $\sum X_t^3 = 0$ if, for example, the times 1991, 1992, 1993 are coded to $-1, 0$, and $1$, respectively. Then $\sum X_t = 0$, and $\sum X_t^3 = (-1)^3 + 0^3 + (1)^3 = 0$. This "coded form" is commonly used in time series analysis.

Normal equations for a second-degree polynomial (quadratic) trend, where $\sum X_t = 0$ by coding, are

| Coded Form | With Solutions | |
|---|---|---|
| $\sum X_t Y = b \sum X_t^2$ | $b = \dfrac{\sum X_t Y}{\sum X_t^2}$ | (14.3) |
| $\left.\begin{array}{l} \sum Y = na \quad + c \sum X_t^2 \\ \sum X_t^2 Y = a \sum X_t^2 + c \sum X_t^4 \end{array}\right\} \longrightarrow$ by algebra | | |

*Note:* Coding that ensures $\sum X_t = 0$ also gives $\sum X_t^3 = 0$.

Odd powers ($X_t^5$, $X_t^7$, etc.) for higher-order polynomials would also sum to zero. The solutions for $a$, $b$, and $c$ in the Equations (14.3) require some algebraic calculations.

Second-degree polynomial graphs have one change in direction that evidences a nonconstant difference in the trend value. Economic data typically fit only a portion of the curve, but that is often sufficient. The procedure for fitting time data to a second-degree polynomial uses three normal equations rather than two. As with a first-degree model, coding so that $\sum X_t = 0$ simplifies the solutions.

We will not discuss procedures for trend analysis for the situations in Figures 14.3(c) and (d). See [11]* for a discussion. Example 14.2 and the following discussion give a good beginning.

**EXAMPLE 14.2**

Use the data of Table 14.2 (a) to develop a second-degree polynomial trend and from it a projection for 1996 and (b) to explain why the quadratic model gives at least as good a trend projection as does the linear least squares trend, which was discussed in Example 14.1.

* Bracketed numbers refer to items listed in "For Further Reading" at the end of the chapter.

*Solution* For a second-degree polynomial trend, the equation takes the form $\hat{Y} = a + bX_t + cX_t^2$ with signature (1987 = 0, $X_t$ = years, $Y$ = \$ millions). Estimating the equation coefficients uses Equation (14.3) and data from Table 14.2. Dividing the $\sum X_t Y$ value by the $\sum X_t^2$, we have

$$b = \frac{\sum X_t Y}{\sum X_t^2} = \frac{394}{110} = 3.5818$$

Solving the following two equations simultaneously uses

$$\sum Y = na + c \sum X_t^2 \qquad\qquad 235 = 11a + 110c$$

$$\sum X_t^2 Y = a \sum X_t^2 + c \sum X_t^4 \qquad 2458 = 110a + 1958c$$

This gives $a = 20.1049$ and $c = -0.1259$. This trend equation is

$$\hat{Y} = 20.1049 + 3.5818X_t - 0.1259X_t^2$$
$$(1987 = 0, X_t = \text{years}, Y = \$ \text{ millions})$$

Then, setting $X_t = 9$ and $X_t^2 = 81$ gives a trend forecast for 1996 of $\hat{Y} = \$62.5$ million.

By least squares regression theory the second-degree polynomial trend equation explains *at least as much* of the variation in dollar investments as does the linear model, which yielded \$53.8 million. Hypotheses tests can determine whether the improvement is statistically significant. The models differ in their trend projection for 1996 investment by \$8.7 million. Figure 14.4 shows the data plot, along with the estimated quadratic trend.

---

**FIGURE 14.4** Investment Amounts with a Second-Degree Polynomial Trend

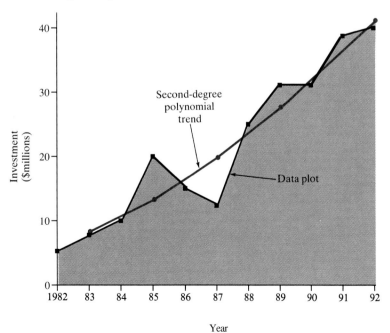

Example 14.2 illustrated the mechanics for estimating a trend by a coded least squares regression model. The resulting equation was "centered" at the middle of the year (July 1, 1987) because there was an odd number of years. When there is an even number of years, the coded $X$-values are . . . , $-1.5$, $-0.5$, $0.5$, $1.5$, $1.5$, $2.5$, . . . . The resulting equation is then centered at the median position, which is the year-end (or year-beginning).

Testing for a "best" polynomial trend model requires use of a procedure developed in Chapter 13. That procedure involves a coding process, where $X_t = X_1$, $X_t^2 = X_2$, $X_t^3 = X_3$, and so on, followed by testing via the extra sum of squares principle. We shall not take up that procedure here. However, we should briefly consider how to shift the origin of a series.

### Shifting the Origin of a Series

Sometimes it is desirable to update a time series by shifting the origin to a later time period. We can accomplish this by adding or subtracting the desired number of periods in the shift from $X_t$ in the original equation.

---

**EXAMPLE 14.3**

Using the following time series equation, shift the origin to 1993.

$$\hat{Y} = 400 + 60X_t \quad (1985 = 0, \ X_t = \text{years}, \ Y = \text{units})$$

*Solution*    The new origin is eight periods forward in time $(+8)$ and the slope is positive. We simply add eight yearly amounts of 60 units to the original intercept and restate the signature to identify the revised origin.

$$\hat{Y} = 400 + 60(X_t + 8) = 400 + 60X_t + 480$$

$$\hat{Y} = 880 + 60X_t \quad (1993 = 0, \ X_t = \text{years}, \ Y = \text{units})$$

Had we wanted to move the origin back in time to 1983, our equation would have been $\hat{Y} = 400 + 60(X_t - 2)$.

---

Modifying trend equations is not restricted to shifting the base period. If we want to express the $X_t$ value in months instead of years, we simply divide the coefficient $X_t$ by 12. This does not affect the trend $(Y)$ units, but the signature should reflect this change by expressing the origin in terms of months rather than years. Example 14.4 illustrates this procedure.

---

**EXAMPLE 14.4**

Using the trend equation shown, express the $X_t$-value in months rather than years. Then shift the origin to mid-September of 1993.

$$\hat{Y} = 880 + 60X_t \quad (\text{July 1, 1993} = 0, \ X_t = \text{years}, \ Y = \text{units})$$

*Solution*    
$$\hat{Y} = 880 + (60/12)X_t = 880 + 5X_t$$
$$(\text{July 1, 1993} = 0, \ X_t = \text{months}, \ Y = \text{units})$$

Shifting the origin 2.5 months (to mid-September), we have

$$\hat{Y} = 880 + 5(X_t + 2.5) = 880 + 5X_t + 12.5$$

$$\hat{Y} = 892.5 + 5X_t \quad (\text{Sept. 15, 1993} = 0, \ X_t = \text{months}, \ Y = \text{units})$$

Trend can explain much of the variation in time series values. But other components, including the business cycle, can be equally important. Our next step in developing a time series forecast considers both trend and cyclical variations.

**Problems 14.3**

5. Over the years, a food-processing firm had sales (in millions of dollars) as shown in the data table.

| Year | Sales | Year | Sales | Year | Sales |
|------|-------|------|-------|------|-------|
| 1982 | $5    | 1986 | $6    | 1990 | $9    |
| 1983 | 3     | 1987 | 7     | 1991 | 13    |
| 1984 | 5     | 1988 | 10    | 1992 | 10    |
| 1985 | 7     | 1989 | 9     |      |       |

   a. Plot the data as a time series and indicate whether a linear fit to the data is appropriate.

   b. Draw a freehand trend line through the data points.

   c. Use your freehand trend to project a value for 1994.

6. a. Use the least squares method to develop a linear trend equation for the data in Problem 5. State an equation complete with signature.

   b. Use your equation to predict a value for 1993. Compare your projection to the actual value, that is 1993 = 8.

7. Data on the gross national product for a country—in billions of (U.S.) dollars—is shown in the table.

| Year | GNP    | Year | GNP    | Year | GNP    |
|------|--------|------|--------|------|--------|
| 1982 | 796.3  | 1986 | 1063.4 | 1990 | 1528.8 |
| 1983 | 868.5  | 1987 | 1171.1 | 1991 | 1706.5 |
| 1984 | 935.5  | 1988 | 1306.6 | 1992 | 1889.6 |
| 1985 | 982.4  | 1989 | 1412.9 |      |        |

   a. Plot the data and fit a freehand trend line to it.

   b. Use the least squares method to develop a linear trend equation.

   c. Based on the trend equation from (b), forecast a value of the GNP for the year 2007.

8. Shift the following equation to an origin of 1993 = 0.

$$\hat{Y} = 1420 + 23X_t \quad (1986 = 0, \ X_t = \text{years}, \ Y = \text{tons})$$

9. Given the time series equation

$$\hat{Y} = 480 + 24X_t \quad (1993 = 0, \ X_t = \text{years}, \ Y = \text{motors/year})$$

rewrite the equation so the signature is

$$(\text{January } 1993 = 0, \ X_t = \text{months}, \ Y = \text{motors/year})$$

10. A time series equation is of the form

$$\hat{Y} = 520 + 36X_t \quad (1988 = 0, \ X_t = \text{years}, \ Y = \text{cartons/year})$$

where $Y$ represents cartons shipped annually.
    a. Shift the origin to 1993 and restate the equation.
    b. Using the original equation, change the time scale of the $X_t$-variable to months and restate the equation.
    c. Using the equation derived in part (b), forecast the rate (trend value/year) of shipments in September 1995.

11. Your company is a supplier of food packaging materials and your supervisor has asked you to attempt to develop a forecasting equation for food consumption in a country that is a major market. You have obtained the data for food consumption (in billions of dollars). Use the normal equations to fit a second-degree (quadratic) curve to the data and calculate trend values for 1983–1992.

| Year | $ | Year | $ | Year | $ |
|------|-------|------|-------|------|-------|
| 1982 | 109.6 | 1986 | 140.6 | 1990 | 209.5 |
| 1983 | 118.3 | 1987 | 150.4 | 1991 | 225.5 |
| 1984 | 126.1 | 1988 | 168.1 | 1992 | 246.2 |
| 1985 | 136.3 | 1989 | 189.8 | | |

## 14.4 CLASSICAL TIME SERIES: TREND BY MOVING AVERAGES METHOD

Moving averages is a procedure for trend and cyclical analysis. With moving averages, we smooth out the irregular or random components. Seasonal variations are of short duration, often 1-year, so a 12-month moving average will smooth out the seasonal effects too. The remaining variations then describe the trend and business cycles over time. This section describes the procedures for using both simple and weighted moving averages to help isolate trend and cyclical effects.

We obtain *moving averages* by repetitively summing and averaging the data values from a fixed number of periods. Each time, the oldest data value is deleted and a new value is added.

A moving average is an average moving through time. The moving average (MA) values are computed by

$$\text{MA}_t = \frac{\sum X_t}{N} \tag{14.4}$$

$$= \text{MA}_{t-1} + \frac{\text{Last value in} - \text{First value out}}{N}$$

where the average is calculated for a number, $N$, of adjacent time periods.

The moving total is adjusted each period by the difference between the last value in and the first value out.

The number of periods used in the average is important to the analyst. However, the average is centered in the middle of the data, so there is no average value available for the endpoints. Averages made up of only two or three periods lose fewer end-time points and follow the actual data more closely but with less smoothing. Moving averages with more periods, from five to ten, tend to have a very significant smoothing effect, filtering more of the seasonal and random factors. We will generally use a moderate number, three to seven periods, for a moving average.

## EXAMPLE 14.5

Compute a three-year moving average for the corporate investment data of Table 14.1 and plot it as a solid graph. Follow this with a dotted graph for a five-year moving average plot on your graph. Use both the three-year and the five-year moving averages procedure to make a forecast for investments in 1993.

*Solution*   The data calculations for both (three-year and five-year) moving averages are described in a table.

| Year | Investment ($ millions) | Three-Year Moving Total | Three-Year Moving Average | Five-Year Moving Average |
|------|------------|------------|------------|------------|
| 1982 | 5 | | | |
| 1983 | 7 | 22 | ÷ 3 =   7.3 | |
| 1984 | 10 | 37 | ÷ 3 =  12.3 | 11.4 |
| 1985 | 20 | 45 | 15.0 | 13.0 |
| 1986 | 15 | 48 | 16.0 | 16.3 |
| 1987 | 13 | 53 | 17.7 | 20.8 |
| 1988 | 25 | 69 | 23.0 | 23.0 |
| 1989 | 31 | 87 | 29.0 | 27.6 |
| 1990 | 31 | 100 | 33.3 | (33.0) |
| 1991 | 38 | 109 | (36.3) | |
| 1992 | 40 | | | |

The calculations summarized in the table are straightforward, *but* a last-value-in, first-value-out process can simplify the arithmetic. For example, the three-year moving average for 1984 uses the 1983 moving average, $7.3 + [(20 - 5)/3] = 12.3$. The plots appear in Figure 14.5. The figure includes a data plot as well as the three-year and five-year moving average trend estimates.

The computed value of the moving average is recorded in the center position of the data it summarizes. Thus the three-year moving average value of 7.3 for 1982, 1983, and 1984 is centered in the middle (July 1) of 1983. Observe the greater smoothing effect of the longer-period five-year average.

The three-year moving average forecast for the 1993 trend-cycle of $36.3 million is circled, as is the five-year moving average forecast of $33.0 million. Much of both graphs resides in the shaded area of Figure 14.5. This suggests that both these estimates are probably too low. This bias occurs when a strictly increasing (or decreasing) trend is present, as appears to be the case here.

Example 14.5 demonstrates that the longer the time period being averaged, the farther the computed data points are from the ends of the time scale. If there is a trend, moving averages tend to lag the latest real data values. So

**FIGURE 14.5** Corporate Investment with a Three-Year Moving Average Trend

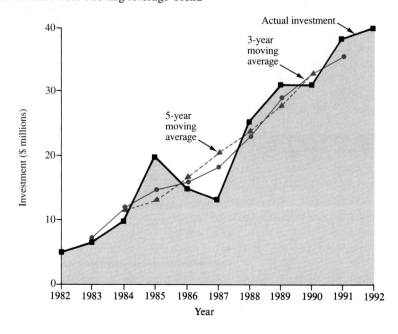

the longer the period for averaging, the longer, too, the lag. With a steadily increasing (or decreasing) moving average, longer-term averages will regularly underestimate (or overestimate) the actual values. The advantage of a shorter period average is in having a value that is closer to the current time.

**Weighted Moving Average**

The *weighted moving average* (MA) is one means of adjusting for lag by placing more weight upon the most recent values.

The weighted moving averages smoothing form is

$$MA = \frac{\sum [(wt) \cdot X_t]}{\sum (wt)} \tag{14.5}$$

Weights can be anything from arbitrarily selected percentages to real numbers based on experience.

---

**EXAMPLE 14.6**

Use a weighted moving average to develop a forecast value for 1993 from the data of Example 14.5. Let the most recent period be weighted 5, the next most recent 3, and the most distant period, 1. Compare the result with that using the simple moving average.

*Solution*

$$MA = \frac{\sum [wt \cdot X_t]}{\sum (wt)} = \frac{(5)(40) + 3(38) + 1(31)}{5 + 3 + 1} = \$38.3 \text{ million}$$

This value 38.3 is higher than the simple moving average value 36.3. The weighting compensates for some of the bias due to lag.

---

A moving average helps identify trend because the average smooths out the effects of short-term variations, that is, the seasonal and random effects.

Similarly, the moving average can help to identify cyclical effects. Growth periods show consecutive points above the smoothed average, whereas recession periods fall below the smoothed average.

A moving average can be used to depict any data—no mathematical curve or equation is involved. Its flexibility and ease of use are strong advantages. Furthermore, the averaging process results in a smoothing out of extremes and incorporates past performance in projecting to the future.

The absence of a forecasting equation is, however, a prime disadvantage of the moving average method. In lieu of an equation, the best prediction of a future period's value is simply the last average value. This may be a better estimate of the future than an individual value would be, but it is still only an average of past data.

The weighted moving average has an advantage over the simple moving average in that it allows greater weight for those values that one chooses to emphasize in time. But the choice of weights is subjective. (This shortcoming can be partially overcome by exponential smoothing, discussed in Section 14.6.) As with a simple moving average, the weighted moving average procedure does not produce an explicit trend curve or equation. So forecasts are essentially limited to one value beyond the last observed time value.

**Problems 14.4**

12. a. Compute a three-year moving average for the data in Problem 5.
    b. Use the three-year moving average to predict a value for 1997.
    c. Use a weighted three-year moving average to predict a value for 1997. Let the weights be 3, 2, and 1, so you give the most weight to the most recent values.

13. Continue using the data from Problem 5.
    a. Do you feel the moving average provides a good estimate of future sales?
    b. Suppose the next six years' sales were 1993 = 8, 1994 = 6, 1995 = 9, 1996 = 9, 1997 = 11. Compared to (a), would you feel more confident using the moving average that covers 1982 through 1997? Why?
    c. What would be the benefit of using a seven-year moving average rather than a three-year moving average?

14. The data in Problem 7 are one country's gross national product, spanning the years 1982 through 1992.
    a. Use a three-year moving average to predict a value for 1993.
    b. Compared to actual, do you expect this value to be low, high, or quite close? Explain your choice.

**14.5
CLASSICAL TIME
SERIES: SEASONAL
VARIATION,
CYCLES, AND
RANDOM
VARIATION**

Time series often exhibit similar patterns of variation at regular short-term intervals. For example, energy demands have well-identified seasonal peaks as a result of heating and cooling loads. Department store sales show a *monthly* pattern, which typically peaks in December. Airlines feature special weekend excursion packages to help level the *weekly* demand. The load on a subway system cycles *daily*, with heaviest demand during rush hours. In each case there is a recurring pattern over time. When the time length is a year or less, we refer to the pattern as a *seasonal variation*.

> A seasonal variation is any regularly recurring pattern where the repetitive time is one year or less.

Many seasonal variations are due to the weather, which affects agriculture, employment, education, and recreation. Other variations stem from weekly and daily occurrences such as regular work-hours. These seasonal variations are all analyzed in the same manner.

Why should a businessperson be interested in a knowledge of seasonal fluctuations? The manager who does a better job of predicting demand cycles can do a better job of scheduling production, managing inventories, anticipating financial needs, minimizing employment fluctuations, and the like. In government work, knowledge of seasonal patterns helps one draw valid conclusions about real growth in the economy, unemployment levels, and changes in consumer prices. For example, an increase in total employment from May to June does not mean the employment picture is rosy. When we take into account seasonal effect of school-age workers entering the labor force, we may find the seasonally adjusted picture to in fact be worse. An understanding of how to derive and use seasonal patterns, or *seasonal indexes*, is important in business and government.

## The Seasonal Index

A seasonal index (SI) is a number describing a seasonal pattern.

A *seasonal index* tells what proportion of the trend value an actual value is. Thus, if December sales are 120% of a trend amount and July sales are only 85%, the seasonal indexes for the two months would be 120 and 85, respectively. We shall illustrate the mechanics of computing seasonal indexes in a moment. But first let us reemphasize the relationship between actual, trend, and seasonal index values.

Suppose you are producing camping equipment and already have a well-established seasonal index—say, from previous work or government statistics. If you could forecast an average or trend value for a month, you could estimate the actual ($Y$) by multiplying the trend value by the seasonal index.

$$\text{Estimated actual} = (\text{Trend}) \cdot (\text{Seasonal index})$$

$$\hat{Y} = T \cdot (\text{SI}) \tag{14.6}$$

---

**EXAMPLE 14.7**

A recreational products firm has used a least squares trend line to project their demand for sailboards for next year. Their trend estimate is 144,000 units for the year, following a highly seasonal pattern in which the April index is 80 and the May index is 170. Estimate the actual demand for April and May.

*Solution*   We shall assume that no adjustment for a cycle effect is necessary.

$$\hat{Y} = T \cdot (\text{SI}) \quad \text{where } T = \frac{144{,}000 \text{ units}}{12 \text{ months}} = 12{,}000 \text{ units/month}$$

April $\hat{Y} = (12{,}000)(80\%) = 9{,}600$ units

May $\hat{Y} = (12{,}000)(170\%) = 20{,}400$ units

---

By simple algebraic manipulation of Equation (14.6), we can derive an expression for a *deseasonalized*, or trend, value given an actual ($Y$) and a seasonal index (SI). Here, SI denotes a seasonal adjustment indexed to a base value and then multiplied by 100.

$$T = \frac{Y \times 100}{\text{SI}} \qquad (14.7)$$

A tremendous amount of government data on employment, new construction, earnings, industrial production, and so on is reported as deseasonalized, or *seasonally adjusted*, data. This means the actual data have been divided by a seasonal index so that seasonal effects have already been compensated for. Then, if the reported (trend) value shows an increase, we know it is a true increase and not simply due to seasonal factors. For example, if there is a (seasonally adjusted) increase in new construction, we know the increase is genuine and is not due only to seasonal factors.

**Ratio-to-Moving-Averages Method**

This brings us to the computation of a seasonal index. In the algebra of decomposition of a time series, the seasonal index results from canceling all the effects except the seasonal ($S$). In algebraic terms this could mean dividing the actual value ($Y = TCSR$) by $TC$ and smoothing out any $R$. If the time period used to compute the $TCR$ is of proper length, 12 months for annual data, the seasonal effect will automatically be negated in the $TCR$ components. The resulting seasonal index is then much like the ratio of an actual value ($Y$) to a moving average of $TCR$. Thus, the seasonal effect is approximated by

$$\text{SI} = \frac{TCRS}{TCR} \times 100 = \frac{Y \times 100}{\text{Moving average}}$$

Assuming further that the moving average does an adequate job of damping out random effects ($R$) and is not significantly influenced by cyclical factors ($C$), we can depict the seasonal index as

$$\text{SI} \cong \frac{Y}{T}(100) = \frac{Y}{\text{MA}}(100) \qquad (14.8)$$

Our computation of a seasonal index will utilize this relationship of actual to trend values. The actual will be an average of several actuals (e.g., several January values), and the trend will be a moving average (MA) centered on the particular period. The result of each individual ratio ($Y/\text{MA}$) is a *specific seasonal* for a particular period (month and year). The specific seasonals are then averaged to obtain a truly representative seasonal index. This technique is called the *ratio-to-moving-averages method* of computing a seasonal index.

> The ratio-to-moving-averages method is a technique for computing a seasonal index that relies upon comparing actual values to the average of surrounding values centered upon each specific period.
>
> A procedure for computing a seasonal index by *ratio-to-moving-averages method*:
> a. Compute a moving average, MA.
> b. Compare the actual to the moving average to obtain a specific seasonal, $Y/\text{MA}$.
> c. Advance one period and repeat.
> d. Average the specific seasonals for like months (or like quarters) to obtain that month's (or quarter's) seasonal index. Multiply by 100 to express as an index number.

**FIGURE 14.6**    Ratio-to-Moving-Averages Illustrated for Month *C*

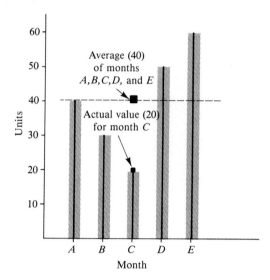

The key relationship is that of the individual monthly value to the average centered upon that month. In Figure 14.6, the actual for month *C* is 20 units, whereas the five-month moving average centered on month *C* has a value of 40. The specific seasonal for month *C* is then the actual (*Y*) divided by the moving average value, $[(20/40) \cdot 100]$, or 50%, reported as 50. The specific seasonals for all *C* months (over multiple years) would then have to be averaged for a month *C* seasonal index.

Having an odd number of periods, such as the five in Figure 14.6, simplifies computations because the moving average is automatically centered directly over one of the periods—period *C* in Figure 14.6. However, for annual data, a 5-month moving average will not suffice. A 12-month moving average works well for annual data because each month is then represented once. This smooths the series and dampens out all seasonal variation. It has a problem, however, in that a 12-month average is centered between months 6 and 7 rather than in the middle of a month. We usually rectify this by averaging two consecutive moving total figures. Thus, if one set of 12 figures is centered on July 1 and the next set on August 1, the center of the two sets will be July 15. Of course, the total must now be divided by 24 instead of by 12 because 24 items of data are now included. The method itself is equivalent to using a 13-period moving average and weighting the two end-periods at 1 each and all the others at 2 each.

We will illustrate the computation of a seasonal index using quarterly rather than monthly data in order to reduce the number of computations by working with 4 quarters rather than 12 months. The methodology is the same.

**EXAMPLE 14.8**

Quarterly sales revenues (in $000) for the Amtex Company are as shown in Table 14.3. Use the ratio-to-moving-average method to develop a seasonal index.

*Solution*    Table 14.4 displays the calculations by the ratio-to-moving-averages procedure. We first calculate a centered moving average by adding two adjacent four-quarter totals (from column (3)) and dividing by 8 (column (5)). Then we find the specific seasonals (column (6)) by dividing actuals, *Y* (column (2)) by the moving average (column (5)).

**TABLE 14.3**   Quarterly Sales of Amtex Company ($000)

| Quarter | 1986 | 1987 | 1988 | 1989 | 1990 | 1991 | 1992 |
|---------|------|------|------|------|------|------|------|
| 1 | 640 | 660 | 690 | 730 | 800 | 850 | 880 |
| 2 | 620 | 650 | 620 | 720 | 800 | 820 | 880 |
| 3 | 640 | 670 | 700 | 760 | 820 | 860 | 900 |
| 4 | 680 | 720 | 740 | 820 | 900 | 890 | 910 |

**TABLE 14.4**   Calculation of Centered Moving Averages and Specific Seasonals

| (1) Quarter | (2) Sales $Y$ (units) | (3) Four-Quarter Moving Total | (4) Eight-Quarter Moving Total | (5) Col (4) ÷ 8 = Centered Moving Average (MA) | (6) Specific Seasonal $(Y \times 100)/\text{MA}$ |
|---|---|---|---|---|---|
| 1986  1 | 640 | | — | | |
| 2 | 620 | | — | | |
| | | 2580 | | | |
| 3 | 640 | | 5180 | 647.50 | 98.8 |
| | | 2600 | | | |
| 4 | 680 | | 5230 | 653.75 | 104.0 |
| | | 2630 | | | |
| 1987  1 | 660 | | 5290 | 661.25 | 99.8 |
| | | 2660 | | | |
| 2 | 650 | | 5360 | 670.00 | 97.0 |
| | | 2700 | | | |
| 3 | 670 | | 5430 | 678.75 | 98.7 |
| | | 2730 | | | |
| 4 | 720 | | 5430 | 678.75 | 106.1 |
| | | 2700 | | | |
| 1988  1 | 690 | | 5430 | 678.75 | 101.7 |
| | | 2730 | | | |
| 2 | 620 | | 5480 | 685.00 | 90.5 |
| | | 2750 | | | |
| 3 | 700 | | 5540 | 692.50 | 101.1 |
| | | 2790 | | | |
| 4 | 740 | | 5680 | 710.00 | 104.2 |
| | | 2890 | | | |
| 1989  1 | 730 | | 5840 | 730.00 | 100.0 |
| | | 2950 | | | |
| 2 | 720 | | 5930 | 747.50 | 96.3 |
| | | 3030 | | | |
| 3 | 760 | | 6130 | 766.25 | 99.2 |
| | | 3100 | | | |
| 4 | 820 | | 6280 | 785.00 | 104.5 |
| | | 3180 | | | |
| 1990  1 | 800 | | 6420 | 802.50 | 99.7 |
| | | 3240 | | | |
| 2 | 800 | | 6560 | 820.00 | 97.6 |
| | | 3320 | | | |
| 3 | 820 | | 6690 | 836.25 | 98.1 |
| | | 3370 | | | |
| 4 | 900 | | 6760 | 845.00 | 106.1 |
| | | 3390 | | | |
| 1991  1 | 850 | | 6820 | 852.50 | 99.7 |
| | | 3430 | | | |
| 2 | 820 | | 6850 | 856.25 | 95.8 |
| | | 3420 | | | |
| 3 | 860 | | 6870 | 858.75 | 100.1 |
| | | 3450 | | | |
| 4 | 890 | | 6960 | 870.00 | 102.3 |
| | | 3510 | | | |
| 1992  1 | 880 | | 7060 | 882.50 | 99.7 |
| | | 3550 | | | |
| 2 | 880 | | 7120 | 890.00 | 98.9 |
| | | 3570 | | | |
| 3 | 900 | | — | | |
| 4 | 910 | | | | |

**TABLE 14.5** Calculation of Seasonal Index Using a Modified Mean

| Year | Quarter | | | |
|------|---------|--------|-------|--------|
| | *First* | *Second* | *Third* | *Fourth* |
| 1986 | — | — | 98.8 | 104.0 |
| 1987 | 99.8 | 97.0 | 98.7 | 106.1 |
| 1988 | H 101.7 | L 90.5 | H 101.1 | 104.2 |
| 1989 | 100.0 | 96.3 | 99.2 | 104.5 |
| 1990 | 99.7 | 97.6 | L 98.1 | H 106.6 |
| 1991 | L 99.7 | 95.8 | 100.1 | L 102.3 |
| 1992 | 99.7 | H 98.9 | — | — |
| Modified total | 399.2 | 386.7 | 396.8 | 418.8 |
| Modified mean | 99.8 | 96.7 | 99.2 | 104.7 |

Total of modified means = 400.4

Correction factor $\dfrac{400}{400.4} = 0.999$

| Seasonal Indexes | 99.7 | 96.6 | 99.1 | 104.6 |
|------|------|------|------|-------|

Next, in Table 14.5, a representative value of the specific seasonal is selected for each quarter. It is common practice to use a *modified mean* here in order to eliminate the effect of random or irregular variations. This is accomplished by deleting the highest (H) and lowest (L) specific seasonals for each quarter. The modifed mean of the remaining figures is then computed and corrected to ensure the four quarterly indexes equal 400. The correction factor is defined in (14.9):

$$\text{SI correction factor} = \frac{\text{Expected sum of means}}{\text{Actual sum of means}} \qquad \textbf{(14.9)}$$

The correction factor is then multiplied by each modified mean. Since the actual sample total mean is greater than 400, the correction factor is less than 1. If the actual total of modified means were less than 400, the correction would be greater than 1.

Our interpretation of the seasonal index is that the first-quarter sales are quite close (99.7%) to trend values. Second-quarter sales are typically low (96.6%), and fourth-quarter sales average 104.6% of trend. Third-quarter sales follow much the same pattern as the first quarter.

The seasonal pattern of an economic series sometimes undergoes a gradual change over time as technologies develop, demand shifts, resources are consumed, and so on. For example, the increasing use of electrical energy for air conditioning has generated a summer peak that was nonexistent many years ago. Special techniques are available for developing seasonal indexes under conditions of changing seasonal patterns (see [15]). However, one of the most important points for an analyst to remember is to update the seasonal index frequently so that it makes use of the most recent data available.

**Including Business-Cycle and Random Variations**

Our discussion thus far has concerned trends and the ratio-to-moving-average method of deriving a seasonal index. The remaining components of the classical model of a time series, cyclical and irregular movements, are much more difficult to isolate and predict. Let us review what we have done thus far.

We began with the classical model

$$Y = T \cdot S \cdot C \cdot R$$

Our first step in the decomposition of actual data from a time series was to isolate the trend. Next we chose one specific estimate of trend (a moving average) and divided it into the actual $Y$-value multiplied by 100 to obtain specific seasonal indexes. In essence we have isolated a "seasonal effect." By dropping the highest and lowest specific seasonals, we were attempting to eliminate some of the irregular or random effects. Another method of doing this would have been to use the median of the specific seasonals rather than a modified mean. We should realize, however, that random effects are, by definition, unpredictable. We cannot forecast them—the best we can do is *allow* for them.

Thus far, then, we have developed an estimate of the trend of a series and a seasonal index that is at least partially free of business-cycle and irregular values. Given some historical data, we project a trend value and multiply it by the SI; we then have a forecast that takes some account of everything except cyclical factors.

*Cyclical variation* in a time series is a recurring wavelike pattern.

**Cyclical variation (e.g., business cycles) is the long-term (multiple-year) repetitive pattern that can be irregular in duration and in the range of high-to-low time series values.**

The classical example is the business cycle or state of the economy—prosperity, recession, depression, or recovery. The time period of a cycle is usually not fixed. And the amplitude (size) can vary greatly because a business cycle describes the economic conditions at a point in time, and the conditions change over time. This irregularity makes business cycles difficult to describe and even more difficult to predict.

Cyclical factors can sometimes be isolated in a quantitative fashion, but the data are typically much more difficult to obtain and interpret than are seasonal data. For this reason, business forecasters often use only rough adjustments for cyclical effects. For example, if a company's business is closely tied to housing construction, and if economic indicators on housing starts suggest a 15% drop in that activity over the next year, then the company may want to apply a 0.85 multiplier to its forecast of sales to that industry.

Business cycles tend to be difficult to predict, both in timing and in magnitude of fluctuation. It can be difficult to separate the trend and seasonal effects from the business cycle. Yet when the effects are singled out one at a time, it may be possible to quantify each component. When valid data exist, they can be used by a skilled analyst for a decomposition to develop forecasts. An example shows a forecast for Amtex Company sales.

**EXAMPLE 14.9**

Using the Amtex data of Table 14.3, develop a forecast for fourth-quarter 1993 sales by the decomposition of the classical model, $Y = T \cdot C \cdot S \cdot R$. Assume that random variations will effectively cancel out and so can be ignored. Also, leading economists foresee an improving economy. This suggests a cyclical component of 1.10 for this industry.

*Solution*   Refer to the data in Tables 14.1 through 14.5 for this decomposition. First we should confirm that a linear trend will adequately represent the data. The fourth-period centered moving average tabulated in column (5) of Table 14.4 suggests a steadily increasing trend (and a graph would support it). So we elect to use a least-squares analysis to describe the trend. Coding the data around an origin of July 1, 1989, we have coded $X$-values of $-0.5, -1.5, -2.5, \ldots$ back in time and $0.5, 1.5, 2.5, \ldots$ ahead. For the $n = 28$ data points, $\sum Y = 21{,}370$, so

$$a = (\sum Y)/n = 763.2143$$

$$b = (\sum X_t Y)/\sum X_t^2 = 20{,}655/1827 = 11.3054$$

The trend equation is thus

$$\hat{Y} = 763.21 + 11.31 X_t \quad \text{(July 1, 1989} = 0, \; X_t = \text{quarters}, \; Y = \text{units)}$$

Using this equation, we can now forecast the trend value for the fourth quarter, 1993, the midpoint of which is 17.5 quarters away from July 1, 1989. The trend forecast is as follows:

$$\hat{Y} = 763.21 + 11.31(17.5) = 961.14 \text{ units}$$

Now we must modify the trend forecast for the seasonal and cyclical effects. From Table 14.5 we have the fourth-quarter seasonal index of 104.6 for this data. Also, the cyclical effect was estimated to raise values to 110% of what they would otherwise be. A forecast can now be made.

$$\text{Seasonal adjustment to trend} = \hat{Y}_{sz} = \hat{Y} \cdot (S)$$
$$= 961.14(1.046) = 1005.35 \text{ units}$$

$$\text{Cyclical adjustment to trend} = \hat{Y}_{cl} = \hat{Y}_{sz} \cdot (\text{cyclical value})$$
$$= 1005.35(1.10) = 1105.88 \text{ units}$$

The projected sales for the fourth-quarter of 1993 are 1106 units.

---

Example 14.9 has incorporated the (4) components for a time series analysis and (point) forecast by classical decomposition. Next, this process is summarized to a general forecasting procedure.

**Summary for Forecasting by Classical Decomposition**

In summary, the procedure for forecasting by the classical decomposition method is to first obtain a trend value and then apply the seasonal index and cyclical relative to incorporate business cycle and seasonal variation.

### PROCEDURE FOR FORECASTING BY CLASSICAL-MODEL DECOMPOSITION

1. Calculate a trend value for the specified time.
2. Obtain a seasonal index adjustment value. Multiply the trend value by the seasonal adjustment.
3. Multiply the seasonally adjusted trend value (Step 2 result) by a cyclical adjustment value. This result is the point forecast.

This assumes that random variations are not significant and can be ignored. So a component for the random error is not included in the procedures, except as was used to compute the seasonal index.

The techniques of classical time series, including description of trend, seasonal, and cyclical variations, can help business analysts to understand the past action of a series. Visual pictures from time data plots can provide additional information for business planning. However, the classical approach falls short of modern business forecasts. The next section briefly describes some techniques that include a measure of the goodness of the forecast. This allows a statistical assessment of the error in the forecast, so it provides the decision maker with a measure of confidence in the forecast. This can lead to better business planning with the aid of time series data.

**Problems 14.5**

15. Blue Mountain Wine Company has forecast the production of 9600 gal of wine in 1995. Quarterly seasonal indexes are as follows:

| First quarter | 40 | Third quarter | 180 |
|---|---|---|---|
| Second quarter | 70 | Fourth quarter | 110 |

a. What is the trend value of production for the third quarter?
b. Estimate the actual production during the second quarter of 1995.

 16. Quarterly sales data for Silicon Valley Products Company of San Jose are as shown below (in ten thousands of units). Use the ratio-to-moving-average method and compute seasonal indexes for each quarter. (*Note:* Use a modified arithmetic mean method and eliminate the highest and lowest values of the specific seasonals for each quarter.)

| Quarter | Year 1 | Year 2 | Year 3 | Year 4 | Year 5 |
|---|---|---|---|---|---|
| 1 | 10 | 20 | 30 | 60 | 120 |
| 2 | 50 | 60 | 80 | 110 | 200 |
| 3 | 30 | 50 | 70 | 120 | 180 |
| 4 | 30 | 40 | 70 | 150 | 180 |

17. The table below gives *specific seasonals* for sales of a furniture manufacturer in North Carolina. Compute the quarterly seasonal index using the ratio-to-moving-average method by using the median value of each quarter (rather than a modified mean).

| | Specific Seasonal | | | |
|---|---|---|---|---|
| Year | Q1 | Q2 | Q3 | Q4 |
| 1986 | 98.0 | 137.3 | 85.7 | 78.7 |
| 1987 | 89.3 | 142.9 | 88.5 | 89.2 |
| 1988 | 92.3 | 144.2 | 76.4 | 84.6 |
| 1989 | 102.2 | 130.6 | 77.5 | 79.5 |
| 1990 | 98.2 | 135.6 | 74.6 | 88.0 |
| 1991 | 93.1 | 139.8 | 83.2 | 82.2 |
| 1992 | 95.0 | 133.0 | 75.6 | 80.2 |

18. Use the linear trend equation given in Example 14.9 [$\hat{Y} = 763.21 + 11.31X_t$, (1989 = 0, $X_t$ = quarter, $Y$ = units)]. Then use it and the second-quarter seasonal index modified mean of 96.6 from Table 14.5 for seasonal along with a cycle relative of 0.955 to forecast Amtex Co. sales for second quarter of 1992. Compare your forecast to the actual sales. See Table 14.3.

19. Develop forecasts for the (a) first, (b) second, and (c) third quarters of 1993 for Amtex Co. sales using the information given in Problem 18. Let business-cycle relatives be 1.01, 1.04, and 1.06, respectively.

## 14.6 EXPONENTIAL SMOOTHING FORECASTING METHODS

An extension of time series methods has gained widespread use in business today. Exponential smoothing is especially useful because it incorporates a measure of error into each forecast. Moreover, subsequent forecasts can be adjusted to take account of the error that has occurred in the past.

A moving average was defined earlier as a method for smoothing a time series. This procedure is appropriate for a series that follows reasonably close to a linear or curvilinear trend. Weighted moving averages give unequal weights to the time observations. With exponential smoothing, usually the largest weights go to the most recent data points. The remaining data points receive weights that diminish exponentially with time. This procedure is especially useful for short-term—daily, weekly, monthly—forecasts that show strong irregular variations.

The *exponential smoothing model* is essentially a modified moving average model.

> *Exponential smoothing* is a forecasting technique that uses the weighted average of (1) the series actual values for the previous period and (2) the forecast for that period.

In symbols, a first-order exponential smoothing model can be described as follows:

### FIRST-ORDER EXPONENTIAL SMOOTHING MODEL

$$\hat{Y}_t = \alpha Y_{t-1} + (1 - \alpha)\hat{Y}_{t-1} \qquad \text{with} \quad 0 < \alpha < 1 \qquad (14.10)$$
$$= \hat{Y}_{t-1} + \alpha(Y_{t-1} - \hat{Y}_{t-1}) \qquad t = 2, 3, \ldots$$

where  $\hat{Y}_t$ = forecast amount for the current time period, $t$

$Y_{t-1}$ = actual values of the series at the previous time period, $t - 1$

$\hat{Y}_{t-1}$ = forecast for time period $t - 1$

The model is defined by one parameter, alpha, which is the *smoothing constant*. By Equation 14.10 the forecast at time $t$ is based on (1) the forecast at time $t - 1$ plus (2) an adjustment for the error in the forecast made at time $t - 1$. Since the forecast for time $t - 1$ is based on comparable data from time period $t - 2$ and so on, the effect on a current time forecast (at time $t$) diminishes as the time period is further removed. Simply put, what happens at time $t - 1$ has more effect on the series at time $t$ than does the response at time

$t - 2$, than at time $t - 3$, and so on. The effect is exponential and is smoothed (by the $Y_{t-1} - \hat{Y}_{t-1}$ differences).

The key to good forecasts by this model is to establish a most reasonable weighting constant. For example, if the forecast is to be quite responsive to the previous period actual value, the analyst would use an $\alpha$-value near 1. In contrast, an $\alpha$-value near zero would result in considerable smoothing. The $\alpha$-values used in industry often lie in the range of 0.05 to 0.3 or 0.4. In the extreme case, where $\alpha = 1$, the forecast is simply the actual value at time $t - 1$. See Equation (14.10).

Applying the model is a straightforward, repetitive process that requires minimal data storage, so *it is economical for computer memory space*. The process begins with an initial value, $Y$. Then the forecasts begin with time period $t = 2$. For $t = 2$, $\hat{Y}_2 = Y_1$, because $\hat{Y}_{2-1} = \hat{Y}_1 = Y_1$ in (14.10). That is, the first forecast is the second-period smoothed value. Each succeeding forecast adjusts the preceding period smoothed value with a correction for the error in the last forecast value,

$$\hat{Y}_3 = \hat{Y}_2 + \alpha(Y_2 - \hat{Y}_2)$$
$$\hat{Y}_4 = \hat{Y}_3 + \alpha(Y_3 - \hat{Y}_3)$$
$$\vdots$$

As the process continues, the error incorporates a diminishing effect from each preceding period; hence the name *exponential*. Again, for $\alpha$ near 1 the greatest weight is on the most recent values, whereas an $\alpha$ near zero affords past-period values an increased influence. We illustrate with some production-demand data.

---

**EXAMPLE 14.10**

Use exponential smoothing on the data in Table 14.6 to forecast the demand for the NR transmissions. The distributor wants to be able to establish inventory supply procedures on a weekly basis. How satisfactory is the exponentially smoothed model?

*Solution*

Initially we let $\alpha = 0.5$, a middle value. Later we will explore using other values and whether the forecasts can be improved. *For the first period*, we let $\hat{Y}_{81} = Y_{81} = 24$. Then the forecast for week 82 is

$$\hat{Y}_{82} = \hat{Y}_{81} + \alpha(Y_{81} - \hat{Y}_{81})$$
$$= 24 + 0.5(24 - 24) = 24$$

**TABLE 14.6** Weekly Demand for NR Transmissions

| Week $t$ | Demand $Y_t$ | Week $t$ | Demand $Y_t$ |
|---|---|---|---|
| 81 | 24 | 86 | 38 |
| 82 | 32 | 87 | 17 |
| 83 | 25 | 88 | 44 |
| 84 | 8 | 89 | 23 |
| 85 | 33 | 90 | 11 |

**TABLE I4.7** Comparison of Exponentially Smoothed Forecast Values against Actual Demands, $\alpha = 0.5$

| Week $t$ | Forecast $\hat{Y}_t$ | Actual $Y_t$ | Error $\hat{Y}_t - Y_t$ |
|---|---|---|---|
| 81 | — | 24 | — |
| 82 | 24.00 | 32 | −8.00 |
| 83 | 28.00 | 25 | 3.00 |
| 84 | 26.50 | 8 | 18.50 |
| 85 | 17.25 | 33 | −15.75 |
| 86 | 25.12 | 38 | −12.88 |
| 87 | 31.56 | 17 | 14.56 |
| 88 | 24.28 | 44 | −19.77 |
| 89 | 34.14 | 23 | 11.14 |
| 90 | 28.50 | 11 | 18.57 |

Continuing, $\hat{Y}_{83} = 24 + 0.5(32 - 24) = 28$, while $Y_{83} = 25$. So the error of forecast for week 83 is (Forecast − Actual value) $= 28 - 25 = +3$, or 3 units above the actual values. Table 14.7 displays forecast, actual, and error values for NR transmissions for weeks 82 through 90.

Some of the errors are substantial. Figure 14.7 provides a visual display of the goodness of the smoothed to actual values. The fit is only moderately good. It can be improved.

You might well wonder what is the effect of assigning a high or a low $\alpha$-value. Simply stated, as $\alpha$ increases, more weight is given the $t - 1$ (most recent) value. Thus, for $\alpha$ near 1, the forecast reflects quick changes in the series. So $\alpha$-values near 1 are appropriate for series in which management wants a

**FIGURE I4.7** A Plot of Demand and an Exponential Smoothed Forecast for NR Transmissions Using $\alpha = 0.5$

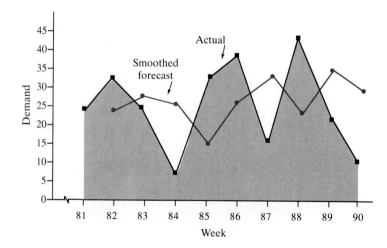

strong response to recent variations. Conversely, if more smoothing is desired, a lower α-value is appropriate.

### GUIDE FOR CHOOSING A VALUE FOR α

Larger values for α, near 1, give quick response to the changes in a time series, whereas values nearer to 0 give more stable, but slower, responding smoothed forecasts.

An overall guide is helpful, but an objective measure for choosing the value is more useful. The *mean square error*, MSE, is an objective measure for comparing alternate values for α. That value that gives the *least MSE* is the best value for making forecasts.

The mean square error, MSE, for a forecast is

$$\text{MSE} = \frac{\sum (Y_t - \hat{Y}_t)^2}{n - 1} \tag{14.11}$$

Our next example calls for applying different exponential smoothing coefficients to the data in Table 14.6. The least MSE gives a guide for choosing an α-value.

---

**EXAMPLE 14.11**

What value for α is most suitable for an exponential smoothing forecast of the NR transmission demands in Table 14.6? Choose from alternatives α = 0.9, 0.5, 0.2, and 0.1. Use your choice to forecast demand for NR transmissions for week 91.

*Solution*

We have selected values for α ranging from 0.9 down to 0.1. The smoothed forecasts are displayed in Table 14.8. The actual demand series is quite erratic. For α = 0.9 the forecast responds mostly to the preceding-time actual demand, so this α value gives relatively unstable forecasts. Similarly, comparing forecast entries line by line to actual demand in Table 14.8 shows that the α = 0.05 estimates are also unstable. The α = 0.01 forecasts are nearly always the closest to actual demand. Our intuition is that α = 0.01 gives the least error and hence the best fit to the actual demands.

**TABLE 14.8**  Smoothed Forecast for Alternative Values for α

| Week | Actual Demand | Smoothed Forecast | | | |
|------|--------|-----------|-----------|-----------|-----------|
|      |        | α = 0.9 | α = 0.5 | α = 0.2 | α = 0.1 |
| 81 | 24 | — | — | — | — |
| 82 | 32 | 24 | 24 | 24 | 24 |
| 83 | 25 | 31.2 | 28 | 25.6 | 24.8 |
| 84 | 8 | 25.62 | 26.5 | 25.48 | 24.8 |
| 85 | 33 | 9.76 | 17.25 | 21.98 | 23.12 |
| 86 | 38 | 30.68 | 25.12 | 24.18 | 24.11 |
| 87 | 17 | 37.19 | 31.56 | 26.94 | 25.50 |
| 88 | 44 | 12.51 | 24.28 | 24.95 | 24.65 |
| 89 | 23 | 40.85 | 34.17 | 28.76 | 26.59 |
| 90 | 11 | 26.32 | 28.58 | 27.61 | 24.29 |

The MSE provides an objective measure for the error in the forecast. The calculations use numbers from Table 14.8. For

$$\alpha = 0.9, \text{MSE} = \frac{(32 - 24)^2 + (25 - 31.2)^2 + \cdots + (11 - 26.32)^2}{9} = 328.80$$

$$\alpha = 0.5, \text{MSE} = \qquad\qquad\qquad\qquad\qquad = 207.10$$

$$\alpha = 0.2, \text{MSE} = \qquad\qquad\qquad\qquad\qquad = 161.46$$

$$\alpha = 0.1, \text{MSE} = \qquad\qquad\qquad\qquad\qquad = 141.45$$

For these choices the best alternative is $\alpha = 0.1$. It has the least MSE. Then the forecast for week 91 is $\hat{Y}_{91} = 24.29 + 0.1 \cdot (11 - 24.29) = 22.95$, or 23 units. The actual demand was 39 units. By Figure 14.7 the actual demand is quite erratic. Exponential smoothing (with $\alpha = 0.05$) gives smoothed values; it may contain substantial errors in some forecasts.

From Example 14.11 we can see the difference in choosing a low versus a high value for $\alpha$. When the series is quite volatile, then a low $\alpha$-value is appropriate. Yet an $\alpha$-value near 1 will follow a series more quickly. This may be important if the series tends to deviate more consistently in one direction or another for several time periods before it returns to the average.

From Figure 14.7 the series for NR transmission demand was erratic with no obvious trend pattern. A trend analysis of the data might result in a horizontal trend line. So we used what is referred to as *single, or first-order, exponential smoothing*. If a series has a trend over time, then a double-smoothed statistic is appropriate. This smooths the single-smoothed values. If the series has a strong second-degree polynomial trend, then triple smoothing can provide a better forecast. As you would guess, the triple smooths the double-smoothed values. See Mendenhall et al. [10] for more on this topic.

Exponential smoothing has several advantages over other forecasting models. It provides a simple approach that can accommodate both irregular fluctuations (single-exponential smoothing) and systematic trend patterns (double smoothing). Again, this procedure is especially useful for forecasting data over short time periods (days, weeks, or months), wherein business cycles can be ignored. Also it is easily programmed on a computer. The calculations are relatively simple. They require only minimal data storage, including the past-period forecast and actual values.

The exponential smoothing procedure also has several limitations. First, it does not provide a fixed model. There is no guarantee that an $\alpha$-value will be reasonable for future series values. So the $\alpha$-value should be checked and may need to be reset. Second, exponential smoothing does not provide long-term predictions. Forecasts are limited to one period into the future. Finally, it does not provide a strong conceptual base for statistical inference.

This completes our brief excursion into some of the forecasting applications of exponential smoothing. In the next section, we examine the use of index numbers for condensing time series data.

**Problems 14.6**

20. The data show demand for NR transmissions for weeks 1–10. Consider the possibility of making a forecast for week 11 from a first-order exponential smoothing model.

a. Plot the data as a time series using demand (vertical axis) against weeks (horizontal axis).
b. Scrutinize this data plot. Does it show an irregular pattern with little trend, a linear pattern, or a curvilinear pattern?
c. Explain why you would or would not use first-order exponential smoothing for a week-11 forecast.

| Week $t$ | Demand $Y_t$ | Week $t$ | Demand $Y_t$ |
|---|---|---|---|
| 1 | 41 | 6 | 21 |
| 2 | 31 | 7 | 33 |
| 3 | 27 | 8 | 29 |
| 4 | 35 | 9 | 39 |
| 5 | 7 | 10 | 49 |

21. Repeat parts (a), (b), and (c) of Problem 20 using new data as shown in the table below. Make the forecast for week 51. First answering Problem 23 can help you make the forecast.

| Week $t$ | Demand $Y_t$ | Week $t$ | Demand $Y_t$ |
|---|---|---|---|
| 41 | 36 | 46 | 37 |
| 42 | 28 | 47 | 22 |
| 43 | 20 | 48 | 54 |
| 44 | 39 | 49 | 32 |
| 45 | 15 | 50 | 13 |

22. Use the data in Problem 21 to complete the table below. By comparing the smoothed forecast values to actual demands, make a visual estimate of which $\alpha$-value (0.1, 0.5, or 0.9) appears by your judgment to give the best fit to this data.

| Week | Demand | Smoothed Forecast | | |
|---|---|---|---|---|
| | | $\alpha = 0.1$ | $\alpha = 0.5$ | $\alpha = 0.9$ |
| 41 | 36 | — | — | — |
| 42 | 28 | 36 | 36 | 36 |
| 43 | 20 | 35.20 | ⋮ | ⋮ |
| ⋮ | ⋮ | | | |
| 50 | 13 | ⋮ | ⋮ | ⋮ |

23. Use your table from Problem 22 to determine which $\alpha$-value provides the best fit. *Note:* Some useful data appear on the next page.

| Week | Demand | Error for Smoothed Forecast | | |
|------|--------|------------|------------|------------|
|      |        | $\alpha = 0.1$ | $\alpha = 0.5$ | $\alpha = 0.9$ |
| 41   | 36     | —          | —          | —          |
| 42   | 28     | −8         | −8         | −8         |
| 43   | 20     | −15.20     |            |            |
| ⋮    | ⋮      |            |            |            |
| 50   | 13     |            |            |            |

a. Complete the table of error values for $\alpha = 0.1$.
b. Complete the values for $\alpha = 0.5$ and $\alpha = 0.9$.
c. Use the mean square error calculation to decide which $\alpha$-value (0.1, 0.5, or 0.9) gives the best fit to this data. Compare this to your judgment choice in Problem 22.

## 14.7 FORMULATION OF INDEX NUMBERS

In many ways our very lives are indexed. Indexes that affect us appear in daily papers, magazines, and television news broadcasts. Index numbers express a price, quantity, or value relative to a base measure. This means they convey the ratio of a price (or quantity or value) in one period divided by the price (or quantity or value) in another period. Section 14.7 describes how index numbers are formulated; Section 14.8 tells how they are used.

> Index numbers are ratios that measure the relative change in a variable or mix of variables over a reference dimension such as time. The relative change is for prices, or quantities, or values of commodities at some point in time to the same mix of commodities at a base time.

We shall deal only with price ($P$), quantity ($Q$), and value ($V$) indexes. These are related in that value is the price times quantity ($V = P \times Q$). The year designated as the *base*, or *ordinary period*, is denoted by the subscript "$o$," and the *given*, or *comparative period*, is denoted by subscript "$n$."

> The base period for an index number is the specific time period used as a basis of comparison.

> The given period for an index number is the time period of interest, and is often a current period.

Thus $P_o$ signifies the price of an item in the base year, and $\sum P_n$ is a summation of the prices of several items in the given year. All index numbers are expressed in percentage form—with calculations simplified by stating indexes as 100 times their decimal value. Thus an index of 100 is average; 110 is 10% above average, just as 90 is 10% below average.

The wholesale price index and the consumer price index are widely used indexes. You may have seen references to them in discussions of the cost of living or union wage settlements. The consumer price index (CPI), for example, expresses a cost of goods and services to urban consumers relative to a national norm. It is so widely used that the wages of over 8 million workers are automatically adjusted to it.

**TABLE 14.9** Hypothetical Prices ($) of Various Fuels for Years 1 through 6

| | | Year | | | | | |
|---|---|---|---|---|---|---|---|
| Fuel | Unit | *1* | *2* | *3* | *4* | *5* | *6* |
| Coal | ton | 24 | 26 | 28 | 29 | 25 | 26 |
| Oil | barrel | 28 | 29 | 28 | 25 | 21 | 22 |
| Natural gas | 1000 ft$^3$ | 4 | 4 | 5 | 6 | 6 | 6 |
| Uranium oxide | pound | 25 | 32 | 38 | 38 | 32 | 31 |

To illustrate different types of indexes in a manner that both efficiently and distinctively brings out the salient features of each index, we shall work consistently with the data in Table 14.9, which reflects some hypothetical differences in the price of various fuels used by an electric utility to generate electric power. The figures are illustrative only and do not necessarily represent actual prices. The period covered is six years, so if year 1 represented 1988, year 6 would then represent 1993. Index numbers for fuel costs such as these could help an electric utility manager decide which of several available energy resources to plan on using for power generation in the future.

**The Simple Price Index**

The simplest form of an index is a ratio that compares the price of an item in a given year to the price of the same quantity of the same item in a base year. We refer to this as a *simple price index*.

A simple price index $[(P_n/P_o) \cdot 100]$ is an index number that expresses a single price in a given year relative to the price in the base year.

Thus a simple price index "relates" two prices. A shorthand notation is $P_n/P_o$. For example, the simple price index for coal in year 6 with base-year 1 is $[P_6/P_1 \cdot 100] = [(26/24) \cdot 100] = 108$. This tells us that the price of coal rose 8% over the five years. A comparable calculation for uranium would show it has risen by 24% (that is, $[(31/25) \cdot 100] = 124$, or a 24% increase). Simple price indexes are easy to compute and interpret. However, they concern only one item over the time interval. If the firm used both coal and uranium, each index should be computed and analyzed separately.

**The Simple Quantity Index**

When we are interested in relative change in the quantity of production, sales, or some other variable, a *simple quantity index* may be appropriate.

A simple quantity index $[(Q_n/Q_o) \cdot 100]$ is an index number that expresses a single quantity in a given year relative to the quantity in the base year.

For example, the electric utility manager, faced with limited supplies of coal, may wish to know the quantity of coal consumed this year relative to that burned five years earlier. The simple quantity index for coal in year 6 with base-year 1 would be $[(Q_6/Q_1) \cdot 100]$. As can be seen, the computation of a simple quantity index is similar in form to that for a simple price index except we use quantities instead of prices.

**The Simple Value Index**   When used separately, both the simple price index and the simple quantity index share the same problem of not taking the relative economic importance of the items into account. However, a simple value index $[(V_n/V_o) \cdot 100]$ does reflect economic importance because it measures the relative value of an item in both the base year $(V_o)$ and the given year $(V_n)$. Value is measured as the price $(P)$ times the quantity $(Q)$.

**SIMPLE VALUE INDEX**

$$V_r = \frac{V_n}{V_o} \cdot 100 = \left(\frac{P_n}{P_o} \cdot \frac{Q_n}{Q_o}\right) \cdot 100 \qquad (14.12)$$

$$= \text{(Simple price)} \cdot \text{(Simple quantity)} \cdot 100$$

As you can observe from Equation (14.12), a value index is the product of a simple price times a simple quantity index. Note that the index is still classified as *simple* because only relative changes in value are measured and the prices are not weighted by a constant quantity.

---

**EXAMPLE 14.12**

In 1987 an electrical manufacturer purchased 2800 pounds of insulation costing $1.50 per pound, whereas in 1992 the firm ordered 2000 pounds of insulation at $3.00 per pound. Compute the simple value relative for 1992 with 1987 as the base year.

*Solution*

$$V_r = \left(\frac{P_n}{P_o} \cdot \frac{Q_n}{Q_o}\right) \cdot 100 = \left(\frac{(\$3.00/\text{lb})}{(\$1.50/\text{lb})} \cdot \frac{(2000 \text{ lb})}{(2800 \text{ lb})}\right) \cdot 100$$

$$\rightarrow [(2.0 \cdot 0.71) \cdot 100] = 142\%$$

Note that the price index (200) reveals a doubling of price, but the quantity index (71) shows that consumption in 1992 is down to 71% of the 1987 consumption volume. In total, the value of insulation purchased has increased by 42%.

---

Value indexes permit us to focus upon the total cost change for goods or services over the respective time periods. Managers who must operate within a fixed budget quickly learn that good control cannot depend upon price or quantity alone. The value index takes both factors into account. However, when both price and quantity change simultaneously, it becomes difficult to use the value index for control purposes. This is because the value index lacks a common reference such as common prices or common quantities.

**Simple Aggregate Indexes: Some Cautions**   A major weakness of the indexes we have seen so far is that they do not reflect the relative importance of the various components in a consistent way. For example, the simple price, quantity, and value indexes do not accommodate the fact that an electric utility may be using three or four different sources of fuel. With simple indexes, if the firm used both coal and uranium, each index would have to be computed and analyzed separately.

An average consisting of the prices of several items may be a more useful basis for comparisons over time. However, we must be careful when combining index numbers.

*Price relatives* cannot simply be added and averaged. For example, from the data in Table 14.9, we saw that the simple price relative for coal in year 6 with base-year 1 was 1.08, and that for uranium was 1.24. Suppose you wanted to obtain an average for all four price relatives, so you included the values for oil (0.79) and natural gas (1.50) and divided by 4. You would have an average of price relatives, $P_r$, of

$$P_r = \frac{1.08 + 1.24 + 0.79 + 1.50}{4} = 1.15$$

This has a numerical result suggesting an average increase in fuel prices of 15% over the six-year period. But is that a reasonable interpretation of the data? Upon closer examination we would have to conclude that it is open to serious question. Note that the units that went into our average of the relative prices for coal, oil, etc., were themselves dimensionless (%) values. Thus our average of 1.15 is independent of the units of measurement of the components of the index. And when summed via the arithmetic mean, each fuel is given the same weight, even though the fuels surely are not of equal importance. If we had used the median as the average, or the geometric mean, we would have obtained different answers. So we must conclude that a simple average of price relatives has serious shortcomings.

A *simple aggregate price index* also has some shortcomings. It expresses the combined prices of the aggregate of items in a given year relative to the combined prices of the same items in a base year. It is equally suspect. For example, suppose we added the prices in year 1 of a ton of coal ($24), a barrel of oil ($28), one thousand cubic feet of natural gas ($4), and a pound of uranium oxide ($25) to get a total of $81. The comparable price total for year 6 is $85, so the simple aggregate price index $P_{6/1}$ is, from Table 14.9, $[85/81]100 = 105$. But note that the summations of prices involve different units of measurement (dollars/ton and dollars/barrel), so the result is not meaningful.

In addition, the relative importance of the various components depends upon the quantity in which they happen to be quoted. Thus, if uranium oxide were quoted in tons (like coal) instead of pounds, the simple aggregate price index for $P_{6/1}$ would be 124 instead of 105. This discrepancy exists because weights are not consciously assigned to prices. Furthermore, it does no good to convert everything to common units, such as tons. Beyond that, a simple aggregate index includes no recognition of the economic importance of the various items in the index. For example, if the electric utility had 20 coal-fired plants and only 1 natural gas-fired plant, the index would not reflect its heavier dependence upon coal.

## Weighted Aggregate Indexes

If an electric utility uses 85% coal and 15% oil for generating power, a representative price index for that firm should weight the coal and oil quantities in that proportion. The value index was an approach to weighting, but it did not use consistent weights. That is, both prices and quantities in the base year and the given year were allowed to vary. This left us with no standard or reference point from which to gauge either price or quantity changes. We then noted that simple aggregate indexes were not the answer to our dilemma. We turn now to *weighted aggregate indexes*.

A weighted aggregate index utilizes quantities or prices in a specified time period as weights in order to make the index more representative of the changes under study.

Weighted aggregate price indexes reflect the importance of various components by weighting the component price by the quantity (or volume) used in either the base year, the given year, or some other typical year. We discuss four alternatives. These differ in the expression used as weights. For consistency we restrict this discussion to price indexes with quantity weights. The *Laspeyres weighted aggregate index* uses base-year quantities as weights. The *Paasche index* differs only in that it uses given year quantities as weights.

The remaining two forms are attempts to overcome a bias of overestimating (Laspeyres) or underestimating (Paasche) price changes if quantities do in fact change according to economic theory. The *typical-year method* assumes that an average, or typical-year, quantity is a more representative amount than that of either the base year or the current year. Finally, we introduce the *Fisher ideal index*, which was designed to overcome the shortcomings of the Laspeyres and Paasche indexes.* The Fisher ideal index is the geometric mean of the Laspeyres and Paasche indexes and gives the same result regardless of the direction in time in which it is computed. It thus satisfies a criteria called the *time reversal test*. It also satisfies a *factor reversal test* in that the price ratios multiplied by the quantity ratios always equal the value ratios.

These four weighted aggregate index forms are summarized below. Your close attention will single out their differences.

### WEIGHTED AGGREGATE PRICE INDEXES

$$\text{Laspeyres Index} \quad L_{n|o} = \frac{\sum (P_n \cdot Q_o)}{\sum (P_o \cdot Q_o)} \times 100 \qquad (14.13)$$

$$\text{Paasche Index} \quad P_{n|o} = \frac{\sum (P_n \cdot Q_n)}{\sum (P_o \cdot Q_n)} \times 100 \qquad (14.14)$$

$$\text{Typical period} \quad T_{n|o} = \frac{\sum (P_n \cdot Q_t)}{\sum (P_o \cdot Q_t)} \times 100 \qquad (14.15)$$

$$\text{Fisher's ideal index} \quad F_{n|o} = \sqrt{L_{n|o} \cdot P_{n|o}} \qquad (14.16)$$

The subscripts are $o$ = base year, $n$ = given year, $t$ = typical year.

We illustrate the differences among the four index forms with data for an auto parts supplier in Example 14.13.

---

**EXAMPLE 14.13**

Records from an automotive parts supplier show the prices and inventory levels for three years listed in Table 14.10. Compute price indexes for the year 1992 with 1982 as the base year. Let 1987 represent a typical year. What is the increase in price levels for this

---

\* Standard usage of Laspeyres and Paasche indexes does not include an apostrophe and letter *s* to indicate the possessive case.

**TABLE 14.10**  Inventory Levels for an Automotive Parts Supplier

|  | Price ($ per unit) | | | Quantity (units) | | |
|---|---|---|---|---|---|---|
| Part | *1982* | *1987* | *1992* | *1982* | *1987* | *1992* |
| Batteries | 38 | 35 | 29 | 60 | 62 | 65 |
| Mufflers | 40 | 32 | 32 | 130 | 130 | 110 |
| Oil filters | 2 | 3 | 4 | 1500 | 1200 | 500 |

10-year period as defined by (a) $L_{n|o}$, (b) $P_{n|o}$, (c) $F_{n|o}$, and (d) $T_{n|o}$? Compare the four indexes.

Solution   The calculations refer to Table 14.10. For these indexes, the prices of individual items are multiplied by quantities, then summed.

**a.** Laspeyres   $L_{n|o} = \dfrac{\sum [(\text{Prices in 1992}) \cdot (\text{Quantities in 1982})]}{\sum [(\text{Prices in 1982}) \cdot (\text{Quantities in 1982})]} \cdot 100$

$= \dfrac{11{,}900}{10{,}480} \cdot 100 = 114\%$

The price level has risen 14%.

**b.** Paasche   $P_{n|o} = \dfrac{\sum [(\text{Prices in 1992}) \cdot (\text{Quantities in 1992})]}{\sum [(\text{Prices in 1982}) \cdot (\text{Quantities in 1992})]} \cdot 100$

$= 94\%$

The price level is *reduced* by 6%.

**c.** Fisher's   $F_{n|o} = \sqrt{(114)(94)} = 103.5\%$

The change is an increase of 3.5%.

**d.** Typical year $= \dfrac{\sum [(\text{Prices in 1992}) \cdot (\text{Quantities in 1987})]}{\sum [(\text{Prices in 1982}) \cdot (\text{Quantities in 1987})]} \cdot 100$

$= \dfrac{10{,}758}{9{,}956} = 108\%$

The price-level change is an increase of 8%.

So, based on personal preference, these procedures show a price range from a 6% decrease to a 14% increase over the 10 years from 1982 to 1992.

---

Relative to the other measures, the Laspeyres value is an overestimate of the price increase. This occurs because the Laspeyres index ignores the changes in quantities. For example, the sharp increase in oil filter prices in Table 14.10 was accompanied by a decline in quantities purchased. However, since the Laspeyres index does not take given-year quantities into account, this effect went unrecognized. As a result, the price index is higher than it might otherwise be. An opposite effect results from emphasizing the (reduced) given-year quantities. The Paasche index leads to a conclusion of an aggregate price decline in this case. The Fisher index and the typical index give values between these. In Example 14.13 *the range of values—94 to 114—illustrates how statistics can be used (by unknowing or perhaps deceptive individuals) to make different and even contradictory claims.*

Selecting the most appropriate index sometimes requires a knowledge of the underlying economic conditions. Laspeyres index compares the total value (Price · Quantity) of a mix of goods in the current, or given, year, $\sum (P_n \cdot Q_o)$ with the same quantity of the mix of goods in the base year, $\sum (P_o \cdot Q_o)$. The quantity is assumed to be fixed at the base-year amount. Economists point out a weakness in this, observing that the law of supply and demand suggests that as prices increase, consumers tend to reduce their consumption of those items. This is illustrated in Example 14.13. By not reflecting the reduced quantities purchased in later periods, the Laspeyres index tends to overstate the values (i.e., price changes) in a period of rising prices. It is most reasonable with a level economy.

Although a steady consumption pattern has often been a poor assumption in the past, the rate of consumption of vital resources is changing in many countries as shortages develop. Numerous business and government organizations have initiated programs to stabilize or reduce energy consumption. As we see more such constraints on consumption in the future, combined with continued price inflation, we may expect to find increased use of Laspeyres index.

As does Laspeyres index, the Paasche also contains a bias—but in the opposite direction. With given-year quantities as weights, the Paasche index tends to underestimate price changes if quantities do in fact change according to economic theory. Another difficulty with the Paasche index lies in the continual need to update the quantities of all items each period. This results in increased data collection costs.

A typical-year index provides a value somewhere between the Laspeyres and Paasche indexes. But the typical period is a free choice for the analyst. The worth of subsequent results depends on the analyst's ability or luck in choosing a typical year.

Fisher's index gives an appraisal that overcomes some of the problems noted in the Laspeyres and Paasche procedures. It also accommodates the irregular pattern in price and quantity changes that discredits a typical-year calculation. Being the geometric mean of the Laspeyres and Paasche indexes, Fisher's index takes on a value between these two. The meaning of the index value is somewhat obscure, however, and for this reason Fisher's ideal index is not as widely used as might be expected.

**Problems 14.7**

24. The prices of a chocolate bar in the years 1988 and 1992 were $0.70 and $1.50, respectively.
    a. Using 1988 as the base year and 1992 as the given year, find the simple price index.
    b. Interpret your result.
    c. Using 1992 as the base year, find the price index.
    d. Interpret your result from (c).

25. The German division of a U.S. firm has labor costs in thousands of dollars (for a constant number of workers) as shown for years 1 through 8.

a. Using year 1 as a base, find the simple price indexes corresponding to all the given years.
b. Using years 4 and 5 as the base period, compute the price index for year 8. (*Hint:* Average the year 4 and year 5 prices.)

| Year | 1 | 2 | 3 | 4 | 5 | 6 | 7 | 8 |
|---|---|---|---|---|---|---|---|---|
| Labor cost ($000) | 11,500 | 12,450 | 13,140 | 14,070 | 15,310 | 16,630 | 16,240 | 17,080 |

26. The Farwell Manufacturing Company has developed the following indexes of manufacturing productivity, labor costs, and related measures for the years 1987–1991, with the base year being 1981 = 100.

| | 1987 | 1988 | 1989 | 1990 | 1991 |
|---|---|---|---|---|---|
| Output per hour | 119.4 | 112.8 | 116.3 | 124.2 | 126.9 |
| Hourly compensation | 147.0 | 161.4 | 179.4 | 194.8 | 212.0 |
| Unit labor costs | 123.2 | 143.1 | 154.3 | 156.9 | 167.0 |
| Employment | 103.1 | 103.0 | 94.4 | 97.6 | 100.7 |

a. How much higher was manufacturing productivity, in terms of output per hour, in 1991 than in 1981?
b. By how much did output per hour rise from 1990 to 1991?
c. Using 1987 as the base year, determine how much unit labor costs rose over the four years to 1991.
d. Compare the increase in unit labor costs computed in (c) to the hourly compensation figures for the same years, 1987 and 1991. Which rate appears to be rising faster?

27. Suppose unit labor costs based upon a country's own national currency were as shown. Assume further that all used year 1 as the base year. Compute the simple average of price relatives for year 11 with year 1 as the base year.

| Country | Unit Labor Cost (Year 1 = 100) | |
|---|---|---|
| | *Year 3* | *Year 11* |
| U.S. | 116.5 | 167.0 |
| Canada | 108.1 | 181.3 |
| Japan | 112.2 | 232.7 |
| France | 111.1 | 205.4 |
| Germany | 115.0 | 170.8 |
| Sweden | 105.6 | 223.7 |
| U.K. | 121.7 | 314.8 |

28. Use the data of Problem 27 to compute the simple average of price relatives for year 3 with year 1 as the base year.

29. The prices of related office supplies used by a firm over the years 1987–1992 are shown in the table below.

|  |  | Price ($) | |
|---|---|---|---|
| Product | Unit | *1987* | *1992* |
| Paper | Ream | 2.20 | 4.50 |
| Pencils | Carton | 10.80 | 13.90 |
| Phone | Per day | 9.00 | 10.10 |
| Calculators | Each | 178.00 | 21.50 |

   a. Compute the simple price index for paper in 1992, using 1987 as the base year.
   b. Compute the simple average of price relatives for 1992 with 1987 as the base year.

30. Compute the simple aggregate price index for the office supplies given in Problem 29 and discuss the deficiency of your resultant index. (Use 1987 as the base year).

31. A small business supplying bearings to industrial and agricultural users had sales of a one-inch motor bearing as listed in the table. Find the simple quantity index for 1991 and 1992 with 1988 as the base year.

| Year | 1988 | 1989 | 1990 | 1991 | 1992 |
|---|---|---|---|---|---|
| Price each | $12 | $14 | $18 | $24 | $24 |
| Quantity sold | 3000 | 3200 | 4200 | 4500 | 3000 |

32. Refer to the data from Problem 26.
   a. Did the quantity of manufacturing employment increase from 1987 to 1989?
   b. Use the simple quantity index to determine how many workers were employed in manufacturing in 1991 for every 100 workers employed in 1989.

33. Use the data from Problem 31 to compute the value indexes for all the years, using 1988 as the base year.

34. Suppose the simple quantity index for lumber required for a building products firm for 1993 is 140 and the actual quantity required is 12 million board feet. The base year is 1983, and the 1988 simple quantity index is 120. What was the actual quantity in 1988?

35. Mercy General Hospital had 80 employees in 1983 and an annual payroll cost of $1.76 million. By 1993 it had added 40 employees and the payroll cost increased to $2.70 million per year. Use 1983 as the base year.
   a. Find the simple quantity index for number of employees for 1993.
   b. Recognizing that the total payroll figures represent "value" amounts, find the value index for 1993 with 1983 as the base year.
   c. Using knowledge of the relationship expressed in Equation 14.12, determine the simple price index.

36. The quantities produced and prices received for products produced by a chemical company were as shown.

|        | Price ($/pound) | | | Quantity (pounds) | | |
|--------|------|------|------|------|------|------|
|        | *1988* | *1990* | *1992* | *1988* | *1990* | *1992* |
| Clox-2 | 0.80 | 0.81 | 0.82 | 600  | 800  | 700  |
| D-Alloy | 1.10 | 1.40 | 2.20 | 500  | 1000 | 3000 |
| Sutex  | 0.50 | 0.40 | 0.30 | 4000 | 6000 | 9000 |

   a. Compute Laspeyres price index for 1992 with 1988 as the base year.
   b. Compute the Paasche index for 1992 with 1988 as the base year.
   c. Compute Fisher's ideal price index with 1988 as the base year.
   d. Explain why the Paasche index has a higher value than Laspeyres index.

37. Given $\sum (P_o \cdot Q_o) = 4000$  $\sum (P_n \cdot Q_o) = 8800$
    $\sum (P_n \cdot Q_n) = 6800$  $\sum (P_o \cdot Q_n) = 3400$

   a. find the Laspeyres index.
   b. find the Paasche index.
   c. find Fisher's ideal index.
   d. in what ways does Fisher's ideal index provide a "better" estimate of price changes than either the Laspeyres or the Paasche index?

## 14.8
## APPLICATIONS OF
## INDEX NUMBERS

Price indexes chart the trend of our food and clothing prices as well as that of the construction materials that go into our homes. Other indexes range from unemployment in California to steel production in Pennsylvania. Knowledge of the relative change in variables is useful not only for making adjustments in wage and unemployment benefits but also for helping managers forecast their market potential and assign prices to their products. Production managers use indexes to help set quantity levels of production, and accountants use them to place dollar values on their firm's inventory.

In this section we first discuss how to update an index. Then we concentrate on the uses of a few indexes, principally the consumer price index. This will provide you with knowledge of some common applications. Then you can find out more about specific indexes from the government documents in your local or university library.

## Shifting the Base of an Index

In order to retain the usefulness of an index over time, its base period must occasionally be updated. The wholesale price index, for example, has been continuously tracking the prices received by manufacturers and producers since 1890. However, both the composition and the prices of goods in the index have changed drastically since the 1890s.

For business purposes, the base periods should be fairly recent points in time that represent typical or normal conditions—periods that exhibit some economic stability.

In many cases, the base period is designated as the average over two or more years, such as 1984–1986. The U.S. government exercises a significant influence over the selection of a base period for economic data because it is the world's largest producer of economic data.

The procedure for shifting the base period of index numbers to a new base is quite simple.

### RULE FOR SHIFTING THE BASE

**Divide all the index numbers of the old series by the index number corresponding to the new base, and express each result as a percentage.**

This procedure makes the new base equal to 100%.

## EXAMPLE 14.14

An index of producer prices for 1982–1990 is shown in the table, with 1982 = 100. Change the base to 1986–1988.

| Year | 1982 | 1983 | 1984 | 1985 | 1986 | 1987 | 1988 | 1989 | 1990 |
|---|---|---|---|---|---|---|---|---|---|
| Producer price (1982 = 100) | 100.0 | 101.6 | 103.7 | 104.7 | 103.2 | 105.4 | 108.0 | 113.6 | 118.0 |

Source: *Economic Indicators*, Sept. 1990, by the Council of Economic Advisors (1990 value estimated).

*Solution*    The arithmetic mean for 1986–1988 is

$$\text{Base} = \frac{103.2 + 105.4 + 108.0}{3} = 105.5$$

Dividing each index by this new base of 105.5 results in the new table below.

| Year | 1982 | 1983 | 1984 | 1985 | 1986 | 1987 | 1988 | 1989 | 1990 |
|---|---|---|---|---|---|---|---|---|---|
| Producer price (1986–1988 = 100) | 94.8 | 96.3 | 98.3 | 99.2 | 97.8 | 99.9 | 102.4 | 107.7 | 111.8 |

Note that no specific year has an index value of exactly 100. This is because the base year is an average. You should always indicate the years upon which the base is averaged.

## Using Index Numbers to Deflate a Time Series

One of the most troublesome aspects of managing modern economic systems is that of dealing with inflation. Increased costs for a fixed level of goods and services resulting from higher raw-material costs, new labor contracts, deficit government spending, and the like cause commodity prices to rise. The supply of money is then increased (by the Federal Reserve Board) to accommodate financial transactions, but the purchasing power of an individual monetary unit (i.e., a dollar) decreases. This is a situation of inflation. Individual wages may appear to be rising over the years, but when adjusted for the increased cost of living, real income may, in fact, be declining. The adjustment of a series of economic data to account for the increases in the price level is referred to as *statistical deflation*, or *deflation of time series* data. The time series data are observations taken at periodic intervals, such as the measurements of the U.S. gross national product. By deflating the time series, we filter out the phantom or unreal increase and focus in on the real effects, which are devoid of artificial results due to prices.

> The procedure for deflation of an economic time series consists simply of dividing the parent, or subject series, by an appropriate price deflator.

This procedure is illustrated by Example 14.15.

## EXAMPLE 14.15

Table 14.11 shows the U.S. gross national product figures in billions of current dollars for the years 1982–1990. Included are the implicit price deflators for gross national product (with 1982 = 100) determined by the U.S. Department of Commerce, Bureau of Economic Analysis.

a. Deflate the GNP to obtain the "real" GNP figures in terms of a 1982 base.
b. Compare the reported GNP growth with the real growth in the GNP for the eight-year period 1983–1990 as adjusted for price level changes.

**TABLE 14.11**   Gross National Product (GNP), with a Price Deflator

| Year | GNP ($ billion) | Price Deflator (1982 = 100) |
|------|------|------|
| 1982 | 3166.0 | 100.0 |
| 1983 | 3405.7 | 103.9 |
| 1984 | 3772.2 | 107.7 |
| 1985 | 4014.9 | 110.9 |
| 1986 | 4231.6 | 113.8 |
| 1987 | 4515.6 | 117.4 |
| 1988 | 4873.7 | 121.3 |
| 1989 | 5200.8 | 126.3 |
| 1990 | 5300.0 | 131.0 |

Source: *Economic Indicators*, Sept. 1990, by the Council of Economic Advisors (1990 value estimated).

a. Divide the actual GNP figure for each year by the implicit price deflator and multiply by 100 (see table).

| Year | 1982 | 1983 | $\cdots$ | 1989 | 1990 |
|---|---|---|---|---|---|
| Deflated GNP ($ billion) | 3166.0 | 3277.9 | $\cdots$ | 4117.8 | 4045.8 |

b. For the eight-year period 1983–1990, the reported growth in the GNP is from $3405.7 to $5300.0 (billion).

$$\text{Percent growth} = \left(\frac{\$5300 - \$3405.7}{3405.7}\right)100 = 55.6\%$$

For the same time, 1983–1990, the growth in deflated GNP is from $3277.9 to $4045.8 (billion), giving

$$\text{Percent growth (deflated)} = \left(\frac{\$4045.8 - \$3277.9}{3277.9}\right)100 = 23.4\%$$

The *real* output of goods and services over the years 1983–1990 increased only 23.4%. Most of the reported growth was due to price inflation.

## The Consumer Price Index and Other Indicators

The *consumer price index* is the most widely used economic statistic published in the United States.

> **The consumer price index (CPI) is a weighted aggregate index of the prices of selected goods and services used by consumers in the United States.**

The primary purpose of the CPI is to measure goods and services purchased by urban consumers. The CPI is important for several reasons [9].

1. It is recognized as perhaps the best measure of inflation in the United States. Thus, it impacts heavily upon government economic policy.
2. It is used as an escalator for numerous income payments and contracts. As such, it is used to adjust the wages of approximately 8 million workers and affects the pensions and incomes of approximately 50 million Social Security recipients as well as numerous private contracts, rents, royalties, insurance and alimony payments, and so on.
3. It is used as a deflator for other economic series. Thus it affects economic decisions pertaining to retail sales, personal income, and the like.

In 1978 the CPI was revised to include two indexes: CPI-U for all urban households and CPI-W for wage and clerical workers. The CPI-U reflects the spending patterns of about 80% of the total population and includes self-employed, professional, and other salaried workers, and retired persons and the unemployed. It is the CPI used by the U.S. government and published monthly in *Economic Indicators* by the Council of Economic Advisors in Washington, D.C.

The CPI includes several thousand items sampled from 85 urban areas. Major expenditure groups in the new index are weighted to give different emphasis. The following percentages illustrate this: (1) food and alcoholic beverages = 20.1%, (2) housing = 37.7%, (3) apparel = 5.2%, (4) transportation = 21.8%, (5) medical care = 6.0%, (6) entertainment = 4.2%, and (7) other goods and services = 5.5%. These weights are adjusted appropriately as changes occur in the consumption patterns of the focus group. The weights also acknowledge product changes, such as the inclusion of portable hair dryers and small computers. Being a *fixed weight aggregate price index*, the CPI does not automatically reflect changes in the consumption pattern of existing items or the introduction of new items. It reflects only the price changes of items included in the index when it is designed. In this respect, it is not really a good "cost-of-living" index, even though many people rely upon it as such.

Two applications of the CPI as a price deflator are central in management–labor relations and for consumer marketing. The first application is the use of the CPI to adjust wages according to increases in the cost of living. For example, if the CPI is 232 in month 1 and rises to 238 in month 2, then an average manufacturing wage of $9.81 per hour under a contract that contains an *escalator clause* would automatically increase to [($9.81)(238/232)] = $10.06 per hour. This type of automatic cost-of-living wage increase is often a part of long-term labor–management contracts.

In a second application the CPI can be used to answer the question, "Relative to a fixed base period, how much goods can be purchased by a current dollar?" Suppose the CPI rating for this month is 218 with base 1967 = 100. The *buying power* of your dollars relative to the 1967 base is [(100/218) · 100] = 46%. That is, today your dollars will buy less than half the consumer goods that they bought in 1967! Another way to say this is that the cost of goods has increased by 118%. Note that the *buying power index* is simply the reciprocal of the CPI. We used a related application in the discussion on time series wherein seasonal indexes were used to deseasonalize a time series.

**Problems 14.8**

38. a. Use the data in Example 14.14 with 1987–1989 = 100 to develop a table of producer price indexes for the years 1987 through 1990.
    b. If the producer prices were increased 5% in each of the next three years, what would be the index for 1993?

39. Use the data in Table 14.11 to deflate the GNP for 1989 relative to 1986 as base. That is, compute the real GNP for 1989 based on 1986 GNP = 100.

40. Use the information you have developed in Problem 39 to calculate each of the following for 1989 relative to 1986.
    a. Percent growth
    b. Percent real growth

 41. Suppose you are currently employed at a wage of $6.00 per hour. You are advised that you will receive a $0.75 per hour wage increase after one year of employment. The relevant price index is 167 this year and 184 when you get the wage increase.

    a. Compute your deflated wage rate (1) currently and (2) after the year of employment.

    b. For the price index as an escalator index, would you be better off with (1) the $0.75 per hour increase or (2) an increase based on the price index inflation? Explain.

42. Suppose you work in the housing construction industry. Two indexes are under consideration for an escalator for housing construction (union) labor wage increases (see the table). Consider the perspective of (1) labor or (2) management. Choose, for each, the preferred index as a base for a wage escalator. Explain each choice.

| Year | 1988 | 1989 | 1990 | 1991 | 1992 |
|---|---|---|---|---|---|
| Housing starts | 112 | 115 | 120 | 125 | 132 |
| Consumer costs | 196 | 215 | 240 | 268 | 300 |

## SUMMARY

A time series describes the action of some variable over time. The *classical model* components are trend, cyclical, seasonal, and random. Linear trends can be isolated by several techniques, the most scientific being the least squares lines of best fit. Coding time series data, $\sum X_t = 0$, simplifies the computation of the least squares line. A linear trend equation takes the form $\hat{Y} = a + bX_t$ (signature), where the signature describes the origin and the units of $X_t$ and $Y$. Numerous models also exist for describing nonlinear trends.

Seasonal variation is measured in a variety of ways, one of the most common being via the *ratio-to-moving-average* seasonal index (SI). The seasonal index describes what proportion of the trend value an actual value is. It is used extensively in government statistics and is essential to forecasting periodic values. The averaging process itself helps filter out extreme irregularities, leaving the cyclical component as the least predictable. When some knowledge of business-cycle forces is available, however, it can also be accommodated in the time series model.

*Exponential smoothing* incorporates trend and irregular measures in one simple equation. This procedure can be useful for series that have a sufficient number of previous data periods to go by. It is basically only an "averaging" technique, but it is very useful for forecasting inventories in which records must be continuously updated for thousands of items. Table 14.12 shows key equations in time series.

*Index numbers* measure the change in a variable, or variables, over time relative to some base point. A *simple price index* is a ratio of the price in a given period, $P_n$, to the price in the base period, $P_o$. The *average of* several *price relatives* can be useful but requires great caution. Also, *a simple aggregate of prices* for items that do not have the same units is mathematically incorrect.

Quantity indexes are similar in concept to price indexes, whereas value indexes represent a combination of the two. Value equals price times quantity. However, only relative changes in value are measured, because prices are not weighted by a constant quantity.

The two *weighted aggregate price indexes* that use constant quantities as weights are the *Laspeyres* and *Paasche indexes*. The Laspeyres index uses base-year quantities as weights; the Paasche index uses given-year quantities. Each contains a bias, which

**TABLE 14.12**  Key Equations in Time Series

Classical model for time series

$$Y = T \cdot C \cdot S \cdot R \qquad (14.1)$$

Linear trend coefficient estimates, for $\sum X_t = 0$

$$a = \bar{Y}, \quad b = \frac{\sum X_t Y}{\sum X_t^2} \qquad (14.2)$$

Quadratic (second-degree polynomial) trend, for $\sum X_t = 0$

$$b = \sum X_t Y / \sum X_t^2$$

$$\sum Y = na + c \sum X_t^2 \qquad (14.3)$$

$$\sum X_t^2 Y = a \sum X_t^2 + c \sum X_t^4$$

Estimated actual = Trend · Seasonal index

$$\hat{Y} = T \cdot (\mathrm{SI}) \qquad (14.6)$$

First-order exponential smoothing model

$$\hat{Y}_t = \alpha Y_{t-1} + (1 - \alpha)\hat{Y}_{t-1} \qquad 0 < \alpha \leqslant 1$$

$$= \hat{Y}_{t-1} + \alpha(Y_{t-1} - \hat{Y}_{t-1}) \quad t = 2, 3, 4, \ldots \qquad (14.10)$$

Mean square error for a forecast

$$\mathrm{MSE} = \sum (Y_t - \hat{Y}_t)^2 / (n - 1) \qquad (14.11)$$

*Fisher's ideal* price *index* endeavors to correct. Fisher's index is the geometric mean of the others and always lies somewhere between the two.

The most widely used index number is the *consumer price index* (CPI). Other index numbers, such as the gross national product (GNP), are used by our federal government as economic indicators.

Over time, indexes need to be updated, which involves a procedure called *shifting the (index) base*. A *price deflator* is an index that is used to adjust an economic time series. The GNP and CPI indexes are used to deflate economic data.

The key equations for this chapter appear in Table 14.13.

**TABLE 14.13**  Key Equations for Index Numbers

**Simple Value Index**
$$V_r = \frac{V_n}{V_o} \times 100 = \left(\frac{P_n \times Q_n}{P_o \times Q_o}\right) \times 100 \qquad (14.12)$$

**Weighted Aggregate Price Indexes**

Laspeyres index
$$L_{n|o} = \left(\frac{\sum P_n Q_o}{\sum P_o Q_o}\right) \times 100 \qquad (14.13)$$

Paasche index
$$P_{n|o} = \left(\frac{\sum P_n Q_n}{\sum P_o Q_n}\right) \times 100 \qquad (14.14)$$

Typical period
$$T_{n|o} = \left(\frac{\sum P_n Q_t}{\sum P_o Q_t}\right) \times 100 \qquad (14.15)$$

Fisher's ideal index
$$F_{n|o} = \sqrt{L_{n|o} \cdot P_{n|o}} \qquad (14.16)$$

## KEY TERMS

base period, 684
business cycle, 657
classical model, 657
classical model decomposition, 676
consumer price index (CPI), 696
cyclical variations, 657, 675
deflation of an economic time series, 695
deseasonalized (or seasonally adjusted)
   value, 670
escalator clause, 697
exponential smoothing, 678
Fisher's ideal index, 688
given period, 684
index numbers, 684
Laspeyres index, 688
least squares (trend), 660
mean square error for a forecast, 681
modified mean, 674
moving averages (trend), 666

Paasche index, 688
random (or irregular) variation, 657
ratio-to-moving-average method, 671
seasonal index, 670
seasonal variation, 657, 669
shifting the base, 694
signature, 661
simple price index, 685
simple quantity index, 685
simple value index, 686
smoothing constant, 678
specific seasonal, 671
time series, 656
trend, 657, 658
typical period index, 688
weighted aggregate index, 688
weighted moving averages smoothing
   form, 668

## CHAPTER 14 PROBLEMS

*Time Series*

43. Using the least squares method of trend calculation, determine the intercept and slope coefficients for a linear equation centered on 1985.

| Year | 1983 | 1984 | 1985 | 1986 | 1987 | 1988 | 1989 | 1990 | 1991 | 1992 |
|------|------|------|------|------|------|------|------|------|------|------|
| Units | 4 | 7 | 10 | 5 | 9 | 15 | 12 | 13 | 9 | 16 |

44. What three important characteristics of a least squares line make it a line of best fit?

45. Which of the following choices is the principal value of a plot of annual data points (a scatter diagram) of a time series?
    a. To determine if there is a changing seasonal pattern
    b. To suggest an approximate value for the seasonal index
    c. To identify the type of relationship (linear, second-order, etc.) between the $X$- and $Y$-variables
    d. To predict an $X$-value on the basis of a given value of $Y$

46. The data in the table represent annual production of chemical reactors at Chemtek, Inc.

| Year | Reactors $Y$ |
|------|--------------|
| 1988 | 20 |
| 1989 | 30 |
| 1990 | 25 |
| 1991 | 40 |
| 1992 | 35 |

a. Develop a linear least squares line of best fit.
b. Use the equation to project reactor production in 1998.

47. A firm has developed a time series forecasting equation of the form $\hat{Y} = 144 + 72X_t$ (origin 1985, $X_t$ unit = 1 year, $Y$ = annual sales in $000).
    a. Estimate the trend value for 1995.
    b. Change the $X_t$ unit to months, and state the resultant equation complete with signature.
    c. Using the equation developed in part (b), estimate the annual sales rate at one year from the origin data.

48. Given the time series equation $\hat{Y} = 144 + 72X_t$ (origin = 1993, $X_t$ unit = 1 year, $Y$ = annual demand), answer the following questions.
    a. On what day is the equation now centered?
    b. State the equation that would correspond with the following signature: (Origin = July 1, 1993, $X_t$ unit = 1 month, $Y$ = annual demand)
    c. Modify the equation as required and use it to compute the *annual* demand rate ($Y$ = annual demand) during September 1993.

49. Coldboy International Foods received the following number of export orders (in hundreds) in the years 1984–1989: 32, 28, 30, 24, 26.
    a. Plot the data and draw a freehand trend line.
    b. Derive a least squares trend equation and state it, complete with signature.
    c. Shift the origin of the equation to 1988.

50. Wheatland Foods, Inc., had the following average hourly wages over a nine-year period.

| Year | 1982 | 1983 | 1984 | 1985 | 1986 | 1987 | 1988 | 1989 | 1990 |
|------|------|------|------|------|------|------|------|------|------|
| Wage | $3.70 | 3.80 | 4.30 | 4.40 | 4.80 | 5.40 | 6.50 | 6.70 | 7.80 |

a. Graph the data and draw a freehand curve through the data graph.
b. Use the method of least squares to fit a quadratic trend to the data. State the equation complete with signature.
c. Forecast the average hourly wage at Wheatland in 1992.

51. At National Mart Discount Stores, sales in November and December were $4,700,000 and $5,620,000, respectively. The seasonal indexes for November and December are 120.0 and 140.0, respectively. Is there any real (deseasonalized) increase in sales from November to December?

52. Monthly demand for an electronics component was as shown.

| Jan. | 100 | Apr. | 150 | July | 300 |
|------|-----|------|-----|------|-----|
| Feb. | 90 | May | 240 | Aug. | 280 |
| Mar. | 80 | June | 320 | Sept. | 220 |

a. Plot the data as recorded (which can be considered a one-month moving average).
b. Plot a five-month moving average as a dotted line.

c. What conclusion can you draw with respect to the length of the moving average versus the smoothing effect?

d. Assume the 12-month moving average centered on June was 260 and you are using this, along with other data, to compute a seasonal index via the ratio-to-moving-average method. What value of the specific seasonal would this yield?

*Index Numbers*

53. Indicate whether the following are true or false.

a. A simple price index of 120 indicates that an item costs 20% more in the given year than in the base year.

b. The simple average of price relatives entails division of the summation of base-year prices, $\sum P_0$, by the summation of the given-year prices, $\sum P_n$.

c. The simple average of price relatives does not use quantities as weights.

d. The simple aggregate price index requires that all items be expressed in the same quantity (barrel, ton, etc.).

e. The simple quantity index weights the given-year and base-year quantities by the base-year prices.

54. Assume that expenditures (in billions of dollars) and volume of construction (in millions of square feet) for a region of the country are as shown in the table. Using 1985 as the base year and 1991 as the given year and the cost and volume figures as indicative of price and quantity, find each of the following:

a. Simple price index

b. Simple quantity index

c. Simple value index

| Year | 1985 | 1986 | 1987 | 1988 | 1989 | 1990 | 1991 |
|---|---|---|---|---|---|---|---|
| Cost | 151.1 | 173.8 | 205.6 | 330.4 | 230.7 | 239.4 | 232 |
| Volume | 592 | 739 | 977 | 1059 | 904 | 906 | 685 |

55. The Mack Metals Co. uses a price index compiled from the average prices and quantities of two products, $X$ and $Y$. Find the weighted aggregate price index (Laspeyres) for 1993 with 1988 base.

| | 1988 | | 1993 | |
|---|---|---|---|---|
| | Price, $P_o$ | Quantity, $Q_o$ | Price, $P_n$ | Quantity, $Q_n$ |
| X | 20 | 10 | 30 | 7 |
| Y | 5 | 50 | 10 | 70 |

56. The data in the table show 1987 as the base year.

| | 1987 | | 1992 | |
|---|---|---|---|---|
| | $P_o$ | $Q_o$ | $P_n$ | $Q_n$ |
| Commodity X | $1.00 | 10 | 0.90 | 20 |
| Commodity Y | 2.00 | 20 | 2.12 | 40 |
| Commodity Z | 5.00 | 100 | 5.50 | 70 |

a. Find the simple average of relatives for prices.
b. Compute the Laspeyres index for prices.

57. Given the price and quantity data shown, find the following price indexes.

| $P_o$ | $Q_o$ | $P_n$ | $Q_n$ |
|-------|-------|-------|-------|
| 20    | 8     | 24    | 9     |
| 15    | 12    | 22    | 8     |
| 40    | 5     | 30    | 10    |

a. Laspeyres
b. Paasche
c. Fisher's ideal

*Note:* Before you work Problem 57, estimate which index will be highest and which will be lowest. Then compute the index numbers to see if your estimates were correct.

58. A machinery price index had a base year of 1983. In 1992, the index value was 120.0. The base was then shifted so that 1992 became the base of the new series. Under the new base, what would be the value of the index for 1983?

59. In 1980 the average wage in a farm equipment manufacturing plant was $260 per week. By 1990, it had risen to $400 per week. A cost-of-living index relevant to employees in this plant had changed from 121 in 1980 to 230 in 1990. What was the percent change in "real," or deflated, wages over the period 1980–1990?

60. In a strong inflationary period, prices increased 22% over a two-year period. By how much has the purchasing power of money declined during the two years?

61. A manufacturing firm had the following price, quantity, and index data for an appliance product, where the simple value index $(V_{1992}/V_{1989}) \cdot 100 = 109$.

|                  | 1989        | 1992     |
|------------------|-------------|----------|
| Price            | $14.00 each | $13.50   |
| Quantity         | 4500 units  |          |
| Production index | 125.1       | 134.9    |

a. Using the simple value index, compute the quantity produced in 1992.
b. Did the increase in dollar value of this product keep pace with the production index change?

**STATISTICS IN ACTION: CHALLENGE PROBLEMS**

62. All of us are aware of the CPI, which provides a national measure of the cost of living. But are you aware that there is a subnational-level price index? The ACCRA cost-of-living index measures relative price levels for consumer goods and services in participating cities compared to a national city average. The number of participating cities varies but is in the range of 225–300 for this quarterly measure. A partial listing is shown below. Complete survey results may be available through your local Chamber of Commerce or at a city library. The survey data are presented in two sections. The first section is an all-cities index report that shows the index for each

ACCRA Cost-of-Living Index, First Quarter 1990

| Component Index Weights | 100% Composite Index | 17% Grocery Items | 22% Housing | 11% Utilities | 13% Transportation | 7% Health Care | 30% Misc. Goods and Services |
|---|---|---|---|---|---|---|---|
| MSA/PMSA | | | | | | | |
| URBAN AREA AND STATE | | | | | | | |
| Albany, GA MSA | | | | | | | |
| Albany, GA | 93.6 | 96.4 | 84.0 | 105.8 | 85.4 | 86.8 | 99.6 |
| Atlanta, GA MSA | | | | | | | |
| Atlanta, GA | 101.8 | 101.3 | 94.5 | 115.5 | 100.9 | 117.9 | 99.0 |
| Augusta, GA-SC MSA | | | | | | | |
| Augusta, GA | 99.3 | 96.2 | 88.8 | 115.6 | 102.3 | 94.0 | 102.6 |
| Macon-Warner Robins, GA MSA | | | | | | | |
| Macon, GA | 95.5 | 93.8 | 91.9 | 108.5 | 94.1 | 85.4 | 97.4 |
| Savannah, GA MSA | | | | | | | |
| Savannah, GA | 94.4 | 93.4 | 80.5 | 94.7 | 103.6 | 101.3 | 99.6 |
| Nonmetropolitan Areas | | | | | | | |
| Americus, GA | 98.1 | 99.7 | 84.2 | 116.2 | 106.8 | 79.8 | 101.2 |
| Carrollton, GA | 100.5 | 104.1 | 82.9 | 121.2 | 89.3 | 102.6 | 108.3 |
| Rome, GA | 95.0 | 97.1 | 79.5 | 118.0 | 95.0 | 87.4 | 98.5 |
| Boise City, ID MSA | | | | | | | |
| Boise, ID | 100.3 | 96.4 | 108.6 | 73.7 | 99.8 | 110.4 | 104.0 |
| Nonmetropolitan Areas | | | | | | | |
| Pocatello, ID | 91.6 | 94.1 | 83.0 | 81.0 | 89.0 | 107.2 | 97.7 |
| Champaign-Urbana-Rantoul, IL MSA | | | | | | | |
| Champaign-Urbana, IL | 104.6 | 100.8 | 108.6 | 102.9 | 107.0 | 112.7 | 101.4 |
| Davenport-Rock Island-Moline, IL-IA MSA | | | | | | | |
| Quad-Cities, IL-IA | 97.6 | 99.6 | 105.9 | 93.4 | 94.9 | 93.0 | 94.0 |
| Decatur, IL MSA | | | | | | | |
| Decatur, IL | 95.1 | 97.5 | 83.7 | 102.8 | 100.8 | 89.3 | 98.0 |
| Peoria, IL MSA | | | | | | | |
| Peoria, IL | 105.4 | 107.2 | 115.9 | 95.4 | 99.9 | 102.3 | 103.3 |
| Rockford, IL MSA | | | | | | | |
| Rockford, IL | 104.0 | 96.0 | 98.2 | 129.1 | 106.2 | 100.0 | 103.6 |

city's standing in an all-items index that includes groceries, housing, utilities, transportation, health care, and miscellaneous goods and services. The second section is a price report by city for the individual items that were included in the survey. These reports allow relative comparisons of, for example, housing costs in participating cities. This might be used as one base for cost-of-living adjustments by a national firm in relocating its employees to different parts of the country.

SOURCE: "The Cost of Living Index." Published quarterly by the American Chamber of Commerce Researcher's Association (ACCRA), 1100 Milam Building, 25th Floor, Houston, TX 77002.

### Questions

a. From the partial ACCRA City Composite Index provided, compare any three cities. Which one (of the three) has the highest all-items index value? the highest grocery items index? the highest housing index? Is one city always highest?
b. In the table, the data in the second column from the left are referred to as the Composite Index. How does it differ from a simple average? Which element has the highest weighting? Which has the least?

ACCRA Cost-of-Living Index, First Quarter 1990   (*continued*)

| Component Index Weights | 100% Composite Index | 17% Grocery Items | 22% Housing | 11% Utilities | 13% Transportation | 7% Health Care | 30% Misc. Goods and Services |
|---|---|---|---|---|---|---|---|
| **Springfield, IL MSA** | | | | | | | |
| Springfield, IL | 99.5 | 99.4 | 100.0 | 84.9 | 105.5 | 106.5 | 100.3 |
| Nonmetropolitan Areas | | | | | | | |
| Freeport, IL | 101.2 | 99.1 | 96.7 | 133.4 | 99.3 | 97.1 | 95.7 |
| Quincy, IL | 95.7 | 90.9 | 91.1 | 110.8 | 102.9 | 90.3 | 94.5 |
| **Provo-Orem, UT MSA** | | | | | | | |
| Provo-Orem, UT | 88.5 | 88.2 | 77.7 | 85.7 | 94.5 | 85.9 | 95.7 |
| **Salt Lake City-Ogden, UT MSA** | | | | | | | |
| Salt Lake City, UT | 92.9 | 90.0 | 82.3 | 86.4 | 95.2 | 95.0 | 103.3 |
| Nonmetropolitan Areas | | | | | | | |
| Cedar City, UT | 88.8 | 98.5 | 68.9 | 80.8 | 94.9 | 89.1 | 98.2 |
| St. George, UT | 90.3 | 95.7 | 83.3 | 64.2 | 98.9 | 94.4 | 97.1 |
| **Nonmetropolitan Areas** | | | | | | | |
| Montpelier-Barre, VT | 119.0 | 105.8 | 147.8 | 161.7 | 104.3 | 98.9 | 100.8 |
| **Olympia, WA MSA** | | | | | | | |
| Olympia-Lacey-Tumwater, WA | 96.5 | 98.6 | 88.9 | 74.0 | 91.2 | 138.9 | 101.6 |
| **Richland-Kennewick-Pasco, WA MSA** | | | | | | | |
| Richland-Kennewick-Pasco, WA | 98.3 | 102.7 | 82.0 | 94.6 | 96.7 | 130.3 | 102.2 |
| **Seattle, WA PMSA** | | | | | | | |
| Seattle, WA | 113.2 | 110.4 | 137.7 | 64.0 | 117.6 | 136.5 | 107.6 |
| **Spokane, WA MSA** | | | | | | | |
| Spokane, WA MSA | 94.3 | 97.6 | 94.4 | 67.9 | 88.5 | 116.4 | 99.5 |
| **Tacoma, WA PMSA** | | | | | | | |
| Tacoma, WA | 101.7 | 107.5 | 96.8 | 78.6 | 97.2 | 135.5 | 104.7 |
| **Yakima, WA MSA** | | | | | | | |
| Yakima, WA | 97.9 | 104.8 | 88.4 | 90.5 | 98.9 | 121.9 | 97.6 |
| **Nonmetropolitan Areas** | | | | | | | |
| Walla Walla, WA | 95.5 | 106.3 | 83.9 | 74.1 | 95.7 | 112.4 | 101.8 |

63. Study the following economic data as reported in various U.S. government publications.

| Period | Civilian Employment (in thousands) | Unemployment Total (in thousands) | Gross National Product (in billions) | Prime Rate Charged by Banks (Interest Rate) |
|---|---|---|---|---|
| 1982 | 99,500 | 10,700 | 3,200 | 14.86 |
| 1983 | 100,800 | 10,700 | 3,400 | 10.79 |
| 1984 | 105,000 | 8,500 | 3,800 | 12.04 |
| 1985 | 107,100 | 8,300 | 4,000 | 9.93 |
| 1986 | 109,600 | 8,200 | 4,200 | 8.33 |
| 1987 | 112,400 | 7,400 | 4,500 | 8.21 |
| 1988 | 115,000 | 6,700 | 4,900 | 9.32 |
| 1989 | 117,300 | 6,500 | 5,200 | 10.87 |
| 1990 | 117,200 | 7,600 | 5,300 | 10.00 |

a. Using 1982–1983 as the base period, compute the unemployment index for (a) 1986 and (b) 1990.

b. Compare the percentage change in the GNP with the changes in the unemployment from 1982 to 1990. What conclusion can you draw from this comparison?

c. Graph the interest rates and the GNP for the years 1982–1990. Do you observe any trends in the interest rates and/or the GNP?

---

**BUSINESS CASE**

*Which Way to Colorado?*

If Thomas Henderson was going to gain top management support for building a plant in Colorado, he first had to convince Janis Malichovich that this was a good move. Tom Henderson was the general manager of manufacturing, and he strongly favored a move to the more centralized Colorado location. Janis, however, was not yet convinced of the wisdom of such a move. And as corporate comptroller she kept a tight rein on the purse strings. She did not look forward to moving to a location that might have high operating costs, even if the fixed costs of the initial facility were reasonably low. Costs of raw material supplies at the firm's plant in Florida had been quite reasonable over the past few years, but she had no evidence that operations at a Colorado site would fare as well.

**Tom**   What's it going to take to convince you, Janis? Are we going to have to fly you up there so you can spend some time in the area and just get a better feel for the place?

**Janis**   You know that's not the problem, Tom. I know just as well as you do that we need to move into a bigger plant—there's no question about that. I just don't want us to make a $12 million investment and then find we can't afford to buy the parts we need to put our electronic gadgets together. From what I can tell, some of the raw material items we use are more expensive in Colorado than they are here in Florida.

**Tom**   That may be true for some of the components, Janis, but they are the exceptions, not the rule. Suppose I pull together some comparative data for you. Then we can look them over tomorrow. OK?

**Janis**   That's fine with me, but let's not get bogged down in too many specific items. Can you include some summary figures so we don't lose sight of the big picture?

**Tom**   I think so. I'll put Carla Anderson on it this morning. She's pretty good at this sort of thing. See you tomorrow.

Back in his office, Tom went over his strategy with Carla. They decided to select four key raw material items. Then Carla would go back into purchasing records to find price and usage data for the past five years. In the meantime, Tom would call the assistant manager in the Denver sales office and have him contact potential suppliers of those same items in Colorado. The suppliers there would be asked to check their old catalogs to find out what prices would have prevailed for those same items if they had been purchased for shipment to the proposed Colorado location.

It took several phone calls and some digging into old files, but the data were available. Moreover, the two locations were networked by computer, so the data transfer was really quite simple. The Denver office even activated the printer in the Florida plant and printed out the requested prices. By 11:00 A.M. the next day, Carla had the materials together in table form for Tom. The top sheet of his packet contained the comparison shown in Table 14.14.

**Tom**   Thanks very much, Carla. I don't have much time to go over the figures now, but they don't look all that good, do they? Some of the items are lower, but the Colorado prices of some items are quite high too. What do you make of this?

**TABLE 14.14**   Price and Volume Comparisons: Florida Site vs. Colorado Site

### Florida

| Item | Price per Unit ($) | | | | | Q = No. of Items Used | | | | |
|------|-----|-----|-----|-----|-----|-----|-----|-----|-----|-----|
| | *1* | *2* | *Year 3* | *4* | *5* | *1* | *2* | *Year 3* | *4* | *5* |
| A (electronic item) | 12 | 12 | 23 | 27 | 29 | 13 | 15 | 28 | 50 | 48 |
| B (mechanical item) | 125 | 122 | 98 | 88 | 82 | 20 | 18 | 9 | 12 | 8 |
| C (chemical item) | 82 | 124 | 105 | 144 | 142 | 5 | 8 | 14 | 32 | 50 |
| D (electronic item) | 53 | 55 | 43 | 35 | 28 | 34 | 34 | 20 | 16 | 10 |

### Colorado

| Item | Price per Unit ($) | | | | | Q = No. of Items Used | | | | |
|------|-----|-----|-----|-----|-----|-----|-----|-----|-----|-----|
| | *1* | *2* | *Year 3* | *4* | *5* | *1* | *2* | *Year 3* | *4* | *5* |
| A (electronic item) | 19 | 18 | 15 | 14 | 13 | 13 | 15 | 28 | 50 | 48 |
| B (mechanical item) | 121 | 118 | 105 | 128 | 138 | 20 | 18 | 9 | 12 | 8 |
| C (chemical item) | 112 | 105 | 94 | 98 | 102 | 5 | 8 | 14 | 32 | 50 |
| D (electronic item) | 38 | 50 | 53 | 65 | 78 | 34 | 34 | 20 | 16 | 10 |

**Carla**   Yes, I know, there is a real difference. But I think you can make a case. I wish there were time to go over them in detail with you. However, I did put some more comparative materials inside. If you have time, you might look at them on the way to your meeting.

On his way to Janis' office, Tom went over the second page of Carla's report, where she had done a calculation of a Laspeyres price index comparison for the Florida and Colorado sites. His heart sank as he scanned the calculations to focus upon the resulting price index numbers (see Table 14.15), which were 75.6 for Florida and 125.4 for Colorado. Laspeyres price index—Florida:

$$L_{5|1} = \frac{\sum (P_n Q_o)}{\sum (P_o Q_o)} (100) = \frac{3679}{4868} (100) = 75.6$$

Laspeyres price index—Colorado:

$$L_{5|1} = \frac{\sum (P_n Q_o)}{\sum (P_o Q_o)} \, 100 = \left(\frac{6091}{4859}\right) 100 = 125.4$$

Not having enough time to review any more of the materials beforehand, Tom resigned himself to the hope that there was something more in the folder than what his cursory examination of the data seemed to indicate. So he reluctantly presented it to Janis. After she looked over the data, Janis came back with a surprising comment.

**Janis**   I thought I might have to give a little on this one, Tom, but you do make a good case.

**Tom**   I do?

**Janis**   Yes. It's the *Paasche index* that convinces me. I can see there has been a definite shift in our raw material usage over the past few years. And the Colorado site will enable us to capitalize on that trend. You can count on my support for that location!

**TABLE 14.15**   Laspeyres Price Index Calculation for Raw Materials

### Florida

| | Base | Price per Unit ($) | | | | | Price · Base-Year Quantity | | | | |
|---|---|---|---|---|---|---|---|---|---|---|---|
| | | Year | | | | | Year | | | | |
| Item | $Q_1$ | 1 | 2 | 3 | 4 | 5 | 1 | 2 | 3 | 4 | 5 |
| A | 13 | 12 | | | | 29 | 156 | | | | 377 |
| B | 20 | 125 | | | | 82 | 2500 | | | | 1640 |
| C | 5 | 82 | | | | 142 | 410 | | | | 2760 |
| D | 34 | 53 | | | | 28 | 1802 | | | | 952 |
| Totals | | | | | | | 4868 | | | | 3679 |

### Colorado

| | Base | Price per Unit ($) | | | | | Price · Base-Year Quantity | | | | |
|---|---|---|---|---|---|---|---|---|---|---|---|
| | | Year | | | | | Year | | | | |
| Item | $Q_1$ | 1 | 2 | 3 | 4 | 5 | 1 | 2 | 3 | 4 | 5 |
| A | 13 | 19 | | | | 13 | 247 | | | | 169 |
| B | 20 | 121 | | | | 138 | 2420 | | | | 2760 |
| C | 5 | 112 | | | | 102 | 560 | | | | 510 |
| D | 34 | 48 | | | | 78 | 1632 | | | | 2652 |
| Totals | | | | | | | 4859 | | | | 6091 |

**Tom**   This is great! By the way, did I ever tell you how much I appreciate having Carla Anderson on my staff?

**Janis**   Yes, many times. And I can see why.

**Business Case Questions**

64. Explain the (philosophical) difference between the Laspeyres and Paasche approaches to index number calculations. Which would be more appropriate for this case? Why?

65. Compute the Paasche price index calculation for the case data, for both the Florida and Colorado locations. Use the year-1 and year-5 figures only.

66. Using the results from Problems 64 and 65, explain why you think Janis agreed that the Colorado location was appropriate.

67. Compute Fisher's ideal index for the case data for both locations. Use the year-1 and year-5 data only. (Problem 65 results can be used.)

68. Define the quantities for a typical year as the simple average for the quantities of years 4 and 5. Now compute the typical period index for the case data for both locations, again using the year-1 and year-5 prices.

**COMPUTER APPLICATION**   Previous computer applications have considered targeting households for the sale of personal computers. This market identification interests local computer stores, department stores, and discount firms that sell computers through direct mail. This application concerns targeting household types that already have personal

**TABLE 14.16**  Mediamark Research, Inc., Survey Summary by Age of Owners of Home/Personal Computers—Decision-Makers

| Base: Adult Decision-makers | Total U.S. (000) | A Responders (000) | B % Down | C % Across | D Index |
|---|---|---|---|---|---|
| ALL ADULTS | 172957 | 14271[a] | 100.0 | 8.3 | 100[b] |
| *Age subcategories* | | | | | |
| 18–24 | 27409 | 1853[c] | 13.0 | 6.8 | 82 |
| 25–34 | 41340 | 3787 | 26.5 | 9.2 | 111 |
| 35–44 | 31803 | 5015[d] | 35.1 | 15.8 | 191 |
| 45–54 | 22672 | 2043 | 14.3 | 9.0 | 109 |
| 55–64 | 22564 | 917 | 6.4 | 4.1 | 49 |
| 65 or over | 27170 | 656 | 4.6 | 2.4 | 29 |

[a,b] $\dfrac{14{,}271}{172{,}957} = 0.0825$ or $8.3\%\ (= 100\ \text{index})$

[c] $\dfrac{1{,}853}{27{,}409} = 0.0676$ or $6.8\%$

[d] $\dfrac{5{,}015}{31{,}803} = 0.1577$ or $15.8\%$

computers. The data source was a national survey with over 14 million estimated owners who are decision-makers.* Profiling the existing owners makes sense if we may assume that future owners will have traits like the current owners.

Table 14.16 is an excerpt on age from a past summary.

The total U.S. adult population is projected as 172,957,000—including 14,271,000 adults, or 8.3%, who say they own a personal computer and are decision-makers. These items are summarized in the display on the top row under Total and under the A and C column headings.

Owner percents for various adult ages are given in column C. For example, the ownership is 6.8% in the age span for 18–24 years, that for 25–34 years is 9.2%, and so on. Using a simple quantity index with 8.3% (= 100) as base gives the following:

$$18\text{–}24 \text{ years old: } (6.8/8.3) \cdot 100 = 81.9 \text{ or } 82 \text{ index}$$
$$25\text{–}34 \text{ years old: } (9.2/8.3) \cdot 100 = 111 \text{ index}$$

These and more index numbers appear in column D of Table 14.16. We can obtain the INDEX numbers by dividing a column-C entry (before rounding for the display by 8.25(%).

From the index column (D), the highest indices on age occur for 35–44 years (191 index), then 25–34 years (111), and 45–54 years (109 index). The older

* Mediamark Research, Inc., New York, NY. The results from approximately 20,000 total responders are weighted to estimate national numbers.

**TABLE 14.17** Mediamark Research, Inc., Survey Summary of Demographics for Owners of Home/Personal Computers—Decision-Makers

| Base: Adult Decision-Makers | Total U.S. (000) | A Responders (000) | Own Any B % Down | Own Any C % Across | D Index |
|---|---|---|---|---|---|
| ALL ADULTS | 172957 | 14271 | 100.0 | 8.3 | 100 |
| MEN | 82458 | 8324 | 58.3 | 10.1 | 122 |
| WOMEN | 90499 | 5947 | 41.7 | 6.6 | 80 |
| HOUSEHOLD HEADS | 94362 | 8272 | 58.0 | 8.8 | 106 |
| HOMEMAKERS | 95700 | 6935 | 48.6 | 7.2 | 88 |
| GRADUATED COLLEGE | 30178 | 4527 | 31.7 | 15.0 | 182 |
| ATTENDED COLLEGE | 31384 | 3167 | 22.2 | 10.1 | 122 |
| GRADUATED HIGH SCHOOL | 67356 | 4823 | 33.8 | 7.2 | 87 |
| DID NOT GRADUATE HIGH SCHOOL | 44039 | 1755 | 12.3 | 4.0 | 48 |
| 18–24 | 27409 | 1853 | 13.0 | 6.8 | 82 |
| 25–34 | 41340 | 3787 | 26.5 | 9.2 | 111 |
| 35–44 | 31803 | 5015 | 35.1 | 15.8 | 191 |
| 45–54 | 22672 | 2043 | 14.3 | 9.0 | 109 |
| 55–64 | 22564 | 917 | 6.4 | 4.1 | 49 |
| 65 OR OVER | 27170 | 656 | 4.6 | 2.4 | 29 |
| 18–34 | 68749 | 5640 | 39.5 | 8.2 | 99 |
| 18–49 | 112153 | 11868 | 83.2 | 10.6 | 128 |
| 25–54 | 95815 | 10645 | 76.0 | 11.3 | 137 |
| EMPLOYED FULL TIME | 95802 | 10288 | 72.1 | 10.7 | 130 |
| PART-TIME | 11463 | 1049 | 7.4 | 9.2 | 111 |
| NOT EMPLOYED | 65692 | 2933 | 20.6 | 4.5 | 54 |
| PROFESSIONAL | 14053 | 2076 | 14.5 | 14.8 | 179 |
| EXECUTIVE/ADMIN./ MANAGERIAL | 12766 | 2006 | 14.1 | 15.7 | 190 |
| CLERICAL/SALES/ TECHNICAL | 33232 | 3473 | 24.3 | 10.5 | 127 |
| PRECISION/CRAFTS/ REPAIR | 13410 | 1037 | 7.3 | 7.7 | 94 |
| OTHER EMPLOYED | 33804 | 2745 | 19.2 | 8.1 | 98 |
| H/D INCOME $50,000 OR MORE | 36223 | 4826 | 33.8 | 13.3 | 161 |
| $40,000–49,999 | 24296 | 3208 | 22.5 | 13.2 | 160 |
| $35,000–39,999 | 14270 | 1509 | 10.6 | 10.6 | 128 |
| $25,000–34,999 | 32350 | 2243 | 15.7 | 6.9 | 84 |
| $15,000–24,999 | 32084 | 1333 | 9.3 | 4.2 | 50 |
| LESS THAN $15,000 | 33734 | 1151 | 8.1 | 3.4 | 41 |
| CENSUS REGION: | | | | | |
| NORTH EAST | 37206 | 3242 | 22.7 | 8.7 | 106 |
| NORTH CENTRAL | 42935 | 3028 | 21.2 | 7.1 | 85 |
| SOUTH | 59451 | 4574 | 32.1 | 7.7 | 93 |
| WEST | 33364 | 3427 | 24.0 | 10.3 | 124 |

**TABLE 14.17**   (*continued*)

| Base: Adult Decision-Makers | Total U.S. (000) | A Responders (000) | Own Any B % Down | Own Any C % Across | D Index |
|---|---|---|---|---|---|
| MARKETING REG.: | | | | | |
| NEW ENGLAND | 9694 | 961 | 6.9 | 10.1 | 123 |
| MIDDLE ATLANTIC | 30810 | 2448 | 17.2 | 7.9 | 96 |
| EAST CENTRAL | 24387 | 2068 | 14.6 | 8.6 | 104 |
| WEST CENTRAL | 27301 | 1804 | 12.6 | 6.6 | 80 |
| SOUTH EAST | 31257 | 2394 | 16.8 | 7.7 | 93 |
| SOUTH WEST | 20210 | 1543 | 10.8 | 7.6 | 93 |
| PACIFIC | 29297 | 3013 | 21.1 | 10.3 | 125 |
| COUNTY SIZE A | 71708 | 6498 | 45.5 | 9.1 | 110 |
| COUNTY SIZE B | 51235 | 4484 | 31.4 | 8.8 | 106 |
| COUNTY SIZE C | 26364 | 1878 | 13.2 | 7.1 | 86 |
| COUNTY SIZE D | 23650 | 1411 | 9.9 | 6.0 | 72 |
| MSA CENTRAL CITY | 64765 | 5535 | 38.8 | 8.5 | 104 |
| MSA SUBURBAN | 67877 | 6325 | 44.3 | 9.3 | 113 |
| NON-MSA | 40315 | 2411 | 16.9 | 6.0 | 72 |
| SINGLE | 36543 | 2416 | 16.9 | 6.6 | 80 |
| MARRIED | 105233 | 10329 | 72.4 | 9.8 | 119 |
| OTHER | 31181 | 1526 | 10.7 | 4.9 | 59 |
| PARENTS | 60647 | 7856 | 55.0 | 13.0 | 157 |
| WORKING PARENTS | 45208 | 6477 | 45.4 | 14.3 | 174 |
| HOUSEHOLD SIZE: | | | | | |
| 1 PERSON | 20592 | 729 | 5.1 | 3.5 | 43 |
| 2 PERSONS | 51147 | 2934 | 20.6 | 5.7 | 70 |
| 3 OR MORE | 101218 | 10806 | 74.3 | 10.5 | 127 |
| ANY CHILD IN HOUSEHOLD | 73678 | 8999 | 63.1 | 12.2 | 148 |
| UNDER 2 YEARS | 12820 | 913 | 6.4 | 7.1 | 86 |
| 2–5 YEARS | 26707 | 2476 | 17.3 | 9.3 | 112 |
| 6–11 YEARS | 32448 | 4413 | 30.9 | 13.6 | 165 |
| 12–17 YEARS | 35162 | 5146 | 36.1 | 14.6 | 177 |
| WHITE | 150344 | 12827 | 89.9 | 8.5 | 103 |
| BLACK | 19088 | 1196 | 8.4 | 6.3 | 76 |
| HOME OWNED | 120315 | 11073 | 77.6 | 9.2 | 112 |
| DAILY NEWSPAPERS: | | | | | |
| READ ANY | 104108 | 9345 | 65.5 | 9.0 | 109 |
| READ ONE DAILY | 79194 | 6680 | 46.9 | 8.4 | 102 |
| READ TWO OR MORE DAILIES | 24825 | 2656 | 18.6 | 10.7 | 130 |
| SUNDAY NEWSPAPERS: | | | | | |
| READ ANY | 110422 | 10070 | 70.6 | 9.1 | 111 |
| READ ONE SUNDAY | 95678 | 8457 | 59.3 | 8.8 | 107 |
| READ TWO OR MORE SUNDAYS | 14744 | 1613 | 11.3 | 10.9 | 133 |

and the young adult responders had lower indices of ownership. (Recall our usage for 35–50 years as a target population to include families with high-school and college-age persons.)

Table 14.17 is from the product-use summary on *Audio and Video Equipment, Leisure Activities.*\*

Table 14.17 gives indices for individual and demographic characteristics, including age, just discussed. Other key characteristics for this product are education, occupation type, income, marital status, household size, and the presence of children by age group.

This kind of survey data is used by businesses to estimate the market potential of products. This research estimated that 8.3% of all adults owned a home/personal computer and were decision-makers. Then the 91.7% of adult decision-maker households who did not own computers are viewed as potential buyers. A word of caution is necessary in applying these results. Because home computers are a fast-growing consumer product, market profiles that are six months or one year apart can be different. You should use the most recent results that are available.

Below are some questions about profiles for personal computer owners. Although these problems do not require a computer for their solutions, the questions are designed to give you another perspective on computer applications.

## COMPUTER PROBLEMS

Here are some questions that relate to building profiles for home/personal computer owners. Suppose you are a marketer with a firm that sells microcomputers. Use Table 14.17 to answer the next questions.

69. How would you respond if the marketing manager proposed to advertise to household (H/D) with incomes $25,000 and higher? Consider the H/D INCOME subtable, and then explain why you would agree with the manager *or* suggest an alternate income span.

70. Explain which attributes for *families* were most descriptive of personal computer owners. Choose, as appropriate, characteristics from marital status, parents or working parents, household size, and presence of children by age. Consider only attributes with an index greater than 110.

71. The indexes presented here are simple quantity indexes. Explain how you would use the INDEXES to obtain a composite index for H/D INCOME $35,000 or more. *Suggestion:* Use a weighted average where TOTAL U.S. column numbers are the weights.

72. Explain why it is *not* appropriate to combine the indexes for age and income, for instance, as an average of quantity relatives to condense several attributes into one index measure.

73. Agrichem, Inc., experienced the following production pattern for a pesticide used in the western United States. Use the ratio-to-moving-average method (with modified mean) to develop a monthly seasonal index for the pesticide. (*Note:* This problem involves considerable calculation and is best handled via computer.)

\* Ibid, p. 138.

| Month | 1985 | 1986 | 1987 | 1988 | 1989 | 1990 | 1991 | 1992 |
|-------|------|------|------|------|------|------|------|------|
| Jan. | 251 | 264 | 262 | 288 | 276 | 250 | 306 | 287 |
| Feb. | 245 | 255 | 253 | 274 | 268 | 286 | 256 | 274 |
| Mar. | 275 | 285 | 304 | 317 | 297 | 295 | 307 | 323 |
| Apr. | 298 | 309 | 327 | 301 | 328 | 317 | 342 | 333 |
| May | 329 | 320 | 342 | 326 | 356 | 345 | 344 | 383 |
| June | 354 | 379 | 365 | 406 | 395 | 411 | 380 | 444 |
| July | 341 | 354 | 386 | 352 | 384 | 323 | 302 | 402 |
| Aug. | 240 | 237 | 238 | 267 | 188 | 224 | 238 | 202 |
| Sept. | 208 | 217 | 189 | 220 | 204 | 144 | 225 | 230 |
| Oct. | 154 | 160 | 144 | 125 | 183 | 153 | 165 | 182 |
| Nov. | 187 | 198 | 171 | 200 | 214 | 146 | 163 | 177 |
| Dec. | 206 | 217 | 180 | 208 | 200 | 223 | 243 | 260 |

74. Using the seasonal index developed in Problem 73, deseasonalize the 1992 production figures for Agrichem, Inc.

75. Harmon Plywood Co. derived the following *specific seasonals* representing lumber sales. Use the modified mean approach to compute a seasonal index and make whatever corrections are necessary so that the sum of the 12 values equals 1200.

| Month | 1985 | 1986 | 1987 | 1988 | 1989 | 1990 | 1991 |
|-------|------|------|------|------|------|------|------|
| Jan. | 85.1 | 85.3 | 84.9 | 84.6 | 84.7 | 85.1 | 84.9 |
| Feb. | 74.1 | 74.9 | 76.2 | 77.6 | 75.0 | 74.1 | 76.7 |
| Mar. | 89.0 | 88.8 | 90.2 | 92.4 | 91.7 | 86.2 | 91.5 |
| Apr. | 100.6 | 101.8 | 101.4 | 98.9 | 100.1 | 99.4 | 102.5 |
| May | 109.5 | 111.1 | 110.5 | 112.0 | 108.7 | 114.2 | 110.1 |
| June | 141.2 | 143.5 | 142.0 | 137.0 | 129.0 | 137.0 | 141.6 |
| July | 121.1 | 119.3 | 118.7 | 117.6 | 114.3 | 124.2 | 122.1 |
| Aug. | 102.4 | 100.7 | 96.8 | 105.8 | 111.2 | 101.5 | 96.7 |
| Sept. | 94.2 | 93.6 | 95.3 | 97.7 | 93.3 | 90.2 | 95.6 |
| Oct. | 96.4 | 98.7 | 101.3 | 90.2 | 92.1 | 98.8 | 96.6 |
| Nov. | 102.7 | 103.2 | 99.7 | 98.7 | 102.3 | 100.1 | 99.8 |
| Dec. | 89.2 | 90.0 | 88.6 | 91.3 | 95.2 | 88.3 | 87.0 |

**FOR FURTHER READING**

[1] Albright, S. Christian. *Statistics for Business and Economics.* New York: Macmillan, 1987.
[2] *Business Conditions Digest.* Published monthly by the U.S. Department of Commerce, Bureau of Economic Analysis, Washington, D.C.
[3] *Current Population Report*, U.S. Department of Commerce, Bureau of the Census, Washington, D.C.
[4] Dixon, Wilfred J. (ed.). *BMDP Statistical Software.* Los Angeles: University of California Press, 1990.
[5] *Economic Indicators.* Published monthly by the Council of Economic Advisors, Washington, D.C.

[6] Hamburg, Morris. *Statistical Analysis for Decision Making*. 5th ed. Orlando, FL: Harcourt Brace Jovanovich, 1991.

[7] Iman, Ronald L., and W. J. Conover. *Modern Business Statistics*, 2nd ed. New York: Wiley, 1989.

[8] "Inter-City Cost of Living Indicators." Published quarterly by the American Chamber of Commerce Researcher's Association (ACCRA), 1100 Milan Building, 25th Floor, Houston, TX, 77002.

[9] Levin, Richard I. *Statistics for Management*, 4th ed. Englewood Cliffs, NJ: Prentice Hall, 1987.

[10] Mendenhall, William, James E. Reinmuth, Robert Beaver, and Dale Durham. *Statistics for Management and Economics*, 6th ed. Boston: PWS-Kent, 1989.

[11] Miller, Robert B., and Dean W. Wichern. *Intermediate Business Statistics*. New York: Holt, Rinehart, and Winston, 1977.

[12] Neter, John, William Wasserman, and G. A. Whitmore. *Applied Statistics*, 3rd ed. Boston: Allyn and Bacon, 1988.

[13] *Survey of Current Business*. Published monthly by the U.S. Department of Commerce, Bureau of Economic Analysis, Washington, D.C.

[14] Pfaffenberger, Roger C., and James H. Patterson. *Statistical Methods for Business and Economics*, 3rd ed. Homewood, IL: Irwin, 1987.

[15] Richmond, Samuel B. *Statistical Analysis*, 2nd ed. New York: Wiley, 1964.

[16] *The Wall Street Journal*. Published by Dow Jones and Co., New York.

# TOTAL QUALITY CONTROL

From the beginning of the Industrial Revolution, around 1900, to the present day, virtually every manufacturer of goods and services has attached importance to producing and marketing "high-quality" products. But it was not until the 1940s and 1950s (around World War II) that quality-control methodology was really implemented in the United States on a large scale. This innovation was motivated largely by the defense needs of the Allied forces—to ensure that military equipment could be relied upon to perform to its design standards. And this attention to quality helped the Allied forces win the war.

At the end of World War II, Japanese industry was in disarray, and its products were typically perceived as "shoddy." As part of the U.S. reconstruction efforts, Dr. W. Edwards Deming (a quality-control specialist) was invited to Tokyo to help Japanese industry "get back on its feet." His success in using statistical quality-control techniques was so phenomenal that today, about fifty years later, Japanese firms and financial institutions dominate the world economy. And one of the most coveted awards one can receive in Japanese industry is the prestigious "Deming" prize for outstanding progress in the area of quality control.

> After World War II, most Japanese companies had to start literally from the ground up. Every day brought new challenges to managers and workers alike, and every day meant progress. Simply staying in business has required unending progress, and KAIZEN (continual improvement) has become a way of life. It was fortunate that the various tools that helped elevate this KAIZEN concept to new heights were introduced to Japan in the late 1950s and early 1960s by such experts as (Professor) W. E. Deming and J. M. Juran. However, most new concepts, systems, and tools that are widely used in Japan today have been developed in Japan and represent qualitative improvements upon the statistical quality control and total quality control of the 1960s.
>
> Masaaki Imai [5]*

* Bracketed numbers refer to items listed in "For Further Reading" at the end of the chapter.

**FIGURE 15.1** The General Motors Quality Network Model

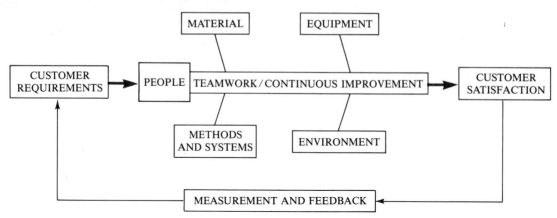

A TOOL TO ANALYZE, UNDERSTAND, AND IMPROVE ANY PROCESS

Today American firms are concerned with being able to compete in the market-place for the sale of consumer products. Certainly in the areas of electronics and automotive engineering, it is reasonable to say that the Japanese dominate world markets, a state of affairs that has resulted in a new motivation to improve the quality of American-made products. This improvement comes about through extending traditional management techniques to focus upon *total quality control*. One example is General Motors Corporation's "Quality Network Process Model," as shown in Figure 15.1.

Figure 15.1 is a fishbone process diagram that has people at its center. It suggests that any plan to improve quality must focus first on people. Quality improvement involves a high level of teamwork—cooperation among all the people who are associated with the company and its products. And the processes of improvement are a focus for the company.

Statistical quality-control methods are described by three general levels of process control. The first level is simple identification or screening of low-quality products. The objective here is to identify low-grade raw materials or poor-quality finished products, and reject these production "lots" with a high degree of accuracy. This is the traditional (lot) acceptance sampling, for which numerous "acceptance sampling plans" exist in quality-control texts. The second (higher) level of control is continuous monitoring of a process to check that the level of production is being maintained at a standard. This level uses specially designed control charts to detect undesirable shifts or movements in the production process before problems get out of hand. The third level of control concerns the continual checking and improvement of quality to ensure (1) that a standard is maintained and (2) that ways to continually improve the production are identified. This is *total quality control*.

**Total quality control is the continual improvement of people, products, and the work environment.**

Total quality control goes beyond the product and the production line to include management action about people. This is being implemented in U.S. industry by "quality circles," "task forces," and other management–worker team efforts aimed at improving the product and the working environment.

Because an understanding of basic concepts is so important, we concentrate on the second level—on monitoring techniques—and most specifically on the use of control charts. In addition, we will illustrate some techniques for quality improvement. See references [1] and [3] for a discussion of acceptance sampling.

Implementing total quality control requires using trained observers who are aware that change is continuous, rather than intermittent or ending. A commonplace statement is that quality control is 90% understanding systems and only 10% statistics (that is, gathering, analyzing, and interpreting relevant data). People who have designed and maintained the systems understand the operation. Statisticians understand how to use and interpret data. Total quality control is accomplished through individuals who can work together for *continuous process improvement*.

> **Continuous process improvement requires a thorough knowledge of production systems and an understanding sufficient to improve those processes and systems.**

This definition is not restricted to the production of goods; it also includes services. Statistics and data analysis permeate the production of both goods and services. The Chapter 15 Business Case describes a use of total quality control in a business service setting.

The overall objective for Chapter 15 is to study the fundamental statistics of total quality control. This includes application of Deming's PDCA cycle and the use of control charts.

> **Upon completing Chapter 15, you should be able to**
>
> 1. construct a control chart;
> 2. read and use with some discernment *p*-charts for the proportion defective, *C*-charts for the number defective per item, and $\bar{X}$-charts and *R*-charts for variables data;
> 3. use the basic tools for total quality control, including control charts, Pareto diagrams, system flowcharting, and cause-and-effect diagrams;
> 4. apply the elements of Deming's PDCA cycle to a basic process-control problem;
> 5. apply the runs test to determine if process sample data are a random sample

Chapter 9 provided some foundation for the concepts of this chapter, including the idea of control (confidence) limits. Here, we also use the concepts of normal distribution and random sampling. However, the techniques for applying these differ, and total quality control requires a new kind of problem-solving skill.

## 15.2
## A PLAN FOR TOTAL QUALITY CONTROL: THE PDCA CYCLE

The plan for this chapter requires some new terminology. Several ideas are embodied in the following example: Consider a production operation that is a direct-mail lettershop. Its purpose is to affix a mailing label to each envelope, insert the advertising pieces into the envelopes, and then mail the assembled "package." From a management viewpoint, the lettershop is one of many *systems* in the total organization.

> **A system is a set of processes working together, according to a plan, to fulfill some function of the organization.**

---

**FIGURE 15.2**    Deming's PDCA Cycle

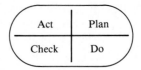

Other systems are the computer operations, purchasing, marketing, sales, and more. Each system is made up of components, called *processes.*

> A process is any action that produces an end result.

Thus the system of a lettershop includes one process of attaching address labels, another of assembling the mailing pieces, another of bagging the mail, and still another process of moving the mail in mailbags to a warehouse for shipping.

Professor Deming has provided us an overall plan for studying the quality aspects of operations that include various processes. His plan is sometimes referred to as the PDCA cycle, and is described in Figure 15.2.

The PDCA diagram shows four recurring steps as a guide for continuous process improvement. The *Plan stage* is important in that it involves describing the system under study. What are the processes that make up the system? It can be informative to develop a flowchart of the processes to show how they work together to define a system. The Plan stage also requires identifying quality characteristics or measures of performance for the system. Planning includes identifying those system characteristics (those that can be measured) that are probable causes of problems—or sources for improvement. Given the overall plan, one can then develop a plan for a specific quality-improvement activity. Every plan needs a statement of objectives, which may result in specific hypotheses to test. Here, it can be helpful to have a clear set of *operational definitions.*

> Operational definitions are written (versus verbal) statements of the requirements for a product to meet standards.

The *operational definitions* should include a specific decision criterion by which results can be judged as acceptable or not acceptable. Such definitions may include measurement procedures, sampling techniques, and test methods.

The Plan also requires a definition of the data and the data collection process. We will view four process measures: for discrete variables—(1) attributes and (2) number of defects; for continuous variables—(3) the mean and (4) the range.

The *Do stage* of the PDCA cycle is simply the portion in which workers carry out the plan that was developed in the Plan stage. This involves developing, and evaluating, test data. The Do stage is where the results are produced. Here it can be crucial to include those persons who are responsible for gathering and analyzing the data as well as at least some of the people who were involved in the Plan stage.

In the *Check stage*, the data are summarized and compared to an existing, or baseline, condition.

> A baseline standard is the expected level of operational performance.

The *baseline standard* is established by experience. It can be directed by customer requirements or industry standards. For example, if the work experience is that 1000 mailable direct-mail pieces can be produced per hour per production line, then this defines a baseline. By comparing the baseline to the current data, we can verify the extent or kind of change needed to achieve the standard. In this way, improvement can be measured.

The *Act stage* is a reaction to the data evaluation. What changes should be made to the system? If problems are identified, then, we hope, their causes can be identified and corrected. The results need to be evaluated by management in order to effect planning and evaluation for the next cycle. If the change is positive, then it should become a part of the system standards and be incorporated into a new baseline.

Most of our discussion will center on the Do and Check stages of the PDCA cycle, because these are the points where statistics enters the plan. Professor Deming's observation is "Reduce variation and you improve quality" [2]. How do we recognize variation, and its sources, and then react to make improvements? This requires a study of process variation. A control chart is a place to begin.

A *control chart* is a graphic device that is used to measure possible sources of variation. Figure 15.3, illustrates a generalized control chart and classifies three common sources of variation that can be observed.

The measurements of a variable change in time because the process itself is changing. We can expect small, irregular changes to occur due to conditions in the work environment, differences in production inputs, and differences in people. These random variations are described by the small, chance variations of data points around a process centerline. Any process will have some *common cause* for variation.

> A common cause for variation is any expected change arising from normal operations.

The benchmark rate of production should incorporate any variation attributable to common causes. When the process variation is simply due to random occurrences, and the data points reside within a small distance from the process mean (with no systematic pattern), we say the process is stable and *in control*.

> A process is stable and in control when its production is predictable and remains within acceptable limits.

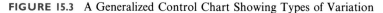

**FIGURE 15.3** A Generalized Control Chart Showing Types of Variation

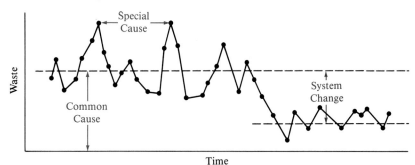

**FIGURE 15.4**   Statistics in Quality Improvement

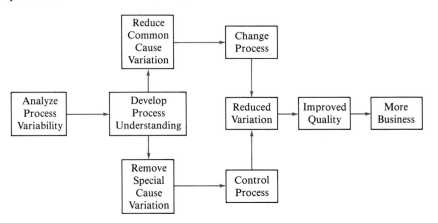

Although these conditions are somewhat nebulous now, in Section 15.3 we will define explicit conditions for a process to be declared in control. Otherwise, variation can occur from system changes. Machines wear down and processes become less efficient over time. These can lead to system changes or shifts in a process. The system change in Figure 15.3 might be the result of replacing old machinery with new. More often the change is an erratic shift toward more waste, which indicates an *identifiable cause*.

> An identifiable cause is an unusual influence that shifts a process
> beyond its normal operating range.

This can make the process unstable, unpredictable, and out of control. Examples of a special cause are a slipping arm on a mail-piece inserter and letter jams due to increased wear on the machine. Special causes can be difficult to isolate. However, a control chart can provide insights that will identify a recurring, specific problem due to special causes, as illustrated in Figure 15.4.

Control charts are useful graphics for checking that a process is in control. The first objective is to identify and remove assignable causes for variation—whether they are system changes, common causes, or special causes. When the process is in control, a process characteristic can be monitored with control charts. The role of process control is described in Figure 15.4, which shows a statistics application of the PDCA cycle.

A control chart gives a snapshot of some process characteristic at intermittent, and usually equally spaced, points in time. It displays data. If the characteristic is an attribute or binomial-scaled measure, like the number of defective items, we use a *p*-type control chart. When the process measure counts the number of defects per item described by a Poisson process, we use a *C*-chart. And for *variables or continuous data*, analysts use an $\bar{X}$-chart, which describes the process mean. These statistical tools are topics of the successive sections of this chapter.

## 15.3 PROPORTION DEFECTIVE: *p*-CHARTS

We focus attention now on control charts, beginning with charts showing the proportion defective—*p*-charts. They are used when quality assessment classifies the item in only two ways, that is, good or bad, pass or fail, defective or nondefective, and so on.

Analyzing the percent defective in production is a common application of quality control. This type of control is frequently referred to as control of attributes—that is, when the item is either good or it is defective. An example of attributes quality control is suggested by the label you might find in the pocket of a new item of clothing. It asserts "This product was inspected by quality inspector #1234," so we can assume it is a "good" item as opposed to a "defective" one.

How do producers control a process so that an excessive number of low-quality items are not produced or distributed? And how can process improvement be measured? This is referred to as process control, and in production it is accomplished by means of control charts. *Control charts* facilitate the continuous study of variation, which leads to identifying its causes, taking action to remove those causes, and reviewing the result of the action.

> A control chart is a graph used to monitor the quality characteristics of a production process over time. It typically shows the upper (UCL) and lower (LCL) control limits, within which sample statistics are expected to lie if the process remains in control.

Control charts are used to monitor a process. Control (chart) limits are set up on the assumption that small (random) changes are to be expected but that the real changes in a process will be investigated in an attempt to find the assignable causes. For example, assume oil-drilling shaft diameters are supposed to be 3.125 inches. We could expect some variation in diameters due to factors such as hidden (random) effects of the cutting speed used, temperature of the lubricants, and inaccuracies of the measuring instruments used by the quality-control inspector. But suppose that in production the shaft diameter evidenced a significant deviation or trend away from the 3.125-inch specification. An investigation might reveal that a different grade of steel was used or that the operator forgot to start a lubricating pump. These are *assignable causes* that shift the production output to unacceptable levels.

Control charts rely upon periodic sampling—usually of fixed sample sizes. As long as the process is in control, the sample statistics will nearly always remain within acceptable limits. Even though random deviations occur, the process is governed by a relatively constant system of causes.

One basic concern of a statistical quality-control program is to determine, with a prespecified risk, when a process is out of control and to find the assignable causes so the problems can be corrected. Hence with control charts our null hypothesis is that the process is in control.

Process control can be used for data to control either attributes ($\pi$) or variables ($\mu$). For an attributes process control the data are classified by a dichotomy such as good or bad or defective or nondefective. *Control charts for attributes* typically track the percent (or number) of defective items. The acceptable limits for attributes are based upon the *standard error of proportion* $\sigma_p$.

Variables are measurable data (interval or ratio level) such as diameter, height, weight, or time. *Control charts for variables* typically track the average (mean) measurement of fixed-size samples of the variable. The acceptable limits for variables are based upon the *standard error of the mean* $\sigma_{\bar{x}}$. Both attributes charts (called *p*-charts) and variables charts (called $\bar{X}$-charts) commonly use control limits that are three standard errors distance from the central value or centerline.

## p-Control Charts

Control charts for attributes are called *p-charts* because the fraction defective of the sample (or *p*) is used to determine whether the process is in control. For attributes control, where items are classed as either defective or nondefective we let *p* = the sample fraction defective. In this case, a defective is any item that does not meet quality specifications. Then, for $\pi$ = the true fraction defective, we define a "$3\sigma$ control chart" having upper and lower *control limits* (UCL and LCL) as follows:

**Process control limits for the fraction defective as defined by $3\sigma_p$ control limits are**

$$\text{UCL}_p = \bar{p} + 3\hat{\sigma}_p \qquad (15.1)$$

$$\text{LCL}_p = \bar{p} - 3\hat{\sigma}_p$$

**where $\text{LCL}_p = 0$ if the preceding calculation results in a negative value and**

$$\bar{p} = \frac{\sum\limits^{k} p_i}{kn_i}$$

**defines the centerline with**

$$\hat{\sigma}_p = \sqrt{\frac{\bar{p}(1 - \bar{p})}{k \cdot n_i}}$$

We consider only cases in which all samples are the same size, so $\sum^k n_i = kn_i$.

A key assumption is that the sampling distribution of the fraction defective is approximately normal when the sample size is sufficiently large. Then, with control limits at three standard errors from the central value, we would expect that 99.7% of the sample proportions would be within the $3\hat{\sigma}_p$ limits. For the process to be in control means that the percent defective is distributed around the centerline with no evident pattern or shift and with sufficiently small variations in the percent defective that no more than 2.5% of the samples exceed the $3\sigma$ limits. Here is an example.

---

**EXAMPLE 15.1**

A system for preparing boxes of fresh oranges for distribution includes processes of initial inspection, washing, waxing, grading, and boxing, followed by a final inspection. Figure 15.5 is a system flowchart.

In the flowchart in Figure 15.5 there are three points where the fruit is inspected and, if found defective, should be discarded. These are (1) at entry and (2) after the fruit is washed, dried, waxed, and sorted; (3) finally, a quality-control inspection is made after the fruit is put into boxes.

The quality-control plan requires first that some workers visually and by touch screen the fruit as it enters the production line on a conveyor belt. Any fruit with low-quality characteristics (poor color, soft spots for rot, hardness for unripe, or spongy and light for frozen) are discarded. This occurs at Inspection 1 and again at Inspection 2, as illustrated in Figure 15.5. This is highly dependent upon the skills of the sorters, who usually are experienced. Another quality-control inspection occurs at the end of the system, and makes use of electronic equipment to check a random sample of 200 oranges per

**FIGURE 15.5**   A System Flowchart for the Preparation of Fresh Oranges for
Distribution

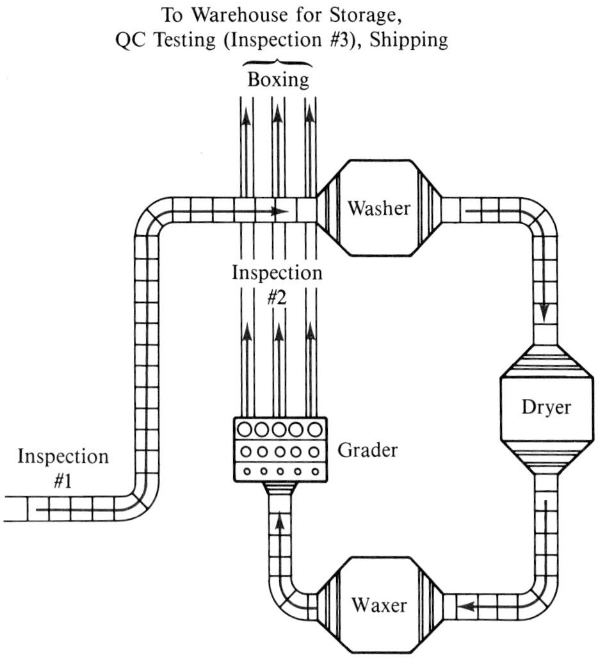

hour for color, water content, and firmness. The distributor has found that with more
than 8% bad fruit the buyers, who wholesale-distribute to grocers, will cut back on their
purchases. This is the customer standard. Table 15.1 displays sample inspection data
on oranges for 28 lots of 200 oranges each. Use these data to establish a $3\hat{\sigma}$ control
chart for the distribution of defective oranges—that is, bad fruit.

**TABLE 15.1**   Fraction Defective in Processing Oranges for Distribution, Lot Sample
Size = 200

| Sample Lot Number | Number of Defectives | Fraction Defective | Sample Lot Number | Number of Defectives | Fraction Defective |
|---|---|---|---|---|---|
| 1 | 4 | 0.020 | 15 | 15 | 0.075 |
| 2 | 3 | 0.015 | 16 | 14 | 0.070 |
| 3 | 2 | 0.010 | 17 | 8 | 0.040 |
| 4 | 3 | 0.015 | 18 | 6 | 0.030 |
| 5 | 1 | 0.005 | 19 | 4 | 0.020 |
| 6 | 4 | 0.020 | 20 | 21 | 0.105 |
| 7 | 13 | 0.065 | 21 | 11 | 0.055 |
| 8 | 5 | 0.025 | 22 | 7 | 0.035 |
| 9 | 7 | 0.035 | 23 | 10 | 0.050 |
| 10 | 8 | 0.040 | 24 | 14 | 0.070 |
| 11 | 10 | 0.050 | 25 | 4 | 0.020 |
| 12 | 6 | 0.030 | 26 | 3 | 0.015 |
| 13 | 12 | 0.060 | 27 | 10 | 0.050 |
| 14 | 5 | 0.025 | 28 | 14 | 0.070 |
| | | | | Total defectives = 224 | |

**FIGURE 15.6**   A *p*-Control Chart for the Defectives in Sample Lots of 200 Oranges

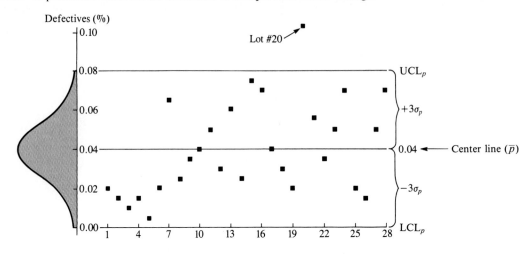

*Solution*   Table 15.1 displays the results of 28 samples each of a lot of 200 oranges. We observe that the fraction defective ranges from 1% (0.001) to 7.5% (shaded in Table 15.1).
The statistics for a control chart are

$$\bar{p} = \frac{\text{Total number of defective oranges}}{\text{Total number in samples}} = \frac{224}{28 \cdot 200} = 0.040$$

$$\hat{\sigma}_p = \sqrt{\frac{0.040 \cdot 0.960}{200}} = 0.0139$$

Then,

$$\text{LCL}_p = \bar{p} - 3\hat{\sigma}_p = 0.040 - 3(0.0139) = 0.040 - 0.042 = -0.002 \quad (\text{use LCL}_p = 0)$$

$$\text{UCL}_p = \bar{p} + 3\hat{\sigma}_p = 0.040 + 0.042 \qquad\qquad = 0.082$$

Since the lower control limit is below zero, which is not possible , that limit is replaced with zero. This, along with the data in Table 15.1, provides the information for a *p*-control chart as shown in Figure 15.6.

We observe in Figure 15.6 that the control limits of $\text{UCL}_p = 0.082$ and $\text{LCL}_p = 0$ encompass all samples except that from Lot 20. This lot would be evaluated and, if an assignable cause were determined, a correction made for it. The upper control limit, rounded, just meets the standard (with more than 8% defective the distributors will reject shipments). By viewing the fraction defective values in Figure 15.6 by sample lot number, which reflects time, we can detect a possible trend toward a higher percent defective. The system should be reviewed for possible system change.

Although the production in Example 15.1 raises some questions, this gives some standard for comparison for future production. Another cycle may be required to ensure that the process is in control. Then the limits would be placed on sheets and used by quality inspectors to plot the percentage of defectives on a periodic basis (10 minutes, or half-hour, or hour). By this it would become evident within one inspection if the process were out of control (by a point appearing beyond the upper control limit).

When a process is in control we expect values to be somewhat randomly dispersed between the upper and lower control limits. But how does a control chart appear when a process is *not* in control? A shift out of control can appear in several ways:

1. As a gradual sloping trend, indicative of process wear or deterioration of some kind
2. As a jump in *p*-values, to either a higher or a lower level, with one or more values outside the control limits
3. As a predominance of values on either the high or the low side of the centerline. This indicates that the process is no longer centered about the initial process mean.

Control charts are an alternative to testing. They are appropriate when repeated sampling is done at relatively short intervals in time; this is common in manufacturing industries. Their use is based upon the belief that it is better to monitor a process regularly to ensure it does not go out of control than to wait until after products are finished to determine whether they are acceptable or not.

**Problems I5.3**

1. Following are three control charts. From among the following choices, indicate the condition or conditions that describe(s) each chart.
   a. in control
   b. system change
   c. common cause
   d. special cause
   e. out of control
   f. variation

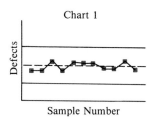
Chart 1

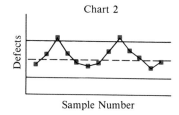
Chart 2

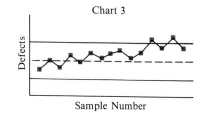
Chart 3

2. Explain whether the process appears to be in control or out of control for each part of the figure below. If out of control, then explain what kind of pattern might underlie the problem. (For example, "Part (a) is out of control and the process is making continually more defectives.")

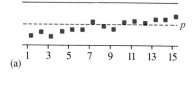
(a)

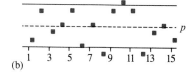
(b)

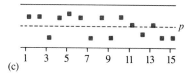
(c)

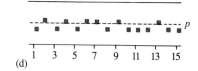
(d)

Sequence number-observations

3. A bank analyst is checking on the rate of excessive overdrafts (as defectives) for sample lots of 400 overdraft checks. An excessive overdraft is one of more than $25.00. For the 18 sample lots listed in the table
   a. calculate $\bar{p}$ and estimate $\sigma_p$.
   b. establish the $UCL_p$ and $LCL_p$ at the $\pm 3\hat{\sigma}_p$ range.
   c. draw the control chart and determine whether the process is in control.

$$\text{\textit{Hint:} Let } \bar{p} = \frac{159}{18 \cdot 400} = 0.0221.$$

Bank Overdrafts in 18 Samples Each of $n = 400$

| Sample ($n = 400$) | Number of Excessive Overdrafts | Fraction Excessive | Sample ($n = 400$) | Number of Excessive Overdrafts | Fraction Excessive |
|---|---|---|---|---|---|
| 1 | 4 | 0.0100 | 10 | 10 | 0.0250 |
| 2 | 2 | 0.0050 | 11 | 7 | 0.0175 |
| 3 | 10 | 0.0250 | 12 | 15 | 0.0375 |
| 4 | 12 | 0.0300 | 13 | 3 | 0.0075 |
| 5 | 3 | 0.0075 | 14 | 18 | 0.0450 |
| 6 | 1 | 0.0025 | 15 | 9 | 0.0225 |
| 7 | 19 | 0.0475 | 16 | 8 | 0.0200 |
| 8 | 14 | 0.0350 | 17 | 7 | 0.0175 |
| 9 | 5 | 0.0125 | 18 | 12 | 0.0300 |
| | | | | Total = 159 | |

4. A manufacturing manager in charge of producing men's blue jeans wants to characterize the performance of the production line. A sample of 30 hours of production includes random samples of 100 jeans per day. The objective is to identify "seconds," as defined by usual inspection standards: any discoloration, mis-stiching, pockets not square, rough zipper, uneven hem, missing belt loop, and more. The table gives the results.

| Hour | Number of Seconds Jeans | Hour | Number of Seconds Jeans | Hour | Number of Seconds Jeans |
|---|---|---|---|---|---|
| 1 | 4 | 11 | 5 | 21 | 6 |
| 2 | 6 | 12 | 3 | 22 | 4 |
| 3 | 4 | 13 | 5 | 23 | 3 |
| 4 | 5 | 14 | 5 | 24 | 2 |
| 5 | 3 | 15 | 4 | 25 | 5 |
| 6 | 7 | 16 | 3 | 26 | 12 |
| 7 | 2 | 17 | 2 | 27 | 14 |
| 8 | 3 | 18 | 6 | 28 | 4 |
| 9 | 7 | 19 | 0 | 29 | 1 |
| 10 | 2 | 20 | 1 | 30 | 6 |
| | | | | Total | 134 |

Establish a $3\hat{\sigma}_p$ control chart on the fraction of this production that is "seconds"-quality jeans. Discuss whether the process is in control. If it is not in control, then what are the indictors of a cause (problem)?

5. In the preceding Problem, assume the manager suspects a problem so is looking for an assignable cause. His investigation finds that the supplier of zippers was having a problem with sticky zippers, which were a major cause of the excessive number of "seconds" for hours 26 and 27. In fact, the production-line supervisor had called for a new batch of zippers and the line started hour 28 using the new batch. Since the hour-26 and hour-27 results were determined to have a special cause that was subsequently corrected, these hours can be eliminated from the control chart, and the limits recalculated. Reset the $LCL_p$ and $UCL_p$ and determine if the process is in control.

## 15.4 THE NUMBER OF DEFECTS PER ITEM: C-CHARTS

In Section 5.7 we learned that the Poisson random variable is appropriate to use when the number of occurrences of an event is unknown, but the probability is high that in many trials it will happen only a few times, or not at all. Some examples for which the Poisson distribution is applicable are the number of visitors per minute entering a CEO's office, the number of fax messages that enter a data message center in a 10-second period, the number of spelling errors per page for a word-processing pool, and the number of days of absence due to sickness in a firm of 10 employees. The Poisson distribution is applicable to business situations that require counting unusual events for a fixed time, a fixed space, or some other constant unit.

A unique characteristic of the Poisson probability distribution is that the mean and variance are equal values. Thus, control charts for the number of defects per item have a very simple form: Mean $\pm 3\sqrt{\text{Mean}}$

A *C*-chart is a control chart for the count of the number of defects per unit, where a defect is a rare event.

Except for the random variable, the construction of the control chart is the same as before.

Process control limits for the number of defects per item as defined by $3\sigma$ control limits are

$$LCL_C = \bar{C} - 3\sqrt{\bar{C}}$$
$$UCL_C = \bar{C} + 3\sqrt{\bar{C}}$$

(15.2)

and

$$\bar{C} = \frac{\sum\limits_{i}^{k} C_i}{k} \quad \text{defines the centerline}$$

where  $k = $ the number of samples

$LCL_C = 0$ if the preceding calculation results in a negative value

Here is an example.

**FIGURE 15.7** A Process for Merging Old and New Listings into One Updated List

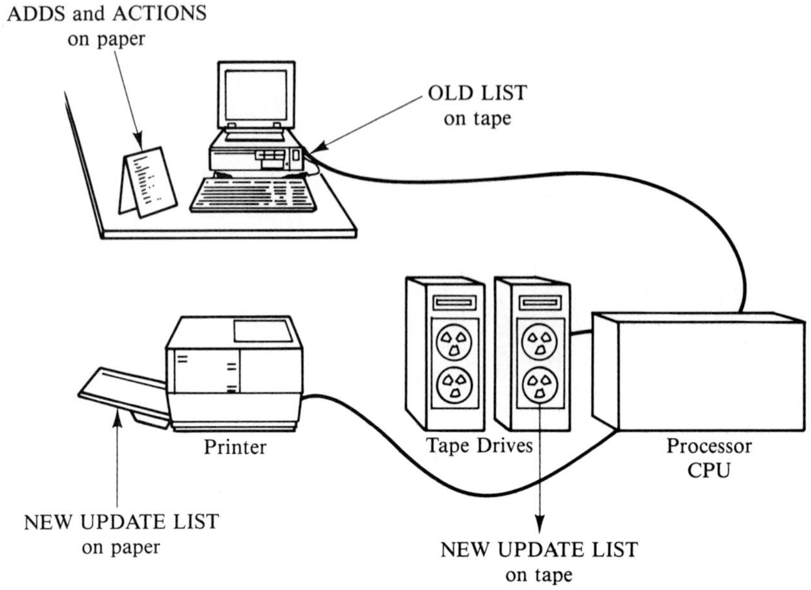

---

**EXAMPLE 15.2**

The system for compiling a telephone directory includes a number of steps, or processes. The current directory is acquired for the listing area. Then an update source is brought in, with deletes, adds, and changes, and the two are compared. Through the comparison, deletes are dropped and new listings or adds are set in place. The process requires repetitive and detailed checking—for example, for names to be deleted in one place but retained for other addresses. After the new and the old are "updated" into one, the listing has to be prepared for printing. This introduces more opportunity for editing errors.

Our example considers only one process in the total system of compiling a directory: the process of editing the list after the old and the new corrected listings are combined. The process is described in the flowchart of Figure 15.7.

From Figure 15.7 it is evident that errors can occur at several steps. The editing process is manual and requires the editor to view panels individually on a computer screen—the old—and "adds and additions" on paper—the new. Errors can occur in omission of adds, in misspellings, incorrect addresses, improper ordering of entries, and more. Any of these editing errors is considered a defect. Still, the operational definition of an acceptable product is zero defects.

This high standard is dictated by a highly competitive industry. In some cases compilers prepare directories at little or no cost to advertisers (that is, yellow-page listings) just to try to get into this profitable market! In spite of the numerous ways errors can occur, the team effort is highly efficient and the error rate is quite small per unit produced. So it is reasonable to assume that a Poisson process applies. Management wishes to develop a control chart for monitoring the number of defects (errors) per signature (a signature is a sequence of 16 pages) unit. A sample of 50 signatures appears in Table 15.2. Use this data and a $C$-chart to describe the quality performance for this update editing process.

*Solution*  The operational definition of this compiling process is zero defects. A defect can be a misspelling, an omission, a wrong address or telephone number, or other problem. Zero defects is an ideal, which obviously will not always be achieved. So the sample data can

**TABLE 15.2**   Number of Defects, $C$, per Directory Signature ($k = 50$ signatures)

| Unit | Defects | Unit | Defects | Unit | Defects | Unit | Defects | Unit | Defects |
|------|---------|------|---------|------|---------|------|---------|------|---------|
| 1 | 0 | 11 | 1 | 21 | 1 | 31 | 0 | 41 | 0 |
| 2 | 1 | 12 | 0 | 22 | 0 | 32 | 1 | 42 | 0 |
| 3 | 0 | 13 | 1 | 23 | 0 | 33 | 0 | 43 | 2 |
| 4 | 0 | 14 | 0 | 24 | 2 | 34 | 1 | 44 | 1 |
| 5 | 1 | 15 | 2 | 25 | 0 | 35 | 0 | 45 | 0 |
| 6 | 0 | 16 | 1 | 26 | 1 | 36 | 0 | 46 | 0 |
| 7 | 0 | 17 | 2 | 27 | 0 | 37 | 1 | 47 | 1 |
| 8 | 2 | 18 | 0 | 28 | 0 | 38 | 0 | 48 | 0 |
| 9 | 0 | 19 | 0 | 29 | 1 | 39 | 0 | 49 | 0 |
| 10 | 0 | 20 | 1 | 30 | 0 | 40 | 1 | 50 | 0 |

identify an existing process level. Then the process goal will be to move toward the (zero defect) standard. A control chart is defined from the defective rate.

$$\bar{C} = \frac{\sum\limits_{i}^{50} C_i}{50} = \frac{25}{50} = 0.5$$

$$\text{LCL}_C = \bar{C} - 3\sqrt{\bar{C}} = 0.5 - 3\sqrt{0.5} = 0.5 - 2.1 = -1.6 \qquad (\text{Therefore, } \text{LCL}_C = 0.)$$

$$\text{UCL}_C = \bar{C} + 3\sqrt{\bar{C}} = 0.5 + 2.1 \qquad\qquad = 2.6$$

A $C$-chart defined by these limits appears as Figure 15.8.

**FIGURE 15.8**   A $C$-Chart for Defects per Signature in a Directory List Update Process

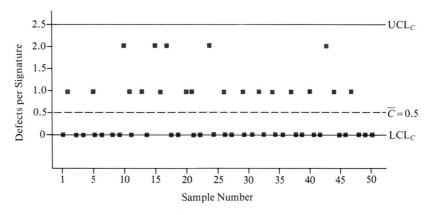

This process is in control, as evidenced by a fairly regular pattern of defects per item. The $C$-chart graph is steady and it appears predictable. It can be used as a guide for comparison of future production. However, this does not achieve the industry standard of zero defects. So, attention should be given to improving the quality of the performance. This is a case for the PDCA cycle.

In the preceding example, the process is in control but a higher level of performance is still warranted. Now, total-quality-control techniques can be put to use. The Pareto chart and cause-and-effect diagram are two procedures that can help to identify sources of process variation.

**A Pareto chart is a histogram that displays a category of the most common source of defects with a count, followed by the next most common source, and so on.**

The purpose of a Pareto chart is to identify sources of variation and to obtain some measure of the relative quantity of errors that can be associated with each source. This is a tool for identifying the special causes of process problems. It can suggest where improvement in quality would be most productive. A Pareto chart is displayed in the next example.

A cause-and-effect diagram identifies possible causes for reduced quantity and isolates them into subgroups. By associating possible causes to types, or subgroups, we can segment the data into more homogeneous subgroupings so that the variation can be analyzed more effectively. Again, the objective is to identify potential sources for high variation and then to analyze for specific problems with control charts.

The next example extends the directory-compiling data to an analysis by both a Pareto chart and a cause-and-effect diagram.

---

**EXAMPLE 15.3**

The compiling manager for the update activity described in Example 15.2 has decided to investigate the sources for errors in the editing process. He will prepare a cause-and-effect diagram and a Pareto chart. What sources of variation should he include?

*Solution*    The manager started with a list of all the activities that relate to the editing activity. He classified them into natural groupings of the distinguishable entities of environment, methods, machines and equipment, materials, and people.

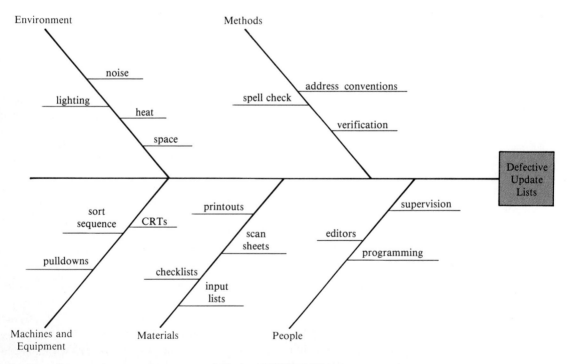

A Cause-and-Effect Diagram

Having the cause-and-effect diagram made it fairly easy to categorize the kinds of errors that appeared most often on marked copies of the update lists (described in Example 15.2, Figure 15.7).

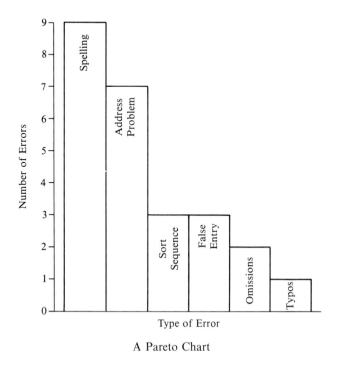

A Pareto Chart

The Pareto chart displays the primary causes, with the highest-frequency item listed first. This Pareto chart shows two dominant error sources: spelling—nine cases, and address problems—seven cases.

   Management concluded from these two graphics that the street-address operation and the causes for spelling errors should be investigated further. Several of the compiling editors agreed that the most common problems they experienced were having incomplete street addresses and being unable to check spelling while keeping a high level of production. From the cause-and-effect diagram, then, improvements are needed in Methods.

   The action that followed was to develop rules for street-coding that the computer could use in its processing before the human edit process. And the editors were provided a spell-check function in their computer editing capabilities. More control charts would indicate whether these changes had helped move closer to zero defects.

---

   The examples of this section have touched upon the elements of the PDCA cycle. Total quality improvement is an ongoing activity. In the compiling activity there will always remain the possibility for improvement. And the isolation of process-problem causes may not be as simple as in the previous examples. The section problems give several other opportunities for identifying process-problem sources using control charts.

**Problems 15.4**

6. Explain the conditions that dictate when a C-chart is appropriate as compared to a p-chart. What random variable distribution is the basis for each one?

7. Considering the amount of space available on a piece of paper, it is spectacular that the paper industry can produce sheets of paper with so few blemishes. These imperfections can include discolorations, spots, lines, waves, holes, and more. The distribution of imperfections per sheet of paper is reasonably described by the Poisson distribution. Given the following samples with number of "blemishes" (imperfections),

| Sheet | Blemishes | Sheet | Blemishes | Sheet | Blemishes | Sheet | Blemishes |
|-------|-----------|-------|-----------|-------|-----------|-------|-----------|
| 1 | 6 | 6 | 3 | 11 | 1 | 16 | 2 |
| 2 | 7 | 7 | 10 | 12 | 4 | 17 | 4 |
| 3 | 4 | 8 | 3 | 13 | 5 | 18 | 2 |
| 4 | 9 | 9 | 1 | 14 | 7 | 19 | 8 |
| 5 | 2 | 10 | 4 | 15 | 2 | 20 | 5 |

a. develop a control chart to check for process control.
b. what number of blemishes would you expect for the next sheet produced in this process?

8. Consider a fast-food restaurant, where the process of serving an order includes someone taking the order; the order being placed "on-call" to be prepared; someone putting the order together—cooking, pouring, mixing, etc.; and finally someone serving the order. Define a defective order as one that is incorrectly served—i.e., it does not include all of the items that were placed on the order, or some were improperly served. Suppose this experiment concerns the number of improperly served orders per 100 orders. Answer the following.
a. Define the process variable.
b. What is the operational definition for a nondefective order?
c. What additional information is required to define an appropriate control chart procedure?

9. In Problem 8, the process can fail at a number of places. Use that description, and your knowledge of fast-food service, to construct a cause-and-effect diagram for defective (food) orders in a fast-food hamburger shop. Use Figure 15.1 as a place to begin. Expand it to include at least two items for each category—materials, equipment, methods, environment, people.

10. There appears to be a problem with a process in which the page edges are trimmed from bound, but not yet cover-bound, copies of yellow-pages directories. The process is fairly simple: The operator places a book in a well-marked position, then presses a release button for a knife to drop down and cut the uneven page ends. Given the process numbers for defective cuts, below, indicate if there is a problem and suggest some possible causes. Assume that a Poisson distribution is reasonable to describe the number of defects (improper cuts) per 100 cuts. *Suggestion:* Begin with a control chart.

| Sample | 1 | 2 | 3 | 4 | 5 | 6 | 7 | 8 | 9 | 10 | 11 | 12 | 13 | 14 | 15 | 16 | 17 | 18 | 19 | 20 |
|---|---|---|---|---|---|---|---|---|---|---|---|---|---|---|---|---|---|---|---|---|
| Defects | 1 | 0 | 2 | 1 | 0 | 0 | 1 | 1 | 0 | 1 | 4 | 2 | 3 | 3 | 4 | 3 | 5 | 4 | 3 | 4 |

| Sample | 21 | 22 | 23 | 24 | 25 | 26 | 27 | 28 | 29 | 30 | 31 | 32 | |
|---|---|---|---|---|---|---|---|---|---|---|---|---|---|
| Defects | 6 | 5 | 7 | 6 | 6 | 7 | 6 | 10 | 12 | 12 | 11 | 12 | Total 142 |

11. The process in Problem 10, in fact, was not in control. A problem was identified and corrected. Then, after test runs for one hour, the process was sampled again. Thirty samples, each of 100 books, gave a total of 61 defects. Use this information, and the knowledge that the process is a Poisson distribution process, to establish a working standard, including
   a. a process mean.
   b. the upper control limit.
   c. a response if the process shows that any sample has seven or more defects.

## 15.5 CONTROL CHARTS FOR VARIABLES DATA: THE $\bar{X}$-CHART AND THE $R$-CHART

So far our study has been of discrete variables, including $p$-charts for binomial data and $C$-charts for Poisson distributed number of defects per item data. Other control situations—for example, production weights for loaves of bread, the linear feet of sheet aluminum on a spool produced in a fixed time, and the length of car radiator hoses molded per hour—require a continuous measure. Variables data assume that the control measurement is from a continuous distribution, or at least that the data are amenable to the central limit theorem, so that the distribution of the sample means is normal. In the latter case it is appropriate to maintain samples of size larger than 30.

### $\bar{X}$-Charts

The formulation of $\bar{X}$-control limits is, again, based on control limits at three standard errors from the mean. These are $\bar{X} \pm 3\sigma_{\bar{x}}$ where $\sigma_{\bar{x}}$ can be estimated using $s$, the sample standard deviation, and the sample size. However, to simplify calculations, control limits have been developed that use the range as an estimator of $\sigma$. The reason for using the range is convenience: It is easy to compute (and understand), and it requires only simple hand-calculations from the two extreme values. This uses

$$\hat{\sigma}_{\bar{x}} = \frac{\bar{R}}{d_2} = \frac{\frac{\sum_{i}^{k} R_i}{k}}{d_2}$$

where   $k$ = the number of samples used to estimate the mean

$d_2$ = a number required to make $\bar{R}/d_2$ an unbiased estimator for $\sigma_{\bar{x}}$

Equation 15.3 first requires us to calculate the range $R$ for each sample. Next we calculate the mean, $\bar{R}$, for $k$ = the number of samples. Then,

$$3\sigma_{\bar{x}} = 3\frac{\sigma}{\sqrt{n}} \sim \frac{3\bar{R}}{d_2\sqrt{n}} = A_2\bar{R} \qquad (15.3)$$

Values for $d_2$ and $A_2$ appear in Appendix B, Table 12. The control limits are defined by Equation 15.4.

For each sample,

$$\text{Range} = \text{Highest value} - \text{Lowest value}$$

with

$$\bar{X} = \frac{\text{Sum of the measurements in a sample}}{\text{Number of items in the sample}} \quad \text{(sample means)}$$

$$\bar{\bar{X}} = \frac{\text{Sum of the } \bar{X}\text{-values for the samples}}{\text{Number of samples}} = \frac{\sum\limits_{i}^{k} \bar{X}_i}{k} \quad \text{(process mean)}$$

$$\text{LCL}_{\bar{x}} = \bar{\bar{X}} - A_2\bar{R}$$

$$\text{UCL}_{\bar{x}} = \bar{\bar{X}} + A_2\bar{R}$$

(15.4)

where $A_2 = \dfrac{3}{d_2\sqrt{n}}$ and $d_2$ are defined in Table B.12 for $2 \leq k \leq 25$.

We indicated earlier that variables data control charts can provide more information than attributes charts. With variables one can control for both the process mean and the process variation. We first analyze an $\bar{X}$-chart for control of the process mean, and then analyze an $R$-chart for process variation.

---

**EXAMPLE 15.4**

A direct-mail production operation prepares mailing pieces, including an advertising piece and a return mail envelope, addressed and placed in mailbags. The operational definition is to fill mailbags as near to 65 pounds in weight as possible. This requirement is designed to meet a mailing standard that penalizes for heavy bags. It also penalizes for too many light bags, because there is a handling charge per mailbag. Thus it is economical to have weights that are as near as possible to the 65-pound standard. Figure 15.9 illustrates a process flow diagram for filling the mailbags.

As the diagram indicates, the bags are attached to mailbag chutes at spouts on an envelope inserter machine. Return envelopes and advertising pieces are stacked and inserted in a mail envelope. This is sealed at the end of the inserter line and drops from the conveyor into a mailbag. Each inserter operator is required to tie and mark the mailbags as they become full. This process does not allow time for weighing the bags

**FIGURE 15.9** Schematic for an Inserter Machine in a Direct-Mail Operation

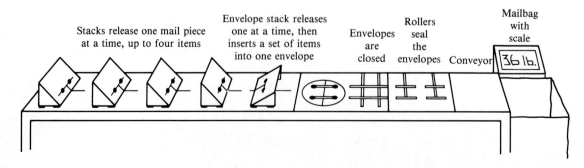

and adding or removing envelopes, but each chute has a weighing device that turns on a light to warn the operator when a bag reaches a preset weight. Because the mailbags go to one distribution center, there is no need to underfill any of the bags.

The data in Table 15.3 are weights for 30 observations each of samples of five mailbags randomly selected from the output for two operators. Establish a $\bar{X}$-chart. Is this production within the specifications for mean weight?

Table 15.3 provides information to develop an $\bar{X}$-control chart.

**Solution**

$$\bar{\bar{X}} = \frac{\bar{X}_1 + \bar{X}_2 + \bar{X}_3 + \cdots + \bar{X}_{30}}{30} = \frac{67.54 + 63.04 + 72.64 + \cdots + 62.84}{30} = 65.23$$

which defines the centerline as $\bar{\bar{X}} = 65.23$.

$$\bar{R} = \frac{R_1 + R_2 + R_3 + \cdots + R_{30}}{30} = \frac{4.9 + 6.8 + 13.0 + \cdots + 4.4}{30} = 8.19$$

**TABLE 15.3**   Weights for Mailbags Produced by Two Inserters in a Direct-Mail Production Operation

| Sample Observation | Mailbag Weight (pounds) | | | | | $\bar{X}$ | $R$ |
| | 1 | 2 | 3 | 4 | 5 | | |
|---|---|---|---|---|---|---|---|
| 1 | 67.9 | 68.3 | 70.1 | 66.2 | 65.2 | 67.54 | 4.9 |
| 2 | 63.3 | 58.9 | 65.7 | 64.6 | 62.7 | 63.04 | 6.8 |
| 3 | 80.8 | 70.5 | 75.4 | 67.8 | 68.7 | 72.64 | 13.0 |
| 4 | 63.2 | 66.2 | 61.9 | 67.6 | 59.9 | 63.76 | 7.7 |
| 5 | 64.3 | 68.3 | 67.0 | 70.3 | 69.7 | 67.92 | 6.0 |
| 6 | 61.2 | 62.1 | 63.4 | 60.5 | 71.0 | 63.64 | 10.5 |
| 7 | 59.6 | 65.7 | 71.2 | 63.8 | 65.2 | 65.10 | 11.6 |
| 8 | 59.4 | 54.2 | 64.7 | 62.5 | 57.6 | 59.68 | 10.5 |
| 9 | 70.1 | 66.6 | 61.0 | 71.2 | 68.6 | 67.50 | 10.2 |
| 10 | 69.3 | 64.2 | 69.4 | 65.4 | 66.9 | 67.04 | 5.2 |
| 11 | 61.1 | 71.6 | 62.1 | 63.4 | 59.5 | 63.54 | 12.1 |
| 12 | 60.2 | 59.2 | 67.3 | 67.3 | 61.7 | 63.14 | 8.1 |
| 13 | 65.2 | 63.3 | 64.8 | 67.0 | 58.0 | 63.66 | 9.0 |
| 14 | 64.9 | 59.8 | 65.4 | 64.8 | 63.9 | 63.76 | 5.6 |
| 15 | 61.2 | 60.3 | 65.8 | 67.2 | 69.7 | 64.84 | 9.4 |
| 16 | 60.5 | 64.7 | 68.3 | 69.9 | 64.6 | 65.60 | 9.4 |
| 17 | 68.6 | 69.5 | 68.1 | 72.0 | 70.4 | 69.72 | 3.9 |
| 18 | 60.1 | 58.8 | 60.6 | 54.9 | 60.0 | 58.88 | 5.7 |
| 19 | 71.0 | 73.0 | 69.0 | 74.1 | 69.2 | 71.26 | 5.1 |
| 20 | 66.9 | 60.3 | 64.1 | 56.4 | 59.8 | 62.10 | 10.7 |
| 21 | 63.9 | 60.4 | 64.2 | 65.4 | 66.5 | 64.08 | 6.1 |
| 22 | 65.3 | 70.9 | 69.3 | 66.4 | 61.5 | 66.68 | 9.4 |
| 23 | 70.5 | 62.1 | 68.6 | 68.3 | 68.1 | 67.52 | 8.4 |
| 24 | 63.4 | 66.4 | 67.4 | 69.1 | 57.7 | 64.80 | 11.4 |
| 25 | 67.1 | 61.3 | 66.0 | 69.6 | 68.1 | 66.42 | 8.3 |
| 26 | 59.8 | 70.1 | 64.2 | 65.5 | 62.5 | 64.42 | 10.3 |
| 27 | 64.5 | 69.0 | 60.3 | 66.8 | 61.5 | 64.42 | 8.7 |
| 28 | 68.1 | 67.0 | 62.6 | 63.4 | 69.5 | 66.12 | 6.9 |
| 29 | 62.6 | 66.7 | 66.1 | 62.4 | 68.7 | 65.30 | 6.3 |
| 30 | 62.5 | 60.6 | 65.0 | 63.0 | 63.1 | 62.84 | 4.4 |

|  |  |  |  |  |  | 1956.96 | 245.6 |
|---|---|---|---|---|---|---|---|
|  |  |  |  |  | Totals | $\bar{\bar{X}} = 65.23$ | |
|  |  |  |  |  |  | $\bar{R} =$ | 8.19 |

Then,

$$\text{LCL}_X = \overline{\overline{X}} - A_2\overline{R} = 65.23 - 0.577(8.19) = 65.23 - 4.73 = 60.50 \text{ pounds}$$

$$\text{UCL}_X = \overline{\overline{X}} + A_2\overline{R} = 65.23 + 4.73 \qquad\qquad\quad = 69.96 \text{ pounds}$$

with $A_2 = 0.577$ by Table B.12 for $k = 5$ samples.

This provides the necessary information to construct the $\overline{X}$-control chart.

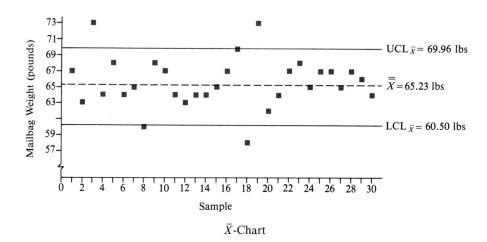

$\overline{X}$-Chart

The control chart demonstrates that this process is not in control. There are four points beyond the control limits. Because two of the points are above the upper limit and two others reside below the lower limit, one could question whether there is a problem with excessive process variability. This is investigated in the next example. Potential causes need to be investigated, and corrections made when the causes are identified.

---

It would be appropriate to prepare a cause-and-effect diagram for the preceding example. An assignable cause is suspected because, except for the four points, the sample shows a stable pattern. Some of the possible special causes might be the operator (there were two) or an out-of-adjustment inserter arm. Or there may be a nonrandom pattern in the data. This possibility is examined later.

**R-Charts**

When the mean process is out of control, as in the preceding example, it can be either in control or out of control for the variation. In fact, *because the $\overline{X}$-chart control limits depend on R, it is appropriate to first check the R-chart for control*. If the R-chart for variation is not in control, then it makes no sense to proceed with an $\overline{X}$-chart. So the limits are now defined for an R-chart.

$$R. = \text{Range for each sample} = \text{Highest value} - \text{Lowest value}$$

$$\overline{R} = \frac{\text{Sum of the ranges for the samples}}{\text{Number of samples}} = \frac{\sum\limits_{i}^{k} R_i}{k} = \text{the centerline}$$

$$\text{LCL}_R = D_3 \cdot \overline{R}$$

$$\text{UCL}_R = D_4 \cdot \overline{R} \qquad\qquad\qquad\qquad (15.5)$$

where                              $k$ = the number of samples

and where $D_3$ and $D_4$ values are based on random sampling from a normal distribution that appear in Appendix B, Table 12 for values of $k$ ranging from 2 to 25. Here is an example.

---

## EXAMPLE 15.5

Use the data in Example 15.4 and Table 15.3 to develop an $R$-chart to check the process variation. Is the variance in mailbag weights (that is, the process variation) in control?

*Solution*   The data for an $R$-chart are provided in Table 15.3. This gives

$$\bar{R} = \frac{\sum\limits_{}^{30} R_i}{30} = \frac{245.6}{30} = 8.19 \text{ pounds}$$

Then, from Table B.12, for $k = 5$ weights per sample $D_3 = 0$ and $D_4 = 2.115$. The $D_3 = 0$ value is established to recognize that the lower bound for the range can be no lower than zero. Then the bounds are

$$\text{LCL}_R = D_3 \cdot \bar{R} = \quad 0(8.19) = \quad 0 \quad \text{pounds}$$

$$\text{UCL}_R = D_4 \cdot \bar{R} = 2.115(8.19) = 17.32 \text{ pounds}$$

and the centerline is $\bar{R} = 8.19$. The control chart limits are as defined below.

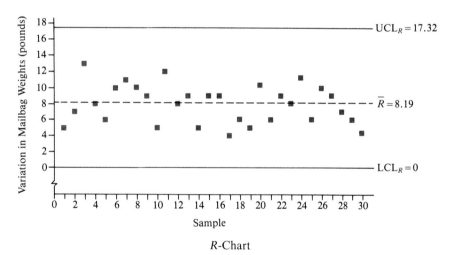

R-Chart

From the $R$-chart we see that the process variation is in control. The chart is stable. All points are within the limits and show no obvious pattern in the sequence of observations. The range, and therefore the variance, in weights is predictable. This chart can be used as a standard for future control.

---

The finding that the process variation is in control assures that the process mean $\bar{X}$-chart limits are using a valid value in $R$ for the calculation of the mean limits. Examples 15.4 and 15.5 illustrate a case in which the variability is in control but the process mean is not in control. In such cases, causes need to be identified and explored to bring the process mean in control.

## A Test for Random Observations

Chapter 17, Section 3 contains a runs test for randomness that can be applied equally well to quality processes. The runs test, being a nonparametric procedure, does not require the normality assumption. The procedure is used here for runs above and below the centerline. (You are invited to read Section 17.3 for background on the runs test before beginning this exercise.) We use the letter A for data points that reside above the centerline and the letter B for data points that reside below the centerline. We consider only the case for $n_A \geq 10$ and $n_B \geq 10$, which enables us to use a $Z$-test:

$H_0$:    The sequence of A's and B's is produced by a random process.

$H_A$:    The sequence of A's and B's is not produced by a random process.

The test statistic

$$Z = \frac{r - \mu_r}{\sigma_r} \quad \text{repeats (17.3)} \tag{15.6}$$

$$\mu_r = \frac{2n_1 n_2}{n_1 + n_2} + 1$$

and

$$\sigma_r = \sqrt{\frac{2n_1 n_2 (2n_1 n_2 - n_1 - n_2)}{(n_1 + n_2)^2 (n_1 + n_2 - 1)}}$$

where $r$ denotes the number of runs requires $n_1 \geq 10$ and $n_2 \geq 10$ with rejection region $|\text{calc } Z| > Z_{\alpha/2}$.

The test is two-tailed because either too few or too many runs will indicate that the sequence is not produced by a random process. The test applies to $\bar{X}$-charts, $C$-charts, $p$-charts, and $R$-charts.

The runs test is applied to the $R$-chart data of Example 15.5 to check if that data is random. If so, then this will add credibility to the Example 15.5 conclusion that the variation of mailbag weights was in control.

## EXAMPLE 15.6

The data from Table 15.3 are used to determine whether the range values in the mailbag-weights process can be considered a random sequence in the order they were observed. Only the pertinent data are presented here.

| Sample | 1 | 2 | 3 | 4 | 5 | 6 | 7 | 8 | 9 | 10 | 11 | 12 | 13 | 14 | 15 |
|--------|-----|-----|------|-----|-----|------|------|------|------|-----|------|-----|-----|-----|-----|
| R-Value | 4.9 | 6.8 | 13.0 | 7.7 | 6.0 | 10.5 | 11.6 | 10.5 | 10.2 | 5.2 | 12.1 | 8.1 | 9.0 | 5.6 | 9.4 |

| Sample | 16 | 17 | 18 | 19 | 20 | 21 | 22 | 23 | 24 | 25 | 26 | 27 | 28 | 29 | 30 |
|--------|-----|-----|-----|-----|------|-----|-----|-----|------|-----|------|-----|-----|-----|-----|
| R-Value | 9.4 | 3.9 | 5.7 | 5.1 | 10.7 | 6.1 | 9.4 | 8.4 | 11.4 | 8.3 | 10.3 | 8.7 | 6.9 | 6.3 | 4.4 |

Test the hypothesis of a random sequence against a two-tailed alternative hypothesis. Let $\alpha = 0.05$.

*Solution*   Here $r$ = the number of runs above, A, or below, B, the centerline $\bar{R}$. From Example 15.5 we know that $\bar{R} = 8.19$. Then the range $R$-values give the sequence

BB, A, BB, AAAA, B, A, B, A, B, AA, BBB, A, B, AAAAAA, BBB

So $r = 15$ runs with $A = 16$ and $B = 14$. Then,

$H_0$:   The distribution of range values for mailbag weights is a random sequence about their mean value, $\bar{R}$.

$H_A$:   The distribution of range values for mailbag weights shows either too few or too many runs to be considered a random sequence.

Test: Reject $H$ if $|Z| > 1.96$.

$$\text{Calculated } Z = \frac{r - \mu_r}{\sigma_r}$$

$$= \frac{15 - 15.93}{2.68} = -0.35$$

where   $\mu_r = \dfrac{2(16)(14)}{30} + 1$   $= 15.93$

$$\sigma_r = \sqrt{\frac{2 \cdot 16 \cdot 14 \cdot (2 \cdot 16 \cdot 14 - 14 - 16)}{(16 + 14)^2 (16 + 14 - 1)}} = 2.68$$

Conclusion: Do not reject the null hypothesis.

The data evidence a random sequence about the range mean. This supports the Example 15.5 conclusion that the $R$-chart is in control with random variation.

---

Although we have barely touched on the techniques for total quality control, this chapter introduces the tools that are used to develop a continuous quality-improvement program. Duncan [3] provides a more complete coverage of this topic, including a "length of runs" and "runs up and down" procedures. The following section Problems provide a variety of applications for $\bar{X}$- and $R$-charts.

**Problems 15.5**

12. Relative to the process mean and its variation, there are four possible conditions for control in a variables situation, as illustrated in the four graphs (a)–(d).

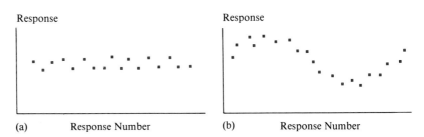

(a)          Response Number          (b)          Response Number

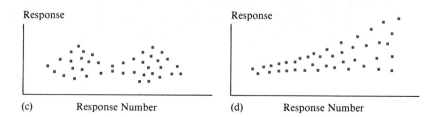

(c)    Response Number        (d)    Response Number

Match (a)–(d) each with one of the following:
 i. changing mean and changing variation
 ii. constant mean and constant variation
 iii. changing mean and constant variation
 iv. constant mean and changing variation

13. Given: the response chart for calls answered per minute, averaged over 5- minute time periods, for three operators who work for a telephone answering service. $\triangle$ = person 1, $\bigcirc$ = person 2, $\blacksquare$ = person 3.

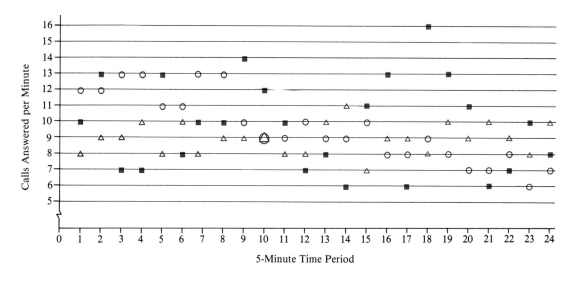

This is a variables measure. For example, in the first 5-minute time period operator 1 ($\triangle$) answered 39 calls, for $39/5 = 7.8$ calls answered per minute. This is rounded to 8 calls per minute for a more convenient use. Construct $\bar{\bar{X}}$ and $R$ control charts for these data. Is the process in control for
a. the mean?                    b. the variation?

14. For the data in Problem 13, the answering service manager suggests that a trend might exist in the mean process. Perform a runs test on the data, using $\bar{\bar{X}} = 9.5$ and $\alpha = 0.05$ with a one-tailed test. (The manager feels there will be too few runs.)

15. Again referring to Problem 13, the service manager has moved person 3 ($\blacksquare$) to a less stressful position. What data in that problem would support this action? And, more generally, how does the control chart relate to measuring individual performance? *Suggestion:* Compute the mean and standard deviation for the performance for each person, and compare.

16. A business manager for a company with over 500 employees wants to investigate the usage of the company's five fax machines. Is the process of fax-machine usage in control with respect to time required to send a fax message? The study is restricted to one-page messages for the five machines. The data represent message-use time in seconds.

Time (seconds) to Send a One-Page Message on a Fax Machine

| Sample | Machine | | | | | $\bar{X}$ | R |
| | 1 | 2 | 3 | 4 | 5 | | |
|---|---|---|---|---|---|---|---|
| 1 | 30.9 | 30.5 | 32.0 | 30.8 | 28.5 | 30.5 | 3.5 |
| 2 | 31.9 | 28.5 | 31.2 | 31.9 | 31.4 | 31.0 | 3.4 |
| 3 | 29.7 | 30.1 | 29.1 | 28.6 | 29.5 | 29.4 | 1.5 |
| 4 | 29.2 | 29.3 | 29.5 | 29.8 | 29.4 | 29.4 | 0.6 |
| 5 | 30.3 | 30.6 | 30.5 | 30.2 | 31.5 | 30.6 | 1.3 |
| 6 | 30.2 | 30.5 | 28.8 | 29.5 | 29.6 | 29.7 | 1.7 |
| 7 | 29.2 | 29.5 | 29.3 | 29.4 | 29.9 | 29.5 | 0.7 |
| 8 | 30.2 | 29.5 | 28.9 | 28.5 | 31.6 | 29.7 | 3.1 |
| 9 | 28.9 | 31.6 | 30.5 | 29.4 | 30.6 | 30.2 | 2.7 |
| 10 | 28.6 | 31.6 | 30.3 | 29.8 | 31.2 | 30.3 | 3.0 |
| 11 | 31.2 | 29.2 | 29.8 | 28.6 | 28.9 | 29.5 | 2.6 |
| 12 | 30.2 | 30.8 | 29.9 | 28.8 | 29.5 | 29.8 | 2.0 |
| 13 | 30.5 | 30.9 | 32.0 | 29.8 | 28.8 | 30.4 | 3.2 |
| 14 | 30.8 | 29.9 | 30.2 | 29.2 | 30.5 | 30.1 | 1.6 |
| 15 | 29.8 | 28.9 | 28.5 | 30.2 | 29.5 | 29.4 | 1.7 |
| 16 | 28.8 | 28.9 | 29.2 | 28.5 | 29.0 | 28.9 | 0.7 |
| 17 | 30.2 | 28.5 | 28.9 | 29.5 | 29.9 | 29.4 | 1.7 |
| 18 | 31.9 | 30.2 | 30.8 | 28.5 | 30.9 | 30.5 | 3.4 |
| 19 | 29.9 | 30.2 | 29.2 | 28.6 | 28.9 | 29.4 | 1.6 |
| 20 | 30.2 | 31.8 | 32.1 | 29.6 | 30.5 | 30.8 | 2.5 |

Totals $\bar{X}$ = 598.5    42.5

$\bar{\bar{X}} = 29.9$

$\bar{R} = 2.1$

**SUMMARY**

Total quality control is a continuous improvement process requiring that everyone involved gain an understanding of the systems and processes. It uses a model, the *PDCA* (*Plan–Do–Check–Act*) *cycle*, introduced by Professor W. Edwards Deming. This defines a recurring cycle of action for continual improvement. It is the guide that we have used for this introduction to total quality control.

Quality improvement efforts have traditionally focused on industrial production processes. A *process* is any action that produces a result. *Total quality control* concerns the improvement of *systems*, which are groups of processes working together to fulfill some function of an organization.

Professor Deming has pointed out that *reducing variation will result in improved quality*. Statistics and data analysis are applied to processes to help personnel understand and reduce variation. Chapter 15 emphasizes the Check and Do stages of the PDCA cycle.

The control chart is the principal tool used to analyze variation. A *control chart* is a well-defined graph used to monitor variation by comparison of a response (a process attribute or measure) to upper and lower control limits. A process is *in control* if the sample response remains within the control limits and does not show a pattern or trend

**TABLE 15.4** Key Equations for Quality Control

| Measure | Level | Process Distribution | Centerline | Control Limits |
|---|---|---|---|---|
| p-chart, for attributes | Discrete | Binomial | $\bar{p} = \dfrac{\sum_{i}^{k} p_i}{k n_i}$ (15.1) | $LCL_p = \bar{p} - 3\hat{\sigma}_p$ ($\geq 0$) <br> $UCL_p = \bar{p} + 3\hat{\sigma}_p$ <br> where $\hat{\sigma}_p = \sqrt{\dfrac{\bar{p}(1 - \bar{p})}{k n_i}}$ |
| C-chart, for defects per unit | Discrete | Poisson | $C = \dfrac{\sum_{i}^{k} C_i}{k}$ (15.2) | $LCL_C = \bar{C} - 3\sigma_C$ ($\geq 0$) <br> $UCL_C = \bar{C} + 3\sigma_C$ <br> where $\sigma_C = \sqrt{\bar{C}}$ |
| $\bar{X}$-chart, for variables data | Continuous | Normal | $\bar{\bar{X}} = \dfrac{\sum_{i}^{k} \bar{X}_i}{n}$ (15.4) | $LCL_{\bar{x}} = \bar{\bar{X}} - A_2 \bar{R}$ ($\geq 0$) <br> $UCL_{\bar{x}} = \bar{\bar{X}} + A_2 \bar{R}$ <br> and $A_2$ appears in Table B.12 |
| R-chart, for variables data | Continuous | Normal | $\bar{R} = \dfrac{\sum_{i}^{k} R_i}{k}$ (15.5) | $LCL_R = D_3 R$ ($\geq 0$) <br> $UCL_R = D_4 R$ <br> and $D_3, D_4$ appear in Table B.12 |

with time. Four kinds of control charts were studied: *p-charts for attributes, C-charts for defects per item*, and *$\bar{X}$- and R-charts for variables (continuous) data*. The chapter key equations for upper and lower control limits and the centerline appear in Table 15.4.

Another tool introduced was the *system flowchart*, which is an engineering diagram that helps managers to visualize the system or to view selected processes. It shows sources for variation. A *cause-and-effect diagram* is a data analysis tool that is used to identify causes for substantial variation and sources for subgroups. The *Pareto table* is a simple graphic that can point out high rates of defectives, and so enable managers to identify possible sources of process variation.

Total quality control has applications beyond industry. The Business Case, following, illustrates total quality control and the PDCA cycle used in a service process.

## KEY TERMS

**CHAPTER PROBLEMS**

17. A newspaper printing press operates at high speeds, so is subject to substantial lost production if it stops. A printer manager suspects a problem with the slack in the paper feeder, a side view of which appears in the schematic below:

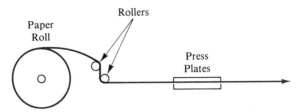

The production measure is the number of paper "breaks" per 1000 feet of print paper run. Under normal operations, there is a small number of breaks, so this can be considered a Poisson process. Use the data provided to establish a control chart.

| Run | Breaks | Run | Breaks | Run | Breaks |
|-----|--------|-----|--------|-----|--------|
| 1 | 4 | 11 | 5 | 21 | 6 |
| 2 | 3 | 12 | 7 | 22 | 3 |
| 3 | 9 | 13 | 6 | 23 | 8 |
| 4 | 6 | 14 | 4 | 24 | 4 |
| 5 | 2 | 15 | 3 | 25 | 5 |
| 6 | 7 | 16 | 7 | 26 | 5 |
| 7 | 5 | 17 | 5 | 27 | 7 |
| 8 | 6 | 18 | 8 | 28 | 6 |
| 9 | 4 | 19 | 4 | 29 | 7 |
| 10 | 8 | 20 | 4 | 30 | 4 |
|   |   |   |   |   | 160  Total |

Is this process in control? Do you think the print manager would be satisfied with this as a resolution to her query? Explain.

18. Establish a cause-and-effect diagram for the production process described in Problem 17. Use the general schematic in Figure 15.1 to begin your diagram. Select one or two sources that you think are most likely to be a cause for a high count of paper "breaks." One source might be the paper strength or quality. Give six to eight more.

19. Perform the runs test for a random sequence on the data in Problem 17. Does this result raise any questions about that process? Use a two-tailed alternative hypothesis and $\alpha = 0.10$ for the runs test.

20. For the paper feed diagram in Problem 17, the plant engineer has suggested placing an idler roller on springs to take up the slack that was occurring at the arrow marked idler roller.

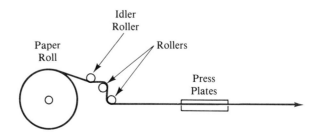

After the idler roller was added, another sample of 30 readings gave the following $C$-chart. How effective was this system change? That is, indicate the process changes you find by comparing this chart with the one you prepared in Problem 17.

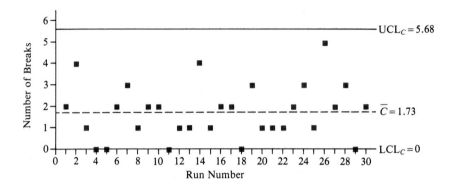

21. A molding process for manufacturing electric-light covers is controlled by weight. A production problem involves cleaning off the edges of the mold to remove excess plastic. Both the mean weight for strength and the variation in the amount of waste are of concern to the production manager. The industry standard for the mean weight is 50 grams $\pm 5$ grams, and for the range is 5 grams $\pm 5$ grams. Because an abrasive is used to clean the excess plastic, the weight is controlled by the amount of abrasive used. The use of one amount of the abrasive resulted in the following plastic weights.

Plastic Mold Weights for Electric-Light Covers

| Sample | Molding Machine | | | | $\bar{X}$ | $R$ |
|--------|----|----|----|----|-------|-----|
| | 1 | 2 | 3 | 4 | | |
| 1 | 46 | 51 | 53 | 50 | 50.00 | 7 |
| 2 | 52 | 51 | 52 | 49 | 51.00 | 3 |
| 3 | 51 | 47 | 49 | 46 | 48.25 | 5 |
| 4 | 51 | 56 | 50 | 48 | 51.25 | 8 |
| 5 | 54 | 50 | 52 | 53 | 52.25 | 4 |
| 6 | 48 | 50 | 55 | 46 | 49.75 | 9 |
| 7 | 49 | 50 | 52 | 49 | 50.00 | 3 |
| 8 | 51 | 47 | 56 | 45 | 49.75 | 11 |
| 9 | 50 | 54 | 52 | 53 | 52.25 | 4 |
| 10 | 55 | 47 | 50 | 49 | 50.25 | 8 |
| 11 | 50 | 48 | 50 | 48 | 49.00 | 2 |
| 12 | 46 | 53 | 49 | 54 | 50.25 | 8 |
| 13 | 47 | 46 | 52 | 49 | 48.50 | 6 |
| 14 | 52 | 51 | 56 | 48 | 51.75 | 8 |
| 15 | 53 | 55 | 54 | 50 | 53.00 | 5 |
| 16 | 50 | 49 | 53 | 49 | 50.25 | 4 |
| 17 | 50 | 55 | 54 | 48 | 51.75 | 7 |
| 18 | 45 | 55 | 58 | 52 | 52.50 | 13 |
| 19 | 54 | 54 | 55 | 52 | 53.75 | 3 |
| 20 | 51 | 49 | 53 | 47 | 50.00 | 6 |
| | | | | Totals | 1015.50 | 124 |
| | | | | | $\bar{\bar{X}} = 50.8$ | |
| | | | | | | $\bar{R} = 6.2$ |

a. Develop an $\bar{X}$ control chart and check for the mean weight standard. Is the process in control for the process mean?

b. Develop an $R$-chart and check for process control in variation.

22. In the molding process described in Problem 21, explain the effect that (a) and (b) would have on (i) the process mean and (ii) the process variation.
    a. Increasing the amount of abrasive used to clean the plastic mold
    b. Decreasing the amount of abrasive used to clean the plastic mold

23. A lawn service provides a chemical spray application designed to control many kinds of weeds. The spray is applied with a pressure hose, much like you would water grass with a garden hose. In addition to the chemical tank, usually mounted on a truck, the equipment includes a pressure pump and wet gear (boots, gloves, and coat) for the service person. Because it schedules 50 lawns for service each day, this company considers as defects any lawns that are on the schedule but are not serviced on the same day. The table displays the number of lawns not serviced as scheduled for a sample of 25 workdays.

| Day | No. Not Serviced | Day | No. Not Serviced | Day | No. Not Serviced | Day | No. Not Serviced | Day | No. Not Serviced |
|---|---|---|---|---|---|---|---|---|---|
| 1 | 2 | 6 | 0 | 11 | 5 | 16 | 0 | 21 | 4 |
| 2 | 6 | 7 | 24 | 12 | 2 | 17 | 3 | 22 | 3 |
| 3 | 2 | 8 | 3 | 13 | 2 | 18 | 5 | 23 | 8 |
| 4 | 7 | 9 | 1 | 14 | 1 | 19 | 2 | 24 | 4 |
| 5 | 2 | 10 | 2 | 15 | 4 | 20 | 4 | 25 | 0 |

Day 7 was a special case in that it rained at mid-day, and work was stopped for the remainder of that day. Day 7 should be excluded from a control chart construction. Develop a chart for the percent of lawns not serviced on the same day as scheduled. Discuss the control chart for any indications of system change, common cause, and assignable cause.

24. Perform a runs tests for randomness on the lawn service data from Problem 23. Is this process in control and predictable?

25. Because every process is subject to some variations, production can always be improved. Identify possible sources for process variation for the service described in Problem 23 by constructing a cause-and-effect diagram. As before, begin with the categories materials, equipment, methods, people, and environment. Then fill in two or more sources for each category.

**STATISTICS IN ACTION: CHALLENGE PROBLEMS**

26. The use of box plots in place of single points in a control chart provides an alternative display that can simplify the evaluation for $\bar{X}$- and $R$-charts. Iglewicz and Hoaglin have prepared an example that shows the basic elements of box plots with application to process control. The process under study is glue joint gaps of corrugated boxes, measured in units of 1/128th inch. The known standard is 48 units. The displays are Figure 1, $\bar{X}$-charts "Before" and "After" a process adjustment and Figure 2, $R$-charts, again "Before" and "After" the process adjustment.

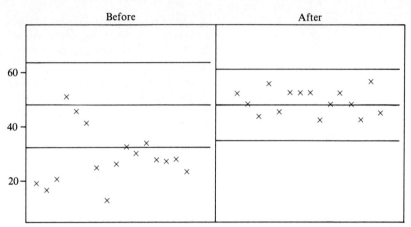

Figure 1. $\overline{X}$ charts for size of gap in glue joints before and after process change

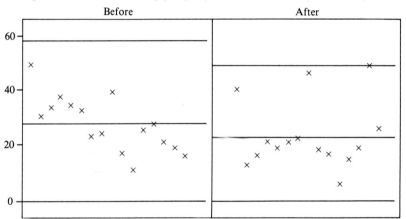

Figure 2. $R$ charts for size of gap in glue joints before and after process change

Questions    The "Before" charts describe measurements on 80 boxes, in groups of 5, for 16 $\overline{X}$ and $R$ data points. The "After" charts include only 15 readings.

a. Explain the conditions, either in or out of control, for each of the $\overline{X}$ control charts. What effect has the change had on the process mean?

b. Explain the condition, either in or out of control, for each of the $R$-charts. What effect has the process change had on the variation?

SOURCE: Iglewicz, Boris, and David C. Hoaglin. "Use of Box Plots for Process Evaluation. "*Journal of Quality Technology* 19, no. 4 (Oct. 1987): 180–190.

27. (Refer to Problem 26) Iglewicz and Hoaglin define box plots to represent four conditions ( #5 repeats two conditions):

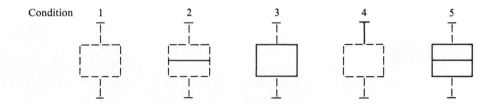

A dashed line indicates normal conditions, while a solid, unbroken line or figure indicates a problem. Condition #1 describes a control chart with both location and variation measures in control, #2 is not in control for the location (the authors use a $F$-mean location measure), #3 is out of control for variation, #4 indicates outliers, and #5 has both location and dispersion measures out of control, but no outlier points.

Figures 4 and 5 from Iglewicz and Hoaglin are reproduced here. These are "Before" and "After" box plot control charts that include both location and dispersion measures.

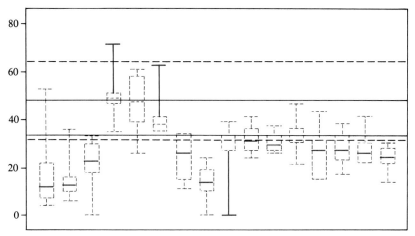

Figure 4. Boxplot control chart for size of gap in glue joints *before* process change

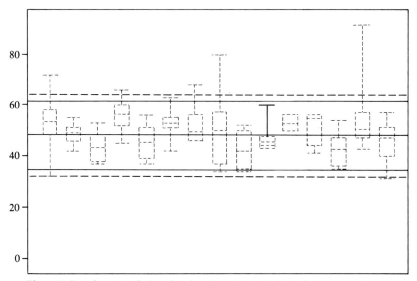

Figure 5. Boxplot control chart for size of gap in glue joints *after* process change

a. Compare the "Before" box plot control chart with the regular control chart ("Before") of Problem 26. Are the in-control conditions the same for location (as for $\bar{X}$) and for dispersion (as for $R$)? What extra information is provided by this box plot control chart?

b. Answer the questions of part (a), but now compare the "After" box plot control chart to the regular "After" control chart.

28. Although this chapter has not discussed probabilities per se, there are useful probability questions in quality control. For example, we address oil spills—which are considered a rare event—with a Poisson distribution. Suppose two oil-tanker delivery companies—A and B—that have an equal chance for the same types of spills, had 3 and 10 spills, or 10 and 3 spills, respectively. Is 3 statistically different from 10? An answer, which we start and ask you to finish, comes from concepts in Chapters 5 and 9.

SOURCE: Nelson, Lloyd S. "Is 3 different from 10?" *Journal of Quality Technology* 18, no. 1 (Jan. 1986): 74–75.

Questions    Consider this an experiment with $3 + 10 = 13$ items set into two groups, with equal chance (i.e., $H_0$: $\lambda_A = \lambda_B$ being equal Poisson process rates, so $p = 1/2$). What is the chance of obtaining occurrences as different as 3 and 10, or 10 and 3, or wider (11 and 2, etc.)? To answer, compute the probability for a binomial distribution, with $n = 13$, $p = 1/2$. Then test $H_0$: $p = 1/2$ ($\lambda_A = \lambda_B$) against $H_A$: $p \neq 1/2$ ($\lambda_A \neq \lambda_B$), using a *p*-value approach. Are 3 oil spills significantly different from 10?

---

## BUSINESS CASE

*Improving Pizza Delivery Performance*

Paul England, the owner of Pizza, Inc., had called this meeting with his two store managers to discuss a growing problem: The time elapsed between receiving an order for pizza and delivering it seemed to be increasing. This was troublesome since Pizza, Inc., had advertised that "Your pizza will be delivered within 45 minutes from the time that you call in your order, or it's free." Paul knew it was good business (and more profitable) to meet his company's advertising claims. Anyway, there had been a few times that the advertised time was not met. And delivery was a big part of the total business. If it got worse, it could cost Pizza, Inc., some customers!

Jerry Erickson managed the Southside and West locations. There had been few payouts here. However, most of the freebies were from the East and Northside locations, where Peg Thompson was manager. Paul had asked for work flowcharts—a description of the total system from entry of orders to their delivery. And he had directed Jerry and Peg to make some delivery time studies in preparation for this meeting.

Paul had a good understanding of his business, but he was not a numbers person. So he had previously sent Peg for some training in a short course on quality control. Jerry was innovative, and they were both good managers. Paul felt that with this background, the three of them would at least come up with some ideas. And they might get lucky and find some solutions.

Paul    Well, business has been fairly good. We still have an edge on our competitors, the three national pizza chains that operate here in town.

Peg    Yes, but my receipts aren't growing as I would expect. The sitdown business is about as usual. But there seems to be some drop in the delivery business. I think Jerry has experienced some drop there too.

Jerry    Peg is right. Our delivery is down slightly. Maybe it's because of the new advertising that the chain stores are putting on radio and TV.

Paul    I rather believe the answer is in our service. It's a known fact that we make the best pizza in town. Possibly your delivery study will give us some answers. Let's have a look at it.

Peg and Jerry had worked together to develop the information Paul had requested, including a job flowchart (Figure 15.10) and four (4) control charts, one for each store (Figure 15.11). The charts show the mean delivery time for each of the past 30 days.

**FIGURE 15.10** A Pizza Order Job Flow

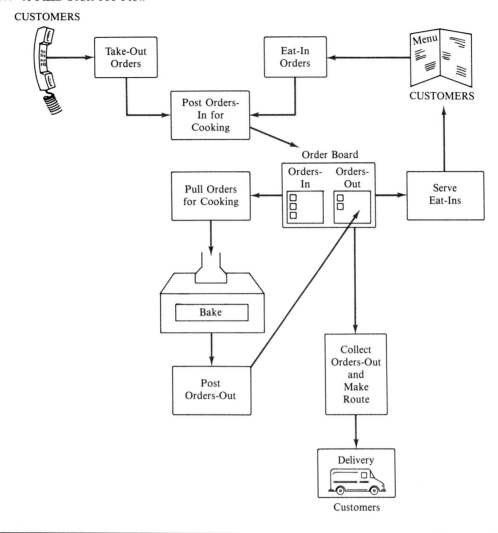

**FIGURE 15.11** Mean Delivery-Time Control Charts

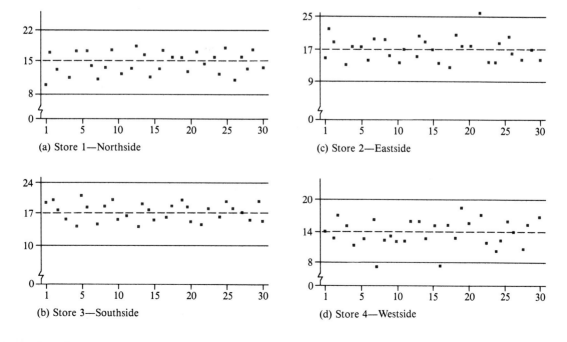

(a) Store 1—Northside

(b) Store 3—Southside

(c) Store 2—Eastside

(d) Store 4—Westside

| Paul | You two have gone to some trouble. Tell me what these charts show. |
|---|---|
| Jerry | Well, I can talk about the pizza order job flowcharts, because this is my procedure. And Peg uses the same system. But she knows about the numbers and so will have to explain the delivery-time charts. |
| Paul | (Looking at the pizza order job flow) Tell me, do your workers have much trouble deciding which in-orders to bake first: the deliveries or the sit-down orders? |
| Peg | I don't think that's a problem. We take the first order that comes to the order board. It gets cooked first. And so on. |
| Jerry | Yes. Southside and West stores do the same. I agree that this is not a problem. They post the orders immediately, and we have enough bakers. The orders get quick attention. There are seldom more than three unfilled orders on the board. We could do a time study on this process, though. |
| Paul | No—not yet, anyway. You've convinced me that this is not a major problem area. Let's look at Peg's delivery numbers. (He points to the control charts, Figure 15.11.) |
| Peg | Jerry and I worked together on these control charts. What they show is the average time for deliveries from each store. The centerline gives an overall average for the 30 days of the study. The outside lines are upper and lower limits for the expected delivery times, and the points are the actual average time it took for deliveries each day. |
| Paul | You mean your drivers had to carry a stopwatch along their routes? |
| Jerry | Well, every driver already has a watch, and in only a few cases did they forget to wear it. But all they had to do was mark the hour and minute each time they left their car to make a delivery. For motivation we gave a bonus for complete records. |
| Paul | That makes good sense. Those folks can always use a few more dollars. Then what do the charts show? |
| Peg | Each of the charts is a little different. Northside shows a fairly regular pattern, with times pretty evenly scattered about the centerline. That's close to an ideal chart. That delivery is okay, and the overall average was 15 minutes. That's the centerline. |
| Paul | Okay. I understand. Then what about Eastside (Store 2)? There is one point outside your lines there. What does that mean? |
| Jerry | That time is beyond a bound. It was determined to be the result of a special cause. |
| Peg | Yes. I asked the delivery driver about that time. He had minor car trouble and had to call in for a backup. Cars aren't 100% dependable, so those things will continue to happen. Otherwise, the Eastside chart looks good. It's in control, and the one special-cause value was explained. |
| Paul | Then what about Westside? (Pointing at the two outside points) Those two were short times. Is that good? |
| Peg | Yes and no. They do indicate short delivery times, and those can both be explained by unusually close delivery distances. On the other hand, when all the points are in bounds, that means delivery times can be predicted. |
| Paul | Well, give me short delivery times. That's what our customers want. |
| Peg | Yours is a fair point. In fact, we made control charts that show the range for the most extremes, the range of time for the deliveries on each day for each store. |
| Paul | Should we look at these, too? |
| Jerry | No. There's no need. They are all in control. But we made copies for you anyway. |
| Paul | Well, this all makes good sense. Peg, you have made good use of the quality-control course, but it hasn't solved our problem, . . . yet. So, we need to determine where to go from here. |

After careful analysis of the flowchart, the three decided to study the process used for "posting orders" for takeout. Peg's stores, where there had been more late deliveries, posted orders when they finished baking. Jerry's stores posted the order, but his bakers also called out the delivery order numbers that were still on the order-out board. This was announced over the work-delivery intercom, so it was announced to the available delivery drivers. Since this was another process in the product delivery system, the group decided to do a study with control charts on this process. They would meet again to evaluate this as a source for wasted time (variation).

Although they might not recognize it as such, Paul, Peg, and Jerry were using the PDCA cycle in their attempt to improve a part of their business.

**Business Case Questions**

29. For the Pizza, Inc., Business Case, identify where each of the PDCA cycle components was used.
    a. Plan          b. Do          c. Check          d. Act

30. Give an example from the Business Case for the use of each of the following statistics in quality improvement. (Refer to Figure 15.3.)
    a. Analyze process variability
    b. Develop process understanding          i. Operational definitions
    c. Identify special cause for variation          j. A system flowchart
    d. A change in a process          k. A process
    e. Reduce variation          l. A tool for measuring process
    f. Process control          variation
    g. Improved quality          m. A measure of variation
    h. More business          n. A statistic

**COMPUTER APPLICATION AND COMPUTER QUESTIONS**

A computer application and questions are not presented for this chapter because there are limited program procedures available for quality-control analysis in Minitab, SPSS, and other widely distributed statistics software packages.

**FOR FURTHER READING**

[1] Burr, Irving. *Statistical Quality Control Methods.* New York: Dekker, 1976.
[2] Deming, W. E. *Quality, Productivity, and Competitive Position.* Cambridge, MA: MIT Center for Advanced Engineering Study, 1982.
[3] Duncan, Acheson J. *Quality Control and Industrial Statistics*, 4th ed. Homewood, Il: Irwin, 1974.
[4] Grant, E. L., and R. S. Leavenworth. *Statistical Quality Control*, 5th ed. New York: McGraw-Hill, 1980.
[5] Imai, Masaaki. *Kaizen (Continuing Improvement).* New York: McGraw-Hill, 1986.
[6] Ishikawa, K. *A Guide to Quality Control.* Tokyo: Asian Productivity Association, 1976.
[7] Snee, Ronald D. "Statistical Thinking and Its Contribution to Total Quality." *The American Statistican* 44, no. 2 (May 1990): 116–121.
[8] "The Refined Focus of Automative Quality." *Quality Progress* (Oct. 1989): 47–50.
[9] Wadsworth, H. M., K. S. Stephens, and A. B. Godfrey. *Modern Methods for Quality Control and Improvement.* New York: Wiley, 1986.

# CHAPTER

# 16

## CHI-SQUARE TESTS OF FREQUENCIES AND PROPORTIONS

In earlier chapters we learned how to test whether a sample mean or proportion was significantly different from a hypothesized value. We also tested hypotheses concerning the equality of two means ($H_0$: $\mu_1 = \mu_2$) and of two proportions ($H_0$: $\pi_1 = \pi_2$). Later we extended the analysis to test for the equality of *more than two population means* by use of an *F*-ratio statistic.

We now study a method for testing the equality of *more than two population proportions*. The test we use is called chi-square (pronounced "kī-square" and symbolized by $\chi^2$) because it utilizes a test statistic obtained from the $\chi^2$ probability distribution. The $\chi^2$ test uses computations of frequencies to infer about the equality of population proportions.

Chi-square tests have extensive applications to managerial decision situations. In particular, they are useful in two areas: (1) tests of goodness of fit and (2) tests of independence of classification. The following situations show where these two tests are appropriate.

**Goodness-of-Fit Application**

The marketing manager of Cleaver Products Co., a firm producing chemicals, soaps, and cosmetics, is evaluating a proposed commercial for television. The commercial is based upon data obtained from a sample of consumers who were each given four boxes of laundry detergent. Each box contained a different brand but was unmarked except for a code number, as shown in Figure 16.1. The box with code #7226 was Cleaver's soap. But this was known only to the independent advertising agency that designed the survey.

After each consumer used all four detergents, he or she wrote the number of the preferred box on a preaddressed card and mailed it to an inde-

FIGURE 16.1 Are Customer Preferences Uniformly Distributed?

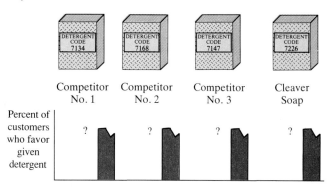

pendent market research company. Results were tallied and given to Cleaver Products. The preferences were fairly close, but the Cleaver Products detergent did emerge with the most preferences. The proposed commercial will identify all four competitive detergents and will claim that there is a real difference in customer preference. However, the marketing manager is very leary of making a false advertising claim, because it is possible that the difference might be due to the chance selection of households for this sample. A valid method is needed to show that the proportions of consumers favoring the various brands are not all equal. A chi-square test of the goodness of fit of the data to a uniform (equal) distribution will satisfy this need. With proper statistical testing, the manager may be able to reject the null hypothesis that the proportions favoring the various brands are equal.

**Independence of Classification Application**

A large motor hotel chain has new hotels in Houston, Denver, Portland, and Spokane. The company's operations manager wishes to verify that the level of customer satisfaction in each of these locations is similar. To measure this, guests at the four locations are randomly selected and asked to indicate their levels of satisfaction according to a five-step scale ranging from highly dissatisfied through neutral to highly satisfied.

Five-Step Satisfaction Scale

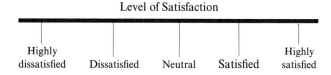

After sufficient data have been collected, the manager can use a $\chi^2$ test to determine whether the level of service is independent of the location. To do this the data are arranged in a table with four rows (locations) and five columns (levels of satisfaction), as illustrated in Table 16.1. These data will be used to compute a $\chi^2$ statistic, with calculations based upon the frequencies, or counts, for the $(4 \times 5 =) 20$ groups resulting from the two-way classification.

Unlike much of our previous study, $\chi^2$ tests are *based on frequency (count) data*. Further, they do *not* require assumptions about the parameters of the

**TABLE 16.1**    Data for a Test to Determine if Level of Satisfaction is Independent of Location

| Location | Level of Satisfaction | | | | |
| | Highly Dissatisfied | Dissatisfied | Neutral | Satisfied | Highly Satisfied |
|---|---|---|---|---|---|
| Houston | 18 | 37 | 130 | 135 | 80 |
| Denver | 12 | 25 | 88 | 40 | 35 |
| Portland | 9 | 27 | 110 | 94 | 60 |
| Spokane | 11 | 11 | 22 | 31 | 25 |

distribution. This makes them especially useful in marketing and other studies in which behavioral data are classified into qualitative groups. Examples of such classifications include the preferences of customers, skill levels of employees, and attitudes of personnel toward working conditions. In these situations, nominal-scaled data are adequate for using $\chi^2$ tests.

We begin our study of $\chi^2$ statistics by gaining a basic understanding of the $\chi^2$ distribution and of how to use the $\chi^2$ tables in Appendix B. Then we discuss tests of goodness of fit and independence of classification. We end the chapter with some optional material that shows how to use $\chi^2$ for inference in binomial and multinomial situations.

Upon completing Chapter 16, you should be able to

1. find and interpret values in the $\chi^2$ probability distribution table;
2. use the $\chi^2$ goodness-of-fit test to determine whether empirical data conform to a common theoretical distribution, such as a uniform, normal, or binomial distribution;
3. formulate contingency tables and use the $\chi^2$ statistic to test for independence of classification;
*4. use the $\chi^2$ statistic to make inferences in binomial and multinomial experiments.

## 16.2 THE CHI-SQUARE STATISTIC

Much of our statistical inference thus far has been based on measures of the deviation of values from their average, or hypothesized, value. This concept forms the basis for the calculation of standard deviations, standard errors, and numerous tests of hypothesis.

Like previous distributions we have worked with, the $\chi^2$ distribution is linked to the measurement of deviations. In fact, the theoretical $\chi^2$-distribution is a distribution of the sum of squared deviations of values drawn from a standard normal distribution. As discussed earlier, the standard normal distribution has a mean of 0 and a standard deviation of 1. We can define $\chi^2$ as follows.

$\chi^2$ is a sampling distribution of the sum of squared deviations from a normal $N(0,1)$ distribution.

★ Star indicates supplementary material for enhanced (optional) coursework.

Tabulated values of $\chi^2$ are given in Table 3 of Appendix B, which expresses sample size in terms of degrees of freedom (df). Because $\chi^2$ is the sum of squared deviations, there is a different set of $\chi^2$-values for each degree of freedom (or sample size) designation. Values in the body of Table B.3 increase as the degrees of freedom increase, going down the page. Values across the top of the table show the areas in the right tail (side) of the distribution. These areas correspond to $\alpha$-levels and designate the probability of obtaining calculated $\chi^2$-values greater than those shown in the respective columns below.

Figure 16.2 illustrates the sampling distribution for the $\chi^2$-statistic for a sample of size $n = 5$, which would correspond to $5 - 1 = 4$ df. It has been marked using the values from Table 3 of Appendix B for 4 df. Thus we see that whereas 90% of the $\chi^2$ sampling distribution values are greater than 1.064 (lightest shading), only 5% are greater than 9.488, only 2.5% are greater than 11.143, and only 1% are greater than 13.277 (darkest shading).

Tabulated values generally represent the percentage of the sum of squared deviation values we might expect to obtain from random samples of a given size drawn from a normal $N(0,1)$ distribution.

The logic of $\chi^2$-tests on qualitative data is based on comparisons of observed sum of squared deviations with values obtained from the $\chi^2$-distribution table. If the actual data follow a specified pattern, then the differences between actual and expected values will be small. Furthermore, the distribution of differences will tend to be normally distributed and the sum of the squared deviations will tend to the $\chi^2$-distribution.

The procedure for using $\chi^2$ for both goodness of fit and independence of classification tests begins with the formulation of null and alternative hypotheses. For goodness-of-fit tests, the null hypothesis states that the data are distributed in an expected way (such as uniformly, or normally). For tests of independence, the null hypothesis states that the classifications are statistically independent.

**FIGURE 16.2**   $\chi^2$-Distribution for 4 df

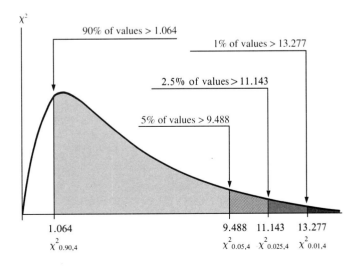

Values of $\chi^2$ for 4 df

This means that the proportions shown for each category of a variable are not affected by a second classification. The computing procedure is basically the same for both tests. We preset an $\alpha$-level of significance. Next, we determine observed ($O_i$) and expected ($E_i$) frequencies. The *observed frequencies* come from the sample data; the *expected frequencies* are computed using the null hypothesis.

> Observed frequencies, $O_i$, are frequencies of sample data that fall into various classes or categories.
>
> Expected frequencies, $E_i$, are the frequencies in each class or category that would be expected if the null hypothesis were true.

Then the difference between the observed and expected frequencies for each class is found, squared, and divided by the expected frequency $(O_i - E_i)^2/E_i$. These values are then summed, giving a $\chi^2$-value. These steps are summarized in Table 16.2. The $\chi^2$ calculated value describes the $\chi^2$-statistic for tests on qualitative data. Equation (16.1) summarizes the statistic calculation.

The $\chi^2$-statistic for qualitative (nominal) variables is

$$\chi^2 = \sum_{i=1}^{K} \left[ \frac{(O_i - E_i)^2}{E_i} \right] \qquad (16.1)$$

where $K$ = the number of classes

$O_i$ = observed frequency in class $i$

$E_i$ = expected frequency in class $i$

Because all deviations in Equation (16.1) are squared, their sum is a positive value. For tests of hypothesis the rejection region is in the upper tail of the $\chi^2$-distribution. Small calculated $\chi^2$-values indicate agreement between the sample data and the null hypothesis. Large values suggest that the differences are excessive, and cause us to reject the null hypothesis.

To use the $\chi^2$-statistic, we assume the data can be divided into some logical classification that describes a variable that has nominal measurement. In addition, we need enough observations in each class to ensure that the value calculated using Equation (16.1) takes on an appropriate $\chi^2$-distribution. If the

**TABLE 16.2** Procedure for Calculating $\chi^2$-Value from Frequency Data

| Observed Frequency | Expected Frequency | Observed − Expected | $\dfrac{(\text{Observed} - \text{Expected})^2}{\text{Expected}}$ |
|---|---|---|---|
| $O_i$ | $E_i$ | $O_i - E_i$ | $\dfrac{(O_i - E_i)^2}{E_i}$ |
| $O_1$ | $E_1$ | $O_1 - E_1$ | xxx |
| $O_2$ | $E_2$ | $O_2 - E_2$ | xxx |
| $\vdots$ | $\vdots$ | $\vdots$ | $\vdots$ |
| $O_K$ | $E_K$ | $O_K - E_K$ | xxx |
| | | | Total   $\chi^2$ calculated value |

**FIGURE 16.3**   The $\chi^2$-Distribution for Various Degrees of Freedom

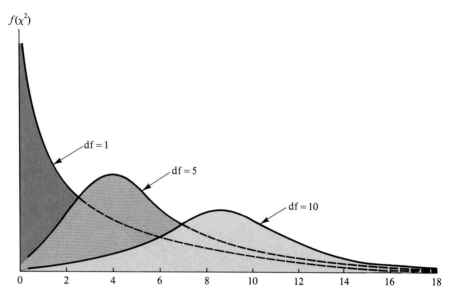

expected frequencies in any class are too small (say, less than five) our statistical inference is less valid. One way to avoid this problem is simply to combine adjacent classes so the theoretical frequencies are at least 5.

For further understanding of $\chi^2$, refer to Figure 16.3. The specific shape of the curve depends on the number of degrees of freedom, which in turn is a function of the number of classes. For small degrees of freedom, the probability curve is quite positively skewed.

As the degrees of freedom increase, the $\chi^2$-distribution approaches the familiar bell shape of a normal distribution, although it remains slightly skewed to the right.

**FIGURE 16.4**   The Chi-Square Probability Distribution for 10 df, Where $\alpha = 0.05$

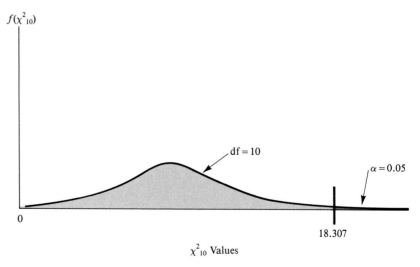

Table 3 of Appendix B contains areas under the $\chi^2$ curve. The column heading indicates the probability that the $\chi^2$-value in that column will be exceeded, given the respective degrees of freedom. See Figure 16.4 for an interpretation of this. For example, the probability that a $\chi^2$-value with 10 df will exceed 18.307 is only 0.05 (dark-shaded area).

Examples 16.1 and 16.2 illustrate how to use Table B.3 for testing and/or estimation purposes.

---

**EXAMPLE 16.1**

What is the critical chi-square value to test a hypothesis at the $\alpha = 0.05$ level using a $\chi^2$-statistic with 6 degrees of freedom?

*Solution*    First we designate the $\chi^2$-value for $\alpha = 0.05$ and 6 df as $\chi^2_{0.05,6}$. Then Table B.3 gives us $\chi^2_{0.05,6} = 12.592$. Thus a calculated $\chi^2$-value larger than 12.592 would be required to reject the null hypothesis.

---

Example 16.2 uses an estimation situation to illustrate the use of both upper- and lower-tail $\chi^2$-distribution percentage points.

---

**EXAMPLE 16.2**

An estimation problem calls for a 90% confidence interval about $\chi^2_{15}$ where 15 represents the df. That is, determine $a$ and $b$ such that $P(a < \chi^2_{15} < b) = 0.90$.

*Solution*    Although the $\chi^2$-distribution is *not* symmetric, we can still allocate equal probabilities to each of the two tails—in this case, $1 - 0.90 = 0.10$ is divided into equal parts, each of 0.05—and find the corresponding $a$ and $b$ values. From Table B.3 we observe that

$$a = \chi^2_{0.95,15} = 7.26 \quad \text{(lower-tail)}$$

$$b = \chi^2_{0.05,15} = 25.00 \quad \text{(upper-tail)}$$

That is, $P(7.26 < \chi^2_{15} < 25.00) = 0.90$. Note that the $\chi^2$-distribution is skewed to the right, so that confidence intervals are not symmetrical.

---

Given this introduction to the $\chi^2$-distribution, we turn next to its application in goodness-of-fit tests. These tests are used to give managers assurance that their data follow a certain probability distribution.

**Problems 16.2**

1. Use Table B.3 to answer the following questions concerning the $\chi^2$ probability distribution.
   a. Determine $P(\chi^2_{12} < 4.404)$, where the subscript 12 is the number of degrees of freedom.
   b. If $P(\chi^2_{20} > 23.828) = \alpha$, find $\alpha$.
   c. For $P(\chi^2_3 > C) = 0.01$, find $C$.

2. Show your understanding of Table B.3 by answering the following questions.
   a. Find $C$ and $D$ for $P(C < \chi^2_5 < D) = 0.95$, so that 2.5% of the lack of confidence comes from either tail.
   b. $P(\chi^2_{11} > 3.053) = \alpha$; what is $\alpha$?
   c. Find the 99th percentile point on the distribution of $\chi^2_{19}$.

3. For a $\chi^2$-test using $\alpha = 0.01$ and with df $= 13$, the calculated value of $\chi^2$ was 18.3. Assume the $\chi^2$-distribution described in this section, so that the rejection region is entirely in the upper-tail.
   a. Should the null hypothesis be rejected?
   b. Another $\chi^2$-test uses $\alpha = 0.025$ and 27 df. What must the calculated $\chi^2$-value be in order to reject the (null) hypothesis of no difference?

4. Recall that $Z_{0.025} = 1.96$. Now $Z_{0.025}^2 = (1.96)^2 = \chi_{\alpha,1}^2$. With the aid of Table B.3, if needed, identify the value of $\alpha$.

## 16.3
## TESTS OF
## GOODNESS OF FIT

We sometimes assume that certain random variables follow a normal, or perhaps binomial or uniform, distribution. A $\chi^2$ *goodness-of-fit* test permits us to test that assumption.

> Goodness-of-fit tests are $\chi^2$-tests used to determine whether observed data are distributed according to some specified probability distribution.

Goodness-of-fit tests can be applied to any frequency data. The observed frequencies are compared with the frequencies expected under the specified distribution using the same $\chi^2$ equation (16.1).

$$\chi_{k-m-1}^2 = \sum_{i=1}^{k} \left( \frac{(O_i - E_i)^2}{E_i} \right) \qquad \text{(16.1 repeated)}$$

where   $k$ = the number of classes

   $m$ = the number of unspecified parameters whose values must be approximated using the sample data

The degrees of freedom for the common distributions that we have studied are as follows.

1. Uniform:   $m = 0$, so df $= k - 1$ since no parameters are estimated.
2. Binomial:   $m = 0$, so df $= k - 1$ when $\pi$ is known.
   $m = 1$, so df $= k - 2$ when $\pi$ is estimated by $p$.
3. Normal:   $m = 2$, so df $= k - 3$ when $\mu$ and $\sigma^2$ are estimated.
4. Poisson:   $m = 1$, so df $= k - 2$ since $\mu$ is usually estimated.

In Example 16.3 we illustrate a goodness-of-fit test on data that are hypothesized to have a uniform distribution.

## EXAMPLE 16.3

The manager of a drive-in bank is concerned about having the correct number of tellers needed and so has had customer arrivals counted for three work periods. The counts are as follows:

| Period | 10–12 A.M. | 12–2 P.M. | 2–4 P.M. | Total |
|---|---|---|---|---|
| Arrivals | 96 | 120 | 84 | 300 |

Test the hypothesis that the arrivals are uniformly distributed over each of the three periods. Use $\alpha = 0.05$.

*Solution*    The solution follows the standard steps that we have used to test a hypothesis.

$H_0$: Arrival data are uniformly distributed. (The number of arrivals in each period is the same.)

$H_A$: Arrival data are not uniform. (The number of arrivals in some period (or periods) is significantly greater than the numbers in other periods.)

$\alpha = 0.05$ (that is, the probability of rejecting a (real) uniform distribution will occur by chance for 5% of all possible random samples.)

Test:    Reject $H_0$ if calculated $\chi^2 > \chi^2_{0.05,2} = 5.991$.

Calculation:    The expected frequencies per period under the assumption of a uniform distribution would be (Total observed/Number of time periods) = 300/3 = 100. Equation (16.1) applied to the arrival counts gives the results in the table.

| Period ($i$) | Number of Arrivals | | | | |
|:---:|:---:|:---:|:---:|:---:|:---:|
| | $O_i$ | $E_i$ | $(O_i - E_i)$ | $(O_i - E_i)^2$ | $\dfrac{(O_i - E_i)^2}{E_i}$ |
| 1 | 96 | 100 | $-4$ | 16 | 0.16 |
| 2 | 120 | 100 | 20 | 400 | 4.00 |
| 3 | 84 | 100 | $-16$ | 256 | 2.56 |
| Totals | 300 | 300 | | | 6.72 |

$$\text{Calculated } \chi^2 = \sum \left[ \frac{(O_i - E_i)^2}{E_i} \right] = 6.72$$

Conclusion:    Reject $H_0$, since $6.72 > 5.991$.

Conclude that the arrivals in some periods are significantly greater than in other periods. The arrivals do not appear to fit a uniform distribution for these work periods. (In this simplified example the inequalities appeared evident from the observed data.)

---

The goodness-of-fit test is unique. While previous tests have concerned the value of population parameters, this test addresses the shape of the underlying probability distribution.

Another example illustrates the goodness-of-fit test for a binomial situation.

**EXAMPLE 16.4**    To check for defectives, random samples of 25 production units were taken each day for a period of 220 days. The observed number of days for the various defective levels is as shown. Do these data fit a binomial distribution that has 0.05 probability of being defective?

| $X_i$<br>No. of<br>Defectives/25 Units | $O_i$<br>Observed<br>No. of Days |
|:---:|:---:|
| 0 | 57 |
| 1 | 76 |
| 2 | 52 |
| 3 | 22 |
| $\geqslant 4$ | 13 |
| Total | 220 |

**Solution**

$H_0$: Data are distributed binomially, with $n = 25$, $\pi = 0.05$.

$H_A$: Data are not distributed binomially, with $\pi = 0.05$.

$\alpha = 0.05$

Test:    Reject $H_0$ if calculated $\chi^2 > \chi^2_{0.05,4} = 9.489$.

(Since no parameters are estimated from sample data, $m = 0$ and $k - m - 1 = 5 - 0 - 1 = 4$ df.)

Calculation:    The binomial probabilities are read from Table B.5, where

$$P(X = 0 \mid n = 25, \pi = 0.05) = 0.277$$

$$P(X = 1 \mid n = 25, \pi = 0.05) = 0.365$$

and so on. For $n = 220$ trials, we determine the expected values ($E_i$) by multiplying the binomial probabilities by the $n_i$ value of 220, as shown in the table.

| Expected Frequencies for a<br>Goodness of Fit to a Binomial Distribution<br>$E_i = n_i P_i(X)$ |
|:---|
| $E_1$: $n_1 P_1 = (220) \cdot P(X = 0 \mid n = 25, \pi = 0.05) = 220(0.277) = \quad 60.94$ |
| $E_2$: $n_2 P_2 = (220) \cdot P(X = 1 \mid n = 25, \pi = 0.05) = 220(0.365) = \quad 80.30$ |
| $E_3$: $n_3 P_3 = (220) \cdot P(X = 2 \mid n = 25, \pi = 0.05) = 220(0.231) = \quad 50.82$ |
| $E_4$: $n_4 P_4 = (220) \cdot P(X = 3 \mid n = 25, \pi = 0.05) = 220(0.093) = \quad 20.46$ |
| $E_5$: $n_5 P_5 = (220) \cdot P(X \geqslant 4 \mid n = 25, \pi = 0.05) = 220(0.034) = \quad 7.48^*$ |
| Total $\sum E_i = 220.00$ |

\* Probabilities must sum to 1 and $\sum E_i = 220$.

$$\text{Calculated } \chi^2 = \frac{(57 - 60.94)^2}{60.94} + \frac{(76 - 80.30)^2}{80.30} + \cdots + \frac{(13 - 7.48)^2}{7.48}$$

$$= 5.0$$

Conclusion:    Do not reject $H_0$, since $5.0 < 9.49$. With $\alpha = 0.05$, we conclude that the distribution of defectives in samples of $n = 25$ is adequately described by a binomial distribution with $\alpha = 0.05$.

The test for the fit of data to a Poisson distribution follows a similar procedure, except that (1) we compute the expected frequencies by using the Poisson distribution, and (2) we compute the degrees of freedom as $k - 1 - 1$. Although the process is the same, the fit of data to a normal distribution is more complicated because we must compute expected frequencies for the respective areas under the normal curve. We illustrate the general approach using an example of the weights of male corporate executives.

**EXAMPLE 16.5**

A weight therapist is interested in the actual distribution of weights of male corporate executives. The therapist estimates that the expected percentages, as determined by the area *under a normal distribution*, would be as shown in column (2) of the data table.

| (1)<br>Class<br>(lb) | (2)<br>Normal<br>Distribution<br>(%) | (3)<br>Actual<br>Frequency<br>Distribution<br>of Weights |
|---|---|---|
| <150 | 6.68 | 40 |
| 150 < 160 | 24.17 | 40 |
| 160 < 170 | 38.30 | 60 |
| 170 < 180 | 24.17 | 30 |
| ≥180 | 6.68 | 30 |
| | Total | 200 |

The actual distribution of weights of a sample of 200 male corporate executives is shown as a frequency distribution (see graph). A normal distribution is shown (dotted) approximating the bar graph. Do the data support the hypothesis that the weights come from a normal distribution? Use $\alpha = 0.05$ and test $H_0$: The distribution of executive weights is normal [with $(\mu, \sigma^2)$ not specified].

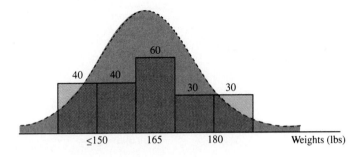

*Solution*   The expected percentages under the assumption of a normal distribution are given, so we need only multiply the percentage values by 200 to get expected frequency counts. For a normal distribution with unknown mean and variance, 2 df are lost because both $\mu$ and $\sigma^2$ must be estimated. The computed $\chi^2$-statistic for a fit to the normal distribution thus has df $= k - 2 - 1 = 5 - 3 = 2$.

| Class | $O_i$ Observed | $E_i$ Expected | $(O_i - E_i)$ | $\dfrac{(O_i - E_i)^2}{E_i}$ |
|---|---|---|---|---|
| $< 150$ | 40 | 13.36 | 26.64 | 53.12 |
| $150 < 160$ | 40 | 48.34 | $-8.34$ | 1.44 |
| $160 < 170$ | 60 | 76.60 | $-16.60$ | 3.60 |
| $170 < 180$ | 30 | 48.34 | $-18.34$ | 6.96 |
| $\geqslant 180$ | 30 | 13.36 | 16.64 | 20.73 |
| | | | Total | 85.85 |

$H_0$: Weights are normally distributed.

$H_A$: Weights are not normally distributed.

$\alpha = 0.05$

Test:   Reject $H_0$ if calculated $\chi^2 > \chi^2_{0.05,2} = 5.991$.

$$\text{Calculated } \chi^2 = \sum \left[ \frac{(O_i - E_i)^2}{E_i} \right] = 85.85$$

Conclusion:   Reject $H_0$. The weights are not normally distributed.

---

If the expected percentages had not been given in Example 16.5, we would have had to use an estimate for the standard deviation ($s = 10$) and compute the area under the normal curve that would fall within each class. The computation is not difficult—just time-consuming. For the distribution given, the estimate for $\mu$ is $\bar{X} = 165$.

Note that the same general concepts apply to the fit of data to the binomial, normal, Poisson, and other distributions. In conclusion, the $\chi^2$-statistic provides one means of testing the goodness of fit to a specific probability distribution. The rejection of a null hypothesis simply tells us that the data do not satisfactorily fit the hypothesized distribution. The fit may, however, be close enough to use for some decision purposes, so the realities of the situation should always be evaluated. Also, an alternative goodness-of-fit test, the Kolmogorov–Smirnov test, is described in Chapter 17.

**Problems 16.3**

5. Assuming you have $k$ groups of data, how many degrees of freedom do you lose when testing the fit of data to (a) a uniform distribution, (b) a normal distribution, and (c) a binomial distribution where $\pi = 0.05$?

6. In a goodness-of-fit test, what is the meaning of your results if your calculated $\chi^2$ is (a) equal to zero, (b) greater than the table $\chi^2$-value for the 0.05 level and proper degrees of freedom?

7. Indicate whether each of the following is true or false.
   a. A $\chi^2$-test inherently assumes that observed data are normally distributed.
   b. A $\chi^2$-test of whether the heights of employees are normally distributed as expected would be a goodness-of-fit test.

    c. A test of whether six sales areas are equal with respect to sales potential could be a $\chi^2$-test of the goodness of fit to a binomial distribution.

    d. A test of whether the absence rate on each of five work days is uniform would utilize a $\chi^2$-test statistic with 3 degrees of freedom.

8. In a $\chi^2$ goodness-of-fit test of data to a uniform distribution with ten classes, the calculated $\chi^2$ value was 12.68. Is this sufficient to reject the hypothesis of a uniform distribution at the 0.10 level? Why or why not?

9. A manufacturing firm's safety engineer feels the accident rate on Monday is significantly greater than on other days. Data for the past year are shown. Test, at the 10% level, whether the data are uniformly distributed.

| Day | No. of Accidents |
|-----|-----|
| Monday | 15 |
| Tuesday | 8 |
| Wednesday | 11 |
| Thursday | 9 |
| Friday | 7 |

10. A quality-control engineer routinely takes daily samples of ten electronic panels for spacecraft instruments and checks them for imperfections. Over 200 consecutive working days, the engineer obtained 180 samples with 0 defectives, 19 samples with 1 defective, and 1 sample with 2 defectives. Test at $\alpha = 0.05$ whether these samples can be looked upon as coming from a binomial distribution having $\pi = 0.05$. That is, make a goodness-of-fit test to the binomial distribution given

$$P(0 \mid 10, 0.05) = 0.599$$

$$P(1 \mid 10, 0.05) = 0.315$$

$$P(\text{At least } 2 \mid 10, 0.05) = 0.086$$

11. Given the tabulated data, do you feel that this sample could have come from a normal distribution? Use $\alpha = 0.05$.

| Class Boundaries | Observed | Expected |
|-----|-----|-----|
| 7.5 < 10.5 | 12 | 10.2 |
| 10.5 < 13.5 | 10 | 10.5 |
| 13.5 < 16.5 | 15 | 15.1 |
| 16.5 < 19.5 | 19 | 17.1 |
| 19.5 < 22.5 | 12 | 15.4 |
| 22.5 < 25.5 | 14 | 10.8 |
| 25.5 < 28.5 | 8 | 10.9 |
| Total | 90 | 90.0 |

*Note:* df = 4

12. A sample of grades for 79 students in a statistic class is given in columns (1) and (2). The expected numbers of students for each grade class in column (3) were obtained from normal curve areas. The expected numbers are based on the sample mean, $\bar{X}$, and the sample standard deviation. Using $\alpha = 0.05$, determine whether or not the normal curve accurately describes the sample data.

| (1)<br>Grade Classes | (2)<br>Actual Number | (3)<br>Theoretical Number |
|---|---|---|
| 49 or less | 13 | 11 |
| 50 < 60 | 7 | 13 |
| 60 < 70 | 10 | 17 |
| 70 < 80 | 29 | 16 |
| 80 < 90 | 12 | 12 |
| 90 and above | 8 | 10 |
| Total | 79 | 79 |

## 16.4 TESTS OF INDEPENDENCE OF CLASSIFICATION

In Section 16.3 we used the $\chi^2$-statistic to test whether actual data were distributed in accordance with some specified distribution. In this section, we classify data according to subcategories. The data can be nominal-scale (qualitative) or higher-order (quantitative). The two variables used in Table 16.1 at the beginning of this chapter were qualitative: (1) location of hotel, and (2) level of satisfaction of the customer. Each of the categories forms the basis for a frequency distribution. The distributions arising from data in the illustrative example might appear as shown in Figure 16.5.

**FIGURE 16.5**   Frequency Distributions of Number of Customers Surveyed

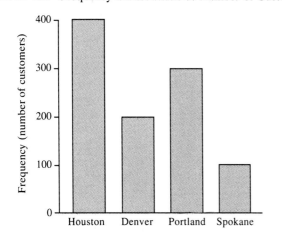

(a) Classification by location

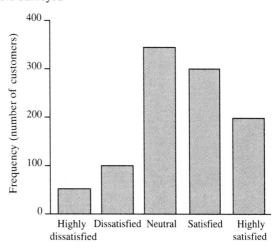

(b) Classification by level of satisfaction

**TABLE 16.3** Contingency Table (Showing Observed Frequencies)

| Location | Observed Levels of Customer Satisfaction | | | | | Totals |
| | *Highly Dissatisfied* | *Dissatisfied* | *Neutral* | *Satisfied* | *Highly Satisfied* | |
|---|---|---|---|---|---|---|
| Houston | 18 | 37 | 130 | 135 | 80 | 400 |
| Denver | 12 | 25 | 88 | 40 | 35 | 200 |
| Portland | 9 | 27 | 110 | 94 | 60 | 300 |
| Spokane | 11 | 11 | 22 | 31 | 25 | 100 |
| Totals | 50 | 100 | 350 | 300 | 200 | 1000 |

Our intent here is to work with both distributions at the same time. Data that are cross-classified in the form of a two-way frequency distribution are referred to as a *contingency table*.

> A contingency table is a table of frequency data that has one mutually exclusive basis of classification down the rows and another across the columns. The table is referred to as an *r* by *c* table ($r \cdot c$), where *r* = number of rows and *c* = number of columns.

Table 16.3 illustrates the form of a $4 \times 5$ contingency table for the two-way classification of the data in Table 16.1. Actual numbers in the cells of the table are not as important, at this point, as the purpose and methodology of analysis.

The question of concern in evaluating contingency tables is whether the two methods of classifying the data are independent.

> Tests of independence of classification are $\chi^2$-tests used to determine whether two methods of classification of data, as shown in a contingency table, are independent of each other.

The null hypothesis assumes the two classifications are independent. We then set the level of significance and compute the $\chi^2$-value in accordance with Equation (16.1). A large calculated $\chi^2$-value evidences a significant difference between the observed and expected values and causes us to reject the hypothesis of independent classifications.

Suppose we were testing a hypothesis of independence of classification of the data shown in Table 16.3. In addition to collecting the "observed" frequencies, we would have to derive some "expected" frequencies. Table 16.3 suggests what these expected frequencies should be. If the null hypothesis of independence is true, the proportions of customers that are highly dissatisfied, and so on, should be independent of location. For example, only 50 of the 1000 customers surveyed were highly dissatisfied. If the variables are independent, we should expect the percentage $50/1000 = 5\%$ to hold across all locations. Thus the expected values for the various locations are

Expected Number of Highly Dissatisfied
(under the assumption of independence of classification)

$$\text{Houston} \quad \left(\frac{50}{1000}\right) \cdot (400) = 20$$

$$\text{Denver} \quad \left(\frac{50}{1000}\right) \cdot (200) = 10$$

$$\text{Portland} \quad \left(\frac{50}{1000}\right) \cdot (300) = 15$$

$$\text{Spokane} \quad \left(\frac{50}{1000}\right) \cdot (100) = \phantom{0}5$$

Note that we can also arrive at the expected frequencies by working across the columns rather than down the rows. Thus, if Houston supplies 400 of the 1000 customers (i.e., 40%), then under the null hypothesis the expected frequency of *highly dissatisfied* customers would be $[(\frac{400}{1000}) \cdot 50] = 20$, of *dissatisfied* would be $[(\frac{400}{1000}) \cdot 100] = 40$, and so on. This leads us to a useful generalization for computing *expected frequencies*.

### RULE FOR EXPECTED FREQUENCIES

Under the assumption of independence, the expected cell frequencies in a two-way contingency table are estimated as the product of the appropriate row total times the column total divided by the sample size:

$$E_{ij} = \frac{\sum R_i \sum C_j}{n} \tag{16.2}$$

$$\chi^2 = \sum_j^c \sum_i^r \left[\frac{(O_{ij} - E_{ij})^2}{E_{ij}}\right]$$

The rule for expected frequencies gives us a procedure for estimating the $E_{ij}$ values necessary for Equation (16.1). The $\chi^2$ equation for contingency tables simply directs that we sum the individual cell calculated $\chi^2$-values both down the columns and across the rows. In a cross-classification, if the $E_{ij}$ values exhibit good agreement with observed counts $(O_{ij})$ for each cell, the computed $\chi^2$-value will be relatively small and will support the (null) hypothesis of independence.

**Degrees of Freedom for Contingency Tables**

The critical $\chi^2$-value is dependent upon the number of classes, just as with goodness-of-fit tests. However, when we have cross-classified data, the degrees of freedom are determined by the number of rows and number of columns in the contingency table. Refer to (the 4 × 5) Table 16.4. Given that the row and column totals must be satisfied, we are free $(F)$ in Table 16.4 to specify only four values in each of three rows and not free $(N)$ to specify any others. This puts one restriction on rows and another on columns. Thus there remain $[(4 - 1)(5 - 1)] = 12$ degrees of freedom. See Table 16.4. Using $r$ = number of

**TABLE 16.4** Degrees of Freedom in a 4 × 5 Contingency Table

| | Row | Col. | 1 | 2 | 3 | 4 | 5 | Totals |
|---|---|---|---|---|---|---|---|---|
| | | | Customer Satisfaction Level | | | | | |
| Location | 1 | | F | F | F | F | N | 400 |
| | 2 | | F | F | F | F | N | 200 |
| | 3 | | F | F | F | F | N | 300 |
| | 4 | | N | N | N | N | N | 100 |
| | | | 50 | 100 | 350 | 300 | 200 | 1000 |

$F$ = value can be a *free* choice

$N$ = value is *not* free, being determined by the row and columns fixed totals and choices for other free values

rows and $c$ = number of columns, we can generalize this as a method for determining degrees of freedom for contingency tables.

$$\text{Contingency table df} = (r - 1)(c - 1) \qquad (16.3)$$

Now that we have some procedures and rules for conducting $\chi^2$-tests of independence of classification, let us follow through with an illustration in a step-by-step fashion.

**EXAMPLE 16.6**

A human resource management analyst hypothesizes that there is some relation between hours worked and a worker's opinion about forming a labor union. The population is the workforce in a small factory town in Pennsylvania. The workforce was classified according to the average number of hours worked per week over the past month. Some members of the workforce were not employed and others worked as much as 60 hours. The opinions from a random sample of 700 people were classified as shown in the table. Test whether the workers' opinions about forming a labor union are independent of the number of hours worked per week. Use $\alpha = 0.05$.

| Work Hours per Week | Observed Frequencies, $O_{ij}$ Opinion of Forming a Labor Union | | | |
|---|---|---|---|---|
| | *In Favor* | *Against* | *No Opinion* | Totals |
| None | 80 | 124 | 16 | 220 |
| 1 < 20 | 75 | 95 | 10 | 180 |
| 20 < 40 | 61 | 73 | 6 | 140 |
| 40 or more | 74 | 78 | 8 | 160 |
| Total | 290 | 370 | 40 | 700 |

*Solution* The hypotheses are

$H_0$: Number of hours of work and opinion concerning a labor union are statistically independent.

$H_A$: Work hours and opinion concerning a labor union are *not* statistically independent. (Opinion concerning a labor union depends upon the number of hours worked.)

$\alpha = 0.05$

Test:   Use a $\chi^2$-test with

$$df = [(r - 1)(c - 1)] = [(4 - 1)(3 - 1)] = 6$$

Reject $H_0$ if calculated $\chi^2 > \chi^2_{0.05,6} = 12.592$.

Calculation:   First, compute the expected frequencies. From column 1, the proportion "in favor" is $290/700 = 0.414$. Cell expectations for "in favor" for the various work groups are as follows:

$$\text{None:} \quad \frac{(290)(220)}{700} = 91.1$$

$$1 < 20: \quad \frac{(290)(180)}{700} = 74.6$$

$$20 < 40: \quad \frac{(290)(140)}{700} = 58.0$$

$$40 \text{ or more:} \quad 290 - (91.1 + 74.6 + 58.0) = 66.3$$

Similarly, expected cell frequencies are computed for the "against" and "no opinion" cells. In all, 12 values are needed. The resulting values are shown in the expected frequencies table.

| Work Hours per Week | Expected Frequencies, $E_{ij}$ Opinion of Forming a Labor Union | | | Totals |
|---|---|---|---|---|
| | *In Favor* | *Against* | *No Opinion* | |
| None | 91.1 | 116.3 | 12.6 | 220 |
| 1 < 20 | 74.6 | 95.1 | 10.3 | 180 |
| 20 < 40 | 58.0 | 74.0 | 8.0 | 140 |
| 40 or more | 66.3 | 84.6 | 9.1 | 160 |
| Totals | 290 | 370 | 40 | 700 |

$$\text{Calculated } \chi^2 = \sum_{j}^{3} \sum_{i}^{4} \left[ \frac{(O_{ij} - E_{ij})^2}{E_{ij}} \right]$$

$$= \frac{(80 - 91.1)^2}{91.1} + \frac{(75 - 74.6)^2}{74.6} + \cdots + \frac{(8 - 9.1)^2}{9.1} = 5.0$$

Conclusion:   Do not reject $H_0$ of independence since calculated $\chi^2 < \chi^2_{0.05,6}$ $(5.0 < 12.592)$.

The frequencies of those who favor, oppose, or have no opinion are not significantly different from what are expected on the basis of the row and column totals only. The differences are not significant at the 5% level, so we have no statistical evidence that work hours affect people's opinions concerning the formation of a labor union in this community.

---

In the example, we chose not to multiply the values out for the "40 or more" row, since these remaining values could be obtained by subtraction from a row or column total.

Experienced statisticians add two additional notes of caution in interpreting $\chi^2$. *First*, if an expected cell frequency becomes less than five, the calculation of $\chi^2$ from Equation (16.1) is less accurate as an approximation to the sampling distribution of $\chi^2$. Methods of handling this situation are discussed in Section 16.5. *Second*, complete agreement of observed and expected total values rarely occurs. Thus, whenever the calculated $\chi^2$-value approaches zero, we should suspect a bias in the data, and further investigation may be in order.

**Problems 16.4**

13. Indicate whether the following statements are true or false.
    a. Chi-square can be used to test for equality of proportions but not for equality of means.
    b. Chi-square tests assume underlying population data are distributed according to the normal distribution.
    c. A (5 × 8) contingency table has 5 columns and 8 rows.
    d. A (5 × 8) contingency table has 28 df.

14. Indicate whether the following statements are true or false.
    a. The theoretical $\chi^2$-distribution utilizes the sum of squared deviations from a normal (0,1) distribution.
    b. Chi-square tests of independence of classification are 1 df tests.
    c. For a $\chi^2_{0.05,1}$ test, we should expect calculated $\chi^2$-values to exceed 5.025 about 5% of the time.
    d. A chi-square rejection value of 32.00 (from Table B.3) could be used for testing a square (25-cell) frequency table at the 0.01 level.

15. Management and labor employees of a large firm were randomly sampled with respect to their attitudes toward a proposed union contract (see table). Use $\chi^2$ to test the hypothesis, at the 5% level, that attitude toward the contract is independent of position (management or labor) by answering the following.
    a. State the null hypothesis.
    b. How many degrees of freedom are appropriate for this test?
    c. What is the (test) value of $\chi^2$ beyond which the hypothesis will be rejected?
    d. Compute the expected frequencies under the null hypothesis and show them by completing the table of expected results.
    e. What is your calculated $\chi^2$-value?
    f. State your decision and conclusions.

| (a) Observed Results | Favor | Opposed | Totals |
|---|---|---|---|
| Management | 75 | 25 | 100 |
| Labor | 65 | 35 | 100 |
| Totals | 140 | 60 | 200 |

| (b) Expected Results | Favor | Opposed | Totals |
|---|---|---|---|
| Management | | | 100 |
| Labor | | | 100 |
| Totals | 140 | 60 | 200 |

16. The market research department of a dairy products company has obtained the following data from a survey of 100 people. The survey was designed to determine whether a person's age has any influence on his or her taste for yogurt. Set up and test, at the 5% level, the hypothesis that a taste for yogurt is independent of age.

| | Observed Frequency | | |
|---|---|---|---|
| | *16–25 yrs* | *26–35 yrs* | *36–45 yrs* |
| Like yogurt | 10 | 20 | 30 |
| Dislike yogurt | 20 | 15 | 5 |

17. An appliance manufacturer obtained the data shown below in a preliminary phone survey designed to evaluate its service nationwide. Use the two-way contingency table to answer these questions:
   a. What is the expected number of consumers in the West who have a good concept of the firm's service (assuming independence)?
   b. What is the number of degrees of freedom for testing the null hypothesis $H_0$: Location and concept of service are independent?
   c. Suppose that the value for the usual statistic that is computed for this type of problem is 6.53. Do you reject the hypothesis given in part (b) at the 5% level of significance? Why or why not?

| Concept of Service | Location | | |
|---|---|---|---|
| | *East* | *Midwest* | *West* |
| Poor | 14 | 16 | 10 |
| Good | 16 | 14 | 30 |

18. For the sample data given, test for independence of income and car style preference. Use $\alpha = 0.05$ and state your conclusion.

| Income | Car Style Preference | | |
|---|---|---|---|
| | *Four-door* | *Hardtop* | *Station Wagon* |
| Low | 30 | 10 | 20 |
| High | 20 | 10 | 10 |

19. The production supervisor at Spokane Products Co. wishes to determine if one or more press operators produce a higher-quality product than others. A selection of four skilled workers were each instructed to run a single press intermittently during one day. Each item produced was inspected and graded, as shown in the table. Test the hypothesis that the proportion of high-, medium-, and low-grade products is the same for each operator. That is, test that the quality is independent of the operator— $H_0: \pi_1 = \pi_2 = \pi_3 = \pi_4$—at the 5% level of significance.

| Product | Grade | Operator Number | | | | Totals |
|---|---|---|---|---|---|---|
| | | *1* | *2* | *3* | *4* | |
| High | 1 | 61 | 75 | 74 | 80 | 290 |
| Medium | 2 | 73 | 95 | 78 | 124 | 370 |
| Low | 3 | 6 | 10 | 8 | 10 | 40 |
| | Totals | 140 | 180 | 160 | 220 | 700 |

20. A production supervisor at the Gould Factory feels that there is a dependence between the hour and the day during which individual work-delays (due to breakdowns, defective products production, etc.) occur. Do you agree? Use $\alpha = 0.05$ and the following record on the number of stops. Data in the table show both observed and expected values. For example, Monday from 9 to 10 A.M. had *15 observed* and *9.7 expected*.

| Day | Observed and Expected Frequencies for Selected Hours | | | |
|---|---|---|---|---|
| | *9–10* | *11–12* | *2–3* | *4–5* |
| Monday | 15 | 12 | 6 | 9 |
| | 9.7 | 10.3 | 9.1 | 12.9 |
| Wednesday | 8 | 10 | 4 | 12 |
| | 7.8 | 8.4 | 7.3 | 10.5 |
| Friday | 7 | 10 | 18 | 19 |
| | 12.5 | 13.3 | 11.6 | 16.6 |

21. A random sample of 350 executives is selected and classified according to their drinking and smoking habits. Are drinking and smoking habits independent for this group? Use $\alpha = 0.01$ and calculated $\chi^2 = 57.25$.

| Drinking Habit | Smoking Habit | | | |
|---|---|---|---|---|
| | *None* | *Light* | *Moderate* | *Heavy* |
| None | 51 | 38 | 37 | 14 |
| Social | 28 | 42 | 11 | 10 |
| Heavy | 21 | 40 | 12 | 46 |

22. A market research firm questions whether including a small prize in a questionnaire would affect the number of responses to its survey. In a preliminary study, 300 questionnaires, half with a prize and half without, were sent to a random selection of persons. This produced the values shown. Is there sufficient evidence to conclude that response to the questionnaire depends upon the inclusion of a prize? Use $\alpha = 0.05$, and state the null hypothesis in terms of proportions.

| Prize | Response | No Response | Totals |
|---|---|---|---|
| Included | 97 | 53 | 150 |
| Not included | 80 | 70 | 150 |
| Totals | 177 | 123 | 300 |

★16.5
INFERENCE IN
BINOMIAL AND
MULTINOMIAL
EXPERIMENTS

The binomial distribution was introduced in Chapter 5; inference concerning $\pi$, the true proportion of successes, was discussed in Chapter 9. We required $n\pi \geqslant 5$ and $n(1 - \pi) \geqslant 5$ as conditions suitable for the use of the normal approximation. In this section we introduce a $\chi^2$-test and discuss its comparability to the binomial and normal distribution tests of proportions.

**Binomial Experiments**

The discussion will separately utilize the binomial, normal, and $\chi^2$ statistics. Recall that for binomially distributed data, the appropriate degrees of freedom for $\chi^2$-tests are the number of classes ($k$) minus 1: If the population proportion ($\pi$) must first be estimated from $p$, df = $k - 2$.

**EXAMPLE 16.7**

A test of the advertising effectiveness of two displays was made as follows: Each display was placed at the checkout counter of a single store and remained there for the same time period. During this time, a record (see table) was kept of the frequency of sales of the advertised product. Assuming that the stores handle comparable trade, are sales statistically equal? Use $\alpha = 0.05$.

★ Star indicates enhanced-study material that some instructors may consider optional.

| Display | 1 | 2 | Total |
|---|---|---|---|
| Sales (number of units) | 82 | 118 | 200 |

*Solution*    The null hypothesis is that the two displays have equal sales of the product. Either a $Z$-statistic or the $\chi^2$-statistic can be applied. The $\chi^2$ procedure is discussed first. Since the test is for a difference in sales, a does-not-equal alternative is appropriate. We let $\pi$ = true percentage of sales when display 1 is used.

$$H_0: \pi = 1/2 \qquad H_A: \pi \neq 1/2 \qquad \alpha = 0.05$$

a. As a $\chi^2$-test:

Test:   Reject $H_0$ if calculated $\chi^2 > \chi^2_{0.05,1} = 3.841$. (There are two classes and we lose 1 df. Hence df $= 2 - 1 = 1$.)

| Observed, $O_i$ | 82 | 118 |
|---|---|---|
| Expected, $E_i$ | 100 | 100 |

Use the given data with the single restriction $\sum O_i = n = 200$. Then, calculated

$$\chi^2 = \sum [(O_i - E_i)^2 / E_i] = [(82 - 100)^2 / 100] + [(118 - 100)^2 / 100]$$
$$= 3.24 + 3.24 = 6.48$$

Conclusion:   Since $6.48 > 3.841$, we reject $H_0$ for $\alpha = 0.05$.

b. As a $Z$-test of proportions:

$$H_0: \pi = 0.50 \qquad H_A: \pi \neq 0.50 \qquad \alpha = 0.05$$

Test:   Reject $H_0$ if |Calculated $Z$| $> 1.96$.

The comparable calculation for $X$ = number of sales under display 1 is

$$Z = \frac{p - \pi_0}{\sigma_p} \qquad \text{where} \quad p = \frac{82}{200} = 0.41, \quad \pi_0 = 0.50$$

$$= \frac{0.41 - 0.50}{0.0354} \qquad \sigma_p = \sqrt{\frac{\pi_0(1 - \pi_0)}{n}} = \sqrt{\frac{(0.5)(0.5)}{200}} = 0.0354$$

$$= -2.55$$

Conclusion:   Again, reject $H_0$.

Notice in Example 16.7 that the $\chi^2$ test is equivalent to using either tail of a two-tailed test using the normal ($Z$) distribution. This is illustrated in Figure 16.6. From the figure we can observe that, for two-tailed tests with df $= 1$,

**FIGURE 16.6**  Equivalence of $\chi^2$- and Z-Tests of Proportions

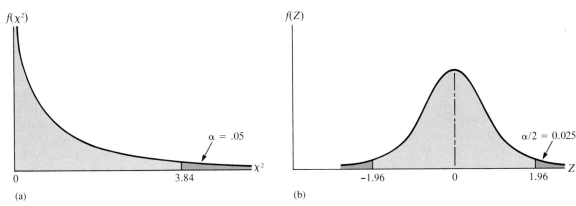

(a)
(b)

$$\chi^2 \text{ Rejection value} = (Z \text{ Rejection value})^2$$

That is,
$$\chi^2 = (Z_{\alpha/2})^2$$
$$3.84 = (1.96)^2$$

Similarly, for the calculated value,

$$6.48 = (-2.55)^2 \quad \text{(except for rounding)}$$

In both cases, the data indicate a significant difference in sales for the two displays at the $\alpha = 0.05$ level for these displays.

The conclusion is the same whether you are using the Z or the $\chi^2$ procedure. The $\chi^2$ approach, being a nonparametric procedure, is slightly more flexible in that we need not assume the data are distributed according to any specific (normal) distribution.

For a *two-tailed test* of equal proportions in comparing two populations, we have a choice of using either the $\chi^2$- or the Z-test. For *one-tailed tests*, the $\chi^2$ procedure is not applicable. Here, for samples less than 30 (where exact binomial probability tables are available), when either $(n\pi) < 5$ or $n(1-\pi) < 5$ we suggest using exact binomial probabilities. (Even this has shortcomings; with a limited number of observations, it is not possible to set the $\alpha$-risk arbitrarily.) For $n\pi \geqslant 5$ and $n(1-\pi) \geqslant 5$, a normal approximation (Chapter 6) is recommended.

Whenever frequency data have cells with *expected* frequency values less than 5, some form of *short-cell* correction should be made in order to obtain meaningful statistical inference.*

Any class with *expected* frequency less than 5 is called a short cell.

---

* Yates' correction is a long-standing procedure for tables with a short cell and 1 df that adds 0.5 to each observed − expected difference before squaring. Numerous statistics researchers contend that an expected value of 5 is an excessive restriction on the $\chi^2$ procedure; for example, see [5]. (Bracketed numbers refer to items listed in "For Further Reading" at the end of the chapter.) The point is that at some low *expected* value there is insufficient information to allow reasonable inference. Then some adjustment becomes necessary.

### METHODS OF DEALING WITH A SHORT-CELL SITUATION

1. Increasing the sample size in order to overcome the low expectation.
2. Combining the low-frequency cells with other cells via logical combinations. (This is usually the preferred method.)
3. Dropping the short cell and others that are no longer meaningful.

Option 1 is our first preference because it maintains the original conditions. Marketing research and opinion polls often have short cells. Example 16.8 illustrates such a case in a multinomial situation. Also, Problems 41, 48, and 49 require your consideration of short cells.

## Multinomial Experiments

*Multinomial experiments* provide us with further insight into the potential use of the $\chi^2$-statistic.

A multinomial experiment is one in which frequency data are classified into more than two categories or classes and each trial of the experiment can result in one of $K$ outcomes, with the expected number of outcomes of a type, $i$, in $n$ trials being $n\pi_i$, where $\pi_i$ is the fixed probability that any trial results in the outcome $i$.

Specifically, the conditions for a multinomial experiment are as follows:

1. The experiment consists of $n$ identical and independent trials.
2. The probability that any trial will result in an outcome $i$ is $\pi_i$, for $i = 1, 2, \ldots, K$, and $\pi_i$ is constant for all trials.
3. Each trial results in one of $K$ outcomes, with $\sum^k \pi_i = 1$.
4. The random variable is $n_i$, $i = 1, 2, \ldots, K$, the number of trials resulting in outcome $i$, with $\sum^k n_i = n$

Then the *multinomial probability distribution*, which describes the number of observations resulting in each of the $K$ outcomes, is described by

$$P(n_1, n_2, \ldots, n_K) = \frac{n!}{n_1! n_2! \cdots n_K!} \pi_1^{n_1} \pi_2^{n_2} \cdots \pi_K^{n_K} \qquad (16.4)$$

A simple example of a multinomial experiment is rolling a fair die $n$ times. A probability question is "What is the probability that 60 rolls will result in ten occurrences of each face, for $K = 1, 2, \ldots, 6$ as the upturned face?" From our experience with binomial experiments, we know that an answer could require extensive calculations. So researchers have developed a more practical way to obtain a solution.

An important assumption here is independence of observations. This condition is ensured as long as the die is reshaken, so that an outcome does not depend on what has happened before. For a fair die and with independent tosses, the expected frequencies would be the same, 10, for all (six) multinomial possibilities.

If we were to rename the faces so we have just two possibilities, such as (1) = the 1 face and (2) = other than the 1 face, then the preceding reduces to a simple binomial experiment. Thus, other than counting for more numerous classes, the multinomial requires little more calculation than in binomial experiments.

The statistical tests of interest concern the probabilities $\pi_1, \pi_2, \ldots, \pi_K$. For this we will hypothesize specific values for the $\pi$'s, and then determine whether the sample data agree with the hypothesized values. We make the test by examining the observed number of occurrences of each type and comparing this to the expected number of occurrences as specified by a null hypothesis.

About 1900, Karl Pearson proposed a procedure using the chi-square statistic to test for specified probabilities:

$$\chi^2_{K-1} = \sum^K \left[ \frac{(n_i - E_i)^2}{E_i} \right] \qquad \text{(16.5)}$$

with $E_i = n\pi_{i_0}$ and where $\pi_{i_0}$ = probabilities specified under the null hypothesis.

The following example illustrates a multinomial experiment, where the binomial form is extended to cover a multinomial situation. In it, we assume independence among the test subjects (that preferences of one did not influence another). We begin the problem with a short-cell situation.

---

**EXAMPLE 16.8**

Suppose 40 potential consumers selected at random gave their preference for three package designs for a product. We hypothesize that a clear plastic wrap (design C) has only 10% of the preference, with the remainder evenly split between two others. In other words, we hypothesize that design A = 0.45, design B = 0.45, and design C = 0.10. Alternately, $H_A$ is that they are not 0.45, 0.45, and 0.10. Test the hypothesis on design preference at the $\alpha = 0.05$ level.

| Package Design | Observed Number Who Favor |
|----------------|---------------------------|
| A | 21 |
| B | 13 |
| C | 6 |
| Total | 40 |

---

*Solution*

The source of the problems, a low expected value, is in the plastic wrap design C, $E_C = n \cdot \pi_C = [(40)(0.1)] = 4.0$. This low expected value indicates a short cell. The correct approach for overcoming the short cell is a matter of opinion and should be determined by the most logical approach under the circumstances. Certainly the preferred approach would be to obtain at least ten more opinions before an analysis is made. That is, $40 + 10 = 50$ opinions would yield $E_C = n \cdot \pi_C = [(50)(10)] = 5$; and this is minimal! In this case the cost of additional information, i.e., for interviewing ten or more potential consumers, could be reasonably low.

The table for extended design preferences assumes this was done, and so includes 20 more opinions, for a total of 60 consumers.

| Package Design | Observed Number Who Favor |
|:--------------:|:-------------------------:|
| A | 32 |
| B | 20 |
| C | 8 |
| | Total 60 |

Again,

$$H_0: \pi_A = 0.45 : \pi_B = 0.45 : \pi_C = 0.10$$
$$H_A: \pi_A : \pi_B : \pi_C \neq 0.45 : 0.45 : 0.10$$

Test:   Reject $H_0$ if calculated $\chi^2 > 5.991$ with df $= 3 - 1 = 2$.

The calculation of approximate $\chi^2$ extends the preceding:

$$\text{Calculated } \chi^2 = \frac{[32 - (60)(0.45)]^2}{27} + \frac{[20 - (60)(0.45)]^2}{27}$$
$$+ \frac{[8 - (60)(0.10)]^2}{6} = 3.41$$

For df $= 3 - 1 = 2$, tabular $\chi^2_{0.05, 2} = 5.991$.

Conclusion:   Since $3.41 < 5.991$, we cannot reject $H_0$, at $\alpha = 0.05$.

The sample shows that at this time there is no statistical difference in preference between package designs A and B, assuming the plastic wrap has only 10% of the preference.

As with other small-sample tests, we should view the preceding "cannot-reject-the-hypothesis" conclusion with caution. There exists a strong possibility of a type 2 error, another reason that we should take larger samples.

In conclusion, caution is the best attitude toward treating short cells. Attempts to combine classes are very suspect if subjective placement is made. For example, it would be difficult to justify combining the frequency for the plastic wrap preference with any other class without good logical support for such a move. Also, we should consider leaving out meaningful information *only* if everything else fails.

**Problems 16.5**     23. A binomial experiment including $n = 50$ trials produces $X = 24$ successes. Use the $\chi^2$ procedure to test the null hypothesis that $\pi = 0.40$. Let $\alpha = 0.05$ and use a does-not-equal alternative hypothesis. Use df $= 1$.

24. A marketing expert is asked to test the hypothesis that more than 65% of all consumers will choose a suburban over a downtown shopping center. If 70 of 100 randomly chosen consumers expressed preference for the suburban center, establish and carry out a suitable test on the preceding contention. Use $\alpha = 0.10$. In deciding which test statistic to use, consider that the alternative hypothesis is one-tailed.

25. A random sample of 200 workers who have taken a safety course contained 62 who had been involved in one or more accidents in the past five years. Can we reject the hypothesis that the proportion of accident-free workers in this group is equal to 0.75? Use $\alpha = 0.10$ and your choice of either a $Z$- or a $\chi^2$-test.

26. In a survey made at a power company's annual meeting, six randomly selected stockholders were asked if they favored the addition of coal-fueled generating plants to their system. Three answered no, and the remainder answered yes. Does the sample show that opinion is evenly divided on this issue?

   a. Why is the $\chi^2$-test inappropriate?

   b. If an increased sample gave a total of 31 out of 50 who were against the plants, how would you answer the original questions? Use $\alpha = 0.025$.

## SUMMARY

Theoretical *chi-square* is a sampling distribution of the sum of squared deviations from a normal (0,1) distribution. We use a test statistic from the $\chi^2$ distribution to test (1) whether observed frequencies conform to a given pattern (goodness of fit) and (2) for independence of classification. When calculated values of $\chi^2$ exceed table values, the difference between observed ($O_i$) and expected ($E_i$) frequencies is too large to attribute to chance, given the $\alpha$-level of significance. This causes us to reject the null hypothesis.

*Goodness-of-fit tests* can be used to determine whether observed data are distributed according to some specified distribution, such as the uniform, binomial, normal, or Poisson. No underlying assumptions are required. However, there must be *five or more expected frequency* in each class to yield accurate results. If there is little difference between the actual and expected values, the calculated $\chi^2$-value will be small. This, in turn, will support the null hypothesis that the actual data closely "fits" the hypothesized distribution.

Tests of *independence-of-classification* rest upon the null hypothesis that two methods of classifying data are independent of each other. For example, a test may be conducted to determine whether the level of customer satisfaction with hotel service is independent of location of the hotel. Observed data are first arranged in a two-way classification referred to as a contingency table. Expected frequencies are determined with Equation (16.2).

The calculated $\chi^2$-value is computed and compared with a $\chi^2$-test statistic (Table B.3) value for the appropriate $\alpha$-level and df. We compute the degrees of freedom for contingency tables by multiplying $[(r - 1)(c - 1)]$. If calculated $\chi^2 > \chi^2_{\alpha,df}$, then the variables are not independent and the null hypothesis is rejected.

Optional material in Section 16.5 deals with the use of $\chi^2$ for inference in binomial and multinomial experiments. It shows that for tests of proportions where $H_A$ is of a does-not-equal type, the $\chi^2$- and $Z$-tests are equivalent. It also suggests methods for short-cell correction (that is, when a cell frequency is less than 5).

Table 16.5 lists the key equations for $\chi^2$-tests.

**TABLE 16.5**   Key Equations for Chi-Square Tests

For goodness of fit

$$\chi^2 = \sum_{i=1}^{K} \left[ \frac{(O_i - E_i)^2}{E_i} \right] \tag{16.1}$$

For tests of independence of classification

$$\chi^2 = \sum_{j}^{c} \sum_{i}^{r} \left[ \frac{(O_{ij} - E_{ij})^2}{E_{ij}} \right] \tag{16.2}$$

with expected frequencies

$$E_{ij} = \frac{\sum R_i \cdot \sum C_j}{n}$$

and

$$df = [(r - 1)(c - 1)] \tag{16.3}$$

For multinomial experiments

$$\chi^2 = \sum_{i=1}^{K} \left[ \frac{(n_i - E_i)^2}{E_i} \right] \tag{16.5}$$

with   $E_1 = n\pi_{1_0}, E_2 = n\pi_{2_0}, \ldots, E_K = n\pi_{K_0}$

---

**KEY TERMS**

chi-square ($\chi^2$), 752, 754
contingency table, 766
expected frequency, 756
goodness-of-fit test, 759
multinomial experiment, 776

observed frequency, 756
rule for expected frequencies, 767
short cells in contingency tables, 775
test of independence, 766

**CHAPTER 16 PROBLEMS**

27. The number of business conventions held in a large city was recorded over many months. Test the claim that the numbers have been evenly distributed over the week, that is, the goodness of fit of the data to a uniform distribution. Use $\alpha = 0.05$. Notice that df $= 7 - 1 = 6$.

| Day | Sun. | Mon. | Tues. | Wed. | Thur. | Fri. | Sat. | Total |
|-----|------|------|-------|------|-------|------|------|-------|
| Number of Meetings | 88 | 84 | 70 | 100 | 90 | 72 | 56 | 560 |

28. Answer the following by using the $\chi^2$-distribution table (Table B.3):
   a. $P(\chi_5^2 < c) = 0.05$; find $c$.
   b. $P(\chi_{13}^2 > 15.984) = \alpha$; find $\alpha$.
   c. If the $\chi^2$-test has 23 df and we want $\alpha = 0.01$, what calculated value is required to reject equality of proportions?

29. A hospital administrator has collected the following data on the number of admissions for a randomly chosen week. Test the goodness of fit of this data to a uniform distribution. That is, test $H_0: \pi_{Su} = \pi_M = \cdots = \pi_{Sa} = 1/7$. Use $\alpha = 0.025$.

| Day | Sun. | Mon. | Tues. | Wed. | Thur. | Fri. | Sat. | Total |
|---|---|---|---|---|---|---|---|---|
| Admissions | 27 | 16 | 15 | 15 | 14 | 25 | 28 | 140 |

30. Test whether the proportion of sales was dependent upon a seasonal factor. Units are cases. Let $\alpha = 0.01$.

| Number of Cases | Nov. | Dec. | Jan. |
|---|---|---|---|
| Retail | 33 | 32 | 20 |
| Wholesale | 106 | 128 | 80 |

31. A certain drug is claimed to be effective in curing the common cold. In an experiment on 106 people with colds, half of them were given the drug and the others were given sugar pills. The patients were all told they were given something that, it was hoped, would cure their colds. If 50 of those given the drug were helped and 44 of those given the sugar felt they were helped, make a test for independence of effect (helped/not helped) and treatment. Use $\alpha = 0.05$. State practical conclusions.

32. A home center carries two major appliances—ranges and refrigerators. A 1-month record of sales shows the following number of units of the major appliances sold by two salespeople. Does this record support the claim that each salesperson's ability to sell depends on the product he or she is selling? State recommendations at $\alpha = 0.025$.

| Salesperson | Ranges | Refrigerators |
|---|---|---|
| Ann | 26 | 8 |
| Steve | 14 | 18 |

33. A manufacturer of sportswear has designed a new package for its product and has run the following test: Two stores were selected, one in a higher-income area and the other in a middle-income neighborhood. A display of sportswear was arranged, consisting of packages with the old design and others with the new one. At the end of the test, 350 packages had been sold in store A. Of these, 298 were the new design. In store B, 150 packages were sold, of which 102 were the new design. Is there a significant difference between the proportion of new design packages sold in the two stores? Use $\alpha = 0.05$.

34. A safety engineer for a manufacturer feels that there is a dependence between the hour and the day during which accidents occur. Do you agree? Use $\alpha = 0.05$ and the record of the number and day of accidents (see table). Note 15 accidents were observed and 9.6 accidents were expected for Monday from 9 to 10 A.M., and so on. State your recommendations.

| | Hour | | | |
|---|---|---|---|---|
| Day | *9–10* | *11–12* | *2–3* | *4–5* |
| Monday | $15^{9.6}$ | $6^{9.0}$ | $12^{10.3}$ | $9^{13.2}$ |
| Wednesday | $8^{7.8}$ | $4^{7.3}$ | $10^{8.3}$ | $12^{10.6}$ |
| Friday | $7^{12.6}$ | $18^{11.8}$ | $10^{13.4}$ | $20^{17.2}$ |

★35. A binomial experiment of 400 independent trials showed 250 successes. Does this indicate other than $\pi = 0.50$? Use $\alpha = 0.05$.

★36. The following is a two-way classification by sex and employment status for a sample of persons at a stockholders' meeting. Test the hypothesis that 50% of the people in this group are now self-employed. Use $\alpha = 0.05$. (*Suggestion:* Use subclasses *self-employed* and *not self-employed* for employment status.)

| Sex | Employed by Another | Self-Employed | Retired |
|-----|---------------------|---------------|---------|
| Male | 11 | 22 | 7 |
| Female | 5 | 21 | 14 |

★37. The number of arrivals were recorded at a drive-in apothecary. Test the hypothesis that 40% of the arrivals are during 12 to 2 P.M. Use $\alpha = 0.01$.

| Hours | 10–12 A.M. | 12–2 P.M. | 2–4 P.M. |
|-------|-----------|-----------|----------|
| Arrivals | 96 | 118 | 86 |

★38. A national retailing firm is studying the advertising media that most influence its customers. An analysis is made of findings from interviews with 180 randomly selected customers. Test for independence of media and income bracket. Use $\alpha = 0.01$. Would you make any further tests? As a preliminary procedure, check for short cells. Note that $O_{11} = 10$ while $E_{11} = 8.1$ and so on.

| | Medium | | | |
|---|---|---|---|---|
| Income | TV | Local News | Magazines | Personal Habit |
| Over $30,000 | $10^{8.1}$ | $17^{13.9}$ | $16^{13.3}$ | 7 |
| $20,000 to $30,000 | $7^{12.9}$ | $13^{22.2}$ | $24^{21.3}$ | 36 |
| Under $20,000 | 12 | 20 | 8 | 10 |

★39. Suppose you are a purchasing agent and have received parts from two suppliers, A and B. Assuming that your random sample is adequate, which company would you choose to continue as sole supplier of the parts? Use $\alpha = 0.05$ and test $H_0: \pi_A = \pi_B$, where $\pi_A$ = percent defective received from company A, and so on. Assume that all other relevant factors, such as cost and delivery, are equal. Begin with a check for short cells.

★ Starred problems require the optional material in Section 16.5; however, some can be solved using a standard Z-test.

| Company | Quantity Received | |
| | Defective | Nondefective |
| --- | --- | --- |
| A | 18 | 182 |
| B | 12 | 288 |

40. Tat investigated differences between black female fashion-opinion leaders and non-leaders, based on media exposure and demographics. Data were obtained from a survey of shoppers at a southeastern city's shopping center. The community had a high concentration of black people. The sample included 241 interviews, resulting in 212 usable responses. Opinion leadership was measured by the answer to the question "People come to me more often than I go to them for information about fashion." Table 2, on page 16 of Tat's article "Opinion Leadership and the Search for Fashion Information," includes the following questions:

| Question | Opinion Leaders | Nonleaders | Calculated Chi-Square | $p$-Value |
| --- | --- | --- | --- | --- |
| To find out the latest fashion, I always | | | | |
| 1. go to the stores selling fashion. | | | | |
| Agree | 64 | 93 | 8.21 | <0.01 |
| Disagree | 10 | 45 | | |
| 2. look at newspaper ads or articles about fashion. | | | | |
| Agree | 69 | 124 | 0.606 | >0.10 |
| Disagree | 5 | 14 | | |
| 3. read fashion magazines such as *Vogue* and *Essence*. | | | | |
| Agree | 67 | 105 | 6.091 | <0.05 |
| Disagree | 7 | 33 | | |
| 4. go to fashion shows. | | | | |
| Agree | 47 | 58 | 7.411 | <0.01 |
| Disagree | 27 | 80 | | |
| 5. watch television. | | | | |
| Agree | 64 | 83 | 15.863 | <0.01 |
| Disagree | 10 | 55 | | |
| 6. listen to radio. | | | | |
| Agree | 42 | 46 | 8.486 | <0.01 |
| Disagree | 32 | 92 | | |
| 7. discuss fashion with other women. | | | | |
| Agree | 70 | 116 | 5.217 | <0.05 |
| Disagree | 4 | 22 | | |
| 8. observe what other women wear. | | | | |
| Agree | 68 | 126 | 0.006 | >0.10 |
| Disagree | 6 | 12 | | |

SOURCE: Tat, Peter K. "Opinion Leadership in Black Female Fashion Buying Behavior." *The Mid-Atlantic Journal of Business* 23, no. 1 (Winter 1984–85): 11–20.

**Questions** Based on the results displayed in the above table, identify those media most useful for distinguishing fashion information used by black female fashion leaders versus nonleaders. Justify your choices with chi-square calculated results and their *p*-values.

41. What is the impact of television advertising on children? Gorn and Florsheim indicate that even if children are not consumers of a product type, exposure to television may influence how they view being an adult. The study included (70) nine- to ten-year-old girls for the effect of exposure to television commercials for adult-oriented products—here, specifically lipstick.

    An experimental group was allowed to watch two 30-second TV commercials that advertised lipstick. The control group was not allowed to see the lipstick commercials. Afterward, in one question the girls were told there was a new group of teachers coming to their school, and then were asked to choose a gift for their new teacher's birthday. Choices were a book, a scarf, lipstick, or a plant. The results were as follows:

| Gift Preference | Control Group | Experimental Group |
|---|---|---|
| Lipstick | 0 | 7 |
| Other | 36 | 27 |

with significance test $\chi_1^2 = 6.11$ and *p*-value $< 0.01$.
*Note:* A Yates' correction (adds 0.5 to each difference, observed − expected, before squaring) was used if the table had a short cell and 1 df.

SOURCE: Gorn, Gerald J., and Renée Florsheim. "The Effects of Commercials for Adult Products on Children." *Journal of Consumer Research* 11, no. 4 (March 1985): 962–967.

**Questions** Using the researchers' results, state an appropriate test conclusion. Then discuss their choice for a short-cell situation test.

---

**BUSINESS CASE**

*The State Lottery— Was It Rigged?*

Tim Sullivan burst into Sharon Moore's office with a startled look on his face.

**Tim** We could be in big trouble, Mrs. Moore! Look at this article.

**Mrs. M.** What article? Calm down, Tim. What are you talking about?

**Tim** Here, in yesterday's *Gazette*. (Tim hands Mrs. Moore a portion of a newspaper, with an article circled in red.) Rob Alexander claims our state lottery might be rigged! And he has some figures to back up what he says!

**Mrs. M.** It's just politics, Tim. He's in a tight race for the state house seat, and he's just playing for public attention. What kind of figures does he have?

Tim Sullivan and his boss, Sharon Moore, were both employed by the state. Sharon was the supervisor of the state lottery project that had been voted in two years ago. Since its inception, the lottery had generated over $100 million in revenues for the state. But it had also been a controversial money-raiser. The most recent debate had centered around use of the (surprisingly large) revenues. This latest controversy suggested that the lottery was not being run carefully and that some winning numbers were occurring much more frequently than others. The political candidate who made the allegation, Robert Alexander, had observed that a well-to-do businessperson had won significant sums in the past four months. He even suggested that there might be some profiting at the expense of the "ordinary citizens" by knowing that the lottery results were biased and selecting number combinations that had higher-than-usual probabilities of winning.

The lottery worked like this: Anyone could buy tickets, at $1 each. The purchaser marked off three digits on his or her ticket. Every day a drawing was held, and ticket holders with the correct three numbers received a small monetary prize. They were also advanced to a winner's round for the week. Weekly winners were then carried forward to compete for much larger monthly amounts.

The lottery drawings were conducted in front of TV cameras at 6:25 each evening. There were three machines, each used to yield one digit. Ten ping pong balls, each with one number from 0 through 9, were released into the (glass) air chamber of each machine. A strong airflow kept the balls mixed in mid-air. Then a representative of a certified public accounting firm opened a vent and one ball immediately popped into the opening. One ball was selected from each machine in this way, giving the winning digits for the particular day.

Mr. Alexander complained that the selection procedure was unfair because two digits had been occuring more frequently than others. For proof, he listed data in a table that showed how many times each digit occurred over the past year (330 days of drawings).

| Digit | 0 | 1 | 2 | 3 | 4 | 5 | 6 | 7 | 8 | 9 |
|---|---|---|---|---|---|---|---|---|---|---|
| Frequency | 93 | 96 | 89 | 98 | 99 | 88 | 120 | 95 | 94 | 118 |

Mr. Alexander went on to suggest that the lottery officials either were not attentive to what was happening or were conveniently overlooking the problem because they had nothing to lose by letting it continue. He felt that anyone with a sense of figures would pick up on the problem if he saw the data. It had, in his words, "less than a hundred-to-one shot" of occurring without some kind of bias.

**Tim**    So what do we do now, Mrs. Moore? His figures do look pretty convincing. What really bugs me, though, are the insinuations that we might be allowing something crooked to go on.

**Mrs. M.**    That irks me too, Tim. We're going to have to counteract his insinuations with some factual analysis of our own. Do me a favor. (Sharon pointed to a statistics book on her bookshelf.) Take that book and do a little homework tonight by studying up on the chi-square distribution. I'm sure you covered that in your statistics course at the university.

**Tim**    Oh, yes. I remember the name—but that's about all. You're always pulling that book out on me. Last month it was to compute the various probabilities of the different winning lottery combinations.

**Mrs. M.**    Well, we are supposed to do things in a professional way around here, aren't we?

**Tim**    Oh, sure. Don't get me wrong. I agree with your approach. Only Mr. Alexander appears to have some pretty solid data. It's going to be hard to refute.

The next morning Tim was waiting for Mrs. Moore when she arrived at 8:00 A.M.

Tim     Like I said, Mrs. Moore, it'll be a breeze!

Mrs. M.  Tim . . . you've been using your calculator again!

Tim     It only took 5 minutes once I got back into $\chi^2$ again. Here, look at this.

Tim showed Mrs. Moore a sheet of paper with the $\chi^2$ calculations (see table).

**Chi-Square Calculations for the Lottery Example**

| Digit | 0 | 1 | 2 | 3 | 4 | 5 | 6 | 7 | 8 | 9 |
|---|---|---|---|---|---|---|---|---|---|---|
| Observed, $O$ | 93 | 96 | 89 | 98 | 99 | 88 | 120 | 95 | 94 | 118 |
| Expected, $E$ | 99 | 99 | 99 | 99 | 99 | 99 | 99 | 99 | 99 | 99 |
| $(O - E)^2$ | 36 | 9 | 100 | 1 | 0 | 121 | 441 | 16 | 25 | 361 |
| $(O - E)^2/E$ | 0.36 | 0.09 | 1.01 | 0.01 | 0 | 1.22 | 4.45 | 0.16 | 0.25 | 3.65 |

$$H_0: \pi_0 = \pi_1 = \cdots = \pi_9$$

$H_A$: The proportions are not all equal.

$$\alpha = 0.10$$

Use $\chi^2$, and reject if calculated $\chi^2 > \chi^2_{0.10,9} = 14.684$.

$$\text{Calculated } \chi^2 = \sum \left[ \frac{(O - E)^2}{E} \right]$$

$$= 0.36 + 0.09 + \cdots + 3.65 = 11.2$$

Conclusion:    Do not reject $H_0$.

Tim     See, we can't reject the hypothesis of equal proportions of digits. So the high frequency of the digits 6 and 9 is probably just due to chance. Right?

Mrs. M.  I'd say so. Given that your calculated $\chi^2$ of 11.2 is somewhat below the test criterion of 14.684, we could expect occurrences like this more than 10% of the time. It's nothing like the one in a hundred Mr. Alexander is babbling on about. Do you think you can write up a short factual statement for the press? I'm sure they'll be calling us within an hour or so.

Tim     I think I can handle that.

Mrs. M.  Be nice. Just because Mr. Alexander doesn't know his stats doesn't mean he's all bad. We may be working for him if he gets elected, you know!

Tim     Okay—but heaven forbid!

**Business Case Questions**

42. Discuss the chi-square goodness-of-fit test on the lottery data that are given in this application. State your conclusion for $\alpha = 0.05$, for $\alpha = 0.01$.

43. How should Tim and Mrs. Moore respond if asked, "Is there truth to Mr. Alexander's claim that the lottery is biased and favors some digits over others?" Use your answer to Problem 42 to write a short but factual statement for the press.

## COMPUTER APPLICATION

*Chi-Square Tests on Proportions*

We return to the application of marketing personal computers that target households with 3+ persons and with head of household 35–50 years old. The marketing manager asks, "Aside from income, what variables can we determine that will describe the target population?" The firm's practice is to restrict its mail advertising to variables that satisfy an 80% efficiency criterion. That is, for a variable to be included, its categories that define a selection condition must describe 80% or more of the target population. What measurable characteristics are available from the PUMS data?

The variables selected for this analysis included the metropolitan statistical area, MSA; the home ownership status, or TENURE; the dwelling size, DUSIZE; and the number of CARS. An SPSS setup follows.

```
1   NUMBERED
2   TITLE FREQUENCIES ON COUNT DATA
3   DATA LIST FILE = INFILE/
4       MSA 8-11 PERSONS 12-13 TENURE 14
5       DUSIZE 15-16 CARS 19 AGE1 43-44
6   SET BLANKS = -1
7   SELECT IF ((PERSONS GE 3) AND (AGE1 GE 35) AND (AGE1 LE 50))          ─── 1
8   RECODE MSA (-1 0 = 0) (ELSE = 1)◄─────
9   RECODE TENURE (1 = 1) (ELSE = 0)
10  RECODE DUSIZE (2 3 = 1) (ELSE = 0)
11  RECODE CARS (2 = 1) (ELSE = 0)
12  SET SEED = 3236
13  SAMPLE 175 FROM 3757
14  LIST VARIABLES = MSA TENURE DUSIZE CARS/
15           FORMAT = NUMBERED/ CASES FROM 1 TO 175
16  FREQUENCIES VARIABLES = MSA TENURE DUSIZE CARS/
17           STATISTICS = ALL/                                            ─── 2
18  NPAR TESTS CHISQUARE = MSA TENURE DUSIZE CARS (0, 1)/◄─────
19           EXPECTED = 20, 80
20  FINISH
```

Observe that the SELECT IF statement, as used earlier, reduces the file to the target population of 3757 records. We have excluded the LIST VARIABLES and FREQUENCIES outputs simply to save space. Because most of the procedures have been used in earlier exercises, only two need to be discussed.

**1** At steps 8 through 11 the variables are RECODED to the values 0 and 1. This allows us to test if a specific category includes 80% of the count. In each case, logic dictates the split. For example, TENURE is split into renters (0) and owners (1). As required by SPSS, the higher code, 1, is assigned to owners because that category is expected to have the higher count.

**2** The chi-square test is performed on each variable with the predefined null hypothesis that specifies 80% classification in the category with code 1. The NPAR procedure, at steps 18 and 19, requires an EXPECTED split specification. Here 20% (for the lesser-count category) or 80% (for the higher-count category) describes the relative expectation. The chi-square test compares OBSERVED CASES against the EXPECTED count.

The information in Table 16.6 summarizes the results of the SPSS computer run for the target population of 3+ person households with householder 35 to 50 years old.

**TABLE 16.6**  SPSS Output of Chi-Square Tests for High-Efficiency (80%) Demographic Categories for a Target Population

CHI-SQUARE TEST

MSA

| CATEGORY | CASES OBSERVED | EXPECTED | RESIDUAL |
|---|---|---|---|
| NO — 0 | 36 | 35 | 1 |
| YES — 1 | 139 | 140 | -1 |
| TOTAL | 175 | | |

| CHI-SQUARE | DF | SIGNIFICANCE (*p*-value) |
|---|---|---|
| .036 | 1 | .850 |

CHI-SQUARE TEST

TENURE

| CATEGORY | CASES OBSERVED | EXPECTED | RESIDUAL |
|---|---|---|---|
| RENTER — 0 | 36 | 35 | 1 |
| OWNER  — 1 | 139 | 140 | -1 |
| TOTAL | 175 | | |

| CHI-SQUARE | DF | SIGNIFICANCE (*p*-value) |
|---|---|---|
| .036 | 1 | .850 |

CHI-SQUARE TEST

DUSIZE

| CATEGORY | CASES OBSERVED | EXPECTED | RESIDUAL |
|---|---|---|---|
| NON-SFDU — 0 | 29 | 35 | -6 |
| SFDU — 1 | 146 | 140 | 6 |
| TOTAL | 175 | | |

| CHI-SQUARE | DF | SIGNIFICANCE (*p*-value) |
|---|---|---|
| 1.286 | 1 | .257 |

CHI-SQUARE TEST

CARS

| CATEGORY | CASES OBSERVED | EXPECTED | RESIDUAL |
|---|---|---|---|
| 0 OR 1 — 0 | 66 | 35 | 31 |
| 2+  — 1 | 109 | 140 | -31 |
| TOTAL | 175 | | |

| CHI-SQUARE | DF | SIGNIFICANCE (*p*-value) |
|---|---|---|
| 258.036 | 1 | .000 |

We have experienced the RESIDUAL term before. Here it simply shows the difference for OBSERVED − EXPECTED. The chi-square tests use the $p$-value (SIGNIFI-CANCE) approach. The quick summaries, following, will indicate an appropriate decision relative to the use of each variable.

1. For MSA:

$$H_0: \text{non-MSA residence}:\text{MSA residence} = 0.2:0.8$$

$$H_A: \text{non-MSA residence}:\text{MSA residence} \neq 0.2:0.8$$

From Table 16.6 the calculated $\chi^2 = 0.036$ with $p$-value $= 0.85$.
Summary: Essentially 80% of the target population live in metropolitan statistical areas. The profile will be restricted to residents within MSA's.

2. For TENURE:

$$H_0: \text{renters}:\text{owners} = 20\%:80\%$$

$$H_A: \text{renters}:\text{owners} \neq 20\%:80\%$$

Again, from Table 16.6, the calculated $\chi^2 = 0.036$ with $p$-value $= 0.85$
Summary: We can achieve an 80% efficiency by restricting the mailing to only households that are owner-occupied.

3. For DUSIZE:

$$H_0: \text{non-SFDU}:\text{SFDU} = 20:80$$

$$H_A: \text{non-SFDU}:\text{SFDU} \neq 20:80$$

The calculated $\chi^2 = 1.286$ has $p$-value $= 0.257$.
Summary: The rate of non-SFDU to SFDU (single-family) residences is not statistically different from 20% for non-SFDU's to 80% for SFDU's. Of the 3+ person households with householder 35–50 years old, 80% are single-family dwellings.

4. For CARS:

$$H_0: \text{0 or 1 car}:\text{2+ cars} = 0.2:0.8$$

$$H_A: \text{0 or 1 car}:\text{2+ cars} \neq 0.2:0.8$$

This chi-square test has calculated $\chi^2 = 258.036$ with $p$-value $= 0.000$.
Summary: Other than 80% of the target population records had 2+ cars. The sample showed OBSERVED 109 of 175 for 62% had 2+ cars. This variable does not meet the 80% criterion. It is excluded.

After considering these factors, the manager decided to select households from the target population based on the criteria

Resides in a metropolitan statistical area (MSA)
Owns (includes "is buying") a single-family dwelling residence

These conditions were added to the mailing-selection requirements because (1) they seemed timely to the personal computer marketing study and (2) data were available for coding each record with these variables. A word of caution: Several variables imposed simultaneously can reduce the count of qualifying records, so it is appropriate to follow up with a qualified population inventory or frequency run. Accordingly, more qualifiers can greatly reduce the count availability—and thereby reduce, rather than increase, the total revenues!

**Computer Application Questions**

44. In some ways duplex residents are more like SFDU occupants than those in 3+ units. Given that one random sample of 175 records from the target population had 146 SFDU's, 5 duplexes, and 24 other dwellings, test

$$H_0: \text{single-family:duplex:other} = 80:5:15$$

$$H_A: \text{single-family:duplex:other} \neq 80:5:15$$

Use a *p*-value test approach and a $\chi^2$ test statistic.

45. No mention is made of using a personal computer response variable in the profiling. Assume the preceding variables are available along with survey data on personal computer ownership. Explain ways of using this information to identify the variables to use for the profiling.

**COMPUTER PROBLEMS**

*Problems Using the PUMS Data Base*

46. Select an *n*th systematic sample of 100 records from the PUMS 20,116-record file. Record the last position of the record number (column 50) and list these numbers in a frequency distribution. Use the $\chi^2$ goodness-of-fit test for a uniform distribution on the values $0, 1, 2, \ldots, 9$. Use a 0.05 test level.

SPSS users, try

```
 1  NUMBERED
 2  TITLE        NTH SAMPLE
 3  DATA LIST    FILE = INFILE/
 4               RECNO 1-5  RECUNITS 5
 5  SET BLANKS = -1
 6  COMPUTE #CASENUM = #CASENUM + 1
 7  SELECT IF  (MOD(#CASENUM, 200) = 0)
 8  LIST VARIABLES = RECNO RECUNITS/
 9          FORMAT = NUMBERED/ CASES FROM 1 TO 110
10  FREQUENCIES VARIABLES = RECUNITS  (-1 9)/ BARCHART
11  NPAR TESTS CHISQUARE = RECUNITS (0 9)/
                          EXPECTED = 10*10
12  FINISH
```

*Note:* The Student Minitab does not have a procedure comparable to NPAR TEST CHISQUARE.

47. Repeat the preceding problem, except now select a *random sample* of 100 records from the PUMS 20,116-record file. Again, perform a chi-square goodness-of-fit test to a uniform distribution on the values $0, 1, 2, \ldots, 9$. State test conclusions. Then compare the benefits of this random sampling to those

of the systematic sampling procedure for (1) ease of drawing the sample and (2) randomness of the record unit numbers.

SPSS users can use the routine outlined in the preceding problem. Simply replace lines 6 and 7 of that routine with

```
SAMPLE 100 FROM 20116
```

Otherwise, the routine is the same.

48. Is the number of bathrooms in a home independent of the number of bedrooms? These variables are ordinal level, but we use only the nominal classes. Select a sample of 200 records from the PUMS data base. Use only those records with # bathroom codes 1, 2, 3, 4 or more (where 4, 5, 6 . . . are combined in the "4 or more" code). Make a test of independence for $\alpha = 0.05$.

*Suggestion:* With SPSS and the 20,116-record PUMS file, use

```
TEMPORARY
SELECT IF ((BATHS GT 0) AND (BEDRMS GT 0))
RECODE BEDRMS (4 THRU HIGH = 4)
CROSSTABS VARIABLES = BATHS (1 4) BEDRMS (1 4)/
    TABLES = BATHS BY BEDRMS
STATISTICS = 1
```

Interpret the meaning of the calculated $\chi^2$-value. (*Note:* You can use the same sample procedures to select the variables for Problem 49.) Upon viewing the two-way table, you will find some short cells. Use logic to collapse the table and then perform a $\chi^2$-test.

49. Are the variables "highest year of school" (columns 48–49) and "income from all sources" independent for $\alpha = 0.05$? Note that these variables are ordinal and interval scale, but the test requires only nominal scale. Select a random sample of 200 records from the PUMS data base. Exclude records with school code = 00 *or* income code = 00000. Otherwise, recode as shown in the table and perform a $\chi^2$ test of independence. State your test conclusions. Be prepared to experience short cells. If so, you may have to combine some adjacent classes to assure reasonable expected values. Use information for heads of household (person 1) only.

| Recode Values for School | Recode Values for Income |
|---|---|
| (01 through 10) = 1 | (under $10,000) = 1 |
| (11 through 14) = 2 | ($10,000 to $19,999) = 2 |
| (15 through 17) = 3 | ($20,000 to $39,999) = 3 |
| (18 or more) = 4 | ($40,000 and higher) = 4 |

One sample gives the following table, which needs to be collapsed to give reasonable (> 5) expected cell counts. Collapse the data by combining INCOMES $20,000 and higher. Also, combine the elementary-school and high-school categories into a "High School or Less" category. This gives a 3-rows by 3-columns matrix. Now test $H_0$: Income is independent of highest year of schooling.

| Education | Income | | | |
|---|---|---|---|---|
| | Under $10,000 | $10,000–$19,999 | $20,000–$39,999 | $40,000 and Higher |
| Elementary school | 15 | 8 | 1 | 0 |
| High school | 57 | 28 | 18 | 3 |
| 1–3 years college | 11 | 9 | 7 | 3 |
| 4+ years college | 10 | 8 | 13 | 2 |

For Minitab users, here is a setup to perform the $\chi^2$-test on the collapsed table:

```
MTB > READ C1-C3
DATA> 72 36 22
DATA> 11  9 10
DATA> 10  8 15
DATA> END
MTB > PRINT C1-C3
MTB > CHISQUARE C1-C3
```

50. In the interest of completeness, this problem provides a computer solution for a goodness-of-fit test to a normal distribution. The procedure is the one-sample Kolmogorov–Smirnov test as discussed in Chapter 17. A chi-square procedure can be used but has shortcomings. For example, it requires extensive precalculations to determine the expected values before beginning the computer procedure. See Section 17.7 for other advantages of the K–S procedure. The data are read as individual values. This Problem uses the customer demand numbers from the Chapter 3 Computer Application (repeated here).

| Demand, units | | | | | |
|---|---|---|---|---|---|
| 2 | 18 | 26 | 32 | 37 | 43 |
| 3 | 18 | 26 | 32 | 37 | 43 |
| 7 | 20 | 26 | 32 | 38 | 44 |
| 8 | 21 | 26 | 33 | 38 | 45 |
| 9 | 21 | 27 | 33 | 38 | 45 |
| 11 | 21 | 27 | 33 | 38 | 45 |
| 12 | 21 | 27 | 33 | 38 | 46 |
| 12 | 22 | 28 | 33 | 38 | 47 |
| 12 | 22 | 28 | 34 | 39 | 47 |
| 13 | 23 | 29 | 34 | 39 | 47 |
| 13 | 23 | 29 | 34 | 39 | 48 |
| 14 | 23 | 29 | 35 | 39 | 48 |
| 14 | 23 | 29 | 35 | 39 | 48 |
| 15 | 23 | 29 | 35 | 39 | 49 |
| 15 | 24 | 30 | 35 | 40 | 49 |
| 15 | 24 | 31 | 36 | 41 | 54 |
| 16 | 25 | 31 | 36 | 41 | 56 |
| 17 | 25 | 31 | 36 | 42 | 58 |
| 17 | 25 | 31 | 36 | 42 | 61 |
| 17 | 26 | 31 | 36 | 42 | 67 |

Here is an SPSS routine:

```
NUMBERED
TITLE GOODNESS OF FIT TO A NORMAL DISTRIBUTION
DATA LIST FILE = INFILE/    DEMAND  1-2
SET BLANKS = -1
FREQUENCIES VARIABLES = DEMAND/HIST = NORMAL
NPAR TESTS K-S (NORMAL) = DEMAND
FINISH
```

Use the $p$-value test procedure to decide if we should accept $H_0$: The demand data have a normal distribution. The procedure assumes a two-tailed alternative hypothesis.

51. Is the average U.S. worker working the traditional 40 hours per week? Test the hypothesis that "usual hours worked per week" (columns 53–54) were in the ratio of $H_0$: under 30:30–50:over 50 hours = 1:8:1.
   a. Draw a random sample of 200 records from the PUMS.
   b. Exclude records with hours code = 00.
   c. Otherwise, recode as described in $H_0$ (under 30 = 1, 30 < 50 = 2, over 50 = 3) and prepare a frequency distribution. Here is a table for one sample:

| Hours Worked | Frequency |
|:---:|:---:|
| 1–29 | 17 |
| 30–50 | 137 |
| 51+ | 8 |

   *Note:* This excludes records with hours code = 00, so the count is less than 200.
   d. Use the procedures and calculations as outlined in Section 16.5 to perform the test. Let $\alpha = 0.05$.

## FOR FURTHER READING

[1] Albright, S. Christian. *Statistics for Business and Economics.* New York: Macmillan, 1987.
[2] Becker, William E., and Donald L. Harnett. *Business and Economic Statistics.* Menlo Park, CA: Addison-Wesley, 1987.
[3] Levin, Richard J. *Statistics for Management*, 4th ed. Englewood Cliffs, NJ: Prentice-Hall, 1987.
[4] Plane, Donald R., and Edward B. Oppermann. *Business and Economic Statistics.* Plano, TX: Business Publications, 1986.
[5] Richards, Larry E., and Jerry L. LaCava. *Business Statistics: Why and When*, 2nd ed. New York: McGraw-Hill, 1983.
[6] Schaefer, Robert L., and Richard B. Anderson. *The Student Edition of Minitab.* Reading, MA: Addison-Wesley, 1989.
[7] *SPSS Reference Guide* (version 4.0). Chicago: SPSS, Inc., 1990.

# 17

## NONPARAMETRIC STATISTICS

In the preceding chapters, our inference has related primarily to population parameters. The parametric methods we used relied upon data that are typically described in terms of some measurement, such as average expenditures, percent defective, or the difference in mean sales. In general, we apply parametric methods to data that have one or more unknown parameters. To draw conclusions, we make assumptions about an underlying probability distribution. Specifically, we have made certain assumptions about the underlying probability distribution—such as normality—and have gone on to make inferences concerning the mean, $\mu$; difference in means, $\mu_1 - \mu_2$; and so on.

Nonparametric statistics take a different viewpoint. Business-decision situations often arise in which restrictive assumptions, such as normal distribution and fixed variance, are not met. Moreover, the data may simply convey a frequency or ranking, with no measure of the magnitude of differences in the observations. In these situations *nonparametric*, or *distribution-free*, *methods* are appropriate.

> **Nonparametric methods are statistical procedures that require fewer assumptions about population parameters and are less restrictive (in distribution requirements) than their parametric counterparts.**

Nonparametric tests are said to be distribution-free because they do not assume that the observations come from a population distributed in a specific way, such as normally. We can also use nonparametric techniques on observations that can be ordered in some way but cannot be further quantified. Many social and behavioral science data fall into this category. This has led to numerous applications in the fields of personnel and organization management.

**TABLE 17.1** Frequently Used Nonparametric Procedures

| Level of Data | Single Sample | Multiple Samples (of $K$ groups) | | Tests of Association |
| --- | --- | --- | --- | --- |
| | | *Dependent* | *Independent* | |
| Nominal | Chi-square | McNemar ($K = 2$) Cochran $Q$ ($K > 2$) | Chi-square ($K = 2$) | Contingency coefficient |
| Ordinal | Runs test | Sign test ($K = 2$) | Median test ($K = 2$) | Spearman's rank correlation |
| | Kolomogorov–Smirnov | Signed-rank (Wilcoxon) test ($K = 2$) | Mann–Whitney $U$ test ($K = 2$) | |
| | Sign test ($K = 1$) | Friedman (two-way ANOVA by ranks) ($K > 2$) | Kruskal–Wallis (1-way ANOVA) ($K > 2$) | |

This chapter includes selected nonparametric methods that are useful to business analysts, market researchers, economists, and managers in general. Both the number of these methods and their uses have grown rapidly in recent years. This is not only because of the freedom they enjoy with respect to minimal assumptions but also because nonparametric procedures are easy to apply. In fact, to the novice the problem is more often which procedure applies than how to apply to given procedure. Table 17.1 presents an overview of some of the most useful nonparametric methods, classified according to the level of data used and the type of sample from which the data are taken.

Although these methods are not all discussed in this chapter, descriptions of those not explained here can be found in standard reference texts [2, 3].* We have, in fact, already used one of the nonparametric methods in Chapter 16 (when discussing $\chi^2$), although we did not elaborate on nonparametric characteristics in general at the time. The bibliography at the end of the chapter lists several useful reference books.

As is suggested in Table 17.1, the selection of any analytical procedure is a function of the level of data, which we shall discuss in the next section. Given the level of data, the appropriate procedure depends upon the number ($K$) of groups involved, that is, whether we are dealing with a single sample ($K = 1$) or multiple samples ($K \geqslant 2$). If two or more samples are being tested, the next classification concerns whether the samples are dependent or independent. Dependent samples are often concerned with changes in a subject variable according to some type of "before" and "after" measurements, whereas independent samples are not influenced by each other.

We offer a word of caution about the nonparametric methods described in Table 17.1. In the past, nonparametric methods have not been as widely used as parametric tests, partly because they do not afford as much information. This is because when the distribution type and possibly some of its parameters are known, then $t$-, $Z$-, and other parametric tests are typically more powerful than their nonparametric counterparts. That is, for a given sample size and $\alpha$-risk

---

* Bracketed numerals refer to items listed in "For Further Reading" at the end of the chapter.

level, and with a known distribution, parametric tests commonly have a lower probability for type 2 error than do comparable nonparametric tests. In such situations nonparametric tests are less likely to detect real differences from a hypothesized situation and thus are less discriminating when it comes to rejecting a false hypothesis. However, they can be more powerful than parametric tests for highly skewed data.

The major advantage of nonparametric methods is that they have less restrictive assumptions and so are applicable to a wider variety of data than are their parametric counterparts. In addition, they can serve as a quick, and often easy, check on the reasonableness of parametric results. For these reasons, selected nonparametric procedures are becoming increasingly popular as statistical tools for use by business managers. Applications in the marketing and human resource management areas are especially noteworthy.

> Upon completion of Chapter 17, you should have an understanding of when specific nonparametric methods are most appropriate and how to apply selected methods. In particular, you should be able to
>
> 1. distinguish between nominal, ordinal, interval, and ratio-scale data;
> 2. recommend a nonparametric test method for a given situation;
> 3. given appropriate data, conduct the following tests:
>    a. Runs test
>    b. Mann–Whitney $U$ test
>    c. Kruskal–Wallis test
>    d. Sign test
>    e. Kolmogorov–Smirnov (one sample) test
>    f. Rank correlation test

## 17.2 DATA CLASSIFICATION SCALES AND NONPARAMETRIC TESTS

Measurement is the process of assigning numbers or descriptors to observations. The kind of measurement depends on the rules by which the values are assigned. In this section we review four measurement scales for data: (1) nominal, (2) ordinal, (3) interval, and (4) ratio. We also identify some analytical methods appropriate for working with each kind of data.

### Nominal Data

The *nominal*, or name, classification scale is the weakest level of measurement.

> Nominal data are data classified into categories by virtue of a category name only.

Data from nominal-scaled variables typically consist of frequencies of observations in the various classes. For example, a survey of users of laundry soap may reveal that 40 prefer cold-water soap, 50 prefer an all-temperature brand, and 80 prefer a hot-water soap. Other examples could be drawn from the surveys used to find out what TV channel is drawing the most viewers and from voter classification according to two or more political parties.

The most common information available from nominal data is a frequency count of the number of observations in a given class, which can yield a mode as a measure of central tendency. However, nominal data can also be coded in such a way that numbers serve as substitutes for names (such as $1 =$ good and $0 =$ defective). In these cases, some form of average of the 1's and 0's can be computed and used as a measure of central tendency. For the most part, however, nominal data conveys only classification information. This section identifies several nonparametric methods that can be applied to nominal data.

*Chi-square* is the first nonparametric procedure listed in Table 17.1. Recall that in working with $\chi^2$-tests of independence of classification, we utilized frequency data in categories, or name classes. This $\chi^2$ procedure is rightfully a nonparametric method because it requires no assumptions concerning the parameters of an underlying distribution. Furthermore, it needs only nominal (frequency) data in the analysis. The only requirements for the data are that the observations must be independently selected and the *expected* number of observations in each cell must be greater than or equal to 5.

As shown in Table 17.1, the *McNemar, Cochran, $\chi^2$ (K-group)*, and *contingency coefficient tests* are also useful for understanding nominal data. These procedures make use of the $\chi^2$-distribution. The McNemar test is particularly applicable in pretest–posttest situations where the data are classified in a dichotomy (two categories) and the analyst wishes to measure some characteristic behavior of a sample before and after the effect of another variable. The purpose of this is to determine whether a significant change has occurred. For example, a personnel manager may wish to determine whether a seminar on communication has had a significant effect on a management group by recording the number of managers having "good" and "poor" communications skill ratings before the seminar and again after it is completed. The McNemar test would be appropriate.

The *Cochran Q-test* is an extension of the *McNemar test* to more than two related groups, again on a "before" and "after" basis. Whereas the McNemar test involves calculating a $\chi^2$-statistic, the Cochran test rests upon calculation of a $Q$-statistic. For large samples, however, the $Q$-distribution is approximately a $\chi^2$-distribution. See [2] for more discussion and examples.

The *contingency coefficient* measures the degree of association between two sets of attributes and is an extension of $\chi^2$. Ordering of the data is not required, and at least one measure should be nominal. The degree of association is measured by a contingency coefficient, $C$, which incorporates a $\chi^2$-value. The calculated $C$-value tends to differ significantly from zero if a strong association exists between the two variables. See [6] for a discussion of this measure.

## Ordinal Data

The *ordinal*, or ranking, level is appropriate when the observations can be scaled from high to low or vice versa.

> Ordinal data use numbers to identify the positions, or ranks, of observations and give a relevant ordering to the data.

For example, consider an attitude scale of $1 =$ Strongly agree, $2 =$ Agree, $3 =$ Disagree, $4 =$ Strongly disagree. Ordinal data possess the property of *order*, or relative standing, so that a complete rank-ordering of the data can be made. The central tendency of ordinal data can be described by either the mode or the median. Note, however, that the intervals of differences in position are not meaningful in themselves, since the ordering does not necessarily imply equality of the intervals. Thus, in the scale example, the interval between Agree and

Strongly agree $(2 - 1)$ is not necessarily equal to the interval between Disagree and Agree $(3 - 2)$.

Tests designed for ordinal data are more powerful than tests for nominal data *because they make use of the ordered position of the observations.** The *runs test* is one of the simplest tests useful for ordinal data as well as nominal data. This test permits inference based upon the order of occurrence of outcomes from a single sample. This test requires only that the order or sequence in which the individual observations are obtained is preserved. It enables us to detect either too few or too many sequences of identical occurrences (e.g., colors, sizes) such that an assumption of randomness might be questioned. The Kolmogorov–Smirnov test is a test of differences between observed and expected frequencies (like the $\chi^2$-test), except that it takes advantage of the ordered ranking of the data, whereas the $\chi^2$-test does not.

Ordinal data tests of multiple groups for *dependent samples* include the sign and signed-rank tests (which apply to two samples) and the Friedman analysis by variance by ranks (which applies to more than two samples). The sign test involves assigning positive or negatives signs to the difference in a before–after or yes–no type situation. Inference is based upon a count of the number of signs (either $+$ or $-$). The signed-rank test, also called the Wilcoxon $T$-test, is the counterpart to the Mann–Whitney $U$-test for independent samples and applies when samples are dependent, or "matched." It is a test of no difference in a before–after situation that bases inference on the sums of the positive and negative ranks of the two groups. The Friedman analysis of variance by ranks is an extension to more than two sets of related samples. It is, for ordinal data, analogous to the repeated measures in ANOVA as used on interval data. The null hypothesis is one of no difference in scores under the various groupings (e.g., over time), and the calculated test statistic is a special ranked version of $\chi^2$.

Ordinal data tests of multiple groups of *independent samples* include the median test, the Mann–Whitney (M–W) $U$-test, and the Kruskal–Wallis (K–W) test. The median test can determine whether two samples are from a population with the same median, whereas the Mann–Whitney $U$-test examines the hypothesis that two populations have identical distributions. Mann–Whitney involves use of the sum of the rankings (in each of two groups) for calculation of a $U$-statistic. This is then compared to a tabled critical value of the $U$-distribution to determine if the hypothesis should be rejected. The Kruskal–Wallis test is similar to the Mann—Whitney $U$-test in that it uses the sums of ranks, but Kruskal–Wallis can accommodate three or more samples. Much like a one-way ANOVA, this is a test of the hypothesis of no difference among groups. The test requires calculation of an $H$-statistic, which is compared to a tabled chi-squared value.

Spearman's rank correlation coefficient is a simple means of testing the level of correlation in two sets of rank ordered data. The test is based on a summation of the actual differences in rank as compared to what would have occurred if random samples were used.

## Interval Data

The *interval scale* is a stronger level of measurement than a nominal or ordinal scale because it is characterized by a constant unit of measure.

---

* The Wald–Wolfowitz test is a two-sample runs test for independent samples that is not mentioned in Table 17.1. See [3].

> Interval data are classified according to a scale that uses unit distances to express the differences between observation values.

With an interval scale, the *differences* between any two intervals on the same scale can be compared. For example, the thermometer we use to measure our body temperature has an interval scale. Normal body temperature should be about 98.6°F. On this scale, the difference between 98.6° and 102.6° (a 4° difference) is twice the difference between 98.6° and 100.6° (a 2° difference).

For interval data, the zero point and the basic unit of measurement are, however, arbitrary. For example, we can measure and work with temperatures on either the Fahrenheit or the Celsius scale, even though both have different zero points and different intervals of hotness. With interval data, the mode, median, and mean are all legitimate measures of central tendency. Furthermore, as we have seen earlier in the text, parametric methods utilizing the normal, $t$-, and $F$-distributions can be used for inference.

## Ratio Data

When interval data also possess a natural or absolute zero point, we refer to them as *ratio data*.

> Ratio data are data that satisfy both ordinal- and interval-scale criteria; in addition, the ratio between measurements is anchored to a fixed or natural zero point.

The ratio scale is a higher level than the interval scale in that it permits comparison of the absolute magnitude of numbers. Thus a voltage of 220 V is twice as high as one of 110 V (ratio data), but a temperature of 220°F is not twice as hot as 110°F (interval data). This is because the voltage scale has an absolute zero, but zero on the Fahrenheit scale does not correspond to absolute zero. Other ratio-scale data include dollar sales, distances, hours (minutes, seconds) of labor, cars per household, and median age.

## Nonparametric Tests Illustrated in Chapter 17

We now move on to a more detailed description of selected nonparametric tests. Insofar as the $\chi^2$-test has already been discussed in Chapters 10 and 16 and has wide applicability for nominal data, we concentrate here on nonparametric procedures appropriate for ordinal data. In particular, this chapter includes applications of the runs test (for single samples), the Mann–Whitney $U$-test (for two independent samples), the sign test (for two dependent samples), the Kruskal–Wallis (nonparametric one-way ANOVA) procedure, the Kolmogorov–Smirnov test (of goodness of fit), and Spearman's rank correlation (for tests of association).

## 17.3 THE RUNS TEST: A TEST OF RANDOMNESS FOR SAMPLES

Throughout this textbook we have indicated that random selection of sample items is critical to making valid statistical inference. Numerous analyses require randomness. In regression, a line of best fit is characterized by random deviations above and below the regression line. Acceptable industrial production is characterized by a random fluctuation of percent defectives about a control chart centerline. Practically every test, whether parametric or nonparametric, is based on the assumption of random samples.

A nonparametric procedure that evaluates randomness uses *runs*.

> A run is a succession of identical symbols that are followed and preceded by different symbols or by no symbols at all.

For example, the succession *aaa bb aa bb a* contains five runs. The first run—*aaa*–has length three, the second—*bb*—has length two, and so on.

The runs test is a method for detecting nonrandomness in a sample. This test is based on the order of occurrence for the sample outcomes. The data are frequently time-ordered. The test requires counting the number of runs observed in a single sample. An extreme number of runs (either too few or too many) indicates a lack of randomness. Too few runs or bunching could indicate a time trend or general dependence. Too many runs indicates systematic short-period (cyclic) fluctuations.

When both $n_1$ and $n_2$ are 10 or smaller, and a probability function based on the number of possible combinations is used, then an exact runs test is available [8]. Fortunately the test can be reduced to an evaluation of the number of runs observed, $r$, compared to critical points from the runs distribution in Appendix B, Table 11. For runs of two kinds, Table B.11 requires only values for $n_1$ and $n_2$, the count of each kind, with $n_1 \leqslant 10$ and $n_2 \leqslant 10$. Here is an example.

---

**EXAMPLE 17.1**

A human resources department has just implemented a new employee dental care program for the company. The numbers of production and nonproduction employees are essentially the same. The director of human resources has requested a test that the sequence of applications for dental care payments is random, at a level of significance of 10%.

*Solution*

The null hypothesis is that the following data are a random sample. Since we cannot focus on a particular reason for a lack of randomness, we use a two-tailed test:

$H_0$: This is a random sample.

$H_A$: This is not a random sample (by too many, or by too few, runs).

The sample data are

PP, N, P, NN, PP, NN, PPP, NN, PP, N

The commas separate $r = 10$ runs consisting of $n_1 = 8$ N's representing nonproduction employee requests and $n_2 = 10$ P's for production employee requests for dental payments. Observe that the smaller count is defined as $n_1$. This is a matter of convenience, which uses the symmetry of Table B.11 to restrict the listings to values below the diagonal.

The test statistics are defined in Table B.11 for the line $(n_1, n_2) = (8, 10)$. We want a test with probability for rejecting the null hypothesis in the tails that (1) is symmetrical and (2) totals as near as possible 10%. Then, $\alpha/2 = 0.10/2 = 0.05$ gives $r_{01} = 6$ and $r_{02} = 13$ for actual $\alpha = P(r \leqslant 6 \text{ or } r \geqslant 13) = 0.048 + (1 - 0.964) = 0.084$. The $\alpha = 0.084$ is as close to the proposed 10% as this two-tailed test can be defined, with $n_1 = 8$ and $n_2 = 10$.

Conclusion: Since $r = 10$ runs were observed, and $(r_{01} =) 6 < 10 < 13 (= r_{02})$ is within the acceptance region, we cannot reject the null hypothesis of randomness for $\alpha = 0.084$. The number of runs is neither too small nor too large to cause us to question the randomness of dental care payments for this company's production and nonproduction employees.

This illustrates the procedure for small samples that uses the runs distribution and Table B.11. For larger samples, we revert to a normal approximation test and use Table B.1. This requires knowing the runs distribution mean and standard deviation.

To determine the meaning of an extreme number of runs—that is, too few or too many—we first compute the mean and standard deviation of the sampling distribution for the statistic that describes the total number of runs. We will consider only runs of two symbols, such as $S$, $F$ (success–failure); $+$, $-$ (plus–minus); or $M$, $F$ (male–female). For elements of two kinds, where $n_1 + n_2$ sample observations are involved, let $n_1$ denote the number of elements of the first kind and $n_2$ the number of the second kind. Then the sampling distribution of $r$, the number of runs, has a mean and a standard deviation as follows:

### PARAMETERS FOR $r$, THE RUNS STATISTIC

$$\text{Mean} \qquad \mu_r = \frac{2n_1 n_2}{n_1 + n_2} + 1 \qquad\qquad (17.1)$$

$$\text{Standard deviation } \sigma_r = \sqrt{\frac{2n_1 n_2 (2n_1 n_2 - n_1 - n_2)}{(n_1 + n_2)^2 (n_1 + n_2 - 1)}} \qquad (17.2)$$

The normal distribution gives a reasonable approximation to the sampling distribution of $r$, when $n_1 > 10$ and $n_2 > 10$. We obtain the approximation by "standardizing $r$" in the usual manner.

### NORMAL APPROXIMATION TO THE DISTRIBUTION OF $r$, THE RUNS STATISTIC

$$Z \doteq \frac{r - \mu_r}{\sigma_r} \qquad\qquad (17.3)$$

The null hypothesis for the runs test is $H_0$: This is a random sample. If the researcher suspects a trend dependence in the observations (for example, a tendency to increase with time), the alternative hypothesis should indicate a one-sided test. If there exists a possibility for either an excess or a few runs, then a two-sided test is appropriate.

A second example uses a normal approximation to compare the numbers of observations above and below the median. If the ordered scores are a random sample, we should expect a moderate mix of magnitudes in the order of the observations. We ignore any observations that fall at the median value.

---

**EXAMPLE 17.2**

A new field-worker for a state welfare agency was asked to take a random sample of 46 heads of households from a poverty area and determine the numbers of days of their unemployment during the past six months. The responses were 36, 34, 18, 25, 0, 12, 28, 84, 33, 68, 10, 20, 39, 60, 42, 57, 34, 0, 4, 20, 7, 12, 26, 75, 33, 60, 4, 22, 40, 60, 41, 56, 38, 5, 0, 10, 0, 14, 0, 60, 35, 64, 10, 18, 38, 40. Is this sample, in the order received, a random sample? Use $\alpha = 0.10$.

*Solution*   Our standard, the median, is $(28 + 33)/2 = 30.5$ days. (Recall that the scores must be ordered to get this value). The total days unemployed is either less than the median value, L, or greater than the median value, G. The sample, as received, shows GG, LLLLL, GGG, LL, GGGGG, LLLLLL, GGG, LL, GGGGG, LLLLLL, GGG, LL, GG. There are $r = 13$ runs, and $n_1 = n_2 = 23$. The test is as follows.

1. $H_0$: The sample of households is random.
   $H_A$: The sample is not random.
2. $\alpha = 0.10$
3. Use a $Z$-test. Reject $H_0$ if calculated $|Z| > 1.645$.

4. $\mu_r = \dfrac{2n_1 n_2}{n_1 + n_2} + 1 = \dfrac{2(23)(23)}{23 + 23} + 1 = 24$

   $\sigma_r = \sqrt{\dfrac{2n_1 n_2 (2n_1 n_2 - n_1 - n_2)}{(n_1 + n_2)^2 (n_1 + n_2 - 1)}}$

   $\quad = \sqrt{\dfrac{2(23)(23) \cdot [2(23)(23) - 23 - 23]}{(23 + 23)^2 (23 + 23 - 1)}} = 3.35$

5. $Z \doteq \dfrac{r - \mu_r}{\sigma_r} = \dfrac{13 - 24}{3.35} = -3.28$

6. Conclusion: Reject $H_0$ since $|-3.28| > 1.645$. The chance is less than 10% that a random sample of 46 households would result in 13 or fewer runs. The sample is not random, so the supervisor should question the manner in which it was collected.

---

The exercises that follow give you an opportunity to use the runs test in both large and small sample situations.

**Problems 17.3**

1. a. Test for randomness in the order of males, M, and females, F, standing in a cafeteria line, given M, F, MM, FFF, M, F, M, F, MM, FF, M, FFF, M, F, M, FF, M, F, M, FFFF, M, F, MM, FF, M, F, M, FF, M, F. Use $\alpha = 0.10$ and a two-tailed test.
   b. Repeat the runs test of part (a); only use the first 15 labels. This test uses Table B.11. State a test conclusion. Then explain which risk—$\alpha$ or $\beta$—is a concern and why it is a concern.

2. Suppose you work in the market research department of a manufacturer of women's apparel and have been asked to do some forecasting for a rainwear item. You have obtained U.S. Weather Bureau records of rainfall in your market area for the last 42 years. Annual precipitation is given chronologically by year. You denote each of the 21 lowest figures with the symbol L and the higher 21 with H. You then count a total of 10 runs of L's and H's. Can you conclude, at the 5% level of significance, that the precipitation pattern is random for years? If not, what conclusion can you draw?

3. The numbers of defective pieces produced in an electronics plant were recorded for successive hours over a 40-hour workweek. Plot this data on a time scale with hours on the horizontal axis and number of defectives on the vertical. In your opinion, is there a pattern? Now apply the test for runs about the median. Use (a) $\alpha = 0.05$ and (b) $\alpha = 0.10$ and a one-tailed test.

| Hour | 1 | 2 | 3 | 4 | 5 | 6 | 7 | 8 | 9 | 10 | 11 | 12 | 13 | 14 | 15 |
|---|---|---|---|---|---|---|---|---|---|---|---|---|---|---|---|
| Number defective | 2 | 3 | 1 | 4 | 2 | 3 | 3 | 5 | 3 | 1 | 3 | 2 | 4 | 4 | 4 |

| Hour | 16 | 17 | 18 | 19 | 20 | 21 | 22 | 23 | 24 | 25 | 26 | 27 | 28 | 29 | 30 |
|---|---|---|---|---|---|---|---|---|---|---|---|---|---|---|---|
| Number defective | 6 | 1 | 3 | 2 | 3 | 2 | 3 | 4 | 5 | 2 | 4 | 3 | 4 | 5 | 6 |

| Hour | 31 | 32 | 33 | 34 | 35 | 36 | 37 | 38 | 39 | 40 |
|---|---|---|---|---|---|---|---|---|---|---|
| Number defective | 5 | 7 | 3 | 4 | 2 | 5 | 6 | 4 | 5 | 7 |

4. Using a random-numbers table (Table 7 in Appendix B) as a source, draw a random selection of 50 single digits. Assume the population median is 4.5 (for digits 0 to 9 inclusive), and test at $\alpha = 0.01$ whether your sample is random. Then ask a friend to name 50 random digits from 0 to 9 inclusive without telling him or her why you want the numbers. Your friend will give several repeats. Did your friend give random numbers? Apply the runs test again at $\alpha = 0.01$.

5. a. A selection of 66 time-ordered observations produced 30 positive (+) observations, 36 negatives (−), and 12 runs. Test at $\alpha = 0.05$ for a trend in the data.
   b. If another, smaller, sample of 11 time-ordered observations gave 5 positives (+) and 6 negatives (−) with 4 runs, then repeat the test in part (a) for a trend in the data. Give conclusions. Discuss the effect of the small sample on (i) setting the $\alpha$-level and on (ii) the size of the $\beta$-risk.

## 17.4 THE MANN–WHITNEY *U*-TEST FOR EQUAL DISTRIBUTIONS FOR TWO INDEPENDENT SAMPLES

As indicated in Table 17.1, the Mann–Whitney *U*-test is an ordinal-level test of the equality of distributions for two independent samples. Whereas the $\chi^2$-test for independence (Chapter 16) is appropriate for (weaker) nominal data and the *t*- and *Z*-tests are appropriate for (stronger) interval data, the Mann–Whitney *U*-test is a relatively powerful test for situations in which both variables can at least be measured on an ordinal scale. It uses ranked data and *is also called the Wilcoxon rank-sum test.*

The null hypothesis for the *U*-test is typically that two populations have the same mean. The test assumes that the samples taken from the two populations are independent of each other and that the distributions themselves are symmetrical (and have the same variance). If the symmetric condition does not hold, the median replaces the mean as the location parameter for a null hypothesis of equal medians. *This test applies to independent samples drawn from continuous variable distributions.* There is no matching of pairs, or "before–after" comparisons.

The procedure for applying the Mann–Whitney *U*-test to situations in which $n_1$ and $n_2$ are both greater than 10 is as follows:

1. Formulate $H_0: \mu_1 = \mu_2$ (or median$_1$ = median$_2$ if the distributions appear skewed) and establish the test level of significance, $\alpha$.
2. Select samples of $n_1$ (from group 1) and $n_2$ (from group 2).
3. Using both samples (together), rank all the observations from highest to lowest. Let the highest observation have rank 1. Find the sum for the ranks associated with each sample.

4. Compute the Mann–Whitney $U$-statistic.

### THE MANN–WHITNEY $U$-STATISTIC

$$U_1 = \left( n_1 n_2 + \frac{n_1(n_1 + 1)}{2} \right) - \sum R_1 \qquad (17.4)$$

where $n_1$ and $n_2$ are the two sample sizes and $R_1$ is the sum of the ranks assigned to the first sample.

Tie scores are given an average rank.

5. Compute the mean and standard deviation of the sampling distribution of the $U$-statistic.

### PARAMETERS FOR THE MANN–WHITNEY $U$-STATISTIC DISTRIBUTION

Mean $\qquad\qquad \mu_U = \dfrac{n_1 n_2}{2} \qquad\qquad\qquad (17.5)$

Standard deviation $\qquad \sigma_U = \sqrt{\dfrac{n_1 n_2 (n_1 + n_2 + 1)}{12}} \qquad (17.6)$

6. When $n_1$ and $n_2$ are both at least 10, the sampling distribution of the $U$-statistic is approximately normal, so the test statistic is then computed as follows.

Normal distribution approximation to the Mann–Whitney $U$-statistic distribution:

$$Z \doteq \frac{U_1 - \mu_U}{\sigma_U} \qquad (17.7)$$

7. Reject $H_0$: $\mu_1 = \mu_2$ at the $\alpha$-level of significance if the calculated $Z$-value from Equation (17.7) is more negative than $-Z_{\alpha/2}$ or is greater than $Z_{\alpha/2}$; otherwise do not reject $H_0$. When either $n_1$ or $n_2$ is less than 10, the test statistic for a Mann–Whitney $U$-test is found in special tables* [2].

Here is an example that uses the normal approximations.

---

**EXAMPLE 17.3**

Two potential suppliers of streetlighting equipment—Supplier A and Supplier B—have included data on life tests of their streetlights (in months) in bids supplied to a city manager. The manager wishes to test the equality of the medians of the two populations using rank sums. Are the median lifetimes of the streetlights equal? (*Note:* Because the condition of symmetric distributions is not specified, so is not known, the test is for equal medians.)

---

\* Some statisticians recommend use of the special tables whenever both samples are less than 20 observations [2]. The modification for small sample tests is discussed later.

| Supplier | Lifetimes of Streetlights (months) |
|---|---|
| A | 35, 66, 58, 46, 42, 40, 49, 59, 32, 58, 68, 41, 29, 75, 30, 53, 60, 63, 58, 56 |
| B | 48, 47, 37, 47, 50, 34, 71, 36, 56, 45, 43, 28, 47, 40, 33, 51, 42, 53, 34, 29 |

*Solution*

1. Let $H_0$: median$_A$ = median$_B$ and $H_A$: median$_A$ $\neq$ median$_B$. Since the $\alpha$-level is unspecified, we let $\alpha = 0.05$. Therefore, reject $H_0$ if calculated $Z < -1.96$ or $> 1.96$.

2. The sample data are given; $n_1 = 20$ and $n_2 = 20$.

3. The scores (in order) and resulting combined ranks are listed in the following table. Note that when ties are involved, we assign a single value to the tied items. This was done for the raw score 29, which occupies ranks 38 and 39 so is assigned a value of 38.5.

| Supplier A | | Supplier B | |
|---|---|---|---|
| *Raw Score* | *Rank* | *Raw Score* | *Rank* |
| 75 | 1 | 71 | 2 |
| 68 | 3 | 56 | 11.5 |
| 66 | 4 | 53 | 13.5 |
| 63 | 5 | 51 | 15 |
| 60 | 6 | 50 | 16 |
| 59 | 7 | 48 | 18 |
| 58 | 9 | 47 | 20 |
| 58 | 9 | 47 | 20 |
| 58 | 9 | 47 | 20 |
| 56 | 11.5 | 45 | 23 |
| 53 | 13.5 | 43 | 24 |
| 49 | 17 | 42 | 25.5 |
| 46 | 22 | 40 | 28.5 |
| 42 | 25.5 | 37 | 30 |
| 41 | 27 | 36 | 31 |
| 40 | 28.5 | 34 | 33.5 |
| 35 | 32 | 34 | 33.5 |
| 32 | 36 | 33 | 35 |
| 30 | 37 | 29 | 38.5 |
| 29 | 38.5 | 28 | 40 |
| $n_1 = 20$ | $\sum R_1 = 341.5$ | $n_2 = 20$ | $\sum R_2 = 478.5$ |

4. $U_1 = n_1 n_2 + \dfrac{n_1(n_1 + 1)}{2} - \sum R_1$

$$= (20)(20) + \frac{20(20 + 1)}{2} - 341.5 = 268.5$$

5. $\mu_U = \dfrac{n_1 n_2}{2} = \dfrac{(20)(20)}{2} = 200$

$$\sigma_U = \sqrt{\frac{(20)(20)(20 + 20 + 1)}{12}} = 36.97$$

6. $Z \doteq \dfrac{U_1 - \mu_U}{\sigma_U} = \dfrac{268.5 - 200}{36.97} = 1.85$

7. Do not reject the hypothesis of equal medians at $\alpha = 0.05$. The difference could be due to chance.

---

Several factors affect the test in Example 17.3. Although the hypothesis of equal medians is not rejected at $\alpha = 0.05$, it would be rejected at $\alpha = 0.10$. That is, $0.05 < p\text{-value} < 0.10$. Thus, if we were willing to be wrong 10% of the time, the difference in lifetimes of the streetlights from the two suppliers could be deemed significant. A correction factor for ties in ranks would tend to make it easier to reject the $H_0$ at the $\alpha = 0.05$ level, but the correction for ties is typically not made.

The Mann–Whitney $U$-test is not as powerful as the standard $t$-test for equality of means. It is, nevertheless, a relatively powerful alternative for ordinal data for which the $t$-test does not apply. In addition, the data in the parent population need not be normally distributed (although for testing the equality of means the distribution should by symmetric).

When either $n_1$ or $n_2$ is less than 10 (or both are less than 20), the test procedure is slightly different. In addition to $U_1$ (from Equation 17.4), a value is computed for $U_2 = n_1 n_2 - U_1$. The $U$-statistic is then defined to be the *smaller* of the $U_1$- or $U_2$-values and it is compared with tabled values of the $U$-distribution. Since the $U$-distribution is a function of both sample sizes, several subtables are required for the various levels of significance (and because of the space required we have not included those tables). The hypothesis is rejected when the calculated $U$-value is *smaller* than the tabled $U$-value. This is because the procedure for calculating $U$ involves *subtraction* of the sum of the ranks from a number related directly to the number of observations. If you are interested, see the chapter references for more information on the small sample procedures.

**Problems 17.4**

6. The following data represent CPA exam test scores by students from two state universities (40 students total). Use the Mann–Whitney $U$-test for the null hypothesis that there is no difference between the median scores of students from the two schools (i.e., two-tailed test). Let $\alpha = 0.05$.

$$n_1 = 18 \qquad \sum R_1 = 315 \qquad n_2 = 22 \qquad \sum R_2 = 505$$

7. An independent laboratory tested two brands of gasoline (A and B) in several different automobile engines to obtain mileage data. Twelve cars used brand A and 16 used brand B. After the test, the mileage figures were merged and arrayed for the two brands (see table). Test, at the 0.01 level of significance, whether there is a difference in the mean mileage from the two brands (use a two-tailed test and assume symmetric distributions).

| Brand A | 1, | 3, | 4, | 5, | 8, | 10, | 13, | 14, | 17, | 19, | 22, | 27 |
|---------|----|----|----|----|----|-----|-----|-----|-----|-----|-----|----|
| Brand B | 2, | 6, | 7, | 9, | 11, | 12, | 15, | 16, | 18, | 20, | 21, | 23, |
|         | 24, | 25, | 26, | 28 | | | | | | | | |

8. The personnel manager for a large corporation has two different development programs (S = systems-emphasis and R = results-oriented) available to train company management. She schedules 20 managers through program S and 10 through program R and ranks their performance one year later. The ranks are as shown in the table. Test, at the 0.05 level, whether there is a significant difference in the performance of managers taking program S versus program R (two-tailed test).

| Program S | 1, 2, 4, 6, 8, 9, 10, 11, 13, 14, 15, 17, 18, 19, 20, 21, 22, 25, 27, 30 |
|---|---|
| Program R | 3, 5, 7, 12, 16, 23, 24, 26, 28, 29 |

9. Independent samples from two distributions follow. These represent the percentage of citrus concentrate in two processors used for mixing whole orange juice. Use the Mann–Whitney *U*-test to determine if the two can be considered the same. Use $\alpha = 0.10$, and the normal approximation for large samples.

| Sample 1 | Sample 2 |
|---|---|
| 10.8 | 12.0 |
| 12.6 | 12.3 |
| 13.3 | 12.7 |
| 12.1 | 13.1 |
| 10.0 | 11.6 |
| 12.2 | 11.3 |
| 9.7 | 13.7 |
| 11.7 | 10.3 |
| 10.3 | 14.5 |
|  | 13.1 |
|  | 11.0 |

10. The purchasing agent for a trucking firm must place a large order for truck tires. He obtained the mileage-until-wearout data from random samples of 15 tires (in thousands of miles) from each of two competing suppliers. Use a 0.10 *p*-value to determine whether there is a significant difference in the median mileage offered by the two brands. Use a two-tailed Mann–Whitney *U*-test.

| Firetone | 26, 28, 24, 30, 18, 22, 17, 24, 28, 32, 21, 34, 22, 25, 27 |
|---|---|
| Goodtone | 27, 23, 29, 21, 15, 18, 23, 17, 21, 14, 19, 15, 31, 16, 26 |

## 17.5
## THE KRUSKAL–WALLIS TEST FOR EQUAL DISTRIBUTIONS FOR TWO OR MORE INDEPENDENT SAMPLES

Chapter 11 provided an introduction to designed experiments and the methods of analysis of variance. There the response variable was assumed to be interval or ratio scale with normal distributions and equal variances. Here we discuss a test for the equality of a location value for three or more groups, where the scale of measurement is ordinal level. For example, do tax accountants, university accounting faculty members, and IRS tax auditors have the same level of acceptance for changes to income tax laws? The Kruskal–Wallis procedure applies as long as the data can be ordered.

The Kruskal–Wallis $H$-statistic is designed to test the hypothesis of identical distributions, versus a shift in centers. It extends the two-sample Mann–Whitney test of the preceding section. Kruskal–Wallis requires minimal assumptions; the necessary conditions are that the samples for each group are independent one from another and are randomly drawn from their respective populations. Also, the observations are required to be ordinal or higher scale.

The Kruskal–Wallis test is analogous to an $F$-test based on a one-way ANOVA. However, this test statistic uses ranks instead of the original observations. Assignment of ranks is made for all ($N$) observations competing across all ($K$) groups. The test statistic is defined by the ranks and group sample sizes.

The Kruskal–Wallis $H$-statistic for a nonparametric location test is

$$H = \frac{12}{N(N + 1)} \sum_{i=1}^{K} \left( \frac{R_i^2}{n_i} \right) - 3(N + 1) \qquad (17.8)$$

where   $N = n_1 + n_2 + \cdots + n_K =$ total sample, summed over all $K$ groups

$n_i =$ the sample size for group $i$ ($i = 1, 2, \ldots, K$)

$R_i =$ the sum of ranks for the observations in group $i$

By Equation (17.8), for a fixed $N$, the value of the $H$-statistic increases if the ranks are different across and uniform within groups.

The null hypothesis states that there is no difference in location measure across the groups. A large value for $H$ indicates that the distributions differ in location. Conversely, mixed ranks, with some high, some low, and some middle values in each group, result in a low value for the $H$-statistic. This case is analogous to low variations across groups relative to the variation within groups. You may recall that this would be descriptive of equal means in the parametric $F$-test of Chapter 11. A low $H$-value supports the null hypothesis of equal location measures across the $K$ groups. Tied observations are assigned an average of ranks. See [6] for an adjustment when there is an excessive number of ties.

The exact sampling distribution for the $H$-statistic is described in [6]. However, the sampling distribution tends toward the $\chi^2$ with df $= K - 1$. Chi-square allows equal or unequal numbers in the groups and is used for tests with either small or large samples.

## EXAMPLE 17.4

Attitudes toward a U.S. Treasury Department proposal for a change in the existing tax rate brackets have been assessed for three groups: tax accountants, university accounting faculty, and IRS tax auditors (see the table below). A score of 100 means perfect agreement, whereas 0 means complete disfavor. Is there a real difference among the groups with respect to attitude toward the proposed change? Let $\alpha = 0.05$.

| Tax Accountants | | Accounting Professors | | IRS Tax Auditors | |
|---|---|---|---|---|---|
| *Score* | *Rank* | *Score* | *Rank* | *Score* | *Rank* |
| 18 | 36 | 93 | 2.5 | 39 | 27 |
| 33 | 29.5 | 21 | 35 | 96 | 1 |
| 36 | 28 | 27 | 33 | 93 | 2.5 |
| 60 | 20 | 33 | 29.5 | 90 | 4 |
| 72 | 12 | 48 | 23 | 84 | 6 |
| 63 | 17 | 62 | 18 | 87 | 5 |
| 54 | 21 | 51 | 22 | 75 | 11 |
| 45 | 24 | 32 | 31 | 78 | 9.5 |
| 42 | 25 | 66 | 16 | 78 | 9.5 |
| 30 | 32 | 69 | 14 | 81 | 7 |
| 24 | 34 | 71 | 13 | 79 | 8 |
| 41 | 26 | 68 | 15 | 61 | 19 |
| | $R_1 = 304.5$ | | $R_2 = 252.0$ | | $R_3 = 109.5$ |

$$\text{Check:} \quad \sum R_i = \frac{N(N+1)}{2} = \frac{36 \cdot 37}{2} = 666$$

*Solution*   Note that the observations are used for ordering, whereas the test statistic is based on ranks. This test uses ordinal-scale measures. The sum of ranks, denoted by $R_i$, with $N = 12 + 12 + 12 = 36$, gives us an $H$-statistic value of

$$H = \frac{12}{36 \cdot 37} \left( \frac{304.5^2 + 252^2 + 109.5^2}{12} \right) - (3 \cdot 37) = 15.29$$

Since tabular $\chi^2_{0.05,2} = 5.99$, we reject the hypothesis of agreement on this issue. The distribution average (rank) opinion is *not* the same for these groups. By the rank-sums ($R_i$), the IRS tax auditors favor the proposal to a higher degree (have lower ranks) than either of the other groups. More testing is required to determine whether the tax-accountant and accounting-professor groups differ in their average opinion of the proposed change.

The Kruskal–Wallis test affords a conclusion that $K$ groups come from distributions with the same location value. The distributions are assumed to be similar in shape and dispersion. Although the Kruskal–Wallis test identifies significant differences among groups—like its parametric test analog, the $F$-distribution—it does not identify the individual groups that are significantly different. For this we can use the Mann–Whitney test (or see [3]).

The Kruskal–Wallis test offers a nonparametric alternative to the normal distribution theory one-way ANOVA and $F$-test. Kruskal–Wallis has simple computations, but it also has some limitations. First, the calculation of ranks can be clumsy with even a moderate number, $N$, of observations. Also, if numerous ties exist, then the $H$-statistic calculation requires an adjustment. Finally, for interval or ratio data the Kruskal–Wallis procedure requires fewer assumptions and so has less power relative to the $F$-test.

The Kruskal–Wallis test provides a way to check the more exacting, but lengthy, parametric $F$-test for equality of means if used with interval or ratio data. It provides a valid approach for testing the equality of distribution medians when the scale of the criterion is of an ordinal measure.

**Problems 17.5**

11. The tabulated data represent ages for junior accountants in three industries. Assuming random samples and an interval scale, test for equal means. Let $\alpha = 0.05$ and apply the Kruskal–Wallis test.

|  | A | B | C |
|---|---|---|---|
|  | 30 | 35 | 28 |
|  | 25 | 42 | 28 |
|  | 31 | 38 | 30 |
|  | 38 | 39 | 24 |
|  | 36 | 31 | 25 |
| Simple means | 32 | 37 | 27 |

12. The data in Problem 11 were used earlier in a one-way ANOVA and $F$-test for equal means. See Chapter 11. What assumptions beyond those given in Problem 11 are required for the parametric $F$-test?

13. The data in the table are known to come from uniform distributions of bread products shipped by a bakery wholesaler (in thousands of pounds).

| Loader 1 | 1.60 | 1.90 | 1.86 | 1.50 | 1.70 |
| Loader 2 | 1.62 | 1.48 | 1.63 | 1.51 | 1.57 |
| Loader 3 | 1.82 | 1.81 | 1.73 | 1.76 | 1.77 |
| Loader 4 | 1.98 | 1.82 | 1.65 | 1.43 | 1.74 |

  a. Are the load averages equal for these four loaders? Let $\alpha = 0.05$.
  b. Explain why the $F$-ratio test for means is not appropriate here.

14. The data represent sales (in millions of dollars) for two metro and three suburban locations of a national department store in a single market. Test the hypothesis of equal medians via the Kruskal–Wallis procedure with $\alpha = 0.10$.

| Met 1 | Met 2 | Sub 1 | Sub 2 | Sub 3 |
|---|---|---|---|---|
| 4.0 | 3.9 | 7.2 | 7.3 | 3.6 |
| 3.2 | 5.2 | 6.9 | 6.5 | 4.2 |
| 3.9 | 4.2 | 8.3 | 6.0 | 3.7 |
| 4.7 | 5.4 | 7.5 | 5.8 | 4.0 |
|  | 4.5 | 6.8 |  |  |

a. Does this data show a real difference in sales for these four locations for this market?

b. Does the present test afford a conclusion specific to which locations have highest and which have lower median sales? Explain.

 15. The years of service at retirement are recorded for random samples of executives from five industries. Test the claim that retiring executives from the population of these samples have the same median work tenure at retirement. Let $\alpha = 0.025$. Can we conclude that banking, stock and bond brokers, and S&L executives retire with fewer years of service than do insurance and real estate executives? Explain how to determine the individual differences for these industry types.

| Years of Service at Retirement (by Industry) | | | | |
|---|---|---|---|---|
| Insurance | Banking | Stocks & Bonds | S&L Institutions | Real Estate |
| 32 | 23 | 19 | 21 | 37 |
| 37 | 24 | 16 | 28 | 30 |
| 29 | 22 | 18 | 26 | 26 |
|  | 10 | 13 | 25 |  |
|  | 16 |  | 20 |  |

## 17.6
## THE SIGN TEST: A NONPARAMETRIC PAIRED-DIFFERENCES (BLOCKING) TEST

The *sign test* is a widely used nonparametric test procedure.

> The sign test is a nonparametric method of testing the null hypothesis that the values of a random variable have an equal chance ($p = 0.5$) of being greater or less than some specified amount.

In the simplest form of the one-sample sign test, values of the test variable are assigned plus ($+$) or minus ($-$) signs according to whether they have a higher or lower value than the hypothesized central value. The hypothesis can concern the mean if the distribution is symmetrical; otherwise, inference should be restricted to the median.

To illustrate, suppose the hypothesized value is $H_0$: median $= 12$ and the observed values are 14, 16, 13, 11, 12, 14, 13, and 10. The signs would be $+$, $+$, $+$, $-$, $+$, $+$, $-$, respectively. All values greater than 12 are assigned a ($+$); those less than 12 are assigned a ($-$), and those equal to 12 are disregarded.

In the sign test, the null hypothesis assumes that the plus and minus signs are distributed randomly above and below the hypothesized value (12) according to the binomial distribution with $p = 0.5$. That is, the *sampling distribution for the sign statistic* is binomial.

> The sampling distribution for the sign statistic is binomial with $p = 0.5$.

$$P(X|n, 0.5) = {}_nC_X \cdot (0.5)^n, \quad X = 0, 1, 2, \ldots, n \qquad \text{(17.9)}$$

where   $X =$ the number of $+$ signs

Chapter 17   Nonparametric Statistics

If the actual proportion of plus signs obtained has a probability of occurrence of less than the specified $\alpha$-level, then the hypothesis must be rejected. In the preceding case, the binomial table (Table 5 in Appendix B) reveals that the probability of 5 or more successes ($+$ signs) in 7 trials is $0.164 + 0.055 + 0.008 = 0.227$, so that it would not be small enough to reject the hypothesis at a 0.10 level. That is to say, the chance of obtaining 5, 6, or 7 pluses in a random sample of $n = 7$, when the population proportion is 0.5, is as large as 0.227. These possibilities are not an unlikely occurrence, so we cannot reject the hypothesis that the median is 0.5. This is a one-tailed test, since all the rejection region is summed from one tail of the distribution.

Our major use of the sign test, however, is for two-sample situations in which we have matched pairs, twins, or before–after values. In these situations the sign records the direction of difference between each pair under study. The hypothesis is still that the plus and minus signs are randomly distributed with binomial probabilities of $p = 0.5$. The hypothesis is supported if about half the differences are positive and rejected if too few differences of either sign exist.

For a one-tailed test of $H_0$: $\pi = 0.5$, the binomial probability is summed, as illustrated, and the null hypothesis is rejected if the probability of occurrence is less than the $\alpha$-value. For a two-tailed test involving a predominance of either $+$ or $-$ signs, the distributions are unequal if the observed count falls below the $\alpha/2$ percentage level. The following example illustrates a two-tailed test.

## EXAMPLE 17.5

A study was made to determine the effect of TV advertising on the attitudes of young people. The readings in the table are scores from 0 through 20 on an attitudinal scale recorded before and after individuals viewed the advertisement. Higher scores reflect a more favorable attitude. Test, at the $\alpha = 0.05$ level, whether this is a significant change (that is, *any* change, up or down, so use a two-tailed test) in attitude.

| | Scores for a TV Advertising Study | | |
|---|---|---|---|
| *Individual* | *Before* | *After* | *Sign of Difference* |
| A | 14 | 14 | |
| B | 16 | 18 | + |
| C | 15 | 16 | + |
| D | 18 | 17 | − |
| E | 15 | 16 | + |
| F | 17 | 19 | + |
| G | 19 | 20 | + |
| H | 17 | 18 | + |
| I | 17 | 19 | + |
| J | 16 | 15 | − |
| K | 19 | 18 | − |
| L | 15 | 16 | + |

*Solution*   The differences are as shown for a two-tailed test of $H_0$: Attitude is unchanged. The first individual showed tie scores and so is eliminated, leaving $n = 11$ observations. The probability of getting only three or fewer negative differences is $P(X \leqslant 3 | n = 11, p = 0.50)$ $= 0.0005 + 0.0054 + 0.0269 + 0.0806 = 0.1134$, which exceeds $\alpha/2 = 0.025$. We cannot reject the null hypothesis. By this criterion, attitude appears to be not affected (neither decreased nor increased) by this advertising. Note that for the two-tailed test, we simply used the $\alpha/2$-value (0.025) as the test criterion.

The $\alpha$-risk is calculated from the binomial table (Table B.5) where, for $n = 11$ and $p = 0.5$, the critical region of $X < 3$ or $X > 8$ gives $\alpha = [2 \cdot (0.0005 + 0.0054 + 0.0269)] = 0.0656$. This is the closest we can get to $\alpha = 0.05$ by summing the values in the tails of the distribution.

---

There is some chance that this conclusion is incorrect, that is, that we have accepted the null hypothesis when it is false. To evaluate this $\beta$-risk, observe that in our sample of $n = 11$, the acceptance region includes anywhere from 3 to 8 plus signs. If the true proportion were really $\pi = 0.6$, the probability of obtaining from 3 to 8 plus signs (a probability value within the acceptance region) would be $P(X = 3) + P(X = 4) + \cdots + P(X = 8) = 0.0234 + 0.0701 + 0.1471 + 0.2207 + 0.2365 + 0.1774 = 0.8752$. This is the $\beta$-risk of accepting a false hypothesis given $H_0: \pi = 0.5$, with $n = 11$. If there is a detrimental consequence to accepting a false hypothesis, then more data should be collected and further testing done with a bigger sample. At 87.5%, the chance is high for accepting a false hypothesis.

Example 17.5 illustrates the use of a two-sample sign test in which data are *paired* according to "before" and "after" values. A second example utilizes the same test for *independent* samples that are randomly paired. In this illustration, $n$ is relatively large and both $np$ and $n(1 - p)$ are at least 5, so we also use the normal approximation to the binomial distribution. In the sign test, the normal approximation can be used for samples as small as $n = 10$ because of the symmetry of the binomial distribution for $p = 0.5$ ($H_0: \pi = 0.5$).

The normal approximation to the sign test uses

$$Z \doteq \frac{X - n\pi}{\sqrt{n\pi(1 - \pi)}} = \frac{X - 0.5n}{0.5\sqrt{n}} \tag{17.10}$$

where   $X =$ the number of $+$ signs

$\pi = 0.5$

Assumption:   $n \geq 10$.

---

**EXAMPLE 17.6**

A delivery company has tested two delivery procedures (procedures X and Y) on separate delivery routes and counted the number of deliveries per 10-minute period (see table). Test, at the $\alpha = 0.1$ level, whether the number of deliveries from the use of the two procedures are equal.

| Procedure X | 3, | 5, | 2, | 4, | 1, | 7, | 4, | 5, | 6, | 8, | 3, | 0, | 5, | 3, | 1, | 2, | 4, | 7, | 10, | 6, | 7, | 4 |
|---|---|---|---|---|---|---|---|---|---|---|---|---|---|---|---|---|---|---|---|---|---|---|
| Procedure Y | 4, | 2, | 1, | 3, | 2, | 5, | 4, | 0, | 3, | 1, | 6, | 4, | 2, | 3, | 5, | 2, | 1, | 4, | 0, | 6, | 4, | 2 |

*Solution*   We randomly pair (or match) the data and determine the sign of the difference as shown in the table. First we delete those two pairs that have zero difference. Then our test is concerned with whether as many as 15 plus signs out of 20 samples would have occurred by chance under a null hypothesis of $H_0: \pi = 0.5$. Since the requirements for the normal approximation are satisfied ($n \geq 10$, $np \geqslant 5$, $nq \geqslant 5$), we can solve for $\mu$ and $\sigma_n$.

| X | Y | Difference | X | Y | Difference |
|---|---|---|---|---|---|
| 3 | 1 | + | 3 | 4 | − |
| 5 | 5 | 0 | 0 | 2 | − |
| 2 | 6 | − | 5 | 4 | + |
| 4 | 3 | + | 3 | 2 | + |
| 1 | 4 | − | 1 | 0 | + |
| 7 | 2 | + | 2 | 1 | + |
| 4 | 1 | + | 4 | 5 | − |
| 5 | 0 | + | 7 | 3 | + |
| 6 | 2 | + | 10 | 6 | + |
| 8 | 4 | + | 6 | 3 | + |
| 7 | 3 | + | | | |
| 4 | 4 | 0 | | | |

$$\mu = np \doteq n\pi = 20(0.5) = 10$$

$$\sigma_n = \sqrt{n(\pi)(1 - \pi)} = \sqrt{20(0.5)(0.5)} = 2.24$$

Thus we have

$$H_0: \text{Mean number of plus signs} = \mu = 10$$

$$H_A: \text{Mean number of plus signs is} > 10 \text{ or} < 10$$

$$\alpha = 0.10 \text{ (for two-tailed we use } \alpha/2 = 0.05)$$

Use the sign test with normal approximation; reject $H_0$ if

$$|\text{Calculated } Z| \text{ is more than } Z_{0.05} = 1.645$$

Because we are using the (continuous) normal approximation to the (discrete) binomial, we will use the continuity correction ($\pm 0.5$) discussed in Chapter 6. Thus an $X$-value of 15 or more plus signs would, in continuous terms, be $15 - 0.5 = 14.5$ and higher.

$$Z = \frac{X - n\pi}{\sigma_n} = \frac{14.5 - 10}{2.24} = 2.01$$

Conclusion: Reject $H_0$, since $2.01 > 1.645$. Relative to delivery procedure $Y$, procedure $X$ appears to result in more deliveries per 10-minute period.

---

The sign test gives good results for small samples, down to $n = 10$, and it is easier to compute than are comparable $Z$- and $t$-tests. However, when applied to interval data, it lacks the power of these parametric procedures. That is, for a fixed sample size and $\alpha$-level, the sign test is more likely to give a false conclusion of accepting the null hypothesis.

**Problems 17.6**

16. The sign test and the runs test are both one-sample nonparametric procedures that use nominal values, such as $+$ or $-$, or $T$ or $F$. Explain ways that these procedures differ.

17. A real estate broker estimates that, on the average, it takes 30 days to close the sale of a piece of property (i.e., $H_0$: median $= 30$ days). A random sample of closing times of 20 transactions revealed the following number of days: 7, 45, 32, 20, 28, 6, 17, 70, 10, 18, 31, 3, 26, 44, 14, 24, 18, 28, 60, 24. Test, at $\alpha = 0.10$, against the alternative that the median is less than 30, using the one-sample sign test.

18. A building materials manufacturer supplies bags of cement that are stamped "Weight 100 pounds." When 100 bags from a production run were weighed, 15 were less than 100 pounds and 85 were more. Using a minus sign to indicate a weight deficiency and a plus sign to indicate an excess, test the hypothesis that the proportion of pluses equals 0.50. Use a two-tailed test at $\alpha = 0.01$.

19. The market research department of a soft-drink producer tested 12 subjects' preference for a new cola before and after a period of strenuous work (see table). Preference was measured on a 0–10 scale, with 10 being the strongest. Conduct a two-sample sign test, at $\alpha = 0.10$, of $H_0$: The preference before equals the preference after, against the following:
    a. The preference after the exercise is different (i.e., two-tailed).
    b. The preference after the exercise is stronger (i.e., one-tailed).

| Subject | Preference Before | Preference After |
|---------|-------------------|------------------|
| A | 4 | 6 |
| B | 8 | 8 |
| C | 2 | 6 |
| D | 4 | 5 |
| E | 5 | 4 |
| F | 3 | 3 |
| G | 6 | 8 |
| H | 3 | 7 |
| I | 7 | 3 |
| J | 6 | 9 |
| K | 2 | 3 |
| L | 4 | 4 |

**17.7
THE KOLMOGOROV–
SMIRNOV TEST FOR
GOODNESS OF FIT**

Several one-sample nonparametric tests are available for inference about a location value of a distribution. However, these procedures generally fall short of completely identifying the distribution. The Kolmogorov–Smirnov (K–S) test is a one-sample procedure that is appropriate when we can completely specify the theoretical distribution and wish to test whether a sampled population follows that distribution. The Kolmogorov–Smirnov test is an alternative to the

Pearson's $\chi^2$ goodness-of-fit procedure that we used in Chapter 16. Comparisons of the two will be made after we describe the K–S test.

The Kolmogorov–Smirnov test assumes that the random variable has a continuous probability distribution. Then the distribution function is labeled as $f(X)$, and the cumulative theoretical distribution function is given the label $F(x) = P(X \leqslant x)$. In this expression, $X$ is a random variable and $x$ represents an individual observation value.

The Kolmogorov–Smirnov test uses a random sample of $n$ observations with a cumulative empirical distribution defined by $S_n(x) = k/n$, for $k = 1, 2, \ldots,$ $n$. $S_n(x)$ is simply the proportion of sample observations that resides at or below $X = x$. For a sample of $n$ observations, $S_n(x)$ takes the values $1/n, 2/n, \ldots, n/n = 1$ for ordered $x$-values. The test statistic, $K$, describes the maximum observed difference of the empirical values to the theoretical cumulative distribution [9]. We provide an outline of the procedures here.

The Kolmogorov–Smirnov test requires that the hypothesized distribution be divided into $n$ nonoverlapping segments, each containing $1/n$ parts of the distribution. The null hypothesis is that the completely specified theoretical distribution is that of the sampled variable. Then applying the $(n)$ sample values to the theoretical distribution provides a comparison of $S_n(x)$ to $F(x)$, at $n$ data points. Since a comparison is made over $n$ ordered values, this is an evaluation of the fit over the total distribution. The test statistic is as follows.

> **Kolmogorov–Smirnov $K$-statistic for testing goodness of fit to a completely specified distribution:**
>
> $$K = \max |S_n(x) - F(x)| \qquad (17.11)$$
>
> where  $S_n(x)$ is the empirical cumulative distribution, $k/n$, for
>   $k = 1, 2, \ldots, n$
>
>   $F(x)$ is the theoretical cumulative distribution applied to the sample observations
>
> Assumptions:  The theoretical distribution is continuous and is completely specified. Random sampling is used.

The test procedure consists of the following steps:

1. Specify the theoretical distribution, $f(X)$, including its parameters.
2. Select a random sample of $n$ observations from the population under test.
3. Order the sample observations, low to high, and construct the cumulative empirical distribution, $S_n(x)$, with values $1/n, 2/n, \ldots, 1$.
4. Using the $(n)$ sample observations, ordered from the lowest value to the highest, calculate the theoretical cumulative distribution values, $F(x)$, using $F(x) = P(X \leqslant x)$.
5. Calculate the difference $S_n(x) - F(x)$ for each $x$. Then note $K$, the maximum (biggest) calculated difference.
6. Compare the calculated value to the $(\alpha)$ upper-tail percentage point for the K–S distribution, $K_{\alpha,n}$ in Table 9 of Appendix B.

The null hypothesis is supported if the empirical cumulative probability distribution is nearly identical to the hypothesized theoretical (cumulative) distribution. Small differences in all $n$-intervals will signal a true null hypothesis. Otherwise, a sizable difference of $K > K_{\alpha,n}$ will signal a rejection of $H_0$ and the hypothesized distribution.

The test does not depend on any parameter conditions; it simply requires that the theoretical distribution be continuous. The sample data can be ordinal scale. However, ties create a problem of classification (see [6]). An example will illustrate the test procedures.

---

**EXAMPLE 17.7**

The computer run-time for a daily audit of a bank's accounts is hypothesized to be normal with a mean 20.0 minutes and standard deviation of 1.0 minutes. This 20.0-minute time is critical because the daily account activity must be checked by bank auditors each morning prior to the start of business activities. After the audit there is a computer update of accounts, which requires 5.0 minutes (this is assumed to be a fixed time). Suppose a random sample from six days revealed times (in minutes) of 20.2, 21.9, 18.6, 17.9, 20.6, and 19.9. Is 8:30 A.M. a sufficiently early start-time to ensure that the auditors can check accounts and be current for the 9:00 A.M. bank opening? Let $\alpha = 0.05$.

*Solution*   We follow the six steps previously outlined.

1. The theoretical distribution is hypothesized to be $N(20.0,1.0^2)$. Then, for $X =$ audit time in minutes,

$$H_0: X \text{ is distributed as } N(20.0,1.0^2).$$

$$H_A: X \text{ is not distributed as } N(20.0,1.0^2).$$

2. The data are audit times, in minutes, for a random sample of six days, given as 20.2, 21.9, 18.6, 17.9, 20.6, 19.9.
3. The ordered data appear as 17.9, 18.6, 19.9, 20.2, 20.6, 21.9 minutes. The empirical cumulative distribution is $S_6(17.9) = 1/6$, $S_6(18.6) = 2/6, \ldots , S_6(21.9) = 1$.
4. Then,

$$F(17.9) = P(X < 17.9 \,|\, N(20.0,1^2))$$

$$= P\left(Z < \frac{17.9 - 20.0}{1.0}\right) = 0.018$$

$F(18.6)$, $F(19.9)$, $F(20.2)$, $F(20.6)$, and $F(21.9)$ are found similarly.

5. The calculations for the $K$-statistic determination are shown in the table.

| Observation Values Ordered | $S_n(x)$ | $F(x)$ | $|S_n(x) - F(x)|$ |
|---|---|---|---|
| 17.9 | 0.167 | 0.018 | 0.119 |
| 18.6 | 0.333 | 0.081 | 0.152 ⟵ $K$ |
| 19.9 | 0.500 | 0.460 | 0.040 |
| 20.2 | 0.667 | 0.579 | 0.088 |
| 20.6 | 0.833 | 0.726 | 0.007 |
| 21.9 | 1.000 | 0.971 | 0.029 |

**6.** From Table 9 in Appendix B, $K_{0.05,6} = 0.519$ (two-tailed test). Since the calculated $K$ (0.152) is less than the value (0.519) from the table, we cannot reject $H_0$. The distribution for audit times is reasonably described by $N(20.0$ minutes, $1.0^2$ minutes). Therefore, 8:30 A.M. is a sufficient start-time for this auditing work for $\alpha = 0.05$. (On average, audit plus account updates should be completed in less than the 30 minutes allowed: 20 minutes for the audit plus 5 minutes for the update, allowing for $\pm 20$ minutes.)

These conditions are not as extensive as those for the $\chi^2$ goodness-of-fit test. The K–S test has several other advantages over the $\chi^2$ procedure. First, the assumptions are minimal, including random sampling and a continuous distribution. An exact distribution for the $K$-statistic is known, and table values are given for alternative sample sizes. This is less restrictive than $\chi^2$, which is theoretically correct only with very large (infinite) sample sizes. Chi-square is only an approximate test for all possible sample sizes, whereas Kolmogorov–Smirnov is an exact test.

The K–S test uses ungrouped data, with every observation representing a point at which goodness of fit is examined. For continuous measure, $\chi^2$ ignores the uniqueness of the data by grouping them into classes. The $\chi^2$ procedure suffers from additional approximations associated with the choice of class interval, start and end points for extreme classes, and uncertainties in estimation of parameters (like using $\bar{X}$ for $\mu$ and $s$ for $\sigma$ in testing a hypothesized normal distribution). Furthermore, for very small samples, the $\chi^2$-test is not applicable at all, but the Kolmogorov–Smirnov test is applicable.

Conversely, the $\chi^2$ procedure has several advantages over the K–S test. Chi-square does not require the hypothesized distribution to be completely specified before sampling. With $\chi^2$ the estimation of parameters is accompanied by one adjustment for degrees of freedom. This ensures a reasonable assignment of the significance level. No such adjustment is possible with the Kolmogorov–Smirnov test.

The $\chi^2$-test is readily applied to discrete variable distributions, whereas the K–S test assumes a continuous variable distribution. Special procedures, including adjustments for ties and approximation for $\alpha$-risk, must be followed in order to apply the Kolmogorov–Smirnov test to discrete distributions.

The problems that follow offer an opportunity for practice with the Kolmogorov–Smirnov test. The Business Case illustrates the treatment of a discrete distribution with classed data.

**Problems 17.7**

20. A drive-in bank has advertised: "On the average, we can serve you within two minutes." Unknown to the bank, a competitor has timed the service for a random sample of eight customers. Times of 2.3, 1.9, 2.6, 2.4, 2.0, 2.1, 1.7, and 1.8 minutes were recorded, with a sample average of 2.1 minutes. Does this evidence satisfy the advertising claim? Make a two-tailed test for goodness of fit to a normal distribution, $N(2.0, 0.2^2)$. Use the Kolmogorov–Smirnov test with $\alpha = 0.05$.

 21. A continuously operating machine has time to failure, in hours of use, that is hypothesized to follow an exponential distribution $f(X) = e^{-X}$, where $X > 0$. Use the theoretical cumulative distribution $F(X) = 1 - e^{-X}$, $X > 0$, and sample failure times of 0.3, 6.1, 1.6, 0.8, 0.1, 3.2, 1.1 to perform the

K–S goodness-of-fit test with $\alpha = 0.05$. Does this distribution adequately describe the failure time measure?

22. The university shuttle between the city and agricultural campuses is set on a 10-minute schedule. The arrival times for a random sample of six trips are 9.6, 10.3, 10.4, 9.5, 10.0, 10.1 minutes. Test the goodness of fit to a continuous uniform distribution $f(X) = 1$, $9.5 < X < 10.5$ minutes. Then test $F(X) = X - 9.5$ for $9.5 < X < 10.5$. Let $\alpha = 0.05$.

23. Given the following distribution, test the hypothesis that the data come from a normal distribution. We have found $\bar{X} = 85.0$ and $s = 29.2$.

| Class | Observed Count |
| --- | --- |
| $25 \leqslant X < 50$ | 15 |
| $50 \leqslant X < 75$ | 25 |
| $75 \leqslant X < 100$ | 30 |
| $100 \leqslant x < 125$ | 20 |
| $125 \leqslant x < 150$ | 10 |

a. Test for a normal distribution $N(90,25^2)$ using the Kolmogorov–Smirnov procedure with $\alpha = 0.05$.

b. Test for normal distribution with $\mu$, $\sigma$, estimated by $\bar{X}$, $s$, and a $\chi^2$ goodness of fit with $\alpha = 0.05$.

## 17.8 (SPEARMAN'S) RANK CORRELATION: A NONPARAMETRIC CORRELATION MEASURE

Rank correlation is an easy and useful way to measure the closeness of association between two sets of ordinal, or ranked, data. Like many other nonparametric methods, the rank correlation method can also be applied to interval- or ratio-scale data, even though it utilizes only the respective rankings of the data. For this reason it is sometimes used on interval or ratio data to get an estimate prior to computing the Pearson correlation. Spearman's rank-order coefficient of correlation, $r_s$, is as follows.

### SPEARMAN'S RANK-ORDER CORRELATION STATISTIC

$$r_s = 1 - \frac{6 \sum d_i^2}{n(n^2 - 1)} \tag{17.12}$$

where   $n$ = the number of paired observations

$d_i = X_i - Y_i$

= difference between ranks of paired observations

Ranks are assigned as follows: 1 = highest to $n$ = lowest rating; ranks are separate for each sample.

The Spearman $r_s$ is comparable to the Pearson correlation coefficient discussed in Chapter 12, except that interval or ratio values are replaced by ranks.

However, Spearman's $r_s$ becomes less accurate than the Pearson coefficient as the number of ties in either variable increases. The same limits of $-1$ to $+1$ apply to the Spearman correlation coefficient $\rho_s$. Generally, the null hypothesis tested is of zero correlation: $H_0: \rho_s = 0$.

Notice that Equation (17.12) calls for subtracting a ratio of the squared differences from 1. If the calculated value of $r_s$ exceeds the appropriate value in Table B.8, then the hypothesis of zero correlation is rejected. In that case the differences between ranks are so small that they signal association that was unlikely to occur by chance.

Table B.8 gives critical values for the rank correlation coefficient for two-tailed tests with $\alpha$-values of 0.01, 0.02, 0.05, and 0.10 and for $n = 5$ to $n = 30$ paired observations. The table is set up for direct use with two-tailed tests.

### RULES FOR USING TABLE B.8 FOR TESTS OF SIGNIFICANCE ON $\rho_s$

1. For two-tailed tests, read the table entry for the specified $\alpha$-level.
2. For one-tailed tests, double the specified $\alpha$ and read from the column with that heading.

For two-tailed tests we read the table directly, using the indicated $\alpha$-risk level. For one-tailed tests, the $\alpha$-risk is half that listed in each column heading. For example, for $n = 10$, a two-tailed test at the $\alpha = 0.10$ level would call for rejection of the hypothesis of zero correlation ($H_0: \rho_s = 0$) if the calculated $|r_s|$ is greater than 0.564. This cuts off 5% of the area in each tail; the percentile is equivalent to that for a one-tailed test at $\alpha = 0.05$. Thus, if you need to use the table for a one-tailed test at the 0.05 $\alpha$-level, you use the column labeled $\alpha = 0.10$.

---

**EXAMPLE 17.8**

A consumer service magazine has rated eight automobiles in terms of two features—style and comfort. The rating scale used was 1 to 8 with 1 being the highest rank. Use the rank correlation coefficient with $\alpha = 0.10$ for a two-tailed test to determine if there is a significant association between the ratings. Notice that autos B and F have the same rank: $(3 + 4)/2 = 3.5$. Ties are assigned the mean of the tied ranks.

|  | Rating of Consumer Magazine | |
| --- | --- | --- |
| Auto | *Rank in Style* | *Rank in Comfort* |
| A | 7.0 | 6 |
| B | 3.5 | 1 |
| C | 5.0 | 5 |
| D | 2.0 | 4 |
| E | 7.0 | 8 |
| F | 3.5 | 3 |
| G | 7.0 | 7 |
| H | 1.0 | 2 |

*Solution*     $H_0: \rho_s = 0$ (i.e., zero correlation)     $H_A: \rho_s \neq 0$     $\alpha = 0.10$

Test: Reject $H_0$ if calculated $|r_s| > 0.643$. Calculation of $r_s$:

| Auto | Rank in Style | Rank in Comfort | Difference $d_i$ | (Difference)$^2$ $d_i^2$ |
|------|------|------|------|------|
| A | 7.0 | 6 | 1.0 | 1.00 |
| B | 3.5 | 1 | 2.5 | 6.25 |
| C | 5.0 | 5 | 0.0 | 0.00 |
| D | 2.0 | 4 | −2.0 | 4.00 |
| E | 7.0 | 8 | −1.0 | 0.25 |
| F | 3.5 | 3 | +0.5 | 0.25 |
| G | 7.0 | 7 | 0.0 | 0.00 |
| H | 1.0 | 2 | −1.0 | 1.00 |

$$\sum d_i^2 = 13.5$$

$$r_s = 1 - \frac{6 \sum d_i^2}{n(n^2 - 1)} = 1 - \frac{6(13.5)}{8(64 - 1)} = 0.84$$

Conclusion: Reject the hypothesis of zero correlation. There is a significant rank association between style and comfort for $\alpha = 0.10$.

In summary, the procedure for using the rank correlation test is as follows:

### STEPS FOR A TEST OF ASSOCIATION USING THE RANK CORRELATION TEST

1. Order the observations separately from 1 to $n$ for each variable, with the highest taking rank 1. Ties take the mean of the ranks that would be assigned had no ties occurred.
2. Compute the difference in ranks ($d_i$), then the rank correlation coefficient $r_s$.
3. For $5 \leqslant n \leqslant 30$, test for a significant association by comparing the calculated $r_s$ with the test values in Table 8, Appendix B. If the calculated $|r_s|$ equals or exceeds the tabled value, then the correlation is significant at the specified $\alpha$-level.

For large samples ($n > 30$), the Z-distribution is an adequate approximation to Table B.8. For a test of $H_0: \rho_s = 0$, we compute the following Z-statistic [3]:

### THE Z-DISTRIBUTION APPROXIMATION TO SPEARMAN'S RANK CORRELATION TEST

$$Z \doteq \frac{r_s - 0}{\sqrt{\dfrac{1}{n - 1}}} = r_s \sqrt{n - 1} \qquad (17.13)$$

Assumption:    $n > 30$.

Here is an example.

**EXAMPLE 17.9**

Sunco Products Corporation, a national manufacturer of cores for spooled products including paper towels and toilet paper, operates numerous small production plants. Sunco management believes in small-city industry, where minimum-wage labor keeps production costs low. The company's comptroller has examined the relation between profits and the level of production capacity for 33 plants. Profits are in thousands of dollars. How strong is the correlation between weekly profits and percent of production capacity? Test for a significant rank correlation using $\alpha = 0.05$. What conclusion can the comptroller make from the test result?

| Weekly Profits | Percent of Capacity | Weekly Profits | Percent of Capacity | Weekly Profits | Percent of Capacity |
|---|---|---|---|---|---|
| 2.8 | 51 | 8.5 | 73 | 13.4 | 94 |
| 3.1 | 57 | 8.5 | 65 | 13.7 | 90 |
| 4.6 | 68 | 8.6 | 71 | 13.8 | 95 |
| 5.9 | 70 | 8.9 | 74 | 14.1 | 87 |
| 6.5 | 65 | 10.3 | 80 | 14.4 | 85 |
| 6.6 | 72 | 10.4 | 86 | 14.6 | 95 |
| 6.7 | 63 | 10.7 | 77 | 15.2 | 83 |
| 7.1 | 50 | 11.3 | 91 | 15.4 | 93 |
| 7.6 | 66 | 11.9 | 79 | 15.5 | 88 |
| 8.0 | 58 | 12.5 | 92 | 16.2 | 99 |
| 8.4 | 69 | 13.0 | 96 | 16.6 | 96 |

*Solution*    Calculation of the ranks, then differences in ranks, results in $\sum d_i^2 = 744.50$, and

$$r_s = 1 - \frac{6 \sum d_i^2}{n(n^2 - 1)} = 1 - \frac{6 \cdot 744.50}{33(33^2 - 1)} = 1 - 0.1244 = 0.8756$$

So, $H_0$: $\rho_s = 0$, $H_A$: $\rho_s \neq 0$ requires the following:

Test: Reject $H_0$ if calculated $|Z| > 1.96$.

Calculated $Z = (0.8756 \sqrt{33 - 1}) = 4.95$

Conclusion: For $\alpha = 0.05$, the Spearman rank correlation is significantly different from zero ($\rho_s \neq 0$).

For this Sunco data there is a high degree of association between the ranks on weekly profits and the ranks of percent of production capacity. The association is strong. This does not, however, mean that the original variables are linearly associated. The comptroller could use this as an indicator that profits are related to the plant's percent of production capacity

The Spearman $r_s$ provides a quick calculation for correlation on the ranks of two variables. For Example 17.9 or other with ordinal or ratio data, this may be preliminary to the more powerful Pearson $r$ calculation and test for zero *linear* correlation.

**Problems 17.8**  24. A strike at Intex Co. has been followed by collective bargaining sessions, where union and management representatives have each ranked the importance of eight key issues as shown in the table.

| Issue | Union Ranking | Management Ranking |
|---|---|---|
| 1. Wages | 3 | 5 |
| 2. Vacation | 5 | 4 |
| 3. Work hours | 4 | 6 |
| 4. Health care | 8 | 7 |
| 5. Insurance | 7 | 8 |
| 6. Pension | 2 | 1 |
| 7. Safety | 1 | 2 |
| 8. Work standards | 6 | 3 |

Determine whether there is significant agreement between union and management representatives on the issues. Use a two-tailed rank correlation test with $\alpha = 0.10$.

25. Given the associated values (see table), the Pearson correlation is $r = -0.85$. Compute the value of the (Spearman's) rank correlation coefficient $r_s$ and test for a significant negative correlation. Let $\alpha = 0.05$.

| X | 2 | 2 | 3 | 3 | 4 | 4 |
|---|---|---|---|---|---|---|
| Y | 7 | 5 | 3 | 5 | 1 | 3 |

26. The dollars spent for meat by a random sample of 15 customers during three days before a consumer boycott are to be compared with their dollar purchases of meat during three days after the boycott. Compute $r_s$ and test for a significant rank correlation. Let $\alpha = 0.10$.

| Customer | Before | After |
|---|---|---|
| 1 | $12.31 | $13.47 |
| 2 | 16.28 | 15.92 |
| 3 | 13.12 | 12.78 |
| 4 | 10.40 | 11.70 |
| 5 | 18.21 | 14.36 |
| 6 | 26.93 | 21.80 |
| 7 | 15.00 | 19.66 |
| 8 | 8.76 | 10.49 |
| 9 | 21.12 | 16.18 |
| 10 | 14.08 | 15.12 |
| 11 | 6.13 | 11.70 |
| 12 | 11.76 | 12.34 |
| 13 | 21.11 | 22.02 |
| 14 | 19.08 | 16.98 |
| 15 | 13.39 | 15.42 |

27. Using the data in Example 17.9, convert the two variables to ranked values and do the computations for $\sum d_i^2$ and $r_s$. Compare your answers with the numbers given in the example.

## SUMMARY

*Nonparametric methods* are statistical techniques that have less restrictive assumptions than corresponding parametric methods. As such, they are applicable to the nominal and ordinal (name or count) data frequently encountered in business and social sciences. They can be used on interval- and ratio-scale data, but then in many cases they are less powerful than their parametric counterparts.

*Nominal data* are classified according to a name, or symbol. This is the weakest scale of measurement. Tests for nominal data include the $\chi^2$-tests we discussed in Chapter 16. Other nominal data tests also relate to the $\chi^2$-statistic. These include the McNemar and Cochran's $Q$-statistic for tests on dependent samples. Another is the contingency coefficient for testing the association of nominal data. This work discusses only $\chi^2$; see [6].

*Ordinal data* possess a means for ordering or ranking the data; so they convey more information than nominal data. Techniques for ordinal data include the *runs test* (e.g., random sequence of exam questions), the *Mann–Whitney U-test* (e.g., life tests on a product from two suppliers), the *Kruskal–Wallis* nonparametric one-way ANOVA (e.g., tests for equal sales among three classes), the *sign test* (e.g., attitudes before versus after viewing a TV ad), the *Kolmogorov–Smirnov test* (e.g., whether waiting times in a service line are normally distributed), and *rank correlation* (e.g., for ratings on style versus comfort).

Nonparametric tests use simple calculations. Some require special sampling distributions; for fairly small sample sizes, most can utilize a normal distribution approximation. Some discussion was made of the relative power of a nonparametric test and its parametric counterpart. In particular, we mentioned the Mann–Whitney $U$-statistic versus a $t$-test for equal means for two independent samples and the Kruskal–Wallis test for equality in three or more independent samples versus an $F$-statistic. In Chapter 9 we assigned the highest power to tests that have the highest probability of rejecting the null hypothesis (1) given that it actually is false and (2) for given conditions of fixed $\alpha$-risk and sample size. When parametric and nonparametric tests are applied to interval or ratio data that satisfy the (parametric) assumptions of the test, then the parametric tests tend to be more powerful. For cases in which the parametric assumptions are not met or the data are nominal or ordinal level, then only the nonparametric test is appropriate.

Table 17.2 provides a summary of the important equations for Chapter 17. The assumptions and examples in specific sections can add to the equation descriptions. Check them carefully for each procedure.

Table 17.3 summarizes the nonparametric methods discussed in this chapter. This table, in conjunction with the introductory remarks of this chapter, can help you select a nonparametric procedure that is most appropriate for a given situation. Any such summary runs the risk of oversimplification, however, and this is no exception. Use this table only as a guide, and then follow up by reviewing the assumptions and examples of the selected technique in more detail.

## KEY TERMS

interval data, 799
Kolmogorov–Smirnov test statistic, 816
Kruskal–Wallis test statistic, 808
Mann–Whitney $U$-test parameters, 804
Mann–Whitney $U$-test statistic, 804
nominal data, 796

nonparametric methods, 794
normal approximations
  for Mann–Whitney $U$-statistic
    distribution, 804
  for runs statistic distribution, 801
  for sign-test statistic distribution, 813

**TABLE 17.2** Key Equations for Nonparametric Statistics

### Runs Statistic

Mean $\quad \mu_r = \dfrac{2n_1 n_2}{n_1 + n_2} + 1$ (17.1)

Standard deviation $\quad \sigma_r = \sqrt{\dfrac{2n_1 n_2 (2n_1 n_2 - n_1 - n_2)}{(n_1 + n_2)^2 (n_1 + n_2 - 1)}}$ (17.2)

Normal distribution approximation $\quad Z \doteq \dfrac{r - \mu_r}{\sigma_r}$ (17.3)

### Mann–Whitney $U$-Statistic

$$U_1 = \left( n_1 n_2 + \frac{n_1(n_1 + 1)}{2} \right) - \sum R_1$$ (17.4)

Mean $\quad \mu_U = n_1 n_2 / 2$ (17.5)

Standard deviation $\quad \sigma_U = \sqrt{\dfrac{n_1 n_2 (n_1 + n_2 - 1)}{12}}$ (17.6)

Normal approximation $\quad Z \doteq \dfrac{\mu_1 - \mu_U}{\sigma_U}$ (17.7)

### Kruskal–Wallis $H$-Statistic

$$H = \frac{12}{N(N + 1)} \sum_{i=1}^{k} \left( \frac{R_i^2}{n_i} \right) - 3(N + 1)$$ (17.8)

### Sign Statistic

$$P(X \,|\, n, 0.5) = {}_nC_X (0.5)^n \quad X = 0, 1, 2, \ldots, n$$ (17.9)

$$Z \doteq \frac{X - 0.5n}{0.5\sqrt{n}}$$ (17.10)

### Kolmogorov–Smirnov $K$-Statistic

$$K = \max |S_n(X) - F(X)|$$ (17.11)

### Spearman's Rank Correlation

$$r_s \doteq 1 - \frac{6 \sum d_i^2}{n(n^2 - 1)}$$ (17.12)

Normal approximation $\quad Z \doteq r_s \sqrt{n - 1} \quad$ for $n > 30$ (17.13)

**TABLE 17.3** Table of Nonparametric Applications

| Application | Level of Data | Where Used | Assumptions | Null Hypothesis | Test Statistic | Advantages (A) and Disadvantages (D) |
|---|---|---|---|---|---|---|
| Chi-square (Independence of classifications) | Nominal (frequencies) | To see if two classifications of data in a contingency table are independent; e.g., "Are grades received by students independent of instructor A, B, and C?" | Samples are independent (i.e., not related). Cell frequencies ≥5. Mutually exclusive categories. | $H_0$: Variables are independent and not affected by 2nd classification; e.g, $H_0$: $\pi_1 = \pi_2 = \pi_3$ | $\chi^2 = \sum \dfrac{(O - E)^2}{E}$ (where $O$ = observed freq. $E$ = expected freq.) | A: Widely used and easily understood. D: Not accurate when cell freq. <5. |
| McNemar | Nominal (frequencies) | In "before–after" designs where data are classified in a table to see if there is a change over time; e.g., "Does behavior of managers change as a result of a seminar?" | Samples are dependent (i.e., related, e.g., by being same individual). Cell frequencies ≥5. | $H_0$: No change in variables over time. | $\chi^2 = \dfrac{(|A - D| - 1)^2}{A + D}$ (where $|A - D|$ = the absolute difference in cell frequencies) | A: Applies to wide variety of data. D: Not appropriate when cell freq. <5. |
| Cochran Q | Nominal (*dichotomous* classification) | To see if one classification has greater proportion than others where >2 groups are involved; e.g., "Do musicians, teachers, and engineers have the same attitude toward nuclear power?" | Samples are dependent. Data can be classified dichotomously (yes = 1, no = 0). | $H_0$: No difference among $k$ groups in proportions having a given characteristic. | Cochran Q (For large $n$, the Q has an approximate $\chi^2$-distribution.) | A: Applies when more than two samples are related. Sample size is not a crucial limitation. |
| Contingency coefficient | Nominal (frequencies) | To measure degree of association between variables; e.g., "How strongly are education and income level related?" | No significant assumptions. | No hypothesis. Result of calculation is an index of association for nominal data. | $C = \sqrt{\dfrac{\chi^2}{N + \chi^2}}$ ($C$ is a measure of the degree of association; $N$ = sample size) | D: C-values are not directly comparable unless sample size and proportions are carefully controlled. |
| Runs test | Ordinal | To see if data obtained in a sample are really random; e.g., "Do trainees' scores differ depending on whether they use classroom or self-paced techniques?" | $n$ should be ≥10. No other significant assumptions (normal distribution gives reasonable approximation). | $H_0$: No difference in scores of two groups (i.e., scores occur randomly). | $Z = \dfrac{r - u_r}{\sigma_r}$ (where $r$ = no. of runs) | A: Many applications and easily applied. |

| Test | Data | Use | Assumptions | $H_0$ | Statistic | Advantages/Disadvantages |
|---|---|---|---|---|---|---|
| Kolmogorov–Smirnov (K–S) | Ordinal | To see if an observed sample distribution is different from a theoretical distribution; e.g., "Is there a difference in preferences for dog food?" | Single sample. Must be able to specify a theoretical cumulative frequency distribution. | $H_0$: There is no difference or preference among the various classes. | $K = \max\lvert S_n(x) - F(x)\rvert$ <br> $S_n(x)$ = empirical cum distribution <br> $F(x)$ = theoretical cum distribution | A: Applies to small samples. More powerful than when data are ordinal. Doesn't require given freq. in cells. Uses ungrouped data. <br> D: Requires a completely specified distribution (under $H_0$) before sampling. |
| Sign test | Ordinal | In "before–after" designs, to see if there is a change over time; e.g., "Does use of a calculator improve students' math skills?" | Samples are dependent (i.e., related and randomly taken). | $H_0$: No difference in level of variable (e.g., math skills) as result of another variable (e.g., calculator usage). | Binomial ($p = 0.5$) for $n \leq 25$ <br> $Z = \dfrac{X - 0.5n}{0.5\sqrt{n}}$ <br> for $n \geq 10$ | A: Simple and easy to apply. <br> D: Not as precise as Wilcoxon test. |
| Signed rank (Wilcoxon) | Ordinal | In pretest–posttest designs where ranks reflect amount of difference between pairs; e.g., "Does safety training program change workers' awareness of safety hazards?" | Samples are dependent (i.e., related or matched). Data approach interval level, e.g., equal-appearing intervals. | $H_0$: No difference in level of variable (e.g., awareness) before and after some event (e.g., training program). | $\sum \tau$ <br> (where $\sum \tau$ = sum of ranks with less frequent sign). Use modified $Z$ (normal) for $n > 25$. | A: Considers both magnitude and direction of difference in values. Easily applied. |
| Friedman (2-way ANOVA by rank) | Ordinal | In multicondition designs, to see if there is a significant difference among $k$ groups; e.g., "Is there a difference in workers' output under three different lighting conditions?" | Samples are dependent under all levels (e.g., same individual). No other assumptions of parametric ANOVA. | $H_0$: No difference in characteristic (e.g., output) under the various conditions (e.g., levels of lighting). | $\chi^2$ (a special ranked $\chi^2$ where $n \leq 9$). $\chi^2$ for $n > 9$, or >4 groups | (A nonparametric equivalent to two-way ANOVA) <br> A: Applies to both large and small samples. <br> D: Not as powerful as ANOVA, with normal distributed variables. |

(continued)

**TABLE 17.3** (*continued*)

| Application | Level of Data | Where Used | Assumptions | Null Hypothesis | Test Statistic | Advantages (A) and Disadvantages (D) |
|---|---|---|---|---|---|---|
| Median test | Ordinal | To see if two samples come from populations with same median; e.g., "Is median expenditure on books by arts & science and business students the same?" | Samples are independent cell frequencies $\geq 5$. | $H_0$: Both groups come from population with same median (e.g., level of expenditure). | $\chi^2 = \sum \frac{(O - E)^2}{E}$ (where this is basically a contingency-table test) | A: Convenient test of central tendency for skewed data (e.g., wealth). Uses the widely understood $\chi^2$-statistic. Easily applied. |
| Mann–Whitney $U$/(rank-sum)-Test | Ordinal | To see if two samples are significantly different on basis of ranked data; e.g., "Do union and non-union personnel attach same relative importance to fringe benefits?" | Samples independent. No assumptions concerning normal distribution. | $H_0$: Both groups have same underlying distribution (e.g., preference of values). | $U_1 = n_1 n_2 + \frac{n_1(n_1 + 1)}{2}$ $- \sum R_1$ $U_2 = n_1 n_2 - U_1$ (where $\sum R_1$ = sum of ranks for first group of $n_1$) | A: A good nonparametric substitute for "$t$" test. Reject when $U$ calc. <tabled value. Works especially well for large samples. Can correct for tied scores. |
| Kruskal–Wallis (1-Way ANOVA) | Ordinal | To see if there is a difference between $k$ samples with respect to some characteristic; e.g., "Do farmers, lawyers, and factory works differ in feeling toward universal health insurance?" | Samples in each group are independent and random. No other assumptions of parametric ANOVA. | $H_0$: No difference among groups with respect to given characteristic (e.g., attitude toward price support). | $H = \frac{12}{N(N + 1)} \sum \left( \frac{R_i^2}{N_i} \right)$ $- 3(N + 1)$ (where $N$ = no. in $k$ samples, $\sum R_i$ = sum of ranks) | (a nonparametric equivalent to one-way ANOVA) A: Applies to both large and small samples. D: Less powerful than Mann–Whitney. |
| Spearman's Rank Correlation | Ordinal | To measure the degree of correlation between two sets of ranks-ordered data; e.g., "Do labor and management agree on collective bargaining issues?" | Treats ordinal (rank) data as interval (i.e., by subtraction). | $H_0$: No difference in ranks as evidenced by the two groups. | $r_s = 1 - \frac{6 \sum d^2}{n(n^2 - 1)}$ (where $\sum d$ = sum of differences, $n$ = no. of items) | A: Widely used for correlation of ordinal data. Easy to apply. |

SOURCES: [2], [3].

28. Using a random-numbers table (Table B.7), take the first 50 digits in column 1 (i.e., 1, 2, 2, ..., 9, 1) as your data points. Label the digits A or B according to whether they are above or below a median of 4.5. Then conduct a runs test to see if the scores represent a random sample of observations about the median. Use a two-tailed test with $\alpha = 0.10$.

29. In tests made on a random sample of 20 lightbulbs from Supplier A and 30 lightbulbs from Supplier B, the bulbs were ranked according to how long (in hours) they operated before failure. The results were $n_A = 20$, $n_B = 30$, $\sum R_A = 546$, and $\sum R_B = 690$. Use a two-tailed Mann–Whitney $U$-test with an $\alpha$-value of 0.05 to test for a difference in mean lifetimes of the bulbs from suppliers A and B.

30. A market research study was designed to determine the influence of advertising on buyer behavior. Consumers in each of two groups were observed and timed with respect to how many seconds it took them to choose between two brands in the supermarket. The experimental group had all been exposed to TV advertising for one product during the week before the test, whereas the control group had not. The given measures (see table) are time in seconds before making the purchase. Use the Mann–Whitney $U$-test with $\alpha = 0.05$ and a one-tailed test.

| Control group | 30.5 | 26.2 | 34.1 | 29.7 | 30.2 | 31.4 | 28.6 |
|---|---|---|---|---|---|---|---|
| | 35.1 | 26.9 | 18.7 | 29.3 | 32.0 | 28.8 | 24.1 |
| (times in seconds) | 33.3 | 30.2 | 30.7 | 34.8 | 28.7 | 26.4 | 30.1 |
| | 37.9 | 33.6 | 26.2 | 29.1 | 30.5 | | |
| Experimental group | 18.6 | 12.4 | 21.9 | 23.7 | 20.2 | 25.6 | 30.9 |
| | 41.0 | 17.3 | 19.9 | 16.7 | 24.6 | 14.1 | 23.7 |
| (times in seconds) | 21.7 | 30.4 | 38.3 | 12.3 | 16.8 | 20.4 | 15.4 |
| | 13.8 | 22.2 | 15.9 | 19.9 | 25.0 | | |

31. Researchers compared two self-paced executive training programs by having two groups (of 11 managers each) participate in the two different programs. One group were line managers (A) and the other were staff managers. (B) Their scores on the common evaluation test are shown in the table. Apply the Mann–Whitney $U$-test to determine if the two groups have comparable training. Use $\alpha = 0.10$.

| Group A | 76 | 82 | 53 | 91 | 87 | 42 | 75 | 88 | 57 | 63 | 83 |
|---|---|---|---|---|---|---|---|---|---|---|---|
| Group B | 93 | 87 | 65 | 81 | 62 | 58 | 43 | 66 | 52 | 60 | 74 |

32. An investment broker claims that her daily stock recommendations average more than a 12% annual rate of return for her clients. A random sample of her recommendations over 12 days revealed the listed rates of return for monies invested that day. Use the one-sample sign test at $\alpha = 0.20$ to test $H_0$: Rate of return = 12% against the alternative that it is greater than 12%.

| 12.5% | 10.2% | 14.0% | 27.3% | 12.4% | 4.2% |
|---|---|---|---|---|---|
| 6.5% | 18.7% | 23.4% | 2.8% | 15.2% | 12.4% |

33. Eleven disposable flashlight batteries from each of two manufacturers were given a battery lifetime test (hours of use), with the results as listed below.

| Company | Battery Lifetime (hours) | | | | | | | | | | |
|---|---|---|---|---|---|---|---|---|---|---|---|
| A | 7 | 6 | 8 | 5 | 8 | 6 | 9 | 4 | 5 | 4 | 5 |
| B | 8 | 10 | 7 | 9 | 8 | 11 | 8 | 6 | 7 | 6 | 6 |

a. Use the sign test with $\alpha = 0.10$ to test whether the mean lifetimes are equal against the alternative that Company B lifetimes are longer.
b. Would the hypothesis be rejected at $\alpha = 0.10$ if the test were two-tailed?

34. In a manufacturing operation involving microprocessors, two assembly lines produce the same component at the same overall rate. The number of acceptable components per day over an eight-day period from each line is as shown in the table. Assume the data given are randomly paired and use the two-sample sign test at $\alpha = 0.05$ to determine if there is a significant difference in the mean number of acceptable components (two-tailed test).
a. Use the binomial table probabilities.
b. Would the normal approximation yield the same test conclusion?

| Line A | 205 | 212 | 193 | 208 | 198 | 214 | 203 | 208 |
|---|---|---|---|---|---|---|---|---|
| Line B | 200 | 212 | 180 | 175 | 220 | 197 | 212 | 201 |

35. Compute the value of $r_s$ for the data in Problem 34. Assuming the values are paired observations, use the $t$-distribution and Table B.2 to test $H_0: \rho_s = 0$ against $H_A: \rho_s < 0$ at $\alpha = 0.025$.

36. An international marketing firm studied whether salaries paid to its employees were actually indicative of performance as measured on a scale of 1 to 10, with 1 being the highest. Ratings, as determined by the employees' supervisors, were as follows:

| Employee | Performance Ranking | Salary Scale |
|---|---|---|
| Murphy | 7 | 9 |
| O'Keefe | 5 | 4 |
| Kennedy | 9 | 7 |
| Byrne | 2 | 1 |
| Kelley | 6 | 5 |
| Hannigan | 4 | 3 |
| O'Harmon | 3 | 2 |
| Sullivan | 10 | 8 |
| Fitzpatrick | 8 | 6 |
| Brajcich | 1 | 10 |

a. Is there a significant correlation between performance and salary in this firm? Use the 90% level of significance and the two-tailed rank correlation test.

b. Review the data and try to determine why your conclusion comes out as it does. Make whatever deletion is necessary and recalculate your answer. (Do not rerank the data.) What conclusion do you draw now?

37. The number of words typed correctly per minute on a computer keyboard is a measure often cited for typing proficiency. The measures for students in first-term keyboarding as trained under three different methods are listed. Test the hypothesis of equal means using the three methods. Use the Kruskal–Wallis test with $\alpha = 0.025$.

| Method | Words per Minute | | | | | |
|--------|----|----|----|----|----|----|
| 1 | 50 | 57 | 52 | 49 | 53 | 46 |
| 2 | 48 | 53 | 44 | 46 | 47 | 51 |
| 3 | 59 | 59 | 55 | 57 | | |

38. A complicated merger plan was devised for testing the ability of business executives to solve a complex problem. The conditions allow only one best solution. A random sample of five executives who solved the problem required 3.89 hours, 2.10 hours, 2.60 hours, 4.01 hours, and 3.63 hours, respectively. Test for goodness of fit to a normal distribution with mean 3.0 hr and standard deviation 1/2 hour. Use the Kolmogorov–Smirnov test with $\alpha = 0.05$.

39. The university business club is active in a morning jogging exercise. Members want to set a time specification for a standard three-mile "around town" course and have clocked individuals from several groups (see table). Is it reasonable to consider these readings as coming from one population that is $N(26.0 \text{ minutes}, 1.0^2 \text{ minutes})$? Test for equal means using the Kruskal–Wallis test with $\alpha = 0.05$.

| Business Students | Business Faculty | Jr. Chamber of Commerce Members |
|-------------------|------------------|---------------------------------|
| 25.7 | 25.4 | 24.8 |
| 24.9 | 26.9 | 25.6 |
| 26.5 | 28.3 | 26.4 |
| 24.2 | 27.2 | 25.1 |

40. A pharmaceutical company employment advertisement was run in different regions of the country on consecutive days and generated the following 42 responses from men (M) and Women (W). Test whether the occurrences are random. MM, F, MMM, FF, MMMM, FFF, MM, FFFF, M, FFF, MMMM, F, MM, FFF, M, FF, MMMM. Use $\alpha = 0.05$ and a two-tailed test.

**STATISTICS IN ACTION: CHALLENGE PROBLEMS**

41. U.S. schools are big users of computers for instruction, with major applications as shown in the table.

|  | Computer Users (%) | |
|---|---|---|
| Type of Instruction | Middle Grades | High School |
| Learning math | 12 | 8 |
| Word processing (how to use) | 14 | 15 |
| Keyboarding (how to) | 15 | 12 |
| Learning English | 11 | 8 |
| Programming | 9 | 12 |
| Recreational use | 9 | 5 |
| Tools, e.g., spreadsheets | 8 | 11 |
| Learning science | 7 | 6 |
| Learning social studies | 5 | 3 |
| Business education | 4 | 10 |
| Industrial arts | 3 | 5 |
| Fine arts | 2 | 2 |
| Learning foreign languages | 2 | 2 |
| Other | 1 | 1 |
| Total | 100 | 100 |

SOURCE: Adapted from U.S. Bureau of the Census. *Statistical Abstract of the United States: 1990* (110th ed.), Table 237. Washington, D.C., 1990.

Questions
a. Compute a Spearman's correlation for usage by middle grades versus high schools.
b. Why would a Pearson correlation (a parametric measure) not be appropriate here?
c. Explain how you would use the Spearman correlation and the data to identify the areas most likely for computer sales in both middle grades and high schools.

42. In the preceding Problem, consider the most appropriate procedure for testing the equality of distributions for the computer usage in the middle grades versus in high schools. Discuss the options of (a) a Mann–Whitney $U$-test, (b) a sign test, and (c) a chi-square test of goodness of fit for the two distributions. *Note:* Calculations are not requested. This is a problem in data analysis—to determine the appropriate technique(s). Explain why each of (a)–(c) is or is not appropriate for the suggested test.

43. Does the U.S. Presidency shorten the office-holder's life expectancy? Paul M. Sommers has gathered data on the former Presidents and compared their life spans against life expectancies from the U.S. Bureau of Census and various volumes of *Vital Statistics of the United States.* His Table 1 is reprinted here.

**TABLE I**    Longevity of U.S. Presidents

| | Age at First Inauguration[a] | Life Expectancy after First Inauguration | Actual Years Lived after First Inauguration |
|---|---|---|---|
| George Washington | 57.1 | 16.3 | 10.6 |
| John Adams | 61.3 | 14.2 | 29.3 |
| Thomas Jefferson | 57.9 | 16.1 | 25.3 |
| James Madison | 58.0 | 16.2 | 27.3 |
| James Monroe | 58.8 | 15.8 | 14.3 |
| John Q. Adams | 57.6 | 16.7 | 23.0 |
| Andrew Jackson | 62.0 | 14.2 | 16.3 |
| Martin Van Buren | 54.2 | 18.9 | 25.4 |
| William H. Harrison | 68.1 | 10.6 | 0.1 |
| John Tyler | 51.0 | 20.9 | 20.8 |
| James K. Polk | 49.8 | 22.0 | 4.3 |
| Zachary Taylor | 64.3 | 12.9 | 1.3 |
| Millard Fillmore | 50.5 | 21.4 | 23.7 |
| Franklin Pierce | 48.3 | 22.2 | 16.6 |
| James Buchanan | 65.9 | 10.8 | 11.2 |
| Abraham Lincoln[b] | 52.1 | ~~19.7~~ | ~~4.1~~ |
| Andrew Johnson (R) | 56.3 | 17.3 | 10.3 |
| Ulysses S. Grant (R) | 46.9 | 23.6 | 16.4 |
| Rutherford B. Hayes (R) | 54.4 | 19.2 | 15.9 |
| James A. Garfield[b] (R) | 49.3 | ~~22.7~~ | ~~0.5~~ |
| Chester A. Arthur (R) | 51.0 | 21.6 | 5.2 |
| Grover Cleveland (D) | 48.0 | 22.9 | 23.3 |
| Benjamin Harrison (R) | 55.5 | ~~17.7~~ | ~~12.0~~ |
| William McKinley[b] (R) | 54.1 | 18.2 | 4.5 |
| Theodore Roosevelt (R) | 42.9 | 25.8 | 17.3 |
| William H. Taft (R) | 51.5 | 19.7 | 21.0 |
| Woodrow Wilson (D) | 56.2 | 17.0 | 10.9 |
| Warren G. Harding (R) | 55.3 | 18.7 | 2.4 |
| Calvin Coolidge (R) | 51.1 | 21.6 | 9.4 |
| Herbert C. Hoover (R) | 54.6 | 18.6 | 35.6 |
| Franklin D. Roosevelt (D) | 51.1 | 21.3 | 12.1 |
| Harry S Truman (D) | 60.8 | 15.0 | 27.7 |
| Dwight D. Eisenhower (R) | 62.3 | 14.6 | 16.2 |
| John F. Kennedy[b] (D) | 43.7 | ~~28.6~~ | ~~2.8~~ |
| Lyndon B. Johnson (D) | 55.2 | 19.1 | 9.2 |
| Richard M. Nixon (R) | 56.0 | 18.7 | — |
| Gerald R. Ford (R) | 61.1 | 15.8 | — |
| James E. Carter (D) | 52.3 | 22.9 | — |
| Ronald W. Reagan (R) | 70.0 | 11.5 | — |

[a] Rounded up or down to the nearest tenth of a year.
[b] Assassinated.

SOURCE: Paul M. Sommers. "Does the American Presidency Shorten Life?" *Journal of Recreational Mathematics* 21, no. 1 (1989): 39–45.

Answer the author's question. You will need to make a sign test using Equation (17.10) for $H_0$: Median difference value is zero, for Difference = Life expectancy − Actual years lived. Your calculations need to exclude Lincoln, Garfield, McKinley, and Kennedy, who were assassinated, and all living former presidents.

44. Is a chi-square goodness of fit an appropriate alternative test? How about the Kolmogorov–Smirnov one-sample test? Explain.

How might this study be modified to investigate the longevity of Fortune 500 business executives?

---

When Tom Heller heard that he was to be given a shot at managing the large new Superfine Supermarket that was under construction in University Plaza, he was elated! Tom had worked hard for promotions since joining the company after receiving his B.S. in marketing four years ago. But he hadn't expected to be manager of a huge supermarket so soon. Furthermore, if after six months he was able to report that customer satisfaction with his operations was above average, the job was his for the next two years for sure—then another promotion! If customers were dissatisfied, he would most likely be reassigned as the produce section manager at his old market on Main and 5th Street.

Tom had now been manager for six months. It was time to make his report. He knew that Mary Halifax, Superfine's vice president, would insist on some hard, cold facts. Thus Tom was in a quandary. He felt that he should generate some reports to demonstrate that business was doing well. But most of the statistics he had worked with in college were either geared to impractical games or burdened with underlying assumptions and conditions that probably didn't exist. Tom's problem seemed less clearly defined here, and furthermore he *had* to come up with some data on customer preference. He *had* to prove himself!

Tom stopped by the university library on his way home from work and picked up a relatively new statistics textbook. Before long he was reviewing various nonparametric methods, formulating some approaches to measure customer satisfaction. He came up with ten hypotheses:

1. The store offers friendly customer service.
2. The store has a convenient location.
3. The store has convenient hours.
4. The store offers competitive prices.
5. The store offers a good product selection.
6. The store is kept orderly and clean.
7. The store offers abundant parking space.
8. The store provides immediate checkout service.
9. Employees perform their work duties accurately.
10. The store is well managed.

His statistics source led to the Kolmogorov–Smirnov test as an appropriate procedure. But this needed some actual data.

Tom arranged to use a questionnaire to collect data from a random sample of 100 customers and developed response categories according to a Likert 5-point scale. For example, question 1 was

1. The store offers friendly customer service.
   ( ) Strongly Agree
   ( ) Agree
   ( ) Neither Agree or Disagree
   ( ) Disagree
   ( ) Strongly Disagree

---

\* This Business Case is adapted from an unpublished market research study by students at Gonzaga University.

**TABLE I7.4**    Kolmogorov–Smirnov Test

| Question 1: | The store offers friendly customer service. | | | | | |
|---|---|---|---|---|---|---|
| (1) | (2) | (3) | (4) Observed Cumulative Proportion | (5) | (6) Theoretical Cumulative Proportion | (7) Difference Between (4) and (6) |
| | Observed Number | Observed Proportion | | Theoretical Proportion | | |
| Strongly agree | 51 | 0.51 | 0.51 | 0.20 | 0.20 | 0.31 |
| Agree | 43 | 0.43 | 0.94 | 0.20 | 0.40 | 0.54* |
| Neither | 3 | 0.03 | 0.97 | 0.20 | 0.60 | 0.37 |
| Disagree | 2 | 0.02 | 0.99 | 0.20 | 0.80 | 0.19 |
| Strongly disagree | 1 | 0.01 | 1.00 | 0.20 | 1.00 | 0.00 |

\* Kolmogorov–Smirnov $D$ = absolute value of the maximum deviation between observed cumulative proportion (Col. 4) and the theoretical cumulative proportion (Col. 6). Here $D = 0.54$.

The Kolmogorov–Smirnov one-sample test would measure shopper attitudes toward the ten store attributes. The Likert scale provided ordinal data. Tom established the null hypothesis as a statement of no difference, or no preference, among the various classes. The test was whether the observed sample distribution is different from a theoretical distribution of 20% in each response category. (See Table 17.4.)

Tom computed differences between observed and expected frequencies, and determined the largest absolute difference between the observed and expected cumulative frequencies. This he designated as $D$. Then he compared the largest absolute difference with a critical value of $D$ for a given level of significance. If the observed $D$ were greater than the table $D$-value, the null hypothesis would be rejected, and Tom could conclude that a preference existed. He used tables in nonparametric reference texts [2, 3].

The various null hypotheses were formulated in such a way that the data, if strong and convincing, would tend to destroy them. Thus, for question 1, the null hypothesis was $H_0$: Customers show no preference in attitude toward the statement that "The store offers *friendly customer service.*" If the null hypothesis were true, approximately equal proportions of the 100 respondents would strongly disagree, agree, neither agree or disagree, and so on. Thus the theoretical proportion remains 0.20 because there are five categories. Table 17.4 shows the relevant calculations at $\alpha = 0.05$; table value of $D = (1.36/n) = (1.36/100) = 0.136$ [3, p. 36].

Conclusion: Since $0.54 > 0.136$, reject $H_0$ of no preference and conclude that respondents *do* feel the store offers friendly customer service.

The same test was applied to each hypothesis. The results were a basis for Tom's report to Ms. Halifax.

Tom Heller's final report provided convincing evidence that his customers looked upon the store with a favorable attitude in all (10) categories *except prices*. Many respondents either felt that prices were higher than in comparable stores or they had no opinion.

To further investigate the importance of prices, Tom conducted a Friedman ANOVA-by-ranks test [3] to determine whether customers have some inherent preference to the relative importance of each attribute. His study revealed that cleanliness ranked higher than any other attribute. Prices rated fifth in relative importance of the attributes. He also found that customers ranked the meat department higher than any others and the deli department lowest.

Using these nonparametric methods, Tom Heller was able to present a factual, accurate summary of customer satisfaction to Ms. Halifax. He concluded (statistically) that Superfine Supermarket customers had a favorable attitude about the store and that they shopped there largely because of factors other than price.

## Business Case Questions

45. Tom's hypothesis (4) was that the store offers competitive prices. The (100) responses summarized by response category were

| | | | |
|---|---|---|---|
| Strongly agree | = 16 | Disagree | = 21 |
| Agree | = 18 | Strongly disagree | = 21 |
| Neither | = 24 | Total | 100 |

Perform the calculations for the Kolmorogov–Smirnov $D$-statistic. Then use the critical value of $D = 0.136$ to test that customers show no perference in attitude toward statement (4), that is, the five response categories will occur equally often.

46. An option to the Likert 5-point response scale could be a 3-point scale including: ( ) Agree ( ) Neither agree nor disagree ( ) Disagree. Discuss why a 5-point scale may be better. (This question asks for a statistical answer. You are not expected to know about Likert and related response scales.)

## COMPUTER APPLICATION

*Selecting a Market to Test by Mail Advertising*

Again, we return to the marketing situation that targets on households with 3+ persons and with head of household 35–50 years old. This application views PUMS data on three Metropolitan Statistical Area (MSA) markets—Birmingham, Alabama; Lansing, Michigan; and Salt Lake City, Utah. The marketing manager wants to select one market to test for mail advertising to sell home computers. The lists include some considerations for a choice—the incomes and ages for a small random sample of households from the three markets. See Table 17.5.

The samples appear to have unequal mean incomes, whereas the mean ages are reasonably close. Do nonparametric statistics tests support these impressions?

Additional statistics are available for comparative tests on the markets. Several nonparametric tests were explored because quick, yet fairly reliable, results were wanted. We use the *p*-value testing approach.

Our first test uses the Kruskal–Wallis (one-way) nonparametric ANOVA for equal income distributions. Another Kruskal–Wallis one-way test is for equal mean ages for heads of household. Here is an SPSS routine that uses a small data set pulled from the 20,116-record PUMS file specifically for this study. Because this is a reduced file, the variable locations are not the same as on the PUMS file.

```
1  NUMBERED
2  TITLE   KRUSKAL-WALLIS TEST: A NONPARAMETRIC ONE-WAY ANOVA
3  DATA LIST   FILE= INFILE/
4  MSA 1   INCOME 3-7  AGE1 9-10   AGE2 12-13   AUTOS 15              1
5  SORT CASES BY MSA                                                  2
6  COMPUTE GROUP = 1
7  IF (MSA EQ 2) GROUP = 2
8  IF (MSA EQ 3) GROUP = 3
9  NPAR TESTS K-W = INCOME AGE1 BY GROUP (1,3)
```

Two notes will provide a brief description of procedures that we have not used before. See [10] for more detail on this and the subsequent nonparametric procedures.

**TABLE 17.5** Data for Computer Application

| Group | 1 | | 2 | | 3 | |
|---|---|---|---|---|---|---|
| | Birmingham, AL ($n = 1000$) Household Income | Age of Head of Household | Lansing, MI ($n = 4040$) Household Income | Age of Head of Household | Salt Lake City, UT ($n = 7160$) Household Income | Age of Head of Household |
| | $ 4,995 | 36 | $26,725 | 48 | $37,020 | 42 |
| | 43,330 | 49 | 30,880 | 36 | 30,005 | 39 |
| | 11,420 | 42 | 69,720 | 36 | 33,065 | 46 |
| | 10,840 | 46 | 29,910 | 36 | 41,810 | 44 |
| | 2,505 | 50 | 40,370 | 38 | 34,710 | 48 |
| | 24,315 | 36 | 56,575 | 42 | 31,975 | 48 |
| | 20,750 | 36 | 15,725 | 36 | 26,995 | 45 |
| | 2,635 | 46 | 71,995 | 46 | 34,610 | 44 |
| | 2,005 | 42 | 35,335 | 48 | 18,005 | 41 |
| | 38,040 | 50 | 27,890 | 44 | 17,230 | 45 |
| | 42,315 | 43 | 27,610 | 37 | 30,080 | 42 |
| | 21,605 | 45 | | | 30,085 | 48 |
| | 11,605 | 35 | | | 12,030 | 48 |
| | 29,840 | 45 | | | | |
| Mean | $19,014.29 | 42.93 | $39,335.91 | 40.64 | $29,047.69 | 44.62 |

1 The Kruskal–Wallis procedure requires that the data be sorted into groups with all MSA 1 (group 1) records ordered first, then MSA 2 records, and then MSA 3 records. This is accomplished by SORT CASES BY MSA at program statement 5.

2 The COMPUTE–IF statements, lines 6 through 8, are actually not required in this case because these group numbers already are 1, 2, and 3. This conforms to the NPAR TESTS procedure, where the variable name following the BY statement defines the GROUP assignment. Cases from the (3) GROUPS are ranked in a single set and the rank sum is reported for each group. The outputs include the number of cases in each group and the mean rank for each group, a calculated chi-square value, a $p$-value, and the same again after correcting for ties.

These test results appear in Summary 1.

Summary 1: The Kruskal–Wallis (K–W) procedure tests whether the data from three independent samples, as defined by a grouping variable, can be considered to come from one population. The K–W one-way nonparametric ANOVA test for equal income distributions by SPSS software produced the results shown in the following table.

Data for Computer Application Summary 1—Income

---

```
              Kruskal-Wallis One-Way ANOVA

 INCOME BY GROUP

 MEAN RANK     CASES

    13.29        14       GROUP = 1    Birmingham
    25.00        11       GROUP = 2    Lansing
    21.54        13       GROUP = 3    Salt Lake City
                 ──
                 38       TOTAL

                                      CORRECTED FOR TIES
    CASES    CHI-SQUARE     SIGNIF    CHI-SQUARE    SIGNIF
     38        7.5094       .0234       7.5094      .0234
```

---

This test indicates that one or more of the markets has a distribution for income that is different from the others, with *p*-value = 0.0234. So the mean incomes are not all equal. *Note:* Because there are no ties, the result "corrected for ties" is the same as not corrected.

Data for Computer Application Summary 1—Age for Head of Household (AGE1)

---

```
              Kruskal-Wallis One-Way ANOVA

 AGE1 BY GROUP

 MEAN RANK     CASES

    19.79        14       GROUP = 1    Birmingham
    15.14        11       GROUP = 2    Lansing
    22.88        13       GROUP = 3    Salt Lake City
                 ──
                 38       TOTAL

                                      CORRECTED FOR TIES
    CASES    CHI-SQUARE     SIGNIF    CHI-SQUARE    SIGNIF
     38        2.9111       .2333       2.9551      .2282
```

---

This K–W test indicates that the AGE1—householder age—distribution is statistically the same for the three MSAs because the *p*-value is 0.2282 for a two-tailed test. This result may seem a bit surprising considering the differences in MEAN RANK. However, the original data, in Table 17.5, show that the means are actually close, relative to the variability for age within each group.

Due to the "no difference" result observed here for the comparison of age of head of household means for the three markets, we will not compute a Mann–Whitney test for AGE1 in Summary 2.

Summary 2:    Because management wants a more affluent market for the initial test, the marketing manager requests a comparison for Lansing, Michigan (with highest mean), against Salt Lake City, Utah. Is the Lansing income distribution higher than that for Salt Lake City? This comparison uses a Mann–Whitney $U$-test.

Since these tests can be performed in one computer run, the SPSS program steps are numbered sequentially, continuing from the above step 8:

```
 9  SORT CASES BY INCOME ←———————————————————————— 3
10  TEMPORARY
11  SELECT IF (MSA GT 1) ←—————————————————————————— 4
12  COMPUTE NUGROUP = 1
13  IF (MSA EQ 3) NUGROUP = 2
14  NPAR TESTS M-W = INCOME BY NUGROUP (1, 2)
```

Again, we provide a few essential comments.

**3** The TEMPORARY statement requires (by SPSS rules) that SORT CASES precede it. Hence the ordering of statements 9 through 11.

**4** The NUGROUPing that occurs at statements 12 and 13 restricts the MSA groups to numbers 2 and 3—the Lansing, Michigan, and Salt Lake City, Utah, markets. Again, there is not a test for AGE1 because the K–W test indicates that the three markets have statistically equal mean age for the householder. The result is Summary 2.

Data for Computer Application Summary 2
———————————————————————————————————————

```
      Mann-Whitney U/Wilcoxon Rank Sum W Test

INCOME BY GROUP

MEAN RANK      CASES
  13.73          11      GROUP = 1.00   Lansing
  11.46          13      GROUP = 3.00   Salt Lake City
                ———
                 24      TOTAL

                        EXACT        CORRECTED FOR TIES
   U      W           2-TAILED P      Z        2-TAILED P
  58.0   151.0          .4585      -.7821         .4341
```

The test uses a normal approximation because both samples exceed ten records. The calculated $Z$, corrected for ties, is $Z = -0.7821$, which is not significant for $\alpha = 0.05$. The income distributions for Lansing and Salt Lake City are not significantly different.

Because Summary 1 indicates that one or more statistical differences exist for mean income, at least the Lansing, Michigan (the highest value), and Birmingham, Alabama (the lowest value), mean incomes are different.

Summary 3:   Additional information is available for each market. This summary includes Spearman correlation values for two variables at a time on AGE1, AGE2 (spouse ages), INCOME, and AUTOS. Because it is at least as affluent as the (2) other markets being tested, the Lansing, Michigan, market will be tested first. The Spearman correlations are displayed.

A Spearman correlation summary requires only one additional step. This compares pairs of values. Here a specific sort order is not a requirement.

15 NONPAR CORR VARIABLES = AGE1 AGE2 INCOME AUTOS/
              PRINT = TWOTAIL SIG SPEARMAN

The line-15 requirements include a two-tailed test. So a *p*-value or SIGNIFICANCE that is 0.025 or smaller indicates a significant nonparametric Spearman correlation. This result appears in Summary 3.

Data for Computer Application Summary 3—
Spearman's Correlation

Spearman Correlation Coefficients

| | | | |
|---|---|---|---|
| AGE2 | .8805 | | |
| | N( 10) | | |
| | SIG .001 | | |
| | | | |
| INCOME | .0816 | .3537 | |
| | N( 10) | N( 10) | |
| | SIG .823 | SIG .316 | |
| | | | |
| AUTOS | .6349 | .6587 | .1970 |
| | N( 10) | N( 10) | N( 10) |
| | SIG .049 | SIG .038 | SIG .585 |
| | | | |
| | AGE1 | AGE2 | INCOME |

*Note:* One record was excluded because it had a person #2 other than the spouse of the householder. So the sample size is $11 - 1 = 10$.

Only AGE1 and AGE2 have a significant Spearman correlation for a 0.05 level of significance. This high value is expected because in Chapter 12 we found a high positive linear (Pearson) correlation for husband/wife ages.

**Computer Application Questions**

47. Use the data presented along with Equation (17.4) to calculate the Mann–Whitney approximate *Z*-value for the test on income distributions for Lansing, Michigan, and Salt Lake City, Utah. Compare your calculated *Z* versus the computer-generated value ($-0.7821$) in Summary 2.

48. Discuss the Spearman correlation values that are given in Summary 3, including the following.
    a. Check the significance for AGE1 and AGE2 using Table 8, Appendix B.
    b. Discuss the low linear correlation for AGE1 and INCOME. *Suggestion:* Review the Lansing, Michigan, data given at the start of this application.

**COMPUTER PROBLEMS**

*Data Base Problems*

The existing software offers a wide variety of options for solving nonparametric test questions. So we have presented the computer problems in two sets. Problems 49 through 55 are designed for hand-calculated solutions with minimal computer assistance. We feel there is much understanding to be gained by working out the solutions by hand with a calculator.

Problems 56 through 61 repeat the same questions as Problems 49–55, but offer alternatives for a computer solution with the SPSS and Minitab packages. The advantage of using a computer to reduce the calculations then becomes readily apparent.

49. Select a random sample of 50 records from the PUMS data base. For each one, record the monthly cost of electricity. Compute their median monthly cost for electricity and list the dollar amounts. Now record, in the observed sequence, whether each record is (+), indicating a monthly cost higher than the median, or (−), for records with this cost below the median. Exclude records that have exactly the median value. Then perform the runs test, with $\alpha = 0.05$, to determine whether this sample, in the order selected, is random with respect to the median cost of electricity. *Suggestion:* For SPSS use

```
SAMPLE 50 FROM 20116
LIST VARIABLES = ELECTRIC/
    FORMAT = NUMBERED/
    CASES FROM 1 TO 50
```

This will provide a listing of 50 electricity costs in the order of selection. Or use the sample, following, that we have drawn.

Monthly Cost of Electricity

| Record | Cost | Record | Cost | Record | Cost | Record | Cost | Record | Cost |
|--------|------|--------|------|--------|------|--------|------|--------|------|
| 1 | $ 0 | 11 | $45 | 21 | $ 0 | 31 | $35 | 41 | $15 |
| 2 | 0 | 12 | 35 | 22 | 18 | 32 | 50 | 42 | 75 |
| 3 | 90 | 13 | 14 | 23 | 32 | 33 | 80 | 43 | 35 |
| 4 | 27 | 14 | 15 | 24 | 25 | 34 | 9 | 44 | 14 |
| 5 | 11 | 15 | 86 | 25 | 0 | 35 | 25 | 45 | 18 |
| 6 | 25 | 16 | 30 | 26 | 12 | 36 | 12 | 46 | 40 |
| 7 | 50 | 17 | 20 | 27 | 30 | 37 | 47 | 47 | 21 |
| 8 | 26 | 18 | 53 | 28 | 11 | 38 | 36 | 48 | 60 |
| 9 | 145 | 19 | 55 | 29 | 33 | 39 | 22 | 49 | 45 |
| 10 | 40 | 20 | 50 | 30 | 28 | 40 | 24 | 50 | 15 |

50. We know that it takes some years for people to peak in their earning power. When can one expect to earn a higher income—during ages 35–44, or 45–54? This test is conducted on ages for male heads of household (00 in cols 43–44 with code 0 in col 42) who have worked 30 hours or more (i.e., cols 53–54 are 30 + ).

a. Select a random sample of 20 records (from 2564) with male head of household 35–44 years old and worked 30+ hours/week, and another 20 (from 2219) who were 45–54 years old and worked 30+ hours/week. Record the income from all sources for both samples.

To use SPSS to draw the sample (after reading in the data), input

```
SET BLANKS = -1
SELECT IF ((AGE1 GE 35) AND (AGE1 LE 44) AND
                (SEX1 EQ 0) AND (HRSWRKD GE 30))
SAMPLE 20 FROM 2564
SORT CASES BY INCOME (D)
LIST VARIABLES = INCOME/ FORMAT = NUMBERED/
                CASES FROM 1 TO 20
```

This gives the sample sorted in descending order—that is, from the highest value, listed first, to the lowest value. Then repeat to select 20 records for ages 45–54 and HRSWRKD = 30 hours, which includes 2219 records in the PUMS. This gives

Incomes for Heads of Household in Specific Age Groups

| Row | For Ages 35–44 | For Ages 45–54 | Row | For Ages 35–44 | For Ages 45–54 |
|-----|-----|-----|-----|-----|-----|
| 1 | 75,000 | 70,120 | 11 | 26,510 | 33,450 |
| 2 | 51,015 | 63,300 | 12 | 26,065 | 31,845 |
| 3 | 48,310 | 62,630 | 13 | 25,510 | 27,365 |
| 4 | 45,915 | 57,010 | 14 | 24,010 | 27,015 |
| 5 | 44,440 | 48,335 | 15 | 20,845 | 26,245 |
| 6 | 38,600 | 39,015 | 16 | 18,990 | 19,850 |
| 7 | 36,615 | 35,805 | 17 | 18,810 | 9,505 |
| 8 | 36,335 | 35,005 | 18 | 10,740 | 9,115 |
| 9 | 32,210 | 34,315 | 19 | 8,110 | 8,910 |
| 10 | 26,815 | 33,880 | 20 | 6,115 | 1,785 |

b. Assign numbers 1–20 for those records with ages 35–44 years, and assign 21–40 for those aged 45–54 years.

c. Sort the 40 incomes as one sample, high to low, and assign ranks with ties receiving the mean rank.

d. Continue to compute the Mann–Whitney $U$-statistic and test for equal distribution of high and low ranks for the two age groups. Let $\alpha = 0.05$.

51. a. Select a random sample of 150 records from the Markets (MSAs): 1920 = Dallas–Ft. Worth; 2080 = Denver; 7320 = San Diego. Record monthly payments-to-lender (positions 27–30). For selecting the sample of 150 records, it can be useful to know that the PUMS file contains 269 records for the Dallas–Ft. Worth MSA, 150 records for the Denver MSA, and 168 records for the San Diego MSA, so you can expect unequal numbers for the three MSAs. *Note:* If one of the choices—payments or rent—is not blank, the other will be blank (or zero). Sometimes both are blank (or zero). Then you would exclude the record from both parts (a) and (b) of this question.

Pull the sample of payments-to-lender, with each value keyed to an MSA. Sort the items from high to low and list them.

Here is a SPSS routine, after the data have been read in, to draw the sample for payments-to-lender:

```
SELECT IF ((MSA EQ 1920) OR (MSA EQ 2080)
           OR (MSA EQ 7320))
SAMPLE 150 FROM 587
SELECT IF (PAYMENT NE 0)
SORT CASES BY MSA
LIST VARIABLES = PAYMENT MSA/ FORMAT = NUMBERED/
                 CASES FROM 1 TO 150
```

This gives payments-to-lender for three samples as shown below. Use the $H$ statistic to test $H_0$: Median monthly payments to lenders are the same.

**Dallas–Ft. Worth (MSA 1920 = 1) payments (24)**

| | | | | |
|---|---|---|---|---|
| 934 | 359 | 325 | 160 | 119 |
| 650 | 359 | 300 | 158 | 112 |
| 608 | 354 | 277 | 150 | 100 |
| 420 | 342 | 276 | 140 | 94 |
| 411 | 330 | 262 | 122 | 77 |

**Denver (MSA 2080 = 2) payments (12)**

| | | |
|---|---|---|
| 847 | 455 | 140 |
| 589 | 430 | 108 |
| 579 | 413 | |
| 550 | 370 | |
| 475 | 180 | |

**San Diego (MSA 7320 = 3) payments (18)**

| | | | |
|---|---|---|---|
| 784 | 494 | 375 | 136 |
| 665 | 484 | 312 | 124 |
| 653 | 474 | 293 | 98 |
| 630 | 420 | 248 | |
| 556 | 415 | 164 | |

b. Repeat (a) but now calculate the $H$-statistic for gross rents to test $H_0$: The median gross rent is the same. This uses gross rent (positions 31–33) from the PUMS.

An SPSS routine for rents is the same as in part (a) except that we put RENT in place of PAYMENT, in two places. Our sample gives

**Dallas–Ft. Worth rents (24)**

| | | | | |
|---|---|---|---|---|
| 651 | 356 | 237 | 212 | 105 |
| 618 | 325 | 237 | 165 | 75 |
| 450 | 278 | 237 | 155 | 62 |
| 375 | 272 | 237 | 155 | 40 |
| 375 | 272 | 212 | 149 | |

**Denver rents (18)**

| | | | |
|---|---|---|---|
| 630 | 339 | 252 | 158 |
| 610 | 337 | 235 | 133 |
| 541 | 303 | 210 | 95 |
| 407 | 295 | 200 | |
| 389 | 257 | 195 | |

**San Diego rents (12)**

| | | |
|---|---|---|
| 442 | 278 | 165 |
| 365 | 265 | 118 |
| 332 | 232 | |
| 317 | 226 | |
| 307 | 223 | |

52. Fifty years ago it was common for a husband to be older than his wife. Does that relation hold for the younger generation of married people? To test this question, select from the PUMS a random sample of 50 records in which the head of household (1) is married and (2) has age in the range of 20–39 years. Record age and sex for both householder and spouse. Now compute the difference, either + or − for husband's age minus wife's age. Let $p$ = the proportion of married couples with husband older than wife, for heads of household 20–39 years old. Exclude ties. Count the number of + differences. Use this in the sign test to test whether the median husband's age equals the median wife's age. Use $\alpha = 0.05$.

Here is an SPSS routine to draw the samples (assumes you have read in the data):

```
SET BLANKS = -1
SELECT IF (((AGE1 GE 20) AND (AGE1 LE 39)) AND
                  (MAR EQ 0) AND (SPOUSE EQ 1))
SAMPLE 50 FROM 4827
LIST VARIABLES = AGE1 AGE2 SEX1 SEX2/
                  CASES FROM 1 TO 50
SELECT IF (AGE1 NE AGE2)
COMPUTE AGEM = AGE1
IF (SEX1 EQ 1) AGEM = AGE2
COMPUTE AGEF = AGE2
IF (SEX1 EQ 1) AGEF = AGE1
```

This gives the following data:

| Couple | AGEM | AGEF | Couple | AGEM | AGEF | Couple | AGEM | AGEF |
|--------|------|------|--------|------|------|--------|------|------|
| 1 | 36 | 32 | 18 | 31 | 32 | 35 | 27 | 25 |
| 2 | 37 | 37 | 19 | 30 | 25 | 36 | 36 | 36 |
| 3 | 26 | 31 | 20 | 31 | 31 | 37* | 37 | 31 |
| 4 | 33 | 31 | 21 | 31 | 26 | 38 | 25 | 21 |
| 5 | 34 | 29 | 22 | 34 | 33 | 39 | 34 | 25 |
| 6 | 22 | 20 | 23 | 25 | 22 | 40 | 32 | 30 |
| 7 | 23 | 24 | 24 | 38 | 32 | 41 | 30 | 26 |
| 8 | 32 | 25 | 25 | 24 | 25 | 42 | 32 | 27 |
| 9 | 39 | 40 | 26 | 35 | 34 | 43 | 20 | 23 |
| 10 | 39 | 44 | 27 | 37 | 37 | 44 | 29 | 21 |
| 11 | 36 | 34 | 28 | 36 | 34 | 45 | 37 | 33 |
| 12 | 33 | 34 | 29 | 30 | 25 | 46 | 38 | 32 |
| 13 | 29 | 25 | 30 | 37 | 34 | 47 | 24 | 23 |
| 14* | 35 | 31 | 31 | 34 | 31 | 48 | 36 | 33 |
| 15 | 23 | 23 | 32 | 31 | 31 | 49 | 29 | 29 |
| 16 | 35 | 38 | 33 | 36 | 39 | 50 | 35 | 36 |
| 17 | 34 | 23 | 34 | 28 | 24 | | | |

AGE1 = Head of household age
AGE2 = Spouse age
AGEM = Male's age
AGEF = Female's age
* *Note:* Head of household is female, so these two ages were interchanged.

53. Correlation is another way to view the age relationship in the data in Problem 52. That is, if husband and wife ages have a high positive correlation, then the ages are closely related. Use the data from Problem 52 and Equation (17.13) to calculate Spearman's $r_s$ and to test for zero correlation. Before you test, estimate the numeric size for $r_s$. Note that in Spearman's test, the ranking for the ages of husbands is separate from the ranking of the ages for wives. Use $\alpha = 0.05$, and include ties.

54. Explain how your conclusions to the preceding two problems support each other. Explain what each one tells that is different from the other.

55. Select a random sample of 20 records from the PUMS. Record the age for head of household. Calculate sample $Z$-score values by hand (calculator), or use a computer program. Based on your sample data, calculate $F(X)$-values to split the distribution for comparison, with $S_1(X) = 2/20$, $S_2(X) = 4/20, \ldots,$ $S_{10}(X) = 1$.

      From SPSS (after data input) we get

```
SAMPLE 20 FROM 20116
LIST VARIABLES = AGE1/CASES FROM 1 TO 20
```

This gives 31, 20, 55, 82, 22, 45, 31, 28, 51, 26, 24, 41, 60, 27, 41, 42, 47, 49, 52, 25. Then calculate cumulative $F(X)$-values for the Kolmogorov–Smirnov $K$-statistic. Recall that the empirical cumulative distribution on the ordered sample values is $S_{20}(\#1) = 1/20$; $S_{20}(\#2) = 2/20, \ldots, S_{20}(\#20) = 1$. Continue to test the hypothesis that the age distribution for head of household is normal with mean $= 45$ and standard deviation $= 12$ years.

The following exercises repeat the earlier computer problems but request preprogrammed nonparametric test calculations. Use sample data from the indicated problem, then compare your results with your earlier findings. That is, compare your answers to Problems 49–55 with your results for Problems 56–61, respectively.

56. Repeat the runs test for randomness on monthly cost of electricity, Problem 49.

      For SPSS users (after data input),

```
SET BLANKS = -1
SAMPLE 50 FROM 20116
LIST VARIABLES = ELECTRIC/FORMAT = NUMBERED/
                      CASES FROM 1 TO 50
NPAR TESTS RUNS (MEDIAN) = ELECTRIC
```

      For Minitab users (after data input),

```
MTB > PRINT 'ELECTRIC'
MTB > MEDIAN 'ELECTRIC'
MTB > RUNS [27.5] 'ELECTRIC'
```

*Note:* The value "[27.5]" is the median for the data presented in Problem 49. A median value needs to be specified, because the Minitab default value is the mean.

      Because this is a large sample, $N = 50$, the calculated statistic is the $Z$-approximation. The SPSS output includes level of significance, that is, a $p$-value. Comparable Minitab output indicates *the test is significant at p-value.*

57. Repeat the Mann–Whitney *U*-test on earning power for males aged 35–44 versus ones 45–54 with your data from Problem 50 (or the data provided there).

    For SPSS users (assumes continuation of the routine started in Problem 50 or reselecting your sample),

```
SORT CASES BY AGE1
COMPUTE GROUP = 1
IF (AGE1 GT 44) GROUP = 2
NPAR TESTS M-W = INCOME BY GROUP (1,2)
```

    For Minitab users (after you have read in the data that appears in Problem 50),

```
MTB > PRINT '35-44' '45-54'
MTB > MANN-WHITNEY [95] '35-44' '45-54'
```

*Note:* The "[95]" is a user-specified value that denotes a 95% probability level. Then the $\alpha$-level is 0.05. Minitab uses 'ETA1' and 'ETA2' to represent the two group medians.

58. Repeat the Kruskal–Wallis *H*-statistic test of Problem 51, using that data and the SPSS or similar procedures.

    a. Test $H_0$: Mean monthly payments-to-lender are equal in these three MSAs.

```
NPAR TESTS K-W = PAYMENT BY MSA (1,3)
```

    Use, from SPSS,

| MSA   | 1920 | 1920 | $\cdots$ | 2080 | 2080 | $\cdots$ | 7320 | 7320 |
|-------|------|------|----------|------|------|----------|------|------|
| Group | 1    | 1    | $\cdots$ | 2    | 2    | $\cdots$ | 3    | 3    |

    For SPSS users (after data are input),

```
SELECT IF ((MSA EQ 1920) OR (MSA EQ 2080)
OR (MSA EQ 7320))
SAMPLE 150 FROM 587
SELECT IF (PAYMENT NE 0)
SORT CASES BY MSA
COMPUTE GROUP = 1
IF (MSA EQ 2080) GROUP = 2
IF (MSA EQ 7320) GROUP = 3
NPAR TESTS K-W PAYMENT BY
GROUP (1,3)/
```

    For Minitab users,

```
MTB > READ C1 C2
DATA> 934 1
DATA> 650 1
  ⋮
DATA>  77 1
```

```
                    DATA>  847  2
                      ⋮
                    DATA>  108  2
                    DATA>  784  3
                      ⋮
                    DATA>   98  3
                    DATA>  END
                    MTB > NAME C1 'PMT' C2 'MSA'
                    MTB > KRUSKAL-WALLIS 'PMT' 'MSA'
```

b. For SPSS and Minitab, use the same steps as for part (a), except read in the rent data and replace PMT with RNT throughout.

59. Perform the sign test of Problem 52 with data from there.
         For SPSS users (after DATA LIST),

```
SELECT IF (((AGE1 GE 20) AND (AGE1 LE 39))
AND (MAR EQ 0) AND (SPOSE EQ 1))
SAMPLE 50 FROM 4827
SELECT IF (AGE1 NE AGE2)
COMPUTE AGEM = AGE1
IF (SEX1 EQ 1) AGEM = AGE2
COMPUTE AGEF = AGE2
IF (SEX1 EQ 1) AGEF = AGE1
NPAR TESTS SIGN = AGEM WITH AGEF/
```

         For Minitab users,

*Line*

```
  1   MTB > READ C1 C2
  2   DATA> 36 32
  3   DATA> 37 37
        ⋮
 51   DATA> 35 36
 52   DATA> END
 53   MTB > NAME C1 'AGEM' C2 'AGEF'
 54   MTB > LET C3 = C1 - C2
 55   MTB > NAME C3 'ADIFF'
 56   MTB > STEST 0 'ADIFF';
 57   SUBC> ALTERNATIVE = 0.
```

60. Repeat Problem 53, a Spearman rank correlation test on husbands' and wives' ages, but now use SPSS (or comparable). For SPSS the proper statements are

```
SELECT IF (((AGE1 GE 20) AND (AGE1 LE 39))
AND (MAR EQ 0) AND (RELN2 EQ 1))
SAMPLE 50 FROM 4827
COMPUTE AGEM = AGE1
IF (SEX1 EQ 1) AGEM = AGE2
COMPUTE AGEF = AGE2
IF (SEX1 EQ 1) AGEF = AGE1
NONPAR CORR VARIABLES = AGEF AGEM/
        PRINT = TWOTAIL SIG SPEARMAN
```

The outputs can include $n$ (# pairs) and the significance ($p$-) level.

For Minitab, use lines 1 through 53 from Problem 59 (or simply continue on with that processing). Then use

```
MTB > RANK C1 C4
MTB > RANK C2 C5
MTB > NAME C4 'RANKM' C5 'RANKF'
MTB > CORRELATE 'RANKM' 'RANKF'
```

This gives the Spearman correlation value, $r_s$, then requires a hand-calculation for $Z_{calc}$ and the test.

61. Test age of head of household for a normal distribution, as in Problem 55, but now use the Kolmogorov–Smirnov procedure. From SPSS,

```
SAMPLE 20 FROM 20116
NPAR TESTS K-S (NORMAL, 45, 12) = AGE
```

*Note:* Minitab does not provide a Kolmogorov–Smirnov goodness-of-fit procedure.

## FOR FURTHER READING

[1] Albright, S. Christian. *Statistics for Business and Economics.* New York: Macmillan Publishing Company, 1987.

[2] Conover, W. J. *Practical Nonparametric Statistics*, 2nd ed. New York: Wiley, 1980.

[3] Gibbons, Jean Dickinson. Methods for Quantitative Analysis, 2nd ed. New York: Am Sciences Press, 1985.

[4] Ingram, John A. *Introductory Statistics.* Menlo Park, CA: Benjamin-Cummings Publishing Co., 1974.

[5] Mansfield, Edwin. *Statistics for Business and Economics*, 3rd ed. New York: W. W. Norton and Company, 1987.

[6] Marascuilo, Leonard A. and Maryellen McSweeney. *Nonparametric and Distribution-Free Methods for the Social Sciences.* Monterey, CA: Brooks/Cole Publishing Co., 1977.

[7] Meek, Gary E., Howard L. Taylor, Kenneth A. Dunning, and Keith A. Klajehn. *Business Statistics.* Boston: Allyn and Bacon, Inc., 1987.

[8] Pfaffenberger, Roger C. and James H. Patterson. *Statistical Methods for Business and Economics*, 3rd ed. Homewood, IL: Irwin, 1987.

[9] Siegel, Sidney. *Nonparametric Statistics.* New York: McGraw-Hill. 1965.

[10] *SPSS Reference Guide.* Chicago: SPSS, Inc., 1990.

# 18

## DECISION THEORY

Until the 1950s business statistics was largely confined to classical approaches. Classical methods emphasize an objective, scientific approach to decision situations. Also, as we have seen, they accommodate uncertainties by allowing the manager to specify a desired level of confidence or an allowable risk of error (type I or II).

These objective methods leave little opportunity for incorporating other factors, such as one's own preferences, subjective experience, or even sample information, into a decision process. For example, it is often impossible to neatly quantify environmental values (e.g., scenic beauty) or risks (e.g., an oil spill). But such values and risks must be recognized when business decisions are made (e.g., locating an oil pipeline).

Decision analysts naturally tend to analyze those factors that can easily be defined in a precise way. The more unclear elements of a decision may get pushed into the background. As a result, the "right answer" using a quantitative approach may seem intuitively "wrong" from a judgmental standpoint. For example, monetary values may assume more importance than human values. This sometimes causes us to lose confidence in (or perhaps even reject) formal statistical analysis. However, the problem lies not in the validity of classical methodology, but rather in its forced application to situations that are more complex than the methodology is designed to handle.

A number of new techniques developed during the last several years have enhanced our abilities to deal with realistic decision problems. These techniques are collectively referred to as *statistical decision theory*. Much of the development rests on an extension of Bayes' theorem, so the methods are also referred to as *Bayesian decision theory*.

**Distinctive features of statistical decision theory include**

1. its emphasis on decisions that must be made under conditions of risk or uncertainty;

2. its capacity for incorporating prior knowledge and economic costs or profits into the decision process;

3. its ability to incorporate individual or group preference in the decision models.

The objective of this chapter is to introduce decision theory methods. We begin with some background concepts and an introduction to decision theory criteria. Then we focus on the expected monetary value and the methodology for evaluating the worth of sample information. The final section of the chapter extends some of these same concepts to nonmonetary values through utility theory. In total, this chapter is an overview of the decision theory area and an introduction to an in-depth study of the subject. A detailed treatment of decision theory is beyond our scope here, so we encourage you to consult the end-of-chapter references for more discussion of these topics.

Upon completion of Chapter 18, you should be able to

1. explain the decision theory criteria for making decisions under uncertainty;

2. use Bayes' theorem to compute revised probabilities;

3. depict a decision situation in the form of a decision tree;

4. compute the expected value of perfect information and the expected net gain from sample information;

5. understand how utility theory concepts might be applied to decision problems.

## 18.2
## THE DECISION-MAKING ENVIRONMENT

A prime function of management is to make and implement decisions. Decision situations arise when the decision maker must choose from among two or more possible courses of action. The difficulty in making this choice typically stems from the fact that the decision maker does not usually know for sure what outcome will result. The decision must be made in an uncertain environment. Figure 18.1 depicts the environment of decisions as ranging from one of extreme uncertainty, where one has no information, to one of complete certainty, where all information is available. The environment is typically uncontrollable, but the better information we have about it, the better decisions we can usually make.

Usually, the more certain we are, the better choices we can make. For example, the more complete information we obtain from weather satellites, the better prepared we can be for storms, hurricanes, and the like. Short of complete information, Figure 18.1 suggests that large samples from objective information tend to give us more certainty than smaller samples or subjective information.

### Decision Making under Complete Certainty

The ends of the continuum of Figure 18.1 represent states of the environment where either all the information is known (complete certainty) or no information is available (extreme uncertainty).

A decision under complete certainty is one for which the outcome from each possible alternative can be known before an alternative is selected.

**FIGURE 18.1**  Uncertainty–Certainty Continuum for Decision Situation

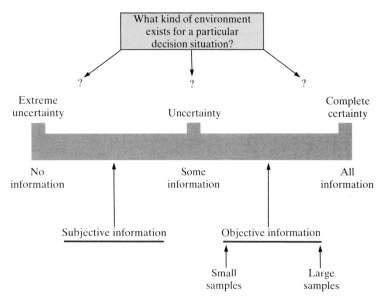

---

**EXAMPLE 18.1**

You have just accepted a job that requires you to have a new car to call on customers. Your company will allow you $16,000 but has specified that you must select a specific four-door model Chevrolet, Toyota, Ford or Honda. The company will also cover all operating and maintenance expenses. Which car should you select if you wish to minimize your initial costs?

*Solution*  Your decision should be quite straightforward. Once you obtain the initial purchase price data from the various dealers, you can select the least expensive model. As long as your criterion is simply cost, you can make the decision in complete certainty.

---

Decision situations are often converted into a decision under complete certainty by either the criteria or the underlying assumptions. In Example 18.1, the lone criterion of least initial cost simplified the decision. If you preferred a more sporty model or if the brand was allowed to influence your decision, the decision would not be so simple.

Decisions under complete certainty (or assumed certainty) utilize deterministic mathematical methods such as algebra, calculus, and linear programming. The values of variables used in these solution techniques either are known, or are assumed to be known, or can be definitively derived. Thus, in the algebraic expression describing profit as the excess of revenues over costs $(P = R - C)$, the analyst assumes that all revenue and cost information is known with absolute certainty.

**Decision Making under Extreme Uncertainty**

At the "no information" end of the spectrum, in Figure 18.1, the decision maker is in complete ignorance about what conditions will prevail.

> A decision under extreme uncertainty is one for which no information is available about the likelihood of a given outcome before an alternative is selected.

Decision theorists have researched the extreme-uncertainty end of the continuum in the context of decision making under conflict, where opponents or competitors are constantly trying to outwit or surprise each other. It is probably not surprising that these approaches are sometimes referred to as *game theory*.

> A game theory approach pits the decision maker against the environment as if it were an opponent and enables the decision maker to select a course of action that will result in the least cost in the long run.

The result of a game theory approach is typically a mixed strategy, in which the decision maker follows one course of action part of the time and another during the remainder of the time. The decision maker determines the appropriate proportion of time by developing an expression for expected profit under each of two states of nature. Then, by setting the profit equations equal and solving for the proportion of time, $p$, the decision maker finds a value that will yield an equal profit regardless of the state of nature. The following example illustrates this.

**EXAMPLE 18.2**

Harmon Mining Company has mineral rights to several sections of property in Arizona, which could contain either gold or oil. Company managers are extremely uncertain which type of mineral exists. Depending upon the exploratory method used, excavation or drilling, Harmon's profits from the land might be estimated as shown in the table. Suppose Harmon Mining intends to make several exploratory trips to the various sections but must choose one exploratory method each time prior to leaving. What percent of the time should they use each method?

|  | Profits ($000) if the Land Contains | |
| --- | --- | --- |
| Method | *Gold* | *Oil* |
| Excavation | $120 | −$80 |
| Drilling | −$100 | $200 |

*Solution*

Let $p$ = the percent of the time they take excavation equipment. Then $(1 − p)$ is the percent of the time they take the drilling equipment. Establish an equation that will provide the firm with the same amount of profit regardless of whether the land contains gold or oil. Thus,

$$\text{If gold, expected profit} = 120(p) − 100(1 − p)$$

$$\text{If oil, expected profit} = −80(p) + 200(1 − p)$$

Setting these profits equal,

$$120(p) − 100(1 − p) = −80(p) + 200(1 − p)$$

$$120p − 100 − 100p = −80p + 200 − 200p$$

$$p = 0.60$$

Harmon Mining Co. should take the excavation equipment 60% of the time and the drilling equipment 100% − 60% = 40% of the time. A random process with outcome $p$ (take excavation equipment) = 0.60 could be used to apportion the times.

The approach in Example 18.2 will yield equal expected profits under both land conditions:

If gold,    $E$(Profit) = \$120(0.6) − \$100(1 − 0.6) = \$32 (thousand)

If oil,      $E$(Profit) = −\$80(0.6) + \$200(1 − 0.6) = \$32 (thousand)

Note that in employing a game theory strategy, we have simply found the percentage of time a given choice of action should be selected in order for the expected outcomes under both environmental conditions to be equal. This technique is theoretically appropriate for situations in which the decision maker has *absolutely no prior knowledge of probabilities* associated with the states of the environment. The objective is to protect ourselves by equalizing the expected outcome regardless of which state occurs. This is an interesting (theoretical) approach, but the use of game theory is very limited. So we will limit our study of it to these basic concepts.

**Decision Making under Uncertainty (DMUU)**

Most of the difficult business decisions fall into the category of *decision making under uncertainty*, which lies between the endpoints in Figure 18.1. These are situations where some, but not all, of the desirable information is available. Decision makers must then rely upon good judgment supported by statistical inference.

> In decision making under uncertainty (DMUU), some information is available about the probability of various outcomes before an alternative is selected.

As in any choice situation, the decision maker must select a course of action that best satisfies some goal, or *criterion*, such as market share, reliability, or profits. The consequence, or *payoff*, depends upon which action is selected and the environmental condition, or event, that occurs following the choice. The environment is sometimes referred to as the *state of nature*. For example, the states of nature for an investment decision might be classified as (1) recession, (2) moderate economic growth, and (3) strong economic expansion.

*Uncertainty* refers to the decision maker's ignorance about which state of nature will prevail, for only one will occur and the decision maker is in no position to control that. For example, a firm developing a new textile fabric may have to make an equipment investment in one of three possible levels of automation (the choice of action). These choices will yield different amounts of profit (the payoff), depending upon whether customer demand for the new fabric is poor, average, or very strong (the states of nature). Whereas management cannot be certain what the level of consumer demand will be before marketing the product, they may be able to assign some chance, or probability, to each of the states on the basis of past experience, market research, or other knowledge of competitive products. These can be objective probabilities (if available) or subjective probabilities. The respective probabilities must, of course, cover all possibilities, so that the collective chance of all states totals 100%.

The elements common to a decision situation thus include

1. Actions (acts)
2. States of nature (events)
   These are the outcomes or states of the environment. They are beyond the control of the decision maker, and he or she is uncertain which one will occur.
3. Payoff table (payoffs)
   This is a matrix showing the consequences or net benefit of each act given each state of nature. Payoffs are often expressed as a monetary criteria, but other criteria (utility values) can be used.
4. Measure of uncertainty (probabilities)
   This is the numerical chance that a given state of nature will prevail.

Uncertainties are usually expressed by assigning (objective or subjective) probabilities to the various states of nature.

## 18.3
## PAYOFF TABLES
## AND DECISION
## TREES

Decision theorists have developed different ways to summarize and express the elements of a decision problem. Two of the most widely used techniques are payoff tables and decision trees. We will illustrate both forms here briefly and then use these aids as we study the underlying theory in the remainder of the chapter.

### Payoff and Opportunity Loss Tables

Payoff tables display actions, states of nature, outcomes, or payoffs that would result from choosing each action given each possible state of nature. Table 18.1 illustrates a payoff table where the values in the cells represent five-year present-value profits (losses) in thousands of dollars.

Note that the payoff values depend upon the action (level of automation) variable and the state-of-nature (demand) variable. Say, for example, that with full automation and strong demand, the estimated profit is $240 (thousand). If the firm purchases and installs the automation equipment but only poor demand materializes, the result would be a $90 (thousand) loss. The decimal values at the head of each column represent probability estimates that the given state of nature will occur. For example, the likelihood of a poor demand is estimated to be 0.4. Using probability values like this enables the decision maker to incorporate uncertainty into the situation.

**TABLE 18.1**   Payoff Table of Profits ($000)

| Probabilities<br>Action<br>(level of automation) | State of Nature (level of demand) | | |
| --- | --- | --- | --- |
| | $\theta_1 = Poor$<br>*(0.4)* | $\theta_2 = Average$<br>*(0.5)* | $\theta_3 = Strong$<br>*(0.1)* |
| $A_1$ = No automation | 0 | 10 | 40 |
| $A_2$ = Some automation | −20 | 60 | 100 |
| $A_3$ = Full automation | −90 | 80 | 240 |

**Payoffs**

**TABLE 18.2** Opportunity Loss Table ($000)

| | State of Nature (level of demand) | | |
|---|---|---|---|
| Action | $\theta_1 = Poor$ (0.4) | $\theta_2 = Average$ (0.5) | $\theta_3 = Strong$ (0.1) |
| $A_1$ = No automation | 0 | 70 | 200 |
| $A_2$ = Some automation | 20 | 20 | 140 |
| $A_3$ = Full automation | 90 | 0 | 0 |

An alternative method of expressing outcomes is in terms of *opportunity losses*.

> An opportunity loss represents the lost benefit or profit loss that would result if some act other than the best act is chosen for a given state of nature.

An opportunity loss table can easily be calculated from a payoff table. In it, the outcomes depict potential losses the decision maker can incur by failing to select the best action under each state of nature. Table 18.2 shows the opportunity loss table that would result from converting the payoffs in Table 18.1 into opportunity losses. First the best (largest) payoff under each state of nature is selected. Then each payoff under that state (column) is subtracted from the largest payoff. The results are opportunity loss values.

For example, consider the payoffs under average demand in Table 18.1. The $80 payoff resulting from a choice of $A_3$ is the largest. Subtracting the $10 ($A_1$) value from this we have $80 − $10 = $70. For $A_2$ we have $80 − $60 = $20 and for $A_3$, $80 − $80 = $0. These values of 70, 20, and 0 are opportunity loss values and are shown under average demand in Table 18.2. Values for poor and strong demand are calculated from their own columns in a similar way. Note that each state of nature will have at least one zero entry.

The intuitive meaning of an opportunity loss value is that a given choice commits the decision maker to a course of action and that action rules out some possibility of gain. For example, from Table 18.2 we see that choosing no automation ($A_1$) means the decision maker forgoes a possible $200 more in profit if the state of nature happens to be a strong demand. That is, the profit will be only $40 with no automation, whereas it would have been $240 had the choice been $A_3$.

Whenever one course of action is better than another course under every state of nature, the lesser alternative is said to be *dominated by the greater one*. Dominated actions will never be selected as best acts and can effectively be removed from consideration if that simplifies the analysis of the problem.

**Decision Trees**

The decision situation of Table 18.1 is depicted in schematic form in the decision tree diagram of Figure 18.2.

> A decision tree is a schematic representation of a multilevel or sequential decision problem in which the decision choices are represented by squares (□) and the chance, or uncontrollable events, by circles (○).

**FIGURE 18.2**    Decision Tree with Payoffs

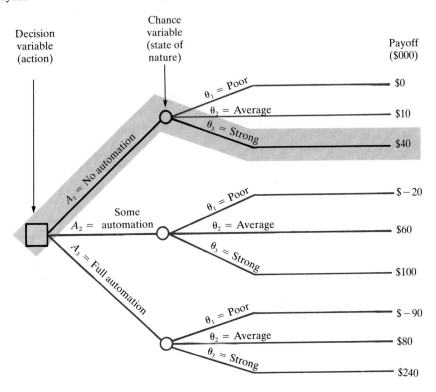

We construct decision trees from left to right by sequentially and systematically outlining each course of action, followed by each of the possible states of nature. Thus if the decision maker chooses $A_1$ (no automation), that choice will be followed by either a poor ($\theta_1$), average ($\theta_2$), or strong ($\theta_3$) demand. The payoff from each demand is shown at the far right of each branch in Figure 18.2. For example, if the decision is to not automate ($A_1$) and demand is strong, the payoff would be $40 (thousand). This path is highlighted in Figure 18.2.

Decision trees help us to structure and more readily visualize a decision situation in an objective way. They force the decision maker to explicitly identify the alternatives and distinguish between the controllable ($\square$) and uncontrollable ($\bigcirc$) variables. Not all variables are easily quantified, however, so decision trees also have limitations. Nevertheless, they are a convenient tool of analysis and, as we shall see later, make it easier to incorporate probabilities into decision situations.

## 18.4 DECISION THEORY CRITERIA

A number of simplified decision criteria are available to the decision maker. Which is best depends upon one's attitude toward risk. Assuming that the decision maker can estimate the payoffs that will occur under the various states and (at least subjectively) assign some chance to each state of nature, he or she might choose an action on the basis of one of the following criteria.

Decision criteria for an action includes the following options:

1. maximax—maximize the maximum possible gain;
2. maximin—maximize the minimum possible gain (or minimize the maximum possible loss—that is, minimax);
3. maximum likelihood—maximize the gain assuming the most likely state of nature prevails;
4. minimax regret—minimize the opportunity loss (or minimize the regret by selecting the act with the least loss).

A decision maker who is willing to gamble on the chance of a big gain would tend to choose the *maximax* criterion, whereas the conservative decision maker would likely prefer the security of a *maximin* criterion. One who "plays the odds" would probably go for the *maximum likelihood* criterion. The *minimax regret* criterion would appeal to one who wished to minimize the maximum possible opportunity loss that could occur. It is not as optimistic as the maximax criterion nor as pessimistic as the maximin criterion. An example will illustrate how these criteria can influence a decision maker's choice.

---

**EXAMPLE 18.3**

For the data in Table 18.1 (page 854), find the best choice of action based upon the decision criteria of (a) maximax, (b) maximin, and (c) maximum likelihood. (d) Finally, use Table 18.2 (page 855) to choose the best action based upon a criterion of minimax regret.

*Solution*

a. *Maximax:* From Table 18.1, choose full automation ($A_3$) in the (optimistic) hope of realizing a strong demand and a maximum payoff of $240 thousand.
b. *Maximin:* Choose no automation ($A_1$) in the (pessimistic) attempt to limit the possible monetary loss to $0. Note that this also minimizes the maximum possible loss (minimax) in this matrix where loss values are involved.
c. *Maximum likelihood:* Consider only the state of average demand (that is, the most likely state, with probability 0.5) and undertake full automation ($A_3$) in hopes of maximizing the payoff of $80 thousand.
d. *Minimax regret:* From Table 18.2, the worst opportunity loss under $A_3$ is only $90, whereas it can run as high as $140 with $A_2$ or $200 with $A_1$. Therefore, choose $A_3$.

---

**18.5
EXPECTED
MONETARY VALUE**

A traditional approach to decisions of the type we have been discussing is, of course, to use a probability weighted average, which we refer to as the expected value. We determine the *expected value* of each course of action, $E(A)$, simply by multiplying the payoff value of each possible outcome ($\theta_{ij}$) by its probability of occurence, $P(\theta_j)$, and summing these products across all possible states, $E(A) = \sum [\theta_{ij} \cdot P(\theta_j)]$. When the payoffs are in monetary terms, the resulting expected values are called *expected monetary values* (EMVs). This approach to decisions essentially says that we obtain the long-run estimate of the outcome from a course of action by weighting the possible outcomes by the chances they will occur. The best course of action is the one with the highest EMV; it is designated by EMV*.

---

* *Asterisked value* is the *best course of action.*

The highest expected monetary value (EMV*) of an uncertainty situation is the expected value of the act having the highest expected value. Under repeated trials (a long-run situation), it is the best course of action from an economic standpoint.

**EXAMPLE 18.4**

Find the EMV* for the data displayed in Table 18.1 (repeated from p. 857).

**TABLE 18.1**    Payoff Table of Profits ($000)

| Probabilities Action (level of automation) | State of Nature (level of demand) | | |
|---|---|---|---|
| | $\theta_1 = Poor$ *(0.4)* | $\theta_2 = Average$ *(0.5)* | $\theta_3 = Strong$ *(0.1)* |
| $A_1$ = No automation | 0 | 10 | 40 |
| $A_2$ = Some automation | $-20$ | 60 | 100 |
| $A_3$ = Full automation | $-90$ | 80 | 240 |

Payoffs

*Solution*    Utilizing the expression $E(A) = \sum [\theta_{ij} \cdot P(\theta_j)]$ gives

$$E(A_1) = 0(0.4) + 10(0.5) + 40(0.1) = 9 \ (\$000)$$

$$E(A_2) = -20(0.4) + 60(0.5) + 100(0.1) = 32 \ (\$000) \leftarrow EMV^*$$

$$E(A_3) = -90(0.4) + 80(0.5) + 240(0.1) = 28 \ (\$000)$$

The optimal action according to an EMV* criterion is therefore to choose some automation ($A_2$) for an expected value of $32 (thousand). Note that the $32 (thousand) *is a long-run monetary expectation* only. Moreover, this exact profit is unlikely to occur. In fact, a return to the payoff table (Table 18.1) reveals that all the payoffs (for $A_2$) are more than $25 (thousand) away from this figure. It is simply a weighted average of the payoffs, where the weights are probabilities of occurrence.

We could have obtained the same result ($A_2$ at $32,000) from the opportunity loss table (Table 18.2) using the same probabilities. However, in keeping with the concept of opportunity loss, we would seek the *minimum expected opportunity loss* (EOL*) rather than the *maximum* expected value gain.

The minimum expected opportunity loss (EOL*) of an uncertainty situation is the expected value of the act, in the opportunity loss table, with the lowest expected value. It is the best course of action from an economic standpoint.

**EXAMPLE 18.5**

Use the same state-of-nature probabilities as in Table 18.1 along with the opportunity loss values of Table 18.2 to find the optimal expected opportunity loss, EOL*.

**TABLE 18.2**   Opportunity Loss Table ($000)

| | State of Nature (level of demand) | | |
| --- | --- | --- | --- |
| Action | $\theta_1 = Poor$ (0.4) | $\theta_2 = Average$ (0.5) | $\theta_3 = Strong$ (0.1) |
| $A_1$ = No automation | 0 | 70 | 200 |
| $A_2$ = Some automation | 20 | 20 | 140 |
| $A_3$ = Full automation | 90 | 0 | 0 |

*Solution*   The expected opportunity loss values are

$$\text{for } A_1 \quad 0(0.4) + 70(0.5) + 200(0.1) = 55 \ (\$000)$$

$$\text{for } A_2 \quad 20(0.4) + 20(0.5) + 140(0.1) = 32 \ (\$000) \leftarrow \text{EOL}^*$$

$$\text{for } A_3 \quad 90(0.4) + 0(0.5) + 0(0.1) = 36 \ (\$000)$$

Thus both the EMV* and EOL* decision criteria result in the same decision—choose some automation, $A_2$.

Note that the value of the EMV* is the same as the value of the EOL*. This can be generalized. That is, maximizing the expected payoff is the same as minimizing the expected opportunity loss. Let us now consider this same problem in the framework of a decision tree.

**EXAMPLE 18.6**

Use a decision tree to display the data from Table 18.1, and identify the tree branch that represents the optimal course of action, by EMV*.

*Solution*   Figure 18.3 depicts the decision problem in the form of a decision tree. Note that probabilities for the various states of nature are shown above the lines. The monetary expected

**FIGURE 18.3**   Decision Tree with Expected Values ($000)

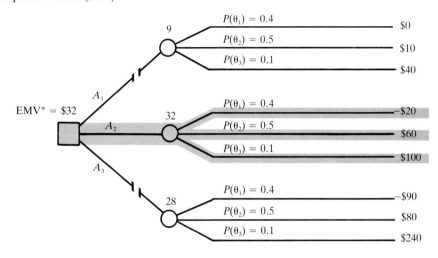

value of each course of action is $\sum [\theta \cdot P(\theta)]$, and the numerical values resulting from this calculation are shown above the chance fork (or node) for each given course of action. The EMV of the optimal course of action (EMV* = $32,000) is then shown above the square representing the decision fork on the left. Slash marks identify branches that are nonoptimal.

Decision trees are constructed from left to right. However, they are analyzed backward, i.e., from right to left. Payoff values are multiplied by the probability of their respective states of nature, summed, and the expected value of that course of action is stated above the node from whence the branches emerge. If the automation decision were itself a branch of an even larger tree, we would continue the process of analysis by computing expected values by moving backward (to the left).

For example, suppose the automation decision were dependent upon the likelihood of the company being awarded a patent on a new production process, as shown in Figure 18.4. The $32 (thousand) expected value would then be treated like a payoff value, and the expected value at the prior node would be computed as $32 \cdot (0.6) + \$50 \cdot (0.4) = \$39.2$ (thousand). This ability to depict sequential (and complex) decision situations schematically is one of the major advantages of a decision tree approach to decision analysis.

**FIGURE 18.4**   Decision Tree Extension

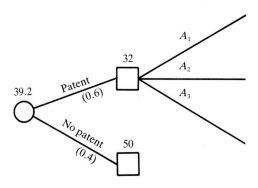

We move now to an additional concept that permits us to place an upper limit on the value of information we might obtain in order to improve our choice. This quantity is referred to as the *expected value of perfect information*.

**18.6**
**EXPECTED VALUE OF PERFECT INFORMATION**

To compute the expected value of perfect information, we must first compute the *expected profit under certainty* (EPC) and the maximum expected monetary value (EMV*).

> The expected profit under certainty (EPC) is the profit that would result if no uncertainty existed with respect to the state of nature (e.g., the demand).

In order to calculate the EPC, we presume that each state of nature exists in proportion to its probability of occurrence. Then the best act under a given state can always be chosen. Calculation of an EPC is a part of the next example.

> The expected value of perfect information (EVPI) is the maximum incremental amount that can rationally be spent to remove uncertainty concerning which state of nature will occur.

For example, uncertainty might be removed by research studies to establish the economic or technological feasibility of a project. The EVPI is the excess of value over the maximum expected monetary value (EMV*) up to what would be the expected profit under certainty (EPC).

> The procedure for computing the EVPI is as follows:
>
> 1. Establish the payoff matrix with actions, states of nature, and payoffs. Assign probabilities to each state.
> 2. Compute the EMV by summing the payoff values times their probability of occurence over each act and identifying the maximum EMV = EMV*.
> 3. Compute the EPC by weighting the highest payoff under each state by the probability that state will occur and summing over the various states.
> 4. Compute the expected value of perfect information, EVPI.

$$EVPI = EPC - EMV^* \qquad \text{(18.1)}$$

The resulting EVPI is the maximum value that perfect information would be worth and, as such, is a measure of the cost of uncertainty. It may represent the maximum expenditure for a market survey or a feasibility study to verify the desirability of a project.

---

**EXAMPLE 18.7**

Find the EVPI for the data displayed in Table 18.1.

*Solution*   Our first step is to compute the EPC, which is the best course of action under each state of nature weighted by the probability of that state. This value, as shown in the table, is EPC = 64.

| (1)<br>If the state<br>is known<br>to be | (2)<br>Then the best<br>action and<br>profit are | (3)<br>And the %<br>of time this<br>occurs is $P(\theta)$ | (4)<br>Expected<br>profit<br>$(2) \cdot (3)$ |
|---|---|---|---|
| $\theta_1$ | $A_1$: \$   0 | 0.4 | 0 |
| $\theta_2$ | $A_2$:    80 | 0.5 | 40 |
| $\theta_3$ | $A_3$:   240 | 0.1 | 24 |
| | | | EPC = 64 |

Subtracting the best act under uncertainty (the EMV* from Example 18.4) from this EPC gives the following result.

| | |
|---|---|
| Expected profit under certainty (EPC) | $64 (000) |
| Expected value of optimal act (EMV*) | − 32 (000) |
| Expected value of perfect information (EVPI) | $32 (000) |

This means that it would be worth up to $32 (thousand) to know whether demand will be poor, average, or strong prior to making the decision about how fully the firm should automate its textile equipment. Such information would (theoretically) enable us to choose the best course of action for each state of nature.

---

The EVPI establishes a maximum worth of information, but perfect information is rarely, if ever, available. The next section will deal with the problem of estimating the value of a limited sample of information. We seek an improvement over what would result from an EMV* choice, but we certainly don't want to incur costs that are greater than the EVPI.

**Problems 18.6**

1. What is the proper term for
   a. an approach to decision making under extreme uncertainty in which the various courses of action are selected on a proportional basis so that the expected outcomes under all states of the environment are equal?
   b. the difference between the expected value under certainty and the expected value of the optimal act?
   c. an uncontrolled variable that represents a part of the background environment within which the decision maker must work?
   d. the condition referring to the decision maker's ignorance about which state of nature will prevail?
   e. the monetary result representing the intersection of an act and event combination?

2. Briefly describe the range of states of nature under which management decisions are made, and indicate the relative location of subjective and objective probabilities on such a continuum.

3. A financial manager must invest $60,000 of pension funds weekly but is extremely uncertain about whether the stock market will go up or not. She must put the money into one of two portfolios and has developed an estimate of (present value) profit from one such investment (see table). Based on a game theory strategy, what percent of the funds should be directed into (a) stocks? (b) bonds?

| Potential Investment | State of the Market | |
|---|---|---|
| | *Up* | *Not up* |
| Portfolio A (primarily stocks) | $100,000 | −$40,000 |
| Portfolio B (primarily bonds) | $40,000 | $60,000 |

4. Given the payoff table shown, where values are monetary amounts and the probabilities for the states of nature $\theta_1$ through $\theta_4$, respectively, are (0.2), (0.5), (0.1), and (0.2), find the best action under each criterion.
   a. Maximax
   b. Maximin
   c. Maximum likelihood
   d. Maximum expected monetary value

|       | $\theta_1$ | $\theta_2$ | $\theta_3$ | $\theta_4$ |
|-------|------------|------------|------------|------------|
| $A_1$ | $-20$      | $-10$      | 5          | 100        |
| $A_2$ | $-10$      | 40         | 50         | 80         |
| $A_3$ | 40         | 30         | 50         | 70         |
| $A_4$ | 20         | 10         | 40         | 20         |

5. For the decision tree shown (see figure), compute the expected values for $A$, $B$, and $C$, and identify the best course of action.

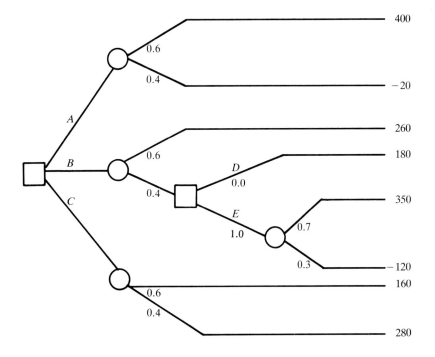

6. Computer Laptop Co. is planning to construct a plant in Chicago to manufacture computers. One alternative is to build a large plant costing $11 million and a second is to build a small plant costing $5 million. The present value of income from the facilities (before deducting plant construction costs) is estimated (in millions of dollars) as shown in the table on the next page. If the firm finds that demand is strong, after having already committed to a small plant, it still has the option of expanding to capitalize on the strong market at an additional cost of $8 million.

| Demand | Income | |
|---|---|---|
| | Small Plant | Large Plant |
| Average (probability = 0.4) | $8 | $12 |
| Strong (probability = 0.6) | $11 | $18 |

   a. Use a decision tree approach to determine the optimal decision based upon a maximum expected monetary value criterion.
   b. State your recommended course of action.
   c. What is the expected payoff?

7. For the payoff table shown,
   a. find the best strategy using game theory.
   b. find the best act according to the (i) maximax and (ii) maximin criteria.

| Act | State of Nature | |
|---|---|---|
| | $E_1$ | $E_2$ |
| A | 50 | −150 |
| B | −100 | 200 |

If $P(E_1) = 0.80$, find the best act using a criterion of
c. maximum probability.
d. maximum expected value.
e. Illustrate your answer to part (d) by means of a decision tree.

8. Find the EVPI for the decision situation described in Problem 7. Note that $P(E_1) = 0.80$ and $P(E_2) = 0.20$.

9. Find the EVPI for the decision situation described in Problem 4.

10. As a part of a national advertising campaign, Southern States Oil Co. (SSOCO) plans to offer customers a "Mystery Package" at a nominal charge. It will contain either a lubrication (L) or gasoline (G) coupon. The coupons are supplied randomly in such a way that $P(L) = 0.3$ and $P(G) = 0.7$. The payoff varies depending on whether the customer uses a credit card from another company (A) or one from SSOCO (B) to claim the package. The value to the customer is shown in the payoff table.

| Act | State of Nature (coupon) | |
|---|---|---|
| | Lube | Gas |
| A = use other credit card | $1.20 | −$.80 |
| B = use SSOCO credit card | −$1.00 | 2.00 |

a. Assuming all customers employed a game theory strategy, what percent of the time would they use a SSOCO credit card?
b. What is the expected value for times when customers use
 (i) credit cards from another company?
 (ii) SSOCO credit cards?
c. Assuming customers could buy perfect information about the state of nature (from the dealer), what would be the expected value of such information? (*Hint:* Compute EVPI = EPC − EMV*.)

<table>
<tr><td>

**18.7
REVISION OF
PROBABILITIES
USING BAYES'
THEOREM**

</td></tr>
</table>

Thus far, our decision-making activities have been concerned with making the best use of the given or prior information available. In the expected-value framework, the probabilities we have associated with the various states of nature are referred to as *prior probabilities*. They are prior in the sense that they facilitate decision making if no more information is available or before additional data are collected.

We turn now to the methodology of incorporating additional information into the structure for decisions. Our vehicle for doing this is *Bayes' theorem*, to which we were introduced in Chapter 4, Equation (4.6).

Bayes' theorem is a rule for finding conditional probabilities of a state of nature or of the environment, $\theta$, given some empirical data $Z$:

$$P(\theta|Z) = \frac{P(\theta \cap Z)}{P(Z)} \qquad \text{(4.6), p. 141}$$

In terms of our current discussion, we could interpret Bayes' theorem as

$P(\text{State of nature}|\text{Empirical or sample data})$

$$= \frac{P(\text{Joint occurrence of state of nature } and \text{ sample result})}{P(\text{Sample result})}$$

Bayes' theorem was expanded in Chapter 4 to illustrate the makeup of the probabilities in the numerator and denominator, respectively. As suggested in Equation (4.7), for a two-state-of-nature situation, the probabilities of the states of nature $\theta_1$ and $\theta_2$ can be written as follows.

**BAYES' THEOREM FOR TWO STATES OF NATURE**

$$P(\theta_1|Z) = \frac{P(\theta_1)P(Z|\theta_1)}{P(\theta_1)P(Z|\theta_1) + P(\theta_2)P(Z|\theta_2)} \qquad \text{(18.2)}$$

$$P(\theta_2|Z) = \frac{P(\theta_2)P(Z|\theta_2)}{P(\theta_1)P(Z|\theta_1) + P(\theta_2)P(Z|\theta_2)} \qquad \text{(18.3)}$$

Although we did not pursue the theory of revision of probabilities in Chapter 4, the groundwork has been fully laid. The probabilities developed using Bayes' theorem are referred to as *revised*, or *posterior*, probabilities. The typical situation is that the prior probability, for example, $P(\theta_1)$, is estimated from experience or subjective judgment. Bayes' theorem then allows the decision maker to develop a revised probability, $P(\theta_1|Z)$, after taking into account additional information, $Z$. In many cases the additional information represents sample evidence, so it is objective, but it could also be derived from subjective information. $P(\theta|Z)$ represents the chance (conditional probability) of obtaining the sample result under each state of nature. After we illustrate the tabular methodology for revising probabilities in this section, we shall consider whether the value of the sample information justifies its cost.

As suggested by Bayes' theorem, the procedure for revising a probability of a state of nature consists of combining the prior $P(\theta)$ and conditional $P(Z|\theta)$ probabilities into a joint probability $P(Z \cap \theta)$ and dividing by the empirical probability of obtaining the sample result, $P(Z)$. We illustrate this for the decision problem involving automation of a textile plant.

## EXAMPLE 18.8

Given a payoff table with prior probabilities $\theta_1 = 0.4$, $\theta_2 = 0.5$, and $\theta_3 = 0.1$,

|  | State of Nature | | |
| --- | --- | --- | --- |
| Act | $\theta_1 = Poor$ <br> *(0.4)* | $\theta_2 = Average$ <br> *(0.5)* | $\theta_3 = Strong$ <br> *(0.1)* |
| $A_1$ = No automation | 0 | 10 | 40 |
| $A_2$ = Some automation | −20 | 60 | 100 |
| $A_3$ = Full automation | −90 | 80 | 240 |

suppose a survey conducted by a market research firm suggests that the market will be strong. Furthermore, on the basis of previous experience, the market research firm adds a statement concerning the reliability of their surveys:

In the past, when actual demand was strong, our surveys have correctly indicated a strong demand 80% of the time. However, when demand turned out to be average, they have incorrectly indicated a strong demand 30% of the time, and when demand was poor, our surveys have incorrectly indicated a strong demand 10% of the time.

Revise the probabilities for $\theta_1$, $\theta_2$, and $\theta_3$ on the basis of the sample evidence and reliability information.

**Solution**    The reliability information constitutes conditional evidence that says, in effect,

$$P(\text{Sample indicates strong demand}|\text{Demand is strong}) = 0.80$$

$$P(\text{Sample indicates strong demand}|\text{Demand is average}) = 0.30$$

$$P(\text{Sample indicates strong demand}|\text{Demand is poor}) = 0.10$$

If we let $Z$ = the sample evidence (of a strong demand), we have

$$P(Z|\theta_3) = 0.8$$

$$P(Z|\theta_2) = 0.3$$

$$P(Z|\theta_1) = 0.1$$

The use of this conditional information to revise prior probabilities is illustrated in Table 18.3.

**TABLE 18.3**   Computation for Revision of Prior Probabilities of Demand Based on Sample Evidence

| (1)<br>State of<br>Nature | (2)<br>Prior<br>Probability | (3)<br>Conditional<br>Probability | (4) = (2) × (3)<br>Joint<br>Probability | (5) = (4)/$\sum$(4)<br>Posterior<br>Probability |
|---|---|---|---|---|
| $\theta_1$, Poor demand | 0.4 | 0.1 | 0.04 | 0.148 |
| $\theta_2$, Average demand | 0.5 | 0.3 | 0.15 | 0.556 |
| $\theta_3$, Strong demand | 0.1 | 0.8 | 0.08 | 0.296 |
| $\sum$ | | | 0.27 | 1.000 |

Note that the prior probability of a strong demand was only $P(\theta_3) = 0.10$, whereas the sample evidence suggested the demand would be strong. Furthermore, the research firm had experienced an 80% accuracy on such occasions. As a result, the revised probability of a strong demand was raised to 0.296. Correspondingly, the probability associated with a poor demand was lowered from 0.40 to 0.148.

Using these revised (posterior) probabilities, we can compute a revised expected value in the same manner as was done in Example 18.4. We shall designate the maximum of the revised expected monetary values as the $\text{EMV}_R$ value.

**EXAMPLE 18.9**

Using the data from Table 18.1 (page 858) and the posterior probabilities of Example 18.8, compute the revised expected profits, designate the largest one as $\text{EMV}_R$, and indicate the best course of action now, after incorporating the sample information.

*Solution*   The revised (posterior) probability values (of 0.148, 0.556, and 0.296) from Table 18.3 are now assigned to the states of nature as shown in the table.

| Act | State of Nature (demand) | | |
|---|---|---|---|
| | $\theta_1 = Poor$<br>*(0.148)* | $\theta_2 = Average$<br>*(0.556)* | $\theta_3 = Strong$<br>*(0.296)* |
| $A_1$ = No automation | 0 | 10 | 40 |
| $A_2$ = Some automation | −20 | 60 | 100 |
| $A_3$ = Full automation | −90 | 80 | 240 |

$E(A_1) = 0(0.148) + 10(0.556) + 40(0.296) = 17.4\ (\$000)$

$E(A_2) = -20(0.148) + 60(0.556) + 100(0.296) = 60.0\ (\$000)$

$E(A_3) = -90(0.148) + 80(0.556) + 240(0.296) = 102.2\ (\$000) \leftarrow \text{EMV}_R$

The best course of action from a revised expected monetary value standpoint is to proceed with full automation, for an expected profit of $102.2 (thousand).

In this example, the sample information has influenced a change in the decision from "some automation" (as determined in Example 18.4) to "full automation," a course of action that now happens to agree with the maximax and maximum-likelihood criteria. This shift arises from the market research, which indicated a stronger demand than was originally anticipated. Note, however, that the revised probability of a strong demand is still only about 0.3 (i.e., 0.296), which is less than one chance in three. And a substantial component of the $102.2 thousand (i.e., $71,000) arises from the $240 payoff *if* demand is strong. A prudent decision maker should never automatically accept the $EMV_R$ criteria, just as he or she should not blindly make a decision based upon prior expected profits, EMV.

Additional sample information could conceivably raise the revised $P(\theta_3)$ even more. On the other hand, it might lower $P(\theta_3)$. Thus the process of incorporating more information can be a continuing one depending upon how much the sample information costs versus how much it is worth. We proceed with that analysis in the next section.

**Problems 18.7**

11. Data were collected by the market research department for an electronics firm that has developed a new telephone-controlled home computer center. Complete the calculation of the revised probabilities $P(\theta|Z)$.

| State of Nature | $P(\theta)$ | $P(Z|\theta)$ | $[P(\theta) \cdot P(Z|\theta)]$ | $P(\theta|Z)$ |
|---|---|---|---|---|
| $\theta_1$ = Captive market | 0.8 | 0.4 | | |
| $\theta_2$ = Competition | 0.2 | 0.8 | | |

12. Superior Building Supply purchases ungraded lumber from a mill at a good discount but is never certain what it will get. The prior probabilities of obtaining the various grades are estimated to be

Economy grade   $P(E_1) = 0.4$     Good grade   $P(E_3) = 0.2$

Shop grade   $P(E_2) = 0.3$   Superior grade   $P(E_4) = 0.1$

If the conditional probabilities associated with a sample observation $Z$ are $P(Z|E_1) = 0.6$, $P(Z|E_2) = 0.3$, $P(Z|E_3) = 0.3$, and $P(Z|E_4) = 0.1$, find the revised probabilities of the various grades.

13. Let $U$ = Electone Co. stock goes up in price, $S$ = it stays the same, and $D$ = it goes down. Also, let $F$ = a favorable recommendation on the stock from a broker. An investor feels there is a good chance the stock will stay the same ($P(S) = 0.6$) and rates the chances of it going up or down as being equal. His broker issues a favorable recommedation. From past history in dealing with this broker, the investor knows the following:

$P(\text{Favorable recommendation}|\text{Stock actually goes up}) = 0.5$

$P(\text{Favorable recommendation}|\text{Stock stays the same}) = 0.7$

$P(\text{Favorable recommendation}|\text{Stock actually goes down}) = 0.8$

Find the revised probability of the stock going up in price given the broker's recommendation and the reliability information.

14. Oxygen valves received by an aircraft company are either satisfactory, *S*, or defective, *D*. (*Note:* These are states of the environment; $P(S) = 0.95$ and $P(D) = 0.05$). For a recent shipment of valves, the receipt inspection department indicated they were satisfactory. In the past, checks on the reliability of the inspection process have revealed that when the valves were actually satisfactory, inspection had correctly indicated they were satisfactory 80% of the time. However, when the valves turned out to be defective, the inspection department had passed them as satisfactory 12% of the time. Revise the probabilities for satisfactory and defective on the basis of this additional evidence from the inspection department.

# 18.8
## VALUE OF SAMPLE INFORMATION

To this point, we have considered the following approaches to a decision process under uncertainty.

1. Follow a strategy such as game theory, maximax, maximin, maximum likelihood, or minimum regret.
2. Compute the expected value of each act and choose that act with the highest EMV (EMV*) or the lowest EOL (EOL*).
3. Follow the computation of EMV* with a calculation of the EVPI. If you do purchase additional information, never pay more than the expected value of perfect information for it.
4. Use sample and/or reliability information to revise prior probabilities and then choose that act with the highest $EMV_R$.

Whereas we have used sample and reliability information to revise prior probabilities, we have not yet fully covered the question of determining whether the cost of sample information is justified. We know it is not worth more than the EVPI, but how much can we afford to spend on additional information? No statistical technique can answer this question with certainty, but we can develop an expected-value answer if we are willing (1) to anticipate the possible sample outcomes and (2) to base our action on the premise of making the best use of the sample information that does result.

The procedure for computing the gain from sample information is

Step A: Determine the EMV* prior to (without) sampling.

Step B: Compute revised (posterior) probabilities that could result from the various sample outcomes.

Step C: Select the best course of action given each sample outcome, and weigh the expected value of this best action by the revised probability of that state of nature occurring. Sum over the various sample outcomes to get an anticipated expected value from sampling, $EMV_s$.

Step D: Compute the difference in expected values before (EMV*) and after ($EMV_s$) sampling. This difference is known as the *expected value of sample information* (EVSI).

Step E: Subtract the cost of sampling from the EVSI. The difference is the *expected net gain from sample information* (ENGS). It must be positive or the sampling is not economically justified on an expected-value basis.

On first reading, the procedure may seem complicated, but an example should help clarify it. We have been through many of the steps before; this simply puts them together and expands slightly on the calculations. The first example deals with the "before sampling" situation. As an aid in understanding the procedure, we shall refer to Steps A through E in Examples 18.10 through 18.12.

**EXAMPLE 18.10**

The inventory manager of Robotic Controls, Inc., is faced with a difficult decision. He must set a stock level for several million dollars worth of inventory of microelectronic parts for next quarter. He has developed the following payoff table depicting net profit results from his alternative courses of action. (All values are in thousands of dollars.) Based on past experience with these products, the manager has assigned prior probabilities of 0.10, 0.70, and 0.20 to the prospects of low, medium, and high demand, respectively.

|  | Profit when demand is | | |
|---|---|---|---|
| Act | Low | Medium | High |
| Stock limited inventory | 500 | 500 | 500 |
| Stock average inventory | 440 | 550 | 550 |
| Stock large inventory | 380 | 490 | 600 |

a. Compute the EMV* and indicate the optimal course of action without sampling (Step A).
b. Find the EVPI.

*Solution*   Step A

a. The table shows the probabilities associated with each state of nature.

| | Payoff Table for Demand | | |
|---|---|---|---|
| Act | $\theta_1 = Low$ Demand *(0.1)* | $\theta_2 = Medium$ Demand *(0.7)* | $\theta_3 = High$ Demand *(0.2)* |
| $A_1$ = Limited stock | 500 | 500 | 500 |
| $A_2$ = Average stock | 440 | 550 | 550 |
| $A_3$ = Large stock | 380 | 490 | 600 |

$$E(A_1) = 500(0.1) + 500(0.7) + 500(0.2) = 500$$

$$E(A_2) = 440(0.1) + 550(0.7) + 550(0.2) = 539$$

$$E(A_3) = 380(0.1) + 490(0.7) + 600(0.2) = 501$$

EMV* = Maximum expected monetary value = 539($000)

b. To compute the EVPI we must first compute the EPC as in the table. This requires EVPI = EPC − EMV*.

| If state is | Best action and profit is | % of time this occurs | Expected profit |
|---|---|---|---|
| $\theta_1$ | $A_1 = 500$ | 0.1 | 50 |
| $\theta_2$ | $A_2 = 550$ | 0.7 | 385 |
| $\theta_3$ | $A_3 = 600$ | 0.2 | 120 |
| | | EPC = | 555 |

Now, the EVPI = EPC − EMV*:

$$\begin{aligned}
\text{EPC} &= \text{expected profit under certainty} &&= 555 \\
\text{EMV*} &= \text{expected value of optional act} &&= 539 \\
\text{EVPI} &= \text{expected value of perfect information} &&= \overline{16}\ (\$000)
\end{aligned}$$

The next example shows how additional information can be used to revise the probabilities of demand.

**EXAMPLE 18.11**

Suppose the inventory manager of Robotic Controls, Inc., in Example 18.10 has the opportunity to survey potential customers in order to improve his prior estimate of the probabilities of demand. He feels the survey results could be classified as being discouraging, moderate, or encouraging. However, from reviewing past records he knows that survey results are not always indicative of the resulting demand. As a matter of fact, the conditional probabilities $P(Z|\theta)$ of getting various survey results Z, given each state of nature $\theta$, are as shown in the table.

Conditional Probabilities of $P(Z|\theta)$ from Inventory Manager

| | State of Nature | | |
|---|---|---|---|
| Survey Results | $\theta_1 = Low$ Demand (0.10) | $\theta_2 = Medium$ Demand (0.70) | $\theta_3 = High$ Demand (0.20) |
| $Z_1 =$ Discouraging | 0.6 | 0.2 | 0.1 |
| $Z_2 =$ Moderate | 0.3 | 0.6 | 0.2 |
| $Z_3 =$ Encouraging | 0.1 | 0.2 | 0.7 |
| | $\overline{1.0}$ | $\overline{1.0}$ | $\overline{1.0}$ |

a. Use this additional experience (from the table) to revise the prior probabilities of demand. That is, develop the complete set of revised (posterior) probabilities that would result under each of the three customer reaction categories. *Note:* These calculations are for *anticipated* outcomes, since the survey has not yet been taken.

b. Interpret the result of your calculations.

*Solution*    Step B

a. The revised probabilities sought are the conditional probabilities: $P(\text{Demand}|\text{Survey results})$. Using Bayes' theorem, we find, for example, the probability of a low demand $\theta_1$, given a potentially discouraging survey result $Z_1$, to be

$$P(\theta_1|Z_1) = \frac{P(\theta_1)P(Z_1|\theta_1)}{P(\theta_1)P(Z_1|\theta_1) + P(\theta_2)P(Z_1|\theta_2) + P(\theta_3)P(Z_1|\theta_3)}$$

$$= \frac{(0.10)(0.60)}{(0.10)(0.60) + (0.70)(0.20) + (0.20)(0.10)} = \frac{0.06}{0.22} = 0.27$$

Note that the denominator is the marginal probability of $Z$—that is, $P(Z)$ from Equation (18.2).

Since we need the revised probabilities for all three states of nature under each anticipated sample outcome, we can solve this more efficiently in tabular form, as in Table 18.4. Using the table format, we can readily find the posterior probabilities given all three results, $Z_1$, $Z_2$, and $Z_3$.

**TABLE 18.4**    Computation for Revision of Conditional Probabilities

| Anticipated Result $Z$ | State of Nature $\theta$ | Prior Probability $P(\theta)$ | Conditional Probability $P(Z|\theta)$ | Joint Probability $P(\theta)P(Z|\theta)$ | Posterior Probability $P(\theta|Z)$ |
|---|---|---|---|---|---|
| $Z_1$ = Discouraging result | | | | | |
| Low | $\theta_1$ | 0.10 | 0.60 | 0.60 | 0.27 |
| Medium | $\theta_2$ | 0.70 | 0.20 | 0.14 | 0.64 |
| High | $\theta_3$ | 0.20 | 0.10 | 0.02 | 0.09 |
| | | | | 0.22 | 1.00 |
| $Z_2$ = Moderate result | | | | | |
| Low | $\theta_1$ | 0.10 | 0.30 | 0.03 | 0.06 |
| Medium | $\theta_2$ | 0.70 | 0.60 | 0.42 | 0.86 |
| High | $\theta_3$ | 0.20 | 0.20 | 0.04 | 0.08 |
| | | | | 0.49 | 1.00 |
| $Z_3$ = Encouraging result | | | | | |
| Low | $\theta_1$ | 0.10 | 0.10 | 0.01 | 0.034 |
| Medium | $\theta_2$ | 0.70 | 0.20 | 0.14 | 0.483 |
| High | $\theta_3$ | 0.20 | 0.70 | 0.14 | 0.483 |
| | | | | 0.29 | 1.000 |

b. As an aid to understanding the computational process, it is helpful to view the probabilities in summary form. We see, for example, that if the survey results turn out to be discouraging, then the revised probabilities can be taken from the matrix in Table 18.5 as

$$P(\theta_1|Z_1) = 0.06/0.22 = 0.27$$

$$P(\theta_2|Z_1) = 0.14/0.22 = 0.64$$

$$P(\theta_3|Z_1) = 0.02/0.22 = 0.09$$

**TABLE 18.5**   New Marginal Probabilities for $Z_1$, $Z_2$, and $Z_3$

| Result | Low $\theta_1$ | Medium $\theta_2$ | High $\theta_3$ | Marginal Total $P(Z)$ |
|---|---|---|---|---|
| $Z_1$ = Discouraging | 0.06 | 0.14 | 0.02 | 0.22 |
| $Z_2$ = Moderate | 0.03 | 0.42 | 0.04 | 0.49 |
| $Z_3$ = Encouraging | 0.01 | 0.14 | 0.14 | 0.29 |
| Old prior $P(\theta)$ | 0.10 | 0.70 | 0.20 | 1.00 |

$$P(\theta_1|Z_1) = 0.06/0.22 = 0.27$$

In other words, if customer reaction turns out to be discouraging, the probabilities change, as shown in the first three lines of Table 18.4. The resultant changes from the prior to posterior probabilities are shown in the table below.

Data for Example 18.11: Change in Probabilities Given Discouraging Sample Result

| For | From prior | To posterior | Change |
|---|---|---|---|
| $\theta_1$ | 0.10 | 0.27 | Strong increase |
| $\theta_2$ | 0.70 | 0.64 | Slight decrease |
| $\theta_3$ | 0.20 | 0.09 | Strong decrease |

If another sampling activity were to take place at this stage, the revised (posterior) probabilities would simply become new priors and the process could be repeated again and again as long as the cost of sampling was justified.

We continue our illustration now by bringing cost factors more explicitly into the picture.

**EXAMPLE 18.12**

From Example 18.10 the EMV* of stocking an average inventory amount was found to be $539 (thousand). The proposed sampling, which can be expected to reveal either discouraging, moderate, or encouraging results, will cost $10 (thousand). Use the revised probabilities of Example 18.11 to
a. calculate the new expected values for the alternative courses of action under each anticipated survey result.
b. combine the appropriate expected values to get an anticipated expected value after sampling.
c. compute the expected value of sample information (EVSI).
d. determine the expected net gain from using the sample information (Steps C, D, E).

*Solution*    Step C

a. Expected values under each survey result would be as follows:

$Z_1$ *Discouraging results:*    Revised $P(\theta_1) = 0.27$,    $P(\theta_2) = 0.64$,
$P(\theta_3) = 0.09$

$E(A_1) = 500(0.27) + 500(0.64) + 500(0.09) = 500.0$

$E(A_2) = 440(0.27) + 550(0.64) + 550(0.09) = 520.30 \leftarrow \text{EMV}|Z_1$

$E(A_3) = 380(0.27) + 490(0.64) + 600(0.09) = 470.20$

$Z_2$ *Moderate results:*    Revised $P(\theta_1)^* = 0.06$,    $P(\theta_2) = 0.86$,
$P(\theta_3) = 0.08$

$E(A_1) = 500(0.06) + 500(0.86) + 500(0.08) = 500.00$

$E(A_2) = 440(0.06) + 550(0.86) + 550(0.08) = 543.40 \leftarrow \text{EMV}|Z_2$

$E(A_3) = 380(0.06) + 490(0.86) + 600(0.08) = 492.20$

$Z_3$ *Encouraging results:*    Revised $P(\theta_1) = 0.034$,    $P(\theta_2) = 0.483$,
$P(\theta_3) = 0.483$

$E(A_1) = 500(0.034) + 500(0.483) + 500(0.483) = 500.00$

$E(A_2) = 440(0.034) + 550(0.483) + 550(0.483) = 546.26 \leftarrow \text{EMV}|Z_3$

$E(A_3) = 380(0.034) + 490(0.483) + 600(0.483) = 539.39$

b. If and when samples are taken, the decision maker is then in a position to choose the optimal course of action depending upon the sample result that occurs. Thus, he or she obtains the anticipated expected value after sampling, $\text{EMV}_s$, by selecting the optimal act under each sample result. That probability is, of course, the marginal $P(Z)$, and it can be obtained directly from the joint probability totals of Table 18.4.

$$\begin{aligned}
\text{EMV}_s &= (\text{EMV}|Z_1) \cdot P(Z_1) + (\text{EMV}|Z_2) \cdot P(Z_2) + (\text{EMV}|Z_3) \cdot P(Z_3) \\
&= (520.30)(0.22) + (543.40)(0.49) + (546.26)(0.29) \\
&= \$539.15
\end{aligned}$$

The computation of $\text{EMV}_s$ is further illustrated in the decision tree in Figure 18.5 on page 875.

Step D

c. The expected value of sample information, EVSI, is the difference between the $\text{EMV}_s$ (after sampling) and the EMV* (before sampling):

| | |
|---|---|
| $\text{EMV}_s$ | $539.15 (thousand) |
| EMV* | $-539.00$ (thousand)    (from Example 18.10) |
| EVSI | $  0.15 (thousand) |

---

* For $Z_2$ moderate results in Table 18.4: Revised $P(\theta_1) = P(\theta_1|Z_2) = 0.03/0.49 = 0.06$, and so on in the same manner as was done for $Z_1$ (see Example 18.11, part (b)).

**FIGURE 18.5**   Decision Tree Using Revised Probabilities

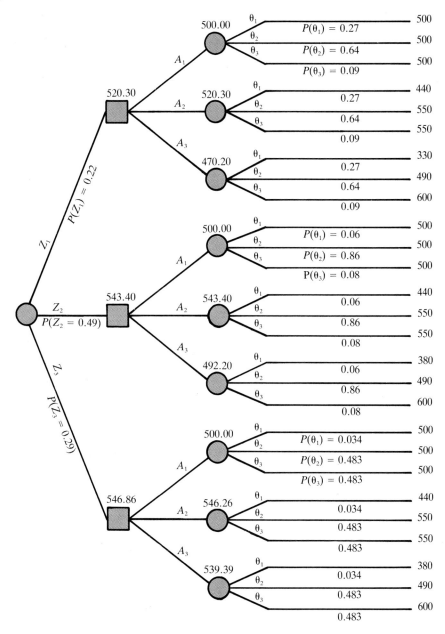

This tells us that the amount of sampling justified is only $150. Relative to the cost of sampling, it is insignificant.

Step E

d. ENGS = EVSI − cost of sampling = $0.15 − $10 = −$9.85 (thousand). Since the EVSI is so small ($150), the cost of $10 (thousand) far outweighs the potential benefits, resulting in an expected net loss of $9850, so the sampling is not economically justified.

There are really two checks to use in evaluating the feasibility of sampling information. First, the cost of sampling should not exceed the expected value of perfect information (EVPI). Since EVPI = $16 (thousand) in Example 18.10 and the cost of sampling = $10 (thousand), this condition is satisfied. The second check is that the expected net gain from sample information ENGS should be positive. In our example it is not, so the sampling cost is not justified.*

This completes our excursion into Bayesian analysis, although we could go into much greater depth if space permitted. References at the end of the chapter consider opportunity loss techniques and expand on the role of the hypergeometric, binomial, and other distributions in supplying appropriate prior and posterior probabilities.

**Problems 18.8**

15. In a certain decision situation, EMV* = $400, EPC = $940, and the cost of sampling is $200. If the anticipated expected value after sampling (EMV$_s$) is $750, what is
   a. the expected value of perfect information, EVPI?
   b. the expected value of sample information, EVSI?
   c. the expected net gain from sample information, ENGS?

16. Given below is the payoff table (shown on the left, with all values in thousand of dollars), where the prior probabilities are $\theta_1 = 0.3$, $\theta_2 = 0.4$, $\theta_3 = 0.3$. Sample information can be obtained for a cost of $0.5 (thousand). The conditional probabilities $P(Z|\theta)$ of the reliability of sample data are shown on the right.

|  | Payoff Table | | |  | $P(Z\|\theta)$, Where $\theta$ Is | | |
|---|---|---|---|---|---|---|---|
|  | $\theta_1$ | $\theta_2$ | $\theta_3$ |  | $\theta_1$ | $\theta_2$ | $\theta_3$ |
| $A_1$ | 40 | 30 | 10 | $Z_1$ | 0.9 | 0.5 | 0.9 |
| $A_2$ | 80 | 50 | −100 | $Z_2$ | 0.1 | 0.5 | 0.1 |

   a. Find the EMV and the best decision without sample information.
   b. Find the EPC and EVPI.
   c. Revise the prior probabilities given $Z_1$ and $Z_2$.
   d. Compute new expected values with sample information and indicate whether the sample information cost is justified.

17. The Country Flavor Cereal Co. is considering spending $200,000 on an advertising campaign to promote their new "environmental" breakfast food. Success will be judged by the incremental gain that results from customer reaction to the campaign. It is expected to be either enthusiastic ($700,000

---

* In this particular case, given the reliability of past surveys, and the prior probabilities, it appears the expected value of the project is likely to be less than it is. This is explained by the fact that the sampling is likely to shift the expected values enough to lower expectations of the whole project. If sampling cannot change a prior decision (no matter what the outcome of the sample), then the sample has no value. This would be the case when EMV* > EMV$_R$ (the revised expected monetary value).

gain before promotional cost), favorable ($300,000 gain), or unfavorable ($200,000 loss) as a result of the promotion. Management's estimate of the outcomes at this time is

$$P(\theta_1) = 0.2 \quad \text{(enthusiastic acceptance)}$$

$$P(\theta_2) = 0.4 \quad \text{(favorable acceptance)}$$

$$P(\theta_3) = 0.4 \quad \text{(unfavorable acceptance)}$$

With no campaign there would be zero incremental gain under all three outcome states.

Before commencing the campaign, the firm's marketing manager proposes to spend $5000 for a research study to assess customers' reactions to the proposed advertising campaign. She knows customer reaction from a survey will not accurately predict success, but based on past experience with studies of this kind, she assesses the chances of various sample outcomes, given the ultimate market outcomes, as shown in the table. Values represent the conditional probability of customer reaction given the state of nature $P(Z|\theta)$—for example, $P(Z_1|\theta) = 0.8$, as shown in the table. Using the information given, determine the answers to parts (a)–(e).

Conditional Probabilities $P(Z|\theta)$

| Customer Reaction | State of Nature | | |
|---|---|---|---|
| | $\theta_1 = Enthusiastic$ | $\theta_2 = Favorable$ | $\theta_3 = Unfavorable$ |
| $Z_1 = $ Positive | 0.8 | 0.3 | 0.1 |
| $Z_2 = $ Neutral | 0.2 | 0.5 | 0.2 |
| $Z_3 = $ Negative | 0.0 | 0.2 | 0.7 |

a. Assume that a decision is to be made without the market research information. What is the best course of action on an expected-value basis? That is, should the firm spend $200,000 on advertising or not?
b. Find the EVPI. Is the cost of the research information less than the EVPI?
c. Revise the probabilities and compute an anticipated expected value after survey research $EMV_s$.
d. What is the EVSI?
e. Compute the expected net gain (or loss) from sample information. Should the proposed study be conducted?

**18.9
A UTILITY
APPROACH TO
DECISION MAKING**

In the earlier parts of this chapter, we have reviewed several approaches to decision making. Most of the analysis was geared toward selecting a course of action based upon some monetary criterion. Our analysis had two inherent assumptions that do not always hold true in practice. First, we assumed that each incremental change in a monetary payoff had a consistent (linear) value to the decision maker. Second, we disregarded the decision maker's attitude toward risk.

In reality, individuals (and organizations) place different values on the same incremental (marginal) amount of money depending upon their current level of wealth. Thus $100 has more value to a penniless person than it does to a millionaire. A firm on the verge of bankruptcy would probably realize a good deal more satisfaction from a $500,000 sale than an economically stable firm would from the same sale. Also, some managers are more willing to risk a firm's resources on new products, production processes, or markets than are others. Individuals differ, and so do organizations. But not many decision criteria are flexible enough to take these differences into account.

Utility theory is one approach that allows us to adjust for the organization's level of assets and its attitude toward risk. And, in its more sophisticated form, utility theory also allows decision makers to incorporate their own values into a decision framework and to base the decision upon multiple criteria rather than just simple monetary values. It represents a truly fascinating and potentially explosive area of decision theory. However, the field is still in its infancy, undergoing growing pains.

Some analysts view *utility* theory as more suited only to describing a decision maker's behavior, whereas others want to use it to *guide*, or suggest, how a manager should behave.

> Utility is a measure of one's preference for some criterion. It is typically measured in relative units such as *utilities* or *utiles*.

> A utility function is a numerical graph depicting the utility value one assigns to different levels of a criterion (often money). It can take various shapes depending upon one's level of assets and attitude toward risk.

In this section we review the assumptions underlying utility theory and explore how one's utility function for monetary amounts can be determined. Then we extend the utility concepts to nonmonetary and multiple-objective criteria. Finally, we see how utility values can be used in the decision process. With this, we are just touching on a topic that is likely to become increasingly important in years to come.

## Assumptions Underlying Utility Theory

The inquiry into utility dates back to Daniel Bernoulli's observation (1738) that people assigned different values to the potential outcomes of a gamble. Bernoulli observed that an item's value was more a function of its utility than its price. Others, such as Von Neumann and Morgenstern (1944), Luce and Raiffa (1957), and Schlaifer (1959), have added substantially to the theory.

Much of the utility theory research has utilized *reference lottery* concepts, wherein an individual's utility is established by reference to a hypothetical gamble between (1) getting a stated certain outcome and (2) having a chance at getting a most preferred outcome along with a complementary chance of getting the least preferred outcome. The decision maker explores the individual's preferences by assigning probability values to the reference lottery. The key value is the one that makes the individual *indifferent* to choosing the reference lottery (with its chance ($P$) of a best outcome and chance ($1 - P$) of a worst outcome) versus choosing the certain outcome, the one that is being evaluated.

Although the theory is still being refined, five basic assumptions are currently recognized. These apply to single individuals who show a consistent behavior in their choices.

1. When confronted with two alternatives, an individual can *preference-rank* them and say whether one act is preferred to another or whether both are equally desirable.
2. An individual can act *transitively*; if *A* is preferred to *B* and *B* to *C*, then *A* is also preferred to *C*.
3. An individual can be indifferent to the *substitution* of acts or consequences upon which he or she places equal value.
4. An individual can identify the utility boundaries of the best outcome (*P* = 1) and worst outcome (*P* = 0). The *continuity-of-preference* assumption holds that there is some gamble representing the likelihood of obtaining the best and worst outcomes (a value of *P*) that is equally preferable to a middle or in-between outcome.
5. For gambles that have identical payoffs, the gamble that has the higher probability for the more attractive payoff will always be preferred.

**Types of Utility Functions**

Three classical utility functions are illustrated in Figure 18.6. An individual who placed a positive and uniformly increasing value on additional amounts of a good would have a linear utility function, such as that of the neutral individual in Figure 18.6. Empirical studies of business executives reveal that they typically have nonlinear functions, evidencing either a positive (risk-seeking) or negative (risk-avoiding) rate of change of marginal utility for money. In Figure 18.6, the risk seeker has a positive utility function with an increasing rate of change in wealth, whereas the risk avoider has one that evidences a smaller increase in utility with each unit increase in wealth.

A risk seeker would be more inclined to gamble an organization's resources on an uncertain venture, whereas a risk avoider would follow a conservative tack. Even though these tendencies exist, some analysts feel that a linear function is sufficiently representative to be useful in most situations. This is especially true when the amounts of money involved are small relative to the total assets of the organization [13].

**FIGURE 18.6**   Utility Functions

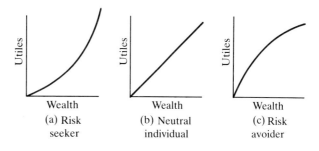

**FIGURE 18.7**    A Utility Function for a Marketing Executive

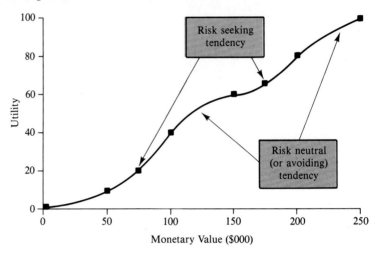

---

## EXAMPLE 18.13

Utility measurement studies of a marketing executive for a computer software firm revealed the preference values shown. Graph the utility function and comment on the individual's characteristics.

| Dollar amount ($000) | 0 | 50 | 75 | 100 | 150 | 175 | 200 | 250 |
|---|---|---|---|---|---|---|---|---|
| Utility value (Utiles) | 0 | 10 | 20 | 40 | 60 | 65 | 80 | 100 |

*Solution*    The utility function, graphed in Figure 18.7, displays a mixture of risk seeking and risk avoidance. The executive appears to have a preference for risk taking, since the utility function shows an increasingly greater utility for higher monetary amounts (the graph is mostly curved up). However, the executive seems to be more neutral in the $100,000–$150,000 range. Just above this, the executive is a risk seeker. This might be explained by the fact that the executive is willing to gamble with smaller amounts, but $100,000–$150,000 represents significant funds to this person individually. However, the executive again appears to be willing to risk corporate funds as larger amounts are involved. (*Note:* The analyst may wish to reaffirm the values in the $100,000–$150,000 range by doing more studies.)

---

## Assigning Utility Values to Monetary Payoffs

Let us simplify the assignment of utility values by first limiting our criteria to monetary amounts. Two methods are in use. One involves using a *reference lottery* that always has a pair of probabilities attached to the larger and smaller sums. In this case, the decision maker specifies the certain monetary amount that he or she feels is equivalent to the lottery. The second method involves having the decision maker *assign a probability value*. We shall begin with the reference lottery involving probabilities because it is a good basis for understanding the theory. Later we will go on to the direct assignment of probability values.

We begin by arbitrarily designating zero utiles as a minimum (worst) outcome and 100 utiles as the maximum (best) outcome. In general, we expect that a higher monetary outcome will yield a higher utility. However, we seek to derive utility values that incorporate an individual's preference and that reflect his or her level of assets and/or attitude toward risk. The procedure can be summarized in six steps.

### A PROCEDURE FOR ASSIGNING UTILITY VALUES TO MONETARY AMOUNTS

1. Identify the minimum ($min) and maximum ($max) monetary values.
2. Assign a utility value ($U$) of zero to $min and 100 to $max. That is,

$$U(\$min) = 0$$

$$U(\$max) = 100$$

3. Compute a third utility value for an amount between $min and $max by determining the amount of money to which the individual is indifferent between (a) receiving that amount ($\$_3$) for certain and (b) participating in a lottery wherein the person will receive $min with a probability of 0.5 and $max with a probability of 0.5.
4. Compute the utility of the lottery amount, $\$_3$, as $U(\$_3) = 0.5U(\$min) + 0.5U(\$max)$.
5. Compute more utility points in a similar manner using amounts whose utilities are known—$U(\$_3)$ and $U(\$min)$. Continue until a sufficient number of consistent points are obtained.
6. Graph the utility function.

---

**EXAMPLE 18.14**

An account executive at a Denver brokerage firm frequently makes investment recommendations to clients. Some bond investments offer almost certain profits, whereas the mining and oil stocks have a potential of larger profits but also the risk of a substantial loss.
a. Using values of zero utiles for $0 profit and 100 utiles for $20,000 profit, show how the endpoints of an investor's utility function might be derived.
b. Supply estimates as necessary to derive five other utility values for parts (b)–(f).
c. Graph the resultant utility function.
(Adapted from [16].)

*Solution*    a. The minimum and maximum values give us two endpoints on the utility function:

$$U(\$0) = 0$$

$$U(\$20,000) = 100$$

b. Compute a third utility value by asking the broker at what amount of guaranteed (bond) profit ($\$_3$) she would be indifferent to receiving it versus taking a chance on some stocks (a lottery) that would yield $0 with a probability of 0.5 and $20,000 with a probability of 0.5. Assume her answer is $\$_3 = \$4000$. This is illustrated in Figure 18.8(a), where the amount for certain is just sufficient to balance the $0 and $20,000 outcomes, which each have a probability of 0.5. Compute the utility of the $\$_3$ amount.

**FIGURE 18.8(a)**    Determining an Intermediate Utility Value for a $0 and $20,000
Combination

$$U(\$4000) = U(\$0) \text{ at } 0.5 \text{ probability} + U(\$20,000) \text{ at } (1 - 0.5) \text{ probability}$$
$$= (0) \cdot (0.5) + (100) \cdot (0.5) = 50$$

c. Compute a fourth utility value by asking the broker at what amount of certain profit
($S_4$) she would be indifferent to receiving it versus entering into a lottery that would
yield $4000 with a probability of 0.5 and $20,000 with a probability of 0.5. This is
shown in Figure 18.8(b). Assume her answer is $S_4 = \$8000$. Then compute the utility
of the $S_4$ amount:

$$U(\$8000) = U(\$4000) \cdot (0.5) + U(20,000) \cdot (0.5)$$
$$U(\$8000) = \quad 50(0.5) \quad + \quad 100(0.5) \quad = 75$$

**FIGURE 18.8(b)**    Determining an Intermediate Utility Value for a $4000 and $20,000
Combination

(Note that the computation of $U(\$8000)$ uses the 50-utile value for $U(\$4000)$ obtained
in the previous step along with the initial value of 100 utiles for $U(\$20,000)$).

d. Compute a fifth utility value. In this case, to illustrate the use of other probabilities,
suppose you ask the broker at what amount of certain profit ($S_5$) she would be indifferent to receiving it versus entering into a lottery that would yield $4000 with a
probability of 0.3 and $20,000 with a probability of 0.7. Assume this answer is $15,000.
This is illustrated in Figure 18.8(c). Then compute the utility of the $S_5$ amount.

$$U(\$15,000) = U(\$4000)(0.3) + U(\$20,000)(0.7)$$
$$= \quad 50(0.3) \quad + \quad 100(0.7) \quad = 85$$

**FIGURE 18.8(c)**    Determining Another Intermediate Utility Value for a $4000 and
$20,000 Combination

Note that the lottery amounts in (c) and (d) are the same, but the higher probability
for the $20,000 results in a higher equivalent amount under certainty ($15,000).

e. More intermediate points could be computed using the values already obtained. For example, assume $2000 is the certainty equivalent of $0 at 0.4 probability and $4000 is the value at $(1 - 0.4) = 0.6$. Then $U(\$2000) = 0(0.4) + 50(0.6) = 30$.

f. In the same manner, the utility of losses can also be ascertained. For example, assume there is a chance the broker could suffer a $3000 loss. Figure 18.8(d) illustrates this.

**FIGURE 18.8(d)**   Determining an Intermediate Utility Value in which an Option
Involves a Loss

Assume the broker is indifferent to receiving $0 for certain versus entering into a lottery that would yield $-\$3000$ with a probability of 0.5 and $15,000 (gain) with a probability of 0.5. Then

$$U(0) = U(-\$3000) \cdot (0.5) + U(\$15{,}000) \cdot (0.5)$$

$$0 = U(-\$3000) \cdot (0.5) + \quad 85 \quad \cdot (0.5)$$

$$U(-\$3000) = \frac{-42.5}{0.5} = -85$$

g. We can now graph the utility function for which we have derived the values shown in the table and in Figure 18.9.

| Dollar amount ($000) | −3000 | 0 | 2000 | 4000 | 8000 | 15,000 | 20,000 |
|---|---|---|---|---|---|---|---|
| Utility values (utiles) | −85 | 0 | 30 | 50 | 75 | 85 | 100 |

**FIGURE 18.9**   A Utility Function for the Brokerage Account Executive in
Example 18.14

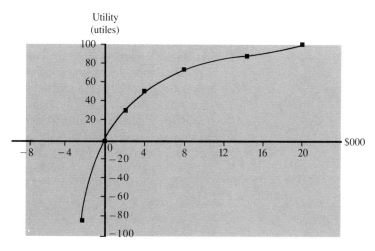

The broker in Example 18.14 has a consistent and risk-avoiding utility function for money.

Once the function has been mapped, we can use it to find utility values for intermediate points on the curve and for expected value calculations using utility values rather than monetary values.

---

**EXAMPLE 18.15**

The broker whose utility function is graphed in Example 18.14 has an opportunity to make a $10,000 profit for certain by means of an investment in bonds.
a. What is her utility value for the $10,000?
b. Suppose she has the option of investing in a stock that will end up with either a $2000 profit or a $20,000 profit. What probability distribution (risk) values must correspond to the stock outcomes in order for the broker to be indifferent to the bond versus the stock alternatives?

*Solution*    a. From the graph in Figure 18.9, the $10,000 profit corresponds to a utility value of approximately 80 utiles.
b. If we let $P$ equal the probability of the $2000 profit and $(1 - P)$ equal the probability of the $20,000 profit, then

$$U(\$10,000) = U(\$2000) \cdot (P) + U(\$20,000) \cdot (1 - P)$$

$$80 = 30P + 100(1 - P)$$

$$80 = 30P + 100 - 100P$$

$$70P = 20$$

$$P = 0.29$$

So the probability of the $2000 profit must be 0.29 and that of the $20,000 profit as high as $(1 - 0.29) = 0.71$.

---

Once an individual's utility function (or selected values from the function) has been determined, it provides a basis for decision. Then decisions can be made on the basis of maximizing the expected utility value (EUV) rather than the expected monetary value (EMV).

**Extensions of Utility Theory to Multi-Attribute Situations**

A major reason why utility theory is becoming so widely recognized today is that utility functions can be developed for decision alternatives that involve outcomes other than money. Money can be included as one component of a package of effects, but it need not be the sole criterion. By using utility theory, the decision maker can incorporate less tangible human and social values into a structured and systematic decision-making process. The individual's utility function is first determined by a questioning technique, wherein the subject is asked to express his or her preference in a series of choices, one of which always contains a risk or chance situation. The choices represent decision outcomes that may very well include financial, marketing, legal, social, and other implications. After the utility values of the various outcomes are determined, a "best" course of action can be determined on an expected utility basis.

To illustrate the potential of a utility-based approach to decisions, let us examine the opportunity faced by the (hypothetical) European Petroleum Company, EPCO. We look first at a Bayesian approach and then view the problem from a utility perspective. This problem also illustrates the use of probability values to derive utility functions.

**EXAMPLE 18.16**

EPCO top management is committed to the prospect of developing nuclear fuel cells for automobiles. The cells would make use of spent nuclear reactor fuel by-products, which would produce an electric current flow by a direct conversion process. This would revolutionize the automobile industry by replacing the internal combustion engine with a clean electric motor. As an alternative to finishing the project by itself, the firm is considering joining with an experienced transportation-oriented firm. Depending upon the success of the project, EPCO estimates the impact upon the company to be as shown in Table 18.6. Projected figures are ten-year present-value amounts in millions of dollars and employment figures represent a two-year change.

Based upon feasibility studies and consultations with various production and market research groups, the project manager has assigned subjective probabilities of success as $P(\theta_1) = 0.2$, $P(\theta_2) = 0.4$, and $P(\theta_3) = 0.4$. She feels that some prototype studies could give a better indication of success and ultimately modify these probabilities. She has been asked to identify the best course of action if a decision had to be taken right now.

a. Use a Bayesian approach to determine the best course of action.

b. Use a utility theory approach to determine the best course of action.

**TABLE 18.6**   Outcomes Depending on Success of Project

| | $\theta_1$<br>Highly Successful | $\theta_2$<br>Moderately Successful | $\theta_3$<br>Not Successful |
|---|---|---|---|
| D: Develop independently on own | Profit = $300 million<br><br>Employment up 150%<br><br>Dominate market<br><br>50% chance of advancing transportation technology<br><br>Favorable effect on environment<br>balance of payments<br>use of nuclear waste | Profit = $40 million<br><br>Employment up 20%<br><br>Strong market position<br><br>30% chance of advancing transportation technology<br><br>Moderate effect on environment<br>balance of payments<br>use of nuclear waste | Profit (loss) = −$60 million<br><br>Employment down 30%<br><br>No market position<br><br>2% chance of advancing transportation technology<br><br>No effect on environment<br>balance of payments<br>use of nuclear waste |
| J: Joint venture | Profit = $200 million<br><br>Employment up 80%<br><br>Strong market position<br><br>75% chance of advancing transportation technology<br><br>Very favorable effect on environment<br>balance of payments<br>use of nuclear waste | Profit = $30 million<br><br>Employment up 10%<br><br>Strong market position<br><br>40% chance of advancing transportation technology<br><br>Moderately favorable effect on environment<br>balance of payments<br>use of nuclear waste | Profit (loss) = −$20 million<br><br>Employment down 7%<br><br>No market position<br><br>4% chance of advancing transportation technology<br><br>No effect on environment<br>balance of payments<br>use of nuclear waste |

*Solution*  a. From a traditional Bayesian standpoint, the best course of action is based totally upon economic optimization by finding EMV*. Thus we have

| | Payoff Values | | |
|---|---|---|---|
| | $\theta_1 = 0.2$ | $\theta_2 = 0.4$ | $\theta_3 = 0.4$ |
| D | 300 | 40 | −60 |
| J | 200 | 30 | −20 |

$$E(D) = \$300(0.2) + \$40(0.4) - \$60(0.5) = \$52 \text{ (million)}$$

$$E(J) = \$200(0.2) + \$30(0.4) - \$20(0.4) = \$44 \text{ (million)}$$

Therefore, the optimal action from a Bayesian viewpoint is for EPCO to develop the fuel cell on its own for an expected profit of $52 million.

b. To obtain a utility-based solution we must first assign utility values to the outcomes on the basis of an individual's or a group's value system. Considering the objectives (criteria) in Table 18.6, let us assume the project manager favors high profits and a relatively stable but rising employment situation (no large fluctuations). For example, she may prefer to see a gradual employment increase of 10% rather than a dramatic, but disruptive, increase of 150%. The manager also favors a reasonably strong market position. From a professional standpoint, however, she is very anxious that this development make a genuine contribution to transportation technology. She also recognizes social values of a cleaner environment, making use of reactor waste products, and trade that would have a favorable effect on the nation's economy and balance of payments.

We can determine the manager's utility value for the various outcomes by first assigning arbitrary values of 100 and 0 to the endpoints of the utility function. Conditions at the endpoints are defined by the manager in terms of what she envisions as best and worst.

**Let $Z_{best}$** be an outcome the manager feels is as good as any combination of outcomes in the situation. (For example, $Z_{best} = \$300$ million profit, employment up 10%, strong market position, 75% chance of advancing transportation technology, and very favorable effect on environment, use of nuclear waste, and balance of payments.) Let $U[Z_{best}] = 100$.

**Let $Z_{worst}$** be an outcome as bad as any combination of outcomes in the situation. (For example, $Z_{worst} = \$60$ million loss, employment down 30%, no market position, only 2% chance of advancing transportation technology, no effect on environment, use of nuclear waste, or balance of payments.) Let $U[Z_{worst}] = 0$.

We now determine the utility value for each of the originally defined outcomes by finding a probability value for a "reference gamble equivalent to the outcome." For example, to evaluate the outcome "develop on own" by "highly successful" (as $D\theta_1$ in Table 18.6), we might employ the following questioning technique.

**Question to Manager**  You have a chance to obtain $Z_b$ ($300 million profit, employment up 10%, etc.) with probability $P$, or $Z_w$ ($60 million loss, employment down 30%, etc.) with probability $1 - P$. What probability would you assign to $P$ such that you are indifferent to that result versus the outcome $D\theta_1$ ($300 million profit, employment up 150%, dominate market, etc.)?

**FIGURE 18.10**   Trade-Offs for Balance in a Utility Function

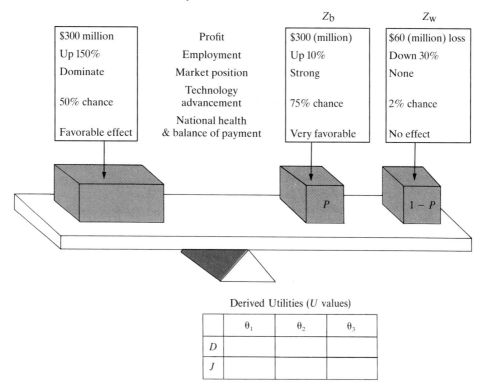

Derived Utilities ($U$ values)

|   | $\theta_1$ | $\theta_2$ | $\theta_3$ |
|---|---|---|---|
| $D$ |   |   |   |
| $J$ |   |   |   |

| | |
|---|---|
| **Response** | I'm not sure I understand. |
| **Question to Manager** | In other words, assume you could be assured of having the outcome $D\theta_1$ (for certain). Would you be willing to give that up for a chance to obtain $Z_b$? |
| **Response** | Probably so, if the chance of obtaining $Z_b$ were high enough. Would that mean I also incur a risk of getting stuck with $Z_w$? |
| **Question to Manager** | Yes, each probability you assign to $Z_b$ means that you are accepting a complementary probability of $Z_w$. You might envision the choice as it is depicted in Figure 18.10, where we achieve the balance point by adjusting the chance of $Z_b$ (and concurrently that of $Z_w$) to a value where you are indifferent toward receiving either the package being evaluated (here, $D\theta_1$) or the $Z_b$ and $Z_w$ package. What chance at $Z_b$ would you feel is equivalent to $D\theta_1$? |
| **Response** | (after some thought)   Probably 85%. <br> Therefore, $P(Z_b) = 0.85$ and $P(Z_w) = 1 - P(Z_b) = 1 - 0.85 = 0.15$. |

The response of $P(Z_b) = 0.85$ is in effect a response that the manager equates the utility of highly successful independent development to an 85% chance of the best conceivable outcome and a 15% chance of the worst. We can now determine the utility value for $D\theta_1$ by setting up an equation expressing this relationship.

$$U[D\theta_1] = 0.85(U[Z_b]) + 0.15(U[Z_w])$$
$$= 0.85(100) + 0.15(0) = 85 \text{ utiles}$$

In a similar manner, we can determine utility values for the other outcomes ($D\theta_2$, $D\theta_3$, $J\theta_1$, etc.) by substituting them for $D\theta_1$ and repeating the process. Assume that the values shown in the table are obtained.

**Utilities for Manager**

| | Derived Utilities | | |
|---|---|---|---|
| | $\theta_1$ | $\theta_2$ | $\theta_3$ |
| D | 85 | 35 | 0 |
| J | 80 | 40 | 10 |

The derived utility values are shown as outcomes in the decision tree of Figure 18.11 for convenience.

**FIGURE 18.11**   Utility Decision Tree

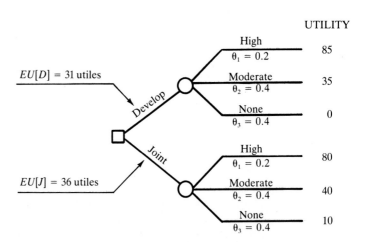

Note that the square box (□) represents the manager's choice and the circles (○) depict the uncontrollable variable of degree of success. We referred to this earlier as the state of nature. We can now determine the optimal course of action on an expected utility basis from the following computation of expected value.

**Expected value of the utilities:**

$$EU[A] = \sum_{j=1}^{n} [(\theta_{i,j})P(\theta_j)] \qquad (18.4)$$

Equation 18.4 is simply the expected value of the utilities. It is another form comparable to an expected monetary value.

$$EU[\text{develop}] = EU[D] = 85(0.2) + 35(0.4) + 0(0.4) = 31 \text{ utiles}$$

$$EU[\text{joint venture}] = EU[j] = 80(0.2) + 40(0.4) + 10(0.4) = 36 \text{ utiles}$$

The best decision is now to proceed with the joint venture, because the manager has a higher expected utility for the outcomes from the joint venture (36 utiles versus 31 utiles). Thus EUV = 36 utiles.

---

In this decision situation, we have monetary values, employee and professional considerations, and social considerations, all evaluated subjectively as a package of effects, even though only some figures were expressed quantitatively. Another approach would be to evaluate such components separately and then combine them all into an overall utility function. A study along these lines is reported in the literature [8].

## Unique Potential of Utility Theory

Utility-based decision making is both ancient and modern. In a historical sense, it incorporates the intuitively personalized experience and value judgments that many modern decision methodologies and mathematical models lack, causing them to be branded as *academic* or *impractical*. On the other hand, while utility theory might also be assigned the same label, it is an analytical approach to decision, utilizing *probability estimation* and consistency checks, certainly worthy of serious consideration.

Two factors make it a potentially powerful force in a decision methodology: (1) its *integrative character* and (2) its *value system base*.

Managerial decisions typically involve numerous decision variables that are *interrelated* within a complex set of organizational goals. Both the decision variables and the goals can be difficult to delineate, especially when intangible values are involved. To facilitate action, sometimes variables are arbitrarily excluded from consideration, or perhaps the objectives are artificially narrowed by a formalized procedure that restricts consideration to one or two oversimplified goals. As a result, the decision process does not incorporate a total system perspective, and suboptimization occurs. Whereas a sophisticated quantitative analysis may clearly recommend that a manager act in one way, his or her intuition may favor another.

Utility theory is one step toward decision making on the basis of a more integrated information base. It takes into account the fact that multiple goals can be sought simultaneously and it capitalizes on the unique capability of the human brain to integrate specific as well as less tangible information. No electronic system has yet been able to match this skill. The brain can utilize quantified information if available, but it can also take intangible objectives into consideration and develop a *preferential ordering of outcomes*, which we have identified as a *utility function*.

When a decision maker assigns a utility value to a potential outcome, he or she is injecting a personal *value system* into the formalized decision structure. This effect of personalizing the decision-making process should tend to enhance the social responsibility of organizational decision making. However, these changes cannot be expected to occur automatically or without side effects on the organization.

Since individual value systems differ, it may be necessary to explore the differences and similarities in an attempt to formulate a common basis for decision. A solid knowledge base would seem to be a good starting point for reducing the variance among individual values and for making the common basis a logical and consistent one for management decisions.

**Problems 18.9**

18. Graph the utility function data points for Managers *A*, *B*, and *C*. Tell what type of risk characteristics each manager has. Describe each function in terms of risk-seeking, neutral, or risk-avoiding tendencies.

**Manager A**

| Dollar amount ($000) | 50 | 100 | 150 | 200 | 250 | 275 |
|---|---|---|---|---|---|---|
| Utility value (utiles) | 5 | 20 | 40 | 60 | 80 | 100 |

**Manager B**

| Dollar amount ($000) | 50 | 100 | 200 | 300 | 400 | 500 |
|---|---|---|---|---|---|---|
| Utility value (utiles) | 20 | 40 | 60 | 80 | 95 | 100 |

**Manager C**

| Dollar amount ($000) | 50 | 100 | 200 | 250 | 300 | 400 |
|---|---|---|---|---|---|---|
| Utility value (utiles) | 10 | 20 | 41 | 50 | 59 | 80 |

19. In a utility theory study of monetary amounts, an analyst used endpoints of $0 (0 utiles) and $80,000 (100 utiles). The manager under study was indifferent to receiving $30,000 for certain versus a lottery with a 0.50 probability of $0 and a 0.5 probability of $80,000. What was the manager's utility value (in utiles) for the $30,000?

20. A utility assessment interview with the manager yielded the following results, where $U(\$0) = 0$ and $U(\$12,000) = 100$. Use the data provided to compute the manager's utility function and graph the function.

| Amount for Certain $ | Lottery of Equivalent Preference | | | |
|---|---|---|---|---|
| | $ | at P | Versus $ | at (1 − P) |
| (a)  5,000 | 0 | 0.5 | 12,000 | 0.5 |
| (b)  4,000 | 0 | 0.6 | 5,000 | 0.4 |
| (c)  8,000 | 5,000 | 0.3 | 12,000 | 0.7 |
| (d)  2,000 | 0 | 0.75 | 4,000 | 0.25 |
| (e)  10,000 | 5,000 | 0.1 | 12,000 | 0.9 |

21. The manager of an electronics firm was found to have a utility score of 60 for a $40,000 amount when her utility function was derived for a range of $0 (0 utiles) to $200,000 (100 utiles). Now suppose she is presented with a choice of $40,000 for certain versus a lottery involving $20,000 at probability *P* and $200,000 at $(1 − P)$. What probability values would correspond to the *P* and $(1 − P)$ amounts if she is indifferent to the choice? The manager's utility value for $20,000 is 25.

22. Suppose you had the following utility choices. Indicate which alternative you would choose and why.

| Receive for Certain | $Z_b$ | $Z_w$ |
|---|---|---|
| (a)  $0 | $5 at $P = 0.6$ | $-$3$ at $(1 - P) = 0.4$ |
| (b)  $0 | $25,000 at $P = 0.6$ | $-$15,000$ at $(1 - P) = 0.4$ |
| (c)  $200,000 | $500,000 at $P = 0.5$ | $0 at $(1 - P) = 0.5$ |

23. In a utility assessment experiment, a decision maker was offered the choice of no-gain or no-loss for certain ($0) versus a 60% chance of gaining $60,000 and a 40% chance of losing $40,000. The $60,000 gain was assigned a utility of $Z_b = 100$ utiles and the $40,000 loss a utility of $Z_w = 0$ utiles. In assessing a third point, the decision maker was indifferent to the $0 for certain versus the monetary amounts when the $P(\$60,000$ gain$) = 0.7$ and the $P(\$40,000$ loss$) = 0.3$.
    a. Graph the decision maker's utility function using this data.
    b. Is the decision maker a risk seeker or a risk avoider?

24. The following utility values for various profit amounts were derived for a manager of a steel fabrication firm.
    a. Plot the utility values for each profit amount and draw a utility curve through the points. (*Note:* Plot utility on the vertical axis and profit on the horizontal axis.)

| Profit | Utility |
|---|---|
| $     0 | 00 |
| 10,000 | 40 |
| 25,000 | 80 |
| 40,000 | 96 |
| 50,000 | 100 |

    b. Indicate whether the manager is a risk seeker, a risk avoider, or neutral.
    c. Suppose a decision has two courses of action, $A$ and $B$, with payoff values as shown, where the probabilities associated with $\theta_1$, $\theta_2$, and $\theta_3$ are 0.3, 0.6, and 0.1, respectively. Determine the EMV for each course of action.

| | State of Nature | | |
|---|---|---|---|
| | $\theta_1$ | $\theta_2$ | $\theta_3$ |
| $A$ | $20,000 | $35,000 | $45,000 |
| $B$ | 10,000 | 30,000 | 50,000 |

    d. Taking approximate values from the utility function you graphed in (a), determine the expected utility for each course of action, and compare the EMV* decision criterion with the EUV decision criterion. Which action should be chosen under an EMV* criterion? an EUV criterion?

e. Construct a utility decision tree for this decision situation.

## SUMMARY

The decision environment ranges from complete certainty to extreme uncertainty about the state of nature. *Uncertainty* refers to the decision maker's ignorance about which state of nature will prevail. When no information is available, a game theory strategy may be followed, but different strategies are usually more useful.

A payoff table shows the potential outcomes of each act given each state of nature. Although maximax, maximin, and maximum likelihood are all feasible criteria, the *expected monetary value* (EMV*) is a more popular approach. EMV* is the largest computed total obtained when payoff values for a given act are multiplied by their probabilities of occurrence and summed over the various states of nature. EMV* is always equal to the smallest *expected opportunity loss* (EOL*). When the EMV* is subtracted from the EPC (*expected profit under certainty*), the EVPI (*expected value of perfect information*) is obtained. The EVPI is the maximum a manager could justify spending based on perfect information that would reveal *in advance* what state of nature prevails at any given time.

*Bayes' theorem* is a logical method of revising the prior probabilities of the state of nature ($\theta$) on the basis of additional (sample) information $Z$. Using Bayes' theorem, the *prior* $P(\theta)$ and conditional $P(Z|\theta)$ probabilities are combined into a joint probability $P(\theta \cap Z)$ and divided by the marginal probability of getting the resultant sample outcome $P(Z)$. This yields revised, or *posterior*, probabilities $P(\theta|Z)$. The revised probabilities can then be used to compute a *revised expected monetary value*, $\text{EMV}_R$, which identifies the best course of action after incorporating sample information.

If the full set of potential sample outcomes can be identified, the probability revisions can be anticipated for all possible outcomes in advance of actually collecting the sample data. When this is done, we can compute the *expected value after sampling*, $\text{EMV}_s$ by weighting the best course of action given each sample outcome, $\text{EMV}|Z$, with the revised probability of that state occurring $P(Z)$, and summing over the various outcomes. The difference between the initial EMV* and the $\text{EMV}_s$, after sampling, is designated as the *expected value of sample information*, EVSI. When we subtract the cost of sampling from EVSI, we obtain the *expected net gain from sample information*, ENGS. It must be positive for the sampling to be economically justified.

*Utility theory* is a decision-making technique that holds much promise for integrating responsibility and value-based considerations into an objective decision framework. By using utility measures, a decision maker can include intangible factors that have previously branded some decisions as "academic" or "unrealistic." From a procedural standpoint, to include these intangibles, we simply maximize an expected utility value (EUV) rather than an expected monetary value (EMV). Table 18.7 contains the key equations for this chapter.

**TABLE 18.7**    Key Equations Used for Decision Theory

$$\text{EVPI} = \text{EPC} - \text{EMV*} \tag{18.1}$$

$$P(\theta_1|Z) = \frac{P(\theta_1)P(Z|\theta_1)}{P(\theta_1)P(Z|\theta_1) + P(\theta_2)P(Z|\theta_2)} \tag{18.2}$$

$$P(\theta_2|Z) = \frac{P(\theta_2)P(Z|\theta_2)}{P(\theta_1)P(Z|\theta_1) + P(\theta_2)P(Z|\theta_2)} \tag{18.3}$$

$$\text{EU}[A] = \sum_{j=i}^{n} (\theta_{ij})P(\theta_j) \tag{18.4}$$

## KEY TERMS

## CHAPTER 18 PROBLEMS

25. Given the payoff matrix shown, where the decision choice is between $X$ and $Y$, what percent of the time should the decision maker choose $X$ if he or she follows a game theory strategy?

|   | State of Nature | |
|---|---|---|
|   | $\theta_1$ | $\theta_2$ |
| $X$ | $60 | $0 |
| $Y$ | $-$60 | $80 |

26. Given the payoff matrix shown, where $P(\theta_1) = 0.3$ and $P(\theta_2) = 0.7$ and the decision maker must choose between $X$ and $Y$,
    a. what is the expected value for each course of action?
    b. what is the EMV*?
    c. what is the expected value of perfect information about the state of nature?

|   | State of Nature | |
|---|---|---|
|   | $\theta_1$ | $\theta_2$ |
| $X$ | $60 | $0 |
| $Y$ | $-$60 | $80 |

27. Given the payoff matrix shown, where $P(\theta_1) = 0.4$ and $P(\theta_2) = 0.6$,
    a. using game theory strategy, what percent of the time would the decision maker choose $A_1$?

b. what is the expected value of each course of action?

c. what is the expected value of perfect information with respect to the state of nature?

| | State of Nature | |
|---|---|---|
| | $\theta_1$ | $\theta_2$ |
| $A_1$ | $-\$30$ | $\$40$ |
| $A_2$ | $\$30$ | $\$0$ |

28. Use the matrix shown to find the appropriate course of action utilizing the following decision criteria:

   a. Maximax

   b. Maximin

   c. Maximum likelihood

   d. Bayesian (expected value). *Note:* Where necessary, you may use prior probability values of $P(\theta_1) = 0.05$, $P(\theta_2) = 0.60$, $P(\theta_3) = 0.35$.

| | $\theta_1$ | $\theta_2$ | $\theta_3$ |
|---|---|---|---|
| $A_1$ | 0 | 0 | 0 |
| $A_2$ | $-3$ | 2 | 2 |
| $A_3$ | $-6$ | $-2$ | 4 |

29. Use the prior probabilities $P(\theta_1) = 0.30$ and $P(\theta_2) = 0.70$ and the conditional probabilities $P(Z|\theta_1) = 0.90$ and $P(Z|\theta_2) = 0.10$. Find $P(\theta_1|Z)$.

30. The Ocean Candy Co. is considering changing a 3/4-ounce chocolate bar to 1 ounce and increasing the price to $1.25. Marketing feels there is a 0.70 probability this will increase profits. However, a limited market test resulted in reduced profits. Marketing research feels the chance this result (reduced profit) would occur given that the change would actually increase profits (on a national basis) is 0.2 and the chance it would occur given that it would not increase profits nationally is 0.8. Find the manager's revised probability of the profitability of increasing the price.

31. Lakeland Fabrics purchases bulk yarn of unknown quality from a Hong Kong supplier. Lakeland managements feels there is a 50% chance the yarn will be of high quality ($\theta_1$), a 30% chance it will be of medium quality ($\theta_2$), and a 20% chance it will be of low quality ($\theta_3$). In order to reduce its uncertainty, Lakeland has paid $200 to a New York agent to sample a shipment from this supplier. The agent feels it is high quality. Having used this agent before, Lakeland management feels that the chance the agent would report high quality Z, given it was really high, is 0.80. Similarly, $P(Z|\theta_2) = 0.40$ and $P(Z|\theta_3) = 0.20$. Revise the prior probabilities on the basis of this additional information.

32. As television sets come off the Quantown Plant assembly line, they are supposed to be in perfect condition. However, experience has shown that due to rough handling the sets are sometimes out of adjustment. As each set leaves the assembly line, the

shipping supervisor can either accept the set (and ship it without further testing) or reject it. The gains and losses associated with accepting or rejecting a set are as follows:

|  | Adjustment | |
|---|---|---|
|  | Good $\theta_1$ | Bad $\theta_2$ |
| $A$ = accept | $5 | −$20 |
| $R$ = reject | −$5 | $30 |

a. Assuming there is no other information to go by (which suggests a game theory strategy), what percent of the time should the supervisor accept a set coming off the assembly line?
b. Experience indicates that the adjustment will be bad on 20% of the sets (and good on 80% of them). Compute the expected value of the alternative courses of action and indicate the optimal action on an expected monetary value basis.
c. Suppose the company can rent a machine that will indicate whether the set is properly adjusted (good or bad). Experience with the machine shows that 80% of the sets that are really in good adjustment will be rated good on the machine, and 70% of those actually out of adjustment (bad) will be rated bad on the machine. Determine the following probabilities.
   i. $P(\text{Good}|\text{Rated good}) = P(\theta_1|Z_1)$
   ii. $P(\text{Bad}|\text{Rated good}) = P(\theta_2|Z_1)$
   iii. $P(\text{Good}|\text{Rated bad}) = P(\theta_1|Z_2)$
   iv. $P(\text{Bad}|\text{Rated bad}) = P(\theta_2|Z_2)$
d. Using the revised probabilities, what is the optimal act and $EMV_R$ if the machine rates the television set as (i) good? (ii) bad?
e. Compute the anticipated expected value after sampling $EMV_s$ and the expected value of sample information EVSI.
f. Suppose the cost to use the machine is $1.20 each time. What is the expected net gain (ENGS) from use of the machine?

33. Given EMV* (without research) = $800, EPC = $1800, and the cost of research (sampling) is $500, if the anticipated expected monetary value after research is $EMV_s$ = $2200, find (a) EVPI, (b) EVSI, (c) ENGS.

34. Payoff table values are as shown and prior probabilities are $P(\theta_1) = 0.3$, $P(\theta_2) = 0.5$, $P(\theta_3) = 0.2$.

|  | Payoff Table | | |
|---|---|---|---|
|  | $\theta_1$ | $\theta_2$ | $\theta_3$ |
| $A_1$ | 50 | 80 | 30 |
| $A_2$ | 70 | 10 | 90 |

The decision maker is considering buying some sample information $Z$ for a cost of 20 (in thousands of dollars). Historical data show that the reliability of this type of information is described by the following conditional probabilities, $P(Z|\theta)$.

Conditional Probabilities $P(Z|\theta)$

|  | $\theta_1$ | $\theta_2$ | $\theta_3$ |
|---|---|---|---|
| $Z_1$ | 0.8 | 0.3 | 0.1 |
| $Z_2$ | 0.2 | 0.7 | 0.9 |

a. Find EMV* and the best decision without sample information.
b. Find EPC and EVPI.
c. Revise the prior probabilities given $Z_1$ and $Z_2$.
d. Compute new expected values with sample information and indicate whether the sample information cost is justified.

35. A new recreational area is being proposed in a national forest. Three locations are under consideration: Cascade Summit, Mountain View, and Big Springs. The cost and income estimates (millions of dollars) have been made in present-value terms for the useful life of the facilities. In addition to the given values, the estimated present value income from the resort lodges is estimated at $6 million from Cascade Summit, $9 million from Mountain View, and $11 million from Big Springs resort. Figure 18.12 shows the beginning of a decision tree depicting the location alterna-

---

**FIGURE 18.12** A Decision Tree for Selection of a New Recreation Area

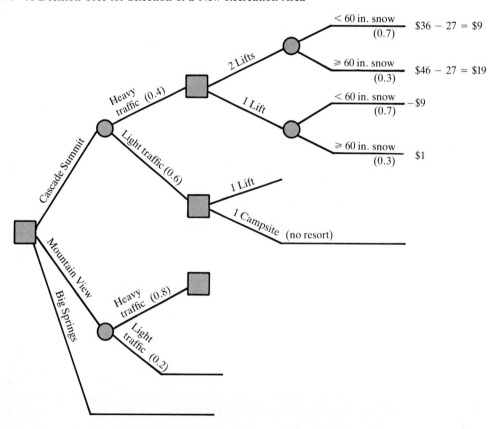

tives. Complete the decision tree and use it to select the best course of action based on the highest expected monetary advantage.

Recreational Area Analysis ($ millions)

| | Road Cost | Ski Lift Costs | | Costs per Campground | Resort Cost |
| --- | --- | --- | --- | --- | --- |
| | | *1 Lift* | *2 Lifts* | | |
| Cascade Summit | $18 | $3 | $5 | $0.6 | $4 |
| Mountain View | 12 | 3 | — | 0.8 | 3 |
| Big Springs | 4 | — | — | — | 5 |

| | Revenues from Ski Lifts | | | | Revenues from Campgrounds | | | |
| --- | --- | --- | --- | --- | --- | --- | --- | --- |
| | *1 Lift* | | *2 Lifts* | | *1 Campground* | | *2 Campgrounds* | |
| (Snow & Traffic) | <60″ | ⩾60″ | <60″ | ⩾60″ | Heavy | Light | Heavy | Light |
| Cascade Summit | $10 | $20 | $30 | $40 | — | $0.6 | — | — |
| Mountain View | 8 | 22 | — | — | 1.0 | 0.8 | 1.8 | — |
| Big Springs | — | — | — | — | — | — | — | — |

36. A management analyst has made a utility function study of an airline executive, Sue Warren, and found that she has a utility value of 20 utiles for a $1000 gain and 70 utiles for a $30,000 gain. In addition, on a lottery assessment she was indifferent to having $10,000 for certain versus a lottery with a 0.5 probability of a $1000 gain and a 0.5 probability of a $30,000 gain.
    a. Find the executive's utility value for $10,000.
    b. Assume that Sue is indifferent to receiving $15,000 for certain versus a lottery with a 0.4 probability of gaining $1000 and a 0.6 chance of gaining $30,000. What is her utility value for $15,000?
    c. Further studies show the executive is indifferent to receiving a $1000 gain for certain versus a lottery that involves a loss of $5000 at $(P = 0.3)$ and a gain of $30,000 (at $1 - P = 0.7$). What is her utility value for the $5000 loss?
    d. Another point of indifference on Sue's utility function is for a certain amount of $30,000 versus a lottery with a 0.9 probability of $50,000 and 0.1 probability of $1000. What is her utility for $50,000?
    e. Graph Sue's utility function and comment upon her attitude toward risk.

37. Use the utility function of Example 18.14 and assume the broker has an opportunity for a certain profit of $8000. What probabilities must go with a reference lottery of $3000 and $12,000 to make the broker indifferent?

38. The marketing manager of a new consumer product doesn't know whether competing companies will copy her company's design and is attempting to develop an appropriate pricing policy. The payoffs for her strategies are estimated to be as shown (in thousands of dollars). An analyst working with the marketing manager has estimated that there is a 70% chance the design will be copied. In attempting to develop a utility function, the analyst has assigned arbitrary utility values of zero and 100 utiles to the payoffs of $10 and $60, respectively. Further analysis revealed that the marketing manager equated a certain $30 payoff with the package of an 80% chance of $60 and a 20% chance of $10.

|  | Payoff Table If Competition | |
|---|---|---|
|  | *Copies* | *Does Not Copy* |
| $A_1$ = Price low | 30 | 30 |
| $A_2$ = Price high | 10 | 60 |

a. Develop a payoff matrix where the payoff amounts are in utility units of utiles.
b. Graph the manager's utility function.
c. Find the optimal strategy on (i) an EMV basis and (ii) an EUV basis.

39. You feel there is a 20% chance your competitor will succeed in research on a fuel cell that could revolutionize your industry. If so, the competitor will introduce the product in three years. If you begin research now you could positively develop a better product, but unless he also introduces his product, your money would be wasted for lack of a market. If you start research and it turns out your competitor introduces the product, you stand to gain $2 million in profits, but if he does not introduce it you will lose $200,000. On the other hand, if you do not undertake the research, you stand to make $400,000 if he does introduce it and $150,000 if he does not.

    You are the decision maker and have to make a decision based on what you know. Evaluate this decision with respect to several criteria, choose a course of action, and defend your choice.

**STATISTICS IN ACTION: CHALLENGE PROBLEMS**

40. Epstein and Wilamowsky point out that whereas the expected-value criterion is correct in the "long run," the number of truly long-run decision situations may, in fact, be very limited. They also use the St. Petersburg paradox to illustrate a case in which the expected-value criterion is not appropriate for analyzing the long-run profit outcome.

    Consider the paradox as follows. A player repeatedly tosses a coin until a tail appears, at which point the game ends. The "house" then pays the player $2^n$, where $n$ is the toss that resulted in the first tail. The authors show that the expected payoff to the player is an infinite number of dollars:

$$E(\text{Payoff}) = \$2^1\left(\frac{1}{2^1}\right) + \$2^2\left(\frac{1}{2^2}\right) + \$2^3\left(\frac{1}{2^3}\right) + \cdots = \sum_{n=1}^{\infty} 2^n\left(\frac{1}{2^n}\right) = \sum_{n=1}^{\infty} 1 = \infty$$

Attempts to resolve the St. Petersburg paradox by the following techniques fail: (1) questioning the possible infinite number of tosses, (2) questioning the assigned probabilities, and (3) disputing the monetary payoffs. The authors argue that the St. Petersburg game is not amenable to using the expected-value concept.

SOURCE: Epstein, Sheldon, and Yonah Wilamowsky. "Expected Value as a Decision-Making Criterion." *The Mid-Atlantic Journal of Business* 19, no. 2 (Summer 1981): 49–55.

Question *An Infinite Length Game* Suppose the game rules were changed so that the player could not continue past 15 trials. Under this rule, what is the expected payoff?

41. *Realistic Payoff Probabilities* Assume that an event (i.e., each individual outcome from an experiment) with a probability of less than one in a million is effectively zero. Under this rule, what is the expected payoff?

42. *Realistic Payoffs*   Arguing that the utility function for money is not linear, another researcher has suggested a utility function of a dollar return, $R$, as $U(R) = \sqrt{R}$. Using this function, we find that the expected utility of the St. Petersburg game is

$$E(u) = \sum_{n=1}^{\infty} \sqrt{2^n}\left(\frac{1}{2^n}\right) = \sum 1\sqrt{2^n} = 2.41$$

This corresponds to a monetary payoff of $\$2.41^2 = \$5.81$. However, still another researcher showed the fallacy of this isolated case by substituting a different payoff amount. Suppose the payoff is $\$2^{2n}$ (instead of $\$2^n$). What is the expected utility?

In summary, while expected value is likely correct in the long run, it has short-run limitations and should not be used indiscriminately.

43. Alemi, Fos, and Lacorte used utility theory extensively in a study of health-care contract negotiations. Their six subjects included three managers who were frequently involved in negotiating contracts (M1, M2, M3) and two physicians plus a physician consultant (P1, P2, P3). First the individuals reached a consensus on what the negotiation issues were to be, and all participants were fully instructed to be as realistic as possible about achieving the best result they could. Then the subjects completed a questionnaire about their priorities in the upcoming negotiations. The researchers then assessed the utility for different resolutions of an issue, $U_{ij}$ by "asking the subjects to specify the best and worst resolution." Between these points, a linear interpolation was used.

An additive Multi-Attribute Utility model (AMAU) was then constructed for each subject's preferences, wherein the utility of a contract was the weighted sum of the preferences concerning the resolution of each issue. Finally, each subject negotiated a contract with each member of the other group. The contracts involved 43 issues and took between 45 and 90 minutes per contract.

Success was defined as the ability of a negotiator to achieve his own high-priority goals, recognizing that this entailed sacrificing some lower-priority goals. The sketch below illustrates a representative success rate for one member of each group.

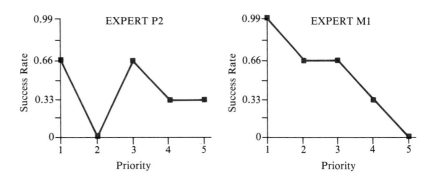

The authors also constructed graphs to show the optimal and negotiated contracts between managers and physicians. In the next sketch, optimal contracts are shown by the top and right perimeter lines. The actual negotiated contract is identified by a point on the graph, ▲, and the shaded area shows how much the negotiation could have been improved. The optimal curve is constructed from three points (one where the physician's preferences are ignored, one where the manager's preferences are ignored, and one where each is treated equally).

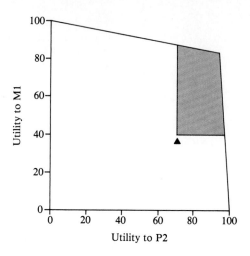

SOURCE: Alemi, Farrokh, Peter Fos, and William Lacorte. "A Demonstration of Methods for Studying Negotiations Between Physicians and Health Care Managers." *Decision Sciences* 21, no. 3 (Summer 1990): 633–641.

Question    Assuming that the success-rate priority graphs are representative of the two groups, which group was better at achieving its (predetermined) priorities?

44. Some reference in the above discussion suggests that the utilities across the different issues could be added to estimate the total utility of the contract. Locate this reference. Do you feel this is a valid assumption?

45. Based on the data above, did manager M1 and physician P2 arrive at an optimal resolution of their negotiation?

46. Identify any concerns you feel might limit the generalizations one could draw about whether managers or physicians are better negotiators.

---

**BUSINESS CASE**

*Constructing Your Own Utility Function*

Plastronics Keyboard manufactures keyboards for computers and has been asked by a major customer to gear up for heavier production. The general manager of operations has identified three possible responses and the payoffs that might be expected over the next five years depending on whether the market demand is weak, moderately good, or very strong. The action alternatives, with payoff amounts in millions of dollars, are shown in the table below.

Data for Plastronics Keyboard

|  | Payoff ($ millions) when market demand is | | |
|---|---|---|---|
|  | $\theta_1 = Weak$ | $\theta_2 = Good$ | $\theta_3 = Strong$ |
| $A_1$ = Retain existing capacity | 100 | 150 | 150 |
| $A_2$ = Expand existing capacity | 90 | 180 | 200 |
| $A_3$ = Build new plant | 0 | 160 | 320 |

This application has been designed to allow you to develop your own utility function for the situation described above. To do this, complete the following steps and fill in the values requested. In Part I, the evaluations will be for the purpose of establishing utility values for your own curve based upon the monetary amounts above. In Part II, other values will be considered in addition to monetary amounts.

*Part I*

47. a. *Deriving your utility values* To derive your utility value, let $Z_w$ = the least preferable outcome $(A_3\theta_1)$ and $Z_b$ = the most preferable outcome $(A_3\theta_3)$.

Arbitrarily, let the utility of $Z_w$ be 0, $[U(A_3\theta_1) = 0]$, and of $Z_b$ be 1, $[U(A_3\theta_3) = 1]$. Then determine a value of each $(A\theta)$ outcome such that you are totally indifferent between receiving that outcome or receiving a gamble that offers you a 50% chance of $Z_b$ along with a 50% chance of $Z_w$. To derive this value, assume you are questioned as follows:

**Question** You have a chance to obtain $Z_b$ ($320) with probability 0.5 and $Z_w$ ($0) with a probability of 0.5. What amount of money to be received for certain would make you indifferent to receiving the certain amount versus the chance of $Z_b$ and $Z_w$?

**Your response** I'm not sure I understand.

**Question** Consider the situation posed (in Figure 18.13). Using that as a reference, assume you have a 50% chance of getting $320 along with a 50% chance of nothing. Would you trade that for some certain amount?

**Your response** Sure, if the amount was enough.

**Question** What would be the minimum amount? That is, at what amount would you consider both alternatives equal?

**Your response** Maybe $125.

**Question** Then $125 is your certainty equivalent, which we will call *M*.

---

**FIGURE 18.13**   Initial Decision Situation in Development of a Utility Function

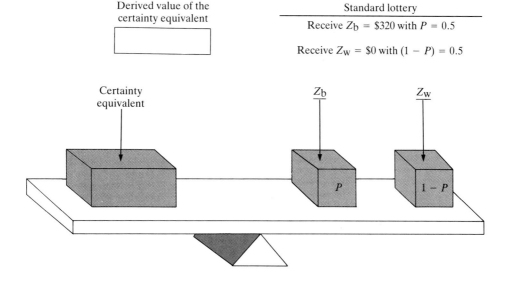

Derived value of the certainty equivalent

Standard lottery

Receive $Z_b$ = $320 with $P = 0.5$

Receive $Z_w$ = $0 with $(1 - P) = 0.5$

Certainty equivalent

$Z_b$

$Z_w$

$P$

$1 - P$

In symbolic form, the amount $M$ is equivalent to the lottery amount, and substituting the utility values of 1.00 for $Z_b$ and 0 for $Z_w$ we can show that the utility value of the certainty equivalent $M$ is 0.50:

$$U[\$M] = PU[\$Z_b] + (1 - P)U[\$Z_w]$$
$$= 0.5U[\$320] + (0.5)U[\$0]$$
$$= 0.5(1.00) + 0.5(0)$$
$$= 0.50$$

The utility of \$_____ is 0.5 utiles (fill in with your own value of $M$).

A step-by-step discussion of the procedure used to establish points $B$ and $C$ is shown in Table 18.8.

b. *Graphing your utility function*  If other intermediate points are necessary, use a similar procedure. Then plot the points and graph your utility function in Figure 18.14.

c. *Using your utility function*  Determine your own preference values for the payoffs of the original decision problem by reading the appropriate utility values from the curve. Show them in the table of utility functions.

d. *Computing expected utility values*  Assume that the probabilities associated with the various states of demand are $\theta_1 = 0.2$, $\theta_2 = 0.5$, and $\theta_3 = 0.3$. Using these values and the utility values from the table you derived in part (c), compute the expected utility value for each action.

$$\text{EMV}[A_1] = \text{\_\_\_\_\_} \qquad \text{EMV}[A_2] = \text{\_\_\_\_\_} \qquad \text{EMV}[A_3] = \text{\_\_\_\_\_}$$

**TABLE 18.8**  Procedure to Establish Points $B$ and $C$

Let us designate the first intermediate point as $B$. To establish point $B$, fill in your value of $M$ and then assess your certainty equivalent, $B$.

We now have $U(Z_w = 0, \boxed{\phantom{XXX}} = 0.5,$ and $\$320 = 1.0)$.

| Standard lottery | Derived value of |
|---|---|
| $Z_b = \$320$ with $P = 0.5$ | certainty equivalent, $B$ |
| $M = \boxed{\$\phantom{XXX}}$ with $(1 - P) = 0.5$ | $\boxed{\$\phantom{XXX}}$ |

Then,  $U[\$B] = PU[Z_b] + (1 - P)U[\$M]$
$$= 0.5(1.0) + (0.5)(0.5)$$
$$= 0.5 + 0.25 = 0.75$$

Let the next intermediate point be $C$. To establish point $C$, fill in the values in the cells provided.

| Standard lottery | Derived value of |
|---|---|
| $Z_w = 0$ with $P = 0.5$ | certainty equivalent, $C$ |
| $M = \boxed{\$\phantom{XXX}}$ with $(1 - P) = 0.50$ | $\boxed{\$\phantom{XXX}}$ |

Then,  $U[\$C] = PU[Z_w] + (1 - P)U[\$M]$
$$U[\$C] = \boxed{\phantom{XXXX}}$$

**FIGURE 18.14**   Graph of an Individual Utility Function

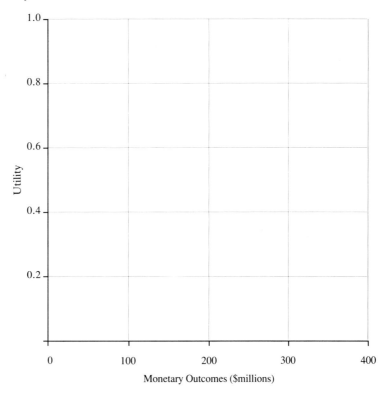

Table of Utility Functions

| Action | $\theta_1$ Weak | $\theta_2$ Good | $\theta_3$ Strong |
|---|---|---|---|
| $A_1$ = retain | | | |
| $A_2$ = expand | | | |
| $A_3$ = build | | | |

Selecting the course with the highest EUV might be different from selecting the course with the highest expected monetary value (EMV). Compute the EMV for the three courses of action and compare the results.

$$\text{EUV}[A_1] = \underline{\hspace{1cm}}$$

$$\text{EUV}[A_2] = \underline{\hspace{1cm}}$$

$$\text{EUV}[A_3] = \underline{\hspace{1cm}}$$

Which course of action would you choose?

*Part II*

For this part of the Business Case, suppose the impacts of the decision alternatives in the Plastronics Keyboard situation are assessed further, with the consequences as summarized in Table 18.9. It shows some additional market demand considerations for the decision under study. Our procedure in this part will be aimed at determining a utility value for each $(A, \theta)$ combination.

**TABLE 18.9** Market Demand Considerations

| | Market Demand | | |
| --- | --- | --- | --- |
| | $\theta_1 = Weak$ | $\theta_2 = Good$ | $\theta_3 = Strong$ |
| $A_1$ = Retain existing capacity | Profit = $100 million<br><br>Employment steady<br><br>Increase market share by 2%<br><br>No labor union; secure and informal work environment<br><br>No relocation of employees<br><br>Older facilities<br><br>Tight operating budget | Profit = $150 million<br><br>Employment up 10%<br><br>Hold current market share<br><br>No labor union; secure and informal work environment<br><br>No relocation of employees<br><br>Older facilities<br><br>Reasonable operating budget | Profit = $150 million<br><br>Employment up 10%<br><br>Lose 4% of market share<br><br>No labor union; secure and informal work environment<br><br>No relocation of employees<br><br>Older facilities<br><br>Reasonable operating budget |
| $A_2$ = Expand existing capacity | Profit = $90 million<br><br>Employment down 10%<br><br>Hold current market share<br><br>No labor union; semiformal work environment<br><br>No relocation of employees<br><br>Old and new facilities<br><br>Tight operating budget | Profit = $180 million<br><br>Employment up 20%<br><br>Increase market share by 2%<br><br>No labor union; semiformal work environment<br><br>No relocation of employees<br><br>Old and new facilities<br><br>Tight operating budget | Profit = $200 million<br><br>Employment up 40%<br><br>Increase market share by 4%<br><br>No labor union; semiformal work environment<br><br>Minor relocation of 10% of employees<br><br>Old and new facilities<br><br>Reasonable operating budget |
| $A_3$ = Build new plant | Profit = $0 million<br><br>Employment down 15%<br><br>Hold current market share<br><br>Unionized workers; formal work environment<br><br>Relocate 40% of employees<br><br>New facilities and equipment<br><br>Very tight operating budget | Profit = $160 million<br><br>Employment up 20%<br><br>Increase market share by 3%<br><br>Unionized workers; formal work environment<br><br>Relocate 50% of employees<br><br>New facilities and equipment<br><br>Tight operating budget | Profit = $320 million<br><br>Employment up 110%<br><br>Increase market share by 12%<br><br>Unionized workers; formal work environment<br><br>Relocate 70% of employees<br><br>New facilities and equipment<br><br>Reasonable operating budget |

**FIGURE 18.15**    Selection of $Z_{\text{best}}$ and $Z_{\text{worst}}$ Conditions to Serve as Endpoints of Utility and Function

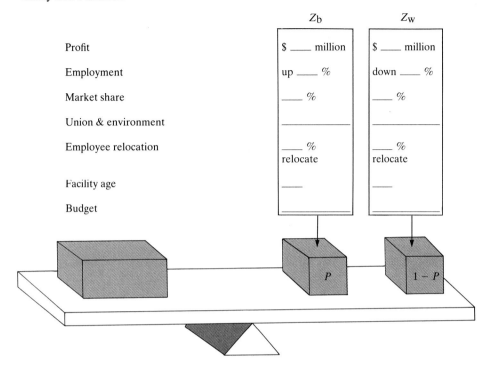

48. a. Follow a procedure similar to the preceding one to assess your utility for the various packages of outcomes. Begin by selecting what you consider to be the $Z_{\text{best}}$ and $Z_{\text{worst}}$ conditions with respect to profit, employment, market share, and so on. Fill in the values in the space provided in Figure 18.15. These values will establish the endpoints for your utility function for this decision.

b. Using a self-questioning and calculation procedure, find the utility values associated with each outcome. In this case, however, it will be easier to find the probability value for a reference gamble equivalent to the outcome. For example, to evaluate the outcome $(A_1\theta_1)$, the approach might be as follows.

**Question** You have a chance to obtain $Z_b$ ($320 million profit, employment up _____%, etc.) with probability $(P)$ or $Z_w$ ($0 profit, employment down _____ %, etc.) with probability $(1 - P)$. What probability would you assign to $P$ such that you are indifferent to that result versus the outcome $(A_1\theta_1)$?

**Response** This assumes I could have the outcome $(A_1\theta_1)$ for certain, versus a chance $P$ of getting $Z_b$ and $(1 - P)$ of getting $Z_w$?

**Question** Yes. Select a value $P$ for which you would be indifferent to receiving either the outcome $A_1\theta_1$ or the package of $Z_b$ and $Z_w$ (together). What chance at $Z_b$ do you feel is the certainty-equivalent to $A_1\theta_1$?

**Response** (Assume a value of 0.80 for illustration purposes.) I would choose $P(Z_b) = 0.80$ and $P(Z_w) = 1 - P(Z_b) = 1.00 - 0.80 = 0.20$.

This response would equate the utility of retaining existing capacity in a weak demand market to an 80% chance of the best conceivable outcome along with a 20% chance of the worst outcome. The utility value for $A_1\theta_1$ can now be determined by setting up an equation expressing this relationship:

$$U(A_1\theta_1) = 0.80(U[Z_b]) + 0.20(U[Z_w])$$
$$= 0.80(100) + 0.20(0) = 80 \text{ utiles}$$

We can determine utility values for the other outcomes by substituting them for $(A_1\theta_1)$ and repeating the process. Determine your utility for each alternative and record it in a table (such as the one shown).

Utility Values for Alternative Outcomes

| | Market Demand | | |
|---|---|---|---|
| | $\theta_1$ *Weak* | $\theta_2$ *Good* | $\theta_3$ *Strong* |
| $A_1$ = Retain | | | |
| $A_2$ = Expand | | | |
| $A_3$ = Build | | | |

c. Assume that we estimate the probabilities associated with the various states of demand as $\theta_1 = 0.2$, $\theta_2 = 0.5$, and $\theta_3 = 0.3$. Use the maximum-likelihood method of choice to arrive at a decision for the Plastronics Keyboard situation and indicate your recommended action. Is your choice any different than when you took *only* monetary values into account? Explain.

## FOR FURTHER READING

[1] Albright, S. Christian. *Statistics for Business and Economics.* New York: Macmillan, 1987.

[2] Edwards, W. "The Theory of Decision Making," in W. Edwards and A. Tversky (eds.), *Decision Making.* Middlesex, England: Penguin Books, 1967.

[3] Hamburg, Morris. *Statistical Analysis for Decision Making*, 5th ed. San Diego: Harcourt Brace Jovanovich, 1991.

[4] Hammond, John S. "Better Decisions with Preference Theory," *Harvard Business Review* (Nov./Dec. 1967): 123–141.

[5] Hampton, J. M., P. G. Moore, and H. Thomas. "Subjective Probability and Its Measurement." *Journal of the Royal Statistical Society A* 136 (1973): 21–42.

[6] Hull, J., P. G. Moore, and H. Thomas. "Utility and Its Measurement." *Journal of the Royal Statistical Society A* 136 (1973): 226–247.

[7] Keeney, Ralph L. "Decision Analysis: An Overview." *Journal of Operations Research* 30 (Sept./Oct. 1982): 803–838.

[8] Lapin, Lawrence L. *Statistics for Modern Business Decisions*, 5th ed. San Diego: Harcourt Brace Jovanovich, 1990.

[9] Magee, John. "Decision Trees for Decision Making." *Harvard Business Review* (July/Aug. 1964): 126–138.

[10] Monks, Joseph G. "A Utility Approach to R&D Decisions." *R&D Management: Manchester Business School* 6, no. 2 (1976): 59–66.

[11] Pain, N. R. "A Useful Approach to the Group Choice Problem." *Decision Sciences* 4, no. 1 (Jan. 1975): 21–30.

[12] Schlaifer, R. *Analysis of Decision Under Uncertainty.* New York: McGraw-Hill, 1969. (Reprint ed., Melbourne, FL: Krieger, 1978.)

[13] Swalm, R. O. "Utility Theory—Insights into Risk Taking." *Harvard Business Review* (Nov./Dec. 1966): 123–126.

[14] Von Neumann, John, and O. Morgenstern. *Theory of Games and Economic Behavior*. Princeton, NJ: Princeton University Press, 1947.

[15] Watson, Collin S., et al. *Statistics for Management and Economics*, 4th ed. Boston: Allyn and Bacon, 1990.

[16] Winkler, R. L. "Research Directions in Decision Making under Uncertainty." *Decision Sciences* 13, no. 4 (Oct. 1982): 517–533.

[17] Zeleny, M. *Multiple Criteria Decision Making*. New York: McGraw-Hill, 1982.

# APPENDIXES

**APPENDIX A**
DATA BASE AND DATA DICTIONARY

**APPENDIX B**
STATISTICAL TABLES

**APPENDIX C**
MATH ESSENTIALS FOR STATISTICS

## APPENDIX A
## DATA BASE AND
## DATA DICTIONARY

The following pages list the variables with their codes defined for the information in our PUMS data base. As much as possible, we have retained the information and codes as they appear in the original Census Bureau publication. Refer to the *Public Use Microdata Samples (PUMS) Technical Documentation* if you wish to relate our data to the source.

**Sample of 100 Records from the PUMS A File**

Table A.1 includes a random sample of 100 records taken from the PUMS A file. The record layout appears in Figure A.1. This is also the format of the 20,116-record total PUMS data base, which is available to instructors on computer tape or cartridge. The 100 records are used to answer some of the computer problems that appear at the end of each chapter in this textbook.

The variables on the record layout are defined beginning in Table A.2. Therein the PUMS report locations are presented to the left of the data base record positions. Thus, if it is needed, you can check back to the source.

The complete data base file of 20,116 records can be obtained from the publisher.

**TABLE A.I** PUMS Data Base—100-Record Sample (Figure A.1, following, gives a detailed data format.)

**FIGURE A.1**    Detailed Layout of Data Format for PUMS Data Base Files
(see Table A.1)

--- Household Information ---

| Posn | Field | Positions |
|------|-------|-----------|
| | RECNUMBR | 1 2 3 4 5 |
| | STATECD | 6 7 |
| | MSA CODE | 8 9 10 11 |
| | #PERS | 12 13 |
| | TENURE | 14 |
| | #UNITS | 15 16 |
| | #BDRMS | 17 18 |
| | #BATHOSS | 19 |
| | YR AMVD IN | 20 |
| | MO COST ELEC | 21 22 23 |
| | MO COST GAS | 24 25 26 |
| | TOTAL MO PMT TO LENDR | 27 28 29 30 |
| | GROSS RENT | 31 32 33 |
| | HSHDTP | 34 |
| | HSHD INCOME | 35 36 37 38 39 |

Person #1

| Field | Positions |
|-------|-----------|
| RELATN | 40 41 |
| AGE | 42 43 |
| MARITL | 44 45 |
| RACE | 46 47 |
| HISP / EDT | 48 49 |
| WKS WRKD | 50 51 |
| HRS WRKD | 53 54 |
| INC FROM ALL SOURCES | 55 56 57 58 59 |

Person #2

| Field | Positions |
|-------|-----------|
| RELATN | 60 61 |
| AGE | 62 63 |
| MARITL | 64 65 |
| HISP / ED | 66 67 |
| WKS WRKD | 68 69 |
| HRS WRKD | 70 71 |
| INC FROM ALL SOURCES | 72 73 74 75 76 77 78 79 |

#3

| Field | Positions |
|-------|-----------|
| RELATN | 78 79 |
| AGE | 80 81 |

#4

| Field | Positions |
|-------|-----------|
| RELATN | 82 83 |
| AGE | 84 85 |

#5

| Field | Positions |
|-------|-----------|
| RELATN | 86 87 |
| AGE | 88 89 |

#6

| Field | Positions |
|-------|-----------|
| RELATN | 93 94 |
| AGE | 95 96 |

**TABLE A.2**  Data Dictionary

*Note:* All PUMS records that are vacant households or group quarters have been excluded from this data base.

| PUMS (source) | Data Base | Variable Descriptions |
|---|---|---|
| | **Positions** | |
| | 1–5 | Record number—00001 through 20,116 was generated for a random sample of 20,116 records selected from the PUMS A file. |
| H4–5 | 6–7 | FIPS* State Code |

|  |  |  |  |  |
|---|---|---|---|---|
| 01 Alabama | 15 Hawaii | 25 Massachusetts | 35 New Mexico | 46 South Dakota |
| 02 Alaska | 16 Idaho | 26 Michigan | 36 New York | 47 Tennessee |
| 04 Arizona | 17 Illinois | 27 Minnesota | 37 North Carolina | 48 Texas |
| 05 Arkansas | 18 Indiana | 28 Mississippi | 38 North Dakota | 49 Utah |
| 06 California | 19 Iowa | 29 Missouri | 39 Ohio | 50 Vermont |
| 08 Colorado | 20 Kansas | 30 Montana | 40 Oklahoma | 51 Virginia |
| 09 Connecticut | 21 Kentucky | 31 Nebraska | 41 Oregon | 53 Washington |
| 10 Delaware | 22 Louisiana | 32 Nevada | 42 Pennsylvania | 54 West Virginia |
| 11 District of Columbia | 23 Maine | 33 New Hampshire | 44 Rhode Island | 55 Wisconsin |
| 12 Florida | 24 Maryland | 34 New Jersey | 45 South Carolina | 56 Wyoming |
| 13 Georgia | | | | |

| PUMS (source) | Data Base | Variable Descriptions |
|---|---|---|
| H10–13 | 8–11 | Metropolitan Statistical Area, MSA, Codes<br>0000–N/A, areas outside of an MSA<br>0040–9340–FIPS MSA Codes<br>9999–county group consisting of two plus MSAs, or a mixed MSA/non-MSA area<br>(See the PUMS Technical Documentation, or a Rand-McNally Atlas, for MSA Codes. Where *this* is required in problems, the MSA Code is given.) |
| H26–27 | 12–13 | Persons – Number of Person Records for this Housing Unit<br>01 – one person<br>02 – 31 |
| H29 | 14 | Tenure – Type of Ownership<br>1 – owner occupied<br>2 – renter occupied, with cash rent<br>3 – renter occupied, no cash rent |
| H36–37 | 15–16 | Units in Structure<br>01 – Mobile home or trailer        06 – building for 5 to 9 families<br>02 – one family, detached       07 – building for 10–19 families<br>03 – one family, attached       08 – building for 20–49 families<br>04 – building for 2 families     09 – building for 50+ families<br>05 – building for 3 or 4 families   10 – boat, tent, van, etc. |
| H45 | 17 | Bedrooms<br>1 – none                4 – three bedrooms<br>2 – one bedroom         5 – four bedrooms<br>3 – two bedrooms        6 – five or more bedrooms |
| H48 | 18 | Bathrooms<br>1 – no bath or a half bath      3 – one complete + half bath(s)<br>2 – one complete bathroom     4 – two or more complete baths |
| H56 | 19 | Automobiles Available<br>1 – none              3 – two<br>2 – one               4 – three or more |

\* Federal Information Processing Standards (FIPS) as developed by the Census Bureau.

**TABLE A.2** (*continued*)

| Positions | | Variable Descriptions |
|---|---|---|
| *PUMS (source)* | *Data Base* | |

| | | |
|---|---|---|
| H59 | 20 | Year Householder Moved into Unit<br>1 – 1979 to March 1980　　　4 – 1960 to 1969<br>2 – 1975 to 1978　　　　　　　5 – 1950 to 1959<br>3 – 1970 to 1974　　　　　　　6 – 1949 or earlier |
| H67–69 | 21–23 | Monthly Cost of Electricity<br>000 – N/A, or no payment for electricity<br>001–199 – cost in dollars<br>200 – $200 or more |
| H71–73 | 24–26 | Monthly Cost of Gas<br>000 – N/A, or no payment for gas<br>001–149 – cost in dollars<br>150 – $150 or more |
| H89–92 | 27–30 | Total Monthly Payment to Lender<br>0000 – no regular payment required, includes renters or N/A<br>0001–1499 – payment in dollars<br>1500 – $1500 or more |
| H101–103 | 31–33 | Gross Rent<br>000–N/A, owner-occupied unit, or no payment required<br>001–998 – gross rent in dollars<br>999 – $999 or more |
| H104 | 34 | Household Type<br>1 – married couple family　　　　　　　　　　　　　　　3 – as 2 but interchange male, female<br>2 – family household with male householder, no female　4 – nonfamily household |
| H107–111 | 35–39 | Household Income in 1979<br>00000 – no income or N/A<br>1–74995 – income in dollars<br>75000 – income of $75,000 or more |
| P2–3 | 40–41 | Relationship<br>−1, or blank – no person (occurs only for persons 2 through 6)<br>0 – householder<br>Family member other than householder:<br>1 – *spouse*　　　　　　　　4 – parent<br>2 – child　　　　　　　　　5 – other relative<br>3 – brother or sister<br>Person not related to householder:<br>6 – roomer or boarder　　　8 – paid employee<br>7 – partner or roommate　　9 – other nonrelative |
| P7 | 42 | Sex<br>−1 – blank – no person<br>0 – male<br>1 – female |
| P8–9 | 43–44 | Age<br>−1, blank – no person<br>00–89 – age in years<br>90–90 or more years |

| Positions | | |
|---|---|---|
| *PUMS (source)* | *Data Base* | Variable Descriptions |

| | | |
|---|---|---|
| P11 | 45 | Marital Status<br>−1, blank – no person<br>0 – now married, except separated<br>1 – widowed      3 – separated<br>2 – divorced      4 – single or N/A (under 15 yrs of age) |
| P12–13 | 46–47 | Race<br>−1, blank – no person      7 – Korean<br>1 – white Caucasian      8 – Asian Indian<br>2 – black      9 – Vietnamese<br>3 – American Indian, Eskimo, Aleut, Asian and Pacific      10 – Hawaiian<br>Islander      11 – other Asian and Pacific Islander<br>4 – Japanese      Other:<br>5 – Chinese      12 – Spanish write-in entry<br>6 – Filipino      13 – other |
| P40 | 48–49 | Highest Year of School Attended<br>−1, blank – no person<br>0 – never attended school or N/A (under 3 years of age)<br>1 – nursery school<br>2 – kindergarten<br>Elementary:<br>03–10 – first grade through eighth grade<br>Highschool:<br>11–14 – ninth grade through twelfth grade<br>College:<br>15–21 – first year through seventh year<br>22 – eighth year or more |
| P81 | 50 | (Labor Force) Work Status<br>−1, blank – no person<br>0 – N/A (under 16 years of age)<br>In labor force:<br>   Civilian labor force:<br>   Employed:<br>1 – at work<br>2 – with a job but not at work<br>3 – unemployed      5 – with a job but not at work<br>   Armed forces:      6 – not in labor force<br>4 – at work |
| P95–96 | 51–52 | Weeks Worked in 1979<br>−1, blank – no person<br>0 – N/A (under 16 years of age or did not work in 1979)<br>1–52 – weeks worked |
| P97–98 | 53–54 | Usual Hours Worked (outside of the home) per week in 1979<br>−1, blank – no person<br>0 – N/A (under 16 years of age or did not work in 1979)<br>1–98 – usual number of hours<br>99 – 99 or more hours per week |
| P134–138 | 55–59 | Income from all Sources in 1979<br>−1, blank – no person<br>00000 – N/A (under 15 years of age or no income from any source)<br>1–74995 – income in dollars<br>75000 – income of $75000 or more |
| | 60–98 | Descriptions for persons 2 through 6 using variables that were previously defined. |

# APPENDIX B
# STATISTICAL TABLES

**TABLE B.I** The Standard Normal (Z) Distribution

| Z | .00 | .01 | .02 | .03 | .04 | .05 | .06 | .07 | .08 | .09 |
|---|-----|-----|-----|-----|-----|-----|-----|-----|-----|-----|
| 0.0 | .0000 | .0040 | .0080 | .0120 | .0160 | .0199 | .0239 | .0279 | .0319 | .0359 |
| 0.1 | .0398 | .0438 | .0478 | .0517 | .0557 | .0596 | .0636 | .0675 | .0714 | .0753 |
| 0.2 | .0793 | .0832 | .0871 | .0910 | .0948 | .0987 | .1026 | .1064 | .1103 | .1141 |
| 0.3 | .1179 | .1217 | .1255 | .1293 | .1331 | .1368 | .1406 | .1443 | .1480 | .1517 |
| 0.4 | .1554 | .1591 | .1628 | .1664 | .1700 | .1736 | .1772 | .1808 | .1844 | .1879 |
| 0.5 | .1915 | .1950 | .1985 | .2019 | .2054 | .2088 | .2123 | .2157 | .2190 | .2224 |
| 0.6 | .2257 | .2291 | .2324 | .2357 | .2389 | .2422 | .2454 | .2486 | .2517 | .2549 |
| 0.7 | .2580 | .2611 | .2642 | .2673 | .2704 | .2734 | .2764 | .2794 | .2823 | .2852 |
| 0.8 | .2881 | .2910 | .2939 | .2967 | .2995 | .3023 | .3051 | .3078 | .3106 | .3133 |
| 0.9 | .3159 | .3186 | .3212 | .3238 | .3264 | .3289 | .3315 | .3340 | .3365 | .3389 |
| 1.0 | .3413 | .3438 | .3461 | .3485 | .3508 | .3531 | .3554 | .3577 | .3599 | .3621 |
| 1.1 | .3643 | .3665 | .3686 | .3708 | .3729 | .3749 | .3770 | .3790 | .3810 | .3830 |
| 1.2 | .3849 | .3869 | .3888 | .3907 | .3925 | .3944 | .3962 | .3980 | .3997 | .4015 |
| 1.3 | .4032 | .4049 | .4066 | .4082 | .4099 | .4115 | .4131 | .4147 | .4162 | .4177 |
| 1.4 | .4192 | .4207 | .4222 | .4236 | .4251 | .4265 | .4279 | .4292 | .4306 | .4319 |
| 1.5 | .4332 | .4345 | .4357 | .4370 | .4382 | .4394 | .4406 | .4418 | .4429 | .4441 |
| 1.6 | .4452 | .4463 | .4474 | .4484 | .4495 | .4505 | .4515 | .4525 | .4535 | .4545 |
| 1.7 | .4554 | .4564 | .4573 | .4582 | .4591 | .4599 | .4608 | .4616 | .4625 | .4633 |
| 1.8 | .4641 | .4649 | .4656 | .4664 | 4671 | .4678 | .4686 | .4693 | .4699 | .4706 |
| 1.9 | .4713 | .4719 | .4726 | .4732 | .4738 | .4744 | .4750 | .4756 | .4761 | .4767 |
| 2.0 | .4772 | .4778 | .4783 | .4788 | .4793 | .4798 | .4803 | .4808 | .4812 | .4817 |
| 2.1 | .4821 | .4826 | .4830 | .4834 | .4838 | .4842 | .4846 | .4850 | .4854 | .4857 |
| 2.2 | .4861 | .4864 | .4868 | .4871 | .4875 | .4878 | .4881 | .4884 | .4887 | .4890 |
| 2.3 | .4893 | .4896 | .4898 | .4901 | .4904 | .4906 | .4909 | .4911 | .4913 | .4916 |
| 2.4 | .4918 | .4920 | .4922 | .4925 | .4927 | .4929 | .4931 | .4932 | .4934 | .4936 |
| 2.5 | .4938 | .4940 | .4941 | .4943 | .4945 | .4946 | .4948 | .4949 | .4951 | .4952 |
| 2.6 | .4953 | .4955 | .4956 | .4957 | .4959 | .4960 | .4961 | .4962 | .4963 | .4964 |
| 2.7 | .4965 | .4966 | .4967 | .4968 | .4969 | .4970 | .4971 | .4972 | .4973 | .4974 |
| 2.8 | .4974 | .4975 | .4976 | .4977 | .4977 | .4978 | .4979 | .4979 | .4980 | .4981 |
| 2.9 | .4981 | .4982 | .4982 | .4983 | .4984 | .4984 | .4985 | .4985 | .4986 | .4986 |
| 3.0 | .4987 | .4987 | .4987 | .4988 | .4988 | .4989 | .4989 | .4989 | .4990 | .4990 |

Abridged from Table II of A. Hald, *Statistical Tables and Formulas* (New York: John Wiley, 1952). Reprinted by permission of John Wiley & Sons, Inc., Copyright ©.

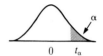

**TABLE B.2**   Student's $t$-Distribution

| Degrees of Freedom | $t_{.100}$ | $t_{.050}$ | $t_{.025}$ | $t_{.010}$ | $t_{.005}$ |
|---|---|---|---|---|---|
| 1 | 3.078 | 6.314 | 12.706 | 31.821 | 63.657 |
| 2 | 1.886 | 2.920 | 4.303 | 6.965 | 9.925 |
| 3 | 1.638 | 2.535 | 3.182 | 4.541 | 5.841 |
| 4 | 1.533 | 2.132 | 2.776 | 3.747 | 4.604 |
| 5 | 1.476 | 2.015 | 2.571 | 3.365 | 4.032 |
| 6 | 1.440 | 1.943 | 2.447 | 3.143 | 3.707 |
| 7 | 1.415 | 1.895 | 2.365 | 2.998 | 3.499 |
| 8 | 1.397 | 1.860 | 2.306 | 2.896 | 3.355 |
| 9 | 1.383 | 1.833 | 2.262 | 2.821 | 3.250 |
| 10 | 1.372 | 1.812 | 2.228 | 2.764 | 3.169 |
| 11 | 1.363 | 1.796 | 2.201 | 2.718 | 3.106 |
| 12 | 1.356 | 1.782 | 2.179 | 2.681 | 3.055 |
| 13 | 1.350 | 1.771 | 2.160 | 2.650 | 3.012 |
| 14 | 1.345 | 1.761 | 2.145 | 2.624 | 2.977 |
| 15 | 1.341 | 1.753 | 2.131 | 2.602 | 2.947 |
| 16 | 1.337 | 1.746 | 2.120 | 2.583 | 2.921 |
| 17 | 1.333 | 1.740 | 2.110 | 2.567 | 2.898 |
| 18 | 1.330 | 1.734 | 2.101 | 2.552 | 2.878 |
| 19 | 1.328 | 1.729 | 2.093 | 2.539 | 2.861 |
| 20 | 1.325 | 1.725 | 2.086 | 2.528 | 2.845 |
| 21 | 1.323 | 1.721 | 2.080 | 2.518 | 2.831 |
| 22 | 1.321 | 1.717 | 2.074 | 2.508 | 2.819 |
| 23 | 1.319 | 1.714 | 2.069 | 2.500 | 2.807 |
| 24 | 1.318 | 1.711 | 2.064 | 2.492 | 2.797 |
| 25 | 1.316 | 1.708 | 2.060 | 2.485 | 2.787 |
| 26 | 1.315 | 1.706 | 2.056 | 2.479 | 2.779 |
| 27 | 1.314 | 1.703 | 2.052 | 2.473 | 2.771 |
| 28 | 1.313 | 1.701 | 2.048 | 2.467 | 2.763 |
| 29 | 1.311 | 1.699 | 2.045 | 2.462 | 2.756 |
| $\infty$ | 1.282 | 1.645 | 1.960 | 2.326 | 2.576 |

**TABLE B.3**  The Chi-Square Distribution (upper-tail area = $\alpha$)

| df | 0.995 | 0.99 | 0.975 | 0.95 | 0.90 | 0.75 | 0.25 | 0.10 | 0.05 | 0.025 | 0.01 | 0.005 |
|----|-------|------|-------|------|------|------|------|------|------|-------|------|-------|
| 1 | — | — | 0.001 | 0.004 | 0.016 | 0.102 | 1.323 | 2.706 | 3.841 | 5.024 | 6.635 | 7.879 |
| 2 | 0.010 | 0.020 | 0.051 | 0.103 | 0.211 | 0.575 | 2.773 | 4.605 | 5.991 | 7.378 | 9.210 | 10.597 |
| 3 | 0.072 | 0.115 | 0.216 | 0.352 | 0.584 | 1.213 | 4.108 | 6.251 | 7.815 | 9.348 | 11.345 | 12.838 |
| 4 | 0.207 | 0.297 | 0.484 | 0.711 | 1.064 | 1.923 | 5.385 | 7.779 | 9.488 | 11.143 | 13.277 | 14.860 |
| 5 | 0.412 | 0.554 | 0.831 | 1.145 | 1.610 | 2.675 | 6.626 | 9.236 | 11.071 | 12.833 | 15.086 | 16.750 |
| 6 | 0.676 | 0.872 | 1.237 | 1.635 | 2.204 | 3.455 | 7.841 | 10.645 | 12.592 | 14.449 | 16.812 | 18.548 |
| 7 | 0.989 | 1.239 | 1.690 | 2.167 | 2.833 | 4.255 | 9.037 | 12.017 | 14.067 | 16.013 | 18.475 | 20.278 |
| 8 | 1.344 | 1.646 | 2.180 | 2.733 | 3.490 | 5.071 | 10.219 | 13.362 | 15.507 | 17.535 | 20.090 | 21.955 |
| 9 | 1.735 | 2.088 | 2.700 | 3.325 | 4.168 | 5.899 | 11.389 | 14.684 | 16.919 | 19.023 | 21.666 | 23.589 |
| 10 | 2.156 | 2.558 | 3.247 | 3.940 | 4.865 | 6.737 | 12.549 | 15.987 | 18.307 | 20.483 | 23.209 | 25.188 |
| 11 | 2.603 | 3.053 | 3.816 | 4.575 | 5.578 | 7.584 | 13.701 | 17.275 | 19.675 | 21.920 | 24.725 | 26.757 |
| 12 | 3.074 | 3.571 | 4.404 | 5.226 | 6.304 | 8.438 | 14.845 | 18.549 | 21.026 | 23.337 | 26.217 | 28.299 |
| 13 | 3.565 | 4.107 | 5.009 | 5.892 | 7.042 | 9.299 | 15.984 | 19.812 | 22.362 | 24.736 | 27.688 | 29.819 |
| 14 | 4.075 | 4.660 | 5.629 | 6.571 | 7.790 | 10.165 | 17.117 | 21.064 | 23.685 | 26.119 | 29.141 | 31.319 |
| 15 | 4.601 | 5.229 | 6.262 | 7.261 | 8.547 | 11.037 | 18.245 | 22.307 | 24.996 | 27.488 | 30.578 | 32.801 |
| 16 | 5.142 | 5.812 | 6.908 | 7.962 | 9.312 | 11.912 | 19.369 | 23.542 | 26.926 | 28.845 | 32.000 | 34.267 |
| 17 | 5.697 | 6.408 | 7.564 | 8.672 | 10.085 | 12.792 | 20.489 | 24.769 | 27.587 | 30.191 | 33.409 | 35.718 |
| 18 | 6.265 | 7.015 | 8.231 | 9.390 | 10.865 | 13.675 | 21.605 | 25.989 | 28.869 | 31.526 | 34.805 | 37.156 |
| 19 | 6.844 | 7.633 | 8.907 | 10.117 | 11.651 | 14.562 | 22.718 | 27.204 | 30.144 | 32.852 | 36.191 | 38.582 |
| 20 | 7.434 | 8.260 | 9.591 | 10.851 | 12.443 | 15.452 | 23.828 | 28.412 | 31.410 | 34.170 | 37.566 | 39.997 |
| 21 | 8.034 | 8.897 | 10.283 | 11.591 | 13.240 | 16.344 | 24.935 | 29.615 | 32.671 | 35.479 | 38.932 | 41.401 |
| 22 | 8.643 | 9.542 | 10.982 | 12.338 | 14.042 | 17.240 | 26.039 | 30.813 | 33.924 | 36.781 | 40.289 | 42.796 |
| 23 | 9.260 | 10.196 | 11.689 | 13.091 | 14.848 | 18.137 | 27.141 | 32.007 | 35.172 | 38.076 | 41.638 | 44.181 |
| 24 | 9.886 | 10.856 | 12.401 | 13.848 | 15.659 | 19.037 | 28.241 | 33.196 | 36.415 | 39.364 | 42.980 | 45.559 |
| 25 | 10.520 | 11.524 | 13.120 | 14.611 | 16.473 | 19.939 | 29.339 | 34.382 | 37.652 | 40.646 | 44.314 | 46.928 |
| 26 | 11.160 | 12.198 | 13.844 | 15.379 | 17.292 | 20.843 | 30.435 | 35.563 | 38.885 | 41.923 | 45.642 | 48.290 |
| 27 | 11.808 | 12.879 | 14.573 | 16.151 | 18.114 | 21.749 | 31.528 | 36.741 | 40.113 | 43.194 | 46.963 | 49.645 |
| 28 | 12.461 | 13.565 | 15.308 | 16.928 | 18.939 | 22.657 | 32.620 | 37.916 | 41.337 | 44.461 | 48.278 | 50.993 |
| 29 | 13.121 | 14.257 | 16.047 | 17.708 | 19.768 | 23.567 | 33.711 | 39.087 | 42.557 | 45.722 | 49.588 | 52.336 |
| 30 | 13.787 | 14.954 | 16.791 | 18.493 | 20.599 | 24.478 | 34.800 | 40.256 | 43.773 | 46.979 | 50.892 | 53.672 |
| 40 | 20.707 | 22.164 | 24.433 | 26.509 | 29.051 | 33.660 | 45.616 | 51.805 | 55.758 | 59.342 | 63.691 | 66.766 |
| 50 | 27.991 | 29.707 | 32.357 | 34.764 | 37.689 | 42.942 | 56.334 | 63.167 | 67.505 | 71.420 | 76.154 | 79.490 |
| 60 | 35.534 | 37.485 | 40.482 | 43.188 | 46.459 | 52.294 | 66.981 | 74.397 | 79.082 | 83.298 | 88.379 | 91.952 |
| 70 | 43.275 | 45.442 | 48.758 | 51.739 | 55.329 | 61.698 | 77.577 | 85.527 | 90.531 | 95.023 | 100.425 | 104.215 |
| 80 | 51.172 | 53.540 | 57.153 | 60.391 | 64.278 | 71.145 | 88.130 | 96.578 | 101.879 | 106.629 | 112.329 | 116.321 |
| 90 | 59.196 | 61.754 | 65.647 | 69.126 | 73.291 | 80.625 | 98.650 | 107.565 | 113.145 | 118.136 | 124.116 | 128.299 |
| 100 | 67.328 | 70.065 | 74.222 | 77.929 | 82.358 | 90.133 | 109.141 | 118.498 | 124.342 | 129.561 | 135.807 | 140.169 |

From *Biometrika Tables for Statisticians*, vol. 2, E. S. Pearson and H. O. Hartley, Cambridge: The University Press, 1972 by permission of the Biometrika trustees.

**TABLE B.4**   F-Distribution: Upper 10 Percent Points

Numerator degrees of freedom

| df₂ \ df₁ | 1 | 2 | 3 | 4 | 5 | 6 | 7 | 8 | 9 | 10 | 12 | 15 | 20 | 24 | 30 | 40 | 60 | 120 | ∞ |
|---|---|---|---|---|---|---|---|---|---|---|---|---|---|---|---|---|---|---|---|
| 1 | 39.863 | 49.500 | 53.593 | 55.833 | 57.240 | 58.204 | 58.906 | 59.439 | 59.858 | 60.195 | 60.705 | 61.220 | 61.740 | 62.002 | 62.265 | 62.529 | 62.794 | 63.061 | 63.328 |
| 2 | 8.5263 | 9.0000 | 9.1618 | 9.2434 | 9.2926 | 9.3255 | 9.3491 | 9.3668 | 9.3805 | 9.3916 | 9.4081 | 9.4247 | 9.4413 | 9.4496 | 9.4579 | 9.4662 | 9.4746 | 9.4829 | 9.4912 |
| 3 | 5.5383 | 5.4624 | 5.3908 | 5.3426 | 5.3092 | 5.2847 | 5.2662 | 5.2517 | 5.2400 | 5.2304 | 5.2156 | 5.2003 | 5.1845 | 5.1764 | 5.1681 | 5.1597 | 5.1512 | 5.1425 | 5.1337 |
| 4 | 4.5448 | 4.3246 | 4.1909 | 4.1072 | 4.0506 | 4.0097 | 3.9790 | 3.9549 | 3.9357 | 3.9199 | 3.8955 | 3.8704 | 3.8443 | 3.8310 | 3.8174 | 3.8036 | 3.7896 | 3.7753 | 3.7607 |
| 5 | 4.0604 | 3.7797 | 3.6195 | 3.5202 | 3.4530 | 3.4045 | 3.3679 | 3.3393 | 3.3163 | 3.2974 | 3.2682 | 3.2380 | 3.2067 | 3.1905 | 3.1741 | 3.1573 | 3.1402 | 3.1228 | 3.1050 |
| 6 | 3.7759 | 3.4633 | 3.2888 | 3.1808 | 3.1075 | 3.0546 | 3.0145 | 2.9830 | 2.9577 | 2.9369 | 2.9047 | 2.8712 | 2.8363 | 2.8183 | 2.8000 | 2.7812 | 2.7620 | 2.7423 | 2.7222 |
| 7 | 3.5894 | 3.2574 | 3.0741 | 2.9605 | 2.8833 | 2.8274 | 2.7849 | 2.7516 | 2.7247 | 2.7025 | 2.6681 | 2.6322 | 2.5947 | 2.5753 | 2.5555 | 2.5351 | 2.5142 | 2.4928 | 2.4708 |
| 8 | 3.4579 | 3.1131 | 2.9238 | 2.8064 | 2.7264 | 2.6683 | 2.6241 | 2.5893 | 2.5612 | 2.5380 | 2.5020 | 2.4642 | 2.4246 | 2.4041 | 2.3830 | 2.3614 | 2.3391 | 2.3162 | 2.2926 |
| 9 | 3.3603 | 3.0065 | 2.8129 | 2.6927 | 2.6106 | 2.5509 | 2.5053 | 2.4694 | 2.4403 | 2.4163 | 2.3789 | 2.3396 | 2.2983 | 2.2768 | 2.2547 | 2.2320 | 2.2085 | 2.1843 | 2.1592 |
| 10 | 3.2850 | 2.9245 | 2.7277 | 2.6053 | 2.5216 | 2.4606 | 2.4140 | 2.3772 | 2.3473 | 2.3226 | 2.2841 | 2.2435 | 2.2007 | 2.1784 | 2.1554 | 2.1317 | 2.1072 | 2.0818 | 2.0554 |
| 11 | 3.2252 | 2.8595 | 2.6602 | 2.5362 | 2.4512 | 2.3891 | 2.3416 | 2.3040 | 2.2735 | 2.2482 | 2.2087 | 2.1671 | 2.1230 | 2.1000 | 2.0762 | 2.0516 | 2.0261 | 1.9997 | 1.9721 |
| 12 | 3.1765 | 2.8068 | 2.6055 | 2.4801 | 2.3940 | 2.3310 | 2.2828 | 2.2446 | 2.2135 | 2.1878 | 2.1474 | 2.1049 | 2.0597 | 2.0360 | 2.0115 | 1.9861 | 1.9597 | 1.9323 | 1.9036 |
| 13 | 3.1362 | 2.7632 | 2.5603 | 2.4337 | 2.3467 | 2.2830 | 2.2341 | 2.1953 | 2.1638 | 2.1376 | 2.0966 | 2.0532 | 2.0070 | 1.9827 | 1.9576 | 1.9315 | 1.9043 | 1.8759 | 1.8462 |
| 14 | 3.1022 | 2.7265 | 2.5222 | 2.3947 | 2.3069 | 2.2426 | 2.1931 | 2.1539 | 2.1220 | 2.0954 | 2.0537 | 2.0095 | 1.9625 | 1.9377 | 1.9119 | 1.8852 | 1.8572 | 1.8280 | 1.7973 |
| 15 | 3.0732 | 2.6952 | 2.4898 | 2.3614 | 2.2730 | 2.2081 | 2.1582 | 2.1185 | 2.0862 | 2.0593 | 2.0171 | 1.9722 | 1.9243 | 1.8990 | 1.8728 | 1.8454 | 1.8168 | 1.7867 | 1.7551 |
| 16 | 3.0481 | 2.6682 | 2.4618 | 2.3327 | 2.2438 | 2.1783 | 2.1280 | 2.0880 | 2.0553 | 2.0281 | 1.9854 | 1.9399 | 1.8913 | 1.8656 | 1.8388 | 1.8108 | 1.7816 | 1.7507 | 1.7182 |
| 17 | 3.0262 | 2.6446 | 2.4374 | 2.3077 | 2.2183 | 2.1524 | 2.1017 | 2.0613 | 2.0284 | 2.0009 | 1.9577 | 1.9117 | 1.8624 | 1.8362 | 1.8090 | 1.7805 | 1.7506 | 1.7191 | 1.6856 |
| 18 | 3.0070 | 2.6239 | 2.4160 | 2.2858 | 2.1958 | 2.1296 | 2.0785 | 2.0379 | 2.0047 | 1.9770 | 1.9333 | 1.8868 | 1.8368 | 1.8103 | 1.7827 | 1.7537 | 1.7232 | 1.6910 | 1.6567 |
| 19 | 2.9899 | 2.6056 | 2.3970 | 2.2663 | 2.1760 | 2.1094 | 2.0580 | 2.0171 | 1.9836 | 1.9557 | 1.9117 | 1.8647 | 1.8142 | 1.7873 | 1.7592 | 1.7298 | 1.6988 | 1.6659 | 1.6308 |
| 20 | 2.9747 | 2.5893 | 2.3801 | 2.2489 | 2.1582 | 2.0913 | 2.0397 | 1.9985 | 1.9649 | 1.9367 | 1.8924 | 1.8449 | 1.7938 | 1.7667 | 1.7382 | 1.7083 | 1.6768 | 1.6433 | 1.6074 |
| 21 | 2.9610 | 2.5746 | 2.3649 | 2.2333 | 2.1423 | 2.0751 | 2.0233 | 1.9819 | 1.9480 | 1.9197 | 1.8750 | 1.8271 | 1.7756 | 1.7481 | 1.7193 | 1.6890 | 1.6569 | 1.6228 | 1.5862 |
| 22 | 2.9486 | 2.5613 | 2.3512 | 2.2193 | 2.1279 | 2.0605 | 2.0084 | 1.9668 | 1.9327 | 1.9043 | 1.8593 | 1.8111 | 1.7590 | 1.7312 | 1.7021 | 1.6714 | 1.6389 | 1.6041 | 1.5668 |
| 23 | 2.9374 | 2.5493 | 2.3387 | 2.2065 | 2.1149 | 2.0472 | 1.9949 | 1.9531 | 1.9189 | 1.8903 | 1.8450 | 1.7964 | 1.7439 | 1.7159 | 1.6864 | 1.6554 | 1.6224 | 1.5871 | 1.5490 |
| 24 | 2.9271 | 2.5383 | 2.3274 | 2.1949 | 2.1030 | 2.0351 | 1.9826 | 1.9407 | 1.9063 | 1.8775 | 1.8319 | 1.7831 | 1.7302 | 1.7019 | 1.6721 | 1.6407 | 1.6073 | 1.5715 | 1.5327 |
| 25 | 2.9177 | 2.5283 | 2.3170 | 2.1842 | 2.0922 | 2.0241 | 1.9714 | 1.9292 | 1.8947 | 1.8658 | 1.8200 | 1.7708 | 1.7175 | 1.6890 | 1.6589 | 1.6272 | 1.5934 | 1.5570 | 1.5176 |
| 26 | 2.9091 | 2.5191 | 2.3075 | 2.1745 | 2.0822 | 2.0139 | 1.9610 | 1.9188 | 1.8841 | 1.8550 | 1.8090 | 1.7596 | 1.7059 | 1.6771 | 1.6468 | 1.6147 | 1.5805 | 1.5437 | 1.5036 |
| 27 | 2.9012 | 2.5106 | 2.2987 | 2.1655 | 2.0730 | 2.0045 | 1.9515 | 1.9091 | 1.8743 | 1.8451 | 1.7989 | 1.7492 | 1.6951 | 1.6662 | 1.6356 | 1.6032 | 1.5686 | 1.5313 | 1.4906 |
| 28 | 2.8938 | 2.5028 | 2.2906 | 2.1571 | 2.0645 | 1.9959 | 1.9427 | 1.9001 | 1.8652 | 1.8359 | 1.7895 | 1.7395 | 1.6852 | 1.6560 | 1.6252 | 1.5925 | 1.5575 | 1.5198 | 1.4784 |
| 29 | 2.8870 | 2.4955 | 2.2831 | 2.1494 | 2.0566 | 1.9878 | 1.9345 | 1.8918 | 1.8568 | 1.8274 | 1.7808 | 1.7306 | 1.6759 | 1.6465 | 1.6155 | 1.5825 | 1.5472 | 1.5090 | 1.4670 |
| 30 | 2.8807 | 2.4887 | 2.2761 | 2.1422 | 2.0492 | 1.9803 | 1.9269 | 1.8841 | 1.8490 | 1.8195 | 1.7727 | 1.7223 | 1.6673 | 1.6377 | 1.6065 | 1.5732 | 1.5376 | 1.4989 | 1.4564 |
| 40 | 2.8354 | 2.4404 | 2.2261 | 2.0909 | 1.9968 | 1.9269 | 1.8725 | 1.8289 | 1.7929 | 1.7627 | 1.7146 | 1.6624 | 1.6052 | 1.5741 | 1.5411 | 1.5056 | 1.4672 | 1.4248 | 1.3769 |
| 60 | 2.7911 | 2.3933 | 2.1774 | 2.0410 | 1.9457 | 1.8747 | 1.8194 | 1.7748 | 1.7380 | 1.7070 | 1.6574 | 1.6034 | 1.5435 | 1.5107 | 1.4755 | 1.4373 | 1.3952 | 1.3476 | 1.2915 |
| 120 | 2.7478 | 2.3473 | 2.1300 | 1.9923 | 1.8959 | 1.8238 | 1.7675 | 1.7220 | 1.6842 | 1.6524 | 1.6012 | 1.5450 | 1.4821 | 1.4472 | 1.4094 | 1.3676 | 1.3203 | 1.2646 | 1.1926 |
| ∞ | 2.7055 | 2.3026 | 2.0838 | 1.9449 | 1.8473 | 1.7741 | 1.7167 | 1.6702 | 1.6315 | 1.5987 | 1.5458 | 1.4871 | 1.4206 | 1.3832 | 1.3419 | 1.2951 | 1.2400 | 1.1686 | 1.0000 |

Denominator degrees of freedom

From Biometrika Tables for Statisticians, vol. 2, E. S. Pearson and H. O. Hartely, Cambridge: The University Press, 1972 by permission of the Biometrika trustees.

**TABLE B.4** *(continued)*
*F*-Distribution: Upper 5 Percent Points

Numerator degrees of freedom

| df₂ \ df₁ | 1 | 2 | 3 | 4 | 5 | 6 | 7 | 8 | 9 | 10 | 12 | 15 | 20 | 24 | 30 | 40 | 60 | 120 | ∞ |
|---|---|---|---|---|---|---|---|---|---|---|---|---|---|---|---|---|---|---|---|
| 1 | 161.45 | 199.50 | 215.71 | 224.58 | 230.16 | 233.99 | 236.77 | 238.88 | 240.54 | 241.88 | 243.91 | 245.95 | 248.01 | 249.05 | 250.10 | 251.14 | 252.20 | 253.25 | 254.31 |
| 2 | 18.513 | 19.000 | 19.164 | 19.247 | 19.296 | 19.330 | 19.353 | 19.371 | 19.385 | 19.396 | 19.413 | 19.429 | 19.446 | 19.454 | 19.462 | 19.471 | 19.479 | 19.487 | 19.496 |
| 3 | 10.128 | 9.5521 | 9.2766 | 9.1172 | 9.0135 | 8.9406 | 8.8867 | 8.8452 | 8.8123 | 8.7855 | 8.7446 | 8.7029 | 8.6602 | 8.6385 | 8.6166 | 8.5944 | 8.5720 | 8.5494 | 8.5264 |
| 4 | 7.7086 | 6.9443 | 6.5914 | 6.3882 | 6.2561 | 6.1631 | 6.0942 | 6.0410 | 5.9988 | 5.9644 | 5.9117 | 5.8578 | 5.8025 | 5.7744 | 5.7459 | 5.7170 | 5.6877 | 5.6581 | 5.6281 |
| 5 | 6.6079 | 5.7861 | 5.4095 | 5.1922 | 5.0503 | 4.9503 | 4.8759 | 4.8183 | 4.7725 | 4.7351 | 4.6777 | 4.6188 | 4.5581 | 4.5272 | 4.4957 | 4.4638 | 4.4314 | 4.3985 | 4.3650 |
| 6 | 5.9874 | 5.1433 | 4.7571 | 4.5337 | 4.3874 | 4.2839 | 4.2067 | 4.1468 | 4.0990 | 4.0600 | 3.9999 | 3.9381 | 3.8742 | 3.8415 | 3.8082 | 3.7743 | 3.7398 | 3.7047 | 3.6689 |
| 7 | 5.5914 | 4.7374 | 4.3468 | 4.1203 | 3.9715 | 3.8660 | 3.7870 | 3.7257 | 3.6767 | 3.6365 | 3.5747 | 3.5107 | 3.4445 | 3.4105 | 3.3758 | 3.3404 | 3.3043 | 3.2674 | 3.2298 |
| 8 | 5.3177 | 4.4590 | 4.0662 | 3.8379 | 3.6875 | 3.5806 | 3.5005 | 3.4381 | 3.3881 | 3.3472 | 3.2839 | 3.2184 | 3.1503 | 3.1152 | 3.0794 | 3.0428 | 3.0053 | 2.9669 | 2.9276 |
| 9 | 5.1174 | 4.2565 | 3.8625 | 3.6331 | 3.4817 | 3.3738 | 3.2927 | 3.2296 | 3.1789 | 3.1373 | 3.0729 | 3.0061 | 2.9365 | 2.9005 | 2.8637 | 2.8259 | 2.7872 | 2.7475 | 2.7067 |
| 10 | 4.9646 | 4.1028 | 3.7083 | 3.4780 | 3.3258 | 3.2172 | 3.1355 | 3.0717 | 3.0204 | 2.9782 | 2.9130 | 2.8450 | 2.7740 | 2.7372 | 2.6996 | 2.6609 | 2.6211 | 2.5801 | 2.5379 |
| 11 | 4.8443 | 3.9823 | 3.5874 | 3.3567 | 3.2039 | 3.0946 | 3.0123 | 2.9480 | 2.8962 | 2.8536 | 2.7876 | 2.7186 | 2.6464 | 2.6090 | 2.5705 | 2.5309 | 2.4901 | 2.4480 | 2.4045 |
| 12 | 4.7472 | 3.8853 | 3.4903 | 3.2592 | 3.1059 | 2.9961 | 2.9134 | 2.8486 | 2.7964 | 2.7534 | 2.6866 | 2.6169 | 2.5436 | 2.5055 | 2.4663 | 2.4259 | 2.3842 | 2.3410 | 2.2962 |
| 13 | 4.6672 | 3.8056 | 3.4105 | 3.1791 | 3.0254 | 2.9153 | 2.8321 | 2.7669 | 2.7144 | 2.6710 | 2.6037 | 2.5331 | 2.4589 | 2.4202 | 2.3803 | 2.3392 | 2.2966 | 2.2524 | 2.2064 |
| 14 | 4.6001 | 3.7389 | 3.3439 | 3.1122 | 2.9582 | 2.8477 | 2.7642 | 2.6987 | 2.6458 | 2.6022 | 2.5342 | 2.4630 | 2.3879 | 2.3487 | 2.3082 | 2.2664 | 2.2229 | 2.1778 | 2.1307 |
| 15 | 4.5431 | 3.6823 | 3.2874 | 3.0556 | 2.9013 | 2.7905 | 2.7066 | 2.6408 | 2.5876 | 2.5437 | 2.4753 | 2.4034 | 2.3275 | 2.2878 | 2.2468 | 2.2043 | 2.1601 | 2.1141 | 2.0658 |
| 16 | 4.4940 | 3.6337 | 3.2389 | 3.0069 | 2.8524 | 2.7413 | 2.6572 | 2.5911 | 2.5377 | 2.4935 | 2.4247 | 2.3522 | 2.2756 | 2.2354 | 2.1938 | 2.1507 | 2.1058 | 2.0589 | 2.0096 |
| 17 | 4.4513 | 3.5915 | 3.1968 | 2.9647 | 2.8100 | 2.6987 | 2.6143 | 2.5480 | 2.4943 | 2.4499 | 2.3807 | 2.3077 | 2.2304 | 2.1898 | 2.1477 | 2.1040 | 2.0584 | 2.0107 | 1.9604 |
| 18 | 4.4139 | 3.5546 | 3.1599 | 2.9277 | 2.7729 | 2.6613 | 2.5767 | 2.5102 | 2.4563 | 2.4117 | 2.3421 | 2.2686 | 2.1906 | 2.1497 | 2.1071 | 2.0629 | 2.0166 | 1.9681 | 1.9168 |
| 19 | 4.3807 | 3.5219 | 3.1274 | 2.8951 | 2.7401 | 2.6283 | 2.5435 | 2.4768 | 2.4227 | 2.3779 | 2.3080 | 2.2341 | 2.1555 | 2.1141 | 2.0712 | 2.0264 | 1.9795 | 1.9302 | 1.8780 |
| 20 | 4.3512 | 3.4928 | 3.0984 | 2.8661 | 2.7109 | 2.5990 | 2.5140 | 2.4471 | 2.3928 | 2.3479 | 2.2776 | 2.2033 | 2.1242 | 2.0825 | 2.0391 | 1.9938 | 1.9464 | 1.8963 | 1.8432 |
| 21 | 4.3248 | 3.4668 | 3.0725 | 2.8401 | 2.6848 | 2.5727 | 2.4876 | 2.4205 | 2.3660 | 2.3210 | 2.2504 | 2.1757 | 2.0960 | 2.0540 | 2.0102 | 1.9645 | 1.9165 | 1.8657 | 1.8117 |
| 22 | 4.3009 | 3.4434 | 3.0491 | 2.8167 | 2.6613 | 2.5491 | 2.4638 | 2.3965 | 2.3419 | 2.2967 | 2.2258 | 2.1508 | 2.0707 | 2.0283 | 1.9842 | 1.9380 | 1.8894 | 1.8380 | 1.7831 |
| 23 | 4.2793 | 3.4221 | 3.0280 | 2.7955 | 2.6400 | 2.5277 | 2.4422 | 2.3748 | 2.3201 | 2.2747 | 2.2036 | 2.1282 | 2.0476 | 2.0050 | 1.9605 | 1.9139 | 1.8648 | 1.8128 | 1.7570 |
| 24 | 4.2597 | 3.4028 | 3.0088 | 2.7763 | 2.6207 | 2.5082 | 2.4226 | 2.3551 | 2.3002 | 2.2547 | 2.1834 | 2.1077 | 2.0267 | 1.9838 | 1.9390 | 1.8920 | 1.8424 | 1.7896 | 1.7330 |
| 25 | 4.2417 | 3.3852 | 2.9912 | 2.7587 | 2.6030 | 2.4904 | 2.4047 | 2.3371 | 2.2821 | 2.2365 | 2.1649 | 2.0889 | 2.0075 | 1.9643 | 1.9192 | 1.8718 | 1.8217 | 1.7684 | 1.7110 |
| 26 | 4.2252 | 3.3690 | 2.9752 | 2.7426 | 2.5868 | 2.4741 | 2.3883 | 2.3205 | 2.2655 | 2.2197 | 2.1479 | 2.0716 | 1.9898 | 1.9464 | 1.9010 | 1.8533 | 1.8027 | 1.7488 | 1.6906 |
| 27 | 4.2100 | 3.3541 | 2.9604 | 2.7278 | 2.5719 | 2.4591 | 2.3732 | 2.3053 | 2.2501 | 2.2043 | 2.1323 | 2.0558 | 1.9736 | 1.9299 | 1.8842 | 1.8361 | 1.7851 | 1.7306 | 1.6717 |
| 28 | 4.1960 | 3.3404 | 2.9467 | 2.7141 | 2.5581 | 2.4453 | 2.3593 | 2.2913 | 2.2360 | 2.1900 | 2.1179 | 2.0411 | 1.9586 | 1.9147 | 1.8687 | 1.8203 | 1.7689 | 1.7138 | 1.6541 |
| 29 | 4.1830 | 3.3277 | 2.9340 | 2.7014 | 2.5454 | 2.4324 | 2.3463 | 2.2783 | 2.2229 | 2.1768 | 2.1045 | 2.0275 | 1.9446 | 1.9005 | 1.8543 | 1.8055 | 1.7537 | 1.6981 | 1.6376 |
| 30 | 4.1709 | 3.3158 | 2.9223 | 2.6896 | 2.5336 | 2.4205 | 2.3343 | 2.2662 | 2.2107 | 2.1646 | 2.0921 | 2.0148 | 1.9317 | 1.8874 | 1.8409 | 1.7918 | 1.7396 | 1.6835 | 1.6223 |
| 40 | 4.0847 | 3.2317 | 2.8387 | 2.6060 | 2.4495 | 2.3359 | 2.2490 | 2.1802 | 2.1240 | 2.0772 | 2.0035 | 1.9245 | 1.8389 | 1.7929 | 1.7444 | 1.6928 | 1.6373 | 1.5766 | 1.5089 |
| 60 | 4.0012 | 3.1504 | 2.7581 | 2.5252 | 2.3683 | 2.2541 | 2.1665 | 2.0970 | 2.0401 | 1.9926 | 1.9174 | 1.8364 | 1.7480 | 1.7001 | 1.6491 | 1.5943 | 1.5343 | 1.4673 | 1.3893 |
| 120 | 3.9201 | 3.0718 | 2.6802 | 2.4472 | 2.2899 | 2.1750 | 2.0868 | 2.0164 | 1.9588 | 1.9105 | 1.8337 | 1.7505 | 1.6587 | 1.6084 | 1.5543 | 1.4952 | 1.4290 | 1.3519 | 1.2539 |
| ∞ | 3.8415 | 2.9957 | 2.6049 | 2.3719 | 2.2141 | 2.0986 | 2.0096 | 1.9384 | 1.8799 | 1.8307 | 1.7522 | 1.6664 | 1.5705 | 1.5173 | 1.4591 | 1.3940 | 1.3180 | 1.2214 | 1.0000 |

Denominator degrees of freedom

**TABLE B.4**  (continued)
F-Distribution: Upper 2½ Percent Points

Numerator degrees of freedom

| df₂ \ df₁ | 1 | 2 | 3 | 4 | 5 | 6 | 7 | 8 | 9 | 10 | 12 | 15 | 20 | 24 | 30 | 40 | 60 | 120 | ∞ |
|---|---|---|---|---|---|---|---|---|---|---|---|---|---|---|---|---|---|---|---|
| 1 | 647.79 | 799.50 | 864.16 | 899.58 | 921.85 | 937.11 | 948.22 | 956.66 | 963.28 | 968.63 | 976.71 | 984.87 | 993.10 | 997.25 | 1001.4 | 1005.6 | 1009.8 | 1014.0 | 1018.3 |
| 2 | 38.506 | 39.000 | 39.165 | 39.248 | 39.298 | 39.331 | 39.335 | 39.373 | 39.387 | 39.398 | 39.415 | 39.431 | 39.448 | 39.456 | 39.465 | 39.473 | 39.481 | 39.490 | 39.498 |
| 3 | 17.443 | 16.044 | 15.439 | 15.101 | 14.885 | 14.735 | 14.624 | 14.540 | 14.473 | 14.419 | 14.337 | 14.253 | 14.167 | 14.124 | 14.081 | 14.037 | 13.992 | 13.947 | 13.902 |
| 4 | 12.218 | 10.649 | 9.9792 | 9.6045 | 9.3645 | 9.1973 | 9.0741 | 8.9796 | 8.9047 | 8.8439 | 8.7512 | 8.6565 | 8.5599 | 8.5109 | 8.4613 | 8.4111 | 8.3604 | 8.3092 | 8.2573 |
| 5 | 10.007 | 8.4336 | 7.7636 | 7.3879 | 7.1464 | 6.9777 | 6.8531 | 6.7572 | 6.6811 | 6.6192 | 6.5245 | 6.4277 | 6.3286 | 6.2780 | 6.2269 | 6.1750 | 6.1225 | 6.0693 | 6.0153 |
| 6 | 8.8131 | 7.2599 | 6.5988 | 6.2272 | 5.9876 | 5.8198 | 5.6955 | 5.5996 | 5.5234 | 5.4613 | 5.3662 | 5.2687 | 5.1684 | 5.1172 | 5.0652 | 5.0125 | 4.9589 | 4.9044 | 4.8491 |
| 7 | 8.0727 | 6.5415 | 5.8898 | 5.5226 | 5.2852 | 5.1186 | 4.9949 | 4.8993 | 4.8232 | 4.7611 | 4.6658 | 4.5678 | 4.4667 | 4.4150 | 4.3624 | 4.3089 | 4.2544 | 4.1989 | 4.1423 |
| 8 | 7.5709 | 6.0595 | 5.4160 | 5.0526 | 4.8173 | 4.6517 | 4.5286 | 4.4333 | 4.3572 | 4.2951 | 4.1997 | 4.1012 | 3.9995 | 3.9472 | 3.8940 | 3.8398 | 3.7844 | 3.7279 | 3.6702 |
| 9 | 7.2093 | 5.7147 | 5.0781 | 4.7181 | 4.4844 | 4.3197 | 4.1970 | 4.1020 | 4.0260 | 3.9639 | 3.8682 | 3.7694 | 3.6669 | 3.6142 | 3.5604 | 3.5055 | 3.4493 | 3.3918 | 3.3329 |
| 10 | 6.9367 | 5.4564 | 4.8256 | 4.4683 | 4.2361 | 4.0721 | 3.9498 | 3.8549 | 3.7790 | 3.7168 | 3.6209 | 3.5217 | 3.4185 | 3.3654 | 3.3110 | 3.2554 | 3.1984 | 3.1399 | 3.0798 |
| 11 | 6.7241 | 5.2559 | 4.6300 | 4.2751 | 4.0440 | 3.8807 | 3.7586 | 3.6638 | 3.5879 | 3.5257 | 3.4296 | 3.3299 | 3.2261 | 3.1725 | 3.1176 | 3.0613 | 3.0035 | 2.9441 | 2.8828 |
| 12 | 6.5538 | 5.0959 | 4.4742 | 4.1212 | 3.8911 | 3.7283 | 3.6065 | 3.5118 | 3.4358 | 3.3736 | 3.2773 | 3.1772 | 3.0728 | 3.0187 | 2.9633 | 2.9063 | 2.8478 | 2.7874 | 2.7249 |
| 13 | 6.4143 | 4.9653 | 4.3472 | 3.9959 | 3.7667 | 3.6043 | 3.4827 | 3.3880 | 3.3120 | 3.2497 | 3.1532 | 3.0527 | 2.9477 | 2.8932 | 2.8372 | 2.7797 | 2.7204 | 2.6590 | 2.5955 |
| 14 | 6.2979 | 4.8567 | 4.2417 | 3.8919 | 3.6634 | 3.5014 | 3.3799 | 3.2853 | 3.2093 | 3.1469 | 3.0502 | 2.9493 | 2.8437 | 2.7888 | 2.7324 | 2.6742 | 2.6142 | 2.5519 | 2.4872 |
| 15 | 6.1995 | 4.7650 | 4.1528 | 3.8043 | 3.5764 | 3.4147 | 3.2934 | 3.1987 | 3.1227 | 3.0602 | 2.9633 | 2.8621 | 2.7559 | 2.7006 | 2.6437 | 2.5850 | 2.5242 | 2.4611 | 2.3953 |
| 16 | 6.1151 | 4.6867 | 4.0168 | 3.7294 | 3.5021 | 3.3406 | 3.2194 | 3.1248 | 3.0488 | 2.9862 | 2.8890 | 2.7875 | 2.6808 | 2.6252 | 2.5678 | 2.5085 | 2.4471 | 2.3831 | 2.3163 |
| 17 | 6.0420 | 4.6189 | 4.0112 | 3.6648 | 3.4379 | 3.2767 | 3.1556 | 3.0610 | 2.9849 | 2.9222 | 2.8249 | 2.7230 | 2.6158 | 2.5598 | 2.5020 | 2.4422 | 2.3801 | 2.3153 | 2.2474 |
| 18 | 5.9781 | 4.5597 | 3.9539 | 3.6083 | 3.3820 | 3.2209 | 3.0999 | 3.0053 | 2.9291 | 2.8664 | 2.7689 | 2.6667 | 2.5590 | 2.5027 | 2.4445 | 2.3842 | 2.3214 | 2.2558 | 2.1869 |
| 19 | 5.9216 | 4.5075 | 3.9034 | 3.5587 | 3.3327 | 3.1718 | 3.0509 | 2.9563 | 2.8801 | 2.8172 | 2.7196 | 2.6171 | 2.5089 | 2.4523 | 2.3937 | 2.3329 | 2.2696 | 2.2032 | 2.1333 |
| 20 | 5.8715 | 4.4613 | 3.8587 | 3.5147 | 3.2891 | 3.1283 | 3.0074 | 2.9128 | 2.8365 | 2.7737 | 2.6758 | 2.5731 | 2.4645 | 2.4076 | 2.3486 | 2.2873 | 2.2234 | 2.1562 | 2.0853 |
| 21 | 5.8266 | 4.4199 | 3.8188 | 3.4754 | 3.2501 | 3.0895 | 2.9686 | 2.8740 | 2.7977 | 2.7348 | 2.6368 | 2.5338 | 2.4247 | 2.3675 | 2.3082 | 2.2465 | 2.1819 | 2.1141 | 2.0422 |
| 22 | 5.7863 | 4.3828 | 3.7829 | 3.4401 | 3.2151 | 3.0546 | 2.9338 | 2.8392 | 2.7628 | 2.6998 | 2.6017 | 2.4984 | 2.3890 | 2.3315 | 2.2718 | 2.2097 | 2.1446 | 2.0760 | 2.0032 |
| 23 | 5.7498 | 4.3492 | 3.7505 | 3.4083 | 3.1835 | 3.0232 | 2.9023 | 2.8077 | 2.7313 | 2.6682 | 2.5699 | 2.4665 | 2.3567 | 2.2989 | 2.2389 | 2.1763 | 2.1107 | 2.0415 | 1.9677 |
| 24 | 5.7166 | 4.3187 | 3.7211 | 3.3794 | 3.1548 | 2.9946 | 2.8738 | 2.7791 | 2.7027 | 2.6396 | 2.5411 | 2.4374 | 2.3273 | 2.2693 | 2.2090 | 2.1460 | 2.0799 | 2.0099 | 1.9353 |
| 25 | 5.6864 | 4.2909 | 3.6943 | 3.3530 | 3.1287 | 2.9685 | 2.8478 | 2.7531 | 2.6766 | 2.6135 | 2.5149 | 2.4110 | 2.3005 | 2.2422 | 2.1816 | 2.1183 | 2.0516 | 1.9811 | 1.9055 |
| 26 | 5.6586 | 4.2655 | 3.6697 | 3.3289 | 3.1048 | 2.9447 | 2.8240 | 2.7293 | 2.6528 | 2.5896 | 2.4908 | 2.3867 | 2.2759 | 2.2174 | 2.1565 | 2.0928 | 2.0257 | 1.9545 | 1.8781 |
| 27 | 5.6331 | 4.2421 | 3.6472 | 3.3067 | 3.0828 | 2.9228 | 2.8021 | 2.7074 | 2.6309 | 2.5676 | 2.4688 | 2.3644 | 2.2533 | 2.1946 | 2.1334 | 2.0693 | 2.0018 | 1.9299 | 1.8527 |
| 28 | 5.6096 | 4.2205 | 3.6264 | 3.2863 | 3.0626 | 2.9027 | 2.7820 | 2.6872 | 2.6106 | 2.5473 | 2.4484 | 2.3438 | 2.2324 | 2.1735 | 2.1121 | 2.0477 | 1.9797 | 1.9072 | 1.8291 |
| 29 | 5.5878 | 4.2006 | 3.6072 | 3.2674 | 3.0438 | 2.8840 | 2.7633 | 2.6686 | 2.5919 | 2.5286 | 2.4295 | 2.3248 | 2.2131 | 2.1540 | 2.0923 | 2.0276 | 1.9591 | 1.8861 | 1.8072 |
| 30 | 5.5675 | 4.1821 | 3.5894 | 3.2499 | 3.0265 | 2.8667 | 2.7460 | 2.6513 | 2.5746 | 2.5112 | 2.4120 | 2.3072 | 2.1952 | 2.1359 | 2.0739 | 2.0089 | 1.9400 | 1.8664 | 1.7867 |
| 40 | 5.4239 | 4.0510 | 3.4633 | 3.1261 | 2.9037 | 2.7444 | 2.6238 | 2.5289 | 2.4519 | 2.3882 | 2.2882 | 2.1819 | 2.0677 | 2.0069 | 1.9429 | 1.8752 | 1.8028 | 1.7242 | 1.6371 |
| 60 | 5.2856 | 3.9253 | 3.3425 | 3.0077 | 2.7863 | 2.6274 | 2.5068 | 2.4117 | 2.3344 | 2.2702 | 2.1692 | 2.0613 | 1.9445 | 1.8817 | 1.8152 | 1.7440 | 1.6668 | 1.5810 | 1.4821 |
| 120 | 5.1523 | 3.8046 | 3.2269 | 2.8943 | 2.6740 | 2.5154 | 2.3948 | 2.2994 | 2.2217 | 2.1570 | 2.0548 | 1.9450 | 1.8249 | 1.7597 | 1.6899 | 1.6141 | 1.5299 | 1.4327 | 1.3104 |
| ∞ | 5.0239 | 3.6889 | 3.1161 | 2.7858 | 2.5665 | 2.4082 | 2.2875 | 2.1918 | 2.1136 | 2.0483 | 1.9447 | 1.8326 | 1.7085 | 1.6402 | 1.5660 | 1.4835 | 1.3883 | 1.2684 | 1.0000 |

Denominator degrees of freedom

.025

$F_{df_1, df_2}$

**TABLE B.4** *(continued)*
*F*-Distribution: Upper 1 Percent Points

Numerator degrees of freedom

| df₂ \ df₁ | 1 | 2 | 3 | 4 | 5 | 6 | 7 | 8 | 9 | 10 | 12 | 15 | 20 | 24 | 30 | 40 | 60 | 120 | ∞ |
|---|---|---|---|---|---|---|---|---|---|---|---|---|---|---|---|---|---|---|---|
| 1 | 4052.2 | 4999.5 | 5403.4 | 5624.6 | 5763.6 | 5859.0 | 5928.4 | 5981.1 | 6022.5 | 6055.8 | 6106.3 | 6157.3 | 6208.7 | 6234.6 | 6260.6 | 6286.8 | 6313.0 | 6339.4 | 6365.9 |
| 2 | 98.503 | 99.000 | 99.166 | 99.249 | 99.299 | 99.333 | 99.356 | 99.374 | 99.388 | 99.399 | 99.416 | 99.433 | 99.449 | 99.458 | 99.466 | 99.474 | 99.482 | 99.491 | 99.499 |
| 3 | 34.116 | 30.817 | 29.457 | 28.710 | 28.237 | 27.911 | 27.672 | 27.489 | 27.345 | 27.229 | 27.052 | 26.872 | 26.690 | 26.598 | 26.505 | 26.411 | 26.316 | 26.221 | 26.125 |
| 4 | 21.198 | 18.000 | 16.694 | 15.977 | 15.522 | 15.207 | 14.976 | 14.799 | 14.659 | 14.546 | 14.374 | 14.198 | 14.020 | 13.929 | 13.838 | 13.745 | 13.652 | 13.558 | 13.463 |
| 5 | 16.258 | 13.274 | 12.060 | 11.392 | 10.967 | 10.672 | 10.456 | 10.289 | 10.158 | 10.051 | 9.8883 | 9.7222 | 9.5526 | 9.4665 | 9.3793 | 9.2912 | 9.2020 | 9.1118 | 9.0204 |
| 6 | 13.745 | 10.925 | 9.7795 | 9.1483 | 8.7459 | 8.4661 | 8.2600 | 8.1017 | 7.9761 | 7.8741 | 7.7183 | 7.5590 | 7.3958 | 7.3127 | 7.2285 | 7.1432 | 7.0567 | 6.9690 | 6.8800 |
| 7 | 12.246 | 9.5466 | 8.4513 | 7.8466 | 7.4604 | 7.1914 | 6.9928 | 6.8400 | 6.7188 | 6.6201 | 6.4691 | 6.3143 | 6.1554 | 6.0743 | 5.9920 | 5.9084 | 5.8236 | 5.7373 | 5.6495 |
| 8 | 11.259 | 8.6491 | 7.5910 | 7.0061 | 6.6318 | 6.3707 | 6.1776 | 6.0289 | 5.9106 | 5.8143 | 5.6667 | 5.5151 | 5.3591 | 5.2793 | 5.1981 | 5.1156 | 5.0316 | 4.9461 | 4.8588 |
| 9 | 10.561 | 8.0215 | 6.9919 | 6.4221 | 6.0569 | 5.8018 | 5.6129 | 5.4671 | 5.3511 | 5.2565 | 5.1114 | 4.9621 | 4.8080 | 4.7290 | 4.6486 | 4.5666 | 4.4831 | 4.3978 | 4.3105 |
| 10 | 10.044 | 7.5594 | 6.5523 | 5.9943 | 5.6363 | 5.3858 | 5.2001 | 5.0567 | 4.9424 | 4.8491 | 4.7059 | 4.5581 | 4.4054 | 4.3269 | 4.2469 | 4.1653 | 4.0819 | 3.9965 | 3.9090 |
| 11 | 9.6460 | 7.2057 | 6.2167 | 5.6683 | 5.3160 | 5.0692 | 4.8861 | 4.7445 | 4.6315 | 4.5393 | 4.3974 | 4.2509 | 4.0990 | 4.0209 | 3.9411 | 3.8596 | 3.7761 | 3.6904 | 3.6024 |
| 12 | 9.3302 | 6.9266 | 5.9525 | 5.4120 | 5.0643 | 4.8206 | 4.6395 | 4.4994 | 4.3875 | 4.2961 | 4.1553 | 4.0096 | 3.8584 | 3.7805 | 3.7008 | 3.6192 | 3.5355 | 3.4494 | 3.3608 |
| 13 | 9.0738 | 6.7010 | 5.7394 | 5.2053 | 4.8616 | 4.6204 | 4.4410 | 4.3021 | 4.1911 | 4.1003 | 3.9603 | 3.8154 | 3.6646 | 3.5868 | 3.5070 | 3.4253 | 3.3413 | 3.2548 | 3.1654 |
| 14 | 8.8616 | 6.5149 | 5.5639 | 5.0354 | 4.6950 | 4.4558 | 4.2779 | 4.1399 | 4.0297 | 3.9394 | 3.8001 | 3.6557 | 3.5052 | 3.4274 | 3.3476 | 3.2656 | 3.1813 | 3.0942 | 3.0040 |
| 15 | 8.6831 | 6.3589 | 5.4170 | 4.8932 | 4.5556 | 4.3183 | 4.1415 | 4.0045 | 3.8948 | 3.8049 | 3.6662 | 3.5222 | 3.3719 | 3.2940 | 3.2141 | 3.1319 | 3.0471 | 2.9595 | 2.8684 |
| 16 | 8.5310 | 6.2262 | 5.2922 | 4.7726 | 4.4374 | 4.2016 | 4.0259 | 3.8896 | 3.7804 | 3.6909 | 3.5527 | 3.4089 | 3.2587 | 3.1808 | 3.1007 | 3.0182 | 2.9330 | 2.8447 | 2.7528 |
| 17 | 8.3997 | 6.1121 | 5.1850 | 4.6690 | 4.3359 | 4.1015 | 3.9267 | 3.7910 | 3.6822 | 3.5931 | 3.4552 | 3.3117 | 3.1615 | 3.0835 | 3.0032 | 2.9205 | 2.8348 | 2.7459 | 2.6530 |
| 18 | 8.2854 | 6.0129 | 5.0919 | 4.5790 | 4.2479 | 4.0146 | 3.8406 | 3.7054 | 3.5971 | 3.5082 | 3.3706 | 3.2273 | 3.0771 | 2.9990 | 2.9185 | 2.8354 | 2.7493 | 2.6597 | 2.5660 |
| 19 | 8.1849 | 5.9259 | 5.0103 | 4.5003 | 4.1708 | 3.9386 | 3.7653 | 3.6305 | 3.5225 | 3.4338 | 3.2965 | 3.1533 | 3.0031 | 2.9249 | 2.8442 | 2.7608 | 2.6742 | 2.5839 | 2.4893 |
| 20 | 8.0960 | 5.8489 | 4.9382 | 4.4307 | 4.1027 | 3.8714 | 3.6987 | 3.5644 | 3.4567 | 3.3682 | 3.2311 | 3.0880 | 2.9377 | 2.8594 | 2.7785 | 2.6947 | 2.6077 | 2.5168 | 2.4212 |
| 21 | 8.0166 | 5.7804 | 4.8740 | 4.3688 | 4.0421 | 3.8117 | 3.6396 | 3.5056 | 3.3981 | 3.3098 | 3.1730 | 3.0300 | 2.8796 | 2.8010 | 2.7200 | 2.6359 | 2.5484 | 2.4568 | 2.3603 |
| 22 | 7.9454 | 5.7190 | 4.8166 | 4.3134 | 3.9880 | 3.7583 | 3.5867 | 3.4530 | 3.3458 | 3.2576 | 3.1209 | 2.9779 | 2.8274 | 2.7488 | 2.6675 | 2.5831 | 2.4951 | 2.4029 | 2.3055 |
| 23 | 7.8811 | 5.6637 | 4.7649 | 4.2636 | 3.9392 | 3.7102 | 3.5390 | 3.4057 | 3.2986 | 3.2106 | 3.0740 | 2.9311 | 2.7805 | 2.7017 | 2.6202 | 2.5355 | 2.4471 | 2.3542 | 2.2558 |
| 24 | 7.8229 | 5.6136 | 4.7181 | 4.2184 | 3.8951 | 3.6667 | 3.4959 | 3.3629 | 3.2560 | 3.1681 | 3.0316 | 2.8887 | 2.7380 | 2.6591 | 2.5773 | 2.4923 | 2.4035 | 2.3100 | 2.2107 |
| 25 | 7.7698 | 5.5680 | 4.6755 | 4.1774 | 3.8550 | 3.6272 | 3.4568 | 3.3239 | 3.2172 | 3.1294 | 2.9931 | 2.8502 | 2.6993 | 2.6203 | 2.5383 | 2.4530 | 2.3637 | 2.2696 | 2.1694 |
| 26 | 7.7213 | 5.5263 | 4.6366 | 4.1400 | 3.8183 | 3.5911 | 3.4210 | 3.2884 | 3.1818 | 3.0941 | 2.9578 | 2.8150 | 2.6640 | 2.5848 | 2.5026 | 2.4170 | 2.3273 | 2.2325 | 2.1315 |
| 27 | 7.6767 | 5.4881 | 4.6009 | 4.1056 | 3.7848 | 3.5580 | 3.3882 | 3.2558 | 3.1494 | 3.0618 | 2.9256 | 2.7827 | 2.6316 | 2.5522 | 2.4699 | 2.3840 | 2.2938 | 2.1985 | 2.0965 |
| 28 | 7.6356 | 5.4529 | 4.5681 | 4.0740 | 3.7539 | 3.5276 | 3.3581 | 3.2259 | 3.1195 | 3.0320 | 2.8959 | 2.7530 | 2.6017 | 2.5223 | 2.4397 | 2.3535 | 2.2629 | 2.1670 | 2.0642 |
| 29 | 7.5977 | 5.4204 | 4.5378 | 4.0449 | 3.7254 | 3.4995 | 3.3303 | 3.1982 | 3.0920 | 3.0045 | 2.8685 | 2.7256 | 2.5742 | 2.4946 | 2.4118 | 2.3253 | 2.2344 | 2.1379 | 2.0342 |
| 30 | 7.5625 | 5.3903 | 4.5097 | 4.0179 | 3.6990 | 3.4735 | 3.3045 | 3.1726 | 3.0665 | 2.9791 | 2.8431 | 2.7002 | 2.5487 | 2.4689 | 2.3860 | 2.2992 | 2.2079 | 2.1108 | 2.0062 |
| 40 | 7.3141 | 5.1785 | 4.3126 | 3.8283 | 3.5138 | 3.2910 | 3.1238 | 2.9930 | 2.8876 | 2.8005 | 2.6648 | 2.5216 | 2.3689 | 2.2880 | 2.2034 | 2.1142 | 2.0194 | 1.9172 | 1.8047 |
| 60 | 7.0771 | 4.9774 | 4.1259 | 3.6490 | 3.3389 | 3.1187 | 2.9530 | 2.8233 | 2.7185 | 2.6318 | 2.4961 | 2.3523 | 2.1978 | 2.1154 | 2.0285 | 1.9360 | 1.8363 | 1.7263 | 1.6006 |
| 120 | 6.8509 | 4.7865 | 3.9491 | 3.4795 | 3.1735 | 2.9559 | 2.7918 | 2.6629 | 2.5586 | 2.4721 | 2.3363 | 2.1915 | 2.0346 | 1.9500 | 1.8600 | 1.7628 | 1.6557 | 1.5330 | 1.3805 |
| ∞ | 6.6349 | 4.6052 | 3.7816 | 3.3192 | 3.0173 | 2.8020 | 2.6393 | 2.5113 | 2.4073 | 2.3209 | 2.1847 | 2.0385 | 1.8783 | 1.7908 | 1.6964 | 1.5923 | 1.4730 | 1.3246 | 1.0000 |

Denominator degrees of freedom

**TABLE B.5**   Binomial Point Probabilities

|   |   | | | | | | | $\pi$ | | | | | | | |
|---|---|------|------|------|------|------|------|------|------|------|------|------|------|------|
| $n$ | $X$ | .05 | .10 | .20 | .25 | .30 | .40 | .50 | .60 | .70 | .75 | .80 | .90 | .95 |
| 1 | 0 | .9500 | .9000 | .8000 | .7500 | .7000 | .6000 | .5000 | .4000 | .3000 | .2500 | .2000 | .1000 | .0500 |
|   | 1 | .0500 | .1000 | .2000 | .2500 | .3000 | .4000 | .5000 | .6000 | .7000 | .7500 | .8000 | .9000 | .9500 |
| 2 | 0 | .9025 | .8100 | .6400 | .5625 | .4900 | .3600 | .2500 | .1600 | .0900 | .0625 | .0400 | .0100 | .0025 |
|   | 1 | .0950 | .1800 | .3200 | .3750 | .4200 | .4800 | .5000 | .4800 | .4200 | .3750 | .3200 | .1800 | .0950 |
|   | 2 | .0025 | .0100 | .0400 | .0625 | .0900 | .1600 | .2500 | .3600 | .4900 | .5625 | .6400 | .8100 | .9025 |
| 3 | 0 | .8574 | .7290 | .5120 | .4219 | .3430 | .2160 | .1250 | .0640 | .0270 | .0156 | .0080 | .0010 | .0001 |
|   | 1 | .1354 | .2430 | .3840 | .4219 | .4410 | .4320 | .3750 | .2880 | .1890 | .1406 | .0960 | .0270 | .0071 |
|   | 2 | .0071 | .0270 | .0960 | .1406 | .1890 | .2880 | .3750 | .4320 | .4410 | .4219 | .3840 | .2430 | .1354 |
|   | 3 | .0001 | .0010 | .0080 | .0156 | .0270 | .0640 | .1250 | .2160 | .3430 | .4219 | .5120 | .7290 | .8574 |
| 4 | 0 | .8145 | .6561 | .4096 | .3164 | .2401 | .1296 | .0625 | .0256 | .0081 | .0039 | .0016 | .0001 |       |
|   | 1 | .1715 | .2916 | .4096 | .4219 | .4116 | .3456 | .2500 | .1536 | .0756 | .0469 | .0256 | .0036 | .0005 |
|   | 2 | .0135 | .0486 | .1536 | .2109 | .2646 | .3456 | .3750 | .3456 | .2646 | .2109 | .1536 | .0486 | .0135 |
|   | 3 | .0005 | .0036 | .0256 | .0469 | .0756 | .1536 | .2500 | .3456 | .4116 | .4219 | .4096 | .2916 | .1715 |
|   | 4 |       | .0001 | .0016 | .0039 | .0081 | .0256 | .0625 | .1296 | .2401 | .3164 | .4096 | .6561 | .8145 |
| 5 | 0 | .7738 | .5905 | .3277 | .2373 | .1681 | .0778 | .0313 | .0102 | .0024 | .0010 | .0003 |       |       |
|   | 1 | .2036 | .3281 | .4096 | .3955 | .3602 | .2592 | .1562 | .0768 | .0284 | .0146 | .0064 | .0004 |       |
|   | 2 | .0214 | .0729 | .2048 | .2637 | .3087 | .3456 | .3125 | .2304 | .1323 | .0879 | .0512 | .0081 | .0011 |
|   | 3 | .0011 | .0081 | .0512 | .0879 | .1323 | .2304 | .3125 | .3456 | .3087 | .2637 | .2048 | .0729 | .0214 |
|   | 4 |       | .0004 | .0064 | .0146 | .0284 | .0768 | .1562 | .2592 | .3602 | .3955 | .4096 | .3281 | .2036 |
|   | 5 |       |       | .0003 | .0010 | .0024 | .0102 | .0313 | .0778 | .1681 | .2373 | .3277 | .5905 | .7738 |
| 6 | 0 | .7351 | .5314 | .2621 | .1780 | .1176 | .0467 | .0156 | .0041 | .0007 | .0002 | .0001 |       |       |
|   | 1 | .2321 | .3543 | .3932 | .3560 | .3025 | .1866 | .0938 | .0369 | .0102 | .0044 | .0015 | .0001 |       |
|   | 2 | .0305 | .0984 | .2458 | .2966 | .3241 | .3110 | .2344 | .1382 | .0595 | .0330 | .0154 | .0012 | .0001 |
|   | 3 | .0021 | .0146 | .0819 | .1318 | .1852 | .2765 | .3125 | .2765 | .1852 | .1318 | .0819 | .0146 | .0021 |
|   | 4 | .0001 | .0012 | .0154 | .0330 | .0595 | .1382 | .2344 | .3110 | .3241 | .2966 | .2458 | .0984 | .0305 |
|   | 5 |       | .0001 | .0015 | .0044 | .0102 | .0369 | .0938 | .1866 | .3025 | .3560 | .3932 | .3543 | .2321 |
|   | 6 |       |       | .0001 | .0002 | .0007 | .0041 | .0156 | .0467 | .1176 | .1780 | .2621 | .5314 | .7351 |
| 7 | 0 | .6983 | .4783 | .2097 | .1335 | .0824 | .0280 | .0078 | .0016 | .0002 | .0001 |       |       |       |
|   | 1 | .2573 | .3720 | .3670 | .3115 | .2471 | .1306 | .0547 | .0172 | .0036 | .0013 | .0004 |       |       |
|   | 2 | .0406 | .1240 | .2753 | .3115 | .3177 | .2613 | .1641 | .0774 | .0250 | .0115 | .0043 | .0002 |       |
|   | 3 | .0036 | .0230 | .1147 | .1730 | .2269 | .2903 | .2734 | .1935 | .0972 | .0577 | .0287 | .0026 | .0002 |
|   | 4 | .0002 | .0026 | .0287 | .0577 | .0972 | .1935 | .2734 | .2903 | .2269 | .1730 | .1147 | .0230 | .0036 |
|   | 5 |       | .0002 | .0043 | .0115 | .0250 | .0774 | .1641 | .2613 | .3177 | .3115 | .2753 | .1240 | .0406 |
|   | 6 |       |       | .0004 | .0013 | .0036 | .0172 | .0547 | .1306 | .2471 | .3115 | .3670 | .3720 | .2573 |
|   | 7 |       |       |       | .0001 | .0002 | .0016 | .0078 | .0280 | .0824 | .1335 | .2097 | .4783 | .6983 |
| 8 | 0 | .6634 | .4305 | .1678 | .1001 | .0576 | .0168 | .0039 | .0007 | .0001 |       |       |       |       |
|   | 1 | .2793 | .3826 | .3355 | .2670 | .1976 | .0896 | .0312 | .0079 | .0012 | .0004 | .001 |       |       |
|   | 2 | .0515 | .1488 | .2936 | .3115 | .2965 | .2090 | .1094 | .0413 | .0100 | .0038 | .0011 |       |       |
|   | 3 | .0054 | .0331 | .1468 | .2076 | .2541 | .2787 | .2188 | .1239 | .0467 | .0231 | .0092 | .0004 |       |
|   | 4 | .0004 | .0046 | .0459 | .0865 | .1361 | .2322 | .2734 | .2322 | .1361 | .0865 | .0459 | .0046 | .0004 |
|   | 5 |       | .0004 | .0092 | .0231 | .0467 | .1238 | .2188 | .2787 | .2541 | .2076 | .1468 | .0331 | .0054 |
|   | 6 |       |       | .0011 | .0038 | .0100 | .0413 | .1094 | .2090 | .2965 | .3115 | .2936 | .1488 | .0515 |
|   | 7 |       |       | .0001 | .0004 | .0012 | .0079 | .0312 | .0896 | .1976 | .2670 | .3355 | .3826 | .2793 |
|   | 8 |       |       |       |       | .0001 | .0007 | .0039 | .0168 | .0576 | .1001 | .1678 | .4305 | .6634 |

**TABLE B.5**  (*continued*)

| n | X | | | | | | | π | | | | | | |
|---|---|-----|-----|-----|-----|-----|-----|-----|-----|-----|-----|-----|-----|-----|
| | | .05 | .10 | .20 | .25 | .30 | .40 | .50 | .60 | .70 | .75 | .80 | .90 | .95 |
| 9 | 0 | .6302 | .3874 | .1342 | .0751 | .0404 | .0101 | .0020 | .0003 | | | | | |
| | 1 | .2985 | .3874 | .3020 | .2253 | .1556 | .0605 | .0176 | .0035 | .0004 | .0001 | | | |
| | 2 | .0628 | .1722 | .3020 | .3003 | .2668 | .1612 | .0703 | .0212 | .0039 | .0012 | .0003 | | |
| | 3 | .0077 | .0446 | .1762 | .2336 | .2668 | .2508 | .1641 | .0743 | .0210 | .0087 | .0028 | .0001 | |
| | 4 | .0006 | .0074 | .0661 | .1168 | .1715 | .2508 | .2461 | .1672 | .0735 | .0389 | .0165 | .0008 | |
| | 5 | | .0008 | .0165 | .0389 | .0735 | .1672 | .2461 | .2508 | .1715 | .1168 | .0661 | .0074 | .0006 |
| | 6 | | .0001 | .0028 | .0087 | .0210 | .0743 | .1641 | .2508 | .2668 | .2336 | .1762 | .0446 | .0077 |
| | 7 | | | .0003 | .0012 | .0039 | .0212 | .0703 | .1612 | .2668 | .3003 | .3020 | .1722 | .0628 |
| | 8 | | | | .0001 | .0004 | .0035 | .0176 | .0605 | .1556 | .2253 | .3020 | .3874 | .2985 |
| | 9 | | | | | | .0003 | .0020 | .0101 | .0404 | .0751 | .1342 | .3874 | .6302 |
| 10 | 0 | .5987 | .3487 | .1074 | .0563 | .0282 | .0060 | .0010 | .0001 | | | | | |
| | 1 | .3151 | .3874 | .2684 | .1877 | .1211 | .0403 | .0098 | .0016 | .0001 | | | | |
| | 2 | .0746 | .1937 | .3020 | .2816 | .2335 | .1209 | .0439 | .0106 | .0014 | .004 | .0001 | | |
| | 3 | .0105 | .0574 | .2013 | .2503 | .2668 | .2150 | .1172 | .0425 | .0090 | .0031 | .0008 | | |
| | 4 | .0010 | .0112 | .0881 | .1460 | .2001 | .2508 | .2051 | .1115 | .0368 | .0162 | .0055 | .0001 | |
| | 5 | .0001 | .0015 | .0264 | .0584 | .1029 | .2007 | .2461 | .2007 | .1029 | .0584 | .0264 | .0015 | .0001 |
| | 6 | | .0001 | .0055 | .0162 | .0368 | .1115 | .2051 | .2508 | .2001 | .1460 | .0881 | .0112 | .0010 |
| | 7 | | | .0008 | .0031 | .0090 | .0425 | .1172 | .2150 | .2668 | .2503 | .2013 | .0574 | .0105 |
| | 8 | | | .0001 | .0004 | .0015 | .0106 | .0439 | .1209 | .2335 | .2816 | .3020 | .1937 | .0746 |
| | 9 | | | | | .0001 | .0016 | .0098 | .0403 | .1211 | .1877 | .2684 | .3874 | .3151 |
| | 10 | | | | | | .0001 | .0010 | .0060 | .0282 | .0563 | .1074 | .3487 | .5987 |
| 11 | 0 | .5688 | .3138 | .0859 | .0422 | .0198 | .0036 | .0005 | | | | | | |
| | 1 | .3293 | .3835 | .2362 | .1549 | .0932 | .0266 | .0054 | .0007 | | | | | |
| | 2 | .0867 | .2131 | .2953 | .2581 | .1998 | .0887 | .0269 | .0052 | .0005 | .0001 | | | |
| | 3 | .0137 | .0710 | .2215 | .2581 | .2568 | .1774 | .0806 | .0234 | .0037 | .0011 | .0002 | | |
| | 4 | .0014 | .0158 | .1107 | .1721 | .2201 | .2365 | .1611 | .0701 | .0173 | .0064 | .0017 | | |
| | 5 | .0001 | .0025 | .0388 | .0803 | .1321 | .2207 | .2256 | .1471 | .0566 | .0268 | .0097 | .0003 | |
| | 6 | | .0003 | .0097 | .0268 | .0566 | .1471 | .2256 | .2207 | .1321 | .0803 | .0388 | .0025 | .0001 |
| | 7 | | | .0017 | .0064 | .0173 | .0701 | .1611 | .2365 | .2201 | .1721 | .1107 | .0158 | .0014 |
| | 8 | | | .0002 | .0011 | .0037 | .0234 | .0806 | .1774 | .2568 | .2581 | .2215 | .0710 | .0137 |
| | 9 | | | | .0001 | .0005 | .0052 | .0269 | .0887 | .1998 | .2581 | .2953 | .2131 | .0867 |
| | 10 | | | | | | .0007 | .0054 | .0266 | .0932 | .1549 | .2362 | .3835 | .3293 |
| | 11 | | | | | | | .0005 | .0036 | .0198 | .0422 | .0859 | .3138 | .5688 |
| 12 | 0 | .5404 | .2824 | .0687 | .0317 | .0138 | .0022 | .0002 | | | | | | |
| | 1 | .3413 | .3766 | .2062 | .1267 | .0712 | .0174 | .0029 | .0003 | | | | | |
| | 2 | .0988 | .2301 | .2835 | .2323 | .1678 | .0639 | .0161 | .0025 | .0002 | | | | |
| | 3 | .0173 | .0852 | .2362 | .2581 | .2397 | .1419 | .0537 | .0125 | .0015 | .0004 | .0001 | | |
| | 4 | .0021 | .0213 | .1329 | .1936 | .2311 | .2128 | .1209 | .0420 | .0078 | .0024 | .0005 | | |
| | 5 | .0002 | .0038 | .0532 | .1032 | .1585 | .2270 | .1934 | .1009 | .0291 | .0115 | .0033 | | |
| | 6 | | .0005 | .0155 | .0402 | .0792 | .1766 | .2256 | .1766 | .0792 | .0402 | .0155 | .0005 | |
| | 7 | | | .0033 | .0115 | .0291 | .1009 | .1934 | .2270 | .1585 | .1032 | .0532 | .0038 | .0002 |
| | 8 | | | .0005 | .0024 | .0078 | .0420 | .1208 | .2128 | .2311 | .1936 | .1329 | .0213 | .0021 |
| | 9 | | | .0001 | .0004 | .0015 | .0125 | .0537 | .1419 | .2397 | .2581 | .2362 | .0852 | .0173 |
| | 10 | | | | | .0002 | .0025 | .0161 | .0639 | .1678 | .2323 | .2835 | .2301 | .0988 |
| | 11 | | | | | | .0003 | .0029 | .0174 | .0712 | .1267 | .2062 | .3766 | .3413 |
| | 12 | | | | | | | .0002 | .0022 | .0138 | .0317 | .0687 | .2824 | .5404 |

**TABLE B.5**  (*continued*)

| | | | | | | | $\pi$ | | | | | | | |
|---|---|---|---|---|---|---|---|---|---|---|---|---|---|---|
| n | X | .05 | .10 | .20 | .25 | .30 | .40 | .50 | .60 | .70 | .75 | .80 | .90 | .95 |
| 13 | 0 | .5133 | .2542 | .0550 | .0238 | .0097 | .0013 | .0001 | | | | | | |
| | 1 | .3512 | .3672 | .1787 | .1029 | .0540 | .0113 | .0016 | .0001 | | | | | |
| | 2 | .1109 | .2448 | .2680 | .2059 | .1388 | .0453 | .0095 | .0012 | .0001 | | | | |
| | 3 | .0214 | .0997 | .2457 | .2517 | .2181 | .1107 | .0349 | .0065 | .0006 | .0001 | | | |
| | 4 | .0028 | .0277 | .1535 | .2097 | .2337 | .1845 | .0873 | .0243 | .0034 | .0009 | .0002 | | |
| | 5 | .0003 | .0055 | .0691 | .1258 | .1803 | .2214 | .1571 | .0656 | .0142 | .0047 | .0011 | | |
| | 6 | | .0008 | .0230 | .0559 | .1030 | .1968 | .2095 | .1312 | .0442 | .0186 | .0058 | .0001 | |
| | 7 | | .0001 | .0058 | .0186 | .0442 | .1312 | .2095 | .1968 | .1030 | .0559 | .0230 | .0008 | |
| | 8 | | | .0011 | .0047 | .0142 | .0656 | .1571 | .2214 | .1803 | .1258 | .0691 | .0055 | .0003 |
| | 9 | | | .0002 | .0009 | .0034 | .0243 | .0873 | .1845 | .2337 | .2097 | .1535 | .0277 | .0028 |
| | 10 | | | | .0001 | .0006 | .0065 | .0349 | .1107 | .2181 | .2517 | .2457 | .0997 | .0214 |
| | 11 | | | | | .0001 | .0012 | .0095 | .0453 | .1388 | .2059 | .2680 | .2448 | .1109 |
| | 12 | | | | | | .0001 | .0016 | .0113 | .0540 | .1029 | .1787 | .3672 | .3512 |
| | 13 | | | | | | | .0001 | .0013 | .0097 | .0238 | .0550 | .2542 | .5133 |
| 14 | 0 | .4877 | .2288 | .0440 | .0178 | .0068 | .0008 | .0001 | | | | | | |
| | 1 | .3593 | .3559 | .1539 | .0832 | .0407 | .0073 | .0009 | .0001 | | | | | |
| | 2 | .1229 | .2570 | .2501 | .1802 | .1134 | .0317 | .0056 | .0006 | | | | | |
| | 3 | .0259 | .1142 | .2501 | .2402 | .1943 | .0845 | .0222 | .0033 | .0002 | | | | |
| | 4 | .0037 | .0349 | .1720 | .2202 | .2290 | .1549 | .0611 | .0136 | .0014 | .0003 | | | |
| | 5 | .0004 | .0078 | .0860 | .1468 | .1963 | .2066 | .1222 | .0408 | .0066 | .0018 | .0003 | | |
| | 6 | | .0013 | .0322 | .0734 | .1262 | .2066 | .1833 | .0918 | .0232 | .0082 | .0020 | | |
| | 7 | | .0002 | .0092 | .0280 | .0618 | .1574 | .2095 | .1574 | .0618 | .0280 | .0092 | .0002 | |
| | 8 | | | .0020 | .0082 | .0232 | .0918 | .1833 | .2066 | .1262 | .0734 | .0322 | .0013 | |
| | 9 | | | .0003 | .0018 | .0066 | .0408 | .1222 | .2066 | .1963 | .1468 | .0860 | .0078 | .0004 |
| | 10 | | | | .0003 | .0014 | .0136 | .0611 | .1549 | .2290 | .2202 | .1720 | .0349 | .0037 |
| | 11 | | | | | .0002 | .0033 | .0222 | .0845 | .1943 | .2402 | .2501 | .1142 | .0259 |
| | 12 | | | | | | .0006 | .0056 | .0317 | .1134 | .1802 | .2501 | .2570 | .1229 |
| | 13 | | | | | | .0001 | .0009 | .0073 | .0407 | .0832 | .1539 | .3559 | .3593 |
| | 14 | | | | | | | .0001 | .0008 | .0068 | .0178 | .0440 | .2288 | .4877 |
| 15 | 0 | .4633 | .2059 | .0352 | .0134 | .0047 | .0005 | | | | | | | |
| | 1 | .3658 | .3432 | .1319 | .0668 | .0305 | .0047 | .0005 | | | | | | |
| | 2 | .1348 | .2669 | .2309 | .1559 | .0916 | .0219 | .0032 | .0003 | | | | | |
| | 3 | .0307 | .1285 | .2501 | .2252 | .1700 | .0634 | .0139 | .0016 | .0001 | | | | |
| | 4 | .0049 | .0428 | .1876 | .2252 | .2186 | .1268 | .0417 | .0074 | .0006 | .0001 | | | |
| | 5 | .0006 | .0105 | .1032 | .1651 | .2061 | .1859 | .0916 | .0245 | .0030 | .0007 | .0001 | | |
| | 6 | | .0019 | .0430 | .0917 | .1472 | .2066 | .1527 | .0612 | .0116 | .0034 | .0007 | | |
| | 7 | | .0003 | .0138 | .0393 | .0811 | .1771 | .1964 | .1181 | .0348 | .0131 | .0035 | | |
| | 8 | | | .0035 | .0131 | .0348 | .1181 | .1964 | .1771 | .0811 | .0393 | .0138 | .0003 | |
| | 9 | | | .0007 | .0034 | .0116 | .0612 | .1527 | .2066 | .1472 | .0917 | .0430 | .0019 | |
| | 10 | | | .0001 | .0007 | .0030 | .0245 | .0916 | .1859 | .2061 | .1651 | .1032 | .0105 | .0006 |
| | 11 | | | | .0001 | .0006 | .0074 | .0417 | .1268 | .2186 | .2252 | .1876 | .0428 | .0049 |
| | 12 | | | | | .0001 | .0016 | .0139 | .0634 | .1700 | .2252 | .2501 | .1285 | .0307 |
| | 13 | | | | | | .0003 | .0032 | .0219 | .0916 | .1559 | .2309 | .2669 | .1348 |
| | 14 | | | | | | | .0005 | .0047 | .0305 | .0668 | .1319 | .3432 | .3658 |
| | 15 | | | | | | | | .0005 | .0047 | .0134 | .0352 | .2059 | .4633 |

**TABLE B.5**  (*continued*)

| n | X | .05 | .10 | .20 | .25 | .30 | .40 | .50 | .60 | .70 | .75 | .80 | .90 | .95 |
|---|---|---|---|---|---|---|---|---|---|---|---|---|---|---|
| 20 | 0 | .3585 | .1216 | .0115 | .0032 | .0008 | | | | | | | | |
| | 1 | .3774 | .2702 | .0576 | .0211 | .0068 | .0005 | | | | | | | |
| | 2 | .1887 | .2852 | .1369 | .0669 | .0278 | .0031 | .0002 | | | | | | |
| | 3 | .0596 | .1901 | .2054 | .1339 | .0716 | .0124 | .0011 | | | | | | |
| | 4 | .0133 | .0898 | .2182 | .1897 | .1304 | .0350 | .0046 | .0003 | | | | | |
| | 5 | .0022 | .0319 | .1746 | .2023 | .1789 | .0746 | .0148 | .0013 | | | | | |
| | 6 | .0003 | .0089 | .1091 | .1686 | .1916 | .1244 | .0370 | .0049 | .0002 | | | | |
| | 7 | | .0020 | .0546 | .1124 | .1643 | .1659 | .0739 | .0146 | .0010 | .0002 | | | |
| | 8 | | .0004 | .0222 | .0609 | .1144 | .1797 | .1201 | .0355 | .0039 | .0008 | .0001 | | |
| | 9 | | .0001 | .0074 | .0271 | .0654 | .1597 | .1602 | .0710 | .0120 | .0030 | .0005 | | |
| | 10 | | | .0020 | .0099 | .0308 | .1171 | .1762 | .1171 | .0308 | .0099 | .0020 | | |
| | 11 | | | .0005 | .0030 | .0120 | .0710 | .1602 | .1597 | .0654 | .0271 | .0074 | .0001 | |
| | 12 | | | .0001 | .0008 | .0039 | .0355 | .1201 | .1797 | .1144 | .0609 | .0222 | .0004 | |
| | 13 | | | | .0002 | .0010 | .0146 | .0739 | .1659 | .1643 | .1124 | .0546 | .0020 | |
| | 14 | | | | | .0002 | .0049 | .0370 | .1244 | .1916 | .1686 | .1091 | .0089 | .0003 |
| | 15 | | | | | | .0013 | .0148 | .0746 | .1789 | .2023 | .1746 | .0319 | .0022 |
| | 16 | | | | | | .0003 | .0046 | .0350 | .1304 | .1897 | .2182 | .0898 | .0133 |
| | 17 | | | | | | | .0011 | .0124 | .0716 | .1339 | .2054 | .1901 | .0596 |
| | 18 | | | | | | | .0002 | .0031 | .0278 | .0669 | .1369 | .2852 | .1887 |
| | 19 | | | | | | | | .0005 | .0068 | .0211 | .0576 | .2702 | .3774 |
| | 20 | | | | | | | | | .0008 | .0032 | .0115 | .1216 | .3585 |
| 25 | 0 | .2774 | .0718 | .0038 | .0008 | .0001 | | | | | | | | |
| | 1 | .3650 | .1994 | .0236 | .0063 | .0014 | | | | | | | | |
| | 2 | .2305 | .2659 | .0708 | .0251 | .0074 | .0004 | | | | | | | |
| | 3 | .0930 | .2265 | .1358 | .0641 | .0243 | .0019 | .0001 | | | | | | |
| | 4 | .0269 | .1384 | .1867 | .1175 | .0572 | .0071 | .0004 | | | | | | |
| | 5 | .0060 | .0646 | .1960 | .1645 | .1030 | .0199 | .0016 | | | | | | |
| | 6 | .0010 | .0239 | .1633 | .1828 | .1472 | .0442 | .0053 | .0002 | | | | | |
| | 7 | .0001 | .0072 | .1108 | .1654 | .1712 | .0800 | .0143 | .0009 | | | | | |
| | 8 | | .0018 | .0623 | .1241 | .1651 | .1200 | .0322 | .0031 | .0001 | | | | |
| | 9 | | .0004 | .0294 | .0781 | .1336 | .1511 | .0609 | .0088 | .0004 | | | | |
| | 10 | | .0001 | .0118 | .0417 | .0916 | .1612 | .0974 | .0212 | .0013 | .0002 | | | |
| | 11 | | | .0040 | .0189 | .0536 | .1465 | .1328 | .0434 | .0042 | .0007 | .0001 | | |
| | 12 | | | .0012 | .0074 | .0268 | .1139 | .1550 | .0760 | .0115 | .0025 | .0003 | | |
| | 13 | | | .0003 | .0025 | .0115 | .0760 | .1550 | .1139 | .0268 | .0074 | .0012 | | |
| | 14 | | | .0001 | .0007 | .0042 | .0434 | .1328 | .1465 | .0536 | .0189 | .0040 | | |
| | 15 | | | | .0002 | .0013 | .0212 | .0974 | .1612 | .0916 | .0417 | .0118 | .0001 | |
| | 16 | | | | | .0004 | .0088 | .0609 | .1511 | .1336 | .0781 | .0294 | .0004 | |
| | 17 | | | | | .0001 | .0031 | .0322 | .1200 | .1651 | .1241 | .0623 | .0018 | |
| | 18 | | | | | | .0009 | .0143 | .0800 | .1712 | .1654 | .1108 | .0072 | .0002 |
| | 19 | | | | | | .0002 | .0053 | .0442 | .1472 | .1828 | .1633 | .0239 | .0010 |
| | 20 | | | | | | | .0016 | .0199 | .1030 | .1645 | .1960 | .0646 | .0060 |
| | 21 | | | | | | | .0004 | .0071 | .0572 | .1175 | .1867 | .1384 | .0269 |
| | 22 | | | | | | | .0001 | .0019 | .0243 | .0641 | .1358 | .2265 | .0930 |
| | 23 | | | | | | | | .0004 | .0074 | .0251 | .0708 | .2659 | .2305 |
| | 24 | | | | | | | | .0014 | .0063 | .0236 | .1994 | .3650 | |
| | 25 | | | | | | | | | .0001 | .0008 | .0038 | .0718 | .2774 |

**TABLE B.6**   Poisson Point Probabilities by Values of $e^{-x}$

Values of $e^{-x}$

| X | $e^{-x}$ | X | $e^{-x}$ | X | $e^{-x}$ | X | $e^{-x}$ | X | $e^{-x}$ | X | $e^{-x}$ | X | $e^{-x}$ | X | $e^{-x}$ | X | $e^{-x}$ | X | $e^{-x}$ |
|---|---|---|---|---|---|---|---|---|---|---|---|---|---|---|---|---|---|---|---|
| 0.00 | 1.00000 | 0.35 | .70469 | 0.70 | .49659 | 1.05 | .34994 | 1.40 | .24660 | 1.75 | .17377 | 2.10 | .12246 | 2.45 | .08629 | 2.80 | .06081 | 3.75 | .02352 |
| 0.01 | .99005 | 0.36 | .69768 | 0.71 | .49164 | 1.06 | .34646 | 1.41 | .24414 | 1.76 | .17204 | 2.11 | .12124 | 2.46 | .08544 | 2.81 | .06020 | 3.80 | .02237 |
| 0.02 | .98020 | 0.37 | .69073 | 0.72 | .48675 | 1.07 | .34301 | 1.42 | .24171 | 1.77 | .17033 | 2.12 | .12003 | 2.47 | .08458 | 2.82 | .05961 | 3.85 | .02128 |
| 0.03 | .97045 | 0.38 | .68386 | 0.73 | .48191 | 1.08 | .33960 | 1.43 | .23931 | 1.78 | .16864 | 2.13 | .11884 | 2.48 | .08374 | 2.83 | .05901 | 3.90 | .02024 |
| 0.04 | .96079 | 0.39 | .67706 | 0.74 | .47711 | 1.09 | .33622 | 1.44 | .23693 | 1.79 | .16696 | 2.14 | .11765 | 2.49 | .08291 | 2.84 | .05843 | 3.95 | .01925 |
| 0.05 | .95123 | 0.40 | .67032 | 0.75 | .47237 | 1.10 | .33287 | 1.45 | .23457 | 1.80 | .16530 | 2.15 | .11648 | 2.50 | .08208 | 2.85 | .05784 | 4.00 | .01832 |
| 0.06 | .94176 | 0.41 | .66365 | 0.76 | .46767 | 1.11 | .32956 | 1.46 | .23224 | 1.81 | .16365 | 2.16 | .11533 | 2.51 | .08127 | 2.86 | .05727 | 4.10 | .01657 |
| 0.07 | .93239 | 0.42 | .65705 | 0.77 | .46301 | 1.12 | .32628 | 1.47 | .22993 | 1.82 | .16203 | 2.17 | .11418 | 2.52 | .08046 | 2.87 | .05670 | 4.20 | .01500 |
| 0.08 | .92312 | 0.43 | .65051 | 0.78 | .45841 | 1.13 | .32303 | 1.48 | .22764 | 1.83 | .16041 | 2.18 | .11304 | 2.53 | .07966 | 2.88 | .05613 | 4.30 | .01357 |
| 0.09 | .91393 | 0.44 | .64404 | 0.79 | .45384 | 1.14 | .31982 | 1.49 | .22537 | 1.84 | .15882 | 2.19 | .11192 | 2.54 | .07887 | 2.89 | .05558 | 4.40 | .01228 |
| 0.10 | .90484 | 0.45 | .63763 | 0.80 | .44933 | 1.15 | .31664 | 1.50 | .22313 | 1.85 | .15724 | 2.20 | .11080 | 2.55 | .07808 | 2.90 | .05502 | 4.50 | .01111 |
| 0.11 | .89583 | 0.46 | .63128 | 0.81 | .44486 | 1.16 | .31349 | 1.51 | .22091 | 1.86 | .15567 | 2.21 | .10970 | 2.56 | .07730 | 2.91 | .05448 | 4.60 | .01005 |
| 0.12 | .88692 | 0.47 | .62500 | 0.82 | .44043 | 1.17 | .31037 | 1.52 | .21871 | 1.87 | .15412 | 2.22 | .10861 | 2.57 | .07654 | 2.92 | .05393 | 4.70 | .00910 |
| 0.13 | .87810 | 0.48 | .61878 | 0.83 | .43605 | 1.18 | .30728 | 1.53 | .21654 | 1.88 | .15259 | 2.23 | .10753 | 2.58 | .07577 | 2.93 | .05340 | 4.80 | .00823 |
| 0.14 | .86936 | 0.49 | .61263 | 0.84 | .43171 | 1.19 | .30422 | 1.54 | .21438 | 1.89 | .15107 | 2.24 | .10646 | 2.59 | .07502 | 2.94 | .05287 | 4.90 | .00745 |
| 0.15 | .86071 | 0.50 | .60653 | 0.85 | .42741 | 1.20 | .30119 | 1.55 | .21225 | 1.90 | .14957 | 2.25 | .10540 | 2.60 | .07427 | 2.95 | .05234 | 5.00 | .00674 |
| 0.16 | .85214 | 0.51 | .60050 | 0.86 | .42316 | 1.21 | .29820 | 1.56 | .21014 | 1.91 | .14808 | 2.26 | .10435 | 2.61 | .07353 | 2.96 | .05182 | 5.10 | .00610 |
| 0.17 | .84366 | 0.52 | .59452 | 0.87 | .41895 | 1.22 | .29523 | 1.57 | .20805 | 1.92 | .14661 | 2.27 | .10331 | 2.62 | .07280 | 2.97 | .05130 | 5.20 | .00552 |
| 0.18 | .83527 | 0.53 | .58860 | 0.88 | .41478 | 1.23 | .29229 | 1.58 | .20598 | 1.93 | .14515 | 2.28 | .10228 | 2.63 | .07208 | 2.98 | .05079 | 5.30 | .00499 |
| 0.19 | .82696 | 0.54 | .58275 | 0.89 | .41066 | 1.24 | .28938 | 1.59 | .20393 | 1.94 | .14370 | 2.29 | .10127 | 2.64 | .07136 | 2.99 | .05029 | 5.40 | .00452 |
| 0.20 | .81873 | 0.55 | .57695 | 0.90 | .40657 | 1.25 | .28650 | 1.60 | .20190 | 1.95 | .14227 | 2.30 | .10026 | 2.65 | .07065 | 3.00 | .04979 | 5.50 | .00409 |
| 0.21 | .81058 | 0.56 | .57121 | 0.91 | .40252 | 1.26 | .28365 | 1.61 | .19989 | 1.96 | .14086 | 2.31 | .09926 | 2.66 | .06995 | 3.05 | .04736 | 5.60 | .00370 |
| 0.22 | .80252 | 0.57 | .56553 | 0.92 | .39852 | 1.27 | .28083 | 1.62 | .19790 | 1.97 | .13946 | 2.32 | .09827 | 2.67 | .06925 | 3.10 | .04505 | 5.70 | .00335 |
| 0.23 | .79453 | 0.58 | .55990 | 0.93 | .39455 | 1.28 | .27804 | 1.63 | .19593 | 1.98 | .13807 | 2.33 | .09730 | 2.68 | .06856 | 3.15 | .04285 | 5.80 | .00303 |
| 0.24 | .78663 | 0.59 | .55433 | 0.94 | .39063 | 1.29 | .27527 | 1.64 | .19389 | 1.99 | .13670 | 2.34 | .09633 | 2.69 | .06788 | 3.20 | .04076 | 5.90 | .00274 |
| 0.25 | .77880 | 0.60 | .54881 | 0.95 | .38674 | 1.30 | .27253 | 1.65 | .19205 | 2.00 | .13534 | 2.35 | .09537 | 2.70 | .06721 | 3.25 | .03877 | 6.00 | .00248 |
| 0.26 | .77105 | 0.61 | .54335 | 0.96 | .38289 | 1.31 | .26982 | 1.66 | .19014 | 2.01 | .13399 | 2.36 | .09442 | 2.71 | .06654 | 3.30 | .03688 | 6.25 | .00193 |
| 0.27 | .76338 | 0.62 | .53794 | 0.97 | .37908 | 1.32 | .26714 | 1.67 | .18825 | 2.02 | .13266 | 2.37 | .09348 | 2.72 | .06587 | 3.35 | .03508 | 6.50 | .00150 |
| 0.28 | .75578 | 0.63 | .53259 | 0.98 | .37531 | 1.33 | .26448 | 1.68 | .18637 | 2.03 | .13134 | 2.38 | .09255 | 2.73 | .06522 | 3.40 | .03337 | 6.75 | .00117 |
| 0.29 | .74826 | 0.64 | .52729 | 0.99 | .37158 | 1.34 | .26185 | 1.69 | .18452 | 2.04 | .13003 | 2.39 | .09163 | 2.74 | .06457 | 3.45 | .03175 | 7.00 | .00091 |
| 0.30 | .74082 | 0.65 | .52205 | 1.00 | .36788 | 1.35 | .25924 | 1.70 | .18268 | 2.05 | .12873 | 2.40 | .09072 | 2.75 | .06393 | 3.50 | .03020 | 7.50 | .00055 |
| 0.31 | .73345 | 0.66 | .51685 | 1.01 | .36422 | 1.36 | .25666 | 1.71 | .18087 | 2.06 | .12745 | 2.41 | .08982 | 2.76 | .06329 | 3.55 | .02872 | 8.00 | .00034 |
| 0.32 | .72615 | 0.67 | .51171 | 1.02 | .36059 | 1.37 | .25411 | 1.72 | .17907 | 2.07 | .12619 | 2.42 | .08892 | 2.77 | .06266 | 3.60 | .02732 | 8.50 | .00020 |
| 0.33 | .71892 | 0.68 | .50662 | 1.03 | .35701 | 1.38 | .25158 | 1.73 | .17728 | 2.08 | .12493 | 2.43 | .08804 | 2.78 | .06204 | 3.65 | .02599 | 9.00 | .00012 |
| 0.34 | .71177 | 0.69 | .50158 | 1.04 | .35345 | 1.39 | .24908 | 1.74 | .17552 | 2.09 | .12369 | 2.44 | .08716 | 2.79 | .06142 | 3.70 | .02472 | 9.50 | .00007 |
|  |  |  |  |  |  |  |  |  |  |  |  |  |  |  |  |  |  | 10.00 | .00005 |

**TABLE B.7**    Random Numbers

A Table of 14,000 Random Units

| Line/Col. | (1) | (2) | (3) | (4) | (5) | (6) | (7) | (8) | (9) | (10) | (11) | (12) | (13) | (14) |
|---|---|---|---|---|---|---|---|---|---|---|---|---|---|---|
| 1 | 10480 | 15011 | 01536 | 02011 | 81647 | 91646 | 69179 | 14194 | 62590 | 36207 | 20969 | 99570 | 91291 | 90700 |
| 2 | 22368 | 46573 | 25595 | 85393 | 30995 | 89198 | 27982 | 53402 | 93965 | 34095 | 52666 | 19174 | 39615 | 99505 |
| 3 | 24130 | 48360 | 22527 | 97265 | 76393 | 64809 | 15179 | 24830 | 49340 | 32081 | 30680 | 19655 | 63348 | 58629 |
| 4 | 42167 | 93093 | 06243 | 61680 | 07856 | 16376 | 39440 | 53537 | 71341 | 57004 | 00849 | 74917 | 97758 | 16379 |
| 5 | 37570 | 39975 | 81837 | 16656 | 06121 | 91782 | 60468 | 81305 | 49684 | 60672 | 14110 | 06927 | 01263 | 54613 |
| 6 | 77921 | 06907 | 11008 | 42751 | 27756 | 53498 | 18602 | 70659 | 90655 | 15053 | 21916 | 81825 | 44394 | 42880 |
| 7 | 99562 | 72905 | 56420 | 69994 | 98872 | 31016 | 71194 | 18738 | 44013 | 48840 | 63213 | 21069 | 10634 | 12952 |
| 8 | 96301 | 91977 | 05463 | 07972 | 18876 | 20922 | 94595 | 56869 | 69014 | 60045 | 18425 | 84903 | 42508 | 32307 |
| 9 | 89579 | 14342 | 63661 | 10281 | 17453 | 18103 | 57740 | 84378 | 25331 | 12566 | 58678 | 44947 | 05585 | 56941 |
| 10 | 85475 | 36857 | 43342 | 53988 | 53060 | 59533 | 38867 | 62300 | 08158 | 17983 | 16439 | 11458 | 18593 | 64952 |
| 11 | 28918 | 69578 | 88231 | 33276 | 70997 | 79936 | 56865 | 05859 | 90106 | 31595 | 01547 | 85590 | 91610 | 78188 |
| 12 | 63553 | 40961 | 48235 | 03427 | 49626 | 69445 | 18663 | 72695 | 52180 | 20847 | 12234 | 90511 | 33703 | 90322 |
| 13 | 09429 | 93969 | 52636 | 92737 | 88974 | 33488 | 36320 | 17617 | 30015 | 08272 | 84115 | 27156 | 30613 | 74952 |
| 14 | 10365 | 61129 | 87529 | 85689 | 48237 | 52267 | 67689 | 93394 | 01511 | 26358 | 85104 | 20285 | 29975 | 89868 |
| 15 | 07119 | 97336 | 71048 | 08178 | 77233 | 13916 | 47564 | 81056 | 97735 | 85977 | 29372 | 74461 | 28551 | 90707 |
| 16 | 51085 | 12765 | 51821 | 51259 | 77452 | 16308 | 60756 | 92144 | 49442 | 53900 | 70960 | 63990 | 75601 | 40719 |
| 17 | 02368 | 21382 | 52404 | 60268 | 89368 | 19885 | 55322 | 44819 | 01188 | 65255 | 64835 | 44919 | 05944 | 55157 |
| 18 | 01011 | 54092 | 33362 | 94904 | 31273 | 04146 | 18594 | 29852 | 71585 | 85030 | 51132 | 01915 | 92747 | 64951 |
| 19 | 52162 | 53916 | 46369 | 58586 | 23216 | 14513 | 83149 | 98736 | 23495 | 64350 | 94738 | 17752 | 35156 | 35749 |
| 20 | 07056 | 97628 | 33787 | 09998 | 42698 | 06691 | 76988 | 13602 | 51851 | 46104 | 88916 | 19509 | 25625 | 58104 |
| 21 | 48663 | 91245 | 85828 | 14346 | 09172 | 30168 | 90229 | 04734 | 59193 | 22178 | 30421 | 61666 | 99904 | 32812 |
| 22 | 54164 | 58492 | 22421 | 74103 | 47070 | 25306 | 76468 | 26384 | 58151 | 06646 | 21524 | 15227 | 96909 | 44592 |
| 23 | 32639 | 32363 | 05597 | 24200 | 13363 | 38005 | 94342 | 28728 | 35806 | 06912 | 17012 | 64161 | 18296 | 22851 |
| 24 | 29334 | 27001 | 87637 | 87308 | 58731 | 00256 | 45834 | 15398 | 46557 | 41135 | 10367 | 07684 | 36188 | 18510 |
| 25 | 02488 | 33062 | 28834 | 07351 | 19731 | 92420 | 60952 | 61280 | 50001 | 67658 | 32586 | 86679 | 50720 | 94953 |
| 26 | 81525 | 72295 | 04839 | 96423 | 24878 | 82651 | 66566 | 14778 | 76797 | 14780 | 13300 | 87074 | 79666 | 95725 |
| 27 | 29676 | 20591 | 68086 | 26432 | 46901 | 20849 | 89768 | 81536 | 86645 | 12659 | 92259 | 57102 | 80428 | 25280 |
| 28 | 00742 | 57392 | 39064 | 66432 | 84673 | 40027 | 32832 | 61362 | 98947 | 96067 | 64760 | 64584 | 96096 | 98253 |
| 29 | 05366 | 04213 | 25669 | 26422 | 44407 | 44048 | 37937 | 63904 | 45766 | 66134 | 75470 | 66520 | 34693 | 90449 |
| 30 | 91921 | 26418 | 64117 | 94305 | 26766 | 25940 | 39972 | 22209 | 71500 | 64568 | 91402 | 42416 | 07844 | 69618 |
| 31 | 00582 | 04711 | 87917 | 77341 | 42206 | 35126 | 74087 | 99547 | 81817 | 42607 | 43808 | 76655 | 62028 | 76630 |
| 32 | 00725 | 69884 | 62797 | 56170 | 86324 | 88072 | 76222 | 36086 | 84637 | 93161 | 76038 | 65855 | 77919 | 88006 |
| 33 | 69011 | 65797 | 95876 | 55293 | 18988 | 27354 | 26575 | 08625 | 40801 | 59920 | 29841 | 80150 | 12777 | 48501 |
| 34 | 25976 | 57948 | 29888 | 88604 | 67917 | 48708 | 18912 | 82271 | 65424 | 69774 | 33611 | 54262 | 85963 | 03547 |
| 35 | 09763 | 83473 | 73577 | 12908 | 30883 | 18317 | 28290 | 35797 | 05998 | 41688 | 34952 | 37888 | 38917 | 88050 |

Reprinted with permission from *Handbook of Tables for Probability and Statistics*, 2nd ed., William H. Beyer (ed.), 1968. Copyright © CRC Press, Inc., Boca Raton, FL.

**TABLE B.7**   (*continued*)

A Table of 14,000 Random Units

| Line/Col. | (1) | (2) | (3) | (4) | (5) | (6) | (7) | (8) | (9) | (10) | (11) | (12) | (13) | (14) |
|-----------|-----|-----|-----|-----|-----|-----|-----|-----|-----|------|------|------|------|------|
| 36 | 91567 | 42595 | 27958 | 30134 | 04024 | 86385 | 29880 | 99730 | 55536 | 84855 | 29080 | 09250 | 79656 | 73211 |
| 37 | 17955 | 56349 | 90999 | 49127 | 20044 | 59931 | 06115 | 20542 | 18059 | 02008 | 73708 | 83517 | 36103 | 42791 |
| 38 | 46503 | 18584 | 18845 | 49618 | 02304 | 51038 | 20655 | 58727 | 28168 | 15475 | 56942 | 53389 | 20562 | 87338 |
| 39 | 92157 | 89634 | 94824 | 78171 | 84610 | 82834 | 09922 | 25417 | 44137 | 48413 | 25555 | 21246 | 35509 | 20468 |
| 40 | 14577 | 62765 | 35605 | 81263 | 39667 | 47358 | 56873 | 56307 | 61607 | 49518 | 89656 | 20103 | 77490 | 18062 |
| 41 | 98427 | 07523 | 33362 | 64270 | 01638 | 92477 | 66969 | 98420 | 04880 | 45585 | 46565 | 04102 | 46880 | 45709 |
| 42 | 34914 | 63976 | 88720 | 82765 | 34476 | 17032 | 87589 | 40836 | 32427 | 70002 | 70663 | 88863 | 77775 | 69348 |
| 43 | 70060 | 28277 | 39475 | 46473 | 23219 | 53416 | 94970 | 25832 | 69975 | 94884 | 19661 | 72828 | 00102 | 66794 |
| 44 | 53976 | 54914 | 06990 | 67245 | 68350 | 82948 | 11398 | 42878 | 80287 | 88267 | 47363 | 46634 | 06541 | 97809 |
| 45 | 76072 | 29515 | 40980 | 07391 | 58745 | 25774 | 22987 | 80059 | 39911 | 96189 | 41151 | 14222 | 60697 | 59583 |
| 46 | 90725 | 52210 | 83974 | 29992 | 65831 | 38857 | 50490 | 83765 | 55657 | 14361 | 31720 | 57375 | 56228 | 41546 |
| 47 | 64364 | 67412 | 33339 | 31926 | 14883 | 24413 | 59744 | 92351 | 97473 | 89286 | 35931 | 04110 | 23726 | 51900 |
| 48 | 08962 | 00358 | 31662 | 25388 | 61642 | 34072 | 81249 | 35648 | 56891 | 69352 | 48373 | 45578 | 78547 | 81788 |
| 49 | 95012 | 68379 | 93526 | 70765 | 10593 | 04542 | 76463 | 54328 | 02349 | 17247 | 28865 | 14777 | 62730 | 92277 |
| 50 | 15664 | 10493 | 20492 | 38391 | 91132 | 21999 | 59516 | 81652 | 27195 | 48223 | 46751 | 22923 | 32261 | 85653 |
| 51 | 16408 | 81899 | 04153 | 53381 | 79401 | 21438 | 83035 | 92350 | 36693 | 31238 | 59649 | 91754 | 72772 | 02338 |
| 52 | 18629 | 81953 | 05520 | 91962 | 04739 | 13092 | 97662 | 24822 | 94730 | 06496 | 35090 | 04822 | 86772 | 98289 |
| 53 | 73115 | 35101 | 47498 | 87637 | 99016 | 71060 | 88824 | 71013 | 18735 | 20286 | 23153 | 72924 | 35165 | 43040 |
| 54 | 57491 | 16703 | 23167 | 49323 | 45021 | 33132 | 12544 | 41035 | 80780 | 45393 | 44812 | 12515 | 98931 | 91202 |
| 55 | 30405 | 83946 | 23792 | 14422 | 15059 | 45799 | 22716 | 19792 | 09983 | 74353 | 68668 | 30429 | 70735 | 25499 |
| 56 | 16631 | 35006 | 85900 | 98275 | 32388 | 52390 | 16815 | 69298 | 82732 | 38480 | 73817 | 32523 | 41961 | 44437 |
| 57 | 96773 | 20206 | 42559 | 78985 | 05300 | 22164 | 24369 | 54224 | 35083 | 19687 | 11052 | 91491 | 60383 | 19746 |
| 58 | 38935 | 64202 | 14349 | 82674 | 66523 | 44133 | 00697 | 35552 | 35970 | 19124 | 63318 | 29686 | 03387 | 59846 |
| 59 | 31624 | 76384 | 17403 | 53363 | 44167 | 64486 | 64758 | 75366 | 76554 | 31601 | 12614 | 33072 | 60332 | 92325 |
| 60 | 78919 | 19474 | 23632 | 27889 | 47914 | 02584 | 37680 | 20801 | 72152 | 39339 | 34806 | 08930 | 85001 | 87820 |
| 61 | 03931 | 33309 | 57047 | 74211 | 63445 | 17361 | 62825 | 39908 | 05607 | 91284 | 68833 | 25570 | 38818 | 46920 |
| 62 | 74426 | 33278 | 43972 | 10119 | 89917 | 15665 | 52872 | 73823 | 73144 | 88662 | 88970 | 74492 | 51805 | 99378 |
| 63 | 09066 | 00903 | 20795 | 95452 | 92648 | 45454 | 09552 | 88815 | 16553 | 51125 | 79375 | 97596 | 16296 | 66092 |
| 64 | 42238 | 12426 | 87025 | 14267 | 20979 | 04508 | 64535 | 31355 | 86064 | 29472 | 47689 | 05974 | 52468 | 16834 |
| 65 | 16153 | 08002 | 26504 | 41744 | 81959 | 65642 | 74240 | 56302 | 00033 | 67107 | 77510 | 70625 | 28725 | 34191 |
| 66 | 21457 | 40742 | 29820 | 96783 | 29400 | 21840 | 15035 | 34537 | 33310 | 06116 | 95240 | 15957 | 16572 | 06004 |
| 67 | 21581 | 57802 | 02050 | 89728 | 17937 | 37621 | 47075 | 42080 | 97403 | 48626 | 68995 | 43805 | 33386 | 21597 |
| 68 | 55612 | 78095 | 83197 | 33732 | 05810 | 24813 | 86902 | 60397 | 16489 | 03264 | 88525 | 42786 | 05269 | 92532 |
| 69 | 44657 | 66999 | 99324 | 51281 | 84463 | 60563 | 79312 | 93454 | 68876 | 25471 | 93911 | 25650 | 12682 | 73572 |
| 70 | 91340 | 84979 | 46949 | 81973 | 37949 | 61023 | 43997 | 15263 | 80644 | 43942 | 89203 | 71795 | 99533 | 50501 |

**TABLE B.7**  (*continued*)

A Table of 14,000 Random Units

| Line/Col. | (1) | (2) | (3) | (4) | (5) | (6) | (7) | (8) | (9) | (10) | (11) | (12) | (13) | (14) |
|---|---|---|---|---|---|---|---|---|---|---|---|---|---|---|
| 71 | 91227 | 21199 | 31935 | 27022 | 84067 | 05462 | 35216 | 14486 | 29891 | 68607 | 41867 | 14951 | 91696 | 85065 |
| 72 | 50001 | 38140 | 66321 | 19924 | 72163 | 09538 | 12151 | 06878 | 91903 | 18749 | 34405 | 56087 | 82790 | 70925 |
| 73 | 65390 | 05224 | 72958 | 28609 | 81406 | 39147 | 25549 | 48542 | 42627 | 45233 | 57202 | 94617 | 23772 | 07896 |
| 74 | 27504 | 96131 | 83944 | 41575 | 10573 | 08619 | 64482 | 73923 | 36152 | 05184 | 94142 | 25299 | 84387 | 34925 |
| 75 | 37169 | 94851 | 39117 | 89632 | 00959 | 16487 | 65536 | 49071 | 39782 | 17095 | 02330 | 74301 | 00275 | 48280 |
| 76 | 11508 | 70225 | 51111 | 38351 | 19444 | 66499 | 71945 | 05422 | 13442 | 78675 | 84081 | 66938 | 93654 | 59894 |
| 77 | 37449 | 30362 | 06694 | 54690 | 04052 | 53115 | 62757 | 95348 | 78662 | 11163 | 81651 | 50245 | 34971 | 52924 |
| 78 | 46515 | 70331 | 85922 | 38329 | 57015 | 15765 | 97161 | 17869 | 45349 | 61796 | 66345 | 81073 | 49106 | 79860 |
| 79 | 30986 | 81223 | 42416 | 58353 | 21532 | 30502 | 32305 | 86482 | 05174 | 07901 | 54339 | 58861 | 74818 | 46942 |
| 80 | 63798 | 64995 | 46583 | 09765 | 44160 | 78128 | 83991 | 42865 | 92520 | 83531 | 80377 | 35909 | 81250 | 54238 |
| 81 | 82486 | 84846 | 99254 | 67632 | 43218 | 50076 | 21361 | 64816 | 51202 | 88124 | 41870 | 52689 | 51275 | 83556 |
| 82 | 21885 | 32906 | 92431 | 09060 | 64297 | 51674 | 64126 | 62570 | 26123 | 05155 | 59194 | 52799 | 28225 | 85762 |
| 83 | 60336 | 98782 | 07408 | 53458 | 13564 | 59089 | 26445 | 29789 | 85205 | 41001 | 12535 | 12133 | 14645 | 23541 |
| 84 | 43937 | 46891 | 24010 | 25560 | 86355 | 33941 | 25786 | 54990 | 71899 | 15475 | 95434 | 98227 | 21824 | 19585 |
| 85 | 97656 | 63175 | 89303 | 16275 | 07100 | 92063 | 21942 | 18611 | 47348 | 20203 | 18534 | 03862 | 78095 | 50136 |
| 86 | 03299 | 01221 | 05418 | 38982 | 55758 | 92237 | 26759 | 86367 | 21216 | 98442 | 08303 | 56613 | 91511 | 75928 |
| 87 | 79626 | 06486 | 03574 | 17668 | 07785 | 76020 | 79924 | 25651 | 83325 | 88428 | 85076 | 72811 | 22717 | 50585 |
| 88 | 85636 | 68335 | 47539 | 03129 | 65651 | 11977 | 02510 | 26113 | 99447 | 68645 | 34327 | 15152 | 55230 | 93448 |
| 89 | 18039 | 14367 | 61337 | 06177 | 12143 | 46609 | 32989 | 74014 | 64708 | 00533 | 35398 | 58408 | 13261 | 47908 |
| 90 | 08362 | 15656 | 60627 | 36478 | 65648 | 16764 | 53412 | 09013 | 07832 | 41574 | 17639 | 82163 | 60859 | 75567 |
| 91 | 79556 | 29068 | 04142 | 16268 | 15387 | 12856 | 66227 | 38358 | 22478 | 73373 | 88732 | 09443 | 82558 | 05250 |
| 92 | 92608 | 82674 | 27072 | 32534 | 17075 | 27698 | 98204 | 64863 | 11951 | 34648 | 88022 | 56148 | 34925 | 57031 |
| 93 | 23982 | 25835 | 40055 | 67006 | 12293 | 02753 | 14827 | 22235 | 35071 | 99704 | 37543 | 11601 | 35503 | 85171 |
| 94 | 09915 | 96306 | 05908 | 97901 | 28395 | 14186 | 00821 | 80703 | 70426 | 75647 | 76310 | 88717 | 37890 | 40129 |
| 95 | 50937 | 33300 | 26695 | 62247 | 69927 | 76123 | 50842 | 43834 | 86654 | 70959 | 79725 | 93872 | 28117 | 19233 |
| 96 | 42488 | 78077 | 69882 | 61657 | 34136 | 79180 | 97526 | 43092 | 04098 | 73571 | 80799 | 76536 | 71255 | 64239 |
| 97 | 46764 | 86273 | 63003 | 93017 | 31204 | 36692 | 40202 | 35275 | 57306 | 55543 | 53203 | 18098 | 47625 | 88684 |
| 98 | 03237 | 45430 | 55417 | 63282 | 90816 | 17349 | 88298 | 90183 | 36600 | 78406 | 06216 | 95787 | 42579 | 90730 |
| 99 | 86591 | 81482 | 52667 | 61583 | 14972 | 90053 | 89534 | 76036 | 49199 | 43716 | 97548 | 04379 | 46370 | 28672 |
| 100 | 38534 | 01715 | 94964 | 87288 | 65680 | 43772 | 39560 | 12918 | 86537 | 62738 | 19636 | 51132 | 25739 | 56947 |

**TABLE B.8**   Critical Values of Spearman's Rank Correlation Coefficient

| $n$ | $\alpha = 0.10$ | $\alpha = 0.05$ | $\alpha = 0.02$ | $\alpha = 0.01$ |
|---|---|---|---|---|
| 5 | 0.900 | — | — | — |
| 6 | 0.829 | 0.886 | 0.943 | — |
| 7 | 0.714 | 0.786 | 0.893 | 0.929 |
| 8 | 0.643 | 0.738 | 0.833 | 0.881 |
| 9 | 0.600 | 0.700 | 0.783 | 0.833 |
| 10 | 0.564 | 0.648 | 0.745 | 0.818 |
| 11 | 0.536 | 0.618 | 0.709 | 0.794 |
| 12 | 0.497 | 0.591 | 0.703 | 0.780 |
| 13 | 0.475 | 0.566 | 0.673 | 0.745 |
| 14 | 0.457 | 0.545 | 0.646 | 0.716 |
| 15 | 0.441 | 0.525 | 0.623 | 0.689 |
| 16 | 0.425 | 0.507 | 0.601 | 0.666 |
| 17 | 0.412 | 0.490 | 0.582 | 0.645 |
| 18 | 0.399 | 0.476 | 0.564 | 0.625 |
| 19 | 0.388 | 0.462 | 0.549 | 0.608 |
| 20 | 0.377 | 0.450 | 0.534 | 0.591 |
| 21 | 0.368 | 0.438 | 0.521 | 0.576 |
| 22 | 0.359 | 0.428 | 0.508 | 0.562 |
| 23 | 0.351 | 0.418 | 0.496 | 0.549 |
| 24 | 0.343 | 0.409 | 0.485 | 0.537 |
| 25 | 0.336 | 0.400 | 0.475 | 0.526 |
| 26 | 0.329 | 0.392 | 0.465 | 0.515 |
| 27 | 0.323 | 0.385 | 0.456 | 0.505 |
| 28 | 0.317 | 0.377 | 0.448 | 0.496 |
| 29 | 0.311 | 0.370 | 0.440 | 0.487 |
| 30 | 0.305 | 0.364 | 0.432 | 0.478 |

**TABLE B.9**  Percentage Points in Kolmogorov–Smirnov Statistics

| One-sided tests α = | .10 | .05 | .025 | .01 | .005 | | | .10 | .05 | .025 | .01 | .005 |
|---|---|---|---|---|---|---|---|---|---|---|---|---|
| Two-sided tests α = | .20 | .10 | .05 | .02 | .01 | | | .20 | .10 | .05 | .02 | .01 |
| n = 1 | .900 | .950 | .975 | .990 | .995 | n = | 31 | .187 | 214 | .238 | .266 | .285 |
| 2 | .684 | .776 | .842 | .900 | .929 | | 32 | .184 | .211 | .234 | .262 | .281 |
| 3 | .565 | .636 | .708 | .785 | .829 | | 33 | .182 | .208 | .231 | .258 | .277 |
| 4 | .493 | .565 | .624 | .689 | .734 | | 34 | .179 | .205 | .227 | .254 | .273 |
| 5 | .447 | .509 | .563 | .627 | .669 | | 35 | .177 | .202 | .224 | .251 | .269 |
| 6 | .410 | .468 | .519 | .577 | .617 | | 36 | .174 | .199 | .221 | .247 | .265 |
| 7 | .381 | .436 | .483 | .538 | .576 | | 37 | .172 | .196 | .218 | .244 | .262 |
| 8 | .358 | .410 | .454 | .507 | .542 | | 38 | .170 | .194 | .215 | .241 | .258 |
| 9 | .339 | .387 | .430 | .480 | .513 | | 39 | .168 | .191 | .213 | .238 | .255 |
| 10 | .323 | .369 | .409 | .457 | .489 | | 40 | .165 | .189 | .210 | .235 | .252 |
| 11 | .308 | .352 | .391 | .437 | .468 | | 41 | .163 | .187 | .208 | .232 | .249 |
| 12 | .296 | .338 | .375 | .419 | .449 | | 42 | .162 | .185 | .205 | .229 | .246 |
| 13 | .285 | .325 | .361 | .404 | .432 | | 43 | .160 | .183 | .203 | .227 | .243 |
| 14 | .275 | .314 | .349 | .390 | .418 | | 44 | .158 | .181 | .201 | .224 | .241 |
| 15 | .266 | .304 | .338 | .377 | .404 | | 45 | .156 | .179 | .198 | .222 | .238 |
| 16 | .258 | .295 | .327 | .366 | .392 | | 46 | .155 | .177 | .196 | .219 | .235 |
| 17 | .250 | .286 | .318 | .355 | .381 | | 47 | .153 | .175 | .194 | .217 | .233 |
| 18 | .244 | .279 | .309 | .346 | .371 | | 48 | .151 | .173 | .192 | .215 | .231 |
| 19 | .237 | .271 | .301 | .337 | .361 | | 49 | .150 | .171 | .190 | .213 | .228 |
| 20 | .232 | .265 | .294 | .329 | .352 | | 50 | .148 | .170 | .188 | .211 | .226 |
| 21 | .226 | .259 | .287 | .321 | .344 | | 55 | .142 | .162 | .180 | .201 | .216 |
| 22 | .221 | .253 | .281 | .314 | .337 | | 60 | .136 | .155 | .172 | .193 | .207 |
| 23 | .216 | .247 | .275 | .307 | .330 | | 65 | .131 | .149 | .166 | .185 | .199 |
| 24 | .212 | .242 | .269 | .301 | .323 | | 70 | .126 | .144 | .160 | .179 | .192 |
| 25 | .208 | .238 | .264 | .295 | .317 | | 75 | .122 | .139 | .154 | .173 | .185 |
| 26 | .204 | .233 | .259 | .290 | .311 | | 80 | .118 | .135 | .150 | .167 | .179 |
| 27 | .200 | .229 | .254 | .284 | .305 | | 85 | .114 | .131 | .145 | .162 | .174 |
| 28 | .197 | .225 | .250 | .279 | .300 | | 90 | .111 | .127 | .141 | .158 | .169 |
| 29 | .193 | .221 | .246 | .275 | .295 | | 95 | .108 | .124 | .137 | .154 | .165 |
| 30 | .190 | .218 | .242 | .270 | .290 | | 100 | .106 | .121 | .134 | .150 | .161 |
| Approximation for large n: | | | | | | | | $\dfrac{1.07}{\sqrt{n}}$ | $\dfrac{1.22}{\sqrt{n}}$ | $\dfrac{1.36}{\sqrt{n}}$ | $\dfrac{1.52}{\sqrt{n}}$ | $\dfrac{1.63}{\sqrt{n}}$ |

From "Table of Percentage Points in Kolmogorov–Smirnov Statistics," by L. H. Miller, *Journal of the American Statistical Association.* 1956, vol. 51, pp. 111–121, with approximation by the asymptotic formula as derived by Smirnov and given on p. 115.

**TABLE B.10**   Durbin–Watson Test Bounds

Level of Significance $\alpha = .05$

| $n$ | $p-1=1$ $d_L$ | $d_U$ | $p-1=2$ $d_L$ | $d_U$ | $p-1=3$ $d_L$ | $d_U$ | $p-1=4$ $d_L$ | $d_U$ | $p-1=5$ $d_L$ | $d_U$ |
|---|---|---|---|---|---|---|---|---|---|---|
| 15 | 1.08 | 1.36 | 0.95 | 1.54 | 0.82 | 1.75 | 0.69 | 1.97 | 0.56 | 2.21 |
| 16 | 1.10 | 1.37 | 0.98 | 1.54 | 0.86 | 1.73 | 0.74 | 1.93 | 0.62 | 2.15 |
| 17 | 1.13 | 1.38 | 1.02 | 1.54 | 0.90 | 1.71 | 0.78 | 1.90 | 0.67 | 2.10 |
| 18 | 1.16 | 1.39 | 1.05 | 1.53 | 0.93 | 1.69 | 0.82 | 1.87 | 0.71 | 2.06 |
| 19 | 1.18 | 1.40 | 1.08 | 1.53 | 0.97 | 1.68 | 0.86 | 1.85 | 0.75 | 2.02 |
| 20 | 1.20 | 1.41 | 1.10 | 1.54 | 1.00 | 1.68 | 0.90 | 1.83 | 0.79 | 1.99 |
| 21 | 1.22 | 1.42 | 1.13 | 1.54 | 1.03 | 1.67 | 0.93 | 1.81 | 0.83 | 1.96 |
| 22 | 1.24 | 1.43 | 1.15 | 1.54 | 1.05 | 1.66 | 0.96 | 1.80 | 0.86 | 1.94 |
| 23 | 1.26 | 1.44 | 1.17 | 1.54 | 1.08 | 1.66 | 0.99 | 1.79 | 0.90 | 1.92 |
| 24 | 1.27 | 1.45 | 1.19 | 1.55 | 1.10 | 1.66 | 1.01 | 1.78 | 0.93 | 1.90 |
| 25 | 1.29 | 1.45 | 1.21 | 1.55 | 1.12 | 1.66 | 1.04 | 1.77 | 0.95 | 1.89 |
| 26 | 1.30 | 1.46 | 1.22 | 1.55 | 1.14 | 1.65 | 1.06 | 1.76 | 0.98 | 1.88 |
| 27 | 1.32 | 1.47 | 1.24 | 1.56 | 1.16 | 1.65 | 1.08 | 1.76 | 1.01 | 1.86 |
| 28 | 1.33 | 1.48 | 1.26 | 1.56 | 1.18 | 1.65 | 1.10 | 1.75 | 1.03 | 1.85 |
| 29 | 1.34 | 1.48 | 1.27 | 1.56 | 1.20 | 1.65 | 1.12 | 1.74 | 1.05 | 1.84 |
| 30 | 1.35 | 1.49 | 1.28 | 1.57 | 1.21 | 1.65 | 1.14 | 1.74 | 1.07 | 1.83 |
| 31 | 1.36 | 1.50 | 1.30 | 1.57 | 1.23 | 1.65 | 1.16 | 1.74 | 1.09 | 1.83 |
| 32 | 1.37 | 1.50 | 1.31 | 1.57 | 1.24 | 1.65 | 1.18 | 1.73 | 1.11 | 1.82 |
| 33 | 1.38 | 1.51 | 1.32 | 1.58 | 1.26 | 1.65 | 1.19 | 1.73 | 1.13 | 1.81 |
| 34 | 1.39 | 1.51 | 1.33 | 1.58 | 1.27 | 1.65 | 1.21 | 1.73 | 1.15 | 1.81 |
| 35 | 1.40 | 1.52 | 1.34 | 1.58 | 1.28 | 1.65 | 1.22 | 1.73 | 1.16 | 1.80 |
| 36 | 1.41 | 1.52 | 1.35 | 1.59 | 1.29 | 1.65 | 1.24 | 1.73 | 1.18 | 1.80 |
| 37 | 1.42 | 1.53 | 1.36 | 1.59 | 1.31 | 1.66 | 1.25 | 1.72 | 1.19 | 1.80 |
| 38 | 1.43 | 1.54 | 1.37 | 1.59 | 1.32 | 1.66 | 1.26 | 1.72 | 1.21 | 1.79 |
| 39 | 1.43 | 1.54 | 1.38 | 1.60 | 1.33 | 1.66 | 1.27 | 1.72 | 1.22 | 1.79 |
| 40 | 1.44 | 1.54 | 1.39 | 1.60 | 1.34 | 1.66 | 1.29 | 1.72 | 1.23 | 1.79 |
| 45 | 1.48 | 1.57 | 1.43 | 1.62 | 1.38 | 1.67 | 1.34 | 1.72 | 1.29 | 1.78 |
| 50 | 1.50 | 1.59 | 1.46 | 1.63 | 1.42 | 1.67 | 1.38 | 1.72 | 1.34 | 1.77 |
| 55 | 1.53 | 1.60 | 1.49 | 1.64 | 1.45 | 1.68 | 1.41 | 1.72 | 1.38 | 1.77 |
| 60 | 1.55 | 1.62 | 1.51 | 1.65 | 1.48 | 1.69 | 1.44 | 1.73 | 1.41 | 1.77 |
| 65 | 1.57 | 1.63 | 1.54 | 1.66 | 1.50 | 1.70 | 1.47 | 1.73 | 1.44 | 1.77 |
| 70 | 1.58 | 1.64 | 1.55 | 1.67 | 1.52 | 1.70 | 1.49 | 1.74 | 1.46 | 1.77 |
| 75 | 1.60 | 1.65 | 1.57 | 1.68 | 1.54 | 1.71 | 1.51 | 1.74 | 1.49 | 1.77 |
| 80 | 1.61 | 1.66 | 1.59 | 1.69 | 1.56 | 1.72 | 1.53 | 1.74 | 1.51 | 1.77 |
| 85 | 1.62 | 1.67 | 1.60 | 1.70 | 1.57 | 1.72 | 1.55 | 1.75 | 1.52 | 1.77 |
| 90 | 1.63 | 1.68 | 1.61 | 1.70 | 1.59 | 1.73 | 1.57 | 1.75 | 1.54 | 1.78 |
| 95 | 1.64 | 1.69 | 1.62 | 1.71 | 1.60 | 1.73 | 1.58 | 1.75 | 1.56 | 1.78 |
| 100 | 1.65 | 1.69 | 1.63 | 1.72 | 1.61 | 1.74 | 1.59 | 1.76 | 1.57 | 1.78 |

Level of Significance $\alpha = .01$

| $n$ | $p-1=1$ $d_L$ | $d_U$ | $p-1=2$ $d_L$ | $d_U$ | $p-1=3$ $d_L$ | $d_U$ | $p-1=4$ $d_L$ | $d_U$ | $p-1=5$ $d_L$ | $d_U$ |
|---|---|---|---|---|---|---|---|---|---|---|
| 15 | 0.81 | 1.07 | 0.70 | 1.25 | 0.59 | 1.46 | 0.49 | 1.70 | 0.39 | 1.96 |
| 16 | 0.84 | 1.09 | 0.74 | 1.25 | 0.63 | 1.44 | 0.53 | 1.66 | 0.44 | 1.90 |
| 17 | 0.87 | 1.10 | 0.77 | 1.25 | 0.67 | 1.43 | 0.57 | 1.63 | 0.48 | 1.85 |
| 18 | 0.90 | 1.12 | 0.80 | 1.26 | 0.71 | 1.42 | 0.61 | 1.60 | 0.52 | 1.80 |
| 19 | 0.93 | 1.13 | 0.83 | 1.26 | 0.74 | 1.41 | 0.65 | 1.58 | 0.56 | 1.77 |
| 20 | 0.95 | 1.15 | 0.86 | 1.27 | 0.77 | 1.41 | 0.68 | 1.57 | 0.60 | 1.74 |
| 21 | 0.97 | 1.16 | 0.89 | 1.27 | 0.80 | 1.41 | 0.72 | 1.55 | 0.63 | 1.71 |
| 22 | 1.00 | 1.17 | 0.91 | 1.28 | 0.83 | 1.40 | 0.75 | 1.54 | 0.66 | 1.69 |
| 23 | 1.02 | 1.19 | 0.94 | 1.29 | 0.86 | 1.40 | 0.77 | 1.53 | 0.70 | 1.67 |
| 24 | 1.04 | 1.20 | 0.96 | 1.30 | 0.88 | 1.41 | 0.80 | 1.53 | 0.72 | 1.66 |
| 25 | 1.05 | 1.21 | 0.98 | 1.30 | 0.90 | 1.41 | 0.83 | 1.52 | 0.75 | 1.65 |
| 26 | 1.07 | 1.22 | 1.00 | 1.31 | 0.93 | 1.41 | 0.85 | 1.52 | 0.78 | 1.64 |
| 27 | 1.09 | 1.23 | 1.02 | 1.32 | 0.95 | 1.41 | 0.88 | 1.51 | 0.81 | 1.63 |
| 28 | 1.10 | 1.24 | 1.04 | 1.32 | 0.97 | 1.41 | 0.90 | 1.51 | 0.83 | 1.62 |
| 29 | 1.12 | 1.25 | 1.05 | 1.33 | 0.99 | 1.42 | 0.92 | 1.51 | 0.85 | 1.61 |
| 30 | 1.13 | 1.26 | 1.07 | 1.34 | 1.01 | 1.42 | 0.94 | 1.51 | 0.88 | 1.61 |
| 31 | 1.15 | 1.27 | 1.08 | 1.34 | 1.02 | 1.42 | 0.96 | 1.51 | 0.90 | 1.60 |
| 32 | 1.16 | 1.28 | 1.10 | 1.35 | 1.04 | 1.43 | 0.98 | 1.51 | 0.92 | 1.60 |
| 33 | 1.17 | 1.29 | 1.11 | 1.36 | 1.05 | 1.43 | 1.00 | 1.51 | 0.94 | 1.59 |
| 34 | 1.18 | 1.30 | 1.13 | 1.36 | 1.07 | 1.43 | 1.01 | 1.51 | 0.95 | 1.59 |
| 35 | 1.19 | 1.31 | 1.14 | 1.37 | 1.08 | 1.44 | 1.03 | 1.51 | 0.97 | 1.59 |
| 36 | 1.21 | 1.32 | 1.15 | 1.38 | 1.10 | 1.44 | 1.04 | 1.51 | 0.99 | 1.59 |
| 37 | 1.22 | 1.32 | 1.16 | 1.38 | 1.11 | 1.45 | 1.06 | 1.51 | 1.00 | 1.59 |
| 38 | 1.23 | 1.33 | 1.18 | 1.39 | 1.12 | 1.45 | 1.07 | 1.52 | 1.02 | 1.58 |
| 39 | 1.24 | 1.34 | 1.19 | 1.39 | 1.14 | 1.45 | 1.09 | 1.52 | 1.03 | 1.58 |
| 40 | 1.25 | 1.34 | 1.20 | 1.40 | 1.15 | 1.46 | 1.10 | 1.52 | 1.05 | 1.58 |
| 45 | 1.29 | 1.38 | 1.24 | 1.42 | 1.20 | 1.48 | 1.16 | 1.53 | 1.11 | 1.58 |
| 50 | 1.32 | 1.40 | 1.28 | 1.45 | 1.24 | 1.49 | 1.20 | 1.54 | 1.16 | 1.59 |
| 55 | 1.36 | 1.43 | 1.32 | 1.47 | 1.28 | 1.51 | 1.25 | 1.55 | 1.21 | 1.59 |
| 60 | 1.38 | 1.45 | 1.35 | 1.48 | 1.32 | 1.52 | 1.28 | 1.56 | 1.25 | 1.60 |
| 65 | 1.41 | 1.47 | 1.38 | 1.50 | 1.35 | 1.53 | 1.31 | 1.57 | 1.28 | 1.61 |
| 70 | 1.43 | 1.49 | 1.40 | 1.52 | 1.37 | 1.55 | 1.34 | 1.58 | 1.31 | 1.61 |
| 75 | 1.45 | 1.50 | 1.42 | 1.53 | 1.39 | 1.56 | 1.37 | 1.59 | 1.34 | 1.62 |
| 80 | 1.47 | 1.52 | 1.44 | 1.54 | 1.42 | 1.57 | 1.39 | 1.60 | 1.36 | 1.62 |
| 85 | 1.48 | 1.53 | 1.46 | 1.55 | 1.43 | 1.58 | 1.41 | 1.60 | 1.39 | 1.63 |
| 90 | 1.50 | 1.54 | 1.47 | 1.56 | 1.45 | 1.59 | 1.43 | 1.61 | 1.41 | 1.64 |
| 95 | 1.51 | 1.55 | 1.49 | 1.57 | 1.47 | 1.60 | 1.45 | 1.62 | 1.42 | 1.64 |
| 100 | 1.52 | 1.56 | 1.50 | 1.58 | 1.48 | 1.60 | 1.46 | 1.63 | 1.44 | 1.65 |

From J. Durbin and G. S. Watson, "Testing for Serial Correlation in Least Squares Regression. II," *Biometrika*, vol. 38 (1951), pp. 159–78, by permission of the Biometrika trustees.

**TABLE B.11**  Cumulative Distribution of the Total Number of Runs, $P(r \leq r_0)$

| $(n_1, n_2)$ | 2 | 3 | 4 | 5 | 6 | 7 | 8 | 9 | 10 | 11 | 12 | 13 | 14 | 15 | 16 | 17 | 18 | 19 | 20 |
|---|---|---|---|---|---|---|---|---|---|---|---|---|---|---|---|---|---|---|---|
| (2,3) | .200 | .500 | .900 | 1.000 | | | | | | | | | | | | | | | |
| (2,4) | .133 | .400 | .800 | 1.000 | | | | | | | | | | | | | | | |
| (2,5) | .095 | .333 | .714 | 1.000 | | | | | | | | | | | | | | | |
| (2,6) | .071 | .286 | .643 | 1.000 | | | | | | | | | | | | | | | |
| (2,7) | .056 | .250 | .583 | 1.000 | | | | | | | | | | | | | | | |
| (2,8) | .044 | .222 | .533 | 1.000 | | | | | | | | | | | | | | | |
| (2,9) | .036 | .200 | .491 | 1.000 | | | | | | | | | | | | | | | |
| (2,10) | .030 | .182 | .455 | 1.000 | | | | | | | | | | | | | | | |
| (3,3) | .100 | .300 | .700 | .900 | 1.000 | | | | | | | | | | | | | | |
| (3,4) | .057 | .200 | .543 | .800 | .971 | 1.000 | | | | | | | | | | | | | |
| (3,5) | .036 | .143 | .429 | .714 | .929 | 1.000 | | | | | | | | | | | | | |
| (3,6) | .024 | .107 | .345 | .643 | .881 | 1.000 | | | | | | | | | | | | | |
| (3,7) | .017 | .083 | .283 | .583 | .833 | 1.000 | | | | | | | | | | | | | |
| (3,8) | .012 | .067 | .236 | .533 | .788 | 1.000 | | | | | | | | | | | | | |
| (3,9) | .009 | .055 | .200 | .491 | .745 | 1.000 | | | | | | | | | | | | | |
| (3,10) | .007 | .045 | .171 | .455 | .706 | 1.000 | | | | | | | | | | | | | |
| (4,4) | .029 | .114 | .371 | .629 | .886 | .971 | 1.000 | | | | | | | | | | | | |
| (4,5) | .016 | .071 | .262 | .500 | .786 | .929 | .992 | 1.000 | | | | | | | | | | | |
| (4,6) | .010 | .048 | .190 | .405 | .690 | .881 | .976 | 1.000 | | | | | | | | | | | |
| (4,7) | .006 | .033 | .142 | .333 | .606 | .833 | .954 | 1.000 | | | | | | | | | | | |
| (4,8) | .004 | .024 | .109 | .279 | .533 | .788 | .929 | 1.000 | | | | | | | | | | | |
| (4,9) | .003 | .018 | .085 | .236 | .471 | .745 | .902 | 1.000 | | | | | | | | | | | |
| (4,10) | .002 | .014 | .068 | .203 | .419 | .706 | .874 | 1.000 | | | | | | | | | | | |

Adapted from Swed, Frieda S., and Eisenhart, C. 1943. "Tables for testing randomness of grouping a sequence of alternatives." *Annals of Mathematical Statistics, 14,* 66–87. Reprinted with the permission of the Institute of Mathematical Statistics.

**TABLE B.11**  (*continued*)

$r_0$

| $(n_1, n_2)$ | 2 | 3 | 4 | 5 | 6 | 7 | 8 | 9 | 10 | 11 | 12 | 13 | 14 | 15 | 16 | 17 | 18 | 19 | 20 |
|---|---|---|---|---|---|---|---|---|---|---|---|---|---|---|---|---|---|---|---|
| (5,5) | .008 | .040 | .167 | .357 | .643 | .833 | .960 | .992 | 1.000 | 1.000 | | | | | | | | | |
| (5,6) | .004 | .024 | .110 | .262 | .522 | .738 | .911 | .976 | .988 | 1.000 | | | | | | | | | |
| (5,7) | .003 | .015 | .076 | .197 | .424 | .652 | .854 | .955 | .992 | 1.000 | | | | | | | | | |
| (5,8) | .002 | .010 | .054 | .152 | .347 | .576 | .793 | .929 | .984 | 1.000 | | | | | | | | | |
| (5,9) | .001 | .007 | .039 | .119 | .287 | .510 | .734 | .902 | .972 | 1.000 | | | | | | | | | |
| (5,10) | .001 | .005 | .029 | .095 | .239 | .455 | .678 | .874 | .958 | 1.000 | | | | | | | | | |
| (6,6) | .002 | .013 | .067 | .175 | .392 | .608 | .825 | .933 | .987 | .998 | 1.000 | | | | | | | | |
| (6,7) | .001 | .008 | .043 | .121 | .296 | .500 | .733 | .879 | .966 | .992 | .999 | 1.000 | | | | | | | |
| (6,8) | .001 | .005 | .028 | .086 | .226 | .413 | .646 | .821 | .937 | .984 | .998 | 1.000 | | | | | | | |
| (6,9) | .000 | .003 | .019 | .063 | .175 | .343 | .566 | .762 | .902 | .972 | .994 | 1.000 | | | | | | | |
| (6,10) | .000 | .002 | .013 | .047 | .137 | .288 | .497 | .706 | .864 | .958 | .990 | 1.000 | | | | | | | |
| (7,7) | .001 | .004 | .025 | .078 | .209 | .383 | .617 | .791 | .922 | .975 | .996 | .999 | 1.000 | | | | | | |
| (7,8) | .000 | .002 | .015 | .051 | .149 | .296 | .514 | .704 | .867 | .949 | .988 | .998 | 1.000 | 1.000 | | | | | |
| (7,9) | .000 | .001 | .010 | .035 | .108 | .231 | .427 | .622 | .806 | .916 | .975 | .994 | .999 | 1.000 | | | | | |
| (7,10) | .000 | .001 | .006 | .024 | .080 | .182 | .355 | .549 | .743 | .879 | .957 | .990 | .998 | 1.000 | | | | | |
| (8,8) | .000 | .001 | .009 | .032 | .100 | .214 | .405 | .595 | .786 | .900 | .968 | .991 | .999 | 1.000 | 1.000 | | | | |
| (8,9) | .000 | .001 | .005 | .020 | .069 | .157 | .319 | .500 | .702 | .843 | .939 | .980 | .996 | .999 | 1.000 | 1.000 | | | |
| (8,10) | .000 | .000 | .003 | .013 | .048 | .117 | .251 | .419 | .621 | .782 | .903 | .964 | .990 | .998 | 1.000 | 1.000 | | | |
| (9,9) | .000 | .000 | .003 | .012 | .044 | .109 | .238 | .399 | .601 | .762 | .891 | .956 | .988 | .997 | 1.000 | 1.000 | 1.000 | | |
| (9,10) | .000 | .000 | .002 | .008 | .029 | .077 | .179 | .319 | .510 | .681 | .834 | .923 | .974 | .992 | .999 | 1.000 | 1.000 | 1.000 | |
| (10,10) | .000 | .000 | .001 | .004 | .019 | .051 | .128 | .242 | .414 | .586 | .758 | .872 | .949 | .981 | .996 | .999 | 1.000 | 1.000 | 1.000 |

**TABLE B.12**   Table of Constants for Control Charts

**Factors for Computing Control Chart Upper and Lower Bounds**

| | For Means | | For Ranges | |
|---|---|---|---|---|
| Sample | $d_2$ | $A_2$ | $D_3$ | $D_4$ |
| 2 | 1.128 | 1.880 | 0 | 3.267 |
| 3 | 1.693 | 1.023 | 0 | 2.575 |
| 4 | 2.059 | 0.729 | 0 | 2.282 |
| 5 | 2.326 | 0.577 | 0 | 2.115 |
| 6 | 2.534 | 0.483 | 0 | 2.004 |
| 7 | 2.704 | 0.419 | 0.076 | 1.924 |
| 8 | 2.847 | 0.373 | 0.136 | 1.864 |
| 9 | 2.970 | 0.337 | 0.184 | 1.816 |
| 10 | 3.078 | 0.308 | 0.223 | 1.777 |
| 11 | 3.173 | 0.285 | 0.256 | 1.744 |
| 12 | 3.258 | 0.266 | 0.284 | 1.716 |
| 13 | 3.336 | 0.249 | 0.308 | 1.692 |
| 14 | 3.407 | 0.235 | 0.329 | 1.671 |
| 15 | 3.472 | 0.223 | 0.348 | 1.652 |
| 16 | 3.532 | 0.212 | 0.364 | 1.636 |
| 17 | 3.588 | 0.203 | 0.379 | 1.621 |
| 18 | 3.640 | 0.194 | 0.392 | 1.608 |
| 19 | 3.689 | 0.187 | 0.404 | 1.596 |
| 20 | 3.735 | 0.180 | 0.414 | 1.586 |
| 21 | 3.778 | 0.173 | 0.425 | 1.575 |
| 22 | 3.819 | 0.167 | 0.434 | 1.566 |
| 23 | 3.858 | 0.162 | 0.443 | 1.557 |
| 24 | 3.895 | 0.157 | 0.452 | 1.548 |
| 25 | 3.931 | 0.153 | 0.459 | 1.541 |

# APPENDIX C
## MATH ESSENTIALS FOR STATISTICS

Since the mathematics prerequisites for a business course in statistics are quite varied, we have collected a selection of concepts that we believe should be common knowledge at the onset of this work. The concepts include mathematical and arithmetic tools about which students often raise questions. They are not restricted to any single area of mathematics but rather include the skill areas that are essential to business statistics.

This material is offered so that you can be assured of having an adequate background for studying business statistics from this book. For some persons this information can serve as a quick review. For others, it may serve as a first introduction to some concepts. Use the problems at the end of this appendix as a review and test of your understanding. The answers appear in the answer section at the end of Appendix C. Table C.1 reviews math symbols and gives examples of their use.

**TABLE C.I**   Symbols

| Symbol | Verbal Equivalent | Example |
|---|---|---|
| $\doteq$ | is approximately equal to | $2.98 \doteq 3.0$ |
| $<$ | is less than | $2 < 3$ |
| $>$ | is greater than | $3 > 2$ |
| $\leqslant (\geqslant)$ | is less (greater) than or equal to | $a \leqslant 5 \; (5 \geqslant a)$ |
| $|a|$ | the absolute (positive) value | $|3| = 3, |-3| = 3$ |
| $\sqrt{b}$ | the square root of $b$ $(>0)$ | $\sqrt{4} = 2$ |
| $c^2$ | the square of $c$ | $3^2 = 9$ |
| $\ldots$ | continuing this pattern | $1, 2, \ldots, 5$ implies $1, 2, 3, 4, 5$ |
| $\approx$ | is similar to | $4.11 \approx 4.1$ |
| $\dfrac{a}{b}$ or $a/b$ | $a$ divided by $b$, requires $b \neq 0$ | $6/2 = 3$ |

## Calculations with Zero and One

1. The product of a number and zero equals zero.

$$6 \cdot 0 = 0 \qquad -100 \cdot 0 = 0$$

2. Zero raised to a positive power has the value zero.

$$0^{1000} = 0^2 = 0^{100} = 0$$

3. Division of zero by a nonzero number yields the quotient zero.

$$0/10 = 0/2 = 0$$

4. Division by zero has no meaning: The result is undefined.

$$3/0 \text{ is undefined}$$

5. The value of a nonzero number raised to power zero is one.

$$-1^0 = 10^0 = 1$$

6. The product of a number and one equals the number.

$$1 \cdot 3 = 3 \qquad 1 \cdot 0 = 0$$

7. One raised to any power has the value one.

$$1^0 = 1^1 = 1^{-10} = 1$$

**8.** A number divided by one has as its quotient the given number.

$$0/1 = 0 \qquad (1/2)/1 = 1/2$$

## Exponents and Roots

In the form $a^k$, $a$ is the base number and $k$ is an exponent (or power). The basic rules regarding exponents are as follows.

**1.** By definition, $a^0 = 1$, $a^1 = a$, and $a^{-1} = \dfrac{1}{a}$ for any $a \neq 0$.

$$10^0 = 1 \qquad 10^1 = 10 \qquad 10^{-1} = \frac{1}{10}$$

**2.** The addition or subtraction of numbers with exponents requires that the separate terms be evaluated first.

$$3^2 - 2^2 = (3 \cdot 3) - (2 \cdot 2) = 9 - 4 = 5$$

**3.** The product of exponential numbers with the same base has that common base raised to the sum of the powers.

$$2^1 \cdot 2^2 \cdot 2^3 = 2^{1+2+3} = 2^6 = 64$$

**4.** The quotient of exponential numbers with the same base is the common base raised to the difference (numerator − denominator) of the powers.

$$\frac{2^3}{2^2} = 2^{3-2} = 2 \qquad \frac{2^2}{2^3} = 2^{2-3} = 2^{-1} = \frac{1}{2}$$

**5.** The product of powers having a different base requires that the separate terms first be evaluated and then the product taken.

$$2^3 \cdot 3^2 = 8 \cdot 9 = 72 \qquad 2^{-1} \cdot 3 = \frac{1}{2} \cdot \frac{3}{1} = \frac{3}{2}$$

**6.** Division of powers with different bases also requires that the separate terms be evaluated first.

$$\frac{3^2}{2^3} = \frac{9}{8} \quad \text{or} \quad 1.125$$

## Rules for Rounding

The accuracy used for tabled decimal numbers should be

1. at most, that given in the table if the number is to be used solely for comparison and will not be used in later calculations.
2. directed by the rules for decimal accuracy that follow if the numbers are to be used in later calculations.

The use of tabled values solely for comparison occurs in Chapters 9 through 18 in the form of test statistics.

Several rules are given for determining decimal accuracy for answers using arithmetic operations.

For addition or subtraction, the answer carries the same accuracy as the least accurate of the combined numbers.

$$\text{Addition} \quad 3.21 + 6.9 + 4.00 = 3.21 + 6.90 + 4.00$$
$$= 14.11 \approx 14.1$$

$$\text{Subtraction} \quad 6.98 - 7.2 = -0.22 \approx 0.2$$

### RULE 1

For multiplication, the answer has the same accuracy as the least accurate of the combined numbers so long as the product exceeds $+1$ or is more negative than $-1$.

$$2361 \cdot 2.1 = 4958.1$$
$$0.36 \cdot 3.6 = 1.296 \approx 1.3$$

### RULE 2

For two or more numbers whose product is less than 1 but greater than $-1$, use one more decimal place in the product than appears in the most accurate of the original numbers.

$$0.2 \cdot 0.3 = 0.06 (0.1)^2 = 0.1 \cdot 0.1 = 0.01 \quad 0.23 \cdot 0.01 = 0.002$$

*Note:* The last two products (0.01) would be 0 (zero) by Rule 1.

The accuracy in a quotient (division) uses the same rule as for addition and subtraction. The accuracy is that of the least accurate of the numbers involved in the division.

$$2 \div 0.2 = 2.00 \div 0.2 = 10.0$$
$$44.5 \div 2.11 = 21.09 \approx 21.1$$

If you doubt a decimal answer, a safe approach is to check your work by using an opposite procedure; for the preceding example, $2.11 \cdot (21.1) = 44.52 \approx 44.5$ gives a check back to the original numbers.

Common practice in rounding decimal numbers is to round up if the digit to the right of the place to which you are rounding is more than 5 and to round down if the digit is less than 5. Decimal numbers ending with 5000 . . . require some convention.

### ENGINEER'S RULE

When the last digits in a decimal number are 5000 . . . , round to the nearest even digit at the desired level of accuracy.

For example, 30.25000 . . . , 600.500 . . . , and 55.99500 . . . are rounded to 30.2, 600, and 56.00, respectively. This procedure eliminates potential rounding biases as it sometimes leads to rounding up and at other times to rounding down. The rule applies only when the last digits are 5000 . . . .

## Squares and Square Roots with Calculators

Squares and square roots are used extensively in statistics. Both can be obtained on calculators or found in tables. We encourage you to use a calculator, where square roots are available with a few key strokes. The square root key is nearly universal on electronic hand calculators. Finding a square root requires only keying a number, pressing the square root button, and then reading the square root from a display. Since this procedure is mechanical, it is essential for you to have a reasonable approximation of the keyboard answer. That is why we recommend a "first approximation" by logic.

For a first approximation, suppose that you want the square root for $n = 90$, that is, $\sqrt{90}$. Since $9^2 = 81$ and $10^2 = 100$, you can estimate $81 < n < 100$, so $9 < \sqrt{n} < 10$. A first approximation is 9. That is, for a first approximation use a number that when squared is less than but near the actual value.

Having obtained a first approximation, including setting the decimal position, you can then use a calculator (or a table) to obtain a more accurate number. For $n = 90$, the square root is found by calculator to be 9.4868 or 9.5. With practice this first approximation is something that you can do fairly quickly. And it gives a check against making calculator entry errors and reading tables incorrectly.

Remember that when you are finding square roots for numbers between 0 and 1, the square root is larger than the number. For example $\sqrt{0.25} = 0.5$; the root 0.5 is greater than the number 0.25.

Some common operations involving square roots are given in Table C.2. In some cases the operations can be simplified. Assume that both $a$ and $b$ are positive numbers.

**TABLE C.2    Common Operations Involving Square Roots**

| Math Operation | Simplification | Example |
|---|---|---|
| $\sqrt{a} + \sqrt{b}$ | None | $\sqrt{4} + \sqrt{9} = 2 + 3 = 5$ |
| $\sqrt{a} - \sqrt{b}$ | None | $\sqrt{4} - \sqrt{9} = 2 - 3 = -1$ |
| $\sqrt{a} \cdot \sqrt{b}$ | $\sqrt{a \cdot b}$ | $\sqrt{2} \cdot \sqrt{8} = \sqrt{2 \cdot 8} = \sqrt{16} = 4$ |
| $\dfrac{\sqrt{a}}{\sqrt{b}}$ | $\sqrt{\dfrac{a}{b}}$ | $\dfrac{\sqrt{8}}{\sqrt{2}} = \sqrt{\dfrac{8}{2}} = \sqrt{4} = 2$ |

## Summation Notation and Rules

The summation symbol is the mathematician's shorthand way of saying "sum a list of numbers." The symbol $\sum$ is the Greek capital letter sigma. It is used to shorten a statement of addition that incorporates numerous terms.

$$\underbrace{3 + 3 + 3 + \cdots + 3}_{\textbf{100 terms}} = \sum^{100} 3 = 100 \cdot 3 = 300$$

$$\sum_{i=1}^{5} X_i = X_1 + X_2 + X_3 + X_4 + X_5$$

In the $\sum X_i$ example, the symbol $i$ is a counter called the index of summation. We assume the counter goes $i = 1, 2, \ldots, n$ (in the example, $n = 5$). We sum a first value, plus a second, . . . , through a last value, $X_n$. Instead of

$$\sum_{i=1}^{n} X_i$$

we use $\sum X$ for an abbreviated notation.

Parentheses ( ), brackets [ ], and braces { } are used as grouping symbols to indicate that whatever is enclosed is to be treated as a single unit. Sometimes two or more of the grouping symbols are used in a single summation. One of the most common uses is shown in the following equation.

For $X_1 = 0$, $X_2 = 1$, $X_3 = 2$,

$$
\begin{aligned}
\sum_{}^{3} [(X_i - 1)^2] &= [(X_1 - 1)^2] + [(X_2 - 1)^2] + [(X_3 - 1)^2] \\
&= [(0 - 1)^2] + [(1 - 1)^2] + [(2 - 1)^2] \\
&= [(-1)^2] + [0^2] + [1^2] = 1 + 0 + 1 = 2
\end{aligned}
$$

In the preceding example we performed subtractions, then squared, and then summed numbers. Observe that the work moved from the inside grouping to the outside grouping symbols.

When symbols are used to show grouping, perform the indicated operations beginning from the inside and working outward.

The following four rules can simplify the process of finding sums. Each one shows equivalent forms for left and right sides. The process in the right member, or side, is equivalent to but is usually computationally shorter than that on the left. The values $X_1 = 0$, $X_2 = 1$, $X_3 = 2$, $Y_1 = -1$, $Y_2 = 1$, and $Y_3 = 2$ are used in the following examples.

1. $\displaystyle\sum^{n}(a) = \overbrace{a + a + \cdots + a}^{n} = na$

   EXAMPLE   $\displaystyle\sum^{10}(2) = 10 \cdot 2 = 20$

2. $\displaystyle\sum^{n}(bX) = bX_1 + bX_2 + \cdots + bX_n$

   $\qquad\qquad = b(X_1 + X_2 + \cdots + X_n) = b\displaystyle\sum^{n}(X)$

   EXAMPLE   $\displaystyle\sum(2X) = 2[\sum X]$

   $\qquad\qquad = 2[0 + 1 + 2] = 6$

3. $\displaystyle\sum(cX + dY) = (cX_1 + dY_2) + (cX_2 + dY_2) + \cdots + (cX_n + dY_n)$

   $\qquad\qquad = cX_1 + cX_2 + \cdots + cX_n + dY_1 + dY_2 + \cdots + dY_n$

   $\qquad\qquad = c\displaystyle\sum^{n}(X) + d\displaystyle\sum^{n}(Y)$

   EXAMPLE   $\displaystyle\sum^{3}(2X + 3Y) = 2\left[\sum^{3} X\right] + 3\left[\sum^{3} Y\right]$

   $\qquad\qquad = 2[0 + 1 + 2] + 3[-1 + 1 + 2]$

   $\qquad\qquad = 2 \cdot [3] + 3 \cdot [2] = 12$

4. $\sum\limits^{n} (X + e) = (X_1 + e) + (X_2 + e) + \cdots + (X_n + e)$

$$= (X_1 + X_2 + \cdots + X_n) + \overbrace{e + e + \cdots + e}^{n}$$

$$= \left[ \sum^{n} X \right] + ne$$

EXAMPLE    $\sum\limits^{3} [X + (-1)] = \left[ \sum^{3} X \right] + 3(-1)$

$$= [0 + 1 + 2] - 3 = 0$$

## Rectangular Cooordinates

Using rectangular coordinates is one way to display points in two dimensions. Figure C.1(a) illustrates the rectangular coordinate system. The horizontal and vertical lines are called the $X$-axis and $Y$-axis, respectively. The axes intersect at right angles; they are perpendicular. The plane of the figure is divided into four parts, called quadrants, which are labeled I, II, III, and IV.

Any position or point is identified by its coordinates (values projected to the coordinate axes). These are written as ordered pairs $(X,Y)$, with the intersection of the axes designating an origin or beginning position $(0,0)$. An ordered pair lists first the abscissa (or $X$-value) and then the ordinate (or $Y$-value).

Observe the position for $(0,2)$ and then that for $(-2,1)$ in Figure C.1(b). The point $(-2,1)$ is appropriately located in Quadrant II. This quadrant contains all points for which the $X$-coordinate (abscissa) is negative but the $Y$-coordinate is positive. Similarly, in Quadrant I, $X$ and $Y$ are both positive.

The line segment $L$ is a representation for many points: all of those in its path. Many other representations of numerous points, called graphs, can be made in two dimensions.

**FIGURE C.1**

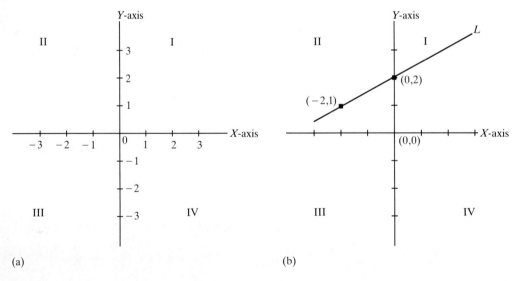

(a)    (b)

## Linear and Curvilinear Equations

A *linear equation* is a relation that can be described on a rectangular coordinate system by a line. We define linear relations using equations like $Y = 2 + (1/2)X$. This specific form is displayed as the line $L$ in Figure C.1(b). In general, a special name—domain—is given to the collection of all values that can be taken by $X$ in a relation. In Figure C.1(b), $X$ can take any real value, so the domain is unrestricted. Similarly, the values taken by $Y$ are called the range. Here too the range is unlimited, but it is restricted by the relation. For $Y = 2 + (1/2)X$, the value $Y = 1$ can occur only when $X = -2$ (see Figure C.1(b)). We say that the $X$-values are independent by choice, but that $Y$-values depend on the choice of a value for $X$. For example, the choice $X = -2$ requires $Y = 2 + 1/2(-2) = 1$.

A linear relation, generalized by $Y = a + bX$, plots as a line on the coordinate axes with $a = $ the point on the graph that has 0 as the value of its abscissa; that is, $(0,a)$ is the $Y$-intercept and $b = $ the slope of the line. The slope is defined as change in $Y$ divided by change in $X$.

**For any two points $(X_1, Y_1)$ and $(X_2, Y_2)$ with $X_1 \neq X_2$,**

$$b = \frac{Y_2 - Y_1}{X_2 - X_1}$$

**is the slope for the line that includes these two points.**

**Positive and Negative Slope**   Lines like "/" have a positive slope, while lines like "\" have a negative slope.

For example, in Figure C.1(b) the line $L$ contains the points $(-2,1)$ and $(0,2)$ and so has slope $b = (2 - 1)/[0 - (-2)] = 1/2$. Since the $Y$-intercept is at $Y = 2$, the relation describing the line is $Y = 2 + (1/2)X$. Observe that the slope $b$ is positive, so the general direction of the line is "/."

**Straight-line Graphs**   The feature that distinguishes a line graph from other graphs is that the slope is *constant* between any two points in its path. The graph is "straight."

For example, the relation $Y = -3X$ (or $Y = 0 - 3X$) describes a line, $\ell$, with a slope $-3$, therefore a line "\." The $Y$-intercept value is zero; this line passes through the origin (see Figure C.2).

You will be asked to evaluate some linear relations, especially in Chapter 12, which requires solving for an unknown. Several examples illustrate the correct procedure.

1. The same number is added to (or subtracted from) each member (side) of an equation:

$$4 = 3 + X \quad \text{so} \quad 4 - 3 = (3 + X) - 3 \quad \text{or} \quad X = 1$$

2. Each member (side) is multiplied by the same nonzero number:

$$\tfrac{1}{2}X = 5 \quad \text{so} \quad (2 \cdot \tfrac{1}{2})X = 2 \cdot 5 \quad \text{or} \quad X = 10$$

**FIGURE C.2**

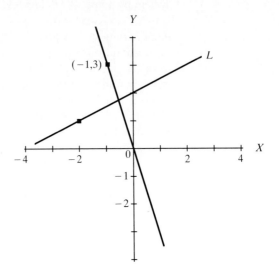

**3.** Each member is divided by the same nonzero number:

$$2X = 8 \quad \text{so} \quad \tfrac{2}{2}X = \tfrac{8}{2} \quad \text{or} \quad X = 4$$

Another procedure allows any added or subtracted term to be moved from one side of an equality to the other side provided that its sign is changed. Zero takes the place of the lost term. This is another form of the first example shown in the list above.

$$X + 2 = 6 \quad \text{so} \quad X + 0 = 6 - 2, \quad X = 4$$

$$Y + 2X - 2 = 3X \quad \text{so} \quad Y + 0 = 3X - (2X - 2)$$
$$= 3X - 2X + 2 = X + 2$$

**Curvilinear Equations**    These produce graphs with other than a straight line. The two that are illustrated in Figure C.3 are most common to our work.

**FIGURE C.3**

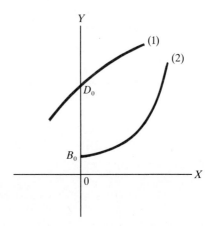

Curve 1 represents a parabolic, or second-degree polynomial, curve. It has nonconstant slope and general equation $Y = D_0 + D_1 X + D_2 X^2$. For points near $X = 0$, the $X$-term has much influence on the curve, while for points further removed from $X = 0$, the $X^2$-term can have a substantial influence on the direction of the curve.

Similarly, curve 2 is nonlinear. This curve shows an exponential growth pattern, for example, as an accumulation of dollars of savings with compound interest. The form is $Y = B_0 \cdot B_1^X$, where $B_0$ is the $Y$-intercept. For $B_1 > 1$ the curve turns upward, whereas for $B_1 < 1$ the curve turns downward. For $B_1 = 1$, the equation becomes $Y = B_0 \cdot 1^X = B_0$. This graphs as a straight line, parallel to the $X$-axis, and cuts the $Y$-axis at height $B_0$.

Appendix C Problems 15, 16, and 17 provide an opportunity for you to explore the graphs of parabolic and exponential curves. Working these problems will give you some feeling for curvilinear plots. Also, answers are provided. This will afford you some experience for data plots and the meaning of linear models for Chapters 12 and 13.

## Coding for Calculations of Mean and Standard Deviation*

We generally would not be interested in using coding for 10 to 50 observations, since calculations of $\bar{X}$ and $s$ are easily handled by the techniques of Chapter 2, by hand or on a calculator. Coding is one alternative for calculating statistics for larger data sets. Even then, if the original observations are available, you may prefer the exact answers that can be obtained with a computer.

Coding establishes a system for simplifying the calculation of statistics. The coding operations we use are (1) adding (or subtracting) a constant and (2) multiplying (or dividing) by a constant. We use a small sample of ungrouped data to show the basic idea of adding a constant in calculating a mean.

For example, in using ungrouped data to average the heights of players on a basketball team whose members are 6'9", 6'5", 7'0", 6'10", and 7'2", we could work with the deviations about an assumed mean, say, 7 feet, rather than summing the given heights and dividing. This coding device is to subtract the constant 7 feet from each observation. We would then have $-3''$, $-7''$, $0''$, $-2''$, and $+2''$, which yield an average difference of $-2$ inches ($-10$ inches/5). Thus the mean height is $7' + (-2'')$, or 6'10". This value is identical to $(6'9'' + 6'5'' + \cdots + 7'2'')/5 = 6'10''$. But coding makes the arithmetic easier.

Coding is practical for 50+ observations when the data are grouped. We illustrate for calculation of the sample mean and standard deviation. The idea of *adding back* for an initial over- (or under-) estimate works for grouped data estimation of the mean.

Computing the Mean Using Coded Data   The coded method for computing the mean of a frequency distribution having equal-length class intervals is similar to the ungrouped illustration just given. We arbitrarily assume a mean and adjust it by a correction for deviations about this "guessed" value. Very simply, if our guess is on target, the correction is zero; if the guess is low, the correction is added; if the guess is high, the correction is subtracted. This is defined in a rule.

---

\* Because we use the notation from Chapter 3 in this discussion, it is best to study this topic after you have studied Chapter 3.

## RULE FOR COMPUTING THE MEAN FROM CODED DATA

$$\bar{X} = \bar{X}_a + \left(\frac{\sum fd}{n}\right)i$$

where   $\bar{X}_a$ = midpoint of any class (an assumed mean)

$f$ = class frequency

$d$ = unit deviation of class midpoint from $\bar{X}_a$

$n$ = number of observations

$i$ = class interval

The term $[(\sum fd)/n]$ represents, in coded units, the average of deviations from the assumed mean. The class interval, $i$, is the length of each coded unit. Note that $\bar{X}_a$ can be taken as the midpoint of any class we choose.

For example, Table C.3 diaplays a frequency distribution of the amounts spent for a business dinner by a sample of 102 bank executives. Code these data, and compute the mean expenditure.

We arbitrarily let $\bar{X}_a$ = \$50, the midpoint of the third class. Using the rule, we get

$$\bar{X} = \bar{X}_a + \left(\frac{\sum fd}{n}\right)i$$

$$= \$50.00 + \left(\frac{60}{102}\right)20.00 = \$61.76$$

In the example, the guessed value of \$50.00 was low, so we must correct it *by adding* the appropriate adjustment.

**Computing the Standard Deviation Using Coded Data**   The coded method of computing the standard deviation of a frequency distribution having equal class intervals uses an equation similar to the expression for standard deviation shown in Chapter 2. The differences are (1) that $\sum X$ and $\sum (X^2)$ are replaced by $\sum (fd)$ and $\sum (fd)^2$ and (2) that the result is multiplied by the class interval $i$ in order to decode the final answer. That is, this procedure uses the coding devices of (1) adding a constant and (2) dividing by another constant.

**TABLE C.3**   Calculation of the Mean and Standard Deviation with Coding

| Cost of Dinner $ | Number of Executives | Class Midpoint M | Unit Deviation from $\bar{X}_a$ d | Weighted Deviation (fd) | Squared and Weighted (fd²) |
|---|---|---|---|---|---|
| 0 under 20.00 | 7 | $10 | −2 | −14 | 28 |
| 20.00 under 40.00 | 10 | 30 | −1 | −10 | 10 |
| 40.00 under 60.00 | 16 | 50 | 0 | 0 | 0 |
| 60.00 under 80.00 | 54 | 70 | 1 | 54 | 54 |
| 80.00 under 100.00 | 15 | 90 | 2 | 30 | 60 |
|  | $n = \overline{102}$ |  |  | $(\sum fd) = \overline{60}$ | $(\sum fd^2) = \overline{152}$ |

## RULES FOR COMPUTING THE STANDARD DEVIATION FROM CODED DATA

$$s = i \sqrt{\frac{n(\sum fd^2) - (\sum fd)^2}{n(n - 1)}}$$

or

$$s = i \sqrt{\frac{(\sum fd^2) - [(\sum fd)^2/n]}{n - 1}}$$

In the next example these constants are (1) $50 and (2) $20. We will find the standard deviation of the sample data given in Table C.3. This is a measure of the variation in the amount spent for a business dinner by bank executives.

$$s = i \cdot \sqrt{\frac{n(\sum fd^2) - (\sum fd)^2}{n(n - 1)}}$$

$$= \$20.00 \cdot \sqrt{\frac{102(152) - (60)^2}{102(101)}} = \$21.50$$

To summarize our use of coding: Adding or subtracting a constant shifts the value of the mean by the amount of the constant. But it has no effect on the variance or standard deviation: All numbers are shifted the same distance so that the amount of dispersion remains unchanged. Thus, coding by adding (or by subtracting) a constant requires decoding the mean by using the opposite operation. Meanwhile, division by the class interval, $i$, reduces class midpoints by a constant amount. This also requires a correction. Then multiplying (or dividing) by a constant also multiplies (or divides) the mean and standard deviation by the constant and multiplies (or divides) the variance by the square of the constant. The necessary decoding is built into the formula (above) for standard deviation calculation.

**Problems for Appendix C**

1. When possible, evaluate each of the following; they all involve calculations with zero or one.

   a. $4 \cdot 0$

   b. $2^0$

   c. $\dfrac{0}{1000}$

   d. $\dfrac{2}{1}$

   e. $0^1$

   f. $1^{-20}$

   g. $\dfrac{1/100}{1}$

   h. $10 \cdot 1 \cdot 0$

   i. $\dfrac{1}{4 \cdot 6 \cdot 0}$

2. Find the answers, with proper decimal location, for the following:

   a. $9.9 \cdot 0.001$

   b. $0.01 \cdot 0.001$

   c. $(1.01)^2$

   d. $(0.03)^2$

   e. $\dfrac{400}{0.0004}$

   f. $\dfrac{0.4}{4.4}$

3. Perform the indicated arithmetic operations on these common (ratio) fractions:

   a. $\dfrac{1}{4} - \dfrac{1}{5}$

   c. $2\dfrac{1}{4} \cdot \dfrac{2}{5}$

   e. $-\dfrac{1}{2}\bigg/\dfrac{1}{6}$

b. $\dfrac{1}{2} + \dfrac{1}{3}$   d. $\dfrac{1}{2}\dfrac{1}{5}\dfrac{1}{2}$   f. $\dfrac{6}{1/3}$

4. Using the concepts of ratios and proportions, show that

   a. $\dfrac{27}{171} = \dfrac{3}{19}$   c. if $\dfrac{4}{7} = \dfrac{2}{X}$, then $X = 3.5$.

   b. if $\dfrac{X}{6} = \dfrac{4}{15}$, then $X = 1.6$.

5. Convert, as indicated, either to a fraction, a decimal, or a percent:

   a. 0.667 to a percent   d. $\dfrac{2}{5}$ to a percent

   b. 1.5% to a decimal   e. $\dfrac{5}{9}$ to a decimal

   c. 0.125 to a ratio fraction   f. 0.5% to a ratio fraction

6. Perform the indicated operations involving exponents or radicals:

   a. $2^3 - 3^2$   c. $\dfrac{3^2}{2^{-1}}$   e. Simplify $\dfrac{\sqrt{9}}{\sqrt{3}}$

   b. $(-1)^2 \cdot (-1)^2$   d. $\dfrac{3^{-2}}{2^1}$   f. Simplify $\dfrac{-\sqrt{9}}{3}$

7. Compute for these inequalities:
   a. Add 2 to both sides of $-3 < -1$.
   b. Which is bigger 5 or $|-6|$? Express your answer as an inequality.
   c. $1/2 > 1/3$ multiplied on both sides by $-1$.
   d. for $3 < 5$, divide each side by $-2$.
   e. Which is bigger, $(1/2)^2$ or $(1/2)^4$? Express as an inequality.

8. Indicate the number of decimals of accuracy in each answer.
   a. 30/15   c. $6.921 - 5.8109$   e. $(0.5)^2$
   b. $3.6 + 1.11 + 4.001$   d. $6 \cdot 0.21$   f. 12.5/10

9. Evaluate each of the following:
   a. $n = 251$, $n^2 = ?$   c. $\sqrt{87}$   e. $\sqrt{8754}$
   b. $n = 0.251$, $n^2 = ?$   d. $\sqrt{875}$   f. $\sqrt{87542}$
   Use a calculator to answer each part.

10. Evaluate these expressions. (*Hint:* All of the radicals can be made perfect squares, so a calculator or square roots table need not be used.)

    a. $\sqrt{9} + \sqrt{16}$   c. $\sqrt{5} \cdot \sqrt{5}$   e. $\dfrac{\sqrt{32}}{\sqrt{8}}$

    b. $\sqrt{9} - \sqrt{16}$   d. $\sqrt{0.8} \cdot \sqrt{0.2}$   f. $\dfrac{\sqrt{0.27}}{\sqrt{3}}$

11. Perform each of the summations. Use the grouping symbols (parentheses) as necessary. Let $X_1 = 0$, $X_2 = 1$, $X_3 = -1$, and $X_4 = 2$.

    a. $\sum\limits^{5} 4$   c. $\sum\limits^{4} [X - (1/2)]$   e. $\sum\limits^{4} (2X)$

    b. $\sum\limits^{4} (3X + 1)$   d. $\sum\limits^{4} [X - (1/2)]^2$   f. $\sum\limits^{4} |X - (1/2)|$

12. Use the rectangular coordinates (as described on p. A-34) to answer the following.
    a. In which quadrant, I, II, III, or IV, is the point $(2, -1)$?
    b. Are the signs $(+ \text{ or } -)$ for both coordinates of points in Quadrant III always negative? (Exclude (0,0) from those points in Quadrant III.)
    c. Are the points $(-2, 1)$ and $(2, -1)$ in the same quadrant? In which quadrant(s) are these points?

13. Perform the designated operations for each linear relation:
    a. Graph on coordinate axes the line defined by $Y = 1 + 2X$.
    b. Graph on coordinate axes the line defined by $Y = 2X$.
    c. Find values for $b_0$ and $b_1$ for each of the lines in the figure.
    d. In $Y = a + bX$, with $a$, $b$ constants, explain how the direction of the line (slope) relates to the sign for $b$.

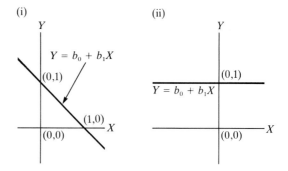

14. Find the unknown, $X$, for each of the linear relations.
    a. $X + 1 = 2$                    b. $3X = 24$

15. Several points that satisfy the relation $Y = 2 + 3X + X^2$ are

| $X$ | $-1$ | 0 | 1 | 2 | 3 | 4 |
|-----|------|---|---|----|---|---|
| $Y$ | 0 | 2 | 6 | 12 | | |

a. Complete the table.
b. Complete the graph.

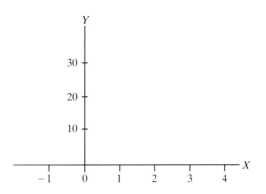

16. Suppose the rate of compound interest for college loans is 10% per year. Graph the accumulation of debt on a $1000 loan for a period of five years. Assume no repayment is made during this time and the debt equation is $D = \$1000(1.10)^X$, where $D$ = accumulated debt and $X$ = number of years after that in which the debt was incurred. Here is a start:

| Year | Accumulated Debt |
|------|------------------|
| 0 | $1,000 |
| 1 | $1,100 |
| 2 | $1210 $[= \$1000(1.10)^2]$ |
| 3 | |
| 4 | |
| 5 | |

17. Evaluate each of the following:
   a. $Y = -2 - X$. Give (i) slope and (ii) $Y$ for $X = 2$.
   b. $Y = 3 - 2X - X^2$. Find $Y$ for $X = 1$, then for $X = -3$, and finally for $X = 0$. Tell what you know about the graph for this relation, but do not graph it.
   c. Write an equation to show debt, $D$, in relation to years, $X$, for an initial debt of $500 taken with 12% annual compound interest.

**Answers to Appendix C Questions**

1. a. 0    b. 1    c. 0    d. 2    e. 0    f. 1    g. 1/100    h. 0
   i. not defined
2. a. 0.0099    b. 0.00001    c. 1.0201    d. 0.0009    e. 1 million    f. 0.09
3. a. 1/20    b. 5/6    c. 1/5    d. 1/20    e. $-3$    f. 18
4. a. $\dfrac{27}{171} = \dfrac{(3 \cdot 9)}{(19 \cdot 9)} = \dfrac{3}{19}$    b. $15X = 6 \cdot 4$, or $X = 24/15 = 1.6$
   c. $4X = 2 \cdot 7$, or $X = 7/2 = 3.5$
5. a. 66.7%    b. 0.015    c. $\dfrac{1}{8}$    d. 40%    e. 0.45    f. $\dfrac{1}{200}$
6. a. $-1$    b. 1    c. 18    d. $\dfrac{1}{18}$    e. $\sqrt{3}$    f. $-1$
7. a. $-1 < +1$    b. $|-6| > 5$    c. $-\frac{1}{2} < -\frac{1}{3}$    d. $-1.5 > -2.5$
   e. $(\frac{1}{2})^2 > (\frac{1}{2})^4$ as $\frac{1}{4} > \frac{1}{16}$
8. a. 2.0    b. 8.7    c. 1.110    d. 1.26    e. 0.25    f. 1.2

9.

| | Table | Calculator | | Table | Calculator |
|---|-------|-----------|---|-------|-----------|
| a. | 63001 | same | d. | 29.7 | 29.6 |
| b. | 0.0625 | 0.0630 | e. | 93.8 | 93.6 |
| c. | 9.3 | same | f. | 296.6 | 295.9 |

where they differ, the calculator answers are more accurate.
10. a. 7    b. $-1$    c. 5    d. 0.4    e. 2    f. 0.3
11. a. 20    b. 10    c. 0    d. 5    e. 4    f. 4
12. a. IV
    b. This is true—assuming that we exclude points on any axis from belonging in the quadrant.
    c. No, II, IV.

13. a.  See figure.                                b.  See figure.

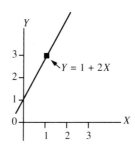                    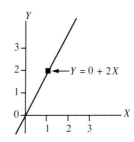

    c.  (i)  $Y = 1 - X$     (ii)  $Y = 1 + 0X = 1$

    d.  The sign indicates the direction of the slope: for $+$ slope, the $Y$-values increase with $X$; for $-$ slope, $Y$-values decrease as $X$ increases.

14. a.  1      b.  8

15. a.

| $X$ | $-1$ | 0 | 1 | 2 | 3 | 4 |
|---|---|---|---|---|---|---|
| $Y$ | 0 | 2 | 6 | 12 | 20 | 30 |

    b.  See figure.

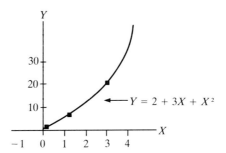

16. See figure.

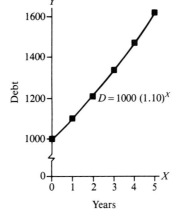

| Year | Accumulated Debt |
|---|---|
| 3 | $1331.00 |
| 4 | $1464.10 |
| 5 | $1610.51 |

17. a.  (i)  $-1$     (ii)  $-4$

    b.  0, 0, 3

       Since $Y = +3$ for $X = 0$ and the graph cuts the $X$-axis at points either side of $X = 0$, then the curve opens downward.

    c.  $D = 500 \ (1.12)^X$, where $D =$ accumulated debt and $X =$ years.

# ANSWERS TO ODD-NUMBERED PROBLEMS

## CHAPTER ONE

1. a. sample  b. descriptive statistic
   c. primary data  d. decision problem
   e. statistic  f. secondary data  g. parameter
   h. statistical inference  i. population
   j. probability  k. data base  l. statistical plan
3. a. sample  b. sample  c. sample
   d. population
5. a. *A* will give a larger (and we hope more accurate) result. However, *B* and *C* might also be sufficient.
   b. If sample is representative, *C* will reflect the entire population at a lower cost. However, it may not be quite as accurate as *A* or *B*.
7. a. The sample count is less.
   b. The sample would be about as descriptively accurate as the census, but a larger sample lends more confidence in the result.
   c. The sample would use statistical inference; the census would constitute actual counted results.
9. provide a description
11. Answers will differ. Numerous uses exist in marketing, operations, accounting, etc.
13. (1) *Problem:* Operations were running in the red.
    (2) *Plan:* Work with labor as much as possible and take decisive action. (3) *Data base:* Labor costs compiled for all Safeway stores. (4) *Sample results:* Several states had high labor costs. (5) *Analysis:* A few key states were excessive. (6) *Resolution:* Differences were brought out and alternatives discussed. Labor refused settlement, so Safeway relocated its stores.

## CHAPTER TWO

1. $\bar{X} = 8$  Median $= 8$  Mode $= 8$
3. Alice: Mean $= 73.5$  Bill: Mode $= 86$
   Clark: Median $= 75$
5. a. $\bar{X}_W = 27.8$ yrs. The average must be weighted by the proportion of men and women.
   b. 45.2 yrs.
7. a. 8  b. 2  c. 2.7  d. 38.14%
9. a. Range $= \$700$  IR $= \$575 - \$225 = \$350$
   MAD $= \$150$  $s = \$200$
   b. Range $= 7$ hundreds, IR $= 3.5$ hundreds,
   MAD $= 1.5$ hundreds, and $s = 2$ hundreds, so the two answers are identical.

11. a. $\bar{X} = -5$ and $\sum (X_i - 5) = 0$
    b. $\sum (X_i - \bar{X})^2 = 14$, so $s^2 = 14/(5 - 1) = 3.5$
    c. No, $s^2 = 0$ when all values are the same.
13. $V_{US} = 13.52\%$, whereas $V_A = 6.67\%$, so U.S. value is about twice that of European value, suggesting tests are not comparable.
15. a. Maximum amounts are Store $1 = 2.8\%$, Store $2 = 11\%$, Store $3 = 100\%$.
    b. Percentages differ because standard deviations differ.
    c. Using Chebyshev's Rule for Store 1, at least 75% of weeks have sales within 2 standard deviations of the mean. So sales could possibly be $<\$100,000$ in $\leq 25\%$ of the weeks.
17. a. 95.5%    b. 81.9%

19.

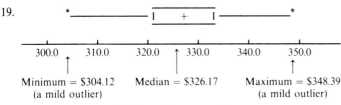

Minimum = $304.12    Median = $326.17    Maximum = $348.39
(a mild outlier)                            (a mild outlier)

Weekly takehome pay ($)

21. a. $s_K = 5.0$, $s_J = 22.3$
    b. $V_K = 6.25\%$, $V_J = 27.9\%$
    c. Kale's inventory investment is more consistently close to its mean.
23. a. at least 75%    b. at most 25%
25. $\bar{X} = 12$    Median = 11.5    Mode = none    $s^2 = 16$
    $s = 4$
27. Using $\bar{X} = 7.1$ and $s = 0.2$, $V = 2.8\%$ so the machine is not satisfying minimum requirements.
29. Mean$_{X,Y} = 44.0$
31. $B$ has a higher % gain at 134% versus $A$ at 124%.
33. At least 88.9% are between 60 and 78 inches tall. If bell-shaped, 99.7% would be between 60 and 78, and 95.4% would be between 63 and 75 inches tall.
35. A box plot for the original data, with a last value of 0.36, is

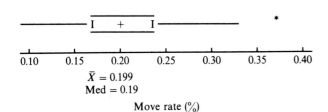

$\bar{X} = 0.199$
Med = 0.19

Move rate (%)

There is a slight positive skew (right) by the $\bar{X} >$ Med, and the median is left of center in the box plot. 0.36

is mild outlier. A 20% move rate is reasonable.
    A box plot for the data with the last value changed from 0.36 to 0.26 is

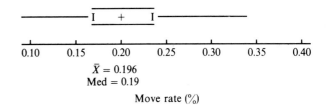

$\bar{X} = 0.196$
Med = 0.19

Move rate (%)

The right skew is lessened and the mean moves closer to the median value. Otherwise the distributions are close. There is no outlier.

## CHAPTER THREE

1. a. 3.00 and 5.99    b. 2.995 and 5.995
   c. 7.495    d. 3.00
   e. 12.00 (for distribution and 11.57 for actual data)
3. See figure.

| Class | $f$ |
|-------|-----|
| 3–7   | 2   |
| 8–12  | 1   |
| 13–17 | 5   |
| 18–22 | 4   |
| 23–27 | 3   |

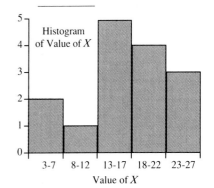

5. See figure. Class boundaries: $329.995 and $339.995. Class midpoint: $334.995.

| Class | $f$ |
|---|---|
| $300.00–309.99 | 1 |
| 310.00–319.99 | 3 |
| 320.00–329.99 | 10 |
| 330.00–339.99 | 4 |
| 340.00–349.99 | 2 |
| | 20 |

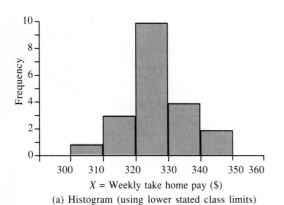

(a) Histogram (using lower stated class limits)

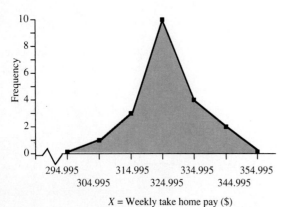

(b) Frequency polygon (using class midpoints)

7. a. 59.5 and 69.5    b. 10    c. 0.5    d. 18.3%
   e. See figure.

| LCB | Freq < LCB |
|---|---|
| 29.5 | 0 |
| 39.5 | 2 |
| 49.5 | 8 |
| 59.5 | 12 |
| 69.5 | 24 |
| 79.5 | 45 |
| 89.5 | 52 |
| 99.5 | 60 |

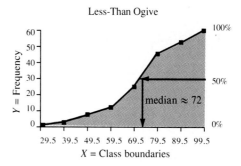

9. a. 76 lbs. appears excessive. Possibly it is a coding error or an incorrect weight. Or maybe it is real.
   b. The bags need to be reweighed. If the discrepancy is not explained by the summing of two weights, then another sample is required. The sample should remain at 25 mailbags.

11. a. Median = Mode = 0%
    b. The number of cases of 0% is very high. These values may be possible for a small number of households, such as appears in the area description table. This is a problem of scale, where more households in each area would yield a more precise measure. Improve the plan by going to a bigger unit of area, redrawing the sample, and recomputing the statistics.

13. a. The personal ledger is missing check #677. The error was found by comparing entries record by record.
    b. bank deposit statement versus check book entry, grocery "tape" list versus items in a grocery bag, accounting credits versus debits, and stock transaction sheet versus monthly brokerage accounting report.

15. a. See figure.

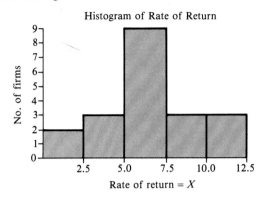

Histogram of Rate of Return

No. of firms

Rate of return $= X$

b. Stem-and-leaf display:

```
 0 | 4
 1 |
 2 | 1 6
 3 | 0
 4 | 5
 5 | 0 0
 6 | 3 3 6 9
 7 | 3 3 4 5 6 7
 8 |
 9 |
10 |
11 | 6
12 | 1 2
```

The stem-and-leaf display retains the identity of individual values. This allows exact calculations for median, $\bar{X}$, and more.

c. Median $= \dfrac{66 + 69}{2} = 67.5$

Percent of cases $= \left(\dfrac{3}{20}\right)100 = 15(\%)$

17. a. 15    b. 15.49    c. 4.47    d. 15    e. 20
19. a. $11.24, $2.31    b. $11.20    c. 74%, 26%, 60%
21. a. See figure.

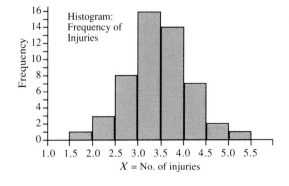

Histogram: Frequency of Injuries

Frequency

$X$ = No. of injuries

b. Mean $= 3.4$    Median $= 3.39$    Mode $= 3.2$
c. $s = 0.69$, $s^2 = 0.47$
23. a. Discrete is countable whereas continuous is measurable.
b. Raw data is as recorded; an array is ordered data.
c. Class limits are stated extremes of the class intervals; class boundaries are carried to one more digit of accuracy.
d. Histograms use adjacent rectangles to depict frequencies within class intervals; frequency polygons are line graphs where frequencies are plotted above the class midpoints.
e. No difference between class midpoint and class mark.
25. a. 24.5    b. 23.56    c. 24.5    d. 6.0    e. 36.0
27. a. See figure.    b. $98.68/day    c. $0.67/day

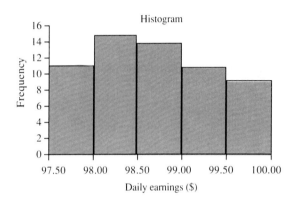

Histogram

Frequency

Daily earnings ($)

29. a. 20,530    b. 84%    c. 50 workers
31. For an increment of 0.1, the stem-and-leaf display is

```
24 | 0
25 | 0 0
26 | 0 0 0 0 0 0 0
27 | 0 0 0 0 0 0 0 0 0 0
28 | 0 0 0 0 0 0 0 0 0 0 0 0 0 0 0 0 0 0 0
29 | 0 0 0 0 0 0 0 0 0
30 | 0
31 | 0
```

The median is 2.8. The exact result is found by counting to $(48 + 1)/2 = 24.5$th position, which has value 2.8.

## CHAPTER FOUR

1. a. $\{1, 2, 3\}$    b. $\{1, 2\}$    c. $\{\varnothing\}$    d. $\{3\}$
   e. $\{\varnothing\}$

3. a. See figure.

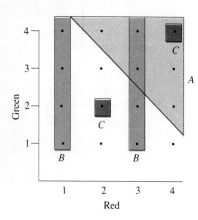

b. $A$ = total spots is 6 or more
   $B$ = red object showed one or three
   $C$ = both red and green showed same number and
        were even
c. $(A \cup B)$ = {(1,1), (1,2), (1,3), (1,4), (3,1), (3,2),
              (3,3), (3,4), (2,4), (4,2), (4,3), (4,4)}
   $(B \cap C)$ = {$\varnothing$}
      $B'$ = {(2,1), (2,2), (2,3), (2,4), (4,1), (4,2), (4,3),
             (4,4)}
   $(A \cap C')$ = {(2,4), (3,3), (3,4), (4,2), (4,3)}
d. $(A \cup B)$ = total number is $\geq 6$ *or* red object showed
              one or three or both
   $(B \cap C)$ = red object showed one or three
              and both red and green objects showed
              same even number. This is impossible,
              so the set is empty.
      $B'$ = red object showed a two or a four
   $(A \cap C')$ = total is $\geq 6$ *and* red and green
              objects are not the same even number
e. no, yes
5. 0.30
7. a. 0.63    b. 0.24    c. 0.39    d. 0.61    e. 1.00
9. a. $A = (P_1,D_1), (P_1,D_2), (P_1,D_3)$
      $B = (P_1,D_3), (P_2,D_3), (P_3,D_3)$
      $C = (P_2,D_3), (P_3,D_3)$
   b. $P(A) = 0.25$
      $P(C) = 0.60$
      $P(A \cap B) = 0.12$
      $P(A \cup C) = 0.85$
   c. $A$ and $C$ are mutually exclusive.
11. 0.30    Events are mutually exclusive.
13. a. 0.40    b. 0.24    c. 0.76    d. 0.24    e. 0.76
15. a. 0.32    b. 0.88    c. 0.12
17. a. not true    b. not true    c. true    d. not true
19. a. marginal 0.27    b. conditional 0.75
    c. joint 0.10    d. 0.30

21. 0.86
23. a. 0.26    b. 0.31
25. 0.26
27. a. 50,625    b. 2,450    c. 45    d. 1000    e. 720
    f. 120
29. a. 21    b. 49    c. 42
31. a. 120    b. 126
33. 18%, which seems reasonable
35. a. 0.10    b. 0.56    c. 0.04    d. 0.41    e. 0.74
    f. 0.26
37. a. See figure.    b. 0.50    c. 1.0    d. 0.40
    e. no, $R$ and $E$ intersect

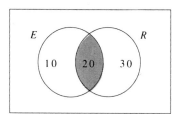

39. Events are independent if $P(A \cap B) = P(A)P(B)$. They
    are mutually exclusive if their intersection is zero.
41. a. 9%    b. 0.80    c. 0.24
    d. Not independent. This means opinion is influenced
       by home ownership.
    e. no, $R$ and $E$ intersect.
43. 98%
45. a. 0.75    b. 0.25
47 a. 0.10    b. 0.67
49. 0.60
51. a. no, class depends upon sex.    b. 0.25    c. 0.39
53. 0.70

## CHAPTER FIVE

1. a.

| X | P(X) |
|---|------|
| 5 | 0.10 |
| 6 | 0.15 |
| 7 | 0.30 |
| 8 | 0.15 |
| 9 | 0.25 |
| 10 | 0.05 |
|   | 1.00 |

b. $E(X) = \sum XP(X) = 5(0.10) + 6(0.15)$
$+ 7(0.30) + 8(0.15) + 9(0.25) + 10(0.05)$
$= 7.45$

c. This is a long-run expectation and would not apply to a specific date.

3. a. 1/2    b. 1/4    c. 1/4
5. 0.282
7. a. 0.357

b.

| X | 0 | 1 | 2 | 3 |
|---|---|---|---|---|
| P(X) | 0.119 | 0.476 | 0.357 | 0.048 |

9. a. 0.333    b. 0.300
11. a. $\sum P(X) = 1$
    b. $\sum P(X) = 1$ and $0 \le P(X) \le 1$
13. a. 1, 1, 1, 1, 1, 1    b. 3, 4, 20, 3, 4, 20
    c. 1, 1, 6, 6, 15, 15,    d. 1, 1    e. n, n
    f. $\dfrac{n!}{X!(n-X)!}$, 12, 1, 153, 15,504

15. a.

| X | 0 | 1 | 2 | 3 | 4 |
|---|---|---|---|---|---|
| P(X) | 0.6561 | 0.2916 | 0.0486 | 0.0036 | 0.0001 |

    b. 0.0523
17. a. 0.0067    b. 0.00512    c. 0.00512
19. 0.4633
21. a. $\mu = 2.0$, Mode $= 2.0$, $\sigma^2 = 1.2$, $\sigma = 1.1$
    b. $\mu = \sum XP(X) = 2.0$, $\sigma^2 = \sum P(X)(X - \mu)^2 = 1.2$
23. For $X = 1$ $P(X) = 0.48$, $XP(X) = 0.48$, $X - 0.80 = 0.20$,
    $(X - 0.80)^2 = 0.04$, and $(X - 0.80)^2 P(X) = 0.0192$, so
    $\sum (X - 0.80)^2 P(X) = 0.48 = \sigma^2$, $\pi = 0.4$
25. $\mu = 112/84 = 1.33$, $\sigma^2 = 0.556$
27. a. $\mu = 3.10$, $\sigma^2 = 1.290$, $\sigma = 1.136$
    b. 0.60, 1.00, 0.80
29. 0.3020
31. a. 0.3658    b. 0.3432
33. 0.0361
35. a. 0.0872    b. 0.0183    c. 0.8641    d. 0.0153
37. 225
39. Net gain = $29,600 - $20,000 = $9,600
41. $\mu = 3.00$, $\sigma^2 = 10.00$, $\sigma = 3.16$, $P(X \le 0) = 0.40$
43. a. 5.50    b. 0.850    c. 0.922
45. 0.9445
47. 0.0244
49. 0.3233
51. a. number of correct answers

b.

| X | 0 | 1 | 2 | 3 | 4 |
|---|---|---|---|---|---|
| P(X) | 0.0625 | 0.2500 | 0.3750 | 0.2500 | 0.0625 |

c. See figure.

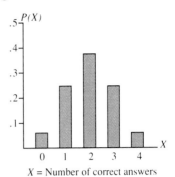

**5-51**    X = Number of correct answers

d. 27 or 28 students

## CHAPTER SIX

1. a. $\mu = 0$, $\sigma^2 = 8.3$, $\sigma = 2.9$    b. 1/2, 0, 1/2
3. a. See figure.    b. 1/2, 1/10, 3/10

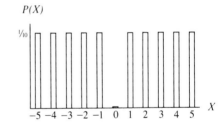

c. Probabilities differ because this is a discrete distribution whereas other is continuous.
5. a. no    b. 1/3 of time    c. 2/7
7. See figure, parts (a) through (f).
    a. 0.3413    b. 0.9500    c. 0.1488    d. 0.0505
    e. 0.0250    f. 0.8641

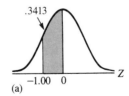

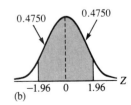

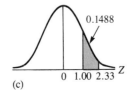

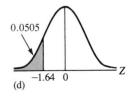

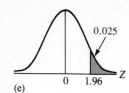

(e)

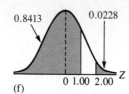

(f)

9. a. $-1.30$    b. $\pm 1.40$    c. $-0.67$    d. $-1.42$
   e. 3.08
11. a. 0.10    b. $\cong 1.00$    c. 0.9803    d. 0.4831
    e. 0.2283
    f. $Z = 0.30$ is larger because it is further to the right and includes more area
13. a. $\cong 1.00$    b. 0.8413    c. $\cong 0.0$    d. 0.6687
    e. 0.7971
15. 0.0228
17. a. 9.5%    b. $.76 per 100 bags
19. 0.7967
21. a. 75.80%    b. 391 lbs.    c. 373 lbs, 399 lbs.
    d. 3rd    e. 0.2119, or 21st percentile
23. a. 0.2025
    b. 0.2144    Approximation is 1.2% high.
25. 0.0838
27. 0.5478
29. 0.1469
31. a. $\mu = -1.17$    b. 2/3
33. 0.0228
35. a. 0.1446    b. 0.0703    c. 2.10    d. $\pm 0.30$
    e. 0.0735    f. 1.04
37. a. 69th    b. 169
39. 0.0228
41. a. 18    b. 3
43. 0.9821
45. a. 0.1170    b. $880
47. a. 0.1937    b. 0.1839    c. 0.241

## CHAPTER SEVEN

1. a. $\dfrac{40!}{2!38!} = 780$    b. $\dfrac{400!}{2!398!} = 79{,}800$

3. a. Mean = 28.4 yrs.    % voter turnout = 62.3%
   b. No. A bias may exist because some were more interested in voting than others.
5. a. 60%    b. 0.956
7. $\mu_{\bar{x}} = 50$ meters    $\sigma_{\bar{x}} = 2.0$ meters
9. $\sigma$ and $\sigma_x$ both refer to the distribution of individual data points and are equivalent terms. $\sigma_{\bar{x}}$ refers to the distribution of sample means.

11. a. There are 20 samples of $n = 3$, from ABC, ABD, ABE, . . . , to DEF. The sample means are as shown:

| $\bar{X}$ | 2.00 | 2.33 | 2.67 | 3.00 | 3.33 | 3.67 | 4.00 |
|---|---|---|---|---|---|---|---|
| $f(\bar{X})$ | 2 | 2 | 4 | 4 | 4 | 2 | 2 |
| $\bar{X}f(\bar{X})$ | 4.00 | 4.67 | 10.67 | 12.00 | 13.33 | 7.33 | 8.00 |

b. $\bar{\bar{X}} = \dfrac{\sum \bar{X}f(\bar{X})}{\text{Number of samples}} = \dfrac{60}{20} = 3$

c. $\sigma_X = 1.29$ gives $\sigma_{\bar{x}} = \dfrac{\sigma_X}{\sqrt{n}} \sqrt{\dfrac{N-n}{N-1}} = 0.577$

d. The means are identical. But the larger sample size reduces $\sigma_{\bar{x}}$.
13. The standard deviation is larger than the standard error by the factor of the square root of $n$.
15. 0.0475
17. 0.9147
19. a. 34.46%    b. 0.9962
21. *Stratification* is the grouping of population elements into relatively homogenous subgroups prior to sampling. It provides good precision for a limited size of sample. *Random sampling* assures every individual unit (and combination) an equal chance of being selected. This constitutes a basis for statistical evaluation. *Convenience samples* are nonprobability samples selected from elements that happen to be available at the time, and *purposeful samples* are consciously arranged to include specific items.
23. The standard error is not dependent upon the population size, unless the sample is $\geq 5\%$ of the population. However, insofar as 100 stores probably constitute $> 5\%$ of the number of stores in a town of 20,000 population, the finite population correction would apply. So the statement may be considered false.
25. 0.1587
27. a. 50%    b. 2.536

## CHAPTER EIGHT

1. 104 to 116 employees
3. $40.86 to $43.40
5. a. 56.3 to 63.7 mph
   Thus 95% of all intervals computed in this way will include the true mean speed.
   b. 55.2 to 64.8 mph
   The 95% confidence interval width is smaller (7.4 mph) than the 99% interval width (9.6 mph).

7. a. $\bar{X} = 0.8$ changes    $s = 0.97$    b. 0.52 to 1.08
9. a. 2.08    b. 95%    c. 1.81
11. a. Go to row labeled 17, column headed $t_{0.05}$. Because 5% of the area is above 1.740 (and 95% is below), this is $P_{95}$.
    b. For a two-tailed test, use column headed 0.025. For row df = 12, this column entry is 2.179.
    c. This is a matter of space. Values for $\alpha = 0.01$, $0.02, \ldots, 0.99$ would take many pages.
13. 5.97 to 6.03 oz.
15. $-5.4$ to $-0.6$ lbs.
17. $1245 to $1355
19. 425
21. a. 139 loads    b. 25.1 to 25.3 tons per load
    c. 75,600 tons
23. a. true    b. false    c. false    d. false    e. true
25. a. 0.64    b. 0.016    c. 0.031    d. 60.9% to 67.1%
27. 2172
29. 65.8% to 84.2%
31. a. No. The sample of $n = 1800$ will cost $2,200.
    b. 20.3% to 24.1%    c. 7203 customers    d. $7703
33. Yes. The interval of 0.33 to 6.87 units/hr does not include zero, so we can conclude that the human workers produce more.
35. a. $-8\%$ to $+18\%$
    b. No. The interval includes zero.
37. $-1.1\%$ to 11.1%
39. d
41. a. T    b. T    c. F    d. T    e. F    f. T
43. 166 stockholders
45. 385 members
47. a. 18.7 to 21.3    b. 144
49. 10.1 to 15.9 beats/min. increase
51. a. 62 households    b. $7.84
53. 271 students
55. $309.88 to $320.12
57. $\bar{X} = 320$, $s = 9.08$, CL = 29.7 to 34.3

## CHAPTER NINE

1. $H_0: \mu = \$14.75$    $H_A: \mu < \$14.75$
3. $Z_{calc} = -1.75$    Decision: Do not reject $H_0$.
5. a. $H_0$: Time to admit a patient $\geq 5$ min.
   b. $H_0$: Time to admit a patient $> 5$ min.
7. a. $H_0: \mu = 10$ points    $H_A: \mu > 10$ points
   b. Reject if $Z_{calc} > 1.645$.
   c. $Z_{calc} = 6.00$
   d. Reject $H_0$ since $6.000 > 1.645$.
9. No. Reject $H_0: \mu = \$88,950$
   because $Z_{calc} = 4.05 > 1.96$.

11. Yes. Reject $H_0: \mu = \$320$
    because $Z_{calc} = -2.80 < -1.645$.
13. a. 1.714    b. 1.761    c. 2.132    d. 2.492
    e. 1.711    f. 1.318
15. a. $-1.721$    b. 1.812    c. $\alpha = 0.01$
17. Reject $H_0: \mu = 4,000$ lbs. since $t_{calc} = -3.16 < -2.821$.
19. Accept $H_0: \pi = 0.65$
    because $Z_{calc} = -0.629 > -2.055$.
21. Accept $H_0: \pi = 0.80$
    because $Z_{calc} = -1.25 > -1.44$.
23. $H_0: \pi = 0.50$    $H_A: \pi > 0.50$
    Reject if $Z_{calc} > 1.645$.
    $$Z_{calc} = \frac{0.57 - 0.50}{0.05} = 1.400 \quad \text{Do not reject } H_0, \text{ because}$$
    $1.400 < 1.645$. Data are not sufficient to suggest that a majority favor the change, at $\alpha = 0.05$.
25. a. no difference
    b. With $p$-value approach, no prior standard need be set.
    c. no difference
    d. The $p$-value is an objective measure of the weight of evidence for rejecting the null hypothesis. Conclusion can be stated at the smallest value of $\alpha$ for which a test result could be considered statistically significant.
27. $p$-value is 10%
29. zero
31. a. Reject $H_0: \mu = 14.0$ min. because $2.00 > 1.645$.
    b. 14.82 min.    c. 0.7389
33. 156
35. a. 0.0918    b. 0.0228, 0.2514, 0.5000, 0.9082, 0.9972
    c. See figure.

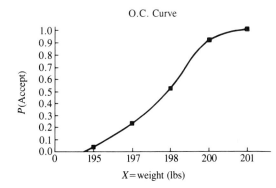

O.C. Curve

37. Power-curve values are based upon hypothetical values of $\mu$, not of $\bar{X}$. If $\mu < 3.0$, the hypothesis is not false, so there is no need of "power to reject a false hypothesis."
39. For $H_0: \pi_1 - \pi_2 = 0$, the hypothesized difference must be zero, and the standard error of difference is a weighted average of the sample values. Under $H_0: \pi_1 = \pi_2 =$ some nonzero amount, we assume the proportions are different, and the standard error of difference is calculated as $\sqrt{(p_1 q_1/n_1) + (p_2 q_2/n_2)}$.

41. Reject $H_0$: $\mu_1 - \mu_2 = 0$
    because $Z_{\text{calc}} = -2.18 < -1.645$.
43. Do not reject $H_0$: $\pi_{\text{new}} - \pi_{\text{old}} = 0$
    because $Z_{\text{calc}} = 0.435 < 1.645$. The new method does not statistically improve on the current rate of selecting ripe (low-acid) fruit.
45. Yes. Reject $H_0$: $\mu = 66.8$ yrs
    because $Z_{\text{calc}} = 2.25 > 1.645$.
47. (1) $H_0$: $\pi = 0.95$   $H_A$: $\pi \neq 0.95$
    (2) $\alpha = 0.10$, reject if $|Z_{\text{calc}}| > 1.96$
    (3) $p = 128/140 = 0.914$
    $$\sigma_p = \sqrt{(0.95)(0.05)/140}$$
    $$= 0.0184$$
    $$Z_{\text{calc}} = \frac{0.9143 - 0.950}{0.0184} = -1.94$$
    (4) Reject $H_0$: $\pi = 0.95$ because $|Z_{\text{calc}}| = |-1.94| = 1.94 > 1.645$.
49. Reject $H_0$: $\pi = 0.5$ because $Z_{\text{calc}} = -2.00 < -1.96$.
51. Reject $H_0$: $\mu = 85$ because $Z_{\text{calc}} = -2.33 < -1.645$. The average is $<85$.
53. Do not reject, because $t_{\text{calc}} = 1.54 < 1.833$.
55. $t_{\text{calc}} = -5.10$
57. Do not reject, because $t_{\text{calc}} = -1.342 > \pm 1.729$.
59. (1) $H_0$: $\mu = 30$ mo.   $H_A$: $\mu \neq 30$ mo.
    (2) $\alpha = 0.01$, reject $H_0$ if $|t_{\text{calc}}| > 3.250$
    (3) $t_{\text{calc}} = \dfrac{\bar{X} - \mu}{s_{\bar{X}}} = \dfrac{25.0 - 30.0}{1.51} = -3.32$
    where $s_{\bar{X}} = \dfrac{s}{\sqrt{n}} \sqrt{\dfrac{N-n}{N-1}} = \dfrac{5}{\sqrt{10}} \sqrt{\dfrac{100-10}{99}}$
    $$= 1.508$$
    (4) Reject $H_0$, since $t_{\text{calc}} = -3.317 < -3.250$. We assume battery lives are normally distributed and random sampling.
61. a. No, because $Z_{\text{calc}} = -1.29 > -1.645$.
    b. 0.1131
63. a. $H_0$: $\mu = \$326$   $H_A$: $\mu > \$326$
    b. $C_1 = \$326 + 1.645\,(10/\sqrt{n})$
    c. $C_1 = \$330 - 1.28\,(10/\sqrt{n})$
    d. $n = 54$, $C_1 = \$328.24$
65. a. 0.500   b. 0.0227   c. $n \cong 189$

## CHAPTER TEN

1. Do not reject $H_0$: $\mu_d = 0$ since $t_{\text{calc}} = 2.31 < 2.821$. The weight gain is not statistically significant.
3. Reject $H_0$: $\mu_d = 0$ since $t_{\text{calc}} = 3.90 > 2.306$. The new station does facilitate a faster work rate.
5. a. 12.592   b. 3.940   c. 36.191
7. a. 28.412   b. CL's = 2.74 to 22.02   c. 165 to 4.69

9. Do not reject $H_0$: $\sigma_1^2 = \sigma_2^2$ because $F_{\text{calc}} = 1.41 < 3.438$. The variances are not significantly different.
11. Do not reject $H_0$: $\sigma_1^2 = \sigma_2^2$ because $F_{\text{calc}} = 2.216 < 5.050$. The variances are not significantly different.
13. Reject $H_0$: $\mu_1 - \mu_2 = 0$ since $t_{\text{calc}} = -3.61 < -2.048$.
15. a. $s_1^2 = 3.073$, $s_2^2 = 0.923$
    b. Calculations are correct (within rounding error).
17. No. The test does not reveal any significant difference because $t_{\text{calc}} = 1.06 < 2.365$. However, due to the small sample size there should be some concern about type 2 error.
19. Reject $H_0$: $\mu_d = 0$ since $t_{\text{calc}} = 2.60 > 1.860$. The first rating is significantly higher.
21. a. Do not reject $H_0$: $\mu_d = 30$ because $t_{\text{calc}} = 1.45 < 2.131$. The improvement in scores is not significantly over 30 points.
    b. 29.1 to 34.9 points
23. Reject $H_0$: $\mu_1 - \mu_2 = 5$ since $t_{\text{calc}} = 2.15 > 1.282$. The time required by team 1 is significantly more than 5 min. longer than that for team 2.
25. a. Do not reject $H_0$: $\sigma_1^2 = \sigma_2^2$ because $1.78 < 2.07$. Therefore we can test for equality of means.
    b. Means are not significantly different, since $1.72 < 1.96$. (The pooled variance $= 50.65$.)
27. Reject $H_0$: $\mu_d = 0$ since $t_{\text{calc}} = 3.30 > 2.262$. Method 1 gives a higher overcharge, so choose Method 2.
29. a. Variances are equal since $F_{\text{calc}} = 1.08 < 3.15$ $(\doteq F_{0.025,13,13})$.
    b. Do not reject equality of means since $-2.056 < t_{\text{calc}}$ of $-1.79 < 2.056$. However, this is a small sample, so be concerned with $\beta$-risk.
    c. Insofar as the test of equality of means (in b.) was not rejected, we should expect that the difference can be zero. The interval of from $-12.9$ to $+0.9$ supports this, and either men or women could have a higher score. To retest with a larger sample would be appropriate because the smaller samples suggest a high $\beta$-risk for reasonable alternative values.

## CHAPTER ELEVEN

1. a. cans of a cola   b. machines A and B
   c. the difference in means of time to seal pop cans
   d. Is the average (mean) time to fill pop cans the same for machines A and B?
3. Variances (in $F$-ratio statistic) are calculated from sums of squares (SS) with all components squared numbers. So for SS $> 0$, the least possible value for the $F$-statistic is zero.
5. a. The data for mailing methods 1 and 2 show different levels, hence different means. So some, possibly all, of the means appear different.

b. We believe the brands A, B, C, means are not statistically different. A subsequent test will allow us to determine if the difference *among* means is large relative to the variation *within* the individual brands.

7. All (3) means appear to be statistically equal, but a test for equal means is needed to make a conclusion.

9. Since $F_{calc} = 12.04$ (given) exceeds the table $F$ of 5.29, reject $H_0$ and conclude that *at least* one treatment mean does not equal the others.

11. $H_0: \mu_1 = \mu_2 = \mu_3$   $H_A$: At least one $\mu_i \neq$ others. $F_{calc} = $ MST/MSE $= 131.67/6.17 = 21.35$
Conclusion: Reject $H_0$ because $21.35 >$ test $F = 5.10$. At least one mean $\neq$ others.

13. $F_{calc} = $ MST/MSE $= 1425/251 = 5.68$   and   $F_{test} = F_{0.05,4,10} = 3.48$
Reject $H_0$ because $F_{calc} = 5.68 > 3.48$. At least one machinist has a mean production rate that is different from the others.

15. a. $H_0: \mu_{PR} = \mu_{No\ PR}$   $H_A: \mu_{PR} \neq \mu_{No\ PR}$
$F_{calc} = $ MST/MSE $= 182.14/14.125 = 12.89$
Reject $H_0$ because $F_{calc} = 12.89 > 4.000$. Conclude that $\mu_{PR}$ does not equal $\mu_{No\ PR}$.
b. $H_0: \sigma_{PR}^2 = \sigma_{No\ PR}^2$   $H_A: \sigma_{PR}^2 \neq \sigma_{No\ PR}^2$
Test: Reject $H_0$ if $F_{calc} > 2.0$ ($\cong F_{\alpha/2,35,35}$).

$$F_{calc} = \frac{S_{No\ PR}^2}{s_{PR}^2} = \frac{(4.0)^2}{(3.5)^2} = 1.31$$

Conclusion: The two variances are statistically equal, $\alpha = 0.05$.

17. MSE $= [s_1^2 + s_2^2 + s_3^2]/3$
After expansion and collection of terms, we have

$$MSE = \frac{1}{3(4-1)} \sum_{}^{3} \sum_{}^{4} (Y_{ij} - \bar{Y}_i)^2.$$

19. a.

| Source of variation | df | Sum of Squares | Mean Squares |
|---|---|---|---|
| Treatments | 2 | 100.11* | 50.055 |
| Error | 69 | 1004.79 | 14.562 |
| Total | 71 | | |

*$1104.90 - 1004.79 = 100.11$

b. $H_0: \mu_1 = \mu_2 = \mu_3$   $H_A$: At least one $\mu_i \neq$ others.
c. Critical (test) $F$-value $= F_{0.05,2,69} \cong 3.15$
d. Calculated $F = $ MST/MSE $= 50.055/14.562 = 3.44$. At least one $\mu_i \neq$ others.

21.

| Source | df | SS | MS | F |
|---|---|---|---|---|
| Treatment groups | 3 | 120.0 | 40.0 | 4.21 |
| Error | 8 | 76.0 | 9.5 | |
| Total | 11 | 196.0 | | |

Test: Reject $H_0$ if $F_{calc} > F_{0.01,3,8} = 7.59$.
Decision: Do not reject. Means are statistically equal.

23.

| Source | df | SS | MS | F |
|---|---|---|---|---|
| Treatment groups | 2 | 9,209.08 | 4,605.54 | 6.04 |
| Error | 15 | 11,433.53 | 762.24 | |
| Total | 17 | 20,642.61 | | |

Test: Reject $H_0$ if calculated $F > 3.68$ ($= F_{0.05,2,15}$).
Conclusion: Reject $H_0$ since $F_{calc}$ (6.04) $> F_{test}$ (3.68). The three means are not all equal, $\alpha = 0.05$.

25. a. For unequal numbers $j = 1, 2, 3, \ldots, n_i$ while for equal numbers $j = 1, 2, 3, \ldots, n$ (same $n$ in each group).

b.

| Equal Numbers | df | Unequal Numbers | df |
|---|---|---|---|
| Treatments | $T - 1$ | Treatments | $T - 1$ |
| Error | $nT - T$ | Error | $\sum_{}^{T} n_i - T$ |
| Total | $nT - 1$ | Total | $\sum_{}^{T} n_i - 1$ |

c. no change
d. The error SS calculation can be simplified with equal numbers, i.e.,

$$MSE = \frac{[(n_1 - 1)s_1^2 + (n_2 - 1)s_2^2 + \cdots + (n_T - 1)s_T^2]}{(n_1 - 1) + (n_2 - 1) + \cdots + (n_T - 1)}$$

with $n_1 = n_2 = \cdots n_T$

$$MSE = \frac{(n - 1)s_1^2 + (n - 1)s_2^2 + \cdots + (n - 1)s_T^2}{(n - 1) + (n - 1) + \cdots + (n - 1)}$$

$$= \frac{s_1^2 + s_2^2 + \cdots + s_T^2}{T}$$

Basically, the arithmetic for MSE calculation is grossly lessened with equal numbers.

27. a. Calculated $F = 38.2/18.6 = 2.05$ exceeds test $F_{0.025,11,17} \cong 2.87$. Conclude $\sigma_{Mex}^2 = \sigma_{Brazil}^2$, for $\alpha = 0.05$.
b. SST $= 141.12$, SSE $= 736.4$
MST $= 141.2/1 = 141.12$,
MSE $= 736.4/28 = 26.30$
$H_0: \mu_{Mex} = \mu_{Brazil}$   $H_A: \mu_{Mex} \neq \mu_{Brazil}$
$F_{calc} = $ MST/MSE $= 5.37 > F_{test} = 4.20$ ($F_{0.05,1,28}$)
Conclusion: Reject $H_0$ and conclude that average performance is *not* the same for these 2 areas.

29. MST $= [13(16 - 14.7)^2 + 10(13 - 14.7)^2]/1 = 50.87$ and MSE $= [(13 - 1)28 + (10 - 1)7]/(13 + 10 - 2) = 19.0$ are correct. Other is correct, so we agree with the conclusion: $\mu_1 = \mu_2$.

31. The tabular $F$-value has incorrect $df_2$. Use $F_{0.05,4,43} = 2.61$ with $F_{calc} = 5.1$, so reject $H_0$.

33. a. For a RCB design, $F_{calc} = MST/MSE = 24/3.2 = 7.5$ exceeds $F_{test} = 3.44$ ($F_{0.05,2,22}$), so reject $H_0$.

 b. In comparison, the RCB design has MSE $= 3.2$ with $df_2 = 22$, while the CRD model has MSE $= 4.1$ with $df_2 = 24$. Relative to the difference in MSE, the $df_2$ values are very close, so for the RCB design, more variations are explained with minimal loss to df. Thus it gives a more precise test.

35. $H_0: \mu_1 = \mu_2 = \mu_3$    $H_A$: At least one $\mu_i \neq$ other.

$$F_{calc} = \frac{\dfrac{53.583}{2}}{\dfrac{13.415}{33}} = 58.18 > 3.33 \; (= F_{0.05,2,29})$$

Conclusion: Reject $H_0$. At least one mean does not equal others.

37. $H_0: \mu_1 = \mu_2$    $H_A: \mu_1 \neq \mu_2$ (where $1 =$ incentive plan and $2 =$ non-incentive)
$F_{calc} = 1{,}821{,}307/(1{,}142{,}756/7) = 11.2$
Conclusion: Reject $H_0$, because $F_{calc} = 11.2 > 5.59$. Means are not equal; the incentive plan group has the higher average.

39. a. 3 levels, 3 levels    b. crossover
 c. nonparallel
 d. Yes. Special outperforms the expanded and regular displays at all given levels.

41. a. $F_{calc} = 4.49$ (given) $> 2.93$ ($F_{0.05,4,18}$)
 There is significant interaction for price and display.
 b. Expanded and regular displays appear to have no interaction. There is a difference of extent (nonparallelism) for special display versus the others.

43. See plot of mean production values. There is a small interaction of Model 1 with Model 3, but it does not appear to be significant.

For $H_0$: No significant interaction, $F_{calc} = 1.607$ is not statistically significant. Interaction can be pooled with the residual for tests on main effects.

For $H_0$: Mean production is equal for the models, $F_{calc} = 25.9$ and exceeds $F_{0.05,2,30}$ of 3.32, so there is a significant difference in mean production for models.

For $H_0$: Mean production is equal for robot lines, $F_{calc} = 15.7$ and exceeds $F_{0.05,3,30}$ of 2.84, so there is a significant difference in mean production for robot lines.

Because interaction effect is not significant, the above results are applicable with no (two-way) interaction conditions.

45. a. $H_0: \mu_{high} = \mu_{low}$
 $F_{calc} = 142.86/3.72 = 38.40 > 3.92$ ($F_{0.05,1,139}$)
 Conclusion: Reject $H_0$. Mean ratings are not the same.
 b. Technically, yes. Practically, no. If two means are significantly different, one is larger and one is smaller. This is displayed.

47. a. $H_0: \mu_1 = \mu_2 = \cdots = \mu_6$
 $H_A$: At least one $\mu_i \neq$ others.
 Test: Reject $H_0$ if $F_{calc} > 2.71$ ($F_{0.05,5,20}$).
 Decision: Reject $H_0$. Some means differ.

 b.

| Brand | 1 | 4 | 3 | 6 | 5 | 2 | |
|-------|-----|------|------|------|------|------|--------|
| Mean | 8.3 | 11.7 | 12.4 | 13.3 | 16.3 | 21.6 | $(\bar{Y}_i)$ |

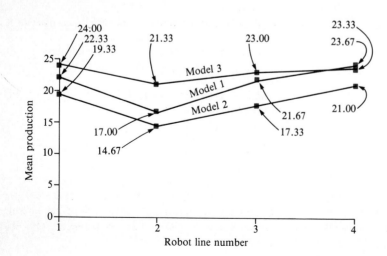

by Equation 11.7:

$$(\bar{Y}_i - \bar{Y}_j) \pm \left[\sqrt{(T-1)F}\ \sqrt{MSE\left(\frac{1}{n_i} + \frac{1}{n_j}\right)}\right]$$

$$(\bar{Y}_i - \bar{Y}_j) \pm \left[\sqrt{(5)(2.71)}\ \sqrt{19.20\left(\frac{1}{5} + \frac{1}{5}\right)}\right]$$

$$(\bar{Y}_i - \bar{Y}_j) \pm 10.20$$

To be significantly different, two means must differ by more than 10.20. Then $2 > 1$ and all other means are *not* significantly different, for $\alpha = 0.05$.

c. The highest mean is for brand 2. For this application use brand 2.

49. a. from Problem 41   MSE = 59.444 and $F_{0.05,4,18} = 2.93$, $n = 12$

$$(\bar{Y}_i - \bar{Y}_j) \pm \left[\sqrt{(T-1)F}\ \sqrt{MSE\left(\frac{1}{n_i} + \frac{1}{n_j}\right)}\right]$$

$$\mu_{reg} - \mu_{red}: (28.17 - 44.42) \pm 7.62$$
$$- 23.87 < \mu_{reg} - \mu_{red} \le 8.63$$

b. Because there is a significant interaction effect, the multiple comparisons do not have a consistent base.

51. a. $H_0: \mu_1 = \mu_2 = \mu_3$   $H_A$: At least one $\mu_i \ne$ others.
b. $df_1 = 2$, $df_2 = 3(20 - 1) = 57$
c. $F_{0.01,2,57} \cong 4.98$
d. Do not test for differences in the (3) individual means, because the test on $\mu_1 = \mu_2 = \mu_3$ does *not* reject equal means.

53.

| Source | df | SS | MS | F |
|---|---|---|---|---|
| Brands | 2 | | 500 | 6.0 |
| Error | 12 | | 83.33 | |
| Total | 14 | | | |

Since $F_{calc} = 6.00 > 3.89$ ($F_{0.05,2,12}$), reject $H_0$ of equal means. So decide that mean lifetimes are not the same for all (3) battery brands.

55. a. $H_0: \sigma_1^2 = \sigma_2^2$   $H_A: \sigma_1^2 \ne \sigma_2^2$
$F_{calc} = 0.70/0.50 = 1.40 < 1.60 \ (\cong \text{test } F_{\alpha/2,100,64})$
Conclusion: Do not reject $H_0$ that $\sigma_1^2 = \sigma_2^2$ at $\alpha = 0.05$ level.

b. $H_0: \mu_A = \mu_B$   $H_A: \mu_A \ne \mu_B$
Test: Reject $H_0$ if $|t_{calc}| > 1.96$ ($t_{0.025,1,62} \approx Z_{0.025}$).

$$t_{calc} = \frac{(1.81 - 1.70) - 0}{\sqrt{\dfrac{0.40}{64} + \dfrac{0.40}{100}}} = 1.09$$

(an equivalent $F_{calc} = 1.09^2 = 1.2$, with test $F = 1.96^2 = 3.81$)

Conclusion: Do not reject $H_0$. The mean elastic strength is statistically equal for these 2 methods (treatments) at $\alpha = 0.05$.

57. $H_0: \mu_1 = \mu_2 = \mu_3$   $H_A$: At least one $\mu_i \ne$ others. ANOVA and F-statistic

| Source | df | SS | MS | F |
|---|---|---|---|---|
| Groups | 2 | 980.58 | 490.29 | 1.48 |
| Error | 42 | 13,839.42 | 329.51 | |
| Total | 44 | | | |

Test: Reject $H_0$ if $F_{calc} > F_{0.05,2,42} = 3.23$.
Conclusion: Do not reject. There is equal achievement on the productivity measurement.

59. By Scheffe's multiple-comparisons procedure, any means that differ by more than
$$\pm\left[\sqrt{(T-1)F_\alpha}\sqrt{MSE(1/n_A + 1/n_B)}\right] = \pm 3.17 \text{ are con-}$$
sidered statistically unequal. Only the Nebraska and Illinois sample means differ by an amount that is greater than 3.17. Illinois has a mean purchasing value that is higher than for Nebraska, $\alpha = 0.025$.

61. By Scheffe's procedure, with $n_a = n_b = n_c = n_d = 6$, then a significant difference in means requires
$$|(Y_A - Y_B)| > \left[\sqrt{3 \cdot 4.15}\ \sqrt{9.9(1/6 + 1/6)}\right]$$
$$> 6.4$$

| Outlet | a | d | c | b |
|---|---|---|---|---|
| Mean | $580 | $599 | $604 | $612 |

Only outlets $d$ and $c$ had equal means. Outlet $b$ has the highest mean sales, outlet $a$ the lowest.

63. For White House apple juice with MSE = 92.467, df = 30, and $F_{0.05,2,30} = 3.32$:
a. regular vs. expanded:
$-4.54 < \mu_{exp} - \mu_{reg} < 15.70$ (cents)
b. regular vs. special:
$9.38 < \mu_{special} - \mu_{reg} < 29.62$ (cents)
c. expanded vs. special:
$14.21 < \mu_{special} - \mu_{exp} < 34.45$ (cents)
Conclusion: The special display has produced a significantly higher unit sales than either the expanded or the regular display.

65. a. $H_0: \mu_{PDM1} = \mu_{PDM2}$
$F_{calc} = 12.28$ (first calculation)
$F_{test} = F_{0.05,1,221} \cong 3.84$
$H_0$: There is no three-way interaction.
$F_{calc} = 4.02$
$F_{test}$ is the same as above (3.84).
b. Don't report the result because the three-way interaction is also significant and this confuses the meaning of significant main effects.

# CHAPTER TWELVE

1. See figure.

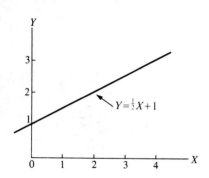

3. See figure.

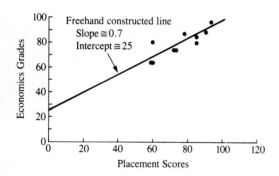

5. a. $A$    b. $A, B$    c. $Y_i = c + dX_i + fX_i^2 + e_i$
   The points display curvature.
7. a. $\hat{Y} = -14 + 4X$
   b. Yes, the point of means $(\bar{X}, \bar{Y})$ is a point on the least squares regression line.
   c. 8
9. See figure.

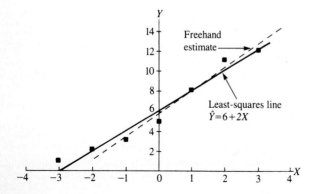

11. $b_0 = -4$, $b_1 = 4.5$
13. a. See figure.

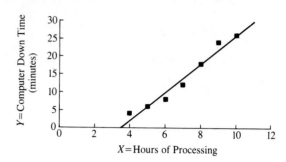

*Note:* The plot is the least squares regression line.
   b. $\hat{Y} = -14 + 4X$ is the same as Problem 7a.
15. See figure. Linear is a good fit. The residuals are flat around $\bar{e} = 0$.

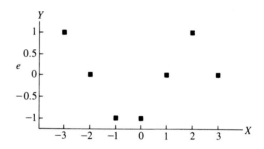

17. 14.14
19. $s_e = 0.89$
21. a. 4    b. $b_1 = 0.53$; $-0.13 < \beta_1 < 1.19$
23. a. This is a $t$-test for zero slope. As $t_{\text{calc}} = 2.82 > t_{0.025,23} = 2.07$, conclude that there is a meaningful linear regression.
    b. $0.5 < \beta_1 < 3.5$   The average increase in production is estimated between $\frac{1}{2}$ and $3\frac{1}{2}$ units for each additional year of experience.
    c. Does not apply. The linear relation is meaningful.
25. $Z_{\text{calc}} = 4.21$ exceeds $|\text{test } Z| = 1.96$.
    Conclude that customer names for this antique collectors list can be used to predict participation in Continental Tours.
27. $90.8 < \mu_{Y|X=10} < 93.2$
29. a. $\mu_{Y|X=60}$: $70.96 \pm 1.96$
    The average production rating for test score = 60 has error $\pm 1.92$ points.
    b. $Y_{X=60}$: $70.96 \pm 9.48$
    Any individual production rating for test score = 60 has error $\pm 9.48$ points. Yes, Susan Ladd's production rating, 75, is within the prediction interval.

31. The error calculation (2.4) is incorrect.

33. $r = -0.5$

   $r^2 = 0.25$ indicates that 25% of the variation in $Y$-values is explained by a linear regression on $X$.

35. $r_1^2 = r_2^2 + 0.09$, so $r_1$ is stronger.

37. $r = 0.77$

   $r^2 = 0.59$ or 59% of the variation in $Y$ is explained by a linear regression on $X$.

39. No, correlation does not mean causation.

41. a. a, c, e    b. b

43. a. $r^2 = 0.90$ so higher sales is positively associated with higher temperatures.

   b. First causal requires $r^2 = 1$, which this is not. Also, all other possible "causes" are not excluded, e.g., advertising.

   c. $t_{calc} = 14.15 >$ test $t = 2.07$, conclude $\rho \neq 0$.

45. Only c is a correct statement.

47. Only b is true.

49. $X_2$ gives a stronger linear description, giving about 10% more explained differences by $X_2$ than by $X_1$.

51. a. $Y$, shipments

   b. $\hat{Y} = 30 + 2X$

      $X$ = weeks, $\hat{Y}$ = estimated shipments

   c. cartons    d. 22 cartons    e. 70 cartons

   f. Confidence limits are $65 < \mu_{Y|X=20} < 75$ cartons.

   g. Prediction limits are $0 < Y_{X=5} < 84$ cartons.

      Observe the wider interval for prediction limits.

53. See figure.

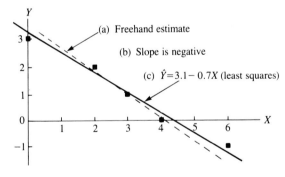

(a) Freehand estimate

(b) Slope is negative

(c) $\hat{Y} = 3.1 - 0.7X$ (least squares)

55. See figure. Yes, a linear trend does appear reasonable.

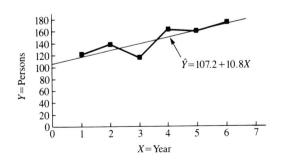

$\hat{Y} = 107.2 + 10.8X$

$X$ = Year

57. a. 33 ovens    b. $30 < \mu_{Y|X=80} < 36$ ovens

   c. $r^2 = 0.5$

   d. 50% of the variation in ovens sold is explained by advertising.

59. $r = 0.91$

   $100r^2 = 83\%$ of the variation in final grades is described by midterm scores.

61. $b_0 = 2.7$, $b_1 = 2.1$

   a. $\hat{Y} = 2.7 + 2.1X$    b. $\hat{Y} = 11.1$

   c. $r^2 = 0.96$, so 96% of variation in heights (trees) is explained by the age of a tree.

# CHAPTER THIRTEEN

1. a. $\beta_0 = 0$, $\beta_1 = 2$, $\beta_3 = 3$

   b. $\beta_0 = 2$, $\beta_1 = \frac{1}{2}$, $\beta_3 = 1$

3. a. $E(Y|X_1X_2) = 3 + 0.5X_1 + 2X_2$

   b. 3 is $Y$ intercept

      0.5 or $50,000 estimated increase in profits per $100,000 of investments

      2.0 or $20,000 estimated increase per $10,000 of payroll

5. a. $b_1 = 2.0$

   b. $b_0 = -6$ is the $Y$-intercept

      $\hat{Y} = -6 + 2X_1 + 5X_2$

      2 is the slope in $X_1Y$ plane

      5 is the slope in $X_2Y$ plane

7. a. $n = 22$    $\sum Y = 569$

      $\sum X_1X_2 = \sum X_2X_1 = 3356.20$

   b. $569 = 22b_0 + 1132b_1$

      $32603 = 1132b_0 + 75570b_1$

   c. $b_0 = -0.9817$

9. a. $b_0 = 2.48$ ($Y$-intercept) has no physical meaning.

      $b_1 = 1.14$ means that for each additional quarter the company is in the red, the expected time to show increased profits goes up 1.14 quarters.

      $b_2 = -0.50$ means that for each additional year the CEO holds office, the expected time to show increased profits goes down by 0.50 quarters.

   b. $\hat{Y} = 4.32$ quarters

      Expect profits to begin in $4+$ quarters.

11. a. $b_2 = 0.208$

   b. $Y = \$12,787$ is the estimated price for this make of van that is 3 years old.

13. a. $b_1$ is dollars response to dollars for advertising expenditures, $X_1$

      $b_2$ is dollars response for worker-hours, $X_2$

      Because they have different units, $b_1$ and $b_2$ give no comparison of the relative strength for $X_1$ and $X_2$.

   b. Use beta coefficients that cancel units giving a pure number. Then compare $beta_1$ and $beta_2$ for $X_1$ and $X_2$, respectively.

15. Use $k$ = number of independent variables.
$n$ = sample size with $n > k \geq 1$
Then $\dfrac{n-1}{n-k} \geq \dfrac{n-1}{n-1} (=1)$ in $R_a^2$ and $R^2$ formulas gives
$R^2 \geq R_a^2$.

17. $\dfrac{MSE_2}{MSE_1} = 0.67$
The scatter about the regression is less for model 2.

19. a. See figure.

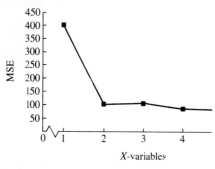

X-variables

b. Check where the MSE begins to level off against computer-generated tests for a reasonable number of $X$-variables in the model. 4 looks best.

21. a. Reject $H_0$ as $F_{calc} = 99.5$ exceeds $F_{test} = 2.92$ ($F_{0.10,3,8}$).
b. 0.97, 0.98
c. $-0.69 < \beta_2 < 3.33$, so both plus and minus values indicate a weak relation ($\beta_2$ could $= 0$) for response on $X_2$.

23. a. $\hat{Y} = 8.9406$ is estimated warp strength.
Given $(X_1, X_2) = (20, 60)$, the estimated average warp strength is within the range 8.10 to 9.78, for $\alpha = 0.02$.
b. Given $(X_1, X_2) = (20, 60)$, the next sheet produced will have warp strength between 7.08 and 10.8, for $\alpha = 0.02$. (This is a wider interval than for part a.)

25. a. $\hat{Y} = \$494.32$ is the estimated dollars liability claims.
b. $b_1 = -\$4.29$ average claim liability decreases for more parcels picked up in Council Bluffs.
$b_2 = +\$18.52$ average claim liability increases for more parcels picked up in Omaha.

27. a. $s_e = 7.54$
b. Yes, because $F_{calc} = 8.26 > 5.14 (= F_{0.05,2,6})$.

29. a. The regression in $X_1$ and $X_2$ is meaningful by the ANOVA and $F$-test using

| Source | df | SS | MS |
|---|---|---|---|
| Regression | 2 | 236.27 | 118.14 |
| Error | 5 | 60.60 | 12.12 |
| Total | 7 | 296.87 | |

Test: $F_{calc} = 9.75 > F_{Tab} = 5.79$.

b. Residual = (Actual − Estimated)
$= (17.00 - 17.03) = -0.03$

31. a. Figure (a) displays nonconstant variance in $X$, there is not a pattern in time. Use weighted least squares.
b. There is a trend in $X$ and a curvilinear pattern in time. Test (to add) terms through $X^3$ and $t^2$.

33. $\beta_2 \neq 0$, for $\alpha = 0.05$, so $X_2$ enters the regression model.
$\beta_{1 \cdot 2} = 0$, for $\alpha = 0.05$ by $F_{calc} = 9.43/112.50 = 0.08$, which is not significant. $X_1$ does not add significant description beyond $X_2$.

35. a. For $D = 1.31$, the test is inconclusive.
b. Take a bigger sample, $n = 15$ is not reliable.

37. Step 1: $H_0: \beta_1 = 0$    $F_{calc} = 47.03$
Reject $H_0$.    $X_1$ enters the model.
Step 2: $H_0: \beta_{2 \cdot 1} = 0$    $F_{calc} = 68.26$
Reject $H_0$.    $X_2$ enters the model.
Step 3: $H_0: \beta_{3 \cdot 21} = 0$    $F_{calc} = 0.10$
Do not reject $H_0$.
$X_3$ does not enter the model.
Best model by forward selection:
$Y = \beta_0 + \beta_1 X_1 + \beta_2 X_2 + \epsilon$

39. $H_0: B_{S \cdot E} = 0$    $F_{calc} = 12.16$    Reject $H_0$.
$H_0: B_{D \cdot SE} = 0$    $F_{calc} = 5.26$    Do not reject $H_0$.
The calculations/results are the same as for Examples 13.8 and 13.9.

41. Observation 8 is *not* a high-leverage point because it is in the middle of the range for both LABOR and ADVT variable values. See the two graphs. It is an outlier. The Minitab output indicates this, and by our rules STRESID $= 2.15 (>2)$ indicates an outlier. This is clearly an influential observation. Without this point (8), the data are highly collinear (labor and advertising costs with sales). This is quantified in R-sq $= 79.6\%$ (with 8), while R-sq $= 97.9\%$ (excluding 8). Other indicators of this difference are big changes in the $p$-value, the $Y$-intercept, and the fit or predicted values. The observation should be investigated for exclusion.

43. a. 8    b. $t_{calc} = \dfrac{\hat{Y} - E(Y|1,8)}{s_{\hat{Y}}} = -0.66$
Conclude that $E(Y|1,8) = 10$.

45. a. Normal distribution is questionable.    b. Yes

47. Step 1: $H_0: \beta_1 = 0$
$F_{calc} = 104.2$ is significant for $\alpha = 0.05$.
Include $X_1$.
Step 2: $H_0: \beta_{2 \cdot 1} = 0$
$F_{calc} = 12.42$ is significant for $\alpha = 0.05$.
Add $X_2$ after $X_1$.
Step 3: $H_0: \beta_{3 \cdot 21} = 0$
$F_{calc} = 0.95$ is not significant for $\alpha = 0.05$.
Exclude $X_3$.
Best linear (first-order) model by forward-stepping procedure with $\alpha = 0.05$ is $Y = \beta_0 + \beta_1 X_1 + \beta_2 X_2 + \varepsilon$.

49. $H_0$: $\beta_4 = 0$   $F_{calc} = 177.86$
    $X_4$ enters the model.
    $H_0$: $\beta_{1 \cdot 4} = 0$   $F_{calc} = 95.34$
    $X_1$ enters after $X_4$.
    $H_0$: $\beta_{2 \cdot 14} = 0$   $F_{calc} = 1.32$
    $X_2$ does not enter after $X_1$, $X_4$.
    A best model by forward-selection is
    $Y = \beta_0 + \beta_1 X_1 + \beta_4 X_4 + \epsilon$.
    $Y$ = gasoline consumption (gallons)
    $X_1$ = kilowatt-hours of electricity
    $X_4$ = total energy (BTU)
    A scree plot displays the $2X$ − variables, $X_1$, $X_4$ and gives a near-minimum MSE. This agrees with the regression testing results.

51. a. $2.71 < E(Y|6,10) < 5.93$
    b. $0.48 < Y < 8.16$
       This would apply to my company that 1 to 8 quarters (a wider interval) is predicted for the time to realize an increase in profits, compared to 3 to 6 quarters on average to realize increased profits. All numbers are rounded here.

## CHAPTER FOURTEEN

1. a. trend   b. seasonal   c. cyclical   d. random
3. See figure. Actual is solid line; cyclical is dashed line. We can observe common patterns in the actual. The business cycle smooths to show a long-term recurring pattern.

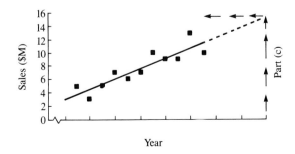

5. a. See figure. Yes, a linear fit is appropriate.
   b. Trend line is shown solid.
   c. Projected value is about $13 M.

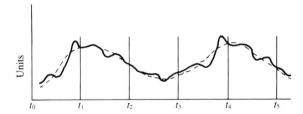

7. a. See figure.

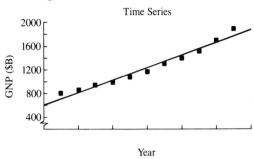

   b. $\hat{Y} = \$1{,}242 + 106.4 X_t$
      (1987 = 0, $X$ = yrs., $Y$ = $B)
   c. $3,370B
9. $\hat{Y} = 493 + 2X_t$ (Jan. 93 = 0, $X$ = mos., $Y$ = motors/yr.)
11. $\hat{Y} = 1570.931 + 13.605 X_t + 8.398 X_t^2$
    (1987 = 0, $X_t$ = yr., $Y$ = $B)

    | | |
    |---|---|
    | 1983 = $1,650.9 | 1988 = $1,592.9 |
    | 1984 = $1,605.7 | 1989 = $1,631.7 |
    | 1985 = $1,577.3 | 1990 = $1,681.3 |
    | 1986 = $1,565.7 | 1991 = $1,759.3 |
    | 1987 = $1570.9 | 1992 = $1,780.9 |

13. a. It may be satisfactory if no trend is present.
    b. Yes, because less evidence of a trend is present.
    c. To obtain more smoothing of the forecast.
15. a. 2,400 gal./quarter   b. 1680 gal./quarter
17. $Q_1 = 96.9$   $Q_2 = 140.1$   $Q_3 = 79.1$   $Q_4 = 83.9$
19. $Q_1 = 933.7$ (thousand)   $Q_2 = 942.9$ (thousand)
    $Q_3 = \$997.8$ (thousand)
21. a. See figure.

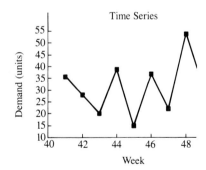

    b. Points appear to be random along a linear pattern.
    c. First-order exponential smoothing would help smooth the random deviations. Week 51 = 32 units.
23. a. For $\alpha = 0.1$, values are $-8.00$, $-15.20$, $5.32$, $-19.21$, $4.71$, $-10.76$, $22.31$, $-1.92$, $-20.73$.
    b. For $\alpha = 0.5$, values are $-8.00$, $-12.00$, $13.00$, $-17.50$, $13.25$, $-8.38$, $27.81$, $-8.09$, $-23.05$.
       For $\alpha = 0.9$, values are $-8.00$, $-8.80$, $18.12$, $-22.19$, $19.78$, $-13.02$, $30.70$, $-18.93$, $-20.89$.

c. MSE values are for

$\alpha = 0.1$  MSE $= 196$;     $\alpha = 0.5$  MSE $= 255$;
$\alpha = 0.9$  MSE $= 360$

Therefore, $\alpha = 0.1$ is smallest error and best coefficient.

25. a. yr. 1 $= 100$  yr. 5 $= 133$

yr. 2 $= 108$  yr. 6 $= 145$

yr. 3 $= 144$  yr. 7 $= 141$

yr. 4 $= 122$  yr. 8 $= 149$

b. 116

27. 213.7

29. a. 205     b. 114.5

31. $Q_{1991/1988} = 150$   $Q_{1992/1988} = 100$

33. $V_{1988/1988} = 100$   $V_{89/88} = 124$   $V_{90/88} = 210$
$V_{91/88} = 300$   $V_{92/88} = 200$

35. a. 150.0     b. 153.4     c. 102.3

37. a. 220     b. 200     c. 209.8

d. Relative to the Laspeyres, the Paasche index is lower. Fisher's, being the geometric mean of the two, helps overcome some of the bias resulting from using either base-year quantities (Laspeyres) or given-year quantities (Paasche).

39. $4,686.2 billion

41. a. Current deflated wage $= $3.95/hr. After one year it $= $3.67/hr.

b. The CPI index increase is 10.2% so the new wage would be ($6.00)(1.102) $= $6.61/hr., which is less than $6.75/hr. You are better off with the 0.75¢ increase.

43. intercept $= 7.52$, slope $= 0.99$

45. c

47. a. $864 (000)

b. $\hat{Y} = 144 + 6X$
(July 1, 1985 $= 0$, $X =$ mo., $Y =$ annual sales in $000)

c. $216 (000)

49. a. See figure.

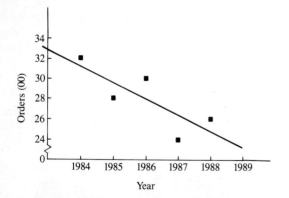

b. $\hat{Y} = 28 - 1.6X_t$
(1986 $= 0$, $X =$ yrs., $Y =$ orders in hundreds)

c. $\hat{Y} = 24.8 - 1.6X_t$
(1988 $= 0$, $X =$ yrs., $Y =$ orders in hundreds)

51. Yes, the deseasonalized trend value is higher.

53. a. T     b. F     c. T     d. F     e. F

55. 177.8

57. a. 112.2     b. 98.9     c. 105.3

59. 19% decrease

61. a. 5,087 units

b. Yes, the actual exceeded the index amount.

## CHAPTER FIFTEEN

1. Chart 1 is in control, is subject to common cause, and shows variation.

Chart 2 is out of control, is subject to common cause, and shows special causes and other variations.

Chart 3 is out of control, is subject to common cause, shows a system change, and shows other variations plus a trend.

3. a. $\bar{p} = 0.0221$   $\hat{\sigma}_p = 0.00735$

b. Control limits $= 0.00005$ to $0.0441$

c. See figure. Process appears to be in control, but with two points outside the limits it is open to question.

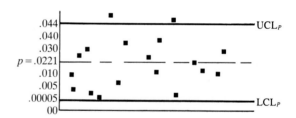

5. $\bar{p} = \dfrac{108}{28 \times 100} = 0.03857\ (3.9\%)$

$\hat{\sigma}_p = \sqrt{\dfrac{0.039 \times 0.961}{100}} = 0.01939\ (1.9\%)$

New control limits are $\text{UCL}_p = 9.6\%$, $\text{LCL}_p = 0$. All observations are within these limits. The process is in control.

7. a. For a Poisson process,

$$\bar{C} = \frac{89}{20} = 4.45$$

$$(\text{LCL}_{\bar{C}}, \text{UCL}_{\bar{C}}) = 4.45 \pm 3\sqrt{4.45} = (0, 10.3)$$

None of the points is outside these limits, so the process is in control.

b. The expected number of imperfections is the process mean, which is estimated by 4.45—so we should expect five imperfections.

9. Here are some categories (others are possible):

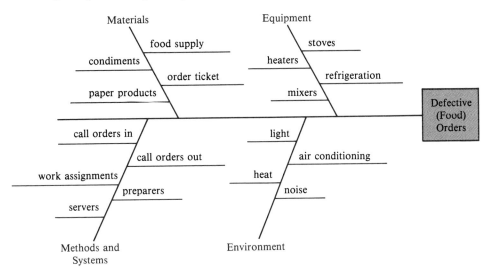

11. a. $\bar{C} = \dfrac{61}{30} = 2.0$ defects (per 100 books)

b. $\text{UCL}_{\bar{C}} = 2.0 \pm 3\sqrt{2} = (0, 6.2)$

c. The process is out of control. First, check the previous cause to be certain it remains corrected. Otherwise, look for another problem source.

13. a. $\bar{X}$-chart should have

$$\text{LCL}_{\bar{X}} = 9.5 - (1.203)(4.125) = 5.3$$
$$\text{UCL}_{\bar{X}} = 9.5 + (1.203)(4.125) = 13.7$$

$R$-chart should have

$$\text{LCL}_R = (4.125)(0) = 0$$
$$\text{UCL}_R = (4.125)(2.575) = 10.6$$

b. All points should lie well within these control limits (when graphed), so both the mean and the variation are in control. (Note, however, that there appears to be some downward drift in the mean.)

15. $\bar{X}_\triangle = 8.96 \quad \bar{X}_\bigcirc = 9.67 \quad \bar{X}_\blacksquare = 9.83$

$s_\triangle = 0.95 \quad s_\bigcirc = 2.16 \quad s_\blacksquare = 2.87$

Individual 3 ($\blacksquare$) had the highest mean and the highest standard deviation. Operator 3's work is quite variable. In this case, the manager wanted stability (as reflected by lower $s$-values), to better plan for scheduling the required number of operators for a work-period. Also, a high mean is not necessarily best, because customer satisfaction may require devoting time to answering questions, taking complete messages, and being personable. Generally the $\bar{X}$- and $R$-charts describe the process mean and variation. Individual variables go beyond the control charts.

17. $\bar{C} = \dfrac{160}{30} = 5.33$

$(\text{LCL}_{\bar{C}}, \text{UCL}_{\bar{C}}) = 5.33 \pm 3\sqrt{5.33} = 5.33 \pm 6.93$
$$= (-1.6, 12.26) \rightarrow (0, 12.26)$$

When graphed, all points should lie within the control limits, so this process is in control. However, the print manager's concern rests with the overall level of defects (breaks), which is too high. It appears that an improvement will require an engineering (systems) change.

19. For runs above A and below B, the process mean is $\bar{C}$ and the number of runs is $r = 22$, with A = 14 points above $C$ and B = 16 below $\bar{C}$. Testing "$H_0$: The process is random" against "$H_A$: The process is not random," we obtain $\mu_r = 15.93$ and $\sigma_r = 2.68$, so $Z_{\text{calc}} = (22 - 15.93)/2.68 = 2.26$. Conclusion: Reject $H_0$ since $2.26 > 1.645$. The process does not appear to be random, and the number of runs is in excess of what should be expected. This suggests a problem is present.

21. a. $\bar{\bar{X}} = 50.8 (\text{LCL}_{\bar{X}}, \text{UCL}_{\bar{X}}) = 50.8 \pm (0.729 \times 6.2)$
$$= (46.3, 55.3)$$

Yes, the process is in control for the mean. It does satisfy the industry standard of 50 grams $\pm$ 5 grams.

b. $\bar{R} = 6.2 (\text{LCL}_R, \text{UCL}_R) = (6.2 \times 0, 6.2 \times 2.282)$
$$= (0, 14.1)$$

The $R$-chart is in control but does not meet the industry standard, which is 5 grams $\pm$ 5 grams.

23. See figure.

$$\bar{p} = \frac{32}{24 \times 50} = 0.06$$

$$(\text{LCL}_p, \text{UCL}_p) = 0.06 \pm 3\sqrt{\frac{0.06 \times 0.94}{50}}$$

$$= 0.06 \pm 0.10 = (0.16)$$

The Day 7 value was deleted due to an assignable cause—rain. The control chart, excluding Day 7, is in control. Day 23, which is on the upper limit, could be checked for an assignable cause. Otherwise, this variation looks pretty much like random variation.

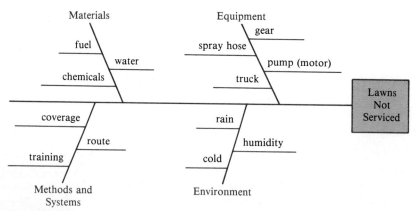

25.

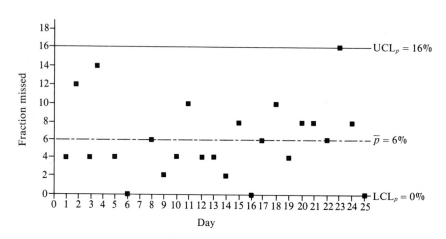

---

## CHAPTER SIXTEEN

1. 0.025    b. 0.250    c. 11.345
3. a. No    b. >43.194
5. a. 1    b. 3    c. 1
7. a. F    b. T    c. F    d. F
9. Do not reject $H_0$ of uniform distribution since $\chi^2_{\text{calc}} = 4.0 < 7.779$.
11. Do not reject $H_0$ of normal distribution since $\chi^2_{\text{calc}} = 3.024 < 9.488$.
13. a. T    b. F    c. F    d. T
15. a. $H_0$: Attitude toward contract is independent of position.

b. 1    c. 3.841

d.

|        | In Favor | Opposed |
|--------|----------|---------|
| Mgt.   | 70       | 30      |
| Labor  | 70       | 30      |

e. 2.381
f. Do not reject hypothesis of independence because 2.381 < 3.841. Attitude appears independent of position.
17. a. 24    b. 2
c. Yes, reject $H_0$ because $\chi^2_{\text{calc}} = 6.54 > 5.991$.

19. Do not reject $H_0$: Quality is independent of operator since $\chi^2_{calc} = 5.0 < 12.592$.
21. No. Reject $H_0$ of independence because $\chi^2_{calc} = 57.246 > 16.812$.
23. Do not reject $H_0$: $\pi = 0.40$ because $\chi^2_{calc} = 1.33 < 3.841$.
25. Yes. Reject $H_0$: $\pi = 0.75$ because $\chi^2_{calc} = 3.84 > 2.706$.
27. Reject $H_0$ of uniform distribution because $\chi^2_{calc} = 16.5 > 12.592$.
29. Do not reject $H_0$ of uniform distribution because $\chi^2_{calc} = 12.0 < 14.449$.
31. Do not reject $H_0$ that cure is independent of treatment because $\chi^2_{calc} = 3.383 < 3.841$. Test does not support use of the drug as a cold remedy.
33. Yes. Reject $H_0$ of equal proportions because $\chi^2_{calc} = 19.29 > 3.841$.
35. Yes. Reject $H_0$: $\pi = 0.50$ because $Z_{calc} = 5.00 > 1.96$.
37. Do not reject $H_0$: $\pi = 0.40$ because $|Z_{calc}|$ of $0.236 < 2.58$.
39. Choose B. Reject $H_0$: $\pi_A = \pi_B$ because $\chi^2_{calc} = 5.319 > 3.841$.

# CHAPTER SEVENTEEN

1. a. Reject $H_0$ of random sample because $Z_{calc} = 2.44 > Z_{table}$ value of 1.96. Order is not random.
   b. Do not reject $H_0$ since $r = 10$ and reject limits are $\leq 5$ and $> 11$. $\beta$-risk is a concern because of small sample.
3. Yes, pattern exists.
   a. Do not reject $H_0$ of random distribution at 5% level because $1.60 < 1.645$.
   b. Reject $H_0$ at 10% level because $1.60 > 1.28$. We can conclude that the number of defectives is increasing if we are willing to be wrong 10% of the time.
5. a. Reject $H_0$ of randomly distributed data because $Z_{calc} = |-5.43| > 1.645$. Data are not random.
   b. Do not reject $H_0$ because observed $r = 4 > r_0 = 3$. If $\alpha$-level were set so that $r_0 = 4$ were used, then the conclusion would be opposite. For such small samples, we can also expect a higher $\beta$-risk.
7. Do not reject $H_0$ of equal mileage because $Z_{calc} = 1.44 < 2.58$. Conclude difference in mileage.
9. Do not reject $H_0$ that percentages are equal because $Z_{calc} = 1.44 < 1.645$.
11. Reject $H_0$ of equal ages in the three industries because $H_{calc} = 8.42 >$ table value of $\chi^2_{0.05,2} = 5.991$.
13. a. Yes, the load averages are not significantly different. $H_{calc} = 5.94 < \chi^2_{0.05,3} = 7.815$
   b. The $F$-ratio test is not appropriate here because the distributions are know to be uniform rather than normal.
15. Reject $H_0$ of equal medians for the five groups of executives because $H_{calc} = 14.4165 >$ table value of $\chi^2_{0.025,4} = 11.143$. The years of service at retirement are significantly different. However, to identify which specific group (or groups) causes rejection of the $H_0$, use a Mann–Whitney test.
17. Reject $H_0$ that median closing time is 30 days and conclude that it is $<30$ days. $P(X \leq 6 | n = 20, p = 0.50) = 0.0577$, which is $<$ the $\alpha$ reject level of 0.10.
19. a. For the two-tailed test, do not reject $H_0$ that preference before $=$ preference after because calculated value of $P(X \geq 7 | n = 9, \pi = 0.50) = 0.0899 >$ the 0.05 in either tail of the distribution.
   b. For the one-tailed test, reject $H_0$ of equal preferences because $0.0899 < 0.10$.
21. Yes. Data supports $H_0$ of exponential distribution as $F(X) = 1 - e^{-x}$. Do not reject $H_0$ because the largest calculated $K$-value of $|-0.1221|$ is $<K_{table}$-value of 0.483.
23. a. Do not reject $H_0$ that data are normally distributed because largest calculated $K$-value of 0.2398 is $<K_{table}$-value of 0.563.
   b. For the chi-square goodness-of-fit test, reject $H_0$ that data are normally distributed because $\chi^2_{calc} = 28.611 >$ table value of $\chi^2_{0.05,2} = 5.991$.
25. Do not reject $H_0$ of zero correlation because $r_s = -0.73 >$ reject value of $-0.829$.
27. $\sum d_i^2 = 744.5$ and $Z_{calc} = 4.95$ as before. Reject $H_0$: $\rho_s = 0$ since $4.95 > 1.96$.
29. For this data $U_1 = 264$, $\mu_U = 300$, and $\sigma_U = 50.50$. Do not reject $H_0$: $\mu_A = \mu_B$ since $|Z_{calc}| = 0.71 < 1.96$.
31. For this data $\sum R_A = 138.5$ and $\sum R_B = 114.5$, so $U_1 = 48.5$ and $\sigma_U = 15.23$. Do not reject $H_0$: $\mu_A = \mu_B$ since $|Z_{calc}| = 0.788 <$ table value of 1.645. The two groups have comparable training.
33. a. Reject $H_0$ of equal lifetimes. $P(X \geq 8 | n = 10, \pi = 0.5) = 0.0547$ is $< \alpha = 0.10$, so conclude that company B's batteries last longer.
   b. For a two-tailed test, we could not reject since $0.0547 > 0.05$, so we would judge the lifetimes to be equal.
35. Do not reject $H_0$ of zero correlation since $|0.13| <$ the table value of 0.738. The correlation is not significantly less than zero at $\alpha = 0.025$ level.
37. Reject $H_0$ that means of the three groups are equal because $H_{calc} = 8.57 >$ table value of $\chi^2_{0.025,2} = 7.378$. The methods do not result in equal (means) number of words typed correctly.
39. Do not reject $H_0$ that group means are equal because $H_{calc} = 4.15 <$ table value of $\chi^2_{0.05,2} = 5.991$. It is reasonable to assume that these readings are from one population.

## CHAPTER EIGHTEEN

1. a. game theory   b. EVPI   c. state-of-nature
   d. uncertainty   e. payoff
3. a. 0.125   b. 0.875
5. $EMV_A = 232.0$   $EMV_B = 239.6$   $EMV_C = 208.0$
   Best choice is B.
7. a. Select A 60% of the time and B 40% of the time.
   b. (i) B   (ii) B   c. A   d. 10
   e. See figure.

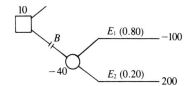

9. 11

11.

| $P(\theta)P(Z\|\theta)$ | $P(\theta\|Z)$ |
|---|---|
| 0.32 | 0.67 |
| 0.16 | 0.33 |
| 0.48 | 1.00 |

13.

| State | $P(\theta)$ | $P(F\|\theta)$ | $P(\theta)P(F\|\theta)$ | $P(\theta\|F)$ |
|---|---|---|---|---|
| U: Stock up | 0.2 | 0.5 | 0.10 | 0.147 |
| S: Stays same | 0.6 | 0.7 | 0.42 | 0.618 |
| D: Stock down | 0.2 | 0.8 | 0.16 | 0.235 |
| | | | 0.68 | 1.000 |

$P$(Stock up|Additional "favorable" info.) is decreased from 0.20 to 0.147 primarily because of poor history of the broker in recommending stocks.
15. a. 540   b. 350   c. 150
17. a. Advertise with an expected value of $180,000.
   b. EVPI = $40.000. Research cost is less than this.
   c. Given $Z_1$, $P(\theta_1|Z_1) = 0.500$, $P(\theta_2|Z_1) = 0.375$,
      $$P(\theta_3|Z_1) = 0.125$$
      Given $Z_2$, $P(\theta_1|Z_2) = 0.125$, $P(\theta_2|Z_2) = 0.625$,
      $$P(\theta_3|Z_2) = 0.250$$
      Given $Z_3$, $P(\theta_1|Z_3) = 0.0$, $P(\theta_2|Z_3) = 0.222$,
      $$P(\theta_3|Z_3) = 0.778$$
   EV (Research)
      $= \sum$ EV (Opt. act$|Z)P(Z)$
      $= 437.5(0.32) + 225(0.32) + 0(0.36)$
      $= \$212(000)$

d. $32,000
e. $27,000   Yes. Study should be conducted.
19. 50 utiles
21. 0.53
23. a. See figure.   b. He is a risk avoider.

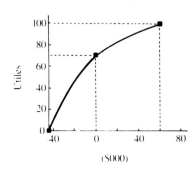

25. 70% of the time
27. a. 30% of the time
   b. $E(A_1) = \$12$   $E(A_2) = \$12$
   c. $24
29. 0.79
31. $P(\theta_1|Z) = 0.7143$   $P(\theta_2|Z) = 0.2143$
   $P(\theta_3|Z) = 0.0714$
33. a. 1,000   b. 1,400   c. 900
35. Best action is to select the Mountain View site, then 1 lift and 1 campsite. That (one) branch of the tree is shown in the figure.

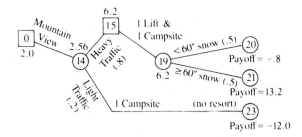

37. $p = 0.17$ and $(1 - p) = 0.83$
39. (1) MAXIMAX: Start research in hopes competitor introduces product and payoff is $2 M.
   (2) MAXIMIN: No research, with hope that competitor introduces product. Even if he doesn't, worst condition is $150,000 gain.
   (3) MAXIMUM PROBABILITY: No research, assuming $B_2$ will prevail for net gain of $150,000.
   (4) MEV: Start research for EMV $(A_1) = \$900,000$. Expected value criterion is inadequate in view of potential $200,000 loss that is the most likely result. With $P(B_2) = 0.8$ the maximum likelihood or maximum criterion may be better. Some measure of utility that takes risk into account would really be helpful here.

# CREDITS

**Text Credits and Acknowledgments**

AMERICAN ACCOUNTING ASSOCIATION    for tables from Raymond C. Dockweiler and Carl G. Willis, "On the Use of Entry Requirements for Undergraduate Accounting Programs," *The Accounting Review* 3 (July 1984): 496–504. Reprinted by permission of the *The Accounting Review*, the authors, and the American Accounting Association.

AMERICAN CHAMBER OF COMMERCE RESEARCHER'S ASSOCIATION    for portion of City Composite Index from ACCRA's *Cost of Living Index*. Reprinted with permission.

AMERICAN DEMOGRAPHICS, INC.    for table from Thomas Exter, "Who Pays for College?" *American Demographics* 12, no. 2 (Feb. 1990): 6; for table from John P. Robinson, "I Love My TV," *American Demographics* 12, no. 9 (Sept. 1990): 24–29; for table from Judith Waldrop, "Up and Down the Income Scale," *American Demographics* 12, no. 7 (July 1990): 24–30. All reprinted with permission © *American Demographics*.

AMERICAN MARKETING ASSOCIATION    for table from John A. Martilla and Dan W. Carvey, "Four Subtle Sins in Marketing Research," *Journal of Marketing* 39 (Jan. 1975): 8–15; for table from J. B. Willkinson, J. Barry Mason, and Christie H. Paksoy, "Assessing the Impact of Short-Term Supermarket Strategy Variables," *Journal of Marketing Research* 19 (Feb. 1982): 72–86. Both reprinted by permission of the American Marketing Association.

AMERICAN SOCIETY OF TESTING MATERIAL    for tables "Formulas for Central Lines and Control Limits" and "Factors for Computing Control Chart Lines — No Standard Given," from *ASTM Manual on Quality Control of Materials*, Special Technical Publication 15-C, January 1951, p. 63. Copyright ASTM. Reprinted with permission.

AMERICAN STATISTICAL ASSOCIATION    for table "Percentage Points in Kolmogorov–Smirnov Statistics," by L. H. Miller, from *Journal of the American Statistical Association* 51 (1956): 111–121. Reprinted by permission of the publisher and the author.

BAYWOOD PUBLISHING COMPANY    for table from Christian T. L. Janssen and R. W. Schutz, "Six Odd Games: An Analysis of the Midseason Slump of the Edmonton Oilers," *Journal of Recreational Mathematics* 17, no. 4 (1984–1985): 275–281; for table from Paul M. Sommers, "Does the American Presidency Shorten Life?" *Journal of Recreational Mathematics* 21, no. 1 (1989): 39–45. Both reprinted by permission of Baywood Publishing Company.

CRC PRESS, INC.    for tables reprinted with permission from *Handbook of Tables for Probability and Statistics*, 2nd ed., William H. Beyer (ed.), 1968. Copyright © CRC Press, Inc., Boca Raton, FL.

ELSEVIER SCIENCE PUBLISHING COMPANY    for table reprinted by permission of the publisher from Warren S. Martin, W. Jack Duncan, Thomas L. Powers, and Jesse C. Sawyer, "Costs and Benefits of Selected Response Inducement Techniques in Mail Survey Research," *Journal of Business Research* 19 (1989): 67–79. Copyright 1989 by Elsevier Science Publishing Co.; for table reprinted by permission of the publisher from Gordon Patzer, "Source Credibility as a Function of Communicator Physical Attractiveness," *Journal of Business Research* 11 (1988): 229–241. Copyright 1988 by Elsevier Science Publishing Co., Inc.

MEDIAMARK RESEARCH INC.    for excerpts from table "Home/Personal Computers: Decision Makers" from Mediamark Research Inc. Spring 1987 survey summary. Reprinted with permission of the publisher.

THE INSTITUTE OF MATHEMATICAL STATISTICS   for table adapted from Frieda S. Swed and C. Eisenhart, "Tables for Testing Randomness of Grouping a Sequence of Alternatives," *Annals of Mathematical Statistics* 14 (1943): 66–87. Reprinted with the permission of the Institute of Mathematical Statistics.

MCB UNIVERSITY PRESS   for table from Bas A. Agbonifoh and Pius E. Edoreh, "Consumer Awareness and Complaining Behavior," *European Journal of Marketing* 20, no. 7 (1986): 43–49. Reprinted by permission of the publisher, MCB University Press.

SETON HALL UNIVERSITY   for table from Hershey H. Friedman and Linda W. Friedman, "More Subtle Sins in Marketing Research," *The Mid-Atlantic Journal of Business* 20, no. 1 (Winter 1981): 21–31; for table from Hershey H. Friedman and Joshua Krausz, "A Portfolio Theory Approach to Solving the Product Elimination Problem," *The Mid-Atlantic Journal of Business* 24, no. 2 (Summer 1986): 43–48; for table from Hershey H. Friedman, Yonah Wilamowsky, and Linda W. Friedman, "A Comparison of Balanced and Unbalanced Rating Scales," *The Mid-Atlantic Journal of Business* 19, no. 2 (1981): 1–7; for table from Peter K. Tat, "Opinion Leadership in Black Female Fashion Buying Behavior," *The Mid-Atlantic Journal of Business* 23, no. 1 (Winter 1984–1985): 11–20. All reprinted by permission of the publisher.

MORRIS J. SLONIM   for excerpts from *Sampling in a Nutshell* (New York: Simon & Schuster, 1971), pp. 19–23. Reprinted by permission of the author.

JOHN WILEY AND SONS, INC.   for table abridged from Table II of A. Hald, *Statistical Tables and Formulas* (New York: John Wiley and Sons, Inc., 1952); for table from Bill Williams, *A Sampler on Sampling* (New York: John Wiley and Sons, Inc., 1978). Both copyright John Wiley and Sons, Inc. Reprinted by permission.

## Figure Credits

Chapter 3, page 100: Reprinted with permission © American Demographics, August 1990.

Figure 15-1: © Copyright 1989 American Society for Quality Control.

Figure 15-2: Reprinted from *Out of the Crisis* by W. Edwards Deming by permission of MIT and W. Edwards Deming. Published by MIT, Center for Advanced Engineering Study, Cambridge, MA 02139. Copyright 1991 by W. Edwards Deming.

Figures 15-3 and 15-4: Reprinted with permission, "Statistical Thinking and Its Contribution to Total Quality Control," *The American Statistician* 44, no. 2 (May 1990): 116–121.

# INDEX